APPLICATIONS

AF861333

Linear Algebra

A Modern Introduction

Fourth Edition

David Poole

CENGAGE

Australia • Brazil • Mexico • Singapore • United Kingdom • United States

CENGAGE

Linear Algebra: A Modern Introduction, Fourth Edition

Linear Algebra: A Modern Introduction, 4th Edition
David Poole

ISBN: 978-1-337-71005-3

Cengage Learning
20 Channel Street
Boston, MA 02210
USA

Cengage Learning is a leading provider of customized learning solutions with employees residing in nearly 40 different countries and sales in more than 125 countries around the world. Find your local representative at: **www.cengage.com.**

Cengage Learning products are represented in Canada by Nelson Education, Ltd.

For your course and learning solutions, visit **www.cengage.com.**

Purchase any of our products at your local college store or at our preferred online store **www.cengagebrain.com.**

Visit our custom book building website at **www.compose.cengage.com.**

Dedicated to the memory of Peter Hilton, who was an exemplary mathematician, educator, and citizen—a unit vector in every sense.

Contents

Chapter 6 Vector Spaces 427

Chapter 7 Distance and Approximation 529

Chapter 8 Codes *Online only* 620

Preface

The last thing one knows when writing a book is what to put first.

—Blaise Pascal
Pensées, 1670

The fourth edition of *Linear Algebra: A Modern Introduction* preserves the approach and features that users found to be strengths of the previous editions. However, I have streamlined the text somewhat, added numerous clarifications, and freshened up the exercises.

I want students to see linear algebra as an exciting subject and to appreciate its tremendous usefulness. At the same time, I want to help them master the basic concepts and techniques of linear algebra that they will need in other courses, both in mathematics and in other disciplines. I also want students to appreciate the interplay of theoretical, applied, and numerical mathematics that pervades the subject.

This book is designed for use in an introductory one- or two-semester course sequence in linear algebra. First and foremost, it is intended for students, and I have tried my best to write the book so that students not only will find it readable but also will *want* to read it. As in the first three editions, I have taken into account the reality that students taking introductory linear algebra are likely to come from a variety of disciplines. In addition to mathematics majors, there are apt to be majors from engineering, physics, chemistry, computer science, biology, environmental science, geography, economics, psychology, business, and education, as well as other students taking the course as an elective or to fulfill degree requirements. Accordingly, the book balances theory and applications, is written in a conversational style yet is fully rigorous, and combines a traditional presentation with concern for student-centered learning.

There is no such thing as a universally best learning style. In any class, there will be some students who work well independently and others who work best in groups; some who prefer lecture-based learning and others who thrive in a workshop setting, doing explorations; some who enjoy algebraic manipulations, some who are adept at numerical calculations (with and without a computer), and some who exhibit strong geometric intuition. In this edition, I continue to present material in a variety of ways—*algebraically, geometrically, numerically*, and *verbally*—so that all types of learners can find a path to follow. I have also attempted to present the theoretical, computational, and applied topics in a flexible yet integrated way. In doing so, it is my hope that all students will be exposed to the many sides of linear algebra.

For more on the recommendations of the Linear Algebra Curriculum Study Group, see *The College Mathematics Journal* **24** (1993), 41–46.

This book is compatible with the recommendations of the Linear Algebra Curriculum Study Group. From a pedagogical point of view, there is no doubt that for most students concrete examples should precede abstraction. I have taken this approach here. I also believe strongly that linear algebra is essentially about vectors and that students need to see vectors first (in a concrete setting) in order to gain some geometric insight. Moreover, introducing vectors early allows students to see how systems of linear equations arise naturally from geometric problems. Matrices then arise equally naturally as coefficient matrices of linear systems and as agents of change (linear transformations). This sets the stage for eigenvectors and orthogonal projections, both of which are best understood geometrically. The dart that appears on the cover of this book symbolizes a vector and reflects my conviction that geometric understanding should precede computational techniques.

I have tried to limit the number of theorems in the text. For the most part, results labeled as theorems either will be used later in the text or summarize preceding work. Interesting results that are not central to the book have been included as exercises or explorations. For example, the cross product of vectors is discussed only in explorations (in Chapters 1 and 4). Unlike most linear algebra textbooks, this book has no chapter on determinants. The essential results are all in Section 4.2, with other interesting material contained in an exploration. The book is, however, comprehensive for an introductory text. Wherever possible, I have included elementary and accessible proofs of theorems in order to avoid having to say, "The proof of this result is beyond the scope of this text." The result is, I hope, a work that is self-contained.

I have not been stingy with the applications: There are many more in the book than can be covered in a single course. However, it is important that students see the impressive range of problems to which linear algebra can be applied. I have included some modern material on finite linear algebra and coding theory that is not normally found in an introductory linear algebra text. There are also several impressive real-world applications of linear algebra and one item of historical, if not practical, interest; these applications are presented as self-contained "vignettes."

I hope that instructors will enjoy teaching from this book. More important, I hope that students using the book will come away with an appreciation of the beauty, power, and tremendous utility of linear algebra and that they will have fun along the way.

What's New in the Fourth Edition

The overall structure and style of *Linear Algebra: A Modern Introduction* remain the same in the fourth edition.

Here is a summary of what is new:

- The applications to coding theory have been moved to the new online Chapter 8.
- To further engage students, five writing projects have been added to the exercise sets. These projects give students a chance to research and write about aspects of the history and development of linear algebra. The explorations, vignettes, and many of the applications provide additional material for student projects.

See pages 49, 82, 283, 301, 443

- There are over 200 new or revised exercises. In response to reviewers' comments, there is now a full proof of the Cauchy-Schwarz Inequality in Chapter 1 in the form of a guided exercise.
- I have made numerous small changes in wording to improve the clarity or accuracy of the exposition. Also, several definitions have been made more explicit by giving them their own definition boxes and a few results have been highlighted by labeling them as theorems.
- All existing ancillaries have been updated.

Features

Clear Writing Style

The text is written is a simple, direct, conversational style. As much as possible, I have used "mathematical English" rather than relying excessively on mathematical notation. However, all proofs that are given are fully rigorous, and Appendix A contains an introduction to mathematical notation for those who wish to streamline their own writing. Concrete examples almost always precede theorems, which are then followed by further examples and applications. This flow—from specific to general and back again—is consistent throughout the book.

Key Concepts Introduced Early

Many students encounter difficulty in linear algebra when the course moves from the computational (solving systems of linear equations, manipulating vectors and matrices) to the theoretical (spanning sets, linear independence, subspaces, basis, and dimension). This book introduces all of the key concepts of linear algebra early, in a concrete setting, before revisiting them in full generality. Vector concepts such as dot product, length, orthogonality, and projection are first discussed in Chapter 1 in the concrete setting of $\mathbb{R}^2$ and $\mathbb{R}^3$ before the more general notions of inner product, norm, and orthogonal projection appear in Chapters 5 and 7. Similarly, spanning sets and linear independence are given a concrete treatment in Chapter 2 prior to their generalization to vector spaces in Chapter 6. The fundamental concepts of subspace, basis, and dimension appear first in Chapter 3 when the row, column, and null spaces of a matrix are introduced; it is not until Chapter 6 that these ideas are given a general treatment. In Chapter 4, eigenvalues and eigenvectors are introduced and explored for 2×2 matrices before their $n \times n$ counterparts appear. By the beginning of Chapter 4, all of the key concepts of linear algebra have been introduced, with concrete, computational examples to support them. When these ideas appear in full generality later in the book, students have had time to get used to them and, hence, are not so intimidated by them.

Emphasis on Vectors and Geometry

In keeping with the philosophy that linear algebra is primarily about vectors, this book stresses geometric intuition. Accordingly, the first chapter is about vectors, and it develops many concepts that will appear repeatedly throughout the text. Concepts such as orthogonality, projection, and linear combination are all found in Chapter 1, as is a comprehensive treatment of lines and planes in $\mathbb{R}^3$ that provides essential insight into the solution of systems of linear equations. This emphasis on vectors, geometry, and visualization is found throughout the text. Linear transformations are introduced as matrix transformations in Chapter 3, with many geometric examples, before general linear transformations are covered in Chapter 6. In Chapter 4, eigenvalues are introduced with "eigenpictures" as a visual aid. The proof of Perron's Theorem is given first heuristically and then formally, in both cases using a geometric argument. The geometry of linear dynamical systems reinforces and summarizes the material on eigenvalues and eigenvectors. In Chapter 5, orthogonal projections, orthogonal complements of subspaces, and the Gram-Schmidt Process are all presented in the concrete setting of $\mathbb{R}^3$ before being generalized to $\mathbb{R}^n$ and, in Chapter 7, to inner product spaces. The nature of the singular value decomposition is also explained informally in Chapter 7 via a geometric argument. Of the more than 300 figures in the text, over 200 are devoted to fostering a geometric understanding of linear algebra.

Explorations

See pages 1, 136, 427, 529

See pages 32, 286, 460, 515, 543, 547

See pages 83, 84, 85, 396, 398

The introduction to each chapter is a guided exploration (Section 0) in which students are invited to discover, individually or in groups, some aspect of the upcoming chapter. For example, "The Racetrack Game" introduces vectors, "Matrices in Action" introduces matrix multiplication and linear transformations, "Fibonacci in (Vector) Space" touches on vector space concepts, and "Taxicab Geometry" sets up generalized norms and distance functions. Additional explorations found throughout the book include applications of vectors and determinants to geometry, an investigation of 3×3 magic squares, a study of symmetry via the tilings of M. C. Escher, an introduction to complex linear algebra, and optimization problems using geometric inequalities. There are also explorations that introduce important numerical considerations and the analysis of algorithms. Having students do some of these explorations is one way of encouraging them to become active learners and to give them "ownership" over a small part of the course.

Applications

See pages 623, 641

See pages 121, 226, 356, 607, 626

The book contains an abundant selection of applications chosen from a broad range of disciplines, including mathematics, computer science, physics, chemistry, engineering, biology, business, economics, psychology, geography, and sociology. Noteworthy among these is a strong treatment of coding theory, from error-detecting codes (such as International Standard Book Numbers) to sophisticated error-correcting codes (such as the Reed-Muller code that was used to transmit satellite photos from space). Additionally, there are five "vignettes" that briefly showcase some very modern applications of linear algebra: the Global Positioning System (GPS), robotics, Internet search engines, digital image compression, and the Codabar System.

Examples and Exercises

There are over 400 examples in this book, most worked in greater detail than is customary in an introductory linear algebra textbook. This level of detail is in keeping with the philosophy that students should want (and be able) to read a textbook. Accordingly, it is not intended that all of these examples be covered in class; many can be assigned for individual or group study, possibly as part of a project. Most examples have at least one counterpart exercise so that students can try out the skills covered in the example before exploring generalizations.

There are over 2000 exercises, more than in most textbooks at a similar level. Answers to most of the computational odd-numbered exercises can be found in the back of the book. Instructors will find an abundance of exercises from which to select homework assignments. The exercises in each section are graduated, progressing from the routine to the challenging. Exercises range from those intended for hand computation to those requiring the use of a calculator or computer algebra system, and from theoretical and numerical exercises to conceptual exercises. Many of the examples and exercises use actual data compiled from real-world situations. For example, there are problems on modeling the growth of caribou and seal populations, radiocarbon dating

See pages 248, 359, 526, 588

of the Stonehenge monument, and predicting major league baseball players' salaries. Working such problems reinforces the fact that linear algebra is a valuable tool for modeling real-life problems.

Additional exercises appear in the form of a review after each chapter. In each set, there are 10 true/false questions designed to test conceptual understanding, followed by 19 computational and theoretical exercises that summarize the main concepts and techniques of that chapter.

Biographical Sketches and Etymological Notes

It is important that students learn something about the history of mathematics and come to see it as a social and cultural endeavor as well as a scientific one. Accordingly, the text contains short biographical sketches about many of the mathematicians who contributed to the development of linear algebra. I hope that these will help to put a human face on the subject and give students another way of relating to the material.

I have found that many students feel alienated from mathematics because the terminology makes no sense to them—it is simply a collection of words to be learned. To help overcome this problem, I have included short etymological notes that give the origins of many of the terms used in linear algebra. (For example, why do we use the word *normal* to refer to a vector that is perpendicular to a plane?)

See page 34

Margin Icons

The margins of the book contain several icons whose purpose is to alert the reader in various ways. Calculus is not a prerequisite for this book, but linear algebra has many interesting and important applications to calculus. The $\frac{dy}{dx}$ icon denotes an example or exercise that requires calculus. (This material can be omitted if not everyone in the class has had at least one semester of calculus. Alternatively, this material can be assigned as projects.) The $a + bi$ icon denotes an example or exercise involving complex numbers. (For students unfamiliar with complex numbers, Appendix C contains all the background material that is needed.) The CAS icon indicates that a computer algebra system (such as Maple, Mathematica, or MATLAB) or a calculator with matrix capabilities (such as almost any graphing calculator) is required—or at least very useful—for solving the example or exercise.

In an effort to help students learn how to read and use this textbook most effectively, I have noted various places where the reader is advised to pause. These may be places where a calculation is needed, part of a proof must be supplied, a claim should be verified, or some extra thought is required. The → icon appears in the margin at such places; the message is "Slow down. Get out your pencil. Think about this."

Technology

This book can be used successfully whether or not students have access to technology. However, calculators with matrix capabilities and computer algebra systems are now commonplace and, properly used, can enrich the learning experience as well as help with tedious calculations. In this text, I take the point of view that students need to master all of the basic techniques of linear algebra by solving by hand examples that are not too computationally difficult. Technology may then be used (in whole or in part) to solve subsequent examples and applications and to apply techniques that rely on earlier ones. For example, when systems of linear equations are first introduced, detailed solutions are provided; later, solutions are simply given, and the reader is expected to verify them. This is a good place to use some form of technology. Likewise, when applications use data that make hand calculation impractical, use technology. All of the numerical methods that are discussed depend on the use of technology.

With the aid of technology, students can explore linear algebra in some exciting ways and discover much for themselves. For example, if one of the coefficients of a linear system is replaced by a parameter, how much variability is there in the solutions? How does changing a single entry of a matrix affect its eigenvalues? This book is not a tutorial on technology, and in places where technology can be used, I have not specified a particular type of technology. The student companion website that accompanies this book offers an online appendix called *Technology Bytes* that gives instructions for solving a selection of examples from each chapter using Maple, Mathematica, and MATLAB. By imitating these examples, students can do further calculations and explorations using whichever CAS they have and exploit the power of these systems to help with the exercises throughout the book, particularly those marked with the CAS icon. The website also contains data sets and computer code in Maple, Mathematica, and MATLAB formats keyed to many exercises and examples in the text. Students and instructors can import these directly into their CAS to save typing and eliminate errors.

Finite and Numerical Linear Algebra

The text covers two aspects of linear algebra that are scarcely ever mentioned together: finite linear algebra and numerical linear algebra. By introducing modular arithmetic early, I have been able to make finite linear algebra (more properly, "linear algebra over finite fields," although I do not use that phrase) a recurring theme throughout the book. This approach provides access to the material on coding theory in Chapter 8 (online). There is also an application to finite linear games in Section 2.4 that students really enjoy. In addition to being exposed to the applications of finite linear algebra, mathematics majors will benefit from seeing the material on finite fields, because they are likely to encounter it in such courses as discrete mathematics, abstract algebra, and number theory.

All students should be aware that in practice, it is impossible to arrive at exact solutions of large-scale problems in linear algebra. Exposure to some of the techniques of numerical linear algebra will provide an indication of how to obtain highly accurate approximate solutions. Some of the numerical topics included in the book are roundoff error and partial pivoting, iterative methods for solving linear systems and computing eigenvalues, the *LU* and *QR* factorizations, matrix norms and condition numbers, least squares approximation, and the singular value decomposition. The inclusion of numerical linear algebra also brings up some interesting and important issues that are completely absent from the *theory* of linear algebra, such as pivoting strategies, the condition of a linear system, and the convergence of iterative methods. This book not only raises these questions but also shows how one might approach them. Gerschgorin disks, matrix norms, and the singular values of a matrix, discussed in Chapters 4 and 7, are useful in this regard.

See pages 83, 84, 124, 180, 311, 392, 555, 561, 568, 590

See pages 319, 563, 600

Appendices

Appendix A contains an overview of mathematical notation and methods of proof, and Appendix B discusses mathematical induction. All students will benefit from these sections, but those with a mathematically oriented major may wish to pay particular attention to them. Some of the examples in these appendices are uncommon (for instance, Example B.6 in Appendix B) and underscore the power of the methods. Appendix C is an introduction to complex numbers. For students familiar with these results, this appendix can serve as a useful reference; for others, this section contains everything they need to know for those parts of the text that use complex numbers. Appendix D is about polynomials. I have found that many students require a refresher about these facts. Most students will be unfamiliar with Descartes's Rule of Signs; it is used in Chapter 4 to explain the behavior of the eigenvalues of Leslie matrices. Exercises to accompany the four appendices can be found on the book's website.

Short answers to most of the odd-numbered computational exercises are given at the end of the book. Exercise sets to accompany Appendixes A, B, C, and D are available on the companion website, along with their odd-numbered answers.

Ancillaries

For Instructors

Enhanced WebAssign® ENHANCED WebAssign
Printed Access Card: 978-1-285-85829-6
Online Access Code: 978-1-285-85827-2
Exclusively from Cengage Learning, Enhanced WebAssign combines the exceptional mathematics content that you know and love with the most powerful online homework solution, WebAssign. Enhanced WebAssign engages students with immediate feedback, rich tutorial content, and interactive, fully customizable eBooks (YouBook), helping students to develop a deeper conceptual understanding of their subject matter. Flexible assignment options give instructors the ability to release assignments conditionally based on students' prerequisite assignment scores. Visit us at **www.cengage.com/ewa** to learn more.

Cengage Learning Testing Powered by Cognero
Cengage Learning Testing Powered by Cognero is a flexible, online system that allows you to author, edit, and manage test bank content from multiple Cengage Learning solutions; create multiple test versions in an instant; and deliver tests from your LMS, your classroom, or wherever you want.

Complete Solutions Manual
The Complete Solutions Manual provides detailed solutions to all exercises in the text, including *Exploration* and *Chapter Review* exercises. The Complete Solutions Manual is available online.

Instructor's Guide
This online guide enhances the text with valuable teaching resources such as group work projects, teaching tips, interesting exam questions, examples and extra

material for lectures, and other items designed to reduce the instructor's preparation time and make linear algebra class an exciting and interactive experience. For each section of the text, the *Instructor's Guide* includes suggested time and emphasis, points to stress, questions for discussion, lecture materials and examples, technology tips, student projects, group work with solutions, sample assignments, and suggested test questions.

Solution Builder
www.cengage.com/solutionbuilder
Solution Builder provides full instructor solutions to all exercises in the text, including those in the explorations and chapter reviews, in a convenient online format. Solution Builder allows instructors to create customized, secure PDF printouts of solutions matched exactly to the exercises assigned for class.

*Access Cognero and additional instructor resources online at **login.cengage.com**.

For Students

Student Solutions Manual (ISBN-13: 978-1-285-84195-3)
The Student Solutions Manual and Study Guide includes detailed solutions to all odd-numbered exercises and selected even-numbered exercises; section and chapter summaries of symbols, definitions, and theorems; and study tips and hints. Complex exercises are explored through a question-and-answer format designed to deepen understanding. Challenging and entertaining problems that further explore selected exercises are also included.

Enhanced WebAssign® ENHANCED WebAssign
Printed Access Card: 978-1-285-85829-6
Online Access Code: 978-1-285-85827-2
Enhanced WebAssign (assigned by the instructor) provides you with instant feedback on homework assignments. This online homework system is easy to use and includes helpful links to textbook sections, video examples, and problem-specific tutorials.

CengageBrain.com
To access additional course materials and companion resources, please visit www.cengagebrain.com. At the CengageBrain.com home page, search for the ISBN of your title (from the back cover of your book) using the search box at the top of the page. This will take you to the product page where free companion resources can be found.

Acknowledgments

The reviewers of the previous edition of this text contributed valuable and often insightful comments about the book. I am grateful for the time each of them took to do this. Their judgement and helpful suggestions have contributed greatly to the development and success of this book, and I would like to thank them personally:
Jamey Bass, City College of San Francisco; Olga Brezhneva, Miami University; Karen Clark, The College of New Jersey; Marek Elzanowski, Portland State University; Christopher Francisco, Oklahoma State University; Brian Jue, California State University, Stanislaus; Alexander Kheyfits, Bronx Community College/CUNY; Henry Krieger, Harvey Mudd College; Rosanna Pearlstein, Michigan State

University; William Sullivan, Portland State University; Matthias Weber, Indiana University.

I am indebted to a great many people who have, over the years, influenced my views about linear algebra and the teaching of mathematics in general. First, I would like to thank collectively the participants in the education and special linear algebra sessions at meetings of the Mathematical Association of America and the Canadian Mathematical Society. I have also learned much from participation in the Canadian Mathematics Education Study Group and the Canadian Mathematics Education Forum.

I especially want to thank Ed Barbeau, Bill Higginson, Richard Hoshino, John Grant McLoughlin, Eric Muller, Morris Orzech, Bill Ralph, Pat Rogers, Peter Taylor, and Walter Whiteley, whose advice and inspiration contributed greatly to the philosophy and style of this book. My gratitude as well to Robert Rogers, who developed the student and instructor solutions, as well as the excellent study guide content. Special thanks go to Jim Stewart for his ongoing support and advice. Joe Rotman and his lovely book *A First Course in Abstract Algebra* inspired the etymological notes in this book, and I relied heavily on Steven Schwartzman's *The Words of Mathematics* when compiling these notes. I thank Art Benjamin for introducing me to the Codabar system and Joe Grcar for clarifying aspects of the history of Gaussian elimination. My colleagues Marcus Pivato and Reem Yassawi provided useful information about dynamical systems. As always, I am grateful to my students for asking good questions and providing me with the feedback necessary to becoming a better teacher.

I sincerely thank all of the people who have been involved in the production of this book. Jitendra Kumar and the team at MPS Limited did an amazing job producing the fourth edition. I thank Christine Sabooni for doing a thorough copyedit. Most of all, it has been a delight to work with the entire editorial, marketing, and production teams at Cengage Learning: Richard Stratton, Molly Taylor, Laura Wheel, Cynthia Ashton, Danielle Hallock, Andrew Coppola, Alison Eigel Zade, and Janay Pryor. They offered sound advice about changes and additions, provided assistance when I needed it, but let me write the book I wanted to write. I am fortunate to have worked with them, as well as the staffs on the first through third editions.

As always, I thank my family for their love, support, and understanding. Without them, this book would not have been possible.

David Poole
dpoole@trentu.ca

To the Instructor

"Would you tell me, please,
which way I ought to go from here?"
"That depends a good deal on where
you want to get to," said the Cat.

—Lewis Carroll
Alice's Adventures in
Wonderland, 1865

This text was written with flexibility in mind. It is intended for use in a one- or two-semester course with 36 lectures per semester. The range of topics and applications makes it suitable for a variety of audiences and types of courses. However, there is more material in the book than can be covered in class, even in a two-semester course. After the following overview of the text are some brief suggestions for ways to use the book.

An Overview of the Text

Chapter 1: Vectors

See page 1

The racetrack game in Section 1.0 serves to introduce vectors in an informal way. (It's also quite a lot of fun to play!) Vectors are then formally introduced from both algebraic and geometric points of view. The operations of addition and scalar multiplication and their properties are first developed in the concrete settings of $\mathbb{R}^2$ and $\mathbb{R}^3$ before being generalized to $\mathbb{R}^n$. Modular arithmetic and finite linear algebra are also introduced. Section 1.2 defines the dot product of vectors and the related notions of length, angle, and orthogonality. The very important concept of (orthogonal) projection is developed here; it will reappear in Chapters 5 and 7. The exploration "Vectors and Geometry" shows how vector methods can be used to prove certain results in Euclidean geometry. Section 1.3 is a basic but thorough introduction to lines and planes in $\mathbb{R}^2$ and $\mathbb{R}^3$. This section is crucial for understanding the geometric significance of the solution of linear systems in Chapter 2. Note that the cross product of vectors in $\mathbb{R}^3$ is left as an exploration. The chapter concludes with an application to force vectors.

See page 32

See page 48

Chapter 2: Systems of Linear Equations

See page 57

The introduction to this chapter serves to illustrate that there is more than one way to think of the solution to a system of linear equations. Sections 2.1 and 2.2 develop the main computational tool for solving linear systems: row reduction of matrices (Gaussian and Gauss-Jordan elimination). Nearly all subsequent computational methods in the book depend on this. The Rank Theorem appears here for the first time; it shows up again, in more generality, in Chapters 3, 5, and 6. Section 2.3 is very important; it introduces the fundamental notions of spanning sets and linear independence of vectors. Do not rush through this material. Section 2.4 contains six applications from which instructors can choose depending on the time available and the interests of the class. The vignette on the Global Positioning System provides another application that students will enjoy. The iterative methods in Section 2.5 will be optional for many courses but are essential for a course with an applied/numerical focus. The three explorations in this chapter are related in that they all deal with aspects of the use of computers to solve linear systems. All students should at least be made aware of these issues.

See pages 72, 205, 386, 486

See page 121

See pages 83, 84, 85

Chapter 3: Matrices

This chapter contains some of the most important ideas in the book. It is a long chapter, but the early material can be covered fairly quickly, with extra time allowed for the crucial material in Section 3.5. Section 3.0 is an exploration that introduces the notion of a linear transformation: the idea that matrices are not just static objects but rather a type of function, transforming vectors into other vectors. All of the basic facts about matrices, matrix operations, and their properties are found in the first two sections. The material on partitioned matrices and the multiple representations of the matrix product is worth stressing, because it is used repeatedly in subsequent sections. The Fundamental Theorem of Invertible Matrices in Section 3.3 is very important and will appear several more times as new characterizations of invertibility are presented. Section 3.4 discusses the very important *LU* factorization of a matrix. If this topic is not covered in class, it is worth assigning as a project or discussing in a workshop. The point of Section 3.5 is to present many of the key concepts of linear algebra (subspace, basis, dimension, and rank) in the concrete setting of matrices before students see them in full generality. Although the examples in this section are all familiar, it is important that students get used to the new terminology and, in particular, understand what the notion of a basis means. The geometric treatment of linear transformations in Section 3.6 is intended to smooth the transition to general linear transformations in Chapter 6. The example of a projection is particularly important because it will reappear in Chapter 5. The vignette on robotic arms is a concrete demonstration of composition of linear (and affine) transformations. There are four applications from which to choose in Section 3.7. Either Markov chains or the Leslie model of population growth should be covered so that they can be used again in Chapter 4, where their behavior will be explained.

See page 136

See pages 172, 206, 296, 512, 605

See page 226

See pages 230, 239

Chapter 4: Eigenvalues and Eigenvectors

The introduction Section 4.0 presents an interesting dynamical system involving graphs. This exploration introduces the notion of an eigenvector and foreshadows the power method in Section 4.5. In keeping with the geometric emphasis of the book, Section 4.1 contains the novel feature of "eigenpictures" as a way of visualizing the eigenvectors of 2×2 matrices. Determinants appear in Section 4.2, motivated by their use in finding the characteristic polynomials of small matrices. This "crash

See page 253

See page 284

See page 286

See pages 325, 330

See page 348

course" in determinants contains all the essential material students need, including an optional but elementary proof of the Laplace Expansion Theorem. The vignette "Lewis Carroll's Condensation Method" presents a historically interesting, alternative method of calculating determinants that students may find appealing. The exploration "Geometric Applications of Determinants" makes a nice project that contains several interesting and useful results. (Alternatively, instructors who wish to give more detailed coverage to determinants may choose to cover some of this exploration in class.) The basic theory of eigenvalues and eigenvectors is found in Section 4.3, and Section 4.4 deals with the important topic of diagonalization. Example 4.29 on powers of matrices is worth covering in class. The power method and its variants, discussed in Section 4.5, are optional, but all students should be aware of the method, and an applied course should cover it in detail. Gerschgorin's Disk Theorem can be covered independently of the rest of Section 4.5. Markov chains and the Leslie model of population growth reappear in Section 4.6. Although the proof of Perron's Theorem is optional, the theorem itself (like the stronger Perron-Frobenius Theorem) should at least be mentioned because it explains *why* we should expect a unique positive eigenvalue with a corresponding positive eigenvector in these applications. The applications on recurrence relations and differential equations connect linear algebra to discrete mathematics and calculus, respectively. The matrix exponential can be covered if your class has a good calculus background. The final topic of discrete linear dynamical systems revisits and summarizes many of the ideas in Chapter 4, looking at them in a new, geometric light. Students will enjoy reading how eigenvectors can be used to help rank sports teams and websites. This vignette can easily be extended to a project or enrichment activity.

Chapter 5: Orthogonality

See page 366

See pages 396, 398

See pages 408, 415

The introductory exploration, "Shadows on a Wall," is mathematics at its best: it takes a known concept (projection of a vector onto another vector) and generalizes it in a useful way (projection of a vector onto a subspace—a plane), while uncovering some previously unobserved properties. Section 5.1 contains the basic results about orthogonal and orthonormal sets of vectors that will be used repeatedly from here on. In particular, orthogonal matrices should be stressed. In Section 5.2, two concepts from Chapter 1 are generalized: the orthogonal complement of a subspace and the orthogonal projection of a vector onto a subspace. The Orthogonal Decomposition Theorem is important here and helps to set up the Gram-Schmidt Process. Also note the quick proof of the Rank Theorem. The Gram-Schmidt Process is detailed in Section 5.3, along with the extremely important *QR* factorization. The two explorations that follow outline how the *QR* factorization is computed in practice and how it can be used to approximate eigenvalues. Section 5.4 on orthogonal diagonalization of (real) symmetric matrices is needed for the applications that follow. It also contains the Spectral Theorem, one of the highlights of the theory of linear algebra. The applications in Section 5.5 are quadratic forms and graphing quadratic equations. I always include at least the second of these in my course because it extends what students already know about conic sections.

Chapter 6: Vector Spaces

See page 427

The Fibonacci sequence reappears in Section 6.0, although it is not important that students have seen it before (Section 4.6). The purpose of this exploration is to show that familiar vector space concepts (Section 3.5) can be used fruitfully in a new setting. Because all of the main ideas of vector spaces have already been introduced in Chapters 1–3, students should find Sections 6.1 and 6.2 fairly familiar. The emphasis here should be on using the vector space axioms to prove properties rather than relying on computational techniques. When discussing change of basis in Section 6.3, it is helpful to show students how to use the notation to remember how the construction works. Ultimately, the Gauss-Jordan method is the most efficient here. Sections 6.4 and 6.5 on linear transformations are important. The examples are related to previous results on matrices (and matrix transformations). In particular, it is important to stress that the kernel and range of a linear transformation generalize the null space and column space of a matrix. Section 6.6 puts forth the notion that (almost) all linear transformations are essentially matrix transformations. This builds on the information in Section 3.6, so students should not find it terribly surprising. However, the examples should be worked carefully. The connection between change of basis and similarity of matrices is noteworthy. The exploration "Tilings, Lattices, and the Crystallographic Restriction" is an impressive application of change of basis. The connection with the artwork of M. C. Escher makes it all the more interesting. The applications in Section 6.7 build on previous ones and can be included as time and interest permit.

See page 515

Chapter 7: Distance and Approximation

Section 7.0 opens with the entertaining "Taxicab Geometry" exploration. Its purpose is to set up the material on generalized norms and distance functions (metrics) that follows. Inner product spaces are discussed in Section 7.1; the emphasis here should be on the examples and using the axioms. The exploration "Vectors and Matrices with Complex Entries" shows how the concepts of dot product, symmetric matrix, orthogonal matrix, and orthogonal diagonalization can be extended from real to complex vector spaces. The following exploration, "Geometric Inequalities and Optimization Problems," is one that students typically enjoy. (They will have fun seeing how many "calculus" problems can be solved without using calculus at all!) Section 7.2 covers generalized vector and matrix norms and shows how the condition number of a matrix is related to the notion of ill-conditioned linear systems explored in Chapter 2. Least squares approximation (Section 7.3) is an important application of linear algebra in many other disciplines. The Best Approximation Theorem and the Least Squares Theorem are important, but their proofs are intuitively clear. Spend time here on the examples—a few should suffice. Section 7.4 presents the singular value decomposition, one of the most impressive applications of linear algebra. If your course gets this far, you will be amply rewarded. Not only does the SVD tie together many notions discussed previously; it also affords some new (and quite powerful) applications. If a CAS is available, the vignette on digital image compression is worth presenting; it is a visually impressive display of the power of linear algebra and a fitting culmination to the course. The further applications in Section 7.5 can be chosen according to the time available and the interests of the class.

See page 529

See page 543

See page 547

See page 607

Chapter 8: Codes

This online chapter contains applications of linear algebra to the theory of codes. Section 8.1 begins with a discussion of how vectors can be used to design

See page 626

error-detecting codes such as the familiar Universal Product Code (UPC) and International Standard Book Number (ISBN). This topic only requires knowledge of Chapter 1. The vignette on the Codabar system used in credit and bank cards is an excellent classroom presentation that can even be used to introduce Section 8.1. Once students are familiar with matrix operations, Section 8.2 describes how codes can be designed to correct as well as detect errors. The Hamming codes introduced here are perhaps the most famous examples of such error-correcting codes. Dual codes, discussed in Section 8.3, are an important way of constructing new codes from old ones. The notion of orthogonal complement, introduced in Chapter 5, is the prerequisite concept here. The most important, and most widely used, class of codes is the class of linear codes that is defined in Section 8.4. The notions of subspace, basis, and dimension are key here. The powerful Reed-Muller codes used by NASA spacecraft are important examples of linear codes. Our discussion of codes concludes in Section 8.5 with the definition of the minimum distance of a code and the role it plays in determining the error-correcting capability of the code.

How to Use the Book

Students find the book easy to read, so I usually have them read a section before I cover the material in class. That way, I can spend class time highlighting the most important concepts, dealing with topics students find difficult, working examples, and discussing applications. I do not attempt to cover all of the material from the assigned reading in class. This approach enables me to keep the pace of the course fairly brisk, slowing down for those sections that students typically find challenging.

In a two-semester course, it is possible to cover the entire book, including a reasonable selection of applications. For extra flexibility, you might omit some of the topics (for example, give only a brief treatment of numerical linear algebra), thereby freeing up time for more in-depth coverage of the remaining topics, more applications, or some of the explorations. In an honors mathematics course that emphasizes proofs, much of the material in Chapters 1–3 can be covered quickly. Chapter 6 can then be covered in conjunction with Sections 3.5 and 3.6, and Chapter 7 can be integrated into Chapter 5. I would be sure to assign the explorations in Chapters 1, 4, 6, and 7 for such a class.

For a one-semester course, the nature of the course and the audience will determine which topics to include. Three possible courses are described below and on the following page. The basic course, described first, has fewer than 36 hours suggested, allowing time for extra topics, in-class review, and tests. The other two courses build on the basic course but are still quite flexible.

A Basic Course

A course designed for mathematics majors and students from other disciplines is outlined on the next page. This course does not mention general vector spaces at all (all concepts are treated in a concrete setting) and is very light on proofs. Still, it is a thorough introduction to linear algebra.

Section	Number of Lectures	Section	Number of Lectures
1.1	1	**3.6**	1–2
1.2	1–1.5	**4.1**	1
1.3	1–1.5	**4.2**	2
2.1	0.5–1	**4.3**	1
2.2	1–2	**4.4**	1–2
2.3	1–2	**5.1**	1–1.5
3.1	1–2	**5.2**	1–1.5
3.2	1	**5.3**	0.5
3.3	2	**5.4**	1
3.5	2	**7.3**	2

Total: 23–30 lectures

Because the students in a course such as this one represent a wide variety of disciplines, I would suggest using much of the remaining lecture time for applications. In my course, I do code vectors in Section 8.1, which students really seem to like, and at least one application from each of Chapters 2–5. Other applications can be assigned as projects, along with as many of the explorations as desired. There is also sufficient lecture time available to cover some of the theory in detail.

A Course with a Computational Emphasis

For a course with a computational emphasis, the basic course outlined on the previous page can be supplemented with the sections of the text dealing with numerical linear algebra. In such a course, I would cover part or all of Sections 2.5, 3.4, 4.5, 5.3, 7.2, and 7.4, ending with the singular value decomposition. The explorations in Chapters 2 and 5 are particularly well suited to such a course, as are almost any of the applications.

A Course for Students Who Have Already Studied Some Linear Algebra

Some courses will be aimed at students who have already encountered the basic principles of linear algebra in other courses. For example, a college algebra course will often include an introduction to systems of linear equations, matrices, and determinants; a multivariable calculus course will almost certainly contain material on vectors, lines, and planes. For students who have seen such topics already, much early material can be omitted and replaced with a quick review. Depending on the background of the class, it may be possible to skim over the material in the basic course up to Section 3.3 in about six lectures. If the class has a significant number of mathematics majors (and especially if this is the only linear algebra course they will take), I would be sure to cover Sections 6.1–6.5, 7.1, and 7.4 and as many applications as time permits. If the course has science majors (but not mathematics majors), I would cover Sections 6.1 and 7.1 and a broader selection of applications, being sure to include the material on differential equations and approximation of functions. If computer science students or engineers are prominently represented, I would try to do as much of the material on codes and numerical linear algebra as I could.

There are many other types of courses that can successfully use this text. I hope that you find it useful for your course and that you enjoy using it.

To the Student

"Where shall I begin, please your Majesty?" he asked.
"Begin at the beginning," the King said, gravely, "and go on till you come to the end: then stop."

—Lewis Carroll
Alice's Adventures in Wonderland, 1865

Linear algebra is an exciting subject. It is full of interesting results, applications to other disciplines, and connections to other areas of mathematics. The *Student Solutions Manual and Study Guide* contains detailed advice on how best to use this book; following are some general suggestions.

Linear algebra has several sides: There are *computational techniques, concepts*, and *applications*. One of the goals of this book is to help you master all of these facets of the subject and to see the interplay among them. Consequently, it is important that you read and understand each section of the text before you attempt the exercises in that section. If you read only examples that are related to exercises that have been assigned as homework, you will miss much. Make sure you understand the definitions of terms and the meaning of theorems. Don't worry if you have to read something more than once before you understand it. Have a pencil and calculator with you as you read. Stop to work out examples for yourself or to fill in missing calculations. The icon in the margin indicates a place where you should pause and think over what you have read so far.

Answers to most odd-numbered computational exercises are in the back of the book. Resist the temptation to look up an answer before you have completed a question. And remember that even if your answer differs from the one in the back, you may still be right; there is more than one correct way to express some of the solutions. For example, a value of $1/\sqrt{2}$ can also be expressed as $\sqrt{2}/2$ and the set of all scalar multiples of the vector $\begin{bmatrix} 3 \\ 1/2 \end{bmatrix}$ is the same as the set of all scalar multiples of $\begin{bmatrix} 6 \\ 1 \end{bmatrix}$.

As you encounter new concepts, try to relate them to examples that you know. Write out proofs and solutions to exercises in a logical, connected way, using complete sentences. Read back what you have written to see whether it makes sense. Better yet, if you can, have a friend in the class read what you have written. If it doesn't make sense to another person, chances are that it doesn't make sense, period.

You will find that a calculator with matrix capabilities or a computer algebra system is useful. These tools can help you to check your own hand calculations and are indispensable for some problems involving tedious computations. Technology also enables you to explore aspects of linear algebra on your own. You can play "what if?" games: What if I change one of the entries in this vector? What if this matrix is of a different size? Can I force the solution to be what I would like it to be by changing something? To signal places in the text or exercises where the use of technology is recommended, I have placed the icon CAS in the margin. The companion website that accompanies this book contains computer code working out selected exercises from the book using Maple, Mathematica, and MATLAB, as well as *Technology Bytes*, an appendix providing much additional advice about the use of technology in linear algebra.

You are about to embark on a journey through linear algebra. Think of this book as your travel guide. Are you ready? Let's go!

1 Vectors

Here they come pouring out of the blue.
Little arrows for me and for you.
—Albert Hammond and
Mike Hazelwood
Little Arrows
Dutchess Music/BMI, 1968

1.0 Introduction: The Racetrack Game

Many measurable quantities, such as length, area, volume, mass, and temperature, can be completely described by specifying their magnitude. Other quantities, such as velocity, force, and acceleration, require both a magnitude and a direction for their description. These quantities are *vectors*. For example, wind velocity is a vector consisting of wind speed and direction, such as 10 km/h southwest. Geometrically, vectors are often represented as arrows or directed line segments.

Although the idea of a vector was introduced in the 19th century, its usefulness in applications, particularly those in the physical sciences, was not realized until the 20th century. More recently, vectors have found applications in computer science, statistics, economics, and the life and social sciences. We will consider some of these many applications throughout this book.

This chapter introduces vectors and begins to consider some of their geometric and algebraic properties. We begin, though, with a simple game that introduces some of the key ideas. [You may even wish to play it with a friend during those (very rare!) dull moments in linear algebra class.]

The game is played on graph paper. A track, with a starting line and a finish line, is drawn on the paper. The track can be of any length and shape, so long as it is wide enough to accommodate all of the players. For this example, we will have two players (let's call them Ann and Bert) who use different colored pens to represent their cars or bicycles or whatever they are going to race around the track. (Let's think of Ann and Bert as cyclists.)

Ann and Bert each begin by drawing a dot on the starting line at a grid point on the graph paper. They take turns moving to a new grid point, subject to the following rules:

1. Each new grid point and the line segment connecting it to the previous grid point must lie entirely within the track.
2. No two players may occupy the same grid point on the same turn. (This is the "no collisions" rule.)
3. Each new move is related to the previous move as follows: If a player moves a units horizontally and b units vertically on one move, then on the next move he or she must move between $a - 1$ and $a + 1$ units horizontally and between

The Irish mathematician William Rowan Hamilton (1805–1865) used vector concepts in his study of complex numbers and their generalization, the quaternions.

$b - 1$ and $b + 1$ units vertically. In other words, if the second move is c units horizontally and d units vertically, then $|a - c| \leq 1$ and $|b - d| \leq 1$. (This is the "acceleration/deceleration" rule.) Note that this rule forces the first move to be 1 unit vertically and/or 1 unit horizontally.

A player who collides with another player or leaves the track is eliminated. The winner is the first player to cross the finish line. If more than one player crosses the finish line on the same turn, the one who goes farthest past the finish line is the winner.

In the sample game shown in Figure 1.1, Ann was the winner. Bert accelerated too quickly and had difficulty negotiating the turn at the top of the track.

To understand rule 3, consider Ann's third and fourth moves. On her third move, she went 1 unit horizontally and 3 units vertically. On her fourth move, her options were to move 0 to 2 units horizontally and 2 to 4 units vertically. (Notice that some of these combinations would have placed her outside the track.) She chose to move 2 units in each direction.

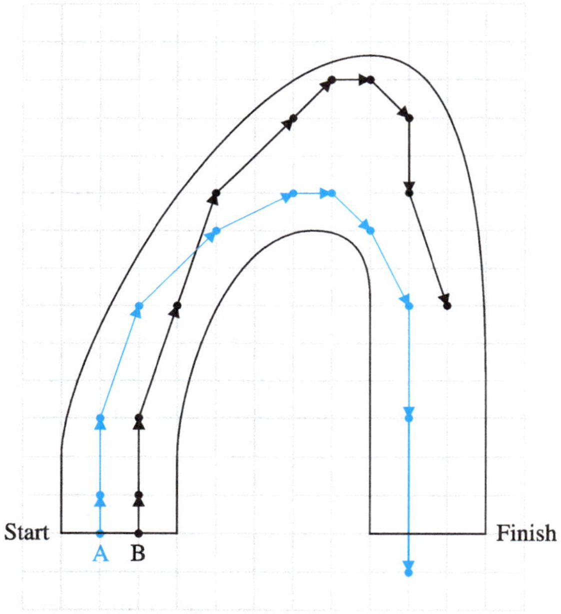

Figure 1.1
A sample game of racetrack

Problem 1 Play a few games of racetrack.

Problem 2 Is it possible for Bert to win this race by choosing a different sequence of moves?

Problem 3 Use the notation $[a, b]$ to denote a move that is a units horizontally and b units vertically. (Either a or b or both may be negative.) If move $[3, 4]$ has just been made, draw on graph paper all the grid points that could possibly be reached on the next move.

Problem 4 What is the net effect of two successive moves? In other words, if you move $[a, b]$ and then $[c, d]$, how far horizontally and vertically will you have moved altogether?

Problem 5 Write out Ann's sequence of moves using the $[a, b]$ notation. Suppose she begins at the origin $(0, 0)$ on the coordinate axes. Explain how you can find the coordinates of the grid point corresponding to each of her moves *without looking at the graph paper*. If the axes were drawn differently, so that Ann's starting point was not the origin but the point $(2, 3)$, what would the coordinates of her final point be?

Although simple, this game introduces several ideas that will be useful in our study of vectors. The next three sections consider vectors from geometric and algebraic viewpoints, beginning, as in the racetrack game, in the plane.

1.1 The Geometry and Algebra of Vectors

Vectors in the Plane

The Cartesian plane is named after the French philosopher and mathematician René Descartes (1596–1650), whose introduction of coordinates allowed *geometric* problems to be handled using *algebraic* techniques.

We begin by considering the Cartesian plane with the familiar x- and y-axes. A ***vector*** is a *directed line segment* that corresponds to a *displacement* from one point A to another point B; see Figure 1.2.

The vector from A to B is denoted by $\overrightarrow{AB}$; the point A is called its ***initial point,*** or ***tail,*** and the point B is called its ***terminal point,*** or ***head***. Often, a vector is simply denoted by a single boldface, lowercase letter such as $\mathbf{v}$.

The set of all points in the plane corresponds to the set of all vectors whose tails are at the origin O. To each point A, there corresponds the vector $\mathbf{a} = \overrightarrow{OA}$; to each vector $\mathbf{a}$ with tail at O, there corresponds its head A. (Vectors of this form are sometimes called *position vectors.*)

The word *vector* comes from the Latin root meaning "to carry." A vector is formed when a point is displaced—or "carried off"—a given distance in a given direction. Viewed another way, vectors "carry" two pieces of information: their length and their direction.

When writing vectors by hand, it is difficult to indicate boldface. Some people prefer to write $\vec{v}$ for the vector denoted in print by $\mathbf{v}$, but in most cases it is fine to use an ordinary lowercase v. It will usually be clear from the context when the letter denotes a vector.

It is natural to represent such vectors using coordinates. For example, in Figure 1.3, $A = (3, 2)$ and we write the vector $\mathbf{a} = \overrightarrow{OA} = [3, 2]$ using square brackets. Similarly, the other vectors in Figure 1.3 are

$$\mathbf{b} = [-1, 3] \quad \text{and} \quad \mathbf{c} = [2, -1]$$

The individual coordinates (3 and 2 in the case of $\mathbf{a}$) are called the ***components*** of the vector. A vector is sometimes said to be an *ordered pair* of real numbers. The order is important since, for example, $[3, 2] \neq [2, 3]$. In general, two vectors are equal if and only if their corresponding components are equal. Thus, $[x, y] = [1, 5]$ implies that $x = 1$ and $y = 5$.

It is frequently convenient to use ***column vectors*** instead of (or in addition to) ***row vectors.*** Another representation of $[3, 2]$ is $\begin{bmatrix} 3 \\ 2 \end{bmatrix}$. (The important point is that the

The word *component* is derived from the Latin words *co,* meaning "together with," and *ponere,* meaning "to put." Thus, a vector is "put together" out of its components.

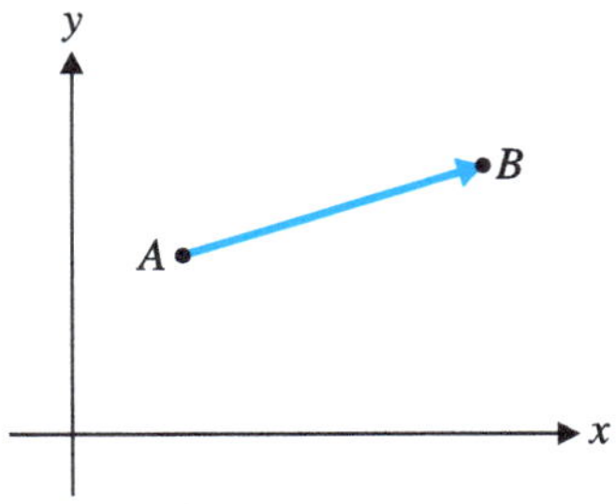

Figure 1.2

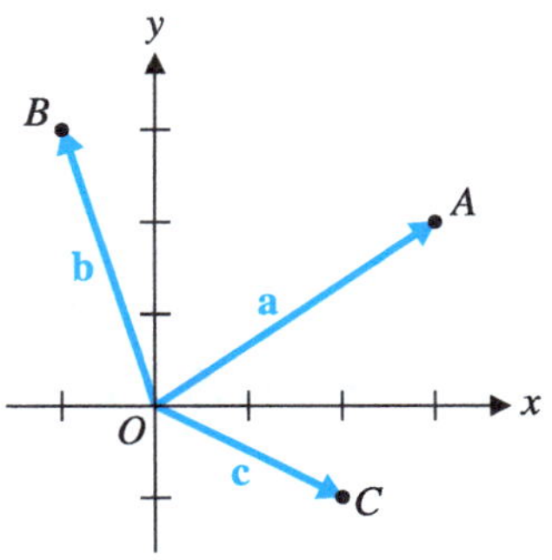

Figure 1.3

components are *ordered.*) In later chapters, you will see that column vectors are somewhat better from a computational point of view; for now, try to get used to both representations.

It may occur to you that we cannot really draw the vector $[0, 0] = \overrightarrow{OO}$ from the origin to itself. Nevertheless, it is a perfectly good vector and has a special name: the ***zero vector***. The zero vector is denoted by $\mathbf{0}$.

$\mathbb{R}^2$ is pronounced "r two."

The set of all vectors with two components is denoted by $\mathbb{R}^2$ (where $\mathbb{R}$ denotes the set of real numbers from which the components of vectors in $\mathbb{R}^2$ are chosen). Thus, $[-1, 3.5]$, $[\sqrt{2}, \pi]$, and $[\frac{5}{3}, 4]$ are all in $\mathbb{R}^2$.

Thinking back to the racetrack game, let's try to connect all of these ideas to vectors whose tails are not at the origin. The etymological origin of the word *vector* in the verb "to carry" provides a clue. The vector $[3, 2]$ may be interpreted as follows: Starting at the origin O, travel 3 units to the right, then 2 units up, finishing at P. The same displacement may be applied with other initial points. Figure 1.4 shows two equivalent displacements, represented by the vectors $\overrightarrow{AB}$ and $\overrightarrow{CD}$.

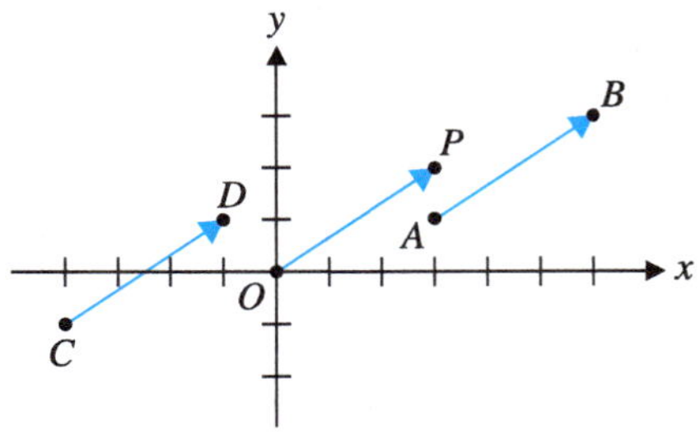

Figure 1.4

When vectors are referred to by their coordinates, they are being considered *analytically*.

We define two vectors as *equal* if they have the same length and the same direction. Thus, $\overrightarrow{AB} = \overrightarrow{CD}$ in Figure 1.4. (Even though they have different initial and terminal points, they represent the same displacement.) Geometrically, two vectors are equal if one can be obtained by sliding (or *translating*) the other parallel to itself until the two vectors coincide. In terms of components, in Figure 1.4 we have $A = (3, 1)$ and $B = (6, 3)$. Notice that the vector $[3, 2]$ that records the displacement is just the difference of the respective components:

$$\overrightarrow{AB} = [3, 2] = [6 - 3, 3 - 1]$$

Similarly,

$$\overrightarrow{CD} = [-1 - (-4), 1 - (-1)] = [3, 2]$$

and thus $\overrightarrow{AB} = \overrightarrow{CD}$, as expected.

A vector such as $\overrightarrow{OP}$ with its tail at the origin is said to be in ***standard position***. The foregoing discussion shows that every vector can be drawn as a vector in standard position. Conversely, a vector in standard position can be redrawn (by translation) so that its tail is at any point in the plane.

Example 1.1

If $A = (-1, 2)$ and $B = (3, 4)$, find $\overrightarrow{AB}$ and redraw it (a) in standard position and (b) with its tail at the point $C = (2, -1)$.

Solution We compute $\overrightarrow{AB} \approx [3 - (-1), 4 - 2] = [4, 2]$. If $\overrightarrow{AB}$ is then translated to $\overrightarrow{CD}$, where $C = (2, -1)$, then we must have $D = (2 + 4, -1 + 2) = (6, 1)$. (See Figure 1.5.)

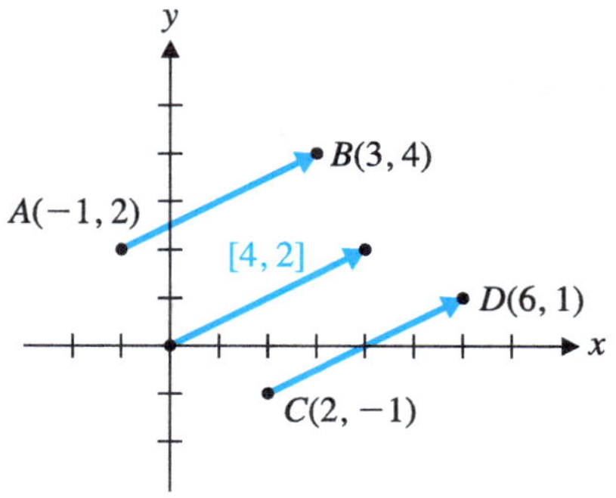

Figure 1.5

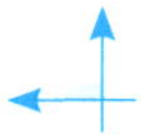

New Vectors from Old

As in the racetrack game, we often want to "follow" one vector by another. This leads to the notion of ***vector addition,*** the first basic vector operation.

If we follow $\mathbf{u}$ by $\mathbf{v}$, we can visualize the total displacement as a third vector, denoted by $\mathbf{u} + \mathbf{v}$. In Figure 1.6, $\mathbf{u} = [1, 2]$ and $\mathbf{v} = [2, 2]$, so the net effect of following $\mathbf{u}$ by $\mathbf{v}$ is

$$[1 + 2, 2 + 2] = [3, 4]$$

which gives $\mathbf{u} + \mathbf{v}$. In general, if $\mathbf{u} = [u_1, u_2]$ and $\mathbf{v} = [v_1, v_2]$, then their ***sum*** $\mathbf{u} + \mathbf{v}$ is the vector

$$\mathbf{u} + \mathbf{v} = [u_1 + v_1, u_2 + v_2]$$

It is helpful to visualize $\mathbf{u} + \mathbf{v}$ geometrically. The following rule is the geometric version of the foregoing discussion.

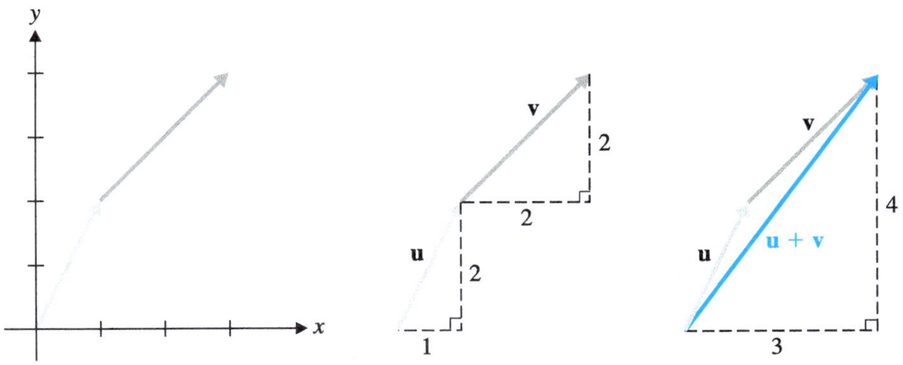

Figure 1.6
Vector addition

The Head-to-Tail Rule

Given vectors **u** and **v** in $\mathbb{R}^2$, translate **v** so that its tail coincides with the head of **u**. The ***sum*** $\mathbf{u} + \mathbf{v}$ of **u** and **v** is the vector from the tail of **u** to the head of **v**. (See Figure 1.7.)

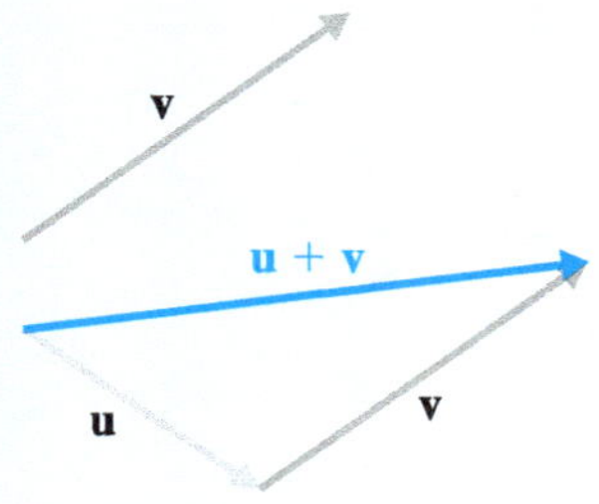

Figure 1.7
The head-to-tail rule

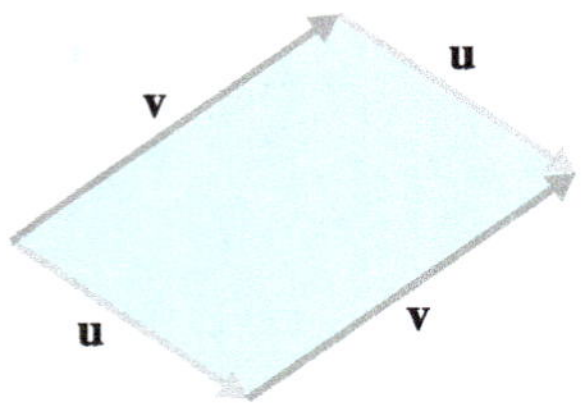

Figure 1.8
The parallelogram determined by **u** and **v**

By translating **u** and **v** parallel to themselves, we obtain a parallelogram, as shown in Figure 1.8. This parallelogram is called the *parallelogram determined by* **u** *and* **v**. It leads to an equivalent version of the head-to-tail rule for vectors in standard position.

The Parallelogram Rule

Given vectors **u** and **v** in $\mathbb{R}^2$ (in standard position), their ***sum*** $\mathbf{u} + \mathbf{v}$ is the vector in standard position along the diagonal of the parallelogram determined by **u** and **v**. (See Figure 1.9.)

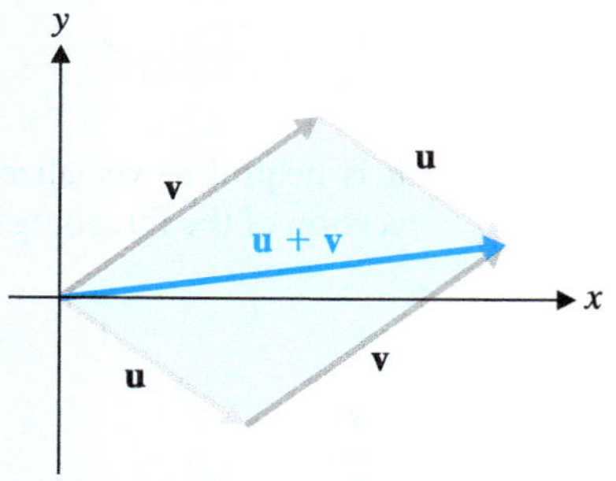

Figure 1.9
The parallelogram rule

Example 1.2

If $\mathbf{u} = [3, -1]$ and $\mathbf{v} = [1, 4]$, compute and draw $\mathbf{u} + \mathbf{v}$.

Solution We compute $\mathbf{u} + \mathbf{v} = [3 + 1, -1 + 4] = [4, 3]$. This vector is drawn using the head-to-tail rule in Figure 1.10(a) and using the parallelogram rule in Figure 1.10(b).

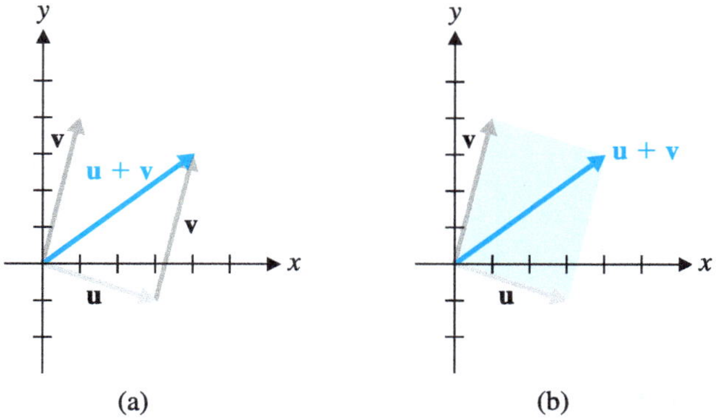

Figure 1.10

The second basic vector operation is ***scalar multiplication.*** Given a vector **v** and a real number c, the ***scalar multiple*** $c\mathbf{v}$ is the vector obtained by multiplying each component of **v** by c. For example, $3[-2, 4] = [-6, 12]$. In general,

$$c\mathbf{v} = c\,[v_1, v_2] = [cv_1, cv_2]$$

Geometrically, $c\mathbf{v}$ is a "scaled" version of **v**.

Example 1.3

If $\mathbf{v} = [-2, 4]$, compute and draw $2\mathbf{v}$, $\frac{1}{2}\mathbf{v}$, and $-2\mathbf{v}$.

Solution We calculate as follows:

$$\begin{aligned} 2\mathbf{v} &= [2(-2), 2(4)] = [-4, 8] \\ \tfrac{1}{2}\mathbf{v} &= [\tfrac{1}{2}(-2), \tfrac{1}{2}(4)] = [-1, 2] \\ -2\mathbf{v} &= [-2(-2), -2(4)] = [4, -8] \end{aligned}$$

These vectors are shown in Figure 1.11.

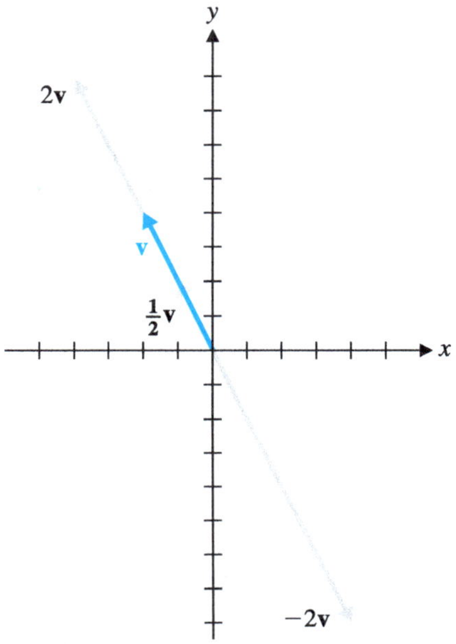

Figure 1.11

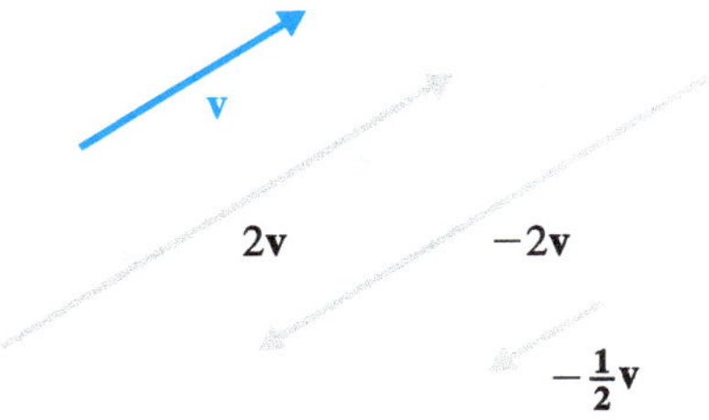

Figure 1.12

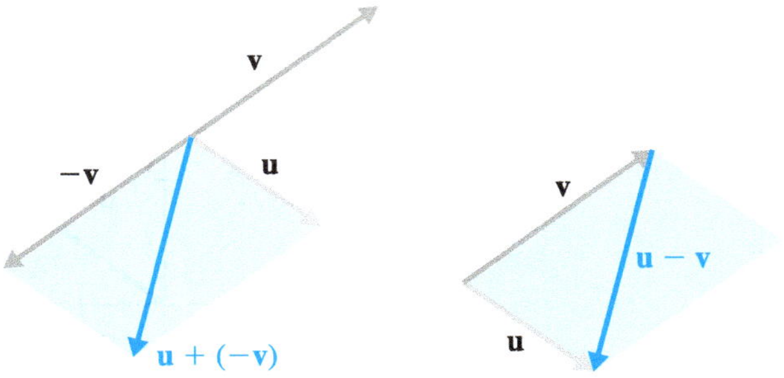

Figure 1.13
Vector subtraction

The term *scalar* comes from the Latin word *scala,* meaning "ladder." The equally spaced rungs on a ladder suggest a scale, and in vector arithmetic, multiplication by a constant changes only the scale (or length) of a vector. Thus, constants became known as scalars.

Observe that $c\mathbf{v}$ has the same direction as $\mathbf{v}$ if $c > 0$ and the opposite direction if $c < 0$. We also see that $c\mathbf{v}$ is $|c|$ times as long as $\mathbf{v}$. For this reason, in the context of vectors, constants (i.e., real numbers) are referred to as ***scalars.*** As Figure 1.12 shows, when translation of vectors is taken into account, two vectors are scalar multiples of each other if and only if they are ***parallel.***

A special case of a scalar multiple is $(-1)\mathbf{v}$, which is written as $-\mathbf{v}$ and is called the ***negative of*** $\mathbf{v}$. We can use it to define ***vector subtraction:*** The ***difference*** of $\mathbf{u}$ and $\mathbf{v}$ is the vector $\mathbf{u} - \mathbf{v}$ defined by

$$\mathbf{u} - \mathbf{v} = \mathbf{u} + (-\mathbf{v})$$

Figure 1.13 shows that $\mathbf{u} - \mathbf{v}$ corresponds to the "other" diagonal of the parallelogram determined by $\mathbf{u}$ and $\mathbf{v}$.

Example 1.4

If $\mathbf{u} = [1, 2]$ and $\mathbf{v} = [-3, 1]$, then $\mathbf{u} - \mathbf{v} = [1 - (-3), 2 - 1] = [4, 1]$.

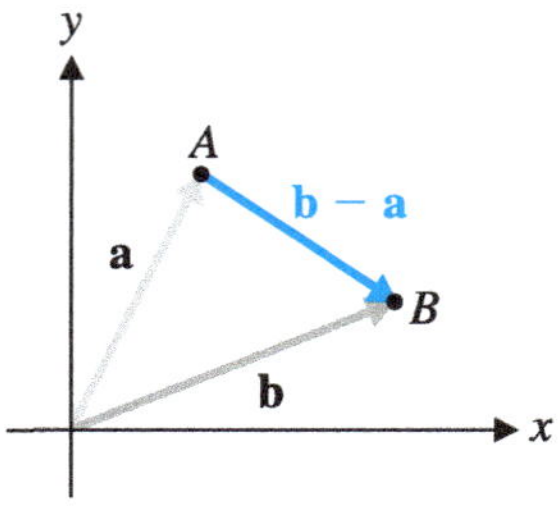

Figure 1.14

The definition of subtraction in Example 1.4 also agrees with the way we calculate a vector such as $\overrightarrow{AB}$. If the points A and B correspond to the vectors $\mathbf{a}$ and $\mathbf{b}$ in standard position, then $\overrightarrow{AB} = \mathbf{b} - \mathbf{a}$, as shown in Figure 1.14. [Observe that the head-to-tail rule applied to this diagram gives the equation $\mathbf{a} + (\mathbf{b} - \mathbf{a}) = \mathbf{b}$. If we had accidentally drawn $\mathbf{b} - \mathbf{a}$ with its head at A instead of at B, the diagram would have read $\mathbf{b} + (\mathbf{b} - \mathbf{a}) = \mathbf{a}$, which is clearly wrong! More will be said about algebraic expressions involving vectors later in this section.]

Vectors in $\mathbb{R}^3$

Everything we have just done extends easily to three dimensions. The set of all *ordered triples* of real numbers is denoted by $\mathbb{R}^3$. Points and vectors are located using three mutually perpendicular coordinate axes that meet at the origin O. A point such as $A = (1, 2, 3)$ can be located as follows: First travel 1 unit along the x-axis, then move 2 units parallel to the y-axis, and finally move 3 units parallel to the z-axis. The corresponding vector $\mathbf{a} = [1, 2, 3]$ is then $\overrightarrow{OA}$, as shown in Figure 1.15.

Another way to visualize vector $\mathbf{a}$ in $\mathbb{R}^3$ is to construct a box whose six sides are determined by the three coordinate planes (the xy-, xz-, and yz-planes) and by three planes through the point $(1, 2, 3)$ parallel to the coordinate planes. The vector $[1, 2, 3]$ then corresponds to the diagonal from the origin to the opposite corner of the box (see Figure 1.16).

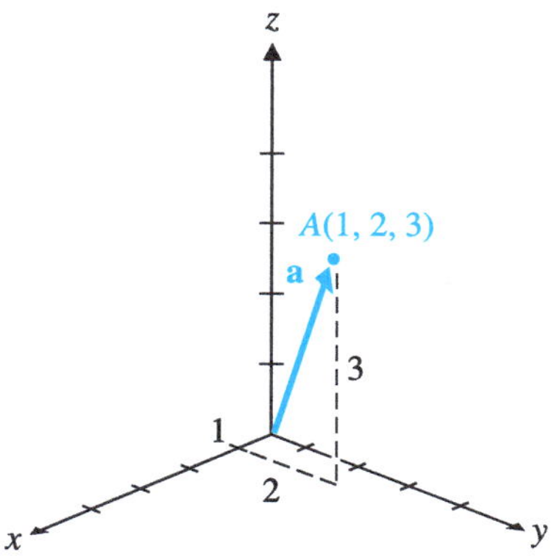

Figure 1.15

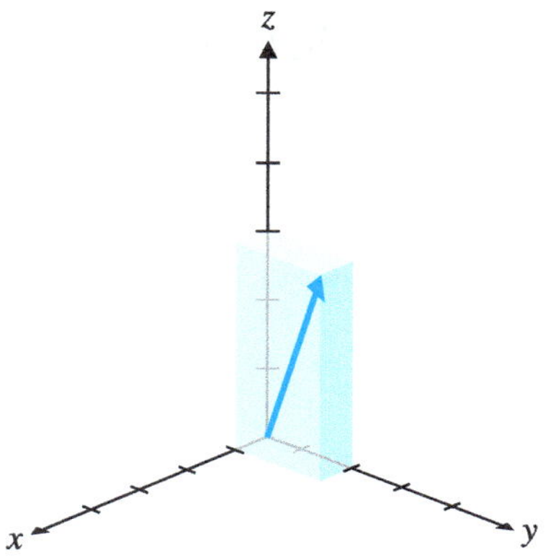

Figure 1.16

The "componentwise" definitions of vector addition and scalar multiplication are extended to $\mathbb{R}^3$ in an obvious way.

Vectors in $\mathbb{R}^n$

In general, we define $\mathbb{R}^n$ as the set of all *ordered n-tuples* of real numbers written as row or column vectors. Thus, a vector $\mathbf{v}$ in $\mathbb{R}^n$ is of the form

$$[v_1, v_2, \ldots, v_n] \quad \text{or} \quad \begin{bmatrix} v_1 \\ v_2 \\ \vdots \\ v_n \end{bmatrix}$$

The individual entries of $\mathbf{v}$ are its components; v_i is called the ith component.

We extend the definitions of vector addition and scalar multiplication to $\mathbb{R}^n$ in the obvious way: If $\mathbf{u} = [u_1, u_2, \ldots, u_n]$ and $\mathbf{v} = [v_1, v_2, \ldots, v_n]$, the ith component of $\mathbf{u} + \mathbf{v}$ is $u_i + v_i$ and the ith component of $c\mathbf{v}$ is just cv_i.

Since in $\mathbb{R}^n$ we can no longer draw pictures of vectors, it is important to be able to calculate with vectors. We must be careful not to assume that vector arithmetic will be similar to the arithmetic of real numbers. Often it is, and the algebraic calculations we do with vectors are similar to those we would do with scalars. But, in later sections, we will encounter situations where vector algebra is quite *unlike* our previous experience with real numbers. So it is important to verify any algebraic properties before attempting to use them.

One such property is *commutativity* of addition: $\mathbf{u} + \mathbf{v} = \mathbf{v} + \mathbf{u}$ for vectors $\mathbf{u}$ and $\mathbf{v}$. This is certainly true in $\mathbb{R}^2$. Geometrically, the head-to-tail rule shows that both $\mathbf{u} + \mathbf{v}$ and $\mathbf{v} + \mathbf{u}$ are the main diagonals of the parallelogram determined by $\mathbf{u}$ and $\mathbf{v}$. (The parallelogram rule also reflects this symmetry; see Figure 1.17.)

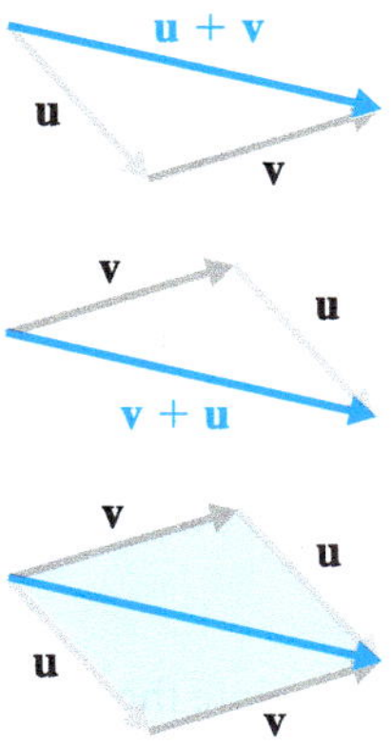

Figure 1.17
$\mathbf{u} + \mathbf{v} = \mathbf{v} + \mathbf{u}$

Note that Figure 1.17 is simply an illustration of the property $\mathbf{u} + \mathbf{v} = \mathbf{v} + \mathbf{u}$. It is not a proof, since it does not cover every possible case. For example, we must also include the cases where $\mathbf{u} = \mathbf{v}$, $\mathbf{u} = -\mathbf{v}$, and $\mathbf{u} = \mathbf{0}$. (What would diagrams for these cases look like?) For this reason, an algebraic proof is needed. However, it is just as easy to give a proof that is valid in $\mathbb{R}^n$ as to give one that is valid in $\mathbb{R}^2$.

The following theorem summarizes the algebraic properties of vector addition and scalar multiplication in $\mathbb{R}^n$. The proofs follow from the corresponding properties of real numbers.

Theorem 1.1 **Algebraic Properties of Vectors in $\mathbb{R}^n$**

Let **u, v,** and **w** be vectors in $\mathbb{R}^n$ and let c and d be scalars. Then

a. $\mathbf{u} + \mathbf{v} = \mathbf{v} + \mathbf{u}$ Commutativity
b. $(\mathbf{u} + \mathbf{v}) + \mathbf{w} = \mathbf{u} + (\mathbf{v} + \mathbf{w})$ Associativity
c. $\mathbf{u} + \mathbf{0} = \mathbf{u}$
d. $\mathbf{u} + (-\mathbf{u}) = \mathbf{0}$
e. $c(\mathbf{u} + \mathbf{v}) = c\mathbf{u} + c\mathbf{v}$ Distributivity
f. $(c + d)\mathbf{u} = c\mathbf{u} + d\mathbf{u}$ Distributivity
g. $c(d\mathbf{u}) = (cd)\mathbf{u}$
h. $1\mathbf{u} = \mathbf{u}$

The word *theorem* is derived from the Greek word *theorema,* which in turn comes from a word meaning "to look at." Thus, a theorem is based on the insights we have when we look at examples and extract from them properties that we try to prove hold in general. Similarly, when we understand something in mathematics—the proof of a theorem, for example—we often say, "I see."

Remarks

- Properties (c) and (d) together with the commutativity property (a) imply that $\mathbf{0} + \mathbf{u} = \mathbf{u}$ and $-\mathbf{u} + \mathbf{u} = \mathbf{0}$ as well.
- If we read the distributivity properties (e) and (f) from right to left, they say that we can *factor* a common scalar or a common vector from a sum.

Proof We prove properties (a) and (b) and leave the proofs of the remaining properties as exercises. Let $\mathbf{u} = [u_1, u_2, \ldots, u_n]$, $\mathbf{v} = [v_1, v_2, \ldots, v_n]$, and $\mathbf{w} = [w_1, w_2, \ldots, w_n]$.

(a)
$$\begin{aligned}\mathbf{u} + \mathbf{v} &= [u_1, u_2, \ldots, u_n] + [v_1, v_2, \ldots, v_n] \\ &= [u_1 + v_1, u_2 + v_2, \ldots, u_n + v_n] \\ &= [v_1 + u_1, v_2 + u_2, \ldots, v_n + u_n] \\ &= [v_1, v_2, \ldots, v_n] + [u_1, u_2, \ldots, u_n] \\ &= \mathbf{v} + \mathbf{u}\end{aligned}$$

The second and fourth equalities are by the definition of vector addition, and the third equality is by the commutativity of addition of real numbers.

(b) Figure 1.18 illustrates associativity in $\mathbb{R}^2$. Algebraically, we have

$$\begin{aligned}(\mathbf{u} + \mathbf{v}) + \mathbf{w} &= ([u_1, u_2, \ldots, u_n] + [v_1, v_2, \ldots, v_n]) + [w_1, w_2, \ldots, w_n] \\ &= [u_1 + v_1, u_2 + v_2, \ldots, u_n + v_n] + [w_1, w_2, \ldots, w_n] \\ &= [(u_1 + v_1) + w_1, (u_2 + v_2) + w_2, \ldots, (u_n + v_n) + w_n] \\ &= [u_1 + (v_1 + w_1), u_2 + (v_2 + w_2), \ldots, u_n + (v_n + w_n)] \\ &= [u_1, u_2, \ldots, u_n] + [v_1 + w_1, v_2 + w_2, \ldots, v_n + w_n] \\ &= [u_1, u_2, \ldots, u_n] + ([v_1, v_2, \ldots, v_n] + [w_1, w_2, \ldots, w_n]) \\ &= \mathbf{u} + (\mathbf{v} + \mathbf{w})\end{aligned}$$

The fourth equality is by the associativity of addition of real numbers. Note the careful use of parentheses.

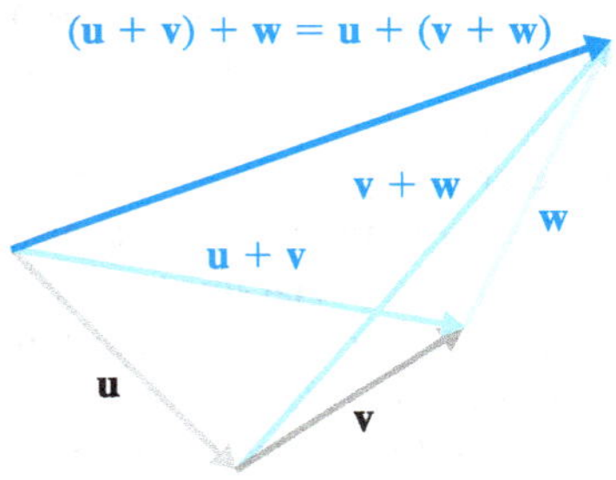

Figure 1.18

By property (b) of Theorem 1.1, we may unambiguously write $\mathbf{u} + \mathbf{v} + \mathbf{w}$ without parentheses, since we may group the summands in whichever way we please. By (a), we may also rearrange the summands—for example, as $\mathbf{w} + \mathbf{u} + \mathbf{v}$—if we choose. Likewise, sums of four or more vectors can be calculated without regard to order or grouping. In general, if $\mathbf{v}_1, \mathbf{v}_2, \ldots, \mathbf{v}_k$ are vectors in $\mathbb{R}^n$, we will write such sums without parentheses:

$$\mathbf{v}_1 + \mathbf{v}_2 + \cdots + \mathbf{v}_k$$

The next example illustrates the use of Theorem 1.1 in performing algebraic calculations with vectors.

Example 1.5

Let **a**, **b**, and **x** denote vectors in $\mathbb{R}^n$.

(a) Simplify $3\mathbf{a} + (5\mathbf{b} - 2\mathbf{a}) + 2(\mathbf{b} - \mathbf{a})$.

(b) If $5\mathbf{x} - \mathbf{a} = 2(\mathbf{a} + 2\mathbf{x})$, solve for **x** in terms of **a**.

Solution We will give both solutions in detail, with reference to all of the properties in Theorem 1.1 that we use. It is good practice to justify all steps the first few times you do this type of calculation. Once you are comfortable with the vector properties, though, it is acceptable to leave out some of the intermediate steps to save time and space.

(a) We begin by inserting parentheses.

$$\begin{aligned}
3\mathbf{a} + (5\mathbf{b} - 2\mathbf{a}) + 2(\mathbf{b} - \mathbf{a}) &= (3\mathbf{a} + (5\mathbf{b} - 2\mathbf{a})) + 2(\mathbf{b} - \mathbf{a}) \\
&= (3\mathbf{a} + (-2\mathbf{a} + 5\mathbf{b})) + (2\mathbf{b} - 2\mathbf{a}) && \text{(a), (e)} \\
&= ((3\mathbf{a} + (-2\mathbf{a})) + 5\mathbf{b}) + (2\mathbf{b} - 2\mathbf{a}) && \text{(b)} \\
&= ((3 + (-2))\mathbf{a} + 5\mathbf{b}) + (2\mathbf{b} - 2\mathbf{a}) && \text{(f)} \\
&= (1\mathbf{a} + 5\mathbf{b}) + (2\mathbf{b} - 2\mathbf{a}) \\
&= ((\mathbf{a} + 5\mathbf{b}) + 2\mathbf{b}) - 2\mathbf{a} && \text{(b), (h)} \\
&= (\mathbf{a} + (5\mathbf{b} + 2\mathbf{b})) - 2\mathbf{a} && \text{(b)} \\
&= (\mathbf{a} + (5 + 2)\mathbf{b}) - 2\mathbf{a} && \text{(f)} \\
&= (7\mathbf{b} + \mathbf{a}) - 2\mathbf{a} && \text{(a)} \\
&= 7\mathbf{b} + (\mathbf{a} - 2\mathbf{a}) && \text{(b)} \\
&= 7\mathbf{b} + (1 - 2)\mathbf{a} && \text{(f), (h)} \\
&= 7\mathbf{b} + (-1)\mathbf{a} \\
&= 7\mathbf{b} - \mathbf{a}
\end{aligned}$$

You can see why we will agree to omit some of these steps! In practice, it is acceptable to simplify this sequence of steps as

$$\begin{aligned}
3\mathbf{a} + (5\mathbf{b} - 2\mathbf{a}) + 2(\mathbf{b} - \mathbf{a}) &= 3\mathbf{a} + 5\mathbf{b} - 2\mathbf{a} + 2\mathbf{b} - 2\mathbf{a} \\
&= (3\mathbf{a} - 2\mathbf{a} - 2\mathbf{a}) + (5\mathbf{b} + 2\mathbf{b}) \\
&= -\mathbf{a} + 7\mathbf{b}
\end{aligned}$$

or even to do most of the calculation mentally.

(b) In detail, we have

$$\begin{aligned}
5\mathbf{x} - \mathbf{a} &= 2(\mathbf{a} + 2\mathbf{x}) && \\
5\mathbf{x} - \mathbf{a} &= 2\mathbf{a} + 2(2\mathbf{x}) && \text{(e)} \\
5\mathbf{x} - \mathbf{a} &= 2\mathbf{a} + (2 \cdot 2)\mathbf{x} && \text{(g)} \\
5\mathbf{x} - \mathbf{a} &= 2\mathbf{a} + 4\mathbf{x} && \\
(5\mathbf{x} - \mathbf{a}) - 4\mathbf{x} &= (2\mathbf{a} + 4\mathbf{x}) - 4\mathbf{x} && \\
(-\mathbf{a} + 5\mathbf{x}) - 4\mathbf{x} &= 2\mathbf{a} + (4\mathbf{x} - 4\mathbf{x}) && \text{(a), (b)} \\
-\mathbf{a} + (5\mathbf{x} - 4\mathbf{x}) &= 2\mathbf{a} + \mathbf{0} && \text{(b), (d)} \\
-\mathbf{a} + (5 - 4)\mathbf{x} &= 2\mathbf{a} && \text{(f), (c)} \\
-\mathbf{a} + (1)\mathbf{x} &= 2\mathbf{a} && \\
\mathbf{a} + (-\mathbf{a} + \mathbf{x}) &= \mathbf{a} + 2\mathbf{a} && \text{(h)} \\
(\mathbf{a} + (-\mathbf{a})) + \mathbf{x} &= (1 + 2)\mathbf{a} && \text{(b), (f)} \\
\mathbf{0} + \mathbf{x} &= 3\mathbf{a} && \text{(d)} \\
\mathbf{x} &= 3\mathbf{a} && \text{(c)}
\end{aligned}$$

Again, we will usually omit most of these steps.

Linear Combinations and Coordinates

A vector that is a sum of scalar multiples of other vectors is said to be a *linear combination* of those vectors. The formal definition follows.

Definition A vector $\mathbf{v}$ is a ***linear combination*** of vectors $\mathbf{v}_1, \mathbf{v}_2, \ldots, \mathbf{v}_k$ if there are scalars $c_1, c_2, \ldots, c_k$ such that $\mathbf{v} = c_1\mathbf{v}_1 + c_2\mathbf{v}_2 + \cdots + c_k\mathbf{v}_k$. The scalars $c_1, c_2, \ldots, c_k$ are called the ***coefficients*** of the linear combination.

Example 1.6

The vector $\begin{bmatrix} 2 \\ -2 \\ -1 \end{bmatrix}$ is a linear combination of $\begin{bmatrix} 1 \\ 0 \\ -1 \end{bmatrix}$, $\begin{bmatrix} 2 \\ -3 \\ 1 \end{bmatrix}$, and $\begin{bmatrix} 5 \\ -4 \\ 0 \end{bmatrix}$, since

$$3\begin{bmatrix} 1 \\ 0 \\ -1 \end{bmatrix} + 2\begin{bmatrix} 2 \\ -3 \\ 1 \end{bmatrix} - \begin{bmatrix} 5 \\ -4 \\ 0 \end{bmatrix} = \begin{bmatrix} 2 \\ -2 \\ -1 \end{bmatrix}$$

Remark Determining whether a given vector is a linear combination of other vectors is a problem we will address in Chapter 2.

In $\mathbb{R}^2$, it is possible to depict linear combinations of two (nonparallel) vectors quite conveniently.

Example 1.7

Let $\mathbf{u} = \begin{bmatrix} 3 \\ 1 \end{bmatrix}$ and $\mathbf{v} = \begin{bmatrix} 1 \\ 2 \end{bmatrix}$. We can use $\mathbf{u}$ and $\mathbf{v}$ to locate a new set of axes (in the same way that $\mathbf{e}_1 = \begin{bmatrix} 1 \\ 0 \end{bmatrix}$ and $\mathbf{e}_2 = \begin{bmatrix} 0 \\ 1 \end{bmatrix}$ locate the standard coordinate axes). We can use

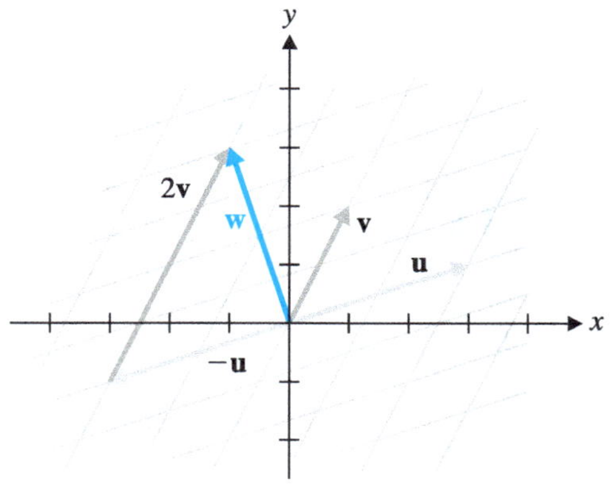

Figure 1.19

these new axes to determine a ***coordinate grid*** that will let us easily locate linear combinations of $\mathbf{u}$ and $\mathbf{v}$.

As Figure 1.19 shows, $\mathbf{w}$ can be located by starting at the origin and traveling $-\mathbf{u}$ followed by $2\mathbf{v}$. That is,

$$\mathbf{w} = -\mathbf{u} + 2\mathbf{v}$$

We say that the coordinates of $\mathbf{w}$ with respect to $\mathbf{u}$ and $\mathbf{v}$ are -1 and 2. (Note that this is just another way of thinking of the coefficients of the linear combination.) It follows that

$$\mathbf{w} = -\begin{bmatrix} 3 \\ 1 \end{bmatrix} + 2\begin{bmatrix} 1 \\ 2 \end{bmatrix} = \begin{bmatrix} -1 \\ 3 \end{bmatrix}$$

(Observe that -1 and 3 are the coordinates of $\mathbf{w}$ with respect to $\mathbf{e}_1$ and $\mathbf{e}_2$.)

Switching from the standard coordinate axes to alternative ones is a useful idea. It has applications in chemistry and geology, since molecular and crystalline structures often do not fall onto a rectangular grid. It is an idea that we will encounter repeatedly in this book.

Binary Vectors and Modular Arithmetic

We will also encounter a type of vector that has no geometric interpretation—at least not using Euclidean geometry. Computers represent data in terms of 0s and 1s (which can be interpreted as off/on, closed/open, false/true, or no/yes). ***Binary vectors*** are vectors each of whose components is a 0 or a 1. As we will see in Chapter 8, such vectors arise naturally in the study of many types of codes.

In this setting, the usual rules of arithmetic must be modified, since the result of each calculation involving scalars must be a 0 or a 1. The modified rules for addition and multiplication are given below.

+	0	1
0	0	1
1	1	0

·	0	1
0	0	0
1	0	1

The only curiosity here is the rule that $1 + 1 = 0$. This is not as strange as it appears; if we replace 0 with the word "even" and 1 with the word "odd," these tables simply summarize the familiar *parity rules* for the addition and multiplication of even and odd integers. For example, $1 + 1 = 0$ expresses the fact that the sum of two odd integers is an even integer. With these rules, our set of scalars $\{0, 1\}$ is denoted by $\mathbb{Z}_2$ and is called the set of ***integers modulo* 2.**

Example 1.8

In $\mathbb{Z}_2$, $1 + 1 + 0 + 1 = 1$ and $1 + 1 + 1 + 1 = 0$. (These calculations illustrate the parity rules again: The sum of three odds and an even is odd; the sum of four odds is even.)

We are using the term *length* differently from the way we used it in $\mathbb{R}^n$. This should not be confusing, since there is no *geometric* notion of length for binary vectors.

With $\mathbb{Z}_2$ as our set of scalars, we now extend the above rules to vectors. The set of all n-tuples of 0s and 1s (with all arithmetic performed modulo 2) is denoted by $\mathbb{Z}_2^n$. The vectors in $\mathbb{Z}_2^n$ are called ***binary vectors of length n.***

Example 1.9

The vectors in $\mathbb{Z}_2^2$ are [0, 0], [0, 1], [1, 0], and [1, 1]. (How many vectors does $\mathbb{Z}_2^n$ contain, in general?)

Example 1.10

Let $\mathbf{u} = [1, 1, 0, 1, 0]$ and $\mathbf{v} = [0, 1, 1, 1, 0]$ be two binary vectors of length 5. Find $\mathbf{u} + \mathbf{v}$.

Solution The calculation of $\mathbf{u} + \mathbf{v}$ takes place over $\mathbb{Z}_2$, so we have

$$\begin{aligned} \mathbf{u} + \mathbf{v} &= [1, 1, 0, 1, 0] + [0, 1, 1, 1, 0] \\ &= [1 + 0, 1 + 1, 0 + 1, 1 + 1, 0 + 0] \\ &= [1, 0, 1, 0, 0] \end{aligned}$$

It is possible to generalize what we have just done for binary vectors to vectors whose components are taken from a finite set $\{0, 1, 2, \ldots, k\}$ for $k \geq 2$. To do so, we must first extend the idea of binary arithmetic.

Example 1.11

The ***integers modulo* 3** is the set $\mathbb{Z}_3 = \{0, 1, 2\}$ with addition and multiplication given by the following tables:

+	0	1	2
0	0	1	2
1	1	2	0
2	2	0	1

·	0	1	2
0	0	0	0
1	0	1	2
2	0	2	1

Observe that the result of each addition and multiplication belongs to the set $\{0, 1, 2\}$; we say that $\mathbb{Z}_3$ is ***closed*** with respect to the operations of addition and multiplication. It is perhaps easiest to think of this set in terms of a 3-hour clock with 0, 1, and 2 on its face, as shown in Figure 1.20.

The calculation $1 + 2 = 0$ translates as follows: 2 hours after 1 o'clock, it is 0 o'clock. Just as 24:00 and 12:00 are the same on a 12-hour clock, so 3 and 0 are equivalent on this 3-hour clock. Likewise, all multiples of 3—positive and negative—are equivalent to 0 here; 1 is equivalent to any number that is 1 more than a multiple of 3 (such as -2, 4, and 7); and 2 is equivalent to any number that is 2 more than a multiple of 3 (such as -1, 5, and 8). We can visualize the number line as wrapping around a circle, as shown in Figure 1.21.

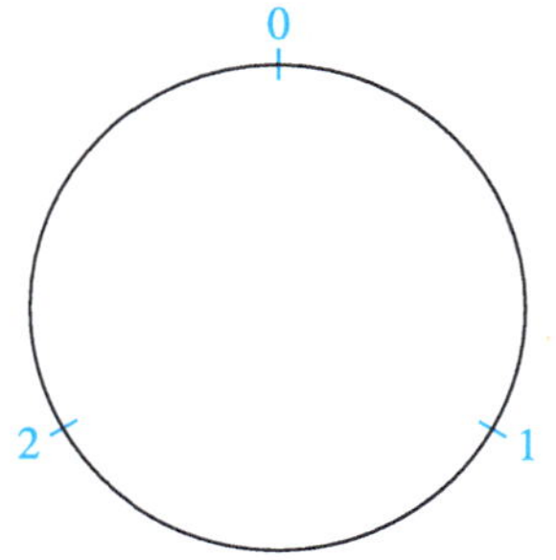

Figure 1.20
Arithmetic modulo 3

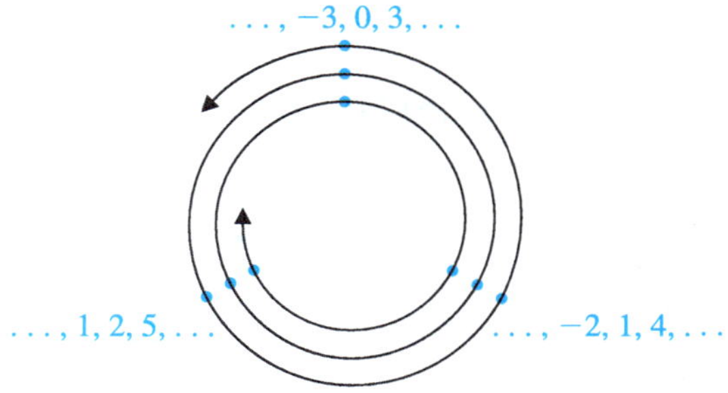

Figure 1.21

Example 1.12

To what is 3548 equivalent in $\mathbb{Z}_3$?

Solution This is the same as asking where 3548 lies on our 3-hour clock. The key is to calculate how far this number is from the nearest (smaller) multiple of 3; that is, we need to know the *remainder* when 3548 is divided by 3. By long division, we find that $3548 = 3 \cdot 1182 + 2$, so the remainder is 2. Therefore, 3548 is equivalent to 2 in $\mathbb{Z}_3$.

In courses in abstract algebra and number theory, which explore this concept in greater detail, the above equivalence is often written as $3548 = 2 \pmod 3$ or $3548 \equiv 2 \pmod 3$, where $\equiv$ is read "is congruent to." We will not use this notation or terminology here.

Example 1.13

In $\mathbb{Z}_3$, calculate $2 + 2 + 1 + 2$.

Solution 1 We use the same ideas as in Example 1.12. The ordinary sum is $2 + 2 + 1 + 2 = 7$, which is 1 more than 6, so division by 3 leaves a remainder of 1. Thus, $2 + 2 + 1 + 2 = 1$ in $\mathbb{Z}_3$.

Solution 2 A better way to perform this calculation is to do it step by step entirely in $\mathbb{Z}_3$.

$$\begin{aligned} 2 + 2 + 1 + 2 &= (2 + 2) + 1 + 2 \\ &= 1 + 1 + 2 \\ &= (1 + 1) + 2 \\ &= 2 + 2 \\ &= 1 \end{aligned}$$

Here we have used parentheses to group the terms we have chosen to combine. We could speed things up by simultaneously combining the first two and the last two terms:

$$\begin{aligned} (2 + 2) + (1 + 2) &= 1 + 0 \\ &= 1 \end{aligned}$$

Repeated multiplication can be handled similarly. The idea is to use the addition and multiplication tables to reduce the result of each calculation to 0, 1, or 2.

Extending these ideas to vectors is straightforward.

Example 1.14

In $\mathbb{Z}_3^5$, let $\mathbf{u} = [2, 2, 0, 1, 2]$ and $\mathbf{v} = [1, 2, 2, 2, 1]$. Then

$$\begin{aligned} \mathbf{u} + \mathbf{v} &= [2, 2, 0, 1, 2] + [1, 2, 2, 2, 1] \\ &= [2 + 1, 2 + 2, 0 + 2, 1 + 2, 2 + 1] \\ &= [0, 1, 2, 0, 0] \end{aligned}$$

Vectors in $\mathbb{Z}_3^5$ are referred to as ***ternary vectors of length* 5.**

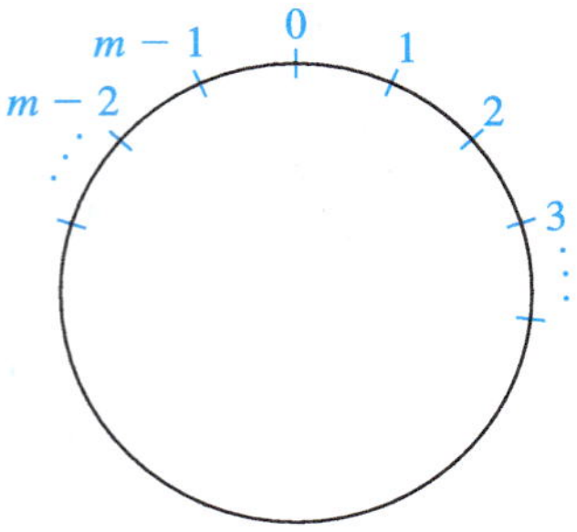

Figure 1.22
Arithmetic modulo m

In general, we have the set $\mathbb{Z}_m = \{0, 1, 2, \ldots, m - 1\}$ of ***integers modulo m*** (corresponding to an m-hour clock, as shown in Figure 1.22). A vector of length n whose entries are in $\mathbb{Z}_m$ is called an ***m-ary vector of length n***. The set of all m-ary vectors of length n is denoted by $\mathbb{Z}_m^n$.

Exercises 1.1

1. Draw the following vectors in standard position in $\mathbb{R}^2$:

(a) $\mathbf{a} = \begin{bmatrix} 3 \\ 0 \end{bmatrix}$ **(b)** $\mathbf{b} = \begin{bmatrix} 2 \\ 3 \end{bmatrix}$

(c) $\mathbf{c} = \begin{bmatrix} -2 \\ 3 \end{bmatrix}$ **(d)** $\mathbf{d} = \begin{bmatrix} 3 \\ -2 \end{bmatrix}$

2. Draw the vectors in Exercise 1 with their tails at the point $(2, -3)$.

3. Draw the following vectors in standard position in $\mathbb{R}^3$:

(a) $\mathbf{a} = [0, 2, 0]$ **(b)** $\mathbf{b} = [3, 2, 1]$
(c) $\mathbf{c} = [1, -2, 1]$ **(d)** $\mathbf{d} = [-1, -1, -2]$

4. If the vectors in Exercise 3 are translated so that their heads are at the point $(3, 2, 1)$, find the points that correspond to their tails.

5. For each of the following pairs of points, draw the vector $\overrightarrow{AB}$. Then compute and redraw $\overrightarrow{AB}$ as a vector in standard position.

(a) $A = (1, -1), B = (4, 2)$
(b) $A = (0, -2), B = (2, -1)$
(c) $A = (2, \frac{3}{2}), B = (\frac{1}{2}, 3)$
(d) $A = (\frac{1}{3}, \frac{1}{3}), B = (\frac{1}{6}, \frac{1}{2})$

6. A hiker walks 4 km north and then 5 km northeast. Draw displacement vectors representing the hiker's trip and draw a vector that represents the hiker's net displacement from the starting point.

Exercises 7–10 refer to the vectors in Exercise 1. Compute the indicated vectors and also show how the results can be obtained geometrically.

7. $\mathbf{a} + \mathbf{b}$ **8.** $\mathbf{b} - \mathbf{c}$
9. $\mathbf{d} - \mathbf{c}$ **10.** $\mathbf{a} + \mathbf{d}$

Exercises 11 and 12 refer to the vectors in Exercise 3. Compute the indicated vectors.

11. $2\mathbf{a} + 3\mathbf{c}$ **12.** $3\mathbf{b} - 2\mathbf{c} + \mathbf{d}$

13. Find the components of the vectors $\mathbf{u}$, $\mathbf{v}$, $\mathbf{u} + \mathbf{v}$, and $\mathbf{u} - \mathbf{v}$, where $\mathbf{u}$ and $\mathbf{v}$ are as shown in Figure 1.23.

14. In Figure 1.24, A, B, C, D, E, and F are the vertices of a regular hexagon centered at the origin.

Express each of the following vectors in terms of $\mathbf{a} = \overrightarrow{OA}$ and $\mathbf{b} = \overrightarrow{OB}$:

(a) $\overrightarrow{AB}$ **(b)** $\overrightarrow{BC}$
(c) $\overrightarrow{AD}$ **(d)** $\overrightarrow{CF}$
(e) $\overrightarrow{AC}$ **(f)** $\overrightarrow{BC} + \overrightarrow{DE} + \overrightarrow{FA}$

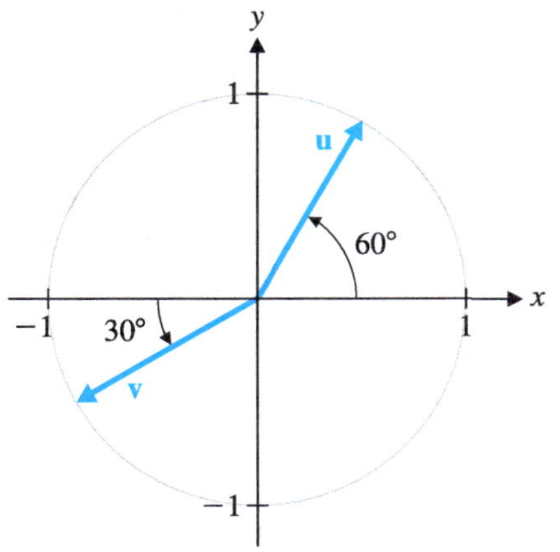

Figure 1.23

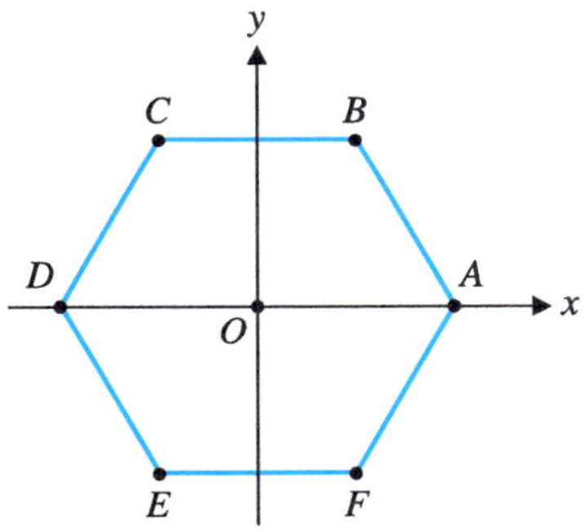

Figure 1.24

In Exercises 15 and 16, simplify the given vector expression. Indicate which properties in Theorem 1.1 you use.

15. $2(\mathbf{a} - 3\mathbf{b}) + 3(2\mathbf{b} + \mathbf{a})$

16. $-3(\mathbf{a} - \mathbf{c}) + 2(\mathbf{a} + 2\mathbf{b}) + 3(\mathbf{c} - \mathbf{b})$

In Exercises 17 and 18, solve for the vector **x** *in terms of the vectors* **a** *and* **b**.

17. $\mathbf{x} - \mathbf{a} = 2(\mathbf{x} - 2\mathbf{a})$

18. $\mathbf{x} + 2\mathbf{a} - \mathbf{b} = 3(\mathbf{x} + \mathbf{a}) - 2(2\mathbf{a} - \mathbf{b})$

In Exercises 19 and 20, draw the coordinate axes relative to **u** *and* **v** *and locate* **w**.

19. $\mathbf{u} = \begin{bmatrix} 1 \\ -1 \end{bmatrix}, \mathbf{v} = \begin{bmatrix} 1 \\ 1 \end{bmatrix}, \mathbf{w} = 2\mathbf{u} + 3\mathbf{v}$

20. $\mathbf{u} = \begin{bmatrix} -2 \\ 1 \end{bmatrix}, \mathbf{v} = \begin{bmatrix} 2 \\ -2 \end{bmatrix}, \mathbf{w} = -\mathbf{u} - 2\mathbf{v}$

In Exercises 21 and 22, draw the standard coordinate axes on the same diagram as the axes relative to **u** *and* **v**. *Use these to find* **w** *as a linear combination of* **u** *and* **v**.

21. $\mathbf{u} = \begin{bmatrix} 1 \\ -1 \end{bmatrix}, \mathbf{v} = \begin{bmatrix} 1 \\ 1 \end{bmatrix}, \mathbf{w} = \begin{bmatrix} 2 \\ 6 \end{bmatrix}$

22. $\mathbf{u} = \begin{bmatrix} -2 \\ 3 \end{bmatrix}, \mathbf{v} = \begin{bmatrix} 2 \\ 1 \end{bmatrix}, \mathbf{w} = \begin{bmatrix} 2 \\ 9 \end{bmatrix}$

23. Draw diagrams to illustrate properties (d) and (e) of Theorem 1.1.

24. Give algebraic proofs of properties (d) through (g) of Theorem 1.1.

In Exercises 25–28, **u** *and* **v** *are binary vectors. Find* $\mathbf{u} + \mathbf{v}$ *in each case.*

25. $\mathbf{u} = \begin{bmatrix} 0 \\ 1 \end{bmatrix}, \mathbf{v} = \begin{bmatrix} 1 \\ 1 \end{bmatrix}$ **26.** $\mathbf{u} = \begin{bmatrix} 1 \\ 1 \\ 0 \end{bmatrix}, \mathbf{v} = \begin{bmatrix} 1 \\ 1 \\ 1 \end{bmatrix}$

27. $\mathbf{u} = [1, 0, 1, 1], \mathbf{v} = [1, 1, 1, 1]$

28. $\mathbf{u} = [1, 1, 0, 1, 0], \mathbf{v} = [0, 1, 1, 1, 0]$

29. Write out the addition and multiplication tables for $\mathbb{Z}_4$.

30. Write out the addition and multiplication tables for $\mathbb{Z}_5$.

In Exercises 31–43, perform the indicated calculations.

31. $2 + 2 + 2$ in $\mathbb{Z}_3$ **32.** $2 \cdot 2 \cdot 2$ in $\mathbb{Z}_3$

33. $2(2 + 1 + 2)$ in $\mathbb{Z}_3$ **34.** $3 + 1 + 2 + 3$ in $\mathbb{Z}_4$

35. $2 \cdot 3 \cdot 2$ in $\mathbb{Z}_4$ **36.** $3(3 + 3 + 2)$ in $\mathbb{Z}_4$

37. $2 + 1 + 2 + 2 + 1$ in $\mathbb{Z}_3$, $\mathbb{Z}_4$, and $\mathbb{Z}_5$

38. $(3 + 4)(3 + 2 + 4 + 2)$ in $\mathbb{Z}_5$

39. $8(6 + 4 + 3)$ in $\mathbb{Z}_9$ **40.** 2^{100} in $\mathbb{Z}_{11}$

41. $[2, 1, 2] + [2, 0, 1]$ in $\mathbb{Z}_3^3$ **42.** $2[2, 2, 1]$ in $\mathbb{Z}_3^3$

43. $2([3, 1, 1, 2] + [3, 3, 2, 1])$ in $\mathbb{Z}_4^4$ and $\mathbb{Z}_5^4$

In Exercises 44–55, solve the given equation or indicate that there is no solution.

44. $x + 3 = 2$ in $\mathbb{Z}_5$ **45.** $x + 5 = 1$ in $\mathbb{Z}_6$

46. $2x = 1$ in $\mathbb{Z}_3$ **47.** $2x = 1$ in $\mathbb{Z}_4$

48. $2x = 1$ in $\mathbb{Z}_5$ **49.** $3x = 4$ in $\mathbb{Z}_5$

50. $3x = 4$ in $\mathbb{Z}_6$ **51.** $6x = 5$ in $\mathbb{Z}_8$

52. $8x = 9$ in $\mathbb{Z}_{11}$ **53.** $2x + 3 = 2$ in $\mathbb{Z}_5$

54. $4x + 5 = 2$ in $\mathbb{Z}_6$ **55.** $6x + 3 = 1$ in $\mathbb{Z}_8$

56. **(a)** For which values of a does $x + a = 0$ have a solution in $\mathbb{Z}_5$?
(b) For which values of a and b does $x + a = b$ have a solution in $\mathbb{Z}_6$?
(c) For which values of a, b, and m does $x + a = b$ have a solution in $\mathbb{Z}_m$?

57. **(a)** For which values of a does $ax = 1$ have a solution in $\mathbb{Z}_5$?
(b) For which values of a does $ax = 1$ have a solution in $\mathbb{Z}_6$?
(c) For which values of a and m does $ax = 1$ have a solution in $\mathbb{Z}_m$?

1.2 Length and Angle: The Dot Product

It is quite easy to reformulate the familiar geometric concepts of length, distance, and angle in terms of vectors. Doing so will allow us to use these important and powerful ideas in settings more general than $\mathbb{R}^2$ and $\mathbb{R}^3$. In subsequent chapters, these simple geometric tools will be used to solve a wide variety of problems arising in applications—even when there is no geometry apparent at all!

The Dot Product

The vector versions of length, distance, and angle can all be described using the notion of the dot product of two vectors.

Definition If

$$\mathbf{u} = \begin{bmatrix} u_1 \\ u_2 \\ \vdots \\ u_n \end{bmatrix} \quad \text{and} \quad \mathbf{v} = \begin{bmatrix} v_1 \\ v_2 \\ \vdots \\ v_n \end{bmatrix}$$

then the ***dot product*** $\mathbf{u} \cdot \mathbf{v}$ of $\mathbf{u}$ and $\mathbf{v}$ is defined by

$$\mathbf{u} \cdot \mathbf{v} = u_1 v_1 + u_2 v_2 + \cdots + u_n v_n$$

In words, $\mathbf{u} \cdot \mathbf{v}$ is the sum of the products of the corresponding components of $\mathbf{u}$ and $\mathbf{v}$. It is important to note a couple of things about this "product" that we have just defined: First, $\mathbf{u}$ and $\mathbf{v}$ must have the same number of components. Second, the dot product $\mathbf{u} \cdot \mathbf{v}$ is a *number,* not another vector. (This is why $\mathbf{u} \cdot \mathbf{v}$ is sometimes called the ***scalar product*** of $\mathbf{u}$ and $\mathbf{v}$.) The dot product of vectors in $\mathbb{R}^n$ is a special and important case of the more general notion of ***inner product,*** which we will explore in Chapter 7.

Example 1.15

Compute $\mathbf{u} \cdot \mathbf{v}$ when $\mathbf{u} = \begin{bmatrix} 1 \\ 2 \\ -3 \end{bmatrix}$ and $\mathbf{v} = \begin{bmatrix} -3 \\ 5 \\ 2 \end{bmatrix}$.

Solution $\mathbf{u} \cdot \mathbf{v} = 1 \cdot (-3) + 2 \cdot 5 + (-3) \cdot 2 = 1$

Notice that if we had calculated $\mathbf{v} \cdot \mathbf{u}$ in Example 1.15, we would have computed

$$\mathbf{v} \cdot \mathbf{u} = (-3) \cdot 1 + 5 \cdot 2 + 2 \cdot (-3) = 1$$

That $\mathbf{u} \cdot \mathbf{v} = \mathbf{v} \cdot \mathbf{u}$ in general is clear, since the individual products of the components commute. This commutativity property is one of the properties of the dot product that we will use repeatedly. The main properties of the dot product are summarized in Theorem 1.2.

Theorem 1.2 Let $\mathbf{u}$, $\mathbf{v}$, and $\mathbf{w}$ be vectors in $\mathbb{R}^n$ and let c be a scalar. Then

a. $\mathbf{u}\cdot\mathbf{v} = \mathbf{v}\cdot\mathbf{u}$ Commutativity
b. $\mathbf{u}\cdot(\mathbf{v}+\mathbf{w}) = \mathbf{u}\cdot\mathbf{v} + \mathbf{u}\cdot\mathbf{w}$ Distributivity
c. $(c\mathbf{u})\cdot\mathbf{v} = c(\mathbf{u}\cdot\mathbf{v})$
d. $\mathbf{u}\cdot\mathbf{u} \geq 0$ and $\mathbf{u}\cdot\mathbf{u} = 0$ if and only if $\mathbf{u} = \mathbf{0}$

Proof We prove (a) and (c) and leave proof of the remaining properties for the exercises.

(a) Applying the definition of dot product to $\mathbf{u}\cdot\mathbf{v}$ and $\mathbf{v}\cdot\mathbf{u}$, we obtain

$$\begin{aligned}\mathbf{u}\cdot\mathbf{v} &= u_1v_1 + u_2v_2 + \cdots + u_nv_n \\ &= v_1u_1 + v_2u_2 + \cdots + v_nu_n \\ &= \mathbf{v}\cdot\mathbf{u}\end{aligned}$$

where the middle equality follows from the fact that multiplication of real numbers is commutative.

(c) Using the definitions of scalar multiplication and dot product, we have

$$\begin{aligned}(c\mathbf{u})\cdot\mathbf{v} &= [cu_1, cu_2, \ldots, cu_n]\cdot[v_1, v_2, \ldots, v_n] \\ &= cu_1v_1 + cu_2v_2 + \cdots + cu_nv_n \\ &= c(u_1v_1 + u_2v_2 + \cdots + u_nv_n) \\ &= c(\mathbf{u}\cdot\mathbf{v})\end{aligned}$$

Remarks

- Property (b) can be read from right to left, in which case it says that we can factor out a common vector $\mathbf{u}$ from a sum of dot products. This property also has a "right-handed" analogue that follows from properties (b) and (a) together: $(\mathbf{v}+\mathbf{w})\cdot\mathbf{u} = \mathbf{v}\cdot\mathbf{u} + \mathbf{w}\cdot\mathbf{u}$.
- Property (c) can be extended to give $\mathbf{u}\cdot(c\mathbf{v}) = c(\mathbf{u}\cdot\mathbf{v})$ (Exercise 58). This extended version of (c) essentially says that in taking a scalar multiple of a dot product of vectors, the scalar can first be combined with whichever vector is more convenient. For example,

$$\left(\tfrac{1}{2}[-1,-3,2]\right)\cdot[6,-4,0] = [-1,-3,2]\cdot\left(\tfrac{1}{2}[6,-4,0]\right) = [-1,-3,2]\cdot[3,-2,0] = 3$$

With this approach we avoid introducing fractions into the vectors, as the original grouping would have.

- The second part of (d) uses the logical connective *if and only if*. Appendix A discusses this phrase in more detail, but for the moment let us just note that the wording signals a *double implication*—namely,

$$\text{if } \mathbf{u} = \mathbf{0}, \text{ then } \mathbf{u}\cdot\mathbf{u} = 0$$

and

$$\text{if } \mathbf{u}\cdot\mathbf{u} = 0, \text{ then } \mathbf{u} = \mathbf{0}$$

Theorem 1.2 shows that aspects of the algebra of vectors resemble the algebra of numbers. The next example shows that we can sometimes find vector analogues of familiar identities.

Example 1.16

Prove that $(\mathbf{u} + \mathbf{v}) \cdot (\mathbf{u} + \mathbf{v}) = \mathbf{u} \cdot \mathbf{u} + 2(\mathbf{u} \cdot \mathbf{v}) + \mathbf{v} \cdot \mathbf{v}$ for all vectors $\mathbf{u}$ and $\mathbf{v}$ in $\mathbb{R}^n$.

Solution

$$\begin{aligned}(\mathbf{u} + \mathbf{v}) \cdot (\mathbf{u} + \mathbf{v}) &= (\mathbf{u} + \mathbf{v}) \cdot \mathbf{u} + (\mathbf{u} + \mathbf{v}) \cdot \mathbf{v} \\ &= \mathbf{u} \cdot \mathbf{u} + \mathbf{v} \cdot \mathbf{u} + \mathbf{u} \cdot \mathbf{v} + \mathbf{v} \cdot \mathbf{v} \\ &= \mathbf{u} \cdot \mathbf{u} + \mathbf{u} \cdot \mathbf{v} + \mathbf{u} \cdot \mathbf{v} + \mathbf{v} \cdot \mathbf{v} \\ &= \mathbf{u} \cdot \mathbf{u} + 2(\mathbf{u} \cdot \mathbf{v}) + \mathbf{v} \cdot \mathbf{v}\end{aligned}$$

(Identify the parts of Theorem 1.2 that were used at each step.)

Length

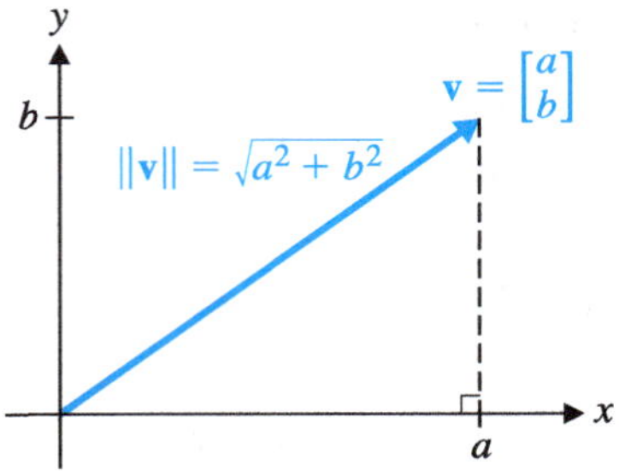

Figure 1.25

To see how the dot product plays a role in the calculation of lengths, recall how lengths are computed in the plane. The Theorem of Pythagoras is all we need.

In $\mathbb{R}^2$, the length of the vector $\mathbf{v} = \begin{bmatrix} a \\ b \end{bmatrix}$ is the distance from the origin to the point (a, b), which, by Pythagoras' Theorem, is given by $\sqrt{a^2 + b^2}$, as in Figure 1.25. Observe that $a^2 + b^2 = \mathbf{v} \cdot \mathbf{v}$. This leads to the following definition.

Definition The ***length*** (or ***norm***) of a vector $\mathbf{v} = \begin{bmatrix} v_1 \\ v_2 \\ \vdots \\ v_n \end{bmatrix}$ in $\mathbb{R}^n$ is the nonnegative scalar $\|\mathbf{v}\|$ defined by

$$\|\mathbf{v}\| = \sqrt{\mathbf{v} \cdot \mathbf{v}} = \sqrt{v_1^2 + v_2^2 + \cdots + v_n^2}$$

In words, the length of a vector is the square root of the sum of the squares of its components. Note that the square root of $\mathbf{v} \cdot \mathbf{v}$ is always defined, since $\mathbf{v} \cdot \mathbf{v} \geq 0$ by Theorem 1.2(d). Note also that the definition can be rewritten to give $\|\mathbf{v}\|^2 = \mathbf{v} \cdot \mathbf{v}$, which will be useful in proving further properties of the dot product and lengths of vectors.

Example 1.17

$$\|[2, 3]\| = \sqrt{2^2 + 3^2} = \sqrt{13}$$

Theorem 1.3 lists some of the main properties of vector length.

Theorem 1.3

Let $\mathbf{v}$ be a vector in $\mathbb{R}^n$ and let c be a scalar. Then

a. $\|\mathbf{v}\| = 0$ if and only if $\mathbf{v} = \mathbf{0}$
b. $\|c\mathbf{v}\| = |c|\,\|\mathbf{v}\|$

Proof Property (a) follows immediately from Theorem 1.2(d). To show (b), we have

$$\|c\mathbf{v}\|^2 = (c\mathbf{v}) \cdot (c\mathbf{v}) = c^2(\mathbf{v} \cdot \mathbf{v}) = c^2\|\mathbf{v}\|^2$$

using Theorem 1.2(c). Taking square roots of both sides, using the fact that $\sqrt{c^2} = |c|$ for any real number c, gives the result.

A vector of length 1 is called a ***unit vector.*** In $\mathbb{R}^2$, the set of all unit vectors can be identified with the *unit circle,* the circle of radius 1 centered at the origin (see Figure 1.26). Given any nonzero vector $\mathbf{v}$, we can always find a unit vector in the same direction as $\mathbf{v}$ by dividing $\mathbf{v}$ by its own length (or, equivalently, *multiplying* by $1/\|\mathbf{v}\|$). We can show this algebraically by using property (b) of Theorem 1.3 above: If $\mathbf{u} = (1/\|\mathbf{v}\|)\mathbf{v}$, then

$$\|\mathbf{u}\| = \|(1/\|\mathbf{v}\|)\mathbf{v}\| = |1/\|\mathbf{v}\||\,\|\mathbf{v}\| = (1/\|\mathbf{v}\|)\|\mathbf{v}\| = 1$$

and $\mathbf{u}$ is in the same direction as $\mathbf{v}$, since $1/\|\mathbf{v}\|$ is a positive scalar. Finding a unit vector in the same direction is often referred to as ***normalizing*** a vector (see Figure 1.27).

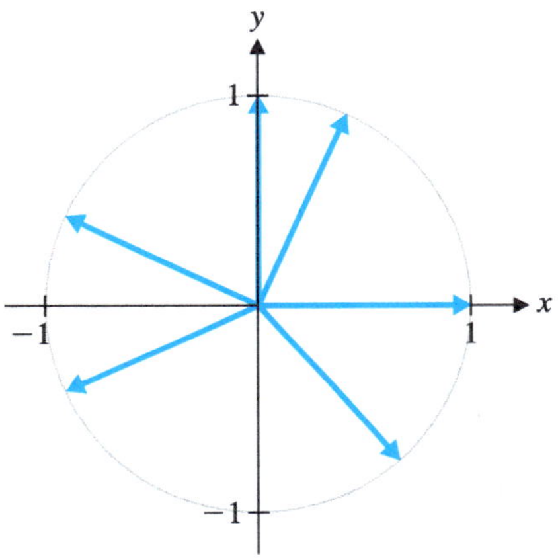

Figure 1.26
Unit vectors in $\mathbb{R}^2$

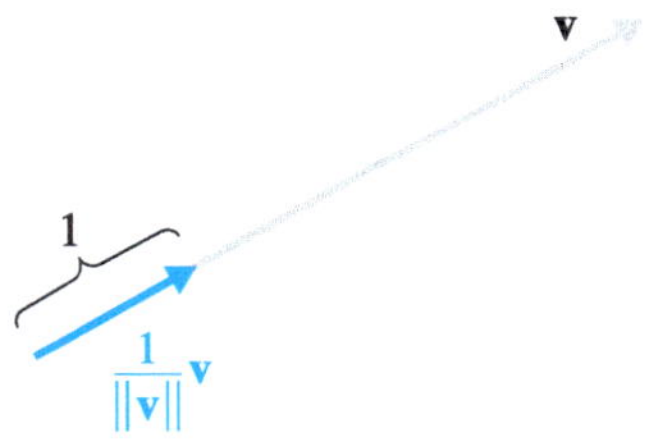

Figure 1.27
Normalizing a vector

Example 1.18

In $\mathbb{R}^2$, let $\mathbf{e}_1 = \begin{bmatrix}1\\0\end{bmatrix}$ and $\mathbf{e}_2 = \begin{bmatrix}0\\1\end{bmatrix}$. Then $\mathbf{e}_1$ and $\mathbf{e}_2$ are unit vectors, since the sum of the squares of their components is 1 in each case. Similarly, in $\mathbb{R}^3$, we can construct unit vectors

$$\mathbf{e}_1 = \begin{bmatrix}1\\0\\0\end{bmatrix}, \quad \mathbf{e}_2 = \begin{bmatrix}0\\1\\0\end{bmatrix}, \quad \text{and} \quad \mathbf{e}_3 = \begin{bmatrix}0\\0\\1\end{bmatrix}$$

Observe in Figure 1.28 that these vectors serve to locate the positive coordinate axes in $\mathbb{R}^2$ and $\mathbb{R}^3$.

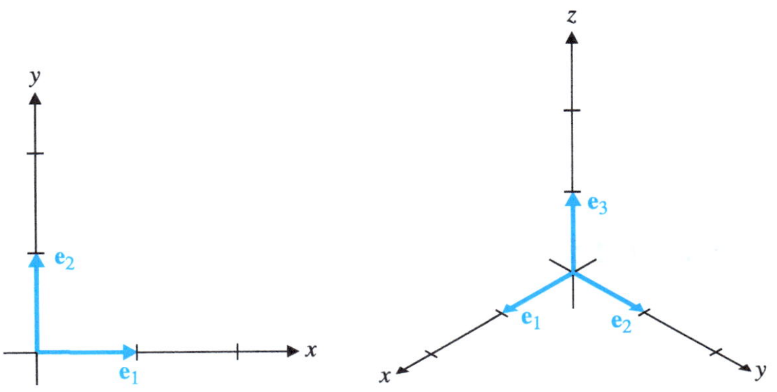

Figure 1.28
Standard unit vectors in $\mathbb{R}^2$ and $\mathbb{R}^3$

In general, in $\mathbb{R}^n$, we define unit vectors $\mathbf{e}_1, \mathbf{e}_2, \ldots, \mathbf{e}_n$, where $\mathbf{e}_i$ has 1 in its ith component and zeros elsewhere. These vectors arise repeatedly in linear algebra and are called the ***standard unit vectors.***

Example 1.19

Normalize the vector $\mathbf{v} = \begin{bmatrix} 2 \\ -1 \\ 3 \end{bmatrix}$.

Solution $\|\mathbf{v}\| = \sqrt{2^2 + (-1)^2 + 3^2} = \sqrt{14}$, so a unit vector in the same direction as $\mathbf{v}$ is given by

$$\mathbf{u} = (1/\|\mathbf{v}\|)\mathbf{v} = (1/\sqrt{14})\begin{bmatrix} 2 \\ -1 \\ 3 \end{bmatrix} = \begin{bmatrix} 2/\sqrt{14} \\ -1/\sqrt{14} \\ 3/\sqrt{14} \end{bmatrix}$$

Since property (b) of Theorem 1.3 describes how length behaves with respect to scalar multiplication, natural curiosity suggests that we ask whether length and vector addition are compatible. It would be nice if we had an identity such as $\|\mathbf{u} + \mathbf{v}\| = \|\mathbf{u}\| + \|\mathbf{v}\|$, but for almost any choice of vectors $\mathbf{u}$ and $\mathbf{v}$ this turns out to be false. [See Exercise 52(a).] However, all is not lost, for it turns out that if we replace the = sign by $\leq$, the resulting inequality is true. The proof of this famous and important result—the Triangle Inequality—relies on another important inequality—the Cauchy-Schwarz Inequality—which we will prove and discuss in more detail in Chapter 7.

Theorem 1.4 **The Cauchy-Schwarz Inequality**

For all vectors $\mathbf{u}$ and $\mathbf{v}$ in $\mathbb{R}^n$,

$$|\mathbf{u} \cdot \mathbf{v}| \leq \|\mathbf{u}\| \, \|\mathbf{v}\|$$

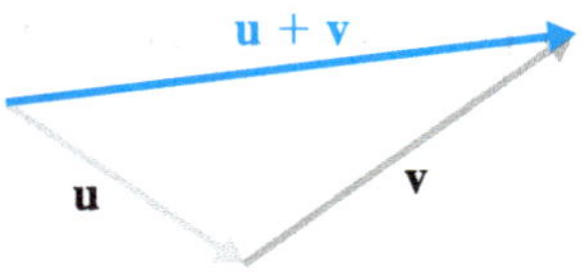

Figure 1.29
The Triangle Inequality

See Exercises 71 and 72 for algebraic and geometric approaches to the proof of this inequality.

In $\mathbb{R}^2$ or $\mathbb{R}^3$, where we can use geometry, it is clear from a diagram such as Figure 1.29 that $\|\mathbf{u} + \mathbf{v}\| \leq \|\mathbf{u}\| + \|\mathbf{v}\|$ for all vectors $\mathbf{u}$ and $\mathbf{v}$. We now show that this is true more generally.

Theorem 1.5 **The Triangle Inequality**

For all vectors $\mathbf{u}$ and $\mathbf{v}$ in $\mathbb{R}^n$,

$$\|\mathbf{u} + \mathbf{v}\| \leq \|\mathbf{u}\| + \|\mathbf{v}\|$$

Proof Since both sides of the inequality are nonnegative, showing that the *square* of the left-hand side is less than or equal to the *square* of the right-hand side is equivalent to proving the theorem. (Why?) We compute

$$\begin{aligned} \|\mathbf{u}+\mathbf{v}\|^2 &= (\mathbf{u}+\mathbf{v})\cdot(\mathbf{u}+\mathbf{v}) \\ &= \mathbf{u}\cdot\mathbf{u} + 2(\mathbf{u}\cdot\mathbf{v}) + \mathbf{v}\cdot\mathbf{v} && \text{By Example 1.9} \\ &\le \|\mathbf{u}\|^2 + 2|\mathbf{u}\cdot\mathbf{v}| + \|\mathbf{v}\|^2 \\ &\le \|\mathbf{u}\|^2 + 2\|\mathbf{u}\|\,\|\mathbf{v}\| + \|\mathbf{v}\|^2 && \text{By Cauchy-Schwarz} \\ &= (\|\mathbf{u}\| + \|\mathbf{v}\|)^2 \end{aligned}$$

as required.

Distance

The distance between two vectors is the direct analogue of the distance between two points on the real number line or two points in the Cartesian plane. On the number line (Figure 1.30), the distance between the numbers a and b is given by $|a-b|$. (Taking the absolute value ensures that we do not need to know which of a or b is larger.) This distance is also equal to $\sqrt{(a-b)^2}$, and its two-dimensional generalization is the familiar formula for the distance d between points (a_1, a_2) and (b_1, b_2)—namely, $d = \sqrt{(a_1-b_1)^2 + (a_2-b_2)^2}$.

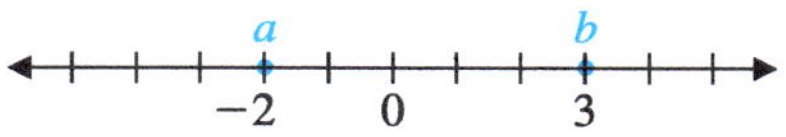

Figure 1.30
$d = |a - b| = |-2 - 3| = 5$

In terms of vectors, if $\mathbf{a} = \begin{bmatrix} a_1 \\ a_2 \end{bmatrix}$ and $\mathbf{b} = \begin{bmatrix} b_1 \\ b_2 \end{bmatrix}$, then d is just the length of $\mathbf{a} - \mathbf{b}$, as shown in Figure 1.31. This is the basis for the next definition.

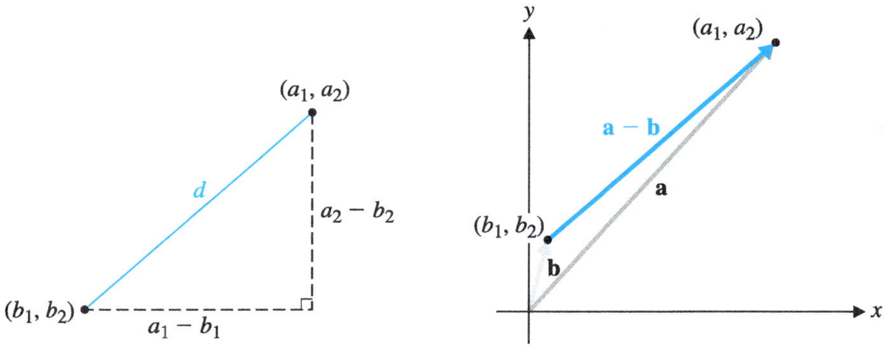

Figure 1.31
$d = \sqrt{(a_1-b_1)^2 + (a_2-b_2)^2} = \|\mathbf{a} - \mathbf{b}\|$

Definition The ***distance*** $d(\mathbf{u}, \mathbf{v})$ between vectors $\mathbf{u}$ and $\mathbf{v}$ in $\mathbb{R}^n$ is defined by

$$d(\mathbf{u}, \mathbf{v}) = \|\mathbf{u} - \mathbf{v}\|$$

Example 1.20

Find the distance between $\mathbf{u} = \begin{bmatrix} \sqrt{2} \\ 1 \\ -1 \end{bmatrix}$ and $\mathbf{v} = \begin{bmatrix} 0 \\ 2 \\ -2 \end{bmatrix}$.

Solution We compute $\mathbf{u} - \mathbf{v} = \begin{bmatrix} \sqrt{2} \\ -1 \\ 1 \end{bmatrix}$, so

$$d(\mathbf{u}, \mathbf{v}) = \|\mathbf{u} - \mathbf{v}\| = \sqrt{(\sqrt{2})^2 + (-1)^2 + 1^2} = \sqrt{4} = 2$$

Angles

The dot product can also be used to calculate the angle between a pair of vectors. In $\mathbb{R}^2$ or $\mathbb{R}^3$, the angle between the nonzero vectors $\mathbf{u}$ and $\mathbf{v}$ will refer to the angle θ determined by these vectors that satisfies $0 \le \theta \le 180°$ (see Figure 1.32).

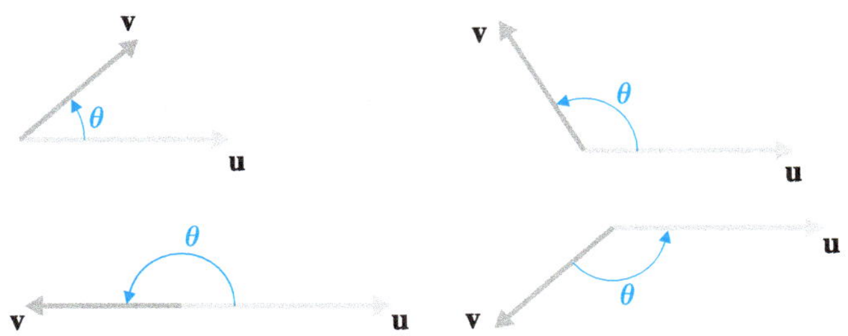

Figure 1.32
The angle between **u** and **v**

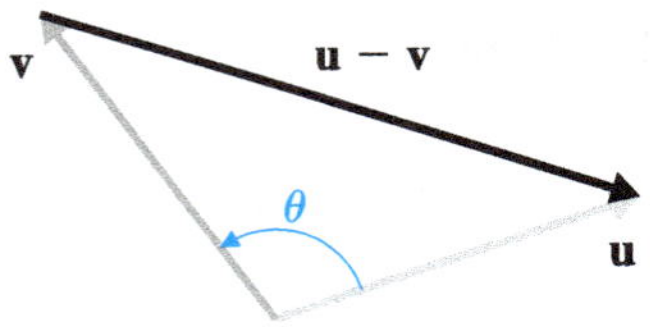

Figure 1.33

In Figure 1.33, consider the triangle with sides $\mathbf{u}$, $\mathbf{v}$, and $\mathbf{u} - \mathbf{v}$, where θ is the angle between $\mathbf{u}$ and $\mathbf{v}$. Applying the law of cosines to this triangle yields

$$\|\mathbf{u} - \mathbf{v}\|^2 = \|\mathbf{u}\|^2 + \|\mathbf{v}\|^2 - 2\|\mathbf{u}\|\,\|\mathbf{v}\| \cos\theta$$

Expanding the left-hand side and using $\|\mathbf{v}\|^2 = \mathbf{v} \cdot \mathbf{v}$ several times, we obtain

$$\|\mathbf{u}\|^2 - 2(\mathbf{u} \cdot \mathbf{v}) + \|\mathbf{v}\|^2 = \|\mathbf{u}\|^2 + \|\mathbf{v}\|^2 - 2\|\mathbf{u}\|\,\|\mathbf{v}\| \cos\theta$$

which, after simplification, leaves us with $\mathbf{u} \cdot \mathbf{v} = \|\mathbf{u}\|\,\|\mathbf{v}\| \cos\theta$. From this we obtain the following formula for the cosine of the angle θ between nonzero vectors $\mathbf{u}$ and $\mathbf{v}$. We state it as a definition.

Definition For nonzero vectors $\mathbf{u}$ and $\mathbf{v}$ in $\mathbb{R}^n$,

$$\cos\theta = \frac{\mathbf{u} \cdot \mathbf{v}}{\|\mathbf{u}\|\,\|\mathbf{v}\|}$$

Example 1.21

Compute the angle between the vectors $\mathbf{u} = [2, 1, -2]$ and $\mathbf{v} = [1, 1, 1]$.

Solution We calculate $\mathbf{u}\cdot\mathbf{v} = 2\cdot 1 + 1\cdot 1 + (-2)\cdot 1 = 1$, $\|\mathbf{u}\| = \sqrt{2^2 + 1^2 + (-2)^2} = \sqrt{9} = 3$, and $\|\mathbf{v}\| = \sqrt{1^2 + 1^2 + 1^2} = \sqrt{3}$. Therefore, $\cos\theta = 1/3\sqrt{3}$, so $\theta = \cos^{-1}(1/3\sqrt{3}) \approx 1.377$ radians, or 78.9°.

Example 1.22

Compute the angle between the diagonals on two adjacent faces of a cube.

Solution The dimensions of the cube do not matter, so we will work with a cube with sides of length 1. Orient the cube relative to the coordinate axes in $\mathbb{R}^3$, as shown in Figure 1.34, and take the two side diagonals to be the vectors [1, 0, 1] and [0, 1, 1]. Then angle θ between these vectors satisfies

$$\cos\theta = \frac{1\cdot 0 + 0\cdot 1 + 1\cdot 1}{\sqrt{2}\sqrt{2}} = \frac{1}{2}$$

from which it follows that the required angle is $\pi/3$ radians, or 60°.

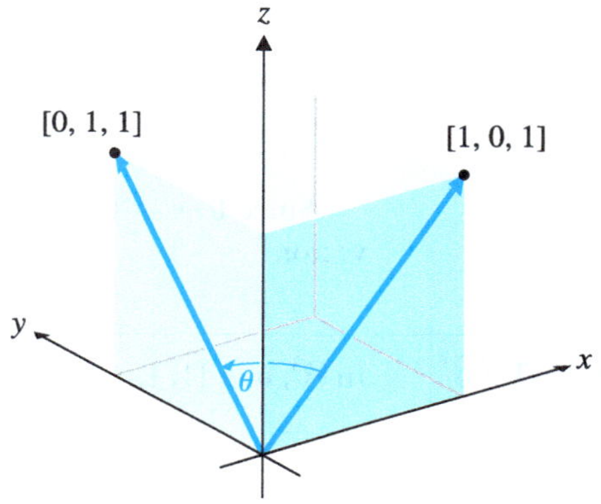

Figure 1.34

(Actually, we don't need to do any calculations at all to get this answer. If we draw a third side diagonal joining the vertices at (1, 0, 1) and (0, 1, 1), we get an equilateral triangle, since all of the side diagonals are of equal length. The angle we want is one of the angles of this triangle and therefore measures 60°. Sometimes, a little insight can save a lot of calculation; in this case, it gives a nice check on our work!)

Remarks

- As this discussion shows, we usually will have to settle for an approximation to the angle between two vectors. However, when the angle is one of the so-called special angles (0°, 30°, 45°, 60°, 90°, or an integer multiple of these), we should be able to recognize its cosine (Table 1.1) and thus give the corresponding angle exactly. In all other cases, we will use a calculator or computer to approximate the desired angle by means of the inverse cosine function.

Table 1.1 Cosines of Special Angles

θ	0°	30°	45°	60°	90°
$\cos\theta$	$\frac{\sqrt{4}}{2} = 1$	$\frac{\sqrt{3}}{2}$	$\frac{\sqrt{2}}{2} = \frac{1}{\sqrt{2}}$	$\frac{\sqrt{1}}{2} = \frac{1}{2}$	$\frac{\sqrt{0}}{2} = 0$

- The derivation of the formula for the cosine of the angle between two vectors is valid only in $\mathbb{R}^2$ or $\mathbb{R}^3$, since it depends on a geometric fact: the law of cosines. In $\mathbb{R}^n$, for $n > 3$, the formula can be taken as a *definition* instead. This makes sense, since the Cauchy-Schwarz Inequality implies that $\left|\dfrac{\mathbf{u}\cdot\mathbf{v}}{\|\mathbf{u}\|\,\|\mathbf{v}\|}\right| \le 1$, so $\dfrac{\mathbf{u}\cdot\mathbf{v}}{\|\mathbf{u}\|\,\|\mathbf{v}\|}$ ranges from -1 to 1, just as the cosine function does.

Orthogonal Vectors

The word *orthogonal* is derived from the Greek words *orthos,* meaning "upright," and *gonia,* meaning "angle." Hence, orthogonal literally means "right-angled." The Latin equivalent is *rectangular.*

The concept of perpendicularity is fundamental to geometry. Anyone studying geometry quickly realizes the importance and usefulness of right angles. We now generalize the idea of perpendicularity to vectors in $\mathbb{R}^n$, where it is called ***orthogonality.***

In $\mathbb{R}^2$ or $\mathbb{R}^3$, two nonzero vectors $\mathbf{u}$ and $\mathbf{v}$ are perpendicular if the angle θ between them is a right angle—that is, if $\theta = \pi/2$ radians, or 90°. Thus, $\dfrac{\mathbf{u}\cdot\mathbf{v}}{\|\mathbf{u}\|\|\mathbf{v}\|} = \cos 90° = 0$, and it follows that $\mathbf{u}\cdot\mathbf{v} = 0$. This motivates the following definition.

Definition Two vectors $\mathbf{u}$ and $\mathbf{v}$ in $\mathbb{R}^n$ are ***orthogonal*** to each other if $\mathbf{u}\cdot\mathbf{v} = 0$.

Since $\mathbf{0}\cdot\mathbf{v} = 0$ for every vector $\mathbf{v}$ in $\mathbb{R}^n$, the zero vector is orthogonal to every vector.

Example 1.23

In $\mathbb{R}^3$, $\mathbf{u} = [1, 1, -2]$ and $\mathbf{v} = [3, 1, 2]$ are orthogonal, since $\mathbf{u}\cdot\mathbf{v} = 3 + 1 - 4 = 0$.

Using the notion of orthogonality, we get an easy proof of Pythagoras' Theorem, valid in $\mathbb{R}^n$.

Theorem 1.6 **Pythagoras' Theorem**

For all vectors $\mathbf{u}$ and $\mathbf{v}$ in $\mathbb{R}^n$, $\|\mathbf{u} + \mathbf{v}\|^2 = \|\mathbf{u}\|^2 + \|\mathbf{v}\|^2$ if and only if $\mathbf{u}$ and $\mathbf{v}$ are orthogonal.

Proof From Example 1.16, we have $\|\mathbf{u} + \mathbf{v}\|^2 = \|\mathbf{u}\|^2 + 2(\mathbf{u}\cdot\mathbf{v}) + \|\mathbf{v}\|^2$ for all vectors $\mathbf{u}$ and $\mathbf{v}$ in $\mathbb{R}^n$. It follows immediately that $\|\mathbf{u} + \mathbf{v}\|^2 = \|\mathbf{u}\|^2 + \|\mathbf{v}\|^2$ if and only if $\mathbf{u}\cdot\mathbf{v} = 0$. See Figure 1.35.

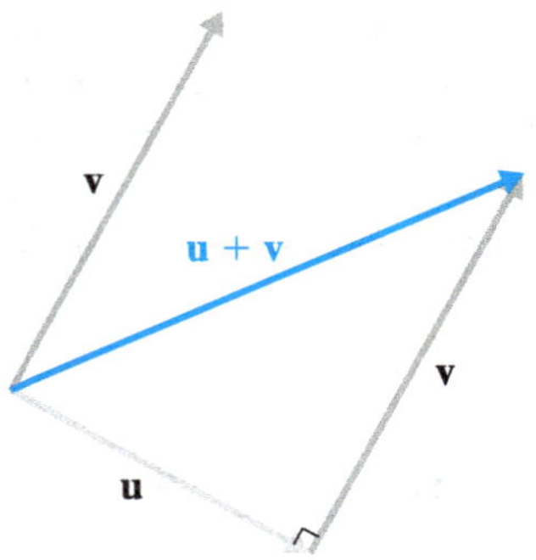

Figure 1.35

The concept of orthogonality is one of the most important and useful in linear algebra, and it often arises in surprising ways. Chapter 5 contains a detailed treatment of the topic, but we will encounter it many times before then. One problem in which it clearly plays a role is finding the distance from a point to a line, where "dropping a perpendicular" is a familiar step.

Projections

We now consider the problem of finding the distance from a point to a line in the context of vectors. As you will see, this technique leads to an important concept: the projection of a vector onto another vector.

As Figure 1.36 shows, the problem of finding the distance from a point B to a line ℓ (in $\mathbb{R}^2$ or $\mathbb{R}^3$) reduces to the problem of finding the length of the perpendicular line segment $\overline{PB}$ or, equivalently, the length of the vector $\overrightarrow{PB}$. If we choose a point A on ℓ, then, in the right-angled triangle ΔAPB, the other two vectors are the leg $\overrightarrow{AP}$ and the hypotenuse $\overrightarrow{AB}$. $\overrightarrow{AP}$ is called the *projection* of $\overrightarrow{AB}$ onto the line ℓ. We will now look at this situation in terms of vectors.

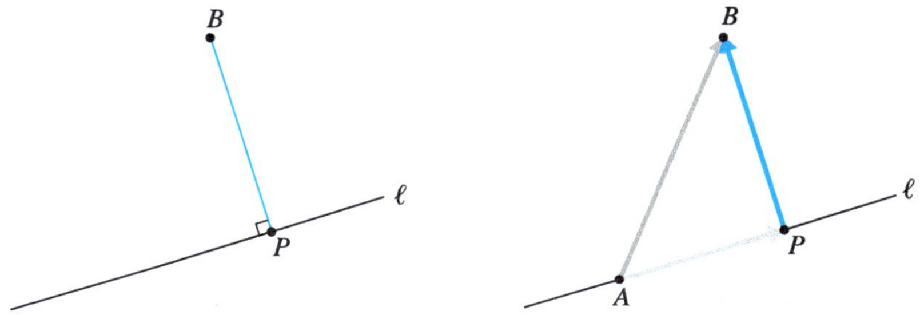

Figure 1.36
The distance from a point to a line

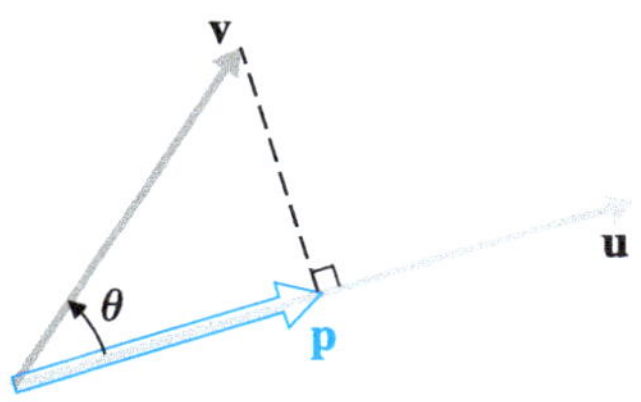

Figure 1.37
The projection of **v** onto **u**

Consider two nonzero vectors $\mathbf{u}$ and $\mathbf{v}$. Let $\mathbf{p}$ be the vector obtained by dropping a perpendicular from the head of $\mathbf{v}$ onto $\mathbf{u}$ and let θ be the angle between $\mathbf{u}$ and $\mathbf{v}$, as shown in Figure 1.37. Then clearly $\mathbf{p} = \|\mathbf{p}\|\hat{\mathbf{u}}$, where $\hat{\mathbf{u}} = (1/\|\mathbf{u}\|)\mathbf{u}$ is the unit vector in the direction of $\mathbf{u}$. Moreover, elementary trigonometry gives $\|\mathbf{p}\| = \|\mathbf{v}\| \cos\theta$, and we know that $\cos\theta = \dfrac{\mathbf{u}\cdot\mathbf{v}}{\|\mathbf{u}\|\,\|\mathbf{v}\|}$. Thus, after substitution, we obtain

$$\begin{aligned}\mathbf{p} &= \|\mathbf{v}\|\left(\frac{\mathbf{u}\cdot\mathbf{v}}{\|\mathbf{u}\|\,\|\mathbf{v}\|}\right)\left(\frac{1}{\|\mathbf{u}\|}\right)\mathbf{u}\\ &= \left(\frac{\mathbf{u}\cdot\mathbf{v}}{\|\mathbf{u}\|^2}\right)\mathbf{u}\\ &= \left(\frac{\mathbf{u}\cdot\mathbf{v}}{\mathbf{u}\cdot\mathbf{u}}\right)\mathbf{u}\end{aligned}$$

This is the formula we want, and it is the basis of the following definition for vectors in $\mathbb{R}^n$.

Definition If $\mathbf{u}$ and $\mathbf{v}$ are vectors in $\mathbb{R}^n$ and $\mathbf{u} \neq \mathbf{0}$, then the ***projection of* v *onto* u** is the vector $\text{proj}_{\mathbf{u}}(\mathbf{v})$ defined by

$$\text{proj}_{\mathbf{u}}(\mathbf{v}) = \left(\frac{\mathbf{u}\cdot\mathbf{v}}{\mathbf{u}\cdot\mathbf{u}}\right)\mathbf{u}$$

An alternative way to derive this formula is described in Exercise 73.

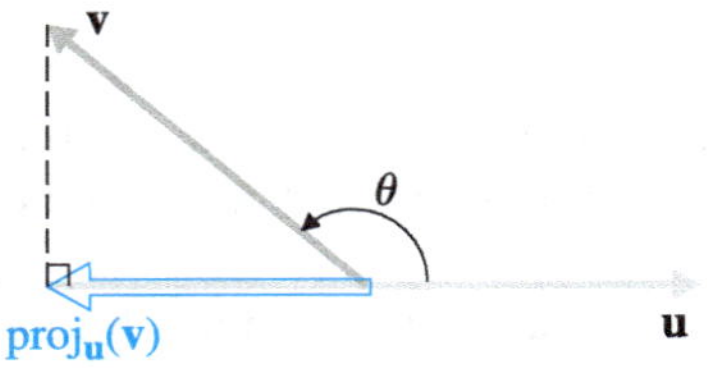

Figure 1.38

Remarks

- The term *projection* comes from the idea of projecting an image onto a wall (with a slide projector, for example). Imagine a beam of light with rays parallel to each other and perpendicular to **u** shining down on **v**. The projection of **v** onto **u** is just the shadow cast, or projected, by **v** onto **u**.
- It may be helpful to think of $\text{proj}_{\mathbf{u}}(\mathbf{v})$ as a function with variable **v**. Then the variable **v** occurs only once on the right-hand side of the definition. Also, it is helpful to remember Figure 1.38, which reminds us that $\text{proj}_{\mathbf{u}}(\mathbf{v})$ is a scalar multiple of the vector **u** (*not* **v**).
- Although in our derivation of the definition of $\text{proj}_{\mathbf{u}}(\mathbf{v})$ we required **v** as well as **u** to be nonzero (why?), it is clear from the geometry that the projection of the zero vector onto **u** is **0**. The definition is in agreement with this, since $\left(\frac{\mathbf{u}\cdot\mathbf{0}}{\mathbf{u}\cdot\mathbf{u}}\right)\mathbf{u} = 0\mathbf{u} = \mathbf{0}$.
- If the angle between **u** and **v** is obtuse, as in Figure 1.38, then $\text{proj}_{\mathbf{u}}(\mathbf{v})$ will be in the opposite direction from **u**; that is, $\text{proj}_{\mathbf{u}}(\mathbf{v})$ will be a *negative* scalar multiple of **u**.
- If **u** is a unit vector then $\text{proj}_{\mathbf{u}}(\mathbf{v}) = (\mathbf{u}\cdot\mathbf{v})\mathbf{u}$. (Why?)

Example 1.24

Find the projection of **v** onto **u** in each case.

(a) $\mathbf{v} = \begin{bmatrix} -1 \\ 3 \end{bmatrix}$ and $\mathbf{u} = \begin{bmatrix} 2 \\ 1 \end{bmatrix}$ (b) $\mathbf{v} = \begin{bmatrix} 1 \\ 2 \\ 3 \end{bmatrix}$ and $\mathbf{u} = \mathbf{e}_3$

(c) $\mathbf{v} = \begin{bmatrix} 1 \\ 2 \\ 3 \end{bmatrix}$ and $\mathbf{u} = \begin{bmatrix} 1/2 \\ 1/2 \\ 1/\sqrt{2} \end{bmatrix}$

Solution

(a) We compute $\mathbf{u}\cdot\mathbf{v} = \begin{bmatrix} 2 \\ 1 \end{bmatrix}\cdot\begin{bmatrix} -1 \\ 3 \end{bmatrix} = 1$ and $\mathbf{u}\cdot\mathbf{u} = \begin{bmatrix} 2 \\ 1 \end{bmatrix}\cdot\begin{bmatrix} 2 \\ 1 \end{bmatrix} = 5$, so

$$\text{proj}_{\mathbf{u}}(\mathbf{v}) = \left(\frac{\mathbf{u}\cdot\mathbf{v}}{\mathbf{u}\cdot\mathbf{u}}\right)\mathbf{u} = \frac{1}{5}\begin{bmatrix} 2 \\ 1 \end{bmatrix} = \begin{bmatrix} 2/5 \\ 1/5 \end{bmatrix}$$

(b) Since $\mathbf{e}_3$ is a unit vector,

$$\text{proj}_{\mathbf{e}_3}(\mathbf{v}) = (\mathbf{e}_3\cdot\mathbf{v})\mathbf{e}_3 = 3\mathbf{e}_3 = \begin{bmatrix} 0 \\ 0 \\ 3 \end{bmatrix}$$

(c) We see that $\|\mathbf{u}\| = \sqrt{\frac{1}{4} + \frac{1}{4} + \frac{1}{2}} = 1$. Thus,

$$\text{proj}_{\mathbf{u}}(\mathbf{v}) = (\mathbf{u}\cdot\mathbf{v})\mathbf{u} = \left(\frac{1}{2} + 1 + \frac{3}{\sqrt{2}}\right)\begin{bmatrix} 1/2 \\ 1/2 \\ 1/\sqrt{2} \end{bmatrix} = \frac{3(1+\sqrt{2})}{2}\begin{bmatrix} 1/2 \\ 1/2 \\ 1/\sqrt{2} \end{bmatrix}$$

$$= \frac{3(1+\sqrt{2})}{4}\begin{bmatrix} 1 \\ 1 \\ \sqrt{2} \end{bmatrix}$$

Exercises 1.2

In Exercises 1–6, find $\mathbf{u} \cdot \mathbf{v}$.

1. $\mathbf{u} = \begin{bmatrix} -1 \\ 2 \end{bmatrix}, \mathbf{v} = \begin{bmatrix} 3 \\ 1 \end{bmatrix}$ **2.** $\mathbf{u} = \begin{bmatrix} 3 \\ -2 \end{bmatrix}, \mathbf{v} = \begin{bmatrix} 4 \\ 6 \end{bmatrix}$

3. $\mathbf{u} = \begin{bmatrix} 1 \\ 2 \\ 3 \end{bmatrix}, \mathbf{v} = \begin{bmatrix} 2 \\ 3 \\ 1 \end{bmatrix}$ CAS **4.** $\mathbf{u} = \begin{bmatrix} 3.2 \\ -0.6 \\ -1.4 \end{bmatrix}, \mathbf{v} = \begin{bmatrix} 1.5 \\ 4.1 \\ -0.2 \end{bmatrix}$

5. $\mathbf{u} = [1, \sqrt{2}, \sqrt{3}, 0], \mathbf{v} = [4, -\sqrt{2}, 0, -5]$

CAS **6.** $\mathbf{u} = [1.12, -3.25, 2.07, -1.83]$,
$\mathbf{v} = [-2.29, 1.72, 4.33, -1.54]$

In Exercises 7–12, find $\|\mathbf{u}\|$ *for the given exercise, and give a unit vector in the direction of* $\mathbf{u}$.

7. Exercise 1 **8.** Exercise 2 **9.** Exercise 3

CAS **10.** Exercise 4 **11.** Exercise 5 CAS **12.** Exercise 6

In Exercises 13–16, find the distance $d(\mathbf{u}, \mathbf{v})$ *between* $\mathbf{u}$ *and* $\mathbf{v}$ *in the given exercise.*

13. Exercise 1 **14.** Exercise 2

15. Exercise 3 CAS **16.** Exercise 4

17. If $\mathbf{u}$, $\mathbf{v}$, and $\mathbf{w}$ are vectors in $\mathbb{R}^n$, $n \geq 2$, and c is a scalar, explain why the following expressions make no sense:

(a) $\|\mathbf{u} \cdot \mathbf{v}\|$ **(b)** $\mathbf{u} \cdot \mathbf{v} + \mathbf{w}$

(c) $\mathbf{u} \cdot (\mathbf{v} \cdot \mathbf{w})$ **(d)** $c \cdot (\mathbf{u} + \mathbf{w})$

In Exercises 18–23, determine whether the angle between $\mathbf{u}$ *and* $\mathbf{v}$ *is acute, obtuse, or a right angle.*

18. $\mathbf{u} = \begin{bmatrix} 3 \\ 0 \end{bmatrix}, \mathbf{v} = \begin{bmatrix} -1 \\ 1 \end{bmatrix}$ **19.** $\mathbf{u} = \begin{bmatrix} 2 \\ -1 \\ 1 \end{bmatrix}, \mathbf{v} = \begin{bmatrix} 1 \\ -2 \\ -1 \end{bmatrix}$

20. $\mathbf{u} = [4, 3, -1], \mathbf{v} = [1, -1, 1]$

CAS **21.** $\mathbf{u} = [0.9, 2.1, 1.2], \mathbf{v} = [-4.5, 2.6, -0.8]$

22. $\mathbf{u} = [1, 2, 3, 4], \mathbf{v} = [-3, 1, 2, -2]$

23. $\mathbf{u} = [1, 2, 3, 4], \mathbf{v} = [5, 6, 7, 8]$

In Exercises 24–29, find the angle between $\mathbf{u}$ *and* $\mathbf{v}$ *in the given exercise.*

24. Exercise 18 **25.** Exercise 19

26. Exercise 20 CAS **27.** Exercise 21

CAS **28.** Exercise 22 CAS **29.** Exercise 23

30. Let $A = (-3, 2)$, $B = (1, 0)$, and $C = (4, 6)$. Prove that ΔABC is a right-angled triangle.

31. Let $A = (1, 1, -1)$, $B = (-3, 2, -2)$, and $C = (2, 2, -4)$. Prove that ΔABC is a right-angled triangle.

CAS **32.** Find the angle between a diagonal of a cube and an adjacent edge.

33. A cube has four diagonals. Show that no two of them are perpendicular.

34. A parallelogram has diagonals determined by the vectors

$$\mathbf{d}_1 = \begin{bmatrix} 2 \\ 2 \\ 0 \end{bmatrix}, \text{ and } \mathbf{d}_2 = \begin{bmatrix} 1 \\ -1 \\ 3 \end{bmatrix}$$

Show that the parallelogram is a rhombus (all sides of equal length) and determine the side length.

35. The rectangle $ABCD$ has vertices at $A = (1, 2, 3)$, $B = (3, 6, -2)$, and $C = (0, 5, -4)$. Determine the coordinates of vertex D.

36. An airplane heading due east has a velocity of 200 miles per hour. A wind is blowing from the north at 40 miles per hour. What is the resultant velocity of the airplane?

37. A boat heads north across a river at a rate of 4 miles per hour. If the current is flowing east at a rate of 3 miles per hour, find the resultant velocity of the boat.

38. Ann is driving a motorboat across a river that is 2 km wide. The boat has a speed of 20 km/h in still water, and the current in the river is flowing at 5 km/h. Ann heads out from one bank of the river for a dock directly across from her on the opposite bank. She drives the boat in a direction perpendicular to the current.

(a) How far downstream from the dock will Ann land?

(b) How long will it take Ann to cross the river?

39. Bert can swim at a rate of 2 miles per hour in still water. The current in a river is flowing at a rate of 1 mile per hour. If Bert wants to swim across the river to a point directly opposite, at what angle to the bank of the river must he swim?

In Exercises 40–45, find the projection of **v** *onto* **u**. *Draw a sketch in Exercises 40 and 41.*

40. $\mathbf{u} = \begin{bmatrix} -1 \\ 1 \end{bmatrix}, \mathbf{v} = \begin{bmatrix} -2 \\ 4 \end{bmatrix}$ **41.** $\mathbf{u} = \begin{bmatrix} 3/5 \\ -4/5 \end{bmatrix}, \mathbf{v} = \begin{bmatrix} 1 \\ 2 \end{bmatrix}$

42. $\mathbf{u} = \begin{bmatrix} 1/2 \\ -1/4 \\ -1/2 \end{bmatrix}, \mathbf{v} = \begin{bmatrix} 2 \\ 2 \\ -2 \end{bmatrix}$ **43.** $\mathbf{u} = \begin{bmatrix} 1 \\ -1 \\ 1 \\ -1 \end{bmatrix}, \mathbf{v} = \begin{bmatrix} 2 \\ -3 \\ -1 \\ -2 \end{bmatrix}$

CAS **44.** $\mathbf{u} = \begin{bmatrix} 0.5 \\ 1.5 \end{bmatrix}, \mathbf{v} = \begin{bmatrix} 2.1 \\ 1.2 \end{bmatrix}$

CAS **45.** $\mathbf{u} = \begin{bmatrix} 3.01 \\ -0.33 \\ 2.52 \end{bmatrix}, \mathbf{v} = \begin{bmatrix} 1.34 \\ 4.25 \\ -1.66 \end{bmatrix}$

Figure 1.39 suggests two ways in which vectors may be used to compute the area of a triangle. The area $\mathcal{A}$ of

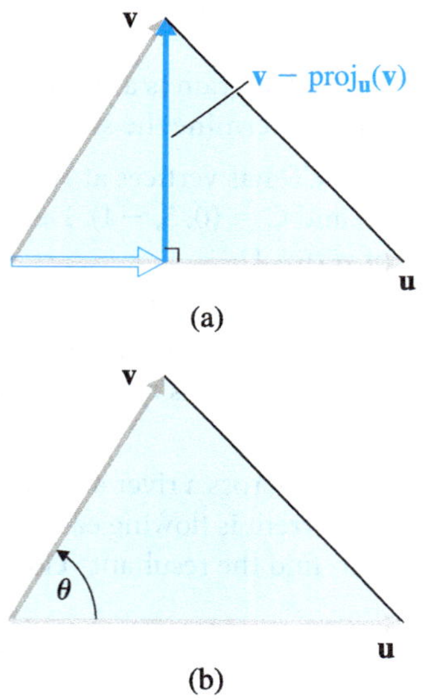

Figure 1.39

the triangle in part (a) is given by $\frac{1}{2}\|\mathbf{u}\| \; \|\mathbf{v} - \text{proj}_{\mathbf{u}}(\mathbf{v})\|$, and part (b) suggests the trigonometric form of the area of a triangle: $\mathcal{A} = \frac{1}{2}\|\mathbf{u}\| \; \|\mathbf{v}\|\sin\theta$ (We can use the identity $\sin\theta = \sqrt{1 - \cos^2\theta}$ to find $\sin\theta$.)

In Exercises 46 and 47, compute the area of the triangle with the given vertices using both methods.

46. $A = (1, -1), B = (2, 2), C = (4, 0)$

47. $A = (3, -1, 4), B = (4, -2, 6), C = (5, 0, 2)$

In Exercises 48 and 49, find all values of the scalar k for which the two vectors are orthogonal.

48. $\mathbf{u} = \begin{bmatrix} 2 \\ 3 \end{bmatrix}, \mathbf{v} = \begin{bmatrix} k+1 \\ k-1 \end{bmatrix}$ **49.** $\mathbf{u} = \begin{bmatrix} 1 \\ -1 \\ 2 \end{bmatrix}, \mathbf{v} = \begin{bmatrix} k^2 \\ k \\ -3 \end{bmatrix}$

50. Describe all vectors $\mathbf{v} = \begin{bmatrix} x \\ y \end{bmatrix}$ that are orthogonal to $\mathbf{u} = \begin{bmatrix} 3 \\ 1 \end{bmatrix}$.

51. Describe all vectors $\mathbf{v} = \begin{bmatrix} x \\ y \end{bmatrix}$ that are orthogonal to $\mathbf{u} = \begin{bmatrix} a \\ b \end{bmatrix}$.

52. Under what conditions are the following true for vectors **u** and **v** in $\mathbb{R}^2$ or $\mathbb{R}^3$?

(a) $\|\mathbf{u} + \mathbf{v}\| = \|\mathbf{u}\| + \|\mathbf{v}\|$ **(b)** $\|\mathbf{u} + \mathbf{v}\| = \|\mathbf{u}\| - \|\mathbf{v}\|$

53. Prove Theorem 1.2(b).

54. Prove Theorem 1.2(d).

In Exercises 55–57, prove the stated property of distance between vectors.

55. $d(\mathbf{u}, \mathbf{v}) = d(\mathbf{v}, \mathbf{u})$ for all vectors **u** and **v**

56. $d(\mathbf{u}, \mathbf{w}) \le d(\mathbf{u}, \mathbf{v}) + d(\mathbf{v}, \mathbf{w})$ for all vectors **u**, **v**, and **w**

57. $d(\mathbf{u}, \mathbf{v}) = 0$ if and only if $\mathbf{u} = \mathbf{v}$

58. Prove that $\mathbf{u} \cdot c\mathbf{v} = c(\mathbf{u} \cdot \mathbf{v})$ for all vectors **u** and **v** in $\mathbb{R}^n$ and all scalars c.

59. Prove that $\|\mathbf{u} - \mathbf{v}\| \ge \|\mathbf{u}\| - \|\mathbf{v}\|$ for all vectors **u** and **v** in $\mathbb{R}^n$. [*Hint:* Replace **u** by $\mathbf{u} - \mathbf{v}$ in the Triangle Inequality.]

60. Suppose we know that $\mathbf{u} \cdot \mathbf{v} = \mathbf{u} \cdot \mathbf{w}$. Does it follow that $\mathbf{v} = \mathbf{w}$? If it does, give a proof that is valid in $\mathbb{R}^n$; otherwise, give a *counterexample* (i.e., a *specific* set of vectors **u**, **v**, and **w** for which $\mathbf{u} \cdot \mathbf{v} = \mathbf{u} \cdot \mathbf{w}$ but $\mathbf{v} \ne \mathbf{w}$).

61. Prove that $(\mathbf{u} + \mathbf{v}) \cdot (\mathbf{u} - \mathbf{v}) = \|\mathbf{u}\|^2 - \|\mathbf{v}\|^2$ for all vectors **u** and **v** in $\mathbb{R}^n$.

62. **(a)** Prove that $\|\mathbf{u} + \mathbf{v}\|^2 + \|\mathbf{u} - \mathbf{v}\|^2 = 2\|\mathbf{u}\|^2 + 2\|\mathbf{v}\|^2$ for all vectors **u** and **v** in $\mathbb{R}^n$.

(b) Draw a diagram showing **u**, **v**, $\mathbf{u} + \mathbf{v}$, and $\mathbf{u} - \mathbf{v}$ in $\mathbb{R}^2$ and use (a) to deduce a result about parallelograms.

63. Prove that $\mathbf{u} \cdot \mathbf{v} = \frac{1}{4}\|\mathbf{u} + \mathbf{v}\|^2 - \frac{1}{4}\|\mathbf{u} - \mathbf{v}\|^2$ for all vectors **u** and **v** in $\mathbb{R}^n$.

64. **(a)** Prove that $\|\mathbf{u} + \mathbf{v}\| = \|\mathbf{u} - \mathbf{v}\|$ if and only if $\mathbf{u}$ and $\mathbf{v}$ are orthogonal.
(b) Draw a diagram showing $\mathbf{u}$, $\mathbf{v}$, $\mathbf{u} + \mathbf{v}$, and $\mathbf{u} - \mathbf{v}$ in $\mathbb{R}^2$ and use (a) to deduce a result about parallelograms.

65. **(a)** Prove that $\mathbf{u} + \mathbf{v}$ and $\mathbf{u} - \mathbf{v}$ are orthogonal in $\mathbb{R}^n$ if and only if $\|\mathbf{u}\| = \|\mathbf{v}\|$.
(b) Draw a diagram showing $\mathbf{u}$, $\mathbf{v}$, $\mathbf{u} + \mathbf{v}$, and $\mathbf{u} - \mathbf{v}$ in $\mathbb{R}^2$ and use (a) to deduce a result about parallelograms.

66. If $\|\mathbf{u}\| = 2$, $\|\mathbf{v}\| = \sqrt{3}$, and $\mathbf{u} \cdot \mathbf{v} = 1$, find $\|\mathbf{u} + \mathbf{v}\|$.

67. Show that there are no vectors $\mathbf{u}$ and $\mathbf{v}$ such that $\|\mathbf{u}\| = 1$, $\|\mathbf{v}\| = 2$, and $\mathbf{u} \cdot \mathbf{v} = 3$.

68. **(a)** Prove that if $\mathbf{u}$ is orthogonal to both $\mathbf{v}$ and $\mathbf{w}$, then $\mathbf{u}$ is orthogonal to $\mathbf{v} + \mathbf{w}$.
(b) Prove that if $\mathbf{u}$ is orthogonal to both $\mathbf{v}$ and $\mathbf{w}$, then $\mathbf{u}$ is orthogonal to $s\mathbf{v} + t\mathbf{w}$ for all scalars s and t.

69. Prove that $\mathbf{u}$ is orthogonal to $\mathbf{v} - \text{proj}_{\mathbf{u}}(\mathbf{v})$ for all vectors $\mathbf{u}$ and $\mathbf{v}$ in $\mathbb{R}^n$, where $\mathbf{u} \neq \mathbf{0}$.

70. **(a)** Prove that $\text{proj}_{\mathbf{u}}(\text{proj}_{\mathbf{u}}(\mathbf{v})) = \text{proj}_{\mathbf{u}}(\mathbf{v})$.
(b) Prove that $\text{proj}_{\mathbf{u}}(\mathbf{v} - \text{proj}_{\mathbf{u}}(\mathbf{v})) = \mathbf{0}$.
(c) Explain (a) and (b) geometrically.

71. The Cauchy-Schwarz Inequality $|\mathbf{u} \cdot \mathbf{v}| \leq \|\mathbf{u}\| \|\mathbf{v}\|$ is equivalent to the inequality we get by squaring both sides: $(\mathbf{u} \cdot \mathbf{v})^2 \leq \|\mathbf{u}\|^2 \|\mathbf{v}\|^2$.

(a) In $\mathbb{R}^2$, with $\mathbf{u} = \begin{bmatrix} u_1 \\ u_2 \end{bmatrix}$ and $\mathbf{v} = \begin{bmatrix} v_1 \\ v_2 \end{bmatrix}$, this becomes

$$(u_1v_1 + u_2v_2)^2 \leq (u_1^2 + u_2^2)(v_1^2 + v_2^2)$$

Prove this algebraically. [*Hint:* Subtract the left-hand side from the right-hand side and show that the difference must necessarily be nonnegative.]
(b) Prove the analogue of (a) in $\mathbb{R}^3$.

72. Figure 1.40 shows that, in $\mathbb{R}^2$ or $\mathbb{R}^3$, $\|\text{proj}_{\mathbf{u}}(\mathbf{v})\| \leq \|\mathbf{v}\|$.

(a) Prove that this inequality is true in general. [*Hint:* Prove that $\text{proj}_{\mathbf{u}}(\mathbf{v})$ is orthogonal to $\mathbf{v} - \text{proj}_{\mathbf{u}}(\mathbf{v})$ and use Pythagoras' Theorem.]

(b) Prove that the inequality $\|\text{proj}_{\mathbf{u}}(\mathbf{v})\| \leq \|\mathbf{v}\|$ is equivalent to the Cauchy-Schwarz Inequality.

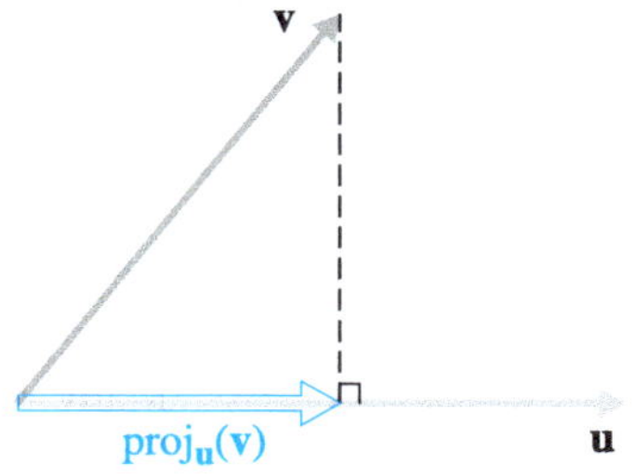

Figure 1.40

73. Use the fact that $\text{proj}_{\mathbf{u}}(\mathbf{v}) = c\mathbf{u}$ for some scalar c, together with Figure 1.41, to find c and thereby derive the formula for $\text{proj}_{\mathbf{u}}(\mathbf{v})$.

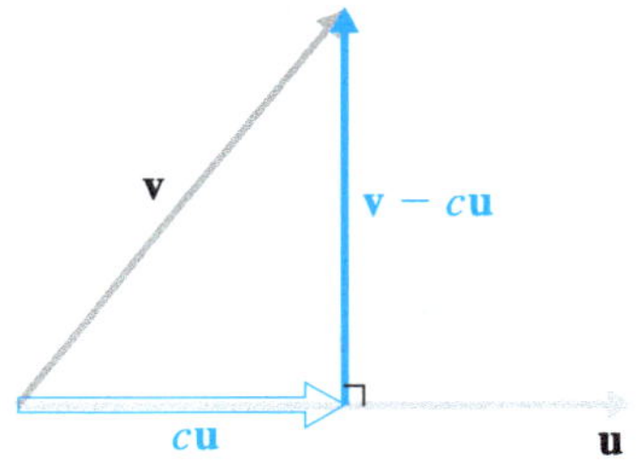

Figure 1.41

74. Using mathematical induction, prove the following generalization of the Triangle Inequality:

$$\|\mathbf{v}_1 + \mathbf{v}_2 + \cdots + \mathbf{v}_n\| \leq \|\mathbf{v}_1\| + \|\mathbf{v}_2\| + \cdots + \|\mathbf{v}_n\|$$

for all $n \geq 1$.

Exploration

Vectors and Geometry

Many results in plane Euclidean geometry can be proved using vector techniques. For example, in Example 1.24, we used vectors to prove Pythagoras' Theorem. In this exploration, we will use vectors to develop proofs for some other theorems from Euclidean geometry.

As an introduction to the notation and the basic approach, consider the following easy example.

Example 1.25 Give a vector description of the midpoint M of a line segment $\overline{AB}$.

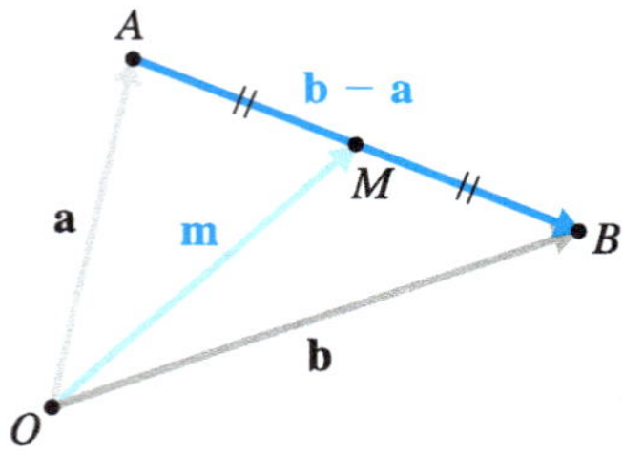

Figure 1.42
The midpoint of $\overline{AB}$

Solution We first convert everything to vector notation. If O denotes the origin and P is a point, let $\mathbf{p}$ be the vector $\overrightarrow{OP}$. In this situation, $\mathbf{a} = \overrightarrow{OA}$, $\mathbf{b} = \overrightarrow{OB}$, $\mathbf{m} = \overrightarrow{OM}$, and $\overrightarrow{AB} = \overrightarrow{OB} - \overrightarrow{OA} = \mathbf{b} - \mathbf{a}$ (Figure 1.42).

Now, since M is the midpoint of $\overline{AB}$, we have

$$\mathbf{m} - \mathbf{a} = \overrightarrow{AM} = \tfrac{1}{2}\overrightarrow{AB} = \tfrac{1}{2}(\mathbf{b} - \mathbf{a})$$

so

$$\mathbf{m} = \mathbf{a} + \tfrac{1}{2}(\mathbf{b} - \mathbf{a}) = \tfrac{1}{2}(\mathbf{a} + \mathbf{b})$$

1. Give a vector description of the point P that is one-third of the way from A to B on the line segment $\overline{AB}$. Generalize.

2. Prove that the line segment joining the midpoints of two sides of a triangle is parallel to the third side and half as long. (In vector notation, prove that $\overrightarrow{PQ} = \frac{1}{2}\overrightarrow{AB}$ in Figure 1.43.)

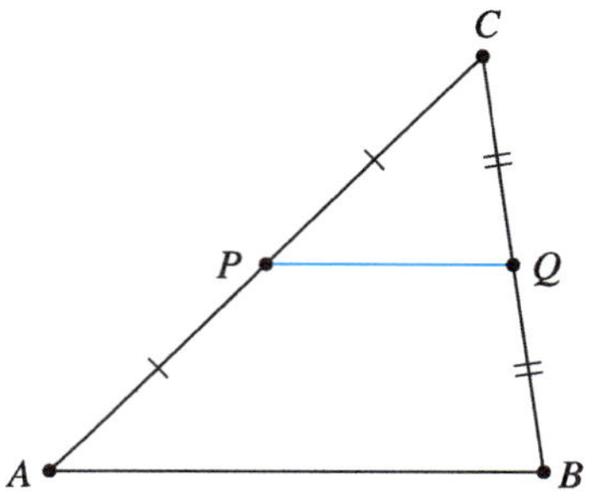

Figure 1.43

3. Prove that the quadrilateral $PQRS$ (Figure 1.44), whose vertices are the midpoints of the sides of an arbitrary quadrilateral $ABCD$, is a parallelogram.

4. A ***median*** of a triangle is a line segment from a vertex to the midpoint of the opposite side (Figure 1.45). Prove that the three medians of any triangle are *concurrent* (i.e., they have a common point of intersection) at a point G that is two-thirds of the distance from each vertex to the midpoint of the opposite side. [*Hint:* In Figure 1.46, show that the point that is two-thirds of the distance from A to P is given by $\frac{1}{3}(\mathbf{a} + \mathbf{b} + \mathbf{c})$. Then show that $\frac{1}{3}(\mathbf{a} + \mathbf{b} + \mathbf{c})$ is two-thirds of the distance from B to Q and two-thirds of the distance from C to R.] The point G in Figure 1.46 is called the ***centroid*** of the triangle.

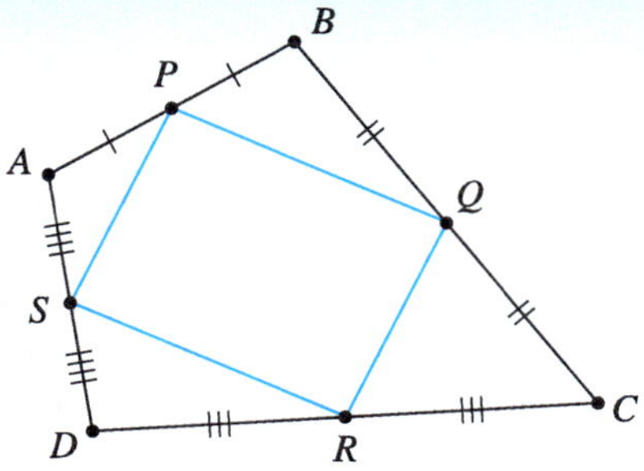

Figure 1.44

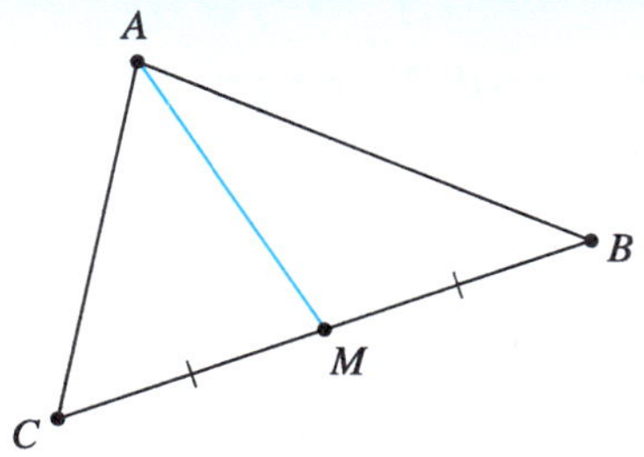

Figure 1.45
A median

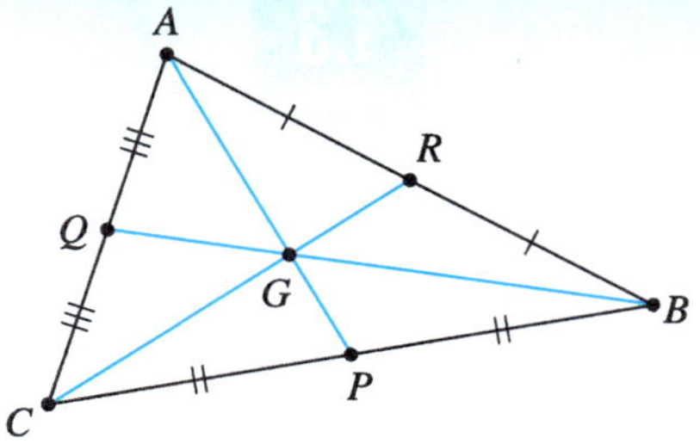

Figure 1.46
The centroid

5. An ***altitude*** of a triangle is a line segment from a vertex that is perpendicular to the opposite side (Figure 1.47). Prove that the three altitudes of a triangle are concurrent. [*Hint:* Let H be the point of intersection of the altitudes from A and B in Figure 1.48. Prove that $\overrightarrow{CH}$ is orthogonal to $\overrightarrow{AB}$.] The point H in Figure 1.48 is called the ***orthocenter*** of the triangle.

6. A ***perpendicular bisector*** of a line segment is a line through the midpoint of the segment, perpendicular to the segment (Figure 1.49). Prove that the perpendicular bisectors of the three sides of a triangle are concurrent. [*Hint:* Let K be the point of intersection of the perpendicular bisectors of $\overline{AC}$ and $\overline{BC}$ in Figure 1.50. Prove that $\overrightarrow{RK}$ is orthogonal to $\overrightarrow{AB}$.] The point K in Figure 1.50 is called the ***circumcenter*** of the triangle.

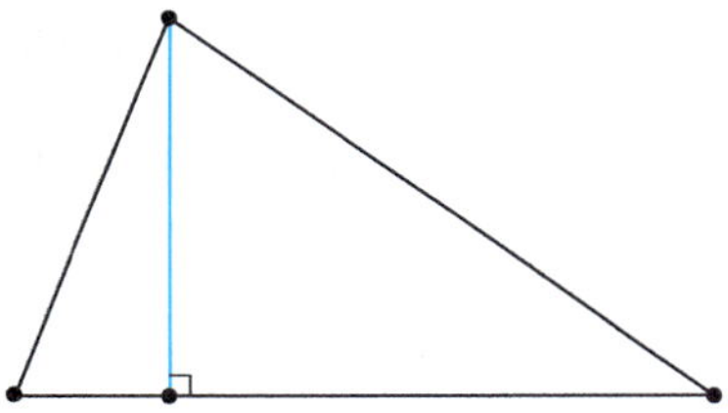

Figure 1.47
An altitude

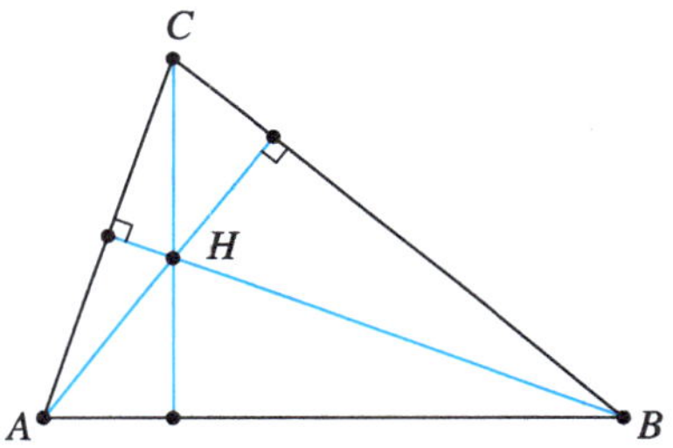

Figure 1.48
The orthocenter

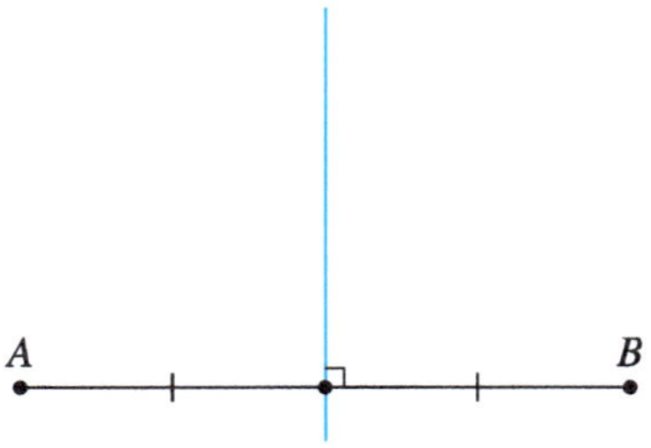

Figure 1.49
A perpendicular bisector

7. Let A and B be the endpoints of a diameter of a circle. If C is any point on the circle, prove that $\angle ACB$ is a right angle. [*Hint:* In Figure 1.51, let O be the center of the circle. Express everything in terms of **a** and **c** and show that $\overrightarrow{AC}$ is orthogonal to $\overrightarrow{BC}$.]

8. Prove that the line segments joining the midpoints of opposite sides of a quadrilateral bisect each other (Figure 1.52).

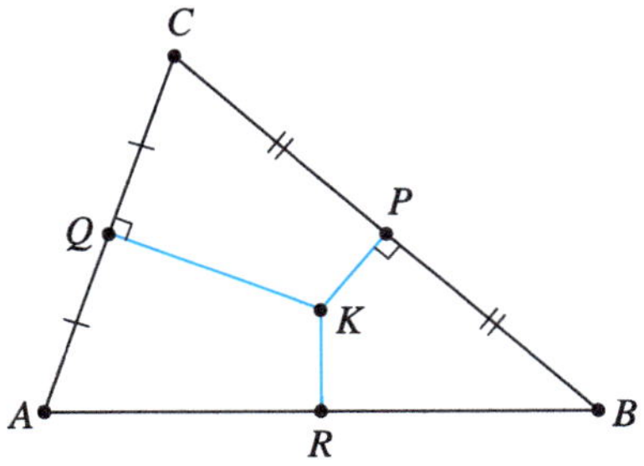

Figure 1.50
The circumcenter

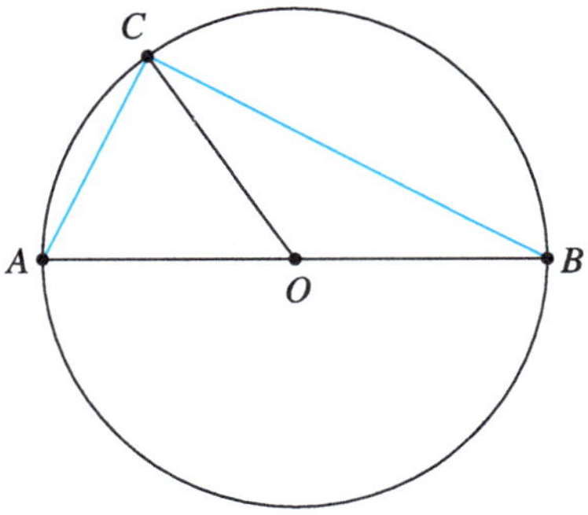

Figure 1.51

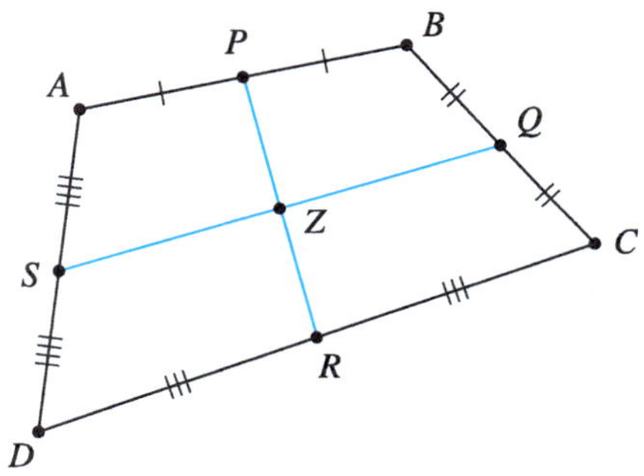

Figure 1.52

1.3 Lines and Planes

We are all familiar with the equation of a line in the Cartesian plane. We now want to consider lines in $\mathbb{R}^2$ from a vector point of view. The insights we obtain from this approach will allow us to generalize to lines in $\mathbb{R}^3$ and then to planes in $\mathbb{R}^3$. Much of the linear algebra we will consider in later chapters has its origins in the simple geometry of lines and planes; the ability to visualize these and to think geometrically about a problem will serve you well.

Lines in $\mathbb{R}^2$ and $\mathbb{R}^3$

In the xy-plane, the general form of the equation of a line is $ax + by = c$. If $b \neq 0$, then the equation can be rewritten as $y = -(a/b)x + c/b$, which has the form $y = mx + k$. [This is the slope-intercept form; m is the slope of the line, and the point with coordinates $(0, k)$ is its y-intercept.] To get vectors into the picture, let's consider an example.

Example 1.26

The line ℓ with equation $2x + y = 0$ is shown in Figure 1.53. It is a line with slope -2 passing through the origin. The left-hand side of the equation is in the form of a dot product; in fact, if we let $\mathbf{n} = \begin{bmatrix} 2 \\ 1 \end{bmatrix}$ and $\mathbf{x} = \begin{bmatrix} x \\ y \end{bmatrix}$, then the equation becomes $\mathbf{n} \cdot \mathbf{x} = 0$. The vector $\mathbf{n}$ is perpendicular to the line—that is, it is *orthogonal* to any vector $\mathbf{x}$ that is parallel to the line (Figure 1.54)—and it is called a ***normal vector*** to the line. The equation $\mathbf{n} \cdot \mathbf{x} = 0$ is the *normal form* of the equation of ℓ.

Another way to think about this line is to imagine a particle moving along the line. Suppose the particle is initially at the origin at time $t = 0$ and it moves along the line in such a way that its x-coordinate changes 1 unit per second. Then at $t = 1$ the particle is at $(1, -2)$, at $t = 1.5$ it is at $(1.5, -3)$, and, if we allow negative values of t (i.e., we consider where the particle was in the past), at $t = -2$ it is (or was) at $(-2, 4)$.

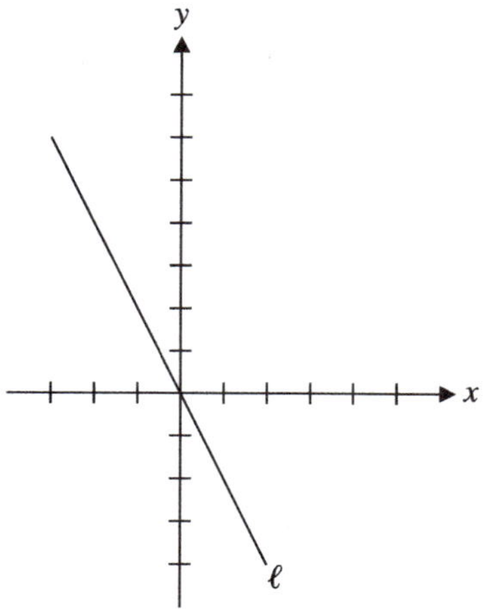

Figure 1.53
The line $2x + y = 0$

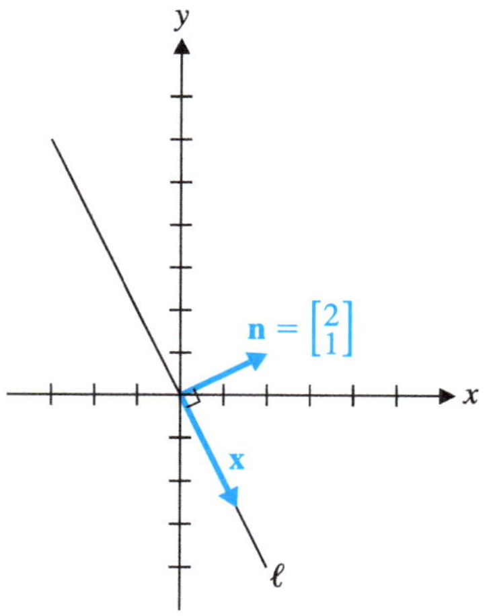

Figure 1.54
A normal vector $\mathbf{n}$

The Latin word *norma* refers to a carpenter's square, used for drawing right angles. Thus, a *normal* vector is one that is perpendicular to something else, usually a plane.

This movement is illustrated in Figure 1.55. In general, if $x = t$, then $y = -2t$, and we may write this relationship in vector form as

$$\begin{bmatrix} x \\ y \end{bmatrix} = \begin{bmatrix} t \\ -2t \end{bmatrix} = t\begin{bmatrix} 1 \\ -2 \end{bmatrix}$$

What is the significance of the vector $\mathbf{d} = \begin{bmatrix} 1 \\ -2 \end{bmatrix}$? It is a particular vector parallel to ℓ, called a ***direction vector*** for the line. As shown in Figure 1.56, we may write the equation of ℓ as $\mathbf{x} = t\mathbf{d}$. This is the *vector form* of the equation of the line.

If the line does not pass through the origin, then we must modify things slightly.

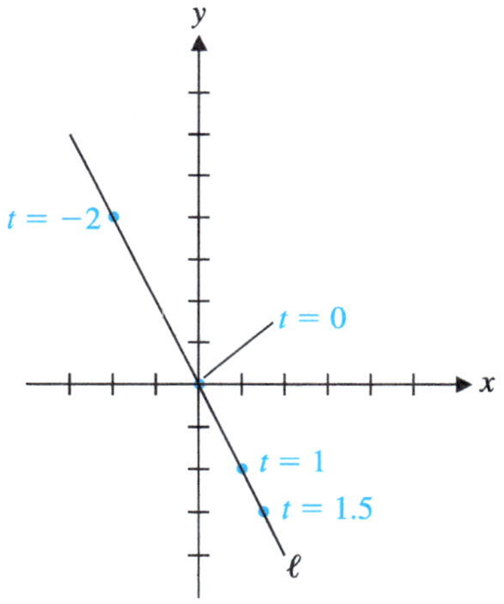

Figure 1.55

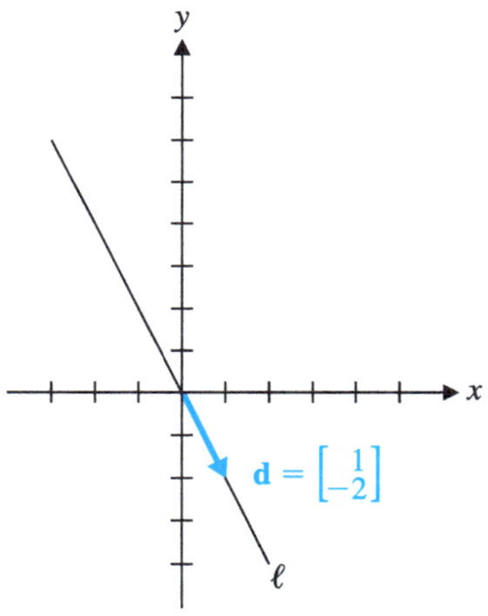

Figure 1.56
A direction vector **d**

Example 1.27

Consider the line ℓ with equation $2x + y = 5$ (Figure 1.57). This is just the line from Example 1.26 shifted upward 5 units. It also has slope -2, but its y-intercept is the point $(0, 5)$. It is clear that the vectors **d** and **n** from Example 1.26 are, respectively, a direction vector and a normal vector for this line too.

Thus, **n** is orthogonal to every vector that is parallel to ℓ. The point $P = (1, 3)$ is on ℓ. If $X = (x, y)$ represents a general point on ℓ, then the vector $\overrightarrow{PX} = \mathbf{x} - \mathbf{p}$ is parallel to ℓ and $\mathbf{n} \cdot (\mathbf{x} - \mathbf{p}) = 0$ (see Figure 1.58). Simplified, we have $\mathbf{n} \cdot \mathbf{x} = \mathbf{n} \cdot \mathbf{p}$. As a check, we compute

$$\mathbf{n}\cdot\mathbf{x} = \begin{bmatrix} 2 \\ 1 \end{bmatrix}\cdot\begin{bmatrix} x \\ y \end{bmatrix} = 2x + y \quad \text{and} \quad n\cdot p = \begin{bmatrix} 2 \\ 1 \end{bmatrix}\cdot\begin{bmatrix} 1 \\ 3 \end{bmatrix} = 5$$

Thus, the normal form $\mathbf{n} \cdot \mathbf{x} = \mathbf{n} \cdot \mathbf{p}$ is just a different representation of the general form of the equation of the line. (Note that in Example 1.26, **p** was the zero vector, so $\mathbf{n} \cdot \mathbf{p} = 0$ gave the right-hand side of the equation.)

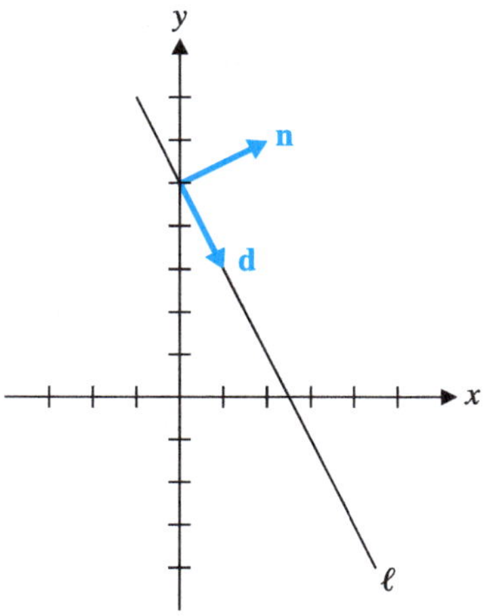

Figure 1.57
The line $2x + y = 5$

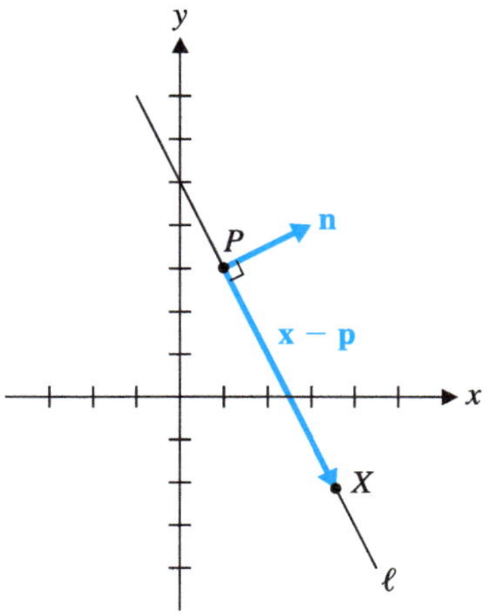

Figure 1.58
$\mathbf{n} \cdot (\mathbf{x} - \mathbf{p}) = 0$

These results lead to the following definition.

Definition The ***normal form of the equation of a line*** ℓ in $\mathbb{R}^2$ is

$$\mathbf{n} \cdot (\mathbf{x} - \mathbf{p}) = 0 \quad \text{or} \quad \mathbf{n} \cdot \mathbf{x} = \mathbf{n} \cdot \mathbf{p}$$

where $\mathbf{p}$ is a specific point on ℓ and $\mathbf{n} \neq \mathbf{0}$ is a normal vector for ℓ.

The ***general form of the equation of*** ℓ is $ax + by = c$, where $\mathbf{n} = \begin{bmatrix} a \\ b \end{bmatrix}$ is a normal vector for ℓ.

Continuing with Example 1.27, let us now find the vector form of the equation of ℓ. Note that, for each choice of $\mathbf{x}$, $\mathbf{x} - \mathbf{p}$ must be parallel to—and thus a multiple of—the direction vector $\mathbf{d}$. That is, $\mathbf{x} - \mathbf{p} = t\mathbf{d}$ or $\mathbf{x} = \mathbf{p} + t\mathbf{d}$ for some scalar t. In terms of components, we have

$$\begin{bmatrix} x \\ y \end{bmatrix} = \begin{bmatrix} 1 \\ 3 \end{bmatrix} + t\begin{bmatrix} 1 \\ -2 \end{bmatrix} \tag{1}$$

or

$$\begin{aligned} x &= 1 + t \\ y &= 3 - 2t \end{aligned} \tag{2}$$

Equation (1) is the vector form of the equation of ℓ, and the componentwise Equations (2) are called *parametric equations* of the line. The variable t is called a *parameter*.

How does all of this generalize to $\mathbb{R}^3$? Observe that the vector and parametric forms of the equations of a line carry over perfectly. The notion of the slope of a line in $\mathbb{R}^2$—which is difficult to generalize to three dimensions—is replaced by the more convenient notion of a direction vector, leading to the following definition.

The word *parameter* and the corresponding adjective *parametric* come from the Greek words *para*, meaning "alongside," and *metron*, meaning "measure." Mathematically speaking, a parameter is a variable in terms of which other variables are expressed—a new "measure" placed alongside old ones.

Definition The ***vector form of the equation of a line*** ℓ in $\mathbb{R}^2$ or $\mathbb{R}^3$ is

$$\mathbf{x} = \mathbf{p} + t\mathbf{d}$$

where $\mathbf{p}$ is a specific point on ℓ and $\mathbf{d} \neq \mathbf{0}$ is a direction vector for ℓ.

The equations corresponding to the components of the vector form of the equation are called ***parametric equations*** of ℓ.

We will often abbreviate this terminology slightly, referring simply to the general, normal, vector, and parametric equations of a line or plane.

Example 1.28

Find vector and parametric equations of the line in $\mathbb{R}^3$ through the point $P = (1, 2, -1)$, parallel to the vector $\mathbf{d} = \begin{bmatrix} 5 \\ -1 \\ 3 \end{bmatrix}$.

Solution The vector equation $\mathbf{x} = \mathbf{p} + t\mathbf{d}$ is

$$\begin{bmatrix} x \\ y \\ z \end{bmatrix} = \begin{bmatrix} 1 \\ 2 \\ -1 \end{bmatrix} + t\begin{bmatrix} 5 \\ -1 \\ 3 \end{bmatrix}$$

The parametric form is

$$\begin{aligned} x &= 1 + 5t \\ y &= 2 - t \\ z &= -1 + 3t \end{aligned}$$

Remarks

- The vector and parametric forms of the equation of a given line ℓ are not unique—in fact, there are infinitely many, since we may use any point on ℓ to determine $\mathbf{p}$ and any direction vector for ℓ. However, all direction vectors are clearly multiples of each other.

In Example 1.28, $(6, 1, 2)$ is another point on the line (take $t = 1$), and $\begin{bmatrix} 10 \\ -2 \\ 6 \end{bmatrix}$ is another direction vector. Therefore,

$$\begin{bmatrix} x \\ y \\ z \end{bmatrix} = \begin{bmatrix} 6 \\ 1 \\ 2 \end{bmatrix} + s\begin{bmatrix} 10 \\ -2 \\ 6 \end{bmatrix}$$

gives a different (but equivalent) vector equation for the line. The relationship between the two parameters s and t can be found by comparing the parametric equations: For a given point (x, y, z) on ℓ, we have

$$\begin{aligned} x &= 1 + 5t = 6 + 10s \\ y &= 2 - t = 1 - 2s \\ z &= -1 + 3t = 2 + 6s \end{aligned}$$

implying that

$$\begin{aligned} -10s + 5t &= 5 \\ 2s - t &= -1 \\ -6s + 3t &= 3 \end{aligned}$$

Each of these equations reduces to $t = 1 + 2s$.

- Intuitively, we know that a line is a *one-dimensional* object. The idea of "dimension" will be clarified in Chapters 3 and 6, but for the moment observe that this idea appears to agree with the fact that the vector form of the equation of a line requires *one* parameter.

Example 1.29 One often hears the expression "two points determine a line." Find a vector equation of the line ℓ in $\mathbb{R}^3$ determined by the points $P = (-1, 5, 0)$ and $Q = (2, 1, 1)$.

Solution We may choose any point on ℓ for $\mathbf{p}$, so we will use P (Q would also be fine).

A convenient direction vector is $\mathbf{d} = \overrightarrow{PQ} = \begin{bmatrix} 3 \\ -4 \\ 1 \end{bmatrix}$ (or any scalar multiple of this). Thus, we obtain

$$\begin{aligned} \mathbf{x} &= \mathbf{p} + t\mathbf{d} \\ &= \begin{bmatrix} -1 \\ 5 \\ 0 \end{bmatrix} + t\begin{bmatrix} 3 \\ -4 \\ 1 \end{bmatrix} \end{aligned}$$

Planes in $\mathbb{R}^3$

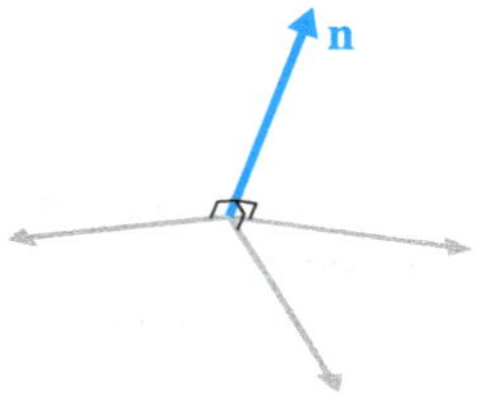

Figure 1.59
n is orthogonal to infinitely many vectors

The next question we should ask ourselves is, How does the general form of the equation of a line generalize to $\mathbb{R}^3$? We might reasonably guess that if $ax + by = c$ is the general form of the equation of a line in $\mathbb{R}^2$, then $ax + by + cz = d$ might represent a line in $\mathbb{R}^3$. In normal form, this equation would be $\mathbf{n} \cdot \mathbf{x} = \mathbf{n} \cdot \mathbf{p}$, where $\mathbf{n}$ is a normal vector to the line and $\mathbf{p}$ corresponds to a point on the line.

To see if this is a reasonable hypothesis, let's think about the special case of the equation $ax + by + cz = 0$. In normal form, it becomes $\mathbf{n} \cdot \mathbf{x} = 0$, where $\mathbf{n} = \begin{bmatrix} a \\ b \\ c \end{bmatrix}$. However, the set of all vectors $\mathbf{x}$ that satisfy this equation is the set of all vectors orthogonal to $\mathbf{n}$. As shown in Figure 1.59, vectors in infinitely many directions have this property, determining a family of parallel *planes*. So our guess was incorrect: It appears that $ax + by + cz = d$ is the equation of a plane—not a line—in $\mathbb{R}^3$.

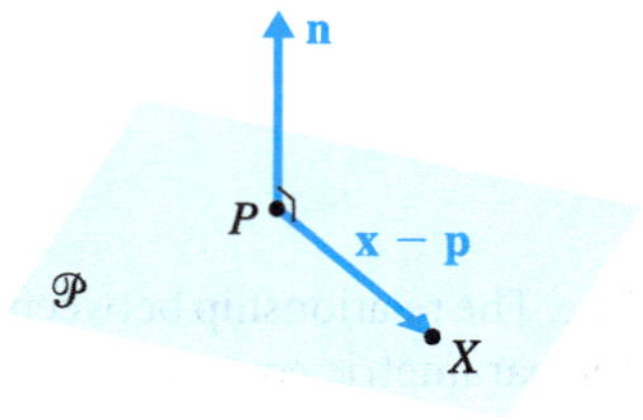

Figure 1.60
$\mathbf{n} \cdot (\mathbf{x} - \mathbf{p}) = 0$

Let's make this finding more precise. Every plane $\mathcal{P}$ in $\mathbb{R}^3$ can be determined by specifying a point $\mathbf{p}$ on $\mathcal{P}$ and a nonzero vector $\mathbf{n}$ normal to $\mathcal{P}$ (Figure 1.60). Thus, if $\mathbf{x}$ represents an arbitrary point on $\mathcal{P}$, we have $\mathbf{n} \cdot (\mathbf{x} - \mathbf{p}) = 0$ or $\mathbf{n} \cdot \mathbf{x} = \mathbf{n} \cdot \mathbf{p}$. If $\mathbf{n} = \begin{bmatrix} a \\ b \\ c \end{bmatrix}$ and $\mathbf{x} = \begin{bmatrix} x \\ y \\ z \end{bmatrix}$, then, in terms of components, the equation becomes $ax + by + cz = d$ (where $d = \mathbf{n} \cdot \mathbf{p}$).

Definition The ***normal form of the equation of a plane*** $\mathcal{P}$ in $\mathbb{R}^3$ is

$$\mathbf{n} \cdot (\mathbf{x} - \mathbf{p}) = 0 \quad \text{or} \quad \mathbf{n} \cdot \mathbf{x} = \mathbf{n} \cdot \mathbf{p}$$

where $\mathbf{p}$ is a specific point on $\mathcal{P}$ and $\mathbf{n} \neq \mathbf{0}$ is a normal vector for $\mathcal{P}$.

The ***general form of the equation of*** $\mathcal{P}$ is $ax + by + cz = d$, where $\mathbf{n} = \begin{bmatrix} a \\ b \\ c \end{bmatrix}$ is a normal vector for $\mathcal{P}$.

Note that any scalar multiple of a normal vector for a plane is another normal vector.

Example 1.30 Find the normal and general forms of the equation of the plane that contains the point $P = (6, 0, 1)$ and has normal vector $\mathbf{n} = \begin{bmatrix} 1 \\ 2 \\ 3 \end{bmatrix}$.

Solution With $\mathbf{p} = \begin{bmatrix} 6 \\ 0 \\ 1 \end{bmatrix}$ and $\mathbf{x} = \begin{bmatrix} x \\ y \\ z \end{bmatrix}$, we have $\mathbf{n} \cdot \mathbf{p} = 1 \cdot 6 + 2 \cdot 0 = 3 \cdot 1 = 9$, so the normal equation $\mathbf{n} \cdot \mathbf{x} = \mathbf{n} \cdot \mathbf{p}$ becomes the general equation $x + 2y + 3z = 9$.

Geometrically, it is clear that parallel planes have the same normal vector(s). Thus, their general equations have left-hand sides that are multiples of each other. So, for example, $2x + 4y + 6z = 10$ is the general equation of a plane that is parallel to the plane in Example 1.30, since we may rewrite the equation as $x + 2y + 3z = 5$—from which we see that the two planes have the same normal vector $\mathbf{n}$. (Note that the planes do not coincide, since the right-hand sides of their equations are distinct.)

We may also express the equation of a plane in vector or parametric form. To do so, we observe that a plane can also be determined by specifying one of its points P (by the vector $\mathbf{p}$) and *two* direction vectors $\mathbf{u}$ and $\mathbf{v}$ parallel to the plane (but not parallel to each other). As Figure 1.61 shows, given any point X in the plane (located

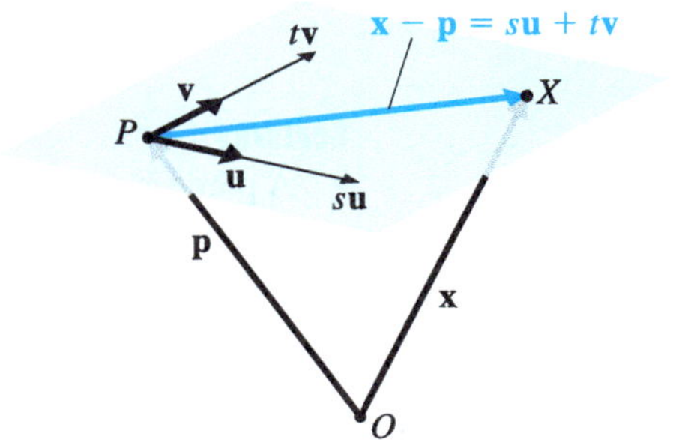

Figure 1.61
$\mathbf{x} - \mathbf{p} = s\mathbf{u} + t\mathbf{v}$

by $\mathbf{x}$), we can always find appropriate multiples $s\mathbf{u}$ and $t\mathbf{v}$ of the direction vectors such that $\mathbf{x} - \mathbf{p} = s\mathbf{u} + t\mathbf{v}$ or $\mathbf{x} = \mathbf{p} + s\mathbf{u} + t\mathbf{v}$. If we write this equation componentwise, we obtain parametric equations for the plane.

Definition The ***vector form of the equation of a plane*** $\mathcal{P}$ in $\mathbb{R}^3$ is

$$\mathbf{x} = \mathbf{p} + s\mathbf{u} + t\mathbf{v}$$

where $\mathbf{p}$ is a point on $\mathcal{P}$ and $\mathbf{u}$ and $\mathbf{v}$ are direction vectors for $\mathcal{P}$ ($\mathbf{u}$ and $\mathbf{v}$ are nonzero and parallel to $\mathcal{P}$, but not parallel to each other).

The equations corresponding to the components of the vector form of the equation are called ***parametric equations*** of $\mathcal{P}$.

Example 1.31 Find vector and parametric equations for the plane in Example 1.30.

Solution We need to find two direction vectors. We have one point $P = (6, 0, 1)$ in the plane; if we can find two other points Q and R in $\mathcal{P}$, then the vectors $\overrightarrow{PQ}$ and $\overrightarrow{PR}$ can serve as direction vectors (unless by bad luck they happen to be parallel!). By trial and error, we observe that $Q = (9, 0, 0)$ and $R = (3, 3, 0)$ both satisfy the general equation $x + 2y + 3z = 9$ and so lie in the plane. Then we compute

$$\mathbf{u} = \overrightarrow{PQ} = \mathbf{q} - \mathbf{p} = \begin{bmatrix} 3 \\ 0 \\ -1 \end{bmatrix} \text{ and } \mathbf{v} = \overrightarrow{PR} = \mathbf{r} - \mathbf{p} = \begin{bmatrix} -3 \\ 3 \\ -1 \end{bmatrix}$$

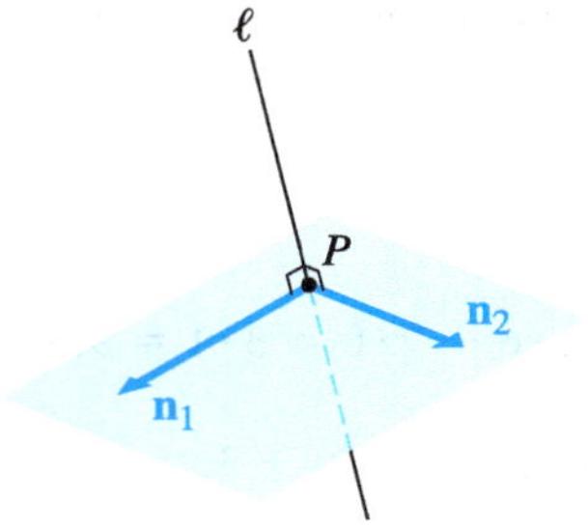

Figure 1.62
Two normals determine a line

which, since they are not scalar multiples of each other, will serve as direction vectors. Therefore, we have the vector equation of $\mathcal{P}$,

$$\begin{bmatrix} x \\ y \\ z \end{bmatrix} = \begin{bmatrix} 6 \\ 0 \\ 1 \end{bmatrix} + s\begin{bmatrix} 3 \\ 0 \\ -1 \end{bmatrix} + t\begin{bmatrix} -3 \\ 3 \\ -1 \end{bmatrix}$$

and the corresponding parametric equations,

$$\begin{aligned} x &= 6 + 3s - 3t \\ y &= \quad\; 3t \\ z &= 1 - \;s - \;t \end{aligned}$$

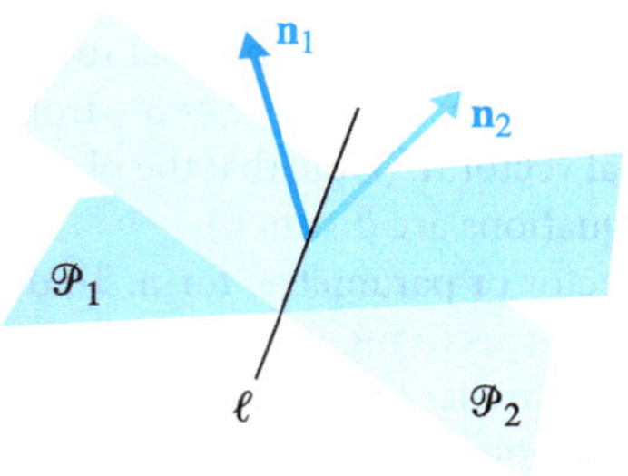

Figure 1.63
The intersection of two planes is a line

[What would have happened had we chosen $R = (0, 0, 3)$?]

Remarks

- A plane is a two-dimensional object, and its equation, in vector or parametric form, requires *two* parameters.
- As Figure 1.59 shows, given a point P and a nonzero vector $\mathbf{n}$ in $\mathbb{R}^3$, there are infinitely many lines through P with $\mathbf{n}$ as a normal vector. However, P and two nonparallel normal vectors $\mathbf{n}_1$ and $\mathbf{n}_2$ do serve to locate a line ℓ uniquely, since ℓ must then be the line through P that is perpendicular to the plane with equation $\mathbf{x} = \mathbf{p} + s\mathbf{n}_1 + t\mathbf{n}_2$ (Figure 1.62). Thus, a line in $\mathbb{R}^3$ can also be specified by a pair of equations

$$\begin{aligned} a_1x + b_1y + c_1z &= d_1 \\ a_2x + b_2y + c_2z &= d_2 \end{aligned}$$

one corresponding to each normal vector. But since these equations correspond to a pair of nonparallel planes (why nonparallel?), this is just the description of a line as the intersection of two nonparallel planes (Figure 1.63). Algebraically, the line consists of all points (x, y, z) that simultaneously satisfy both equations. We will explore this concept further in Chapter 2 when we discuss the solution of systems of linear equations.

Tables 1.2 and 1.3 summarize the information presented so far about the equations of lines and planes.

Observe once again that a single (general) equation describes a line in $\mathbb{R}^2$ but a plane in $\mathbb{R}^3$. [In higher dimensions, an object (line, plane, etc.) determined by a single equation of this type is usually called a ***hyperplane***.] The relationship among

Table 1.2 Equations of Lines in $\mathbb{R}^2$

Normal Form	General Form	Vector Form	Parametric Form
$\mathbf{n}\cdot\mathbf{x}=\mathbf{n}\cdot\mathbf{p}$	$ax+by=c$	$\mathbf{x}=\mathbf{p}+t\mathbf{d}$	$\begin{cases}x=p_1+td_1\\y=p_2+td_2\end{cases}$

Table 1.3 Lines and Planes in $\mathbb{R}^3$

	Normal Form	General Form	Vector Form	Parametric Form
Lines	$\begin{cases}\mathbf{n}_1\cdot\mathbf{x}=\mathbf{n}_1\cdot\mathbf{p}_1\\\mathbf{n}_2\cdot\mathbf{x}=\mathbf{n}_2\cdot\mathbf{p}_2\end{cases}$	$\begin{cases}a_1x+b_1y+c_1z=d_1\\a_2x+b_2y+c_2z=d_2\end{cases}$	$\mathbf{x}=\mathbf{p}+t\mathbf{d}$	$\begin{cases}x=p_1+td_1\\y=p_2+td_2\\z=p_3+td_3\end{cases}$
Planes	$\mathbf{n}\cdot\mathbf{x}=\mathbf{n}\cdot\mathbf{p}$	$ax+by+cz=d$	$\mathbf{x}=\mathbf{p}+s\mathbf{u}+t\mathbf{v}$	$\begin{cases}x=p_1+su_1+tv_1\\y=p_2+su_2+tv_2\\z=p_3+su_3+tv_3\end{cases}$

the dimension of the object, the number of equations required, and the dimension of the space is given by the "balancing formula":

(*dimension* of the object) + (*number* of general equations) = dimension of the space

The higher the dimension of the object, the fewer equations it needs. For example, a plane in $\mathbb{R}^3$ is two-dimensional, requires one general equation, and lives in a three-dimensional space: $2 + 1 = 3$. A line in $\mathbb{R}^3$ is one-dimensional and so needs $3 - 1 = 2$ equations. Note that the dimension of the object also agrees with the number of parameters in its vector or parametric form. Notions of "dimension" will be clarified in Chapters 3 and 6, but for the time being, these intuitive observations will serve us well.

We can now find the distance from a point to a line or a plane by combining the results of Section 1.2 with the results from this section.

Example 1.32

Find the distance from the point $B = (1, 0, 2)$ to the line ℓ through the point $A = (3, 1, 1)$ with direction vector $\mathbf{d} = \begin{bmatrix}-1\\1\\0\end{bmatrix}$.

Solution As we have already determined, we need to calculate the length of $\overrightarrow{PB}$, where P is the point on ℓ at the foot of the perpendicular from B. If we label $\mathbf{v} = \overrightarrow{AB}$, then $\overrightarrow{AP} = \text{proj}_{\mathbf{d}}(\mathbf{v})$ and $\overrightarrow{PB} = \mathbf{v} - \text{proj}_{\mathbf{d}}(\mathbf{v})$ (see Figure 1.64). We do the necessary calculations in several steps.

Step 1: $\mathbf{v} = \overrightarrow{AB} = \mathbf{b} - \mathbf{a} = \begin{bmatrix}1\\0\\2\end{bmatrix} - \begin{bmatrix}3\\1\\1\end{bmatrix} = \begin{bmatrix}-2\\-1\\1\end{bmatrix}$

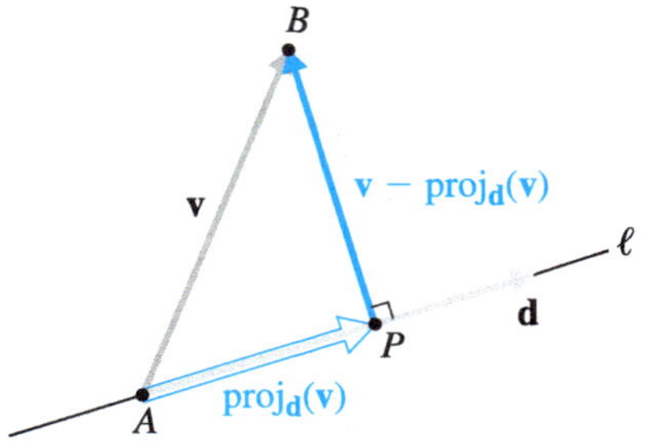

Figure 1.64
$d(B, \ell) = \|\mathbf{v} - \text{proj}_{\mathbf{d}}(\mathbf{v})\|$

Step 2: The projection of **v** onto **d** is

$$\begin{aligned}\text{proj}_{\mathbf{d}}(\mathbf{v}) &= \left(\frac{\mathbf{d}\cdot\mathbf{v}}{\mathbf{d}\cdot\mathbf{d}}\right)\mathbf{d} \\ &= \left(\frac{(-1)\cdot(-2) + 1\cdot(-1) + 0\cdot 1}{(-1)^2 + 1 + 0}\right)\begin{bmatrix}-1\\1\\0\end{bmatrix} \\ &= \tfrac{1}{2}\begin{bmatrix}-1\\1\\0\end{bmatrix} \\ &= \begin{bmatrix}-\frac{1}{2}\\ \frac{1}{2}\\ 0\end{bmatrix}\end{aligned}$$

Step 3: The vector we want is

$$\mathbf{v} - \text{proj}_{\mathbf{d}}(\mathbf{v}) = \begin{bmatrix}-2\\-1\\1\end{bmatrix} - \begin{bmatrix}-\frac{1}{2}\\ \frac{1}{2}\\ 0\end{bmatrix} = \begin{bmatrix}-\frac{3}{2}\\ -\frac{3}{2}\\ 1\end{bmatrix}$$

Step 4: The distance $d(B, \ell)$ from B to ℓ is

$$\|\mathbf{v} - \text{proj}_{\mathbf{d}}(\mathbf{v})\| = \left\|\begin{bmatrix}-\frac{3}{2}\\ -\frac{3}{2}\\ 1\end{bmatrix}\right\|$$

Using Theorem 1.3(b) to simplify the calculation, we have

$$\begin{aligned}\|\mathbf{v} - \text{proj}_{\mathbf{d}}(\mathbf{v})\| &= \tfrac{1}{2}\left\|\begin{bmatrix}-3\\-3\\2\end{bmatrix}\right\| \\ &= \tfrac{1}{2}\sqrt{9 + 9 + 4} \\ &= \tfrac{1}{2}\sqrt{22}\end{aligned}$$

Note

- In terms of our earlier notation, $d(B, \ell) = d(\mathbf{v}, \text{proj}_{\mathbf{d}}(\mathbf{v}))$.

In the case where the line ℓ is in $\mathbb{R}^2$ and its equation has the general form $ax + by = c$, the distance $d(B, \ell)$ from $B = (x_0, y_0)$ is given by the formula

$$d(B, \ell) = \frac{|ax_0 + by_0 - c|}{\sqrt{a^2 + b^2}} \tag{3}$$

You are invited to prove this formula in Exercise 39.

Example 1.33

Find the distance from the point $B = (1, 0, 2)$ to the plane $\mathcal{P}$ whose general equation is $x + y - z = 1$.

Solution In this case, we need to calculate the length of $\overrightarrow{PB}$, where P is the point on $\mathcal{P}$ at the foot of the perpendicular from B. As Figure 1.65 shows, if A is any point on $\mathcal{P}$ and we situate the normal vector $\mathbf{n} = \begin{bmatrix} 1 \\ 1 \\ -1 \end{bmatrix}$ of $\mathcal{P}$ so that its tail is at A, then we need to find the length of the projection of $\overrightarrow{AB}$ onto $\mathbf{n}$. Again we do the necessary calculations in steps.

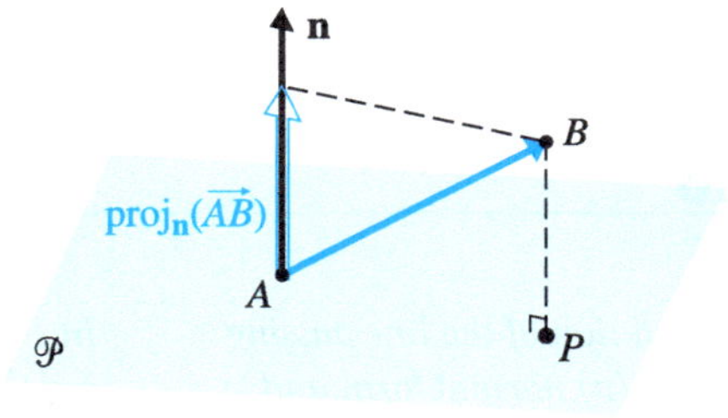

Figure 1.65
$d(B, \mathcal{P}) = \|\text{proj}_{\mathbf{n}}(\overrightarrow{AB})\|$

Step 1: By trial and error, we find any point whose coordinates satisfy the equation $x + y - z = 1$. $A = (1, 0, 0)$ will do.

Step 2: Set

$$\mathbf{v} = \overrightarrow{AB} = \mathbf{b} - \mathbf{a} = \begin{bmatrix} 1 \\ 0 \\ 2 \end{bmatrix} - \begin{bmatrix} 1 \\ 0 \\ 0 \end{bmatrix} = \begin{bmatrix} 0 \\ 0 \\ 2 \end{bmatrix}$$

Step 3: The projection of $\mathbf{v}$ onto $\mathbf{n}$ is

$$\begin{aligned} \text{proj}_{\mathbf{n}}(\mathbf{v}) &= \left(\frac{\mathbf{n} \cdot \mathbf{v}}{\mathbf{n} \cdot \mathbf{n}}\right)\mathbf{n} \\ &= \left(\frac{1 \cdot 0 + 1 \cdot 0 - 1 \cdot 2}{1 + 1 + (-1)^2}\right)\begin{bmatrix} 1 \\ 1 \\ -1 \end{bmatrix} \\ &= -\frac{2}{3}\begin{bmatrix} 1 \\ 1 \\ -1 \end{bmatrix} = \begin{bmatrix} -\frac{2}{3} \\ -\frac{2}{3} \\ \frac{2}{3} \end{bmatrix} \end{aligned}$$

Step 4: The distance $d(B, \mathcal{P})$ from B to $\mathcal{P}$ is

$$\|\text{proj}_{\mathbf{n}}(\mathbf{v})\| = \left|-\tfrac{2}{3}\right| \left\| \begin{bmatrix} 1 \\ 1 \\ -1 \end{bmatrix} \right\|$$

$$= \tfrac{2}{3} \left\| \begin{bmatrix} 1 \\ 1 \\ -1 \end{bmatrix} \right\|$$

$$= \tfrac{2}{3}\sqrt{3}$$

In general, the distance $d(B, \mathcal{P})$ from the point $B = (x_0, y_0, z_0)$ to the plane whose general equation is $ax + by + cz = d$ is given by the formula

$$d(B, \mathcal{P}) = \frac{|ax_0 + by_0 + cz_0 - d|}{\sqrt{a^2 + b^2 + c^2}} \tag{4}$$

You will be asked to derive this formula in Exercise 40.

Exercises 1.3

In Exercises 1 and 2, write the equation of the line passing through P with normal vector **n** *in (a) normal form and (b) general form.*

1. $P = (0, 0), \mathbf{n} = \begin{bmatrix} 3 \\ 2 \end{bmatrix}$ **2.** $P = (1, 2), \mathbf{n} = \begin{bmatrix} 3 \\ -4 \end{bmatrix}$

In Exercises 3–6, write the equation of the line passing through P with direction vector **d** *in (a) vector form and (b) parametric form.*

3. $P = (1, 0), \mathbf{d} = \begin{bmatrix} -1 \\ 3 \end{bmatrix}$ **4.** $P = (-4, 4), \mathbf{d} = \begin{bmatrix} 1 \\ 1 \end{bmatrix}$

5. $P = (0, 0, 0), \mathbf{d} = \begin{bmatrix} 1 \\ -1 \\ 4 \end{bmatrix}$ **6.** $P = (3, 0, -2), \mathbf{d} = \begin{bmatrix} 2 \\ 5 \\ 0 \end{bmatrix}$

In Exercises 7 and 8, write the equation of the plane passing through P with normal vector **n** *in (a) normal form and (b) general form.*

7. $P = (0, 1, 0), \mathbf{n} = \begin{bmatrix} 3 \\ 2 \\ 1 \end{bmatrix}$ **8.** $P = (3, 0, -2), \mathbf{n} = \begin{bmatrix} 2 \\ 5 \\ 0 \end{bmatrix}$

In Exercises 9 and 10, write the equation of the plane passing through P with direction vectors **u** *and* **v** *in (a) vector form and (b) parametric form.*

9. $P = (0, 0, 0), \mathbf{u} = \begin{bmatrix} 2 \\ 1 \\ 2 \end{bmatrix}, \mathbf{v} = \begin{bmatrix} -3 \\ 2 \\ 1 \end{bmatrix}$

10. $P = (6, -4, -3), \mathbf{u} = \begin{bmatrix} 0 \\ 1 \\ 1 \end{bmatrix}, \mathbf{v} = \begin{bmatrix} -1 \\ 1 \\ 1 \end{bmatrix}$

In Exercises 11 and 12, give the vector equation of the line passing through P and Q.

11. $P = (1, -2), Q = (3, 0)$

12. $P = (0, 1, -1), Q = (-2, 1, 3)$

In Exercises 13 and 14, give the vector equation of the plane passing through P, Q, and R.

13. $P = (1, 1, 1), Q = (4, 0, 2), R = (0, 1, -1)$

14. $P = (1, 1, 0), Q = (1, 0, 1), R = (0, 1, 1)$

15. Find parametric equations and an equation in vector form for the lines in $\mathbb{R}^2$ with the following equations:

(a) $y = 3x - 1$ **(b)** $3x + 2y = 5$

16. Consider the vector equation $\mathbf{x} = \mathbf{p} + t(\mathbf{q} - \mathbf{p})$, where $\mathbf{p}$ and $\mathbf{q}$ correspond to distinct points P and Q in $\mathbb{R}^2$ or $\mathbb{R}^3$.

(a) Show that this equation describes the line segment $\overline{PQ}$ as t varies from 0 to 1.

(b) For which value of t is $\mathbf{x}$ the midpoint of $\overline{PQ}$, and what is $\mathbf{x}$ in this case?

(c) Find the midpoint of $\overline{PQ}$ when $P = (2, -3)$ and $Q = (0, 1)$.

(d) Find the midpoint of $\overline{PQ}$ when $P = (1, 0, 1)$ and $Q = (4, 1, -2)$.

(e) Find the two points that divide $\overline{PQ}$ in part (c) into three equal parts.

(f) Find the two points that divide $\overline{PQ}$ in part (d) into three equal parts.

17. Suggest a "vector proof" of the fact that, in $\mathbb{R}^2$, two lines with slopes m_1 and m_2 are perpendicular if and only if $m_1 m_2 = -1$.

18. The line ℓ passes through the point $P = (1, -1, 1)$ and has direction vector $\mathbf{d} = \begin{bmatrix} 2 \\ 3 \\ -1 \end{bmatrix}$. For each of the following planes $\mathscr{P}$, determine whether ℓ and $\mathscr{P}$ are parallel, perpendicular, or neither:

(a) $2x + 3y - z = 1$ **(b)** $4x - y + 5z = 0$

(c) $x - y - z = 3$ **(d)** $4x + 6y - 2z = 0$

19. The plane $\mathscr{P}_1$ has the equation $4x - y + 5z = 2$. For each of the planes $\mathscr{P}$ in Exercise 18, determine whether $\mathscr{P}_1$ and $\mathscr{P}$ are parallel, perpendicular, or neither.

20. Find the vector form of the equation of the line in $\mathbb{R}^2$ that passes through $P = (2, -1)$ and is perpendicular to the line with general equation $2x - 3y = 1$.

21. Find the vector form of the equation of the line in $\mathbb{R}^2$ that passes through $P = (2, -1)$ and is parallel to the line with general equation $2x - 3y = 1$.

22. Find the vector form of the equation of the line in $\mathbb{R}^3$ that passes through $P = (-1, 0, 3)$ and is perpendicular to the plane with general equation $x - 3y + 2z = 5$.

23. Find the vector form of the equation of the line in $\mathbb{R}^3$ that passes through $P = (-1, 0, 3)$ and is parallel to the line with parametric equations

$$\begin{aligned} x &= 1 - t \\ y &= 2 + 3t \\ z &= -2 - t \end{aligned}$$

24. Find the normal form of the equation of the plane that passes through $P = (0, -2, 5)$ and is parallel to the plane with general equation $6x - y + 2z = 3$.

25. A cube has vertices at the eight points (x, y, z), where each of x, y, and z is either 0 or 1. (See Figure 1.34.)

(a) Find the general equations of the planes that determine the six faces (sides) of the cube.

(b) Find the general equation of the plane that contains the diagonal from the origin to $(1, 1, 1)$ and is perpendicular to the xy-plane.

(c) Find the general equation of the plane that contains the side diagonals referred to in Example 1.22.

26. Find the equation of the set of all points that are equidistant from the points $P = (1, 0, -2)$ and $Q = (5, 2, 4)$.

In Exercises 27 and 28, find the distance from the point Q to the line ℓ.

27. $Q = (2, 2)$, ℓ with equation $\begin{bmatrix} x \\ y \end{bmatrix} = \begin{bmatrix} -1 \\ 2 \end{bmatrix} + t\begin{bmatrix} 1 \\ -1 \end{bmatrix}$

28. $Q = (0, 1, 0)$, ℓ with equation $\begin{bmatrix} x \\ y \\ z \end{bmatrix} = \begin{bmatrix} 1 \\ 1 \\ 1 \end{bmatrix} + t\begin{bmatrix} -2 \\ 0 \\ 3 \end{bmatrix}$

In Exercises 29 and 30, find the distance from the point Q to the plane $\mathscr{P}$.

29. $Q = (2, 2, 2)$, $\mathscr{P}$ with equation $x + y - z = 0$

30. $Q = (0, 0, 0)$, $\mathscr{P}$ with equation $x - 2y + 2z = 1$

Figure 1.66 suggests a way to use vectors to locate the point R on ℓ that is closest to Q.

31. Find the point R on ℓ that is closest to Q in Exercise 27.

32. Find the point R on ℓ that is closest to Q in Exercise 28.

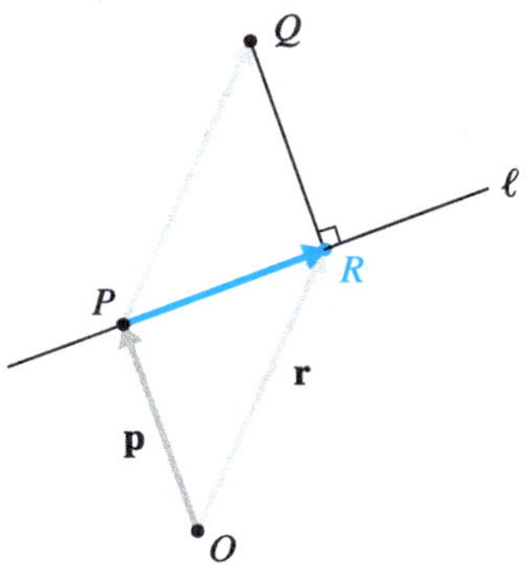

Figure 1.66

$\mathbf{r} = \mathbf{p} + \overrightarrow{PR}$

Figure 1.67 suggests a way to use vectors to locate the point R on $\mathcal{P}$ that is closest to Q.

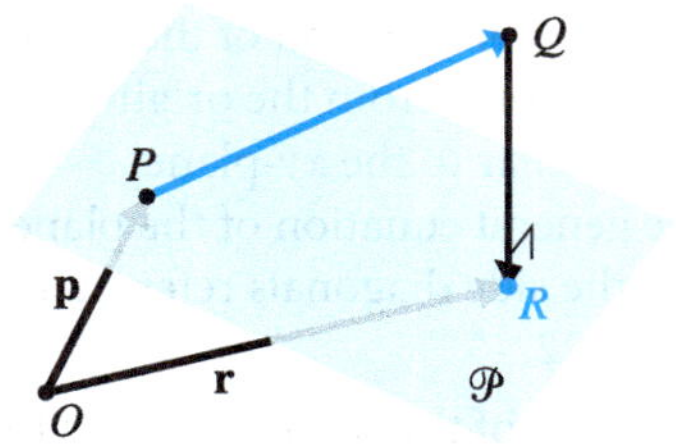

Figure 1.67
$\mathbf{r} = \mathbf{p} + \overrightarrow{PQ} + \overrightarrow{QR}$

33. Find the point R on $\mathcal{P}$ that is closest to Q in Exercise 29.

34. Find the point R on $\mathcal{P}$ that is closest to Q in Exercise 30.

In Exercises 35 and 36, find the distance between the parallel lines.

35. $\begin{bmatrix} x \\ y \end{bmatrix} = \begin{bmatrix} 1 \\ 1 \end{bmatrix} + s\begin{bmatrix} -2 \\ 3 \end{bmatrix}$ and $\begin{bmatrix} x \\ y \end{bmatrix} = \begin{bmatrix} 5 \\ 4 \end{bmatrix} + t\begin{bmatrix} -2 \\ 3 \end{bmatrix}$

36. $\begin{bmatrix} x \\ y \\ z \end{bmatrix} = \begin{bmatrix} 1 \\ 0 \\ -1 \end{bmatrix} + s\begin{bmatrix} 1 \\ 1 \\ 1 \end{bmatrix}$ and $\begin{bmatrix} x \\ y \\ z \end{bmatrix} = \begin{bmatrix} 0 \\ 1 \\ 1 \end{bmatrix} + t\begin{bmatrix} 1 \\ 1 \\ 1 \end{bmatrix}$

In Exercises 37 and 38, find the distance between the parallel planes.

37. $2x + y - 2z = 0$ and $2x + y - 2z = 5$

38. $x + y + z = 1$ and $x + y + z = 3$

39. Prove Equation (3) on page 43.

40. Prove Equation (4) on page 44.

41. Prove that, in $\mathbb{R}^2$, the distance between parallel lines with equations $\mathbf{n} \cdot \mathbf{x} = c_1$ and $\mathbf{n} \cdot \mathbf{x} = c_2$ is given by $\dfrac{|c_1 - c_2|}{\|\mathbf{n}\|}$.

42. Prove that the distance between parallel planes with equations $\mathbf{n} \cdot \mathbf{x} = d_1$ and $\mathbf{n} \cdot \mathbf{x} = d_2$ is given by $\dfrac{|d_1 - d_2|}{\|\mathbf{n}\|}$.

If two nonparallel planes $\mathcal{P}_1$ and $\mathcal{P}_2$ have normal vectors $\mathbf{n}_1$ and $\mathbf{n}_2$ and θ is the angle between $\mathbf{n}_1$ and $\mathbf{n}_2$, then we define the angle between $\mathcal{P}_1$ and $\mathcal{P}_2$ to be either θ or $180° - \theta$, whichever is an acute angle. (Figure 1.68)

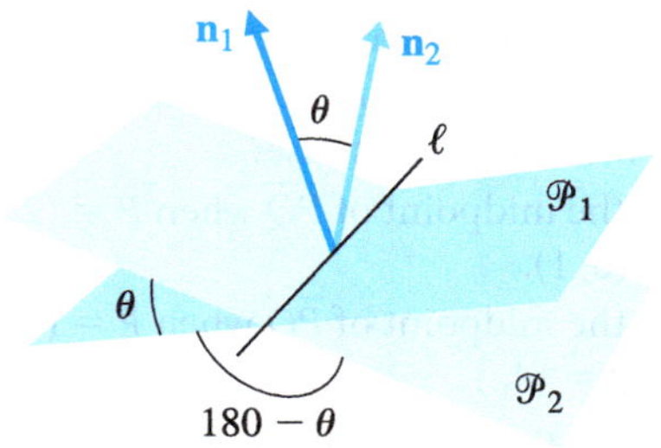

Figure 1.68

In Exercises 43–44, find the acute angle between the planes with the given equations.

43. $x + y + z = 0$ and $2x + y - 2z = 0$

44. $3x - y + 2z = 5$ and $x + 4y - z = 2$

In Exercises 45–46, show that the plane and line with the given equations intersect, and then find the acute angle of intersection between them.

45. The plane given by $x + y + 2z = 0$ and the line given by
$$\begin{aligned} x &= 2 + t \\ y &= 1 - 2t \\ z &= 3 + t \end{aligned}$$

46. The plane given by $4x - y - z = 6$ and the line given by
$$\begin{aligned} x &= t \\ y &= 1 + 2t \\ z &= 2 + 3t \end{aligned}$$

Exercises 47–48 explore one approach to the problem of finding the projection of a vector onto a plane. As Figure 1.69 shows, if $\mathcal{P}$ is a plane through the origin in $\mathbb{R}^3$ with normal vector $\mathbf{n}$, and $\mathbf{v}$ is a vector in $\mathbb{R}^3$, then $\mathbf{p} = \mathrm{proj}_{\mathcal{P}}(\mathbf{v})$ is a vector in $\mathcal{P}$ such that $\mathbf{v} - c\mathbf{n} = \mathbf{p}$ for some scalar c.

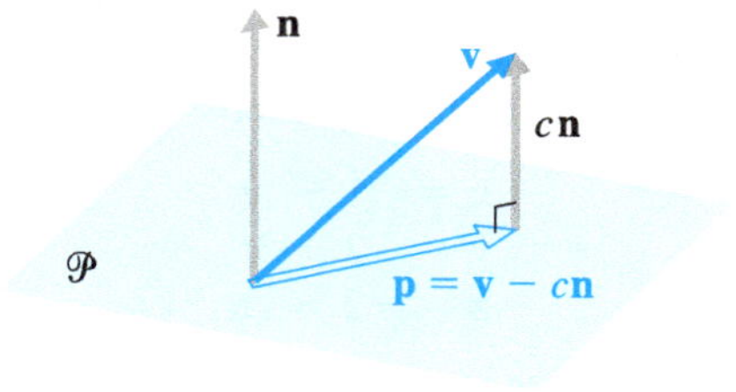

Figure 1.69
Projection onto a plane

47. Using the fact that **n** is orthogonal to every vector in $\mathcal{P}$ (and hence to **p**), solve for c and thereby find an expression for **p** in terms of **v** and **n**.

48. Use the method of Exercise 43 to find the projection of

$$\mathbf{v} = \begin{bmatrix} 1 \\ 0 \\ -2 \end{bmatrix}$$

onto the planes with the following equations:

(a) $x + y + z = 0$ **(b)** $3x - y + z = 0$

(c) $x - 2z = 0$ **(d)** $2x - 3y + z = 0$

Exploration

The Cross Product

It would be convenient if we could easily convert the vector form $\mathbf{x} = \mathbf{p} + s\mathbf{u} + t\mathbf{v}$ of the equation of a plane to the normal form $\mathbf{n} \cdot \mathbf{x} = \mathbf{n} \cdot \mathbf{p}$. What we need is a process that, given two nonparallel vectors $\mathbf{u}$ and $\mathbf{v}$, produces a third vector $\mathbf{n}$ that is orthogonal to both $\mathbf{u}$ and $\mathbf{v}$. One approach is to use a construction known as the ***cross product*** of vectors. *Only valid in* $\mathbb{R}^3$, it is defined as follows:

Definition The *cross product* of $\mathbf{u} = \begin{bmatrix} u_1 \\ u_2 \\ u_3 \end{bmatrix}$ and $\mathbf{v} = \begin{bmatrix} v_1 \\ v_2 \\ v_3 \end{bmatrix}$ is the vector $\mathbf{u} \times \mathbf{v}$ defined by

$$\mathbf{u} \times \mathbf{v} = \begin{bmatrix} u_2v_3 - u_3v_2 \\ u_3v_1 - u_1v_3 \\ u_1v_2 - u_2v_1 \end{bmatrix}$$

A shortcut that can help you remember how to calculate the cross product of two vectors is illustrated below. Under each complete vector, write the first two components of that vector. Ignoring the two components on the top line, consider each block of four: Subtract the products of the components connected by dashed lines from the products of the components connected by solid lines. (It helps to notice that the first component of $\mathbf{u} \times \mathbf{v}$ has no 1s as subscripts, the second has no 2s, and the third has no 3s.)

$$\begin{array}{ccc} u_1 & v_1 & \\ u_2 & v_2 & \\ u_3 & v_3 & u_2v_3 - u_3v_2 \\ u_1 & v_1 & u_3v_1 - u_1v_3 \\ u_2 & v_2 & u_1v_2 - u_2v_1 \end{array}$$

The following problems briefly explore the cross product.

1. Compute $\mathbf{u} \times \mathbf{v}$.

(a) $\mathbf{u} = \begin{bmatrix} 0 \\ 1 \\ 1 \end{bmatrix}, \mathbf{v} = \begin{bmatrix} 3 \\ -1 \\ 2 \end{bmatrix}$ (b) $\mathbf{u} = \begin{bmatrix} 3 \\ -1 \\ 2 \end{bmatrix}, \mathbf{v} = \begin{bmatrix} 0 \\ 1 \\ 1 \end{bmatrix}$

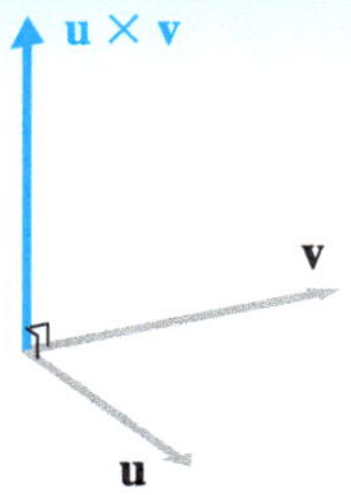

Figure 1.70

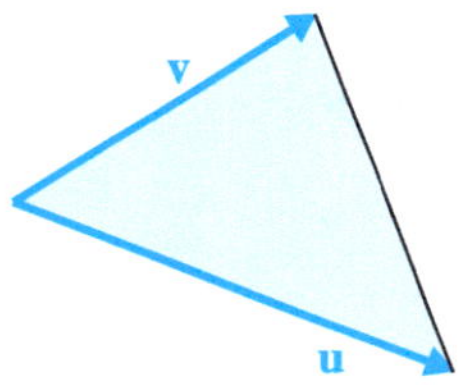

Figure 1.71

(c) $\mathbf{u} = \begin{bmatrix} -1 \\ 2 \\ 3 \end{bmatrix}, \mathbf{v} = \begin{bmatrix} 2 \\ -4 \\ -6 \end{bmatrix}$ (d) $\mathbf{u} = \begin{bmatrix} 1 \\ 1 \\ 1 \end{bmatrix}, \mathbf{v} = \begin{bmatrix} 1 \\ 2 \\ 3 \end{bmatrix}$

2. Show that $\mathbf{e}_1 \times \mathbf{e}_2 = \mathbf{e}_3$, $\mathbf{e}_2 \times \mathbf{e}_3 = \mathbf{e}_1$, and $\mathbf{e}_3 \times \mathbf{e}_1 = \mathbf{e}_2$.
3. Using the definition of a cross product, prove that $\mathbf{u} \times \mathbf{v}$ (as shown in Figure 1.70) is orthogonal to **u** and **v**.
4. Use the cross product to help find the normal form of the equation of the plane.
   (a) The plane passing through $P = (1, 0, -2)$, parallel to $\mathbf{u} = \begin{bmatrix} 0 \\ 1 \\ 1 \end{bmatrix}$ and $\mathbf{v} = \begin{bmatrix} 3 \\ -1 \\ 2 \end{bmatrix}$
   (b) The plane passing through $P = (0, -1, 1)$, $Q = (2, 0, 2)$, and $R = (1, 2, -1)$
5. Prove the following properties of the cross product:
   (a) $\mathbf{v} \times \mathbf{u} = -(\mathbf{u} \times \mathbf{v})$ (b) $\mathbf{u} \times \mathbf{0} = \mathbf{0}$
   (c) $\mathbf{u} \times \mathbf{u} = \mathbf{0}$ (d) $\mathbf{u} \times k\mathbf{v} = k(\mathbf{u} \times \mathbf{v})$
   (e) $\mathbf{u} \times k\mathbf{u} = \mathbf{0}$ (f) $\mathbf{u} \times (\mathbf{v} + \mathbf{w}) = \mathbf{u} \times \mathbf{v} + \mathbf{u} \times \mathbf{w}$
6. Prove the following properties of the cross product:
   (a) $\mathbf{u} \cdot (\mathbf{v} \times \mathbf{w}) = (\mathbf{u} \times \mathbf{v}) \cdot \mathbf{w}$ (b) $\mathbf{u} \times (\mathbf{v} \times \mathbf{w}) = (\mathbf{u} \cdot \mathbf{w})\mathbf{v} - (\mathbf{u} \cdot \mathbf{v})\mathbf{w}$
   (c) $\|\mathbf{u} \times \mathbf{v}\|^2 = \|\mathbf{u}\|^2 \|\mathbf{v}\|^2 - (\mathbf{u} \cdot \mathbf{v})^2$
7. Redo Problems 2 and 3, this time making use of Problems 5 and 6.
8. Let **u** and **v** be vectors in $\mathbb{R}^3$ and let θ be the angle between **u** and **v**.
   (a) Prove that $\|\mathbf{u} \times \mathbf{v}\| = \|\mathbf{u}\| \, \|\mathbf{v}\| \sin \theta$. [*Hint:* Use Problem 6(c).]
   (b) Prove that the area $\mathcal{A}$ of the triangle determined by **u** and **v** (as shown in Figure 1.71) is given by

$$\mathcal{A} = \tfrac{1}{2} \|\mathbf{u} \times \mathbf{v}\|$$

   (c) Use the result in part (b) to compute the area of the triangle with vertices $A = (1, 2, 1)$, $B = (2, 1, 0)$, and $C = (5, -1, 3)$.

Writing Project

The Origins of the Dot Product and Cross Product

The notations for dot and cross product that we use today were introduced in the late 19th century by Josiah Willard Gibbs, a professor of mathematical physics at Yale University. Edwin B. Wilson was a graduate student in Gibbs's class, and he later wrote up his class notes, expanded upon them, and had them published in 1901, with Gibbs's blessing, as *Vector Analysis: A Text-Book for the Use of Students of Mathematics and Physics*. However, the concepts of dot and cross product arose earlier and went by various other names and notations.

Write a report on the evolution of the names and notations for the dot product and cross product.

1. Florian Cajori, *A History of Mathematical Notations* (New York: Dover, 1993).
2. J. Willard Gibbs and Edwin Bidwell Wilson, *Vector Analysis: A Text-Book for the Use of Students of Mathematics and Physics* (New York: Charles Scribner's Sons, 1901). Available online at http://archive.org/details/117714283.
3. Ivor Grattan-Guinness, *Companion Encyclopedia of the History and Philosophy of the Mathematical Sciences* (London: Routledge, 2013).

1.4 Applications

Force Vectors

Force is defined as the product of mass and acceleration due to gravity (which, on Earth, is 9.8 m/s^2). Thus, a 1 kg mass exerts a downward force of 1 kg $\times$ 9.8 m/s^2 or 9.8 kg • m/s^2. This unit of measurement is a ***newton*** (N). So the force exerted by a 1 kg mass is 9.8 N.

We can use vectors to model force. For example, a wind blowing at 30 km/h in a westerly direction or the Earth's gravity acting on a 1 kg mass with a force of 9.8 newtons downward are each best represented by vectors since they each consist of a magnitude and a direction.

It is often the case that multiple forces act on an object. In such situations, the net result of all the forces acting together is a single force called the ***resultant***, which is simply the vector sum of the individual forces (Figure 1.72). When several forces act on an object, it is possible that the resultant force is zero. In this case, the object is clearly not moving in any direction and we say that it is in ***equilibrium***. When an object is in equilibrium and the force vectors acting on it are arranged head-to-tail, the result is a closed polygon (Figure 1.73).

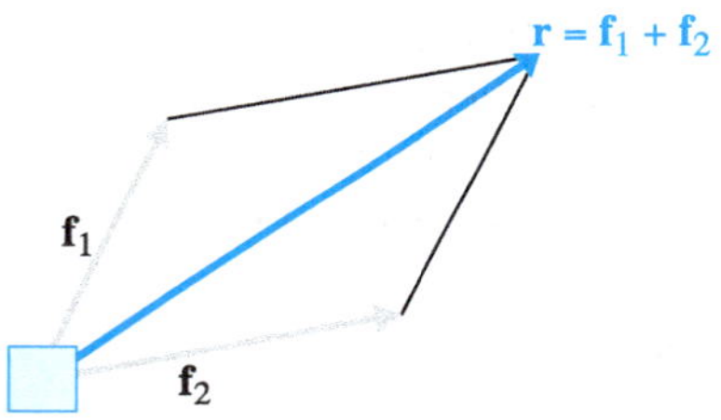

Figure 1.72
The resultant of two forces

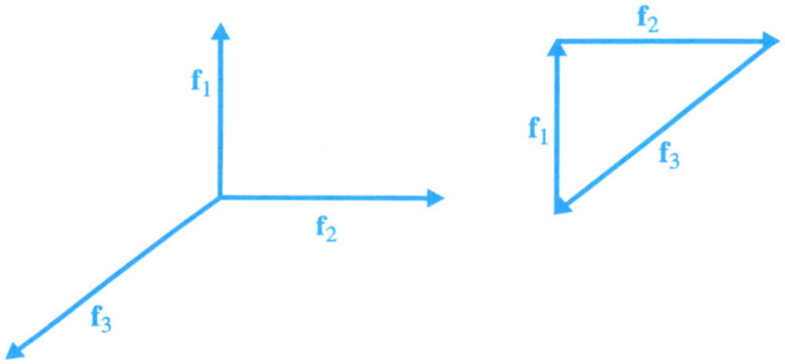

Figure 1.73
Equilibrium

Example 1.34

Ann and Bert are trying to roll a rock out of the way. Ann pushes with a force of 20 N in a northerly direction while Bert pushes with a force of 40 N in an easterly direction.

(a) What is the resultant force on the rock?

(b) Carla is trying to prevent Ann and Bert from moving the rock. What force must Carla apply to keep the rock in equilibrium?

Solution (a) Figure 1.74 shows the position of the two forces. Using the parallelogram rule, we add the two forces to get the resultant $\mathbf{r}$ as shown. By Pythagoras'

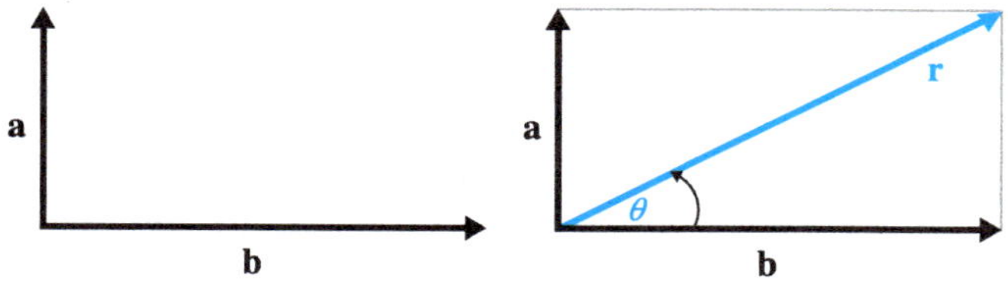

Figure 1.74
The resultant of two forces

Theorem, we see that $\|\mathbf{r}\| = \sqrt{20^2 + 40^2} = \sqrt{2000} \approx 44.72$ N. For the direction of $\mathbf{r}$, we calculate the angle θ between $\mathbf{r}$ and Bert's easterly force. We find that $\sin\theta = 20/\|\mathbf{r}\| \approx 0.447$, so $\theta \approx 26.57°$.

(b) If we denote the forces exerted by Ann, Bert, and Carla by $\mathbf{a}$, $\mathbf{b}$, and $\mathbf{c}$, respectively, then we require $\mathbf{a} + \mathbf{b} + \mathbf{c} = \mathbf{0}$. Therefore $\mathbf{c} = -(\mathbf{a} + \mathbf{b}) = -\mathbf{r}$, so Carla needs to exert a force of 44.72 N in the direction opposite to $\mathbf{r}$.

Often, we are interested in decomposing a force vector into other vectors whose resultant is the given vector. This process is called ***resolving a vector into components***. In two dimensions, we wish to resolve a vector into two components. However, there are infinitely many ways to do this; the most useful will be to resolve the vector into two *orthogonal* components. (Chapters 5 and 7 explore this idea more generally.) This is usually done by introducing coordinate axes and by choosing the components so that one is parallel to the x-axis and the other to the y-axis. These components are usually referred to as the horizontal and vertical components, respectively. In Figure 1.75, $\mathbf{f}$ is the given vector and $\mathbf{f}_x$ and $\mathbf{f}_y$ are its horizontal and vertical components.

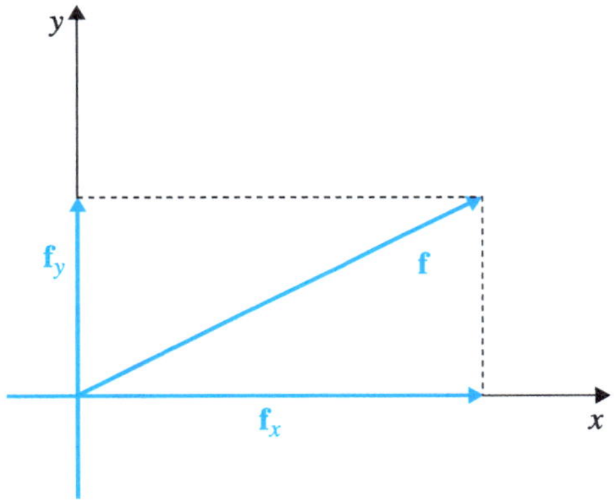

Figure 1.75
Resolving a vector into components

Example 1.35

Ann pulls on the handle of a wagon with a force of 100 N. If the handle makes an angle of 20° with the horizontal, what is the force that tends to pull the wagon forward and what force tends to lift it off the ground?

Solution Figure 1.76 shows the situation and the vector diagram that we need to consider.

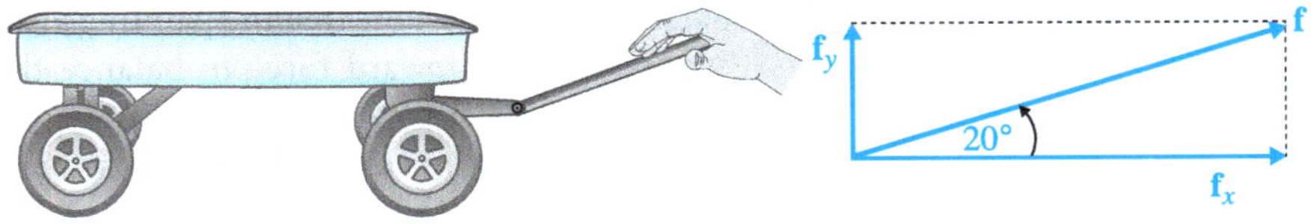

Figure 1.76

We see that

$$\|\mathbf{f}_x\| = \|\mathbf{f}\| \cos 20° \text{ and } \|\mathbf{f}_y\| = \|\mathbf{f}\| \sin 20°$$

Thus, $\|\mathbf{f}_x\| \approx 100(0.9397) \approx 93.97$ and $\|\mathbf{f}_y\| \approx 100(0.3420) \approx 34.20$. So the wagon is pulled forward with a force of approximately 93.97 N and it tends to lift off the ground with a force of approximately 34.20 N.

We solve the next example using two different methods. The first solution considers a triangle of forces in equilibrium; the second solution uses resolution of forces into components.

Example 1.36 Figure 1.77 shows a painting that has been hung from the ceiling by two wires. If the painting has a mass of 5 kg and if the two wires make angles of 45 and 60 degrees with the ceiling, determine the tension in each wire.

Figure 1.77

Solution 1 We assume that the painting is in equilibrium. Then the two wires must supply enough upward force to balance the downward force of gravity. Gravity exerts a downward force of $5 \times 9.8 = 49$ N on the painting, so the two wires must collectively pull upward with 49 N of force. Let $\mathbf{f}_1$ and $\mathbf{f}_2$ denote the tensions in the wires and let $\mathbf{r}$ be their resultant (Figure 1.78). It follows that $\|\mathbf{r}\| = 49$ since we are in equilibrium.

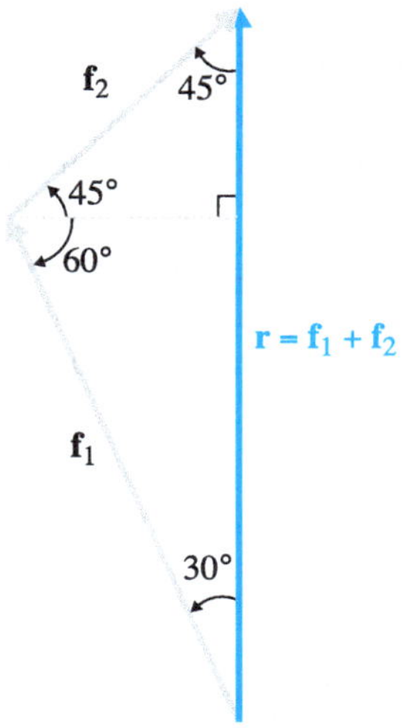

Figure 1.78

Using the law of sines, we have

$$\frac{\|\mathbf{f}_1\|}{\sin 45°} = \frac{\|\mathbf{f}_2\|}{\sin 30°} = \frac{\|\mathbf{r}\|}{\sin 105°}$$

so

$$\|\mathbf{f}_1\| = \frac{\|\mathbf{r}\| \sin 45°}{\sin 105°} \approx \frac{49(0.7071)}{0.9659} \approx 35.87 \text{ and } \|\mathbf{f}_2\| = \frac{\|\mathbf{r}\| \sin 30°}{\sin 105°} \approx \frac{49(0.5)}{0.9659} \approx 25.36$$

Therefore, the tensions in the wires are approximately 35.87 N and 25.36 N.

Solution 2 We resolve $\mathbf{f}_1$ and $\mathbf{f}_2$ into horizontal and vertical components, say, $\mathbf{f}_1 = \mathbf{h}_1 + \mathbf{v}_1$ and $\mathbf{f}_2 = \mathbf{h}_2 + \mathbf{v}_2$, and note that, as above, there is a downward force of 49 N (Figure 1.79).

It follows that

$$\|\mathbf{h}_1\| = \|\mathbf{f}_1\| \cos 60° = \frac{\|\mathbf{f}_1\|}{2}, \quad \|\mathbf{v}_1\| = \|\mathbf{f}_1\| \sin 60° = \frac{\sqrt{3}\|\mathbf{f}_1\|}{2},$$

$$\|\mathbf{h}_2\| = \|\mathbf{f}_2\| \cos 45° = \frac{\|\mathbf{f}_2\|}{\sqrt{2}}, \quad \|\mathbf{v}_2\| = \|\mathbf{f}_2\| \sin 45° = \frac{\|\mathbf{f}_2\|}{\sqrt{2}}$$

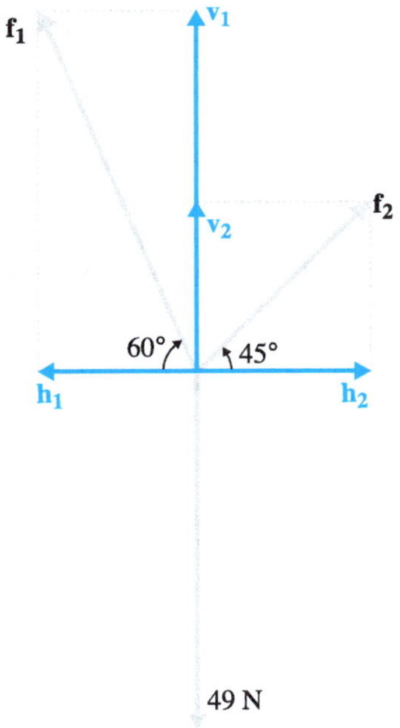

Figure 1.79

Since the painting is in equilibrium, the horizontal components must balance, as must the vertical components. Therefore, $\|\mathbf{h}_1\| = \|\mathbf{h}_2\|$ and $\|\mathbf{v}_1\| + \|\mathbf{v}_2\| = 49$, from which it follows that

$$\|\mathbf{f}_1\| = \frac{2\|\mathbf{f}_2\|}{\sqrt{2}} = \sqrt{2}\|\mathbf{f}_2\| \quad \text{and} \quad \frac{\sqrt{3}\|\mathbf{f}_1\|}{2} + \frac{\|\mathbf{f}_2\|}{\sqrt{2}} = 49$$

Substituting the first of these equations into the second equation yields

$$\frac{\sqrt{3}\|\mathbf{f}_2\|}{\sqrt{2}} + \frac{\|\mathbf{f}_2\|}{\sqrt{2}} = 49, \quad \text{or} \quad \|\mathbf{f}_2\| = \frac{49\sqrt{2}}{1 + \sqrt{3}} \approx 25.36$$

Thus, $\|\mathbf{f}_1\| = \sqrt{2}\|\mathbf{f}_2\| \approx 1.4142(25.36) \approx 35.87$, so the tensions in the hires are approximately 35.87 N and 25.36 N, as before.

Exercises 1.4

Force Vectors

In Exercises 1–6, determine the resultant of the given forces.

1. $\mathbf{f}_1$ acting due north with a magnitude of 12 N and $\mathbf{f}_2$ acting due east with a magnitude of 5 N

2. $\mathbf{f}_1$ acting due west with a magnitude of 15 N and $\mathbf{f}_2$ acting due south with a magnitude of 20 N

3. $\mathbf{f}_1$ acting with a magnitude of 8 N and $\mathbf{f}_2$ acting at an angle of 60° to $\mathbf{f}_1$ with a magnitude of 8 N

4. $\mathbf{f}_1$ acting with a magnitude of 4 N and $\mathbf{f}_2$ acting at an angle of 135° to $\mathbf{f}_1$ with a magnitude of 6 N

5. $\mathbf{f}_1$ acting due east with a magnitude of 2 N, $\mathbf{f}_2$ acting due west with a magnitude of 6 N, and $\mathbf{f}_3$ acting at an angle of 60° to $\mathbf{f}_1$ with a magnitude of 4 N

6. $\mathbf{f}_1$ acting due east with a magnitude of 10 N, $\mathbf{f}_2$ acting due north with a magnitude of 13 N, $\mathbf{f}_3$ acting due west with a magnitude of 5 N, and $\mathbf{f}_4$ acting due south with a magnitude of 8 N

7. Resolve a force of 10 N into two forces perpendicular to each other so that one component makes an angle of 60° with the 10 N force.

8. A 10 kg block lies on a ramp that is inclined at an angle of 30° (Figure 1.80). Assuming there is no friction, what force, parallel to the ramp, must be applied to keep the block from sliding down the ramp?

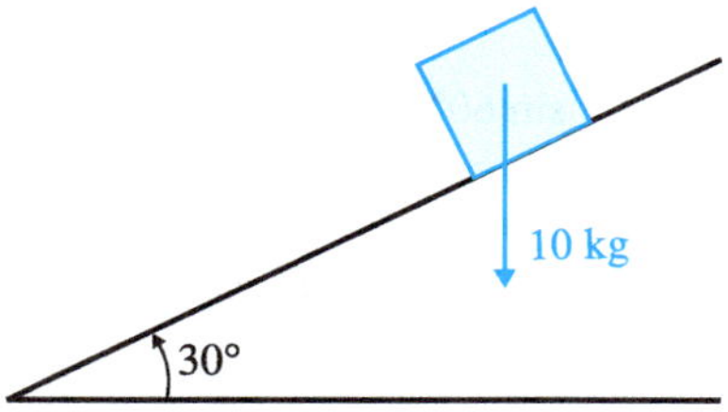

Figure 1.80

9. A tow truck is towing a car. The tension in the tow cable is 1500 N and the cable makes a 45° with the horizontal, as shown in Figure 1.81. What is the vertical force that tends to lift the car off the ground?

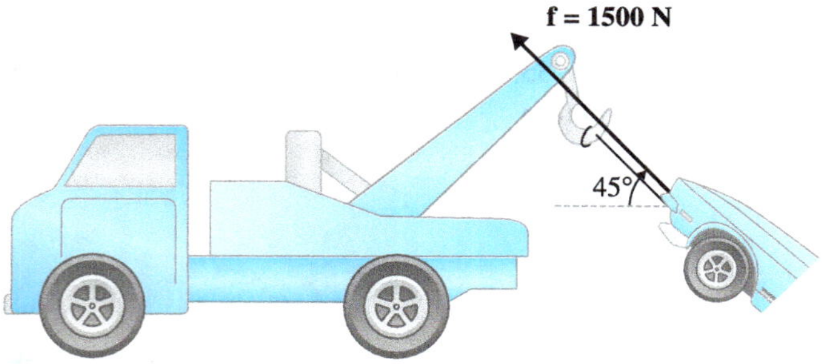

Figure 1.81

10. A lawn mower has a mass of 30 kg. It is being pushed with a force of 100 N. If the handle of the lawn mower makes an angle of 45° with the ground, what is the horizontal component of the force that is causing the mower to move forward?

11. A sign hanging outside Joe's Diner has a mass of 50 kg (Figure 1.82). If the supporting cable makes an angle of 60° with the wall of the building, determine the tension in the cable.

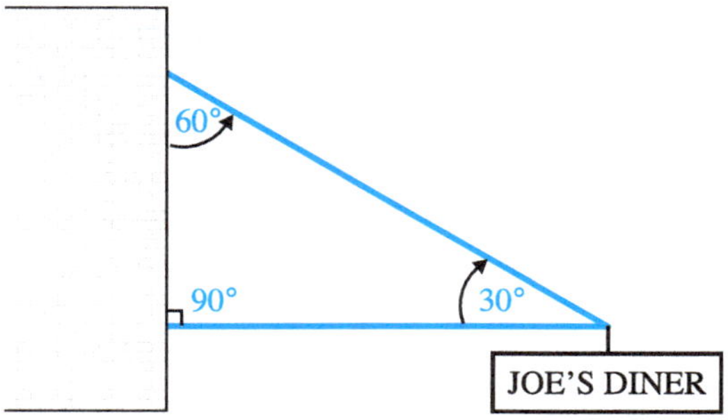

Figure 1.82

12. A sign hanging in the window of Joe's Diner has a mass of 1 kg. If the supporting strings each make an angle of 45° with the sign and the supporting hooks are at the same height (Figure 1.83), find the tension in each string.

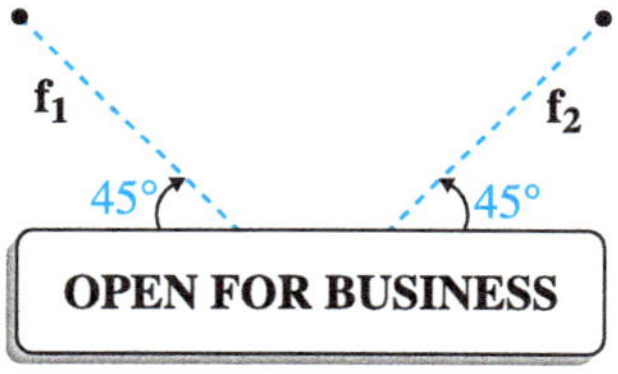

Figure 1.83

13. A painting with a mass of 15 kg is suspended by two wires from hooks on a ceiling. If the wires have lengths of 15 cm and 20 cm and the distance between the hooks is 25 cm, find the tension in each wire.

14. A painting with a mass of 20 kg is suspended by two wires from a ceiling. If the wires make angles of 30° and 45° with the ceiling, find the tension in each wire.

Chapter Review

Key Definitions and Concepts

algebraic properties of vectors, 10
angle between vectors, 24
binary vector, 13
Cauchy-Schwarz Inequality, 22
cross product, 48
direction vector, 35
distance between vectors, 23
dot product, 18
equation of a line, 36
equation of a plane, 38–39
head-to-tail rule, 6
integers modulo m ($\mathbb{Z}_m$), 14–16
length (norm) of a vector, 20
linear combination of vectors, 12
normal vector, 34, 38
m-ary vector 16
orthogonal vectors, 26
parallel vectors, 8
parallelogram rule, 6
projection of a vector onto a vector, 27
Pythagoras' Theorem, 26
scalar multiplication, 7
standard unit vectors, 22
Triangle Inequality, 22
unit vector, 21
vector, 3
vector addition, 5
zero vector, 4

Review Questions

1. Mark each of the following statements true or false:

(a) For vectors **u**, **v**, and **w** in $\mathbb{R}^n$, if $\mathbf{u} + \mathbf{w} = \mathbf{v} + \mathbf{w}$, then $\mathbf{u} = \mathbf{v}$.

(b) For vectors **u**, **v**, and **w** in $\mathbb{R}^n$, if $\mathbf{u} \cdot \mathbf{w} = \mathbf{v} \cdot \mathbf{w}$, then $\mathbf{u} = \mathbf{v}$.

(c) For vectors **u**, **v**, and **w** in $\mathbb{R}^3$, if **u** is orthogonal to **v**, and **v** is orthogonal to **w**, then **u** is orthogonal to **w**.

(d) In $\mathbb{R}^3$, if a line ℓ is parallel to a plane $\mathcal{P}$, then a direction vector **d** for ℓ is parallel to a normal vector **n** for $\mathcal{P}$.

(e) In $\mathbb{R}^3$, if a line ℓ is perpendicular to a plane $\mathcal{P}$, then a direction vector **d** for ℓ is a parallel to a normal vector **n** for $\mathcal{P}$.

(f) In $\mathbb{R}^3$, if two planes are not parallel, then they must intersect in a line.

(g) In $\mathbb{R}^3$, if two lines are not parallel, then they must intersect in a point.

(h) If **v** is a binary vector such that $\mathbf{v} \cdot \mathbf{v} = 0$, then $\mathbf{v} = \mathbf{0}$.

(i) In $\mathbb{Z}_5$, if $ab = 0$ then either $a = 0$ or $b = 0$.

(j) In $\mathbb{Z}_6$, if $ab = 0$ then either $a = 0$ or $b = 0$.

2. If $\mathbf{u} = \begin{bmatrix} -1 \\ 5 \end{bmatrix}$, $\mathbf{v} = \begin{bmatrix} 3 \\ 2 \end{bmatrix}$, and the vector $4\mathbf{u} + \mathbf{v}$ is drawn with its tail at the point $(10, -10)$, find the coordinates of the point at the head of $4\mathbf{u} + \mathbf{v}$.

3. If $\mathbf{u} = \begin{bmatrix} -1 \\ 5 \end{bmatrix}$, $\mathbf{v} = \begin{bmatrix} 3 \\ 2 \end{bmatrix}$, and $2\mathbf{x} + \mathbf{u} = 3(\mathbf{x} - \mathbf{v})$, solve for **x**.

4. Let A, B, C, and D be the vertices of a square centered at the origin O, labeled in clockwise order. If $\mathbf{a} = \overrightarrow{OA}$ and $\mathbf{b} = \overrightarrow{OB}$, find $\overrightarrow{BC}$ in terms of **a** and **b**.

5. Find the angle between the vectors $[-1, 1, 2]$ and $[2, 1, -1]$.

6. Find the projection of $\mathbf{v} = \begin{bmatrix} 1 \\ 1 \\ 1 \end{bmatrix}$ onto $\mathbf{u} = \begin{bmatrix} 1 \\ -2 \\ 2 \end{bmatrix}$.

7. Find a unit vector in the xy-plane that is orthogonal to $\begin{bmatrix} 1 \\ 2 \\ 3 \end{bmatrix}$.

8. Find the general equation of the plane through the point $(1, 1, 1)$ that is perpendicular to the line with parametric equations

$$\begin{aligned} x &= 2 - t \\ y &= 3 + 2t \\ z &= -1 + t \end{aligned}$$

9. Find the general equation of the plane through the point $(3, 2, 5)$ that is parallel to the plane whose general equation is $2x + 3y - z = 0$.

10. Find the general equation of the plane through the points $A(1, 1, 0)$, $B(1, 0, 1)$, and $C(0, 1, 2)$.

11. Find the area of the triangle with vertices $A(1, 1, 0)$, $B(1, 0, 1)$, and $C(0, 1, 2)$.

12. Find the midpoint of the line segment between $A = (5, 1, -2)$ and $B = (3, -7, 0)$.

13. Why are there no vectors **u** and **v** in $\mathbb{R}^n$ such that $\|\mathbf{u}\| = 2$, $\|\mathbf{v}\| = 3$, and $\mathbf{u} \cdot \mathbf{v} = -7$?

14. Find the distance from the point (3, 2, 5) to the plane whose general equation is $2x + 3y - z = 0$.

15. Find the distance from the point (3, 2, 5) to the line with parametric equations $x = t$, $y = 1 + t$, $z = 2 + t$.

16. Compute $3 - (2 + 4)^3(4 + 3)^2$ in $\mathbb{Z}_5$.

17. If possible, solve $3(x + 2) = 5$ in $\mathbb{Z}_7$.

18. If possible, solve $3(x + 2) = 5$ in $\mathbb{Z}_9$.

19. Compute $[2, 1, 3, 3] \cdot [3, 4, 4, 2]$ in $\mathbb{Z}_5^4$.

20. Let $\mathbf{u} = [1, 1, 1, 0]$ in $\mathbb{Z}_2^4$. How many binary vectors **v** satisfy $\mathbf{u} \cdot \mathbf{v} = 0$?

2 Systems of Linear Equations

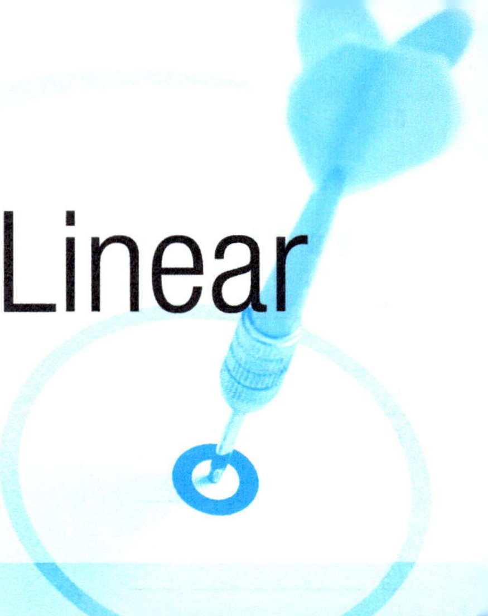

The world was full of equations . . . .
There must be an answer for everything,
if only you knew how to set forth
the questions.

—Anne Tyler
The Accidental Tourist
Alfred A. Knopf, 1985, p. 235

2.0 Introduction: Triviality

The word *trivial* is derived from the Latin root *tri* ("three") and the Latin word *via* ("road"). Thus, speaking literally, a triviality is a place where three roads meet. This common meeting point gives rise to the other, more familiar meaning of *trivial*—commonplace, ordinary, or insignificant. In medieval universities, the *trivium* consisted of the three "common" subjects (grammar, rhetoric, and logic) that were taught before the *quadrivium* (arithmetic, geometry, music, and astronomy). The "three roads" that made up the trivium were the beginning of the liberal arts.

In this section, we begin to examine systems of linear equations. The same system of equations can be viewed in three different, yet equally important, ways—these will be our three roads, all leading to the same solution. You will need to get used to this threefold way of viewing systems of linear equations, so that it becomes commonplace (trivial!) for you.

The system of equations we are going to consider is

$$\begin{aligned} 2x + y &= 8 \\ x - 3y &= -3 \end{aligned}$$

Problem 1 Draw the two lines represented by these equations. What is their point of intersection?

Problem 2 Consider the vectors $\mathbf{u} = \begin{bmatrix} 2 \\ 1 \end{bmatrix}$ and $\mathbf{v} = \begin{bmatrix} 1 \\ -3 \end{bmatrix}$. Draw the coordinate grid determined by $\mathbf{u}$ and $\mathbf{v}$. [*Hint:* Lightly draw the standard coordinate grid first and use it as an aid in drawing the new one.]

Problem 3 On the u-v grid, find the coordinates of $\mathbf{w} = \begin{bmatrix} 8 \\ -3 \end{bmatrix}$.

Problem 4 Another way to state Problem 3 is to ask for the coefficients x and y for which $x\mathbf{u} + y\mathbf{v} = \mathbf{w}$. Write out the two equations to which this vector equation is equivalent (one for each component). What do you observe?

Problem 5 Return now to the lines you drew for Problem 1. We will refer to the line whose equation is $2x + y = 8$ as line 1 and the line whose equation is $x - 3y = -3$ as line 2. Plot the point (0, 0) on your graph from Problem 1 and label it P_0. Draw a

Table 2.1

Point	x	y
P_0	0	0
P_1		
P_2		
P_3		
P_4		
P_5		
P_6		

horizontal line segment from P_0 to line 1 and label this new point P_1. Next draw a *vertical* line segment from P_1 to line 2 and label this point P_2. Now draw a *horizontal* line segment from P_2 to line 1, obtaining point P_3. Continue in this fashion, drawing vertical segments to line 2 followed by horizontal segments to line 1. What appears to be happening?

Problem 6 Using a calculator with two-decimal-place accuracy, find the (approximate) coordinates of the points $P_1, P_2, P_3, \ldots, P_6$. (You will find it helpful to first solve the first equation for x in terms of y and the second equation for y in terms of x.) Record your results in Table 2.1, writing the x- and y-coordinates of each point separately.

The results of these problems show that the task of "solving" a system of linear equations may be viewed in several ways. Repeat the process described in the problems with the following systems of equations:

$$\begin{array}{llll} \text{(a)}\ \begin{aligned} 4x - 2y &= 0 \\ x + 2y &= 5 \end{aligned} & \text{(b)}\ \begin{aligned} 3x + 2y &= 9 \\ x + 3y &= 10 \end{aligned} & \text{(c)}\ \begin{aligned} x + y &= 5 \\ x - y &= 3 \end{aligned} & \text{(d)}\ \begin{aligned} x + 2y &= 4 \\ 2x - y &= 3 \end{aligned} \end{array}$$

Are all of your observations from Problems 1–6 still valid for these examples? Note any similarities or differences. In this chapter, we will explore these ideas in more detail.

2.1 Introduction to Systems of Linear Equations

Recall that the general equation of a line in $\mathbb{R}^2$ is of the form

$$ax + by = c$$

and that the general equation of a plane in $\mathbb{R}^3$ is of the form

$$ax + by + cz = d$$

Equations of this form are called ***linear equations.***

Definition A ***linear equation*** in the n variables $x_1, x_2, \ldots, x_n$ is an equation that can be written in the form

$$a_1x_1 + a_2x_2 + \cdots + a_nx_n = b$$

where the ***coefficients*** $a_1, a_2, \ldots, a_n$ and the ***constant term*** b are constants.

Example 2.1

The following equations are linear:

$$3x - 4y = -1 \qquad r - \tfrac{1}{2}s - \tfrac{15}{3}t = 9 \qquad x_1 + 5x_2 = 3 - x_3 + 2x_4$$

$$\sqrt{2}x + \frac{\pi}{4}y - \left(\sin\frac{\pi}{5}\right)z = 1 \qquad 3.2x_1 - 0.01x_2 = 4.6$$

Observe that the third equation is linear because it can be rewritten in the form $x_1 + 5x_2 + x_3 - 2x_4 = 3$. It is also important to note that, although in these examples (and in most applications) the coefficients and constant terms are real numbers, in some examples and applications they will be complex numbers or members of $\mathbb{Z}_p$ for some prime number p.

The following equations are not linear:

$$xy + 2z = 1 \qquad x_1^2 - x_2^3 = 3 \qquad \frac{x}{y} + z = 2$$

$$\sqrt{2x} + \frac{\pi}{4}y - \sin\left(\frac{\pi}{5}z\right) = 1 \qquad \sin x_1 - 3x_2 + 2^{x_3} = 0$$

Thus, linear equations do not contain products, reciprocals, or other functions of the variables; the variables occur only to the first power and are multiplied only by constants. Pay particular attention to the fourth example in each list: Why is it that the fourth equation in the first list is linear but the fourth equation in the second list is not?

A ***solution*** of a linear equation $a_1x_1 + a_2x_2 + \cdots + a_nx_n = b$ is a vector $[s_1, s_2, \ldots, s_n]$ whose components satisfy the equation when we substitute $x_1 = s_1$, $x_2 = s_2, \ldots, x_n = s_n$.

Example 2.2

(a) $[5, 4]$ is a solution of $3x - 4y = -1$ because, when we substitute $x = 5$ and $y = 4$, the equation is satisfied: $3(5) - 4(4) = -1$. $[1, 1]$ is another solution. In general, the solutions simply correspond to the points on the line determined by the given equation. Thus, setting $x = t$ and solving for y, we see that the complete set of solutions can be written in the parametric form $[t, \frac{1}{4} + \frac{3}{4}t]$. (We could also set y equal to some parameter—say, s—and solve for x instead; the two parametric solutions would look different but would be equivalent. Try this.)

(b) The linear equation $x_1 - x_2 + 2x_3 = 3$ has $[3, 0, 0]$, $[0, 1, 2]$, and $[6, 1, -1]$ as specific solutions. The complete set of solutions corresponds to the set of points in the plane determined by the given equation. If we set $x_2 = s$ and $x_3 = t$, then a parametric solution is given by $[3 + s - 2t, s, t]$. (Which values of s and t produce the three specific solutions above?)

A ***system of linear equations*** is a finite set of linear equations, each with the same variables. A ***solution*** of a system of linear equations is a vector that is *simultaneously* a solution of each equation in the system. The ***solution set*** of a system of linear equations is the set of *all* solutions of the system. We will refer to the process of finding the solution set of a system of linear equations as "solving the system."

Example 2.3

The system

$$\begin{aligned} 2x - y &= 3 \\ x + 3y &= 5 \end{aligned}$$

has $[2, 1]$ as a solution, since it is a solution of both equations. On the other hand, $[1, -1]$ is not a solution of the system, since it satisfies only the first equation.

Example 2.4

Solve the following systems of linear equations:

(a) $\begin{aligned} x - y &= 1 \\ x + y &= 3 \end{aligned}$ (b) $\begin{aligned} x - y &= 2 \\ 2x - 2y &= 4 \end{aligned}$ (c) $\begin{aligned} x - y &= 1 \\ x - y &= 3 \end{aligned}$

Solution

(a) Adding the two equations together gives $2x = 4$, so $x = 2$, from which we find that $y = 1$. A quick check confirms that $[2, 1]$ is indeed a solution of both equations. That this is the *only* solution can be seen by observing that this solution corresponds to the (unique) point of intersection $(2, 1)$ of the lines with equations $x - y = 1$ and $x + y = 3$, as shown in Figure 2.1(a). Thus, $[2, 1]$ is a *unique solution.*

(b) The second equation in this system is just twice the first, so the solutions are the solutions of the first equation alone—namely, the points on the line $x - y = 2$. These can be represented parametrically as $[2 + t, t]$. Thus, this system has *infinitely many solutions* [Figure 2.1(b)].

(c) Two numbers x and y cannot simultaneously have a difference of 1 and 3. Hence, this system has *no solutions.* (A more algebraic approach might be to subtract the second equation from the first, yielding the absurd conclusion $0 = -2$.) As Figure 2.1(c) shows, the lines for the equations are parallel in this case.

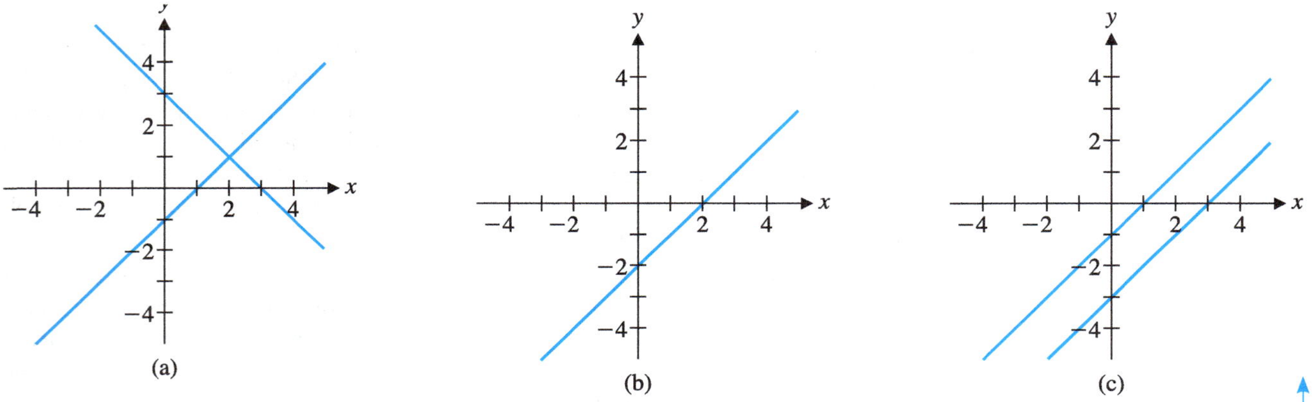

Figure 2.1

A system of linear equations is called ***consistent*** if it has at least one solution. A system with no solutions is called ***inconsistent.*** Even though they are small, the three systems in Example 2.4 illustrate the only three possibilities for the number of solutions of a system of linear equations with real coefficients. We will prove later that these same three possibilities hold for *any* system of linear equations over the real numbers.

A system of linear equations with real coefficients has either

(a) a unique solution (a consistent system) or
(b) infinitely many solutions (a consistent system) or
(c) no solutions (an inconsistent system).

Solving a System of Linear Equations

Two linear systems are called ***equivalent*** if they have the same solution sets. For example,

$$\begin{aligned} x - y &= 1 \\ x + y &= 3 \end{aligned} \qquad \text{and} \qquad \begin{aligned} x - y &= 1 \\ y &= 1 \end{aligned}$$

are equivalent, since they both have the unique solution $[2, 1]$. (Check this.)

Our approach to solving a system of linear equations is to transform the given system into an equivalent one that is easier to solve. The triangular pattern of the second example above (in which the second equation has one less variable than the first) is what we will aim for.

Example 2.5

Solve the system

$$\begin{aligned} x - y - z &= 2 \\ y + 3z &= 5 \\ 5z &= 10 \end{aligned}$$

Solution Starting from the last equation and working backward, we find successively that $z = 2$, $y = 5 - 3(2) = -1$, and $x = 2 + (-1) + 2 = 3$. So the unique solution is $[3, -1, 2]$.

The procedure used to solve Example 2.5 is called ***back substitution.***

We now turn to the general strategy for transforming a given system into an equivalent one that can be solved easily by back substitution. This process will be described in greater detail in the next section; for now, we will simply observe it in action in a single example.

Example 2.6

Solve the system

$$\begin{aligned} x - y - z &= 2 \\ 3x - 3y + 2z &= 16 \\ 2x - y + z &= 9 \end{aligned}$$

Solution To transform this system into one that exhibits the triangular structure of Example 2.5, we first need to eliminate the variable x from Equations 2 and 3. Observe that subtracting appropriate multiples of equation 1 from Equations 2 and 3 will do the trick. Next, observe that we are operating on the coefficients, not on the variables, so we can save ourselves some writing if we record the coefficients and constant terms in the *matrix*

$$\left[\begin{array}{rrr|r} 1 & -1 & -1 & 2 \\ 3 & -3 & 2 & 16 \\ 2 & -1 & 1 & 9 \end{array}\right]$$

The word *matrix* is derived from the Latin word *mater*, meaning "mother." When the suffix *-ix* is added, the meaning becomes "womb." Just as a womb surrounds a fetus, the brackets of a matrix surround its entries, and just as the womb gives rise to a baby, a matrix gives rise to certain types of functions called *linear transformations*. A matrix with m rows and n columns is called an $m \times n$ matrix (pronounced "m by n"). The plural of *matrix* is *matrices*, not "matrixes."

where the first three columns contain the coefficients of the variables in order, the final column contains the constant terms, and the vertical bar serves to remind us of the equal signs in the equations. This matrix is called the ***augmented matrix*** of the system.

There are various ways to convert the given system into one with the triangular pattern we are after. The steps we will use here are closest in spirit to the more general method described in the next section. We will perform the sequence of operations on the given system and simultaneously on the corresponding augmented matrix. We begin by eliminating x from Equations 2 and 3.

$$\begin{aligned} x - y - z &= 2 \\ 3x - 3y + 2z &= 16 \\ 2x - y + z &= 9 \end{aligned} \qquad \left[\begin{array}{rrr|r} 1 & -1 & -1 & 2 \\ 3 & -3 & 2 & 16 \\ 2 & -1 & 1 & 9 \end{array}\right]$$

Subtract 3 times the first equation from the second equation:

$$\begin{aligned} x - y - z &= 2 \\ 5z &= 10 \\ 2x - y + z &= 9 \end{aligned}$$

Subtract 3 times the first row from the second row:

$$\left[\begin{array}{rrr|r} 1 & -1 & -1 & 2 \\ 0 & 0 & 5 & 10 \\ 2 & -1 & 1 & 9 \end{array}\right]$$

Subtract 2 times the first equation from the third equation:

$$\begin{aligned} x - y - z &= 2 \\ 5z &= 10 \\ y + 3z &= 5 \end{aligned}$$

Subtract 2 times the first row from the third row:

$$\left[\begin{array}{rrr|r} 1 & -1 & -1 & 2 \\ 0 & 0 & 5 & 10 \\ 0 & 1 & 3 & 5 \end{array}\right]$$

Interchange Equations 2 and 3:

$$\begin{aligned} x - y - z &= 2 \\ y + 3z &= 5 \\ 5z &= 10 \end{aligned}$$

Interchange rows 2 and 3:

$$\left[\begin{array}{rrr|r} 1 & -1 & -1 & 2 \\ 0 & 1 & 3 & 5 \\ 0 & 0 & 5 & 10 \end{array}\right]$$

This is the same system that we solved using back substitution in Example 2.5, where we found that the solution was $[3, -1, 2]$. This is therefore also the solution to the system given in this example. Why? The calculations above show that *any solution of the given system is also a solution of the final one*. But since the steps we just performed are *reversible*, we could recover the original system, starting with the final system. (How?) So *any solution of the final system is also a solution of the given one*. Thus, the systems are equivalent (as are all of the ones obtained in the intermediate steps above). Moreover, we might just as well work with matrices instead of equations, since it is a simple matter to reinsert the variables before proceeding with the back substitution. (Working with matrices is the subject of the next section.)

Remark Calculators with matrix capabilities and computer algebra systems can facilitate solving systems of linear equations, particularly when the systems are large or have coefficients that are not "nice," as is often the case in real-life applications. As always, though, you should do as many examples as you can with pencil and paper until you are comfortable with the techniques. Even if a calculator or CAS is called for, think about *how* you would do the calculations manually before doing anything. After you have an answer, be sure to think about whether it is reasonable.

Do not be misled into thinking that technology will always give you the answer faster or more easily than calculating by hand. Sometimes it may not give you the answer at all! Roundoff errors associated with the floating-point arithmetic used by calculators and computers can cause serious problems and lead to wildly wrong answers to some problems. See Exploration: Lies My Computer Told Me for a glimpse of the problem. (You've been warned!)

Exercises 2.1

In Exercises 1–6, determine which equations are linear equations in the variables x, y, and z. If any equation is not linear, explain why not.

1. $x - \pi y + \sqrt[3]{5}z = 0$

2. $x^2 + y^2 + z^2 = 1$

3. $x^{-1} + 7y + z = \sin\left(\frac{\pi}{9}\right)$

4. $2x - xy - 5z = 0$

5. $3\cos x - 4y + z = \sqrt{3}$

6. $(\cos 3)x - 4y + z = \sqrt{3}$

In Exercises 7–10, find a linear equation that has the same solution set as the given equation (possibly with some restrictions on the variables).

7. $2x + y = 7 - 3y$

8. $\dfrac{x^2 - y^2}{x - y} = 1$

9. $\dfrac{1}{x} + \dfrac{1}{y} = \dfrac{4}{xy}$

10. $\log_{10} x - \log_{10} y = 2$

In Exercises 11–14, find the solution set of each equation.

11. $3x - 6y = 0$

12. $2x_1 + 3x_2 = 5$

13. $x + 2y + 3z = 4$

14. $4x_1 + 3x_2 + 2x_3 = 1$

In Exercises 15–18, draw graphs corresponding to the given linear systems. Determine geometrically whether each system has a unique solution, infinitely many solutions, or no solution. Then solve each system algebraically to confirm your answer.

15. $\begin{aligned} x + y &= 0 \\ 2x + y &= 3 \end{aligned}$

16. $\begin{aligned} x - 2y &= 7 \\ 3x + y &= 7 \end{aligned}$

17. $\begin{aligned} 3x - 6y &= 3 \\ -x + 2y &= 1 \end{aligned}$

18. $\begin{aligned} 0.10x - 0.05y &= 0.20 \\ -0.06x + 0.03y &= -0.12 \end{aligned}$

In Exercises 19–24, solve the given system by back substitution.

19. $\begin{aligned} x - 2y &= 1 \\ y &= 3 \end{aligned}$

20. $\begin{aligned} 2u - 3v &= 5 \\ 2v &= 6 \end{aligned}$

21. $\begin{aligned} x - y + z &= 0 \\ 2y - z &= 1 \\ 3z &= -1 \end{aligned}$

22. $\begin{aligned} x_1 + 2x_2 + 3x_3 &= 0 \\ -5x_2 + 2x_3 &= 0 \\ 4x_3 &= 0 \end{aligned}$

23. $\begin{aligned} x_1 + x_2 - x_3 - x_4 &= 1 \\ x_2 + x_3 + x_4 &= 0 \\ x_3 - x_4 &= 0 \\ x_4 &= 1 \end{aligned}$

24. $\begin{aligned} x - 3y + z &= 5 \\ y - 2z &= -1 \end{aligned}$

The systems in Exercises 25 and 26 exhibit a "lower triangular" pattern that makes them easy to solve by forward substitution. (We will encounter forward substitution again in Chapter 3.) Solve these systems.

25. $\begin{aligned} x &= 2 \\ 2x + y &= -3 \\ -3x - 4y + z &= -10 \end{aligned}$

26. $\begin{aligned} x_1 &= -1 \\ -\tfrac{1}{2}x_1 + x_2 &= 5 \\ \tfrac{3}{2}x_1 + 2x_2 + x_3 &= 7 \end{aligned}$

Find the augmented matrices of the linear systems in Exercises 27–30.

27. $\begin{aligned} x - y &= 0 \\ 2x + y &= 3 \end{aligned}$

28. $\begin{aligned} 2x_1 + 3x_2 - x_3 &= 1 \\ x_1 + x_3 &= 0 \\ -x_1 + 2x_2 - 2x_3 &= 0 \end{aligned}$

29. $\begin{aligned} x + 5y &= -1 \\ -x + y &= -5 \\ 2x + 4y &= 4 \end{aligned}$

30. $\begin{aligned} a - 2b + d &= 2 \\ -a + b - c - 3d &= 1 \end{aligned}$

In Exercises 31 and 32, find a system of linear equations that has the given matrix as its augmented matrix.

31. $\left[\begin{array}{rrr|r} 0 & 1 & 1 & 1 \\ 1 & -1 & 0 & 1 \\ 2 & -1 & 1 & 1 \end{array}\right]$

32. $\left[\begin{array}{rrrrr|r} 1 & -1 & 0 & 3 & 1 & 2 \\ 1 & 1 & 2 & 1 & -1 & 4 \\ 0 & 1 & 0 & 2 & 3 & 0 \end{array}\right]$

For Exercises 33–38, solve the linear systems in the given exercises.

33. Exercise 27

34. Exercise 28

35. Exercise 29

36. Exercise 30

37. Exercise 31

38. Exercise 32

39. (a) Find a system of two linear equations in the variables x and y whose solution set is given by the parametric equations $x = t$ and $y = 3 - 2t$.

(b) Find another parametric solution to the system in part (a) in which the parameter is s and $y = s$.

40. (a) Find a system of two linear equations in the variables x_1, x_2, and x_3 whose solution set is given by the parametric equations $x_1 = t$, $x_2 = 1 + t$, and $x_3 = 2 - t$.

(b) Find another parametric solution to the system in part (a) in which the parameter is s and $x_3 = s$.

In Exercises 41–44, the systems of equations are nonlinear. Find substitutions (changes of variables) that convert each system into a linear system and use this linear system to help solve the given system.

41. $\dfrac{2}{x} + \dfrac{3}{y} = 0$
$\dfrac{3}{x} + \dfrac{4}{y} = 1$

42. $x^2 + 2y^2 = 6$
$x^2 - y^2 = 3$

43. $\tan x - 2\sin y = 2$
$\tan x - \sin y + \cos z = 2$
$\sin y - \cos z = -1$

44. $-2^a + 2(3^b) = 1$
$3(2^a) - 4(3^b) = 1$

2.2 Direct Methods for Solving Linear Systems

In this section, we will look at a general, systematic procedure for solving a system of linear equations. This procedure is based on the idea of reducing the augmented matrix of the given system to a form that can then be solved by back substitution. The method is *direct* in the sense that it leads directly to the solution (if one exists) in a finite number of steps. In Section 2.5, we will consider some *indirect* methods that work in a completely different way.

Matrices and Echelon Form

There are two important matrices associated with a linear system. The ***coefficient matrix*** contains the coefficients of the variables, and the ***augmented matrix*** (which we have already encountered) is the coefficient matrix augmented by an extra column containing the constant terms.

For the system

$$\begin{aligned} 2x + y - z &= 3 \\ x \qquad + 5z &= 1 \\ -x + 3y - 2z &= 0 \end{aligned}$$

the coefficient matrix is

$$\begin{bmatrix} 2 & 1 & -1 \\ 1 & 0 & 5 \\ -1 & 3 & -2 \end{bmatrix}$$

and the augmented matrix is

$$\left[\begin{array}{rrr|r} 2 & 1 & -1 & 3 \\ 1 & 0 & 5 & 1 \\ -1 & 3 & -2 & 0 \end{array}\right]$$

Note that if a variable is missing (as y is in the second equation), its coefficient 0 is entered in the appropriate position in the matrix. If we denote the coefficient matrix of a linear system by A and the column vector of constant terms by $\mathbf{b}$, then the form of the augmented matrix is $[A \mid \mathbf{b}]$.

In solving a linear system, it will not always be possible to reduce the coefficient matrix to triangular form, as we did in Example 2.6. However, we can always achieve a staircase pattern in the nonzero entries of the final matrix.

The word echelon comes from the Latin word *scala*, meaning "ladder" or "stairs." The French word for "ladder," *échelle*, is also derived from this Latin base. A matrix in echelon form exhibits a staircase pattern.

Definition A matrix is in ***row echelon form*** if it satisfies the following properties:

1. Any rows consisting entirely of zeros are at the bottom.
2. In each nonzero row, the first nonzero entry (called the ***leading entry***) is in a column to the left of any leading entries below it.

Note that these properties guarantee that the leading entries form a staircase pattern. In particular, in any column containing a leading entry, all entries below the leading entry are zero, as the following examples illustrate.

Example 2.7

The following matrices are in row echelon form:

$$\begin{bmatrix} 2 & 4 & 1 \\ 0 & -1 & 2 \\ 0 & 0 & 0 \end{bmatrix} \quad \begin{bmatrix} 1 & 0 & 1 \\ 0 & 1 & 5 \\ 0 & 0 & 4 \end{bmatrix} \quad \begin{bmatrix} 1 & 1 & 2 & 1 \\ 0 & 0 & 1 & 3 \\ 0 & 0 & 0 & 0 \end{bmatrix} \quad \begin{bmatrix} 0 & 2 & 0 & 1 & -1 & 3 \\ 0 & 0 & -1 & 1 & 2 & 2 \\ 0 & 0 & 0 & 0 & 4 & 0 \\ 0 & 0 & 0 & 0 & 0 & 5 \end{bmatrix}$$

If a matrix in row echelon form is actually the augmented matrix of a linear system, the system is quite easy to solve by back substitution alone.

Example 2.8

Assuming that each of the matrices in Example 2.7 is an augmented matrix, write out the corresponding systems of linear equations and solve them.

Solution We first remind ourselves that the last column in an augmented matrix is the vector of constant terms. The first matrix then corresponds to the system

$$\begin{aligned} 2x_1 + 4x_2 &= 1 \\ -x_2 &= 2 \end{aligned}$$

(Notice that we have dropped the last equation $0 = 0$, or $0x_1 + 0x_2 = 0$, which is clearly satisfied for any values of x_1 and x_2.) Back substitution gives $x_2 = -2$ and then $2x_1 = 1 - 4(-2) = 9$, so $x_1 = \frac{9}{2}$. The solution is $[\frac{9}{2}, -2]$.

The second matrix has the corresponding system

$$\begin{aligned} x_1 &= 1 \\ x_2 &= 5 \\ 0 &= 4 \end{aligned}$$

The last equation represents $0x_1 + 0x_2 = 4$, which clearly has no solutions. Therefore, the system has no solutions. Similarly, the system corresponding to the fourth matrix has no solutions. For the system corresponding to the third matrix, we have

$$\begin{aligned} x_1 + x_2 + 2x_3 &= 1 \\ x_3 &= 3 \end{aligned}$$

so $x_1 = 1 - 2(3) - x_2 = -5 - x_2$. There are infinitely many solutions, since we may assign x_2 any value t to get the parametric solution $[-5 - t, t, 3]$.

Elementary Row Operations

We now describe the procedure by which any matrix can be reduced to a matrix in row echelon form. The allowable operations, called ***elementary row operations,*** correspond to the operations that can be performed on a system of linear equations to transform it into an equivalent system.

Definition The following ***elementary row operations*** can be performed on a matrix:

1. Interchange two rows.
2. Multiply a row by a nonzero constant.
3. Add a multiple of a row to another row.

Observe that dividing a row by a nonzero constant is implied in the above definition, since, for example, dividing a row by 2 is the same as multiplying it by $\frac{1}{2}$. Similarly, subtracting a multiple of one row from another row is the same as adding a negative multiple of one row to another row.

We will use the following shorthand notation for the three elementary row operations:

1. $R_i \leftrightarrow R_j$ means interchange rows i and j.
2. kR_i means multiply row i by k.
3. $R_i + kR_j$ means add k times row j to row i (and replace row i with the result).

The process of applying elementary row operations to bring a matrix into row echelon form, called ***row reduction,*** is used to reduce a matrix to echelon form.

Example 2.9 Reduce the following matrix to echelon form:

$$\begin{bmatrix} 1 & 2 & -4 & -4 & 5 \\ 2 & 4 & 0 & 0 & 2 \\ 2 & 3 & 2 & 1 & 5 \\ -1 & 1 & 3 & 6 & 5 \end{bmatrix}$$

Solution We work column by column, from left to right and from top to bottom. The strategy is to create a leading entry in a column and then use it to create zeros below it. The entry chosen to become a leading entry is called a ***pivot,*** and this phase of the process is called ***pivoting.*** Although not strictly necessary, it is often convenient to use the second elementary row operation to make each leading entry a 1.

We begin by introducing zeros into the first column below the leading 1 in the first row:

$$\begin{bmatrix} 1 & 2 & -4 & -4 & 5 \\ 2 & 4 & 0 & 0 & 2 \\ 2 & 3 & 2 & 1 & 5 \\ -1 & 1 & 3 & 6 & 5 \end{bmatrix} \xrightarrow[R_4 + R_1]{\substack{R_2 - 2R_1 \\ R_3 - 2R_1}} \begin{bmatrix} 1 & 2 & -4 & -4 & 5 \\ 0 & 0 & 8 & 8 & -8 \\ 0 & -1 & 10 & 9 & -5 \\ 0 & 3 & -1 & 2 & 10 \end{bmatrix}$$

The first column is now as we want it, so the next thing to do is to create a leading entry in the second row, aiming for the staircase pattern of echelon form. In this case, we do this by interchanging rows. (We could also add row 3 or row 4 to row 2.)

$$\xrightarrow{R_2 \leftrightarrow R_3} \begin{bmatrix} 1 & 2 & -4 & -4 & 5 \\ 0 & -1 & 10 & 9 & -5 \\ 0 & 0 & 8 & 8 & -8 \\ 0 & 3 & -1 & 2 & 10 \end{bmatrix}$$

The pivot this time was -1. We now create a zero at the bottom of column 2, using the leading entry -1 in row 2:

$$\xrightarrow{R_4 + 3R_2} \begin{bmatrix} 1 & 2 & -4 & -4 & 5 \\ 0 & -1 & 10 & 9 & -5 \\ 0 & 0 & 8 & 8 & -8 \\ 0 & 0 & 29 & 29 & -5 \end{bmatrix}$$

Column 2 is now done. Noting that we already have a leading entry in column 3, we just pivot on the 8 to introduce a zero below it. This is easiest if we first divide row 3 by 8:

$$\xrightarrow{\frac{1}{8}R_3} \begin{bmatrix} 1 & 2 & -4 & -4 & 5 \\ 0 & -1 & 10 & 9 & -5 \\ 0 & 0 & 1 & 1 & -1 \\ 0 & 0 & 29 & 29 & -5 \end{bmatrix}$$

Now we use the leading 1 in row 3 to create a zero below it:

$$\xrightarrow{R_4 - 29R_3} \begin{bmatrix} 1 & 2 & -4 & -4 & 5 \\ 0 & -1 & 10 & 9 & -5 \\ 0 & 0 & 1 & 1 & -1 \\ 0 & 0 & 0 & 0 & 24 \end{bmatrix}$$

With this final step, we have reduced our matrix to echelon form.

Remarks

- The row echelon form of a matrix is not unique. (Find a different row echelon form for the matrix in Example 2.9.)

- The leading entry in each row is used to create the zeros below it.
- The pivots are not necessarily the entries that are originally in the positions eventually occupied by the leading entries. In Example 2.9, the pivots were 1, −1, 8, and 24. The original matrix had 1, 4, 2, and 5 in those positions on the "staircase."
- Once we have pivoted and introduced zeros below the leading entry in a column, that column does not change. In other words, the row echelon form emerges from left to right, top to bottom.

Elementary row operations are reversible—that is, they can be "undone." Thus, if some elementary row operation converts A into B, there is also an elementary row operation that converts B into A. (See Exercises 15 and 16.)

Definition Matrices A and B are ***row equivalent*** if there is a sequence of elementary row operations that converts A into B.

The matrices in Example 2.9,

$$\begin{bmatrix} 1 & 2 & -4 & -4 & 5 \\ 2 & 4 & 0 & 0 & 2 \\ 2 & 3 & 2 & 1 & 5 \\ -1 & 1 & 3 & 6 & 5 \end{bmatrix} \quad \text{and} \quad \begin{bmatrix} 1 & 2 & -4 & -4 & 5 \\ 0 & -1 & 10 & 9 & -5 \\ 0 & 0 & 1 & 1 & -1 \\ 0 & 0 & 0 & 0 & 24 \end{bmatrix}$$

are row equivalent. In general, though, how can we tell whether two matrices are row equivalent?

Theorem 2.1 Matrices A and B are row equivalent if and only if they can be reduced to the same row echelon form.

Proof If A and B are row equivalent, then further row operations will reduce B (and therefore A) to the (same) row echelon form.

Conversely, if A and B have the same row echelon form R, then, via elementary row operations, we can convert A into R and B into R. Reversing the latter sequence of operations, we can convert R into B, and therefore the sequence $A \to R \to B$ achieves the desired effect.

Remark In practice, Theorem 2.1 is easiest to use if R is the *reduced* row echelon form of A and B, as defined on page 73. See Exercises 17 and 18.

Gaussian Elimination

When row reduction is applied to the augmented matrix of a system of linear equations, we create an equivalent system that can be solved by back substitution. The entire process is known as ***Gaussian elimination.***

Gaussian Elimination

1. Write the augmented matrix of the system of linear equations.
2. Use elementary row operations to reduce the augmented matrix to row echelon form.
3. Using back substitution, solve the equivalent system that corresponds to the row-reduced matrix.

Remark When performed by hand, step 2 of Gaussian elimination allows quite a bit of choice. Here are some useful guidelines:

(a) Locate the leftmost column that is not all zeros.

(b) Create a leading entry at the top of this column. (It will usually be easiest if you make this a leading 1. See Exercise 22.)

(c) Use the leading entry to create zeros below it.

(d) Cover up the row containing the leading entry, and go back to step (a) to repeat the procedure on the remaining submatrix. Stop when the entire matrix is in row echelon form.

Example 2.10 Solve the system

$$\begin{aligned} 2x_2 + 3x_3 &= 8 \\ 2x_1 + 3x_2 + x_3 &= 5 \\ x_1 - x_2 - 2x_3 &= -5 \end{aligned}$$

Solution The augmented matrix is

$$\left[\begin{array}{ccc|c} 0 & 2 & 3 & 8 \\ 2 & 3 & 1 & 5 \\ 1 & -1 & -2 & -5 \end{array}\right]$$

We proceed to reduce this matrix to row echelon form, following the guidelines given for step 2 of the process. The first nonzero column is column 1. We begin by creating

Bettmann/CORBIS

Carl Friedrich Gauss (1777–1855) is generally considered to be one of the three greatest mathematicians of all time, along with Archimedes and Newton. He is often called the "prince of mathematicians," a nickname that he richly deserves. A child prodigy, Gauss reportedly could do arithmetic before he could talk. At the age of 3, he corrected an error in his father's calculations for the company payroll, and as a young student, he found the formula $n(n + 1)/2$ for the sum of the first n natural numbers. When he was 19, he proved that a 17-sided polygon could be constructed using only a straightedge and a compass, and at the age of 21, he proved, in his doctoral dissertation, that every polynomial of degree n with real or complex coefficients has exactly n zeros, counting multiple zeros—the Fundamental Theorem of Algebra.

Gauss's 1801 publication *Disquisitiones Arithmeticae* is generally considered to be the foundation of modern number theory, but he made contributions to nearly every branch of mathematics as well as to statistics, physics, astronomy, and surveying. Gauss did not publish all of his findings, probably because he was too critical of his own work. He also did not like to teach and was often critical of other mathematicians, perhaps because he discovered—but did not publish—their results before they did.

The method called Gaussian elimination was known to the Chinese in the third century B.C. and was well known by Gauss's time. The method bears Gauss's name because of his use of it in a paper in which he solved a system of linear equations to describe the orbit of an asteroid.

a leading entry at the top of this column; interchanging rows 1 and 3 is the best way to achieve this.

$$\left[\begin{array}{rrr|r} 0 & 2 & 3 & 8 \\ 2 & 3 & 1 & 5 \\ 1 & -1 & -2 & -5 \end{array}\right] \xrightarrow{R_1 \leftrightarrow R_3} \left[\begin{array}{rrr|r} 1 & -1 & -2 & -5 \\ 2 & 3 & 1 & 5 \\ 0 & 2 & 3 & 8 \end{array}\right]$$

We now create a second zero in the first column, using the leading 1:

$$\xrightarrow{R_2 - 2R_1} \left[\begin{array}{rrr|r} 1 & -1 & -2 & -5 \\ 0 & 5 & 5 & 15 \\ 0 & 2 & 3 & 8 \end{array}\right]$$

We now cover up the first row and repeat the procedure. The second column is the first nonzero column of the submatrix. Multiplying row 2 by $\frac{1}{5}$ will create a leading 1.

$$\left[\begin{array}{rrr|r} 1 & -1 & -2 & -5 \\ 0 & 5 & 5 & 15 \\ 0 & 2 & 3 & 8 \end{array}\right] \xrightarrow{\frac{1}{5}R_2} \left[\begin{array}{rrr|r} 1 & -1 & -2 & -5 \\ 0 & 1 & 1 & 3 \\ 0 & 2 & 3 & 8 \end{array}\right]$$

We now need another zero at the bottom of column 2:

$$\xrightarrow{R_3 - 2R_2} \left[\begin{array}{rrr|r} 1 & -1 & -2 & -5 \\ 0 & 1 & 1 & 3 \\ 0 & 0 & 1 & 2 \end{array}\right]$$

The augmented matrix is now in row echelon form, and we move to step 3. The corresponding system is

$$\begin{aligned} x_1 - x_2 - 2x_3 &= -5 \\ x_2 + x_3 &= 3 \\ x_3 &= 2 \end{aligned}$$

and back substitution gives $x_3 = 2$, then $x_2 = 3 - x_3 = 3 - 2 = 1$, and finally $x_1 = -5 + x_2 + 2x_3 = -5 + 1 + 4 = 0$. We write the solution in vector form as

$$\begin{bmatrix} 0 \\ 1 \\ 2 \end{bmatrix}$$

(We are going to write the vector solutions of linear systems as column vectors from now on. The reason for this will become clear in Chapter 3.)

Example 2.11 Solve the system

$$\begin{aligned} w - x - y + 2z &= 1 \\ 2w - 2x - y + 3z &= 3 \\ -w + x - y \qquad &= -3 \end{aligned}$$

Solution The augmented matrix is

$$\left[\begin{array}{rrrr|r} 1 & -1 & -1 & 2 & 1 \\ 2 & -2 & -1 & 3 & 3 \\ -1 & 1 & -1 & 0 & -3 \end{array}\right]$$

which can be row reduced as follows:

$$\left[\begin{array}{rrrr|r} 1 & -1 & -1 & 2 & 1 \\ 2 & -2 & -1 & 3 & 3 \\ -1 & 1 & -1 & 0 & -3 \end{array}\right] \xrightarrow[R_3+R_1]{R_2-2R_1} \left[\begin{array}{rrrr|r} 1 & -1 & -1 & 2 & 1 \\ 0 & 0 & 1 & -1 & 1 \\ 0 & 0 & -2 & 2 & -2 \end{array}\right]$$

$$\xrightarrow{R_3+2R_2} \left[\begin{array}{rrrr|r} 1 & -1 & -1 & 2 & 1 \\ 0 & 0 & 1 & -1 & 1 \\ 0 & 0 & 0 & 0 & 0 \end{array}\right]$$

The associated system is now

$$\begin{aligned} w - x - y + 2z &= 1 \\ y - z &= 1 \end{aligned}$$

which has infinitely many solutions. There is more than one way to assign parameters, but we will proceed to use back substitution, writing the variables corresponding to the leading entries (the ***leading variables***) in terms of the other variables (the ***free variables***).

In this case, the leading variables are w and y, and the free variables are x and z. Thus, $y = 1 + z$, and from this we obtain

$$\begin{aligned} w &= 1 + x + y - 2z \\ &= 1 + x + (1 + z) - 2z \\ &= 2 + x - z \end{aligned}$$

If we assign parameters $x = s$ and $z = t$, the solution can be written in vector form as

$$\begin{bmatrix} w \\ x \\ y \\ z \end{bmatrix} = \begin{bmatrix} 2 + s - t \\ s \\ 1 + t \\ t \end{bmatrix} = \begin{bmatrix} 2 \\ 0 \\ 1 \\ 0 \end{bmatrix} + s\begin{bmatrix} 1 \\ 1 \\ 0 \\ 0 \end{bmatrix} + t\begin{bmatrix} -1 \\ 0 \\ 1 \\ 1 \end{bmatrix}$$

Example 2.11 highlights a very important property: In a consistent system, the free variables are just the variables that are not leading variables. Since the number of leading variables is the number of nonzero rows in the row echelon form of the coefficient matrix, we can predict the number of free variables (parameters) before we find the explicit solution using back substitution. In Chapter 3, we will prove that, although the row echelon form of a matrix is not unique, the number of nonzero rows is the same in *all* row echelon forms of a given matrix. Thus, it makes sense to give a name to this number.

Definition The ***rank*** of a matrix is the number of nonzero rows in its row echelon form.

We will denote the rank of a matrix A by rank(A). In Example 2.10, the rank of the coefficient matrix is 3, and in Example 2.11, the rank of the coefficient matrix is 2. The observations we have just made justify the following theorem, which we will prove in more generality in Chapters 3 and 6.

Theorem 2.2 **The Rank Theorem**

Let A be the coefficient matrix of a system of linear equations with n variables. If the system is consistent, then

$$\text{number of free variables} = n - \text{rank}(A)$$

Thus, in Example 2.10, we have $3 - 3 = 0$ free variables (in other words, a *unique* solution), and in Example 2.11, we have $4 - 2 = 2$ free variables, as we found.

Example 2.12 Solve the system

$$\begin{aligned} x_1 - x_2 + 2x_3 &= 3 \\ x_1 + 2x_2 - x_3 &= -3 \\ 2x_2 - 2x_3 &= 1 \end{aligned}$$

Solution When we row reduce the augmented matrix, we have

$$\left[\begin{array}{rrr|r} 1 & -1 & 2 & 3 \\ 1 & 2 & -1 & -3 \\ 0 & 2 & -2 & 1 \end{array}\right] \xrightarrow{R_2 - R_1} \left[\begin{array}{rrr|r} 1 & -1 & 2 & 3 \\ 0 & 3 & -3 & -6 \\ 0 & 2 & -2 & 1 \end{array}\right]$$

$$\xrightarrow{\frac{1}{3}R_2} \left[\begin{array}{rrr|r} 1 & -1 & 2 & 3 \\ 0 & 1 & -1 & -2 \\ 0 & 2 & -2 & 1 \end{array}\right]$$

$$\xrightarrow{R_3 - 2R_2} \left[\begin{array}{rrr|r} 1 & -1 & 2 & 3 \\ 0 & 1 & -1 & -2 \\ 0 & 0 & 0 & 5 \end{array}\right]$$

leading to the impossible equation $0 = 5$. (We could also have performed $R_3 - \frac{2}{3}R_2$ as the second elementary row operation, which would have given us the same contradiction but a different row echelon form.) Thus, the system has no solutions—it is inconsistent.

Wilhelm Jordan (1842–1899) was a German professor of geodesy whose contribution to solving linear systems was a systematic method of back substitution closely related to the method described here.

Gauss-Jordan Elimination

A modification of Gaussian elimination greatly simplifies the back substitution phase and is particularly helpful when calculations are being done by hand on a system with infinitely many solutions. This variant, known as ***Gauss-Jordan elimination,*** relies on reducing the augmented matrix even further.

Definition A matrix is in ***reduced row echelon form*** if it satisfies the following properties:

1. It is in row echelon form.
2. The leading entry in each nonzero row is a 1 (called a ***leading 1***).
3. Each column containing a leading 1 has zeros everywhere else.

The following matrix is in reduced row echelon form:

$$\begin{bmatrix} 1 & 2 & 0 & 0 & -3 & 1 & 0 \\ 0 & 0 & 1 & 0 & 4 & -1 & 0 \\ 0 & 0 & 0 & 1 & 3 & -2 & 0 \\ 0 & 0 & 0 & 0 & 0 & 0 & 1 \\ 0 & 0 & 0 & 0 & 0 & 0 & 0 \end{bmatrix}$$

For 2×2 matrices, the possible reduced row echelon forms are

$$\begin{bmatrix} 1 & 0 \\ 0 & 1 \end{bmatrix}, \begin{bmatrix} 1 & * \\ 0 & 0 \end{bmatrix}, \begin{bmatrix} 0 & 1 \\ 0 & 0 \end{bmatrix}, \text{ and } \begin{bmatrix} 0 & 0 \\ 0 & 0 \end{bmatrix}$$

where * can be any number.

It is clear that after a matrix has been reduced to echelon form, further elementary row operations will bring it to reduced row echelon form. What is not clear (although intuition may suggest it) is that, unlike the row echelon form, the reduced row echelon form of a matrix is *unique.*

In Gauss-Jordan elimination, we proceed as in Gaussian elimination but reduce the augmented matrix to *reduced* row echelon form.

For a short proof that the reduced row echelon form of a matrix is unique, see the article by Thomas Yuster, "The Reduced Row Echelon Form of a Matrix Is Unique: A Simple Proof," in the March 1984 issue of *Mathematics Magazine* (vol. 57, no. 2, pp. 93–94).

Gauss-Jordan Elimination

1. Write the augmented matrix of the system of linear equations.
2. Use elementary row operations to reduce the augmented matrix to reduced row echelon form.
3. If the resulting system is consistent, solve for the leading variables in terms of any remaining free variables.

Example 2.13 Solve the system in Example 2.11 by Gauss-Jordan elimination.

Solution The reduction proceeds as it did in Example 2.11 until we reach the echelon form:

$$\left[\begin{array}{cccc|c} 1 & -1 & -1 & 2 & 1 \\ 0 & 0 & 1 & -1 & 1 \\ 0 & 0 & 0 & 0 & 0 \end{array}\right]$$

We now must create a zero above the leading 1 in the second row, third column. We do this by adding row 2 to row 1 to obtain

$$\left[\begin{array}{rrrr|r} 1 & -1 & 0 & 1 & 2 \\ 0 & 0 & 1 & -1 & 1 \\ 0 & 0 & 0 & 0 & 0 \end{array}\right]$$

The system has now been reduced to

$$\begin{aligned} w - x \quad\quad + z &= 2 \\ y - z &= 1 \end{aligned}$$

It is now much easier to solve for the leading variables:

$$w = 2 + x - z \qquad \text{and} \qquad y = 1 + z$$

If we assign parameters $x = s$ and $z = t$ as before, the solution can be written in vector form as

$$\begin{bmatrix} w \\ x \\ y \\ z \end{bmatrix} = \begin{bmatrix} 2 + s - t \\ s \\ 1 + t \\ t \end{bmatrix}$$

Remark From a computational point of view, it is more efficient (in the sense that it requires fewer calculations) to first reduce the matrix to row echelon form and then, working from *right to left,* make each leading entry a 1 and create zeros above these leading 1s. However, for manual calculation, you will find it easier to just work from left to right and create the leading 1s and the zeros in their columns as you go.

Let's return to the geometry that brought us to this point. Just as systems of linear equations in two variables correspond to lines in $\mathbb{R}^2$, so linear equations in three variables correspond to planes in $\mathbb{R}^3$. In fact, many questions about lines and planes can be answered by solving an appropriate linear system.

Example 2.14 Find the line of intersection of the planes $x + 2y - z = 3$ and $2x + 3y + z = 1$.

Solution First, observe that there *will* be a line of intersection, since the normal vectors of the two planes—$[1, 2, -1]$ and $[2, 3, 1]$—are not parallel. The points that lie in the intersection of the two planes correspond to the points in the solution set of the system

$$\begin{aligned} x + 2y - z &= 3 \\ 2x + 3y + z &= 1 \end{aligned}$$

Gauss-Jordan elimination applied to the augmented matrix yields

$$\left[\begin{array}{ccc|c} 1 & 2 & -1 & 3 \\ 2 & 3 & 1 & 1 \end{array}\right] \xrightarrow{R_2 - 2R_1} \left[\begin{array}{ccc|c} 1 & 2 & -1 & 3 \\ 0 & -1 & 3 & -5 \end{array}\right]$$

$$\xrightarrow[-R_2]{R_1 + 2R_2} \left[\begin{array}{ccc|c} 1 & 0 & 5 & -7 \\ 0 & 1 & -3 & 5 \end{array}\right]$$

Replacing variables, we have

$$\begin{aligned} x \qquad + 5z &= -7 \\ y - 3z &= 5 \end{aligned}$$

We set the free variable z equal to a parameter t and thus obtain the parametric equations of the line of intersection of the two planes:

$$\begin{aligned} x &= -7 - 5t \\ y &= 5 + 3t \\ z &= t \end{aligned}$$

In vector form, the equation is

$$\begin{bmatrix} x \\ y \\ z \end{bmatrix} = \begin{bmatrix} -7 \\ 5 \\ 0 \end{bmatrix} + t \begin{bmatrix} -5 \\ 3 \\ 1 \end{bmatrix}$$

See Figure 2.2.

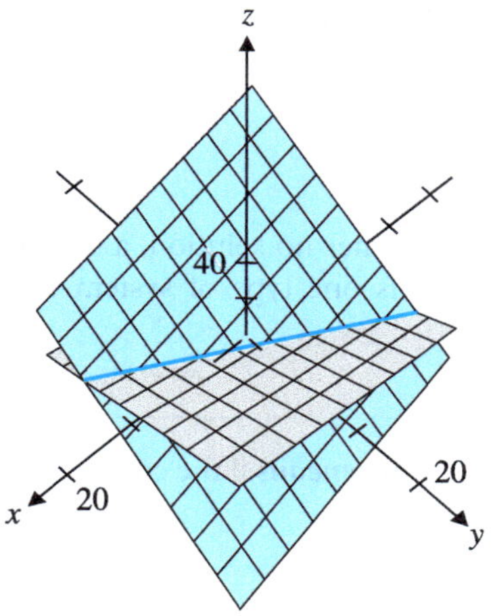

Figure 2.2
The intersection of two planes

Example 2.15

Let $\mathbf{p} = \begin{bmatrix} 1 \\ 0 \\ -1 \end{bmatrix}$, $\mathbf{q} = \begin{bmatrix} 0 \\ 2 \\ 1 \end{bmatrix}$, $\mathbf{u} = \begin{bmatrix} 1 \\ 1 \\ 1 \end{bmatrix}$, and $\mathbf{v} = \begin{bmatrix} 3 \\ -1 \\ -1 \end{bmatrix}$. Determine whether the lines $\mathbf{x} = \mathbf{p} + t\mathbf{u}$ and $\mathbf{x} = \mathbf{q} + t\mathbf{v}$ intersect and, if so, find their point of intersection.

Solution We need to be careful here. Although t has been used as the parameter in the equations of both lines, the lines are independent and therefore so are their parameters. Let's use a different parameter—say, s—for the first line, so its equation becomes $\mathbf{x} = \mathbf{p} + s\mathbf{u}$. If the lines intersect, then we want to find an $\mathbf{x} = \begin{bmatrix} x \\ y \\ z \end{bmatrix}$ that satisfies both equations simultaneously. That is, we want $\mathbf{x} = \mathbf{p} + s\mathbf{u} = \mathbf{q} + t\mathbf{v}$ or $s\mathbf{u} - t\mathbf{v} = \mathbf{q} - \mathbf{p}$.

Substituting the given $\mathbf{p}$, $\mathbf{q}$, $\mathbf{u}$, and $\mathbf{v}$, we obtain the equations

$$\begin{aligned} s - 3t &= -1 \\ s + t &= 2 \\ s + t &= 2 \end{aligned}$$

whose solution is easily found to be $s = \frac{5}{4}$, $t = \frac{3}{4}$. The point of intersection is therefore

$$\begin{bmatrix} x \\ y \\ z \end{bmatrix} = \begin{bmatrix} 1 \\ 0 \\ -1 \end{bmatrix} + \tfrac{5}{4}\begin{bmatrix} 1 \\ 1 \\ 1 \end{bmatrix} = \begin{bmatrix} \frac{9}{4} \\ \frac{5}{4} \\ \frac{1}{4} \end{bmatrix}$$

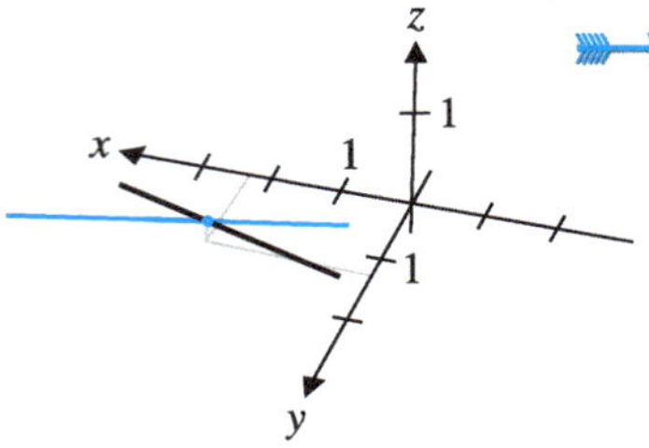

Figure 2.3
Two intersecting lines

See Figure 2.3. (Check that substituting $t = \frac{3}{4}$ in the other equation gives the same point.)

Remark In $\mathbb{R}^3$, it is possible for two lines to intersect in a point, to be parallel, or to do neither. Nonparallel lines that do not intersect are called *skew lines.*

Homogeneous Systems

We have seen that every system of linear equations has either no solution, a unique solution, or infinitely many solutions. However, there is one type of system that always has at least one solution.

Definition A system of linear equations is called ***homogeneous*** if the constant term in each equation is zero.

In other words, a homogeneous system has an augmented matrix of the form $[A \mid \mathbf{0}]$. The following system is homogeneous:

$$\begin{aligned} 2x + 3y - z &= 0 \\ -x + 5y + 2z &= 0 \end{aligned}$$

Since a homogeneous system cannot have no solution (forgive the double negative!), it will have either a unique solution (namely, the zero, or trivial, solution) or infinitely many solutions. The next theorem says that the latter case *must* occur if the number of variables is greater than the number of equations.

Theorem 2.3 If $[A \mid \mathbf{0}]$ is a homogeneous system of m linear equations with n variables, where $m < n$, then the system has infinitely many solutions.

Proof Since the system has at least the zero solution, it is consistent. Also, $\text{rank}(A) \le m$ (why?). By the Rank Theorem, we have

$$\text{number of free variables} = n - \text{rank}(A) \ge n - m > 0$$

So there is at least one free variable and, hence, there are infinitely many solutions.

Note Theorem 2.3 says nothing about the case where $m \ge n$. Exercise 44 asks you to give examples to show that, in this case, there can be either a unique solution or infinitely many solutions.

$\mathbb{R}$ and $\mathbb{Z}_p$ are examples of *fields*. The set of rational numbers $\mathbb{Q}$ and the set of complex numbers $\mathbb{C}$ are other examples. Fields are covered in detail in courses in abstract algebra.

Linear Systems over $\mathbb{Z}_p$

Thus far, all of the linear systems we have encountered have involved real numbers, and the solutions have accordingly been vectors in some $\mathbb{R}^n$. We have seen how other number systems arise—notably, $\mathbb{Z}_p$. When p is a prime number, $\mathbb{Z}_p$ behaves in many respects like $\mathbb{R}$; in particular, we can add, subtract, multiply, and divide (by nonzero numbers). Thus, we can also solve systems of linear equations when the variables and coefficients belong to some $\mathbb{Z}_p$. In such instances, we refer to solving a system *over* $\mathbb{Z}_p$.

For example, the linear equation $x_1 + x_2 + x_3 = 1$, when viewed as an equation over $\mathbb{Z}_2$, has exactly four solutions:

$$\begin{bmatrix} x_1 \\ x_2 \\ x_3 \end{bmatrix} = \begin{bmatrix} 1 \\ 0 \\ 0 \end{bmatrix}, \begin{bmatrix} x_1 \\ x_2 \\ x_3 \end{bmatrix} = \begin{bmatrix} 0 \\ 1 \\ 0 \end{bmatrix}, \begin{bmatrix} x_1 \\ x_2 \\ x_3 \end{bmatrix} = \begin{bmatrix} 0 \\ 0 \\ 1 \end{bmatrix}, \text{ and } \begin{bmatrix} x_1 \\ x_2 \\ x_3 \end{bmatrix} = \begin{bmatrix} 1 \\ 1 \\ 1 \end{bmatrix}$$

(where the last solution arises because $1 + 1 + 1 = 1$ in $\mathbb{Z}_2$).

When we view the equation $x_1 + x_2 + x_3 = 1$ over $\mathbb{Z}_3$, the solutions $\begin{bmatrix} x_1 \\ x_2 \\ x_3 \end{bmatrix}$ are

$$\begin{bmatrix} 1 \\ 0 \\ 0 \end{bmatrix}, \begin{bmatrix} 0 \\ 1 \\ 0 \end{bmatrix}, \begin{bmatrix} 0 \\ 0 \\ 1 \end{bmatrix}, \begin{bmatrix} 2 \\ 2 \\ 0 \end{bmatrix}, \begin{bmatrix} 0 \\ 2 \\ 2 \end{bmatrix}, \begin{bmatrix} 2 \\ 0 \\ 2 \end{bmatrix}, \begin{bmatrix} 1 \\ 1 \\ 2 \end{bmatrix}, \begin{bmatrix} 1 \\ 2 \\ 1 \end{bmatrix}, \begin{bmatrix} 2 \\ 1 \\ 1 \end{bmatrix}$$

(Check these.)

But we need not use trial-and-error methods; row reduction of augmented matrices works just as well over $\mathbb{Z}_p$ as over $\mathbb{R}$.

Example 2.16

Solve the following system of linear equations over $\mathbb{Z}_3$:

$$\begin{aligned} x_1 + 2x_2 + x_3 &= 0 \\ x_1 \qquad + x_3 &= 2 \\ x_2 + 2x_3 &= 1 \end{aligned}$$

Solution The first thing to note in examples like this one is that subtraction and division are not needed; we can accomplish the same effects using addition and multiplication. (This, however, requires that we be working over $\mathbb{Z}_p$, where p is a prime; see Exercise 60 at the end of this section and Exercise 57 in Section 1.1.)

We row reduce the augmented matrix of the system, using calculations modulo 3.

$$\left[\begin{array}{ccc|c} 1 & 2 & 1 & 0 \\ 1 & 0 & 1 & 2 \\ 0 & 1 & 2 & 1 \end{array}\right] \xrightarrow{R_2 + 2R_1} \left[\begin{array}{ccc|c} 1 & 2 & 1 & 0 \\ 0 & 1 & 0 & 2 \\ 0 & 1 & 2 & 1 \end{array}\right]$$

$$\xrightarrow[R_3 + 2R_2]{R_1 + R_2} \left[\begin{array}{ccc|c} 1 & 0 & 1 & 2 \\ 0 & 1 & 0 & 2 \\ 0 & 0 & 2 & 2 \end{array}\right]$$

$$\xrightarrow[2R_3]{R_1+R_3} \left[\begin{array}{ccc|c} 1 & 0 & 0 & 1 \\ 0 & 1 & 0 & 2 \\ 0 & 0 & 1 & 1 \end{array}\right]$$

Thus, the solution is $x_1 = 1, x_2 = 2, x_3 = 1$.

Example 2.17

Solve the following system of linear equations over $\mathbb{Z}_2$:

$$\begin{aligned} x_1 + x_2 + x_3 + x_4 &= 1 \\ x_1 + x_2 \phantom{{}+ x_3 + x_4} &= 1 \\ x_2 + x_3 \phantom{{}+ x_4} &= 0 \\ x_3 + x_4 &= 0 \\ x_1 \phantom{{}+ x_2 + x_3} + x_4 &= 1 \end{aligned}$$

Solution The row reduction proceeds as follows:

$$\left[\begin{array}{cccc|c} 1 & 1 & 1 & 1 & 1 \\ 1 & 1 & 0 & 0 & 1 \\ 0 & 1 & 1 & 0 & 0 \\ 0 & 0 & 1 & 1 & 0 \\ 1 & 0 & 0 & 1 & 1 \end{array}\right] \xrightarrow[R_5+R_1]{R_2+R_1} \left[\begin{array}{cccc|c} 1 & 1 & 1 & 1 & 1 \\ 0 & 0 & 1 & 1 & 0 \\ 0 & 1 & 1 & 0 & 0 \\ 0 & 0 & 1 & 1 & 0 \\ 0 & 1 & 1 & 0 & 0 \end{array}\right]$$

$$\xrightarrow[\substack{R_1 + R_2 \\ R_5 + R_2}]{R_2 \leftrightarrow R_3} \left[\begin{array}{cccc|c} 1 & 0 & 0 & 1 & 1 \\ 0 & 1 & 1 & 0 & 0 \\ 0 & 0 & 1 & 1 & 0 \\ 0 & 0 & 1 & 1 & 0 \\ 0 & 0 & 0 & 0 & 0 \end{array}\right]$$

$$\xrightarrow[R_4 + R_3]{R_2 + R_3} \left[\begin{array}{cccc|c} 1 & 0 & 0 & 1 & 1 \\ 0 & 1 & 0 & 1 & 0 \\ 0 & 0 & 1 & 1 & 0 \\ 0 & 0 & 0 & 0 & 0 \\ 0 & 0 & 0 & 0 & 0 \end{array}\right]$$

Therefore, we have

$$\begin{aligned} x_1 \phantom{{}+ x_2 + x_3} + x_4 &= 1 \\ x_2 \phantom{{}+ x_3} + x_4 &= 0 \\ x_3 + x_4 &= 0 \end{aligned}$$

Setting the free variable $x_4 = t$ yields

$$\begin{bmatrix} x_1 \\ x_2 \\ x_3 \\ x_4 \end{bmatrix} = \begin{bmatrix} 1+t \\ t \\ t \\ t \end{bmatrix} = \begin{bmatrix} 1 \\ 0 \\ 0 \\ 0 \end{bmatrix} + t\begin{bmatrix} 1 \\ 1 \\ 1 \\ 1 \end{bmatrix}$$

Since t can take on the two values 0 and 1, there are exactly two solutions:

$$\begin{bmatrix} 1 \\ 0 \\ 0 \\ 0 \end{bmatrix} \text{ and } \begin{bmatrix} 0 \\ 1 \\ 1 \\ 1 \end{bmatrix}$$

Remark For linear systems over $\mathbb{Z}_p$, there can never be infinitely many solutions. (Why not?) Rather, when there is more than one solution, the number of solutions is finite and is a function of the number of free variables and p. (See Exercise 59.)

Exercises 2.2

In Exercises 1–8, determine whether the given matrix is in row echelon form. If it is, state whether it is also in reduced row echelon form.

1. $\begin{bmatrix} 1 & 0 & 1 \\ 0 & 0 & 3 \\ 0 & 1 & 0 \end{bmatrix}$

2. $\begin{bmatrix} 7 & 0 & 1 & 0 \\ 0 & 1 & -1 & 4 \\ 0 & 0 & 0 & 0 \end{bmatrix}$

3. $\begin{bmatrix} 0 & 1 & 3 & 0 \\ 0 & 0 & 0 & 1 \end{bmatrix}$

4. $\begin{bmatrix} 0 & 0 & 0 \\ 0 & 0 & 0 \\ 0 & 0 & 0 \end{bmatrix}$

5. $\begin{bmatrix} 1 & 0 & 3 & -4 & 0 \\ 0 & 0 & 0 & 0 & 0 \\ 0 & 1 & 5 & 0 & 1 \end{bmatrix}$

6. $\begin{bmatrix} 0 & 0 & 1 \\ 0 & 1 & 0 \\ 1 & 0 & 0 \end{bmatrix}$

7. $\begin{bmatrix} 1 & 2 & 3 \\ 1 & 0 & 0 \\ 0 & 1 & 1 \\ 0 & 0 & 1 \end{bmatrix}$

8. $\begin{bmatrix} 2 & 1 & 3 & 5 \\ 0 & 0 & 1 & -1 \\ 0 & 0 & 0 & 3 \\ 0 & 0 & 0 & 0 \end{bmatrix}$

In Exercises 9–14, use elementary row operations to reduce the given matrix to (a) row echelon form and (b) reduced row echelon form.

9. $\begin{bmatrix} 0 & 0 & 1 \\ 0 & 1 & 1 \\ 1 & 1 & 1 \end{bmatrix}$

10. $\begin{bmatrix} 4 & 3 \\ 2 & 1 \end{bmatrix}$

11. $\begin{bmatrix} 3 & 5 \\ 5 & -2 \\ 2 & 4 \end{bmatrix}$

12. $\begin{bmatrix} 2 & -4 & -2 & 6 \\ 3 & 1 & 6 & 6 \end{bmatrix}$

13. $\begin{bmatrix} 3 & -2 & -1 \\ 2 & -1 & -1 \\ 4 & -3 & -1 \end{bmatrix}$

14. $\begin{bmatrix} -2 & -4 & 7 \\ -3 & -6 & 10 \\ 1 & 2 & -3 \end{bmatrix}$

15. Reverse the elementary row operations used in Example 2.9 to show that we can convert

$$\begin{bmatrix} 1 & 2 & -4 & -4 & 5 \\ 0 & -1 & 10 & 9 & -5 \\ 0 & 0 & 1 & 1 & -1 \\ 0 & 0 & 0 & 0 & 24 \end{bmatrix} \text{ into}$$

$$\begin{bmatrix} 1 & 2 & -4 & -4 & 5 \\ 2 & 4 & 0 & 0 & 2 \\ 2 & 3 & 2 & 1 & 5 \\ -1 & 1 & 3 & 6 & 5 \end{bmatrix}$$

16. In general, what is the elementary row operation that "undoes" each of the three elementary row operations $R_i \leftrightarrow R_j$, kR_i, and $R_i + kR_j$?

In Exercises 17 and 18, show that the given matrices are row equivalent and find a sequence of elementary row operations that will convert A into B.

17. $A = \begin{bmatrix} 1 & 2 \\ 3 & 4 \end{bmatrix}$, $B = \begin{bmatrix} 3 & -1 \\ 1 & 0 \end{bmatrix}$

18. $A = \begin{bmatrix} 2 & 0 & -1 \\ 1 & 1 & 0 \\ -1 & 1 & 1 \end{bmatrix}$, $B = \begin{bmatrix} 3 & 1 & -1 \\ 3 & 5 & 1 \\ 2 & 2 & 0 \end{bmatrix}$

19. What is wrong with the following "proof" that every matrix with at least two rows is row equivalent to a matrix with a zero row?

Perform $R_2 + R_1$ and $R_1 + R_2$. Now rows 1 and 2 are identical. Now perform $R_2 - R_1$ to obtain a row of zeros in the second row.

20. What is the net effect of performing the following sequence of elementary row operations on a matrix (with at least two rows)?

$$R_2 + R_1, R_1 - R_2, R_2 + R_1, -R_1$$

21. Students frequently perform the following type of calculation to introduce a zero into a matrix:

$$\begin{bmatrix} 3 & 1 \\ 2 & 4 \end{bmatrix} \xrightarrow{3R_2 - 2R_1} \begin{bmatrix} 3 & 1 \\ 0 & 10 \end{bmatrix}$$

However, $3R_2 - 2R_1$ is *not* an elementary row operation. Why not? Show how to achieve the same result using elementary row operations.

22. Consider the matrix $A = \begin{bmatrix} 3 & 2 \\ 1 & 4 \end{bmatrix}$. Show that any of the three types of elementary row operations can be used to create a leading 1 at the top of the first column. Which do you prefer and why?

23. What is the rank of each of the matrices in Exercises 1–8?

24. What are the possible reduced row echelon forms of 3×3 matrices?

In Exercises 25–34, solve the given system of equations using either Gaussian or Gauss-Jordan elimination.

25. $\begin{aligned} x_1 + 2x_2 - 3x_3 &= 9 \\ 2x_1 - x_2 + x_3 &= 0 \\ 4x_1 - x_2 + x_3 &= 4 \end{aligned}$

26. $\begin{aligned} x - y + z &= 0 \\ -x + 3y + z &= 5 \\ 3x + y + 7z &= 2 \end{aligned}$

27. $\begin{aligned} x_1 - 3x_2 - 2x_3 &= 0 \\ -x_1 + 2x_2 + x_3 &= 0 \\ 2x_1 + 4x_2 + 6x_3 &= 0 \end{aligned}$

28. $\begin{aligned} 2w + 3x - y + 4z &= 1 \\ 3w - x + z &= 1 \\ 3w - 4x + y - z &= 2 \end{aligned}$

29. $\begin{aligned} 2r + s &= 3 \\ 4r + s &= 7 \\ 2r + 5s &= -1 \end{aligned}$

30. $\begin{aligned} -x_1 + 3x_2 - 2x_3 + 4x_4 &= 0 \\ 2x_1 - 6x_2 + x_3 - 2x_4 &= -3 \\ x_1 - 3x_2 + 4x_3 - 8x_4 &= 2 \end{aligned}$

31. $\begin{aligned} \tfrac{1}{2}x_1 + x_2 - x_3 - 6x_4 &= 2 \\ \tfrac{1}{6}x_1 + \tfrac{1}{2}x_2 - 3x_4 + x_5 &= -1 \\ \tfrac{1}{3}x_1 - 2x_3 - 4x_5 &= 8 \end{aligned}$

32. $\begin{aligned} \sqrt{2}x + y + 2z &= 1 \\ \sqrt{2}y - 3z &= -\sqrt{2} \\ -y + \sqrt{2}z &= 1 \end{aligned}$

33. $\begin{aligned} w + x + 2y + z &= 1 \\ w - x - y + z &= 0 \\ x + y &= -1 \\ w + x + z &= 2 \end{aligned}$

34. $\begin{aligned} a + b + c + d &= 4 \\ a + 2b + 3c + 4d &= 10 \\ a + 3b + 6c + 10d &= 20 \\ a + 4b + 10c + 20d &= 35 \end{aligned}$

In Exercises 35–38, determine by inspection (i.e., without performing any calculations) whether a linear system with the given augmented matrix has a unique solution, infinitely many solutions, or no solution. Justify your answers.

35. $\left[\begin{array}{ccc|c} 0 & 0 & 1 & 2 \\ 0 & 1 & 3 & 1 \\ 1 & 0 & 1 & 1 \end{array}\right]$

36. $\left[\begin{array}{cccc|c} 3 & -2 & 0 & 1 & 1 \\ 1 & 2 & -3 & 1 & -1 \\ 2 & 4 & -6 & 2 & 0 \end{array}\right]$

37. $\left[\begin{array}{cccc|c} 1 & 2 & 3 & 4 & 0 \\ 5 & 6 & 7 & 8 & 0 \\ 9 & 10 & 11 & 12 & 0 \end{array}\right]$

38. $\left[\begin{array}{ccccc|c} 1 & 2 & 3 & 4 & 5 & 6 \\ 6 & 5 & 4 & 3 & 2 & 1 \\ 7 & 7 & 7 & 7 & 7 & 7 \end{array}\right]$

39. Show that if $ad - bc \neq 0$, then the system

$$\begin{aligned} ax + by &= r \\ cx + dy &= s \end{aligned}$$

has a unique solution.

In Exercises 40–43, for what value(s) of k, if any, will the systems have (a) no solution, (b) a unique solution, and (c) infinitely many solutions?

40. $\begin{aligned} kx + 2y &= 3 \\ 2x - 4y &= -6 \end{aligned}$

41. $\begin{aligned} x + ky &= 1 \\ kx + y &= 1 \end{aligned}$

42. $\begin{aligned} x - 2y + 3z &= 2 \\ x + y + z &= k \\ 2x - y + 4z &= k^2 \end{aligned}$

43. $\begin{aligned} x + y + kz &= 1 \\ x + ky + z &= 1 \\ kx + y + z &= -2 \end{aligned}$

44. Give examples of homogeneous systems of m linear equations in n variables with $m = n$ and with $m > n$ that have (a) infinitely many solutions and (b) a unique solution.

In Exercises 45 and 46, find the line of intersection of the given planes.

45. $3x + 2y + z = -1$ and $2x - y + 4z = 5$

46. $4x + y + z = 0$ and $2x - y + 3z = 2$

47. **(a)** Give an example of three planes that have a common line of intersection (Figure 2.4).

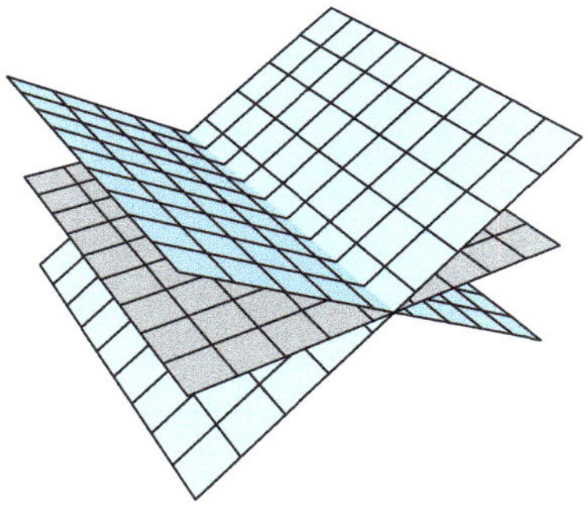

Figure 2.4

(b) Give an example of three planes that intersect in pairs but have no common point of intersection (Figure 2.5).

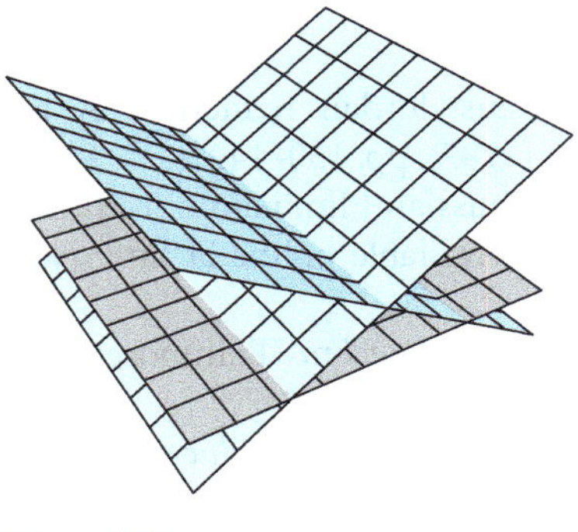

Figure 2.5

(c) Give an example of three planes, exactly two of which are parallel (Figure 2.6).

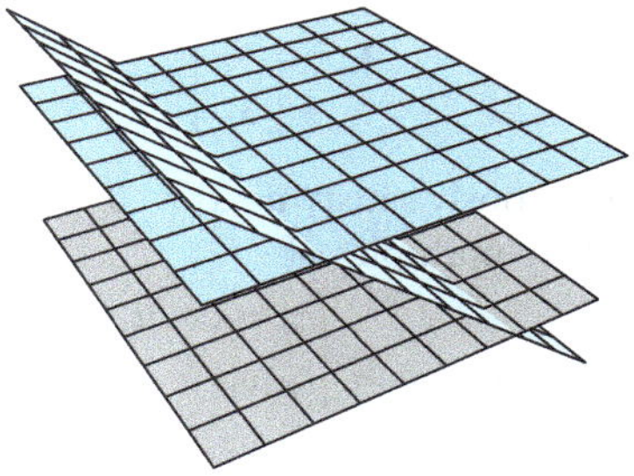

Figure 2.6

(d) Give an example of three planes that intersect in a single point (Figure 2.7).

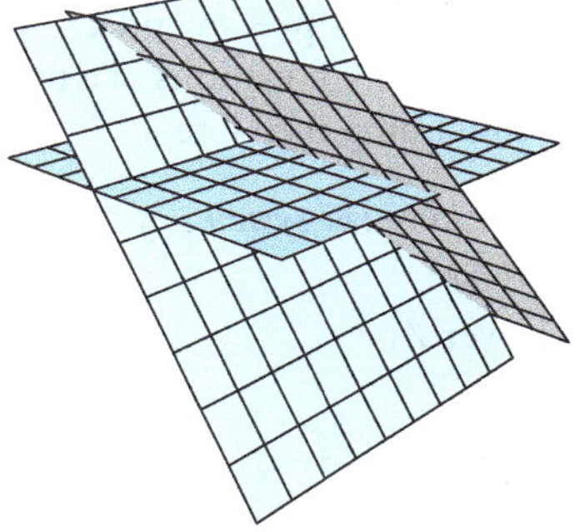

Figure 2.7

In Exercises 48 and 49, determine whether the lines $\mathbf{x} = \mathbf{p} + s\mathbf{u}$ *and* $\mathbf{x} = \mathbf{q} + t\mathbf{v}$ *intersect and, if they do, find their point of intersection.*

48. $\mathbf{p} = \begin{bmatrix} -1 \\ 2 \\ 1 \end{bmatrix}, \mathbf{q} = \begin{bmatrix} 2 \\ 2 \\ 0 \end{bmatrix}, \mathbf{u} = \begin{bmatrix} 1 \\ 2 \\ -1 \end{bmatrix}, \mathbf{v} = \begin{bmatrix} -1 \\ 1 \\ 0 \end{bmatrix}$

49. $\mathbf{p} = \begin{bmatrix} 3 \\ 1 \\ 0 \end{bmatrix}, \mathbf{q} = \begin{bmatrix} -1 \\ 1 \\ -1 \end{bmatrix}, \mathbf{u} = \begin{bmatrix} 1 \\ 0 \\ 1 \end{bmatrix}, \mathbf{v} = \begin{bmatrix} 2 \\ 3 \\ 1 \end{bmatrix}$

50. Let $\mathbf{p} = \begin{bmatrix} 1 \\ 2 \\ 3 \end{bmatrix}, \mathbf{u} = \begin{bmatrix} 1 \\ 1 \\ -1 \end{bmatrix}$, and $\mathbf{v} = \begin{bmatrix} 2 \\ 1 \\ 0 \end{bmatrix}$. Describe all points $Q = (a, b, c)$ such that the line through Q with direction vector $\mathbf{v}$ intersects the line with equation $\mathbf{x} = \mathbf{p} + s\mathbf{u}$.

51. Recall that the cross product of vectors $\mathbf{u}$ and $\mathbf{v}$ is a vector $\mathbf{u} \times \mathbf{v}$ that is orthogonal to both $\mathbf{u}$ and $\mathbf{v}$. (See Exploration: The Cross Product in Chapter 1.) If

$$\mathbf{u} = \begin{bmatrix} u_1 \\ u_2 \\ u_3 \end{bmatrix} \quad \text{and} \quad \mathbf{v} = \begin{bmatrix} v_1 \\ v_2 \\ v_3 \end{bmatrix}$$

show that there are infinitely many vectors

$$\mathbf{x} = \begin{bmatrix} x_1 \\ x_2 \\ x_3 \end{bmatrix}$$

that simultaneously satisfy $\mathbf{u} \cdot \mathbf{x} = 0$ and $\mathbf{v} \cdot \mathbf{x} = 0$ and that all are multiples of

$$\mathbf{u} \times \mathbf{v} = \begin{bmatrix} u_2v_3 - u_3v_2 \\ u_3v_1 - u_1v_3 \\ u_1v_2 - u_2v_1 \end{bmatrix}$$

52. Let $\mathbf{p} = \begin{bmatrix} 1 \\ 1 \\ 0 \end{bmatrix}, \mathbf{q} = \begin{bmatrix} 0 \\ 1 \\ -1 \end{bmatrix}, \mathbf{u} = \begin{bmatrix} 2 \\ -3 \\ 1 \end{bmatrix}$, and $\mathbf{v} = \begin{bmatrix} 0 \\ 6 \\ -1 \end{bmatrix}$.

Show that the lines $\mathbf{x} = \mathbf{p} + s\mathbf{u}$ and $\mathbf{x} = \mathbf{q} + t\mathbf{v}$ are skew lines. Find vector equations of a pair of parallel planes, one containing each line.

In Exercises 53–58, solve the systems of linear equations over the indicated $\mathbb{Z}_p$.

53. $\begin{aligned} x + 2y &= 1 \text{ over } \mathbb{Z}_3 \\ x + y &= 2 \end{aligned}$

54. $\begin{aligned} x + y \quad &= 1 \text{ over } \mathbb{Z}_2 \\ y + z &= 0 \\ x \quad + z &= 1 \end{aligned}$

55. $\begin{aligned} x + y \quad &= 1 \text{ over } \mathbb{Z}_3 \\ y + z &= 0 \\ x \quad + z &= 1 \end{aligned}$

56. $\begin{aligned} 3x + 2y &= 1 \text{ over } \mathbb{Z}_5 \\ x + 4y &= 1 \end{aligned}$

57. $\begin{aligned} 3x + 2y &= 1 \text{ over } \mathbb{Z}_7 \\ x + 4y &= 1 \end{aligned}$

58. $\begin{aligned} x_1 \quad\quad\quad + 4x_4 &= 1 \text{ over } \mathbb{Z}_5 \\ x_1 + 2x_2 + 4x_3 \quad &= 3 \\ 2x_1 + 2x_2 \quad + x_4 &= 1 \\ x_1 \quad + 3x_3 \quad &= 2 \end{aligned}$

59. Prove the following corollary to the Rank Theorem: Let A be an $m \times n$ matrix with entries in $\mathbb{Z}_p$. Any consistent system of linear equations with coefficient matrix A has exactly $p^{n-\text{rank}(A)}$ solutions over $\mathbb{Z}_p$.

60. When p is not prime, extra care is needed in solving a linear system (or, indeed, any equation) over $\mathbb{Z}_p$. Using Gaussian elimination, solve the following system over $\mathbb{Z}_6$. What complications arise?

$$\begin{aligned} 2x + 3y &= 4 \\ 4x + 3y &= 2 \end{aligned}$$

Writing Project

A History of Gaussian Elimination

As noted in the biographical sketch of Gauss in this section, Gauss did not actually "invent" the method known as Gaussian elimination. It was known in some form as early as the third century B.C. and appears in the mathematical writings of cultures throughout Europe and Asia.

Write a report on the history of elimination methods for solving systems of linear equations. What role did Gauss actually play in this history, and why is his name attached to the method?

1. S. Athloen and R. McLaughlin, Gauss-Jordan reduction: A brief history, *American Mathematical Monthly* 94 (1987), pp. 130–142.
2. Joseph F. Grcar, Mathematicians of Gaussian Elimination, *Notices of the AMS*, Vol. 58, No. 6 (2011), pp. 782–792. (Available online at http://www.ams.org/notices/201106/index.html)
3. Roger Hart, *The Chinese Roots of Linear Algebra* (Baltimore: Johns Hopkins University Press, 2011).
4. Victor J. Katz, *A History of Mathematics: An Introduction* (Third Edition) (Reading, MA: Addison Wesley Longman, 2008).

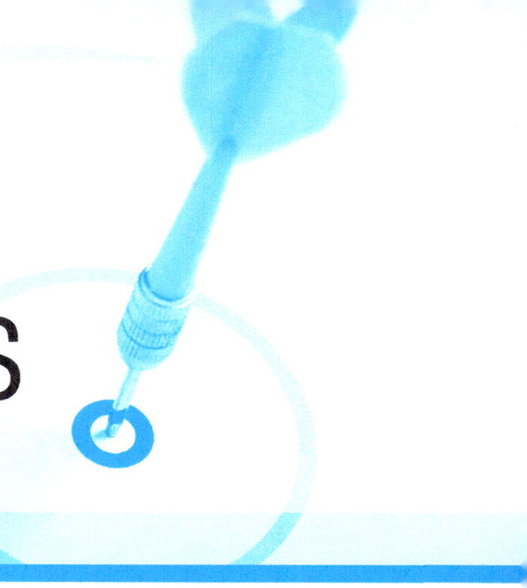

Explorations

CAS Lies My Computer Told Me

Computers and calculators store real numbers in *floating-point form*. For example, 2001 is stored as 0.2001×10^4, and -0.00063 is stored as -0.63×10^{-3}. In general, the floating-point form of a number is $\pm M \times 10^k$, where k is an integer and the *mantissa* M is a (decimal) real number that satisfies $0.1 \le M < 1$.

The maximum number of decimal places that can be stored in the mantissa depends on the computer, calculator, or computer algebra system. If the maximum number of decimal places that can be stored is d, we say that there are d *significant digits*. Many calculators store 8 or 12 significant digits; computers can store more but still are subject to a limit. Any digits that are not stored are either omitted (in which case we say that the number has been *truncated*) or used to *round* the number to d significant digits.

For example, $\pi \approx 3.141592654$, and its floating-point form is 0.3141592654×10^1. In a computer that truncates to five significant digits, π would be stored as 0.31415×10^1 (and displayed as 3.1415); a computer that rounds to five significant digits would store π as 0.31416×10^1 (and display 3.1416). When the dropped digit is a solitary 5, the last remaining digit is rounded so that it becomes even. Thus, rounded to two significant digits, 0.735 becomes 0.74 while 0.725 becomes 0.72.

Whenever truncation or rounding occurs, a *roundoff error* is introduced, which can have a dramatic effect on the calculations. The more operations that are performed, the more the error accumulates. Sometimes, unfortunately, there is nothing we can do about this. This exploration illustrates this phenomenon with very simple systems of linear equations.

1. Solve the following system of linear equations *exactly* (that is, work with rational numbers throughout the calculations).

$$\begin{aligned} x + \phantom{\tfrac{801}{800}}y &= 0 \\ x + \tfrac{801}{800}y &= 1 \end{aligned}$$

2. As a decimal, $\frac{801}{800} = 1.00125$, so, rounded to five significant digits, the system becomes

$$\begin{aligned} x + y &= 0 \\ x + 1.0012y &= 1 \end{aligned}$$

Using your calculator or CAS, solve this system, rounding the result of every calculation to five significant digits.

3. Solve the system two more times, rounding first to four significant digits and then to three significant digits. What happens?

4. Clearly, a very small roundoff error (less than or equal to 0.00125) can result in very large errors in the solution. Explain why geometrically. (Think about the graphs of the various linear systems you solved in Problems 1–3.)

Systems such as the one you just worked with are called ***ill-conditioned.*** They are extremely sensitive to roundoff errors, and there is not much we can do about it. We will encounter ill-conditioned systems again in Chapters 3 and 7. Here is another example to experiment with:

$$\begin{aligned} 4.552x + 7.083y &= 1.931 \\ 1.731x + 2.693y &= 2.001 \end{aligned}$$

Play around with various numbers of significant digits to see what happens, starting with eight significant digits (if you can).

Partial Pivoting

In Exploration: Lies My Computer Told Me, we saw that ill-conditioned linear systems can cause trouble when roundoff error occurs. In this exploration, you will discover another way in which linear systems are sensitive to roundoff error and see that very small changes in the coefficients can lead to huge inaccuracies in the solution. Fortunately, there is something that can be done to minimize or even eliminate this problem (unlike the problem with ill-conditioned systems).

1. (a) Solve the single linear equation $0.00021x = 1$ for x.

   (b) Suppose your calculator can carry only four decimal places. The equation will be rounded to $0.0002x = 1$. Solve this equation.

The difference between the answers in parts (a) and (b) can be thought of as the effect of an error of 0.00001 on the solution of the given equation.

2. Now extend this idea to a system of linear equations.

   (a) With Gaussian elimination, solve the linear system

$$\begin{aligned} 0.400x + 99.6y &= 100 \\ 75.3x - 45.3y &= 30.0 \end{aligned}$$

   using three significant digits. Begin by pivoting on 0.400 and take each calculation to three significant digits. You should obtain the "solution" $x = -1.00$, $y = 1.01$. Check that the actual solution is $x = 1.00$, $y = 1.00$. This is a huge error—200% in the x value! Can you discover what caused it?

   (b) Solve the system in part (a) again, this time interchanging the two equations (or, equivalently, the two rows of its augmented matrix) and pivoting on 75.3. Again, take each calculation to three significant digits. What is the solution this time?

The moral of the story is that, when using Gaussian or Gauss-Jordan elimination to obtain a numerical solution to a system of linear equations (i.e., a decimal approximation), you should choose the pivots with care. Specifically, at each pivoting step, choose from among all possible pivots in a column the entry with the largest absolute value. Use row interchanges to bring this element into the correct position and use it to create zeros where needed in the column. This strategy is known as ***partial pivoting.***

3. Solve the following systems by Gaussian elimination, first without and then with partial pivoting. Take each calculation to three significant digits. (The exact solutions are given.)

(a)
$$\begin{aligned} 0.001x + 0.995y &= 1.00 \\ -10.2x + 1.00y &= -50.0 \end{aligned}$$

Exact solution: $\begin{bmatrix} x \\ y \end{bmatrix} = \begin{bmatrix} 5.00 \\ 1.00 \end{bmatrix}$

(b)
$$\begin{aligned} 10x - 7y \quad &= 7 \\ -3x + 2.09y + 6z &= 3.91 \\ 5x - y + 5z &= 6 \end{aligned}$$

Exact solution: $\begin{bmatrix} x \\ y \\ z \end{bmatrix} = \begin{bmatrix} 0.00 \\ -1.00 \\ 1.00 \end{bmatrix}$

Counting Operations: An Introduction to the Analysis of Algorithms

Gaussian and Gauss-Jordan elimination are examples of ***algorithms:*** systematic procedures designed to implement a particular task—in this case, the row reduction of the augmented matrix of a system of linear equations. Algorithms are particularly well suited to computer implementation, but not all algorithms are created equal. Apart from the speed, memory, and other attributes of the computer system on which they are running, some algorithms are faster than others. One measure of the so-called *complexity* of an algorithm (a measure of its efficiency, or ability to perform its task in a reasonable number of steps) is the number of basic operations it performs as a function of the number of variables that are input.

Let's examine this proposition in the case of the two algorithms we have for solving a linear system: Gaussian and Gauss-Jordan elimination. For our purposes, the basic operations are multiplication and division; we will assume that all other operations are performed much more rapidly and can be ignored. (This is a reasonable assumption, but we will not attempt to justify it.) We will consider only systems of equations with *square* coefficient matrices, so, if the coefficient matrix is $n \times n$, the number of input variables is n. Thus, our task is to find the number of operations performed by Gaussian and Gauss-Jordan elimination as a function of n. Furthermore, we will not worry about special cases that may arise, but rather establish the *worst case* that can arise—when the algorithm takes as long as possible. Since this will give us an estimate of the time it will take a computer to perform the algorithm (if we know how long it takes a computer to perform a single operation), we will denote the number of operations performed by an algorithm by $T(n)$. We will typically be interested in $T(n)$ for large values of n, so comparing this function for different algorithms will allow us to determine which will take less time to execute.

Abu Ja'far Muhammad ibn Musa al-Khwarizmi (c. 780–850) was a Persian mathematician whose book *Hisab al-jabr w'al muqabalah* (c. 825) described the use of Hindu-Arabic numerals and the rules of basic arithmetic. The second word of the book's title gives rise to the English word *algebra*, and the word *algorithm* is derived from al-Khwarizmi's name.

1. Consider the augmented matrix

$$[A \mid \mathbf{b}] = \left[\begin{array}{rrr|r} 2 & 4 & 6 & 8 \\ 3 & 9 & 6 & 12 \\ -1 & 1 & -1 & 1 \end{array}\right]$$

Count the number of operations required to bring $[A \mid \mathbf{b}]$ to the row echelon form

$$\left[\begin{array}{rrr|r} 1 & 2 & 3 & 4 \\ 0 & 1 & -1 & 0 \\ 0 & 0 & 1 & 1 \end{array}\right]$$

(By "operation" we mean a multiplication or a division.) Now count the number of operations needed to complete the back substitution phase of Gaussian elimination. Record the total number of operations.

2. Count the number of operations needed to perform Gauss-Jordan elimination—that is, to reduce $[A \mid \mathbf{b}]$ to its reduced row echelon form

$$\left[\begin{array}{rrr|r} 1 & 0 & 0 & -1 \\ 0 & 1 & 0 & 1 \\ 0 & 0 & 1 & 1 \end{array}\right]$$

(where the zeros are introduced into each column immediately after the leading 1 is created in that column). What do your answers suggest about the relative efficiency of the two algorithms?

We will now attempt to analyze the algorithms in a general, systematic way. Suppose the augmented matrix $[A \mid \mathbf{b}]$ arises from a linear system with n equations and n variables; thus, $[A \mid \mathbf{b}]$ is $n \times (n + 1)$:

$$[A \mid \mathbf{b}] = \left[\begin{array}{cccc|c} a_{11} & a_{12} & \cdots & a_{1n} & b_1 \\ a_{21} & a_{22} & \cdots & a_{2n} & b_2 \\ \vdots & \vdots & \ddots & \vdots & \vdots \\ a_{n1} & a_{n2} & \cdots & a_{nn} & b_n \end{array}\right]$$

We will assume that row interchanges are never needed—that we can always create a leading 1 from a pivot by dividing by the pivot.

3. (a) Show that n operations are needed to create the first leading 1:

$$\left[\begin{array}{cccc|c} a_{11} & a_{12} & \cdots & a_{1n} & b_1 \\ a_{21} & a_{22} & \cdots & a_{2n} & b_2 \\ \vdots & \vdots & \ddots & \vdots & \vdots \\ a_{n1} & a_{n2} & \cdots & a_{nn} & b_n \end{array}\right] \longrightarrow \left[\begin{array}{cccc|c} 1 & * & \cdots & * & * \\ a_{21} & a_{22} & \cdots & a_{2n} & b_2 \\ \vdots & \vdots & \ddots & \vdots & \vdots \\ a_{n1} & a_{n2} & \cdots & a_{nn} & b_n \end{array}\right]$$

(Why don't we need to count an operation for the creation of the leading 1?) Now show that n operations are needed to obtain the first zero in column 1:

$$\left[\begin{array}{cccc|c} 1 & * & \cdots & * & * \\ 0 & * & \cdots & * & * \\ \vdots & \vdots & \ddots & \vdots & \vdots \\ a_{n1} & a_{n2} & \cdots & a_{nn} & b_n \end{array}\right]$$

(Why don't we need to count an operation for the creation of the zero itself?) When the first column has been "swept out," we have the matrix

$$\left[\begin{array}{cccc|c} 1 & * & \cdots & * & * \\ 0 & * & \cdots & * & * \\ \vdots & \vdots & \ddots & \vdots & \vdots \\ 0 & * & \cdots & * & * \end{array}\right]$$

Show that the total number of operations needed up to this point is $n + (n - 1)n$.

(b) Show that the total number of operations needed to reach the row echelon form

$$\left[\begin{array}{cccc|c} 1 & * & \cdots & * & * \\ 0 & 1 & \cdots & * & * \\ \vdots & \vdots & \ddots & \vdots & \vdots \\ 0 & 0 & \cdots & 1 & * \end{array}\right]$$

is

$$[n + (n - 1)n] + [(n - 1) + (n - 2)(n - 1)] + [(n - 2) + (n - 3)(n - 2)] + \cdots + [2 + 1 \cdot 2] + 1$$

which simplifies to

$$n^2 + (n - 1)^2 + \cdots + 2^2 + 1^2$$

(c) Show that the number of operations needed to complete the back substitution phase is

$$1 + 2 + \cdots + (n - 1)$$

(d) Using summation formulas for the sums in parts (b) and (c) (see Exercises 51 and 52 in Section 2.4 and Appendix B), show that the total number of operations, $T(n)$, performed by Gaussian elimination is

$$T(n) = \tfrac{1}{3}n^3 + n^2 - \tfrac{1}{3}n$$

Since every polynomial function is dominated by its leading term for large values of the variable, we see that $T(n) \approx \frac{1}{3}n^3$ for large values of n.

4. Show that Gauss-Jordan elimination has $T(n) \approx \frac{1}{2}n^3$ total operations if we create zeros above and below the leading 1s as we go. (This shows that, for large systems of linear equations, Gaussian elimination is faster than this version of Gauss-Jordan elimination.)

2.3 Spanning Sets and Linear Independence

The second of the three roads in our "trivium" is concerned with linear combinations of vectors. We have seen that we can view solving a system of linear equations as asking whether a certain vector is a linear combination of certain other vectors. We explore this idea in more detail in this section. It leads to some very important concepts, which we will encounter repeatedly in later chapters.

Spanning Sets of Vectors

We can now easily answer the question raised in Section 1.1: When is a given vector a linear combination of other given vectors?

Example 2.18

(a) Is the vector $\begin{bmatrix} 1 \\ 2 \\ 3 \end{bmatrix}$ a linear combination of the vectors $\begin{bmatrix} 1 \\ 0 \\ 3 \end{bmatrix}$ and $\begin{bmatrix} -1 \\ 1 \\ -3 \end{bmatrix}$?

(b) Is $\begin{bmatrix} 2 \\ 3 \\ 4 \end{bmatrix}$ a linear combination of the vectors $\begin{bmatrix} 1 \\ 0 \\ 3 \end{bmatrix}$ and $\begin{bmatrix} -1 \\ 1 \\ -3 \end{bmatrix}$?

Solution

(a) We want to find scalars x and y such that

$$x\begin{bmatrix} 1 \\ 0 \\ 3 \end{bmatrix} + y\begin{bmatrix} -1 \\ 1 \\ -3 \end{bmatrix} = \begin{bmatrix} 1 \\ 2 \\ 3 \end{bmatrix}$$

Expanding, we obtain the system

$$\begin{aligned} x - y &= 1 \\ y &= 2 \\ 3x - 3y &= 3 \end{aligned}$$

whose augmented matrix is

$$\left[\begin{array}{rr|r} 1 & -1 & 1 \\ 0 & 1 & 2 \\ 3 & -3 & 3 \end{array}\right]$$

(Observe that the columns of the augmented matrix are just the given vectors; notice the order of the vectors—in particular, which vector is the constant vector.)

The reduced echelon form of this matrix is

$$\left[\begin{array}{rr|r} 1 & 0 & 3 \\ 0 & 1 & 2 \\ 0 & 0 & 0 \end{array}\right]$$

(Verify this.) So the solution is $x = 3$, $y = 2$, and the corresponding linear combination is

$$3\begin{bmatrix} 1 \\ 0 \\ 3 \end{bmatrix} + 2\begin{bmatrix} -1 \\ 1 \\ -3 \end{bmatrix} = \begin{bmatrix} 1 \\ 2 \\ 3 \end{bmatrix}$$

(b) Utilizing our observation in part (a), we obtain a linear system whose augmented matrix is

$$\left[\begin{array}{rr|r} 1 & -1 & 2 \\ 0 & 1 & 3 \\ 3 & -3 & 4 \end{array}\right]$$

which reduces to

$$\left[\begin{array}{rr|r} 1 & 0 & 5 \\ 0 & 1 & 3 \\ 0 & 0 & -2 \end{array}\right]$$

revealing that the system has no solution. Thus, in this case, $\begin{bmatrix} 2 \\ 3 \\ 4 \end{bmatrix}$ is not a linear combination of $\begin{bmatrix} 1 \\ 0 \\ 3 \end{bmatrix}$ and $\begin{bmatrix} -1 \\ 1 \\ -3 \end{bmatrix}$.

The notion of a spanning set is intimately connected with the solution of linear systems. Look back at Example 2.18. There we saw that a system with augmented matrix $[A \mid \mathbf{b}]$ has a solution precisely when $\mathbf{b}$ is a linear combination of the columns of A. This is a general fact, summarized in the next theorem.

Theorem 2.4 A system of linear equations with augmented matrix $[A \mid \mathbf{b}]$ is consistent if and only if $\mathbf{b}$ is a linear combination of the columns of A.

Let's revisit Example 2.4, interpreting it in light of Theorem 2.4.

(a) The system

$$\begin{aligned} x - y &= 1 \\ x + y &= 3 \end{aligned}$$

has the unique solution $x = 2$, $y = 1$. Thus,

$$2\begin{bmatrix} 1 \\ 1 \end{bmatrix} + \begin{bmatrix} -1 \\ 1 \end{bmatrix} = \begin{bmatrix} 1 \\ 3 \end{bmatrix}$$

See Figure 2.8(a).

(b) The system

$$\begin{aligned} x - y &= 2 \\ 2x - 2y &= 4 \end{aligned}$$

has infinitely many solutions of the form $x = 2 + t$, $y = t$. This implies that

$$(2 + t)\begin{bmatrix} 1 \\ 2 \end{bmatrix} + t\begin{bmatrix} -1 \\ -2 \end{bmatrix} = \begin{bmatrix} 2 \\ 4 \end{bmatrix}$$

for all values of t. Geometrically, the vectors $\begin{bmatrix} 1 \\ 2 \end{bmatrix}$, $\begin{bmatrix} -1 \\ -2 \end{bmatrix}$, and $\begin{bmatrix} 2 \\ 4 \end{bmatrix}$ are all parallel and so all lie along the same line through the origin [see Figure 2.8(b)].

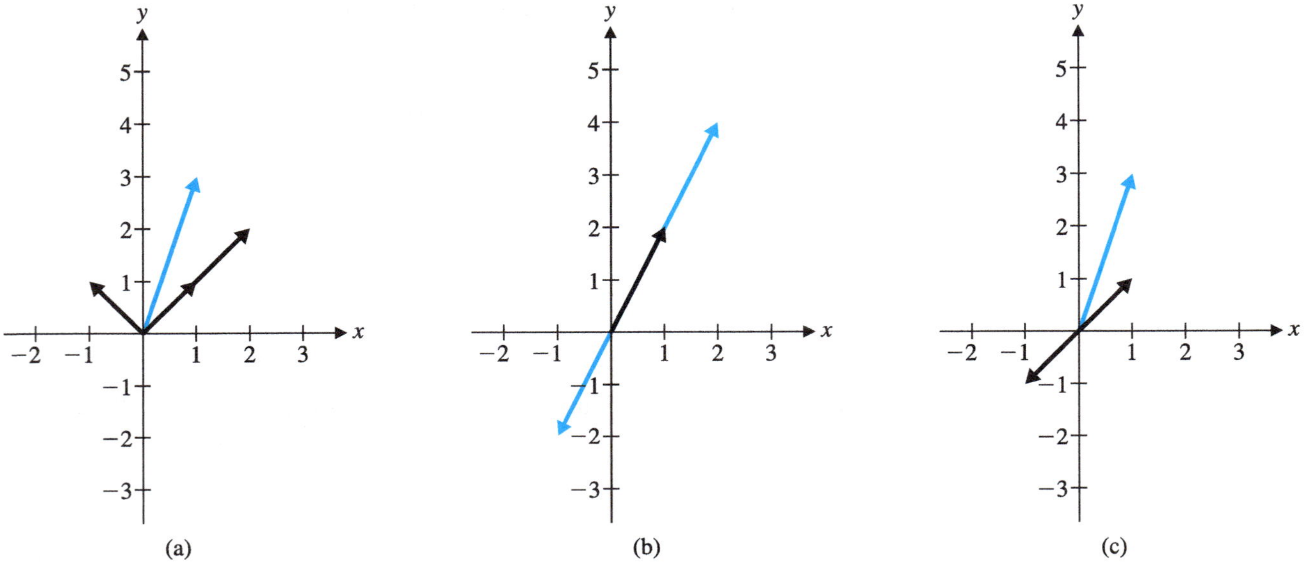

Figure 2.8

(c) The system

$$\begin{aligned} x - y &= 1 \\ x - y &= 3 \end{aligned}$$

has no solutions, so there are no values of x and y that satisfy

$$x\begin{bmatrix} 1 \\ 1 \end{bmatrix} + y\begin{bmatrix} -1 \\ -1 \end{bmatrix} = \begin{bmatrix} 1 \\ 3 \end{bmatrix}$$

In this case, $\begin{bmatrix} 1 \\ 1 \end{bmatrix}$ and $\begin{bmatrix} -1 \\ -1 \end{bmatrix}$ are parallel, but $\begin{bmatrix} 1 \\ 3 \end{bmatrix}$ does not lie along the same line through the origin [see Figure 2.8(c)].

We will often be interested in the collection of *all* linear combinations of a given set of vectors.

Definition If $S = \{\mathbf{v}_1, \mathbf{v}_2, \ldots, \mathbf{v}_k\}$ is a set of vectors in $\mathbb{R}^n$, then the set of all linear combinations of $\mathbf{v}_1, \mathbf{v}_2, \ldots, \mathbf{v}_k$ is called the ***span*** of $\mathbf{v}_1, \mathbf{v}_2, \ldots, \mathbf{v}_k$ and is denoted by $\text{span}(\mathbf{v}_1, \mathbf{v}_2, \ldots, \mathbf{v}_k)$ or $\text{span}(S)$. If $\text{span}(S) = \mathbb{R}^n$, then S is called a ***spanning set*** for $\mathbb{R}^n$.

Example 2.19

Show that $\mathbb{R}^2 = \text{span}\left(\begin{bmatrix} 2 \\ -1 \end{bmatrix}, \begin{bmatrix} 1 \\ 3 \end{bmatrix}\right)$.

Solution We need to show that an arbitrary vector $\begin{bmatrix} a \\ b \end{bmatrix}$ can be written as a linear combination of $\begin{bmatrix} 2 \\ -1 \end{bmatrix}$ and $\begin{bmatrix} 1 \\ 3 \end{bmatrix}$; that is, we must show that the equation $x\begin{bmatrix} 2 \\ -1 \end{bmatrix} + y\begin{bmatrix} 1 \\ 3 \end{bmatrix} = \begin{bmatrix} a \\ b \end{bmatrix}$ can always be solved for x and y (in terms of a and b), regardless of the values of a and b.

The augmented matrix is $\left[\begin{array}{rr|r} 2 & 1 & a \\ -1 & 3 & b \end{array}\right]$, and row reduction produces

$$\left[\begin{array}{rr|r} 2 & 1 & a \\ -1 & 3 & b \end{array}\right] \xrightarrow{R_1 \leftrightarrow R_2} \left[\begin{array}{rr|r} -1 & 3 & b \\ 2 & 1 & a \end{array}\right] \xrightarrow{R_2 + 2R_1} \left[\begin{array}{rr|c} -1 & 3 & b \\ 0 & 7 & a + 2b \end{array}\right]$$

at which point it is clear that the system has a (unique) solution. (Why?) If we continue, we obtain

$$\xrightarrow{\frac{1}{7}R_2} \left[\begin{array}{rr|c} -1 & 3 & b \\ 0 & 1 & (a + 2b)/7 \end{array}\right] \xrightarrow{R_1 - 3R_2} \left[\begin{array}{rr|c} -1 & 0 & (b - 3a)/7 \\ 0 & 1 & (a + 2b)/7 \end{array}\right]$$

from which we see that $x = (3a - b)/7$ and $y = (a + 2b)/7$. Thus, for any choice of a and b, we have

$$\left(\frac{3a - b}{7}\right)\begin{bmatrix} 2 \\ -1 \end{bmatrix} + \left(\frac{a + 2b}{7}\right)\begin{bmatrix} 1 \\ 3 \end{bmatrix} = \begin{bmatrix} a \\ b \end{bmatrix}$$

(Check this.)

Remark It is also true that $\mathbb{R}^2 = \text{span}\left(\begin{bmatrix} 2 \\ -1 \end{bmatrix}, \begin{bmatrix} 1 \\ 3 \end{bmatrix}, \begin{bmatrix} 5 \\ 7 \end{bmatrix}\right)$: If, given $\begin{bmatrix} a \\ b \end{bmatrix}$, we can find x and y such that $x\begin{bmatrix} 2 \\ -1 \end{bmatrix} + y\begin{bmatrix} 1 \\ 3 \end{bmatrix} = \begin{bmatrix} a \\ b \end{bmatrix}$, then we also have $x\begin{bmatrix} 2 \\ -1 \end{bmatrix} + y\begin{bmatrix} 1 \\ 3 \end{bmatrix} + 0\begin{bmatrix} 5 \\ 7 \end{bmatrix} = \begin{bmatrix} a \\ b \end{bmatrix}$. In fact, any set of vectors that *contains* a spanning set for $\mathbb{R}^2$ will also be a spanning set for $\mathbb{R}^2$ (see Exercise 20).

The next example is an important (easy) case of a spanning set. We will encounter versions of this example many times.

Example 2.20

Let $\mathbf{e}_1$, $\mathbf{e}_2$, and $\mathbf{e}_3$ be the standard unit vectors in $\mathbb{R}^3$. Then for any vector $\begin{bmatrix} x \\ y \\ z \end{bmatrix}$, we have

$$\begin{bmatrix} x \\ y \\ z \end{bmatrix} = x\begin{bmatrix} 1 \\ 0 \\ 0 \end{bmatrix} + y\begin{bmatrix} 0 \\ 1 \\ 0 \end{bmatrix} + z\begin{bmatrix} 0 \\ 0 \\ 1 \end{bmatrix} = x\mathbf{e}_1 + y\mathbf{e}_2 + z\mathbf{e}_3$$

Thus, $\mathbb{R}^3 = \text{span}(\mathbf{e}_1, \mathbf{e}_2, \mathbf{e}_3)$.

You should have no difficulty seeing that, in general, $\mathbb{R}^n = \text{span}(\mathbf{e}_1, \mathbf{e}_2, \ldots, \mathbf{e}_n)$.

When the span of a set of vectors in $\mathbb{R}^n$ is not all of $\mathbb{R}^n$, it is reasonable to ask for a description of the vectors' span.

Example 2.21

Find the span of $\begin{bmatrix} 1 \\ 0 \\ 3 \end{bmatrix}$ and $\begin{bmatrix} -1 \\ 1 \\ -3 \end{bmatrix}$. (See Example 2.18.)

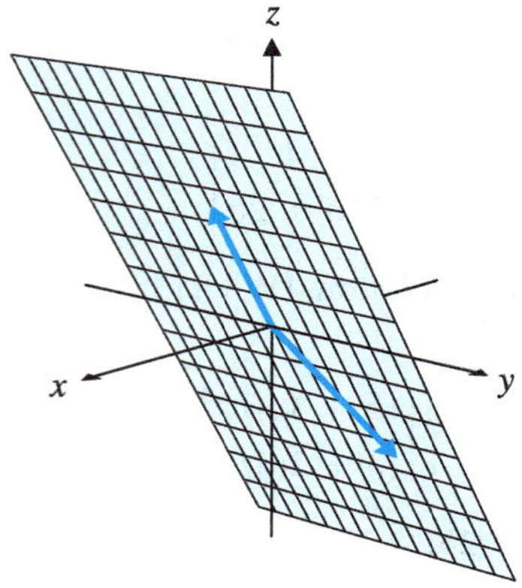

Figure 2.9
Two nonparallel vectors span a plane

Solution Thinking geometrically, we can see that the set of all linear combinations of $\begin{bmatrix}1\\0\\3\end{bmatrix}$ and $\begin{bmatrix}-1\\1\\-3\end{bmatrix}$ is just the plane through the origin with $\begin{bmatrix}1\\0\\3\end{bmatrix}$ and $\begin{bmatrix}-1\\1\\-3\end{bmatrix}$ as direction vectors (Figure 2.9). The vector equation of this plane is $\begin{bmatrix}x\\y\\z\end{bmatrix} = s\begin{bmatrix}1\\0\\3\end{bmatrix} + t\begin{bmatrix}-1\\1\\-3\end{bmatrix}$, which is just another way of saying that $\begin{bmatrix}x\\y\\z\end{bmatrix}$ is in the span of $\begin{bmatrix}1\\0\\3\end{bmatrix}$ and $\begin{bmatrix}-1\\1\\-3\end{bmatrix}$.

Suppose we want to obtain the general equation of this plane. There are several ways to proceed. One is to use the fact that the equation $ax + by + cz = 0$ must be satisfied by the points $(1, 0, 3)$ and $(-1, 1, -3)$ determined by the direction vectors. Substitution then leads to a system of equations in a, b, and c. (See Exercise 17.)

Another method is to use the system of equations arising from the vector equation:

$$\begin{aligned} s - t &= x \\ t &= y \\ 3s - 3t &= z \end{aligned}$$

If we row reduce the augmented matrix, we obtain

$$\left[\begin{array}{cc|c} 1 & -1 & x \\ 0 & 1 & y \\ 3 & -3 & z \end{array}\right] \xrightarrow{R_3 - 3R_1} \left[\begin{array}{cc|c} 1 & -1 & x \\ 0 & 1 & y \\ 0 & 0 & z - 3x \end{array}\right]$$

Now we know that this system is consistent, since $\begin{bmatrix}x\\y\\z\end{bmatrix}$ *is* in the span of $\begin{bmatrix}1\\0\\3\end{bmatrix}$ and $\begin{bmatrix}-1\\1\\-3\end{bmatrix}$ by assumption. So we *must* have $z - 3x = 0$ (or $3x - z = 0$, in more standard form), giving us the general equation we seek.

Remark A normal vector to the plane in this example is also given by the cross product

$$\begin{bmatrix}1\\0\\3\end{bmatrix} \times \begin{bmatrix}-1\\1\\-3\end{bmatrix} = \begin{bmatrix}-3\\0\\1\end{bmatrix}$$

Linear Independence

In Example 2.18, we found that $3\begin{bmatrix}1\\0\\3\end{bmatrix} + 2\begin{bmatrix}-1\\1\\-3\end{bmatrix} = \begin{bmatrix}1\\2\\3\end{bmatrix}$. Let's abbreviate this equation as $3\mathbf{u} + 2\mathbf{v} = \mathbf{w}$. The vector $\mathbf{w}$ "depends" on $\mathbf{u}$ and $\mathbf{v}$ in the sense that it is a linear combination of them. We say that a set of vectors is ***linearly dependent*** if one of them can be written as a linear combination of the others. Note that we also have $\mathbf{u} = -\frac{2}{3}\mathbf{v} + \frac{1}{3}\mathbf{w}$ and $\mathbf{v} = -\frac{3}{2}\mathbf{u} + \frac{1}{2}\mathbf{w}$. To get around the question of which vector to express in terms of the rest, the formal definition is stated as follows:

Definition A set of vectors $\mathbf{v}_1, \mathbf{v}_2, \ldots, \mathbf{v}_k$ is ***linearly dependent*** if there are scalars $c_1, c_2, \ldots, c_k$, *at least one of which is not zero*, such that

$$c_1\mathbf{v}_1 + c_2\mathbf{v}_2 + \cdots + c_k\mathbf{v}_k = \mathbf{0}$$

A set of vectors that is not linearly dependent is called ***linearly independent.***

Remarks

- In the definition of linear dependence, the requirement that at least one of the scalars $c_1, c_2, \ldots, c_k$ must be nonzero allows for the possibility that some may be zero. In the example above, $\mathbf{u}$, $\mathbf{v}$, and $\mathbf{w}$ are linearly dependent, since $3\mathbf{u} + 2\mathbf{v} - \mathbf{w} = \mathbf{0}$ and, in fact, *all* of the scalars are nonzero. On the other hand,

$$\begin{bmatrix} 2 \\ 6 \end{bmatrix} - 2\begin{bmatrix} 1 \\ 3 \end{bmatrix} + 0\begin{bmatrix} 4 \\ 1 \end{bmatrix} = \begin{bmatrix} 0 \\ 0 \end{bmatrix}$$

so $\begin{bmatrix} 2 \\ 6 \end{bmatrix}$, $\begin{bmatrix} 1 \\ 3 \end{bmatrix}$, and $\begin{bmatrix} 4 \\ 1 \end{bmatrix}$ are linearly dependent, since at least one (in fact, two) of the three scalars 1, -2, and 0 is nonzero. (Note that the actual dependence arises simply from the fact that the first two vectors are multiples.) (See Exercise 44.)

- Since $0\mathbf{v}_1 + 0\mathbf{v}_2 + \cdots + 0\mathbf{v}_k = \mathbf{0}$ for *any* vectors $\mathbf{v}_1, \mathbf{v}_2, \ldots, \mathbf{v}_k$, linear dependence essentially says that the zero vector can be expressed as a *nontrivial* linear combination of $\mathbf{v}_1, \mathbf{v}_2, \ldots, \mathbf{v}_k$. Thus, linear independence means that the zero vector can be expressed as a linear combination of $\mathbf{v}_1, \mathbf{v}_2, \ldots, \mathbf{v}_k$ *only* in the trivial way: $c_1\mathbf{v}_1 + c_2\mathbf{v}_2 + \cdots + c_k\mathbf{v}_k = \mathbf{0}$ only if $c_1 = 0, c_2 = 0, \ldots, c_k = 0$.

The relationship between the intuitive notion of dependence and the formal definition is given in the next theorem. Happily, the two notions are equivalent!

Theorem 2.5 Vectors $\mathbf{v}_1, \mathbf{v}_2, \ldots, \mathbf{v}_m$ in $\mathbb{R}^n$ are linearly dependent if and only if at least one of the vectors can be expressed as a linear combination of the others.

Proof If one of the vectors—say, $\mathbf{v}_1$—is a linear combination of the others, then there are scalars $c_2, \ldots, c_m$ such that $\mathbf{v}_1 = c_2\mathbf{v}_2 + \cdots + c_m\mathbf{v}_m$. Rearranging, we obtain $\mathbf{v}_1 - c_2\mathbf{v}_2 - \cdots - c_m\mathbf{v}_m = \mathbf{0}$, which implies that $\mathbf{v}_1, \mathbf{v}_2, \ldots, \mathbf{v}_m$ are linearly dependent, since at least one of the scalars (namely, the coefficient 1 of $\mathbf{v}_1$) is nonzero.

Conversely, suppose that $\mathbf{v}_1, \mathbf{v}_2, \ldots, \mathbf{v}_m$ are linearly dependent. Then there are scalars $c_1, c_2, \ldots, c_m$, not all zero, such that $c_1\mathbf{v}_1 + c_2\mathbf{v}_2 + \cdots + c_m\mathbf{v}_m = \mathbf{0}$. Suppose $c_1 \neq 0$. Then

$$c_1\mathbf{v}_1 = -c_2\mathbf{v}_2 - \cdots - c_m\mathbf{v}_m$$

and we may multiply both sides by $1/c_1$ to obtain $\mathbf{v}_1$ as a linear combination of the other vectors:

$$\mathbf{v}_1 = -\left(\frac{c_2}{c_1}\right)\mathbf{v}_2 - \cdots - \left(\frac{c_m}{c_1}\right)\mathbf{v}_m$$

Note It may appear as if we are cheating a bit in this proof. After all, we cannot be sure that $\mathbf{v}_1$ is a linear combination of the other vectors, nor that c_1 is nonzero. However, the argument is analogous for some other vector $\mathbf{v}_i$ or for a different scalar c_j. Alternatively, we can just relabel things so that they work out as in the above proof. In a situation like this, a mathematician might begin by saying, "without loss of generality, we may assume that $\mathbf{v}_1$ is a linear combination of the other vectors" and then proceed as above.

Example 2.22

Any set of vectors containing the zero vector is linearly dependent. For if $\mathbf{0}, \mathbf{v}_2, \ldots, \mathbf{v}_m$ are in $\mathbb{R}^n$, then we can find a nontrivial combination of the form $c_1\mathbf{0} + c_2\mathbf{v}_2 + \cdots + c_m\mathbf{v}_m = \mathbf{0}$ by setting $c_1 = 1$ and $c_2 = c_3 = \cdots = c_m = 0$.

Example 2.23

Determine whether the following sets of vectors are linearly independent:

(a) $\begin{bmatrix} 1 \\ 4 \end{bmatrix}$ and $\begin{bmatrix} -1 \\ 2 \end{bmatrix}$ (b) $\begin{bmatrix} 1 \\ 1 \\ 0 \end{bmatrix}, \begin{bmatrix} 0 \\ 1 \\ 1 \end{bmatrix}$, and $\begin{bmatrix} 1 \\ 0 \\ 1 \end{bmatrix}$

(c) $\begin{bmatrix} 1 \\ -1 \\ 0 \end{bmatrix}, \begin{bmatrix} 0 \\ 1 \\ -1 \end{bmatrix}$, and $\begin{bmatrix} -1 \\ 0 \\ 1 \end{bmatrix}$ (d) $\begin{bmatrix} 1 \\ 2 \\ 0 \end{bmatrix}, \begin{bmatrix} 1 \\ 1 \\ -1 \end{bmatrix}$, and $\begin{bmatrix} 1 \\ 4 \\ 2 \end{bmatrix}$

Solution In answering any question of this type, it is a good idea to see if you can determine by inspection whether one vector is a linear combination of the others. A little thought may save a lot of computation!

(a) The only way two vectors can be linearly dependent is if one is a multiple of the other. (Why?) These two vectors are clearly not multiples, so they are linearly independent.

(b) There is no obvious dependence relation here, so we try to find scalars c_1, c_2, c_3 such that

$$c_1\begin{bmatrix} 1 \\ 1 \\ 0 \end{bmatrix} + c_2\begin{bmatrix} 0 \\ 1 \\ 1 \end{bmatrix} + c_3\begin{bmatrix} 1 \\ 0 \\ 1 \end{bmatrix} = \begin{bmatrix} 0 \\ 0 \\ 0 \end{bmatrix}$$

The corresponding linear system is

$$\begin{aligned} c_1 + \phantom{c_2 + {}} c_3 &= 0 \\ c_1 + c_2 \phantom{{} + c_3} &= 0 \\ c_2 + c_3 &= 0 \end{aligned}$$

and the augmented matrix is

$$\left[\begin{array}{ccc|c} 1 & 0 & 1 & 0 \\ 1 & 1 & 0 & 0 \\ 0 & 1 & 1 & 0 \end{array}\right]$$

Once again, we make the fundamental observation that the columns of the coefficient matrix are just the vectors in question!

The reduced row echelon form is

$$\left[\begin{array}{ccc|c} 1 & 0 & 0 & 0 \\ 0 & 1 & 0 & 0 \\ 0 & 0 & 1 & 0 \end{array}\right]$$

(check this), so $c_1 = 0, c_2 = 0, c_3 = 0$. Thus, the given vectors are linearly independent.

(c) A little reflection reveals that

$$\begin{bmatrix} 1 \\ -1 \\ 0 \end{bmatrix} + \begin{bmatrix} 0 \\ 1 \\ -1 \end{bmatrix} + \begin{bmatrix} -1 \\ 0 \\ 1 \end{bmatrix} = \begin{bmatrix} 0 \\ 0 \\ 0 \end{bmatrix}$$

so the three vectors are linearly dependent. [Set up a linear system as in part (b) to check this algebraically.]

(d) Once again, we observe no obvious dependence, so we proceed directly to reduce a homogeneous linear system whose augmented matrix has as its columns the given vectors:

$$\left[\begin{array}{ccc|c} 1 & 1 & 1 & 0 \\ 2 & 1 & 4 & 0 \\ 0 & -1 & 2 & 0 \end{array}\right] \xrightarrow{R_2 - 2R_1} \left[\begin{array}{ccc|c} 1 & 1 & 1 & 0 \\ 0 & -1 & 2 & 0 \\ 0 & -1 & 2 & 0 \end{array}\right] \xrightarrow[\substack{R_3 - R_2 \\ -R_2}]{R_1 + R_2} \left[\begin{array}{ccc|c} 1 & 0 & 3 & 0 \\ 0 & 1 & -2 & 0 \\ 0 & 0 & 0 & 0 \end{array}\right]$$

If we let the scalars be c_1, c_2, and c_3, we have

$$\begin{aligned} c_1 + \quad 3c_3 &= 0 \\ c_2 - 2c_3 &= 0 \end{aligned}$$

from which we see that the system has infinitely many solutions. In particular, there must be a nonzero solution, so the given vectors are linearly dependent.

If we continue, we can describe these solutions exactly: $c_1 = -3c_3$ and $c_2 = 2c_3$. Thus, for any nonzero value of c_3, we have the linear dependence relation

$$-3c_3\begin{bmatrix} 1 \\ 2 \\ 0 \end{bmatrix} + 2c_3\begin{bmatrix} 1 \\ 1 \\ -1 \end{bmatrix} + c_3\begin{bmatrix} 1 \\ 4 \\ 2 \end{bmatrix} = \begin{bmatrix} 0 \\ 0 \\ 0 \end{bmatrix}$$

(Once again, check that this is correct.)

We summarize this procedure for testing for linear independence as a theorem.

Theorem 2.6 Let $\mathbf{v}_1, \mathbf{v}_2, \ldots, \mathbf{v}_m$ be (column) vectors in $\mathbb{R}^n$ and let A be the $n \times m$ matrix $[\mathbf{v}_1\ \mathbf{v}_2 \cdots \mathbf{v}_m]$ with these vectors as its columns. Then $\mathbf{v}_1, \mathbf{v}_2, \ldots, \mathbf{v}_m$ are linearly dependent if and only if the homogeneous linear system with augmented matrix $[A \mid \mathbf{0}]$ has a nontrivial solution.

Proof $\mathbf{v}_1, \mathbf{v}_2, \ldots, \mathbf{v}_m$ are linearly dependent if and only if there are scalars $c_1, c_2, \ldots, c_m$, not all zero, such that $c_1\mathbf{v}_1 + c_2\mathbf{v}_2 + \cdots + c_m\mathbf{v}_m = \mathbf{0}$. By Theorem 2.4, this is equivalent to saying that the nonzero vector $\begin{bmatrix} c_1 \\ c_2 \\ \vdots \\ c_m \end{bmatrix}$ is a solution of the system whose augmented matrix is $[\mathbf{v}_1\ \mathbf{v}_2 \cdots \mathbf{v}_m \mid \mathbf{0}]$.

Example 2.24

The standard unit vectors $\mathbf{e}_1$, $\mathbf{e}_2$, and $\mathbf{e}_3$ are linearly independent in $\mathbb{R}^3$, since the system with augmented matrix $[\mathbf{e}_1\ \mathbf{e}_2\ \mathbf{e}_3 \mid \mathbf{0}]$ is already in the reduced row echelon form

$$\left[\begin{array}{ccc|c} 1 & 0 & 0 & 0 \\ 0 & 1 & 0 & 0 \\ 0 & 0 & 1 & 0 \end{array}\right]$$

and so clearly has only the trivial solution. In general, we see that $\mathbf{e}_1, \mathbf{e}_2, \ldots, \mathbf{e}_n$ will be linearly independent in $\mathbb{R}^n$.

Performing elementary row operations on a matrix constructs linear combinations of the rows. We can use this fact to come up with another way to test vectors for linear independence.

Example 2.25

Consider the three vectors of Example 2.23(d) as *row* vectors:

$$[1, 2, 0], \quad [1, 1, -1], \quad \text{and} \quad [1, 4, 2]$$

We construct a matrix with these vectors as its rows and proceed to reduce it to echelon form. Each time a row changes, we denote the new row by adding a prime symbol:

$$\begin{bmatrix} 1 & 2 & 0 \\ 1 & 1 & -1 \\ 1 & 4 & 2 \end{bmatrix} \xrightarrow[R_3' = R_3 - R_1]{R_2' = R_2 - R_1} \begin{bmatrix} 1 & 2 & 0 \\ 0 & -1 & -1 \\ 0 & 2 & 2 \end{bmatrix} \xrightarrow{R_3'' = R_3' + 2R_2'} \begin{bmatrix} 1 & 2 & 0 \\ 0 & -1 & -1 \\ 0 & 0 & 0 \end{bmatrix}$$

From this we see that

$$\mathbf{0} = R_3'' = R_3' + 2R_2' = (R_3 - R_1) + 2(R_2 - R_1) = -3R_1 + 2R_2 + R_3$$

or, in terms of the original vectors,

$$-3[1, 2, 0] + 2[1, 1, -1] + [1, 4, 2] = [0, 0, 0]$$

[Notice that this approach corresponds to taking $c_3 = 1$ in the solution to Example 2.23(d).]

Thus, the rows of a matrix will be linearly dependent if elementary row operations can be used to create a zero row. We summarize this finding as follows:

Theorem 2.7

Let $\mathbf{v}_1, \mathbf{v}_2, \ldots, \mathbf{v}_m$ be (row) vectors in $\mathbb{R}^n$ and let A be the $m \times n$ matrix $\begin{bmatrix} \mathbf{v}_1 \\ \mathbf{v}_2 \\ \vdots \\ \mathbf{v}_m \end{bmatrix}$ with these vectors as its rows. Then $\mathbf{v}_1, \mathbf{v}_2, \ldots, \mathbf{v}_m$ are linearly dependent if and only if $\operatorname{rank}(A) < m$.

Proof Assume that $\mathbf{v}_1, \mathbf{v}_2, \ldots, \mathbf{v}_m$ are linearly dependent. Then, by Theorem 2.2, at least one of the vectors can be written as a linear combination of the others.

We relabel the vectors, if necessary, so that we can write $\mathbf{v}_m = c_1\mathbf{v}_1 + c_2\mathbf{v}_2 + \cdots + c_{m-1}\mathbf{v}_{m-1}$. Then the elementary row operations $R_m - c_1R_1$, $R_m - c_2R_2, \ldots, R_m - c_{m-1}R_{m-1}$ applied to A will create a zero row in row m. Thus, $\text{rank}(A) < m$.

Conversely, assume that $\text{rank}(A) < m$. Then there is some sequence of row operations that will create a zero row. A successive substitution argument analogous to that used in Example 2.25 can be used to show that $\mathbf{0}$ is a nontrivial linear combination of $\mathbf{v}_1, \mathbf{v}_2, \ldots, \mathbf{v}_m$. Thus, $\mathbf{v}_1, \mathbf{v}_2, \ldots, \mathbf{v}_m$ are linearly dependent.

In some situations, we can deduce that a set of vectors is linearly dependent without doing any work. One such situation is when the zero vector is in the set (as in Example 2.22). Another is when there are "too many" vectors to be independent. The following theorem summarizes this case. (We will see a sharper version of this result in Chapter 6.)

Theorem 2.8 Any set of m vectors in $\mathbb{R}^n$ is linearly dependent if $m > n$.

Proof Let $\mathbf{v}_1, \mathbf{v}_2, \ldots, \mathbf{v}_m$ be (column) vectors in $\mathbb{R}^n$ and let A be the $n \times m$ matrix $[\mathbf{v}_1\ \mathbf{v}_2 \cdots \mathbf{v}_m]$ with these vectors as its columns. By Theorem 2.6, $\mathbf{v}_1, \mathbf{v}_2, \ldots, \mathbf{v}_m$ are linearly dependent if and only if the homogeneous linear system with augmented matrix $[A \mid \mathbf{0}]$ has a nontrivial solution. But, according to Theorem 2.6, this will always be the case if A has more columns than rows; it is the case here, since number of columns m is greater than number of rows n.

Example 2.26

The vectors $\begin{bmatrix}1\\3\end{bmatrix}, \begin{bmatrix}2\\4\end{bmatrix}$, and $\begin{bmatrix}3\\1\end{bmatrix}$ are linearly dependent, since there cannot be more than two linearly independent vectors in $\mathbb{R}^2$. (Note that if we want to find the actual dependence relation among these three vectors, we must solve the homogeneous system whose coefficient matrix has the given vectors as columns. Do this!)

Exercises 2.3

In Exercises 1–6, determine if the vector $\mathbf{v}$ *is a linear combination of the remaining vectors.*

1. $\mathbf{v} = \begin{bmatrix}1\\2\end{bmatrix}, \mathbf{u}_1 = \begin{bmatrix}1\\-1\end{bmatrix}, \mathbf{u}_2 = \begin{bmatrix}2\\-1\end{bmatrix}$

2. $\mathbf{v} = \begin{bmatrix}2\\1\end{bmatrix}, \mathbf{u}_1 = \begin{bmatrix}4\\-2\end{bmatrix}, \mathbf{u}_2 = \begin{bmatrix}-2\\1\end{bmatrix}$

3. $\mathbf{v} = \begin{bmatrix}1\\2\\3\end{bmatrix}, \mathbf{u}_1 = \begin{bmatrix}1\\1\\0\end{bmatrix}, \mathbf{u}_2 = \begin{bmatrix}0\\1\\1\end{bmatrix}$

4. $\mathbf{v} = \begin{bmatrix}3\\2\\-1\end{bmatrix}, \mathbf{u}_1 = \begin{bmatrix}1\\1\\0\end{bmatrix}, \mathbf{u}_2 = \begin{bmatrix}0\\1\\1\end{bmatrix}$

5. $\mathbf{v} = \begin{bmatrix}1\\2\\3\end{bmatrix}, \mathbf{u}_1 = \begin{bmatrix}1\\1\\0\end{bmatrix}, \mathbf{u}_2 = \begin{bmatrix}0\\1\\1\end{bmatrix}$,

$\mathbf{u}_3 = \begin{bmatrix}1\\0\\1\end{bmatrix}$

CAS **6.** $\mathbf{v} = \begin{bmatrix}3.2\\2.0\\-2.6\end{bmatrix}, \mathbf{u}_1 = \begin{bmatrix}1.0\\0.4\\4.8\end{bmatrix}, \mathbf{u}_2 = \begin{bmatrix}3.4\\1.4\\-6.4\end{bmatrix}$,

$\mathbf{u}_3 = \begin{bmatrix}-1.2\\0.2\\-1.0\end{bmatrix}$

In Exercises 7 and 8, determine if the vector **b** *is in the span of the columns of the matrix A.*

7. $A = \begin{bmatrix} 1 & 2 \\ 3 & 4 \end{bmatrix}, \mathbf{b} = \begin{bmatrix} 5 \\ 6 \end{bmatrix}$

8. $A = \begin{bmatrix} 1 & 2 & 3 \\ 5 & 6 & 7 \\ 9 & 10 & 11 \end{bmatrix}, \mathbf{b} = \begin{bmatrix} 4 \\ 8 \\ 12 \end{bmatrix}$

9. Show that $\mathbb{R}^2 = \text{span}\left(\begin{bmatrix} 1 \\ 1 \end{bmatrix}, \begin{bmatrix} 1 \\ -1 \end{bmatrix}\right)$.

10. Show that $\mathbb{R}^2 = \text{span}\left(\begin{bmatrix} 3 \\ -2 \end{bmatrix}, \begin{bmatrix} 0 \\ 1 \end{bmatrix}\right)$.

11. Show that $\mathbb{R}^3 = \text{span}\left(\begin{bmatrix} 1 \\ 0 \\ 1 \end{bmatrix}, \begin{bmatrix} 1 \\ 1 \\ 0 \end{bmatrix}, \begin{bmatrix} 0 \\ 1 \\ 1 \end{bmatrix}\right)$.

12. Show that $\mathbb{R}^3 = \text{span}\left(\begin{bmatrix} 1 \\ 1 \\ 0 \end{bmatrix}, \begin{bmatrix} 1 \\ 2 \\ 3 \end{bmatrix}, \begin{bmatrix} 2 \\ 1 \\ -1 \end{bmatrix}\right)$.

In Exercises 13–16, describe the span of the given vectors (a) geometrically and (b) algebraically.

13. $\begin{bmatrix} 2 \\ -4 \end{bmatrix}, \begin{bmatrix} -1 \\ 2 \end{bmatrix}$

14. $\begin{bmatrix} 0 \\ 0 \end{bmatrix}, \begin{bmatrix} 3 \\ 4 \end{bmatrix}$

15. $\begin{bmatrix} 1 \\ 2 \\ 0 \end{bmatrix}, \begin{bmatrix} 3 \\ 2 \\ -1 \end{bmatrix}$

16. $\begin{bmatrix} 1 \\ 0 \\ -1 \end{bmatrix}, \begin{bmatrix} -1 \\ 1 \\ 0 \end{bmatrix}, \begin{bmatrix} 0 \\ -1 \\ 1 \end{bmatrix}$

17. The general equation of the plane that contains the points $(1, 0, 3)$, $(-1, 1, -3)$, and the origin is of the form $ax + by + cz = 0$. Solve for a, b, and c.

18. Prove that $\mathbf{u}$, $\mathbf{v}$, and $\mathbf{w}$ are all in $\text{span}(\mathbf{u}, \mathbf{v}, \mathbf{w})$.

19. Prove that $\mathbf{u}$, $\mathbf{v}$, and $\mathbf{w}$ are all in $\text{span}(\mathbf{u}, \mathbf{u} + \mathbf{v}, \mathbf{u} + \mathbf{v} + \mathbf{w})$.

20. **(a)** Prove that if $\mathbf{u}_1, \ldots, \mathbf{u}_m$ are vectors in $\mathbb{R}^n$, $S = \{\mathbf{u}_1, \mathbf{u}_2, \ldots, \mathbf{u}_k\}$, and $T = \{\mathbf{u}_1, \ldots, \mathbf{u}_k, \mathbf{u}_{k+1}, \ldots, \mathbf{u}_m\}$, then $\text{span}(S) \subseteq \text{span}(T)$. [*Hint:* Rephrase this question in terms of linear combinations.]

(b) Deduce that if $\mathbb{R}^n = \text{span}(S)$, then $\mathbb{R}^n = \text{span}(T)$ also.

21. **(a)** Suppose that vector $\mathbf{w}$ is a linear combination of vectors $\mathbf{u}_1, \ldots, \mathbf{u}_k$ and that each $\mathbf{u}_i$ is a linear combination of vectors $\mathbf{v}_1, \ldots, \mathbf{v}_m$. Prove that $\mathbf{w}$ is a linear combination of $\mathbf{v}_1, \ldots, \mathbf{v}_m$ and therefore $\text{span}(\mathbf{u}_1, \ldots, \mathbf{u}_k) \subseteq \text{span}(\mathbf{v}_1, \ldots, \mathbf{v}_m)$.

(b) In part (a), suppose in addition that each $\mathbf{v}_j$ is also a linear combination of $\mathbf{u}_1, \ldots, \mathbf{u}_k$. Prove that $\text{span}(\mathbf{u}_1, \ldots, \mathbf{u}_k) = \text{span}(\mathbf{v}_1, \ldots, \mathbf{v}_m)$.

(c) Use the result of part (b) to prove that

$$\mathbb{R}^3 = \text{span}\left(\begin{bmatrix} 1 \\ 0 \\ 0 \end{bmatrix}, \begin{bmatrix} 1 \\ 1 \\ 0 \end{bmatrix}, \begin{bmatrix} 1 \\ 1 \\ 1 \end{bmatrix}\right)$$

[*Hint:* We know that $\mathbb{R}^3 = \text{span}(\mathbf{e}_1, \mathbf{e}_2, \mathbf{e}_3)$.]

Use the method of Example 2.23 and Theorem 2.6 to determine if the sets of vectors in Exercises 22–31 are linearly independent. If, for any of these, the answer can be determined by inspection (i.e., without calculation), state why. For any sets that are linearly dependent, find a dependence relationship among the vectors.

22. $\begin{bmatrix} 2 \\ -1 \\ 3 \end{bmatrix}, \begin{bmatrix} -1 \\ 2 \\ 3 \end{bmatrix}$

23. $\begin{bmatrix} 1 \\ 1 \\ 1 \end{bmatrix}, \begin{bmatrix} 1 \\ 2 \\ 3 \end{bmatrix}, \begin{bmatrix} 1 \\ -1 \\ 2 \end{bmatrix}$

24. $\begin{bmatrix} 2 \\ 2 \\ 1 \end{bmatrix}, \begin{bmatrix} 3 \\ 1 \\ 2 \end{bmatrix}, \begin{bmatrix} 1 \\ -5 \\ 2 \end{bmatrix}$

25. $\begin{bmatrix} 0 \\ 1 \\ 2 \end{bmatrix}, \begin{bmatrix} 2 \\ 1 \\ 3 \end{bmatrix}, \begin{bmatrix} 2 \\ 0 \\ 1 \end{bmatrix}$

26. $\begin{bmatrix} -2 \\ 3 \\ 7 \end{bmatrix}, \begin{bmatrix} 4 \\ -1 \\ 5 \end{bmatrix}, \begin{bmatrix} 3 \\ 1 \\ 3 \end{bmatrix}, \begin{bmatrix} 5 \\ 0 \\ 2 \end{bmatrix}$

27. $\begin{bmatrix} 3 \\ 4 \\ 5 \end{bmatrix}, \begin{bmatrix} 6 \\ 7 \\ 8 \end{bmatrix}, \begin{bmatrix} 0 \\ 0 \\ 0 \end{bmatrix}$

28. $\begin{bmatrix} -1 \\ 1 \\ 2 \\ 1 \end{bmatrix}, \begin{bmatrix} 3 \\ 2 \\ 2 \\ 4 \end{bmatrix}, \begin{bmatrix} 2 \\ 3 \\ 1 \\ -1 \end{bmatrix}$

29. $\begin{bmatrix} 1 \\ -1 \\ 1 \\ 0 \end{bmatrix}, \begin{bmatrix} -1 \\ 1 \\ 0 \\ 1 \end{bmatrix}, \begin{bmatrix} 1 \\ 0 \\ 1 \\ -1 \end{bmatrix}, \begin{bmatrix} 0 \\ 1 \\ -1 \\ 1 \end{bmatrix}$

30. $\begin{bmatrix} 0 \\ 0 \\ 0 \\ 1 \end{bmatrix}, \begin{bmatrix} 0 \\ 0 \\ 2 \\ 1 \end{bmatrix}, \begin{bmatrix} 0 \\ 3 \\ 2 \\ 1 \end{bmatrix}, \begin{bmatrix} 4 \\ 3 \\ 2 \\ 1 \end{bmatrix}$

31. $\begin{bmatrix} 3 \\ -1 \\ 1 \\ -1 \end{bmatrix}, \begin{bmatrix} -1 \\ 3 \\ 1 \\ -1 \end{bmatrix}, \begin{bmatrix} 1 \\ 1 \\ 3 \\ 1 \end{bmatrix}, \begin{bmatrix} -1 \\ -1 \\ 1 \\ 3 \end{bmatrix}$

In Exercises 32–41, determine if the sets of vectors in the given exercise are linearly independent by converting the vectors to row vectors and using the method of Example 2.25 and Theorem 2.7. For any sets that are linearly dependent, find a dependence relationship among the vectors.

32. Exercise 22 **33.** Exercise 23

34. Exercise 24 **35.** Exercise 25

36. Exercise 26 **37.** Exercise 27

38. Exercise 28 **39.** Exercise 29

40. Exercise 30 **41.** Exercise 31

42. **(a)** If the columns of an $n \times n$ matrix A are linearly independent as vectors in $\mathbb{R}^n$, what is the rank of A? Explain.

(b) If the rows of an $n \times n$ matrix A are linearly independent as vectors in $\mathbb{R}^n$, what is the rank of A? Explain.

43. **(a)** If vectors $\mathbf{u}$, $\mathbf{v}$, and $\mathbf{w}$ are linearly independent, will $\mathbf{u} + \mathbf{v}$, $\mathbf{v} + \mathbf{w}$, and $\mathbf{u} + \mathbf{w}$ also be linearly independent? Justify your answer.

(b) If vectors $\mathbf{u}$, $\mathbf{v}$, and $\mathbf{w}$ are linearly independent, will $\mathbf{u} - \mathbf{v}$, $\mathbf{v} - \mathbf{w}$, and $\mathbf{u} - \mathbf{w}$ also be linearly independent? Justify your answer.

44. Prove that two vectors are linearly dependent if and only if one is a scalar multiple of the other. [*Hint:* Separately consider the case where one of the vectors is $\mathbf{0}$.]

45. Give a "row vector proof" of Theorem 2.8.

46. Prove that every subset of a linearly independent set is linearly independent.

47. Suppose that $S = \{\mathbf{v}_1, \ldots, \mathbf{v}_k, \mathbf{v}\}$ is a set of vectors in some $\mathbb{R}^n$ and that $\mathbf{v}$ is a linear combination of $\mathbf{v}_1, \ldots, \mathbf{v}_k$. If $S' = \{\mathbf{v}_1, \ldots, \mathbf{v}_k\}$, prove that $\text{span}(S) = \text{span}(S')$. [*Hint:* Exercise 21(b) is helpful here.]

48. Let $\{\mathbf{v}_1, \ldots, \mathbf{v}_k\}$ be a linearly independent set of vectors in $\mathbb{R}^n$, and let $\mathbf{v}$ be a vector in $\mathbb{R}^n$. Suppose that $\mathbf{v} = c_1\mathbf{v}_1 + c_2\mathbf{v}_2 + \cdots + c_k\mathbf{v}_k$ with $c_1 \neq 0$. Prove that $\{\mathbf{v}, \mathbf{v}_2, \ldots, \mathbf{v}_k\}$ is linearly independent.

2.4 Applications

There are too many applications of systems of linear equations to do them justice in a single section. This section will introduce a few applications, to illustrate the diverse settings in which they arise.

Allocation of Resources

A great many applications of systems of linear equations involve allocating limited resources subject to a set of constraints.

Example 2.27

A biologist has placed three strains of bacteria (denoted I, II, and III) in a test tube, where they will feed on three different food sources (A, B, and C). Each day 2300 units of A, 800 units of B, and 1500 units of C are placed in the test tube, and each bacterium consumes a certain number of units of each food per day, as shown in Table 2.2. How many bacteria of each strain can coexist in the test tube and consume all of the food?

Table 2.2

	Bacteria Strain I	Bacteria Strain II	Bacteria Strain III
Food A	2	2	4
Food B	1	2	0
Food C	1	3	1

Solution Let x_1, x_2, and x_3 be the numbers of bacteria of strains I, II, and III, respectively. Since each of the x_1 bacteria of strain I consumes 2 units of A per day, strain I consumes a total of $2x_1$ units per day. Similarly, strains II and III consume a total of $2x_2$ and $4x_3$ units of food A daily. Since we want to use up all of the 2300 units of A, we have the equation

$$2x_1 + 2x_2 + 4x_3 = 2300$$

Likewise, we obtain equations corresponding to the consumption of B and C:

$$\begin{aligned} x_1 + 2x_2 \phantom{{}+x_3} &= 800 \\ x_1 + 3x_2 + x_3 &= 1500 \end{aligned}$$

Thus, we have a system of three linear equations in three variables. Row reduction of the corresponding augmented matrix gives

$$\left[\begin{array}{ccc|c} 2 & 2 & 4 & 2300 \\ 1 & 2 & 0 & 800 \\ 1 & 3 & 1 & 1500 \end{array}\right] \longrightarrow \left[\begin{array}{ccc|c} 1 & 0 & 0 & 100 \\ 0 & 1 & 0 & 350 \\ 0 & 0 & 1 & 350 \end{array}\right]$$

Therefore, $x_1 = 100$, $x_2 = 350$, and $x_3 = 350$. The biologist should place 100 bacteria of strain I and 350 of each of strains II and III in the test tube if she wants all the food to be consumed.

Example 2.28

Repeat Example 2.27, using the data on daily consumption of food (units per day) shown in Table 2.3. Assume this time that 1500 units of A, 3000 units of B, and 4500 units of C are placed in the test tube each day.

Table 2.3

	Bacteria Strain I	Bacteria Strain II	Bacteria Strain III
Food A	1	1	1
Food B	1	2	3
Food C	1	3	5

Solution Let x_1, x_2, and x_3 again be the numbers of bacteria of each type. The augmented matrix for the resulting linear system and the corresponding reduced echelon form are

$$\left[\begin{array}{ccc|c} 1 & 1 & 1 & 1500 \\ 1 & 2 & 3 & 3000 \\ 1 & 3 & 5 & 4500 \end{array}\right] \longrightarrow \left[\begin{array}{ccc|c} 1 & 0 & -1 & 0 \\ 0 & 1 & 2 & 1500 \\ 0 & 0 & 0 & 0 \end{array}\right]$$

We see that in this case we have more than one solution, given by

$$\begin{aligned} x_1 \phantom{{}+2x_2} - x_3 &= 0 \\ x_2 + 2x_3 &= 1500 \end{aligned}$$

Letting $x_3 = t$, we obtain $x_1 = t$, $x_2 = 1500 - 2t$, and $x_3 = t$. In any applied problem, we must be careful to interpret solutions properly. Certainly the number of bacteria cannot be negative. Therefore, $t \geq 0$ and $1500 - 2t \geq 0$. The latter inequality implies that $t \leq 750$, so we have $0 \leq t \leq 750$. Presumably the number of bacteria must be a whole number, so there are exactly 751 values of t that satisfy the inequality. Thus, our 751 solutions are of the form

$$\begin{bmatrix} x_1 \\ x_2 \\ x_3 \end{bmatrix} = \begin{bmatrix} t \\ 1500 - 2t \\ t \end{bmatrix} = \begin{bmatrix} 0 \\ 1500 \\ 0 \end{bmatrix} + t\begin{bmatrix} 1 \\ -2 \\ 1 \end{bmatrix}$$

one for each integer value of t such that $0 \leq t \leq 750$. (So, although mathematically this system has infinitely many solutions, *physically* there are only finitely many.)

Balancing Chemical Equations

When a chemical reaction occurs, certain molecules (the *reactants*) combine to form new molecules (the *products*). A ***balanced chemical equation*** is an algebraic equation that gives the relative numbers of reactants and products in the reaction and has the same number of atoms of each type on the left- and right-hand sides. The equation is usually written with the reactants on the left, the products on the right, and an arrow in between to show the direction of the reaction.

For example, for the reaction in which hydrogen gas (H_2) and oxygen (O_2) combine to form water (H_2O), a balanced chemical equation is

$$2H_2 + O_2 \longrightarrow 2H_2O$$

indicating that two molecules of hydrogen combine with one molecule of oxygen to form two molecules of water. Observe that the equation is balanced, since there are four hydrogen atoms and two oxygen atoms on each side. Note that there will never be a unique balanced equation for a reaction, since any positive integer multiple of a balanced equation will also be balanced. For example, $6H_2 + 3O_2 \longrightarrow 6H_2O$ is also balanced. Therefore, we usually look for the *simplest* balanced equation for a given reaction.

While trial and error will often work in simple examples, the process of balancing chemical equations really involves solving a homogeneous system of linear equations, so we can use the techniques we have developed to remove the guesswork.

Example 2.29

The combustion of ammonia (NH_3) in oxygen produces nitrogen (N_2) and water. Find a balanced chemical equation for this reaction.

Solution If we denote the numbers of molecules of ammonia, oxygen, nitrogen, and water by w, x, y, and z, respectively, then we are seeking an equation of the form

$$wNH_3 + xO_2 \longrightarrow yN_2 + zH_2O$$

Comparing the numbers of nitrogen, hydrogen, and oxygen atoms in the reactants and products, we obtain three linear equations:

$$\begin{aligned} \text{Nitrogen:} \quad w &= 2y \\ \text{Hydrogen:} \quad 3w &= 2z \\ \text{Oxygen:} \quad 2x &= z \end{aligned}$$

Rewriting these equations in standard form gives us a homogeneous system of three linear equations in four variables. [Notice that Theorem 2.3 guarantees that such a system will have (infinitely many) nontrivial solutions.] We reduce the corresponding augmented matrix by Gauss-Jordan elimination.

$$\begin{aligned} w \quad - 2y \quad &= 0 \\ 3w \quad - 2z &= 0 \\ 2x \quad - z &= 0 \end{aligned} \longrightarrow \left[\begin{array}{cccc|c} 1 & 0 & -2 & 0 & 0 \\ 3 & 0 & 0 & -2 & 0 \\ 0 & 2 & 0 & -1 & 0 \end{array}\right] \longrightarrow \left[\begin{array}{cccc|c} 1 & 0 & 0 & -\frac{2}{3} & 0 \\ 0 & 1 & 0 & -\frac{1}{2} & 0 \\ 0 & 0 & 1 & -\frac{1}{3} & 0 \end{array}\right]$$

Thus, $w = \frac{2}{3}z$, $x = \frac{1}{2}z$, and $y = \frac{1}{3}z$. The smallest positive value of z that will produce *integer* values for all four variables is the least common denominator of the fractions $\frac{2}{3}$, $\frac{1}{2}$, and $\frac{1}{3}$—namely, 6—which gives $w = 4$, $x = 3$, $y = 2$, and $z = 6$. Therefore, the balanced chemical equation is

$$4NH_3 + 3O_2 \longrightarrow 2N_2 + 6H_2O$$

Network Analysis

Many practical situations give rise to networks: transportation networks, communications networks, and economic networks, to name a few. Of particular interest are the possible *flows* through networks. For example, vehicles flow through a network of roads, information flows through a data network, and goods and services flow through an economic network.

For us, a ***network*** will consist of a finite number of ***nodes*** (also called ***junctions*** or ***vertices***) connected by a series of directed edges known as ***branches*** or ***arcs***. Each branch will be labeled with a ***flow*** that represents the amount of some commodity that can flow along or through that branch in the indicated direction. (Think of cars traveling along a network of one-way streets.) The fundamental rule governing flow through a network is ***conservation of flow:***

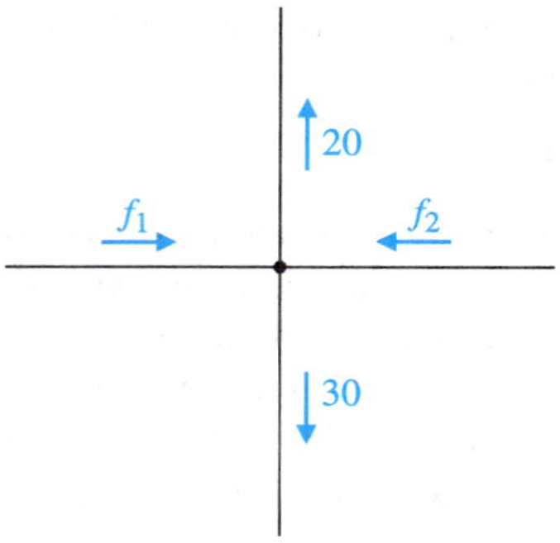

Figure 2.10
Flow at a node: $f_1 + f_2 = 50$

At each node, the flow in equals the flow out.

Figure 2.10 shows a portion of a network, with two branches entering a node and two leaving. The conservation of flow rule implies that the total incoming flow, $f_1 + f_2$ units, must match the total outgoing flow, $20 + 30$ units. Thus, we have the linear equation $f_1 + f_2 = 50$ corresponding to this node.

We can analyze the flow through an entire network by constructing such equations and solving the resulting system of linear equations.

Example 2.30

Describe the possible flows through the network of water pipes shown in Figure 2.11, where flow is measured in liters per minute.

Solution At each node, we write out the equation that represents the conservation of flow there. We then rewrite each equation with the variables on the left and the constant on the right, to get a linear system in standard form.

$$\begin{aligned} \text{Node } A&: \quad 15 = f_1 + f_4 \\ \text{Node } B&: \quad f_1 = f_2 + 10 \\ \text{Node } C&: \quad f_2 + f_3 + 5 = 30 \\ \text{Node } D&: \quad f_4 + 20 = f_3 \end{aligned} \longrightarrow \begin{aligned} f_1 \qquad\qquad + f_4 &= 15 \\ f_1 - f_2 \qquad\qquad &= 10 \\ f_2 + f_3 \qquad &= 25 \\ f_3 - f_4 &= 20 \end{aligned}$$

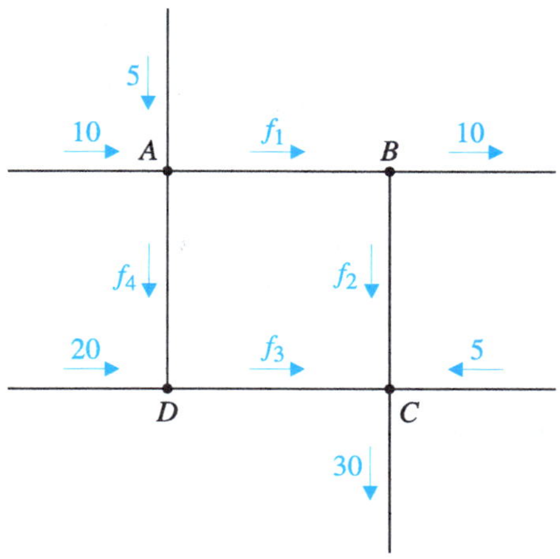

Figure 2.11

Using Gauss-Jordan elimination, we reduce the augmented matrix:

$$\left[\begin{array}{rrrr|r} 1 & 0 & 0 & 1 & 15 \\ 1 & -1 & 0 & 0 & 10 \\ 0 & 1 & 1 & 0 & 25 \\ 0 & 0 & 1 & -1 & 20 \end{array}\right] \longrightarrow \left[\begin{array}{rrrr|r} 1 & 0 & 0 & 1 & 15 \\ 0 & 1 & 0 & 1 & 5 \\ 0 & 0 & 1 & -1 & 20 \\ 0 & 0 & 0 & 0 & 0 \end{array}\right]$$

(Check this.) We see that there is one free variable, f_4, so we have infinitely many solutions. Setting $f_4 = t$ and expressing the leading variables in terms of f_4, we obtain

$$\begin{aligned} f_1 &= 15 - t \\ f_2 &= 5 - t \\ f_3 &= 20 + t \\ f_4 &= t \end{aligned}$$

These equations describe all possible flows and allow us to analyze the network. For example, we see that if we control the flow on branch AD so that $t = 5$ L/min, then the other flows are $f_1 = 10, f_2 = 0$, and $f_3 = 25$.

We can do even better: We can find the minimum and maximum possible flows on each branch. Each of the flows must be nonnegative. Examining the first and second equations in turn, we see that $t \le 15$ (otherwise f_1 would be negative) and $t \le 5$ (otherwise f_2 would be negative). The second of these inequalities is more restrictive than the first, so we must use it. The third equation contributes no further restrictions on our parameter t, so we have deduced that $0 \le t \le 5$. Combining this result with the four equations, we see that

$$\begin{aligned} 10 \le f_1 &\le 15 \\ 0 \le f_2 &\le 5 \\ 20 \le f_3 &\le 25 \\ 0 \le f_4 &\le 5 \end{aligned}$$

We now have a complete description of the possible flows through this network.

Electrical Networks

Electrical networks are a specialized type of network providing information about power sources, such as batteries, and devices powered by these sources, such as light bulbs or motors. A power source "forces" a current of electrons to flow through the network, where it encounters various *resistors,* each of which requires that a certain amount of force be applied in order for the current to flow through it.

The fundamental law of electricity is Ohm's law, which states exactly how much force E is needed to drive a current I through a resistor with resistance R.

Ohm's Law

$$\text{force} = \text{resistance} \times \text{current}$$

or

$$E = RI$$

Force is measured in *volts,* resistance in *ohms,* and current in *amperes* (or *amps,* for short). Thus, in terms of these units, Ohm's law becomes "volts = ohms × amps," and it tells us what the "voltage drop" is when a current passes through a resistor—that is, how much voltage is used up.

Current flows out of the positive terminal of a battery and flows back into the negative terminal, traveling around one or more closed circuits in the process. In a diagram of an electrical network, batteries are represented by  (where the positive terminal is the longer vertical bar) and resistors are represented by . The following two laws, whose discovery we owe to Kirchhoff, govern electrical networks. The first is a "conservation of flow" law at each node; the second is a "balancing of voltage" law around each circuit.

Kirchhoff's Laws

Current Law (nodes)
The sum of the currents flowing into any node is equal to the sum of the currents flowing out of that node.

Voltage Law (circuits)
The sum of the voltage drops around any circuit is equal to the total voltage around the circuit (provided by the batteries).

Figure 2.12 illustrates Kirchhoff's laws. In part (a), the current law gives $I_1 = I_2 + I_3$ (or $I_1 - I_2 - I_3 = 0$, as we will write it); part (b) gives $4I = 10$, where we have used Ohm's law to compute the voltage drop $4I$ at the resistor. Using Kirchhoff's laws, we can set up a system of linear equations that will allow us to determine the currents in an electrical network.

Example 2.31

Determine the currents I_1, I_2, and I_3 in the electrical network shown in Figure 2.13.

Solution This network has two batteries and four resistors. Current I_1 flows through the top branch BCA, current I_2 flows across the middle branch AB, and current I_3 flows through the bottom branch BDA.

At node A, the current law gives $I_1 + I_3 = I_2$, or

$$I_1 - I_2 + I_3 = 0$$

(Observe that we get the same equation at node B.)

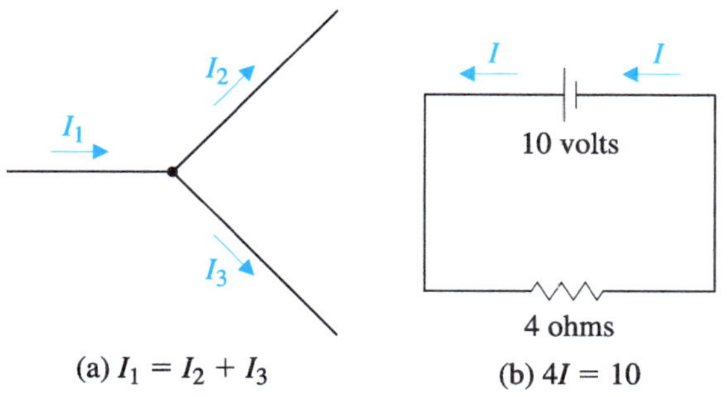

Figure 2.12

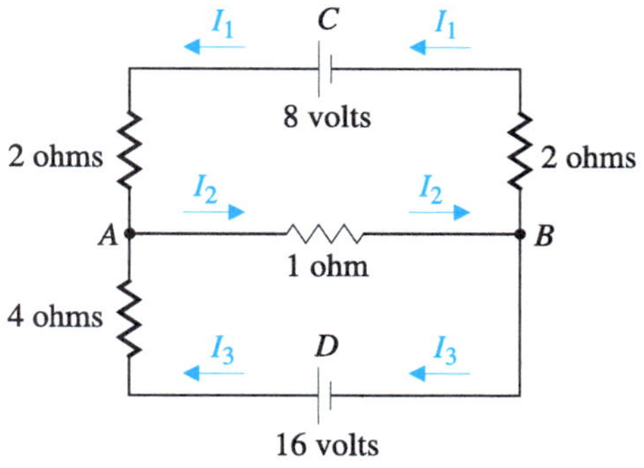

Figure 2.13

Next we apply the voltage law for each circuit. For the circuit *CABC*, the voltage drops at the resistors are $2I_1$, I_2, and $2I_1$. Thus, we have the equation

$$4I_1 + I_2 = 8$$

Similarly, for the circuit *DABD*, we obtain

$$I_2 + 4I_3 = 16$$

(Notice that there is actually a third circuit, *CADBC*, if we "go against the flow." In this case, we must treat the voltages and resistances on the "reversed" paths as negative. Doing so gives $2I_1 + 2I_1 - 4I_3 = 8 - 16 = -8$ or $4I_1 - 4I_3 = -8$, which we observe is just the difference of the voltage equations for the other two circuits. Thus, we can omit this equation, as it contributes no new information. On the other hand, including it does no harm.)

We now have a system of three linear equations in three variables:

$$\begin{aligned} I_1 - I_2 + I_3 &= 0 \\ 4I_1 + I_2 &= 8 \\ I_2 + 4I_3 &= 16 \end{aligned}$$

Gauss-Jordan elimination produces

$$\left[\begin{array}{ccc|c} 1 & -1 & 1 & 0 \\ 4 & 1 & 0 & 8 \\ 0 & 1 & 4 & 16 \end{array}\right] \longrightarrow \left[\begin{array}{ccc|c} 1 & 0 & 0 & 1 \\ 0 & 1 & 0 & 4 \\ 0 & 0 & 1 & 3 \end{array}\right]$$

Hence, the currents are $I_1 = 1$ amp, $I_2 = 4$ amps, and $I_3 = 3$ amps.

Remark In some electrical networks, the currents may have fractional values or may even be negative. A negative value simply means that the current in the corresponding branch flows in the direction opposite that shown on the network diagram.

CAS

Example 2.32

The network shown in Figure 2.14 has a single power source *A* and five resistors. Find the currents $I, I_1, \ldots, I_5$. This is an example of what is known in electrical engineering as a *Wheatstone bridge circuit*.

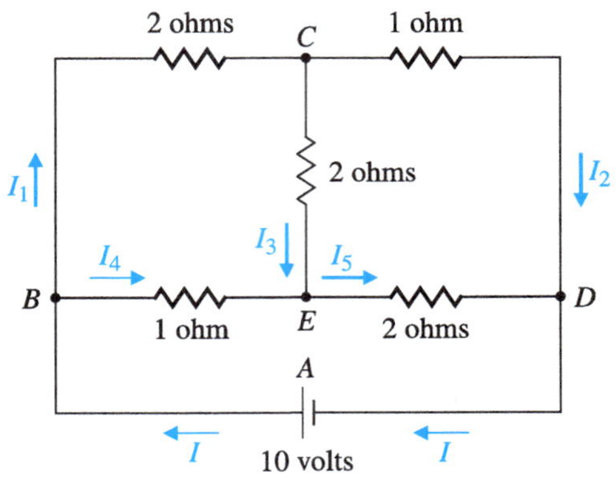

Figure 2.14
A bridge circuit

Solution Kirchhoff's current law gives the following equations at the four nodes:

$$\begin{aligned} \text{Node } B: &\quad I - I_1 - I_4 = 0 \\ \text{Node } C: &\quad I_1 - I_2 - I_3 = 0 \\ \text{Node } D: &\quad I - I_2 - I_5 = 0 \\ \text{Node } E: &\quad I_3 + I_4 - I_5 = 0 \end{aligned}$$

For the three basic circuits, the voltage law gives

$$\begin{aligned} \text{Circuit } ABEDA: &\quad I_4 + 2I_5 = 10 \\ \text{Circuit } BCEB: &\quad 2I_1 + 2I_3 - I_4 = 0 \\ \text{Circuit } CDEC: &\quad I_2 - 2I_5 - 2I_3 = 0 \end{aligned}$$

(Observe that branch DAB has no resistor and therefore no voltage drop; thus, there is no I term in the equation for circuit $ABEDA$. Note also that we had to change signs three times because we went "against the current." This poses no problem, since we will let the sign of the answer determine the direction of current flow.)

We now have a system of seven equations in six variables. Row reduction gives

$$\left[\begin{array}{rrrrrr|r} 1 & -1 & 0 & 0 & -1 & 0 & 0 \\ 0 & 1 & -1 & -1 & 0 & 0 & 0 \\ 1 & 0 & -1 & 0 & 0 & -1 & 0 \\ 0 & 0 & 0 & 1 & 1 & -1 & 0 \\ 0 & 0 & 0 & 0 & 1 & 2 & 10 \\ 0 & 2 & 0 & 2 & -1 & 0 & 0 \\ 0 & 0 & 1 & -2 & 0 & -2 & 0 \end{array}\right] \longrightarrow \left[\begin{array}{rrrrrr|r} 1 & 0 & 0 & 0 & 0 & 0 & 7 \\ 0 & 1 & 0 & 0 & 0 & 0 & 3 \\ 0 & 0 & 1 & 0 & 0 & 0 & 4 \\ 0 & 0 & 0 & 1 & 0 & 0 & -1 \\ 0 & 0 & 0 & 0 & 1 & 0 & 4 \\ 0 & 0 & 0 & 0 & 0 & 1 & 3 \\ 0 & 0 & 0 & 0 & 0 & 0 & 0 \end{array}\right]$$

(Use your calculator or CAS to check this.) Thus, the solution (in amps) is $I = 7$, $I_1 = I_5 = 3$, $I_2 = I_4 = 4$, and $I_3 = -1$. The significance of the negative value here is that the current through branch CE is flowing in the direction opposite that marked on the diagram.

Remark There is only one power source in this example, so the single 10-volt battery sends a current of 7 amps through the network. If we substitute these values into

Ohm's law, $E = RI$, we get $10 = 7R$ or $R = \frac{10}{7}$. Thus, the entire network behaves as if there were a single $\frac{10}{7}$-ohm resistor. This value is called the *effective resistance* (R_{eff}) of the network.

Linear Economic Models

An economy is a very complex system with many interrelationships among the various sectors of the economy and the goods and services they produce and consume. Determining optimal prices and levels of production subject to desired economic goals requires sophisticated mathematical models. Linear algebra has proven to be a powerful tool in developing and analyzing such economic models.

In this section, we introduce two models based on the work of Harvard economist Wassily Leontief in the 1930s. His methods, often referred to as ***input-output analysis***, are now standard tools in mathematical economics and are used by cities, corporations, and entire countries for economic planning and forecasting.

We begin with a simple example.

Example 2.33

The economy of a region consists of three industries, or sectors: service, electricity, and oil production. For simplicity, we assume that each industry produces a single commodity (goods or services) in a given year and that ***income (output)*** is generated from the sale of this commodity. Each industry purchases commodities from the other industries, including itself, in order to generate its output. No commodities are purchased from outside the region and no output is sold outside the region. Furthermore, for each industry, we assume that production exactly equals consumption (output equals input, income equals expenditure). In this sense, this is a ***closed economy*** that is in ***equilibrium***. Table 2.4 summarizes how much of each industry's output is consumed by each industry.

Bettmann/CORBIS

Wassily Leontief (1906–1999) was born in St. Petersburg, Russia. He studied at the University of Leningrad and received his Ph.D. from the University of Berlin. He emigrated to the United States in 1931, teaching at Harvard University and later at New York University. In 1932, Leontief began compiling data for the monumental task of conducting an input-output analysis of the United States economy, the results of which were published in 1941. He was also an early user of computers, which he needed to solve the large-scale linear systems in his models. For his pioneering work, Leontief was awarded the Nobel Prize in Economics in 1973.

Table 2.4

		Produced by (output)		
		Service	**Electricity**	**Oil**
Consumed by (input)	**Service**	1/4	1/3	1/2
	Electricity	1/4	1/3	1/4
	Oil	1/2	1/3	1/4

From the first column of the table, we see that the service industry consumes 1/4 of its own output, electricity consumes another 1/4, and the oil industry uses 1/2 of the service industry's output. The other two columns have similar interpretations. Notice that the sum of each column is 1, indicating that all of the output of each industry is consumed.

Let x_1, x_2, and x_3 denote the annual output (income) of the service, electricity, and oil industries, respectively, in millions of dollars. Since consumption corresponds to expenditure, the service industry spends $\frac{1}{4}x_1$ on its own commodity, $\frac{1}{3}x_2$ on electricity, and $\frac{1}{2}x_3$ on oil. This means that the service industry's total annual expenditure is $\frac{1}{4}x_1 + \frac{1}{3}x_2 + \frac{1}{2}x_3$. Since the economy is in equilibrium, the service industry's expenditure must equal its annual income x_1. This gives the first of the following equations; the other two equations are obtained by analyzing the expenditures of the electricity and oil industries.

$$\begin{aligned} &\text{Service:} && \tfrac{1}{4}x_1 + \tfrac{1}{3}x_2 + \tfrac{1}{2}x_3 = x_1 \\ &\text{Electricity:} && \tfrac{1}{4}x_1 + \tfrac{1}{3}x_2 + \tfrac{1}{4}x_3 = x_2 \\ &\text{Oil:} && \tfrac{1}{2}x_1 + \tfrac{1}{3}x_2 + \tfrac{1}{4}x_3 = x_3 \end{aligned}$$

Rearranging each equation, we obtain a homogeneous system of linear equations, which we then solve. (Check this!)

$$\begin{aligned} -\tfrac{3}{4}x_1 + \tfrac{1}{3}x_2 + \tfrac{1}{2}x_3 &= 0 \\ \tfrac{1}{4}x_1 - \tfrac{2}{3}x_2 - \tfrac{1}{4}x_3 &= 0 \\ \tfrac{1}{2}x_1 + \tfrac{1}{3}x_2 - \tfrac{3}{4}x_3 &= 0 \end{aligned} \longrightarrow \left[\begin{array}{ccc|c} -\frac{3}{4} & \frac{1}{3} & \frac{1}{2} & 0 \\ \frac{1}{4} & -\frac{2}{3} & \frac{1}{4} & 0 \\ \frac{1}{2} & \frac{1}{3} & -\frac{3}{4} & 0 \end{array}\right] \longrightarrow \left[\begin{array}{ccc|c} 1 & 0 & -1 & 0 \\ 0 & 1 & -\frac{3}{4} & 0 \\ 0 & 0 & 0 & 0 \end{array}\right]$$

Setting $x_3 = t$, we find that $x_1 = t$ and $x_2 = \frac{3}{4}t$. Thus, we see that the *relative* outputs of the service, electricity, and oil industries need to be in the ratios $x_1 : x_2 : x_3 = 4 : 3 : 4$ for the economy to be in equilibrium.

Remarks

- The last example illustrates what is commonly called the ***Leontief closed model.***
- Since output corresponds to income, we can also think of x_1, x_2, and x_3 as the *prices* of the three commodities.

We now modify the model in Example 2.33 to accommodate an ***open economy***, one in which there is an *external* as well as an internal demand for the commodities that are produced. Not surprisingly, this version is called the ***Leontief open model.***

Example 2.34

Consider the three industries of Example 2.33 but with consumption given by Table 2.5. We see that, of the commodities produced by the service industry, 20% are consumed by the service industry, 40% by the electricity industry, and 10% by the oil industry. Thus, only 70% of the service industry's output is consumed by this economy. The implication of this calculation is that there is an excess of output (income) over input (expenditure) for the service industry. We say that the service industry is ***productive.*** Likewise, the oil industry is productive but the electricity industry is ***non-productive.*** (This is reflected in the fact that the sums of the first and third columns are less than 1 but the sum of the second column is equal to 1). The excess output may be applied to satisfy an external demand.

Table 2.5

		Produced by (output)		
		Service	**Electricity**	**Oil**
Consumed by (input)	**Service**	0.20	0.50	0.10
	Electricity	0.40	0.20	0.20
	Oil	0.10	0.30	0.30

For example, suppose there is an annual external demand (in millions of dollars) for 10, 10, and 30 from the service, electricity, and oil industries, respectively. Then, equating expenditures (internal demand and external demand) with income (output), we obtain the following equations:

	output	internal demand	external demand
Service	x_1	$= 0.2x_1 + 0.5x_2 + 0.1x_3$	$+ 10$
Electricity	x_2	$= 0.4x_1 + 0.2x_2 + 0.2x_3$	$+ 10$
Oil	x_3	$= 0.1x_1 + 0.3x_2 + 0.3x_3$	$+ 30$

Rearranging, we obtain the following linear system and augmented matrix:

$$\begin{aligned} 0.8x_1 - 0.5x_2 - 0.1x_3 &= 10 \\ -0.4x_1 + 0.8x_2 - 0.2x_3 &= 10 \\ -0.1x_1 - 0.3x_2 + 0.7x_3 &= 30 \end{aligned} \longrightarrow \left[\begin{array}{ccc|c} 0.8 & -0.5 & -0.1 & 10 \\ -0.4 & 0.8 & -0.2 & 10 \\ -0.1 & -0.3 & 0.7 & 30 \end{array}\right]$$

CAS Row reduction yields

$$\left[\begin{array}{ccc|c} 1 & 0 & 0 & 61.74 \\ 0 & 1 & 0 & 63.04 \\ 0 & 0 & 1 & 78.70 \end{array}\right]$$

from which we see that the service, electricity, and oil industries must have an annual production of \$61.74, \$63.04, and \$78.70 (million), respectively, in order to meet both the internal and external demand for their commodities.

We will revisit these models in Section 3.7.

Finite Linear Games

There are many situations in which we must consider a physical system that has only a finite number of *states*. Sometimes these states can be altered by applying certain processes, each of which produces finitely many outcomes. For example, a light bulb can be on or off and a switch can change the state of the light bulb from on to off and vice versa. Digital systems that arise in computer science are often of this type. More frivolously, many computer games feature puzzles in which a certain device must be manipulated by various switches to produce a desired outcome. The finiteness of such situations is perfectly suited to analysis using modular arithmetic, and often linear systems over some $\mathbb{Z}_p$ play a role. Problems involving this type of situation are often called ***finite linear games.***

Example 2.35

A row of five lights is controlled by five switches. Each switch changes the state (on or off) of the light directly above it and the states of the lights immediately adjacent to the left and right. For example, if the first and third lights are on, as in Figure 2.15(a), then pushing switch A changes the state of the system to that shown in Figure 2.15(b). If we next push switch C, then the result is the state shown in Figure 2.15(c).

(a)

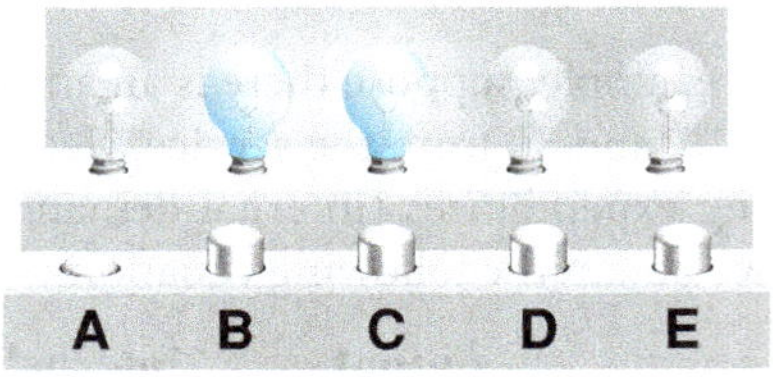

(b)

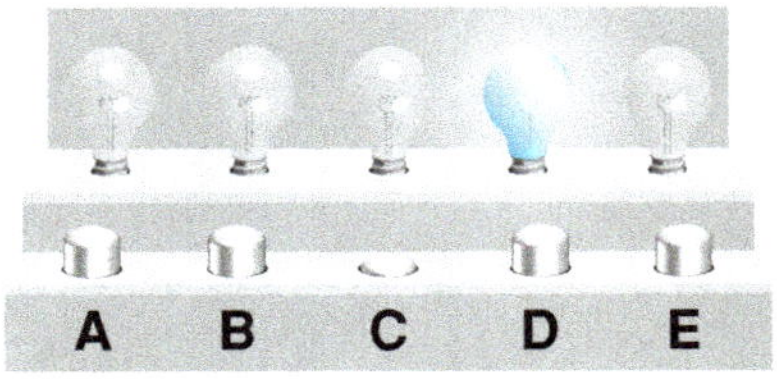

(c)

Figure 2.15

Suppose that initially all the lights are off. Can we push the switches in some order so that only the first, third, and fifth lights will be on? Can we push the switches in some order so that only the first light will be on?

Solution The on/off nature of this problem suggests that binary notation will be helpful and that we should work with $\mathbb{Z}_2$. Accordingly, we represent the states of the five lights by a vector in $\mathbb{Z}_2^5$, where 0 represents off and 1 represents on. Thus, for example, the vector

$$\begin{bmatrix} 0 \\ 1 \\ 1 \\ 0 \\ 0 \end{bmatrix}$$

corresponds to Figure 2.15(b).

We may also use vectors in $\mathbb{Z}_2^5$ to represent the action of each switch. If a switch changes the state of a light, the corresponding component is a 1; otherwise, it is 0. With this convention, the actions of the five switches are given by

$$\mathbf{a} = \begin{bmatrix} 1 \\ 1 \\ 0 \\ 0 \\ 0 \end{bmatrix}, \mathbf{b} = \begin{bmatrix} 1 \\ 1 \\ 1 \\ 0 \\ 0 \end{bmatrix}, \mathbf{c} = \begin{bmatrix} 0 \\ 1 \\ 1 \\ 1 \\ 0 \end{bmatrix}, \mathbf{d} = \begin{bmatrix} 0 \\ 0 \\ 1 \\ 1 \\ 1 \end{bmatrix}, \mathbf{e} = \begin{bmatrix} 0 \\ 0 \\ 0 \\ 1 \\ 1 \end{bmatrix}$$

The situation depicted in Figure 2.15(a) corresponds to the initial state

$$\mathbf{s} = \begin{bmatrix} 1 \\ 0 \\ 1 \\ 0 \\ 0 \end{bmatrix}$$

followed by

$$\mathbf{a} = \begin{bmatrix} 1 \\ 1 \\ 0 \\ 0 \\ 0 \end{bmatrix}$$

It is the vector sum (in $\mathbb{Z}_2^5$)

$$\mathbf{s} + \mathbf{a} = \begin{bmatrix} 0 \\ 1 \\ 1 \\ 0 \\ 0 \end{bmatrix}$$

Observe that this result agrees with Figure 2.15(b).

Starting with any initial configuration **s**, suppose we push the switches in the order A, C, D, A, C, B. This corresponds to the vector sum $\mathbf{s} + \mathbf{a} + \mathbf{c} + \mathbf{d} + \mathbf{a} + \mathbf{c} + \mathbf{b}$. But in $\mathbb{Z}_2^5$, addition is commutative, so we have

$$\begin{aligned} \mathbf{s} + \mathbf{a} + \mathbf{c} + \mathbf{d} + \mathbf{a} + \mathbf{c} + \mathbf{b} &= \mathbf{s} + 2\mathbf{a} + \mathbf{b} + 2\mathbf{c} + \mathbf{d} \\ &= \mathbf{s} + \mathbf{b} + \mathbf{d} \end{aligned}$$

where we have used the fact that $2 = 0$ in $\mathbb{Z}_2$. Thus, we would achieve the same result by pushing only B and D—and the order does not matter. (Check that this is correct.) Hence, in this example, we do not need to push any switch more than once.

So, to see if we can achieve a target configuration **t** starting from an initial configuration **s**, we need to determine whether there are scalars $x_1, \ldots, x_5$ in $\mathbb{Z}_2$ such that

$$\mathbf{s} + x_1\mathbf{a} + x_2\mathbf{b} + \cdots + x_5\mathbf{e} = \mathbf{t}$$

In other words, we need to solve (if possible) the linear system over $\mathbb{Z}_2$ that corresponds to the vector equation

$$x_1\mathbf{a} + x_2\mathbf{b} + \cdots + x_5\mathbf{e} = \mathbf{t} - \mathbf{s} = \mathbf{t} + \mathbf{s}$$

In this case, $\mathbf{s} = \mathbf{0}$ and our first target configuration is

$$\mathbf{t} = \begin{bmatrix} 1 \\ 0 \\ 1 \\ 0 \\ 1 \end{bmatrix}$$

The augmented matrix of this system has the given vectors as columns:

$$\left[\begin{array}{ccccc|c} 1 & 1 & 0 & 0 & 0 & 1 \\ 1 & 1 & 1 & 0 & 0 & 0 \\ 0 & 1 & 1 & 1 & 0 & 1 \\ 0 & 0 & 1 & 1 & 1 & 0 \\ 0 & 0 & 0 & 1 & 1 & 1 \end{array}\right]$$

We reduce it over $\mathbb{Z}_2$ to obtain

$$\left[\begin{array}{ccccc|c} 1 & 0 & 0 & 0 & 1 & 0 \\ 0 & 1 & 0 & 0 & 1 & 1 \\ 0 & 0 & 1 & 0 & 0 & 1 \\ 0 & 0 & 0 & 1 & 1 & 1 \\ 0 & 0 & 0 & 0 & 0 & 0 \end{array}\right]$$

Thus, x_5 is a free variable. Hence, there are exactly two solutions (corresponding to $x_5 = 0$ and $x_5 = 1$). Solving for the other variables in terms of x_5, we obtain

$$\begin{aligned} x_1 &= \phantom{1 +{}} x_5 \\ x_2 &= 1 + x_5 \\ x_3 &= 1 \\ x_4 &= 1 + x_5 \end{aligned}$$

So, when $x_5 = 0$ and $x_5 = 1$, we have the solutions

$$\begin{bmatrix} x_1 \\ x_2 \\ x_3 \\ x_4 \\ x_5 \end{bmatrix} = \begin{bmatrix} 0 \\ 1 \\ 1 \\ 1 \\ 0 \end{bmatrix} \quad \text{and} \quad \begin{bmatrix} x_1 \\ x_2 \\ x_3 \\ x_4 \\ x_5 \end{bmatrix} = \begin{bmatrix} 1 \\ 0 \\ 1 \\ 0 \\ 1 \end{bmatrix}$$

respectively. (Check that these both work.)

Similarly, in the second case, we have

$$\mathbf{t} = \begin{bmatrix} 1 \\ 0 \\ 0 \\ 0 \\ 0 \end{bmatrix}$$

The augmented matrix reduces as follows:

$$\left[\begin{array}{ccccc|c} 1 & 1 & 0 & 0 & 0 & 1 \\ 1 & 1 & 1 & 0 & 0 & 0 \\ 0 & 1 & 1 & 1 & 0 & 0 \\ 0 & 0 & 1 & 1 & 1 & 0 \\ 0 & 0 & 0 & 1 & 1 & 0 \end{array}\right] \longrightarrow \left[\begin{array}{ccccc|c} 1 & 0 & 0 & 0 & 1 & 0 \\ 0 & 1 & 0 & 0 & 1 & 1 \\ 0 & 0 & 1 & 0 & 0 & 1 \\ 0 & 0 & 0 & 1 & 1 & 1 \\ 0 & 0 & 0 & 0 & 0 & 1 \end{array}\right]$$

showing that there is no solution in this case; that is, it is impossible to start with all of the lights off and turn only the first light on.

Example 2.35 shows the power of linear algebra. Even though we might have found out by trial and error that there was no solution, checking all possible ways to push the switches would have been extremely tedious. We might also have missed the fact that no switch need ever be pushed more than once.

Example 2.36

Consider a row with only three lights, each of which can be off, light blue, or dark blue. Below the lights are three switches, A, B, and C, each of which changes the states of particular lights to the *next* state, in the order shown in Figure 2.16. Switch A changes the states of the first two lights, switch B all three lights, and switch C the last two

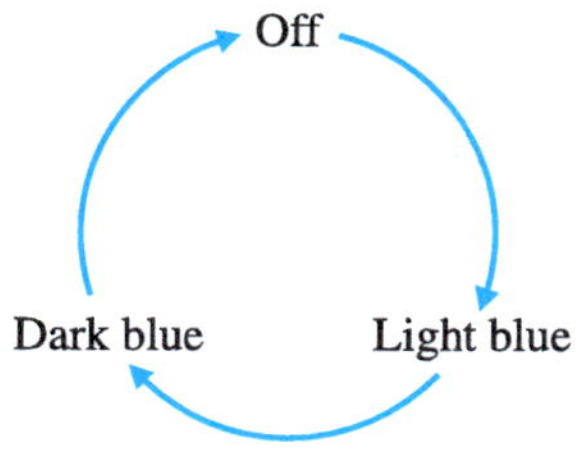

Figure 2.16

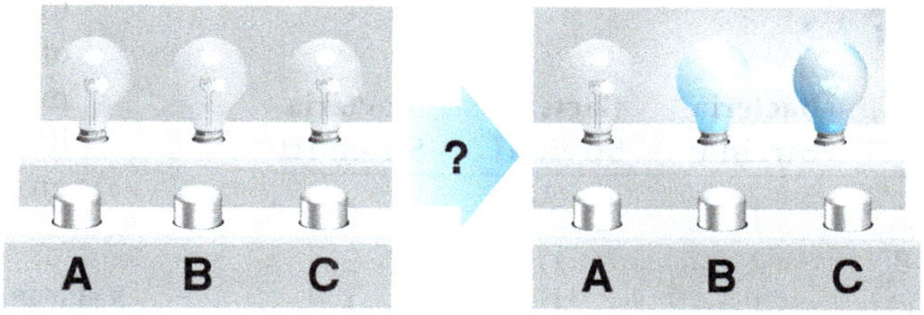

Figure 2.17

lights. If all three lights are initially off, is it possible to push the switches in some order so that the lights are off, light blue, and dark blue, in that order (as in Figure 2.17)?

Solution Whereas Example 2.35 involved $\mathbb{Z}_2$, this one clearly (is it clear?) involves $\mathbb{Z}_3$. Accordingly, the switches correspond to the vectors

$$\mathbf{a} = \begin{bmatrix} 1 \\ 1 \\ 0 \end{bmatrix}, \ \mathbf{b} = \begin{bmatrix} 1 \\ 1 \\ 1 \end{bmatrix}, \ \mathbf{c} = \begin{bmatrix} 0 \\ 1 \\ 1 \end{bmatrix}$$

in $\mathbb{Z}_3^3$, and the final configuration we are aiming for is $\mathbf{t} = \begin{bmatrix} 0 \\ 1 \\ 2 \end{bmatrix}$. (Off is 0, light blue is 1, and dark blue is 2.) We wish to find scalars x_1, x_2, x_3 in $\mathbb{Z}_3$ such that

$$x_1\mathbf{a} + x_2\mathbf{b} + x_3\mathbf{c} = \mathbf{t}$$

(where x_i represents the number of times the ith switch is pushed). This equation gives rise to the augmented matrix $[\mathbf{a}\ \mathbf{b}\ \mathbf{c} \mid \mathbf{t}]$, which reduces over $\mathbb{Z}_3$ as follows:

$$\left[\begin{array}{ccc|c} 1 & 1 & 0 & 0 \\ 1 & 1 & 1 & 1 \\ 0 & 1 & 1 & 2 \end{array}\right] \longrightarrow \left[\begin{array}{ccc|c} 1 & 0 & 0 & 2 \\ 0 & 1 & 0 & 1 \\ 0 & 0 & 1 & 1 \end{array}\right]$$

Hence, there is a unique solution: $x_1 = 2, x_2 = 1, x_3 = 1$. In other words, we must push switch A twice and the other two switches once each. (Check this.)

Exercises 2.4

Allocation of Resources

1. Suppose that, in Example 2.27, 400 units of food A, 600 units of B, and 600 units of C are placed in the test tube each day and the data on daily food consumption by the bacteria (in units per day) are as shown in Table 2.6. How many bacteria of each strain can coexist in the test tube and consume all of the food?

2. Suppose that in Example 2.27, 400 units of food A, 500 units of B, and 600 units of C are placed in the test tube each day and the data on daily food consumption by the bacteria (in units per day) are as shown in Table 2.7. How many bacteria of each

Table 2.6

	Bacteria Strain I	Bacteria Strain II	Bacteria Strain III
Food A	1	2	0
Food B	2	1	1
Food C	1	1	2

Table 2.7

	Bacteria Strain I	Bacteria Strain II	Bacteria Strain III
Food A	1	2	0
Food B	2	1	3
Food C	1	1	1

strain can coexist in the test tube and consume all of the food?

3. A florist offers three sizes of flower arrangements containing roses, daisies, and chrysanthemums. Each small arrangement contains one rose, three daisies, and three chrysanthemums. Each medium arrangement contains two roses, four daisies, and six chrysanthemums. Each large arrangement contains four roses, eight daisies, and six chrysanthemums. One day, the florist noted that she used a total of 24 roses, 50 daisies, and 48 chrysanthemums in filling orders for these three types of arrangements. How many arrangements of each type did she make?

4. **(a)** In your pocket you have some nickels, dimes, and quarters. There are 20 coins altogether and exactly twice as many dimes as nickels. The total value of the coins is \$3.00. Find the number of coins of each type.
   (b) Find *all* possible combinations of 20 coins (nickels, dimes, and quarters) that will make exactly \$3.00.

5. A coffee merchant sells three blends of coffee. A bag of the house blend contains 300 grams of Colombian beans and 200 grams of French roast beans. A bag of the special blend contains 200 grams of Colombian beans, 200 grams of Kenyan beans, and 100 grams of French roast beans. A bag of the gourmet blend contains 100 grams of Colombian beans, 200 grams of Kenyan beans, and 200 grams of French roast beans. The merchant has on hand 30 kilograms of Colombian beans, 15 kilograms of Kenyan beans, and 25 kilograms of French roast beans. If he wishes to use up all of the beans, how many bags of each type of blend can be made?

6. Redo Exercise 5, assuming that the house blend contains 300 grams of Colombian beans, 50 grams of Kenyan beans, and 150 grams of French roast beans and the gourmet blend contains 100 grams of Colombian beans, 350 grams of Kenyan beans, and 50 grams of French roast beans. This time the merchant has on hand 30 kilograms of Colombian beans, 15 kilograms of Kenyan beans, and 15 kilograms of French roast beans. Suppose one bag of the house blend produces a profit of \$0.50, one bag of the special blend produces a profit of \$1.50, and one bag of the gourmet blend produces a profit of \$2.00. How many bags of each type should the merchant prepare if he wants to use up all of the beans *and* maximize his profit? What is the maximum profit?

Balancing Chemical Equations

In Exercises 7–14, balance the chemical equation for each reaction.

7. $FeS_2 + O_2 \longrightarrow Fe_2O_3 + SO_2$
8. $CO_2 + H_2O \longrightarrow C_6H_{12}O_6 + O_2$ (This reaction takes place when a green plant converts carbon dioxide and water to glucose and oxygen during photosynthesis.)
9. $C_4H_{10} + O_2 \longrightarrow CO_2 + H_2O$ (This reaction occurs when butane, C_4H_{10}, burns in the presence of oxygen to form carbon dioxide and water.)
10. $C_7H_6O_2 + O_2 \longrightarrow H_2O + CO_2$
11. $C_5H_{11}OH + O_2 \longrightarrow H_2O + CO_2$ (This equation represents the combustion of amyl alcohol.)
12. $HClO_4 + P_4O_{10} \longrightarrow H_3PO_4 + Cl_2O_7$
13. $Na_2CO_3 + C + N_2 \longrightarrow NaCN + CO$
14. CAS $C_2H_2Cl_4 + Ca(OH)_2 \longrightarrow C_2HCl_3 + CaCl_2 + H_2O$

Network Analysis

15. Figure 2.18 shows a network of water pipes with flows measured in liters per minute.
    (a) Set up and solve a system of linear equations to find the possible flows.
    (b) If the flow through AB is restricted to 5 L/min, what will the flows through the other two branches be?
    (c) What are the minimum and maximum possible flows through each branch?
    (d) We have been assuming that flow is always *positive*. What would *negative* flow mean, assuming we allowed it? Give an illustration for this example.

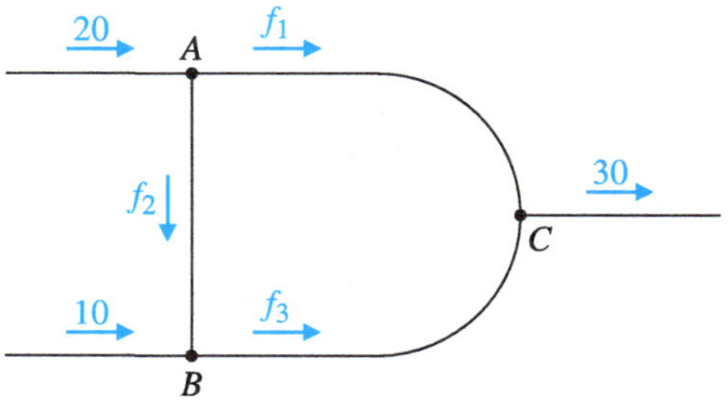

Figure 2.18

16. The downtown core of Gotham City consists of one-way streets, and the traffic flow has been measured at each intersection. For the city block shown in Figure 2.19, the numbers represent the average numbers of vehicles per minute entering and leaving intersections A, B, C, and D during business hours.

- **(a)** Set up and solve a system of linear equations to find the possible flows $f_1, \ldots, f_4$.
- **(b)** If traffic is regulated on CD so that $f_4 = 10$ vehicles per minute, what will the average flows on the other streets be?
- **(c)** What are the minimum and maximum possible flows on each street?
- **(d)** How would the solution change if *all* of the directions were reversed?

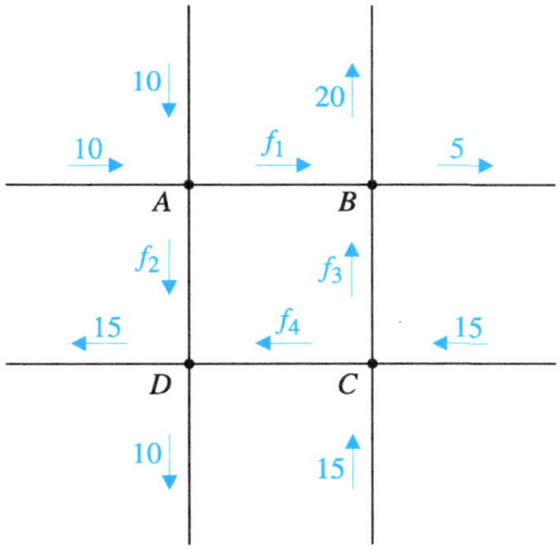

Figure 2.19

17. A network of irrigation ditches is shown in Figure 2.20, with flows measured in thousands of liters per day.

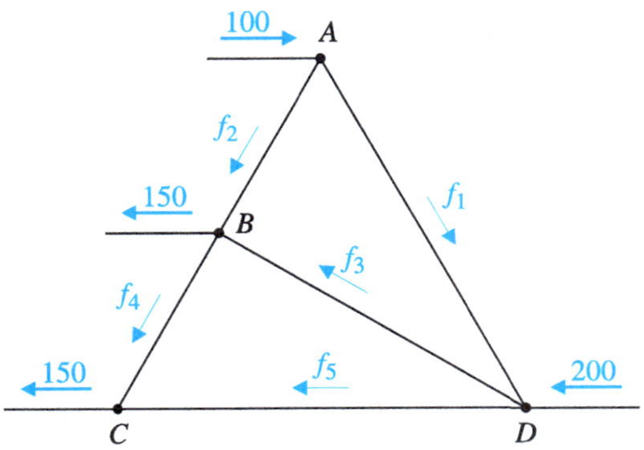

Figure 2.20

- **(a)** Set up and solve a system of linear equations to find the possible flows $f_1, \ldots, f_5$.
- **(b)** Suppose DC is closed. What range of flow will need to be maintained through DB?
- **(c)** From Figure 2.20 it is clear that DB cannot be closed. (Why not?) How does your solution in part (a) show this?
- **(d)** From your solution in part (a), determine the minimum and maximum flows through DB.

18. **(a)** Set up and solve a system of linear equations to find the possible flows in the network shown in Figure 2.21.

- **(b)** Is it possible for $f_1 = 100$ and $f_6 = 150$? [Answer this question first with reference to your solution in part (a) and then directly from Figure 2.21.]
- **(c)** If $f_4 = 0$, what will the range of flow be on each of the other branches?

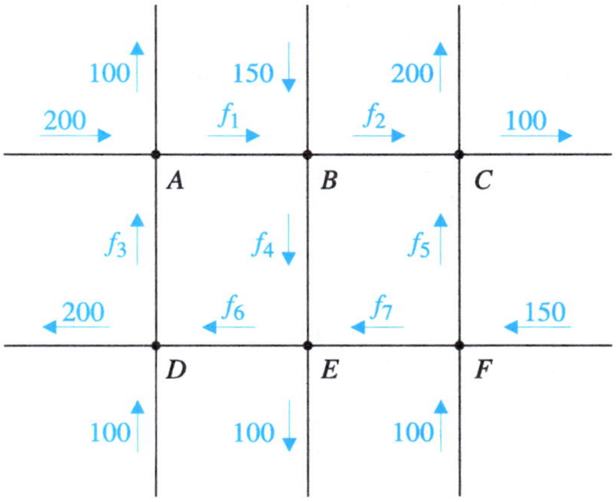

Figure 2.21

Electrical Networks

For Exercises 19 and 20, determine the currents for the given electrical networks.

19.

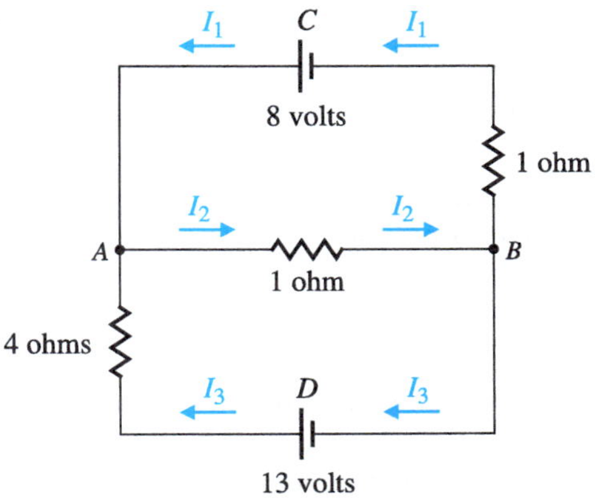

20.

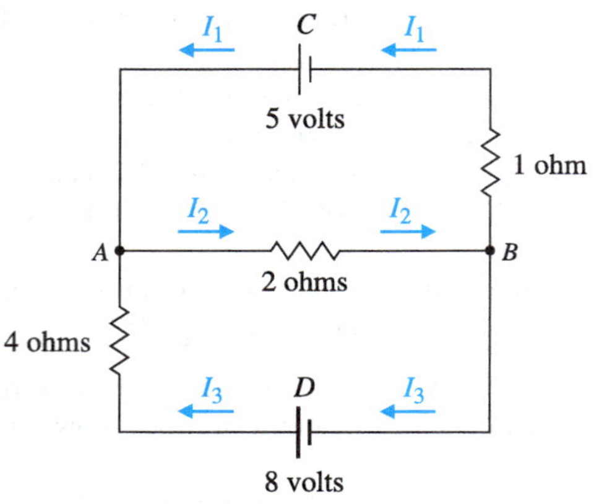

21. (a) Find the currents $I, I_1, \ldots, I_5$ in the bridge circuit in Figure 2.22.

(b) Find the effective resistance of this network.

(c) Can you change the resistance in branch BC (but leave everything else unchanged) so that the current through branch CE becomes 0?

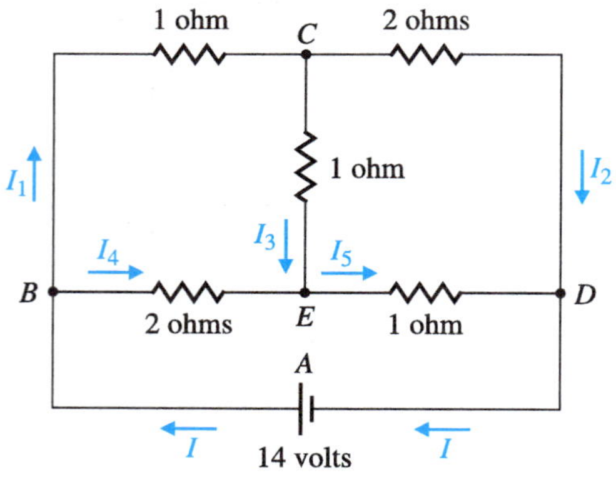

Figure 2.22

22. The networks in parts (a) and (b) of Figure 2.23 show two resistors coupled in *series* and in *parallel*, respectively. We wish to find a general formula for the effective resistance of each network—that is, find R_{eff} such that $E = R_{\text{eff}}I$.

(a) Show that the effective resistance R_{eff} of a network with two resistors coupled in series [Figure 2.23(a)] is given by

$$R_{\text{eff}} = R_1 + R_2$$

(b) Show that the effective resistance R_{eff} of a network with two resistors coupled in parallel [Figure 2.23(b)] is given by

$$R_{\text{eff}} = \frac{1}{\dfrac{1}{R_1} + \dfrac{1}{R_2}}$$

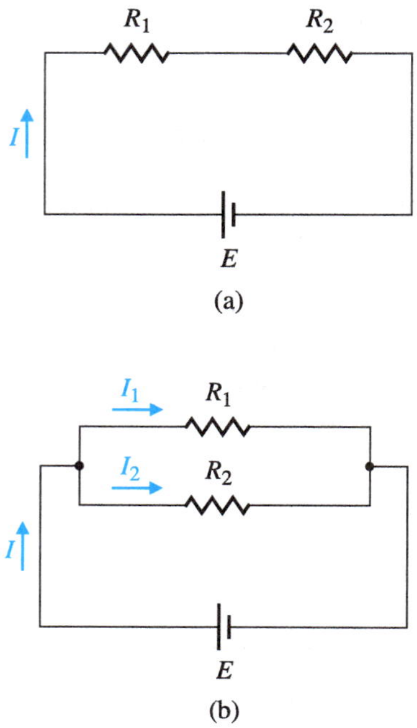

Figure 2.23
Resistors in series and in parallel

Linear Economic Models

23. Consider a simple economy with just two industries: farming and manufacturing. Farming consumes 1/2 of the food and 1/3 of the manufactured goods. Manufacturing consumes 1/2 of the food and 2/3 of the manufactured goods. Assuming the economy is closed and in equilibrium, find the relative outputs of the farming and manufacturing industries.

24. Suppose the coal and steel industries form a closed economy. Every \$1 produced by the coal industry requires \$0.30 of coal and \$0.70 of steel. Every \$1 produced by steel requires \$0.80 of coal and \$0.20 of steel. Find the annual production (output) of coal and steel if the total annual production is \$20 million.

25. A painter, a plumber, and an electrician enter into a cooperative arrangement in which each of them agrees to work for himself/herself and the other two for a total of 10 hours per week according to the schedule shown in Table 2.8. For tax purposes, each person must establish a value for his/her services. They agree to do this so that they each come out even—that is, so that the total amount paid out by each person equals the amount he/she receives. What hourly rate should each person charge if the rates are all whole numbers between \$30 and \$60 per hour?

Table 2.8

		Supplier		
		Painter	Plumber	Electrician
	Painter	2	1	5
Consumer	Plumber	4	5	1
	Electrician	4	4	4

26. Four neighbors, each with a vegetable garden, agree to share their produce. One will grow beans (**B**), one will grow lettuce (**L**), one will grow tomatoes (**T**), and one will grow zucchini (**Z**). Table 2.9 shows what fraction of each crop each neighbor will receive. What prices should the neighbors charge for their crops if each person is to break even and the lowest-priced crop has a value of \$50?

Table 2.9

		Producer			
		B	L	T	Z
	B	0	1/4	1/8	1/6
Consumer	L	1/2	1/4	1/4	1/6
	T	1/4	1/4	1/2	1/3
	Z	1/4	1/4	1/8	1/3

27. Suppose the coal and steel industries form an open economy. Every \$1 produced by the coal industry requires \$0.15 of coal and \$0.20 of steel. Every \$1 produced by steel requires \$0.25 of coal and \$0.10 of steel. Suppose that there is an annual outside demand for \$45 million of coal and \$124 million of steel.

(a) How much should each industry produce to satisfy the demands?

(b) If the demand for coal decreases by \$5 million per year while the demand for steel increases by \$6 million per year, how should the coal and steel industries adjust their production?

28. In Gotham City, the departments of Administration (A), Health (H), and Transportation (T) are interdependent. For every dollar's worth of services they produce, each department uses a certain amount of the services produced by the other departments and itself, as shown in Table 2.10. Suppose that, during the year, other city departments require \$1 million in Administrative services, \$1.2 million in Health services, and \$0.8 million in Transportation services. What does the annual dollar value of the services produced by each department need to be in order to meet the demands?

Table 2.10

		Department		
		A	H	T
	A	\$0.20	0.10	0.20
Buy	H	0.10	0.10	0.20
	T	0.20	0.40	0.30

Finite Linear Games

29. **(a)** In Example 2.35, suppose all the lights are initially off. Can we push the switches in some order so that only the second and fourth lights will be on?

(b) Can we push the switches in some order so that only the second light will be on?

30. **(a)** In Example 2.35, suppose the fourth light is initially on and the other four lights are off. Can we push the switches in some order so that only the second and fourth lights will be on?

(b) Can we push the switches in some order so that only the second light will be on?

31. In Example 2.35, describe all possible configurations of lights that can be obtained if we start with all the lights off.

32. **(a)** In Example 2.36, suppose that all of the lights are initially off. Show that it is possible to push the switches in some order so that the lights are off, dark blue, and light blue, in that order.

(b) Show that it is possible to push the switches in some order so that the lights are light blue, off, and light blue, in that order.

(c) Prove that *any* configuration of the three lights can be achieved.

33. Suppose the lights in Example 2.35 can be off, light blue, or dark blue and the switches work as described in Example 2.36. (That is, the switches control the same lights as in Example 2.35 but cycle through the colors as in Example 2.36.) Show that it is possible to start with all of the lights off and push the switches in some order so that the lights are dark blue, light blue, dark blue, light blue, and dark blue, in that order.

34. For Exercise 33, describe all possible configurations of lights that can be obtained, starting with all the lights off.

CAS **35.** Nine squares, each one either black or white, are arranged in a 3×3 grid. Figure 2.24 shows one possible arrangement. When touched, each square changes its own state and the states of some of its neighbors (black → white and white → black). Figure 2.25 shows how the state changes work. (Touching the square whose number is circled causes the states of the squares marked * to change.) The object of the game is to turn all nine squares black. [Exercises 35 and 36 are adapted from puzzles that can be found in the interactive CD-ROM game *The Seventh Guest* (Trilobyte Software/Virgin Games, 1992).]

Figure 2.24
The nine squares puzzle

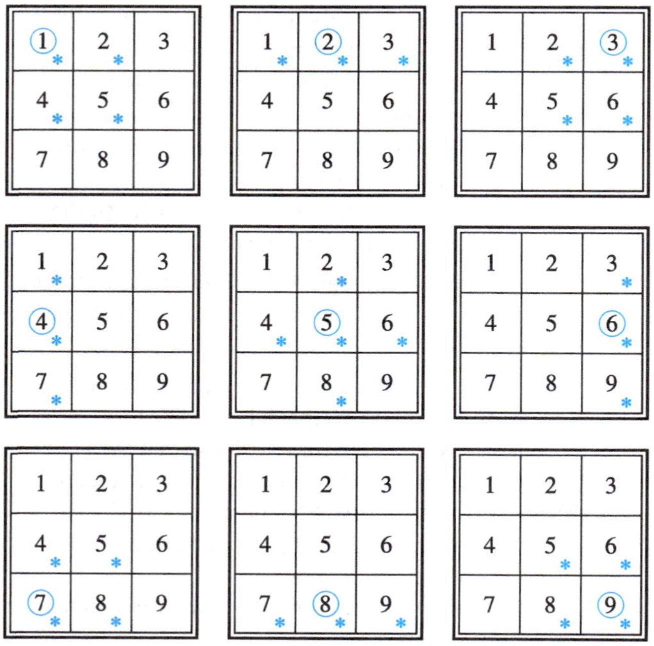

Figure 2.25
State changes for the nine squares puzzle

(a) If the initial configuration is the one shown in Figure 2.24, show that the game can be won and describe a winning sequence of moves.

(b) Prove that the game can always be won, no matter what the initial configuration.

CAS **36.** Consider a variation on the nine squares puzzle. The game is the same as that described in Exercise 35 except that there are three possible states for each square: white, gray, or black. The squares change as shown in Figure 2.25, but now the state changes follow the cycle white → gray → black → white. Show how the winning all-black configuration can be achieved from the initial configuration shown in Figure 2.26.

Figure 2.26
The nine squares puzzle with more states

Miscellaneous Problems

In Exercises 37–53, set up and solve an appropriate system of linear equations to answer the questions.

37. Grace is three times as old as Hans, but in 5 years she will be twice as old as Hans is then. How old are they now?

38. The sum of Annie's, Bert's, and Chris's ages is 60. Annie is older than Bert by the same number of years that Bert is older than Chris. When Bert is as old as Annie is now, Annie will be three times as old as Chris is now. What are their ages?

The preceding two problems are typical of those found in popular books of mathematical puzzles. However, they have their origins in antiquity. A Babylonian clay tablet that survives from about 300 B.C. contains the following problem.

39. There are two fields whose total area is 1800 square yards. One field produces grain at the rate of $\frac{2}{3}$ bushel per square yard; the other field produces grain at the rate of $\frac{1}{2}$ bushel per square yard. If the total yield is 1100 bushels, what is the size of each field?

Over 2000 years ago, the Chinese developed methods for solving systems of linear equations, including a version of Gaussian elimination that did not become well known in Europe until the 19th century. (There is no evidence that Gauss was aware of the Chinese methods when he developed what we now call Gaussian elimination. However, it is clear that the Chinese knew the essence of the method, even though they did not justify its use.) The following problem is taken from the Chinese text Jiuzhang suanshu (Nine Chapters in the Mathematical Art), *written during the early Han Dynasty, about 200* B.C.

40. There are three types of corn. Three bundles of the first type, two of the second, and one of the third make 39 measures. Two bundles of the first type, three of the second, and one of the third make 34 measures. And one bundle of the first type, two of the second, and three of the third make 26 measures. How many measures of corn are contained in one bundle of each type?

41. Describe all possible values of a, b, c, and d that will make each of the following a valid addition table. [Problems 41–44 are based on the article "An Application of Matrix Theory" by Paul Glaister in *The Mathematics Teacher*, 85 (1992), pp. 220–223.]

(a)

+	a	b
c	2	3
d	4	5

(b)

+	a	b
c	3	6
d	4	5

42. What conditions on w, x, y, and z will guarantee that we can find a, b, c, and d so that the following is a valid addition table?

+	a	b
c	w	x
d	y	z

43. Describe all possible values of a, b, c, d, e, and f that will make each of the following a valid addition table.

(a)

+	a	b	c
d	3	2	1
e	5	4	3
f	4	3	1

(b)

+	a	b	c
d	1	2	3
e	3	4	5
f	4	5	6

44. Generalizing Exercise 42, find conditions on the entries of a 3×3 addition table that will guarantee that we can solve for a, b, c, d, e, and f as previously.

45. From elementary geometry we know that there is a unique straight line through any two points in a plane. Less well known is the fact that there is a unique parabola through any *three* noncollinear points in a plane. For each set of points below, find a parabola with an equation of the form $y = ax^2 + bx + c$ that passes through the given points. (Sketch the resulting parabola to check the validity of your answer.)

(a) $(0, 1)$, $(-1, 4)$, and $(2, 1)$
(b) $(-3, 1)$, $(-2, 2)$, and $(-1, 5)$

46. Through any three noncollinear points there also passes a unique circle. Find the circles (whose general equations are of the form $x^2 + y^2 + ax + by + c = 0$) that pass through the sets of points in Exercise 45. (To check the validity of your answer, find the center and radius of each circle and draw a sketch.)

The process of adding rational functions (ratios of polynomials) by placing them over a common denominator is the analogue of adding rational numbers. The reverse process of taking a rational function apart by writing it as a sum of simpler rational functions is useful in several areas of mathematics; for example, it arises in calculus when we need to integrate a rational function and in discrete mathematics when we use generating functions to solve recurrence relations. The decomposition of a rational function as a sum of partial fractions leads to a system of linear equations. In Exercises 47–50, find the partial fraction decomposition of the given form. (The capital letters denote constants.)

47. $\dfrac{3x + 1}{x^2 + 2x - 3} = \dfrac{A}{x - 1} + \dfrac{B}{x + 3}$

48. $\dfrac{x^2 - 3x + 3}{x^3 + 2x^2 + x} = \dfrac{A}{x} + \dfrac{B}{x + 1} + \dfrac{C}{(x + 1)^2}$

CAS **49.** $\dfrac{x - 1}{(x + 1)(x^2 + 1)(x^2 + 4)}$

$$= \frac{A}{x + 1} + \frac{Bx + C}{x^2 + 1} + \frac{Dx + E}{x^2 + 4}$$

CAS **50.** $\dfrac{x^3 + x + 1}{x(x - 1)(x^2 + x + 1)(x^2 + 1)^3} = \dfrac{A}{x} + \dfrac{B}{x - 1}$

$$+ \frac{Cx + D}{x^2 + x + 1} + \frac{Ex + F}{x^2 + 1} + \frac{Gx + H}{(x^2 + 1)^2} + \frac{Ix + J}{(x^2 + 1)^3}$$

Following are two useful formulas for the sums of powers of consecutive natural numbers:

$$1 + 2 + \cdots + n = \frac{n(n + 1)}{2}$$

and

$$1^2 + 2^2 + \cdots + n^2 = \frac{n(n + 1)(2n + 1)}{6}$$

The validity of these formulas for all values of $n \geq 1$ (or even $n \geq 0$) can be established using mathematical induction (see Appendix B). One way to make an educated guess as to what the formulas are, though, is to observe that we can rewrite the two formulas above as

$$\tfrac{1}{2}n^2 + \tfrac{1}{2} \quad \text{and} \quad \tfrac{1}{3}n^3 + \tfrac{1}{2}n^2 + \tfrac{1}{6}n$$

respectively. This leads to the conjecture that the sum of pth powers of the first n natural numbers is a polynomial of degree $p + 1$ in the variable n.

51. Assuming that $1 + 2 + \cdots + n = an^2 + bn + c$, find a, b, and c by substituting three values for n and thereby obtaining a system of linear equations in a, b, and c.

52. Assume that $1^2 + 2^2 + \cdots + n^2 = an^3 + bn^2 + cn + d$. Find a, b, c, and d. [*Hint:* It is legitimate to use $n = 0$. What is the left-hand side in that case?]

53. Show that $1^3 + 2^3 + \cdots + n^3 = (n(n + 1)/2)^2$.

Vignette

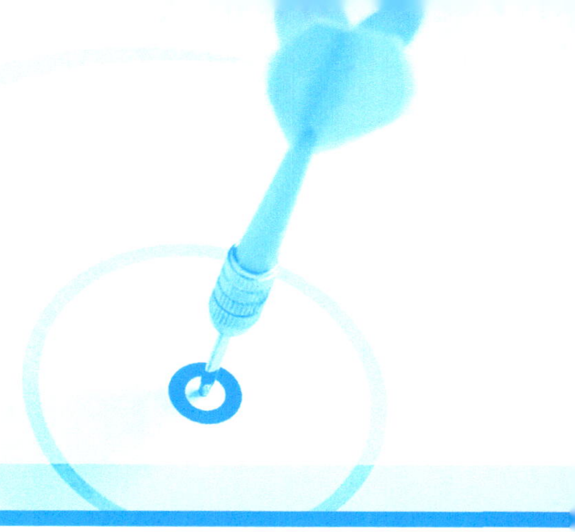

The Global Positioning System

The Global Positioning System (GPS) is used in a variety of situations for determining geographical locations. The military, surveyors, airlines, shipping companies, and hikers all make use of it. GPS technology is becoming so commonplace that some automobiles, cellular phones, and various handheld devices are now equipped with it.

The basic idea of GPS is a variant on three-dimensional triangulation: A point on Earth's surface is uniquely determined by knowing its distances from three other points. Here the point we wish to determine is the location of the GPS receiver, the other points are satellites, and the distances are computed using the travel times of radio signals from the satellites to the receiver.

We will assume that Earth is a sphere on which we impose an xyz-coordinate system with Earth centered at the origin and with the positive z-axis running through the north pole and fixed relative to Earth.

For simplicity, let's take one unit to be equal to the radius of Earth. Thus Earth's surface becomes the unit sphere with equation $x^2 + y^2 + z^2 = 1$. Time will be measured in hundredths of a second. GPS finds distances by knowing how long it takes a radio signal to get from one point to another. For this we need to know the speed of light, which is approximately equal to 0.47 (Earth radii per hundredths of a second).

Let's imagine that you are a hiker lost in the woods at point (x, y, z) at some time t. You don't know where you are, and furthermore, you have no watch, so you don't know what time it is. However, you have your GPS device, and it receives simultaneous signals from four satellites, giving their positions and times as shown in Table 2.11. (Distances are measured in Earth radii and time in hundredths of a second past midnight.)

This application is based on the article "An Underdetermined Linear System for GPS" by Dan Kalman in *The College Mathematics Journal*, 33 (2002), pp. 384–390. For a more in-depth treatment of the ideas introduced here, see G. Strang and K. Borre, *Linear Algebra, Geodesy, and GPS* (Wellesley-Cambridge Press, MA, 1997).

Table 2.11 Satellite Data

Satellite	Position	Time
1	(1.11, 2.55, 2.14)	1.29
2	(2.87, 0.00, 1.43)	1.31
3	(0.00, 1.08, 2.29)	2.75
4	(1.54, 1.01, 1.23)	4.06

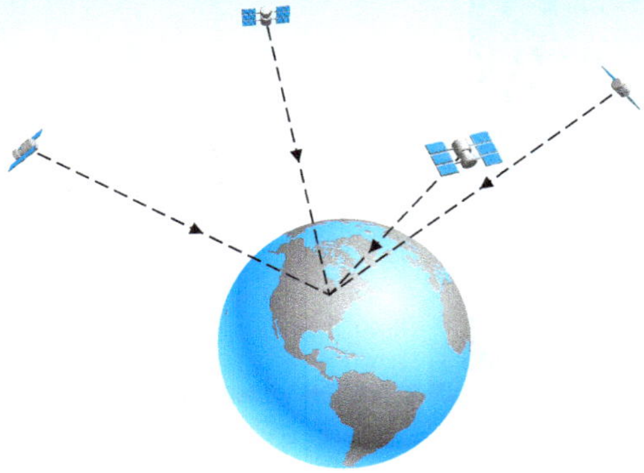

Let (x, y, z) be your position, and let t be the time when the signals arrive. The goal is to solve for x, y, z, and t. Your distance from Satellite 1 can be computed as follows. The signal, traveling at a speed of 0.47 Earth radii/10^{-2} sec, was sent at time 1.29 and arrived at time t, so it took $t - 1.29$ hundredths of a second to reach you. Distance equals velocity multiplied by (elapsed) time, so

$$d = 0.47(t - 1.29)$$

We can also express d in terms of (x, y, z) and the satellite's position (1.11, 2.55, 2.14) using the distance formula:

$$d = \sqrt{(x - 1.11)^2 + (y - 2.55)^2 + (z - 2.14)^2}$$

Combining these results leads to the equation

$$(x - 1.11)^2 + (y - 2.55)^2 + (z - 2.14)^2 = 0.47^2(t - 1.29)^2 \qquad (1)$$

Expanding, simplifying, and rearranging, we find that Equation (1) becomes

$$2.22x + 5.10y + 4.28z - 0.57t = x^2 + y^2 + z^2 - 0.22t^2 + 11.95$$

Similarly, we can derive a corresponding equation for each of the other three satellites. We end up with a system of four equations in x, y, z, and t:

$$\begin{aligned}
2.22x + 5.10y + 4.28z - 0.57t &= x^2 + y^2 + z^2 - 0.22t^2 + 11.95 \\
5.74x \qquad\qquad + 2.86z - 0.58t &= x^2 + y^2 + z^2 - 0.22t^2 + 9.90 \\
2.16y + 4.58z - 1.21t &= x^2 + y^2 + z^2 - 0.22t^2 + 4.74 \\
3.08x + 2.02y + 2.46z - 1.79t &= x^2 + y^2 + z^2 - 0.22t^2 + 1.26
\end{aligned}$$

These are not linear equations, but the nonlinear terms are the same in each equation. If we subtract the first equation from each of the other three equations, we obtain a linear system:

$$\begin{aligned}
3.52x - 5.10y - 1.42z - 0.01t &= 2.05 \\
-2.22x - 2.94y + 0.30z - 0.64t &= 7.21 \\
0.86x - 3.08y - 1.82z - 1.22t &= -10.69
\end{aligned}$$

The augmented matrix row reduces as

$$\left[\begin{array}{rrrr|r} 3.52 & -5.10 & -1.42 & -0.01 & -2.05 \\ -2.22 & -2.94 & 0.30 & -0.64 & -7.21 \\ 0.86 & -3.08 & -1.82 & -1.22 & -10.69 \end{array}\right] \longrightarrow \left[\begin{array}{cccc|c} 1 & 0 & 0 & 0.36 & 2.97 \\ 0 & 1 & 0 & 0.03 & 0.81 \\ 0 & 0 & 1 & 0.79 & 5.91 \end{array}\right]$$

from which we see that

$$
\begin{aligned}
x &= 2.97 - 0.36t \\
y &= 0.81 - 0.03t \\
z &= 5.91 - 0.79t
\end{aligned} \qquad (2)
$$

with t free. Substituting these equations into (1), we obtain

$$
\begin{aligned}
&(2.97 - 0.36t - 1.11)^2 + (0.81 - 0.03t - 2.55)^2 \\
&\quad + (5.91 - 0.79t - 2.14)^2 = 0.47^2(t - 1.29)^2
\end{aligned}
$$

which simplifies to the quadratic equation

$$0.54t^2 - 6.65t + 20.32 = 0$$

There are two solutions:

$$t = 6.74 \quad \text{and} \quad t = 5.60$$

Substituting into (2), we find that the first solution corresponds to $(x, y, z) = (0.55, 0.61, 0.56)$ and the second solution to $(x, y, z) = (0.96, 0.65, 1.46)$. The second solution is clearly not on the unit sphere (Earth), so we reject it. The first solution produces $x^2 + y^2 + z^2 = 0.99$, so we are satisfied that, within acceptable roundoff error, we have located your coordinates as $(0.55, 0.61, 0.56)$.

In practice, GPS takes significantly more factors into account, such as the fact that Earth's surface is not exactly spherical, so additional refinements are needed involving such techniques as least squares approximation (see Chapter 7). In addition, the results of the GPS calculation are converted from rectangular (Cartesian) coordinates into latitude and longitude, an interesting exercise in itself and one involving yet other branches of mathematics.

CAS

2.5 Iterative Methods for Solving Linear Systems

The direct methods for solving linear systems, using elementary row operations, lead to exact solutions in many cases but are subject to errors due to roundoff and other factors, as we have seen. The third road in our "trivium" takes us down quite a different path indeed. In this section, we explore methods that proceed *iteratively* by successively generating sequences of vectors that approach a solution to a linear system. In many instances (such as when the coefficient matrix is *sparse*—that is, contains many zero entries), iterative methods can be faster and more accurate than direct methods. Also, iterative methods can be stopped whenever the approximate solution they generate is sufficiently accurate. In addition, iterative methods often *benefit* from inaccuracy: Roundoff error can actually accelerate their convergence toward a solution.

We will explore two iterative methods for solving linear systems: ***Jacobi's method*** and a refinement of it, the ***Gauss-Seidel method.*** In all examples, we will be considering linear systems with the same number of variables as equations, and we will assume that there is a unique solution. Our interest is in finding this solution using iterative methods.

Example 2.37

Consider the system

$$\begin{aligned} 7x_1 - x_2 &= 5 \\ 3x_1 - 5x_2 &= -7 \end{aligned}$$

Jacobi's method begins with solving the first equation for x_1 and the second equation for x_2, to obtain

$$\begin{aligned} x_1 &= \frac{5 + x_2}{7} \\ x_2 &= \frac{7 + 3x_1}{5} \end{aligned} \qquad (1)$$

We now need an ***initial approximation*** to the solution. It turns out that it does not matter what this initial approximation is, so we might as well take $x_1 = 0, x_2 = 0$. We use these values in Equations (1) to get new values of x_1 and x_2:

$$\begin{aligned} x_1 &= \frac{5 + 0}{7} = \frac{5}{7} \approx 0.714 \\ x_2 &= \frac{7 + 3 \cdot 0}{5} = \frac{7}{5} = 1.400 \end{aligned}$$

Now we substitute these values into (1) to get

$$\begin{aligned} x_1 &= \frac{5 + 1.4}{7} \approx 0.914 \\ x_2 &= \frac{7 + 3 \cdot \frac{5}{7}}{5} \approx 1.829 \end{aligned}$$

(written to three decimal places). We repeat this process (using the old values of x_2 and x_1 to get the new values of x_1 and x_2), producing the sequence of approximations given in Table 2.12.

Courtesy of the Smithsonian Institution Libraries, Washington, D.C.

Carl Gustav Jacobi (1804–1851) was a German mathematician who made important contributions to many fields of mathematics and physics, including geometry, number theory, analysis, mechanics, and fluid dynamics. Although much of his work was in applied mathematics, Jacobi believed in the importance of doing mathematics for its own sake. A fine teacher, he held positions at the Universities of Berlin and Königsberg and was one of the most famous mathematicians in Europe.

Table 2.12

n	0	1	2	3	4	5	6
x_1	0	0.714	0.914	0.976	0.993	0.998	0.999
x_2	0	1.400	1.829	1.949	1.985	1.996	1.999

The successive vectors $\begin{bmatrix} x_1 \\ x_2 \end{bmatrix}$ are called ***iterates,*** so, for example, when $n = 4$, the fourth iterate is $\begin{bmatrix} 0.993 \\ 1.985 \end{bmatrix}$. We can see that the iterates in this example are approaching $\begin{bmatrix} 1 \\ 2 \end{bmatrix}$, which is the exact solution of the given system. (Check this.) We say in this case that Jacobi's method ***converges.***

Jacobi's method calculates the successive iterates in a two-variable system according to the crisscross pattern shown in Table 2.13.

Table 2.13

n	0	1	2	3
x_1				
x_2				

The Gauss-Seidel method is named after C. F. Gauss and Philipp Ludwig von Seidel (1821–1896). Seidel worked in analysis, probability theory, astronomy, and optics. Unfortunately, he suffered from eye problems and retired at a young age. The paper in which he described the method now known as Gauss-Seidel was published in 1874. Gauss, it seems, was unaware of the method!

Before we consider Jacobi's method in the general case, we will look at a modification of it that often converges faster to the solution. The *Gauss-Seidel method* is the same as the Jacobi method except that we use each new value *as soon as we can.* So in our example, we begin by calculating $x_1 = (5 + 0)/7 = \frac{5}{7} \approx 0.714$ as before, but we now use this value of x_1 to get the next value of x_2:

$$x_2 = \frac{7 + 3 \cdot \frac{5}{7}}{5} \approx 1.829$$

We then use this value of x_2 to recalculate x_1, and so on. The iterates this time are shown in Table 2.14.

We observe that the Gauss-Seidel method has converged faster to the solution. The iterates this time are calculated according to the zigzag pattern shown in Table 2.15.

Table 2.14

n	0	1	2	3	4	5
x_1	0	0.714	0.976	0.998	1.000	1.000
x_2	0	1.829	1.985	1.999	2.000	2.000

Table 2.15

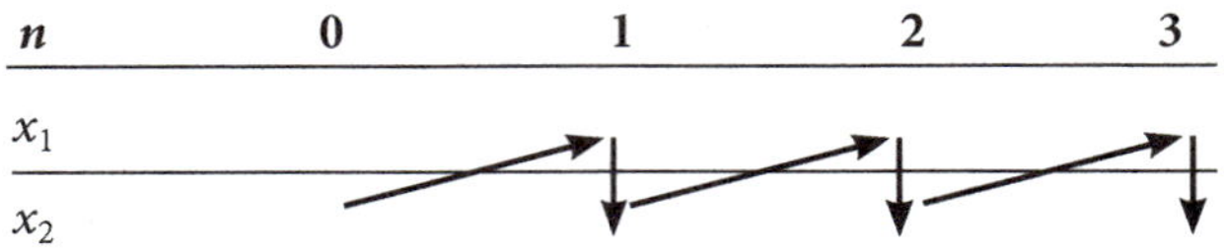

The Gauss-Seidel method also has a nice geometric interpretation in the case of two variables. We can think of x_1 and x_2 as the coordinates of points in the plane. Our starting point is the point corresponding to our initial approximation, (0, 0). Our first calculation gives $x_1 = \frac{5}{7}$, so we move to the point $(\frac{5}{7}, 0) \approx (0.714, 0)$. Then we compute $x_2 = \frac{64}{35} \approx 1.829$, which moves us to the point $(\frac{5}{7}, \frac{64}{35}) \approx (0.714, 1.829)$. Continuing in this fashion, our calculations from the Gauss-Seidel method give rise to a sequence of points, each one differing from the preceding point in exactly one coordinate. If we plot the lines $7x_1 - x_2 = 5$ and $3x_1 - 5x_2 = -7$ corresponding to the two given equations, we find that the points calculated above fall alternately on the two lines, as shown in Figure 2.27. Moreover, they approach the point of intersection of the lines, which corresponds to the solution of the system of equations. This is what *convergence* means!

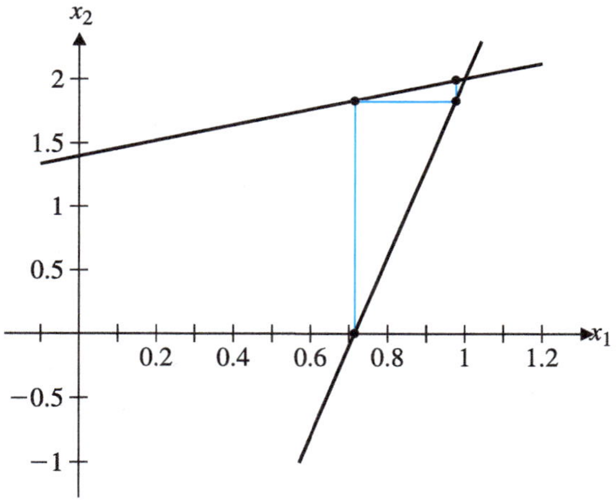

Figure 2.27
Converging iterates

The general cases of the two methods are analogous. Given a system of n linear equations in n variables,

$$\begin{aligned} a_{11}x_1 + a_{12}x_2 + \cdots + a_{1n}x_n &= b_1 \\ a_{21}x_1 + a_{22}x_2 + \cdots + a_{2n}x_n &= b_2 \\ &\vdots \\ a_{n1}x_1 + a_{n2}x_2 + \cdots + a_{nn}x_n &= b_n \end{aligned} \tag{2}$$

we solve the first equation for x_1, the second for x_2, and so on. Then, beginning with an initial approximation, we use these new equations to iteratively update each variable. Jacobi's method uses *all* of the values at the kth iteration to compute the $(k + 1)$st iterate, whereas the Gauss-Seidel method always uses the *most recent* value of each variable in every calculation. Example 2.39 later illustrates the Gauss-Seidel method in a three-variable problem.

At this point, you should have some questions and concerns about these iterative methods. (Do you?) Several come to mind: Must these methods converge? If not, when *do* they converge? *If* they converge, must they converge to the solution? The answer to the first question is no, as Example 2.38 illustrates.

Example 2.38

Apply the Gauss-Seidel method to the system

$$\begin{aligned} x_1 - x_2 &= 1 \\ 2x_1 + x_2 &= 5 \end{aligned}$$

with initial approximation $\begin{bmatrix} 0 \\ 0 \end{bmatrix}$.

Solution We rearrange the equations to get

$$\begin{aligned} x_1 &= 1 + x_2 \\ x_2 &= 5 - 2x_1 \end{aligned}$$

The first few iterates are given in Table 2.16. (Check these.)

The actual solution to the given system is $\begin{bmatrix} x_1 \\ x_2 \end{bmatrix} = \begin{bmatrix} 2 \\ 1 \end{bmatrix}$. Clearly, the iterates in Table 2.16 are not approaching this point, as Figure 2.28 makes graphically clear in an example of ***divergence.***

Table 2.16

n	0	1	2	3	4	5
x_1	0	1	4	−2	10	−14
x_2	0	3	−3	9	−15	33

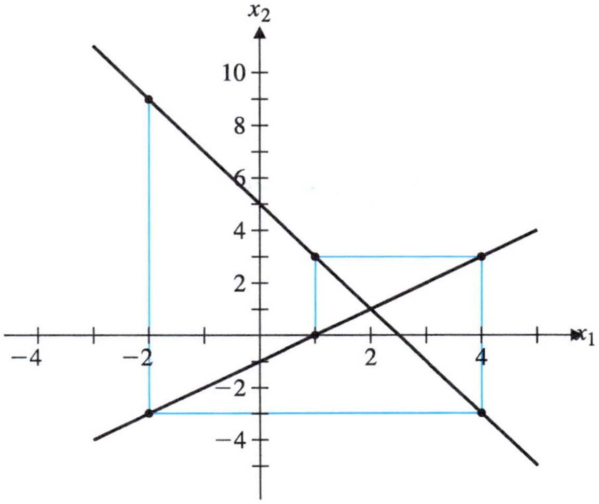

Figure 2.28
Diverging iterates

So when do these iterative methods converge? Unfortunately, the answer to this question is rather tricky. We will answer it completely in Chapter 7, but for now we will give a partial answer, without proof.

Let A be the $n \times n$ matrix

$$A = \begin{bmatrix} a_{11} & a_{12} & \cdots & a_{1n} \\ a_{21} & a_{22} & \cdots & a_{2n} \\ \vdots & \vdots & \ddots & \vdots \\ a_{n1} & a_{n2} & \cdots & a_{nn} \end{bmatrix}$$

We say that A is ***strictly diagonally dominant*** if

$$\begin{aligned} |a_{11}| &> |a_{12}| + |a_{13}| + \cdots + |a_{1n}| \\ |a_{22}| &> |a_{21}| + |a_{23}| + \cdots + |a_{2n}| \\ &\vdots \\ |a_{nn}| &> |a_{n1}| + |a_{n2}| + \cdots + |a_{n,n-1}| \end{aligned}$$

That is, the absolute value of each diagonal entry $a_{11}, a_{22}, \ldots, a_{nn}$ is greater than the sum of the absolute values of the remaining entries in that row.

Theorem 2.9 If a system of n linear equations in n variables has a strictly diagonally dominant coefficient matrix, then it has a unique solution and both the Jacobi and the Gauss-Seidel method converge to it.

Remark Be warned! This theorem is a one-way implication. The fact that a system is *not* strictly diagonally dominant does *not* mean that the iterative methods diverge. They may or may not converge. (See Exercises 15–19.) Indeed, there are examples in which one of the methods converges and the other diverges. However, *if* either of these methods converges, then it must converge to the solution—it cannot converge to some other point.

Theorem 2.10 If the Jacobi or the Gauss-Seidel method converges for a system of n linear equations in n variables, then it must converge to the solution of the system.

Proof We will illustrate the idea behind the proof by sketching it out for the case of Jacobi's method, using the system of equations in Example 2.37. The general proof is similar.

Convergence means that "as iterations increase, the values of the iterates get closer and closer to a limiting value." This means that x_1 and x_2 converge to r and s, respectively, as shown in Table 2.17.

We must prove that $\begin{bmatrix} x_1 \\ x_2 \end{bmatrix} = \begin{bmatrix} r \\ s \end{bmatrix}$ is the solution of the system of equations. In other words, at the $(k+1)$st iteration, the values of x_1 and x_2 must stay the same as at

Table 2.17

n	$\cdots$	k	$k+1$	$k+2$	$\cdots$
x_1	$\cdots$	r	r	r	$\cdots$
x_2	$\cdots$	s	s	s	$\cdots$

the kth iteration. But the calculations give $x_1 = (5 + x_2)/7 = (5 + s)/7$ and $x_2 = (7 + 3x_1)/5 = (7 + 3r)/5$. Therefore,

$$\frac{5+s}{7} = r \quad \text{and} \quad \frac{7+3r}{5} = s$$

Rearranging, we see that

$$\begin{aligned} 7r - s &= 5 \\ 3r - 5s &= -7 \end{aligned}$$

Thus, $x_1 = r$, $x_2 = s$ satisfy the original equations, as required.

By now you may be wondering: If iterative methods don't always converge to the solution, what good are they? Why don't we just use Gaussian elimination? First, we have seen that Gaussian elimination is sensitive to roundoff errors, and this sensitivity can lead to inaccurate or even wildly wrong answers. Also, even if Gaussian elimination does not go astray, we cannot improve on a solution once we have found it. For example, if we use Gaussian elimination to calculate a solution to two decimal places, there is no way to obtain the solution to four decimal places except to start over again and work with increased accuracy.

In contrast, we can achieve additional accuracy with iterative methods simply by doing more iterations. For large systems, particularly those with sparse coefficient matrices, iterative methods are much faster than direct methods when implemented on a computer. In many applications, the systems that arise are strictly diagonally dominant, and thus iterative methods are guaranteed to converge. The next example illustrates one such application.

Example 2.39

Suppose we heat each edge of a metal plate to a constant temperature, as shown in Figure 2.29.

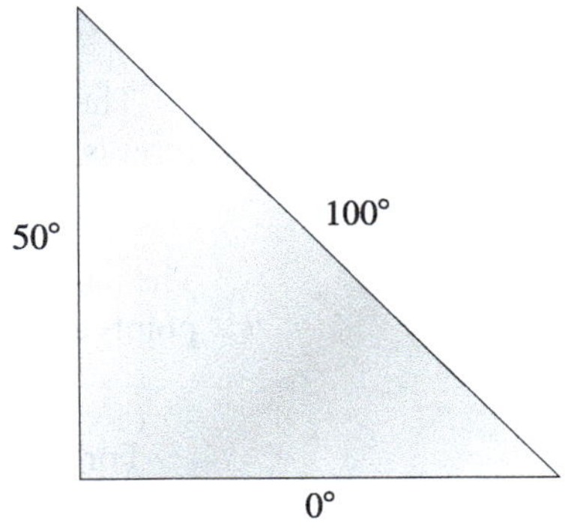

Figure 2.29
A heated metal plate

Eventually the temperature at the interior points will reach *equilibrium,* where the following property can be shown to hold:

The temperature at each interior point P on a plate is the average of the temperatures on the circumference of any circle centered at P inside the plate (Figure 2.30).

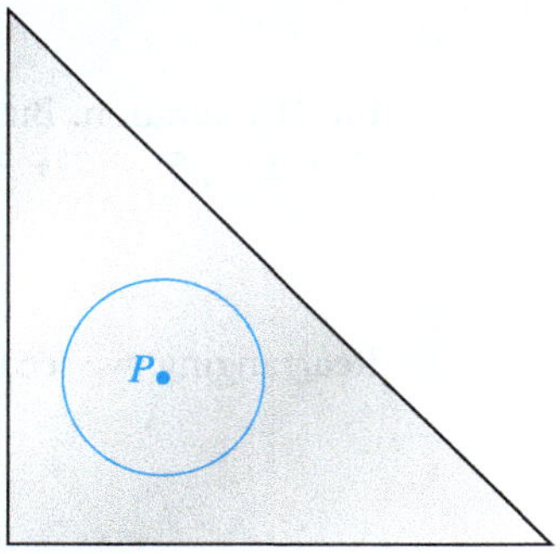

Figure 2.30

To apply this property in an actual example requires techniques from calculus. As an alternative, we can approximate the situation by overlaying the plate with a grid, or mesh, that has a finite number of interior points, as shown in Figure 2.31.

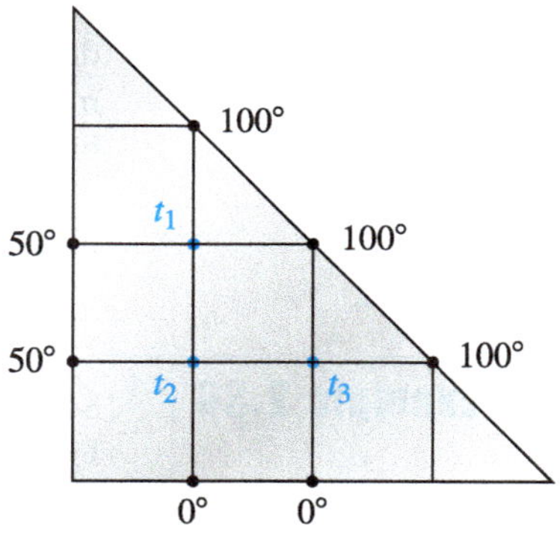

Figure 2.31
The discrete version of the heated plate problem

The discrete analogue of the averaging property governing equilibrium temperatures is stated as follows:

The temperature at each interior point P is the average of the temperatures at the points adjacent to P.

For the example shown in Figure 2.31, there are three interior points, and each is adjacent to four other points. Let the equilibrium temperatures of the interior points be t_1, t_2, and t_3, as shown. Then, by the temperature-averaging property, we have

$$\begin{aligned} t_1 &= \frac{100 + 100 + t_2 + 50}{4} \\ t_2 &= \frac{t_1 + t_3 + 0 + 50}{4} \\ t_3 &= \frac{100 + 100 + 0 + t_2}{4} \end{aligned} \tag{3}$$

or

$$\begin{aligned} 4t_1 - t_2 \qquad &= 250 \\ -t_1 + 4t_2 - t_3 &= 50 \\ - t_2 + 4t_3 &= 200 \end{aligned}$$

Notice that this system is strictly diagonally dominant. Notice also that Equations (3) are in the form required for Jacobi or Gauss-Seidel iteration. With an initial approximation of $t_1 = 0$, $t_2 = 0$, $t_3 = 0$, the Gauss-Seidel method gives the following iterates.

Iteration 1:

$$\begin{aligned} t_1 &= \frac{100 + 100 + 0 + 50}{4} = 62.5 \\ t_2 &= \frac{62.5 + 0 + 0 + 50}{4} = 28.125 \\ t_3 &= \frac{100 + 100 + 0 + 28.125}{4} = 57.031 \end{aligned}$$

Iteration 2:

$$\begin{aligned} t_1 &= \frac{100 + 100 + 28.125 + 50}{4} = 69.531 \\ t_2 &= \frac{69.531 + 57.031 + 0 + 50}{4} = 44.141 \\ t_3 &= \frac{100 + 100 + 0 + 44.141}{4} = 61.035 \end{aligned}$$

Continuing, we find the iterates listed in Table 2.18. We work with five-significant-digit accuracy and stop when two successive iterates agree within 0.001 in all variables.

Thus, the equilibrium temperatures at the interior points are (to an accuracy of 0.001) $t_1 = 74.108$, $t_2 = 46.430$, and $t_3 = 61.607$. (Check these calculations.)

By using a finer grid (with more interior points), we can get as precise information as we like about the equilibrium temperatures at various points on the plate.

Table 2.18

n	0	1	2	3	...	7	8
t_1	0	62.500	69.531	73.535	...	74.107	74.107
t_2	0	28.125	44.141	46.143	...	46.429	46.429
t_3	0	57.031	61.035	61.536	...	61.607	61.607

Exercises 2.5

CAS

In Exercises 1–6, apply Jacobi's method to the given system. Take the zero vector as the initial approximation and work with four-significant-digit accuracy until two successive iterates agree within 0.001 in each variable. In each case, compare your answer with the exact solution found using any direct method you like.

1. $\begin{aligned} 7x_1 - x_2 &= 6 \\ x_1 - 5x_2 &= -4 \end{aligned}$

2. $\begin{aligned} 2x_1 + x_2 &= 5 \\ x_1 - x_2 &= 1 \end{aligned}$

3. $\begin{aligned} 4.5x_1 - 0.5x_2 &= 1 \\ x_1 - 3.5x_2 &= -1 \end{aligned}$

4. $\begin{aligned} 20x_1 + x_2 - x_3 &= 17 \\ x_1 - 10x_2 + x_3 &= 13 \\ -x_1 + x_2 + 10x_3 &= 18 \end{aligned}$

5. $\begin{aligned} 3x_1 + x_2 &= 1 \\ x_1 + 4x_2 + x_3 &= 1 \\ x_2 + 3x_3 &= 1 \end{aligned}$

6. $\begin{aligned} 3x_1 - x_2 &= 1 \\ -x_1 + 3x_2 - x_3 &= 0 \\ -x_2 + 3x_3 - x_4 &= 1 \\ -x_3 + 3x_4 &= 1 \end{aligned}$

In Exercises 7–12, repeat the given exercise using the Gauss-Seidel method. Take the zero vector as the initial approximation and work with four-significant-digit accuracy until two successive iterates agree within 0.001 in each variable. Compare the number of iterations required by the Jacobi and Gauss-Seidel methods to reach such an approximate solution.

7. Exercise 1

8. Exercise 2

9. Exercise 3

10. Exercise 4

11. Exercise 5

12. Exercise 6

In Exercises 13 and 14, draw diagrams to illustrate the convergence of the Gauss-Seidel method with the given system.

13. The system in Exercise 1

14. The system in Exercise 2

In Exercises 15 and 16, compute the first four iterates, using the zero vector as the initial approximation, to show that the Gauss-Seidel method diverges. Then show that the equations can be rearranged to give a strictly diagonally dominant coefficient matrix, and apply the Gauss-Seidel method to obtain an approximate solution that is accurate to within 0.001.

15. $\begin{aligned} x_1 - 2x_2 &= 3 \\ 3x_1 + 2x_2 &= 1 \end{aligned}$

16. $\begin{aligned} x_1 - 4x_2 + 2x_3 &= 2 \\ 2x_2 + 4x_3 &= 1 \\ 6x_1 - x_2 - 2x_3 &= 1 \end{aligned}$

17. Draw a diagram to illustrate the divergence of the Gauss-Seidel method in Exercise 15.

In Exercises 18 and 19, the coefficient matrix is not strictly diagonally dominant, nor can the equations be rearranged to make it so. However, both the Jacobi and the Gauss-Seidel method converge anyway. Demonstrate that this is true of the Gauss-Seidel method, starting with the zero vector as the initial approximation and obtaining a solution that is accurate to within 0.01.

18. $\begin{aligned} -4x_1 + 5x_2 &= 14 \\ x_1 - 3x_2 &= -7 \end{aligned}$

19. $\begin{aligned} 5x_1 - 2x_2 + 3x_3 &= -8 \\ x_1 + 4x_2 - 4x_3 &= 102 \\ -2x_1 - 2x_2 + 4x_3 &= -90 \end{aligned}$

20. Continue performing iterations in Exercise 18 to obtain a solution that is accurate to within 0.001.

21. Continue performing iterations in Exercise 19 to obtain a solution that is accurate to within 0.001.

In Exercises 22–24, the metal plate has the constant temperatures shown on its boundaries. Find the equilibrium temperature at each of the indicated interior points by setting up a system of linear equations and applying either the Jacobi or the Gauss-Seidel method. Obtain a solution that is accurate to within 0.001.

22.

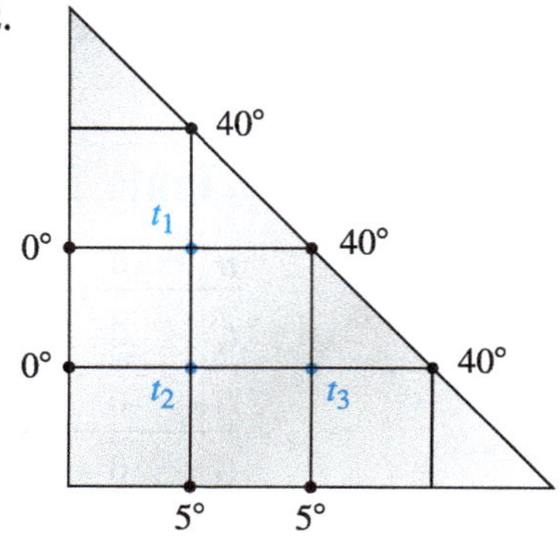

23.

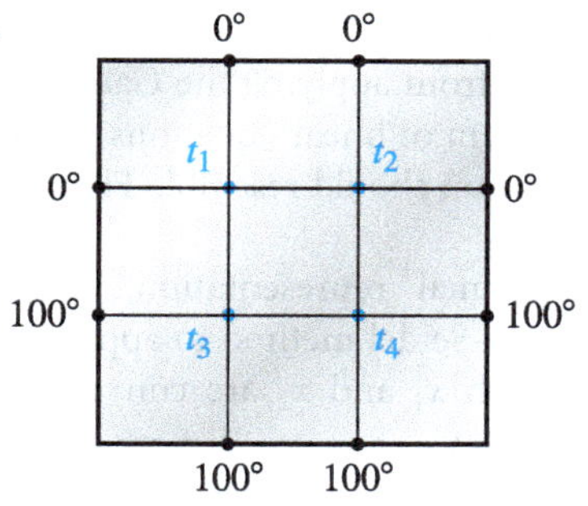

24.

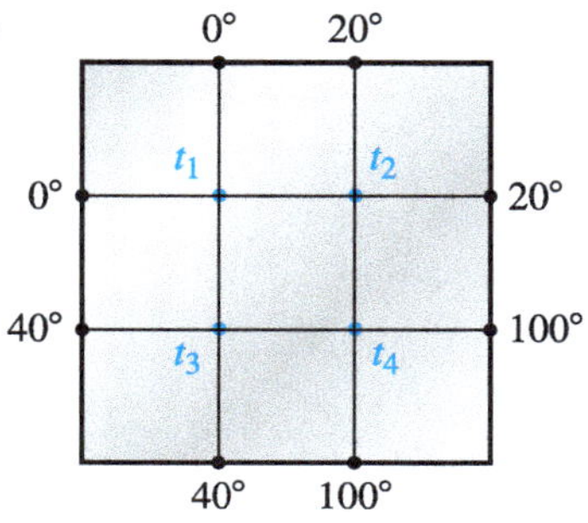

In Exercises 25 and 26, we refine the grids used in Exercises 22 and 24 to obtain more accurate information about the equilibrium temperatures at interior points of the plates. Obtain solutions that are accurate to within 0.001, using either the Jacobi or the Gauss-Seidel method.

25.

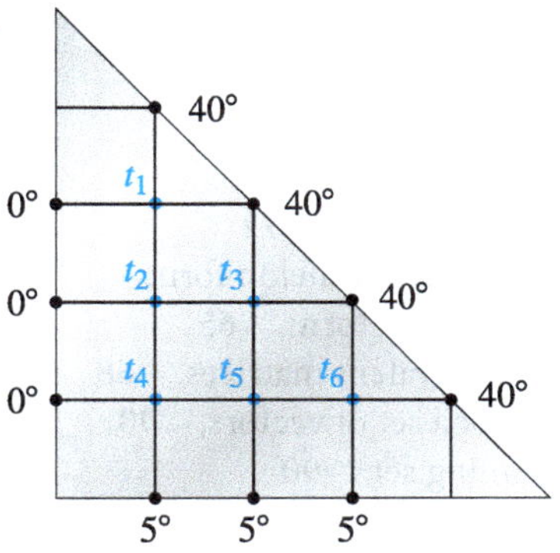

26.

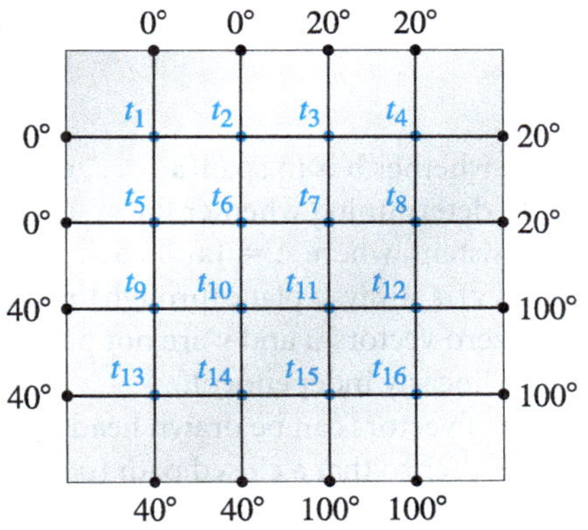

Exercises 27 and 28 demonstrate that sometimes, if we are lucky, the form of an iterative problem may allow us to use a little insight to obtain an exact solution.

27. A narrow strip of paper 1 unit long is placed along a number line so that its ends are at 0 and 1. The paper is folded in half, right end over left, so that its ends are now at 0 and $\frac{1}{2}$. Next, it is folded in half again, this time left end over right, so that its ends are at $\frac{1}{4}$ and $\frac{1}{2}$. Figure 2.32 shows this process. We continue folding the paper in half, alternating right-over-left and left-over-right. If we could continue indefinitely, it is clear that the ends of the paper would converge to a point. It is this point that we want to find.

(a) Let x_1 correspond to the left-hand end of the paper and x_2 to the right-hand end. Make a table with the first six values of $[x_1, x_2]$ and plot the corresponding points on x_1, x_2 coordinate axes.

(b) Find two linear equations of the form $x_2 = ax_1 + b$ and $x_1 = cx_2 + d$ that determine the new values of the endpoints at each iteration. Draw the corresponding lines on your coordinate axes and show that this diagram would result from applying the Gauss-Seidel method to the system of linear equations you have found. (Your diagram should resemble Figure 2.27 on page 126.)

(c) Switching to decimal representation, continue applying the Gauss-Seidel method to approximate the point to which the ends of the paper are converging to within 0.001 accuracy.

(d) Solve the system of equations exactly and compare your answers.

28. An ant is standing on a number line at point A. It walks halfway to point B and turns around. Then it walks halfway back to point A, turns around again, and walks halfway to point B. It continues to do this indefinitely. Let point A be at 0 and point B be at 1. The ant's walk is made up of a sequence of overlapping line segments. Let x_1 record the positions of the left-hand endpoints of these segments and x_2 their right-hand endpoints. (Thus, we begin with $x_1 = 0$ and $x_2 = \frac{1}{2}$. Then we have $x_1 = \frac{1}{4}$ and $x_2 = \frac{1}{2}$, and so on.) Figure 2.33 shows the start of the ant's walk.

(a) Make a table with the first six values of $[x_1, x_2]$ and plot the corresponding points on x_1, x_2 coordinate axes.

(b) Find two linear equations of the form $x_2 = ax_1 + b$ and $x_1 = cx_2 + d$ that determine the new values of the endpoints at each iteration. Draw the corresponding

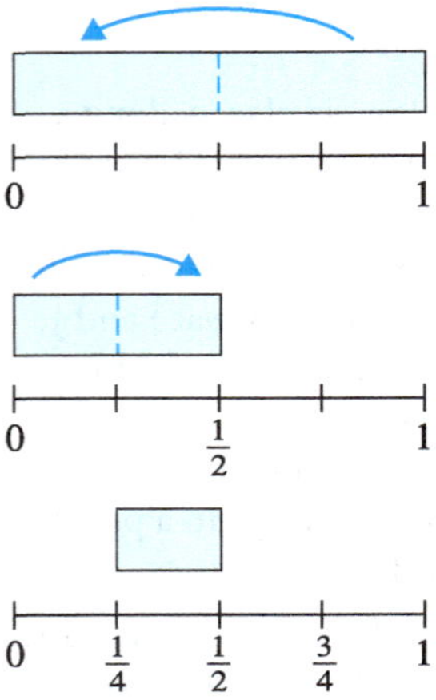

Figure 2.32
Folding a strip of paper

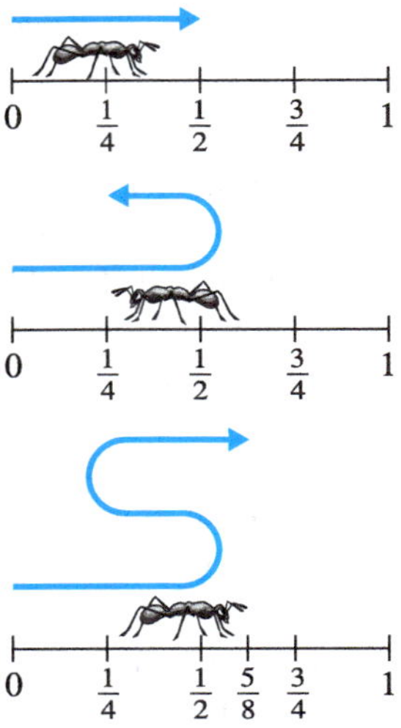

Figure 2.33
The ant's walk

lines on your coordinate axes and show that this diagram would result from applying the Gauss-Seidel method to the system of linear equations you have found. (Your diagram should resemble Figure 2.27 on page 126.)

(c) Switching to decimal representation, continue applying the Gauss-Seidel method to approximate the values to which x_1 and x_2 are converging to within 0.001 accuracy.

(d) Solve the system of equations exactly and compare your answers. Interpret your results.

Chapter Review

Key Definitions and Concepts

Review Questions

1. Mark each of the following statements true or false:

(a) Every system of linear equations has a solution.

(b) Every homogeneous system of linear equations has a solution.

(c) If a system of linear equations has more variables than equations, then it has infinitely many solutions.

(d) If a system of linear equations has more equations than variables, then it has no solution.

(e) Determining whether $\mathbf{b}$ is in span$(\mathbf{a}_1, \ldots, \mathbf{a}_n)$ is equivalent to determining whether the system $[A \mid \mathbf{b}]$ is consistent, where $A = [\mathbf{a}_1 \cdots \mathbf{a}_n]$.

(f) In $\mathbb{R}^3$, span$(\mathbf{u}, \mathbf{v})$ is always a plane through the origin.

(g) In $\mathbb{R}^3$, if nonzero vectors $\mathbf{u}$ and $\mathbf{v}$ are not parallel, then they are linearly independent.

(h) In $\mathbb{R}^3$, if a set of vectors can be drawn head to tail, one after the other so that a closed path (polygon) is formed, then the vectors are linearly dependent.

(i) If a set of vectors has the property that no two vectors in the set are scalar multiples of one another, then the set of vectors is linearly independent.

(j) If there are more vectors in a set of vectors than the number of entries in each vector, then the set of vectors is linearly dependent.

2. Find the rank of the matrix $\begin{bmatrix} 1 & -2 & 0 & 3 & 2 \\ 3 & -1 & 1 & 3 & 4 \\ 3 & 4 & 2 & -3 & 2 \\ 0 & -5 & -1 & 6 & 2 \end{bmatrix}$.

3. Solve the linear system

$$\begin{aligned} x + \ \ y - 2z &= 4 \\ x + 3y - \ \ z &= 7 \\ 2x + \ \ y - 5z &= 7 \end{aligned}$$

4. Solve the linear system

$$\begin{aligned} 3w + 8x - 18y + z &= 35 \\ w + 2x - \ \ 4y \qquad &= 11 \\ w + 3x - \ \ 7y + z &= 10 \end{aligned}$$

5. Solve the linear system

$$\begin{aligned} 2x + 3y &= 4 \\ x + 2y &= 3 \end{aligned}$$

over $\mathbb{Z}_7$.

6. Solve the linear system

$$\begin{aligned} 3x + 2y &= 1 \\ x + 4y &= 2 \end{aligned}$$

over $\mathbb{Z}_5$.

7. For what value(s) of k is the linear system with augmented matrix $\left[\begin{array}{cc|c} k & 2 & 1 \\ 1 & 2k & 1 \end{array}\right]$ inconsistent?

8. Find parametric equations for the line of intersection of the planes $x + 2y + 3z = 4$ and $5x + 6y + 7z = 8$.

9. Find the point of intersection of the following lines, if it exists.

$$\begin{bmatrix} x \\ y \\ z \end{bmatrix} = \begin{bmatrix} 1 \\ 2 \\ 3 \end{bmatrix} + s\begin{bmatrix} 1 \\ -1 \\ 2 \end{bmatrix} \quad \text{and} \quad \begin{bmatrix} x \\ y \\ z \end{bmatrix} = \begin{bmatrix} 5 \\ -2 \\ -4 \end{bmatrix} + t\begin{bmatrix} -1 \\ 1 \\ 1 \end{bmatrix}$$

10. Determine whether $\begin{bmatrix} 3 \\ 5 \\ -1 \end{bmatrix}$ is in the span of $\begin{bmatrix} 1 \\ 1 \\ 3 \end{bmatrix}$ and $\begin{bmatrix} 1 \\ 2 \\ -2 \end{bmatrix}$.

11. Find the general equation of the plane spanned by $\begin{bmatrix} 1 \\ 1 \\ 1 \end{bmatrix}$ and $\begin{bmatrix} 3 \\ 2 \\ 1 \end{bmatrix}$.

12. Determine whether $\begin{bmatrix} 2 \\ 1 \\ -3 \end{bmatrix}, \begin{bmatrix} 1 \\ -1 \\ -2 \end{bmatrix}, \begin{bmatrix} 3 \\ 9 \\ -2 \end{bmatrix}$ are linearly independent.

13. Determine whether $\mathbb{R}^3 = \text{span}(\mathbf{u}, \mathbf{v}, \mathbf{w})$ if:

(a) $\mathbf{u} = \begin{bmatrix} 1 \\ 1 \\ 0 \end{bmatrix}, \mathbf{v} = \begin{bmatrix} 1 \\ 0 \\ 1 \end{bmatrix}, \mathbf{w} = \begin{bmatrix} 0 \\ 1 \\ 1 \end{bmatrix}$

(b) $\mathbf{u} = \begin{bmatrix} 1 \\ -1 \\ 0 \end{bmatrix}, \mathbf{v} = \begin{bmatrix} -1 \\ 0 \\ 1 \end{bmatrix}, \mathbf{w} = \begin{bmatrix} 0 \\ -1 \\ 1 \end{bmatrix}$

14. Let $\mathbf{a}_1, \mathbf{a}_2, \mathbf{a}_3$ be linearly independent vectors in $\mathbb{R}^3$, and let $A = [\mathbf{a}_1 \ \mathbf{a}_2 \ \mathbf{a}_3]$. Which of the following statements are true?

(a) The reduced row echelon form of A is I_3.

(b) The rank of A is 3.

(c) The system $[A \mid \mathbf{b}]$ has a unique solution for any vector $\mathbf{b}$ in $\mathbb{R}^3$.

(d) (a), (b), and (c) are all true.

(e) (a) and (b) are both true, but not (c).

15. Let $\mathbf{a}_1, \mathbf{a}_2, \mathbf{a}_3$ be linearly dependent vectors in $\mathbb{R}^3$, not all zero, and let $A = [\mathbf{a}_1 \ \mathbf{a}_2 \ \mathbf{a}_3]$. What are the possible values of the rank of A?

16. What is the maximum rank of a 5×3 matrix? What is the minimum rank of a 5×3 matrix?

17. Show that if $\mathbf{u}$ and $\mathbf{v}$ are linearly independent vectors, then so are $\mathbf{u} + \mathbf{v}$ and $\mathbf{u} - \mathbf{v}$.

18. Show that $\text{span}(\mathbf{u}, \mathbf{v}) = \text{span}(\mathbf{u}, \mathbf{u} + \mathbf{v})$ for any vectors $\mathbf{u}$ and $\mathbf{v}$.

19. In order for a linear system with augmented matrix $[A \mid \mathbf{b}]$ to be consistent, what must be true about the ranks of A and $[A \mid \mathbf{b}]$?

20. Are the matrices $\begin{bmatrix} 1 & 1 & 1 \\ 2 & 3 & -1 \\ -1 & 4 & 1 \end{bmatrix}$ and $\begin{bmatrix} 1 & 0 & -1 \\ 1 & 1 & 1 \\ 0 & 1 & 3 \end{bmatrix}$ row equivalent? Why or why not?

3 Matrices

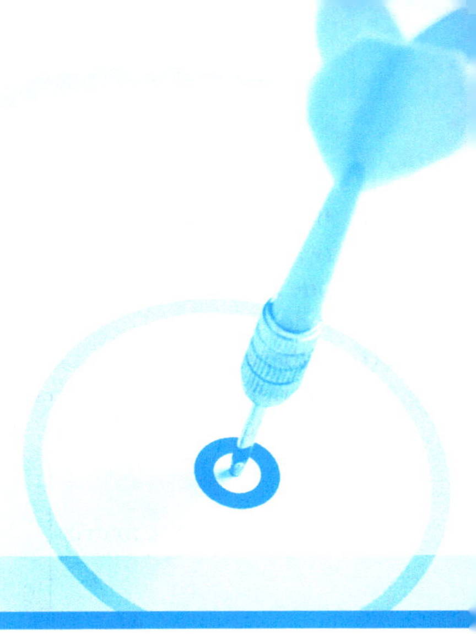

We [Halmos and Kaplansky] share a philosophy about linear algebra: we think basis-free, we write basis-free, but when the chips are down we close the office door and compute with matrices like fury.

—Irving Kaplansky
In *Paul Halmos: Celebrating 50 Years of Mathematics*
J. H. Ewing and F. W. Gehring, eds. Springer-Verlag, 1991, p. 88

3.0 Introduction: Matrices in Action

In this chapter, we will study matrices in their own right. We have already used matrices—in the form of augmented matrices—to record information about and to help streamline calculations involving systems of linear equations. Now you will see that matrices have algebraic properties of their own, which enable us to calculate with them, subject to the rules of matrix algebra. Furthermore, you will observe that matrices are not static objects, recording information and data; rather, they represent certain types of functions that "act" on vectors, transforming them into other vectors. These "matrix transformations" will begin to play a key role in our study of linear algebra and will shed new light on what you have already learned about vectors and systems of linear equations. Furthermore, matrices arise in many forms other than augmented matrices; we will explore some of the many applications of matrices at the end of this chapter.

In this section, we will consider a few simple examples to illustrate how matrices can transform vectors. In the process, you will get your first glimpse of "matrix arithmetic."

Consider the equations

$$\begin{aligned} y_1 &= x_1 + 2x_2 \\ y_2 &= \phantom{x_1 + {}} 3x_2 \end{aligned} \qquad (1)$$

We can view these equations as describing a transformation of the vector $\mathbf{x} = \begin{bmatrix} x_1 \\ x_2 \end{bmatrix}$ into the vector $\mathbf{y} = \begin{bmatrix} y_1 \\ y_2 \end{bmatrix}$. If we denote the matrix of coefficients of the right-hand side by F, then $F = \begin{bmatrix} 1 & 2 \\ 0 & 3 \end{bmatrix}$, and we can rewrite the transformation as

$$\begin{bmatrix} y_1 \\ y_2 \end{bmatrix} = \begin{bmatrix} 1 & 2 \\ 0 & 3 \end{bmatrix} \begin{bmatrix} x_1 \\ x_2 \end{bmatrix}$$

or, more succinctly, $\mathbf{y} = F\mathbf{x}$. [Think of this expression as analogous to the functional notation $y = f(x)$ you are used to: $\mathbf{x}$ is the independent "variable" here, $\mathbf{y}$ is the dependent "variable," and F is the name of the "function."]

Thus, if $\mathbf{x} = \begin{bmatrix} -2 \\ 1 \end{bmatrix}$, then the Equations (1) give

$$\begin{aligned} y_1 &= -2 + 2 \cdot 1 = 0 \\ y_2 &= \phantom{-2 + {}} 3 \cdot 1 = 3 \end{aligned} \quad \text{or} \quad \mathbf{y} = \begin{bmatrix} 0 \\ 3 \end{bmatrix}$$

We can write this expression as $\begin{bmatrix} 0 \\ 3 \end{bmatrix} = \begin{bmatrix} 1 & 2 \\ 0 & 3 \end{bmatrix} \begin{bmatrix} -2 \\ 1 \end{bmatrix}$.

Problem 1 Compute $F\mathbf{x}$ for the following vectors $\mathbf{x}$:

(a) $\mathbf{x} = \begin{bmatrix} 1 \\ 1 \end{bmatrix}$ (b) $\mathbf{x} = \begin{bmatrix} 1 \\ -1 \end{bmatrix}$ (c) $\mathbf{x} = \begin{bmatrix} -1 \\ -1 \end{bmatrix}$ (d) $\mathbf{x} = \begin{bmatrix} -1 \\ 1 \end{bmatrix}$

Problem 2 The heads of the four vectors $\mathbf{x}$ in Problem 1 locate the four corners of a square in the x_1x_2 plane. Draw this square and label its corners A, B, C, and D, corresponding to parts (a), (b), (c), and (d) of Problem 1.

On separate coordinate axes (labeled y_1 and y_2), draw the four points determined by $F\mathbf{x}$ in Problem 1. Label these points A', B', C', and D'. Let's make the (reasonable) assumption that the line segment $\overline{AB}$ is transformed into the line segment $\overline{A'B'}$, and likewise for the other three sides of the square $ABCD$. What geometric figure is represented by $A'B'C'D'$?

Problem 3 The center of square $ABCD$ is the origin $\mathbf{0} = \begin{bmatrix} 0 \\ 0 \end{bmatrix}$. What is the center of $A'B'C'D'$? What algebraic calculation confirms this?

Now consider the equations

$$\begin{aligned} z_1 &= y_1 - y_2 \\ z_2 &= -2y_1 \end{aligned} \tag{2}$$

that transform a vector $\mathbf{y} = \begin{bmatrix} y_1 \\ y_2 \end{bmatrix}$ into the vector $\mathbf{z} = \begin{bmatrix} z_1 \\ z_2 \end{bmatrix}$. We can abbreviate this transformation as $\mathbf{z} = G\mathbf{y}$, where

$$G = \begin{bmatrix} 1 & -1 \\ -2 & 0 \end{bmatrix}$$

Problem 4 We are going to find out how G transforms the figure $A'B'C'D'$. Compute $G\mathbf{y}$ for each of the four vectors $\mathbf{y}$ that you computed in Problem 1. [That is, compute $\mathbf{z} = G(F\mathbf{x})$. You may recognize this expression as being analogous to the composition of functions with which you are familiar.] Call the corresponding points A'', B'', C'', and D'', and sketch the figure $A''B''C''D''$ on z_1z_2 coordinate axes.

Problem 5 By substituting Equations (1) into Equations (2), obtain equations for z_1 and z_2 in terms of x_1 and x_2. If we denote the matrix of these equations by H, then we have $\mathbf{z} = H\mathbf{x}$. Since we also have $\mathbf{z} = GF\mathbf{x}$, it is reasonable to write

$$H = GF$$

Can you see how the entries of H are related to the entries of F and G?

Problem 6 Let's do the above process the other way around: First transform the square $ABCD$, using G, to obtain figure $A^*B^*C^*D^*$. Then transform the resulting figure, using F, to obtain $A^{**}B^{**}C^{**}D^{**}$. [*Note:* Don't worry about the "variables" $\mathbf{x}$,

y, and **z** here. Simply substitute the coordinates of A, B, C, and D into Equations (2) and then substitute the results into Equations (1).] Are $A^{**}B^{**}C^{**}D^{**}$ and $A''B''C''D''$ the same? What does this tell you about the order in which we perform the transformations F and G?

Problem 7 Repeat Problem 5 with general matrices

$$F = \begin{bmatrix} f_{11} & f_{12} \\ f_{21} & f_{22} \end{bmatrix}, \quad G = \begin{bmatrix} g_{11} & g_{12} \\ g_{21} & g_{22} \end{bmatrix}, \quad \text{and} \quad H = \begin{bmatrix} h_{11} & h_{12} \\ h_{21} & h_{22} \end{bmatrix}$$

That is, if Equations (1) and Equations (2) have coefficients as specified by F and G, find the entries of H in terms of the entries of F and G. The result will be a formula for the "product" $H = GF$.

Problem 8 Repeat Problems 1–6 with the following matrices. (Your formula from Problem 7 may help to speed up the algebraic calculations.) Note any similarities or differences that you think are significant.

$$\text{(a)}\ F = \begin{bmatrix} 0 & -1 \\ 1 & 0 \end{bmatrix}, G = \begin{bmatrix} 2 & 0 \\ 0 & 3 \end{bmatrix} \qquad \text{(b)}\ F = \begin{bmatrix} 1 & 1 \\ 1 & 2 \end{bmatrix}, G = \begin{bmatrix} 2 & 1 \\ 1 & 1 \end{bmatrix}$$

$$\text{(c)}\ F = \begin{bmatrix} 1 & 1 \\ 1 & 2 \end{bmatrix}, G = \begin{bmatrix} 2 & -1 \\ -1 & 1 \end{bmatrix} \qquad \text{(d)}\ F = \begin{bmatrix} 1 & -2 \\ -2 & 4 \end{bmatrix}, G = \begin{bmatrix} 2 & 1 \\ 1 & 1 \end{bmatrix}$$

3.1 Matrix Operations

Although we have already encountered matrices, we begin by stating a formal definition and recording some facts for future reference.

Definition A ***matrix*** is a rectangular array of numbers called the ***entries,*** or ***elements,*** of the matrix.

The following are all examples of matrices:

Although numbers will usually be chosen from the set $\mathbb{R}$ of real numbers, they may also be taken from the set $\mathbb{C}$ of complex numbers or from $\mathbb{Z}_p$, where p is prime.

$$\begin{bmatrix} 1 & 2 \\ 0 & 3 \end{bmatrix}, \quad \begin{bmatrix} \sqrt{5} & -1 & 0 \\ 2 & \pi & \frac{1}{2} \end{bmatrix}, \quad \begin{bmatrix} 2 \\ 4 \\ 17 \end{bmatrix}, \quad \begin{bmatrix} 1 & 1 & 1 & 1 \end{bmatrix}, \quad \begin{bmatrix} 5.1 & 1.2 & -1 \\ 6.9 & 0 & 4.4 \\ -7.3 & 9 & 8.5 \end{bmatrix}, \quad [7]$$

The ***size*** of a matrix is a description of the numbers of rows and columns it has. A matrix is called $m \times n$ (pronounced "m by n") if it has m rows and n columns. Thus, the examples above are matrices of sizes 2×2, 2×3, 3×1, 1×4, 3×3, and 1×1, respectively. A $1 \times m$ matrix is called a ***row matrix*** (or ***row vector***), and an $n \times 1$ matrix is called a ***column matrix*** (or ***column vector***).

Technically, there is a distinction between row/column matrices and vectors, but we will not belabor this distinction. We *will,* however, distinguish between *row* matrices/vectors and *column* matrices/vectors. This distinction is important—at the very least—for algebraic computations, as we will demonstrate.

We use *double-subscript* notation to refer to the entries of a matrix A. The entry of A in row i and column j is denoted by a_{ij}. Thus, if

$$A = \begin{bmatrix} 3 & 9 & -1 \\ 0 & 5 & 4 \end{bmatrix}$$

then $a_{13} = -1$ and $a_{22} = 5$. (The notation A_{ij} is sometimes used interchangeably with a_{ij}.) We can therefore compactly denote a matrix A by $[a_{ij}]$ (or $[a_{ij}]_{m \times n}$ if it is important to specify the size of A, although the size will usually be clear from the context).

With this notation, a general $m \times n$ matrix A has the form

$$A = \begin{bmatrix} a_{11} & a_{12} & \cdots & a_{1n} \\ a_{21} & a_{22} & \cdots & a_{2n} \\ \vdots & \vdots & \ddots & \vdots \\ a_{m1} & a_{m2} & \cdots & a_{mn} \end{bmatrix}$$

If the columns of A are the vectors $\mathbf{a}_1, \mathbf{a}_2, \ldots, \mathbf{a}_n$, then we may represent A as

$$A = [\mathbf{a}_1 \quad \mathbf{a}_2 \quad \cdots \quad \mathbf{a}_n]$$

If the rows of A are $\mathbf{A}_1, \mathbf{A}_2, \ldots, \mathbf{A}_m$, then we may represent A as

$$A = \begin{bmatrix} \mathbf{A}_1 \\ \mathbf{A}_2 \\ \vdots \\ \mathbf{A}_m \end{bmatrix}$$

The ***diagonal entries*** of A are $a_{11}, a_{22}, a_{33}, \ldots$, and if $m = n$ (that is, if A has the same number of rows as columns), then A is called a ***square matrix.*** A square matrix whose nondiagonal entries are all zero is called a ***diagonal matrix.*** A diagonal matrix all of whose diagonal entries are the same is called a ***scalar matrix.*** If the scalar on the diagonal is 1, the scalar matrix is called an ***identity matrix.***

For example, let

$$A = \begin{bmatrix} 2 & 5 & 0 \\ -1 & 4 & 1 \end{bmatrix}, \quad B = \begin{bmatrix} 3 & 1 \\ 4 & 5 \end{bmatrix}, \quad C = \begin{bmatrix} 3 & 0 & 0 \\ 0 & 6 & 0 \\ 0 & 0 & 2 \end{bmatrix}, \quad \text{and} \quad D = \begin{bmatrix} 1 & 0 & 0 \\ 0 & 1 & 0 \\ 0 & 0 & 1 \end{bmatrix}$$

The diagonal entries of A are 2 and 4, but A is not square; B is a square matrix of size 2×2 with diagonal entries 3 and 5; C is a diagonal matrix; D is a 3×3 identity matrix. The $n \times n$ identity matrix is denoted by I_n (or simply I if its size is understood).

Since we can view matrices as generalizations of vectors (and, indeed, matrices can and should be thought of as being made up of both row and column vectors), many of the conventions and operations for vectors carry through (in an obvious way) to matrices.

Two matrices are ***equal*** if they have the same size and if their corresponding entries are equal. Thus, if $A = [a_{ij}]_{m\times n}$ and $B = [b_{ij}]_{r\times s}$, then $A = B$ if and only if $m = r$ and $n = s$ and $a_{ij} = b_{ij}$ for all i and j.

Example 3.1

Consider the matrices

$$A = \begin{bmatrix} a & b \\ c & d \end{bmatrix}, \quad B = \begin{bmatrix} 2 & 0 \\ 5 & 3 \end{bmatrix}, \quad \text{and} \quad C = \begin{bmatrix} 2 & 0 & x \\ 5 & 3 & y \end{bmatrix}$$

Neither A nor B can be equal to C (no matter what the values of x and y), since A and B are 2×2 matrices and C is 2×3. However, $A = B$ if and only if $a = 2$, $b = 0$, $c = 5$, and $d = 3$.

Example 3.2

Consider the matrices

$$R = [1 \quad 4 \quad 3] \quad \text{and} \quad C = \begin{bmatrix} 1 \\ 4 \\ 3 \end{bmatrix}$$

Despite the fact that R and C have the same entries in the same order, $R \neq C$ since R is 1×3 and C is 3×1. (If we read R and C aloud, they both sound the same: "one, four, three.") Thus, our distinction between row matrices/vectors and column matrices/vectors is an important one.

Matrix Addition and Scalar Multiplication

Generalizing from vector addition, we define matrix addition *componentwise*. If $A = [a_{ij}]$ and $B = [b_{ij}]$ are $m \times n$ matrices, their ***sum*** $A + B$ is the $m \times n$ matrix obtained by adding the corresponding entries. Thus,

$$A + B = [a_{ij} + b_{ij}]$$

[We could equally well have defined $A + B$ in terms of vector addition by specifying that each column (or row) of $A + B$ is the sum of the corresponding columns (or rows) of A and B.] If A and B are not the same size, then $A + B$ is not defined.

Example 3.3

Let

$$A = \begin{bmatrix} 1 & 4 & 0 \\ -2 & 6 & 5 \end{bmatrix}, \quad B = \begin{bmatrix} -3 & 1 & -1 \\ 3 & 0 & 2 \end{bmatrix}, \quad \text{and} \quad C = \begin{bmatrix} 4 & 3 \\ 2 & 1 \end{bmatrix}$$

Then

$$A + B = \begin{bmatrix} -2 & 5 & -1 \\ 1 & 6 & 7 \end{bmatrix}$$

but neither $A + C$ nor $B + C$ is defined.

The componentwise definition of scalar multiplication will come as no surprise. If A is an $m \times n$ matrix and c is a scalar, then the ***scalar multiple*** cA is the $m \times n$ matrix obtained by multiplying each entry of A by c. More formally, we have

$$cA = c\,[a_{ij}] = [ca_{ij}]$$

[In terms of vectors, we could equivalently stipulate that each column (or row) of cA is c times the corresponding column (or row) of A.]

Example 3.4

For matrix A in Example 3.3,

$$2A = \begin{bmatrix} 2 & 8 & 0 \\ -4 & 12 & 10 \end{bmatrix}, \quad \tfrac{1}{2}A = \begin{bmatrix} \frac{1}{2} & 2 & 0 \\ -1 & 3 & \frac{5}{2} \end{bmatrix}, \quad \text{and} \quad (-1)A = \begin{bmatrix} -1 & -4 & 0 \\ 2 & -6 & -5 \end{bmatrix}$$

The matrix $(-1)A$ is written as $-A$ and called the ***negative*** of A. As with vectors, we can use this fact to define the ***difference*** of two matrices: If A and B are the same size, then

$$A - B = A + (-B)$$

Example 3.5

For matrices A and B in Example 3.3,

$$A - B = \begin{bmatrix} 1 & 4 & 0 \\ -2 & 6 & 5 \end{bmatrix} - \begin{bmatrix} -3 & 1 & -1 \\ 3 & 0 & 2 \end{bmatrix} = \begin{bmatrix} 4 & 3 & 1 \\ -5 & 6 & 3 \end{bmatrix}$$

A matrix all of whose entries are zero is called a ***zero matrix*** and denoted by O (or $O_{m \times n}$ if it is important to specify its size). It should be clear that if A is any matrix and O is the zero matrix of the same size, then

$$A + O = A = O + A$$

and

$$A - A = O = -A + A$$

Matrix Multiplication

Mathematicians are sometimes like Lewis Carroll's Humpty Dumpty: "When *I* use a word," Humpty Dumpty said, "it means just what I choose it to mean—neither more nor less" (from *Through the Looking Glass*).

The Introduction in Section 3.0 suggested that there is a "product" of matrices that is analogous to the composition of functions. We now make this notion more precise. The definition we are about to give generalizes what you should have discovered in Problems 5 and 7 in Section 3.0. Unlike the definitions of matrix addition and scalar multiplication, the definition of the product of two matrices is *not* a componentwise definition. Of course, there is nothing to stop us from defining a product of matrices in a componentwise fashion; unfortunately such a definition has few applications and is not as "natural" as the one we now give.

Definition If A is an $m \times n$ matrix and B is an $n \times r$ matrix, then the ***product*** $C = AB$ is an $m \times r$ matrix. The (i, j) entry of the product is computed as follows:

$$c_{ij} = a_{i1}b_{1j} + a_{i2}b_{2j} + \cdots + a_{in}b_{nj}$$

Remarks

- Notice that A and B need not be the same size. However, the number of *columns* of A must be the same as the number of *rows* of B. If we write the sizes of A, B, and AB in order, we can see at a glance whether this requirement is satisfied. Moreover, we can predict the size of the product before doing any calculations, since the number of *rows* of AB is the same as the number of rows of A, while the number of *columns* of AB is the same as the number of columns of B, as shown below:

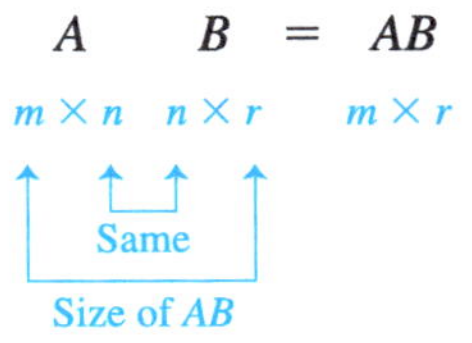

- The formula for the entries of the product looks like a dot product, and indeed it is. It says that the (i, j) entry of the matrix AB is the dot product of the ith row of A and the jth column of B:

$$\begin{bmatrix} a_{11} & a_{12} & \cdots & a_{1n} \\ \vdots & \vdots & & \vdots \\ a_{i1} & a_{i2} & \cdots & a_{in} \\ \vdots & \vdots & & \vdots \\ a_{m1} & a_{m2} & \cdots & a_{mn} \end{bmatrix} \begin{bmatrix} b_{11} & \cdots & b_{1j} & \cdots & b_{1r} \\ b_{21} & \cdots & b_{2j} & \cdots & b_{2r} \\ \vdots & & \vdots & & \vdots \\ b_{n1} & \cdots & b_{nj} & \cdots & b_{nr} \end{bmatrix}$$

Notice that, in the expression $c_{ij} = a_{i1}b_{1j} + a_{i2}b_{2j} + \cdots + a_{in}b_{nj}$, the "outer subscripts" on each ab term in the sum are always i and j whereas the "inner subscripts" always agree and increase from 1 to n. We see this pattern clearly if we write c_{ij} using summation notation:

$$c_{ij} = \sum_{k=1}^{n} a_{ik}b_{kj}$$

Example 3.6

Compute AB if

$$A = \begin{bmatrix} 1 & 3 & -1 \\ -2 & -1 & 1 \end{bmatrix} \quad \text{and} \quad B = \begin{bmatrix} -4 & 0 & 3 & -1 \\ 5 & -2 & -1 & 1 \\ -1 & 2 & 0 & 6 \end{bmatrix}$$

Solution Since A is 2×3 and B is 3×4, the product AB is defined and will be a 2×4 matrix. The first row of the product $C = AB$ is computed by taking the dot product of the first row of A with each of the columns of B in turn. Thus,

$$\begin{aligned} c_{11} &= 1(-4) + 3(5) + (-1)(-1) = 12 \\ c_{12} &= 1(0) + 3(-2) + (-1)(2) = -8 \\ c_{13} &= 1(3) + 3(-1) + (-1)(0) = 0 \\ c_{14} &= 1(-1) + 3(1) + (-1)(6) = -4 \end{aligned}$$

The second row of C is computed by taking the dot product of the second row of A with each of the columns of B in turn:

$$\begin{aligned} c_{21} &= (-2)(-4) + (-1)(5) + (1)(-1) = 2 \\ c_{22} &= (-2)(0) + (-1)(-2) + (1)(2) = 4 \\ c_{23} &= (-2)(3) + (-1)(-1) + (1)(0) = -5 \\ c_{24} &= (-2)(-1) + (-1)(1) + (1)(6) = 7 \end{aligned}$$

Thus, the product matrix is given by

$$AB = \begin{bmatrix} 12 & -8 & 0 & -4 \\ 2 & 4 & -5 & 7 \end{bmatrix}$$

(With a little practice, you should be able to do these calculations mentally without writing out all of the details as we have done here. For more complicated examples, a calculator with matrix capabilities or a computer algebra system is preferable.)

Before we go further, we will consider two examples that justify our chosen definition of matrix multiplication.

Example 3.7

Ann and Bert are planning to go shopping for fruit for the next week. They each want to buy some apples, oranges, and grapefruit, but in differing amounts. Table 3.1 lists what they intend to buy. There are two fruit markets nearby—Sam's and Theo's—and their prices are given in Table 3.2. How much will it cost Ann and Bert to do their shopping at each of the two markets?

Table 3.1

	Apples	Grapefruit	Oranges
Ann	6	3	10
Bert	4	8	5

Table 3.2

	Sam's	Theo's
Apple	\$0.10	\$0.15
Grapefruit	\$0.40	\$0.30
Orange	\$0.10	\$0.20

Solution If Ann shops at Sam's, she will spend

$$6(0.10) + 3(0.40) + 10(0.10) = \$2.80$$

If she shops at Theo's, she will spend

$$6(0.15) + 3(0.30) + 10(0.20) = \$3.80$$

Bert will spend

$$4(0.10) + 8(0.40) + 5(0.10) = \$4.10$$

at Sam's and

$$4(0.15) + 8(0.30) + 5(0.20) = \$4.00$$

at Theo's. (Presumably, Ann will shop at Sam's while Bert goes to Theo's.)

The "dot product form" of these calculations suggests that matrix multiplication is at work here. If we organize the given information into a demand matrix D and a price matrix P, we have

$$D = \begin{bmatrix} 6 & 3 & 10 \\ 4 & 8 & 5 \end{bmatrix} \quad \text{and} \quad P = \begin{bmatrix} 0.10 & 0.15 \\ 0.40 & 0.30 \\ 0.10 & 0.20 \end{bmatrix}$$

The calculations above are equivalent to computing the product

$$DP = \begin{bmatrix} 6 & 3 & 10 \\ 4 & 8 & 5 \end{bmatrix} \begin{bmatrix} 0.10 & 0.15 \\ 0.40 & 0.30 \\ 0.10 & 0.20 \end{bmatrix} = \begin{bmatrix} 2.80 & 3.80 \\ 4.10 & 4.00 \end{bmatrix}$$

Thus, the product matrix DP tells us how much each person's purchases will cost at each store (Table 3.3).

Table 3.3

	Sam's	Theo's
Ann	\$2.80	\$3.80
Bert	\$4.10	\$4.00

Example 3.8 Consider the linear system

$$\begin{aligned} x_1 - 2x_2 + 3x_3 &= 5 \\ -x_1 + 3x_2 + x_3 &= 1 \\ 2x_1 - x_2 + 4x_3 &= 14 \end{aligned} \qquad (1)$$

Observe that the left-hand side arises from the matrix product

$$\begin{bmatrix} 1 & -2 & 3 \\ -1 & 3 & 1 \\ 2 & -1 & 4 \end{bmatrix}\begin{bmatrix} x_1 \\ x_2 \\ x_3 \end{bmatrix}$$

so the system (1) can be written as

$$\begin{bmatrix} 1 & -2 & 3 \\ -1 & 3 & 1 \\ 2 & -1 & 4 \end{bmatrix}\begin{bmatrix} x_1 \\ x_2 \\ x_3 \end{bmatrix} = \begin{bmatrix} 5 \\ 1 \\ 14 \end{bmatrix}$$

or $A\mathbf{x} = \mathbf{b}$, where A is the coefficient matrix, $\mathbf{x}$ is the (column) vector of variables, and $\mathbf{b}$ is the (column) vector of constant terms.

You should have no difficulty seeing that *every* linear system can be written in the form $A\mathbf{x} = \mathbf{b}$. In fact, the notation $[A \mid \mathbf{b}]$ for the augmented matrix of a linear system is just shorthand for the matrix equation $A\mathbf{x} = \mathbf{b}$. This form will prove to be a tremendously useful way of expressing a system of linear equations, and we will exploit it often from here on.

Combining this insight with Theorem 2.4, we see that $A\mathbf{x} = \mathbf{b}$ has a solution if and only if $\mathbf{b}$ is a linear combination of the columns of A.

There is another fact about matrix operations that will also prove to be quite useful: Multiplication of a matrix by a standard unit vector can be used to "pick out" or "reproduce" a column or row of a matrix. Let $A = \begin{bmatrix} 4 & 2 & 1 \\ 0 & 5 & -1 \end{bmatrix}$ and consider the products $A\mathbf{e}_3$ and $\mathbf{e}_2A$, with the unit vectors $\mathbf{e}_3$ and $\mathbf{e}_2$ chosen so that the products make sense. Thus,

$$A\mathbf{e}_3 = \begin{bmatrix} 4 & 2 & 1 \\ 0 & 5 & -1 \end{bmatrix}\begin{bmatrix} 0 \\ 0 \\ 1 \end{bmatrix} = \begin{bmatrix} 1 \\ -1 \end{bmatrix} \quad \text{and} \quad \mathbf{e}_2A = \begin{bmatrix} 0 & 1 \end{bmatrix}\begin{bmatrix} 4 & 2 & 1 \\ 0 & 5 & -1 \end{bmatrix} = \begin{bmatrix} 0 & 5 & -1 \end{bmatrix}$$

Notice that $A\mathbf{e}_3$ gives us the third column of A and $\mathbf{e}_2A$ gives us the second row of A. We record the general result as a theorem.

Theorem 3.1 Let A be an $m \times n$ matrix, $\mathbf{e}_i$ a $1 \times m$ standard unit vector, and $\mathbf{e}_j$ an $n \times 1$ standard unit vector. Then

a. $\mathbf{e}_i A$ is the ith row of A and
b. $A\mathbf{e}_j$ is the jth column of A.

Proof We prove (b) and leave proving (a) as Exercise 41. If $\mathbf{a}_1, \ldots, \mathbf{a}_n$ are the columns of A, then the product $A\mathbf{e}_j$ can be written

$$A\mathbf{e}_j = 0\mathbf{a}_1 + 0\mathbf{a}_2 + \cdots + 1\mathbf{a}_j + \cdots + 0\mathbf{a}_n = \mathbf{a}_j$$

We could also prove (b) by direct calculation:

$$A\mathbf{e}_j = \begin{bmatrix} a_{11} & \cdots & a_{1j} & \cdots & a_{1n} \\ a_{21} & \cdots & a_{2j} & \cdots & a_{2n} \\ \vdots & & \vdots & & \vdots \\ a_{m1} & \cdots & a_{mj} & \cdots & a_{mn} \end{bmatrix} \begin{bmatrix} 0 \\ \vdots \\ 1 \\ \vdots \\ 0 \end{bmatrix} = \begin{bmatrix} a_{1j} \\ a_{2j} \\ \vdots \\ a_{mj} \end{bmatrix}$$

since the 1 in $\mathbf{e}_j$ is the jth entry.

Partitioned Matrices

It will often be convenient to regard a matrix as being composed of a number of smaller ***submatrices.*** By introducing vertical and horizontal lines into a matrix, we can ***partition*** it into ***blocks.*** There is a natural way to partition many matrices, particularly those arising in certain applications. For example, consider the matrix

$$A = \begin{bmatrix} 1 & 0 & 0 & 2 & -1 \\ 0 & 1 & 0 & 1 & 3 \\ 0 & 0 & 1 & 4 & 0 \\ 0 & 0 & 0 & 1 & 7 \\ 0 & 0 & 0 & 7 & 2 \end{bmatrix}$$

It seems natural to partition A as

$$\left[\begin{array}{ccc:cc} 1 & 0 & 0 & 2 & -1 \\ 0 & 1 & 0 & 1 & 3 \\ 0 & 0 & 1 & 4 & 0 \\ \hdashline 0 & 0 & 0 & 1 & 7 \\ 0 & 0 & 0 & 7 & 2 \end{array}\right] = \begin{bmatrix} I & B \\ O & C \end{bmatrix}$$

where I is the 3×3 identity matrix, B is 3×2, O is the 2×3 zero matrix, and C is 2×2. In this way, we can view A as a 2×2 matrix whose entries are themselves matrices.

When matrices are being multiplied, there is often an advantage to be gained by viewing them as partitioned matrices. Not only does this frequently reveal underlying structures, but it often speeds up computation, especially when the matrices are large and have many blocks of zeros. It turns out that the multiplication of partitioned matrices is just like ordinary matrix multiplication.

We begin by considering some special cases of partitioned matrices. Each gives rise to a different way of viewing the product of two matrices.

Suppose A is $m \times n$ and B is $n \times r$, so the product AB exists. If we partition B in terms of its column vectors, as $B = [\mathbf{b}_1 \,\vdots\, \mathbf{b}_2 \,\vdots\, \cdots \,\vdots\, \mathbf{b}_r]$, then

$$AB = A[\mathbf{b}_1 \,\vdots\, \mathbf{b}_2 \,\vdots\, \cdots \,\vdots\, \mathbf{b}_r] = [A\mathbf{b}_1 \,\vdots\, A\mathbf{b}_2 \,\vdots\, \cdots \,\vdots\, A\mathbf{b}_r]$$

This result is an immediate consequence of the definition of matrix multiplication. The form on the right is called the ***matrix-column representation*** of the product.

Example 3.9

If

$$A = \begin{bmatrix} 1 & 3 & 2 \\ 0 & -1 & 1 \end{bmatrix} \quad \text{and} \quad B = \begin{bmatrix} 4 & -1 \\ 1 & 2 \\ 3 & 0 \end{bmatrix}$$

then

$$A\mathbf{b}_1 = \begin{bmatrix} 1 & 3 & 2 \\ 0 & -1 & 1 \end{bmatrix}\begin{bmatrix} 4 \\ 1 \\ 3 \end{bmatrix} = \begin{bmatrix} 13 \\ 2 \end{bmatrix} \quad \text{and} \quad A\mathbf{b}_2 = \begin{bmatrix} 1 & 3 & 2 \\ 0 & -1 & 1 \end{bmatrix}\begin{bmatrix} -1 \\ 2 \\ 0 \end{bmatrix} = \begin{bmatrix} 5 \\ -2 \end{bmatrix}$$

Therefore, $AB = [A\mathbf{b}_1 \,\vdots\, A\mathbf{b}_2] = \left[\begin{array}{c:c} 13 & 5 \\ 2 & -2 \end{array}\right]$. (Check by ordinary matrix multiplication.)

Remark Observe that the matrix-column representation of AB allows us to write each column of AB as a linear combination of the columns of A with entries from B as the coefficients. For example,

$$\begin{bmatrix} 13 \\ 2 \end{bmatrix} = \begin{bmatrix} 1 & 3 & 2 \\ 0 & -1 & 1 \end{bmatrix}\begin{bmatrix} 4 \\ 1 \\ 3 \end{bmatrix} = 4\begin{bmatrix} 1 \\ 0 \end{bmatrix} + \begin{bmatrix} 3 \\ -1 \end{bmatrix} + 3\begin{bmatrix} 2 \\ 1 \end{bmatrix}$$

(See Exercises 23 and 26.)

Suppose A is $m \times n$ and B is $n \times r$, so the product AB exists. If we partition A in terms of its row vectors, as

$$A = \left[\begin{array}{c} \mathbf{A}_1 \\ \hdashline \mathbf{A}_2 \\ \hdashline \vdots \\ \hdashline \mathbf{A}_m \end{array}\right]$$

then

$$AB = \left[\begin{array}{c} \mathbf{A}_1 \\ \hdashline \mathbf{A}_2 \\ \hdashline \vdots \\ \hdashline \mathbf{A}_m \end{array}\right] B = \left[\begin{array}{c} \mathbf{A}_1 B \\ \hdashline \mathbf{A}_2 B \\ \hdashline \vdots \\ \hdashline \mathbf{A}_m B \end{array}\right]$$

Once again, this result is a direct consequence of the definition of matrix multiplication. The form on the right is called the ***row-matrix representation*** of the product.

Example 3.10

Using the row-matrix representation, compute AB for the matrices in Example 3.9.

Solution We compute

$$\mathbf{A}_1B = [1 \quad 3 \quad 2]\begin{bmatrix} 4 & -1 \\ 1 & 2 \\ 3 & 0 \end{bmatrix} = [13 \quad 5] \quad \text{and} \quad \mathbf{A}_2B = [0 \quad -1 \quad 1]\begin{bmatrix} 4 & -1 \\ 1 & 2 \\ 3 & 0 \end{bmatrix}$$
$$= [2 \quad -2]$$

Therefore, $AB = \begin{bmatrix} \mathbf{A}_1B \\ \hline \mathbf{A}_2B \end{bmatrix} = \begin{bmatrix} 13 & 5 \\ \hline 2 & -2 \end{bmatrix}$, as before.

The definition of the matrix product AB uses the natural partition of A into rows and B into columns; this form might well be called the ***row-column representation*** of the product. We can also partition A into columns and B into rows; this form is called the ***column-row representation*** of the product.

In this case, we have

$$A = [\mathbf{a}_1 \vdots \mathbf{a}_2 \vdots \cdots \vdots \mathbf{a}_n] \quad \text{and} \quad B = \begin{bmatrix} \mathbf{B}_1 \\ \hline \mathbf{B}_2 \\ \hline \vdots \\ \hline \mathbf{B}_n \end{bmatrix}$$

so

$$AB = [\mathbf{a}_1 \vdots \mathbf{a}_2 \vdots \cdots \vdots \mathbf{a}_n]\begin{bmatrix} \mathbf{B}_1 \\ \hline \mathbf{B}_2 \\ \hline \vdots \\ \hline \mathbf{B}_n \end{bmatrix} = \mathbf{a}_1\mathbf{B}_1 + \mathbf{a}_2\mathbf{B}_2 + \cdots + \mathbf{a}_n\mathbf{B}_n \qquad (2)$$

Notice that the sum resembles a dot product expansion; the difference is that the individual terms are matrices, not scalars. Let's make sure that this makes sense. Each term $\mathbf{a}_i\mathbf{B}_i$ is the product of an $m \times 1$ and a $1 \times r$ matrix. Thus, each $\mathbf{a}_i\mathbf{B}_i$ is an $m \times r$ matrix—the same size as AB. The products $\mathbf{a}_i\mathbf{B}_i$ are called ***outer products,*** and (2) is called the ***outer product expansion*** of AB.

Example 3.11

Compute the outer product expansion of AB for the matrices in Example 3.9.

Solution We have

$$A = [\mathbf{a}_1 \vdots \mathbf{a}_2 \vdots \mathbf{a}_3] = \left[\begin{array}{c:c:c} 1 & 3 & 2 \\ 0 & -1 & 1 \end{array}\right] \quad \text{and} \quad B = \begin{bmatrix} \mathbf{B}_1 \\ \hline \mathbf{B}_2 \\ \hline \mathbf{B}_3 \end{bmatrix} = \begin{bmatrix} 4 & -1 \\ \hline 1 & 2 \\ \hline 3 & 0 \end{bmatrix}$$

The outer products are

$$\mathbf{a}_1\mathbf{B}_1 = \begin{bmatrix} 1 \\ 0 \end{bmatrix}[4 \quad -1] = \begin{bmatrix} 4 & -1 \\ 0 & 0 \end{bmatrix}, \quad \mathbf{a}_2\mathbf{B}_2 = \begin{bmatrix} 3 \\ -1 \end{bmatrix}[1 \quad 2] = \begin{bmatrix} 3 & 6 \\ -1 & -2 \end{bmatrix},$$

and

$$\mathbf{a}_3\mathbf{B}_3 = \begin{bmatrix} 2 \\ 1 \end{bmatrix}[3 \quad 0] = \begin{bmatrix} 6 & 0 \\ 3 & 0 \end{bmatrix}$$

(Observe that computing each outer product is exactly like filling in a multiplication table.) Therefore, the outer product expansion of AB is

$$\mathbf{a}_1\mathbf{B}_1 + \mathbf{a}_2\mathbf{B}_2 + \mathbf{a}_3\mathbf{B}_3 = \begin{bmatrix} 4 & -1 \\ 0 & 0 \end{bmatrix} + \begin{bmatrix} 3 & 6 \\ -1 & -2 \end{bmatrix} + \begin{bmatrix} 6 & 0 \\ 3 & 0 \end{bmatrix} = \begin{bmatrix} 13 & 5 \\ 2 & -2 \end{bmatrix} = AB$$

We will make use of the outer product expansion in Chapters 5 and 7 when we discuss the Spectral Theorem and the singular value decomposition, respectively.

Each of the foregoing partitions is a special case of partitioning in general. A matrix A is said to be partitioned if horizontal and vertical lines have been introduced, subdividing A into submatrices called blocks. Partitioning allows A to be written as a matrix whose entries are its blocks.

For example,

$$A = \left[\begin{array}{ccc|cc} 1 & 0 & 0 & 2 & -1 \\ 0 & 1 & 0 & 1 & 3 \\ 0 & 0 & 1 & 4 & 0 \\ \hline 0 & 0 & 0 & 1 & 7 \\ 0 & 0 & 0 & 7 & 2 \end{array}\right] \quad \text{and} \quad B = \left[\begin{array}{cc|cc|c} 4 & 3 & 1 & 2 & 1 \\ -1 & 2 & 2 & 1 & 1 \\ 1 & -5 & 3 & 3 & 1 \\ \hline 1 & 0 & 0 & 0 & 2 \\ 0 & 1 & 0 & 0 & 3 \end{array}\right]$$

are partitioned matrices. They have the block structures

$$A = \begin{bmatrix} A_{11} & A_{12} \\ A_{21} & A_{22} \end{bmatrix} \quad \text{and} \quad B = \begin{bmatrix} B_{11} & B_{12} & B_{13} \\ B_{21} & B_{22} & B_{23} \end{bmatrix}$$

If two matrices are the same size and have been partitioned in the same way, it is clear that they can be added and multiplied by scalars block by block. Less obvious is the fact that, with suitable partitioning, matrices can be multiplied blockwise as well. The next example illustrates this process.

Example 3.12

Consider the matrices A and B above. If we ignore for the moment the fact that their entries are matrices, then A appears to be a 2×2 matrix and B a 2×3 matrix. Their product should thus be a 2×3 matrix given by

$$\begin{aligned} AB &= \begin{bmatrix} A_{11} & A_{12} \\ A_{21} & A_{22} \end{bmatrix} \begin{bmatrix} B_{11} & B_{12} & B_{13} \\ B_{21} & B_{22} & B_{23} \end{bmatrix} \\ &= \begin{bmatrix} A_{11}B_{11} + A_{12}B_{21} & A_{11}B_{12} + A_{12}B_{22} & A_{11}B_{13} + A_{12}B_{23} \\ A_{21}B_{11} + A_{22}B_{21} & A_{21}B_{12} + A_{22}B_{22} & A_{21}B_{13} + A_{22}B_{23} \end{bmatrix} \end{aligned}$$

But all of the products in this calculation are actually *matrix* products, so we need to make sure that they are all defined. A quick check reveals that this is indeed the case, since the numbers of *columns* in the blocks of A (3 and 2) match the numbers of *rows* in the blocks of B. The matrices A and B are said to be ***partitioned conformably for block multiplication.***

Carrying out the calculations indicated gives us the product AB in partitioned form:

$$A_{11}B_{11} + A_{12}B_{21} = I_3B_{11} + A_{12}I_2 = B_{11} + A_{12} = \begin{bmatrix} 4 & 3 \\ -1 & 2 \\ 1 & -5 \end{bmatrix} + \begin{bmatrix} 2 & -1 \\ 1 & 3 \\ 4 & 0 \end{bmatrix} = \begin{bmatrix} 6 & 2 \\ 0 & 5 \\ 5 & -5 \end{bmatrix}$$

(When some of the blocks are zero matrices or identity matrices, as is the case here, these calculations can be done quite quickly.) The calculations for the other five blocks of AB are similar. Check that the result is

$$\left[\begin{array}{cc|cc|c} 6 & 2 & 1 & 2 & 2 \\ 0 & 5 & 2 & 1 & 12 \\ 5 & -5 & 3 & 3 & 9 \\ \hline 1 & 7 & 0 & 0 & 23 \\ 7 & 2 & 0 & 0 & 20 \end{array}\right]$$

(Observe that the block in the upper-left corner is the result of our calculations above.) Check that you obtain the same answer by multiplying A by B in the usual way.

Matrix Powers

When A and B are two $n \times n$ matrices, their product AB will also be an $n \times n$ matrix. A special case occurs when $A = B$. It makes sense to define $A^2 = AA$ and, in general, to define A^k as

$$A^k = \underbrace{AA \cdots A}_{k \text{ factors}}$$

if k is a positive integer. Thus, $A^1 = A$, and it is convenient to define $A^0 = I_n$.

Before making too many assumptions, we should ask ourselves to what extent matrix powers behave like powers of real numbers. The following properties follow immediately from the definitions we have just given and are the matrix analogues of the corresponding properties for powers of real numbers.

If A is a square matrix and r and s are nonnegative integers, then

1. $A^r A^s = A^{r+s}$
2. $(A^r)^s = A^{rs}$

In Section 3.3, we will extend the definition and properties to include negative integer powers.

Example 3.13

(a) If $A = \begin{bmatrix} 1 & 1 \\ 1 & 1 \end{bmatrix}$, then

$$A^2 = \begin{bmatrix} 1 & 1 \\ 1 & 1 \end{bmatrix}\begin{bmatrix} 1 & 1 \\ 1 & 1 \end{bmatrix} = \begin{bmatrix} 2 & 2 \\ 2 & 2 \end{bmatrix}, \quad A^3 = A^2A = \begin{bmatrix} 2 & 2 \\ 2 & 2 \end{bmatrix}\begin{bmatrix} 1 & 1 \\ 1 & 1 \end{bmatrix} = \begin{bmatrix} 4 & 4 \\ 4 & 4 \end{bmatrix}$$

and, in general,

$$A^n = \begin{bmatrix} 2^{n-1} & 2^{n-1} \\ 2^{n-1} & 2^{n-1} \end{bmatrix} \quad \text{for all } n \geq 1$$

The above statement can be proved by mathematical induction, since it is an *infinite* collection of statements, one for each natural number n. (Appendix B gives a brief review of mathematical induction.) The basis step is to prove that the formula holds for $n = 1$. In this case,

$$A^1 = \begin{bmatrix} 2^{1-1} & 2^{1-1} \\ 2^{1-1} & 2^{1-1} \end{bmatrix} = \begin{bmatrix} 2^0 & 2^0 \\ 2^0 & 2^0 \end{bmatrix} = \begin{bmatrix} 1 & 1 \\ 1 & 1 \end{bmatrix} = A$$

as required.

The induction hypothesis is to assume that

$$A^k = \begin{bmatrix} 2^{k-1} & 2^{k-1} \\ 2^{k-1} & 2^{k-1} \end{bmatrix}$$

for some integer $k \geq 1$. The induction step is to prove that the formula holds for $n = k + 1$. Using the definition of matrix powers and the induction hypothesis, we compute

$$\begin{aligned} A^{k+1} = A^kA &= \begin{bmatrix} 2^{k-1} & 2^{k-1} \\ 2^{k-1} & 2^{k-1} \end{bmatrix}\begin{bmatrix} 1 & 1 \\ 1 & 1 \end{bmatrix} \\ &= \begin{bmatrix} 2^{k-1} + 2^{k-1} & 2^{k-1} + 2^{k-1} \\ 2^{k-1} + 2^{k-1} & 2^{k-1} + 2^{k-1} \end{bmatrix} \\ &= \begin{bmatrix} 2^k & 2^k \\ 2^k & 2^k \end{bmatrix} \\ &= \begin{bmatrix} 2^{(k+1)-1} & 2^{(k+1)-1} \\ 2^{(k+1)-1} & 2^{(k+1)-1} \end{bmatrix} \end{aligned}$$

Thus, the formula holds for all $n \geq 1$ by the principle of mathematical induction.

(b) If $B = \begin{bmatrix} 0 & -1 \\ 1 & 0 \end{bmatrix}$, then $B^2 = \begin{bmatrix} 0 & -1 \\ 1 & 0 \end{bmatrix}\begin{bmatrix} 0 & -1 \\ 1 & 0 \end{bmatrix} = \begin{bmatrix} -1 & 0 \\ 0 & -1 \end{bmatrix}$. Continuing, we find

$$B^3 = B^2B = \begin{bmatrix} -1 & 0 \\ 0 & -1 \end{bmatrix}\begin{bmatrix} 0 & -1 \\ 1 & 0 \end{bmatrix} = \begin{bmatrix} 0 & 1 \\ -1 & 0 \end{bmatrix}$$

and

$$B^4 = B^3B = \begin{bmatrix} 0 & 1 \\ -1 & 0 \end{bmatrix}\begin{bmatrix} 0 & -1 \\ 1 & 0 \end{bmatrix} = \begin{bmatrix} 1 & 0 \\ 0 & 1 \end{bmatrix}$$

Thus, $B^5 = B$, and the sequence of powers of B repeats in a cycle of four:

$$\begin{bmatrix} 0 & -1 \\ 1 & 0 \end{bmatrix}, \begin{bmatrix} -1 & 0 \\ 0 & -1 \end{bmatrix}, \begin{bmatrix} 0 & 1 \\ -1 & 0 \end{bmatrix}, \begin{bmatrix} 1 & 0 \\ 0 & 1 \end{bmatrix}, \begin{bmatrix} 0 & -1 \\ 1 & 0 \end{bmatrix}, \ldots$$

The Transpose of a Matrix

Thus far, all of the matrix operations we have defined are analogous to operations on real numbers, although they may not always behave in the same way. The next operation has no such analogue.

Definition The ***transpose*** of an $m \times n$ matrix A is the $n \times m$ matrix A^T obtained by interchanging the rows and columns of A. That is, the ith column of A^T is the ith row of A for all i.

Example 3.14

Let

$$A = \begin{bmatrix} 1 & 3 & 2 \\ 5 & 0 & 1 \end{bmatrix}, \quad B = \begin{bmatrix} a & b \\ c & d \end{bmatrix}, \quad \text{and} \quad C = \begin{bmatrix} 5 & -1 & 2 \end{bmatrix}$$

Then their transposes are

$$A^T = \begin{bmatrix} 1 & 5 \\ 3 & 0 \\ 2 & 1 \end{bmatrix}, \quad B^T = \begin{bmatrix} a & c \\ b & d \end{bmatrix}, \quad \text{and} \quad C^T = \begin{bmatrix} 5 \\ -1 \\ 2 \end{bmatrix}$$

The transpose is sometimes used to give an alternative definition of the dot product of two vectors in terms of matrix multiplication. If

$$\mathbf{u} = \begin{bmatrix} u_1 \\ u_2 \\ \vdots \\ u_n \end{bmatrix} \quad \text{and} \quad \mathbf{v} = \begin{bmatrix} v_1 \\ v_2 \\ \vdots \\ v_n \end{bmatrix}$$

then

$$\begin{aligned} \mathbf{u} \cdot \mathbf{v} &= u_1v_1 + u_2v_2 + \cdots + u_nv_n \\ &= \begin{bmatrix} u_1 & u_2 & \cdots & u_n \end{bmatrix} \begin{bmatrix} v_1 \\ v_2 \\ \vdots \\ v_n \end{bmatrix} \\ &= \mathbf{u}^T\mathbf{v} \end{aligned}$$

A useful alternative definition of the transpose is given componentwise:

$$(A^T)_{ij} = A_{ji} \quad \text{for all } i \text{ and } j$$

In words, the entry in row i and column j of A^T is the same as the entry in row j and column i of A.

The transpose is also used to define a very important type of square matrix: a symmetric matrix.

Definition A square matrix A is ***symmetric*** if $A^T = A$—that is, if A is equal to its own transpose.

Example 3.15

Let

$$A = \begin{bmatrix} 1 & 3 & 2 \\ 3 & 5 & 0 \\ 2 & 0 & 4 \end{bmatrix} \quad \text{and} \quad B = \begin{bmatrix} 1 & 2 \\ -1 & 3 \end{bmatrix}$$

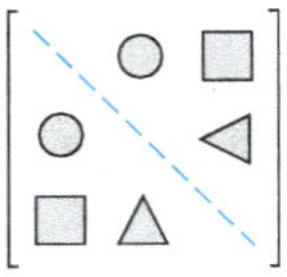

Figure 3.1
A symmetric matrix

Then A is symmetric, since $A^T = A$; but B is not symmetric, since $B^T = \begin{bmatrix} 1 & -1 \\ 2 & 3 \end{bmatrix} \neq B$.

A symmetric matrix has the property that it is its own "mirror image" across its main diagonal. Figure 3.1 illustrates this property for a 3×3 matrix. The corresponding shapes represent equal entries; the diagonal entries (those on the dashed line) are arbitrary.

A componentwise definition of a symmetric matrix is also useful. It is simply the algebraic description of the "reflection" property.

A square matrix A is symmetric if and only if $A_{ij} = A_{ji}$ for all i and j.

Exercises 3.1

Let

$$A = \begin{bmatrix} 3 & 0 \\ -1 & 5 \end{bmatrix}, \quad B = \begin{bmatrix} 4 & -2 & 1 \\ 0 & 2 & 3 \end{bmatrix}, \quad C = \begin{bmatrix} 1 & 2 \\ 3 & 4 \\ 5 & 6 \end{bmatrix},$$

$$D = \begin{bmatrix} 0 & -3 \\ -2 & 1 \end{bmatrix}, \quad E = [4 \quad 2], \quad F = \begin{bmatrix} -1 \\ 2 \end{bmatrix}$$

In Exercises 1–16, compute the indicated matrices (if possible).

1. $A + 2D$

2. $3D - 2A$

3. $B - C$

4. $C - B^T$

5. AB

6. BD

7. $D + BC$

8. BB^T

9. $E(AF)$

10. $F(DF)$

11. FE

12. EF

13. $B^TC^T - (CB)^T$

14. $DA - AD$

15. A^3

16. $(I_2 - D)^2$

17. Give an example of a nonzero 2×2 matrix A such that $A^2 = O$.

18. Let $A = \begin{bmatrix} 2 & 1 \\ 6 & 3 \end{bmatrix}$. Find 2×2 matrices B and C such that $AB = AC$ but $B \neq C$.

19. A factory manufactures three products (doohickies, gizmos, and widgets) and ships them to two warehouses for storage. The number of units of each product shipped to each warehouse is given by the matrix

$$A = \begin{bmatrix} 200 & 75 \\ 150 & 100 \\ 100 & 125 \end{bmatrix}$$

(where a_{ij} is the number of units of product i sent to warehouse j and the products are taken in alphabetical order). The cost of shipping one unit of each product by truck is \$1.50 per doohickey, \$1.00 per gizmo, and \$2.00 per widget. The corresponding unit costs to ship by train are \$1.75, \$1.50, and \$1.00. Organize these costs into a matrix B and then use matrix multiplication to show how the factory can compare the cost of shipping its products to each of the two warehouses by truck and by train.

20. Referring to Exercise 19, suppose that the unit cost of distributing the products to stores is the same for each product but varies by warehouse because of the distances involved. It costs \$0.75 to distribute one unit from warehouse 1 and \$1.00 to distribute one unit from warehouse 2. Organize these costs into a matrix C and then use matrix multiplication to compute the total cost of distributing each product.

In Exercises 21–22, write the given system of linear equations as a matrix equation of the form $A\mathbf{x} = \mathbf{b}$.

21. $$\begin{aligned} x_1 - 2x_2 + 3x_3 &= 0 \\ 2x_1 + x_2 - 5x_3 &= 4 \end{aligned}$$

22. $$\begin{aligned} -x_1 \quad\quad + 2x_3 &= 1 \\ x_1 - x_2 \quad\quad &= -2 \\ x_2 + x_3 &= -1 \end{aligned}$$

In Exercises 23–28, let

$$A = \begin{bmatrix} 1 & 0 & -2 \\ -3 & 1 & 1 \\ 2 & 0 & -1 \end{bmatrix}$$

and $$B = \begin{bmatrix} 2 & 3 & 0 \\ 1 & -1 & 1 \\ -1 & 6 & 4 \end{bmatrix}$$

23. Use the matrix-column representation of the product to write each column of AB as a linear combination of the columns of A.

24. Use the row-matrix representation of the product to write each row of AB as a linear combination of the rows of B.

25. Compute the outer product expansion of AB.

26. Use the matrix-column representation of the product to write each column of BA as a linear combination of the columns of B.

27. Use the row-matrix representation of the product to write each row of BA as a linear combination of the rows of A.

28. Compute the outer product expansion of BA.

In Exercises 29 and 30, assume that the product AB makes sense.

29. Prove that if the columns of B are linearly dependent, then so are the columns of AB.

30. Prove that if the rows of A are linearly dependent, then so are the rows of AB.

In Exercises 31–34, compute AB by block multiplication, using the indicated partitioning.

31. $A = \left[\begin{array}{cc|cc} 1 & -1 & 0 & 0 \\ 0 & 1 & 0 & 0 \\ \hline 0 & 0 & 2 & 3 \end{array}\right]$, $B = \left[\begin{array}{cc|c} 2 & 3 & 0 \\ -1 & 1 & 0 \\ \hline 0 & 0 & 1 \\ 0 & 0 & 1 \end{array}\right]$

32. $A = \left[\begin{array}{cc|cc} 2 & 3 & 1 & 0 \\ 4 & 5 & 0 & 1 \end{array}\right]$, $B = \left[\begin{array}{c|cc} 0 & 1 & 0 \\ 0 & 0 & 1 \\ \hline 1 & 5 & 4 \\ -2 & 3 & 2 \end{array}\right]$

33. $A = \left[\begin{array}{cc|cc} 1 & 2 & 1 & 0 \\ 3 & 4 & 0 & 1 \\ \hline 1 & 0 & -1 & 1 \\ 0 & 1 & 1 & -1 \end{array}\right]$, $B = \left[\begin{array}{cc|cc} 1 & 0 & 0 & 1 \\ 0 & 1 & 1 & 0 \\ \hline 0 & 0 & 0 & -1 \\ 0 & 0 & 1 & 0 \end{array}\right]$

34. $A = \left[\begin{array}{ccc|c} 1 & 0 & 0 & 1 \\ 0 & 1 & 0 & 2 \\ 0 & 0 & 1 & 3 \\ \hline 0 & 0 & 0 & 4 \end{array}\right]$, $B = \left[\begin{array}{ccc|c} 1 & 2 & 3 & 1 \\ 0 & 1 & 4 & 1 \\ 0 & 0 & 1 & 1 \\ \hline 1 & 1 & 1 & -1 \end{array}\right]$

35. Let $A = \begin{bmatrix} 0 & 1 \\ -1 & 1 \end{bmatrix}$.

(a) Compute $A^2, A^3, \ldots, A^7$.
(b) What is A^{2015}? Why?

36. Let $B = \begin{bmatrix} \frac{1}{\sqrt{2}} & -\frac{1}{\sqrt{2}} \\ \frac{1}{\sqrt{2}} & \frac{1}{\sqrt{2}} \end{bmatrix}$. Find, with justification, B^{2015}.

37. Let $A = \begin{bmatrix} 1 & 1 \\ 0 & 1 \end{bmatrix}$. Find a formula for A^n $(n \geq 1)$ and verify your formula using mathematical induction.

38. Let $A = \begin{bmatrix} \cos\theta & -\sin\theta \\ \sin\theta & \cos\theta \end{bmatrix}$.

(a) Show that $A^2 = \begin{bmatrix} \cos 2\theta & -\sin 2\theta \\ \sin 2\theta & \cos 2\theta \end{bmatrix}$.

(b) Prove, by mathematical induction, that

$$A^n = \begin{bmatrix} \cos n\theta & -\sin n\theta \\ \sin n\theta & \cos n\theta \end{bmatrix} \text{ for } n \geq 1$$

39. In each of the following, find the 4×4 matrix $A = [a_{ij}]$ that satisfies the given condition:

(a) $a_{ij} = (-1)^{i+j}$ **(b)** $a_{ij} = j - i$

(c) $a_{ij} = (i - 1)^j$ **(d)** $a_{ij} = \sin\left(\dfrac{(i + j - 1)\pi}{4}\right)$

40. In each of the following, find the 6×6 matrix $A = [a_{ij}]$ that satisfies the given condition:

(a) $a_{ij} = \begin{cases} i + j & \text{if } i \leq j \\ 0 & \text{if } i > j \end{cases}$ **(b)** $a_{ij} = \begin{cases} 1 & \text{if } |i - j| \leq 1 \\ 0 & \text{if } |i - j| > 1 \end{cases}$

(c) $a_{ij} = \begin{cases} 1 & \text{if } 6 \leq i + j \leq 8 \\ 0 & \text{otherwise} \end{cases}$

41. Prove Theorem 3.1(a).

3.2 Matrix Algebra

In some ways, the arithmetic of matrices generalizes that of vectors. We do not expect any surprises with respect to addition and scalar multiplication, and indeed there are none. This will allow us to extend to matrices several concepts that we are already familiar with from our work with vectors. In particular, linear combinations, spanning sets, and linear independence carry over to matrices with no difficulty.

However, matrices have other operations, such as matrix multiplication, that vectors do not possess. We should not expect matrix multiplication to behave like multiplication of real numbers unless we can prove that it does; in fact, it does not. In this section, we summarize and prove some of the main properties of matrix operations and begin to develop an algebra of matrices.

Properties of Addition and Scalar Multiplication

All of the algebraic properties of addition and scalar multiplication for vectors (Theorem 1.1) carry over to matrices. For completeness, we summarize these properties in the next theorem.

Theorem 3.2 **Algebraic Properties of Matrix Addition and Scalar Multiplication**

Let A, B, and C be matrices of the same size and let c and d be scalars. Then

a. $A + B = B + A$ — Commutativity
b. $(A + B) + C = A + (B + C)$ — Associativity
c. $A + O = A$
d. $A + (-A) = O$
e. $c(A + B) = cA + cB$ — Distributivity
f. $(c + d)A = cA + dA$ — Distributivity
g. $c(dA) = (cd)A$
h. $1A = A$

The proofs of these properties are direct analogues of the corresponding proofs of the vector properties and are left as exercises. Likewise, the comments following Theorem 1.1 are equally valid here, and you should have no difficulty using these properties to perform algebraic manipulations with matrices. (Review Example 1.5 and see Exercises 17 and 18 at the end of this section.)

The associativity property allows us to unambiguously combine scalar multiplication and addition without parentheses. If A, B, and C are matrices of the same size, then

$$(2A + 3B) - C = 2A + (3B - C)$$

and so we can simply write $2A + 3B - C$. Generally, then, if $A_1, A_2, \dots, A_k$ are matrices of the same size and $c_1, c_2, \dots, c_k$ are scalars, we may form the ***linear combination***

$$c_1A_1 + c_2A_2 + \cdots + c_kA_k$$

We will refer to $c_1, c_2, \dots, c_k$ as the ***coefficients*** of the linear combination. We can now ask and answer questions about linear combinations of matrices.

Example 3.16

Let $A_1 = \begin{bmatrix} 0 & 1 \\ -1 & 0 \end{bmatrix}$, $A_2 = \begin{bmatrix} 1 & 0 \\ 0 & 1 \end{bmatrix}$, and $A_3 = \begin{bmatrix} 1 & 1 \\ 1 & 1 \end{bmatrix}$.

(a) Is $B = \begin{bmatrix} 1 & 4 \\ 2 & 1 \end{bmatrix}$ a linear combination of A_1, A_2, and A_3?

(b) Is $C = \begin{bmatrix} 1 & 2 \\ 3 & 4 \end{bmatrix}$ a linear combination of A_1, A_2, and A_3?

Solution

(a) We want to find scalars c_1, c_2, and c_3 such that $c_1A_1 + c_2A_2 + c_3A_3 = B$. Thus,

$$c_1\begin{bmatrix} 0 & 1 \\ -1 & 0 \end{bmatrix} + c_2\begin{bmatrix} 1 & 0 \\ 0 & 1 \end{bmatrix} + c_3\begin{bmatrix} 1 & 1 \\ 1 & 1 \end{bmatrix} = \begin{bmatrix} 1 & 4 \\ 2 & 1 \end{bmatrix}$$

The left-hand side of this equation can be rewritten as

$$\begin{bmatrix} c_2 + c_3 & c_1 + c_3 \\ -c_1 + c_3 & c_2 + c_3 \end{bmatrix}$$

Comparing entries and using the definition of matrix equality, we have four linear equations:

$$\begin{aligned} c_2 + c_3 &= 1 \\ c_1 \qquad + c_3 &= 4 \\ -c_1 \qquad + c_3 &= 2 \\ c_2 + c_3 &= 1 \end{aligned}$$

Gauss-Jordan elimination easily gives

$$\left[\begin{array}{rrr|r} 0 & 1 & 1 & 1 \\ 1 & 0 & 1 & 4 \\ -1 & 0 & 1 & 2 \\ 0 & 1 & 1 & 1 \end{array}\right] \longrightarrow \left[\begin{array}{rrr|r} 1 & 0 & 0 & 1 \\ 0 & 1 & 0 & -2 \\ 0 & 0 & 1 & 3 \\ 0 & 0 & 0 & 0 \end{array}\right]$$

(check this!), so $c_1 = 1$, $c_2 = -2$, and $c_3 = 3$. Thus, $A_1 - 2A_2 + 3A_3 = B$, which can be easily checked.

(b) This time we want to solve

$$c_1\begin{bmatrix} 0 & 1 \\ -1 & 0 \end{bmatrix} + c_2\begin{bmatrix} 1 & 0 \\ 0 & 1 \end{bmatrix} + c_3\begin{bmatrix} 1 & 1 \\ 1 & 1 \end{bmatrix} = \begin{bmatrix} 1 & 2 \\ 3 & 4 \end{bmatrix}$$

Proceeding as in part (a), we obtain the linear system

$$\begin{aligned} c_2 + c_3 &= 1 \\ c_1 \qquad + c_3 &= 2 \\ -c_1 \qquad + c_3 &= 3 \\ c_2 + c_3 &= 4 \end{aligned}$$

Row reduction gives

$$\left[\begin{array}{rrr|r} 0 & 1 & 1 & 1 \\ 1 & 0 & 1 & 2 \\ -1 & 0 & 1 & 3 \\ 0 & 1 & 1 & 4 \end{array}\right] \xrightarrow{R_4 - R_1} \left[\begin{array}{rrr|r} 0 & 1 & 1 & 1 \\ 1 & 0 & 1 & 2 \\ -1 & 0 & 1 & 3 \\ 0 & 0 & 0 & 3 \end{array}\right]$$

We need go no further: The last row implies that there is no solution. Therefore, in this case, C is not a linear combination of A_1, A_2, and A_3.

Remark Observe that the columns of the augmented matrix contain the entries of the matrices we are given. If we read the entries of each matrix from left to right and top to bottom, we get the order in which the entries appear in the columns of the augmented matrix. For example, we read A_1 as "0, 1, −1, 0," which corresponds to the first column of the augmented matrix. It is as if we simply "straightened out" the given matrices into column vectors. Thus, we would have ended up with exactly the same system of linear equations as in part (a) if we had asked

$$\text{Is } \begin{bmatrix} 1 \\ 4 \\ 2 \\ 1 \end{bmatrix} \text{ a linear combination of } \begin{bmatrix} 0 \\ 1 \\ -1 \\ 0 \end{bmatrix}, \begin{bmatrix} 1 \\ 0 \\ 0 \\ 1 \end{bmatrix}, \text{ and } \begin{bmatrix} 1 \\ 1 \\ 1 \\ 1 \end{bmatrix}?$$

We will encounter such parallels repeatedly from now on. In Chapter 6, we will explore them in more detail.

We can define the ***span*** of a set of matrices to be the set of all linear combinations of the matrices.

Example 3.17

Describe the span of the matrices A_1, A_2, and A_3 in Example 3.16.

Solution One way to do this is simply to write out a general linear combination of A_1, A_2, and A_3. Thus,

$$\begin{aligned} c_1A_1 + c_2A_2 + c_3A_3 &= c_1\begin{bmatrix} 0 & 1 \\ -1 & 0 \end{bmatrix} + c_2\begin{bmatrix} 1 & 0 \\ 0 & 1 \end{bmatrix} + c_3\begin{bmatrix} 1 & 1 \\ 1 & 1 \end{bmatrix} \\ &= \begin{bmatrix} c_2 + c_3 & c_1 + c_3 \\ -c_1 + c_3 & c_2 + c_3 \end{bmatrix} \end{aligned}$$

(which is analogous to the parametric representation of a plane). But suppose we want to know when the matrix $\begin{bmatrix} w & x \\ y & z \end{bmatrix}$ is in span(A_1, A_2, A_3). From the representation above, we know that it is when

$$\begin{bmatrix} c_2 + c_3 & c_1 + c_3 \\ -c_1 + c_3 & c_2 + c_3 \end{bmatrix} = \begin{bmatrix} w & x \\ y & z \end{bmatrix}$$

for some choice of scalars c_1, c_2, c_3. This gives rise to a system of linear equations whose left-hand side is exactly the same as in Example 3.16 but whose right-hand side is general. The augmented matrix of this system is

$$\left[\begin{array}{ccc|c} 0 & 1 & 1 & w \\ 1 & 0 & 1 & x \\ -1 & 0 & 1 & y \\ 0 & 1 & 1 & z \end{array}\right]$$

and row reduction produces

$$\left[\begin{array}{ccc|c} 0 & 1 & 1 & w \\ 1 & 0 & 1 & x \\ -1 & 0 & 1 & y \\ 0 & 1 & 1 & z \end{array}\right] \longrightarrow \left[\begin{array}{ccc|c} 1 & 0 & 0 & \frac{1}{2}x - \frac{1}{2}y \\ 0 & 1 & 0 & -\frac{1}{2}x - \frac{1}{2}y + w \\ 0 & 0 & 1 & \frac{1}{2}x + \frac{1}{2}y \\ 0 & 0 & 0 & w - z \end{array}\right]$$

(Check this carefully.) The only restriction comes from the last row, where clearly we must have $w - z = 0$ in order to have a solution. Thus, the span of A_1, A_2, and A_3 consists of all matrices $\begin{bmatrix} w & x \\ y & z \end{bmatrix}$ for which $w = z$. That is, $\text{span}(A_1, A_2, A_3) = \left\{\begin{bmatrix} w & x \\ y & w \end{bmatrix}\right\}$.

Note If we had known this *before* attempting Example 3.16, we would have seen immediately that $B = \begin{bmatrix} 1 & 4 \\ 2 & 1 \end{bmatrix}$ is a linear combination of A_1, A_2, and A_3, since it has the necessary form (take $w = 1, x = 4$, and $y = 2$), but $C = \begin{bmatrix} 1 & 2 \\ 3 & 4 \end{bmatrix}$ cannot be a linear combination of A_1, A_2, and A_3, since it does not have the proper form ($1 \neq 4$).

Linear independence also makes sense for matrices. We say that matrices $A_1, A_2, \ldots, A_k$ of the same size are ***linearly independent*** if the only solution of the equation

$$c_1A_1 + c_2A_2 + \cdots + c_kA_k = O \tag{1}$$

is the trivial one: $c_1 = c_2 = \cdots = c_k = 0$. If there are nontrivial coefficients that satisfy (1), then $A_1, A_2, \ldots, A_k$ are called ***linearly dependent.***

Example 3.18

Determine whether the matrices A_1, A_2, and A_3 in Example 3.16 are linearly independent.

Solution We want to solve the equation $c_1A_1 + c_2A_2 + c_3A_3 = O$. Writing out the matrices, we have

$$c_1\begin{bmatrix} 0 & 1 \\ -1 & 0 \end{bmatrix} + c_2\begin{bmatrix} 1 & 0 \\ 0 & 1 \end{bmatrix} + c_3\begin{bmatrix} 1 & 1 \\ 1 & 1 \end{bmatrix} = \begin{bmatrix} 0 & 0 \\ 0 & 0 \end{bmatrix}$$

This time we get a *homogeneous* linear system whose left-hand side is the same as in Examples 3.16 and 3.17. (Are you starting to spot a pattern yet?) The augmented matrix row reduces to give

$$\left[\begin{array}{ccc|c} 0 & 1 & 1 & 0 \\ 1 & 0 & 1 & 0 \\ -1 & 0 & 1 & 0 \\ 0 & 1 & 1 & 0 \end{array}\right] \longrightarrow \left[\begin{array}{ccc|c} 1 & 0 & 0 & 0 \\ 0 & 1 & 0 & 0 \\ 0 & 0 & 1 & 0 \\ 0 & 0 & 0 & 0 \end{array}\right]$$

Thus, $c_1 = c_2 = c_3 = 0$, and we conclude that the matrices A_1, A_2, and A_3 are linearly independent.

Properties of Matrix Multiplication

Whenever we encounter a new operation, such as matrix multiplication, we must be careful not to assume too much about it. It would be nice if matrix multiplication behaved like multiplication of real numbers. Although in many respects it does, there are some significant differences.

Example 3.19 Consider the matrices

$$A = \begin{bmatrix} 2 & 4 \\ -1 & -2 \end{bmatrix} \quad \text{and} \quad B = \begin{bmatrix} 1 & 0 \\ 1 & 1 \end{bmatrix}$$

Multiplying gives

$$AB = \begin{bmatrix} 2 & 4 \\ -1 & -2 \end{bmatrix}\begin{bmatrix} 1 & 0 \\ 1 & 1 \end{bmatrix} = \begin{bmatrix} 6 & 4 \\ -3 & -2 \end{bmatrix} \quad \text{and} \quad BA = \begin{bmatrix} 1 & 0 \\ 1 & 1 \end{bmatrix}\begin{bmatrix} 2 & 4 \\ -1 & -2 \end{bmatrix} = \begin{bmatrix} 2 & 4 \\ 1 & 2 \end{bmatrix}$$

Thus, $AB \neq BA$. So, in contrast to multiplication of real numbers, matrix multiplication is *not commutative*—the order of the factors in a product matters!

It is easy to check that $A^2 = \begin{bmatrix} 0 & 0 \\ 0 & 0 \end{bmatrix}$ (do so!). So, for matrices, the equation $A^2 = O$ does not imply that $A = O$ (unlike the situation for real numbers, where the equation $x^2 = 0$ has only $x = 0$ as a solution).

However gloomy things might appear after the last example, the situation is not really bad at all—you just need to get used to working with matrices and to constantly remind yourself that they are not numbers. The next theorem summarizes the main properties of matrix multiplication.

Theorem 3.3 **Properties of Matrix Multiplication**

Let A, B, and C be matrices (whose sizes are such that the indicated operations can be performed) and let k be a scalar. Then

a. $A(BC) = (AB)C$ — Associativity
b. $A(B + C) = AB + AC$ — Left distributivity
c. $(A + B)C = AC + BC$ — Right distributivity
d. $k(AB) = (kA)B = A(kB)$
e. $I_mA = A = AI_n$ if A is $m \times n$ — Multiplicative identity

Proof We prove (b) and half of (e). We defer the proof of property (a) until Section 3.6. The remaining properties are considered in the exercises.

(b) To prove $A(B + C) = AB + AC$, we let the rows of A be denoted by $\mathbf{A}_i$ and the columns of B and C by $\mathbf{b}_j$ and $\mathbf{c}_j$. Then the jth column of $B + C$ is $\mathbf{b}_j + \mathbf{c}_j$ (since addition is defined componentwise), and thus

$$\begin{aligned}[A(B + C)]_{ij} &= \mathbf{A}_i \cdot (\mathbf{b}_j + \mathbf{c}_j) \\ &= \mathbf{A}_i \cdot \mathbf{b}_j + \mathbf{A}_i \cdot \mathbf{c}_j \\ &= (AB)_{ij} + (AC)_{ij} \\ &= (AB + AC)_{ij}\end{aligned}$$

Since this is true for all i and j, we must have $A(B + C) = AB + AC$.

(e) To prove $AI_n = A$, we note that the identity matrix I_n can be column-partitioned as

$$I_n = [\mathbf{e}_1 \vdots \mathbf{e}_2 \vdots \cdots \vdots \mathbf{e}_n]$$

where $\mathbf{e}_i$ is a standard unit vector. Therefore,

$$\begin{aligned}AI_n &= [A\mathbf{e}_1 \vdots A\mathbf{e}_2 \vdots \cdots \vdots A\mathbf{e}_n] \\ &= [\mathbf{a}_1 \vdots \mathbf{a}_2 \vdots \cdots \vdots \mathbf{a}_n] \\ &= A\end{aligned}$$

by Theorem 3.1(b).

We can use these properties to further explore how closely matrix multiplication resembles multiplication of real numbers.

Example 3.20

If A and B are square matrices of the same size, is $(A + B)^2 = A^2 + 2AB + B^2$?

Solution Using properties of matrix multiplication, we compute

$$\begin{aligned}(A + B)^2 &= (A + B)(A + B) \\ &= (A + B)A + (A + B)B && \text{by left distributivity} \\ &= A^2 + BA + AB + B^2 && \text{by right distributivity}\end{aligned}$$

Therefore, $(A + B)^2 = A^2 + 2AB + B^2$ if and only if $A^2 + BA + AB + B^2 = A^2 + 2AB + B^2$. Subtracting A^2 and B^2 from both sides gives $BA + AB = 2AB$. Subtracting AB from both sides gives $BA = AB$. Thus, $(A + B)^2 = A^2 + 2AB + B^2$ if and only if A and B commute. (Can you give an example of such a pair of matrices? Can you find two matrices that do not satisfy this property?)

Properties of the Transpose

Theorem 3.4

Properties of the Transpose

Let A and B be matrices (whose sizes are such that the indicated operations can be performed) and let k be a scalar. Then

a. $(A^T)^T = A$ b. $(A + B)^T = A^T + B^T$

c. $(kA)^T = k(A^T)$ d. $(AB)^T = B^T A^T$

e. $(A^r)^T = (A^T)^r$ for all nonnegative integers r

Proof Properties (a)–(c) are intuitively clear and straightforward to prove (see Exercise 30). Proving property (e) is a good exercise in mathematical induction (see Exercise 31). We will prove (d), since it is not what you might have expected. [Would you have suspected that $(AB)^T = A^TB^T$ might be true?]

First, if A is $m \times n$ and B is $n \times r$, then B^T is $r \times n$ and A^T is $n \times m$. Thus, the product B^TA^T is defined and is $r \times m$. Since AB is $m \times r$, $(AB)^T$ is $r \times m$, and so $(AB)^T$ and B^TA^T have the same size. We must now prove that their corresponding entries are equal.

We denote the ith row of a matrix X by $\text{row}_i(X)$ and its jth column by $\text{col}_j(X)$. Using these conventions, we see that

$$\begin{aligned} [(AB)^T]_{ij} &= (AB)_{ji} \\ &= \text{row}_j(A) \cdot \text{col}_i(B) \\ &= \text{col}_j(A^T) \cdot \text{row}_i(B^T) \\ &= \text{row}_i(B^T) \cdot \text{col}_j(A^T) = [B^TA^T]_{ij} \end{aligned}$$

(Note that we have used the definition of matrix multiplication, the definition of the transpose, and the fact that the dot product is commutative.) Since i and j are arbitrary, this result implies that $(AB)^T = B^TA^T$.

Remark Properties (b) and (d) of Theorem 3.4 can be generalized to sums and products of finitely many matrices:

$$(A_1 + A_2 + \cdots + A_k)^T = A_1^T + A_2^T + \cdots + A_k^T \quad \text{and} \quad (A_1A_2\cdots A_k)^T = A_k^T \cdots A_2^TA_1^T$$

assuming that the sizes of the matrices are such that all of the operations can be performed. You are asked to prove these facts by mathematical induction in Exercises 32 and 33.

Example 3.21

Let

$$A = \begin{bmatrix} 1 & 2 \\ 3 & 4 \end{bmatrix} \quad \text{and} \quad B = \begin{bmatrix} 4 & -1 & 0 \\ 2 & 3 & 1 \end{bmatrix}$$

Then $A^T = \begin{bmatrix} 1 & 3 \\ 2 & 4 \end{bmatrix}$, so $A + A^T = \begin{bmatrix} 2 & 5 \\ 5 & 8 \end{bmatrix}$, a symmetric matrix.

We have

$$B^T = \begin{bmatrix} 4 & 2 \\ -1 & 3 \\ 0 & 1 \end{bmatrix}$$

so

$$BB^T = \begin{bmatrix} 4 & -1 & 0 \\ 2 & 3 & 1 \end{bmatrix} \begin{bmatrix} 4 & 2 \\ -1 & 3 \\ 0 & 1 \end{bmatrix} = \begin{bmatrix} 17 & 5 \\ 5 & 14 \end{bmatrix}$$

and

$$B^TB = \begin{bmatrix} 4 & 2 \\ -1 & 3 \\ 0 & 1 \end{bmatrix} \begin{bmatrix} 4 & -1 & 0 \\ 2 & 3 & 1 \end{bmatrix} = \begin{bmatrix} 20 & 2 & 2 \\ 2 & 10 & 3 \\ 2 & 3 & 1 \end{bmatrix}$$

Thus, both BB^T and B^TB are symmetric, even though B is not even square! (Check that AA^T and A^TA are also symmetric.)

The next theorem says that the results of Example 3.21 are true in general.

Theorem 3.5

a. If A is a square matrix, then $A + A^T$ is a symmetric matrix.
b. For any matrix A, AA^T and A^TA are symmetric matrices.

Proof We prove (a) and leave proving (b) as Exercise 34. We simply check that

$$(A + A^T)^T = A^T + (A^T)^T = A^T + A = A + A^T$$

(using properties of the transpose and the commutativity of matrix addition). Thus, $A + A^T$ is equal to its own transpose and so, by definition, is symmetric.

Exercises 3.2

In Exercises 1–4, solve the equation for X, given that $A = \begin{bmatrix} 1 & 2 \\ 3 & 4 \end{bmatrix}$ *and* $B = \begin{bmatrix} -1 & 0 \\ 1 & 1 \end{bmatrix}$.

1. $X - 2A + 3B = O$

2. $2X = A - B$

3. $2(A + 2B) = 3X$

4. $2(A - B + X) = 3(X - A)$

In Exercises 5–8, write B as a linear combination of the other matrices, if possible.

5. $B = \begin{bmatrix} 2 & 5 \\ 0 & 3 \end{bmatrix}$, $A_1 = \begin{bmatrix} 1 & 2 \\ -1 & 1 \end{bmatrix}$, $A_2 = \begin{bmatrix} 0 & 1 \\ 2 & 1 \end{bmatrix}$

6. $B = \begin{bmatrix} 2 & 3 \\ -4 & 2 \end{bmatrix}$, $A_1 = \begin{bmatrix} 1 & 0 \\ 0 & 1 \end{bmatrix}$, $A_2 = \begin{bmatrix} 0 & -1 \\ 1 & 0 \end{bmatrix}$, $A_3 = \begin{bmatrix} 1 & 1 \\ 0 & 1 \end{bmatrix}$

7. $B = \begin{bmatrix} 3 & 1 & 1 \\ 0 & 1 & 0 \end{bmatrix}$, $A_1 = \begin{bmatrix} 1 & 0 & -1 \\ 0 & 1 & 0 \end{bmatrix}$, $A_2 = \begin{bmatrix} -1 & 2 & 0 \\ 0 & 1 & 0 \end{bmatrix}$, $A_3 = \begin{bmatrix} 1 & 1 & 1 \\ 0 & 0 & 0 \end{bmatrix}$

8. $B = \begin{bmatrix} 2 & -2 & 3 \\ 0 & 0 & -2 \\ 0 & 0 & 2 \end{bmatrix}$, $A_1 = \begin{bmatrix} 1 & 0 & 0 \\ 0 & 1 & 0 \\ 0 & 0 & 1 \end{bmatrix}$, $A_2 = \begin{bmatrix} 0 & 1 & 1 \\ 0 & 0 & 1 \\ 0 & 0 & 0 \end{bmatrix}$, $A_3 = \begin{bmatrix} -1 & 0 & -1 \\ 0 & 1 & 0 \\ 0 & 0 & -1 \end{bmatrix}$,

$A_4 = \begin{bmatrix} 1 & -1 & 1 \\ 0 & -1 & -1 \\ 0 & 0 & 1 \end{bmatrix}$

In Exercises 9–12, find the general form of the span of the indicated matrices, as in Example 3.17.

9. $\text{span}(A_1, A_2)$ in Exercise 5

10. $\text{span}(A_1, A_2, A_3)$ in Exercise 6

11. $\text{span}(A_1, A_2, A_3)$ in Exercise 7

12. $\text{span}(A_1, A_2, A_3, A_4)$ in Exercise 8

In Exercises 13–16, determine whether the given matrices are linearly independent.

13. $\begin{bmatrix} 1 & 2 \\ 3 & 4 \end{bmatrix}, \begin{bmatrix} 4 & 3 \\ 2 & 1 \end{bmatrix}$

14. $\begin{bmatrix} 1 & 2 \\ 4 & 3 \end{bmatrix}, \begin{bmatrix} 2 & 1 \\ -1 & 0 \end{bmatrix}, \begin{bmatrix} 1 & 1 \\ 1 & 1 \end{bmatrix}$

15. $\begin{bmatrix} 0 & 1 \\ 5 & 2 \\ -1 & 0 \end{bmatrix}, \begin{bmatrix} 1 & 0 \\ 2 & 3 \\ 1 & 1 \end{bmatrix}, \begin{bmatrix} -2 & -1 \\ 0 & 1 \\ 0 & 2 \end{bmatrix}, \begin{bmatrix} -1 & -3 \\ 1 & 9 \\ 4 & 5 \end{bmatrix}$

16. $\begin{bmatrix} 1 & -1 & 0 \\ 0 & 2 & 0 \\ 0 & 2 & 6 \end{bmatrix}, \begin{bmatrix} 2 & 1 & 0 \\ 0 & 3 & 0 \\ 0 & 4 & 9 \end{bmatrix}, \begin{bmatrix} 1 & 2 & 0 \\ 0 & 1 & 0 \\ 0 & 3 & 5 \end{bmatrix}, \begin{bmatrix} -1 & 1 & 0 \\ 0 & -1 & 0 \\ 0 & 0 & -4 \end{bmatrix}$

17. Prove Theorem 3.2(a)–(d).

18. Prove Theorem 3.2(e)–(h).

19. Prove Theorem 3.3(c).

20. Prove Theorem 3.3(d).

21. Prove the half of Theorem 3.3(e) that was not proved in the text.

22. Prove that, for square matrices A and B, $AB = BA$ if and only if $(A - B)(A + B) = A^2 - B^2$.

In Exercises 23–25, if $B = \begin{bmatrix} a & b \\ c & d \end{bmatrix}$, *find conditions on a, b, c, and d such that* $AB = BA$.

23. $A = \begin{bmatrix} 1 & 1 \\ 0 & 1 \end{bmatrix}$ **24.** $A = \begin{bmatrix} 1 & -1 \\ -1 & 1 \end{bmatrix}$ **25.** $A = \begin{bmatrix} 1 & 2 \\ 3 & 4 \end{bmatrix}$

26. Find conditions on a, b, c, and d such that $B = \begin{bmatrix} a & b \\ c & d \end{bmatrix}$ commutes with both $\begin{bmatrix} 1 & 0 \\ 0 & 0 \end{bmatrix}$ and $\begin{bmatrix} 0 & 0 \\ 0 & 1 \end{bmatrix}$.

27. Find conditions on a, b, c, and d such that $B = \begin{bmatrix} a & b \\ c & d \end{bmatrix}$ commutes with every 2×2 matrix.

28. Prove that if AB and BA are both defined, then AB and BA are both square matrices.

A square matrix is called ***upper triangular*** *if all of the entries below the main diagonal are zero. Thus, the form of an upper triangular matrix is*

$$\begin{bmatrix} * & * & \cdots & * & * \\ 0 & * & \cdots & * & * \\ 0 & 0 & \ddots & \vdots & \vdots \\ \vdots & \vdots & & * & * \\ 0 & 0 & \cdots & 0 & * \end{bmatrix}$$

where the entries marked $*$ *are arbitrary. A more formal definition of such a matrix* $A = [a_{ij}]$ *is that* $a_{ij} = 0$ *if* $i > j$.

29. Prove that the product of two upper triangular $n \times n$ matrices is upper triangular.

30. Prove Theorem 3.4(a)–(c).

31. Prove Theorem 3.4(e).

32. Using induction, prove that for all $n \geq 1$, $(A_1 + A_2 + \cdots + A_n)^T = A_1^T + A_2^T + \cdots + A_n^T$.

33. Using induction, prove that for all $n \geq 1$, $(A_1 A_2 \cdots A_n)^T = A_n^T \cdots A_2^T A_1^T$.

34. Prove Theorem 3.5(b).

35. **(a)** Prove that if A and B are symmetric $n \times n$ matrices, then so is $A + B$.
(b) Prove that if A is a symmetric $n \times n$ matrix, then so is kA for any scalar k.

36. **(a)** Give an example to show that if A and B are symmetric $n \times n$ matrices, then AB need not be symmetric.
(b) Prove that if A and B are symmetric $n \times n$ matrices, then AB is symmetric if and only if $AB = BA$.

A square matrix is called ***skew-symmetric*** *if* $A^T = -A$.

37. Which of the following matrices are skew-symmetric?

(a) $\begin{bmatrix} 1 & 2 \\ -2 & 3 \end{bmatrix}$ **(b)** $\begin{bmatrix} 0 & -1 \\ 1 & 0 \end{bmatrix}$

(c) $\begin{bmatrix} 0 & 3 & -1 \\ -3 & 0 & 2 \\ 1 & -2 & 0 \end{bmatrix}$ **(d)** $\begin{bmatrix} 0 & 1 & 2 \\ -1 & 0 & 5 \\ 2 & 5 & 0 \end{bmatrix}$

38. Give a componentwise definition of a skew-symmetric matrix.

39. Prove that the main diagonal of a skew-symmetric matrix must consist entirely of zeros.

40. Prove that if A and B are skew-symmetric $n \times n$ matrices, then so is $A + B$.

41. If A and B are skew-symmetric 2×2 matrices, under what conditions is AB skew-symmetric?

42. Prove that if A is an $n \times n$ matrix, then $A - A^T$ is skew-symmetric.

43. **(a)** Prove that any square matrix A can be written as the sum of a symmetric matrix and a skew-symmetric matrix. [*Hint:* Consider Theorem 3.5 and Exercise 42.]
(b) Illustrate part (a) for the matrix $A = \begin{bmatrix} 1 & 2 & 3 \\ 4 & 5 & 6 \\ 7 & 8 & 9 \end{bmatrix}$.

The ***trace*** *of an* $n \times n$ *matrix* $A = [a_{ij}]$ *is the sum of the entries on its main diagonal and is denoted by* $\text{tr}(A)$. *That is,*

$$\text{tr}(A) = a_{11} + a_{22} + \cdots + a_{nn}$$

44. If A and B are $n \times n$ matrices, prove the following properties of the trace:
(a) $\text{tr}(A + B) = \text{tr}(A) + \text{tr}(B)$
(b) $\text{tr}(kA) = k\text{tr}(A)$, where k is a scalar

45. Prove that if A and B are $n \times n$ matrices, then $\text{tr}(AB) = \text{tr}(BA)$.

46. If A is any matrix, to what is $\text{tr}(AA^T)$ equal?

47. Show that there are no 2×2 matrices A and B such that $AB - BA = I_2$.

3.3 The Inverse of a Matrix

In this section, we return to the matrix description $A\mathbf{x} = \mathbf{b}$ of a system of linear equations and look for ways to use matrix algebra to solve the system. By way of analogy, consider the equation $ax = b$, where a, b, and x represent real numbers and we want to solve for x. We can quickly figure out that we want $x = b/a$ as the solution, but we must remind ourselves that this is true only if $a \neq 0$. Proceeding more slowly, assuming that $a \neq 0$, we will reach the solution by the following sequence of steps:

$$ax = b \Rightarrow \frac{1}{a}(ax) = \frac{1}{a}(b) \Rightarrow \left(\frac{1}{a}(a)\right)x = \frac{b}{a} \Rightarrow 1 \cdot x = \frac{b}{a} \Rightarrow x = \frac{b}{a}$$

(This example shows how much we do in our head and how many properties of arithmetic and algebra we take for granted!)

To imitate this procedure for the matrix equation $A\mathbf{x} = \mathbf{b}$, what do we need? We need to find a matrix A' (analogous to $1/a$) such that $A'A = I$, an identity matrix (analogous to 1). *If* such a matrix exists (analogous to the requirement that $a \neq 0$), then we can do the following sequence of calculations:

$$A\mathbf{x} = \mathbf{b} \Rightarrow A'(A\mathbf{x}) = A'\mathbf{b} \Rightarrow (A'A)\mathbf{x} = A'\mathbf{b} \Rightarrow I\mathbf{x} = A'\mathbf{b} \Rightarrow \mathbf{x} = A'\mathbf{b}$$

(Why would each of these steps be justified?)

Our goal in this section is to determine precisely when we can find such a matrix A'. In fact, we are going to insist on a bit more: We want not only $A'A = I$ but also $AA' = I$. This requirement forces A and A' to be square matrices. (Why?)

Definition If A is an $n \times n$ matrix, an ***inverse*** of A is an $n \times n$ matrix A' with the property that

$$AA' = I \quad \text{and} \quad A'A = I$$

where $I = I_n$ is the $n \times n$ identity matrix. If such an A' exists, then A is called ***invertible.***

Example 3.22

If $A = \begin{bmatrix} 2 & 5 \\ 1 & 3 \end{bmatrix}$, then $A' = \begin{bmatrix} 3 & -5 \\ -1 & 2 \end{bmatrix}$ is an inverse of A, since

$$AA' = \begin{bmatrix} 2 & 5 \\ 1 & 3 \end{bmatrix}\begin{bmatrix} 3 & -5 \\ -1 & 2 \end{bmatrix} = \begin{bmatrix} 1 & 0 \\ 0 & 1 \end{bmatrix} \text{ and } A'A = \begin{bmatrix} 3 & -5 \\ -1 & 2 \end{bmatrix}\begin{bmatrix} 2 & 5 \\ 1 & 3 \end{bmatrix} = \begin{bmatrix} 1 & 0 \\ 0 & 1 \end{bmatrix}$$

Example 3.23

Show that the following matrices are not invertible:

(a) $O = \begin{bmatrix} 0 & 0 \\ 0 & 0 \end{bmatrix}$ (b) $B = \begin{bmatrix} 1 & 2 \\ 2 & 4 \end{bmatrix}$

Solution

(a) It is easy to see that the zero matrix O does not have an inverse. If it did, then there would be a matrix O' such that $OO' = I = O'O$. But the product of the zero matrix with any other matrix is the zero matrix, and so OO' could never equal the identity matrix I. (Notice that this proof makes no reference to the size of the matrices and so is true for $n \times n$ matrices in general.)

(b) Suppose B has an inverse $B' = \begin{bmatrix} w & x \\ y & z \end{bmatrix}$. The equation $BB' = I$ gives

$$\begin{bmatrix} 1 & 2 \\ 2 & 4 \end{bmatrix}\begin{bmatrix} w & x \\ y & z \end{bmatrix} = \begin{bmatrix} 1 & 0 \\ 0 & 1 \end{bmatrix}$$

from which we get the equations

$$\begin{aligned} w \qquad + 2y \qquad &= 1 \\ x \qquad + 2z &= 0 \\ 2w \qquad + 4y \qquad &= 0 \\ 2x \qquad + 4z &= 1 \end{aligned}$$

Subtracting twice the first equation from the third yields $0 = -2$, which is clearly absurd. Thus, there is no solution. (Row reduction gives the same result but is not really needed here.) We deduce that no such matrix B' exists; that is, B is not invertible. (In fact, it does not even have an inverse that works on one side, let alone two!)

Remarks

- Even though we have seen that matrix multiplication is not, in general, commutative, A' (if it exists) must satisfy $A'A = AA'$.
- The examples above raise two questions, which we will answer in this section:

(1) How can we know when a matrix has an inverse?
(2) If a matrix does have an inverse, how can we find it?

- We have not ruled out the possibility that a matrix A might have more than one inverse. The next theorem assures us that this cannot happen.

Theorem 3.6 If A is an invertible matrix, then its inverse is unique.

Proof In mathematics, a standard way to show that there is just one of something is to show that there cannot be more than one. So, suppose that A has two inverses—say, A' and A''. Then

$$AA' = I = A'A \quad \text{and} \quad AA'' = I = A''A$$

Thus,
$$A' = A'I = A'(AA'') = (A'A)A'' = IA'' = A''$$

Hence, $A' = A''$, and the inverse is unique.

Thanks to this theorem, we can now refer to *the* inverse of an invertible matrix. From now on, when A is invertible, we will denote its (unique) inverse by A^{-1} (pronounced "A inverse").

Warning Do not be tempted to write $A^{-1} = \dfrac{1}{A}$! There is no such operation as "division by a matrix." Even if there were, how on earth could we divide the *scalar* 1 by the *matrix A*? If you ever feel tempted to "divide" by a matrix, what you really want to do is multiply by its inverse.

We can now complete the analogy that we set up at the beginning of this section.

Theorem 3.7 If A is an invertible $n \times n$ matrix, then the system of linear equations given by $A\mathbf{x} = \mathbf{b}$ has the unique solution $\mathbf{x} = A^{-1}\mathbf{b}$ for any $\mathbf{b}$ in $\mathbb{R}^n$.

Proof Theorem 3.7 essentially formalizes the observation we made at the beginning of this section. We will go through it again, a little more carefully this time. We are asked to prove two things: that $A\mathbf{x} = \mathbf{b}$ *has* a solution and that it has *only one* solution. (In mathematics, such a proof is called an "existence and uniqueness" proof.)

To show that a solution exists, we need only verify that $\mathbf{x} = A^{-1}\mathbf{b}$ works. We check that

$$A(A^{-1}\mathbf{b}) = (AA^{-1})\mathbf{b} = I\mathbf{b} = \mathbf{b}$$

So $A^{-1}\mathbf{b}$ satisfies the equation $A\mathbf{x} = \mathbf{b}$, and hence there is *at least* this solution.

To show that this solution is unique, suppose $\mathbf{y}$ is another solution. Then $A\mathbf{y} = \mathbf{b}$, and multiplying both sides of the equation by A^{-1} on the left, we obtain the chain of implications

$$A^{-1}(A\mathbf{y}) = A^{-1}\mathbf{b} \Rightarrow (A^{-1}A)\mathbf{y} = A^{-1}\mathbf{b} \Rightarrow I\mathbf{y} = A^{-1}\mathbf{b} \Rightarrow \mathbf{y} = A^{-1}\mathbf{b}$$

Thus, $\mathbf{y}$ is the same solution as before, and therefore the solution is unique.

So, returning to the questions we raised in the Remarks before Theorem 3.6, how can we tell if a matrix is invertible and how can we find its inverse when it is invertible? We will give a general procedure shortly, but the situation for 2×2 matrices is sufficiently simple to warrant being singled out.

Theorem 3.8 If $A = \begin{bmatrix} a & b \\ c & d \end{bmatrix}$, then A is invertible if $ad - bc \neq 0$, in which case

$$A^{-1} = \frac{1}{ad - bc}\begin{bmatrix} d & -b \\ -c & a \end{bmatrix}$$

If $ad - bc = 0$, then A is not invertible.

The expression $ad - bc$ is called the ***determinant*** of A, denoted $\det A$. The formula for the inverse of $\begin{bmatrix} a & b \\ c & d \end{bmatrix}$ (when it exists) is thus $\dfrac{1}{\det A}$ times the matrix obtained by interchanging the entries on the main diagonal and changing the signs on the other two entries. In addition to giving this formula, Theorem 3.8 says that a 2×2 matrix A is invertible if and only if $\det A \neq 0$. We will see in Chapter 4 that the determinant can be defined for all square matrices and that this result remains true, although there is no simple formula for the inverse of larger square matrices.

Proof Suppose that $\det A = ad - bc \neq 0$. Then

$$\begin{bmatrix} a & b \\ c & d \end{bmatrix}\begin{bmatrix} d & -b \\ -c & a \end{bmatrix} = \begin{bmatrix} ad - bc & -ab + ba \\ cd - dc & -cb + da \end{bmatrix} = \begin{bmatrix} ad - bc & 0 \\ 0 & ad - bc \end{bmatrix} = \det A\begin{bmatrix} 1 & 0 \\ 0 & 1 \end{bmatrix}$$

Similarly,

$$\begin{bmatrix} d & -b \\ -c & a \end{bmatrix}\begin{bmatrix} a & b \\ c & d \end{bmatrix} = \det A\begin{bmatrix} 1 & 0 \\ 0 & 1 \end{bmatrix}$$

Since $\det A \neq 0$, we can multiply both sides of each equation by $1/\det A$ to obtain

$$\begin{bmatrix} a & b \\ c & d \end{bmatrix}\left(\frac{1}{\det A}\begin{bmatrix} d & -b \\ -c & a \end{bmatrix}\right) = \begin{bmatrix} 1 & 0 \\ 0 & 1 \end{bmatrix}$$

and
$$\left(\frac{1}{\det A}\begin{bmatrix} d & -b \\ -c & a \end{bmatrix}\right)\begin{bmatrix} a & b \\ c & d \end{bmatrix} = \begin{bmatrix} 1 & 0 \\ 0 & 1 \end{bmatrix}$$

[Note that we have used property (d) of Theorem 3.3.] Thus, the matrix

$$\frac{1}{\det A}\begin{bmatrix} d & -b \\ -c & a \end{bmatrix}$$

satisfies the definition of an inverse, so A is invertible. Since the inverse of A is unique, by Theorem 3.6, we must have

$$A^{-1} = \frac{1}{\det A}\begin{bmatrix} d & -b \\ -c & a \end{bmatrix}$$

Conversely, assume that $ad - bc = 0$. We will consider separately the cases where $a \neq 0$ and where $a = 0$. If $a \neq 0$, then $d = bc/a$, so the matrix can be written as

$$A = \begin{bmatrix} a & b \\ c & d \end{bmatrix} = \begin{bmatrix} a & b \\ ac/a & bc/a \end{bmatrix} = \begin{bmatrix} a & b \\ ka & kb \end{bmatrix}$$

where $k = c/a$. In other words, the second row of A is a multiple of the first. Referring to Example 3.23(b), we see that if A has an inverse $\begin{bmatrix} w & x \\ y & z \end{bmatrix}$, then

$$\begin{bmatrix} a & b \\ ka & kb \end{bmatrix}\begin{bmatrix} w & x \\ y & z \end{bmatrix} = \begin{bmatrix} 1 & 0 \\ 0 & 1 \end{bmatrix}$$

and the corresponding system of linear equations

$$\begin{aligned} aw \qquad\quad + by \qquad\quad &= 1 \\ ax \qquad\quad + bz &= 0 \\ kaw \qquad\quad + kby \qquad\quad &= 0 \\ kax \qquad\quad + kbz &= 1 \end{aligned}$$

has no solution. (Why?)

If $a = 0$, then $ad - bc = 0$ implies that $bc = 0$, and therefore either b or c is 0. Thus, A is of the form

$$\begin{bmatrix} 0 & 0 \\ c & d \end{bmatrix} \quad \text{or} \quad \begin{bmatrix} 0 & b \\ 0 & d \end{bmatrix}$$

In the first case, $\begin{bmatrix} 0 & 0 \\ c & d \end{bmatrix}\begin{bmatrix} w & x \\ y & z \end{bmatrix} = \begin{bmatrix} 0 & 0 \\ * & * \end{bmatrix} \neq \begin{bmatrix} 1 & 0 \\ 0 & 1 \end{bmatrix}$. Similarly, $\begin{bmatrix} 0 & b \\ 0 & d \end{bmatrix}$ cannot have an inverse. (Verify this.)

Consequently, if $ad - bc = 0$, then A is not invertible.

Example 3.24

Find the inverses of $A = \begin{bmatrix} 1 & 2 \\ 3 & 4 \end{bmatrix}$ and $B = \begin{bmatrix} 12 & -15 \\ 4 & -5 \end{bmatrix}$, if they exist.

Solution We have $\det A = 1(4) - 2(3) = -2 \neq 0$, so A is invertible, with

$$A^{-1} = \frac{1}{-2}\begin{bmatrix} 4 & -2 \\ -3 & 1 \end{bmatrix} = \begin{bmatrix} -2 & 1 \\ \frac{3}{2} & -\frac{1}{2} \end{bmatrix}$$

(Check this.)

On the other hand, $\det B = 12(-5) - (-15)(4) = 0$, so B is not invertible.

Example 3.25

Use the inverse of the coefficient matrix to solve the linear system

$$\begin{aligned} x + 2y &= 3 \\ 3x + 4y &= -2 \end{aligned}$$

Solution The coefficient matrix is the matrix $A = \begin{bmatrix} 1 & 2 \\ 3 & 4 \end{bmatrix}$, whose inverse we computed in Example 3.24. By Theorem 3.7, $A\mathbf{x} = \mathbf{b}$ has the unique solution $\mathbf{x} = A^{-1}\mathbf{b}$. Here we have $\mathbf{b} = \begin{bmatrix} 3 \\ -2 \end{bmatrix}$; thus, the solution to the given system is

$$\mathbf{x} = \begin{bmatrix} -2 & 1 \\ \frac{3}{2} & -\frac{1}{2} \end{bmatrix}\begin{bmatrix} 3 \\ -2 \end{bmatrix} = \begin{bmatrix} -8 \\ \frac{11}{2} \end{bmatrix}$$

Remark Solving a linear system $A\mathbf{x} = \mathbf{b}$ via $\mathbf{x} = A^{-1}\mathbf{b}$ would appear to be a good method. Unfortunately, except for 2×2 coefficient matrices and matrices with certain special forms, it is almost always faster to use Gaussian or Gauss-Jordan elimination to find the solution directly. (See Exercise 13.) Furthermore, the technique of Example 3.25 works only when the coefficient matrix is square and invertible, while elimination methods can always be applied.

Properties of Invertible Matrices

The following theorem records some of the most important properties of invertible matrices.

Theorem 3.9

a. If A is an invertible matrix, then A^{-1} is invertible and

$$(A^{-1})^{-1} = A$$

b. If A is an invertible matrix and c is a nonzero scalar, then cA is an invertible matrix and

$$(cA)^{-1} = \frac{1}{c}A^{-1}$$

c. If A and B are invertible matrices of the same size, then AB is invertible and

$$(AB)^{-1} = B^{-1}A^{-1}$$

d. If A is an invertible matrix, then A^T is invertible and

$$(A^T)^{-1} = (A^{-1})^T$$

e. If A is an invertible matrix, then A^n is invertible for all nonnegative integers n and

$$(A^n)^{-1} = (A^{-1})^n$$

Proof We will prove properties (a), (c), and (e), leaving properties (b) and (d) to be proven in Exercises 14 and 15.

(a) To show that A^{-1} is invertible, we must argue that there is a matrix X such that

$$A^{-1}X = I = XA^{-1}$$

But A certainly satisfies these equations in place of X, so A^{-1} is invertible and A is *an* inverse of A^{-1}. Since inverses are unique, this means that $(A^{-1})^{-1} = A$.

(c) Here we must show that there is a matrix X such that

$$(AB)X = I = X(AB)$$

The claim is that substituting $B^{-1}A^{-1}$ for X works. We check that

$$(AB)(B^{-1}A^{-1}) = A(BB^{-1})A^{-1} = AIA^{-1} = AA^{-1} = I$$

where we have used associativity to shift the parentheses. Similarly, $(B^{-1}A^{-1})(AB) = I$ (check!), so AB is invertible and its inverse is $B^{-1}A^{-1}$.

(e) The basic idea here is easy enough. For example, when $n = 2$, we have

$$A^2(A^{-1})^2 = AAA^{-1}A^{-1} = AIA^{-1} = AA^{-1} = I$$

Similarly, $(A^{-1})^2A^2 = I$. Thus, $(A^{-1})^2$ is the inverse of A^2. It is not difficult to see that a similar argument works for any higher integer value of n. However, mathematical induction is the way to carry out the proof.

The basis step is when $n = 0$, in which case we are being asked to prove that A^0 is invertible and that

$$(A^0)^{-1} = (A^{-1})^0$$

This is the same as showing that I is invertible and that $I^{-1} = I$, which is clearly true. (Why? See Exercise 16.)

Now we assume that the result is true when $n = k$, where k is a specific nonnegative integer. That is, the induction hypothesis is to assume that A^k is invertible and that

$$(A^k)^{-1} = (A^{-1})^k$$

The induction step requires that we prove that A^{k+1} is invertible and that $(A^{k+1})^{-1} = (A^{-1})^{k+1}$. Now we know from (c) that $A^{k+1} = A^kA$ is invertible, since A and (by hypothesis) A^k are both invertible. Moreover,

$$\begin{aligned}
(A^{-1})^{k+1} &= (A^{-1})^kA^{-1} \\
&= (A^k)^{-1}A^{-1} && \text{by the induction hypothesis} \\
&= (AA^k)^{-1} && \text{by property (c)} \\
&= (A^{k+1})^{-1}
\end{aligned}$$

Therefore, A^n is invertible for all nonnegative integers n, and $(A^n)^{-1} = (A^{-1})^n$ by the principle of mathematical induction.

Remarks

- While all of the properties of Theorem 3.9 are useful, (c) is the one you should highlight. It is perhaps the most important algebraic property of matrix inverses. It is also the one that is easiest to get wrong. In Exercise 17, you are asked to give a counterexample to show that, contrary to what we might like, $(AB)^{-1} \neq A^{-1}B^{-1}$ in general. The correct property, $(AB)^{-1} = B^{-1}A^{-1}$, is sometimes called the socks-and-shoes rule, because, although we put our socks on before our shoes, we take them off in the reverse order.
- Property (c) generalizes to products of finitely many invertible matrices: If $A_1, A_2, \ldots, A_n$ are invertible matrices of the same size, then $A_1A_2 \cdots A_n$ is invertible and

$$(A_1A_2\cdots A_n)^{-1} = A_n^{-1}\cdots A_2^{-1}A_1^{-1}$$

(See Exercise 18.) Thus, we can state:

> The inverse of a product of invertible matrices is the product of their inverses in the reverse order.

- Since, for real numbers, $\dfrac{1}{a+b} \neq \dfrac{1}{a} + \dfrac{1}{b}$, we should not expect that, for square matrices, $(A + B)^{-1} = A^{-1} + B^{-1}$ (and, indeed, this is not true in general; see Exercise 19). In fact, except for special matrices, there is no formula for $(A + B)^{-1}$.
- Property (e) allows us to define negative integer powers of an invertible matrix:

Definition If A is an invertible matrix and n is a positive integer, then A^{-n} is defined by

$$A^{-n} = (A^{-1})^n = (A^n)^{-1}$$

With this definition, it can be shown that the rules for exponentiation, $A^rA^s = A^{r+s}$ and $(A^r)^s = A^{rs}$, hold for all integers r and s, provided A is invertible.

One use of the algebraic properties of matrices is to help solve equations involving matrices. The next example illustrates the process. Note that we must pay particular attention to the order of the matrices in the product.

Example 3.26

Solve the following matrix equation for X (assuming that the matrices involved are such that all of the indicated operations are defined):

$$A^{-1}(BX)^{-1} = (A^{-1}B^3)^2$$

Solution There are many ways to proceed here. One solution is

$$\begin{aligned} A^{-1}(BX)^{-1} = (A^{-1}B^3)^2 &\Rightarrow ((BX)A)^{-1} = (A^{-1}B^3)^2 \\ &\Rightarrow [((BX)A)^{-1}]^{-1} = [(A^{-1}B^3)^2]^{-1} \\ &\Rightarrow (BX)A = [(A^{-1}B^3)(A^{-1}B^3)]^{-1} \\ &\Rightarrow (BX)A = B^{-3}(A^{-1})^{-1}B^{-3}(A^{-1})^{-1} \\ &\Rightarrow BXA = B^{-3}AB^{-3}A \\ &\Rightarrow B^{-1}BXAA^{-1} = B^{-1}B^{-3}AB^{-3}AA^{-1} \\ &\Rightarrow IXI = B^{-4}AB^{-3}I \\ &\Rightarrow X = B^{-4}AB^{-3} \end{aligned}$$

(Can you justify each step?) Note the careful use of Theorem 3.9(c) and the expansion of $(A^{-1}B^3)^2$. We have also made liberal use of the associativity of matrix multiplication to simplify the placement (or elimination) of parentheses.

Elementary Matrices

We are going to use matrix multiplication to take a different perspective on the row reduction of matrices. In the process, you will discover many new and important insights into the nature of invertible matrices.

If

$$E = \begin{bmatrix} 1 & 0 & 0 \\ 0 & 0 & 1 \\ 0 & 1 & 0 \end{bmatrix} \quad \text{and} \quad A = \begin{bmatrix} 5 & 7 \\ -1 & 0 \\ 8 & 3 \end{bmatrix}$$

we find that

$$EA = \begin{bmatrix} 5 & 7 \\ 8 & 3 \\ -1 & 0 \end{bmatrix}$$

In other words, multiplying A by E (on the left) has the same effect as interchanging rows 2 and 3 of A. What is significant about E? It is simply the matrix we obtain by applying the same elementary row operation, $R_2 \leftrightarrow R_3$, to the identity matrix I_3. It turns out that this always works.

Definition An ***elementary matrix*** is any matrix that can be obtained by performing an elementary row operation on an identity matrix.

Since there are three types of elementary row operations, there are three corresponding types of elementary matrices. Here are some more elementary matrices.

Example 3.27 Let

$$E_1 = \begin{bmatrix} 1 & 0 & 0 & 0 \\ 0 & 3 & 0 & 0 \\ 0 & 0 & 1 & 0 \\ 0 & 0 & 0 & 1 \end{bmatrix}, \quad E_2 = \begin{bmatrix} 0 & 0 & 1 & 0 \\ 0 & 1 & 0 & 0 \\ 1 & 0 & 0 & 0 \\ 0 & 0 & 0 & 1 \end{bmatrix}, \quad \text{and} \quad E_3 = \begin{bmatrix} 1 & 0 & 0 & 0 \\ 0 & 1 & 0 & 0 \\ 0 & 0 & 1 & 0 \\ 0 & -2 & 0 & 1 \end{bmatrix}$$

Each of these matrices has been obtained from the identity matrix I_4 by applying a single elementary row operation. The matrix E_1 corresponds to $3R_2$, E_2 to $R_1 \leftrightarrow R_3$, and E_3 to $R_4 - 2R_2$. Observe that when we left-multiply a $4 \times n$ matrix by one of these elementary matrices, the corresponding elementary row operation is performed on the matrix. For example, if

$$A = \begin{bmatrix} a_{11} & a_{12} & a_{13} \\ a_{21} & a_{22} & a_{23} \\ a_{31} & a_{32} & a_{33} \\ a_{41} & a_{42} & a_{43} \end{bmatrix}$$

then

$$E_1A = \begin{bmatrix} a_{11} & a_{12} & a_{13} \\ 3a_{21} & 3a_{22} & 3a_{23} \\ a_{31} & a_{32} & a_{33} \\ a_{41} & a_{42} & a_{43} \end{bmatrix}, \quad E_2A = \begin{bmatrix} a_{31} & a_{32} & a_{33} \\ a_{21} & a_{22} & a_{23} \\ a_{11} & a_{12} & a_{13} \\ a_{41} & a_{42} & a_{43} \end{bmatrix},$$

and

$$E_3A = \begin{bmatrix} a_{11} & a_{12} & a_{13} \\ a_{21} & a_{22} & a_{23} \\ a_{31} & a_{32} & a_{33} \\ a_{41} - 2a_{21} & a_{42} - 2a_{22} & a_{43} - 2a_{23} \end{bmatrix}$$

Example 3.27 and Exercises 24–30 should convince you that *any* elementary row operation on *any* matrix can be accomplished by left-multiplying by a suitable elementary matrix. We record this fact as a theorem, the proof of which is omitted.

Theorem 3.10 Let E be the elementary matrix obtained by performing an elementary row operation on I_n. If the same elementary row operation is performed on an $n \times r$ matrix A, the result is the same as the matrix EA.

Remark From a computational point of view, it is not a good idea to use elementary matrices to perform elementary row operations—just do them directly. However, elementary matrices can provide some valuable insights into invertible matrices and the solution of systems of linear equations.

We have already observed that every elementary row operation can be "undone," or "reversed." This same observation applied to elementary matrices shows us that they are invertible.

Example 3.28 Let

$$E_1 = \begin{bmatrix} 1 & 0 & 0 \\ 0 & 0 & 1 \\ 0 & 1 & 0 \end{bmatrix}, \quad E_2 = \begin{bmatrix} 1 & 0 & 0 \\ 0 & 4 & 0 \\ 0 & 0 & 1 \end{bmatrix}, \quad \text{and} \quad E_3 = \begin{bmatrix} 1 & 0 & 0 \\ 0 & 1 & 0 \\ -2 & 0 & 1 \end{bmatrix}$$

Then E_1 corresponds to $R_2 \leftrightarrow R_3$, which is undone by doing $R_2 \leftrightarrow R_3$ again. Thus, $E_1^{-1} = E_1$. (Check by showing that $E_1^2 = E_1E_1 = I$.) The matrix E_2 comes from $4R_2$, which is undone by performing $\frac{1}{4}R_2$. Thus,

$$E_2^{-1} = \begin{bmatrix} 1 & 0 & 0 \\ 0 & \frac{1}{4} & 0 \\ 0 & 0 & 1 \end{bmatrix}$$

which can be easily checked. Finally, E_3 corresponds to the elementary row operation $R_3 - 2R_1$, which can be undone by the elementary row operation $R_3 + 2R_1$. So, in this case,

$$E_3^{-1} = \begin{bmatrix} 1 & 0 & 0 \\ 0 & 1 & 0 \\ 2 & 0 & 1 \end{bmatrix}$$

(Again, it is easy to check this by confirming that the product of this matrix and E_3, in both orders, is I.)

Notice that not only is each elementary matrix invertible, but its inverse is another elementary matrix of the same type. We record this finding as the next theorem.

Theorem 3.11 Each elementary matrix is invertible, and its inverse is an elementary matrix of the same type.

The Fundamental Theorem of Invertible Matrices

We are now in a position to prove one of the main results in this book—a set of equivalent characterizations of what it means for a matrix to be invertible. In a sense, much of linear algebra is connected to this theorem, either in the development of these characterizations or in their application. As you might expect, given this introduction, we will use this theorem a great deal. Make it your friend!

We refer to Theorem 3.12 as the first version of the Fundamental Theorem, since we will add to it in subsequent chapters. You are reminded that, when we say that a set of statements about a matrix A are equivalent, we mean that, for a given A, the statements are either all true or all false.

Theorem 3.12 **The Fundamental Theorem of Invertible Matrices: Version 1**

Let A be an $n \times n$ matrix. The following statements are equivalent:

a. A is invertible.
b. $A\mathbf{x} = \mathbf{b}$ has a unique solution for every $\mathbf{b}$ in $\mathbb{R}^n$.
c. $A\mathbf{x} = \mathbf{0}$ has only the trivial solution.
d. The reduced row echelon form of A is I_n.
e. A is a product of elementary matrices.

Proof We will establish the theorem by proving the circular chain of implications

$$(a) \Rightarrow (b) \Rightarrow (c) \Rightarrow (d) \Rightarrow (e) \Rightarrow (a)$$

(a) $\Rightarrow$ (b) We have already shown that if A is invertible, then $A\mathbf{x} = \mathbf{b}$ has the unique solution $\mathbf{x} = A^{-1}\mathbf{b}$ for any $\mathbf{b}$ in $\mathbb{R}^n$ (Theorem 3.7).

(b) $\Rightarrow$ (c) Assume that $A\mathbf{x} = \mathbf{b}$ has a unique solution for any $\mathbf{b}$ in $\mathbb{R}^n$. This implies, in particular, that $A\mathbf{x} = \mathbf{0}$ has a unique solution. But a homogeneous system $A\mathbf{x} = \mathbf{0}$ always has $\mathbf{x} = \mathbf{0}$ as *one* solution. So in this case, $\mathbf{x} = \mathbf{0}$ must be *the* solution.

(c) $\Rightarrow$ (d) Suppose that $A\mathbf{x} = \mathbf{0}$ has only the trivial solution. The corresponding system of equations is

$$\begin{aligned} a_{11}x_1 + a_{12}x_2 + \cdots + a_{1n}x_n &= 0 \\ a_{21}x_1 + a_{22}x_2 + \cdots + a_{2n}x_n &= 0 \\ &\vdots \\ a_{n1}x_1 + a_{n2}x_2 + \cdots + a_{nn}x_n &= 0 \end{aligned}$$

and we are assuming that its solution is

$$\begin{aligned} x_1 \qquad\qquad\qquad &= 0 \\ x_2 \qquad\quad &= 0 \\ \ddots \qquad & \\ x_n &= 0 \end{aligned}$$

In other words, Gauss-Jordan elimination applied to the augmented matrix of the system gives

$$[A \mid \mathbf{0}] = \left[\begin{array}{cccc|c} a_{11} & a_{12} & \cdots & a_{1n} & 0 \\ a_{21} & a_{22} & \cdots & a_{2n} & 0 \\ \vdots & \vdots & \ddots & \vdots & \vdots \\ a_{n1} & a_{n2} & \cdots & a_{nn} & 0 \end{array}\right] \longrightarrow \left[\begin{array}{cccc|c} 1 & 0 & \cdots & 0 & 0 \\ 0 & 1 & \cdots & 0 & 0 \\ \vdots & \vdots & \ddots & \vdots & \vdots \\ 0 & 0 & \cdots & 1 & 0 \end{array}\right] = [I_n \mid \mathbf{0}]$$

Thus, the reduced row echelon form of A is I_n.

(d) $\Rightarrow$ (e) If we assume that the reduced row echelon form of A is I_n, then A can be reduced to I_n using a finite sequence of elementary row operations. By Theorem 3.10, each one of these elementary row operations can be achieved by left-multiplying by an appropriate elementary matrix. If the appropriate sequence of elementary matrices is $E_1, E_2, \ldots, E_k$ (in that order), then we have

$$E_k \cdots E_2E_1A = I_n$$

According to Theorem 3.11, these elementary matrices are all invertible. Therefore, so is their product, and we have

$$A = (E_k \cdots E_2E_1)^{-1}I_n = (E_k \cdots E_2E_1)^{-1} = E_1^{-1}E_2^{-1} \cdots E_k^{-1}$$

Again, each E_i^{-1} is another elementary matrix, by Theorem 3.11, so we have written A as a product of elementary matrices, as required.

(e) $\Rightarrow$ (a) If A is a product of elementary matrices, then A is invertible, since elementary matrices are invertible and products of invertible matrices are invertible.

Example 3.29

If possible, express $A = \begin{bmatrix} 2 & 3 \\ 1 & 3 \end{bmatrix}$ as a product of elementary matrices.

Solution We row reduce A as follows:

$$A = \begin{bmatrix} 2 & 3 \\ 1 & 3 \end{bmatrix} \xrightarrow{R_1 \leftrightarrow R_2} \begin{bmatrix} 1 & 3 \\ 2 & 3 \end{bmatrix} \xrightarrow{R_2 - 2R_1} \begin{bmatrix} 1 & 3 \\ 0 & -3 \end{bmatrix}$$

$$\xrightarrow{R_1 + R_2} \begin{bmatrix} 1 & 0 \\ 0 & -3 \end{bmatrix} \xrightarrow{-\frac{1}{3}R_2} \begin{bmatrix} 1 & 0 \\ 0 & 1 \end{bmatrix} = I_2$$

Thus, the reduced row echelon form of A is the identity matrix, so the Fundamental Theorem assures us that A is invertible and can be written as a product of elementary matrices. We have $E_4E_3E_2E_1A = I$, where

$$E_1 = \begin{bmatrix} 0 & 1 \\ 1 & 0 \end{bmatrix}, \quad E_2 = \begin{bmatrix} 1 & 0 \\ -2 & 1 \end{bmatrix}, \quad E_3 = \begin{bmatrix} 1 & 1 \\ 0 & 1 \end{bmatrix}, \quad E_4 = \begin{bmatrix} 1 & 0 \\ 0 & -\frac{1}{3} \end{bmatrix}$$

are the elementary matrices corresponding to the four elementary row operations used to reduce A to I. As in the proof of the theorem, we have

$$A = (E_4E_3E_2E_1)^{-1} = E_1^{-1}E_2^{-1}E_3^{-1}E_4^{-1} = \begin{bmatrix} 0 & 1 \\ 1 & 0 \end{bmatrix}\begin{bmatrix} 1 & 0 \\ 2 & 1 \end{bmatrix}\begin{bmatrix} 1 & -1 \\ 0 & 1 \end{bmatrix}\begin{bmatrix} 1 & 0 \\ 0 & -3 \end{bmatrix}$$

as required.

Remark Because the sequence of elementary row operations that transforms A into I is not unique, neither is the representation of A as a product of elementary matrices. (Find a different way to express A as a product of elementary matrices.)

"Never bring a cannon on stage in Act I unless you intend to fire it by the last act." –Anton Chekhov

The Fundamental Theorem is surprisingly powerful. To illustrate its power, we consider two of its consequences. The first is that, although the definition of an invertible matrix states that a matrix A is invertible if there is a matrix B such that *both* $AB = I$ *and* $BA = I$ are satisfied, we need only check *one* of these equations. Thus, we can cut our work in half!

Theorem 3.13

Let A be a square matrix. If B is a square matrix such that either $AB = I$ or $BA = I$, then A is invertible and $B = A^{-1}$.

Proof Suppose $BA = I$. Consider the equation $A\mathbf{x} = \mathbf{0}$. Left-multiplying by B, we have $BA\mathbf{x} = B\mathbf{0}$. This implies that $\mathbf{x} = I\mathbf{x} = \mathbf{0}$. Thus, the system represented by $A\mathbf{x} = \mathbf{0}$ has the unique solution $\mathbf{x} = \mathbf{0}$. From the equivalence of (c) and (a) in the Fundamental Theorem, we know that A is invertible. (That is, A^{-1} exists and satisfies $AA^{-1} = I = A^{-1}A$.)

If we now right-multiply both sides of $BA = I$ by A^{-1}, we obtain

$$BAA^{-1} = IA^{-1} \Rightarrow BI = A^{-1} \Rightarrow B = A^{-1}$$

(The proof in the case of $AB = I$ is left as Exercise 41.)

The next consequence of the Fundamental Theorem is the basis for an efficient method of computing the inverse of a matrix.

Theorem 3.14 Let A be a square matrix. If a sequence of elementary row operations reduces A to I, then the same sequence of elementary row operations transforms I into A^{-1}.

Proof If A is row equivalent to I, then we can achieve the reduction by left-multiplying by a sequence $E_1, E_2, \dots, E_k$ of elementary matrices. Therefore, we have $E_k \cdots E_2E_1A = I$. Setting $B = E_k \cdots E_2E_1$ gives $BA = I$. By Theorem 3.13, A is invertible and $A^{-1} = B$. Now applying the same sequence of elementary row operations to I is equivalent to left-multiplying I by $E_k \cdots E_2E_1 = B$. The result is

$$E_k \cdots E_2E_1I = BI = B = A^{-1}$$

Thus, I is transformed into A^{-1} by the same sequence of elementary row operations.

The Gauss-Jordan Method for Computing the Inverse

We can perform row operations on A and I simultaneously by constructing a "super-augmented matrix" $[A \mid I]$. Theorem 3.14 shows that if A is row equivalent to I [which, by the Fundamental Theorem (d) $\Leftrightarrow$ (a), means that A is invertible], then elementary row operations will yield

$$[A \mid I] \longrightarrow [I \mid A^{-1}]$$

If A cannot be reduced to I, then the Fundamental Theorem guarantees us that A is not invertible.

The procedure just described is simply Gauss-Jordan elimination performed on an $n \times 2n$, instead of an $n \times (n+1)$, augmented matrix. Another way to view this procedure is to look at the problem of finding A^{-1} as solving the matrix equation $AX = I_n$ for an $n \times n$ matrix X. (This is sufficient, by the Fundamental Theorem, since a right inverse of A must be a two-sided inverse.) If we denote the columns of X by $\mathbf{x}_1, \dots, \mathbf{x}_n$, then this matrix equation is equivalent to solving for the columns of X, one at a time. Since the columns of I_n are the standard unit vectors $\mathbf{e}_1, \dots, \mathbf{e}_n$, we thus have n systems of linear equations, all with coefficient matrix A:

$$A\mathbf{x}_1 = \mathbf{e}_1, \dots, A\mathbf{x}_n = \mathbf{e}_n$$

Since the same sequence of row operations is needed to bring A to reduced row echelon form in each case, the augmented matrices for these systems, $[A \mid \mathbf{e}_1], \dots, [A \mid \mathbf{e}_n]$, can be combined as

$$[A \mid \mathbf{e}_1\, \mathbf{e}_2 \cdots \mathbf{e}_n] = [A \mid I_n]$$

We now apply row operations to try to reduce A to I_n, which, if successful, will simultaneously solve for the columns of A^{-1}, transforming I_n into A^{-1}.

We illustrate this use of Gauss-Jordan elimination with three examples.

Example 3.30 Find the inverse of

$$A = \begin{bmatrix} 1 & 2 & -1 \\ 2 & 2 & 4 \\ 1 & 3 & -3 \end{bmatrix}$$

if it exists.

Solution Gauss-Jordan elimination produces

$$[A|I] = \left[\begin{array}{ccc|ccc} 1 & 2 & -1 & 1 & 0 & 0 \\ 2 & 2 & 4 & 0 & 1 & 0 \\ 1 & 3 & -3 & 0 & 0 & 1 \end{array}\right]$$

$$\xrightarrow{\substack{R_2 - 2R_1 \\ R_3 - R_1}} \left[\begin{array}{ccc|ccc} 1 & 2 & -1 & 1 & 0 & 0 \\ 0 & -2 & 6 & -2 & 1 & 0 \\ 0 & 1 & -2 & -1 & 0 & 1 \end{array}\right]$$

$$\xrightarrow{(-\frac{1}{2})R_2} \left[\begin{array}{ccc|ccc} 1 & 2 & -1 & 1 & 0 & 0 \\ 0 & 1 & -3 & 1 & -\frac{1}{2} & 0 \\ 0 & 1 & -2 & -1 & 0 & 1 \end{array}\right]$$

$$\xrightarrow{R_3 - R_2} \left[\begin{array}{ccc|ccc} 1 & 2 & -1 & 1 & 0 & 0 \\ 0 & 1 & -3 & 1 & -\frac{1}{2} & 0 \\ 0 & 0 & 1 & -2 & \frac{1}{2} & 1 \end{array}\right]$$

$$\xrightarrow{\substack{R_1 + R_3 \\ R_2 + 3R_3}} \left[\begin{array}{ccc|ccc} 1 & 2 & 0 & -1 & \frac{1}{2} & 1 \\ 0 & 1 & 0 & -5 & 1 & 3 \\ 0 & 0 & 1 & -2 & \frac{1}{2} & 1 \end{array}\right]$$

$$\xrightarrow{R_1 - 2R_2} \left[\begin{array}{ccc|ccc} 1 & 0 & 0 & 9 & -\frac{3}{2} & -5 \\ 0 & 1 & 0 & -5 & 1 & 3 \\ 0 & 0 & 1 & -2 & \frac{1}{2} & 1 \end{array}\right]$$

Therefore,

$$A^{-1} = \begin{bmatrix} 9 & -\frac{3}{2} & -5 \\ -5 & 1 & 3 \\ -2 & \frac{1}{2} & 1 \end{bmatrix}$$

(You should always check that $AA^{-1} = I$ by direct multiplication. By Theorem 3.13, we do not need to check that $A^{-1}A = I$ too.)

Remark Notice that we have used the variant of Gauss-Jordan elimination that first introduces all of the zeros *below* the leading 1s, from left to right and top to bottom, and then creates zeros *above* the leading 1s, from right to left and bottom to top. This approach saves on calculations, as we noted in Chapter 2, but you may find it easier, when working by hand, to create *all* of the zeros in each column as you go. The answer, of course, will be the same.

Example 3.31

Find the inverse of

$$A = \begin{bmatrix} 2 & 1 & -4 \\ -4 & -1 & 6 \\ -2 & 2 & -2 \end{bmatrix}$$

if it exists.

Solution We proceed as in Example 3.30, adjoining the identity matrix to A and then trying to manipulate $[A \mid I]$ into $[I \mid A^{-1}]$.

$$[A \mid I] = \left[\begin{array}{rrr|rrr} 2 & 1 & -4 & 1 & 0 & 0 \\ -4 & -1 & 6 & 0 & 1 & 0 \\ -2 & 2 & -2 & 0 & 0 & 1 \end{array}\right]$$

$$\xrightarrow[R_3+R_1]{R_2+2R_1} \left[\begin{array}{rrr|rrr} 2 & 1 & -4 & 1 & 0 & 0 \\ 0 & 1 & -2 & 2 & 1 & 0 \\ 0 & 3 & -6 & 1 & 0 & 1 \end{array}\right]$$

$$\xrightarrow{R_3-3R_2} \left[\begin{array}{rrr|rrr} 1 & 2 & -1 & 1 & 0 & 0 \\ 0 & 1 & -3 & 2 & 1 & 0 \\ 0 & 0 & 0 & -5 & -3 & 1 \end{array}\right]$$

At this point, we see that it is not possible to reduce A to I, since there is a row of zeros on the left-hand side of the augmented matrix. Consequently, A is not invertible.

As the next example illustrates, everything works the same way over $\mathbb{Z}_p$, where p is prime.

Example 3.32

Find the inverse of

$$A = \begin{bmatrix} 2 & 2 \\ 2 & 0 \end{bmatrix}$$

if it exists, over $\mathbb{Z}_3$.

Solution 1 We use the Gauss-Jordan method, remembering that all calculations are in $\mathbb{Z}_3$.

$$[A \mid I] = \left[\begin{array}{rr|rr} 2 & 2 & 1 & 0 \\ 2 & 0 & 0 & 1 \end{array}\right]$$

$$\xrightarrow{2R_1} \left[\begin{array}{rr|rr} 1 & 1 & 2 & 0 \\ 2 & 0 & 0 & 1 \end{array}\right]$$

$$\xrightarrow{R_2+R_1} \left[\begin{array}{rr|rr} 1 & 1 & 2 & 0 \\ 0 & 1 & 2 & 1 \end{array}\right]$$

$$\xrightarrow{R_1+2R_2} \left[\begin{array}{rr|rr} 1 & 0 & 0 & 2 \\ 0 & 1 & 2 & 1 \end{array}\right]$$

Thus, $A^{-1} = \begin{bmatrix} 0 & 2 \\ 2 & 1 \end{bmatrix}$, and it is easy to check that, over $\mathbb{Z}_3$, $AA^{-1} = I$.

Solution 2 Since A is a 2×2 matrix, we can also compute A^{-1} using the formula given in Theorem 3.8. The determinant of A is

$$\det A = 2(0) - 2(2) = -1 = 2$$

in $\mathbb{Z}_3$ (since $2 + 1 = 0$). Thus, A^{-1} exists and is given by the formula in Theorem 3.8. We must be careful here, though, since the formula introduces the "fraction" $1/\det A$

and there are no fractions in $\mathbb{Z}_3$. We must use multiplicative inverses rather than division.

Instead of $1/\det A = 1/2$, we use 2^{-1}; that is, we find the number x that satisfies the equation $2x = 1$ in $\mathbb{Z}_3$. It is easy to see that $x = 2$ is the solution we want: In $\mathbb{Z}_3$, $2^{-1} = 2$, since $2(2) = 1$. The formula for A^{-1} now becomes

$$A^{-1} = 2^{-1}\begin{bmatrix} 0 & -2 \\ -2 & 2 \end{bmatrix} = 2\begin{bmatrix} 0 & 1 \\ 1 & 2 \end{bmatrix} = \begin{bmatrix} 0 & 2 \\ 2 & 1 \end{bmatrix}$$

which agrees with our previous solution.

Exercises 3.3

In Exercises 1–10, find the inverse of the given matrix (if it exists) using Theorem 3.8.

1. $\begin{bmatrix} 4 & 7 \\ 1 & 2 \end{bmatrix}$ **2.** $\begin{bmatrix} 4 & -2 \\ 2 & 0 \end{bmatrix}$

3. $\begin{bmatrix} 3 & 4 \\ 6 & 8 \end{bmatrix}$ **4.** $\begin{bmatrix} 0 & 1 \\ -1 & 0 \end{bmatrix}$

5. $\begin{bmatrix} \frac{3}{4} & \frac{3}{5} \\ \frac{5}{6} & \frac{2}{3} \end{bmatrix}$ **6.** $\begin{bmatrix} 1/\sqrt{2} & 1/\sqrt{2} \\ -1/\sqrt{2} & 1/\sqrt{2} \end{bmatrix}$

7. $\begin{bmatrix} -1.5 & -4.2 \\ 0.5 & 2.4 \end{bmatrix}$ **8.** $\begin{bmatrix} 3.55 & 0.25 \\ 8.52 & 0.60 \end{bmatrix}$

9. $\begin{bmatrix} a & -b \\ b & a \end{bmatrix}$

10. $\begin{bmatrix} 1/a & 1/b \\ 1/c & 1/d \end{bmatrix}$, where neither a, b, c, nor d is 0

In Exercises 11 and 12, solve the given system using the method of Example 3.25.

11. $\begin{aligned} 2x + y &= -1 \\ 5x + 3y &= 2 \end{aligned}$ **12.** $\begin{aligned} x_1 - x_2 &= 1 \\ 2x_1 + x_2 &= 2 \end{aligned}$

13. Let $A = \begin{bmatrix} 1 & 2 \\ 2 & 6 \end{bmatrix}$, $\mathbf{b}_1 = \begin{bmatrix} 3 \\ 5 \end{bmatrix}$, $\mathbf{b}_2 = \begin{bmatrix} -1 \\ 2 \end{bmatrix}$, and $\mathbf{b}_3 = \begin{bmatrix} 2 \\ 0 \end{bmatrix}$.

(a) Find A^{-1} and use it to solve the three systems $A\mathbf{x} = \mathbf{b}_1$, $A\mathbf{x} = \mathbf{b}_2$, and $A\mathbf{x} = \mathbf{b}_3$.

(b) Solve all three systems at the same time by row reducing the augmented matrix $[A \mid \mathbf{b}_1\ \mathbf{b}_2\ \mathbf{b}_3]$ using Gauss-Jordan elimination.

(c) Carefully count the total number of individual multiplications that you performed in (a) and in (b). You should discover that, even for this 2×2 example, one method uses fewer operations. For larger systems, the difference is even more pronounced, and this explains why computer systems do not use one of these methods to solve linear systems.

14. Prove Theorem 3.9(b).

15. Prove Theorem 3.9(d).

16. Prove that the $n \times n$ identity matrix I_n is invertible and that $I_n^{-1} = I_n$.

17. **(a)** Give a counterexample to show that $(AB)^{-1} \neq A^{-1}B^{-1}$ in general.

(b) Under what conditions on A and B is $(AB)^{-1} = A^{-1}B^{-1}$? Prove your assertion.

18. By induction, prove that if $A_1, A_2, \ldots, A_n$ are invertible matrices of the same size, then the product $A_1A_2 \cdots A_n$ is invertible and $(A_1A_2 \cdots A_n)^{-1} = A_n^{-1} \cdots A_2^{-1}A_1^{-1}$.

19. Give a counterexample to show that $(A + B)^{-1} \neq A^{-1} + B^{-1}$ in general.

In Exercises 20–23, solve the given matrix equation for X. Simplify your answers as much as possible. (In the words of Albert Einstein, "Everything should be made as simple as possible, but not simpler.") Assume that all matrices are invertible.

20. $XA^2 = A^{-1}$ **21.** $AXB = (BA)^2$

22. $(A^{-1}X)^{-1} = A(B^{-2}A)^{-1}$ **23.** $ABXA^{-1}B^{-1} = I + A$

In Exercises 24–30, let

$$A = \begin{bmatrix} 1 & 2 & -1 \\ 1 & 1 & 1 \\ 1 & -1 & 0 \end{bmatrix}, \quad B = \begin{bmatrix} 1 & -1 & 0 \\ 1 & 1 & 1 \\ 1 & 2 & -1 \end{bmatrix},$$

$$C = \begin{bmatrix} 1 & 2 & -1 \\ 1 & 1 & 1 \\ 2 & 1 & -1 \end{bmatrix}, \quad D = \begin{bmatrix} 1 & 2 & -1 \\ -3 & -1 & 3 \\ 2 & 1 & -1 \end{bmatrix}$$

In each case, find an elementary matrix E that satisfies the given equation.

24. $EA = B$ **25.** $EB = A$ **26.** $EA = C$

27. $EC = A$ **28.** $EC = D$ **29.** $ED = C$

30. Is there an elementary matrix E such that $EA = D$? Why or why not?

In Exercises 31–38, find the inverse of the given elementary matrix.

31. $\begin{bmatrix} 3 & 0 \\ 0 & 1 \end{bmatrix}$ **32.** $\begin{bmatrix} 1 & 2 \\ 0 & 1 \end{bmatrix}$

33. $\begin{bmatrix} 0 & 1 \\ 1 & 0 \end{bmatrix}$ **34.** $\begin{bmatrix} 1 & 0 \\ -\frac{1}{2} & 1 \end{bmatrix}$

35. $\begin{bmatrix} 1 & 0 & 0 \\ 0 & 1 & -2 \\ 0 & 0 & 1 \end{bmatrix}$ **36.** $\begin{bmatrix} 0 & 0 & 1 \\ 0 & 1 & 0 \\ 1 & 0 & 0 \end{bmatrix}$

37. $\begin{bmatrix} 1 & 0 & 0 \\ 0 & c & 0 \\ 0 & 0 & 1 \end{bmatrix}, c \neq 0$ **38.** $\begin{bmatrix} 1 & 0 & 0 \\ 0 & 1 & c \\ 0 & 0 & 1 \end{bmatrix}, c \neq 0$

In Exercises 39 and 40, find a sequence of elementary matrices $E_1, E_2, \ldots, E_k$ such that $E_k \cdots E_2E_1A = I$. Use this sequence to write both A and A^{-1} as products of elementary matrices.

39. $A = \begin{bmatrix} 1 & 0 \\ -1 & -2 \end{bmatrix}$ **40.** $A = \begin{bmatrix} 2 & 4 \\ 1 & 1 \end{bmatrix}$

41. Prove Theorem 3.13 for the case of $AB = I$.

42. **(a)** Prove that if A is invertible and $AB = O$, then $B = O$.

(b) Give a counterexample to show that the result in part (a) may fail if A is not invertible.

43. **(a)** Prove that if A is invertible and $BA = CA$, then $B = C$.

(b) Give a counterexample to show that the result in part (a) may fail if A is not invertible.

44. A square matrix A is called ***idempotent*** if $A^2 = A$. (The word idempotent comes from the Latin *idem*, meaning "same," and *potere*, meaning "to have power." Thus, something that is idempotent has the "same power" when squared.)

(a) Find three idempotent 2×2 matrices.

(b) Prove that the only invertible idempotent $n \times n$ matrix is the identity matrix.

45. Show that if A is a square matrix that satisfies the equation $A^2 - 2A + I = O$, then $A^{-1} = 2I - A$.

46. Prove that if a symmetric matrix is invertible, then its inverse is symmetric also.

47. Prove that if A and B are square matrices and AB is invertible, then both A and B are invertible.

In Exercises 48–63, use the Gauss-Jordan method to find the inverse of the given matrix (if it exists).

48. $\begin{bmatrix} 1 & 5 \\ 1 & 4 \end{bmatrix}$ **49.** $\begin{bmatrix} -2 & 4 \\ 3 & -1 \end{bmatrix}$

50. $\begin{bmatrix} 4 & -2 \\ 2 & 0 \end{bmatrix}$ **51.** $\begin{bmatrix} 1 & a \\ -a & 1 \end{bmatrix}$

52. $\begin{bmatrix} 2 & 3 & 0 \\ 1 & -2 & -1 \\ 2 & 0 & -1 \end{bmatrix}$ **53.** $\begin{bmatrix} 1 & -1 & 2 \\ 3 & 1 & 2 \\ 2 & 3 & -1 \end{bmatrix}$

54. $\begin{bmatrix} 1 & 1 & 0 \\ 1 & 0 & 1 \\ 0 & 1 & 1 \end{bmatrix}$ **55.** $\begin{bmatrix} a & 0 & 0 \\ 1 & a & 0 \\ 0 & 1 & a \end{bmatrix}$

56. $\begin{bmatrix} 0 & a & 0 \\ b & 0 & c \\ 0 & d & 0 \end{bmatrix}$ **57.** $\begin{bmatrix} 0 & -1 & 1 & 0 \\ 2 & 1 & 0 & 2 \\ 1 & -1 & 3 & 0 \\ 0 & 1 & 1 & -1 \end{bmatrix}$

58. $\begin{bmatrix} \sqrt{2} & 0 & 2\sqrt{2} & 0 \\ -4\sqrt{2} & \sqrt{2} & 0 & 0 \\ 0 & 0 & 1 & 0 \\ 0 & 0 & 3 & 1 \end{bmatrix}$

59. $\begin{bmatrix} 1 & 0 & 0 & 0 \\ 0 & 1 & 0 & 0 \\ 0 & 0 & 1 & 0 \\ a & b & c & d \end{bmatrix}$ **60.** $\begin{bmatrix} 0 & 1 \\ 1 & 1 \end{bmatrix}$ over $\mathbb{Z}_2$

61. $\begin{bmatrix} 4 & 2 \\ 3 & 4 \end{bmatrix}$ over $\mathbb{Z}_5$ **62.** $\begin{bmatrix} 2 & 1 & 0 \\ 1 & 1 & 2 \\ 0 & 2 & 1 \end{bmatrix}$ over $\mathbb{Z}_3$

63. $\begin{bmatrix} 1 & 5 & 0 \\ 1 & 2 & 4 \\ 3 & 6 & 1 \end{bmatrix}$ over $\mathbb{Z}_7$

Partitioning large square matrices can sometimes make their inverses easier to compute, particularly if the blocks have a nice form. In Exercises 64–68, verify by block multiplication that the inverse of a matrix, if partitioned as shown, is as claimed. (Assume that all inverses exist as needed.)

64. $\begin{bmatrix} A & B \\ O & D \end{bmatrix}^{-1} = \begin{bmatrix} A^{-1} & -A^{-1}BD^{-1} \\ O & D^{-1} \end{bmatrix}$

65. $\begin{bmatrix} O & B \\ C & I \end{bmatrix}^{-1} = \begin{bmatrix} -(BC)^{-1} & (BC)^{-1}B \\ C(BC)^{-1} & I - C(BC)^{-1}B \end{bmatrix}$

66. $\begin{bmatrix} I & B \\ C & I \end{bmatrix}^{-1} = \begin{bmatrix} (I - BC)^{-1} & -(I - BC)^{-1}B \\ -C(I - BC)^{-1} & I + C(I - BC)^{-1}B \end{bmatrix}$

67. $\begin{bmatrix} O & B \\ C & D \end{bmatrix}^{-1}$

$= \begin{bmatrix} -(BD^{-1}C)^{-1} & (BD^{-1}C)^{-1}BD^{-1} \\ D^{-1}C(BD^{-1}C)^{-1} & D^{-1} - D^{-1}C(BD^{-1}C)^{-1}BD^{-1} \end{bmatrix}$

68. $\begin{bmatrix} A & B \\ C & D \end{bmatrix}^{-1} = \begin{bmatrix} P & Q \\ R & S \end{bmatrix}$, where $P = (A - BD^{-1}C)^{-1}$, $Q = -PBD^{-1}$, $R = -D^{-1}CP$, and $S = D^{-1} + D^{-1}CPBD^{-1}$

In Exercises 69–72, partition the given matrix so that you can apply one of the formulas from Exercises 64–68, and then calculate the inverse using that formula.

69. $\begin{bmatrix} 1 & 0 & 0 & 0 \\ 0 & 1 & 0 & 0 \\ 2 & 3 & 1 & 0 \\ 1 & 2 & 0 & 1 \end{bmatrix}$

70. The matrix in Exercise 58

71. $\begin{bmatrix} 0 & 0 & 1 & 1 \\ 0 & 0 & 1 & 0 \\ 0 & -1 & 1 & 0 \\ 1 & 1 & 0 & 1 \end{bmatrix}$

72. $\begin{bmatrix} 0 & 1 & 1 \\ 1 & 3 & 1 \\ -1 & 5 & 2 \end{bmatrix}$

3.4 The *LU* Factorization

Just as it is natural (and illuminating) to factor a natural number into a product of other natural numbers—for example, $30 = 2 \cdot 3 \cdot 5$—it is also frequently helpful to factor matrices as products of other matrices. Any representation of a matrix as a product of two or more other matrices is called a ***matrix factorization.*** For example,

$$\begin{bmatrix} 3 & -1 \\ 9 & -5 \end{bmatrix} = \begin{bmatrix} 1 & 0 \\ 3 & 1 \end{bmatrix}\begin{bmatrix} 3 & -1 \\ 0 & -2 \end{bmatrix}$$

is a matrix factorization.

Needless to say, some factorizations are more useful than others. In this section, we introduce a matrix factorization that arises in the solution of systems of linear equations by Gaussian elimination and is particularly well suited to computer implementation. In subsequent chapters, we will encounter other equally useful matrix factorizations. Indeed, the topic is a rich one, and entire books and courses have been devoted to it.

Consider a system of linear equations of the form $A\mathbf{x} = \mathbf{b}$, where A is an $n \times n$ matrix. Our goal is to show that Gaussian elimination implicitly factors A into a product of matrices that then enable us to solve the given system (and any other system with the same coefficient matrix) easily.

The following example illustrates the basic idea.

Example 3.33 Let

$$A = \begin{bmatrix} 2 & 1 & 3 \\ 4 & -1 & 3 \\ -2 & 5 & 5 \end{bmatrix}$$

Row reduction of A proceeds as follows:

$$A = \begin{bmatrix} 2 & 1 & 3 \\ 4 & -1 & 3 \\ -2 & 5 & 5 \end{bmatrix} \xrightarrow[R_3 + R_1]{R_2 - 2R_1} \begin{bmatrix} 2 & 1 & 3 \\ 0 & -3 & -3 \\ 0 & 6 & 8 \end{bmatrix} \xrightarrow{R_3 + 3R_2} \begin{bmatrix} 2 & 1 & 3 \\ 0 & -3 & -3 \\ 0 & 6 & 8 \end{bmatrix} = U \quad (1)$$

The three elementary matrices E_1, E_2, E_3 that accomplish this reduction of A to echelon form U are (in order):

$$E_1 = \begin{bmatrix} 1 & 0 & 0 \\ -2 & 1 & 0 \\ 0 & 0 & 1 \end{bmatrix}, \quad E_2 = \begin{bmatrix} 1 & 0 & 0 \\ 0 & 1 & 0 \\ 1 & 0 & 1 \end{bmatrix}, \quad E_3 = \begin{bmatrix} 1 & 0 & 0 \\ 0 & 1 & 0 \\ 0 & 2 & 1 \end{bmatrix}$$

Hence,

$$E_3E_2E_1A = U$$

Solving for A, we get

$$A = E_1^{-1}E_2^{-1}E_3^{-1}U = \begin{bmatrix} 1 & 0 & 0 \\ 2 & 1 & 0 \\ 0 & 0 & 1 \end{bmatrix}\begin{bmatrix} 1 & 0 & 0 \\ 0 & 1 & 0 \\ -1 & 0 & 1 \end{bmatrix}\begin{bmatrix} 1 & 0 & 0 \\ 0 & 1 & 0 \\ 0 & -2 & 1 \end{bmatrix}U$$

$$= \begin{bmatrix} 1 & 0 & 0 \\ 2 & 1 & 0 \\ -1 & -2 & 1 \end{bmatrix}U = LU$$

Thus, A can be factored as

$$A = LU$$

where U is an *upper triangular* matrix (see the exercises for Section 3.2), and L is *unit lower triangular.* That is, L has the form

$$L = \begin{bmatrix} 1 & 0 & \cdots & 0 \\ * & 1 & \cdots & 0 \\ \vdots & \vdots & \ddots & \vdots \\ * & * & \cdots & 1 \end{bmatrix}$$

with zeros above and 1s on the main diagonal.

National Portrait Gallery, London

The *LU* factorization was introduced in 1948 by the great English mathematician Alan M. Turing (1912–1954) in a paper entitled "Rounding-off Errors in Matrix Processes" (*Quarterly Journal of Mechanics and Applied Mathematics, 1* (1948), pp. 287–308). During World War II, Turing was instrumental in cracking the German "Enigma" code. However, he is best known for his work in mathematical logic that laid the theoretical groundwork for the development of the digital computer and the modern field of artificial intelligence. The "Turing test" that he proposed in 1950 is still used as one of the benchmarks in addressing the question of whether a computer can be considered "intelligent."

The preceding example motivates the following definition.

Definition Let A be a square matrix. A factorization of A as $A = LU$, where L is unit lower triangular and U is upper triangular, is called an ***LU factorization*** of A.

Remarks

- Observe that the matrix A in Example 3.33 had an LU factorization because *no row interchanges* were needed in the row reduction of A. Hence, all of the elementary matrices that arose were unit lower triangular. Thus, L was guaranteed to be unit lower triangular because inverses and products of unit lower triangular matrices are also unit lower triangular. (See Exercises 29 and 30.)

If a zero had appeared in a pivot position at any step, we would have had to swap rows to get a nonzero pivot. This would have resulted in L no longer being unit lower triangular. We will comment further on this observation below. (Can you find a matrix for which row interchanges will be necessary?)

- The notion of an LU factorization can be generalized to nonsquare matrices by simply requiring U to be a matrix in row echelon form. (See Exercises 13 and 14.)
- Some books define an LU factorization of a square matrix A to be any factorization $A = LU$, where L is lower triangular and U is upper triangular.

The first remark above is essentially a proof of the following theorem.

Theorem 3.15 If A is a square matrix that can be reduced to row echelon form without using any row interchanges, then A has an LU factorization.

To see why the LU factorization is useful, consider a linear system $A\mathbf{x} = \mathbf{b}$, where the coefficient matrix has an LU factorization $A = LU$. We can rewrite the system $A\mathbf{x} = \mathbf{b}$ as $LU\mathbf{x} = \mathbf{b}$ or $L(U\mathbf{x}) = \mathbf{b}$. If we now define $\mathbf{y} = U\mathbf{x}$, then we can solve for $\mathbf{x}$ in two stages:

1. Solve $L\mathbf{y} = \mathbf{b}$ for $\mathbf{y}$ by *forward substitution* (see Exercises 25 and 26 in Section 2.1).
2. Solve $U\mathbf{x} = \mathbf{y}$ for $\mathbf{x}$ by *back substitution*.

Each of these linear systems is straightforward to solve because the coefficient matrices L and U are both triangular. The next example illustrates the method.

Example 3.34

Use an LU factorization of $A = \begin{bmatrix} 2 & 1 & 3 \\ 4 & -1 & 3 \\ -2 & 5 & 5 \end{bmatrix}$ to solve $A\mathbf{x} = \mathbf{b}$, where $\mathbf{b} = \begin{bmatrix} 1 \\ -4 \\ 9 \end{bmatrix}$.

Solution In Example 3.33, we found that

$$A = \begin{bmatrix} 1 & 0 & 0 \\ 2 & 1 & 0 \\ -1 & -2 & 1 \end{bmatrix} \begin{bmatrix} 2 & 1 & 3 \\ 0 & -3 & -3 \\ 0 & 0 & 2 \end{bmatrix} = LU$$

As outlined above, to solve $A\mathbf{x} = \mathbf{b}$ (which is the same as $L(U\mathbf{x}) = \mathbf{b}$), we first solve

$L\mathbf{y} = \mathbf{b}$ for $\mathbf{y} = \begin{bmatrix} y_1 \\ y_2 \\ y_3 \end{bmatrix}$. This is just the linear system

$$\begin{aligned} y_1 \qquad\qquad &= 1 \\ 2y_1 + y_2 \qquad &= -4 \\ -y_1 - 2y_2 + y_3 &= 9 \end{aligned}$$

Forward substitution (that is, working from top to bottom) yields

$$y_1 = 1, y_2 = -4 - 2y_1 = -6, y_3 = 9 + y_1 + 2y_2 = -2$$

Thus $\mathbf{y} = \begin{bmatrix} 1 \\ -6 \\ -2 \end{bmatrix}$ and we now solve $U\mathbf{x} = \mathbf{y}$ for $\mathbf{x} = \begin{bmatrix} x_1 \\ x_2 \\ x_3 \end{bmatrix}$. This linear system is

$$\begin{aligned} 2x_1 + \quad x_2 + 3x_3 &= 1 \\ -3x_2 - 3x_3 &= -6 \\ 2x_3 &= -2 \end{aligned}$$

and back substitution quickly produces

$$\begin{aligned} x_3 &= -1, \\ -3x_2 &= -6 + 3x_3 = -9 \text{ so that } x_2 = 3, \text{ and} \\ 2x_1 &= 1 - x_2 - 3x_3 = 1 \text{ so that } x_1 = \tfrac{1}{2} \end{aligned}$$

Therefore, the solution to the given system $A\mathbf{x} = \mathbf{b}$ is $\mathbf{x} = \begin{bmatrix} \frac{1}{2} \\ 3 \\ -1 \end{bmatrix}$.

An Easy Way to Find *LU* Factorizations

In Example 3.33, we computed the matrix L as a product of elementary matrices. Fortunately, L can be computed directly from the row reduction process without our needing to compute elementary matrices at all. Remember that we are assuming that A can be reduced to row echelon form without using any row interchanges. If this is the case, then the entire row reduction process can be done using only elementary row operations of the form $R_i - kR_j$. (Why do we not need to use the remaining elementary row operation, multiplying a row by a nonzero scalar?) In the operation $R_i - kR_j$, we will refer to the scalar k as the ***multiplier.***

In Example 3.33, the elementary row operations that were used were, in order,

$$\begin{aligned} &R_2 - 2R_1 && \text{(multiplier} = 2) \\ &R_3 + R_1 = R_3 - (-1)R_1 && \text{(multiplier} = -1) \\ &R_3 + 2R_2 = R_3 - (-2)R_2 && \text{(multiplier} = -2) \end{aligned}$$

The multipliers are precisely the entries of L that are below its diagonal! Indeed,

$$L = \begin{bmatrix} 1 & 0 & 0 \\ 2 & 1 & 0 \\ -1 & -2 & 1 \end{bmatrix}$$

and $L_{21} = 2$, $L_{31} = -1$, and $L_{32} = -2$. Notice that the elementary row operation $R_i - kR_j$ has its multiplier k placed in the (i, j) entry of L.

Example 3.35

Find an *LU* factorization of

$$A = \begin{bmatrix} 3 & 1 & 3 & -4 \\ 6 & 4 & 8 & -10 \\ 3 & 2 & 5 & -1 \\ -9 & 5 & -2 & -4 \end{bmatrix}$$

Solution Reducing A to row echelon form, we have

$$A = \begin{bmatrix} 3 & 1 & 3 & -4 \\ 6 & 4 & 8 & -10 \\ 3 & 2 & 5 & -1 \\ -9 & 5 & -2 & -4 \end{bmatrix} \xrightarrow[R_4-(-3)R_1]{\substack{R_2-2R_1 \\ R_3-R_1}} \begin{bmatrix} 3 & 1 & 3 & -4 \\ 0 & 2 & 2 & -2 \\ 0 & 1 & 2 & 3 \\ 0 & 8 & 7 & -16 \end{bmatrix}$$

$$\xrightarrow[R_4-4R_2]{R_3-\frac{1}{2}R_2} \begin{bmatrix} 3 & 1 & 3 & -4 \\ 0 & 2 & 2 & -2 \\ 0 & 0 & 1 & 4 \\ 0 & 0 & -1 & -8 \end{bmatrix}$$

$$\xrightarrow{R_4-(-1)R_3} \begin{bmatrix} 3 & 1 & 3 & -4 \\ 0 & 2 & 2 & -2 \\ 0 & 0 & 1 & 4 \\ 0 & 0 & 0 & -4 \end{bmatrix} = U$$

The first three multipliers are 2, 1, and -3, and these go into the subdiagonal entries of the first column of L. So, thus far,

$$L = \begin{bmatrix} 1 & 0 & 0 & 0 \\ 2 & 1 & 0 & 0 \\ 1 & * & 1 & 0 \\ -3 & * & * & 1 \end{bmatrix}$$

The next two multipliers are $\frac{1}{2}$ and 4, so we continue to fill out L:

$$L = \begin{bmatrix} 1 & 0 & 0 & 0 \\ 2 & 1 & 0 & 0 \\ 1 & \frac{1}{2} & 1 & 0 \\ -3 & 4 & * & 1 \end{bmatrix}$$

The final multiplier, -1, replaces the last * in L to give

$$L = \begin{bmatrix} 1 & 0 & 0 & 0 \\ 2 & 1 & 0 & 0 \\ 1 & \frac{1}{2} & 1 & 0 \\ -3 & 4 & -1 & 1 \end{bmatrix}$$

Thus, an LU factorization of A is

$$A = \begin{bmatrix} 3 & 1 & 3 & -4 \\ 6 & 4 & 8 & -10 \\ 3 & 2 & 5 & -1 \\ -9 & 5 & -2 & -4 \end{bmatrix} = \begin{bmatrix} 1 & 0 & 0 & 0 \\ 2 & 1 & 0 & 0 \\ 1 & \frac{1}{2} & 1 & 0 \\ -3 & 4 & -1 & 1 \end{bmatrix} \begin{bmatrix} 3 & 1 & 3 & -4 \\ 0 & 2 & 2 & -2 \\ 0 & 0 & 1 & 4 \\ 0 & 0 & 0 & -4 \end{bmatrix} = LU$$

as is easily checked.

Remarks

- In applying this method, it is important to note that the elementary row operations $R_i - kR_j$ must be performed from top to bottom within each column (using the diagonal entry as the pivot), and column by column from left to right. To illustrate what can go wrong if we do not obey these rules, consider the following row reduction:

$$A = \begin{bmatrix} 1 & 2 & 2 \\ 1 & 1 & 1 \\ 2 & 2 & 1 \end{bmatrix} \xrightarrow{R_3 - 2R_2} \begin{bmatrix} 1 & 2 & 2 \\ 1 & 1 & 1 \\ 0 & 0 & -1 \end{bmatrix} \xrightarrow{R_2 - R_1} \begin{bmatrix} 1 & 2 & 2 \\ 0 & -1 & -1 \\ 0 & 0 & -1 \end{bmatrix} = U$$

This time the multipliers would be placed in L as follows: $L_{32} = 2$, $L_{21} = 1$. We would get

$$L = \begin{bmatrix} 1 & 0 & 0 \\ 1 & 1 & 0 \\ 0 & 2 & 1 \end{bmatrix}$$

but $A \neq LU$. (Check this! Find a correct LU factorization of A.)

- An alternative way to construct L is to observe that the multipliers can be obtained directly from the matrices obtained at the intermediate steps of the row reduction process. In Example 3.33, examine the pivots and the corresponding columns of the matrices that arise in the row reduction

$$A = \begin{bmatrix} 2 & 1 & 3 \\ 4 & -1 & 3 \\ -2 & 5 & 5 \end{bmatrix} \to A_1 = \begin{bmatrix} 2 & 1 & 3 \\ 0 & -3 & -3 \\ 0 & 6 & 8 \end{bmatrix} \to \begin{bmatrix} 2 & 1 & 3 \\ 0 & -3 & -3 \\ 0 & 0 & 2 \end{bmatrix} = U$$

The first pivot is 2, which occurs in the first column of A. Dividing the entries of this column vector that are on or below the diagonal by the pivot produces

$$\frac{1}{2}\begin{bmatrix} 2 \\ 4 \\ -2 \end{bmatrix} = \begin{bmatrix} 1 \\ 2 \\ -1 \end{bmatrix}$$

The next pivot is -3, which occurs in the second column of A_1. Dividing the entries of this column vector that are on or below the diagonal by the pivot, we obtain

$$\frac{1}{(-3)}\begin{bmatrix} \\ -3 \\ 6 \end{bmatrix} = \begin{bmatrix} \\ 1 \\ -2 \end{bmatrix}$$

The final pivot (which we did not need to use) is 2, in the third column of U. Dividing the entries of this column vector that are on or below the diagonal by the pivot, we obtain

$$\frac{1}{2}\begin{bmatrix} \\ \\ 2 \end{bmatrix} = \begin{bmatrix} \\ \\ 1 \end{bmatrix}$$

If we place the resulting three column vectors side by side in a matrix, we have

$$\begin{bmatrix} 1 & & \\ 2 & 1 & \\ -1 & -2 & 1 \end{bmatrix}$$

which is exactly L once the above-diagonal entries are filled with zeros.

In Chapter 2, we remarked that the row echelon form of a matrix is not unique. However, if an *invertible* matrix A has an LU factorization $A=LU$, then this factorization is unique.

Theorem 3.16 If A is an invertible matrix that has an LU factorization, then L and U are unique.

Proof Suppose $A = LU$ and $A = L_1U_1$ are two LU factorizations of A. Then $LU = L_1U_1$, where L and L_1 are unit lower triangular and U and U_1 are upper triangular. In fact, U and U_1 are two (possibly different) row echelon forms of A.

By Exercise 30, L_1 is invertible. Because A is invertible, its reduced row echelon form is an identity matrix I by the Fundamental Theorem of Invertible Matrices. Hence U also row reduces to I (why?) and so U is invertible also. Therefore,

$$L_1^{-1}(LU)U^{-1} = L_1^{-1}(L_1U_1)U^{-1} \quad \text{so} \quad (L_1^{-1}L)(UU^{-1}) = (L_1^{-1}L_1)(U_1U^{-1})$$

Hence,

$$(L_1^{-1}L)I = I(U_1U^{-1}) \quad \text{so} \quad L_1^{-1}L = U_1U^{-1}$$

But $L_1^{-1}L$ is unit lower triangular by Exercise 29, and U_1U^{-1} is upper triangular. (Why?) It follows that $L_1^{-1}L = U_1U^{-1}$ is *both* unit lower triangular *and* upper triangular. The only such matrix is the identity matrix, so $L_1^{-1}L = I$ and $U_1U^{-1} = I$. It follows that $L = L_1$ and $U = U_1$, so the LU factorization of A is unique.

The P^TLU Factorization

We now explore the problem of adapting the LU factorization to handle cases where row interchanges are necessary during Gaussian elimination. Consider the matrix

$$A = \begin{bmatrix} 1 & 2 & -1 \\ 3 & 6 & 2 \\ -1 & 1 & 4 \end{bmatrix}$$

A straightforward row reduction produces

$$A \rightarrow B = \begin{bmatrix} 1 & 2 & -1 \\ 0 & 0 & 5 \\ 0 & 3 & 3 \end{bmatrix}$$

which is not an upper triangular matrix. However, we can easily convert this into upper triangular form by swapping rows 2 and 3 of B to get

$$U = \begin{bmatrix} 1 & 2 & -1 \\ 0 & 3 & 3 \\ 0 & 0 & 5 \end{bmatrix}$$

Alternatively, we can swap rows 2 and 3 of A first. To this end, let P be the elementary matrix

$$\begin{bmatrix} 1 & 0 & 0 \\ 0 & 0 & 1 \\ 0 & 1 & 0 \end{bmatrix}$$

corresponding to interchanging rows 2 and 3, and let E be the product of the elementary matrices that then reduce PA to U (so that $E^{-1} = L$ is unit lower triangular). Thus $EPA = U$, so $A = (EP)^{-1}U = P^{-1}E^{-1}U = P^{-1}LU$.

Now this handles only the case of a *single* row interchange. In general, P will be the product $P = P_k \cdots P_2P_1$ of all the row interchange matrices $P_1, P_2, \ldots, P_k$ (where P_1 is performed first, and so on). Such a matrix P is called a ***permutation matrix.*** Observe that a permutation matrix arises from permuting the rows of an identity matrix in some order. For example, the following are all permutation matrices:

$$\begin{bmatrix} 0 & 1 \\ 1 & 0 \end{bmatrix}, \begin{bmatrix} 0 & 0 & 1 \\ 1 & 0 & 0 \\ 0 & 1 & 0 \end{bmatrix}, \begin{bmatrix} 0 & 1 & 0 & 0 \\ 0 & 0 & 0 & 1 \\ 1 & 0 & 0 & 0 \\ 0 & 0 & 1 & 0 \end{bmatrix}$$

Fortunately, the inverse of a permutation matrix is easy to compute; in fact, no calculations are needed at all!

Theorem 3.17 If P is a permutation matrix, then $P^{-1} = P^T$.

Proof We must show that $P^TP = I$. But the ith row of P^T is the same as the ith column of P, and these are both equal to the same standard unit vector $\mathbf{e}$, because P is a permutation matrix. So

$$(P^TP)_{ii} = (i\text{th row of } P^T)(i\text{th column of } P) = \mathbf{e}^T\mathbf{e} = \mathbf{e}\cdot\mathbf{e} = 1$$

This shows that diagonal entries of P^TP are all 1s. On the other hand, if $j \neq i$, then the jth column of P is a *different* standard unit vector from $\mathbf{e}$—say $\mathbf{e}'$. Thus, a typical off-diagonal entry of P^TP is given by

$$(P^TP)_{ij} = (i\text{th row of } P^T)(j\text{th column of } P) = \mathbf{e}^T\mathbf{e}' = \mathbf{e}\cdot\mathbf{e}' = 0$$

Hence P^TP is an identity matrix, as we wished to show.

Thus, in general, we can factor a square matrix A as $A = P^{-1}LU = P^TLU$.

Definition Let A be a square matrix. A factorization of A as $A = P^TLU$, where P is a permutation matrix, L is unit lower triangular, and U is upper triangular, is called a ***P^TLU factorization*** of A.

Example 3.36

Find a P^TLU factorization of $A = \begin{bmatrix} 0 & 0 & 6 \\ 1 & 2 & 3 \\ 2 & 1 & 4 \end{bmatrix}$.

Solution First we reduce A to row echelon form. Clearly, we need at least one row interchange.

$$A = \begin{bmatrix} 0 & 0 & 6 \\ 1 & 2 & 3 \\ 2 & 1 & 4 \end{bmatrix} \xrightarrow{R_1 \leftrightarrow R_2} \begin{bmatrix} 1 & 2 & 3 \\ 0 & 0 & 6 \\ 2 & 1 & 4 \end{bmatrix} \xrightarrow{R_3 - 2R_1} \begin{bmatrix} 1 & 2 & 3 \\ 0 & 0 & 6 \\ 0 & -3 & -2 \end{bmatrix}$$

$$\xrightarrow{R_2 \leftrightarrow R_3} \begin{bmatrix} 1 & 2 & 3 \\ 0 & -3 & -2 \\ 0 & 0 & 6 \end{bmatrix}$$

We have used two row interchanges ($R_1 \leftrightarrow R_2$ and then $R_2 \leftrightarrow R_3$), so the required permutation matrix is

$$P = P_2P_1 = \begin{bmatrix} 1 & 0 & 0 \\ 0 & 0 & 1 \\ 0 & 1 & 0 \end{bmatrix}\begin{bmatrix} 0 & 1 & 0 \\ 1 & 0 & 0 \\ 0 & 0 & 1 \end{bmatrix} = \begin{bmatrix} 0 & 1 & 0 \\ 0 & 0 & 1 \\ 1 & 0 & 0 \end{bmatrix}$$

We now find an LU factorization of PA.

$$PA = \begin{bmatrix} 0 & 1 & 0 \\ 0 & 0 & 1 \\ 1 & 0 & 0 \end{bmatrix}\begin{bmatrix} 0 & 0 & 6 \\ 1 & 2 & 3 \\ 2 & 1 & 4 \end{bmatrix} = \begin{bmatrix} 1 & 2 & 3 \\ 2 & 1 & 4 \\ 0 & 0 & 6 \end{bmatrix} \xrightarrow{R_2 - 2R_1} \begin{bmatrix} 1 & 2 & 3 \\ 0 & -3 & -2 \\ 0 & 0 & 6 \end{bmatrix} = U$$

Hence $L_{21} = 2$, and so

$$A = P^TLU = \begin{bmatrix} 0 & 0 & 1 \\ 1 & 0 & 0 \\ 0 & 1 & 0 \end{bmatrix}\begin{bmatrix} 1 & 0 & 0 \\ 2 & 1 & 0 \\ 0 & 0 & 1 \end{bmatrix}\begin{bmatrix} 1 & 2 & 3 \\ 0 & -3 & -2 \\ 0 & 0 & 6 \end{bmatrix}$$

The discussion above justifies the following theorem.

Theorem 3.18 Every square matrix has a P^TLU factorization.

Remark Even for an invertible matrix, the P^TLU factorization is not unique. In Example 3.36, a single row interchange $R_1 \leftrightarrow R_3$ also would have worked, leading to a different P. However, once P has been determined, L and U are unique.

Computational Considerations

If A is $n \times n$, then the total number of operations (multiplications and divisions) required to solve a linear system $A\mathbf{x} = \mathbf{b}$ using an LU factorization of A) is $T(n) \approx n^3/3$, the same as is required for Gaussian elimination. (See the Exploration "Counting Operations," in Chapter 2.) This is hardly surprising since the forward elimination phase produces the LU factorization in $\approx n^3/3$ steps, whereas both forward and backward substitution require $\approx n^2/2$ steps. Therefore, for large values of n, the $n^3/3$ term is dominant. From this point of view, then, Gaussian elimination and the LU factorization are equivalent.

However, the LU factorization has other advantages:

- From a storage point of view, the LU factorization is very compact because we can *overwrite* the entries of A with the entries of L and U *as they are computed.* In Example 3.33, we found that

$$A = \begin{bmatrix} 2 & 1 & 3 \\ 4 & -1 & 3 \\ -2 & 5 & 5 \end{bmatrix} = \begin{bmatrix} 1 & 0 & 0 \\ 2 & 1 & 0 \\ -1 & -2 & 1 \end{bmatrix}\begin{bmatrix} 2 & 1 & 3 \\ 0 & -3 & -3 \\ 0 & 0 & 2 \end{bmatrix} = LU$$

This can be stored as

$$\begin{bmatrix} 2 & -1 & 3 \\ 2 & -3 & -3 \\ -1 & -2 & 2 \end{bmatrix}$$

with the entries placed in the order (1,1), (1,2), (1,3), (2,1), (3,1), (2,2), (2,3), (3,2), (3,3). In other words, the subdiagonal entries of A are replaced by the corresponding multipliers. (Check that this works!)

- Once an LU factorization of A has been computed, it can be used to solve as many linear systems of the form $A\mathbf{x} = \mathbf{b}$ as we like. We just need to apply the method of Example 3.34, varying the vector $\mathbf{b}$ each time.
- For matrices with certain special forms, especially those with a large number of zeros (so-called "sparse" matrices) concentrated off the diagonal, there are methods that will simplify the computation of an LU factorization. In these cases, this method is faster than Gaussian elimination in solving $A\mathbf{x} = \mathbf{b}$.
- For an invertible matrix A, an LU factorization of A can be used to find A^{-1}, if necessary. Moreover, this can be done in such a way that it simultaneously yields a factorization of A^{-1}. (See Exercises 15–18.)

Remark If you have a CAS (such as MATLAB) that has the LU factorization built in, you may notice some differences between your hand calculations and the computer output. This is because most CAS's will automatically try to perform partial pivoting to reduce roundoff errors. (See the Exploration "Partial Pivoting," in Chapter 2.) Turing's paper is an extended discussion of such errors in the context of matrix factorizations.

This section has served to introduce one of the most useful matrix factorizations. In subsequent chapters, we will encounter other equally useful factorizations.

Exercises 3.4

In Exercises 1–6, solve the system $A\mathbf{x} = \mathbf{b}$ using the given LU factorization of A.

1. $A = \begin{bmatrix} -2 & 1 \\ 2 & 5 \end{bmatrix} = \begin{bmatrix} 1 & 0 \\ -1 & 1 \end{bmatrix}\begin{bmatrix} -2 & 1 \\ 0 & 6 \end{bmatrix}$, $\mathbf{b} = \begin{bmatrix} 5 \\ 1 \end{bmatrix}$

2. $A = \begin{bmatrix} 4 & -2 \\ 2 & 3 \end{bmatrix} = \begin{bmatrix} 1 & 0 \\ \frac{1}{2} & 1 \end{bmatrix}\begin{bmatrix} 4 & -2 \\ 0 & 4 \end{bmatrix}$, $\mathbf{b} = \begin{bmatrix} 0 \\ 8 \end{bmatrix}$

3. $A = \begin{bmatrix} 2 & 1 & -2 \\ -2 & 3 & -4 \\ 4 & -3 & 0 \end{bmatrix} = \begin{bmatrix} 1 & 0 & 0 \\ -1 & 1 & 0 \\ 2 & -\frac{5}{4} & 1 \end{bmatrix}$
$\times \begin{bmatrix} 2 & 1 & -2 \\ 0 & 4 & -6 \\ 0 & 0 & -\frac{7}{2} \end{bmatrix}$, $\mathbf{b} = \begin{bmatrix} -3 \\ 1 \\ 0 \end{bmatrix}$

4. $A = \begin{bmatrix} 2 & -4 & 0 \\ 3 & -1 & 4 \\ -1 & 2 & 2 \end{bmatrix} = \begin{bmatrix} 1 & 0 & 0 \\ \frac{3}{2} & 1 & 0 \\ -\frac{1}{2} & 0 & 1 \end{bmatrix}$
$\times \begin{bmatrix} 2 & -4 & 0 \\ 0 & 5 & 4 \\ 0 & 0 & 2 \end{bmatrix}$, $\mathbf{b} = \begin{bmatrix} 2 \\ 0 \\ -5 \end{bmatrix}$

5. $A = \begin{bmatrix} 2 & -1 & 0 & 0 \\ 6 & -4 & 5 & -3 \\ 8 & -4 & 1 & 0 \\ 4 & -1 & 0 & 7 \end{bmatrix} = \begin{bmatrix} 1 & 0 & 0 & 0 \\ 3 & 1 & 0 & 0 \\ 4 & 0 & 1 & 0 \\ 2 & -1 & 5 & 1 \end{bmatrix}$
$\times \begin{bmatrix} 2 & -1 & 0 & 0 \\ 0 & -1 & 5 & -3 \\ 0 & 0 & 1 & 0 \\ 0 & 0 & 0 & 4 \end{bmatrix}$, $\mathbf{b} = \begin{bmatrix} 1 \\ 2 \\ 2 \\ 1 \end{bmatrix}$

6. $A = \begin{bmatrix} 1 & 4 & 3 & 0 \\ -2 & -5 & -1 & 2 \\ 3 & 6 & -3 & -4 \\ -5 & -8 & 9 & 9 \end{bmatrix} = \begin{bmatrix} 1 & 0 & 0 & 0 \\ -2 & 1 & 0 & 0 \\ 3 & -2 & 1 & 0 \\ -5 & 4 & -2 & 1 \end{bmatrix}$
$\times \begin{bmatrix} 1 & 4 & 3 & 0 \\ 0 & 3 & 5 & 2 \\ 0 & 0 & -2 & 0 \\ 0 & 0 & 0 & 1 \end{bmatrix}$, $\mathbf{b} = \begin{bmatrix} 1 \\ -3 \\ -1 \\ 0 \end{bmatrix}$

In Exercises 7–12, find an LU factorization of the given matrix.

7. $\begin{bmatrix} 1 & 2 \\ -3 & -1 \end{bmatrix}$

8. $\begin{bmatrix} 2 & -4 \\ 3 & 1 \end{bmatrix}$

9. $\begin{bmatrix} 1 & 2 & 3 \\ 4 & 5 & 6 \\ 8 & 7 & 9 \end{bmatrix}$ **10.** $\begin{bmatrix} 2 & 2 & -1 \\ 4 & 0 & 4 \\ 3 & 4 & 4 \end{bmatrix}$

11. $\begin{bmatrix} 1 & 2 & 3 & -1 \\ 2 & 6 & 3 & 0 \\ 0 & 6 & -6 & 7 \\ -1 & -2 & -9 & 0 \end{bmatrix}$

12. $\begin{bmatrix} 2 & 2 & 2 & 1 \\ -2 & 4 & -1 & 2 \\ 4 & 4 & 7 & 3 \\ 6 & 9 & 5 & 8 \end{bmatrix}$

Generalize the definition of LU factorization to nonsquare matrices by simply requiring U to be a matrix in row echelon form. With this modification, find an LU factorization of the matrices in Exercises 13 and 14.

13. $\begin{bmatrix} 1 & 0 & 1 & -2 \\ 0 & 3 & 3 & 1 \\ 0 & 0 & 0 & 5 \end{bmatrix}$

14. $\begin{bmatrix} 1 & 2 & 0 & -1 & 1 \\ -2 & -7 & 3 & 8 & -2 \\ 1 & 1 & 3 & 5 & 2 \\ 0 & 3 & -3 & -6 & 0 \end{bmatrix}$

For an invertible matrix with an LU factorization $A = LU$, both L and U will be invertible and $A^{-1} = U^{-1}L^{-1}$. In Exercises 15 and 16, find L^{-1}, U^{-1}, and A^{-1} for the given matrix.

15. A in Exercise 1 **16.** A in Exercise 4

The inverse of a matrix can also be computed by solving several systems of equations using the method of Example 3.34. For an $n \times n$ matrix A, to find its inverse we need to solve $AX = I_n$ for the $n \times n$ matrix X. Writing this equation as $A[\mathbf{x}_1\ \mathbf{x}_2 \cdots \mathbf{x}_n] = [\mathbf{e}_1\ \mathbf{e}_2 \cdots \mathbf{e}_n]$, using the matrix-column form of AX, we see that we need to solve n systems of linear equations: $A\mathbf{x}_1 = \mathbf{e}_1, A\mathbf{x}_2 = \mathbf{e}_2, \ldots, A\mathbf{x}_n = \mathbf{e}_n$. Moreover, we can use the factorization $A = LU$ to solve each one of these systems.

In Exercises 17 and 18, use the approach just outlined to find A^{-1} for the given matrix. Compare with the method of Exercises 15 and 16.

17. A in Exercise 1 **18.** A in Exercise 4

In Exercises 19–22, write the given permutation matrix as a product of elementary (row interchange) matrices.

19. $\begin{bmatrix} 0 & 0 & 1 \\ 1 & 0 & 0 \\ 0 & 1 & 0 \end{bmatrix}$ **20.** $\begin{bmatrix} 0 & 0 & 0 & 1 \\ 0 & 0 & 1 & 0 \\ 0 & 1 & 0 & 0 \\ 1 & 0 & 0 & 0 \end{bmatrix}$

21. $\begin{bmatrix} 0 & 1 & 0 & 0 \\ 0 & 0 & 0 & 1 \\ 1 & 0 & 0 & 0 \\ 0 & 0 & 1 & 0 \end{bmatrix}$ **22.** $\begin{bmatrix} 0 & 0 & 1 & 0 & 0 \\ 1 & 0 & 0 & 0 & 0 \\ 0 & 0 & 0 & 1 & 0 \\ 0 & 0 & 0 & 0 & 1 \\ 0 & 1 & 0 & 0 & 0 \end{bmatrix}$

In Exercises 23–25, find a P^TLU factorization of the given matrix A.

23. $A = \begin{bmatrix} 0 & 1 & 4 \\ -1 & 2 & 1 \\ 1 & 3 & 3 \end{bmatrix}$ **24.** $A = \begin{bmatrix} 0 & 0 & 1 & 2 \\ -1 & 1 & 3 & 2 \\ 0 & 2 & 1 & 1 \\ 1 & 1 & -1 & 0 \end{bmatrix}$

25. $A = \begin{bmatrix} 0 & -1 & 1 & 3 \\ -1 & 1 & 1 & 2 \\ 0 & 1 & -1 & 1 \\ 0 & 0 & 1 & 1 \end{bmatrix}$

26. Prove that there are exactly $n!$ $n \times n$ permutation matrices.

In Exercises 27–28, solve the system $A\mathbf{x} = \mathbf{b}$ using the given factorization $A = P^TLU$. Because $PP^T = I$, $P^TLU\mathbf{x} = \mathbf{b}$ can be rewritten as $LU\mathbf{x} = P\mathbf{b}$. This system can then be solved using the method of Example 3.34.

27. $A = \begin{bmatrix} 0 & 1 & -1 \\ 2 & 3 & 2 \\ 1 & 1 & -1 \end{bmatrix} = \begin{bmatrix} 0 & 1 & 0 \\ 1 & 0 & 0 \\ 0 & 0 & 1 \end{bmatrix}\begin{bmatrix} 1 & 0 & 0 \\ 0 & 1 & 0 \\ \frac{1}{2} & -\frac{1}{2} & 1 \end{bmatrix}$
$\times \begin{bmatrix} 2 & 3 & 2 \\ 0 & 1 & -1 \\ 0 & 0 & -\frac{5}{2} \end{bmatrix} = P^TLU, \ \mathbf{b} = \begin{bmatrix} 1 \\ 1 \\ 5 \end{bmatrix}$

28. $A = \begin{bmatrix} 8 & 3 & 5 \\ 4 & 1 & 2 \\ 4 & 0 & 3 \end{bmatrix} = \begin{bmatrix} 0 & 1 & 0 \\ 0 & 0 & 1 \\ 1 & 0 & 0 \end{bmatrix}\begin{bmatrix} 1 & 0 & 0 \\ 1 & 1 & 0 \\ 2 & -1 & 1 \end{bmatrix}$
$\times \begin{bmatrix} 4 & 1 & 2 \\ 0 & -1 & 1 \\ 0 & 0 & 2 \end{bmatrix} = P^TLU, \ \mathbf{b} = \begin{bmatrix} 16 \\ -4 \\ 4 \end{bmatrix}$

29. Prove that a product of unit lower triangular matrices is unit lower triangular.

30. Prove that every unit lower triangular matrix is invertible and that its inverse is also unit lower triangular.

*An **LDU factorization** of a square matrix A is a factorization $A = LDU$, where L is a unit lower triangular matrix, D is a diagonal matrix, and U is a unit upper triangular matrix (upper triangular with 1s on its diagonal). In Exercises 31 and 32, find an LDU factorization of A.*

31. A in Exercise 1

32. A in Exercise 4

33. If A is symmetric and invertible and has an LDU factorization, show that $U = L^T$.

34. If A is symmetric and invertible and $A = LDL^T$ (with L unit lower triangular and D diagonal), prove that this factorization is unique. That is, prove that if we also have $A = L_1D_1L_1^T$ (with L_1 unit lower triangular and D_1 diagonal), then $L = L_1$ and $D = D_1$.

3.5 Subspaces, Basis, Dimension, and Rank

This section introduces perhaps the most important ideas in the entire book. We have already seen that there is an interplay between geometry and algebra: We can often use geometric intuition and reasoning to obtain algebraic results, and the power of algebra will often allow us to extend our findings well beyond the geometric settings in which they first arose.

In our study of vectors, we have already encountered all of the concepts in this section informally. Here, we will start to become more formal by giving definitions for the key ideas. As you'll see, the notion of a *subspace* is simply an algebraic generalization of the geometric examples of lines and planes through the origin. The fundamental concept of a *basis* for a subspace is then derived from the idea of direction vectors for such lines and planes. The concept of a basis will allow us to give a precise definition of *dimension* that agrees with an intuitive, geometric idea of the term, yet is flexible enough to allow generalization to other settings.

You will also begin to see that these ideas shed more light on what you already know about matrices and the solution of systems of linear equations. In Chapter 6, we will encounter all of these fundamental ideas again, in more detail. Consider this section a "getting to know you" session.

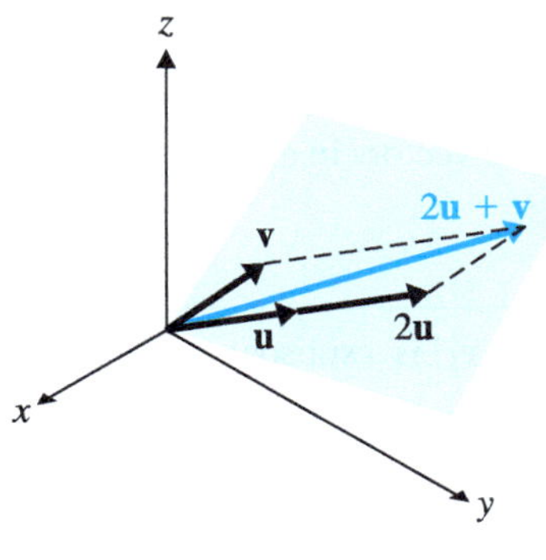

Figure 3.2

A plane through the origin in $\mathbb{R}^3$ "looks like" a copy of $\mathbb{R}^2$. Intuitively, we would agree that they are both "two-dimensional." Pressed further, we might also say that any calculation that can be done with vectors in $\mathbb{R}^2$ can also be done in a plane through the origin. In particular, we can add and take scalar multiples (and, more generally, form linear combinations) of vectors in such a plane, and the results are other vectors *in the same plane*. We say that, like $\mathbb{R}^2$, a plane through the origin is *closed* with respect to the operations of addition and scalar multiplication. (See Figure 3.2.)

But are the vectors in this plane two- or three-dimensional objects? We might argue that they are three-dimensional because they live in $\mathbb{R}^3$ and therefore have three components. On the other hand, they can be described as a linear combination of just two vectors—direction vectors for the plane—and so are two-dimensional objects living in a two-dimensional plane. The notion of a subspace is the key to resolving this conundrum.

Definition A *subspace* of $\mathbb{R}^n$ is any collection S of vectors in $\mathbb{R}^n$ such that:

1. The zero vector $\mathbf{0}$ is in S.
2. If $\mathbf{u}$ and $\mathbf{v}$ are in S, then $\mathbf{u} + \mathbf{v}$ is in S. (S is ***closed under addition.***)
3. If $\mathbf{u}$ is in S and c is a scalar, then $c\mathbf{u}$ is in S. (S is ***closed under scalar multiplication.***)

We could have combined properties (2) and (3) and required, equivalently, that S be ***closed under linear combinations:***

$$\text{If } \mathbf{u}_1, \mathbf{u}_2, \ldots, \mathbf{u}_k \text{ are in } S \text{ and } c_1, c_2, \ldots, c_k \text{ are scalars,}$$
$$\text{then } c_1\mathbf{u}_1 + c_2\mathbf{u}_2 + \cdots + c_k\mathbf{u}_k \text{ is in } S.$$

Example 3.37

Every line and plane through the origin in $\mathbb{R}^3$ is a subspace of $\mathbb{R}^3$. It should be clear geometrically that properties (1) through (3) are satisfied. Here is an algebraic proof in the case of a plane through the origin. You are asked to give the corresponding proof for a line in Exercise 9.

Let $\mathcal{P}$ be a plane through the origin with direction vectors $\mathbf{v}_1$ and $\mathbf{v}_2$. Hence, $\mathcal{P} = \text{span}(\mathbf{v}_1, \mathbf{v}_2)$. The zero vector $\mathbf{0}$ is in $\mathcal{P}$, since $\mathbf{0} = 0\mathbf{v}_1 + 0\mathbf{v}_2$. Now let

$$\mathbf{u} = c_1\mathbf{v}_1 + c_2\mathbf{v}_2 \quad \text{and} \quad \mathbf{v} = d_1\mathbf{v}_1 + d_2\mathbf{v}_2$$

be two vectors in $\mathcal{P}$. Then

$$\mathbf{u} + \mathbf{v} = (c_1\mathbf{v}_1 + c_2\mathbf{v}_2) + (d_1\mathbf{v}_1 + d_2\mathbf{v}_2) = (c_1 + d_1)\mathbf{v}_1 + (c_2 + d_2)\mathbf{v}_2$$

Thus, $\mathbf{u} + \mathbf{v}$ is a linear combination of $\mathbf{v}_1$ and $\mathbf{v}_2$ and so is in $\mathcal{P}$.

Now let c be a scalar. Then

$$c\mathbf{u} = c(c_1\mathbf{v}_1 + c_2\mathbf{v}_2) = (cc_1)\mathbf{v}_1 + (cc_2)\mathbf{v}_2$$

which shows that $c\mathbf{u}$ is also a linear combination of $\mathbf{v}_1$ and $\mathbf{v}_2$ and is therefore in $\mathcal{P}$. We have shown that $\mathcal{P}$ satisfies properties (1) through (3) and hence is a subspace of $\mathbb{R}^3$.

If you look carefully at the details of Example 3.37, you will notice that the fact that $\mathbf{v}_1$ and $\mathbf{v}_2$ were vectors in $\mathbb{R}^3$ played no role at all in the verification of the properties. Thus, the algebraic method we used should generalize beyond $\mathbb{R}^3$ and apply in situations where we can no longer visualize the geometry. It does. Moreover, the method of Example 3.37 can serve as a "template" in more general settings. When we generalize Example 3.37 to the span of an arbitrary set of vectors in any $\mathbb{R}^n$, the result is important enough to be called a theorem.

Theorem 3.19 Let $\mathbf{v}_1, \mathbf{v}_2, \ldots, \mathbf{v}_k$ be vectors in $\mathbb{R}^n$. Then $\text{span}(\mathbf{v}_1, \mathbf{v}_2, \ldots, \mathbf{v}_k)$ is a subspace of $\mathbb{R}^n$.

Proof Let $S = \text{span}(\mathbf{v}_1, \mathbf{v}_2, \ldots, \mathbf{v}_k)$. To check property (1) of the definition, we simply observe that the zero vector $\mathbf{0}$ is in S, since $\mathbf{0} = 0\mathbf{v}_1 + 0\mathbf{v}_2 + \cdots + 0\mathbf{v}_k$.

Now let

$$\mathbf{u} = c_1\mathbf{v}_1 + c_2\mathbf{v}_2 + \cdots + c_k\mathbf{v}_k \quad \text{and} \quad \mathbf{v} = d_1\mathbf{v}_1 + d_2\mathbf{v}_2 + \cdots + d_k\mathbf{v}_k$$

be two vectors in S. Then

$$\begin{aligned}\mathbf{u} + \mathbf{v} &= (c_1\mathbf{v}_1 + c_2\mathbf{v}_2 + \cdots + c_k\mathbf{v}_k) + (d_1\mathbf{v}_1 + d_2\mathbf{v}_2 + \cdots + d_k\mathbf{v}_k)\\ &= (c_1 + d_1)\mathbf{v}_1 + (c_2 + d_2)\mathbf{v}_2 + \cdots + (c_k + d_k)\mathbf{v}_k\end{aligned}$$

Thus, $\mathbf{u} + \mathbf{v}$ is a linear combination of $\mathbf{v}_1, \mathbf{v}_2, \ldots, \mathbf{v}_k$ and so is in S. This verifies property (2).

To show property (3), let c be a scalar. Then

$$\begin{aligned}c\mathbf{u} &= c(c_1\mathbf{v}_1 + c_2\mathbf{v}_2 + \cdots + c_k\mathbf{v}_k)\\ &= (cc_1)\mathbf{v}_1 + (cc_2)\mathbf{v}_2 + \cdots + (cc_k)\mathbf{v}_k\end{aligned}$$

which shows that $c\mathbf{u}$ is also a linear combination of $\mathbf{v}_1, \mathbf{v}_2, \ldots, \mathbf{v}_k$ and is therefore in S. We have shown that S satisfies properties (1) through (3) and hence is a subspace of $\mathbb{R}^n$.

We will refer to span$(\mathbf{v}_1, \mathbf{v}_2, \ldots, \mathbf{v}_k)$ as ***the subspace spanned by*** $\mathbf{v}_1, \mathbf{v}_2, \ldots, \mathbf{v}_k$. We will often be able to save a lot of work by recognizing when Theorem 3.19 can be applied.

Example 3.38

Show that the set of all vectors $\begin{bmatrix} x \\ y \\ z \end{bmatrix}$ that satisfy the conditions $x = 3y$ and $z = -2y$ forms a subspace of $\mathbb{R}^3$.

Solution Substituting the two conditions into $\begin{bmatrix} x \\ y \\ z \end{bmatrix}$ yields

$$\begin{bmatrix} 3y \\ y \\ -2y \end{bmatrix} = y\begin{bmatrix} 3 \\ 1 \\ -2 \end{bmatrix}$$

Since y is arbitrary, the given set of vectors is span$\left(\begin{bmatrix} 3 \\ 1 \\ -2 \end{bmatrix}\right)$ and is thus a subspace of $\mathbb{R}^3$, by Theorem 3.19.

Geometrically, the set of vectors in Example 3.38 represents the line through the origin in $\mathbb{R}^3$ with direction vector $\begin{bmatrix} 3 \\ 1 \\ -2 \end{bmatrix}$.

Example 3.39

Determine whether the set of all vectors $\begin{bmatrix} x \\ y \\ z \end{bmatrix}$ that satisfy the conditions $x = 3y + 1$ and $z = -2y$ is a subspace of $\mathbb{R}^3$.

Solution This time, we have all vectors of the form

$$\begin{bmatrix} 3y + 1 \\ y \\ -2y \end{bmatrix}$$

The zero vector is not of this form. (Why not? Try solving $\begin{bmatrix} 3y + 1 \\ y \\ -2y \end{bmatrix} = \begin{bmatrix} 0 \\ 0 \\ 0 \end{bmatrix}$.) Hence, property (1) does not hold, so this set cannot be a subspace of $\mathbb{R}^3$.

Example 3.40

Determine whether the set of all vectors $\begin{bmatrix} x \\ y \end{bmatrix}$, where $y = x^2$, is a subspace of $\mathbb{R}^2$.

Solution These are the vectors of the form $\begin{bmatrix} x \\ x^2 \end{bmatrix}$—call this set S. This time $\mathbf{0} = \begin{bmatrix} 0 \\ 0 \end{bmatrix}$ belongs to S (take $x = 0$), so property (1) holds. Let $\mathbf{u} = \begin{bmatrix} x_1 \\ x_1^2 \end{bmatrix}$ and $\mathbf{v} = \begin{bmatrix} x_2 \\ x_2^2 \end{bmatrix}$ be in S. Then

$$\mathbf{u} + \mathbf{v} = \begin{bmatrix} x_1 + x_2 \\ x_1^2 + x_2^2 \end{bmatrix}$$

which, in general, is not in S, since it does not have the correct form; that is, $x_1^2 + x_2^2 \neq (x_1 + x_2)^2$. To be specific, we look for a counterexample. If

$$\mathbf{u} = \begin{bmatrix} 1 \\ 1 \end{bmatrix} \quad \text{and} \quad \mathbf{v} = \begin{bmatrix} 2 \\ 4 \end{bmatrix}$$

then both $\mathbf{u}$ and $\mathbf{v}$ are in S, but their sum $\mathbf{u} + \mathbf{v} = \begin{bmatrix} 3 \\ 5 \end{bmatrix}$ is not in S since $5 \neq 3^2$. Thus, property (2) fails and S is not a subspace of $\mathbb{R}^2$.

Remark In order for a set S to be a subspace of some $\mathbb{R}^n$, we must *prove* that properties (1) through (3) hold *in general*. However, for S to *fail* to be a subspace of $\mathbb{R}^n$, it is enough to show that *one* of the three properties fails to hold. The easiest course is usually to find a single, specific *counterexample* to illustrate the failure of the property. Once you have done so, there is no need to consider the other properties.

Subspaces Associated with Matrices

A great many examples of subspaces arise in the context of matrices. We have already encountered the most important of these in Chapter 2; we now revisit them with the notion of a subspace in mind.

Definition Let A be an $m \times n$ matrix.

1. The ***row space*** of A is the subspace row(A) of $\mathbb{R}^n$ spanned by the rows of A.
2. The ***column space*** of A is the subspace col(A) of $\mathbb{R}^m$ spanned by the columns of A.

Remark Observe that, by Example 3.9 and the Remark that follows it, col(A) consists precisely of all vectors of the form $A\mathbf{x}$ where $\mathbf{x}$ is in $\mathbb{R}^n$.

Example 3.41

Consider the matrix

$$A = \begin{bmatrix} 1 & -1 \\ 0 & 1 \\ 3 & -3 \end{bmatrix}$$

(a) Determine whether $\mathbf{b} = \begin{bmatrix} 1 \\ 2 \\ 3 \end{bmatrix}$ is in the column space of A.

(b) Determine whether $\mathbf{w} = [4 \quad 5]$ is in the row space of A.

(c) Describe row(A) and col(A).

Solution

(a) By Theorem 2.4 and the discussion preceding it, $\mathbf{b}$ is a linear combination of the columns of A if and only if the linear system $A\mathbf{x} = \mathbf{b}$ is consistent. We row reduce the augmented matrix as follows:

$$\left[\begin{array}{cc|c} 1 & -1 & 1 \\ 0 & 1 & 2 \\ 3 & -3 & 3 \end{array}\right] \longrightarrow \left[\begin{array}{cc|c} 1 & 0 & 3 \\ 0 & 1 & 2 \\ 0 & 0 & 0 \end{array}\right]$$

Thus, the system is consistent (and, in fact, has a unique solution). Therefore, $\mathbf{b}$ is in col(A). (This example is just Example 2.18, phrased in the terminology of this section.)

(b) As we also saw in Section 2.3, elementary row operations simply create linear combinations of the rows of a matrix. That is, they produce vectors only in the row space of the matrix. If the vector $\mathbf{w}$ is in row(A), then $\mathbf{w}$ is a linear combination of the rows of A, so if we augment A by $\mathbf{w}$ as $\left[\begin{array}{c} A \\ \hline \mathbf{w} \end{array}\right]$, it will be possible to apply elementary row operations to this augmented matrix to reduce it to form $\left[\begin{array}{c} A' \\ \hline \mathbf{0} \end{array}\right]$ using only elementary row operations of the form $R_i + kR_j$, where $i > j$—in other words, *working from top to bottom in each column.* (Why?)

In this example, we have

$$\left[\begin{array}{c} A \\ \hline \mathbf{w} \end{array}\right] = \left[\begin{array}{cc} 1 & -1 \\ 0 & 1 \\ 3 & -3 \\ \hline 4 & 5 \end{array}\right] \xrightarrow[R_4 - 4R_1]{R_3 - 3R_1} \left[\begin{array}{cc} 1 & -1 \\ 0 & 1 \\ 0 & 0 \\ \hline 0 & 9 \end{array}\right] \xrightarrow{R_4 - 9R_2} \left[\begin{array}{cc} 1 & -1 \\ 0 & 1 \\ 0 & 0 \\ \hline 0 & 0 \end{array}\right]$$

Therefore, $\mathbf{w}$ is a linear combination of the rows of A (in fact, these calculations show that $\mathbf{w} = 4[1 \quad -1] + 9[0 \quad 1]$—how?), and thus $\mathbf{w}$ is in row(A).

(c) It is easy to check that, for any vector $\mathbf{w} = [x \quad y]$, the augmented matrix $\left[\begin{array}{c} A \\ \hline \mathbf{w} \end{array}\right]$ reduces to

$$\left[\begin{array}{cc} 1 & 0 \\ 0 & 1 \\ 0 & 0 \\ \hline 0 & 0 \end{array}\right]$$

in a similar fashion. Therefore, every vector in $\mathbb{R}^2$ is in row(A), and so row(A) $= \mathbb{R}^2$.

Finding col(A) is identical to solving Example 2.21, wherein we determined that it coincides with the plane (through the origin) in $\mathbb{R}^3$ with equation $3x - z = 0$. (We will discover other ways to answer this type of question shortly.)

Remark We could also have answered part (b) and the first part of part (c) by observing that any question about the *rows* of A is the corresponding question about the *columns* of A^T. So, for example, $\mathbf{w}$ is in row(A) if and only if $\mathbf{w}^T$ is in col(A^T). This is true if and only if the system $A^T\mathbf{x} = \mathbf{w}^T$ is consistent. We can now proceed as in part (a). (See Exercises 21–24.)

The observations we have made about the relationship between elementary row operations and the row space are summarized in the following theorem.

Theorem 3.20 Let B be any matrix that is row equivalent to a matrix A. Then row(B) $=$ row(A).

Proof The matrix A can be transformed into B by a sequence of row operations. Consequently, the rows of B are linear combinations of the rows of A; hence, linear combinations of the rows of B are linear combinations of the rows of A. (See Exercise 21 in Section 2.3.) It follows that row(B) $\subseteq$ row(A).

On the other hand, reversing these row operations transforms B into A. Therefore, the above argument shows that row(A) $\subseteq$ row(B). Combining these results, we have row(A) $=$ row(B).

There is another important subspace that we have already encountered: the set of solutions of a homogeneous system of linear equations. It is easy to prove that this subspace satisfies the three subspace properties.

Theorem 3.21 Let A be an $m \times n$ matrix and let N be the set of solutions of the homogeneous linear system $A\mathbf{x} = \mathbf{0}$. Then N is a subspace of $\mathbb{R}^n$.

Proof [Note that $\mathbf{x}$ must be a (column) vector in $\mathbb{R}^n$ in order for $A\mathbf{x}$ to be defined and that $\mathbf{0} = \mathbf{0}_m$ is the zero vector in $\mathbb{R}^m$.] Since $A\mathbf{0}_n = \mathbf{0}_m$, $\mathbf{0}_n$ is in N. Now let $\mathbf{u}$ and $\mathbf{v}$ be in N. Therefore, $A\mathbf{u} = \mathbf{0}$ and $A\mathbf{v} = \mathbf{0}$. It follows that

$$A(\mathbf{u} + \mathbf{v}) = A\mathbf{u} + A\mathbf{v} = \mathbf{0} + \mathbf{0} = \mathbf{0}$$

Hence, $\mathbf{u} + \mathbf{v}$ is in N. Finally, for any scalar c,

$$A(c\mathbf{u}) = c(A\mathbf{u}) = c\mathbf{0} = \mathbf{0}$$

and therefore $c\mathbf{u}$ is also in N. It follows that N is a subspace of $\mathbb{R}^n$.

Definition Let A be an $m \times n$ matrix. The ***null space*** of A is the subspace of $\mathbb{R}^n$ consisting of solutions of the homogeneous linear system $A\mathbf{x} = \mathbf{0}$. It is denoted by $\text{null}(A)$.

The fact that the null space of a matrix is a subspace allows us to prove what intuition and examples have led us to understand about the solutions of linear systems: They have either no solution, a unique solution, or infinitely many solutions.

Theorem 3.22 Let A be a matrix whose entries are real numbers. For any system of linear equations $A\mathbf{x} = \mathbf{b}$, exactly one of the following is true:

a. There is no solution.
b. There is a unique solution.
c. There are infinitely many solutions.

At first glance, it is not entirely clear how we should proceed to prove this theorem. A little reflection should persuade you that what we are really being asked to prove is that if (a) and (b) are not true, then (c) is the only other possibility. That is, if there is more than one solution, then there cannot be just two or even finitely many, but there must be infinitely many.

Proof If the system $A\mathbf{x} = \mathbf{b}$ has either no solutions or exactly one solution, we are done. Assume, then, that there are at least two distinct solutions of $A\mathbf{x} = \mathbf{b}$—say, $\mathbf{x}_1$ and $\mathbf{x}_2$. Thus,

$$A\mathbf{x}_1 = \mathbf{b} \quad \text{and} \quad A\mathbf{x}_2 = \mathbf{b}$$

with $\mathbf{x}_1 \neq \mathbf{x}_2$. It follows that

$$A(\mathbf{x}_1 - \mathbf{x}_2) = A\mathbf{x}_1 - A\mathbf{x}_2 = \mathbf{b} - \mathbf{b} = \mathbf{0}$$

Set $\mathbf{x}_0 = \mathbf{x}_1 - \mathbf{x}_2$. Then $\mathbf{x}_0 \neq \mathbf{0}$ and $A\mathbf{x}_0 = \mathbf{0}$. Hence, the null space of A is nontrivial, and since $\text{null}(A)$ is closed under scalar multiplication, $c\mathbf{x}_0$ is in $\text{null}(A)$ for every scalar c. Consequently, the null space of A contains infinitely many vectors (since it contains *at least* every vector of the form $c\mathbf{x}_0$ and there are infinitely many of these).

Now, consider the (infinitely many) vectors of the form $\mathbf{x}_1 + c\mathbf{x}_0$, as c varies through the set of real numbers. We have

$$A(\mathbf{x}_1 + c\mathbf{x}_0) = A\mathbf{x}_1 + cA\mathbf{x}_0 = \mathbf{b} + c\mathbf{0} = \mathbf{b}$$

Therefore, there are infinitely many solutions of the equation $A\mathbf{x} = \mathbf{b}$.

Basis

We can extract a bit more from the intuitive idea that subspaces are generalizations of planes through the origin in $\mathbb{R}^3$. A plane is spanned by any two vectors that are parallel to the plane but are not parallel to each other. In algebraic parlance, two such vectors span the plane and are linearly independent. Fewer than two vectors will not work; more than two vectors is not necessary. This is the essence of a *basis* for a subspace.

Definition A ***basis*** for a subspace S of $\mathbb{R}^n$ is a set of vectors in S that

1. spans S and
2. is linearly independent.

Example 3.42

In Section 2.3, we saw that the standard unit vectors $\mathbf{e}_1, \mathbf{e}_2, \ldots \mathbf{e}_n$ in $\mathbb{R}^n$ are linearly independent and span $\mathbb{R}^n$. Therefore, they form a basis for $\mathbb{R}^n$, called the ***standard basis.***

Example 3.43

In Example 2.19, we showed that $\mathbb{R}^2 = \operatorname{span}\left(\begin{bmatrix} 2 \\ -1 \end{bmatrix}, \begin{bmatrix} 1 \\ 3 \end{bmatrix}\right)$. Since $\begin{bmatrix} 2 \\ -1 \end{bmatrix}$ and $\begin{bmatrix} 1 \\ 3 \end{bmatrix}$ are also linearly independent (as they are not multiples), they form a basis for $\mathbb{R}^2$.

A subspace can (and will) have more than one basis. For example, we have just seen that $\mathbb{R}^2$ has the standard basis $\left\{\begin{bmatrix} 1 \\ 0 \end{bmatrix}, \begin{bmatrix} 0 \\ 1 \end{bmatrix}\right\}$ and the basis $\left\{\begin{bmatrix} 2 \\ -1 \end{bmatrix}, \begin{bmatrix} 1 \\ 3 \end{bmatrix}\right\}$. However, we will prove shortly that the *number* of vectors in a basis for a given subspace will always be the same.

Example 3.44

Find a basis for $S = \operatorname{span}(\mathbf{u}, \mathbf{v}, \mathbf{w})$, where

$$\mathbf{u} = \begin{bmatrix} 3 \\ -1 \\ 5 \end{bmatrix}, \mathbf{v} = \begin{bmatrix} 2 \\ 1 \\ 3 \end{bmatrix}, \text{ and } \mathbf{w} = \begin{bmatrix} 0 \\ -5 \\ 1 \end{bmatrix}$$

Solution The vectors $\mathbf{u}$, $\mathbf{v}$, and $\mathbf{w}$ already span S, so they will be a basis for S if they are also linearly independent. It is easy to determine that they are not; indeed, $\mathbf{w} = 2\mathbf{u} - 3\mathbf{v}$. Therefore, we can ignore $\mathbf{w}$, since any linear combinations involving $\mathbf{u}$, $\mathbf{v}$, and $\mathbf{w}$ can be rewritten to involve $\mathbf{u}$ and $\mathbf{v}$ alone. (Also see Exercise 47 in Section 2.3.) This implies that $S = \operatorname{span}(\mathbf{u}, \mathbf{v}, \mathbf{w}) = \operatorname{span}(\mathbf{u}, \mathbf{v})$, and since $\mathbf{u}$ and $\mathbf{v}$ are certainly linearly independent (why?), they form a basis for S. (Geometrically, this means that $\mathbf{u}$, $\mathbf{v}$, and $\mathbf{w}$ all lie in the same plane and $\mathbf{u}$ and $\mathbf{v}$ can serve as a set of direction vectors for this plane.)

Example 3.45

Find a basis for the row space of

$$A = \begin{bmatrix} 1 & 1 & 3 & 1 & 6 \\ 2 & -1 & 0 & 1 & -1 \\ -3 & 2 & 1 & -2 & 1 \\ 4 & 1 & 6 & 1 & 3 \end{bmatrix}$$

Solution The reduced row echelon form of A is

$$R = \begin{bmatrix} 1 & 0 & 1 & 0 & -1 \\ 0 & 1 & 2 & 0 & 3 \\ 0 & 0 & 0 & 1 & 4 \\ 0 & 0 & 0 & 0 & 0 \end{bmatrix}$$

By Theorem 3.20, $\text{row}(A) = \text{row}(R)$, so it is enough to find a basis for the row space of R. But $\text{row}(R)$ is clearly spanned by its nonzero rows, and it is easy to check that the staircase pattern forces the first three rows of R to be linearly independent. (This is a general fact, one that you will need to establish to prove Exercise 33.) Therefore, a basis for the row space of A is

$$\{[1 \quad 0 \quad 1 \quad 0 \quad -1], [0 \quad 1 \quad 2 \quad 0 \quad 3], [0 \quad 0 \quad 0 \quad 1 \quad 4]\}$$

We can use the method of Example 3.45 to find a basis for the subspace spanned by a given set of vectors.

Example 3.46

Rework Example 3.44 using the method from Example 3.45.

Solution We transpose **u**, **v**, and **w** to get row vectors and then form a matrix with these vectors as its rows:

$$B = \begin{bmatrix} 3 & -1 & 5 \\ 2 & 1 & 3 \\ 0 & -5 & 1 \end{bmatrix}$$

Proceeding as in Example 3.45, we reduce B to its reduced row echelon form

$$\begin{bmatrix} 1 & 0 & \frac{8}{5} \\ 0 & 1 & -\frac{1}{5} \\ 0 & 0 & 0 \end{bmatrix}$$

and use the nonzero row vectors as a basis for the row space. Since we started with column vectors, we must transpose again. Thus, a basis for $\text{span}(\mathbf{u}, \mathbf{v}, \mathbf{w})$ is

$$\left\{ \begin{bmatrix} 1 \\ 0 \\ \frac{8}{5} \end{bmatrix}, \begin{bmatrix} 0 \\ 1 \\ -\frac{1}{5} \end{bmatrix} \right\}$$

Remarks

- In fact, we do not need to go all the way to *reduced* row echelon form—row echelon form is far enough. If U is a row echelon form of A, then the nonzero row vectors of U will form a basis for row(A) (see Exercise 33). This approach has the advantage of (often) allowing us to avoid fractions. In Example 3.46, B can be reduced to

$$U = \begin{bmatrix} 3 & -1 & 5 \\ 0 & -5 & 1 \\ 0 & 0 & 0 \end{bmatrix}$$

which gives us the basis

$$\left\{ \begin{bmatrix} 3 \\ -1 \\ 5 \end{bmatrix}, \begin{bmatrix} 0 \\ -5 \\ 1 \end{bmatrix} \right\}$$

for span($\mathbf{u}, \mathbf{v}, \mathbf{w}$).

- Observe that the methods used in Example 3.44, Example 3.46, and the Remark above will generally produce different bases.

We now turn to the problem of finding a basis for the column space of a matrix A. One method is simply to transpose the matrix. The column vectors of A become the row vectors of A^T, and we can apply the method of Example 3.45 to find a basis for row(A^T). Transposing these vectors then gives us a basis for col(A). (You are asked to do this in Exercises 21–24.) This approach, however, requires performing a new set of row operations on A^T.

Instead, we prefer to take an approach that allows us to use the row reduced form of A that we have already computed. Recall that a product $A\mathbf{x}$ of a matrix and a vector corresponds to a linear combination of the columns of A with the entries of $\mathbf{x}$ as coefficients. Thus, a nontrivial solution to $A\mathbf{x} = \mathbf{0}$ represents a *dependence relation* among the columns of A. Since elementary row operations do not affect the solution set, if A is row equivalent to R, *the columns of A have the same dependence relationships as the columns of R*. This important observation is the basis (no pun intended!) for the technique we now use to find a basis for col(A).

Example 3.47

Find a basis for the column space of the matrix from Example 3.45,

$$A = \begin{bmatrix} 1 & 1 & 3 & 1 & 6 \\ 2 & -1 & 0 & 1 & -1 \\ -3 & 2 & 1 & -2 & 1 \\ 4 & 1 & 6 & 1 & 3 \end{bmatrix}$$

Solution Let $\mathbf{a}_i$ denote a column vector of A and let $\mathbf{r}_i$ denote a column vector of the reduced echelon form

$$R = \begin{bmatrix} 1 & 0 & 1 & 0 & -1 \\ 0 & 1 & 2 & 0 & 3 \\ 0 & 0 & 0 & 1 & 4 \\ 0 & 0 & 0 & 0 & 0 \end{bmatrix}$$

We can quickly see by inspection that $\mathbf{r}_3 = \mathbf{r}_1 + 2\mathbf{r}_2$ and $\mathbf{r}_5 = -\mathbf{r}_1 + 3\mathbf{r}_2 + 4\mathbf{r}_4$. (Check that, as predicted, the corresponding column vectors of A satisfy the same dependence relations.) Thus, $\mathbf{r}_3$ and $\mathbf{r}_5$ contribute nothing to col(R). The remaining column vectors, $\mathbf{r}_1$, $\mathbf{r}_2$, and $\mathbf{r}_4$, are linearly independent, since they are just standard unit vectors. The corresponding statements are therefore true of the column vectors of A.

Thus, among the column vectors of A, we eliminate the dependent ones ($\mathbf{a}_3$ and $\mathbf{a}_5$), and the remaining ones will be linearly independent and hence form a basis for col(A). What is the fastest way to find this basis? Use the columns of A that correspond to the columns of R *containing the leading 1s*. A basis for col(A) is

$$\{\mathbf{a}_1, \mathbf{a}_2, \mathbf{a}_4\} = \left\{ \begin{bmatrix} 1 \\ 2 \\ -3 \\ 4 \end{bmatrix}, \begin{bmatrix} 1 \\ -1 \\ 2 \\ 1 \end{bmatrix}, \begin{bmatrix} 1 \\ 1 \\ -2 \\ 1 \end{bmatrix} \right\}$$

Warning Elementary row operations change the column space! In our example, $\text{col}(A) \neq \text{col}(R)$, since every vector in col(R) has its fourth component equal to 0 but this is certainly not true of col(A). So we must go back to the original matrix A to get the column vectors for a basis of col(A). To be specific, in Example 3.47, $\mathbf{r}_1$, $\mathbf{r}_2$, and $\mathbf{r}_4$ do *not* form a basis for the column space of A.

Example 3.48

Find a basis for the null space of matrix A from Example 3.47.

Solution There is really nothing new here except the terminology. We simply have to find and describe the solutions of the homogeneous system $A\mathbf{x} = \mathbf{0}$. We have already computed the reduced row echelon form R of A, so all that remains to be done in Gauss-Jordan elimination is to solve for the leading variables in terms of the free variables. The final augmented matrix is

$$[R \,|\, \mathbf{0}] = \left[\begin{array}{ccccc|c} 1 & 0 & 1 & 0 & -1 & 0 \\ 0 & 1 & 2 & 0 & 3 & 0 \\ 0 & 0 & 0 & 1 & 4 & 0 \\ 0 & 0 & 0 & 0 & 0 & 0 \end{array}\right]$$

If

$$\mathbf{x} = \begin{bmatrix} x_1 \\ x_2 \\ x_3 \\ x_4 \\ x_5 \end{bmatrix}$$

then the leading 1s are in columns 1, 2, and 4, so we solve for x_1, x_2, and x_4 in terms of the free variables x_3 and x_5. We get $x_1 = -x_3 + x_5$, $x_2 = -2x_3 - 3x_5$, and $x_4 = -4x_5$. Setting $x_3 = s$ and $x_5 = t$, we obtain

$$\mathbf{x} = \begin{bmatrix} x_1 \\ x_2 \\ x_3 \\ x_4 \\ x_5 \end{bmatrix} = \begin{bmatrix} -s + t \\ -2s - 3t \\ s \\ -4t \\ t \end{bmatrix} = s\begin{bmatrix} -1 \\ -2 \\ 1 \\ 0 \\ 0 \end{bmatrix} + t\begin{bmatrix} 1 \\ -3 \\ 0 \\ -4 \\ 1 \end{bmatrix} = s\mathbf{u} + t\mathbf{v}$$

Thus, $\mathbf{u}$ and $\mathbf{v}$ span null(A), and since they are linearly independent, they form a basis for null(A).

Following is a summary of the most effective procedure to use to find bases for the row space, the column space, and the null space of a matrix A.

1. Find the reduced row echelon form R of A.
2. Use the nonzero row vectors of R (containing the leading 1s) to form a basis for row(A).
3. Use the column vectors of A that correspond to the columns of R containing the leading 1s (the pivot columns) to form a basis for col(A).
4. Solve for the leading variables of $R\mathbf{x} = \mathbf{0}$ in terms of the free variables, set the free variables equal to parameters, substitute back into $\mathbf{x}$, and write the result as a linear combination of f vectors (where f is the number of free variables). These f vectors form a basis for null(A).

If we do not need to find the null space, then it is faster to simply reduce A to row echelon form to find bases for the row and column spaces. Steps 2 and 3 above remain valid (with the substitution of the word "pivots" for "leading 1s").

Dimension and Rank

We have observed that although a subspace will have different bases, each basis has the same number of vectors. This fundamental fact will be of vital importance from here on in this book.

Theorem 3.23 **The Basis Theorem**

Let S be a subspace of $\mathbb{R}^n$. Then any two bases for S have the same number of vectors.

Sherlock Holmes noted, "When you have eliminated the impossible, whatever remains, *however improbable,* must be the truth" (from *The Sign of Four* by Sir Arthur Conan Doyle).

Proof Let $\mathcal{B} = \{\mathbf{u}_1, \mathbf{u}_2, \ldots, \mathbf{u}_r\}$ and $\mathcal{C} = \{\mathbf{v}_1, \mathbf{v}_2, \ldots, \mathbf{v}_s\}$ be bases for S. We need to prove that $r = s$. We do so by showing that neither of the other two possibilities, $r < s$ or $r > s$, can occur.

Suppose that $r < s$. We will show that this forces $\mathcal{C}$ to be a linearly dependent set of vectors. To this end, let

$$c_1\mathbf{v}_1 + c_2\mathbf{v}_2 + \cdots + c_s\mathbf{v}_s = 0 \tag{1}$$

Since $\mathcal{B}$ is a basis for S, we can write each $\mathbf{v}_i$ as a linear combination of the elements $\mathbf{u}_j$:

$$\begin{aligned}\mathbf{v}_1 &= a_{11}\mathbf{u}_1 + a_{12}\mathbf{u}_2 + \cdots + a_{1r}\mathbf{u}_r\\ \mathbf{v}_2 &= a_{21}\mathbf{u}_1 + a_{22}\mathbf{u}_2 + \cdots + a_{2r}\mathbf{u}_r\\ &\vdots\\ \mathbf{v}_s &= a_{s1}\mathbf{u}_1 + a_{s2}\mathbf{u}_2 + \cdots + a_{sr}\mathbf{u}_r\end{aligned} \tag{2}$$

Substituting the Equations (2) into Equation (1), we obtain

$$c_1(a_{11}\mathbf{u}_1 + \cdots + a_{1r}\mathbf{u}_r) + c_2(a_{21}\mathbf{u}_1 + \cdots + a_{2r}\mathbf{u}_r) + \cdots + c_s(a_{s1}\mathbf{u}_1 + \cdots + a_{sr}\mathbf{u}_r) = 0$$

Regrouping, we have

$$(c_1a_{11} + c_2a_{21} + \cdots + c_sa_{s1})\mathbf{u}_1 + (c_1a_{12} + c_2a_{22} + \cdots + c_sa_{s2})\mathbf{u}_2 + \cdots + (c_1a_{1r} + c_2a_{2r} + \cdots + c_sa_{sr})\mathbf{u}_r = 0$$

Now, since $\mathcal{B}$ is a basis, the $\mathbf{u}_j$'s are linearly independent. So each of the expressions in parentheses must be zero:

$$\begin{aligned} c_1a_{11} + c_2a_{21} + \cdots + c_sa_{s1} &= 0 \\ c_1a_{12} + c_2a_{22} + \cdots + c_sa_{s2} &= 0 \\ &\vdots \\ c_1a_{1r} + c_2a_{2r} + \cdots + c_sa_{sr} &= 0 \end{aligned}$$

This is a homogeneous system of r linear equations in the s variables $c_1, c_2, \ldots, c_s$. (The fact that the variables appear to the left of the coefficients makes no difference.) Since $r < s$, we know from Theorem 2.3 that there are infinitely many solutions. In particular, there is a nontrivial solution, giving a nontrivial dependence relation in Equation (1). Thus, $\mathcal{C}$ is a linearly dependent set of vectors. But this finding contradicts the fact that $\mathcal{C}$ was given to be a basis and hence linearly independent. We conclude that $r < s$ is not possible. Similarly (interchanging the roles of $\mathcal{B}$ and $\mathcal{C}$), we find that $r > s$ leads to a contradiction. Hence, we must have $r = s$, as desired.

Since all bases for a given subspace must have the same number of vectors, we can attach a name to this number.

Definition If S is a subspace of $\mathbb{R}^n$, then the number of vectors in a basis for S is called the ***dimension*** of S, denoted dim S.

Remark The zero vector $\mathbf{0}$ by itself is always a subspace of $\mathbb{R}^n$. (Why?) Yet any set containing the zero vector (and, in particular, $\{\mathbf{0}\}$) is linearly dependent, so $\{\mathbf{0}\}$ cannot have a basis. We define dim $\{\mathbf{0}\}$ to be 0.

Example 3.49

Since the standard basis for $\mathbb{R}^n$ has n vectors, dim $\mathbb{R}^n = n$. (Note that this result agrees with our intuitive understanding of dimension for $n \leq 3$.)

Example 3.50

In Examples 3.45 through 3.48, we found that row(A) has a basis with three vectors, col(A) has a basis with three vectors, and null(A) has a basis with two vectors. Hence, dim(row(A)) = 3, dim(col(A)) = 3, and dim(null(A)) = 2.

A single example is not enough on which to speculate, but the fact that the row and column spaces in Example 3.50 have the same dimension is no accident. Nor is the fact that the sum of dim(col(A)) and dim(null(A)) is 5, the number of columns of A. We now prove that these relationships are true in general.

Theorem 3.24 The row and column spaces of a matrix A have the same dimension.

Proof Let R be the reduced row echelon form of A. By Theorem 3.20, $\text{row}(A) = \text{row}(R)$, so

$$\begin{aligned}\dim(\text{row}(A)) &= \dim(\text{row}(R)) \\ &= \text{number of nonzero rows of } R \\ &= \text{number of leading 1s of } R\end{aligned}$$

Let this number be called r.

Now $\text{col}(A) \neq \text{col}(R)$, but the columns of A and R have the same dependence relationships. Therefore, $\dim(\text{col}(A)) = \dim(\text{col}(R))$. Since there are r leading 1s, R has r columns that are standard unit vectors, $\mathbf{e}_1, \mathbf{e}_2, \ldots, \mathbf{e}_r$. (These will be vectors in $\mathbb{R}^m$ if A and R are $m \times n$ matrices.) These r vectors are linearly independent, and the remaining columns of R are linear combinations of them. Thus, $\dim(\text{col}(R)) = r$. It follows that $\dim(\text{row}(A)) = r = \dim(\text{col}(A))$, as we wished to prove.

The rank of a matrix was first defined in 1878 by Georg Frobenius (1849–1917), although he defined it using determinants and not as we have done here. (See Chapter 4.) Frobenius was a German mathematician who received his doctorate from and later taught at the University of Berlin. Best known for his contributions to group theory, Frobenius used matrices in his work on group representations.

Definition The ***rank*** of a matrix A is the dimension of its row and column spaces and is denoted by $\text{rank}(A)$.

For Example 3.50, we can thus write $\text{rank}(A) = 3$.

Remarks

- The preceding definition agrees with the more informal definition of rank that was introduced in Chapter 2. The advantage of our new definition is that it is much more flexible.
- The rank of a matrix simultaneously gives us information about linear dependence among the row vectors of the matrix *and* among its column vectors. In particular, it tells us the number of rows and columns that are linearly independent (and this number is the same in each case!).

Since the row vectors of A are the column vectors of A^T, Theorem 3.24 has the following immediate corollary.

Theorem 3.25 For any matrix A,

$$\text{rank}(A^T) = \text{rank}(A)$$

Proof We have

$$\begin{aligned}\text{rank}(A^T) &= \dim(\text{col}(A^T)) \\ &= \dim(\text{row}(A)) \\ &= \text{rank}(A)\end{aligned}$$

Definition The ***nullity*** of a matrix A is the dimension of its null space and is denoted by $\text{nullity}(A)$.

In other words, nullity(A) is the dimension of the solution space of $A\mathbf{x} = \mathbf{0}$, which is the same as the number of free variables in the solution. We can now revisit the Rank Theorem (Theorem 2.2), rephrasing it in terms of our new definitions.

Theorem 3.26 **The Rank Theorem**

If A is an $m \times n$ matrix, then

$$\operatorname{rank}(A) + \operatorname{nullity}(A) = n$$

Proof Let R be the reduced row echelon form of A, and suppose that rank$(A) = r$. Then R has r leading 1s, so there are r leading variables and $n - r$ free variables in the solution to $A\mathbf{x} = \mathbf{0}$. Since dim(null$(A)) = n - r$, we have

$$\begin{aligned}\operatorname{rank}(A) + \operatorname{nullity}(A) &= r + (n - r) \\ &= n\end{aligned}$$

Often, when we need to know the nullity of a matrix, we do not need to know the actual solution of $A\mathbf{x} = \mathbf{0}$. The Rank Theorem is extremely useful in such situations, as the following example illustrates.

Example 3.51 Find the nullity of each of the following matrices:

$$M = \begin{bmatrix} 2 & 3 \\ 1 & 5 \\ 4 & 7 \\ 3 & 6 \end{bmatrix} \quad \text{and}$$

$$N = \begin{bmatrix} 2 & 1 & -2 & -1 \\ 4 & 4 & -3 & 1 \\ 2 & 7 & 1 & 8 \end{bmatrix}$$

Solution Since the two columns of M are clearly linearly independent, rank$(M) = 2$. Thus, by the Rank Theorem, nullity$(M) = 2 - \operatorname{rank}(M) = 2 - 2 = 0$.

There is no obvious dependence among the rows or columns of N, so we apply row operations to reduce it to

$$\begin{bmatrix} 2 & 1 & -2 & -1 \\ 0 & 2 & 1 & 3 \\ 0 & 0 & 0 & 0 \end{bmatrix}$$

We have reduced the matrix far enough (we do not need *reduced* row echelon form here, since we are not looking for a basis for the null space). We see that there are only two nonzero rows, so rank$(N) = 2$. Hence, nullity$(N) = 4 - \operatorname{rank}(N) = 4 - 2 = 2$.

The results of this section allow us to extend the Fundamental Theorem of Invertible Matrices (Theorem 3.12).

Theorem 3.27 **The Fundamental Theorem of Invertible Matrices: Version 2**

Let A be an $n \times n$ matrix. The following statements are equivalent:

a. A is invertible.
b. $A\mathbf{x} = \mathbf{b}$ has a unique solution for every $\mathbf{b}$ in $\mathbb{R}^n$.
c. $A\mathbf{x} = \mathbf{0}$ has only the trivial solution.
d. The reduced row echelon form of A is I_n.
e. A is a product of elementary matrices.
f. $\text{rank}(A) = n$
g. $\text{nullity}(A) = 0$
h. The column vectors of A are linearly independent.
i. The column vectors of A span $\mathbb{R}^n$.
j. The column vectors of A form a basis for $\mathbb{R}^n$.
k. The row vectors of A are linearly independent.
l. The row vectors of A span $\mathbb{R}^n$.
m. The row vectors of A form a basis for $\mathbb{R}^n$.

National Portrait Gallery, London

The nullity of a matrix was defined in 1884 by James Joseph Sylvester (1814–1887), who was interested in *invariants*—properties of matrices that do not change under certain types of transformations. Born in England, Sylvester became the second president of the London Mathematical Society. In 1878, while teaching at Johns Hopkins University in Baltimore, he founded the *American Journal of Mathematics*, the first mathematical journal in the United States.

Proof We have already established the equivalence of (a) through (e). It remains to be shown that statements (f) to (m) are equivalent to the first five statements.

(f) $\Leftrightarrow$ (g) Since $\text{rank}(A) + \text{nullity}(A) = n$ when A is an $n \times n$ matrix, it follows from the Rank Theorem that $\text{rank}(A) = n$ if and only if $\text{nullity}(A) = 0$.

(f) $\Rightarrow$ (d) $\Rightarrow$ (c) $\Rightarrow$ (h) If $\text{rank}(A) = n$, then the reduced row echelon form of A has n leading 1s and so is I_n. From (d) $\Rightarrow$ (c) we know that $A\mathbf{x} = \mathbf{0}$ has only the trivial solution, which implies that the column vectors of A are linearly independent, since $A\mathbf{x}$ is just a linear combination of the column vectors of A.

(h) $\Rightarrow$ (i) If the column vectors of A are linearly independent, then $A\mathbf{x} = \mathbf{0}$ has only the trivial solution. Thus, by (c) $\Rightarrow$ (b), $A\mathbf{x} = \mathbf{b}$ has a unique solution for every $\mathbf{b}$ in $\mathbb{R}^n$. This means that every vector $\mathbf{b}$ in $\mathbb{R}^n$ can be written as a linear combination of the column vectors of A, establishing (i).

(i) $\Rightarrow$ (j) If the column vectors of A span $\mathbb{R}^n$, then $\text{col}(A) = \mathbb{R}^n$ by definition, so $\text{rank}(A) = \dim(\text{col}(A)) = n$. This is (f), and we have already established that (f) $\Rightarrow$ (h). We conclude that the column vectors of A are linearly independent and so form a basis for $\mathbb{R}^n$, since, by assumption, they also span $\mathbb{R}^n$.

(j) $\Rightarrow$ (f) If the column vectors of A form a basis for $\mathbb{R}^n$, then, in particular, they are linearly independent. It follows that the reduced row echelon form of A contains n leading 1s, and thus $\text{rank}(A) = n$.

The above discussion shows that (f) $\Rightarrow$ (d) $\Rightarrow$ (c) $\Rightarrow$ (h) $\Rightarrow$ (i) $\Rightarrow$ (j) $\Rightarrow$ (f) $\Leftrightarrow$ (g). Now recall that, by Theorem 3.25, $\text{rank}(A^T) = \text{rank}(A)$, so what we have just proved gives us the corresponding results about the column vectors of A^T. These are then results about the *row* vectors of A, bringing (k), (l), and (m) into the network of equivalences and completing the proof.

Theorems such as the Fundamental Theorem are not merely of theoretical interest. They are tremendous labor-saving devices as well. The Fundamental Theorem has already allowed us to cut in half the work needed to check that two square matrices are inverses. It also simplifies the task of showing that certain sets of vectors are bases for $\mathbb{R}^n$. Indeed, when we have a set of n vectors in $\mathbb{R}^n$, that set will be a basis for $\mathbb{R}^n$ if *either* of the necessary properties of linear independence or spanning set is true. The next example shows how easy the calculations can be.

Example 3.52 Show that the vectors

$$\begin{bmatrix} 1 \\ 2 \\ 3 \end{bmatrix}, \begin{bmatrix} -1 \\ 0 \\ 1 \end{bmatrix}, \text{ and } \begin{bmatrix} 4 \\ 9 \\ 7 \end{bmatrix}$$

form a basis for $\mathbb{R}^3$.

Solution According to the Fundamental Theorem, the vectors will form a basis for $\mathbb{R}^3$ if and only if a matrix with these vectors as its columns (or rows) has rank 3. We perform just enough row operations to determine this:

$$A = \begin{bmatrix} 1 & -1 & 4 \\ 2 & 0 & 9 \\ 3 & 1 & 7 \end{bmatrix} \longrightarrow \begin{bmatrix} 1 & -1 & 4 \\ 0 & 2 & 1 \\ 0 & 0 & -7 \end{bmatrix}$$

We see that A has rank 3, so the given vectors are a basis for $\mathbb{R}^3$ by the equivalence of (f) and (j).

The next theorem is an application of both the Rank Theorem and the Fundamental Theorem. We will require this result in Chapters 5 and 7.

Theorem 3.28 Let A be an $m \times n$ matrix. Then:

a. $\text{rank}(A^TA) = \text{rank}(A)$
b. The $n \times n$ matrix A^TA is invertible if and only if $\text{rank}(A) = n$.

Proof

(a) Since A^TA is $n \times n$, it has the same number of columns as A. The Rank Theorem then tells us that

$$\text{rank}(A) + \text{nullity}(A) = n = \text{rank}(A^TA) + \text{nullity}(A^TA)$$

Hence, to show that $\text{rank}(A) = \text{rank}(A^TA)$, it is enough to show that $\text{nullity}(A) = \text{nullity}(A^TA)$. We will do so by establishing that the null spaces of A and A^TA are the same.

To this end, let $\mathbf{x}$ be in $\text{null}(A)$ so that $A\mathbf{x} = \mathbf{0}$. Then $A^TA\mathbf{x} = A^T\mathbf{0} = \mathbf{0}$, and thus $\mathbf{x}$ is in $\text{null}(A^TA)$. Conversely, let $\mathbf{x}$ be in $\text{null}(A^TA)$. Then $A^TA\mathbf{x} = \mathbf{0}$, so $\mathbf{x}^TA^TA\mathbf{x} = \mathbf{x}^T\mathbf{0} = 0$. But then

$$(A\mathbf{x}) \cdot (A\mathbf{x}) = (A\mathbf{x})^T(A\mathbf{x}) = \mathbf{x}^TA^TA\mathbf{x} = 0$$

and hence $A\mathbf{x} = \mathbf{0}$, by Theorem 1.2(d). Therefore, $\mathbf{x}$ is in $\text{null}(A)$, so $\text{null}(A) = \text{null}(A^TA)$, as required.

(b) By the Fundamental Theorem, the $n \times n$ matrix A^TA is invertible if and only if $\text{rank}(A^TA) = n$. But, by (a) this is so if and only if $\text{rank}(A) = n$.

Coordinates

We now return to one of the questions posed at the very beginning of this section: How should we view vectors in $\mathbb{R}^3$ that live in a plane through the origin? Are they two-dimensional or three-dimensional? The notions of basis and dimension will help clarify things.

A plane through the origin is a two-dimensional subspace of $\mathbb{R}^3$, with any set of two direction vectors serving as a basis. Basis vectors locate coordinate axes in the plane/subspace, in turn allowing us to view the plane as a "copy" of $\mathbb{R}^2$. Before we illustrate this approach, we prove a theorem guaranteeing that "coordinates" that arise in this way are unique.

Theorem 3.29 Let S be a subspace of $\mathbb{R}^n$ and let $\mathcal{B} = \{\mathbf{v}_1, \mathbf{v}_2, \ldots, \mathbf{v}_k\}$ be a basis for S. For every vector $\mathbf{v}$ in S, there is exactly one way to write $\mathbf{v}$ as a linear combination of the basis vectors in $\mathcal{B}$:

$$\mathbf{v} = c_1\mathbf{v}_1 + c_2\mathbf{v}_2 + \cdots + c_k\mathbf{v}_k$$

Proof Since $\mathcal{B}$ is a basis, it spans S, so $\mathbf{v}$ can be written in *at least one* way as a linear combination of $\mathbf{v}_1, \mathbf{v}_2, \ldots, \mathbf{v}_k$. Let one of these linear combinations be

$$\mathbf{v} = c_1\mathbf{v}_1 + c_2\mathbf{v}_2 + \cdots + c_k\mathbf{v}_k$$

Our task is to show that this is the *only* way to write $\mathbf{v}$ as a linear combination of $\mathbf{v}_1, \mathbf{v}_2, \ldots, \mathbf{v}_k$. To this end, suppose that we also have

$$\mathbf{v} = d_1\mathbf{v}_1 + d_2\mathbf{v}_2 + \cdots + d_k\mathbf{v}_k$$

Then

$$c_1\mathbf{v}_1 + c_2\mathbf{v}_2 + \cdots + c_k\mathbf{v}_k = d_1\mathbf{v}_1 + d_2\mathbf{v}_2 + \cdots + d_k\mathbf{v}_k$$

Rearranging (using properties of vector algebra), we obtain

$$(c_1 - d_1)\mathbf{v}_1 + (c_2 - d_2)\mathbf{v}_2 + \cdots + (c_k - d_k)\mathbf{v}_k = \mathbf{0}$$

Since $\mathcal{B}$ is a basis, $\mathbf{v}_1, \mathbf{v}_2, \ldots, \mathbf{v}_k$ are linearly independent. Therefore,

$$(c_1 - d_1) = (c_2 - d_2) = \cdots = (c_k - d_k) = 0$$

In other words, $c_1 = d_1, c_2 = d_2, \ldots, c_k = d_k$, and the two linear combinations are actually the same. Thus, there is exactly one way to write $\mathbf{v}$ as a linear combination of the basis vectors in $\mathcal{B}$.

Definition Let S be a subspace of $\mathbb{R}^n$ and let $\mathcal{B} = \{\mathbf{v}_1, \mathbf{v}_2, \ldots, \mathbf{v}_k\}$ be a basis for S. Let $\mathbf{v}$ be a vector in S, and write $\mathbf{v} = c_1\mathbf{v}_1 + c_2\mathbf{v}_2 + \cdots + c_k\mathbf{v}_k$. Then $c_1, c_2, \ldots, c_k$ are called the ***coordinates of* v *with respect to* $\mathcal{B}$**, and the column vector

$$[\mathbf{v}]_{\mathcal{B}} = \begin{bmatrix} c_1 \\ c_2 \\ \vdots \\ c_k \end{bmatrix}$$

is called the ***coordinate vector of* v *with respect to* $\mathcal{B}$.**

Example 3.53 Let $\mathcal{E} = \{\mathbf{e}_1, \mathbf{e}_2, \mathbf{e}_3\}$ be the standard basis for $\mathbb{R}^3$. Find the coordinate vector of

$$\mathbf{v} = \begin{bmatrix} 2 \\ 7 \\ 4 \end{bmatrix}$$

with respect to $\mathcal{E}$.

Solution Since $\mathbf{v} = 2\mathbf{e}_1 + 7\mathbf{e}_2 + 4\mathbf{e}_3$,

$$[\mathbf{v}]_{\mathcal{E}} = \begin{bmatrix} 2 \\ 7 \\ 4 \end{bmatrix}$$

It should be clear that the coordinate vector of every (column) vector in $\mathbb{R}^n$ with respect to the standard basis is just the vector itself.

Example 3.54

In Example 3.44, we saw that $\mathbf{u} = \begin{bmatrix} 3 \\ -1 \\ 5 \end{bmatrix}$, $\mathbf{v} = \begin{bmatrix} 2 \\ 1 \\ 3 \end{bmatrix}$, and $\mathbf{w} = \begin{bmatrix} 0 \\ -5 \\ 1 \end{bmatrix}$ are three vectors in the same subspace (plane through the origin) S of $\mathbb{R}^3$ and that $\mathcal{B} = \{\mathbf{u}, \mathbf{v}\}$ is a basis for S. Since $\mathbf{w} = 2\mathbf{u} - 3\mathbf{v}$, we have

$$[\mathbf{w}]_{\mathcal{B}} = \begin{bmatrix} 2 \\ -3 \end{bmatrix}$$

See Figure 3.3.

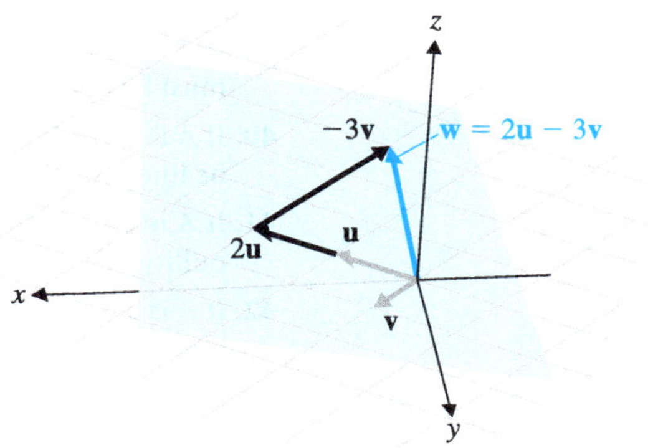

Figure 3.3
The coordinates of a vector with respect to a basis

Exercises 3.5

In Exercises 1–4, let S be the collection of vectors $\begin{bmatrix} x \\ y \end{bmatrix}$ in $\mathbb{R}^2$ that satisfy the given property. In each case, either prove that S forms a subspace of $\mathbb{R}^2$ or give a counterexample to show that it does not.

1. $x = 0$ **2.** $x \geq 0, y \geq 0$

3. $y = 2x$ **4.** $xy \geq 0$

In Exercises 5–8, let S be the collection of vectors $\begin{bmatrix} x \\ y \\ z \end{bmatrix}$ in $\mathbb{R}^3$ that satisfy the given property. In each case, either prove that S forms a subspace of $\mathbb{R}^3$ or give a counterexample to show that it does not.

5. $x = y = z$ **6.** $z = 2x, y = 0$

7. $x - y + z = 1$ **8.** $|x - y| = |y - z|$

9. Prove that every line through the origin in $\mathbb{R}^3$ is a subspace of $\mathbb{R}^3$.

10. Suppose S consists of all points in $\mathbb{R}^2$ that are on the x-axis or the y-axis (or both). (S is called the *union* of the two axes.) Is S a subspace of $\mathbb{R}^2$? Why or why not?

In Exercises 11 and 12, determine whether **b** *is in col(A) and whether* **w** *is in row(A), as in Example 3.41.*

11. $A = \begin{bmatrix} 1 & 0 & -1 \\ 1 & 1 & 1 \end{bmatrix}$, $\mathbf{b} = \begin{bmatrix} 3 \\ 2 \end{bmatrix}$, $\mathbf{w} = [-1 \quad 1 \quad 1]$

12. $A = \begin{bmatrix} 1 & 1 & -3 \\ 0 & 2 & 1 \\ 1 & -1 & -4 \end{bmatrix}$, $\mathbf{b} = \begin{bmatrix} 1 \\ 1 \\ 0 \end{bmatrix}$, $\mathbf{w} = [2 \quad 4 \quad -5]$

13. In Exercise 11, determine whether $\mathbf{w}$ is in row(A), using the method described in the Remark following Example 3.41.

14. In Exercise 12, determine whether $\mathbf{w}$ is in row(A), using the method described in the Remark following Example 3.41.

15. If A is the matrix in Exercise 11, is $\mathbf{v} = \begin{bmatrix} -1 \\ 3 \\ -1 \end{bmatrix}$ in null(A)?

16. If A is the matrix in Exercise 12, is $\mathbf{v} = \begin{bmatrix} 7 \\ -1 \\ 2 \end{bmatrix}$ in null(A)?

In Exercises 17–20, give bases for row(A), col(A), and null(A).

17. $A = \begin{bmatrix} 1 & 0 & -1 \\ 1 & 1 & 1 \end{bmatrix}$ **18.** $A = \begin{bmatrix} 1 & 1 & -3 \\ 0 & 2 & 1 \\ 1 & -1 & -4 \end{bmatrix}$

19. $A = \begin{bmatrix} 1 & 1 & 0 & 1 \\ 0 & 1 & -1 & 1 \\ 0 & 1 & -1 & -1 \end{bmatrix}$

20. $A = \begin{bmatrix} 2 & -4 & 0 & 2 & 1 \\ -1 & 2 & 1 & 2 & 3 \\ 1 & -2 & 1 & 4 & 4 \end{bmatrix}$

In Exercises 21–24, find bases for row(A) and col(A) in the given exercises using A^T.

21. Exercise 17 **22.** Exercise 18

23. Exercise 19 **24.** Exercise 20

25. Explain carefully why your answers to Exercises 17 and 21 are both correct even though there appear to be differences.

26. Explain carefully why your answers to Exercises 18 and 22 are both correct even though there appear to be differences.

In Exercises 27–30, find a basis for the span of the given vectors.

27. $\begin{bmatrix} 1 \\ -1 \\ 0 \end{bmatrix}, \begin{bmatrix} -1 \\ 0 \\ 1 \end{bmatrix}, \begin{bmatrix} 0 \\ 1 \\ -1 \end{bmatrix}$ **28.** $\begin{bmatrix} 1 \\ -1 \\ 1 \end{bmatrix}, \begin{bmatrix} 1 \\ 2 \\ 0 \end{bmatrix}, \begin{bmatrix} 0 \\ 1 \\ 1 \end{bmatrix}, \begin{bmatrix} 2 \\ 1 \\ 2 \end{bmatrix}$

29. $[2 \quad -3 \quad 1], [1 \quad -1 \quad 0], [4 \quad -4 \quad 1]$

30. $[0 \quad 1 \quad -2 \quad 1], [3 \quad 1 \quad -1 \quad 0], [2 \quad 1 \quad 5 \quad 1]$

For Exercises 31 and 32, find bases for the spans of the vectors in the given exercises from among the vectors themselves.

31. Exercise 29 **32.** Exercise 30

33. Prove that if R is a matrix in echelon form, then a basis for row(R) consists of the nonzero rows of R.

34. Prove that if the columns of A are linearly independent, then they must form a basis for col(A).

For Exercises 35–38, give the rank and the nullity of the matrices in the given exercises.

35. Exercise 17

36. Exercise 18

37. Exercise 19

38. Exercise 20

39. If A is a 3×5 matrix, explain why the columns of A must be linearly dependent.

40. If A is a 4×2 matrix, explain why the rows of A must be linearly dependent.

41. If A is a 3×5 matrix, what are the possible values of nullity(A)?

42. If A is a 4×2 matrix, what are the possible values of nullity(A)?

In Exercises 43 and 44, find all possible values of rank(A) as a varies.

43. $A = \begin{bmatrix} 1 & 2 & a \\ -2 & 4a & 2 \\ a & -2 & 1 \end{bmatrix}$ **44.** $A = \begin{bmatrix} a & 2 & -1 \\ 3 & 3 & -2 \\ -2 & -1 & a \end{bmatrix}$

Answer Exercises 45–48 by considering the matrix with the given vectors as its columns.

45. Do $\begin{bmatrix} 1 \\ 1 \\ 0 \end{bmatrix}, \begin{bmatrix} 1 \\ 0 \\ 1 \end{bmatrix}, \begin{bmatrix} 0 \\ 1 \\ 1 \end{bmatrix}$ form a basis for $\mathbb{R}^3$?

46. Do $\begin{bmatrix} 1 \\ -1 \\ 3 \end{bmatrix}, \begin{bmatrix} -1 \\ 5 \\ 1 \end{bmatrix}, \begin{bmatrix} 1 \\ -3 \\ 1 \end{bmatrix}$ form a basis for $\mathbb{R}^3$?

47. Do $\begin{bmatrix} 1 \\ 1 \\ 1 \\ 0 \end{bmatrix}, \begin{bmatrix} 1 \\ 1 \\ 0 \\ 1 \end{bmatrix}, \begin{bmatrix} 1 \\ 0 \\ 1 \\ 1 \end{bmatrix}, \begin{bmatrix} 0 \\ 1 \\ 1 \\ 1 \end{bmatrix}$ form a basis for $\mathbb{R}^4$?

48. Do $\begin{bmatrix} 1 \\ -1 \\ 0 \\ 0 \end{bmatrix}, \begin{bmatrix} 0 \\ 1 \\ 0 \\ -1 \end{bmatrix}, \begin{bmatrix} 0 \\ 0 \\ -1 \\ 1 \end{bmatrix}, \begin{bmatrix} -1 \\ 0 \\ 1 \\ 0 \end{bmatrix}$ form a basis for $\mathbb{R}^4$?

49. Do $\begin{bmatrix} 1 \\ 1 \\ 0 \end{bmatrix}, \begin{bmatrix} 0 \\ 1 \\ 1 \end{bmatrix}, \begin{bmatrix} 1 \\ 0 \\ 1 \end{bmatrix}$ form a basis for $\mathbb{Z}_2^3$?

50. Do $\begin{bmatrix} 1 \\ 1 \\ 0 \end{bmatrix}, \begin{bmatrix} 0 \\ 1 \\ 1 \end{bmatrix}, \begin{bmatrix} 1 \\ 0 \\ 1 \end{bmatrix}$ form a basis for $\mathbb{Z}_3^3$?

In Exercises 51 and 52, show that $\mathbf{w}$ *is in span*($\mathcal{B}$) *and find the coordinate vector* $[\mathbf{w}]_{\mathcal{B}}$.

51. $\mathcal{B} = \left\{ \begin{bmatrix} 1 \\ 2 \\ 0 \end{bmatrix}, \begin{bmatrix} 1 \\ 0 \\ -1 \end{bmatrix} \right\}, \mathbf{w} = \begin{bmatrix} 1 \\ 6 \\ 2 \end{bmatrix}$

52. $\mathcal{B} = \left\{ \begin{bmatrix} 3 \\ 1 \\ 4 \end{bmatrix}, \begin{bmatrix} 5 \\ 1 \\ 6 \end{bmatrix} \right\}, \mathbf{w} = \begin{bmatrix} 1 \\ 3 \\ 4 \end{bmatrix}$

In Exercises 53–56, compute the rank and nullity of the given matrices over the indicated $\mathbb{Z}_p$.

53. $\begin{bmatrix} 1 & 1 & 0 \\ 0 & 1 & 1 \\ 1 & 0 & 1 \end{bmatrix}$ over $\mathbb{Z}_2$

54. $\begin{bmatrix} 1 & 1 & 2 \\ 2 & 1 & 2 \\ 2 & 0 & 0 \end{bmatrix}$ over $\mathbb{Z}_3$

55. $\begin{bmatrix} 1 & 3 & 1 & 4 \\ 2 & 3 & 0 & 1 \\ 1 & 0 & 4 & 0 \end{bmatrix}$ over $\mathbb{Z}_5$

56. $\begin{bmatrix} 2 & 4 & 0 & 0 & 1 \\ 6 & 3 & 5 & 1 & 0 \\ 1 & 0 & 2 & 2 & 5 \\ 1 & 1 & 1 & 1 & 1 \end{bmatrix}$ over $\mathbb{Z}_7$

57. If A is $m \times n$, prove that every vector in null(A) is orthogonal to every vector in row(A).

58. If A and B are $n \times n$ matrices of rank n, prove that AB has rank n.

59. **(a)** Prove that rank(AB) $\leq$ rank(B). [*Hint:* Review Exercise 29 in Section 3.1.]
(b) Give an example in which rank(AB) $<$ rank(B).

60. **(a)** Prove that rank(AB) $\leq$ rank(A). [*Hint:* Review Exercise 30 in Section 3.1 or use transposes and Exercise 59(a).]
(b) Give an example in which rank(AB) $<$ rank(A).

61. **(a)** Prove that if U is invertible, then rank(UA) = rank(A). [*Hint:* $A = U^{-1}(UA)$.]
(b) Prove that if V is invertible, then rank(AV) = rank(A).

62. Prove that an $m \times n$ matrix A has rank 1 if and only if A can be written as the outer product $\mathbf{u}\mathbf{v}^T$ of a vector $\mathbf{u}$ in $\mathbb{R}^m$ and $\mathbf{v}$ in $\mathbb{R}^n$.

63. If an $m \times n$ matrix A has rank r, prove that A can be written as the sum of r matrices, each of which has rank 1. [*Hint:* Find a way to use Exercise 62.]

64. Prove that, for $m \times n$ matrices A and B, rank $(A + B) \leq$ rank(A) + rank(B).

65. Let A be an $n \times n$ matrix such that $A^2 = O$. Prove that rank(A) $\leq n/2$. [*Hint:* Show that col(A) $\subseteq$ null(A) and use the Rank Theorem.]

66. Let A be a skew-symmetric $n \times n$ matrix. (See page 162).
(a) Prove that $\mathbf{x}^T A\mathbf{x} = 0$ for all $\mathbf{x}$ in $\mathbb{R}^n$.
(b) Prove that $I + A$ is invertible. [*Hint:* Show that null($I + A$) = {$\mathbf{0}$}.]

3.6 Introduction to Linear Transformations

In this section, we begin to explore one of the themes from the introduction to this chapter. There we saw that matrices can be used to transform vectors, acting as a type of "function" of the form $\mathbf{w} = T(\mathbf{v})$, where the independent variable $\mathbf{v}$ and the dependent variable $\mathbf{w}$ are vectors. We will make this notion more precise now and look at several examples of such matrix transformations, leading to the concept of a *linear transformation*—a powerful idea that we will encounter repeatedly from here on.

We begin by recalling some of the basic concepts associated with functions. You will be familiar with most of these ideas from other courses in which you encountered functions of the form $f: \mathbb{R} \to \mathbb{R}$ [such as $f(x) = x^2$] that transform real numbers into real numbers. What is new here is that vectors are involved and we are interested only in functions that are "compatible" with the vector operations of addition and scalar multiplication.

Consider an example. Let

$$A = \begin{bmatrix} 1 & 0 \\ 2 & -1 \\ 3 & 4 \end{bmatrix} \quad \text{and} \quad \mathbf{v} = \begin{bmatrix} 1 \\ -1 \end{bmatrix}$$

Then

$$A\mathbf{v} = \begin{bmatrix} 1 & 0 \\ 2 & -1 \\ 3 & 4 \end{bmatrix} \begin{bmatrix} 1 \\ -1 \end{bmatrix} = \begin{bmatrix} 1 \\ 3 \\ -1 \end{bmatrix}$$

This shows that A transforms $\mathbf{v}$ into $\mathbf{w} = \begin{bmatrix} 1 \\ 3 \\ -1 \end{bmatrix}$.

We can describe this transformation more generally. The matrix equation

$$\begin{bmatrix} 1 & 0 \\ 2 & -1 \\ 3 & 4 \end{bmatrix} \begin{bmatrix} x \\ y \end{bmatrix} = \begin{bmatrix} x \\ 2x - y \\ 3x + 4y \end{bmatrix}$$

gives a formula that shows how A transforms an arbitrary vector $\begin{bmatrix} x \\ y \end{bmatrix}$ in $\mathbb{R}^2$ into the vector $\begin{bmatrix} x \\ 2x - y \\ 3x + 4y \end{bmatrix}$ in $\mathbb{R}^3$. We denote this transformation by T_A and write

$$T_A\left(\begin{bmatrix} x \\ y \end{bmatrix}\right) = \begin{bmatrix} x \\ 2x - y \\ 3x + 4y \end{bmatrix}$$

(Although technically sloppy, omitting the parentheses in definitions such as this one is a common convention that saves some writing. The description of T_A becomes

$$T_A\begin{bmatrix} x \\ y \end{bmatrix} = \begin{bmatrix} x \\ 2x - y \\ 3x + 4y \end{bmatrix}$$

with this convention.)

With this example in mind, we now consider some terminology. A ***transformation*** (or ***mapping*** or ***function***) T from $\mathbb{R}^n$ to $\mathbb{R}^m$ is a rule that assigns to each vector $\mathbf{v}$ in $\mathbb{R}^n$ a unique vector $T(\mathbf{v})$ in $\mathbb{R}^m$. The ***domain*** of T is $\mathbb{R}^n$, and the ***codomain*** of T is $\mathbb{R}^m$. We indicate this by writing $T: \mathbb{R}^n \to \mathbb{R}^m$. For a vector $\mathbf{v}$ in the domain of T, the vector $T(\mathbf{v})$ in the codomain is called the ***image*** of $\mathbf{v}$ under (the action of) T. The set of all possible images $T(\mathbf{v})$ (as $\mathbf{v}$ varies throughout the domain of T) is called the ***range*** of T.

In our example, the domain of T_A is $\mathbb{R}^2$ and its codomain is $\mathbb{R}^3$, so we write $T_A: \mathbb{R}^2 \to \mathbb{R}^3$. The image of $\mathbf{v} = \begin{bmatrix} 1 \\ -1 \end{bmatrix}$ is $\mathbf{w} = T_A(\mathbf{v}) = \begin{bmatrix} 1 \\ 3 \\ -1 \end{bmatrix}$. What is the range of

T_A? It consists of all vectors in the codomain $\mathbb{R}^3$ that are of the form

$$T_A\begin{bmatrix} x \\ y \end{bmatrix} = \begin{bmatrix} x \\ 2x - y \\ 3x + 4y \end{bmatrix} = x\begin{bmatrix} 1 \\ 2 \\ 3 \end{bmatrix} + y\begin{bmatrix} 0 \\ -1 \\ 4 \end{bmatrix}$$

which describes the set of all linear combinations of the column vectors $\begin{bmatrix} 1 \\ 2 \\ 3 \end{bmatrix}$ and $\begin{bmatrix} 0 \\ -1 \\ 4 \end{bmatrix}$ of A. In other words, the range of T is the column space of A! (We will have more to say about this later—for now we'll simply note it as an interesting observation.) Geometrically, this shows that the range of T_A is the plane through the origin in $\mathbb{R}^3$ with direction vectors given by the column vectors of A. Notice that the range of T_A is strictly smaller than the codomain of T_A.

Linear Transformations

The example T_A above is a special case of a more general type of transformation called a *linear transformation*. We will consider the general definition in Chapter 6, but the essence of it is that these are the transformations that "preserve" the vector operations of addition and scalar multiplication.

Definition A transformation $T: \mathbb{R}^n \rightarrow \mathbb{R}^m$ is called a ***linear transformation*** if

1. $T(\mathbf{u} + \mathbf{v}) = T(\mathbf{u}) + T(\mathbf{v})$ for all $\mathbf{u}$ and $\mathbf{v}$ in $\mathbb{R}^n$ and
2. $T(c\mathbf{v}) = cT(\mathbf{v})$ for all $\mathbf{v}$ in $\mathbb{R}^n$ and all scalars c.

Example 3.55

Consider once again the transformation $T: \mathbb{R}^2 \rightarrow \mathbb{R}^3$ defined by

$$T\begin{bmatrix} x \\ y \end{bmatrix} = \begin{bmatrix} x \\ 2x - y \\ 3x + 4y \end{bmatrix}$$

Let's check that T is a linear transformation. To verify (1), we let

$$\mathbf{u} = \begin{bmatrix} x_1 \\ y_1 \end{bmatrix} \quad \text{and} \quad \mathbf{v} = \begin{bmatrix} x_2 \\ y_2 \end{bmatrix}$$

Then

$$\begin{aligned} T(\mathbf{u} + \mathbf{v}) &= T\left(\begin{bmatrix} x_1 \\ y_1 \end{bmatrix} + \begin{bmatrix} x_2 \\ y_2 \end{bmatrix}\right) = T\left(\begin{bmatrix} x_1 + x_2 \\ y_1 + y_2 \end{bmatrix}\right) = \begin{bmatrix} x_1 + x_2 \\ 2(x_1 + x_2) - (y_1 + y_2) \\ 3(x_1 + x_2) + 4(y_1 + y_2) \end{bmatrix} \\ &= \begin{bmatrix} x_1 + x_2 \\ 2x_1 + 2x_2 - y_1 - y_2 \\ 3x_1 + 3x_2 + 4y_1 + 4y_2 \end{bmatrix} = \begin{bmatrix} x_1 + x_2 \\ (2x_1 - y_1) + (2x_2 - y_2) \\ (3x_1 + 4y_1) + (3x_2 + 4y_2) \end{bmatrix} \\ &= \begin{bmatrix} x_1 \\ 2x_1 - y_1 \\ 3x_1 + 4y_1 \end{bmatrix} + \begin{bmatrix} x_2 \\ 2x_2 - y_2 \\ 3x_2 + 4y_2 \end{bmatrix} = T\begin{bmatrix} x_1 \\ y_1 \end{bmatrix} + T\begin{bmatrix} x_2 \\ y_2 \end{bmatrix} = T(\mathbf{u}) + T(\mathbf{v}) \end{aligned}$$

To show (2), we let $\mathbf{v} = \begin{bmatrix} x \\ y \end{bmatrix}$ and let c be a scalar. Then

$$\begin{aligned} T(c\mathbf{v}) &= T\left(c\begin{bmatrix} x \\ y \end{bmatrix}\right) = T\left(\begin{bmatrix} cx \\ cy \end{bmatrix}\right) \\ &= \begin{bmatrix} cx \\ 2(cx) - (cy) \\ 3(cx) + 4(cy) \end{bmatrix} = \begin{bmatrix} cx \\ c(2x - y) \\ c(3x + 4y) \end{bmatrix} \\ &= c\begin{bmatrix} x \\ 2x - y \\ 3x + 4y \end{bmatrix} = cT\begin{bmatrix} x \\ y \end{bmatrix} = cT(\mathbf{v}) \end{aligned}$$

Thus, T is a linear transformation.

Remark The definition of a linear transformation can be streamlined by combining (1) and (2) as shown below.

$T: \mathbb{R}^n \to \mathbb{R}^m$ is a linear transformation if

$$T(c_1\mathbf{v}_1 + c_2\mathbf{v}_2) = c_1T(\mathbf{v}_1) + c_2T(\mathbf{v}_2) \quad \text{for all } \mathbf{v}_1, \mathbf{v}_2 \text{ in } \mathbb{R}^n \text{ and scalars } c_1, c_2$$

In Exercise 53, you will be asked to show that the statement above is equivalent to the original definition. In practice, this equivalent formulation can save some writing—try it!

Although the linear transformation T in Example 3.55 originally arose as a *matrix* transformation T_A, it is a simple matter to recover the matrix A from the definition of T given in the example. We observe that

$$T\begin{bmatrix} x \\ y \end{bmatrix} = \begin{bmatrix} x \\ 2x - y \\ 3x + 4y \end{bmatrix} = x\begin{bmatrix} 1 \\ 2 \\ 3 \end{bmatrix} + y\begin{bmatrix} 0 \\ -1 \\ 4 \end{bmatrix} = \begin{bmatrix} 1 & 0 \\ 2 & -1 \\ 3 & 4 \end{bmatrix}\begin{bmatrix} x \\ y \end{bmatrix}$$

so $T = T_A$, where $A = \begin{bmatrix} 1 & 0 \\ 2 & -1 \\ 3 & 4 \end{bmatrix}$. (Notice that when the variables x and y are lined up, the matrix A is just their coefficient matrix.)

Recognizing that a transformation is a matrix transformation is important, since, as the next theorem shows, all matrix transformations are linear transformations.

Theorem 3.30 Let A be an $m \times n$ matrix. Then the matrix transformation $T_A: \mathbb{R}^n \to \mathbb{R}^m$ defined by

$$T_A(\mathbf{x}) = A\mathbf{x} \quad (\text{for } \mathbf{x} \text{ in } \mathbb{R}^n)$$

is a linear transformation.

Proof Let **u** and **v** be vectors in $\mathbb{R}^n$ and let c be a scalar. Then

$$T_A(\mathbf{u}+\mathbf{v}) = A(\mathbf{u}+\mathbf{v}) = A\mathbf{u} + A\mathbf{v} = T_A(\mathbf{u}) + T_A(\mathbf{v})$$

and
$$T_A(c\mathbf{v}) = A(c\mathbf{v}) = c(A\mathbf{v}) = cT_A(\mathbf{v})$$

Hence, T_A is a linear transformation.

Example 3.56

Let $F : \mathbb{R}^2 \to \mathbb{R}^2$ be the transformation that sends each point to its reflection in the x-axis. Show that F is a linear transformation.

Solution From Figure 3.4, it is clear that F sends the point (x, y) to the point $(x, -y)$. Thus, we may write

$$F\begin{bmatrix} x \\ y \end{bmatrix} = \begin{bmatrix} x \\ -y \end{bmatrix}$$

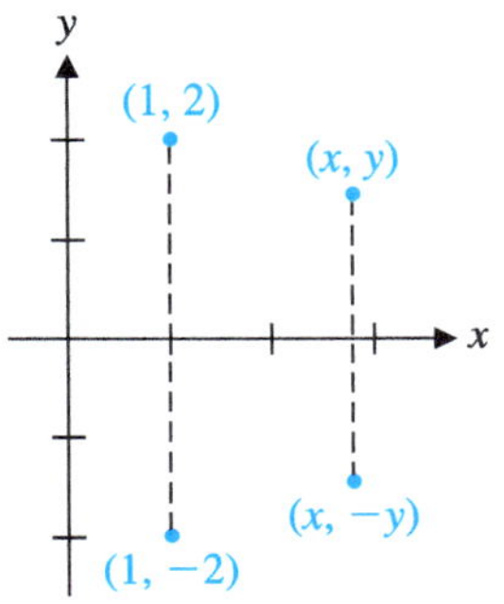

Figure 3.4
Reflection in the x-axis

We could proceed to check that F is linear, as in Example 3.55 (this one is even easier to check!), but it is faster to observe that

$$\begin{bmatrix} x \\ -y \end{bmatrix} = x\begin{bmatrix} 1 \\ 0 \end{bmatrix} + y\begin{bmatrix} 0 \\ -1 \end{bmatrix} = \begin{bmatrix} 1 & 0 \\ 0 & -1 \end{bmatrix}\begin{bmatrix} x \\ y \end{bmatrix}$$

Therefore, $F\begin{bmatrix} x \\ y \end{bmatrix} = A\begin{bmatrix} x \\ y \end{bmatrix}$, where $A = \begin{bmatrix} 1 & 0 \\ 0 & -1 \end{bmatrix}$, so F is a matrix transformation. It now follows, by Theorem 3.30, that F is a linear transformation.

Example 3.57

Let $R : \mathbb{R}^2 \to \mathbb{R}^2$ be the transformation that rotates each point 90° counterclockwise about the origin. Show that R is a linear transformation.

Solution As Figure 3.5 shows, R sends the point (x, y) to the point $(-y, x)$. Thus, we have

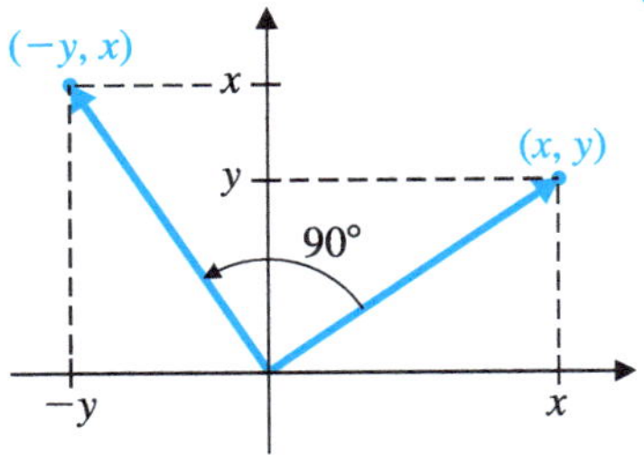

Figure 3.5
A 90° rotation

$$R\begin{bmatrix} x \\ y \end{bmatrix} = \begin{bmatrix} -y \\ x \end{bmatrix} = x\begin{bmatrix} 0 \\ 1 \end{bmatrix} + y\begin{bmatrix} -1 \\ 0 \end{bmatrix} = \begin{bmatrix} 0 & -1 \\ 1 & 0 \end{bmatrix}\begin{bmatrix} x \\ y \end{bmatrix}$$

Hence, R is a matrix transformation and is therefore linear.

Observe that if we multiply a matrix by standard basis vectors, we obtain the columns of the matrix. For example,

$$\begin{bmatrix} a & b \\ c & d \\ e & f \end{bmatrix}\begin{bmatrix} 1 \\ 0 \end{bmatrix} = \begin{bmatrix} a \\ c \\ e \end{bmatrix} \quad \text{and} \quad \begin{bmatrix} a & b \\ c & d \\ e & f \end{bmatrix}\begin{bmatrix} 0 \\ 1 \end{bmatrix} = \begin{bmatrix} b \\ d \\ f \end{bmatrix}$$

We can use this observation to show that *every* linear transformation from $\mathbb{R}^n$ to $\mathbb{R}^m$ arises as a matrix transformation.

Theorem 3.31 Let $T:\mathbb{R}^n \to \mathbb{R}^m$ be a linear transformation. Then T is a matrix transformation. More specifically, $T = T_A$, where A is the $m \times n$ matrix

$$A = [\,T(\mathbf{e}_1) \vdots T(\mathbf{e}_2) \vdots \cdots \vdots T(\mathbf{e}_n)\,]$$

Proof Let $\mathbf{e}_1, \mathbf{e}_2, \ldots, \mathbf{e}_n$ be the standard basis vectors in $\mathbb{R}^n$ and let $\mathbf{x}$ be a vector in $\mathbb{R}^n$. We can write $\mathbf{x} = x_1\mathbf{e}_1 + x_2\mathbf{e}_2 + \cdots + x_n\mathbf{e}_n$ (where the x_i's are the components of $\mathbf{x}$). We also know that $T(\mathbf{e}_1), T(\mathbf{e}_2), \ldots, T(\mathbf{e}_n)$ are (column) vectors in $\mathbb{R}^m$. Let $A = [T(\mathbf{e}_1) \vdots T(\mathbf{e}_2) \vdots \cdots \vdots T(\mathbf{e}_n)]$ be the $m \times n$ matrix with these vectors as its columns. Then

$$\begin{aligned} T(\mathbf{x}) &= T(x_1\mathbf{e}_1 + x_2\mathbf{e}_2 + \cdots + x_n\mathbf{e}_n) \\ &= x_1T(\mathbf{e}_1) + x_2T(\mathbf{e}_2) + \cdots + x_nT(\mathbf{e}_n) \\ &= [\,T(\mathbf{e}_1) \vdots T(\mathbf{e}_2) \vdots \cdots \vdots T(\mathbf{e}_n)\,]\begin{bmatrix} x_1 \\ x_2 \\ \vdots \\ x_n \end{bmatrix} = A\mathbf{x} \end{aligned}$$

as required.

The matrix A in Theorem 3.31 is called the ***standard matrix of the linear transformation T.***

Example 3.58 Show that a rotation about the origin through an angle θ defines a linear transformation from $\mathbb{R}^2$ to $\mathbb{R}^2$ and find its standard matrix.

Solution Let R_θ be the rotation. We will give a geometric argument to establish the fact that R_θ is linear. Let $\mathbf{u}$ and $\mathbf{v}$ be vectors in $\mathbb{R}^2$. If they are not parallel, then Figure 3.6(a) shows the parallelogram rule that determines $\mathbf{u} + \mathbf{v}$. If we now apply R_θ, the entire parallelogram is rotated through the angle θ, as shown in Figure 3.6(b). But the diagonal of this parallelogram must be $R_\theta(\mathbf{u}) + R_\theta(\mathbf{v})$, again by the parallelogram rule. Hence, $R_\theta(\mathbf{u} + \mathbf{v}) = R_\theta(\mathbf{u}) + R_\theta(\mathbf{v})$. (What happens if $\mathbf{u}$ and $\mathbf{v}$ are parallel?)

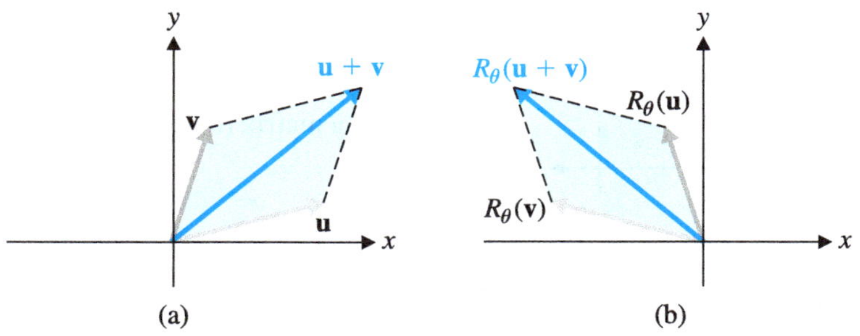

Figure 3.6

Similarly, if we apply R_θ to $\mathbf{v}$ and $c\mathbf{v}$, we obtain $R_\theta(\mathbf{v})$ and $R_\theta(c\mathbf{v})$, as shown in Figure 3.7. But since the rotation does not affect lengths, we must then have $R_\theta(c\mathbf{v}) = cR_\theta(\mathbf{v})$, as required. (Draw diagrams for the cases $0 < c < 1$, $-1 < c < 0$, and $c < -1$.)

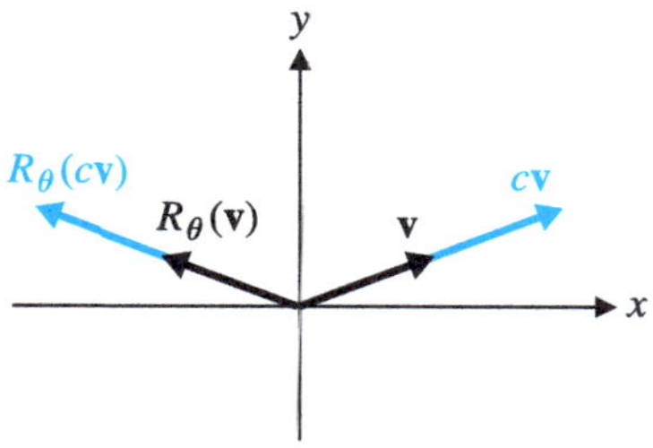

Figure 3.7

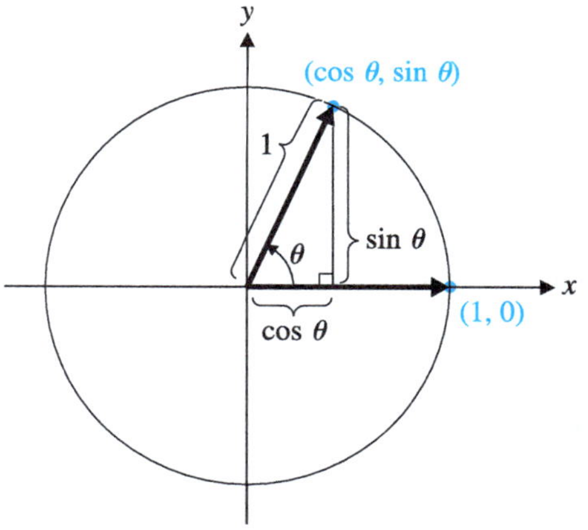

Figure 3.8
$R_\theta(\mathbf{e}_1)$

Therefore, R_θ is a linear transformation. According to Theorem 3.31, we can find its matrix by determining its effect on the standard basis vectors $\mathbf{e}_1$ and $\mathbf{e}_2$ of $\mathbb{R}^2$. Now, as Figure 3.8 shows, $R_\theta\begin{bmatrix}1\\0\end{bmatrix} = \begin{bmatrix}\cos\theta\\ \sin\theta\end{bmatrix}$.

We can find $R_\theta\begin{bmatrix}0\\1\end{bmatrix}$ similarly, but it is faster to observe that $R_\theta\begin{bmatrix}0\\1\end{bmatrix}$ must be perpendicular (counterclockwise) to $R_\theta\begin{bmatrix}1\\0\end{bmatrix}$ and so, by Example 3.57, $R_\theta\begin{bmatrix}0\\1\end{bmatrix} = \begin{bmatrix}-\sin\theta\\ \cos\theta\end{bmatrix}$ (Figure 3.9).

Therefore, the standard matrix of R_θ is $\begin{bmatrix}\cos\theta & -\sin\theta\\ \sin\theta & \cos\theta\end{bmatrix}$.

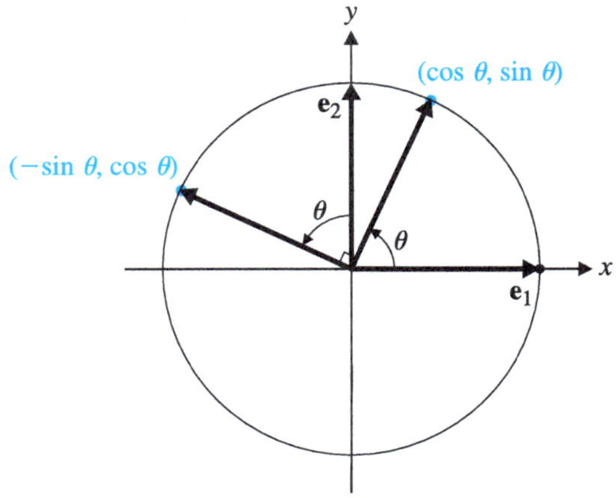

Figure 3.9
$R_\theta(\mathbf{e}_2)$

The result of Example 3.58 can now be used to compute the effect of any rotation. For example, suppose we wish to rotate the point $(2, -1)$ through 60° about the origin. (The convention is that a positive angle corresponds to a counterclockwise

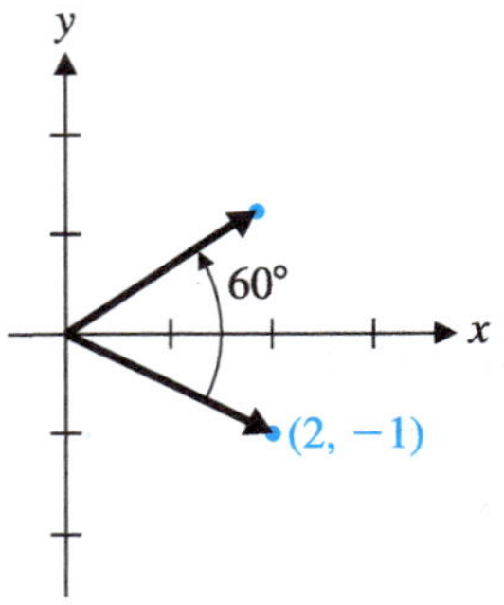

Figure 3.10
A 60° rotation

rotation, while a negative angle is clockwise.) Since $\cos 60° = 1/2$ and $\sin 60° = \sqrt{3}/2$, we compute

$$R_{60}\begin{bmatrix} 2 \\ -1 \end{bmatrix} = \begin{bmatrix} \cos 60° & -\sin 60° \\ \sin 60° & \cos 60° \end{bmatrix}\begin{bmatrix} 2 \\ -1 \end{bmatrix} = \begin{bmatrix} 1/2 & -\sqrt{3}/2 \\ \sqrt{3}/2 & 1/2 \end{bmatrix}\begin{bmatrix} 2 \\ -1 \end{bmatrix}$$
$$= \begin{bmatrix} (2+\sqrt{3})/2 \\ (2\sqrt{3}-1)/2 \end{bmatrix}$$

Thus, the image of the point $(2, -1)$ under this rotation is the point $((2+\sqrt{3})/2, (2\sqrt{3}-1)/2) \approx (1.87, 1.23)$, as shown in Figure 3.10.

Example 3.59

(a) Show that the transformation $P : \mathbb{R}^2 \to \mathbb{R}^2$ that projects a point onto the x-axis is a linear transformation and find its standard matrix.
(b) More generally, if ℓ is a line through the origin in $\mathbb{R}^2$, show that the transformation $P_\ell : \mathbb{R}^2 \to \mathbb{R}^2$ that projects a point onto ℓ is a linear transformation and find its standard matrix.

Solution (a) As Figure 3.11 shows, P sends the point (x, y) to the point $(x, 0)$. Thus,

$$P\begin{bmatrix} x \\ y \end{bmatrix} = \begin{bmatrix} x \\ 0 \end{bmatrix} = x\begin{bmatrix} 1 \\ 0 \end{bmatrix} + y\begin{bmatrix} 0 \\ 0 \end{bmatrix} = \begin{bmatrix} 1 & 0 \\ 0 & 0 \end{bmatrix}\begin{bmatrix} x \\ y \end{bmatrix}$$

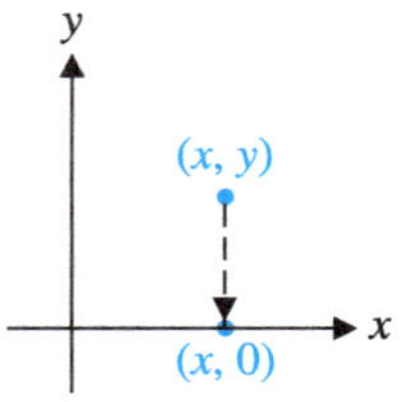

Figure 3.11
A projection

It follows that P is a matrix transformation (and hence a linear transformation) with standard matrix $\begin{bmatrix} 1 & 0 \\ 0 & 0 \end{bmatrix}$.

(b) Let the line ℓ have direction vector $\mathbf{d}$ and let $\mathbf{v}$ be an arbitrary vector. Then P_ℓ is given by $\text{proj}_{\mathbf{d}}(\mathbf{v})$, the projection of $\mathbf{v}$ onto $\mathbf{d}$, which you'll recall from Section 1.2 has the formula

$$\text{proj}_{\mathbf{d}}(\mathbf{v}) = \left(\frac{\mathbf{d}\cdot\mathbf{v}}{\mathbf{d}\cdot\mathbf{d}}\right)\mathbf{d}$$

Thus, to show that P_ℓ is linear, we proceed as follows:

$$P_\ell(\mathbf{u}+\mathbf{v}) = \left(\frac{\mathbf{d}\cdot(\mathbf{u}+\mathbf{v})}{\mathbf{d}\cdot\mathbf{d}}\right)\mathbf{d}$$
$$= \left(\frac{\mathbf{d}\cdot\mathbf{u}+\mathbf{d}\cdot\mathbf{v}}{\mathbf{d}\cdot\mathbf{d}}\right)\mathbf{d}$$
$$= \left(\frac{\mathbf{d}\cdot\mathbf{u}}{\mathbf{d}\cdot\mathbf{d}} + \frac{\mathbf{d}\cdot\mathbf{v}}{\mathbf{d}\cdot\mathbf{d}}\right)\mathbf{d}$$
$$= \left(\frac{\mathbf{d}\cdot\mathbf{u}}{\mathbf{d}\cdot\mathbf{d}}\right)\mathbf{d} + \left(\frac{\mathbf{d}\cdot\mathbf{v}}{\mathbf{d}\cdot\mathbf{d}}\right)\mathbf{d} = P_\ell(\mathbf{u}) + P_\ell(\mathbf{v})$$

Similarly, $P_\ell(c\mathbf{v}) = cP_\ell(\mathbf{v})$ for any scalar c (Exercise 52). Hence, P_ℓ is a linear transformation.

To find the standard matrix of P_ℓ, we apply Theorem 3.31. If we let $\mathbf{d} = \begin{bmatrix} d_1 \\ d_2 \end{bmatrix}$, then

$$P_\ell(\mathbf{e}_1) = \left(\frac{\mathbf{d}\cdot\mathbf{e}_1}{\mathbf{d}\cdot\mathbf{d}}\right)\mathbf{d} = \frac{d_1}{d_1^2 + d_2^2}\begin{bmatrix} d_1 \\ d_2 \end{bmatrix} = \frac{1}{d_1^2 + d_2^2}\begin{bmatrix} d_1^2 \\ d_1 d_2 \end{bmatrix}$$

and
$$P_\ell(\mathbf{e}_2) = \left(\frac{\mathbf{d}\cdot\mathbf{e}_2}{\mathbf{d}\cdot\mathbf{d}}\right)\mathbf{d} = \frac{d_2}{d_1^2 + d_2^2}\begin{bmatrix} d_1 \\ d_2 \end{bmatrix} = \frac{1}{d_1^2 + d_2^2}\begin{bmatrix} d_1 d_2 \\ d_2^2 \end{bmatrix}$$

Thus, the standard matrix of the projection is

$$A = \frac{1}{d_1^2 + d_2^2}\begin{bmatrix} d_1^2 & d_1 d_2 \\ d_1 d_2 & d_2^2 \end{bmatrix} = \begin{bmatrix} d_1^2/(d_1^2 + d_2^2) & d_1 d_2/(d_1^2 + d_2^2) \\ d_1 d_2/(d_1^2 + d_2^2) & d_2^2/(d_1^2 + d_2^2) \end{bmatrix}$$

As a check, note that in part (a) we could take $\mathbf{d} = \mathbf{e}_1$ as a direction vector for the x-axis. Therefore, $d_1 = 1$ and $d_2 = 0$, and we obtain $A = \begin{bmatrix} 1 & 0 \\ 0 & 0 \end{bmatrix}$, as before.

New Linear Transformations from Old

If $T: \mathbb{R}^m \to \mathbb{R}^n$ and $S: \mathbb{R}^n \to \mathbb{R}^p$ are linear transformations, then we may follow T by S to form the ***composition*** of the two transformations, denoted $S \circ T$. Notice that, in order for $S \circ T$ to make sense, the codomain of T and the domain of S must match (in this case, they are both $\mathbb{R}^n$) and the resulting composite transformation $S \circ T$ goes from the domain of T to the codomain of S (in this case, $S \circ T: \mathbb{R}^m \to \mathbb{R}^p$). Figure 3.12 shows schematically how this composition works. The formal definition of composition of transformations is taken directly from this figure and is the same as the corresponding definition of composition of ordinary functions:

$$(S \circ T)(\mathbf{v}) = S(T(\mathbf{v}))$$

Of course, we would like $S \circ T$ to be a linear transformation too, and happily we find that it is. We can demonstrate this by showing that $S \circ T$ satisfies the definition of a linear transformation (which we will do in Chapter 6), but, since for the time being we are assuming that linear transformations and matrix transformations are the same thing, it is enough to show that $S \circ T$ is a matrix transformation. We will use the notation $[T]$ for the standard matrix of a linear transformation T.

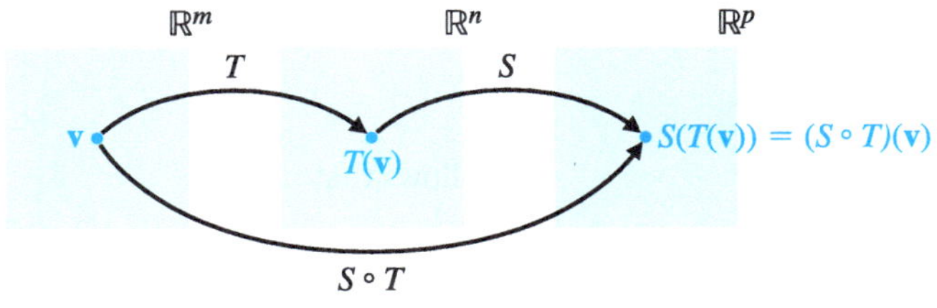

Figure 3.12
The composition of transformations

Theorem 3.32 Let $T: \mathbb{R}^m \to \mathbb{R}^n$ and $S: \mathbb{R}^n \to \mathbb{R}^p$ be linear transformations. Then $S \circ T: \mathbb{R}^m \to \mathbb{R}^p$ is a linear transformation. Moreover, their standard matrices are related by

$$[S \circ T] = [S][T]$$

Proof Let $[S] = A$ and $[T] = B$. (Notice that A is $p \times n$ and B is $n \times m$.) If $\mathbf{v}$ is a vector in $\mathbb{R}^m$, then we simply compute

$$(S \circ T)(\mathbf{v}) = S(T(\mathbf{v})) = S(B\mathbf{v}) = A(B\mathbf{v}) = (AB)\mathbf{v}$$

(Notice here that the dimensions of A and B guarantee that the product AB makes sense.) Thus, we see that the effect of $S \circ T$ is to multiply vectors by AB, from which it follows immediately that $S \circ T$ is a matrix (hence, linear) transformation with $[S \circ T] = [S][T]$.

Isn't this a great result? Say it in words: "The matrix of the composite is the product of the matrices." What a lovely formula!

Example 3.60 Consider the linear transformation $T: \mathbb{R}^2 \to \mathbb{R}^3$ from Example 3.55, defined by

$$T\begin{bmatrix} x_1 \\ x_2 \end{bmatrix} = \begin{bmatrix} x_1 \\ 2x_1 - x_2 \\ 3x_1 + 4x_2 \end{bmatrix}$$

and the linear transformation $S: \mathbb{R}^3 \to \mathbb{R}^4$ defined by

$$S\begin{bmatrix} y_1 \\ y_2 \\ y_3 \end{bmatrix} = \begin{bmatrix} 2y_1 + y_3 \\ 3y_2 - y_3 \\ y_1 - y_2 \\ y_1 + y_2 + y_3 \end{bmatrix}$$

Find $S \circ T: \mathbb{R}^2 \to \mathbb{R}^4$.

Solution We see that the standard matrices are

$$[S] = \begin{bmatrix} 2 & 0 & 1 \\ 0 & 3 & -1 \\ 1 & -1 & 0 \\ 1 & 1 & 1 \end{bmatrix} \quad \text{and} \quad [T] = \begin{bmatrix} 1 & 0 \\ 2 & -1 \\ 3 & 4 \end{bmatrix}$$

so Theorem 3.32 gives

$$[S \circ T] = [S][T] = \begin{bmatrix} 2 & 0 & 1 \\ 0 & 3 & -1 \\ 1 & -1 & 0 \\ 1 & 1 & 1 \end{bmatrix}\begin{bmatrix} 1 & 0 \\ 2 & -1 \\ 3 & 4 \end{bmatrix} = \begin{bmatrix} 5 & 4 \\ 3 & -7 \\ -1 & 1 \\ 6 & 3 \end{bmatrix}$$

It follows that

$$(S \circ T)\begin{bmatrix} x_1 \\ x_2 \end{bmatrix} = \begin{bmatrix} 5 & 4 \\ 3 & -7 \\ -1 & 1 \\ 6 & 3 \end{bmatrix}\begin{bmatrix} x_1 \\ x_2 \end{bmatrix} = \begin{bmatrix} 5x_1 + 4x_2 \\ 3x_1 - 7x_2 \\ -x_1 + x_2 \\ 6x_1 + 3x_2 \end{bmatrix}$$

(In Exercise 29, you will be asked to check this result by setting

$$\begin{bmatrix} y_1 \\ y_2 \\ y_3 \end{bmatrix} = T\begin{bmatrix} x_1 \\ x_2 \end{bmatrix} = \begin{bmatrix} x_1 \\ 2x_1 - x_2 \\ 3x_1 + 4x_2 \end{bmatrix}$$

and substituting these values into the definition of S, thereby calculating $(S \circ T)\begin{bmatrix} x_1 \\ x_2 \end{bmatrix}$ directly.)

Example 3.61

Find the standard matrix of the transformation that first rotates a point 90° counterclockwise about the origin and then reflects the result in the x-axis.

Solution The rotation R and the reflection F were discussed in Examples 3.57 and 3.56, respectively, where we found their standard matrices to be $[R] = \begin{bmatrix} 0 & -1 \\ 1 & 0 \end{bmatrix}$ and $[F] = \begin{bmatrix} 1 & 0 \\ 0 & -1 \end{bmatrix}$. It follows that the composition $F \circ R$ has for its matrix

$$[F \circ R] = [F][R] = \begin{bmatrix} 1 & 0 \\ 0 & -1 \end{bmatrix}\begin{bmatrix} 0 & -1 \\ 1 & 0 \end{bmatrix} = \begin{bmatrix} 0 & -1 \\ -1 & 0 \end{bmatrix}$$

(Check that this result is correct by considering the effect of $F \circ R$ on the standard basis vectors $\mathbf{e}_1$ and $\mathbf{e}_2$. Note the importance of the *order* of the transformations: R is performed before F, but we write $F \circ R$. In this case, $R \circ F$ also makes sense. Is $R \circ F = F \circ R$?)

Inverses of Linear Transformations

Consider the effect of a 90° counterclockwise rotation about the origin followed by a 90° clockwise rotation about the origin. Clearly this leaves every point in $\mathbb{R}^2$ unchanged. If we denote these transformations by R_{90} and R_{-90} (remember that a negative angle measure corresponds to clockwise direction), then we may express this as $(R_{90} \circ R_{-90})(\mathbf{v}) = \mathbf{v}$ for every $\mathbf{v}$ in $\mathbb{R}^2$. Note that, in this case, if we perform the transformations in the other order, we get the same end result: $(R_{-90} \circ R_{90})(\mathbf{v}) = \mathbf{v}$ for every $\mathbf{v}$ in $\mathbb{R}^2$.

Thus, $R_{90} \circ R_{-90}$ (and $R_{-90} \circ R_{90}$ too) is a linear transformation that leaves every vector in $\mathbb{R}^2$ unchanged. Such a transformation is called an ***identity transformation.*** Generally, we have one such transformation for every $\mathbb{R}^n$—namely, $I : \mathbb{R}^n \to \mathbb{R}^n$ such that $I(\mathbf{v}) = \mathbf{v}$ for every $\mathbf{v}$ in $\mathbb{R}^n$. (If it is important to keep track of the dimension of the space, we might write I_n for clarity.)

So, with this notation, we have $R_{90} \circ R_{-90} = I = R_{-90} \circ R_{90}$. A pair of transformations that are related to each other in this way are called ***inverse transformations.***

Definition Let S and T be linear transformations from $\mathbb{R}^n$ to $\mathbb{R}^n$. Then S and T are ***inverse transformations*** if $S \circ T = I_n$ and $T \circ S = I_n$.

Remark Since this definition is symmetric with respect to S and T, we will say that, when this situation occurs, S is the inverse of T and T is the inverse of S. Furthermore, we will say that S and T are ***invertible.***

In terms of matrices, we see immediately that if S and T are inverse transformations, then $[S][T] = [S \circ T] = [I] = I$, where the last I is the identity *matrix*. (Why is the standard matrix of the identity transformation the identity matrix?) We must also have $[T][S] = [T \circ S] = [I] = I$. This shows that $[S]$ and $[T]$ are inverse matrices. It shows something more: If a linear transformation T is invertible, then its standard matrix $[T]$ must be invertible, and since matrix inverses are unique, this means that the inverse of T is also unique. Therefore, we can unambiguously use the notation T^{-1} to refer to *the* inverse of T. Thus, we can rewrite the above equations as $[T][T^{-1}] = I = [T^{-1}][T]$, showing that the matrix of T^{-1} is the inverse matrix of $[T]$. We have just proved the following theorem.

Theorem 3.33 Let $T: \mathbb{R}^n \to \mathbb{R}^n$ be an invertible linear transformation. Then its standard matrix $[T]$ is an invertible matrix, and

$$[T^{-1}] = [T]^{-1}$$

Remark Say this one in words too: "The matrix of the inverse is the inverse of the matrix." Fabulous!

Example 3.62 Find the standard matrix of a 60° clockwise rotation about the origin in $\mathbb{R}^2$.

Solution Earlier we computed the matrix of a 60° counterclockwise rotation about the origin to be

$$[R_{60}] = \begin{bmatrix} 1/2 & -\sqrt{3}/2 \\ \sqrt{3}/2 & 1/2 \end{bmatrix}$$

Since a 60° clockwise rotation is the inverse of a 60° counterclockwise rotation, we can apply Theorem 3.33 to obtain

$$[R_{-60}] = [(R_{60})^{-1}] = \begin{bmatrix} 1/2 & -\sqrt{3}/2 \\ \sqrt{3}/2 & 1/2 \end{bmatrix}^{-1} = \begin{bmatrix} 1/2 & \sqrt{3}/2 \\ -\sqrt{3}/2 & 1/2 \end{bmatrix}$$

(Check the calculation of the matrix inverse. The fastest way is to use the 2×2 shortcut from Theorem 3.8. Also, check that the resulting matrix has the right effect on the standard basis in $\mathbb{R}^2$ by drawing a diagram.)

Example 3.63 Determine whether projection onto the x-axis is an invertible transformation, and if it is, find its inverse.

Solution The standard matrix of this projection P is $\begin{bmatrix} 1 & 0 \\ 0 & 0 \end{bmatrix}$, which is not invertible since its determinant is 0. Hence, P is not invertible either.

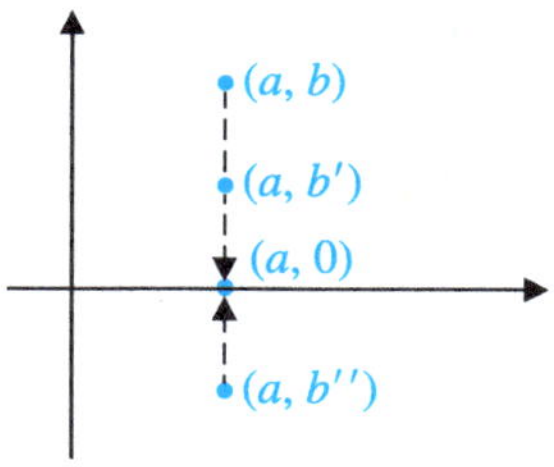

Figure 3.13
Projections are not invertible

Remark Figure 3.13 gives some idea why P in Example 3.63 is not invertible. The projection "collapses" $\mathbb{R}^2$ onto the x-axis. For P to be invertible, we would have to have a way of "undoing" it, to recover the point (a, b) we started with. However, there are infinitely many candidates for the image of $(a, 0)$ under such a hypothetical "inverse." Which one should we use? We cannot simply say that P^{-1} must send $(a, 0)$ to (a, b), since this cannot be a *definition* when we have no way of knowing what b should be. (See Exercise 42.)

Associativity

Theorem 3.3(a) in Section 3.2 stated the associativity property for matrix multiplication: $A(BC) = (AB)C$. (If you didn't try to prove it then, do so now. Even with all matrices restricted 2×2, you will get some feeling for the notational complexity involved in an "elementwise" proof, which should make you appreciate the proof we are about to give.)

Our approach to the proof is via linear transformations. We have seen that every $m \times n$ matrix A gives rise to a linear transformation $T_A : \mathbb{R}^n \to \mathbb{R}^m$; conversely, every linear transformation $T : \mathbb{R}^n \to \mathbb{R}^m$ has a corresponding $m \times n$ matrix $[T]$. The two correspondences are inversely related; that is, given A, $[T_A] = A$, and given T, $T_{[T]} = T$.

Let $R = T_A$, $S = T_B$, and $T = T_C$. Then, by Theorem 3.32,

$$A(BC) = (AB)C \quad \text{if and only if} \quad R \circ (S \circ T) = (R \circ S) \circ T$$

We now prove the latter identity. Let $\mathbf{x}$ be in the domain of T [and hence in the domain of both $R \circ (S \circ T)$ and $(R \circ S) \circ T$—why?]. To prove that $R \circ (S \circ T) = (R \circ S) \circ T$, it is enough to prove that they have the same effect on $\mathbf{x}$. By repeated application of the definition of composition, we have

$$\begin{aligned}(R \circ (S \circ T))(\mathbf{x}) &= R((S \circ T)(\mathbf{x})) \\ &= R(S(T(\mathbf{x}))) \\ &= (R \circ S)(T(\mathbf{x})) = ((R \circ S) \circ T)(\mathbf{x})\end{aligned}$$

as required. (Carefully check how the definition of composition has been used four times.)

This section has served as an introduction to linear transformations. In Chapter 6, we will take a more detailed and more general look at these transformations. The exercises that follow also contain some additional explorations of this important concept.

Exercises 3.6

1. Let $T_A : \mathbb{R}^2 \to \mathbb{R}^2$ be the matrix transformation corresponding to $A = \begin{bmatrix} 2 & -1 \\ 3 & 4 \end{bmatrix}$. Find $T_A(\mathbf{u})$ and $T_A(\mathbf{v})$, where $\mathbf{u} = \begin{bmatrix} 1 \\ 2 \end{bmatrix}$ and $\mathbf{v} = \begin{bmatrix} 3 \\ -2 \end{bmatrix}$.

2. Let $T_A : \mathbb{R}^2 \to \mathbb{R}^3$ be the matrix transformation corresponding to $A = \begin{bmatrix} 3 & -1 \\ 1 & 2 \\ 1 & 4 \end{bmatrix}$. Find $T_A(\mathbf{u})$ and $T_A(\mathbf{v})$, where $\mathbf{u} = \begin{bmatrix} 1 \\ 2 \end{bmatrix}$ and $\mathbf{v} = \begin{bmatrix} 3 \\ -2 \end{bmatrix}$.

In Exercises 3–6, prove that the given transformation is a linear transformation, using the definition (or the Remark following Example 3.55).

3. $T\begin{bmatrix} x \\ y \end{bmatrix} = \begin{bmatrix} x + y \\ x - y \end{bmatrix}$

4. $T\begin{bmatrix} x \\ y \end{bmatrix} = \begin{bmatrix} -y \\ x + 2y \\ 3x - 4y \end{bmatrix}$

5. $T\begin{bmatrix} x \\ y \\ z \end{bmatrix} = \begin{bmatrix} x - y + z \\ 2x + y - 3z \end{bmatrix}$

6. $T\begin{bmatrix} x \\ y \\ z \end{bmatrix} = \begin{bmatrix} x + z \\ y + z \\ x + y \end{bmatrix}$

In Exercises 7–10, give a counterexample to show that the given transformation is not a linear transformation.

7. $T\begin{bmatrix} x \\ y \end{bmatrix} = \begin{bmatrix} y \\ x^2 \end{bmatrix}$

8. $T\begin{bmatrix} x \\ y \end{bmatrix} = \begin{bmatrix} |x| \\ |y| \end{bmatrix}$

9. $T\begin{bmatrix} x \\ y \end{bmatrix} = \begin{bmatrix} xy \\ x + y \end{bmatrix}$

10. $T\begin{bmatrix} x \\ y \end{bmatrix} = \begin{bmatrix} x + 1 \\ y - 1 \end{bmatrix}$

In Exercises 11–14, find the standard matrix of the linear transformation in the given exercise.

11. Exercise 3

12. Exercise 4

13. Exercise 5

14. Exercise 6

In Exercises 15–18, show that the given transformation from $\mathbb{R}^2$ to $\mathbb{R}^2$ is linear by showing that it is a matrix transformation.

15. F reflects a vector in the y-axis.

16. R rotates a vector 45° counterclockwise about the origin.

17. D stretches a vector by a factor of 2 in the x-component and a factor of 3 in the y-component.

18. P projects a vector onto the line $y = x$.

19. The three types of elementary matrices give rise to five types of 2 × 2 matrices with one of the following forms:

$$\begin{bmatrix} k & 0 \\ 0 & 1 \end{bmatrix} \text{ or } \begin{bmatrix} 1 & 0 \\ 0 & k \end{bmatrix}$$

$$\begin{bmatrix} 0 & 1 \\ 1 & 0 \end{bmatrix}$$

$$\begin{bmatrix} 1 & k \\ 0 & 1 \end{bmatrix} \text{ or } \begin{bmatrix} 1 & 0 \\ k & 1 \end{bmatrix}$$

Each of these elementary matrices corresponds to a linear transformation from $\mathbb{R}^2$ to $\mathbb{R}^2$. Draw pictures to illustrate the effect of each one on the unit square with vertices at (0, 0), (1, 0), (0, 1), and (1, 1).

In Exercises 20–25, find the standard matrix of the given linear transformation from $\mathbb{R}^2$ to $\mathbb{R}^2$.

20. Counterclockwise rotation through 120° about the origin

21. Clockwise rotation through 30° about the origin

22. Projection onto the line $y = 2x$

23. Projection onto the line $y = -x$

24. Reflection in the line $y = x$

25. Reflection in the line $y = -x$

26. Let ℓ be a line through the origin in $\mathbb{R}^2$, P_ℓ the linear transformation that projects a vector onto ℓ, and F_ℓ the transformation that reflects a vector in ℓ.

(a) Draw diagrams to show that F_ℓ is linear.

(b) Figure 3.14 suggests a way to find the matrix of F_ℓ, using the fact that the diagonals of a parallelogram bisect each other. Prove that $F_\ell(\mathbf{x}) = 2P_\ell(\mathbf{x}) - \mathbf{x}$, and use this result to show that the standard matrix of F_ℓ is

$$\frac{1}{d_1^2 + d_2^2}\begin{bmatrix} d_1^2 - d_2^2 & 2d_1d_2 \\ 2d_1d_2 & -d_1^2 + d_2^2 \end{bmatrix}$$

(where the direction vector of ℓ is $\mathbf{d} = \begin{bmatrix} d_1 \\ d_2 \end{bmatrix}$).

(c) If the angle between ℓ and the positive x-axis is θ, show that the matrix of F_ℓ is

$$\begin{bmatrix} \cos 2\theta & \sin 2\theta \\ \sin 2\theta & -\cos 2\theta \end{bmatrix}$$

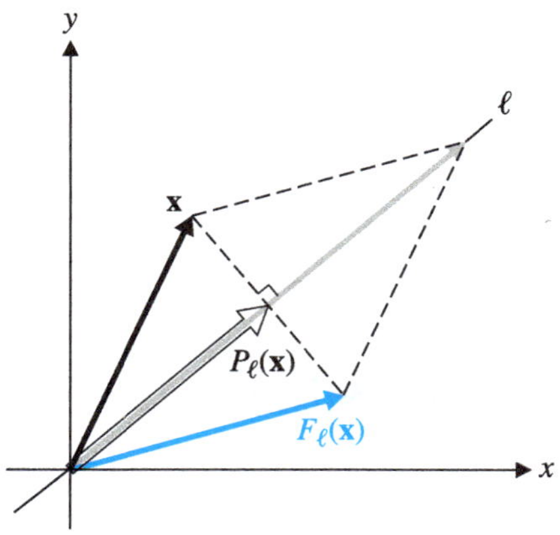

Figure 3.14

In Exercises 27 and 28, apply part (b) or (c) of Exercise 26 to find the standard matrix of the transformation.

27. Reflection in the line $y = 2x$

28. Reflection in the line $y = \sqrt{3}x$

29. Check the formula for $S \circ T$ in Example 3.60, by performing the suggested direct substitution.

In Exercises 30–35, verify Theorem 3.32 by finding the matrix of $S \circ T$ (a) by direct substitution and (b) by matrix multiplication of $[S][T]$.

30. $T\begin{bmatrix} x_1 \\ x_2 \end{bmatrix} = \begin{bmatrix} x_1 - x_2 \\ x_1 + x_2 \end{bmatrix}, S\begin{bmatrix} y_1 \\ y_2 \end{bmatrix} = \begin{bmatrix} 2y_1 \\ -y_2 \end{bmatrix}$

31. $T\begin{bmatrix} x_1 \\ x_2 \end{bmatrix} = \begin{bmatrix} x_1 + 2x_2 \\ -3x_1 + x_2 \end{bmatrix}, S\begin{bmatrix} y_1 \\ y_2 \end{bmatrix} = \begin{bmatrix} y_1 + 3y_2 \\ y_1 - y_2 \end{bmatrix}$

32. $T\begin{bmatrix} x_1 \\ x_2 \end{bmatrix} = \begin{bmatrix} x_2 \\ -x_1 \end{bmatrix}, S\begin{bmatrix} y_1 \\ y_2 \end{bmatrix} = \begin{bmatrix} y_1 + 3y_2 \\ 2y_1 + y_2 \\ y_1 - y_2 \end{bmatrix}$

33. $T\begin{bmatrix} x_1 \\ x_2 \\ x_3 \end{bmatrix} = \begin{bmatrix} x_1 + x_2 - x_3 \\ 2x_1 - x_2 + x_3 \end{bmatrix}, S\begin{bmatrix} y_1 \\ y_2 \end{bmatrix} = \begin{bmatrix} 4y_1 - 2y_2 \\ -y_1 + y_2 \end{bmatrix}$

34. $T\begin{bmatrix} x_1 \\ x_2 \\ x_3 \end{bmatrix} = \begin{bmatrix} x_1 + 2x_2 \\ 2x_2 - x_3 \end{bmatrix}, S\begin{bmatrix} y_1 \\ y_2 \end{bmatrix} = \begin{bmatrix} y_1 - y_2 \\ y_1 + y_2 \\ -y_1 + y_2 \end{bmatrix}$

35. $T\begin{bmatrix} x_1 \\ x_2 \\ x_3 \end{bmatrix} = \begin{bmatrix} x_1 + x_2 \\ x_2 + x_3 \\ x_1 + x_3 \end{bmatrix}, S\begin{bmatrix} y_1 \\ y_2 \\ y_3 \end{bmatrix} = \begin{bmatrix} y_1 - y_2 \\ y_2 - y_3 \\ -y_1 + y_3 \end{bmatrix}$

In Exercises 36–39, find the standard matrix of the composite transformation from $\mathbb{R}^2$ to $\mathbb{R}^2$.

36. Counterclockwise rotation through 60°, followed by reflection in the line $y = x$

37. Reflection in the y-axis, followed by clockwise rotation through 30°

38. Clockwise rotation through 45°, followed by projection onto the y-axis, followed by clockwise rotation through 45°

39. Reflection in the line $y = x$, followed by counterclockwise rotation through 30°, followed by reflection in the line $y = -x$

In Exercises 40–43, use matrices to prove the given statements about transformations from $\mathbb{R}^2$ to $\mathbb{R}^2$.

40. If R_θ denotes a rotation (about the origin) through the angle θ, then $R_\alpha \circ R_\beta = R_{\alpha+\beta}$.

41. If θ is the angle between lines ℓ and m (through the origin), then $F_m \circ F_\ell = R_{+2\theta}$. (See Exercise 26.)

42. **(a)** If P is a projection, then $P \circ P = P$.

(b) The matrix of a projection can never be invertible.

43. If ℓ, m, and n are three lines through the origin, then $F_n \circ F_m \circ F_\ell$ is also a reflection in a line through the origin.

44. Let T be a linear transformation from $\mathbb{R}^2$ to $\mathbb{R}^2$ (or from $\mathbb{R}^3$ to $\mathbb{R}^3$). Prove that T maps a straight line to a straight line or a point. [*Hint:* Use the vector form of the equation of a line.]

45. Let T be a linear transformation from $\mathbb{R}^2$ to $\mathbb{R}^2$ (or from $\mathbb{R}^3$ to $\mathbb{R}^3$). Prove that T maps parallel lines to parallel lines, a single line, a pair of points, or a single point.

In Exercises 46–51, let ABCD be the square with vertices $(-1, 1)$, $(1, 1)$, $(1, -1)$, and $(-1, -1)$. Use the results in Exercises 44 and 45 to find and draw the image of ABCD under the given transformation.

46. T in Exercise 3

47. D in Exercise 17

48. P in Exercise 18

49. The projection in Exercise 22

50. T in Exercise 31

51. The transformation in Exercise 37

52. Prove that $P_\ell(c\mathbf{v}) = cP_\ell(\mathbf{v})$ for any scalar c [Example 3.59(b)].

53. Prove that $T: \mathbb{R}^n \to \mathbb{R}^m$ is a linear transformation if and only if

$$T(c_1\mathbf{v}_1 + c_2\mathbf{v}_2) = c_1T(\mathbf{v}_1) + c_2T(\mathbf{v}_2)$$

for all $\mathbf{v}_1$, $\mathbf{v}_2$ in $\mathbb{R}^n$ and scalars c_1, c_2.

54. Prove that (as noted at the beginning of this section) the range of a linear transformation $T: \mathbb{R}^n \to \mathbb{R}^m$ is the column space of its matrix $[T]$.

55. If A is an invertible 2×2 matrix, what does the Fundamental Theorem of Invertible Matrices assert about the corresponding linear transformation T_A in light of Exercise 19?

Vignette

Robotics

In 1981, the U.S. Space Shuttle *Columbia* blasted off equipped with a device called the Shuttle Remote Manipulator System (SRMS). This robotic arm, known as Canadarm, has proved to be a vital tool in all subsequent space shuttle missions, providing strong, yet precise and delicate handling of its payloads (see Figure 3.15).

Canadarm has been used to place satellites into their proper orbit and to retrieve malfunctioning ones for repair, and it has also performed critical repairs to the shuttle itself. Notably, the robotic arm was instrumental in the successful repair of the *Hubble Space Telescope*. Since 1998, Canadarm has played an important role in the assembly and operation of the *International Space Station*.

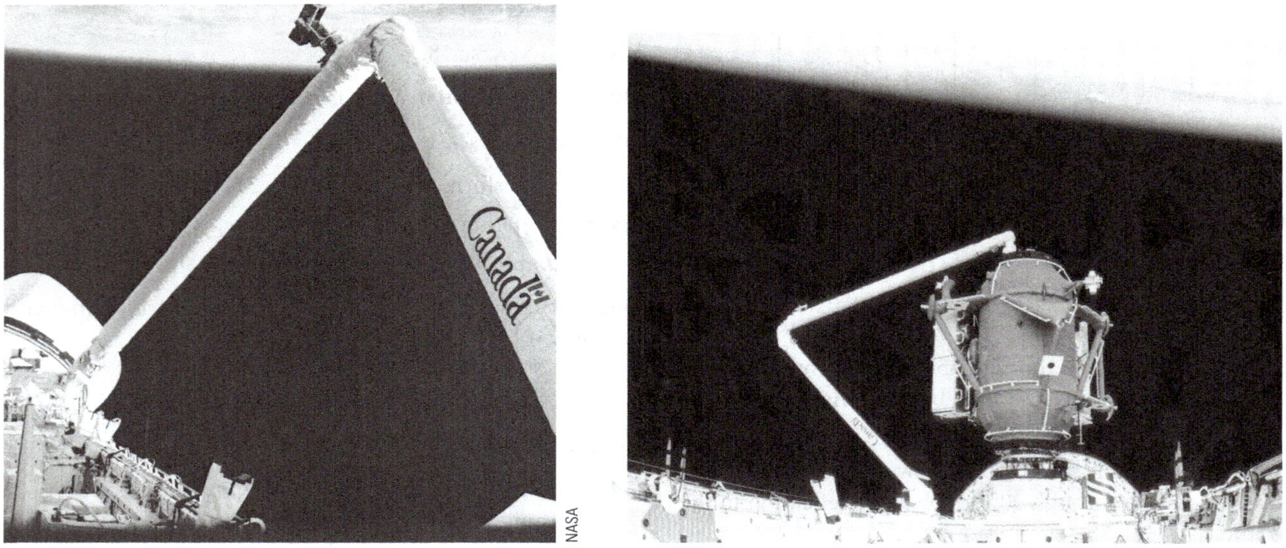

NASA

NASA

Figure 3.15
Canadarm

A robotic arm consists of a series of *links* of fixed length connected at *joints* where they can rotate. Each link can therefore rotate in space, or (through the effect of the other links) be translated parallel to itself, or move by a combination (composition) of rotations and translations. Before we can design a mathematical model for a robotic arm, we need to understand how rotations and translations work in composition. To simplify matters, we will assume that our arm is in $\mathbb{R}^2$.

In Section 3.6, we saw that the matrix of a rotation R about the origin through an angle θ is a linear transformation with matrix $\begin{bmatrix} \cos\theta & -\sin\theta \\ \sin\theta & \cos\theta \end{bmatrix}$ (Figure 3.16(a)). If $\mathbf{v} = \begin{bmatrix} a \\ b \end{bmatrix}$, then a ***translation along* v** is the transformation

$$T(\mathbf{x}) = \mathbf{x} + \mathbf{v} \quad \text{or, equivalently,} \quad T\begin{bmatrix} x \\ y \end{bmatrix} = \begin{bmatrix} x + a \\ y + b \end{bmatrix}$$

(Figure 3.16(b)).

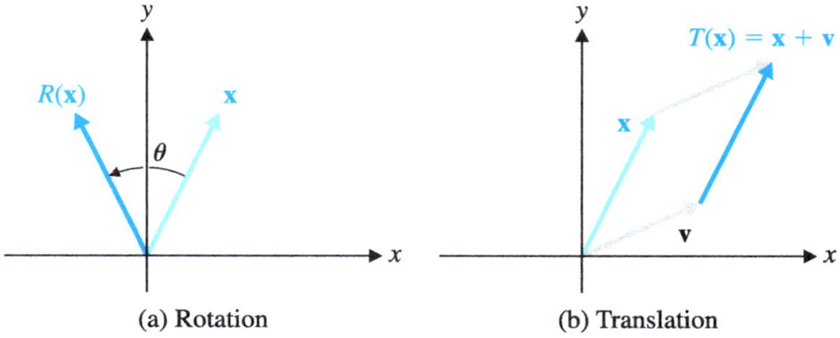

Figure 3.16

Unfortunately, translation is not a linear transformation, because $T(\mathbf{0}) \neq \mathbf{0}$. However, there is a trick that will get us around this problem. We can represent the vector $\mathbf{x} = \begin{bmatrix} x \\ y \end{bmatrix}$ as the vector $\begin{bmatrix} x \\ y \\ 1 \end{bmatrix}$ in $\mathbb{R}^3$. This is called representing $\mathbf{x}$ in ***homogeneous coordinates.*** Then the matrix multiplication

$$\begin{bmatrix} 1 & 0 & a \\ 0 & 1 & b \\ 0 & 0 & 1 \end{bmatrix}\begin{bmatrix} x \\ y \\ 1 \end{bmatrix} = \begin{bmatrix} x + a \\ y + b \\ 1 \end{bmatrix}$$

represents the translated vector $T(\mathbf{x})$ in homogeneous coordinates.

We can treat rotations in homogeneous coordinates too. The matrix multiplication

$$\begin{bmatrix} \cos\theta & -\sin\theta & 0 \\ \sin\theta & \cos\theta & 0 \\ 0 & 0 & 1 \end{bmatrix}\begin{bmatrix} x \\ y \\ 1 \end{bmatrix} = \begin{bmatrix} x\cos\theta - y\sin\theta \\ x\sin\theta + y\cos\theta \\ 1 \end{bmatrix}$$

represents the rotated vector $R(\mathbf{x})$ in homogeneous coordinates. The composition $T \circ R$ that gives the rotation R followed by the translation T is now represented by the product

$$\begin{bmatrix} 1 & 0 & a \\ 0 & 1 & b \\ 0 & 0 & 1 \end{bmatrix}\begin{bmatrix} \cos\theta & -\sin\theta & 0 \\ \sin\theta & \cos\theta & 0 \\ 0 & 0 & 1 \end{bmatrix} = \begin{bmatrix} \cos\theta & -\sin\theta & a \\ \sin\theta & \cos\theta & b \\ 0 & 0 & 1 \end{bmatrix}$$

[Note that $R \circ T \neq T \circ R$.]

To model a robotic arm, we give each link its own coordinate system (called a *frame*) and examine how one link moves in relation to those to which it is directly connected. To be specific, we let the coordinate axes for the link A_i be x_i and y_i, with the x_i-axis aligned with the link. The length of A_i is denoted by a_i, and the angle between x_i and x_{i-1} is denoted by θ_i. The joint between A_i and A_{i-1} is at the point $(0, 0)$ relative to A_i and $(a_{i-1}, 0)$ relative to A_{i-1}. Hence, relative to A_{i-1}, the coordinate system for A_i has been rotated through θ_i and then translated along $\begin{bmatrix} a_{i-1} \\ 0 \end{bmatrix}$ (Figure 3.17). This transformation is represented in homogeneous coordinates by the matrix

$$T_i = \begin{bmatrix} \cos\theta_i & -\sin\theta_i & a_{i-1} \\ \sin\theta_i & \cos\theta_i & 0 \\ 0 & 0 & 1 \end{bmatrix}$$

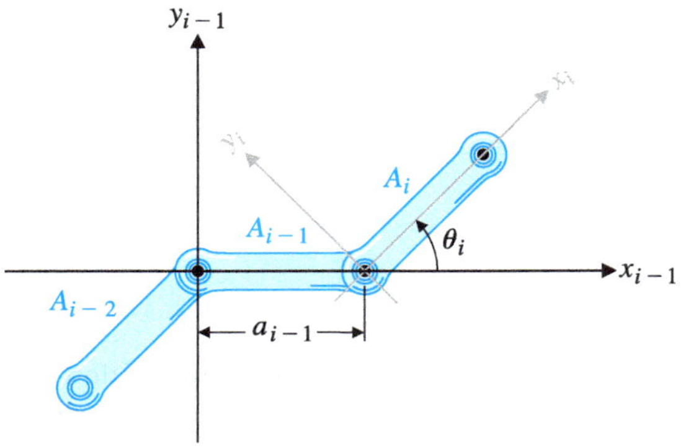

Figure 3.17

To give a specific example, consider Figure 3.18(a). It shows an arm with three links in which A_1 is in its initial position and each of the other two links has been rotated 45° from the previous link. We will take the length of each link to be 2 units. Figure 3.18(b) shows A_3 in its initial frame. The transformation

$$T_3 = \begin{bmatrix} \cos 45 & -\sin 45 & 2 \\ \sin 45 & \cos 45 & 0 \\ 0 & 0 & 1 \end{bmatrix} = \begin{bmatrix} 1/\sqrt{2} & -1/\sqrt{2} & 2 \\ 1/\sqrt{2} & 1/\sqrt{2} & 0 \\ 0 & 0 & 1 \end{bmatrix}$$

causes a rotation of 45° and then a translation by 2 units. As shown in 3.18(c), this places A_3 in its appropriate position relative to A_2's frame. Next, the transformation

$$T_2 = \begin{bmatrix} \cos 45 & -\sin 45 & 2 \\ \sin 45 & \cos 45 & 0 \\ 0 & 0 & 1 \end{bmatrix} = \begin{bmatrix} 1/\sqrt{2} & -1/\sqrt{2} & 2 \\ 1/\sqrt{2} & 1/\sqrt{2} & 0 \\ 0 & 0 & 1 \end{bmatrix}$$

is applied to the previous result. This places both A_3 and A_2 in their correct position relative to A_1, as shown in Figure 3.18(d). Normally, a third transformation T_1 (a rotation) would be applied to the previous result, but in our case, T_1 is the identity transformation because A_1 stays in its initial position.

Typically, we want to know the coordinates of the end (the "hand") of the robotic arm, given the length and angle parameters—this is known as *forward kinematics.* Following the above sequence of calculations and referring to Figure 3.18, we see that

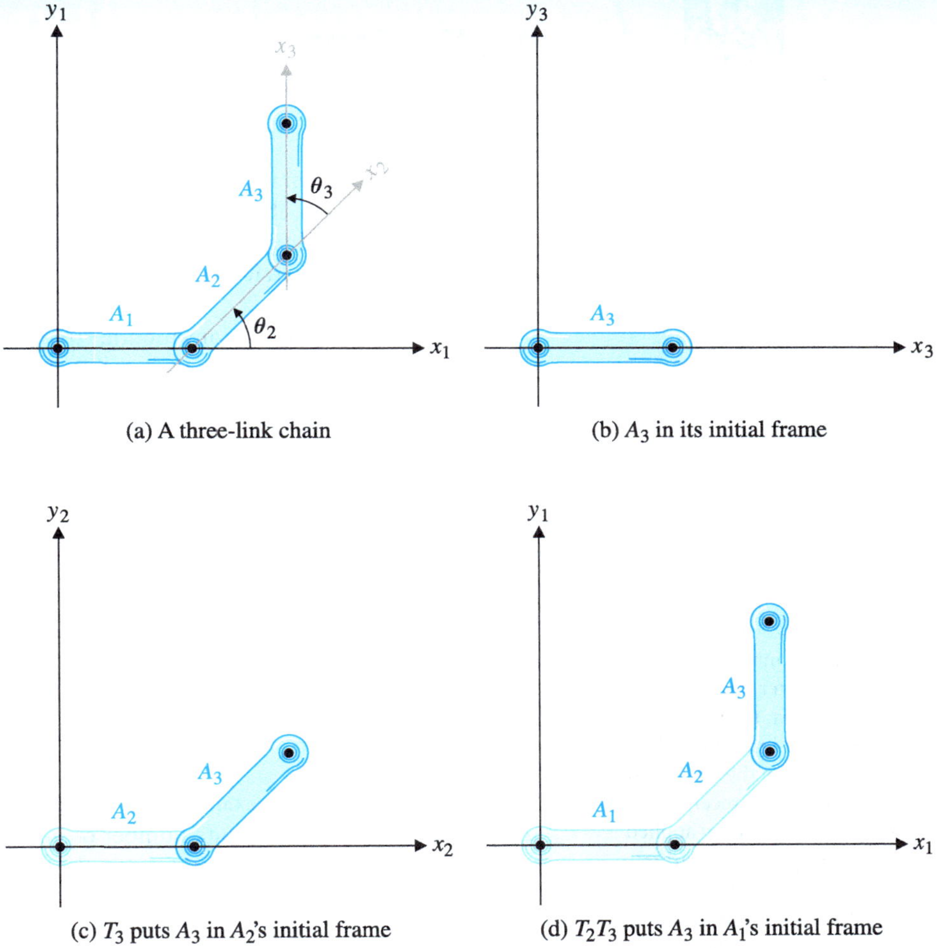

(a) A three-link chain

(b) A_3 in its initial frame

(c) T_3 puts A_3 in A_2's initial frame

(d) T_2T_3 puts A_3 in A_1's initial frame

Figure 3.18

we need to determine where the point (2, 0) ends up after T_3 and T_2 are applied. Thus, the arm's hand is at

$$T_2T_3\begin{bmatrix}2\\0\\1\end{bmatrix} = \begin{bmatrix}1/\sqrt{2} & -1/\sqrt{2} & 2\\ 1/\sqrt{2} & 1/\sqrt{2} & 0\\ 0 & 0 & 1\end{bmatrix}^2\begin{bmatrix}2\\0\\1\end{bmatrix} = \begin{bmatrix}0 & -1 & 2+\sqrt{2}\\ 1 & 0 & \sqrt{2}\\ 0 & 0 & 1\end{bmatrix}\begin{bmatrix}2\\0\\1\end{bmatrix}$$

$$= \begin{bmatrix}2+\sqrt{2}\\ 2+\sqrt{2}\\ 1\end{bmatrix}$$

which represents the point $(2+\sqrt{2}, 2+\sqrt{2})$ in homogeneous coordinates. It is easily checked from Figure 3.18(a) that this is correct.

The methods used in this example generalize to robotic arms in three dimensions, although in $\mathbb{R}^3$ there are more degrees of freedom and hence more variables. The method of homogeneous coordinates is also useful in other applications, notably computer graphics.

3.7 Applications

Markov Chains

A market research team is conducting a controlled survey to determine people's preferences in toothpaste. The sample consists of 200 people, each of whom is asked to try two brands of toothpaste over a period of several months. Based on the responses to the survey, the research team compiles the following statistics about toothpaste preferences.

Of those using Brand A in any month, 70% continue to use it the following month, while 30% switch to Brand B; of those using Brand B in any month, 80% continue to use it the following month, while 20% switch to Brand A. These findings are summarized in Figure 3.19, in which the percentages have been converted into decimals; we will think of them as probabilities.

Andrei A. Markov (1856–1922) was a Russian mathematician who studied and later taught at the University of St. Petersburg. He was interested in number theory, analysis, and the theory of continued fractions, a recently developed field that Markov applied to probability theory. Markov was also interested in poetry, and one of the uses to which he put Markov chains was the analysis of patterns in poems and other literary texts.

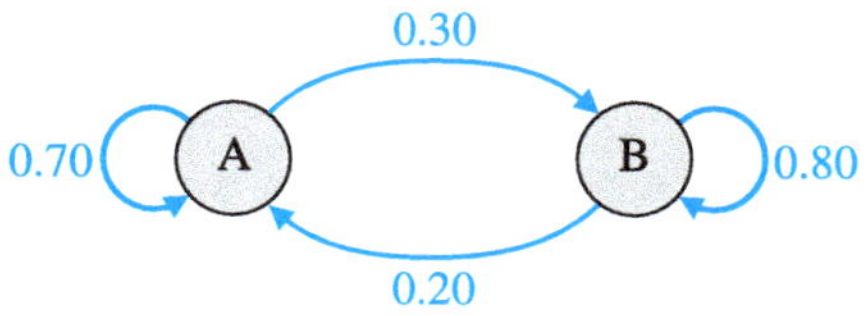

Figure 3.19

Figure 3.19 is a simple example of a (finite) ***Markov chain.*** It represents an evolving process consisting of a finite number of *states*. At each step or point in time, the process may be in any one of the states; at the next step, the process can remain in its present state or switch to one of the other states. The state to which the process moves at the next step and the probability of its doing so depend *only* on the present state and not on the past history of the process. These probabilities are called *transition probabilities* and are assumed to be constants (that is, the probability of moving from state i to state j is always the same).

Example 3.64

In the toothpaste survey described above, there are just two states—using Brand A and using Brand B—and the transition probabilities are those indicated in Figure 3.19. Suppose that, when the survey begins, 120 people are using Brand A and 80 people are using Brand B. How many people will be using each brand 1 month later? 2 months later?

Solution The number of Brand A users after 1 month will be 70% of those initially using Brand A (those who remain loyal to Brand A) plus 20% of the Brand B users (those who switch from B to A):

$$0.70(120) + 0.20(80) = 100$$

Similarly, the number of Brand B users after 1 month will be a combination of those who switch to Brand B and those who continue to use it:

$$0.30(120) + 0.80(80) = 100$$

We can summarize these two equations in a single matrix equation:

$$\begin{bmatrix} 0.70 & 0.20 \\ 0.30 & 0.80 \end{bmatrix} \begin{bmatrix} 120 \\ 80 \end{bmatrix} = \begin{bmatrix} 100 \\ 100 \end{bmatrix}$$

Let's call the matrix P and label the vectors $\mathbf{x}_0 = \begin{bmatrix} 120 \\ 80 \end{bmatrix}$ and $\mathbf{x}_1 = \begin{bmatrix} 100 \\ 100 \end{bmatrix}$. (Note that the components of each vector are the numbers of Brand A and Brand B users, in that order, after the number of months indicated by the subscript.) Thus, we have $\mathbf{x}_1 = P\mathbf{x}_0$.

Extending the notation, let $\mathbf{x}_k$ be the vector whose components record the distribution of toothpaste users after k months. To determine the number of users of each brand after 2 months have elapsed, we simply apply the same reasoning, starting with $\mathbf{x}_1$ instead of $\mathbf{x}_0$. We obtain

$$\mathbf{x}_2 = P\mathbf{x}_1 = \begin{bmatrix} 0.70 & 0.20 \\ 0.30 & 0.80 \end{bmatrix} \begin{bmatrix} 100 \\ 100 \end{bmatrix} = \begin{bmatrix} 90 \\ 110 \end{bmatrix}$$

from which we see that there are now 90 Brand A users and 110 Brand B users.

The vectors $\mathbf{x}_k$ in Example 3.64 are called the ***state vectors*** of the Markov chain, and the matrix P is called its ***transition matrix.*** We have just seen that a Markov chain satisfies the relation

$$\mathbf{x}_{k+1} = P\mathbf{x}_k \quad \text{for } k = 0, 1, 2, \ldots$$

From this result it follows that we can compute an arbitrary state vector *iteratively* once we know $\mathbf{x}_0$ and P. In other words, a Markov chain is *completely determined* by its transition probabilities and its initial state.

Remarks

- Suppose, in Example 3.64, we wanted to keep track of not the *actual* numbers of toothpaste users but, rather, the *relative* numbers using each brand. We could convert the data into percentages or fractions by dividing by 200, the total number of users. Thus, we would start with

$$\mathbf{x}_0 = \begin{bmatrix} \frac{120}{200} \\ \frac{80}{200} \end{bmatrix} = \begin{bmatrix} 0.60 \\ 0.40 \end{bmatrix}$$

to reflect the fact that, initially, the Brand A–Brand B split is 60%–40%. Check by direct calculation that $P\mathbf{x}_0 = \begin{bmatrix} 0.50 \\ 0.50 \end{bmatrix}$, which can then be taken as $\mathbf{x}_1$ (in agreement with the 50–50 split we computed above). Vectors such as these, with nonnegative components that add up to 1, are called ***probability vectors.***

- Observe how the transition probabilities are arranged within the transition matrix P. We can think of the columns as being labeled with the *present* states and the rows as being labeled with the *next* states:

$$\begin{array}{cc} & \textit{Present} \\ & \begin{array}{cc} \text{A} & \text{B} \end{array} \\ \textit{Next} \begin{array}{c} \text{A} \\ \text{B} \end{array} & \begin{bmatrix} 0.70 & 0.20 \\ 0.30 & 0.80 \end{bmatrix} \end{array}$$

The word *stochastic* is derived from the Greek adjective *stokhastikos,* meaning "capable of aiming" (or guessing). It has come to be applied to anything that is governed by the laws of probability in the sense that probability makes predictions about the likelihood of things happening. In probability theory, "stochastic processes" form a generalization of Markov chains.

Note also that the columns of P are probability vectors; any square matrix with this property is called a ***stochastic matrix.***

We can realize the deterministic nature of Markov chains in another way. Note that we can write

$$\mathbf{x}_2 = P\mathbf{x}_1 = P(P\mathbf{x}_0) = P^2\mathbf{x}_0$$

and, in general,

$$\mathbf{x}_k = P^k\mathbf{x}_0 \quad \text{for } k = 0, 1, 2, \dots$$

This leads us to examine the powers of a transition matrix. In Example 3.64, we have

$$P^2 = \begin{bmatrix} 0.70 & 0.20 \\ 0.30 & 0.80 \end{bmatrix}\begin{bmatrix} 0.70 & 0.20 \\ 0.30 & 0.80 \end{bmatrix} = \begin{bmatrix} 0.55 & 0.30 \\ 0.45 & 0.70 \end{bmatrix}$$

What are we to make of the entries of this matrix? The first thing to observe is that P^2 is another stochastic matrix, since its columns sum to 1. (You are asked to prove this in Exercise 14.) Could it be that P^2 is also a transition matrix of some kind? Consider one of its entries—say, $(P^2)_{21} = 0.45$. The tree diagram in Figure 3.20 clarifies where this entry came from.

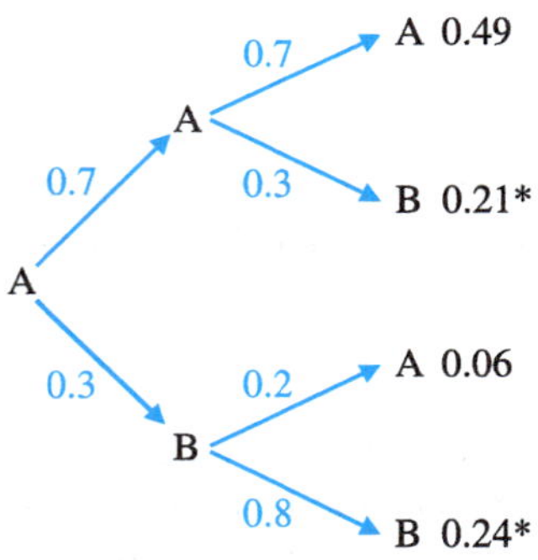

Figure 3.20

There are four possible state changes that can occur over 2 months, and these correspond to the four branches (or paths) of length 2 in the tree. Someone who initially is using Brand A can end up using Brand B 2 months later in two different ways (marked * in the figure): The person can continue to use A after 1 month and then switch to B (with probability $0.7(0.3) = 0.21$), or the person can switch to B after 1 month and then stay with B (with probability $0.3(0.8) = 0.24$). The sum of these probabilities gives an overall probability of 0.45. Observe that these calculations are *exactly* what we do when we compute $(P^2)_{21}$.

It follows that $(P^2)_{21} = 0.45$ represents the probability of moving from state 1 (Brand A) to state 2 (Brand B) in two transitions. (Note that the order of the subscripts is the *reverse* of what you might have guessed.) The argument can be generalized to show that

> $(P^k)_{ij}$ is the probability of moving from state j to state i in k transitions.

In Example 3.64, what will happen to the distribution of toothpaste users in the long run? Let's work with probability vectors as state vectors. Continuing our calculations (rounding to three decimal places), we find

$$\mathbf{x}_0 = \begin{bmatrix} 0.60 \\ 0.40 \end{bmatrix}, \mathbf{x}_1 = \begin{bmatrix} 0.50 \\ 0.50 \end{bmatrix}, \mathbf{x}_2 = P\mathbf{x}_1 = \begin{bmatrix} 0.70 & 0.20 \\ 0.30 & 0.80 \end{bmatrix}\begin{bmatrix} 0.50 \\ 0.50 \end{bmatrix} = \begin{bmatrix} 0.45 \\ 0.55 \end{bmatrix},$$

$$\mathbf{x}_3 = P\mathbf{x}_2 = \begin{bmatrix} 0.70 & 0.20 \\ 0.30 & 0.80 \end{bmatrix}\begin{bmatrix} 0.45 \\ 0.55 \end{bmatrix} = \begin{bmatrix} 0.425 \\ 0.575 \end{bmatrix}, \mathbf{x}_4 = \begin{bmatrix} 0.412 \\ 0.588 \end{bmatrix}, \mathbf{x}_5 = \begin{bmatrix} 0.406 \\ 0.594 \end{bmatrix},$$

$$\mathbf{x}_6 = \begin{bmatrix} 0.403 \\ 0.597 \end{bmatrix}, \mathbf{x}_7 = \begin{bmatrix} 0.402 \\ 0.598 \end{bmatrix}, \mathbf{x}_8 = \begin{bmatrix} 0.401 \\ 0.599 \end{bmatrix}, \mathbf{x}_9 = \begin{bmatrix} 0.400 \\ 0.600 \end{bmatrix}, \mathbf{x}_{10} = \begin{bmatrix} 0.400 \\ 0.600 \end{bmatrix}$$

and so on. It appears that the state vectors approach (or *converge to*) the vector $\begin{bmatrix} 0.4 \\ 0.6 \end{bmatrix}$, implying that eventually 40% of the toothpaste users in the survey will be using Brand A and 60% will be using Brand B. Indeed, it is easy to check that, once this distribution is reached, it will never change. We simply compute

$$\begin{bmatrix} 0.70 & 0.20 \\ 0.30 & 0.80 \end{bmatrix}\begin{bmatrix} 0.4 \\ 0.6 \end{bmatrix} = \begin{bmatrix} 0.4 \\ 0.6 \end{bmatrix}$$

A state vector $\mathbf{x}$ with the property that $P\mathbf{x} = \mathbf{x}$ is called a ***steady state vector.*** In Chapter 4, we will prove that every Markov chain has a unique steady state vector. For now, let's accept this as a fact and see how we can find such a vector without doing any iterations at all.

We begin by rewriting the matrix equation $P\mathbf{x} = \mathbf{x}$ as $P\mathbf{x} = I\mathbf{x}$, which can in turn be rewritten as $(I - P)\mathbf{x} = \mathbf{0}$. Now this is just a homogeneous system of linear equations with coefficient matrix $I - P$, so the augmented matrix is $[I - P \mid \mathbf{0}]$. In Example 3.64, we have

$$[I - P \mid \mathbf{0}] = \left[\begin{array}{cc|c} 1 - 0.70 & -0.20 & 0 \\ -0.30 & 1 - 0.80 & 0 \end{array}\right] = \left[\begin{array}{cc|c} 0.30 & -0.20 & 0 \\ -0.30 & 0.20 & 0 \end{array}\right]$$

which reduces to

$$\left[\begin{array}{cc|c} 1 & -\frac{2}{3} & 0 \\ 0 & 0 & 0 \end{array}\right]$$

So, if our steady state vector is $\mathbf{x} = \begin{bmatrix} x_1 \\ x_2 \end{bmatrix}$, then x_2 is a free variable and the parametric solution is

$$x_1 = \tfrac{2}{3}t, \quad x_2 = t$$

If we require $\mathbf{x}$ to be a probability vector, then we must have

$$1 = x_1 + x_2 = \tfrac{2}{3}t + t = \tfrac{5}{3}t$$

Therefore, $x_2 = t = \frac{3}{5} = 0.6$ and $x_1 = \frac{2}{5} = 0.4$, so $\mathbf{x} = \begin{bmatrix} 0.4 \\ 0.6 \end{bmatrix}$, in agreement with our iterative calculations above. (If we require $\mathbf{x}$ to contain the *actual* distribution, then in this example we must have $x_1 + x_2 = 200$, from which it follows that $\mathbf{x} = \begin{bmatrix} 80 \\ 120 \end{bmatrix}$.)

Example 3.65

A psychologist places a rat in a cage with three compartments, as shown in Figure 3.21. The rat has been trained to select a door at random whenever a bell is rung and to move through it into the next compartment.

(a) If the rat is initially in compartment 1, what is the probability that it will be in compartment 2 after the bell has rung twice? three times?

(b) In the long run, what proportion of its time will the rat spend in each compartment?

Solution Let $P = [p_{ij}]$ be the transition matrix for this Markov chain. Then

$$p_{21} = p_{31} = \tfrac{1}{2},\ p_{12} = p_{13} = \tfrac{1}{3},\ p_{32} = p_{23} = \tfrac{2}{3},\ \text{and}\ p_{11} = p_{22} = p_{33} = 0$$

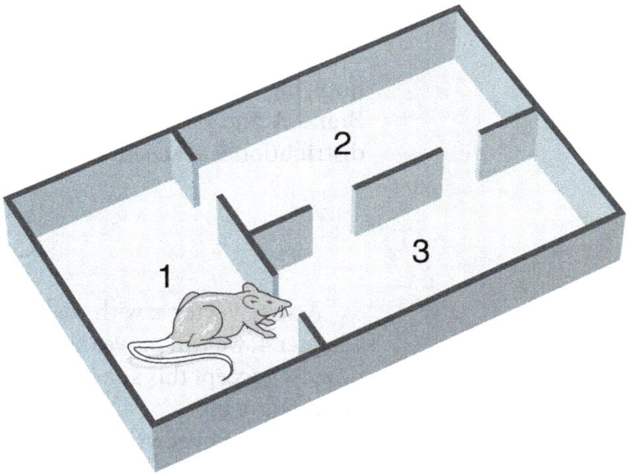

Figure 3.21

(Why? Remember that p_{ij} is the probability of moving from j to i.) Therefore,

$$P = \begin{bmatrix} 0 & \frac{1}{3} & \frac{1}{3} \\ \frac{1}{2} & 0 & \frac{2}{3} \\ \frac{1}{2} & \frac{2}{3} & 0 \end{bmatrix}$$

and the initial state vector is

$$\mathbf{x}_0 = \begin{bmatrix} 1 \\ 0 \\ 0 \end{bmatrix}$$

(a) After one ring of the bell, we have

$$\mathbf{x}_1 = P\mathbf{x}_0 = \begin{bmatrix} 0 & \frac{1}{3} & \frac{1}{3} \\ \frac{1}{2} & 0 & \frac{2}{3} \\ \frac{1}{2} & \frac{2}{3} & 0 \end{bmatrix}\begin{bmatrix} 1 \\ 0 \\ 0 \end{bmatrix} = \begin{bmatrix} 0 \\ \frac{1}{2} \\ \frac{1}{2} \end{bmatrix} = \begin{bmatrix} 0 \\ 0.5 \\ 0.5 \end{bmatrix}$$

Continuing (rounding to three decimal places), we find

$$\mathbf{x}_2 = P\mathbf{x}_1 = \begin{bmatrix} 0 & \frac{1}{3} & \frac{1}{3} \\ \frac{1}{2} & 0 & \frac{2}{3} \\ \frac{1}{2} & \frac{2}{3} & 0 \end{bmatrix}\begin{bmatrix} 0 \\ \frac{1}{2} \\ \frac{1}{2} \end{bmatrix} = \begin{bmatrix} \frac{1}{3} \\ \frac{1}{3} \\ \frac{1}{3} \end{bmatrix} \approx \begin{bmatrix} 0.333 \\ 0.333 \\ 0.333 \end{bmatrix}$$

and

$$\mathbf{x}_3 = P\mathbf{x}_2 = \begin{bmatrix} 0 & \frac{1}{3} & \frac{1}{3} \\ \frac{1}{2} & 0 & \frac{2}{3} \\ \frac{1}{2} & \frac{2}{3} & 0 \end{bmatrix}\begin{bmatrix} \frac{1}{3} \\ \frac{1}{3} \\ \frac{1}{3} \end{bmatrix} = \begin{bmatrix} \frac{2}{9} \\ \frac{7}{18} \\ \frac{7}{18} \end{bmatrix} \approx \begin{bmatrix} 0.222 \\ 0.389 \\ 0.389 \end{bmatrix}$$

Therefore, after two rings, the probability that the rat is in compartment 2 is $\frac{1}{3} \approx 0.333$, and after three rings, the probability that the rat is in compartment 2 is $\frac{7}{18} \approx 0.389$. [Note that these questions could also be answered by computing $(P^2)_{21}$ and $(P^3)_{21}$.]

(b) This question is asking for the steady state vector **x** as a probability vector. As we saw above, **x** must be in the null space of $I - P$, so we proceed to solve the system

$$[I - P \mid \mathbf{0}] = \left[\begin{array}{ccc|c} 1 & -\frac{1}{3} & -\frac{1}{3} & 0 \\ -\frac{1}{2} & 1 & -\frac{2}{3} & 0 \\ -\frac{1}{2} & -\frac{2}{3} & 1 & 0 \end{array}\right] \longrightarrow \left[\begin{array}{ccc|c} 1 & 0 & -\frac{2}{3} & 0 \\ 0 & 1 & -1 & 0 \\ 0 & 0 & 0 & 0 \end{array}\right]$$

Hence, if $\mathbf{x} = \begin{bmatrix} x_1 \\ x_2 \\ x_3 \end{bmatrix}$, then $x_3 = t$ is free and $x_1 = \frac{2}{3}t$, $x_2 = t$. Since **x** must be a probability vector, we need $1 = x_1 + x_2 + x_3 = \frac{8}{3}t$. Thus, $t = \frac{3}{8}$ and

$$\mathbf{x} = \begin{bmatrix} \frac{1}{4} \\ \frac{3}{8} \\ \frac{3}{8} \end{bmatrix}$$

which tells us that, in the long run, the rat spends $\frac{1}{4}$ of its time in compartment 1 and $\frac{3}{8}$ of its time in each of the other two compartments.

Linear Economic Models

We now revisit the economic models that we first encountered in Section 2.4 and recast these models in terms of matrices. Example 2.33 illustrated the Leontief closed model. The system of equations we needed to solve was

$$\begin{aligned} \tfrac{1}{4}x_1 + \tfrac{1}{3}x_2 + \tfrac{1}{2}x_3 &= x_1 \\ \tfrac{1}{4}x_1 + \tfrac{1}{3}x_2 + \tfrac{1}{4}x_3 &= x_2 \\ \tfrac{1}{2}x_1 + \tfrac{1}{3}x_2 + \tfrac{1}{4}x_3 &= x_3 \end{aligned}$$

In matrix form, this is the equation $E\mathbf{x} = \mathbf{x}$, where

$$E = \begin{bmatrix} 1/4 & 1/3 & 1/2 \\ 1/4 & 1/3 & 1/4 \\ 1/2 & 1/3 & 1/4 \end{bmatrix} \text{ and } \mathbf{x} = \begin{bmatrix} x_1 \\ x_2 \\ x_3 \end{bmatrix}$$

The matrix E is called an ***exchange matrix*** and the vector **x** is called a ***price vector***. In general, if $E = [e_{ij}]$, then e_{ij} represents the fraction (or percentage) of industry j's output that is consumed by industry i and x_i is the price charged by industry i for its output.

In a closed economy, the sum of each column of E is 1. Since the entries of E are also nonnegative, E is a stochastic matrix and the problem of finding a solution to the equation

$$E\mathbf{x} = \mathbf{x} \tag{1}$$

is precisely the same as the problem of finding the steady state vector of a Markov chain! Thus, to find a price vector x that satisfies $E\mathbf{x} = \mathbf{x}$, we solve the equivalent homogeneous equation $(I - E)\mathbf{x} = 0$. There will always be infinitely many solutions; we seek a solution where the prices are all nonnegative and at least one price is positive.

The Leontief open model is more interesting. In Example 2.34, we needed to solve the system

$$\begin{aligned} x_1 &= 0.2x_1 + 0.5x_2 + 0.1x_3 + 10 \\ x_2 &= 0.4x_1 + 0.2x_2 + 0.2x_3 + 10 \\ x_3 &= 0.1x_1 + 0.3x_2 + 0.3x_3 + 30 \end{aligned}$$

In matrix form, we have

$$\mathbf{x} = C\mathbf{x} + \mathbf{d} \quad \text{or} \quad (I - C)\mathbf{x} = \mathbf{d} \tag{2}$$

where

$$C = \begin{bmatrix} 0.2 & 0.5 & 0.1 \\ 0.4 & 0.2 & 0.2 \\ 0.1 & 0.3 & 0.3 \end{bmatrix}, \ \mathbf{x} = \begin{bmatrix} x_1 \\ x_2 \\ x_3 \end{bmatrix}, \ \mathbf{d} = \begin{bmatrix} 10 \\ 10 \\ 30 \end{bmatrix}$$

The matrix C is called the ***consumption matrix,*** x is the ***production vector,*** and **d** is the ***demand vector.*** In general, if $C = [c_{ij}]$, $\mathbf{x} = [x_i]$, and $\mathbf{d} = [d_i]$, then c_{ij} represents the dollar value of industry i's output that is needed to produce one dollar's worth of industry j's output, x_i is the dollar value (price) of industry i's output, and d_i is the dollar value of the external demand for industry i's output. Once again, we are interested in finding a production vector **x** with nonnegative entries such that at least one entry is positive. We call such a vector **x** a ***feasible solution.***

Example 3.66

Determine whether there is a solution to the Leontief open model determined by the following consumption matrices:

(a) $C = \begin{bmatrix} 1/4 & 1/3 \\ 1/2 & 1/3 \end{bmatrix}$ (b) $C = \begin{bmatrix} 1/2 & 1/2 \\ 1/2 & 2/3 \end{bmatrix}$

Solution (a) We have

$$I - C = \begin{bmatrix} 1 & 0 \\ 0 & 1 \end{bmatrix} - \begin{bmatrix} 1/4 & 1/3 \\ 1/2 & 1/3 \end{bmatrix} = \begin{bmatrix} 3/4 & -1/3 \\ -1/2 & 2/3 \end{bmatrix}$$

so the equation $(I - C)\mathbf{x} = \mathbf{d}$ becomes

$$\begin{bmatrix} 3/4 & -1/3 \\ -1/2 & 2/3 \end{bmatrix} \begin{bmatrix} x_1 \\ x_2 \end{bmatrix} = \begin{bmatrix} d_1 \\ d_2 \end{bmatrix}$$

In practice, we would row reduce the corresponding augmented matrix to determine a solution. However, in this case, it is instructive to notice that the coefficient matrix $I - C$ is invertible and then to apply Theorem 3.7. We compute

$$\begin{bmatrix} x_1 \\ x_2 \end{bmatrix} = \begin{bmatrix} 3/4 & -1/3 \\ -1/2 & 2/3 \end{bmatrix}^{-1} \begin{bmatrix} d_1 \\ d_2 \end{bmatrix} = \begin{bmatrix} 2 & 1 \\ 3/2 & 9/4 \end{bmatrix} \begin{bmatrix} d_1 \\ d_2 \end{bmatrix}$$

Since d_1, d_2, and all entries of $(I - C)^{-1}$ are nonnegative, so are x_1 and x_2. Thus, we can find a feasible solution for *any* nonzero demand vector.

(b) In this case,

$$I - C = \begin{bmatrix} 1/2 & -1/2 \\ -1/2 & 2/3 \end{bmatrix} \quad \text{and} \quad (I - C)^{-1} = \begin{bmatrix} -4 & -6 \\ -6 & -6 \end{bmatrix}$$

so that

$$\mathbf{x} = (I - C)^{-1}\mathbf{d} = \begin{bmatrix} -4 & -6 \\ -6 & -6 \end{bmatrix}\mathbf{d}$$

Since all entries of $(I - C)^{-1}$ are negative, this will not produce a feasible solution for *any* nonzero demand vector **d**.

Motivated by Example 3.66, we have the following definition. (For two $m \times n$ matrices $A = [a_{ij}]$ and $B = [b_{ij}]$, we will write $A \geq B$ if $a_{ij} \geq b_{ij}$ for all i and j. Similarly, we may define $A > B$, $A \leq B$, and so on. A matrix A is called nonnegative if $A \geq O$ and positive if $A > O$.)

Definition A consumption matrix C is called ***productive*** if $I - C$ is invertible and $(I - C)^{-1} \geq O$.

We now give three results that give criteria for a consumption matrix to be productive.

Theorem 3.34 Let C be a consumption matrix. Then C is productive if and only if there exists a production vector $\mathbf{x} \geq \mathbf{0}$ such that $\mathbf{x} > C\mathbf{x}$.

Proof Assume that C is productive. Then $I - C$ is invertible and $(I - C)^{-1} \geq O$. Let

$$\mathbf{j} = \begin{bmatrix} 1 \\ 1 \\ \vdots \\ 1 \end{bmatrix}$$

Then $\mathbf{x} = (I - C)^{-1}\mathbf{j} \geq 0$ and $(I - C)\mathbf{x} = \mathbf{j} > 0$. Thus, $\mathbf{x} - C\mathbf{x} > 0$ or, equivalently, $\mathbf{x} > C\mathbf{x}$.

Conversely, assume that there exists a vector $\mathbf{x} \geq 0$ such that $\mathbf{x} > C\mathbf{x}$. Since $C \geq O$ and $C \neq O$, we have $\mathbf{x} > 0$ by Exercise 35. Furthermore, there must exist a real number λ with $0 < \lambda < 1$ such that $C\mathbf{x} < \lambda\mathbf{x}$. But then

$$C^2\mathbf{x} = C(C\mathbf{x}) \leq C(\lambda\mathbf{x}) = \lambda(C\mathbf{x}) < \lambda(\lambda\mathbf{x}) = \lambda^2\mathbf{x}$$

By induction, it can be shown that $0 \leq C^n\mathbf{x} < \lambda^n\mathbf{x}$ for all $n \geq 0$. (Write out the details of this induction proof.) Since $0 < \lambda < 1$, λ^n approaches 0 as n gets large. Therefore, as $n \to \infty$, $\lambda^n\mathbf{x} \to 0$ and hence $C^n\mathbf{x} \to 0$. Since $\mathbf{x} > 0$, we must have $C^n \to O$ as $n \to \infty$.

Now consider the matrix equation

$$(I - C)(I + C + C^2 + \cdots + C^{n-1}) = I - C^n$$

As $n \to \infty$, $C^n \to O$, so we have

$$(I - C)(I + C + C^2 + \ldots) = I - O = I$$

Therefore, $I - C$ is invertible, with its inverse given by the infinite matrix series $I + C + C^2 + \ldots$. Since all the terms in this series are nonnegative, we also have

$$(I - C)^{-1} = I + C + C^2 + \ldots \geq O$$

Hence, C is productive.

Remarks

- The infinite series $I + C + C^2 + \ldots$ is the matrix analogue of the geometric series $1 + x + x^2 + \ldots$. You may be familiar with the fact that, for $|x| < 1$, $1 + x + x^2 + \ldots = 1/(1 - x)$.
- Since the vector $C\mathbf{x}$ represents the amounts consumed by each industry, the inequality $\mathbf{x} > C\mathbf{x}$ means that there is some level of production for which each industry is producing more than it consumes.
- For an alternative approach to the first part of the proof of Theorem 3.34, see Exercise 42 in Section 4.6.

Corollary 3.35 Let C be a consumption matrix. If the sum of each row of C is less than 1, then C is productive.

The word *corollary* comes from the Latin word *corollarium,* which refers to a garland given as a reward. Thus, a corollary is a little extra reward that follows from a theorem.

Proof If

$$\mathbf{x} = \begin{bmatrix} 1 \\ 1 \\ \vdots \\ 1 \end{bmatrix}$$

then $C\mathbf{x}$ is a vector consisting of the row sums of C. If each row sum of C is less than 1, then the condition $\mathbf{x} > C\mathbf{x}$ is satisfied. Hence, C is productive.

Corollary 3.36 Let C be a consumption matrix. If the sum of each column of C is less than 1, then C is productive.

Proof If each column sum of C is less than 1, then each row sum of C^T is less than 1. Hence, C^T is productive, by Corollary 3.35. Therefore, by Theorems 3.9(d) and 3.4,

$$((I - C)^{-1})^T = ((I - C)^T)^{-1} = (I^T - C^T)^{-1} = (I - C^T)^{-1} \geq O$$

It follows that $(I - C)^{-1} \geq O$ too and, thus, C is productive.

You are asked to give alternative proofs of Corollaries 3.35 and 3.36 in Exercise 52 of Section 7.2.

It follows from the definition of a consumption matrix that the sum of column j is the total dollar value of all the inputs needed to produce one dollar's worth of industry j's output—that is, industry j's income exceeds its expenditures. We say that such an industry is ***profitable.*** Corollary 3.36 can therefore be rephrased to state that a consumption matrix is productive if all industries are profitable.

Population Growth

P. H. Leslie, "On the Use of Matrices in Certain Population Mathematics," *Biometrika* **33** (1945), pp. 183–212.

One of the most popular models of population growth is a matrix-based model, first introduced by P. H. Leslie in 1945. The *Leslie model* describes the growth of the female portion of a population, which is assumed to have a maximum lifespan. The females are divided into age classes, all of which span an equal number of years. Using data about the average birthrates and survival probabilities of each class, the model is then able to determine the growth of the population over time.

Example 3.67

A certain species of German beetle, the Vollmar-Wasserman beetle (or VW beetle, for short), lives for at most 3 years. We divide the female VW beetles into three age classes of 1 year each: youths (0–1 year), juveniles (1–2 years), and adults (2–3 years). The youths do not lay eggs; each juvenile produces an average of four female beetles; and each adult produces an average of three females.

The survival rate for youths is 50% (that is, the probability of a youth's surviving to become a juvenile is 0.5), and the survival rate for juveniles is 25%. Suppose we begin with a population of 100 female VW beetles: 40 youths, 40 juveniles, and 20 adults. Predict the beetle population for each of the next 5 years.

Solution After 1 year, the number of youths will be the number produced during that year:

$$40 \times 4 + 20 \times 3 = 220$$

The number of juveniles will simply be the number of youths that have survived:

$$40 \times 0.5 = 20$$

Likewise, the number of adults will be the number of juveniles that have survived:

$$40 \times 0.25 = 10$$

We can combine these into a single matrix equation

$$\begin{bmatrix} 0 & 4 & 3 \\ 0.5 & 0 & 0 \\ 0 & 0.25 & 0 \end{bmatrix} \begin{bmatrix} 40 \\ 40 \\ 20 \end{bmatrix} = \begin{bmatrix} 220 \\ 20 \\ 10 \end{bmatrix}$$

or $L\mathbf{x}_0 = \mathbf{x}_1$, where $\mathbf{x}_0 = \begin{bmatrix} 40 \\ 40 \\ 20 \end{bmatrix}$ is the initial population distribution vector and $\mathbf{x}_1 = \begin{bmatrix} 220 \\ 20 \\ 10 \end{bmatrix}$ is the distribution after 1 year. We see that the structure of the equation is exactly the same as for Markov chains: $\mathbf{x}_{k+1} = L\mathbf{x}_k$ for $k = 0, 1, 2, \ldots$ (although the interpretation is quite different). It follows that we can iteratively compute successive population distribution vectors. (It also follows that $\mathbf{x}_k = L^k\mathbf{x}_0$ for $k = 0, 1, 2, \ldots$, as for Markov chains, but we will not use this fact here.)

We compute

$$\mathbf{x}_2 = L\mathbf{x}_1 = \begin{bmatrix} 0 & 4 & 3 \\ 0.5 & 0 & 0 \\ 0 & 0.25 & 0 \end{bmatrix} \begin{bmatrix} 220 \\ 20 \\ 10 \end{bmatrix} = \begin{bmatrix} 110 \\ 110 \\ 5 \end{bmatrix}$$

$$\mathbf{x}_3 = L\mathbf{x}_2 = \begin{bmatrix} 0 & 4 & 3 \\ 0.5 & 0 & 0 \\ 0 & 0.25 & 0 \end{bmatrix} \begin{bmatrix} 110 \\ 110 \\ 5 \end{bmatrix} = \begin{bmatrix} 455 \\ 55 \\ 27.5 \end{bmatrix}$$

$$\mathbf{x}_4 = L\mathbf{x}_3 = \begin{bmatrix} 0 & 4 & 3 \\ 0.5 & 0 & 0 \\ 0 & 0.25 & 0 \end{bmatrix} \begin{bmatrix} 455 \\ 55 \\ 27.5 \end{bmatrix} = \begin{bmatrix} 302.5 \\ 227.5 \\ 13.75 \end{bmatrix}$$

$$\mathbf{x}_5 = L\mathbf{x}_4 = \begin{bmatrix} 0 & 4 & 3 \\ 0.5 & 0 & 0 \\ 0 & 0.25 & 0 \end{bmatrix} \begin{bmatrix} 302.5 \\ 227.5 \\ 13.75 \end{bmatrix} = \begin{bmatrix} 951.2 \\ 151.2 \\ 56.88 \end{bmatrix}$$

Therefore, the model predicts that after 5 years there will be approximately 951 young female VW beetles, 151 juveniles, and 57 adults. (*Note:* You could argue that we should have rounded to the nearest integer at each step—for example, 28 adults after step 3—which would have affected the subsequent iterations. We elected *not* to do this, since the calculations are only approximations anyway and it is much easier to use a calculator or CAS if you do not round as you go.)

The matrix L in Example 3.67 is called a ***Leslie matrix.*** In general, if we have a population with n age classes of equal duration, L will be an $n \times n$ matrix with the following structure:

$$L = \begin{bmatrix} b_1 & b_2 & b_3 & \cdots & b_{n-1} & b_n \\ s_1 & 0 & 0 & \cdots & 0 & 0 \\ 0 & s_2 & 0 & \cdots & 0 & 0 \\ 0 & 0 & s_3 & \cdots & 0 & 0 \\ \vdots & \vdots & \vdots & \ddots & \vdots & \vdots \\ 0 & 0 & 0 & \cdots & s_{n-1} & 0 \end{bmatrix}$$

Here, $b_1, b_2, \ldots$ are the *birth parameters* (b_i = the average numbers of females produced by each female in class i) and $s_1, s_2, \ldots$ are the *survival probabilities* (s_i = the probability that a female in class i survives into class $i + 1$).

What are we to make of our calculations? Overall, the beetle population appears to be increasing, although there are some fluctuations, such as a decrease from 250 to 225 from year 1 to year 2. Figure 3.22 shows the change in the population in each of the three age classes and clearly shows the growth, with fluctuations.

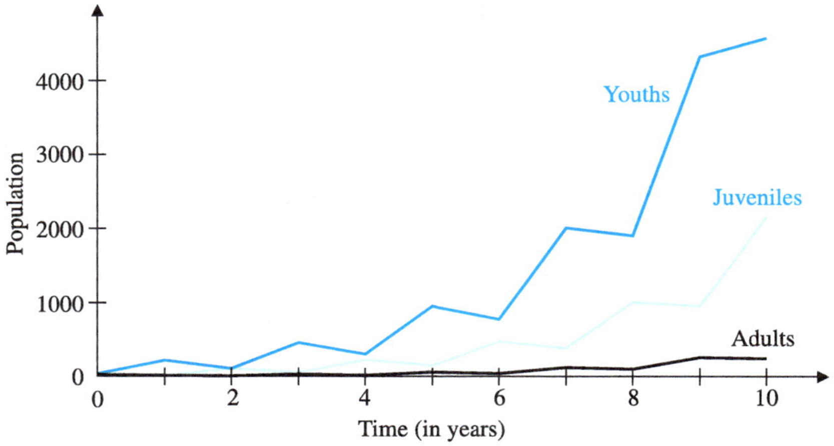

Figure 3.22

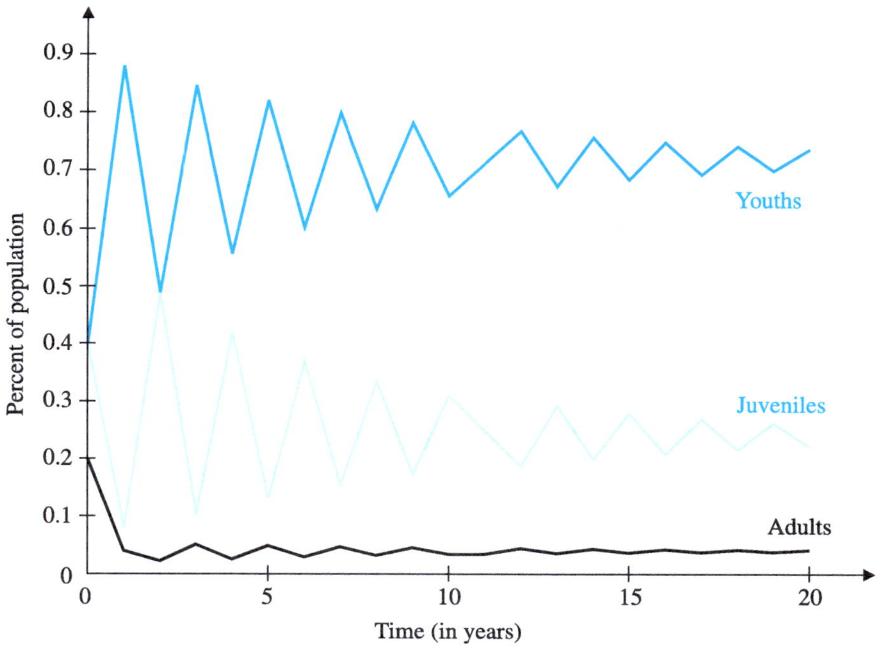

Figure 3.23

If, instead of plotting the *actual* population, we plot the *relative* population in each class, a different pattern emerges. To do this, we need to compute the fraction of the population in each age class in each year; that is, we need to divide each distribution vector by the sum of its components. For example, after 1 year, we have

$$\frac{1}{250}\mathbf{x}_1 = \frac{1}{250}\begin{bmatrix} 220 \\ 20 \\ 10 \end{bmatrix} = \begin{bmatrix} 0.88 \\ 0.08 \\ 0.04 \end{bmatrix}$$

which tells us that 88% of the population consists of youths, 8% is juveniles, and 4% is adults. If we plot this type of data over time, we get a graph like the one in Figure 3.23, which shows clearly that the proportion of the population in each class is approaching a steady state. It turns out that the steady state vector in this example is

$$\begin{bmatrix} 0.72 \\ 0.24 \\ 0.04 \end{bmatrix}$$

That is, in the long run, 72% of the population will be youths, 24% juveniles, and 4% adults. (In other words, the population is distributed among the three age classes in the ratio 18:6:1.) We will see how to determine this ratio exactly in Chapter 4.

Graphs and Digraphs

There are many situations in which it is important to be able to model the interrelationships among a finite set of objects. For example, we might wish to describe various types of networks (roads connecting towns, airline routes connecting cities, communication links connecting satellites, etc.) or relationships among groups or individuals (friendship relationships in a society, predator-prey relationships in

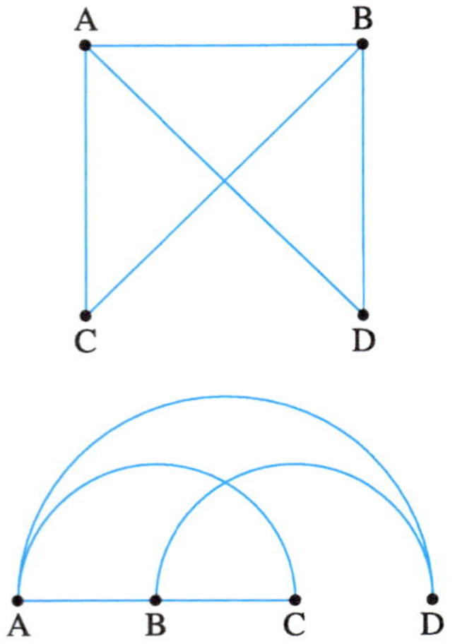

Figure 3.24
Two representations of the same graph

The term *vertex* (*vertices* is the plural) comes from the Latin verb *vertere*, which means "to turn." In the context of graphs (and geometry), a vertex is a corner—a point where an edge "turns" into a different edge.

an ecosystem, dominance relationships in a sport, etc.). Graphs are ideally suited to modeling such networks and relationships, and it turns out that matrices are a useful tool in their study.

A ***graph*** consists of a finite set of points (called *vertices*) and a finite set of *edges*, each of which connects two (not necessarily distinct) vertices. We say that two vertices are *adjacent* if they are the endpoints of an edge. Figure 3.24 shows an example of the same graph drawn in two different ways. The graphs are the "same" in the sense that all we care about are the adjacency relationships that identify the edges.

We can record the essential information about a graph in a matrix and use matrix algebra to help us answer certain questions about the graph. This is particularly useful if the graphs are large, since computers can handle the calculations very quickly.

Definition If G is a graph with n vertices, then its ***adjacency matrix*** is the $n \times n$ matrix A [or $A(G)$] defined by

$$a_{ij} = \begin{cases} 1 & \text{if there is an edge between vertices } i \text{ and } j \\ 0 & \text{otherwise} \end{cases}$$

Figure 3.25 shows a graph and its associated adjacency matrix.

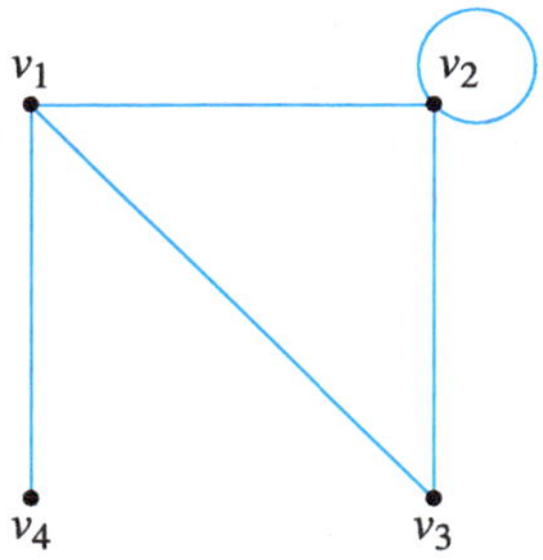

$$A = \begin{bmatrix} 0 & 1 & 1 & 1 \\ 1 & 1 & 1 & 0 \\ 1 & 1 & 0 & 0 \\ 1 & 0 & 0 & 0 \end{bmatrix}$$

Figure 3.25
A graph with adjacency matrix A

Remark Observe that the adjacency matrix of a graph is necessarily a symmetric matrix. (Why?) Notice also that a diagonal entry a_{ii} of A is zero unless there is a loop at vertex i. In some situations, a graph may have more than one edge between a pair of vertices. In such cases, it may make sense to modify the definition of the adjacency matrix so that a_{ij} equals the *number* of edges between vertices i and j.

We define a ***path*** in a graph to be a sequence of edges that allows us to travel from one vertex to another continuously. The ***length*** of a path is the number of edges it contains, and we will refer to a path with k edges as a ***k-path.*** For example, in the graph of Figure 3.25, $v_1v_3v_2v_1$ is a 3-path, and $v_4v_1v_2v_2v_1v_3$ is a 5-path. Notice that the first of these is *closed* (it begins and ends at the same vertex); such a path is called a ***circuit.*** The second uses the edge between v_1 and v_2 twice; a path that does *not* include the same edge more than once is called a ***simple*** path.

We can use the powers of a graph's adjacency matrix to give us information about the paths of various lengths in the graph. Consider the square of the adjacency matrix in Figure 3.25:

$$A^2 = \begin{bmatrix} 3 & 2 & 1 & 0 \\ 2 & 3 & 2 & 1 \\ 1 & 2 & 2 & 1 \\ 0 & 1 & 1 & 1 \end{bmatrix}$$

What do the entries of A^2 represent? Look at the (2, 3) entry. From the definition of matrix multiplication, we know that

$$(A^2)_{23} = a_{21}a_{13} + a_{22}a_{23} + a_{23}a_{33} + a_{24}a_{43}$$

The only way this expression can result in a nonzero number is if at least one of the products $a_{2k}a_{k3}$ that make up the sum is nonzero. But $a_{2k}a_{k3}$ is nonzero if and only if both a_{2k} and a_{k3} are nonzero, which means that there is an edge between v_2 and v_k as well as an edge between v_k and v_3. Thus, there will be a 2-path between vertices 2 and 3 (via vertex k). In our example, this happens for $k = 1$ and for $k = 2$, so

$$\begin{aligned}(A^2)_{23} &= a_{21}a_{13} + a_{22}a_{23} + a_{23}a_{33} + a_{24}a_{43} \\ &= 1 \cdot 1 + 1 \cdot 1 + 1 \cdot 0 + 0 \cdot 0 \\ &= 2\end{aligned}$$

which tells us that there are two 2-paths between vertices 2 and 3. (Check to see that the remaining entries of A^2 correctly give 2-paths in the graph.) The argument we have just given can be generalized to yield the following result, whose proof we leave as Exercise 72.

If A is the adjacency matrix of a graph G, then the (i, j) entry of A^k is equal to the number of k-paths between vertices i and j.

Example 3.68

How many 3-paths are there between v_1 and v_2 in Figure 3.25?

Solution We need the (1, 2) entry of A^3, which is the dot product of row 1 of A^2 and column 2 of A. The calculation gives

$$(A^3)_{12} = 3 \cdot 1 + 2 \cdot 1 + 1 \cdot 1 + 0 \cdot 0 = 6$$

so there are six 3-paths between vertices 1 and 2, which can be easily checked.

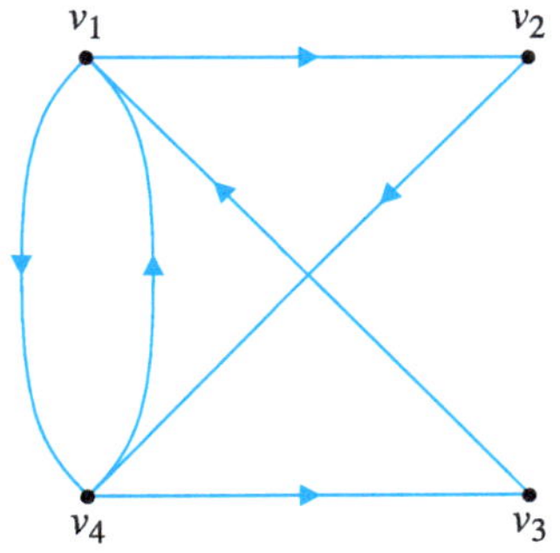

Figure 3.26
A digraph

In many applications that can be modeled by a graph, the vertices are ordered by some type of relation that imposes a direction on the edges. For example, directed edges might be used to represent one-way routes in a graph that models a transportation network or predator-prey relationships in a graph modeling an ecosystem. A graph with directed edges is called a ***digraph.*** Figure 3.26 shows an example.

An easy modification to the definition of adjacency matrices allows us to use them with digraphs.

Definition If G is a digraph with n vertices, then its ***adjacency matrix*** is the $n \times n$ matrix A [or $A(G)$] defined by

$$a_{ij} = \begin{cases} 1 & \text{if there is an edge from vertex } i \text{ to vertex } j \\ 0 & \text{otherwise} \end{cases}$$

Thus, the adjacency matrix for the digraph in Figure 3.26 is

$$A = \begin{bmatrix} 0 & 1 & 0 & 1 \\ 0 & 0 & 0 & 1 \\ 1 & 0 & 0 & 0 \\ 1 & 0 & 1 & 0 \end{bmatrix}$$

Not surprisingly, the adjacency matrix of a digraph is not symmetric in general. (When would it be?) You should have no difficulty seeing that A^k now contains the numbers of *directed* k-paths between vertices, where we insist that all edges along a path flow in the same direction. (See Exercise 72.) The next example gives an application of this idea.

Example 3.69

Five tennis players (Djokovic, Federer, Nadal, Roddick, and Safin) compete in a round-robin tournament in which each player plays every other player once. The digraph in Figure 3.27 summarizes the results. A directed edge from vertex i to vertex j means that player i defeated player j. (A digraph in which there is exactly one directed edge between every pair of vertices is called a ***tournament***.)

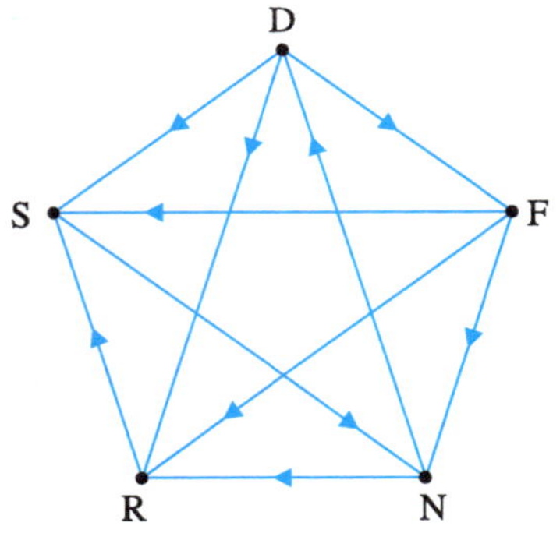

Figure 3.27
A tournament

The adjacency matrix for the digraph in Figure 3.27 is

$$A = \begin{bmatrix} 0 & 1 & 0 & 1 & 1 \\ 0 & 0 & 1 & 1 & 1 \\ 1 & 0 & 0 & 1 & 0 \\ 0 & 0 & 0 & 0 & 1 \\ 0 & 0 & 1 & 0 & 0 \end{bmatrix}$$

where the order of the vertices (and hence the rows and columns of A) is determined alphabetically. Thus, Federer corresponds to row 2 and column 2, for example.

Suppose we wish to rank the five players, based on the results of their matches. One way to do this might be to count the number of wins for each player. Observe that the number of wins each player had is just the sum of the entries in the corresponding row; equivalently, the vector containing all the row sums is given by the product $A\mathbf{j}$, where

$$\mathbf{j} = \begin{bmatrix} 1 \\ 1 \\ 1 \\ 1 \\ 1 \end{bmatrix}$$

In our case, we have

$$A\mathbf{j} = \begin{bmatrix} 0 & 1 & 0 & 1 & 1 \\ 0 & 0 & 1 & 1 & 1 \\ 1 & 0 & 0 & 1 & 0 \\ 0 & 0 & 0 & 0 & 1 \\ 0 & 0 & 1 & 0 & 0 \end{bmatrix} \begin{bmatrix} 1 \\ 1 \\ 1 \\ 1 \\ 1 \end{bmatrix} = \begin{bmatrix} 3 \\ 3 \\ 2 \\ 1 \\ 1 \end{bmatrix}$$

which produces the following ranking:

First: Djokovic, Federer (tie)
Second: Nadal
Third: Roddick, Safin (tie)

Are the players who tied in this ranking equally strong? Djokovic might argue that since he defeated Federer, he deserves first place. Roddick would use the same type of argument to break the tie with Safin. However, Safin could argue that he has two "indirect" victories because he beat Nadal, who defeated *two* others; furthermore, he might note that Roddick has only *one* indirect victory (over Safin, who then defeated Nadal).

Since in a group of ties there may not be a player who defeated all the others in the group, the notion of indirect wins seems more useful. Moreover, an indirect victory corresponds to a 2-path in the digraph, so we can use the square of the adjacency matrix. To compute both wins and indirect wins for each player, we need the row sums of the matrix $A + A^2$, which are given by

$$(A + A^2)\mathbf{j} = \left(\begin{bmatrix} 0 & 1 & 0 & 1 & 1 \\ 0 & 0 & 1 & 1 & 1 \\ 1 & 0 & 0 & 1 & 0 \\ 0 & 0 & 0 & 0 & 1 \\ 0 & 0 & 1 & 0 & 0 \end{bmatrix} + \begin{bmatrix} 0 & 0 & 2 & 1 & 2 \\ 1 & 0 & 1 & 1 & 1 \\ 0 & 1 & 0 & 1 & 2 \\ 0 & 0 & 1 & 0 & 0 \\ 1 & 0 & 0 & 1 & 0 \end{bmatrix} \right) \begin{bmatrix} 1 \\ 1 \\ 1 \\ 1 \\ 1 \end{bmatrix}$$

$$= \begin{bmatrix} 0 & 1 & 2 & 2 & 3 \\ 1 & 0 & 2 & 2 & 2 \\ 1 & 1 & 0 & 2 & 2 \\ 0 & 0 & 1 & 0 & 1 \\ 1 & 0 & 1 & 1 & 0 \end{bmatrix} \begin{bmatrix} 1 \\ 1 \\ 1 \\ 1 \\ 1 \end{bmatrix} = \begin{bmatrix} 8 \\ 7 \\ 6 \\ 2 \\ 3 \end{bmatrix}$$

Thus, we would rank the players as follows: Djokovic, Federer, Nadal, Safin, Roddick. Unfortunately, this approach is not guaranteed to break all ties.

Exercises 3.7

Markov Chains

In Exercises 1–4, let $P = \begin{bmatrix} 0.5 & 0.3 \\ 0.5 & 0.7 \end{bmatrix}$ *be the transition matrix for a Markov chain with two states. Let* $\mathbf{x}_0 = \begin{bmatrix} 0.5 \\ 0.5 \end{bmatrix}$ *be the initial state vector for the population.*

1. Compute $\mathbf{x}_1$ and $\mathbf{x}_2$.

2. What proportion of the state 1 population will be in state 2 after two steps?

3. What proportion of the state 2 population will be in state 2 after two steps?

4. Find the steady state vector.

In Exercises 5–8, let $P = \begin{bmatrix} \frac{1}{2} & \frac{1}{3} & \frac{1}{3} \\ 0 & \frac{1}{3} & \frac{2}{3} \\ \frac{1}{2} & \frac{1}{3} & 0 \end{bmatrix}$ *be the transition matrix for a Markov chain with three states. Let* $\mathbf{x}_0 = \begin{bmatrix} 120 \\ 180 \\ 90 \end{bmatrix}$ *be the initial state vector for the population.*

5. Compute $\mathbf{x}_1$ and $\mathbf{x}_2$.

6. What proportion of the state 1 population will be in state 1 after two steps?

7. What proportion of the state 2 population will be in state 3 after two steps?

8. Find the steady state vector.

9. Suppose that the weather in a particular region behaves according to a Markov chain. Specifically, suppose that the probability that tomorrow will be a wet day is 0.662 if today is wet and 0.250 if today is dry. The probability that tomorrow will be a dry day is 0.750 if today is dry and 0.338 if today is wet. [This exercise is based on an actual study of rainfall in Tel Aviv over a 27-year period. See K. R. Gabriel and J. Neumann, "A Markov Chain Model for Daily Rainfall Occurrence at Tel Aviv," *Quarterly Journal of the Royal Meteorological Society, 88* (1962), pp. 90–95.]

(a) Write down the transition matrix for this Markov chain.

(b) If Monday is a dry day, what is the probability that Wednesday will be wet?

(c) In the long run, what will the distribution of wet and dry days be?

10. Data have been accumulated on the heights of children relative to their parents. Suppose that the probabilities that a tall parent will have a tall, medium-height, or short child are 0.6, 0.2, and 0.2, respectively; the probabilities that a medium-height parent will have a tall, medium-height, or short child are 0.1, 0.7, and 0.2, respectively; and the probabilities that a short parent will have a tall, medium-height, or short child are 0.2, 0.4, and 0.4, respectively.

(a) Write down the transition matrix for this Markov chain.

(b) What is the probability that a short person will have a tall grandchild?

(c) If 20% of the current population is tall, 50% is of medium height, and 30% is short, what will the distribution be in three generations?

(d) What proportion of the population will be tall, of medium height, and short in the long run?

11. A study of piñon (pine) nut crops in the American southwest from 1940 to 1947 hypothesized that nut production followed a Markov chain. [See D. H. Thomas, "A Computer Simulation Model of Great Basin Shoshonean Subsistence and Settlement Patterns," in D. L. Clarke, ed., *Models in Archaeology* (London: Methuen, 1972).] The data suggested that if one year's crop was good, then the probabilities that the following year's crop would be good, fair, or poor were 0.08, 0.07, and 0.85, respectively; if one year's crop was fair, then the probabilities that the following year's crop would be good, fair, or poor were 0.09, 0.11, and 0.80, respectively; if one year's crop was poor, then the probabilities that the following year's crop would be good, fair, or poor were 0.11, 0.05, and 0.84, respectively.

(a) Write down the transition matrix for this Markov chain.

(b) If the piñon nut crop was good in 1940, find the probabilities of a good crop in the years 1941 through 1945.

(c) In the long run, what proportion of the crops will be good, fair, and poor?

12. Robots have been programmed to traverse the maze shown in Figure 3.28 and at each junction randomly choose which way to go.

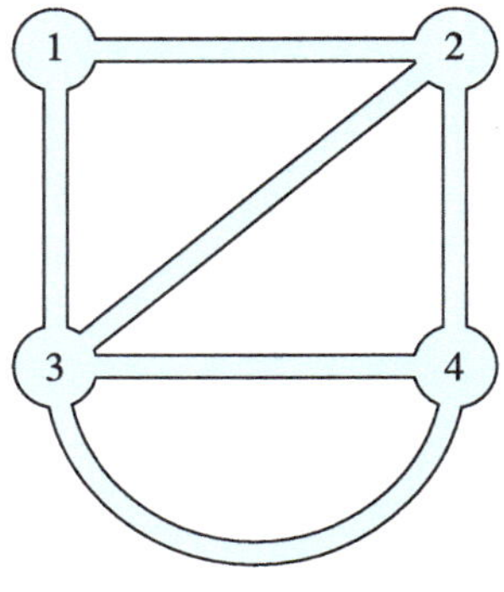

Figure 3.28

(a) Construct the transition matrix for the Markov chain that models this situation.

(b) Suppose we start with 15 robots at each junction. Find the steady state distribution of robots. (Assume that it takes each robot the same amount of time to travel between two adjacent junctions.)

13. Let $\mathbf{j}$ denote a row vector consisting entirely of 1s. Prove that a nonnegative matrix P is a stochastic matrix if and only if $\mathbf{j}P = \mathbf{j}$.

14. (a) Show that the product of two 2×2 stochastic matrices is also a stochastic matrix.
(b) Prove that the product of two $n \times n$ stochastic matrices is also a stochastic matrix.
(c) If a 2×2 stochastic matrix P is invertible, prove that P^{-1} is also a stochastic matrix.

Suppose we want to know the average (or expected) number of steps it will take to go from state i to state j in a Markov chain. It can be shown that the following computation answers this question: Delete the jth row and the jth column of the transition matrix P to get a new matrix Q. (Keep the rows and columns of Q labeled as they were in P.) The expected number of steps from state i to state j is given by the sum of the entries in the column of $(I - Q)^{-1}$ labeled i.

15. In Exercise 9, if Monday is a dry day, what is the expected number of days until a wet day?

16. In Exercise 10, what is the expected number of generations until a short person has a tall descendant?

17. In Exercise 11, if the piñon nut crop is fair one year, what is the expected number of years until a good crop occurs?

18. In Exercise 12, starting from each of the other junctions, what is the expected number of moves until a robot reaches junction 4?

Linear Economic Models

In Exercises 19–26, determine which of the matrices are exchange matrices. For those that are exchange matrices, find a nonnegative price vector that satisfies Equation (1).

19. $\begin{bmatrix} 1/2 & 1/4 \\ 1/2 & 3/4 \end{bmatrix}$

20. $\begin{bmatrix} 1/3 & 2/3 \\ 1/2 & 1/2 \end{bmatrix}$

21. $\begin{bmatrix} 0.4 & 0.7 \\ 0.6 & 0.4 \end{bmatrix}$

22. $\begin{bmatrix} 0.1 & 0.6 \\ 0.9 & 0.4 \end{bmatrix}$

23. $\begin{bmatrix} 1/3 & 0 & 0 \\ 1/3 & 3/2 & 0 \\ 1/3 & -1/2 & 1 \end{bmatrix}$

24. $\begin{bmatrix} 1/2 & 1 & 0 \\ 0 & 0 & 1/3 \\ 1/2 & 0 & 2/3 \end{bmatrix}$

25. $\begin{bmatrix} 0.3 & 0 & 0.2 \\ 0.3 & 0.5 & 0.3 \\ 0.4 & 0.5 & 0.5 \end{bmatrix}$

26. $\begin{bmatrix} 0.50 & 0.70 & 0.35 \\ 0.25 & 0.30 & 0.25 \\ 0.25 & 0 & 0.40 \end{bmatrix}$

In Exercises 27–30, determine whether the given consumption matrix is productive.

27. $\begin{bmatrix} 0.2 & 0.3 \\ 0.5 & 0.6 \end{bmatrix}$

28. $\begin{bmatrix} 0.20 & 0.10 & 0.10 \\ 0.30 & 0.15 & 0.45 \\ 0.15 & 0.30 & 0.50 \end{bmatrix}$

29. $\begin{bmatrix} 0.35 & 0.25 & 0 \\ 0.15 & 0.55 & 0.35 \\ 0.45 & 0.30 & 0.60 \end{bmatrix}$

30. $\begin{bmatrix} 0.2 & 0.4 & 0.1 & 0.4 \\ 0.3 & 0.2 & 0.2 & 0.1 \\ 0 & 0.4 & 0.5 & 0.3 \\ 0.5 & 0 & 0.2 & 0.2 \end{bmatrix}$

*In Exercises 31–34, a consumption matrix C and a demand vector **d** are given. In each case, find a feasible production vector **x** that satisfies Equation (2).*

31. $C = \begin{bmatrix} 1/2 & 1/4 \\ 1/2 & 1/2 \end{bmatrix}, \mathbf{d} = \begin{bmatrix} 1 \\ 3 \end{bmatrix}$

32. $C = \begin{bmatrix} 0.1 & 0.4 \\ 0.3 & 0.2 \end{bmatrix}, \mathbf{d} = \begin{bmatrix} 2 \\ 1 \end{bmatrix}$

33. $C = \begin{bmatrix} 0.5 & 0.2 & 0.1 \\ 0 & 0.4 & 0.2 \\ 0 & 0 & 0.5 \end{bmatrix}, \mathbf{d} = \begin{bmatrix} 3 \\ 2 \\ 4 \end{bmatrix}$

CAS 34. $C = \begin{bmatrix} 0.1 & 0.4 & 0.1 \\ 0 & 0.2 & 0.2 \\ 0.3 & 0.2 & 0.3 \end{bmatrix}, \mathbf{d} = \begin{bmatrix} 1.1 \\ 3.5 \\ 2.0 \end{bmatrix}$

35. Let A be an $n \times n$ matrix, $A \geq O$. Suppose that $A\mathbf{x} < \mathbf{x}$ for some $\mathbf{x}$ in $\mathbb{R}^n$, $\mathbf{x} \geq \mathbf{0}$. Prove that $\mathbf{x} > \mathbf{0}$.

36. Let A, B, C, and D be $n \times n$ matrices and $\mathbf{x}$ and $\mathbf{y}$ vectors in $\mathbb{R}^n$. Prove the following inequalities:
(a) If $A \geq B \geq O$ and $C \geq D \geq O$, then $AC \geq BD \geq O$.
(b) If $A > B$ and $\mathbf{x} \geq \mathbf{0}$, $\mathbf{x} \neq \mathbf{0}$, then $A\mathbf{x} > B\mathbf{x}$.

Population Growth

37. A population with two age classes has a Leslie matrix $L = \begin{bmatrix} 2 & 5 \\ 0.6 & 0 \end{bmatrix}$. If the initial population vector is $\mathbf{x}_0 = \begin{bmatrix} 10 \\ 5 \end{bmatrix}$, compute $\mathbf{x}_1$, $\mathbf{x}_2$, and $\mathbf{x}_3$.

38. A population with three age classes has a Leslie matrix $L = \begin{bmatrix} 0 & 1 & 2 \\ 0.2 & 0 & 0 \\ 0 & 0.5 & 0 \end{bmatrix}$. If the initial population vector is $\mathbf{x}_0 = \begin{bmatrix} 10 \\ 4 \\ 3 \end{bmatrix}$, compute $\mathbf{x}_1$, $\mathbf{x}_2$, and $\mathbf{x}_3$.

39. A population with three age classes has a Leslie matrix $L = \begin{bmatrix} 1 & 1 & 3 \\ 0.7 & 0 & 0 \\ 0 & 0.5 & 0 \end{bmatrix}$. If the initial population vector is $\mathbf{x}_0 = \begin{bmatrix} 100 \\ 100 \\ 100 \end{bmatrix}$, compute $\mathbf{x}_1$, $\mathbf{x}_2$, and $\mathbf{x}_3$.

40. A population with four age classes has a Leslie matrix $L = \begin{bmatrix} 0 & 1 & 2 & 5 \\ 0.5 & 0 & 0 & 0 \\ 0 & 0.7 & 0 & 0 \\ 0 & 0 & 0.3 & 0 \end{bmatrix}$. If the initial population vector is $\mathbf{x}_0 = \begin{bmatrix} 10 \\ 10 \\ 10 \\ 10 \end{bmatrix}$, compute $\mathbf{x}_1$, $\mathbf{x}_2$, and $\mathbf{x}_3$.

41. A certain species with two age classes of 1 year's duration has a survival probability of 80% from class 1 to class 2. Empirical evidence shows that, on average, each female gives birth to five females per year. Thus, two possible Leslie matrices are

$$L_1 = \begin{bmatrix} 0 & 5 \\ 0.8 & 0 \end{bmatrix} \quad \text{and} \quad L_2 = \begin{bmatrix} 4 & 1 \\ 0.8 & 0 \end{bmatrix}$$

(a) Starting with $\mathbf{x}_0 = \begin{bmatrix} 10 \\ 10 \end{bmatrix}$, compute $\mathbf{x}_1, \ldots, \mathbf{x}_{10}$ in each case.

(b) For each case, plot the relative size of each age class over time (as in Figure 3.23). What do your graphs suggest?

42. Suppose the Leslie matrix for the VW beetle is $L = \begin{bmatrix} 0 & 0 & 20 \\ 0.1 & 0 & 0 \\ 0 & 0.5 & 0 \end{bmatrix}$. Starting with an arbitrary $\mathbf{x}_0$, determine the behavior of this population.

43. Suppose the Leslie matrix for the VW beetle is $L = \begin{bmatrix} 0 & 0 & 20 \\ s & 0 & 0 \\ 0 & 0.5 & 0 \end{bmatrix}$. Investigate the effect of varying the survival probability s of the young beetles.

CAS **44.** Woodland caribou are found primarily in the western provinces of Canada and the American northwest. The average lifespan of a female is about 14 years. The birth and survival rates for each age bracket are given in Table 3.4, which shows that caribou cows do not give birth at all during their first 2 years and give birth to about one calf per year during their middle years. The mortality rate for young calves is very high.

Table 3.4

Age (years)	Birth Rate	Survival Rate
0–2	0.0	0.3
2–4	0.4	0.7
4–6	1.8	0.9
6–8	1.8	0.9
8–10	1.8	0.9
10–12	1.6	0.6
12–14	0.6	0.0

The numbers of woodland caribou reported in Jasper National Park in Alberta in 1990 are shown in Table 3.5. Using a CAS, predict the caribou population for 1992 and 1994. Then project the population for the years 2010 and 2020. What do you conclude? (What assumptions does this model make, and how could it be improved?)

Table 3.5 Woodland Caribou Population in Jasper National Park, 1990

Age (years)	Number
0–2	10
2–4	2
4–6	8
6–8	5
8–10	12
10–12	0
12–14	1

Source: World Wildlife Fund Canada

Graphs and Digraphs

In Exercises 45–48, determine the adjacency matrix of the given graph.

45

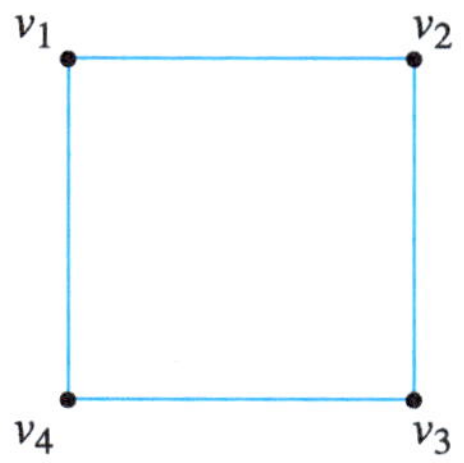

46.

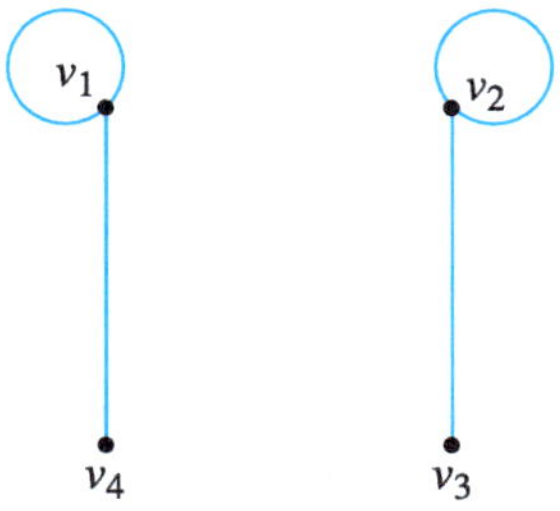

47.

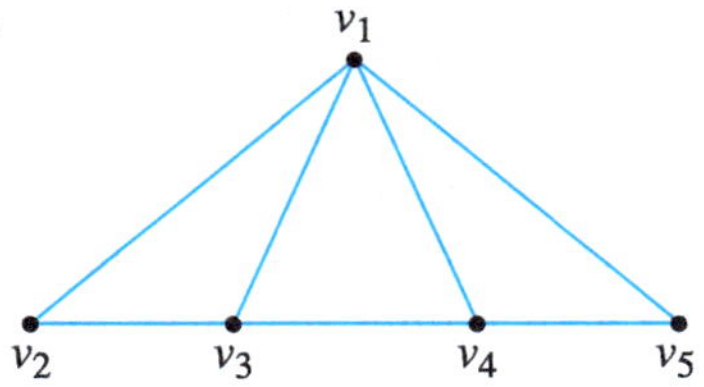

48.

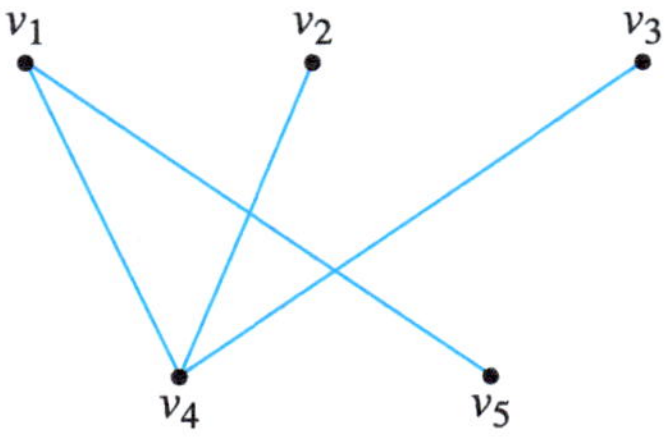

In Exercises 49–52, draw a graph that has the given adjacency matrix.

49. $\begin{bmatrix} 0 & 1 & 1 & 1 \\ 1 & 0 & 0 & 0 \\ 1 & 0 & 0 & 0 \\ 1 & 0 & 0 & 0 \end{bmatrix}$

50. $\begin{bmatrix} 0 & 1 & 0 & 1 \\ 1 & 1 & 1 & 1 \\ 0 & 1 & 0 & 1 \\ 1 & 1 & 1 & 0 \end{bmatrix}$

51. $\begin{bmatrix} 0 & 0 & 1 & 1 & 0 \\ 0 & 0 & 0 & 1 & 1 \\ 1 & 0 & 0 & 0 & 1 \\ 1 & 1 & 0 & 0 & 0 \\ 0 & 1 & 1 & 0 & 0 \end{bmatrix}$

52. $\begin{bmatrix} 0 & 0 & 0 & 1 & 1 \\ 0 & 0 & 0 & 1 & 1 \\ 0 & 0 & 0 & 1 & 1 \\ 1 & 1 & 1 & 0 & 0 \\ 1 & 1 & 1 & 0 & 0 \end{bmatrix}$

In Exercises 53–56, determine the adjacency matrix of the given digraph.

53.

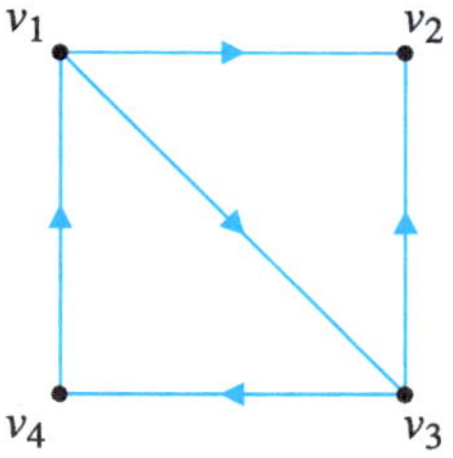

54.

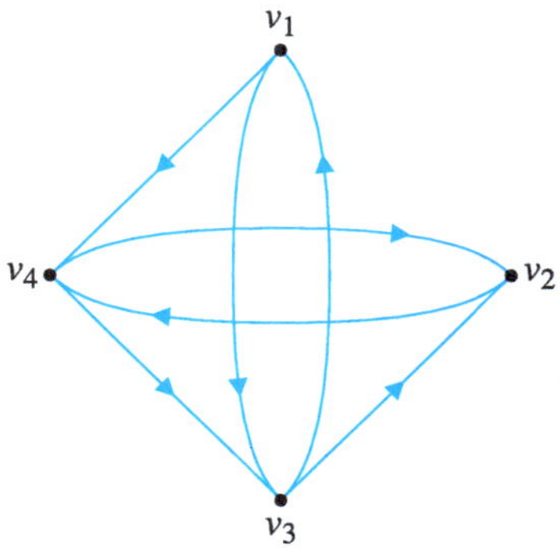

55.

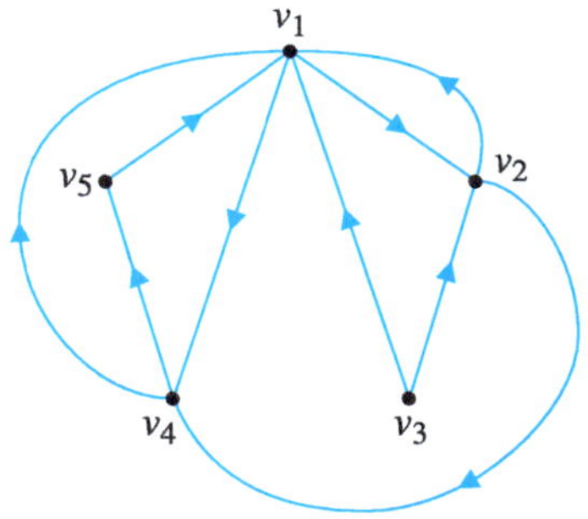

56.

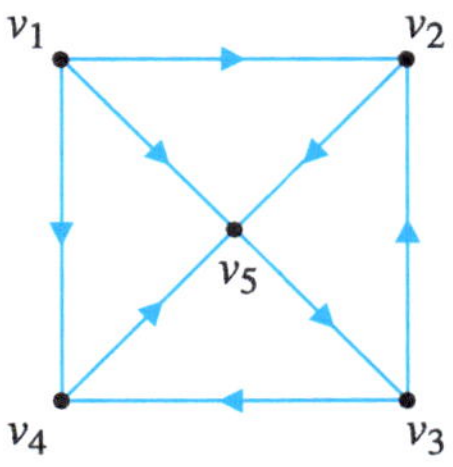

In Exercises 57–60, draw a digraph that has the given adjacency matrix.

57. $\begin{bmatrix} 0 & 1 & 0 & 0 \\ 1 & 0 & 0 & 1 \\ 0 & 1 & 0 & 0 \\ 1 & 0 & 1 & 1 \end{bmatrix}$

58. $\begin{bmatrix} 0 & 1 & 0 & 0 \\ 0 & 0 & 0 & 1 \\ 1 & 0 & 0 & 0 \\ 0 & 0 & 1 & 0 \end{bmatrix}$

59. $\begin{bmatrix} 0 & 0 & 1 & 0 & 1 \\ 1 & 0 & 0 & 1 & 0 \\ 0 & 0 & 0 & 0 & 1 \\ 1 & 0 & 1 & 0 & 0 \\ 0 & 1 & 0 & 1 & 0 \end{bmatrix}$ **60.** $\begin{bmatrix} 0 & 1 & 0 & 0 & 1 \\ 0 & 0 & 0 & 1 & 0 \\ 1 & 0 & 0 & 1 & 1 \\ 1 & 0 & 1 & 0 & 0 \\ 1 & 1 & 0 & 0 & 0 \end{bmatrix}$

In Exercises 61–68, use powers of adjacency matrices to determine the number of paths of the specified length between the given vertices.

61. Exercise 50, length 2, v_1 and v_2

62. Exercise 52, length 2, v_1 and v_2

63. Exercise 50, length 3, v_1 and v_3

64. Exercise 52, length 4, v_2 and v_2

65. Exercise 57, length 2, v_1 to v_3

66. Exercise 57, length 3, v_4 to v_1

67. Exercise 60, length 3, v_4 to v_1

68. Exercise 60, length 4, v_1 to v_4

69. Let A be the adjacency matrix of a graph G.

(a) If row i of A is all zeros, what does this imply about G?

(b) If column j of A is all zeros, what does this imply about G?

70. Let A be the adjacency matrix of a digraph D.

(a) If row i of A^2 is all zeros, what does this imply about D?

(b) If column j of A^2 is all zeros, what does this imply about D?

71. Figure 3.29 is the digraph of a tournament with six players, P_1 to P_6. Using adjacency matrices, rank the players first by determining wins only and then by using the notion of combined wins and indirect wins, as in Example 3.69.

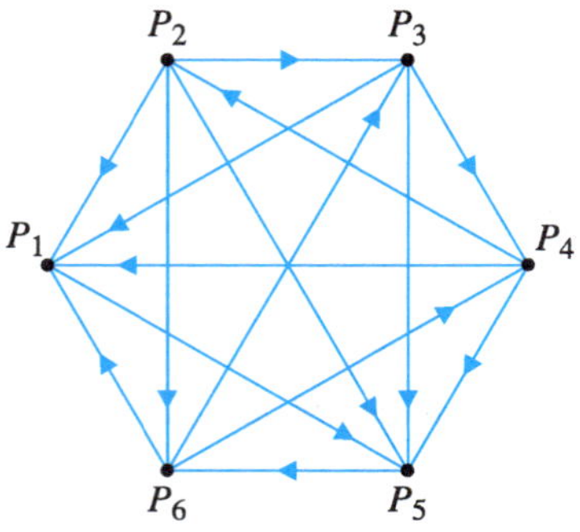

Figure 3.29

72. Figure 3.30 is a digraph representing a food web in a small ecosystem. A directed edge from a to b indicates that a has b as a source of food. Construct the adjacency matrix A for this digraph and use it to answer the following questions.

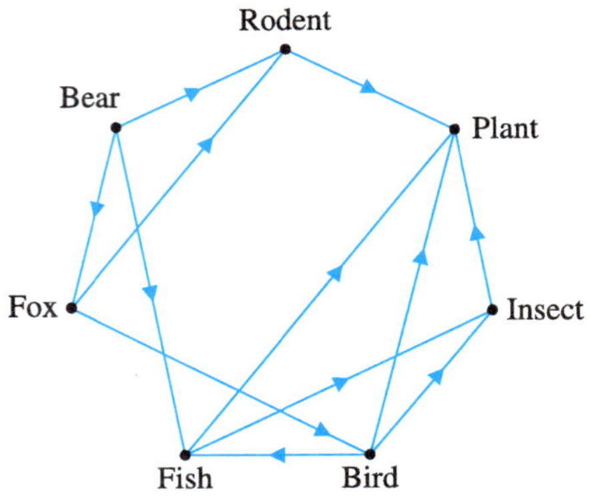

Figure 3.30

(a) Which species has the most direct sources of food? How does A show this?

(b) Which species is a direct source of food for the most other species? How does A show this?

(c) If a eats b and b eats c, we say that a has c as an indirect source of food. How can we use A to determine which species has the most indirect food sources? Which species has the most direct and indirect food sources combined?

(d) Suppose that pollutants kill the plants in this food web, and we want to determine the effect this change will have on the ecosystem. Construct a new adjacency matrix A^* from A by deleting the row and column corresponding to plants. Repeat parts (a) to (c) and determine which species are the most and least affected by the change.

(e) What will the long-term effect of the pollution be? What matrix calculations will show this?

73. Five people are all connected by e-mail. Whenever one of them hears a juicy piece of gossip, he or she passes it along by e-mailing it to someone else in the group according to Table 3.6.

(a) Draw the digraph that models this "gossip network" and find its adjacency matrix A.

Table 3.6

Sender	Recipients
Ann	Carla, Ehaz
Bert	Carla, Dana
Carla	Ehaz
Dana	Ann, Carla
Ehaz	Bert

(b) Define a *step* as the time it takes a person to e-mail everyone on his or her list. (Thus, in one step, gossip gets from Ann to both Carla and Ehaz.) If Bert hears a rumor, how many steps will it take for everyone else to hear the rumor? What matrix calculation reveals this?

(c) If Ann hears a rumor, how many steps will it take for everyone else to hear the rumor? What matrix calculation reveals this?

(d) In general, if A is the adjacency matrix of a digraph, how can we tell if vertex i is connected to vertex j by a path (of some length)?

[The gossip network in this exercise is reminiscent of the notion of "six degrees of separation" (found in the play and film by that name), which suggests that any two people are connected by a path of acquaintances whose length is at most 6. The game "Six Degrees of Kevin Bacon" more frivolously asserts that all actors are connected to the actor Kevin Bacon in such a way.]

74. Let A be the adjacency matrix of a graph G.

(a) By induction, prove that for all $n \geq 1$, the (i, j) entry of A^n is equal to the number of n-paths between vertices i and j.

(b) How do the statement and proof in part (a) have to be modified if G is a digraph?

75. If A is the adjacency matrix of a digraph G, what does the (i, j) entry of AA^T represent if $i \neq j$?

A graph is called ***bipartite*** *if its vertices can be subdivided into two sets U and V such that every edge has one endpoint in U and the other endpoint in V. For example, the graph in Exercise 48 is bipartite with* $U = \{v_1, v_2, v_3\}$ *and* $V = \{v_4, v_5\}$. *In Exercises 76–79, determine whether a graph with the given adjacency matrix is bipartite.*

76. The adjacency matrix in Exercise 49

77. The adjacency matrix in Exercise 52

78. The adjacency matrix in Exercise 51

79. $\begin{bmatrix} 0 & 0 & 1 & 0 & 1 & 1 \\ 0 & 0 & 1 & 0 & 1 & 1 \\ 1 & 1 & 0 & 1 & 0 & 0 \\ 0 & 0 & 1 & 0 & 1 & 1 \\ 1 & 1 & 0 & 1 & 0 & 0 \\ 1 & 1 & 0 & 1 & 0 & 0 \end{bmatrix}$

80. (a) Prove that a graph is bipartite if and only if its vertices can be labeled so that its adjacency matrix can be partitioned as

$$A = \left[\begin{array}{c|c} O & B \\ \hline B^T & O \end{array}\right]$$

(b) Using the result in part (a), prove that a bipartite graph has no circuits of odd length.

Chapter Review

Key Definitions and Concepts

Review Questions

1. Mark each of the following statements true or false:

(a) For any matrix A, both AA^T and A^TA are defined.

(b) If A and B are matrices such that $AB = O$ and $A \neq O$, then $B = O$.

(c) If A, B, and X are invertible matrices such that $XA = B$, then $X = A^{-1}B$.

(d) The inverse of an elementary matrix is an elementary matrix.

(e) The transpose of an elementary matrix is an elementary matrix.

(f) The product of two elementary matrices is an elementary matrix.

(g) If A is an $m \times n$ matrix, then the null space of A is a subspace of $\mathbb{R}^n$.

(h) Every plane in $\mathbb{R}^3$ is a two-dimensional subspace of $\mathbb{R}^3$.

(i) The transformation $T: \mathbb{R}^2 \to \mathbb{R}^2$ defined by $T(\mathbf{x}) = -\mathbf{x}$ is a linear transformation.

(j) If $T: \mathbb{R}^4 \to \mathbb{R}^5$ is a linear transformation, then there is a 4×5 matrix A such that $T(\mathbf{x}) = A\mathbf{x}$ for all $\mathbf{x}$ in the domain of T.

In Exercises 2–7, let $A = \begin{bmatrix} 1 & 2 \\ 3 & 5 \end{bmatrix}$ *and* $B = \begin{bmatrix} 2 & 0 & -1 \\ 3 & -3 & 4 \end{bmatrix}$. *Compute the indicated matrices, if possible.*

2. A^2B **3.** A^2B^2 **4.** $B^TA^{-1}B$

5. $(BB^T)^{-1}$ **6.** $(B^TB)^{-1}$

7. The outer product expansion of AA^T

8. If A is a matrix such that $A^{-1} = \begin{bmatrix} 1/2 & -1 \\ -3/2 & 4 \end{bmatrix}$, find A.

9. If $A = \begin{bmatrix} 1 & 0 & -1 \\ 2 & 3 & -1 \\ 0 & 1 & 1 \end{bmatrix}$ and X is a matrix such that

$$AX = \begin{bmatrix} -1 & -3 \\ 5 & 0 \\ 3 & -2 \end{bmatrix},$$ find X.

10. If possible, express the matrix $A = \begin{bmatrix} 1 & 2 \\ 4 & 6 \end{bmatrix}$ as a product of elementary matrices.

11. If A is a square matrix such that $A^3 = O$, show that $(I - A)^{-1} = I + A + A^2$.

12. Find an LU factorization of $A = \begin{bmatrix} 1 & 1 & 1 \\ 3 & 1 & 1 \\ 2 & -1 & 1 \end{bmatrix}$.

13. Find bases for the row space, column space, and null space of $A = \begin{bmatrix} 2 & -4 & 5 & 8 & 5 \\ 1 & -2 & 2 & 3 & 1 \\ 4 & -8 & 3 & 2 & 6 \end{bmatrix}$.

14. Suppose matrices A and B are row equivalent. Do they have the same row space? Why or why not? Do A and B have the same column space? Why or why not?

15. If A is an invertible matrix, explain why A and A^T must have the same null space. Is this true if A is a noninvertible square matrix? Explain.

16. If A is a square matrix whose rows add up to the zero vector, explain why A cannot be invertible.

17. Let A be an $m \times n$ matrix with linearly independent columns. Explain why A^TA must be an invertible matrix. Must AA^T also be invertible? Explain.

18. Find a linear transformation $T: \mathbb{R}^2 \to \mathbb{R}^2$ such that

$$T\begin{bmatrix} 1 \\ 1 \end{bmatrix} = \begin{bmatrix} 2 \\ 3 \end{bmatrix} \text{ and } T\begin{bmatrix} 1 \\ -1 \end{bmatrix} = \begin{bmatrix} 0 \\ 5 \end{bmatrix}.$$

19. Find the standard matrix of the linear transformation $T: \mathbb{R}^2 \to \mathbb{R}^2$ that corresponds to a counterclockwise rotation of 45° about the origin followed by a projection onto the line $y = -2x$.

20. Suppose that $T: \mathbb{R}^n \to \mathbb{R}^n$ is a linear transformation and suppose that $\mathbf{v}$ is a vector such that $T(\mathbf{v}) \neq \mathbf{0}$ but $T^2(\mathbf{v}) = \mathbf{0}$ (where $T^2 = T \circ T$). Prove that $\mathbf{v}$ and $T(\mathbf{v})$ are linearly independent.

4 Eigenvalues and Eigenvectors

Almost every combination of the adjectives proper, latent, characteristic, eigen *and* secular, *with the nouns* root, number *and* value, *has been used in the literature for what we call a proper value.*

—Paul R. Halmos
Finite Dimensional Vector Spaces (2nd edition)
Van Nostrand, 1958, p. 102

4.0 Introduction: A Dynamical System on Graphs

CAS

We saw in the last chapter that iterating matrix multiplication often produces interesting results. Both Markov chains and the Leslie model of population growth exhibit steady states in certain situations. One of the goals of this chapter is to help you understand such behavior. First we will look at another iterative process, or ***dynamical system,*** that uses matrices. (In the problems that follow, you will find it helpful to use a CAS or a calculator with matrix capabilities to facilitate the computations.)

Our example involves graphs (see Section 3.7). A ***complete graph*** is any graph in which every vertex is adjacent to every other vertex. If a complete graph has n vertices, it is denoted by K_n. For example, Figure 4.1 shows a representation of K_4.

Problem 1 Pick any vector $\mathbf{x}$ in $\mathbb{R}^4$ with nonnegative entries and label the vertices of K_4 with the components of $\mathbf{x}$, so that v_1 is labeled with x_1, and so on. Compute the adjacency matrix A of K_4 and relabel the vertices of the graph with the corresponding components of $A\mathbf{x}$. Try this for several vectors $\mathbf{x}$ and explain, in terms of the graph, how the new labels can be determined from the old labels.

Problem 2 Now *iterate* the process in Problem 1. That is, for a given choice of $\mathbf{x}$, relabel the vertices as described above and then apply A again (and again, and again) until a pattern emerges. Since components of the vectors themselves will get quite large, we will *scale* them by dividing each vector by its *largest component* after each iteration. Thus, if a computation results in the vector

$$\begin{bmatrix} 4 \\ 2 \\ 1 \\ 1 \end{bmatrix}$$

we will replace it by

$$\frac{1}{4}\begin{bmatrix} 4 \\ 2 \\ 1 \\ 1 \end{bmatrix} = \begin{bmatrix} 1 \\ 0.5 \\ 0.25 \\ 0.25 \end{bmatrix}$$

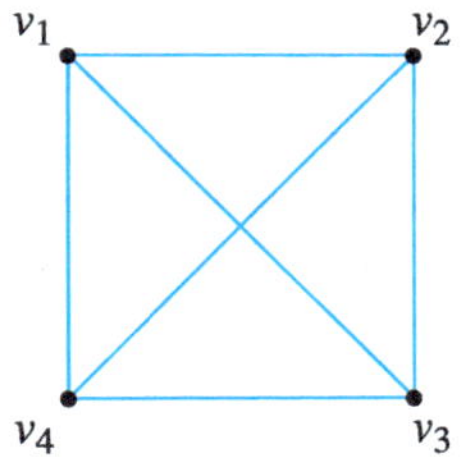

Figure 4.1
K_4

Note that this process guarantees that the largest component of each vector will now be 1. Do this for K_4, then K_3 and K_5. Use at least ten iterations and two-decimal-place accuracy. What appears to be happening?

Problem 3 You should have noticed that, in each case, the labeling vector is approaching a certain vector (a steady state label!). Label the vertices of the complete graphs with this steady state vector and apply the adjacency matrix A one more time (without scaling). What is the relationship between the new labels and the old ones?

Problem 4 Make a conjecture about the general case K_n. What is the steady state label? What happens if we label K_n with the steady state vector and apply the adjacency matrix A without scaling?

Problem 5 The ***Petersen graph*** is shown in Figure 4.2. Repeat the process in Problems 1 through 3 with this graph.

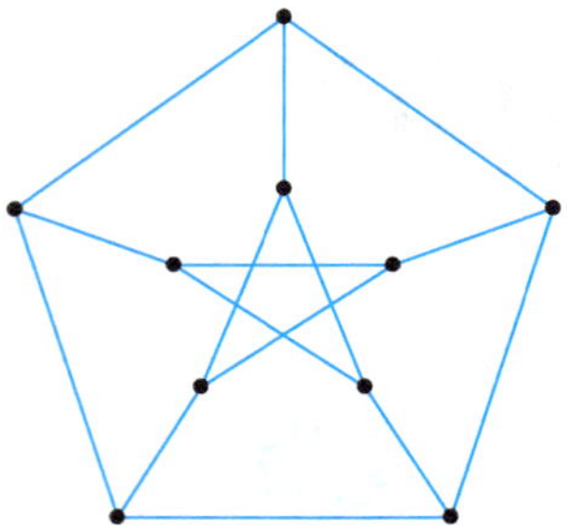

Figure 4.2

We will now explore the process with some other classes of graphs to see if they behave the same way. The ***cycle*** C_n is the graph with n vertices arranged in a cyclic fashion. For example, C_5 is the graph shown in Figure 4.3.

Problem 6 Repeat the process of Problems 1 through 3 with cycles C_n for various *odd* values of n and make a conjecture about the general case.

Problem 7 Repeat Problem 6 with *even* values of n. What happens?

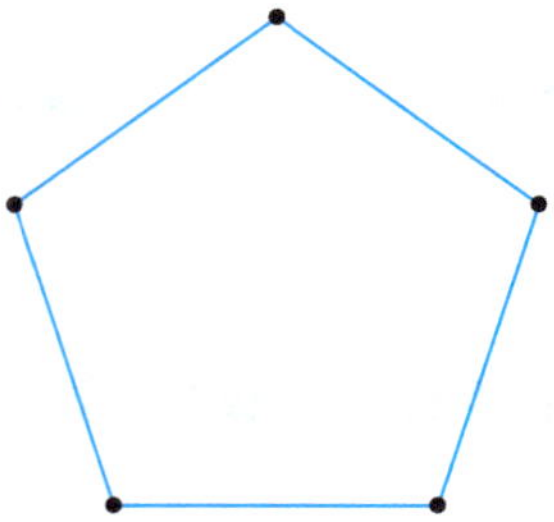

Figure 4.3

A bipartite graph is a ***complete bipartite graph*** (see Exercises 74–78 in Section 3.7) if its vertices can be partitioned into sets U and V such that every vertex in U is adjacent to every vertex in V, and vice versa. If U and V each have n vertices, then the graph is denoted by $K_{n,n}$. For example, $K_{3,3}$ is the graph in Figure 4.4.

Problem 8 Repeat the process of Problems 1 through 3 with complete bipartite graphs $K_{n,n}$ for various values of n. What happens?

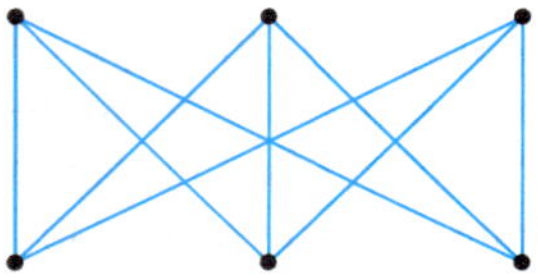

Figure 4.4

By the end of this chapter, you will be in a position to explain the observations you have made in this Introduction.

4.1 Introduction to Eigenvalues and Eigenvectors

In Chapter 3, we encountered the notion of a steady state vector in the context of two applications: Markov chains and the Leslie model of population growth. For a Markov chain with transition matrix P, a steady state vector $\mathbf{x}$ had the property that $P\mathbf{x} = \mathbf{x}$; for a Leslie matrix L, a steady state vector was a population vector $\mathbf{x}$ satisfying $L\mathbf{x} = r\mathbf{x}$, where r represented the steady state growth rate. For example, we saw that

$$\begin{bmatrix} 0.7 & 0.2 \\ 0.3 & 0.8 \end{bmatrix}\begin{bmatrix} 0.4 \\ 0.6 \end{bmatrix} = \begin{bmatrix} 0.4 \\ 0.6 \end{bmatrix} \quad \text{and} \quad \begin{bmatrix} 0 & 4 & 3 \\ 0.5 & 0 & 0 \\ 0 & 0.25 & 0 \end{bmatrix}\begin{bmatrix} 18 \\ 6 \\ 1 \end{bmatrix} = 1.5\begin{bmatrix} 18 \\ 6 \\ 1 \end{bmatrix}$$

The German adjective *eigen* means "own" or "characteristic of." *Eigenvalues* and *eigenvectors* are characteristic of a matrix in the sense that they contain important information about the nature of the matrix. The letter λ (lambda), the Greek equivalent of the English letter L, is used for eigenvalues because at one time they were also known as *latent values*. The prefix *eigen* is pronounced "EYE-gun."

In this chapter, we investigate this phenomenon more generally. That is, for a square matrix A, we ask whether there exist nonzero vectors $\mathbf{x}$ such that $A\mathbf{x}$ is just a scalar multiple of $\mathbf{x}$. This is the *eigenvalue problem*, and it is one of the most central problems in linear algebra. It has applications throughout mathematics and in many other fields as well.

Definition Let A be an $n \times n$ matrix. A scalar λ is called an ***eigenvalue*** of A if there is a nonzero vector $\mathbf{x}$ such that $A\mathbf{x} = \lambda\mathbf{x}$. Such a vector $\mathbf{x}$ is called an ***eigenvector*** of A corresponding to λ.

Example 4.1

Show that $\mathbf{x} = \begin{bmatrix} 1 \\ 1 \end{bmatrix}$ is an eigenvector of $A = \begin{bmatrix} 3 & 1 \\ 1 & 3 \end{bmatrix}$ and find the corresponding eigenvalue.

Solution We compute

$$A\mathbf{x} = \begin{bmatrix} 3 & 1 \\ 1 & 3 \end{bmatrix}\begin{bmatrix} 1 \\ 1 \end{bmatrix} = \begin{bmatrix} 4 \\ 4 \end{bmatrix} = 4\begin{bmatrix} 1 \\ 1 \end{bmatrix} = 4\mathbf{x}$$

from which it follows that $\mathbf{x}$ is an eigenvector of A corresponding to the eigenvalue 4.

Example 4.2

Show that 5 is an eigenvalue of $A = \begin{bmatrix} 1 & 2 \\ 4 & 3 \end{bmatrix}$ and determine all eigenvectors corresponding to this eigenvalue.

Solution We must show that there is a *nonzero* vector $\mathbf{x}$ such that $A\mathbf{x} = 5\mathbf{x}$. But this equation is equivalent to the equation $(A - 5I)\mathbf{x} = \mathbf{0}$, so we need to compute the null space of the matrix $A - 5I$. We find that

$$A - 5I = \begin{bmatrix} 1 & 2 \\ 4 & 3 \end{bmatrix} - \begin{bmatrix} 5 & 0 \\ 0 & 5 \end{bmatrix} = \begin{bmatrix} -4 & 2 \\ 4 & -2 \end{bmatrix}$$

Since the columns of this matrix are clearly linearly dependent, the Fundamental Theorem of Invertible Matrices implies that its null space is nonzero. Thus, $A\mathbf{x} = 5\mathbf{x}$ has a nontrivial solution, so 5 is an eigenvalue of A. We find its eigenvectors by computing the null space:

$$[A - 5I \mid 0] = \left[\begin{array}{cc|c} -4 & 2 & 0 \\ 4 & -2 & 0 \end{array}\right] \longrightarrow \left[\begin{array}{cc|c} 1 & -\frac{1}{2} & 0 \\ 0 & 0 & 0 \end{array}\right]$$

Thus, if $\mathbf{x} = \begin{bmatrix} x_1 \\ x_2 \end{bmatrix}$ is an eigenvector corresponding to the eigenvalue 5, it satisfies $x_1 - \frac{1}{2}x_2 = 0$, or $x_1 = \frac{1}{2}x_2$, so these eigenvectors are of the form

$$\mathbf{x} = \begin{bmatrix} \frac{1}{2}x_2 \\ x_2 \end{bmatrix}$$

That is, they are the nonzero multiples of $\begin{bmatrix} \frac{1}{2} \\ 1 \end{bmatrix}$ (or, equivalently, the nonzero multiples of $\begin{bmatrix} 1 \\ 2 \end{bmatrix}$).

The set of all eigenvectors corresponding to an eigenvalue λ of an $n \times n$ matrix A is just the set of *nonzero* vectors in the null space of $A - \lambda I$. It follows that this set of eigenvectors, together with the zero vector in $\mathbb{R}^n$, *is* the null space of $A - \lambda I$.

Definition Let A be an $n \times n$ matrix and let λ be an eigenvalue of A. The collection of all eigenvectors corresponding to λ, together with the zero vector, is called the ***eigenspace*** of λ and is denoted by E_λ.

Therefore, in Example 4.2, $E_5 = \left\{ t\begin{bmatrix} 1 \\ 2 \end{bmatrix} \right\}$.

Example 4.3

Show that $\lambda = 6$ is an eigenvalue of $A = \begin{bmatrix} 7 & 1 & -2 \\ -3 & 3 & 6 \\ 2 & 2 & 2 \end{bmatrix}$ and find a basis for its eigenspace.

Solution As in Example 4.2, we compute the null space of $A - 6I$. Row reduction produces

$$A - 6I = \begin{bmatrix} 1 & 1 & -2 \\ -3 & -3 & 6 \\ 2 & 2 & -4 \end{bmatrix} \longrightarrow \begin{bmatrix} 1 & 1 & -2 \\ 0 & 0 & 0 \\ 0 & 0 & 0 \end{bmatrix}$$

from which we see that the null space of $A - 6I$ is nonzero. Hence, 6 is an eigenvalue of A, and the eigenvectors corresponding to this eigenvalue satisfy $x_1 + x_2 - 2x_3 = 0$, or $x_1 = -x_2 + 2x_3$. It follows that

$$E_6 = \left\{ \begin{bmatrix} -x_2 + 2x_3 \\ x_2 \\ x_3 \end{bmatrix} \right\} = \left\{ x_2\begin{bmatrix} -1 \\ 1 \\ 0 \end{bmatrix} + x_3\begin{bmatrix} 2 \\ 0 \\ 1 \end{bmatrix} \right\} = \text{span}\left(\begin{bmatrix} -1 \\ 1 \\ 0 \end{bmatrix}, \begin{bmatrix} 2 \\ 0 \\ 1 \end{bmatrix} \right)$$

In $\mathbb{R}^2$, we can give a geometric interpretation of the notion of an eigenvector. The equation $A\mathbf{x} = \lambda\mathbf{x}$ says that the vectors $A\mathbf{x}$ and $\mathbf{x}$ are parallel. Thus, $\mathbf{x}$ is an eigenvector of A if and only if A transforms $\mathbf{x}$ into a parallel vector [or, equivalently, if and only if $T_A(\mathbf{x})$ is parallel to $\mathbf{x}$, where T_A is the matrix transformation corresponding to A].

Example 4.4

Find the eigenvectors and eigenvalues of $A = \begin{bmatrix} 1 & 0 \\ 0 & -1 \end{bmatrix}$ geometrically.

Solution We recognize that A is the matrix of a reflection F in the x-axis (see Example 3.56). The only vectors that F maps parallel to themselves are vectors parallel to the y-axis (i.e., multiples of $\begin{bmatrix} 0 \\ 1 \end{bmatrix}$), which are reversed (eigenvalue -1), and vectors parallel to the x-axis (i.e., multiples of $\begin{bmatrix} 1 \\ 0 \end{bmatrix}$), which are sent to themselves (eigenvalue 1) (see Figure 4.5). Accordingly, $\lambda = -1$ and $\lambda = 1$ are the eigenvalues of A, and the corresponding eigenspaces are

$$E_{-1} = \text{span}\left(\begin{bmatrix} 0 \\ 1 \end{bmatrix} \right) \quad \text{and} \quad E_1 = \text{span}\left(\begin{bmatrix} 1 \\ 0 \end{bmatrix} \right)$$

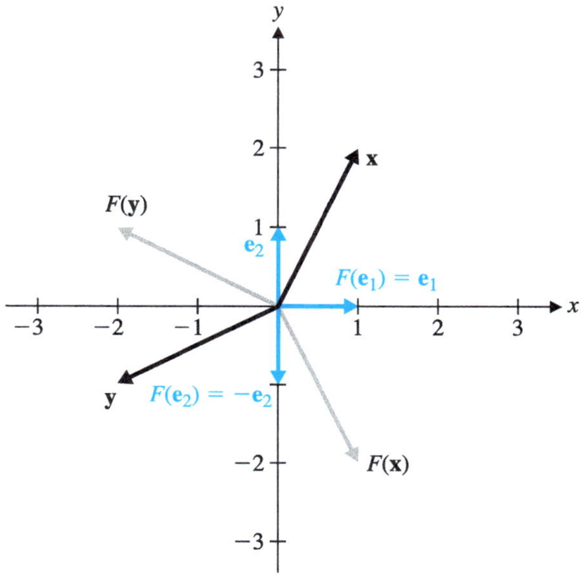

Figure 4.5
The eigenvectors of a reflection

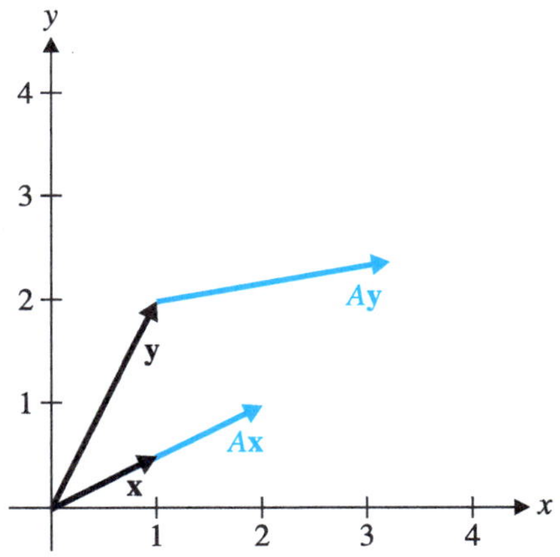

Figure 4.6

The discussion is based on the article "Eigenpictures: Picturing the Eigenvector Problem" by Steven Schonefeld in *The College Mathematics Journal 26* (1996), pp. 316–319.

Another way to think of eigenvectors geometrically is to draw $\mathbf{x}$ and $A\mathbf{x}$ head-to-tail. Then $\mathbf{x}$ will be an eigenvector of A if and only if $\mathbf{x}$ and $A\mathbf{x}$ are aligned in a straight line. In Figure 4.6, $\mathbf{x}$ is an eigenvector of A but $\mathbf{y}$ is not.

If $\mathbf{x}$ is an eigenvector of A corresponding to the eigenvalue λ, then so is any nonzero multiple of $\mathbf{x}$. So, if we want to search for eigenvectors geometrically, we need only consider the effect of A on *unit* vectors. Figure 4.7(a) shows what happens when we transform unit vectors with the matrix $A = \begin{bmatrix} 3 & 1 \\ 1 & 3 \end{bmatrix}$ of Example 4.1 and display the results head-to-tail, as in Figure 4.6. We can see that the vector $\mathbf{x} = \begin{bmatrix} 1/\sqrt{2} \\ 1/\sqrt{2} \end{bmatrix}$ is an eigenvector, but we also notice that there appears to be an eigenvector in the second quadrant. Indeed, this is the case, and it turns out to be the vector $\begin{bmatrix} -1/\sqrt{2} \\ 1/\sqrt{2} \end{bmatrix}$.

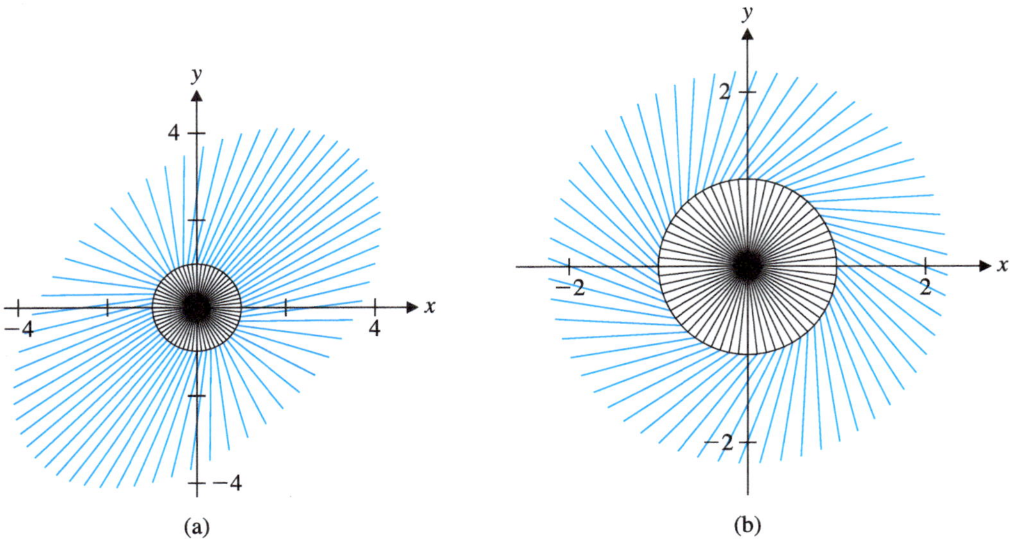

Figure 4.7

In Figure 4.7(b), we see what happens when we use the matrix $A = \begin{bmatrix} 1 & 1 \\ -1 & 1 \end{bmatrix}$. There are no eigenvectors at all!

We now know how to find eigenvectors once we have the corresponding eigenvalues, and we have a geometric interpretation of them—but one question remains: How do we first find the eigenvalues of a given matrix? The key is the observation that λ is an eigenvalue of A if and only if the null space of $A - \lambda I$ is nontrivial.

Recall from Section 3.3 that the determinant of a 2×2 matrix $A = \begin{bmatrix} a & b \\ c & d \end{bmatrix}$ is the expression $\det A = ad - bc$, and A is invertible if and only if $\det A$ is nonzero. Furthermore, the Fundamental Theorem of Invertible Matrices guarantees that a matrix has a nontrivial null space if and only if it is noninvertible—hence, if and only if its determinant is zero. Putting these facts together, we see that (for 2×2 matrices at least) λ is an eigenvalue of A if and only if $\det(A - \lambda I) = 0$. This fact characterizes eigenvalues, and we will soon generalize it to square matrices of arbitrary size. For the moment, though, let's see how to use it with 2×2 matrices.

Example 4.5

Find all of the eigenvalues and corresponding eigenvectors of the matrix $A = \begin{bmatrix} 3 & 1 \\ 1 & 3 \end{bmatrix}$ from Example 4.1.

Solution The preceding remarks show that we must find all solutions λ of the equation $\det(A - \lambda I) = 0$. Since

$$\det(A - \lambda I) = \det\begin{bmatrix} 3 - \lambda & 1 \\ 1 & 3 - \lambda \end{bmatrix} = (3 - \lambda)(3 - \lambda) - 1 = \lambda^2 - 6\lambda + 8$$

we need to solve the quadratic equation $\lambda^2 - 6\lambda + 8 = 0$. The solutions to this equation are easily found to be $\lambda = 4$ and $\lambda = 2$. These are therefore the eigenvalues of A.

To find the eigenvectors corresponding to the eigenvalue $\lambda = 4$, we compute the null space of $A - 4I$. We find

$$[A - 4I\,|\,\mathbf{0}] = \left[\begin{array}{rr|r} -1 & 1 & 0 \\ 1 & -1 & 0 \end{array}\right] \to \left[\begin{array}{rr|r} 1 & -1 & 0 \\ 0 & 0 & 0 \end{array}\right]$$

from which it follows that $\mathbf{x} = \begin{bmatrix} x_1 \\ x_2 \end{bmatrix}$ is an eigenvector corresponding to $\lambda = 4$ if and only if $x_1 - x_2 = 0$ or $x_1 = x_2$. Hence, the eigenspace $E_4 = \left\{\begin{bmatrix} x_2 \\ x_2 \end{bmatrix}\right\} = \left\{x_2\begin{bmatrix} 1 \\ 1 \end{bmatrix}\right\} = \text{span}\left(\begin{bmatrix} 1 \\ 1 \end{bmatrix}\right)$.

Similarly, for $\lambda = 2$, we have

$$[A - 2I\,|\,0] = \left[\begin{array}{rr|r} 1 & 1 & 0 \\ 1 & 1 & 0 \end{array}\right] \to \left[\begin{array}{rr|r} 1 & 1 & 0 \\ 0 & 0 & 0 \end{array}\right]$$

so $\mathbf{y} = \begin{bmatrix} y_1 \\ y_2 \end{bmatrix}$ is an eigenvector corresponding to $\lambda = 2$ if and only if $y_1 + y_2 = 0$ or $y_1 = -y_2$. Thus, the eigenspace $E_2 = \left\{\begin{bmatrix} -y_2 \\ y_2 \end{bmatrix}\right\} = \left\{y_2\begin{bmatrix} -1 \\ 1 \end{bmatrix}\right\} = \text{span}\left(\begin{bmatrix} -1 \\ 1 \end{bmatrix}\right)$.

Figure 4.8 shows graphically how the eigenvectors of A are transformed when multiplied by A: an eigenvector $\mathbf{x}$ in the eigenspace E_4 is transformed into $4\mathbf{x}$, and an eigenvector $\mathbf{y}$ in the eigenspace E_2 is transformed into $2\mathbf{y}$. As Figure 4.7(a) shows, the eigenvectors of A are the *only* vectors in $\mathbb{R}^2$ that are transformed into scalar multiples of themselves when multiplied by A.

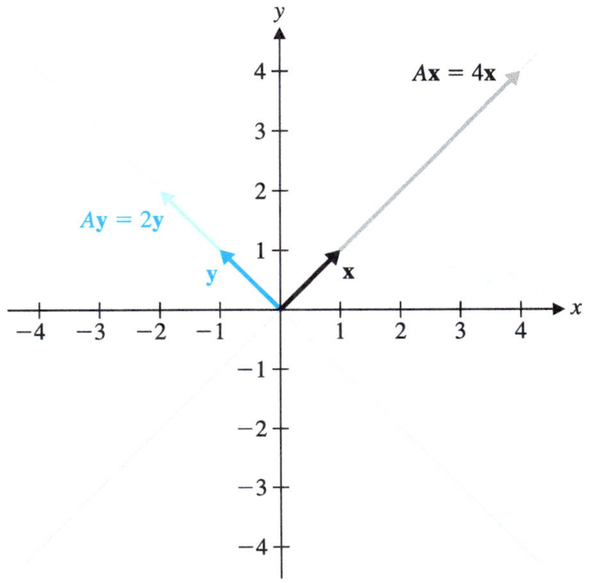

Figure 4.8
How A transforms eigenvectors

Remark You will recall that a polynomial equation with real coefficients (such as the quadratic equation in Example 4.5) need not have real roots; it may have *complex* roots. (See Appendix C.) It is also possible to compute eigenvalues and eigenvectors when the entries of a matrix come from $\mathbb{Z}_p$, where p is prime. Thus, it is important to specify the setting we intend to work in before we set out to compute the eigenvalues of a matrix. However, unless otherwise specified, the eigenvalues of a matrix whose entries are *real* numbers will be assumed to be real as well.

Example 4.6

Interpret the matrix in Example 4.5 as a matrix over $\mathbb{Z}_3$ and find its eigenvalues in that field.

Solution The solution proceeds exactly as above, except we work modulo 3. Hence, the quadratic equation $\lambda^2 - 6\lambda + 8 = 0$ becomes $\lambda^2 + 2 = 0$. This equation is the same as $\lambda^2 = -2 = 1$, giving $\lambda = 1$ and $\lambda = -1 = 2$ as the eigenvalues in $\mathbb{Z}_3$. (Check that the same answer would be obtained by *first* reducing A modulo 3 to obtain $\begin{bmatrix} 0 & 1 \\ 1 & 0 \end{bmatrix}$ and then working with this matrix.)

Example 4.7

Find the eigenvalues of $A = \begin{bmatrix} 0 & -1 \\ 1 & 0 \end{bmatrix}$ (a) over $\mathbb{R}$ and (b) over the complex numbers $\mathbb{C}$.

Solution We must solve the equation

$$0 = \det(A - \lambda I) = \det\begin{bmatrix} -\lambda & -1 \\ 1 & -\lambda \end{bmatrix} = \lambda^2 + 1$$

(a) Over $\mathbb{R}$, there are no solutions, so A has no real eigenvalues.

(b) Over $\mathbb{C}$, the solutions are $\lambda = i$ and $\lambda = -i$. (See Appendix C.)

In the next section, we will extend the notion of determinant from 2×2 to $n \times n$ matrices, which in turn will allow us to find the eigenvalues of arbitrary square matrices. (In fact, this isn't quite true—but we will at least be able to find a polynomial equation that the eigenvalues of a given matrix must satisfy.)

Exercises 4.1

In Exercises 1–6, show that **v** *is an eigenvector of A and find the corresponding eigenvalue.*

1. $A = \begin{bmatrix} 0 & 3 \\ 3 & 0 \end{bmatrix}, \mathbf{v} = \begin{bmatrix} 1 \\ 1 \end{bmatrix}$

2. $A = \begin{bmatrix} 1 & 2 \\ 2 & 1 \end{bmatrix}, \mathbf{v} = \begin{bmatrix} 3 \\ -3 \end{bmatrix}$

3. $A = \begin{bmatrix} -1 & 1 \\ 6 & 0 \end{bmatrix}, \mathbf{v} = \begin{bmatrix} 1 \\ -2 \end{bmatrix}$

4. $A = \begin{bmatrix} 4 & -2 \\ 5 & -7 \end{bmatrix}, \mathbf{v} = \begin{bmatrix} 4 \\ 2 \end{bmatrix}$

5. $A = \begin{bmatrix} 3 & 0 & 0 \\ 0 & 1 & -2 \\ 1 & 0 & 1 \end{bmatrix}, \mathbf{v} = \begin{bmatrix} 2 \\ -1 \\ 1 \end{bmatrix}$

6. $A = \begin{bmatrix} 0 & 1 & -1 \\ 1 & 1 & 1 \\ 1 & 2 & 0 \end{bmatrix}, \mathbf{v} = \begin{bmatrix} -2 \\ 1 \\ 1 \end{bmatrix}$

In Exercises 7–12, show that λ is an eigenvalue of A and find one eigenvector corresponding to this eigenvalue.

7. $A = \begin{bmatrix} 2 & 2 \\ 2 & -1 \end{bmatrix}, \lambda = 3$

8. $A = \begin{bmatrix} 2 & 3 \\ 3 & 2 \end{bmatrix}, \lambda = -1$

9. $A = \begin{bmatrix} 0 & 4 \\ -1 & 5 \end{bmatrix}, \lambda = 1$

10. $A = \begin{bmatrix} 4 & -2 \\ 5 & -7 \end{bmatrix}, \lambda = -6$

11. $A = \begin{bmatrix} 1 & 0 & 2 \\ -1 & 1 & 1 \\ 2 & 0 & 1 \end{bmatrix}, \lambda = -1$

12. $A = \begin{bmatrix} 3 & 1 & -1 \\ 1 & 1 & 1 \\ 4 & 2 & 0 \end{bmatrix}, \lambda = 2$

In Exercises 13–18, find the eigenvalues and eigenvectors of A geometrically.

13. $A = \begin{bmatrix} -1 & 0 \\ 0 & 1 \end{bmatrix}$ (reflection in the y-axis)

14. $A = \begin{bmatrix} 0 & 1 \\ 1 & 0 \end{bmatrix}$ (reflection in the line $y = x$)

15. $A = \begin{bmatrix} 1 & 0 \\ 0 & 0 \end{bmatrix}$ (projection onto the x-axis)

16. $A = \begin{bmatrix} \frac{16}{25} & \frac{12}{25} \\ \frac{12}{25} & \frac{9}{25} \end{bmatrix}$ (projection onto the line through the origin with direction vector $\begin{bmatrix} \frac{4}{5} \\ \frac{3}{5} \end{bmatrix}$)

17. $A = \begin{bmatrix} 2 & 0 \\ 0 & 3 \end{bmatrix}$ (stretching by a factor of 2 horizontally and a factor of 3 vertically)

18. $A = \begin{bmatrix} 0 & -1 \\ 1 & 0 \end{bmatrix}$ (counterclockwise rotation of 90° about the origin)

*In Exercises 19–22, the unit vectors **x** in $\mathbb{R}^2$ and their images $A\mathbf{x}$ under the action of a 2×2 matrix A are drawn head-to-tail, as in Figure 4.7. Estimate the eigenvectors and eigenvalues of A from each "eigenpicture."*

19.

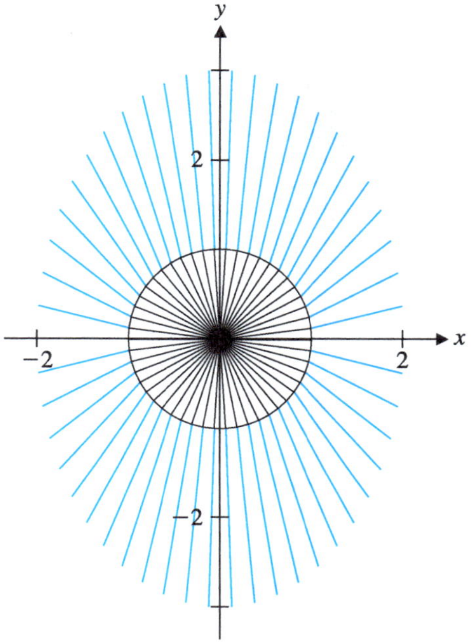

20.

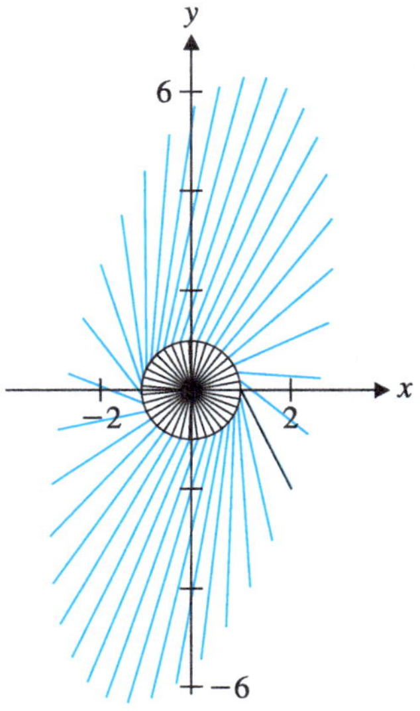

21.

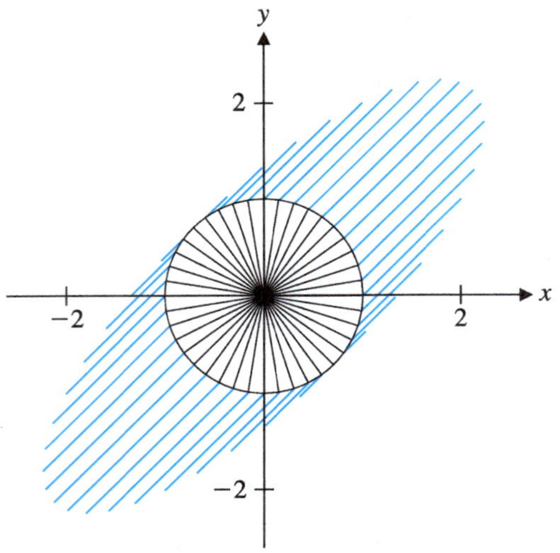

22.

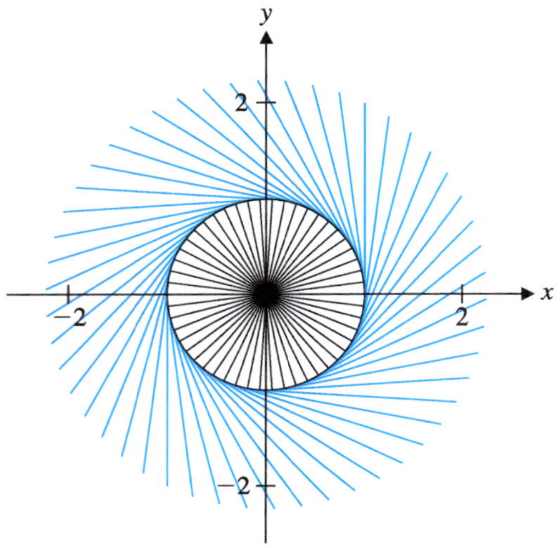

In Exercises 23–26, use the method of Example 4.5 to find all of the eigenvalues of the matrix A. Give bases for each of the corresponding eigenspaces. Illustrate the eigenspaces and the effect of multiplying eigenvectors by A as in Figure 4.8.

23. $A = \begin{bmatrix} 4 & -1 \\ 2 & 1 \end{bmatrix}$ **24.** $A = \begin{bmatrix} 2 & 4 \\ 6 & 0 \end{bmatrix}$

25. $A = \begin{bmatrix} 2 & 5 \\ 0 & 2 \end{bmatrix}$ **26.** $A = \begin{bmatrix} 1 & 2 \\ -2 & 3 \end{bmatrix}$

In Exercises 27–30, find all of the eigenvalues of the matrix A over the complex numbers $\mathbb{C}$. Give bases for each of the corresponding eigenspaces.

27. $A = \begin{bmatrix} 1 & 1 \\ -1 & 1 \end{bmatrix}$ **28.** $A = \begin{bmatrix} 2 & -3 \\ 1 & 0 \end{bmatrix}$

29. $A = \begin{bmatrix} 1 & i \\ i & 1 \end{bmatrix}$ **30.** $A = \begin{bmatrix} 0 & 1+i \\ 1-i & 1 \end{bmatrix}$

In Exercises 31–34, find all of the eigenvalues of the matrix A over the indicated $\mathbb{Z}_p$.

31. $A = \begin{bmatrix} 1 & 0 \\ 1 & 2 \end{bmatrix}$ over $\mathbb{Z}_3$ **32.** $A = \begin{bmatrix} 2 & 1 \\ 1 & 2 \end{bmatrix}$ over $\mathbb{Z}_3$

33. $A = \begin{bmatrix} 3 & 1 \\ 4 & 0 \end{bmatrix}$ over $\mathbb{Z}_5$ **34.** $A = \begin{bmatrix} 1 & 4 \\ 4 & 0 \end{bmatrix}$ over $\mathbb{Z}_5$

35. (a) Show that the eigenvalues of the 2×2 matrix

$$A = \begin{bmatrix} a & b \\ c & d \end{bmatrix}$$

are the solutions of the quadratic equation $\lambda^2 - \operatorname{tr}(A)\lambda + \det A = 0$, where $\operatorname{tr}(A)$ is the trace of A. (See page 162.)

(b) Show that the eigenvalues of the matrix A in part (a) are

$$\lambda = \tfrac{1}{2}\left(a + d \pm \sqrt{(a-d)^2 + 4bc}\right)$$

(c) Show that the trace and determinant of the matrix A in part (a) are given by

$$\operatorname{tr}(A) = \lambda_1 + \lambda_2 \quad \text{and} \quad \det A = \lambda_1\lambda_2$$

where λ_1 and λ_2 are the eigenvalues of A.

36. Consider again the matrix A in Exercise 35. Give conditions on a, b, c, and d such that A has

(a) two distinct real eigenvalues,
(b) one real eigenvalue, and
(c) no real eigenvalues.

37. Show that the eigenvalues of the upper triangular matrix

$$A = \begin{bmatrix} a & b \\ 0 & d \end{bmatrix}$$

are $\lambda = a$ and $\lambda = d$, and find the corresponding eigenspaces.

38. Let a and b be real numbers. Find the eigenvalues and corresponding eigenspaces of

$$A = \begin{bmatrix} a & b \\ -b & a \end{bmatrix}$$

over the complex numbers.

4.2 Determinants

Historically, determinants preceded matrices—a curious fact in light of the way linear algebra is taught today, with matrices before determinants. Nevertheless, determinants arose independently of matrices in the solution of many practical problems, and the theory of determinants was well developed almost two centuries before matrices were deemed worthy of study in and of themselves. A snapshot of the history of determinants is presented at the end of this section.

Recall that the determinant of the 2×2 matrix $A = \begin{bmatrix} a_{11} & a_{12} \\ a_{21} & a_{22} \end{bmatrix}$ is

$$\det A = a_{11}a_{22} - a_{12}a_{21}$$

We first encountered this expression when we determined ways to compute the inverse of a matrix. In particular, we found that

$$\begin{bmatrix} a_{11} & a_{12} \\ a_{21} & a_{22} \end{bmatrix}^{-1} = \frac{1}{a_{11}a_{22} - a_{12}a_{21}} \begin{bmatrix} a_{22} & -a_{12} \\ -a_{21} & a_{11} \end{bmatrix}$$

The determinant of a matrix A is sometimes also denoted by $|A|$, so for the 2×2 matrix $A = \begin{bmatrix} a_{11} & a_{12} \\ a_{21} & a_{22} \end{bmatrix}$ we may also write

$$|A| = \begin{vmatrix} a_{11} & a_{12} \\ a_{21} & a_{22} \end{vmatrix} = a_{11}a_{22} - a_{12}a_{21}$$

Warning This notation for the determinant is reminiscent of absolute value notation. It is easy to mistake $\begin{vmatrix} a_{11} & a_{12} \\ a_{21} & a_{22} \end{vmatrix}$, the notation for determinant, for $\begin{bmatrix} a_{11} & a_{12} \\ a_{21} & a_{22} \end{bmatrix}$, the notation for the matrix itself. Do not confuse these. Fortunately, it will usually be clear from the context which is intended.

We define the determinant of a 1×1 matrix $A = [a]$ to be

$$\det A = |a| = a$$

(Note that we really have to be careful with notation here: $|a|$ does *not* denote the absolute value of a in this case.) How then should we define the determinant of a 3×3 matrix? If you ask your CAS for the inverse of

$$A = \begin{bmatrix} a & b & c \\ d & e & f \\ g & h & i \end{bmatrix}$$

the answer will be equivalent to

$$A^{-1} = \frac{1}{\Delta} \begin{bmatrix} ei - fh & ch - bi & bf - ce \\ fg - di & ai - cg & cd - af \\ dh - eg & bg - ah & ae - bd \end{bmatrix}$$

where $\Delta = aei - afh - bdi + bfg + cdh - ceg$. Observe that

$$\begin{aligned} \Delta &= aei - afh - bdi + bfg + cdh - ceg \\ &= a(ei - fh) - b(di - fg) + c(dh - eg) \\ &= a\begin{vmatrix} e & f \\ h & i \end{vmatrix} - b\begin{vmatrix} d & f \\ g & i \end{vmatrix} + c\begin{vmatrix} d & e \\ g & h \end{vmatrix} \end{aligned}$$

and that each of the entries in the matrix portion of A^{-1} appears to be the determinant of a 2×2 submatrix of A. In fact, this is true, and it is the basis of the definition of the determinant of a 3×3 matrix. The definition is *recursive* in the sense that the determinant of a 3×3 matrix is defined in terms of determinants of 2×2 matrices.

Definition Let $A = \begin{bmatrix} a_{11} & a_{12} & a_{13} \\ a_{21} & a_{22} & a_{23} \\ a_{31} & a_{32} & a_{33} \end{bmatrix}$. Then the ***determinant*** of A is the scalar

$$\det A = |A| = a_{11}\begin{vmatrix} a_{22} & a_{23} \\ a_{32} & a_{33} \end{vmatrix} - a_{12}\begin{vmatrix} a_{21} & a_{23} \\ a_{31} & a_{33} \end{vmatrix} + a_{13}\begin{vmatrix} a_{21} & a_{22} \\ a_{31} & a_{32} \end{vmatrix} \qquad (1)$$

Notice that each of the 2×2 determinants is obtained by deleting the row and column of A that contain the entry the determinant is being multiplied by. For example, the first summand is a_{11} multiplied by the determinant of the submatrix obtained by deleting row 1 and column 1. Notice also that the plus and minus signs alternate in Equation (1). If we denote by A_{ij} the submatrix of a matrix A obtained by deleting row i and column j, then we may abbreviate Equation (1) as

$$\begin{aligned} \det A &= a_{11} \det A_{11} - a_{12} \det A_{12} + a_{13} \det A_{13} \\ &= \sum_{j=1}^{3} (-1)^{1+j} a_{1j} \det A_{1j} \end{aligned}$$

For any square matrix A, $\det A_{ij}$ is called the ***(i, j)-minor*** of A.

Example 4.8 Compute the determinant of

$$A = \begin{bmatrix} 5 & -3 & 2 \\ 1 & 0 & 2 \\ 2 & -1 & 3 \end{bmatrix}$$

Solution We compute

$$\begin{aligned} \det A &= 5\begin{vmatrix} 0 & 2 \\ -1 & 3 \end{vmatrix} - (-3)\begin{vmatrix} 1 & 2 \\ 2 & 3 \end{vmatrix} + 2\begin{vmatrix} 1 & 0 \\ 2 & -1 \end{vmatrix} \\ &= 5(0 - (-2)) + 3(3 - 4) + 2(-1 - 0) \\ &= 5(2) + 3(-1) + 2(-1) = 5 \end{aligned}$$

With a little practice, you should find that you can easily work out 2×2 determinants in your head. Writing out the second line in the above solution is then unnecessary.

Another method for calculating the determinant of a 3×3 matrix is analogous to the method for calculating the determinant of a 2×2 matrix. Copy the first two columns of A to the right of the matrix and take the products of the elements on the six diagonals shown below. Attach plus signs to the products from the downward-sloping diagonals and attach minus signs to the products from the upward-sloping diagonals.

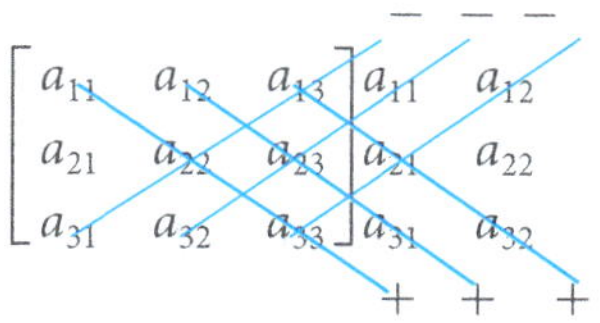

(2)

This method gives

$$a_{11}a_{22}a_{33} + a_{12}a_{23}a_{31} + a_{13}a_{21}a_{32} - a_{31}a_{22}a_{13} - a_{32}a_{23}a_{11} - a_{33}a_{21}a_{12}$$

In Exercise 19, you are asked to check that this result agrees with that from Equation (1) for a 3×3 determinant.

Example 4.9

Calculate the determinant of the matrix in Example 4.8 using the method shown in (2).

Solution We adjoin to A its first two columns and compute the six indicated products:

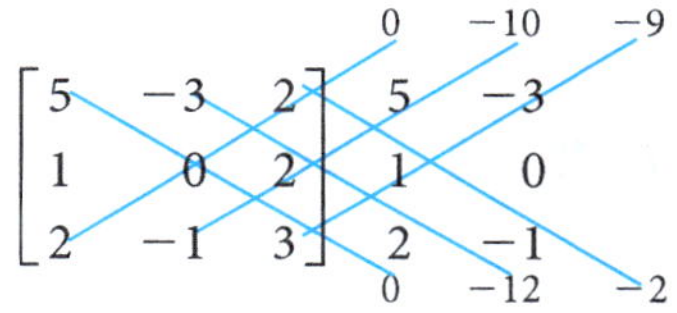

Adding the three products at the bottom and subtracting the three products at the top gives

$$\det A = 0 + (-12) + (-2) - 0 - (-10) - (-9) = 5$$

as before.

Warning We are about to define determinants for arbitrary square matrices. However, there is *no* analogue of the method in Example 4.9 for larger matrices. It is valid *only* for 3×3 matrices.

Determinants of $n \times n$ Matrices

The definition of the determinant of a 3×3 matrix extends naturally to arbitrary square matrices.

Definition Let $A = [a_{ij}]$ be an $n \times n$ matrix, where $n \geq 2$. Then the ***determinant*** of A is the scalar

$$\det A = |A| = a_{11} \det A_{11} - a_{12} \det A_{12} + \cdots + (-1)^{1+n} a_{1n} \det A_{1n}$$
$$= \sum_{j=1}^{n} (-1)^{1+j} a_{1j} \det A_{1j} \qquad (3)$$

It is convenient to combine a minor with its plus or minus sign. To this end, we define the ***(i, j)-cofactor of A*** to be

$$C_{ij} = (-1)^{i+j} \det A_{ij}$$

With this notation, definition (3) becomes

$$\det A = \sum_{j=1}^{n} a_{1j}C_{1j} \tag{4}$$

Exercise 20 asks you to check that this definition correctly gives the formula for the determinant of a 2×2 matrix when $n = 2$.

Definition (4) is often referred to as ***cofactor expansion along the first row.*** It is an amazing fact that we get exactly the same result by expanding along *any* row (or even *any column*)! We summarize this fact as a theorem but defer the proof until the end of this section (since it is somewhat lengthy and would interrupt our discussion if we were to present it here).

Theorem 4.1 **The Laplace Expansion Theorem**

The determinant of an $n \times n$ matrix $A = [a_{ij}]$, where $n \geq 2$, can be computed as

$$\begin{aligned} \det A &= a_{i1}C_{i1} + a_{i2}C_{i2} + \cdots + a_{in}C_{in} \\ &= \sum_{j=1}^{n} a_{ij}C_{ij} \end{aligned} \tag{5}$$

(which is the ***cofactor expansion along the*** **i*th row***) and also as

$$\begin{aligned} \det A &= a_{1j}C_{1j} + a_{2j}C_{2j} + \cdots + a_{nj}C_{nj} \\ &= \sum_{i=1}^{n} a_{ij}C_{ij} \end{aligned} \tag{6}$$

(the ***cofactor expansion along the*** **j*th column***).

Since $C_{ij} = (-1)^{i+j} \det A_{ij}$, each cofactor is plus or minus the corresponding minor, with the correct sign given by the term $(-1)^{i+j}$. A quick way to determine whether the sign is $+$ or $-$ is to remember that the signs form a "checkerboard" pattern:

$$\begin{bmatrix} + & - & + & - & \cdots \\ - & + & - & + & \cdots \\ + & - & + & - & \cdots \\ - & + & - & + & \cdots \\ \vdots & \vdots & \vdots & \vdots & \ddots \end{bmatrix}$$

Example 4.10

Compute the determinant of the matrix

$$A = \begin{bmatrix} 5 & -3 & 2 \\ 1 & 0 & 2 \\ 2 & -1 & 3 \end{bmatrix}$$

by (a) cofactor expansion along the third row and (b) cofactor expansion along the second column.

Solution

(a) We compute

$$\begin{aligned} \det A &= a_{31}C_{31} + a_{32}C_{32} + a_{33}C_{33} \\ &= 2\begin{vmatrix} -3 & 2 \\ 0 & 2 \end{vmatrix} - (-1)\begin{vmatrix} 5 & 2 \\ 1 & 2 \end{vmatrix} + 3\begin{vmatrix} 5 & -3 \\ 1 & 0 \end{vmatrix} \\ &= 2(-6) + 8 + 3(3) \\ &= 5 \end{aligned}$$

(b) In this case, we have

$$\begin{aligned} \det A &= a_{12}C_{12} + a_{22}C_{22} + a_{32}C_{32} \\ &= -(-3)\begin{vmatrix} 1 & 2 \\ 2 & 3 \end{vmatrix} + 0\begin{vmatrix} 5 & 2 \\ 2 & 3 \end{vmatrix} - (-1)\begin{vmatrix} 5 & 2 \\ 1 & 2 \end{vmatrix} \\ &= 3(-1) + 0 + 8 \\ &= 5 \end{aligned}$$

Pierre Simon Laplace (1749–1827) was born in Normandy, France, and was expected to become a clergyman until his mathematical talents were noticed at school. He made many important contributions to calculus, probability, and astronomy. He was an examiner of the young Napoleon Bonaparte at the Royal Artillery Corps and later, when Napoleon was in power, served briefly as Minister of the Interior and then Chancellor of the Senate. Laplace was granted the title of Count of the Empire in 1806 and received the title of Marquis de Laplace in 1817.

Notice that in part (b) of Example 4.10 we needed to do fewer calculations than in part (a) because we were expanding along a column that contained a zero entry—namely, a_{22}; therefore, we did not need to compute C_{22}. It follows that the Laplace Expansion Theorem is most useful when the matrix contains a row or column with lots of zeros, since, by choosing to expand along that row or column, we minimize the number of cofactors we need to compute.

Example 4.11

Compute the determinant of

$$A = \begin{bmatrix} 2 & -3 & 0 & 1 \\ 5 & 4 & 2 & 0 \\ 1 & -1 & 0 & 3 \\ -2 & 1 & 0 & 0 \end{bmatrix}$$

Solution First, notice that column 3 has only one nonzero entry; we should therefore expand along this column. Next, note that the +/− pattern assigns a minus sign to the entry $a_{23} = 2$. Thus, we have

$$\begin{aligned}\det A &= a_{13}C_{13} + a_{23}C_{23} + a_{33}C_{33} + a_{43}C_{43} \\ &= 0(C_{13}) + 2C_{23} + 0(C_{33}) + 0(C_{43}) \\ &= -2\begin{vmatrix} 2 & -3 & 1 \\ 1 & -1 & 3 \\ -2 & 1 & 0 \end{vmatrix}\end{aligned}$$

We now continue by expanding along the third row of the determinant above (the third column would also be a good choice) to get

$$\begin{aligned}\det A &= -2\left(-2\begin{vmatrix} -3 & 1 \\ -1 & 3 \end{vmatrix} - \begin{vmatrix} 2 & 1 \\ 1 & 3 \end{vmatrix}\right) \\ &= -2(-2(-8) - 5) \\ &= -2(11) = -22\end{aligned}$$

(Note that the $+/-$ pattern for the 3×3 minor is not that of the original matrix but that of a 3×3 matrix in general.)

The Laplace expansion is particularly useful when the matrix is (upper or lower) triangular.

Example 4.12 Compute the determinant of

$$A = \begin{bmatrix} 2 & -3 & 1 & 0 & 4 \\ 0 & 3 & 2 & 5 & 7 \\ 0 & 0 & 1 & 6 & 0 \\ 0 & 0 & 0 & 5 & 2 \\ 0 & 0 & 0 & 0 & -1 \end{bmatrix}$$

Solution We expand along the first column to get

$$\det A = 2\begin{vmatrix} 3 & 2 & 5 & 7 \\ 0 & 1 & 6 & 0 \\ 0 & 0 & 5 & 2 \\ 0 & 0 & 0 & -1 \end{vmatrix}$$

(We have omitted all cofactors corresponding to zero entries.) Now we expand along the first column again:

$$\det A = 2 \cdot 3\begin{vmatrix} 1 & 6 & 0 \\ 0 & 5 & 2 \\ 0 & 0 & -1 \end{vmatrix}$$

Continuing to expand along the first column, we complete the calculation:

$$\det A = 2 \cdot 3 \cdot 1\begin{vmatrix} 5 & 2 \\ 0 & -1 \end{vmatrix} = 2 \cdot 3 \cdot 1 \cdot (5(-1) - 2 \cdot 0) = 2 \cdot 3 \cdot 1 \cdot 5 \cdot (-1) = -30$$

Example 4.12 should convince you that the determinant of a triangular matrix is the product of its diagonal entries. You are asked to give a proof of this fact in Exercise 21. We record the result as a theorem.

Theorem 4.2 The determinant of a triangular matrix is the product of the entries on its main diagonal. Specifically, if $A = [a_{ij}]$ is an $n \times n$ triangular matrix, then

$$\det A = a_{11}a_{22}\cdots a_{nn}$$

Note In general (that is, unless the matrix is triangular or has some other special form), computing a determinant by cofactor expansion is not efficient. For example, the determinant of a 3×3 matrix has $6 = 3!$ summands, each requiring two multiplications, and then five additions and subtractions are needed to finish off the calculations. For an $n \times n$ matrix, there will be $n!$ summands, each with $n - 1$ multiplications, and then $n! - 1$ additions and subtractions. The total number of operations is thus

$$T(n) = (n - 1)n! + n! - 1 > n!$$

Even the fastest of supercomputers cannot calculate the determinant of a moderately large matrix using cofactor expansion. To illustrate: Suppose we needed to calculate a 50×50 determinant. (Matrices *much* larger than 50×50 are used to store the data from digital images such as those transmitted over the Internet or taken by a digital camera.) To calculate the determinant directly would require, in general, more than 50! operations, and $50! \approx 3 \times 10^{64}$. If we had a computer that could perform a trillion (10^{12}) operations per second, it would take approximately 3×10^{52} seconds, or almost 10^{45} years, to finish the calculations. To put this in perspective, consider that astronomers estimate the age of the universe to be at least 10 billion (10^{10}) years. Thus, on even a very fast supercomputer, calculating a 50×50 determinant by cofactor expansion would take more than 10^{30} times the age of the universe!

Fortunately, there are better methods—and we now turn to developing more computationally effective means of finding determinants. First, we need to look at some of the properties of determinants.

Properties of Determinants

The most efficient way to compute determinants is to use row reduction. However, not every elementary row operation leaves the determinant of a matrix unchanged. The next theorem summarizes the main properties you need to understand in order to use row reduction effectively.

Theorem 4.3 Let $A = [a_{ij}]$ be a square matrix.

a. If A has a zero row (column), then $\det A = 0$.
b. If B is obtained by interchanging two rows (columns) of A, then $\det B = -\det A$.
c. If A has two identical rows (columns), then $\det A = 0$.
d. If B is obtained by multiplying a row (column) of A by k, then $\det B = k \det A$.
e. If A, B, and C are identical except that the ith row (column) of C is the sum of the ith rows (columns) of A and B, then $\det C = \det A + \det B$.
f. If B is obtained by adding a multiple of one row (column) of A to another row (column), then $\det B = \det A$.

Proof We will prove (b) as Lemma 4.14 at the end of this section. The proofs of properties (a) and (f) are left as exercises. We will prove the remaining properties in terms of rows; the corresponding proofs for columns are analogous.

(c) If A has two identical rows, swap them to obtain the matrix B. Clearly, $B = A$, so $\det B = \det A$. On the other hand, by (b), $\det B = -\det A$. Therefore, $\det A = -\det A$, so $\det A = 0$.

(d) Suppose row i of A is multiplied by k to produce B; that is, $b_{ij} = ka_{ij}$ for $j = 1, \ldots, n$. Since the cofactors C_{ij} of the elements in the ith rows of A and B are identical (why?), expanding along the ith row of B gives

$$\det B = \sum_{j=1}^{n} b_{ij}C_{ij} = \sum_{j=1}^{n} ka_{ij}C_{ij} = k\sum_{j=1}^{n} a_{ij}C_{ij} = k \det A$$

(e) As in (d), the cofactors C_{ij} of the elements in the ith rows of A, B, and C are identical. Moreover, $c_{ij} = a_{ij} + b_{ij}$ for $j = 1, \ldots, n$. We expand along the ith row of C to obtain

$$\det C = \sum_{j=1}^{n} c_{ij}C_{ij} = \sum_{j=1}^{n} (a_{ij} + b_{ij})C_{ij} = \sum_{j=1}^{n} a_{ij}C_{ij} + \sum_{j=1}^{n} b_{ij}C_{ij} = \det A + \det B$$

Notice that properties (b), (d), and (f) are related to elementary row operations. Since the echelon form of a square matrix is necessarily upper triangular, we can combine these properties with Theorem 2 to calculate determinants efficiently. (See Exploration: Counting Operations in Chapter 2, which shows that row reduction of an $n \times n$ matrix uses on the order of n^3 operations, far fewer than the $n!$ needed for cofactor expansion.) The next examples illustrate the computation of determinants using row reduction.

Example 4.13 Compute $\det A$ if

(a) $A = \begin{bmatrix} 2 & 3 & -1 \\ 0 & 5 & 3 \\ -4 & -6 & 2 \end{bmatrix}$

(b) $A = \begin{bmatrix} 0 & 2 & -4 & 5 \\ 3 & 0 & -3 & 6 \\ 2 & 4 & 5 & 7 \\ 5 & -1 & -3 & 1 \end{bmatrix}$

Solution

(a) Using property (f) and then property (a), we have

$$\det A = \begin{vmatrix} 2 & 3 & -1 \\ 0 & 5 & 3 \\ -4 & -6 & 2 \end{vmatrix} \overset{R_3+2R_1}{=} \begin{vmatrix} 2 & 3 & -1 \\ 0 & 5 & 3 \\ 0 & 0 & 0 \end{vmatrix} = 0$$

(b) We reduce A to echelon form as follows (there are other possible ways to do this):

$$\det A = \begin{vmatrix} 0 & 2 & -4 & 5 \\ 3 & 0 & -3 & 6 \\ 2 & 4 & 5 & 7 \\ 5 & -1 & -3 & 1 \end{vmatrix} \overset{R_1 \leftrightarrow R_2}{=} - \begin{vmatrix} 3 & 0 & -3 & 6 \\ 0 & 2 & -4 & 5 \\ 2 & 4 & 5 & 7 \\ 5 & -1 & -3 & 1 \end{vmatrix} \overset{R_1/3}{=} -3 \begin{vmatrix} 1 & 0 & -1 & 2 \\ 0 & 2 & -4 & 5 \\ 2 & 4 & 5 & 7 \\ 5 & -1 & -3 & 1 \end{vmatrix}$$

$$\overset{\substack{R_3-2R_1 \\ R_4-5R_1}}{=} -3 \begin{vmatrix} 1 & 0 & -1 & 2 \\ 0 & 2 & -4 & 5 \\ 0 & 4 & 7 & 3 \\ 0 & -1 & 2 & -9 \end{vmatrix} \overset{R_2 \leftrightarrow R_4}{=} -(-3) \begin{vmatrix} 1 & 0 & -1 & 2 \\ 0 & -1 & 2 & -9 \\ 0 & 4 & 7 & 3 \\ 0 & 2 & -4 & 5 \end{vmatrix}$$

$$\overset{\substack{R_3+4R_2 \\ R_4+2R_2}}{=} 3 \begin{vmatrix} 1 & 0 & -1 & 2 \\ 0 & -1 & 2 & -9 \\ 0 & 0 & 15 & -33 \\ 0 & 0 & 0 & -13 \end{vmatrix}$$

$$= 3 \cdot 1 \cdot (-1) \cdot 15 \cdot (-13) = 585$$

Remark By Theorem 4.3, we can also use elementary column operations in the process of computing determinants, and we can "mix and match" elementary row and column operations. For example, in Example 4.13(a), we could have started by adding column 3 to column 1 to create a leading 1 in the upper left-hand corner. In fact, the method we used was faster, but in other examples column operations may speed up the calculations. Keep this in mind when you work determinants by hand.

Determinants of Elementary Matrices

Recall from Section 3.3 that an elementary matrix results from performing an elementary row operation on an identity matrix. Setting $A = I_n$ in Theorem 4.3 yields the following theorem.

Theorem 4.4 Let E be an $n \times n$ elementary matrix.

a. If E results from interchanging two rows of I_n, then $\det E = -1$.
b. If E results from multiplying one row of I_n by k, then $\det E = k$.
c. If E results from adding a multiple of one row of I_n to another row, then $\det E = 1$.

Proof Since $\det I_n = 1$, applying (b), (d), and (f) of Theorem 4.3 immediately gives (a), (b), and (c), respectively, of Theorem 4.4.

The word *lemma* is derived from the Greek verb *lambanein*, which means "to grasp." In mathematics, a lemma is a "helper theorem" that we "grasp hold of" and use to prove another, usually more important, theorem.

Next, recall that multiplying a matrix B by an elementary matrix *on the left* performs the corresponding elementary row operation on B. We can therefore rephrase (b), (d), and (f) of Theorem 4.3 succinctly as the following lemma, the proof of which is straightforward and is left as Exercise 43.

Lemma 4.5 Let B be an $n \times n$ matrix and let E be an $n \times n$ elementary matrix. Then

$$\det(EB) = (\det E)(\det B)$$

We can use Lemma 4.5 to prove the main theorem of this section: a characterization of invertibility in terms of determinants.

Theorem 4.6 A square matrix A is invertible if and only if $\det A \neq 0$.

Proof Let A be an $n \times n$ matrix and let R be the reduced row echelon form of A. We will show first that $\det A \neq 0$ if and only if $\det R \neq 0$. Let $E_1, E_2, \ldots, E_r$ be the elementary matrices corresponding to the elementary row operations that reduce A to R. Then

$$E_r \cdots E_2 E_1 A = R$$

Taking determinants of both sides and repeatedly applying Lemma 4.5, we obtain

$$(\det E_r) \cdots (\det E_2)(\det E_1)(\det A) = \det R$$

By Theorem 4.4, the determinants of all the elementary matrices are nonzero. We conclude that $\det A \neq 0$ if and only if $\det R \neq 0$.

Now suppose that A is invertible. Then, by the Fundamental Theorem of Invertible Matrices, $R = I_n$, so $\det R = 1 \neq 0$. Hence, $\det A \neq 0$ also. Conversely, if $\det A \neq 0$, then $\det R \neq 0$, so R cannot contain a zero row, by Theorem 4.3(a). It follows that R must be I_n (why?), so A is invertible, by the Fundamental Theorem again.

Determinants and Matrix Operations

Let's now try to determine what relationship, if any, exists between determinants and some of the basic matrix operations. Specifically, we would like to find formulas for $\det(kA)$, $\det(A + B)$, $\det(AB)$, $\det(A^{-1})$, and $\det(A^T)$ in terms of $\det A$ and $\det B$.

Theorem 4.3(d) does *not* say that $\det(kA) = k \det A$. The correct relationship between scalar multiplication and determinants is given by the following theorem.

Theorem 4.7 If A is an $n \times n$ matrix, then

$$\det(kA) = k^n \det A$$

You are asked to give a proof of this theorem in Exercise 44.

Unfortunately, there is no simple formula for $\det(A + B)$, and in general, $\det(A + B) \neq \det A + \det B$. (Find two 2×2 matrices that verify this.) It therefore comes as a pleasant surprise to find out that determinants are quite compatible with matrix multiplication. Indeed, we have the following nice formula due to Cauchy.

Augustin Louis Cauchy (1789–1857) was born in Paris and studied engineering but switched to mathematics because of poor health. A brilliant and prolific mathematician, he published over 700 papers, many on quite difficult problems. His name can be found on many theorems and definitions in differential equations, infinite series, probability theory, algebra, and physics. He is noted for introducing rigor into calculus, laying the foundation for the branch of mathematics known as analysis. Politically conservative, Cauchy was a royalist, and in 1830 he followed Charles X into exile. He returned to France in 1838 but did not return to his post at the Sorbonne until the university dropped its requirement that faculty swear an oath of loyalty to the new king.

Theorem 4.8 If A and B are $n \times n$ matrices, then

$$\det(AB) = (\det A)(\det B)$$

Proof We consider two cases: A invertible and A not invertible.

If A is invertible, then, by the Fundamental Theorem of Invertible Matrices, it can be written as a product of elementary matrices—say,

$$A = E_1E_2\cdots E_k$$

Then $AB = E_1E_2\cdots E_kB$, so k applications of Lemma 4.5 give

$$\det(AB) = \det(E_1E_2\cdots E_kB) = (\det E_1)(\det E_2)\cdots(\det E_k)(\det B)$$

Continuing to apply Lemma 4.5, we obtain

$$\det(AB) = \det(E_1E_2\cdots E_k)\det B = (\det A)(\det B)$$

On the other hand, if A is not invertible, then neither is AB, by Exercise 47 in Section 3.3. Thus, by Theorem 4.6, $\det A = 0$ and $\det(AB) = 0$. Consequently, $\det(AB) = (\det A)(\det B)$, since both sides are zero.

Example 4.14

Applying Theorem 4.8 to $A = \begin{bmatrix} 2 & 1 \\ 2 & 3 \end{bmatrix}$ and $B = \begin{bmatrix} 5 & 1 \\ 2 & 1 \end{bmatrix}$, we find that

$$AB = \begin{bmatrix} 12 & 3 \\ 16 & 5 \end{bmatrix}$$

and that $\det A = 4$, $\det B = 3$, and $\det(AB) = 12 = 4 \cdot 3 = (\det A)(\det B)$, as claimed. (Check these assertions!)

The next theorem gives a nice relationship between the determinant of an invertible matrix and the determinant of its inverse.

Theorem 4.9 If A is invertible, then

$$\det(A^{-1}) = \frac{1}{\det A}$$

Proof Since A is invertible, $AA^{-1} = I$, so $\det(AA^{-1}) = \det I = 1$. Hence, $(\det A)(\det A^{-1}) = 1$, by Theorem 4.8, and since $\det A \neq 0$ (why?), dividing by $\det A$ yields the result.

Example 4.15 Verify Theorem 4.9 for the matrix A of Example 4.14.

Solution We compute

$$A^{-1} = \tfrac{1}{4}\begin{bmatrix} 3 & -1 \\ -2 & 2 \end{bmatrix} = \begin{bmatrix} \frac{3}{4} & -\frac{1}{4} \\ -\frac{1}{2} & \frac{1}{2} \end{bmatrix}$$

so

$$\det A^{-1} = \left(\frac{3}{4}\right)\left(\frac{1}{2}\right) - \left(-\frac{1}{4}\right)\left(-\frac{1}{2}\right) = \frac{3}{8} - \frac{1}{8} = \frac{1}{4} = \frac{1}{\det A}$$

Remark The beauty of Theorem 4.9 is that sometimes we do not need to know *what* the inverse of a matrix is, but only that it *exists*, or to know what its determinant is. For the matrix A in the last two examples, once we know that $\det A = 4 \neq 0$, we immediately can deduce that A is invertible and that $\det A^{-1} = \frac{1}{4}$ without actually computing A^{-1}.

We now relate the determinant of a matrix A to that of its transpose A^T. Since the rows of A^T are just the columns of A, evaluating $\det A^T$ by expanding along the first row is identical to evaluating $\det A$ by expanding along its first column, which the Laplace Expansion Theorem allows us to do. Thus, we have the following result.

Theorem 4.10 For any square matrix A,

$$\det A = \det A^T$$

Cramer's Rule and the Adjoint

Gabriel Cramer (1704–1752) was a Swiss mathematician. The rule that bears his name was published in 1750, in his treatise *Introduction to the Analysis of Algebraic Curves*. As early as 1730, however, special cases of the formula were known to other mathematicians, including the Scotsman Colin Maclaurin (1698–1746), perhaps the greatest of the British mathematicians who were the "successors of Newton."

In this section, we derive two useful formulas relating determinants to the solution of linear systems and the inverse of a matrix. The first of these, ***Cramer's Rule***, gives a formula for describing the solution of certain systems of n linear equations in n variables entirely in terms of determinants. While this result is of little practical use beyond 2×2 systems, it is of great theoretical importance.

We will need some new notation for this result and its proof. For an $n \times n$ matrix A and a vector $\mathbf{b}$ in $\mathbb{R}^n$, let $A_i(\mathbf{b})$ denote the matrix obtained by replacing the ith column of A by $\mathbf{b}$. That is,

$$A_i(\mathbf{b}) = [\mathbf{a}_1 \cdots \underset{\substack{\uparrow \\ \text{Column } i}}{\mathbf{b}} \cdots \mathbf{a}_n]$$

Theorem 1.11 **Cramer's Rule**

Let A be an invertible $n \times n$ matrix and let $\mathbf{b}$ be a vector in $\mathbb{R}^n$. Then the unique solution $\mathbf{x}$ of the system $A\mathbf{x} = \mathbf{b}$ is given by

$$x_i = \frac{\det(A_i(\mathbf{b}))}{\det A} \quad \text{for } i = 1, \ldots, n$$

Proof The columns of the identity matrix $I = I_n$ are the standard unit vectors $\mathbf{e}_1, \mathbf{e}_2, \ldots, \mathbf{e}_n$. If $A\mathbf{x} = \mathbf{b}$, then

$$\begin{aligned} AI_i(\mathbf{x}) &= A[\mathbf{e}_1 \quad \cdots \quad \mathbf{x} \quad \cdots \quad \mathbf{e}_n] = [A\mathbf{e}_1 \quad \cdots \quad A\mathbf{x} \quad \cdots \quad A\mathbf{e}_n] \\ &= [\mathbf{a}_1 \quad \cdots \quad \mathbf{b} \quad \cdots \quad \mathbf{a}_n] = A_i(\mathbf{b}) \end{aligned}$$

Therefore, by Theorem 4.8,

$$(\det A)(\det I_i(\mathbf{x})) = \det(AI_i(\mathbf{x})) = \det(A_i(\mathbf{b}))$$

Now

$$\det I_i(\mathbf{x}) = \begin{vmatrix} 1 & 0 & \cdots & x_1 & \cdots & 0 & 0 \\ 0 & 1 & \cdots & x_2 & \cdots & 0 & 0 \\ \vdots & \vdots & \ddots & \vdots & & \vdots & \vdots \\ 0 & 0 & \cdots & x_i & \cdots & 0 & 0 \\ \vdots & \vdots & & \vdots & \ddots & \vdots & \vdots \\ 0 & 0 & \cdots & x_{n-1} & \cdots & 1 & 0 \\ 0 & 0 & \cdots & x_n & \cdots & 0 & 1 \end{vmatrix} = x_i$$

as can be seen by expanding along the ith row. Thus, $(\det A)x_i = \det(A_i(\mathbf{b}))$, and the result follows by dividing by $\det A$ (which is nonzero, since A is invertible).

Example 4.16

Use Cramer's Rule to solve the system

$$\begin{aligned} x_1 + 2x_2 &= 2 \\ -x_1 + 4x_2 &= 1 \end{aligned}$$

Solution We compute

$$\det A = \begin{vmatrix} 1 & 2 \\ -1 & 4 \end{vmatrix} = 6, \quad \det(A_1(\mathbf{b})) = \begin{vmatrix} 2 & 2 \\ 1 & 4 \end{vmatrix} = 6, \quad \text{and} \quad \det(A_2(\mathbf{b})) = \begin{vmatrix} 1 & 2 \\ -1 & 1 \end{vmatrix} = 3$$

By Cramer's Rule,

$$x_1 = \frac{\det(A_1(\mathbf{b}))}{\det A} = \frac{6}{6} = 1 \quad \text{and} \quad x_2 = \frac{\det(A_2(\mathbf{b}))}{\det A} = \frac{3}{6} = \frac{1}{2}$$

Remark As noted previously, Cramer's Rule is computationally inefficient for all but small systems of linear equations because it involves the calculation of many determinants. The effort expended to compute just one of these determinants, using even the most efficient method, would be better spent using Gaussian elimination to solve the system directly.

The final result of this section is a formula for the inverse of a matrix in terms of determinants. This formula was hinted at by the formula for the inverse of a 3×3 matrix, which was given without proof at the beginning of this section. Thus, we have come full circle.

Let's discover the formula for ourselves. If A is an invertible $n \times n$ matrix, its inverse is the (unique) matrix X that satisfies the equation $AX = I$. Solving for X one column at a time, let $\mathbf{x}_j$ be the jth column of X. That is,

$$\mathbf{x}_j = \begin{bmatrix} x_{1j} \\ \vdots \\ x_{ij} \\ \vdots \\ x_{nj} \end{bmatrix}$$

Therefore, $A\mathbf{x}_j = \mathbf{e}_j$, and by Cramer's Rule,

$$x_{ij} = \frac{\det(A_i(\mathbf{e}_j))}{\det A}$$

However,

$$\det(A_i(\mathbf{e}_j)) = \begin{vmatrix} a_{11} & a_{12} & \cdots & 0 & \cdots & a_{1n} \\ a_{21} & a_{22} & \cdots & 0 & \cdots & a_{2n} \\ \vdots & \vdots & \ddots & \vdots & \iddots & \vdots \\ a_{j1} & a_{j2} & \cdots & 1 & \cdots & a_{jn} \\ \vdots & \vdots & \iddots & \vdots & \ddots & \vdots \\ a_{n1} & a_{n2} & \cdots & 0 & \cdots & a_{nn} \end{vmatrix} = (-1)^{j+i}\det A_{ji} = C_{ji}$$

(with the 0, 0, …, 1, …, 0 entries in the ith column)

which is the (j, i)-cofactor of A.

It follows that $x_{ij} = (1/\det A)C_{ji}$, so $A^{-1} = X = (1/\det A)[C_{ji}] = (1/\det A)[C_{ij}]^T$. In words, the inverse of A is the *transpose* of the matrix of cofactors of A, divided by the determinant of A.

The matrix

$$[C_{ji}] = [C_{ij}]^T = \begin{bmatrix} C_{11} & C_{21} & \cdots & C_{n1} \\ C_{12} & C_{22} & \cdots & C_{n2} \\ \vdots & \vdots & \ddots & \vdots \\ C_{1n} & C_{2n} & \cdots & C_{nn} \end{bmatrix}$$

is called the ***adjoint*** (or ***adjugate***) of A and is denoted by adj A. The result we have just proved can be stated as follows.

Theorem 4.12 Let A be an invertible $n \times n$ matrix. Then

$$A^{-1} = \frac{1}{\det A}\operatorname{adj} A$$

Example 4.17 Use the adjoint method to compute the inverse of

$$A = \begin{bmatrix} 1 & 2 & -1 \\ 2 & 2 & 4 \\ 1 & 3 & -3 \end{bmatrix}$$

Solution We compute $\det A = -2$ and the nine cofactors

$$C_{11} = +\begin{vmatrix} 2 & 4 \\ 3 & -3 \end{vmatrix} = -18 \quad C_{12} = -\begin{vmatrix} 2 & 4 \\ 1 & -3 \end{vmatrix} = 10 \quad C_{13} = +\begin{vmatrix} 2 & 2 \\ 1 & 3 \end{vmatrix} = 4$$

$$C_{21} = -\begin{vmatrix} 2 & -1 \\ 3 & -3 \end{vmatrix} = 3 \quad C_{22} = +\begin{vmatrix} 1 & -1 \\ 1 & -3 \end{vmatrix} = -2 \quad C_{23} = -\begin{vmatrix} 1 & 2 \\ 1 & 3 \end{vmatrix} = -1$$

$$C_{31} = +\begin{vmatrix} 2 & -1 \\ 2 & 4 \end{vmatrix} = 10 \quad C_{32} = -\begin{vmatrix} 1 & -1 \\ 2 & 4 \end{vmatrix} = -6 \quad C_{33} = +\begin{vmatrix} 1 & 2 \\ 2 & 2 \end{vmatrix} = -2$$

The adjoint is the *transpose* of the matrix of cofactors—namely,

$$\operatorname{adj} A = \begin{bmatrix} -18 & 10 & 4 \\ 3 & -2 & -1 \\ 10 & -6 & -2 \end{bmatrix}^T = \begin{bmatrix} -18 & 3 & 10 \\ 10 & -2 & -6 \\ 4 & -1 & -2 \end{bmatrix}$$

Then

$$A^{-1} = \frac{1}{\det A}\operatorname{adj} A = -\frac{1}{2}\begin{bmatrix} -18 & 3 & 10 \\ 10 & -2 & -6 \\ 4 & -1 & -2 \end{bmatrix} = \begin{bmatrix} 9 & -\frac{3}{2} & -5 \\ -5 & 1 & 3 \\ -2 & \frac{1}{2} & 1 \end{bmatrix}$$

which is the same answer we obtained (with less work) in Example 3.30.

Proof of the Laplace Expansion Theorem

Unfortunately, there is no short, easy proof of the Laplace Expansion Theorem. The proof we give has the merit of being relatively straightforward. We break it down into several steps, the first of which is to prove that cofactor expansion along the first row of a matrix is the same as cofactor expansion along the first column.

Lemma 4.13 Let A be an $n \times n$ matrix. Then

$$a_{11}C_{11} + a_{12}C_{12} + \cdots + a_{1n}C_{1n} = \det A = a_{11}C_{11} + a_{21}C_{21} + \cdots + a_{n1}C_{n1} \qquad (7)$$

Proof We prove this lemma by induction on n. For $n = 1$, the result is trivial. Now assume that the result is true for $(n-1) \times (n-1)$ matrices; this is our induction hypothesis. Note that, by the definition of cofactor (or minor), all of the terms containing a_{11} are accounted for by the summand $a_{11}C_{11}$. We can therefore ignore terms containing a_{11}.

The ith summand on the right-hand side of Equation (7) is $a_{i1}C_{i1} = a_{i1}(-1)^{i+1}$ det A_{i1}. Now we expand det A_{i1} along the first row:

$$\begin{vmatrix} a_{12} & a_{13} & \cdots & a_{1j} & \cdots & a_{1n} \\ \vdots & \vdots & \ddots & \vdots & \cdot^{\cdot^{\cdot}} & \vdots \\ a_{i-1,2} & a_{i-1,3} & \cdots & a_{i-1,j} & \cdots & a_{i-1,n} \\ a_{i+1,2} & a_{i+1,3} & \cdots & a_{i+1,j} & \cdots & a_{i+1,n} \\ \vdots & \vdots & \cdot^{\cdot^{\cdot}} & \vdots & \ddots & \vdots \\ a_{n2} & a_{n3} & \cdots & a_{nj} & \cdots & a_{n,n} \end{vmatrix}$$

The jth term in this expansion of det A_{i1} is $a_{1j}(-1)^{1+j-1}$ det $A_{1i,1j}$, where the notation $A_{kl,rs}$ denotes the submatrix of A obtained by deleting rows k and l and columns r and s. Combining these, we see that the term containing $a_{i1}a_{1j}$ on the right-hand side of Equation (7) is

$$a_{i1}(-1)^{i+1}a_{1j}(-1)^{1+j-1} \det A_{1i,1j} = (-1)^{i+j+1}a_{i1}a_{1j} \det A_{1i,1j}$$

What is the term containing $a_{i1}a_{1j}$ on the left-hand side of Equation (7)? The factor a_{1j} occurs in the jth summand, $a_{1j}C_{1j} = a_{1j}(-1)^{1+j}$ det A_{1j}. By the induction hypothesis, we can expand det A_{1j} along its first column:

$$\begin{vmatrix} a_{21} & \cdots & a_{2,j-1} & a_{2,j+1} & \cdots & a_{2n} \\ a_{31} & \cdots & a_{3,j-1} & a_{3,j+1} & \cdots & a_{3n} \\ \vdots & \ddots & \vdots & \vdots & \cdot^{\cdot^{\cdot}} & \vdots \\ a_{i1} & \cdots & a_{i,j-1} & a_{i,j+1} & \cdots & a_{in} \\ \vdots & \cdot^{\cdot^{\cdot}} & \vdots & \vdots & \ddots & \vdots \\ a_{n1} & \cdots & a_{n,j-1} & a_{n,j+1} & \cdots & a_{nn} \end{vmatrix}$$

The ith term in this expansion of det A_{1j} is $a_{i1}(-1)^{(i-1)+1}$ det $A_{1i,1j}$, so the term containing $a_{i1}a_{1j}$ on the left-hand side of Equation (7) is

$$a_{1j}(-1)^{1+j}a_{i1}(-1)^{(i-1)+1} \det A_{1i,1j} = (-1)^{i+j+1}a_{i1}a_{1j} \det A_{1i,1j}$$

which establishes that the left- and right-hand sides of Equation (7) are equivalent.

Next, we prove property (b) of Theorem 4.3.

Lemma 4.14 Let A be an $n \times n$ matrix and let B be obtained by interchanging any two rows (columns) of A. Then

$$\det B = -\det A$$

Proof Once again, the proof is by induction on n. The result can be easily checked when $n = 2$, so assume that it is true for $(n - 1) \times (n - 1)$ matrices. We will prove that the result is true for $n \times n$ matrices. First, we prove that it holds when two adjacent rows of A are interchanged—say, rows r and $r + 1$.

By Lemma 4.13, we can evaluate det B by cofactor expansion along its first column. The ith term in this expansion is $(-1)^{1+i}b_{i1} \det B_{i1}$. If $i \neq r$ and $i \neq r + 1$, then $b_{i1} = a_{i1}$ and B_{i1} is an $(n - 1) \times (n - 1)$ submatrix that is identical to A_{i1} except that two adjacent rows have been interchanged.

$$\begin{vmatrix} a_{11} & a_{12} & \cdots & a_{1n} \\ \vdots & \vdots & & \vdots \\ a_{i1} & a_{i2} & \cdots & a_{in} \\ \vdots & \vdots & & \vdots \\ a_{r+1,1} & a_{r+1,2} & \cdots & a_{r+1,n} \\ a_{r1} & a_{r2} & \cdots & a_{rn} \\ \vdots & \vdots & & \vdots \\ a_{n1} & a_{n2} & \cdots & a_{nn} \end{vmatrix}$$

Thus, by the induction hypothesis, $\det B_{i1} = -\det A_{i1}$ if $i \neq r$ and $i \neq r + 1$.

If $i = r$, then $b_{i1} = a_{r+1,1}$ and $B_{i1} = A_{r+1,1}$.

$$\text{Row } i \rightarrow \begin{vmatrix} a_{11} & a_{12} & \cdots & a_{1n} \\ \vdots & \vdots & & \vdots \\ a_{r+1,1} & a_{r+1,2} & \cdots & a_{r+1,n} \\ a_{r1} & a_{r2} & \cdots & a_{rn} \\ \vdots & \vdots & & \vdots \\ a_{n1} & a_{n2} & \cdots & a_{nn} \end{vmatrix}$$

Therefore, the rth summand in det B is

$$(-1)^{r+1}b_{r1} \det B_{r1} = (-1)^{r+1}a_{r+1,1} \det A_{r+1,1} = -(-1)^{(r+1)+1}a_{r+1,1} \det A_{r+1,1}$$

Similarly, if $i = r + 1$, then $b_{i1} = a_{r1}$, $B_{i1} = A_{r1}$, and the $(r + 1)$st summand in det B is

$$(-1)^{(r+1)+1}b_{r+1,1} \det B_{r+1,1} = (-1)^{r}a_{r1} \det A_{r1} = -(-1)^{r+1}a_{r1} \det A_{r1}$$

In other words, the rth and $(r + 1)$st terms in the first column cofactor expansion of det B are the *negatives* of the $(r + 1)$st and rth terms, respectively, in the first column cofactor expansion of det A.

Substituting all of these results into det B and using Lemma 4.13 again, we obtain

$$\begin{aligned} \det B &= \sum_{i=1}^{n}(-1)^{i+1}b_{i1} \det B_{i1} \\ &= \sum_{\substack{i=1 \\ i\neq r,r+1}}^{n} (-1)^{i+1}b_{i1} \det B_{i1} + (-1)^{r+1}b_{r1} \det B_{r1} + (-1)^{(r+1)+1}b_{r+1,1} \det B_{r+1,1} \\ &= \sum_{\substack{i=1 \\ i\neq r,r+1}}^{n} (-1)^{i+1}a_{i1}(-\det A_{i1}) - (-1)^{(r+1)+1}a_{r+1,1} \det A_{r+1,1} - (-1)^{r+1}a_{r1} \det A_{r1} \\ &= -\sum_{i=1}^{n}(-1)^{i+1}a_{i1} \det A_{i1} \\ &= -\det A \end{aligned}$$

This proves the result for $n \times n$ matrices if adjacent rows are interchanged. To see that it holds for arbitrary row interchanges, we need only note that, for example, rows r and s, where $r < s$, can be swapped by performing $2(s - r) - 1$ interchanges of adjacent rows (see Exercise 67). Since the number of interchanges is *odd* and each one changes the sign of the determinant, the net effect is a change of sign, as desired.

The proof for column interchanges is analogous, except that we expand along row 1 instead of along column 1.

We can now prove the Laplace Expansion Theorem.

Proof of Theorem 4.1 Let B be the matrix obtained by moving row i of A to the top, using $i - 1$ interchanges of adjacent rows. By Lemma 4.14, $\det B = (-1)^{i-1} \det A$. But $b_{1j} = a_{ij}$ and $B_{1j} = A_{ij}$ for $j = 1, \ldots, n$.

$$\det B = \begin{vmatrix} a_{i1} & \cdots & a_{ij} & \cdots & a_{in} \\ a_{11} & \cdots & a_{1j} & \cdots & a_{1n} \\ \vdots & & \vdots & & \vdots \\ a_{i-1,1} & \cdots & a_{i-1,j} & \cdots & a_{i-1,n} \\ a_{i+1,1} & \cdots & a_{i+1,j} & \cdots & a_{i+1,n} \\ \vdots & & \vdots & & \vdots \\ a_{n1} & \cdots & a_{nj} & \cdots & a_{nn} \end{vmatrix}$$

Thus,

$$\begin{aligned} \det A &= (-1)^{i-1} \det B = (-1)^{i-1} \sum_{j=1}^{n} (-1)^{1+j} b_{1j} \det B_{1j} \\ &= (-1)^{i-1} \sum_{j=1}^{n} (-1)^{1+j} a_{ij} \det A_{ij} = \sum_{j=1}^{n} (-1)^{i+j} a_{ij} \det A_{ij} \end{aligned}$$

which gives the formula for cofactor expansion along row i.

The proof for column expansion is similar, invoking Lemma 4.13 so that we can use column expansion instead of row expansion (see Exercise 68).

A Brief History of Determinants

As noted at the beginning of this section, the history of determinants predates that of matrices. Indeed, determinants were first introduced, independently, by Seki in 1683 and Leibniz in 1693. In 1748, determinants appeared in Maclaurin's *Treatise on Algebra*, which included a treatment of Cramer's Rule up to the 4×4 case. In 1750, Cramer himself proved the general case of his rule, applying it to curve fitting, and in 1772, Laplace gave a proof of his expansion theorem.

The term *determinant* was not coined until 1801, when it was used by Gauss. Cauchy made the first use of determinants in the modern sense in 1812. Cauchy, in fact, was responsible for developing much of the early theory of determinants, including several important results that we have mentioned: the product rule for determinants, the characteristic polynomial, and the notion of a diagonalizable matrix. Determinants did not become widely known until 1841, when Jacobi popularized them, albeit in the context of functions of several variables, such as are encountered in a multivariable calculus course. (These types of determinants were called "Jacobians" by Sylvester around 1850, a term that is still used today.)

A self-taught child prodigy, Takakazu Seki Kōwa (1642–1708) was descended from a family of samurai warriors. In addition to discovering determinants, he wrote about diophantine equations, magic squares, and Bernoulli numbers (before Bernoulli) and quite likely made discoveries in calculus.

Gottfried Wilhelm von Leibniz (1646–1716) was born in Leipzig and studied law, theology, philosophy, and mathematics. He is probably best known for developing (with Newton, independently) the main ideas of differential and integral calculus. However, his contributions to other branches of mathematics are also impressive. He developed the notion of a determinant, knew versions of Cramer's Rule and the Laplace Expansion Theorem before others were given credit for them, and laid the foundation for matrix theory through work he did on quadratic forms. Leibniz also was the first to develop the binary system of arithmetic. He believed in the importance of good notation and, along with the familiar notation for derivatives and integrals, introduced a form of subscript notation for the coefficients of a linear system that is essentially the notation we use today.

By the late 19th century, the theory of determinants had developed to the stage that entire books were devoted to it, including Dodgson's *An Elementary Treatise on Determinants* in 1867 and Thomas Muir's monumental five-volume work, which appeared in the early 20th century. While their history is fascinating, today determinants are of theoretical more than practical interest. Cramer's Rule is a hopelessly inefficient method for solving a system of linear equations, and numerical methods have replaced any use of determinants in the computation of eigenvalues. Determinants are used, however, to give students an initial understanding of the characteristic polynomial (as in Sections 4.1 and 4.3).

Exercises 4.2

Compute the determinants in Exercises 1–6 using cofactor expansion along the first row and along the first column.

1. $\begin{vmatrix} 1 & 0 & 3 \\ 5 & 1 & 1 \\ 0 & 1 & 2 \end{vmatrix}$

2. $\begin{vmatrix} 0 & 1 & -1 \\ 2 & 3 & -2 \\ -1 & 3 & 0 \end{vmatrix}$

3. $\begin{vmatrix} 1 & -1 & 0 \\ -1 & 0 & 1 \\ 0 & 1 & -1 \end{vmatrix}$

4. $\begin{vmatrix} 1 & 1 & 0 \\ 1 & 0 & 1 \\ 0 & 1 & 1 \end{vmatrix}$

5. $\begin{vmatrix} 1 & 2 & 3 \\ 2 & 3 & 1 \\ 3 & 1 & 2 \end{vmatrix}$

6. $\begin{vmatrix} 1 & 2 & 3 \\ 4 & 5 & 6 \\ 7 & 8 & 9 \end{vmatrix}$

Compute the determinants in Exercises 7–15 using cofactor expansion along any row or column that seems convenient.

7. $\begin{vmatrix} 5 & 2 & 2 \\ -1 & 1 & 2 \\ 3 & 0 & 0 \end{vmatrix}$

8. $\begin{vmatrix} 1 & 1 & -1 \\ 2 & 0 & 1 \\ 3 & -2 & 1 \end{vmatrix}$

9. $\begin{vmatrix} -4 & 1 & 3 \\ 2 & -2 & 4 \\ 1 & -1 & 0 \end{vmatrix}$

10. $\begin{vmatrix} \cos\theta & \sin\theta & \tan\theta \\ 0 & \cos\theta & -\sin\theta \\ 0 & \sin\theta & \cos\theta \end{vmatrix}$

11. $\begin{vmatrix} a & b & 0 \\ 0 & a & b \\ a & 0 & b \end{vmatrix}$

12. $\begin{vmatrix} 0 & a & 0 \\ b & c & d \\ 0 & e & 0 \end{vmatrix}$

13. $\begin{vmatrix} 1 & -1 & 0 & 3 \\ 2 & 5 & 2 & 6 \\ 0 & 1 & 0 & 0 \\ 1 & 4 & 2 & 1 \end{vmatrix}$

14. $\begin{vmatrix} 2 & 0 & 3 & -1 \\ 1 & 0 & 2 & 2 \\ 0 & -1 & 1 & 4 \\ 2 & 0 & 1 & -3 \end{vmatrix}$

15. $\begin{vmatrix} 0 & 0 & 0 & a \\ 0 & 0 & b & c \\ 0 & d & e & f \\ g & h & i & j \end{vmatrix}$

In Exercises 16–18, compute the indicated 3×3 determinants using the method of Example 4.9.

16. The determinant in Exercise 6

17. The determinant in Exercise 8

18. The determinant in Exercise 11

19. Verify that the method indicated in (2) agrees with Equation (1) for a 3×3 determinant.

20. Verify that definition (4) agrees with the definition of a 2×2 determinant when $n = 2$.

21. Prove Theorem 4.2. [*Hint:* A proof by induction would be appropriate here.]

In Exercises 22–25, evaluate the given determinant using elementary row and/or column operations and Theorem 4.3 to reduce the matrix to row echelon form.

22. The determinant in Exercise 1

23. The determinant in Exercise 9

24. The determinant in Exercise 13

25. The determinant in Exercise 14

In Exercises 26–34, use properties of determinants to evaluate the given determinant by inspection. Explain your reasoning.

26. $\begin{vmatrix} 1 & 1 & 1 \\ 3 & 0 & -2 \\ 2 & 2 & 2 \end{vmatrix}$

27. $\begin{vmatrix} 3 & 1 & 0 \\ 0 & -2 & 5 \\ 0 & 0 & 4 \end{vmatrix}$

28. $\begin{vmatrix} 0 & 0 & 1 \\ 0 & 5 & 2 \\ 3 & -1 & 4 \end{vmatrix}$

29. $\begin{vmatrix} 2 & 3 & -4 \\ 1 & -3 & -2 \\ -1 & 5 & 2 \end{vmatrix}$

30. $\begin{vmatrix} 1 & 2 & 3 \\ 0 & 4 & 1 \\ 1 & 6 & 4 \end{vmatrix}$

31. $\begin{vmatrix} 4 & 1 & 3 \\ -2 & 0 & -2 \\ 5 & 4 & 1 \end{vmatrix}$

32. $\begin{vmatrix} 1 & 0 & 0 & 0 \\ 0 & 0 & 1 & 0 \\ 0 & 1 & 0 & 0 \\ 0 & 0 & 0 & 1 \end{vmatrix}$

33. $\begin{vmatrix} 0 & 2 & 0 & 0 \\ -3 & 0 & 0 & 0 \\ 0 & 0 & 0 & 4 \\ 0 & 0 & 1 & 0 \end{vmatrix}$

34. $\begin{vmatrix} 1 & 0 & 1 & 0 \\ 0 & 1 & 0 & 1 \\ 1 & 1 & 0 & 0 \\ 0 & 0 & 1 & 1 \end{vmatrix}$

Find the determinants in Exercises 35–40, assuming that

$$\begin{vmatrix} a & b & c \\ d & e & f \\ g & h & i \end{vmatrix} = 4$$

35. $\begin{vmatrix} 2a & 2b & 2c \\ d & e & f \\ g & h & i \end{vmatrix}$

36. $\begin{vmatrix} 2a & b/3 & -c \\ 2d & e/3 & -f \\ 2g & h/3 & -i \end{vmatrix}$

37. $\begin{vmatrix} d & e & f \\ a & b & c \\ g & h & i \end{vmatrix}$

38. $\begin{vmatrix} a-c & b & c \\ d-f & e & f \\ g-i & h & i \end{vmatrix}$

39. $\begin{vmatrix} 2c & b & a \\ 2f & e & d \\ 2i & h & g \end{vmatrix}$

40. $\begin{vmatrix} a+2g & b+2h & c+2i \\ 3d+2g & 3e+2h & 3f+2i \\ g & h & i \end{vmatrix}$

41. Prove Theorem 4.3(a).

42. Prove Theorem 4.3(f).

43. Prove Lemma 4.5.

44. Prove Theorem 4.7.

In Exercises 45 and 46, use Theorem 4.6 to find all values of k for which A is invertible.

45. $A = \begin{bmatrix} k & -k & 3 \\ 0 & k+1 & 1 \\ k & -8 & k-1 \end{bmatrix}$

46. $A = \begin{bmatrix} k & k & 0 \\ k^2 & 2 & k \\ 0 & k & k \end{bmatrix}$

In Exercises 47–52, assume that A and B are $n \times n$ matrices with $\det A = 3$ and $\det B = -2$. Find the indicated determinants.

47. $\det(AB)$

48. $\det(A^2)$

49. $\det(B^{-1}A)$

50. $\det(2A)$

51. $\det(3B^T)$

52. $\det(AA^T)$

In Exercises 53–56, A and B are $n \times n$ matrices.

53. Prove that $\det(AB) = \det(BA)$.

54. If B is invertible, prove that $\det(B^{-1}AB) = \det(A)$.

55. If A is idempotent (that is, $A^2 = A$), find all possible values of $\det(A)$.

56. A square matrix A is called ***nilpotent*** if $A^m = O$ for some $m > 1$. (The word nilpotent comes from the Latin *nil,* meaning "nothing," and *potere,* meaning "to have power." A nilpotent matrix is thus one that becomes "nothing"—that is, the zero matrix—when raised to some power.) Find all possible values of $\det(A)$ if A is nilpotent.

In Exercises 57–60, use Cramer's Rule to solve the given linear system.

57. $x + y = 1$
$x - y = 2$

58. $2x - y = 5$
$x + 3y = -1$

59. $2x + y + 3z = 1$
$y + z = 1$
$z = 1$

60. $x + y - z = 1$
$x + y + z = 2$
$x - y = 3$

In Exercises 61–64, use Theorem 4.12 to compute the inverse of the coefficient matrix for the given exercise.

61. Exercise 57

62. Exercise 58

63. Exercise 59

64. Exercise 60

65. If A is an invertible $n \times n$ matrix, show that adj A is also invertible and that

$$(\text{adj } A)^{-1} = \frac{1}{\det A}A = \text{adj } (A^{-1})$$

66. If A is an $n \times n$ matrix, prove that

$$\det(\text{adj } A) = (\det A)^{n-1}$$

67. Verify that if $r < s$, then rows r and s of a matrix can be interchanged by performing $2(s - r) - 1$ interchanges of adjacent rows.

68. Prove that the Laplace Expansion Theorem holds for column expansion along the jth column.

69. Let A be a square matrix that can be partitioned as

$$A = \left[\begin{array}{c|c} P & Q \\ \hline O & S \end{array}\right]$$

where P and S are square matrices. Such a matrix is said to be in ***block (upper) triangular form.*** Prove that

$$\det A = (\det P)(\det S)$$

[*Hint:* Try a proof by induction on the number of rows of P.]

70. (a) Give an example to show that if A can be partitioned as

$$A = \left[\begin{array}{c|c} P & Q \\ \hline R & S \end{array}\right]$$

where P, Q, R, and S are all square, then it is not necessarily true that

$$\det A = (\det P)(\det S) - (\det Q)(\det R)$$

(b) Assume that A is partitioned as in part (a) and that P is invertible. Let

$$B = \left[\begin{array}{c|c} P^{-1} & O \\ \hline -RP^{-1} & I \end{array}\right]$$

Compute $\det(BA)$ using Exercise 69 and use the result to show that

$$\det A = \det P \det(S - RP^{-1}Q)$$

[The matrix $S - RP^{-1}Q$ is called the ***Schur complement*** of P in A, after Issai Schur (1875–1941), who was born in Belarus but spent most of his life in Germany. He is known mainly for his fundamental work on the representation theory of groups, but he also worked in number theory, analysis, and other areas.]

(c) Assume that A is partitioned as in part (a), that P is invertible, and that $PR = RP$. Prove that

$$\det A = \det(PS - RQ)$$

Writing Project

Which Came First: The Matrix or the Determinant?

The way in which matrices and determinants are taught today—matrices before determinants—bears little resemblance to the way these topics developed historically. There is a brief history of determinants at the end of Section 4.2.

Write a report on the history of matrices and determinants. How did the notations used for each evolve over time? Who were some of the key mathematicians involved and what were their contributions?

1. Florian Cajori, *A History of Mathematical Notations* (New York: Dover, 1993).
2. Howard Eves, *An Introduction to the History of Mathematics* (Sixth Edition) (Philadelphia: Saunders College Publishing, 1990).
3. Victor J. Katz, *A History of Mathematics: An Introduction* (Third Edition) (Reading, MA: Addison Wesley Longman, 2008).
4. Eberhard Knobloch, Determinants, in Ivor Grattan-Guinness, ed., *Companion Encyclopedia of the History and Philosophy of the Mathematical Sciences* (London: Routledge, 2013).

Vignette

Lewis Carroll's Condensation Method

Charles Lutwidge Dodgson (1832–1898) is much better known by his pen name, Lewis Carroll, under which he wrote *Alice's Adventures in Wonderland* and *Through the Looking Glass*. He also wrote several mathematics books and collections of logic puzzles.

This vignette is based on the article "Lewis Carroll's Condensation Method for Evaluating Determinants" by Adrian Rice and Eve Torrence in *Math Horizons*, November 2006, pp. 12–15. For further details of the condensation method, see David M. Bressoud, *Proofs and Confirmations: The Story of the Alternating Sign Matrix Conjecture*, MAA Spectrum Series (Cambridge University Press, 1999).

In 1866, Charles Dodgson—better known by his pseudonym Lewis Carroll—published his only mathematical research paper. In it, he described a "new and brief method" for computing determinants, which he called "condensation." Although not well known today and rendered obsolete by numerical methods for evaluating determinants, the condensation method is very useful for hand calculation. When calculators or computer algebra systems are not available, many students find condensation to be their method of choice. It requires only the ability to compute 2×2 determinants.

We require the following terminology.

Definition If A is an $n \times n$ matrix with $n \geq 3$, the ***interior*** of A, denoted $\text{int}(A)$, is the $(n - 2) \times (n - 2)$ matrix obtained by deleting the first row, last row, first column, and last column of A.

We will illustrate the condensation method for the 5×5 matrix

$$A = \begin{bmatrix} 2 & 3 & -1 & 2 & 0 \\ 1 & 2 & 3 & 1 & -4 \\ 2 & -1 & 2 & 1 & 1 \\ 3 & 1 & -1 & 2 & -2 \\ -4 & 1 & 0 & 1 & 2 \end{bmatrix}$$

Begin by setting A_0 equal to the 6×6 matrix all of whose entries are 1. Then, we set $A_1 = A$. It is useful to imagine A_0 as the base of a pyramid with A_1 centered on top of A_0. We are going to add successively smaller and smaller layers to the pyramid until we reach a 1×1 matrix at the top—this will contain $\det A$. (Figure 4.9)

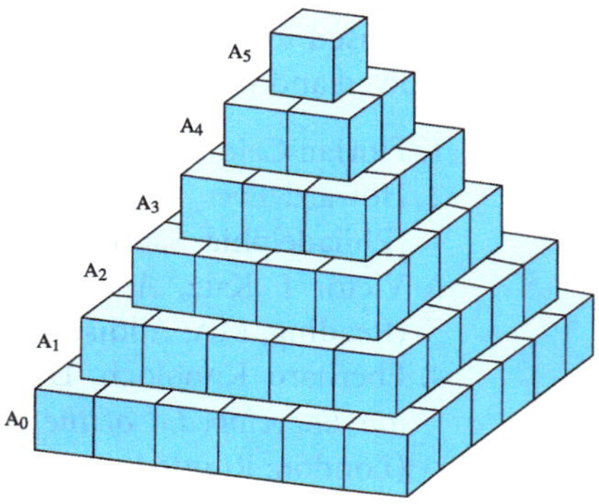

Figure 4.9

Next, we "condense" A_1 into a 4×4 matrix A_2' whose entries are the determinants of all 2×2 submatrices of A_1:

$$A_2' = \begin{bmatrix} \begin{vmatrix} 2 & 3 \\ 1 & 2 \end{vmatrix} & \begin{vmatrix} 3 & -1 \\ 2 & 3 \end{vmatrix} & \begin{vmatrix} -1 & 2 \\ 3 & 1 \end{vmatrix} & \begin{vmatrix} 2 & 0 \\ 1 & -4 \end{vmatrix} \\ \begin{vmatrix} 1 & 2 \\ 2 & -1 \end{vmatrix} & \begin{vmatrix} 2 & 3 \\ -1 & 2 \end{vmatrix} & \begin{vmatrix} 3 & 1 \\ 2 & 1 \end{vmatrix} & \begin{vmatrix} 1 & -4 \\ 1 & 1 \end{vmatrix} \\ \begin{vmatrix} 2 & -1 \\ 3 & 1 \end{vmatrix} & \begin{vmatrix} -1 & 2 \\ 1 & -1 \end{vmatrix} & \begin{vmatrix} 2 & 1 \\ -1 & 2 \end{vmatrix} & \begin{vmatrix} 1 & 1 \\ 2 & -2 \end{vmatrix} \\ \begin{vmatrix} 3 & 1 \\ -4 & 1 \end{vmatrix} & \begin{vmatrix} 1 & -1 \\ 1 & 0 \end{vmatrix} & \begin{vmatrix} -1 & 2 \\ 0 & 1 \end{vmatrix} & \begin{vmatrix} 2 & -2 \\ 1 & 2 \end{vmatrix} \end{bmatrix} = \begin{bmatrix} 1 & 11 & -7 & -8 \\ -5 & 7 & 1 & 5 \\ 5 & -1 & 5 & -4 \\ 7 & 1 & -1 & 6 \end{bmatrix}$$

Now we divide each entry of A_2' by the corresponding entry of $\text{int}(A_0)$ to get matrix A_2. Since A_0 is all 1s, this means $A_2 = A_2'$.

We repeat the procedure, constructing A_3' from the 2×2 submatrices of A_2 and then dividing each entry of A_3' by the corresponding entry of $\text{int}(A_1)$, and so on. We obtain:

$$A_3' = \begin{bmatrix} \begin{vmatrix} 1 & 11 \\ -5 & 7 \end{vmatrix} & \begin{vmatrix} 11 & -7 \\ 7 & 1 \end{vmatrix} & \begin{vmatrix} -7 & -8 \\ 1 & 5 \end{vmatrix} \\ \begin{vmatrix} -5 & 7 \\ 5 & -1 \end{vmatrix} & \begin{vmatrix} 7 & 1 \\ -1 & 5 \end{vmatrix} & \begin{vmatrix} 1 & 5 \\ 5 & -4 \end{vmatrix} \\ \begin{vmatrix} 5 & -1 \\ 7 & 1 \end{vmatrix} & \begin{vmatrix} -1 & 5 \\ 1 & -1 \end{vmatrix} & \begin{vmatrix} 5 & -4 \\ -1 & 6 \end{vmatrix} \end{bmatrix} = \begin{bmatrix} 62 & 60 & -27 \\ -30 & 36 & -29 \\ 12 & -4 & 26 \end{bmatrix},$$

$$A_3 = \begin{bmatrix} 62/2 & 60/3 & -27/1 \\ -30/-1 & 36/2 & -29/1 \\ 12/1 & -4/-1 & 26/2 \end{bmatrix} = \begin{bmatrix} 31 & 20 & -27 \\ 30 & 18 & -29 \\ 12 & 4 & 13 \end{bmatrix},$$

$$A_4' = \begin{bmatrix} \begin{vmatrix} 31 & 20 \\ 30 & 18 \end{vmatrix} & \begin{vmatrix} 20 & -27 \\ 18 & -29 \end{vmatrix} \\ \begin{vmatrix} 30 & 18 \\ 12 & 4 \end{vmatrix} & \begin{vmatrix} 18 & -29 \\ 4 & 13 \end{vmatrix} \end{bmatrix} = \begin{bmatrix} -42 & -94 \\ -96 & 350 \end{bmatrix},$$

$$A_4 = \begin{bmatrix} -42/7 & -94/1 \\ -96/(-1) & 350/5 \end{bmatrix} = \begin{bmatrix} -6 & -94 \\ 96 & 70 \end{bmatrix}$$

$$A_5' = \begin{bmatrix} \begin{vmatrix} -6 & -94 \\ 96 & 70 \end{vmatrix} \end{bmatrix} = [8604],$$

$$A_5 = [8604/18] = [478]$$

As can be checked by other methods, $\det A = 478$. In general, for an $n \times n$ matrix A, the condensation method will produce a 1×1 matrix A_n containing $\det A$.

Clearly, the method breaks down if the interior of any of the A_i' s contains a zero, since we would then be trying to divide by zero to construct A_{i+1}. However, careful use of elementary row and column operations can be used to eliminate the zeros so that we can proceed.

Exploration

Geometric Applications of Determinants

This exploration will reveal some of the amazing applications of determinants to geometry. In particular, we will see that determinants are closely related to area and volume formulas and can be used to produce the equations of lines, planes, and certain other curves. Most of these ideas arose when the theory of determinants was being developed as a subject in its own right.

The Cross Product

Recall from Exploration: The Cross Product in Chapter 1 that the cross product of $\mathbf{u} = \begin{bmatrix} u_1 \\ u_2 \\ u_3 \end{bmatrix}$ and $\mathbf{v} = \begin{bmatrix} v_1 \\ v_2 \\ v_3 \end{bmatrix}$ is the vector $\mathbf{u} \times \mathbf{v}$ defined by

$$\mathbf{u} \times \mathbf{v} = \begin{bmatrix} u_2v_3 - u_3v_2 \\ u_3v_1 - u_1v_3 \\ u_1v_2 - u_2v_1 \end{bmatrix}$$

If we write this cross product as $(u_2v_3 - u_3v_2)\mathbf{e}_1 - (u_1v_3 - u_3v_1)\mathbf{e}_2 + (u_1v_2 - u_2v_1)\mathbf{e}_3$, where $\mathbf{e}_1$, $\mathbf{e}_2$, and $\mathbf{e}_3$ are the standard basis vectors, then we see that the *form* of this formula is

$$\mathbf{u} \times \mathbf{v} = \det \begin{bmatrix} \mathbf{e}_1 & u_1 & v_1 \\ \mathbf{e}_2 & u_2 & v_2 \\ \mathbf{e}_3 & u_3 & v_3 \end{bmatrix}$$

if we expand along the first column. (This is not a proper determinant, of course, since $\mathbf{e}_1$, $\mathbf{e}_2$, and $\mathbf{e}_3$ are vectors, not scalars; however, it gives a useful way of remembering the somewhat awkward cross product formula. It also lets us use properties of determinants to verify some of the properties of the cross product.)

Now let's revisit some of the exercises from Chapter 1.

1. Use the determinant version of the cross product to compute $\mathbf{u} \times \mathbf{v}$.

(a) $\mathbf{u} = \begin{bmatrix} 0 \\ 1 \\ 1 \end{bmatrix}, \mathbf{v} = \begin{bmatrix} 3 \\ -1 \\ 2 \end{bmatrix}$ (b) $\mathbf{u} = \begin{bmatrix} 3 \\ -1 \\ 2 \end{bmatrix}, \mathbf{v} = \begin{bmatrix} 0 \\ 1 \\ 1 \end{bmatrix}$

(c) $\mathbf{u} = \begin{bmatrix} -1 \\ 2 \\ 3 \end{bmatrix}, \mathbf{v} = \begin{bmatrix} 2 \\ -4 \\ -6 \end{bmatrix}$ (d) $\mathbf{u} = \begin{bmatrix} 1 \\ 1 \\ 1 \end{bmatrix}, \mathbf{v} = \begin{bmatrix} 1 \\ 2 \\ 3 \end{bmatrix}$

2. If $\mathbf{u} = \begin{bmatrix} u_1 \\ u_2 \\ u_3 \end{bmatrix}, \mathbf{v} = \begin{bmatrix} v_1 \\ v_2 \\ v_3 \end{bmatrix}$, and $\mathbf{w} = \begin{bmatrix} w_1 \\ w_2 \\ w_3 \end{bmatrix}$, show that

$$\mathbf{u} \cdot (\mathbf{v} \times \mathbf{w}) = \det \begin{bmatrix} u_1 & v_1 & w_1 \\ u_2 & v_2 & w_2 \\ u_3 & v_3 & w_3 \end{bmatrix}$$

3. Use properties of determinants (and Problem 2 above, if necessary) to prove the given property of the cross product.

(a) $\mathbf{v} \times \mathbf{u} = -(\mathbf{u} \times \mathbf{v})$ (b) $\mathbf{u} \times \mathbf{0} = \mathbf{0}$

(c) $\mathbf{u} \times \mathbf{u} = \mathbf{0}$ (d) $\mathbf{u} \times k\mathbf{v} = k(\mathbf{u} \times \mathbf{v})$

(e) $\mathbf{u} \times (\mathbf{v} + \mathbf{w}) = \mathbf{u} \times \mathbf{v} + \mathbf{u} \times \mathbf{w}$ (f) $\mathbf{u} \cdot (\mathbf{u} \times \mathbf{v}) = 0$ and $\mathbf{v} \cdot (\mathbf{u} \times \mathbf{v}) = 0$

(g) $\mathbf{u} \cdot (\mathbf{v} \times \mathbf{w}) = (\mathbf{u} \times \mathbf{v}) \cdot \mathbf{w}$ (the ***triple scalar product identity***)

Area and Volume

We can now give a geometric interpretation of the determinants of 2×2 and 3×3 matrices. Recall that if $\mathbf{u}$ and $\mathbf{v}$ are vectors in $\mathbb{R}^3$, then the area $\mathcal{A}$ of the parallelogram determined by these vectors is given by $\mathcal{A} = \|\mathbf{u} \times \mathbf{v}\|$. (See Exploration: The Cross Product in Chapter 1.)

4. Let $\mathbf{u} = \begin{bmatrix} u_1 \\ u_2 \end{bmatrix}$ and $\mathbf{v} = \begin{bmatrix} v_1 \\ v_2 \end{bmatrix}$. Show that the area $\mathcal{A}$ of the parallelogram determined by $\mathbf{u}$ and $\mathbf{v}$ is given by

$$\mathcal{A} = \left| \det \begin{bmatrix} u_1 & v_1 \\ u_2 & v_2 \end{bmatrix} \right|$$

[*Hint:* Write $\mathbf{u}$ and $\mathbf{v}$ as $\begin{bmatrix} u_1 \\ u_2 \\ 0 \end{bmatrix}$ and $\begin{bmatrix} v_1 \\ v_2 \\ 0 \end{bmatrix}$.]

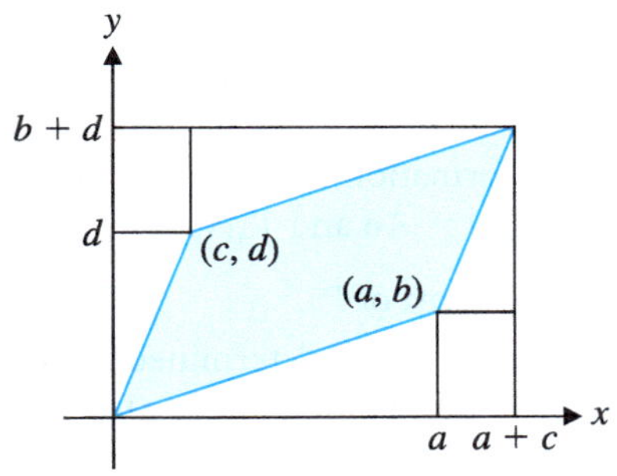

Figure 4.10

5. Derive the area formula in Problem 4 geometrically, using Figure 4.10 as a guide. [*Hint:* Subtract areas from the large rectangle until the parallelogram remains.] Where does the absolute value sign come from in this case?

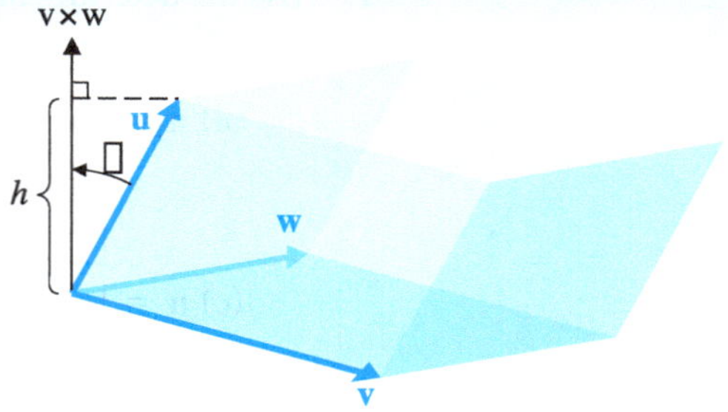

Figure 4.11

6. Find the area of the parallelogram determined by **u** and **v**.

$$\text{(a) } \mathbf{u} = \begin{bmatrix} 2 \\ 3 \end{bmatrix}, \mathbf{v} = \begin{bmatrix} -1 \\ 4 \end{bmatrix} \quad \text{(b) } \mathbf{u} = \begin{bmatrix} 3 \\ 4 \end{bmatrix}, \mathbf{v} = \begin{bmatrix} 5 \\ 5 \end{bmatrix}$$

Generalizing from Problems 4–6, consider a ***parallelepiped,*** a three-dimensional solid resembling a "slanted" brick, whose six faces are all parallelograms with opposite faces parallel and congruent (Figure 4.11). Its volume is given by the area of its base times its height.

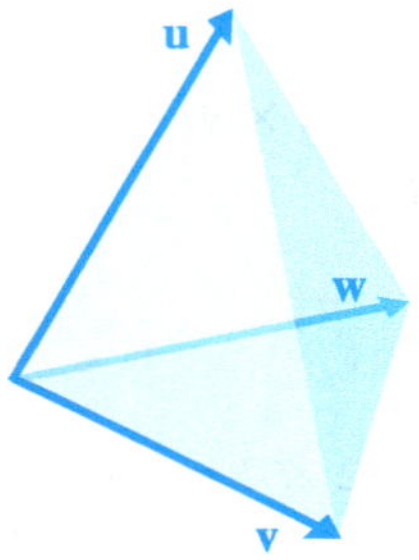

Figure 4.12

7. Prove that the volume $\mathcal{V}$ of the parallelepiped determined by **u**, **v**, and **w** is given by the absolute value of the determinant of the 3×3 matrix $[\mathbf{u} \;\; \mathbf{v} \;\; \mathbf{w}]$ with **u**, **v**, and **w** as its columns. [*Hint:* From Figure 4.11 you can see that the height h can be expressed as $h = \|\mathbf{u}\| \cos\theta$, where θ is the angle between **u** and $\mathbf{v} \times \mathbf{w}$. Use this fact to show that $\mathcal{V} = |\mathbf{u} \cdot (\mathbf{v} \times \mathbf{w})|$ and apply the result of Problem 2.]

8. Show that the volume $\mathcal{V}$ of the tetrahedron determined by **u**, **v**, and **w** (Figure 4.12) is given by

$$\mathcal{V} = \tfrac{1}{6}|\mathbf{u} \cdot (\mathbf{v} \times \mathbf{w})|$$

[*Hint:* From geometry, we know that the volume of such a solid is $\mathcal{V} = \frac{1}{3}$(area of the base)(height).]

Now let's view these geometric interpretations from a transformational point of view. Let A be a 2×2 matrix and let P be the parallelogram determined by the vectors **u** and **v**. We will consider the effect of the matrix transformation T_A on the area of P. Let $T_A(P)$ denote the parallelogram determined by $T_A(\mathbf{u}) = A\mathbf{u}$ and $T_A(\mathbf{v}) = A\mathbf{v}$.

9. Prove that the area of $T_A(P)$ is given by $|\det A|$(area of P).

10. Let A be a 3×3 matrix and let P be the parallelepiped determined by the vectors **u**, **v**, and **w**. Let $T_A(P)$ denote the parallelepiped determined by $T_A(\mathbf{u}) = A\mathbf{u}$, $T_A(\mathbf{v}) = A\mathbf{v}$, and $T_A(\mathbf{w}) = A\mathbf{w}$. Prove that the volume of $T_A(P)$ is given by $|\det A|$(volume of P).

The preceding problems illustrate that the determinant of a matrix captures what the corresponding matrix transformation does to the area or volume of figures upon which the transformation acts. (Although we have considered only certain types of figures, the result is perfectly general and can be made rigorous. We will not do so here.)

Lines and Planes

Suppose we are given two distinct points (x_1, y_1) and (x_2, y_2) in the plane. There is a unique line passing through these points, and its equation is of the form

$$ax + by + c = 0$$

Since the two given points are on this line, their coordinates satisfy this equation. Thus,

$$ax_1 + by_1 + c = 0$$
$$ax_2 + by_2 + c = 0$$

The three equations together can be viewed as a system of linear equations in the variables a, b, and c. Since there *is* a nontrivial solution (i.e., the line exists), the coefficient matrix

$$\begin{bmatrix} x & y & 1 \\ x_1 & y_1 & 1 \\ x_2 & y_2 & 1 \end{bmatrix}$$

cannot be invertible, by the Fundamental Theorem of Invertible Matrices. Consequently, its determinant must be zero, by Theorem 4.6. Expanding this determinant gives the equation of the line.

The equation of the line through the points (x_1, y_1) and (x_2, y_2) is given by

$$\begin{vmatrix} x & y & 1 \\ x_1 & y_1 & 1 \\ x_2 & y_2 & 1 \end{vmatrix} = 0$$

11. Use the method described above to find the equation of the line through the given points.

(a) $(2, 3)$ and $(-1, 0)$ (b) $(1, 2)$ and $(4, 3)$

12. Prove that the three points (x_1, y_1), (x_2, y_2), and (x_3, y_3) are collinear (lie on the same line) if and only if

$$\begin{vmatrix} x_1 & y_1 & 1 \\ x_2 & y_2 & 1 \\ x_3 & y_3 & 1 \end{vmatrix} = 0$$

13. Show that the equation of the plane through the three noncollinear points (x_1, y_1, z_1), (x_2, y_2, z_2), and (x_3, y_3, z_3) is given by

$$\begin{vmatrix} x & y & z & 1 \\ x_1 & y_1 & z_1 & 1 \\ x_2 & y_2 & z_2 & 1 \\ x_3 & y_3 & z_3 & 1 \end{vmatrix} = 0$$

What happens if the three points *are* collinear? [*Hint:* Explain what happens when row reduction is used to evaluate the determinant.]

14. Prove that the four points (x_1, y_1, z_1), (x_2, y_2, z_2), (x_3, y_3, z_3), and (x_4, y_4, z_4) are coplanar (lie in the same plane) if and only if

$$\begin{vmatrix} x_1 & y_1 & z_1 & 1 \\ x_2 & y_2 & z_2 & 1 \\ x_3 & y_3 & z_3 & 1 \\ x_4 & y_4 & z_4 & 1 \end{vmatrix} = 0$$

Curve Fitting

When data arising from experimentation take the form of points (x, y) that can be plotted in the plane, it is often of interest to find a relationship between the variables x and y. Ideally, we would like to find a function whose graph passes through all of the points. Sometimes all we want is an approximation (see Section 7.3), but exact results are also possible in certain situations.

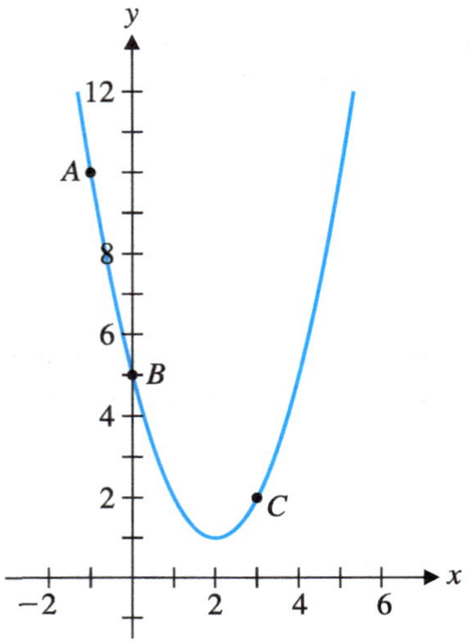

Figure 4.13

15. From Figure 4.13 it appears as though we may be able to find a parabola passing through the points $A(-1, 10)$, $B(0, 5)$, and $C(3, 2)$. The equation of such a parabola is of the form $y = a + bx + cx^2$. By substituting the given points into this equation, set up a system of three linear equations in the variables a, b, and c. *Without solving the system,* use Theorem 4.6 to argue that it must have a unique solution. Then solve the system to find the equation of the parabola in Figure 4.13.

16. Use the method of Problem 15 to find the polynomials of degree at most 2 that pass through the following sets of points.

(a) $A(1, -1)$, $B(2, 4)$, $C(3, 3)$ (b) $A(-1, -3)$, $B(1, -1)$, $C(3, 1)$

17. Generalizing from Problems 15 and 16, suppose a_1, a_2, and a_3 are distinct real numbers. For any real numbers b_1, b_2, and b_3, we want to show that there is a unique quadratic with equation of the form $y = a + bx + cx^2$ passing through the points (a_1, b_1), (a_2, b_2), and (a_3, b_3). Do this by demonstrating that the coefficient matrix of the associated linear system has the determinant

$$\begin{vmatrix} 1 & a_1 & a_1^2 \\ 1 & a_2 & a_2^2 \\ 1 & a_3 & a_3^2 \end{vmatrix} = (a_2 - a_1)(a_3 - a_1)(a_3 - a_2)$$

which is necessarily nonzero. (Why?)

18. Let a_1, a_2, a_3, and a_4 be distinct real numbers. Show that

$$\begin{vmatrix} 1 & a_1 & a_1^2 & a_1^3 \\ 1 & a_2 & a_2^2 & a_2^3 \\ 1 & a_3 & a_3^2 & a_3^3 \\ 1 & a_4 & a_4^2 & a_4^3 \end{vmatrix} = (a_2 - a_1)(a_3 - a_1)(a_4 - a_1)(a_3 - a_2)(a_4 - a_2)(a_4 - a_3) \neq 0$$

For any real numbers b_1, b_2, b_3, and b_4, use this result to prove that there is a unique cubic with equation $y = a + bx + cx^2 + dx^3$ passing through the four points (a_1, b_1), (a_2, b_2), (a_3, b_3), and (a_4, b_4). (Do *not* actually solve for a, b, c, and d.)

19. Let $a_1, a_2, \ldots, a_n$ be n real numbers. Prove that

$$\begin{vmatrix} 1 & a_1 & a_1^2 & \cdots & a_1^{n-1} \\ 1 & a_2 & a_2^2 & \cdots & a_2^{n-1} \\ 1 & a_3 & a_3^2 & \cdots & a_3^{n-1} \\ \vdots & \vdots & \vdots & \ddots & \vdots \\ 1 & a_n & a_n^2 & \cdots & a_n^{n-1} \end{vmatrix} = \prod_{1 \le i < j \le n} (a_j - a_i)$$

where $\prod_{1 \le i < j \le n} (a_j - a_i)$ means the product of all terms of the form $(a_j - a_i)$, where $i < j$ and both i and j are between 1 and n. [The determinant of a matrix of this form (or its transpose) is called a ***Vandermonde determinant***, named after the French mathematician A. T. Vandermonde (1735–1796).]

Deduce that for any n points in the plane whose x-coordinates are all distinct, there is a unique polynomial of degree $n - 1$ whose graph passes through the given points.

4.3 Eigenvalues and Eigenvectors of $n \times n$ Matrices

Now that we have defined the determinant of an $n \times n$ matrix, we can continue our discussion of eigenvalues and eigenvectors in a general context. Recall from Section 4.1 that λ is an eigenvalue of A if and only if $A - \lambda I$ is noninvertible. By Theorem 4.6, this is true if and only if $\det(A - \lambda I) = 0$. To summarize:

> The eigenvalues of a square matrix A are precisely the solutions λ of the equation
>
> $$\det(A - \lambda I) = 0$$

When we expand $\det(A - \lambda I)$, we get a polynomial in λ, called the ***characteristic polynomial*** of A. The equation $\det(A - \lambda I) = 0$ is called the ***characteristic equation*** of A. For example, if $A = \begin{bmatrix} a & b \\ c & d \end{bmatrix}$, its characteristic polynomial is

$$\det(A - \lambda I) = \begin{vmatrix} a - \lambda & b \\ c & d - \lambda \end{vmatrix} = (a - \lambda)(d - \lambda) - bc = \lambda^2 - (a + d)\lambda + (ad - bc)$$

If A is $n \times n$, its characteristic polynomial will be of degree n. According to the Fundamental Theorem of Algebra (see Appendix D), a polynomial of degree n with real or complex coefficients has at most n distinct roots. Applying this fact to the characteristic polynomial, we see that an $n \times n$ matrix with real or complex entries has at most n distinct eigenvalues.

Let's summarize the procedure we will follow (for now) to find the eigenvalues and eigenvectors (eigenspaces) of a matrix.

> Let A be an $n \times n$ matrix.
>
> 1. Compute the characteristic polynomial $\det(A - \lambda I)$ of A.
> 2. Find the eigenvalues of A by solving the characteristic equation $\det(A - \lambda I) = 0$ for λ.
> 3. For each eigenvalue λ, find the null space of the matrix $A - \lambda I$. This is the eigenspace E_λ, the nonzero vectors of which are the eigenvectors of A corresponding to λ.
> 4. Find a basis for each eigenspace.

Example 4.18

Find the eigenvalues and the corresponding eigenspaces of

$$A = \begin{bmatrix} 0 & 1 & 0 \\ 0 & 0 & 1 \\ 2 & -5 & 4 \end{bmatrix}$$

Solution We follow the procedure outlined previously. The characteristic polynomial is

$$\begin{aligned}\det(A - \lambda I) &= \begin{vmatrix} -\lambda & 1 & 0 \\ 0 & -\lambda & 1 \\ 2 & -5 & 4-\lambda \end{vmatrix} \\ &= -\lambda \begin{vmatrix} -\lambda & 1 \\ -5 & 4-\lambda \end{vmatrix} - \begin{vmatrix} 0 & 1 \\ 2 & 4-\lambda \end{vmatrix} \\ &= -\lambda(\lambda^2 - 4\lambda + 5) - (-2) \\ &= -\lambda^3 + 4\lambda^2 - 5\lambda + 2\end{aligned}$$

To find the eigenvalues, we need to solve the characteristic equation $\det(A - \lambda I) = 0$ for λ. The characteristic polynomial factors as $-(\lambda - 1)^2(\lambda - 2)$. (The Factor Theorem is helpful here; see Appendix D.) Thus, the characteristic equation is $-(\lambda - 1)^2(\lambda - 2) = 0$, which clearly has solutions $\lambda = 1$ and $\lambda = 2$. Since $\lambda = 1$ is a multiple root and $\lambda = 2$ is a simple root, let us label them $\lambda_1 = \lambda_2 = 1$ and $\lambda_3 = 2$.

To find the eigenvectors corresponding to $\lambda_1 = \lambda_2 = 1$, we find the null space of

$$A - 1I = \begin{bmatrix} -1 & 1 & 0 \\ 0 & -1 & 1 \\ 2 & -5 & 4-1 \end{bmatrix} = \begin{bmatrix} -1 & 1 & 0 \\ 0 & -1 & 1 \\ 2 & -5 & 3 \end{bmatrix}$$

Row reduction produces

$$[A - I \mid \mathbf{0}] = \left[\begin{array}{ccc|c} -1 & 1 & 0 & 0 \\ 0 & -1 & 1 & 0 \\ 2 & -5 & 3 & 0 \end{array}\right] \longrightarrow \left[\begin{array}{ccc|c} 1 & 0 & -1 & 0 \\ 0 & 1 & -1 & 0 \\ 0 & 0 & 0 & 0 \end{array}\right]$$

(We knew in advance that we *must* get at least one zero row. Why?) Thus, $\mathbf{x} = \begin{bmatrix} x_1 \\ x_2 \\ x_3 \end{bmatrix}$ is in the eigenspace E_1 if and only if $x_1 - x_3 = 0$ and $x_2 - x_3 = 0$. Setting the free variable $x_3 = t$, we see that $x_1 = t$ and $x_2 = t$, from which it follows that

$$E_1 = \left\{ \begin{bmatrix} t \\ t \\ t \end{bmatrix} \right\} = \left\{ t \begin{bmatrix} 1 \\ 1 \\ 1 \end{bmatrix} \right\} = \text{span}\left(\begin{bmatrix} 1 \\ 1 \\ 1 \end{bmatrix} \right)$$

To find the eigenvectors corresponding to $\lambda_3 = 2$, we find the null space of $A - 2I$ by row reduction:

$$[A - 2I \mid \mathbf{0}] = \left[\begin{array}{ccc|c} -2 & 1 & 0 & 0 \\ 0 & -2 & 1 & 0 \\ 2 & -5 & 2 & 0 \end{array}\right] \longrightarrow \left[\begin{array}{ccc|c} 1 & 0 & -\frac{1}{4} & 0 \\ 0 & 1 & -\frac{1}{2} & 0 \\ 0 & 0 & 0 & 0 \end{array}\right]$$

So $\mathbf{x} = \begin{bmatrix} x_1 \\ x_2 \\ x_3 \end{bmatrix}$ is in the eigenspace E_2 if and only if $x_1 = \frac{1}{4}x_3$ and $x_2 = \frac{1}{2}x_3$. Setting the free variable $x_3 = t$, we have

$$E_2 = \left\{ \begin{bmatrix} \frac{1}{4}t \\ \frac{1}{2}t \\ t \end{bmatrix} \right\} = \left\{ t \begin{bmatrix} \frac{1}{4} \\ \frac{1}{2} \\ 1 \end{bmatrix} \right\} = \text{span}\left(\begin{bmatrix} \frac{1}{4} \\ \frac{1}{2} \\ 1 \end{bmatrix} \right) = \text{span}\left(\begin{bmatrix} 1 \\ 2 \\ 4 \end{bmatrix} \right)$$

where we have cleared denominators in the basis by multiplying through by the least common denominator 4. (Why is this permissible?)

Remark Notice that in Example 4.18, A is a 3×3 matrix but has only two distinct eigenvalues. However, if we count multiplicities, A has exactly three eigenvalues ($\lambda = 1$ twice and $\lambda = 2$ once). This is what the Fundamental Theorem of Algebra guarantees. Let us define the ***algebraic multiplicity*** of an eigenvalue to be its multiplicity as a root of the characteristic equation. Thus, $\lambda = 1$ has algebraic multiplicity 2 and $\lambda = 2$ has algebraic multiplicity 1.

Next notice that each eigenspace has a basis consisting of just one vector. In other words, $\dim E_1 = \dim E_2 = 1$. Let us define the ***geometric multiplicity*** of an eigenvalue λ to be $\dim E_\lambda$, the dimension of its corresponding eigenspace. As you will see in Section 4.4, a comparison of these two notions of multiplicity is important.

Example 4.19

Find the eigenvalues and the corresponding eigenspaces of

$$A = \begin{bmatrix} -1 & 0 & 1 \\ 3 & 0 & -3 \\ 1 & 0 & -1 \end{bmatrix}$$

Solution The characteristic equation is

$$0 = \det(A - \lambda I) = \begin{vmatrix} -1-\lambda & 0 & 1 \\ 3 & -\lambda & -3 \\ 1 & 0 & -1-\lambda \end{vmatrix} = -\lambda \begin{vmatrix} -1-\lambda & 1 \\ 1 & -1-\lambda \end{vmatrix}$$

$$= -\lambda(\lambda^2 + 2\lambda) = -\lambda^2(\lambda + 2)$$

Hence, the eigenvalues are $\lambda_1 = \lambda_2 = 0$ and $\lambda_3 = -2$. Thus, the eigenvalue 0 has algebraic multiplicity 2 and the eigenvalue -2 has algebraic multiplicity 1.

For $\lambda_1 = \lambda_2 = 0$, we compute

$$[A - 0I \mid 0] = [A \mid 0] = \left[\begin{array}{ccc|c} -1 & 0 & 1 & 0 \\ 3 & 0 & -3 & 0 \\ 1 & 0 & -1 & 0 \end{array}\right] \longrightarrow \left[\begin{array}{ccc|c} 1 & 0 & -1 & 0 \\ 0 & 0 & 0 & 0 \\ 0 & 0 & 0 & 0 \end{array}\right]$$

from which it follows that an eigenvector $\mathbf{x} = \begin{bmatrix} x_1 \\ x_2 \\ x_3 \end{bmatrix}$ in E_0 satisfies $x_1 = x_3$. Therefore, both x_2 and x_3 are free. Setting $x_2 = s$ and $x_3 = t$, we have

$$E_0 = \left\{ \begin{bmatrix} t \\ s \\ t \end{bmatrix} \right\} = \left\{ s\begin{bmatrix} 0 \\ 1 \\ 0 \end{bmatrix} + t\begin{bmatrix} 1 \\ 0 \\ 1 \end{bmatrix} \right\} = \operatorname{span}\left(\begin{bmatrix} 0 \\ 1 \\ 0 \end{bmatrix}, \begin{bmatrix} 1 \\ 0 \\ 1 \end{bmatrix} \right)$$

For $\lambda_3 = -2$,

$$[A - (-2)I \mid 0] = [A + 2I \mid 0] = \left[\begin{array}{ccc|c} 1 & 0 & 1 & 0 \\ 3 & 2 & -3 & 0 \\ 1 & 0 & 1 & 0 \end{array}\right] \longrightarrow \left[\begin{array}{ccc|c} 1 & 0 & 1 & 0 \\ 0 & 1 & -3 & 0 \\ 0 & 0 & 0 & 0 \end{array}\right]$$

so $x_3 = t$ is free and $x_1 = -x_3 = -t$ and $x_2 = 3x_3 = 3t$. Consequently,

$$E_{-2} = \left\{\begin{bmatrix} -t \\ 3t \\ t \end{bmatrix}\right\} = \left\{t\begin{bmatrix} -1 \\ 3 \\ 1 \end{bmatrix}\right\} = \text{span}\left(\begin{bmatrix} -1 \\ 3 \\ 1 \end{bmatrix}\right)$$

It follows that $\lambda_1 = \lambda_2 = 0$ has geometric multiplicity 2 and $\lambda_3 = -2$ has geometric multiplicity 1. (Note that the algebraic multiplicity equals the geometric multiplicity for each eigenvalue.)

In some situations, the eigenvalues of a matrix are very easy to find. If A is a triangular matrix, then so is $A - \lambda I$, and Theorem 4.2 says that $\det(A - \lambda I)$ is just the product of the diagonal entries. This implies that the characteristic equation of a triangular matrix is

$$(a_{11} - \lambda)(a_{22} - \lambda)\cdots(a_{nn} - \lambda) = 0$$

from which it follows immediately that the eigenvalues are $\lambda_1 = a_{11}, \lambda_2 = a_{22}, \ldots, \lambda_n = a_{nn}$. We summarize this result as a theorem and illustrate it with an example.

Theorem 4.15 The eigenvalues of a triangular matrix are the entries on its main diagonal.

Example 4.20 The eigenvalues of

$$A = \begin{bmatrix} 2 & 0 & 0 & 0 \\ -1 & 1 & 0 & 0 \\ 3 & 0 & 3 & 0 \\ 5 & 7 & 4 & -2 \end{bmatrix}$$

are $\lambda_1 = 2$, $\lambda_2 = 1$, $\lambda_3 = 3$, and $\lambda_4 = -2$, by Theorem 4.15. Indeed, the characteristic polynomial is just $(2 - \lambda)(1 - \lambda)(3 - \lambda)(-2 - \lambda)$.

Note that diagonal matrices are a special case of Theorem 4.15. In fact, a diagonal matrix is both upper and lower triangular.

Eigenvalues capture much important information about the behavior of a matrix. Once we know the eigenvalues of a matrix, we can deduce a great many things without doing any more work. The next theorem is one of the most important in this regard.

Theorem 4.16 A square matrix A is invertible if and only if 0 is *not* an eigenvalue of A.

Proof Let A be a square matrix. By Theorem 4.6, A is invertible if and only if $\det A \neq 0$. But $\det A \neq 0$ is equivalent to $\det(A - 0I) \neq 0$, which says that 0 is not a root of the characteristic equation of A (i.e., 0 is not an eigenvalue of A).

We can now extend the Fundamental Theorem of Invertible Matrices to include results we have proved in this chapter.

Theorem 4.17 **The Fundamental Theorem of Invertible Matrices: Version 3**

Let A be an $n \times n$ matrix. The following statements are equivalent:

a. A is invertible.
b. $A\mathbf{x} = \mathbf{b}$ has a unique solution for every $\mathbf{b}$ in $\mathbb{R}^n$.
c. $A\mathbf{x} = \mathbf{0}$ has only the trivial solution.
d. The reduced row echelon form of A is I_n.
e. A is a product of elementary matrices.
f. $\text{rank}(A) = n$
g. $\text{nullity}(A) = 0$
h. The column vectors of A are linearly independent.
i. The column vectors of A span $\mathbb{R}^n$.
j. The column vectors of A form a basis for $\mathbb{R}^n$.
k. The row vectors of A are linearly independent.
l. The row vectors of A span $\mathbb{R}^n$.
m. The row vectors of A form a basis for $\mathbb{R}^n$.
n. $\det A \neq 0$
o. 0 is not an eigenvalue of A.

Proof The equivalence (a) $\Leftrightarrow$ (n) is Theorem 4.6, and we just proved (a) $\Leftrightarrow$ (o) in Theorem 4.16.

There are nice formulas for the eigenvalues of the powers and inverses of a matrix.

Theorem 4.18 Let A be a square matrix with eigenvalue λ and corresponding eigenvector $\mathbf{x}$.

a. For any positive integer n, λ^n is an eigenvalue of A^n with corresponding eigenvector $\mathbf{x}$.
b. If A is invertible, then $1/\lambda$ is an eigenvalue of A^{-1} with corresponding eigenvector $\mathbf{x}$.
c. If A is invertible, then for any integer n, λ^n is an eigenvalue of A^n with corresponding eigenvector $\mathbf{x}$.

Proof We are given that $A\mathbf{x} = \lambda\mathbf{x}$.

(a) We proceed by induction on n. For $n = 1$, the result is just what has been given. Assume the result is true for $n = k$. That is, assume that, for some positive integer k, $A^k\mathbf{x} = \lambda^k\mathbf{x}$. We must now prove the result for $n = k + 1$. But

$$A^{k+1}\mathbf{x} = A(A^k\mathbf{x}) = A(\lambda^k\mathbf{x})$$

by the induction hypothesis. Using property (d) of Theorem 3.3, we have

$$A(\lambda^k\mathbf{x}) = \lambda^k(A\mathbf{x}) = \lambda^k(\lambda\mathbf{x}) = \lambda^{k+1}\mathbf{x}$$

Thus, $A^{k+1}\mathbf{x} = \lambda^{k+1}\mathbf{x}$, as required. By induction, the result is true for all integers $n \geq 1$.

(b) You are asked to prove this property in Exercise 13.

(c) You are asked to prove this property in Exercise 14.

The next example shows one application of this theorem.

Example 4.21

Compute $\begin{bmatrix} 0 & 1 \\ 2 & 1 \end{bmatrix}^{10} \begin{bmatrix} 5 \\ 1 \end{bmatrix}$.

Solution Let $A = \begin{bmatrix} 0 & 1 \\ 2 & 1 \end{bmatrix}$ and $\mathbf{x} = \begin{bmatrix} 5 \\ 1 \end{bmatrix}$; then what we want to find is $A^{10}\mathbf{x}$. The eigenvalues of A are $\lambda_1 = -1$ and $\lambda_2 = 2$, with corresponding eigenvectors $\mathbf{v}_1 = \begin{bmatrix} 1 \\ -1 \end{bmatrix}$ and $\mathbf{v}_2 = \begin{bmatrix} 1 \\ 2 \end{bmatrix}$. That is,

$$A\mathbf{v}_1 = -\mathbf{v}_1 \quad \text{and} \quad A\mathbf{v}_2 = 2\mathbf{v}_2$$

(Check this.) Since $\{\mathbf{v}_1, \mathbf{v}_2\}$ forms a basis for $\mathbb{R}^2$ (why?), we can write $\mathbf{x}$ as a linear combination of $\mathbf{v}_1$ and $\mathbf{v}_2$. Indeed, as is easily checked, $\mathbf{x} = 3\mathbf{v}_1 + 2\mathbf{v}_2$.

Therefore, using Theorem 4.18(a), we have

$$\begin{aligned} A^{10}\mathbf{x} &= A^{10}(3\mathbf{v}_1 + 2\mathbf{v}_2) = 3(A^{10}\mathbf{v}_1) + 2(A^{10}\mathbf{v}_2) \\ &= 3(\lambda_1^{10})\mathbf{v}_1 + 2(\lambda_2^{10})\mathbf{v}_2 \\ &= 3(-1)^{10}\begin{bmatrix} 1 \\ -1 \end{bmatrix} + 2(2^{10})\begin{bmatrix} 1 \\ 2 \end{bmatrix} = \begin{bmatrix} 3 + 2^{11} \\ -3 + 2^{12} \end{bmatrix} = \begin{bmatrix} 2051 \\ 4093 \end{bmatrix} \end{aligned}$$

This is certainly a lot easier than computing A^{10} first; in fact, there are no matrix multiplications at all!

When it can be used, the method of Example 4.21 is quite general. We summarize it as the following theorem, which you are asked to prove in Exercise 42.

Theorem 4.19

Suppose the $n \times n$ matrix A has eigenvectors $\mathbf{v}_1, \mathbf{v}_2, \ldots, \mathbf{v}_m$ with corresponding eigenvalues $\lambda_1, \lambda_2, \ldots, \lambda_m$. If $\mathbf{x}$ is a vector in $\mathbb{R}^n$ that can be expressed as a linear combination of these eigenvectors—say,

$$\mathbf{x} = c_1\mathbf{v}_1 + c_2\mathbf{v}_2 + \cdots + c_m\mathbf{v}_m$$

then, for any integer k,

$$A^k\mathbf{x} = c_1\lambda_1^k\mathbf{v}_1 + c_2\lambda_2^k\mathbf{v}_2 + \cdots + c_m\lambda_m^k\mathbf{v}_m$$

Warning The catch here is the "if" in the second sentence. There is absolutely no guarantee that such a linear combination is possible. The best possible situation would be if there were a basis of $\mathbb{R}^n$ consisting of eigenvectors of A; we will explore this possibility further in the next section. As a step in that direction, however, we have the following theorem, which states that eigenvectors corresponding to *distinct* eigenvalues are linearly independent.

Theorem 4.20

Let A be an $n \times n$ matrix and let $\lambda_1, \lambda_2, \ldots, \lambda_m$ be distinct eigenvalues of A with corresponding eigenvectors $\mathbf{v}_1, \mathbf{v}_2, \ldots, \mathbf{v}_m$. Then $\mathbf{v}_1, \mathbf{v}_2, \ldots, \mathbf{v}_m$ are linearly independent.

Proof The proof is indirect. We will assume that $\mathbf{v}_1, \mathbf{v}_2, \ldots, \mathbf{v}_m$ are linearly *dependent* and show that this assumption leads to a contradiction.

If $\mathbf{v}_1, \mathbf{v}_2, \ldots, \mathbf{v}_m$ are linearly dependent, then one of these vectors must be expressible as a linear combination of the previous ones. Let $\mathbf{v}_{k+1}$ be the first of the vectors $\mathbf{v}_i$ that can be so expressed. In other words, $\mathbf{v}_1, \mathbf{v}_2, \ldots, \mathbf{v}_k$ are linearly independent, but there are scalars $c_1, c_2, \ldots, c_k$ such that

$$\mathbf{v}_{k+1} = c_1\mathbf{v}_1 + c_2\mathbf{v}_2 + \cdots + c_k\mathbf{v}_k \tag{1}$$

Multiplying both sides of Equation (1) by A from the left and using the fact that $A\mathbf{v}_i = \lambda_i\mathbf{v}_i$ for each i, we have

$$\begin{aligned}\lambda_{k+1}\mathbf{v}_{k+1} = A\mathbf{v}_{k+1} &= A(c_1\mathbf{v}_1 + c_2\mathbf{v}_2 + \cdots + c_k\mathbf{v}_k)\\ &= c_1A\mathbf{v}_1 + c_2A\mathbf{v}_2 + \cdots + c_kA\mathbf{v}_k \qquad (2)\\ &= c_1\lambda_1\mathbf{v}_1 + c_2\lambda_2\mathbf{v}_2 + \cdots + c_k\lambda_k\mathbf{v}_k\end{aligned}$$

Now we multiply both sides of Equation (1) by λ_{k+1} to get

$$\lambda_{k+1}\mathbf{v}_{k+1} = c_1\lambda_{k+1}\mathbf{v}_1 + c_2\lambda_{k+1}\mathbf{v}_2 + \cdots + c_k\lambda_{k+1}\mathbf{v}_k \tag{3}$$

When we subtract Equation (3) from Equation (2), we obtain

$$0 = c_1(\lambda_1 - \lambda_{k+1})\mathbf{v}_1 + c_2(\lambda_2 - \lambda_{k+1})\mathbf{v}_2 + \cdots + c_k(\lambda_k - \lambda_{k+1})\mathbf{v}_k$$

The linear independence of $\mathbf{v}_1, \mathbf{v}_2, \ldots, \mathbf{v}_k$ implies that

$$c_1(\lambda_1 - \lambda_{k+1}) = c_2(\lambda_2 - \lambda_{k+1}) = \cdots = c_k(\lambda_k - \lambda_{k+1}) = 0$$

Since the eigenvalues λ_i are all distinct, the terms in parentheses $(\lambda_i - \lambda_{k+1})$, $i = 1, \ldots, k$, are all nonzero. Hence, $c_1 = c_2 = \cdots = c_k = 0$. This implies that

$$\mathbf{v}_{k+1} = c_1\mathbf{v}_1 + c_2\mathbf{v}_2 + \cdots + c_k\mathbf{v}_k = 0\mathbf{v}_1 + 0\mathbf{v}_2 + \cdots + 0\mathbf{v}_k = \mathbf{0}$$

which is impossible, since the eigenvector $\mathbf{v}_{k+1}$ cannot be zero. Thus, we have a contradiction, which means that our assumption that $\mathbf{v}_1, \mathbf{v}_2, \ldots, \mathbf{v}_m$ are linearly dependent is false. It follows that $\mathbf{v}_1, \mathbf{v}_2, \ldots, \mathbf{v}_m$ must be linearly independent.

Exercises 4.3

In Exercises 1–12, compute (a) the characteristic polynomial of A, (b) the eigenvalues of A, (c) a basis for each eigenspace of A, and (d) the algebraic and geometric multiplicity of each eigenvalue.

1. $A = \begin{bmatrix} 1 & 3 \\ -2 & 6 \end{bmatrix}$

2. $A = \begin{bmatrix} 2 & 1 \\ -1 & 0 \end{bmatrix}$

3. $A = \begin{bmatrix} 1 & 1 & 0 \\ 0 & -2 & 1 \\ 0 & 0 & 3 \end{bmatrix}$

4. $A = \begin{bmatrix} 1 & 0 & 1 \\ 0 & 1 & 1 \\ 1 & 1 & 0 \end{bmatrix}$

5. $A = \begin{bmatrix} 1 & 2 & 0 \\ -1 & -1 & 1 \\ 0 & 1 & 1 \end{bmatrix}$

6. $A = \begin{bmatrix} 1 & 0 & 2 \\ 3 & -1 & 3 \\ 2 & 0 & 1 \end{bmatrix}$

7. $A = \begin{bmatrix} 4 & 0 & 1 \\ 2 & 3 & 2 \\ -1 & 0 & 2 \end{bmatrix}$

8. $A = \begin{bmatrix} 1 & -1 & -1 \\ 0 & 2 & 0 \\ -1 & -1 & 1 \end{bmatrix}$

9. $A = \begin{bmatrix} 3 & 1 & 0 & 0 \\ -1 & 1 & 0 & 0 \\ 0 & 0 & 1 & 4 \\ 0 & 0 & 1 & 1 \end{bmatrix}$

10. $A = \begin{bmatrix} 2 & 1 & 1 & 0 \\ 0 & 1 & 4 & 5 \\ 0 & 0 & 3 & 1 \\ 0 & 0 & 0 & 2 \end{bmatrix}$

11. $A = \begin{bmatrix} 1 & 0 & 0 & 0 \\ 0 & 1 & 0 & 0 \\ 1 & 1 & 3 & 0 \\ -2 & 1 & 2 & -1 \end{bmatrix}$

12. $A = \begin{bmatrix} 4 & 0 & 1 & 0 \\ 0 & 4 & 1 & 1 \\ 0 & 0 & 1 & 2 \\ 0 & 0 & 3 & 0 \end{bmatrix}$

13. Prove Theorem 4.18(b).

14. Prove Theorem 4.18(c). [*Hint:* Combine the proofs of parts (a) and (b) and see the fourth Remark following Theorem 3.9 (page 169).]

In Exercises 15 and 16, A is a 2×2 matrix with eigenvectors $\mathbf{v}_1 = \begin{bmatrix} 1 \\ -1 \end{bmatrix}$ and $\mathbf{v}_2 = \begin{bmatrix} 1 \\ 1 \end{bmatrix}$ corresponding to eigenvalues $\lambda_1 = \frac{1}{2}$ and $\lambda_2 = 2$, respectively, and $\mathbf{x} = \begin{bmatrix} 5 \\ 1 \end{bmatrix}$.

15. Find $A^{10}\mathbf{x}$.

16. Find $A^k\mathbf{x}$. What happens as k becomes large (i.e., $k \to \infty$)?

In Exercises 17 and 18, A is a 3×3 matrix with eigenvectors $\mathbf{v}_1 = \begin{bmatrix} 1 \\ 0 \\ 0 \end{bmatrix}$, $\mathbf{v}_2 = \begin{bmatrix} 1 \\ 1 \\ 0 \end{bmatrix}$, and $\mathbf{v}_3 = \begin{bmatrix} 1 \\ 1 \\ 1 \end{bmatrix}$ corresponding to eigenvalues $\lambda_1 = -\frac{1}{3}$, $\lambda_2 = \frac{1}{3}$, and $\lambda_3 = 1$, respectively, and $\mathbf{x} = \begin{bmatrix} 2 \\ 1 \\ 2 \end{bmatrix}$.

17. Find $A^{20}\mathbf{x}$.

18. Find $A^k\mathbf{x}$. What happens as k becomes large (i.e., $k \to \infty$)?

19. **(a)** Show that, for any square matrix A, A^T and A have the same characteristic polynomial and hence the same eigenvalues.

(b) Give an example of a 2×2 matrix A for which A^T and A have different eigenspaces.

20. Let A be a nilpotent matrix (that is, $A^m = O$ for some $m > 1$). Show that $\lambda = 0$ is the only eigenvalue of A.

21. Let A be an idempotent matrix (that is, $A^2 = A$). Show that $\lambda = 0$ and $\lambda = 1$ are the only possible eigenvalues of A.

22. If $\mathbf{v}$ is an eigenvector of A with corresponding eigenvalue λ and c is a scalar, show that $\mathbf{v}$ is an eigenvector of $A - cI$ with corresponding eigenvalue $\lambda - c$.

23. **(a)** Find the eigenvalues and eigenspaces of

$$A = \begin{bmatrix} 3 & 2 \\ 5 & 0 \end{bmatrix}$$

(b) Using Theorem 4.18 and Exercise 22, find the eigenvalues and eigenspaces of A^{-1}, $A - 2I$, and $A + 2I$.

24. Let A and B be $n \times n$ matrices with eigenvalues λ and μ, respectively.

(a) Give an example to show that $\lambda + \mu$ need not be an eigenvalue of $A + B$.

(b) Give an example to show that $\lambda\mu$ need not be an eigenvalue of AB.

(c) Suppose λ and μ correspond to the *same* eigenvector $\mathbf{x}$. Show that, in this case, $\lambda + \mu$ is an eigenvalue of $A + B$ and $\lambda\mu$ is an eigenvalue of AB.

25. If A and B are two row equivalent matrices, do they necessarily have the same eigenvalues? Either prove that they do or give a counterexample.

Let $p(x)$ be the polynomial

$$p(x) = x^n + a_{n-1}x^{n-1} + \cdots + a_1x + a_0$$

*The **companion matrix** of $p(x)$ is the $n \times n$ matrix*

$$C(p) = \begin{bmatrix} -a_{n-1} & -a_{n-2} & \cdots & -a_1 & -a_0 \\ 1 & 0 & \cdots & 0 & 0 \\ 0 & 1 & \ddots & \vdots & \vdots \\ 0 & 0 & \cdots & 0 & 0 \\ 0 & 0 & \cdots & 1 & 0 \end{bmatrix} \quad (4)$$

26. Find the companion matrix of $p(x) = x^2 - 7x + 12$ and then find the characteristic polynomial of $C(p)$.

27. Find the companion matrix of $p(x) = x^3 + 3x^2 - 4x + 12$ and then find the characteristic polynomial of $C(p)$.

28. **(a)** Show that the companion matrix $C(p)$ of $p(x) = x^2 + ax + b$ has characteristic polynomial $\lambda^2 + a\lambda + b$.

(b) Show that if λ is an eigenvalue of the companion matrix $C(p)$ in part (a), then $\begin{bmatrix} \lambda \\ 1 \end{bmatrix}$ is an eigenvector of $C(p)$ corresponding to λ.

29. **(a)** Show that the companion matrix $C(p)$ of $p(x) = x^3 + ax^2 + bx + c$ has characteristic polynomial $-(\lambda^3 + a\lambda^2 + b\lambda + c)$.

(b) Show that if λ is an eigenvalue of the companion matrix $C(p)$ in part (a), then $\begin{bmatrix} \lambda^2 \\ \lambda \\ 1 \end{bmatrix}$ is an eigenvector of $C(p)$ corresponding to λ.

30. Construct a nontriangular 2×2 matrix with eigenvalues 2 and 5. [*Hint:* Use Exercise 28.]

31. Construct a nontriangular 3×3 matrix with eigenvalues -2, 1, and 3. [*Hint:* Use Exercise 29.]

32. **(a)** Use mathematical induction to prove that, for $n \geq 2$, the companion matrix $C(p)$ of $p(x) = x^n + a_{n-1}x^{n-1} + \cdots + a_1x + a_0$ has characteristic polynomial $(-1)^n p(\lambda)$. [*Hint:* Expand by cofactors along the last column. You may find it helpful to introduce the polynomial $q(x) = (p(x) - a_0)/x$.]

(b) Show that if λ is an eigenvalue of the companion matrix $C(p)$ in Equation (4), then an eigenvector corresponding to λ is given by

$$\begin{bmatrix} \lambda^{n-1} \\ \lambda^{n-2} \\ \vdots \\ \lambda \\ 1 \end{bmatrix}$$

If $p(x) = x^n + a_{n-1}x^{n-1} + \cdots + a_1x + a_0$ and A is a square matrix, we can define a square matrix $p(A)$ by

$$p(A) = A^n + a_{n-1}A^{n-1} + \cdots + a_1A + a_0I$$

An important theorem in advanced linear algebra says that if $c_A(\lambda)$ is the characteristic polynomial of the matrix A, then $c_A(A) = O$ (in words, every matrix satisfies its characteristic equation). This is the celebrated ***Cayley-Hamilton Theorem,*** *named after Arthur Cayley (1821–1895), pictured below, and Sir William Rowan Hamilton (see page 2). Cayley proved this theorem in 1858. Hamilton discovered it, independently, in his work on quaternions, a generalization of the complex numbers.*

Bettmann/CORBIS

33. Verify the Cayley-Hamilton Theorem for $A = \begin{bmatrix} 1 & -1 \\ 2 & 3 \end{bmatrix}$.

That is, find the characteristic polynomial $c_A(\lambda)$ of A and show that $c_A(A) = O$.

34. Verify the Cayley-Hamilton Theorem for

$$A = \begin{bmatrix} 1 & 1 & 0 \\ 1 & 0 & 1 \\ 0 & 1 & 1 \end{bmatrix}.$$

The Cayley-Hamilton Theorem can be used to calculate powers and inverses of matrices. For example, if A is a 2×2 matrix with characteristic polynomial $c_A(\lambda) = \lambda^2 + a\lambda + b$, then $A^2 + aA + bI = O$, so

$$A^2 = -aA - bI$$

and

$$\begin{aligned} A^3 &= AA^2 = A(-aA - bI) \\ &= -aA^2 - bA \\ &= -a(-aA - bI) - bA \\ &= (a^2 - b)A + abI \end{aligned}$$

It is easy to see that by continuing in this fashion we can express any positive power of A as a linear combination of I and A. From $A^2 + aA + bI = O$, we also obtain $A(A + aI) = -bI$, so

$$A^{-1} = -\frac{1}{b}A - \frac{a}{b}I$$

provided $b \neq 0$.

35. For the matrix A in Exercise 33, use the Cayley-Hamilton Theorem to compute A^2, A^3, and A^4 by expressing each as a linear combination of I and A.

36. For the matrix A in Exercise 34, use the Cayley-Hamilton Theorem to compute A^3 and A^4 by expressing each as a linear combination of I, A, and A^2.

37. For the matrix A in Exercise 33, use the Cayley-Hamilton Theorem to compute A^{-1} and A^{-2} by expressing each as a linear combination of I and A.

38. For the matrix A in Exercise 34, use the Cayley-Hamilton Theorem to compute A^{-1} and A^{-2} by expressing each as a linear combination of I, A, and A^2.

39. Show that if the square matrix A can be partitioned as

$$A = \left[\begin{array}{c:c} P & Q \\ \hdashline O & S \end{array}\right]$$

where P and S are square matrices, then the characteristic polynomial of A is $c_A(\lambda) = c_P(\lambda)c_S(\lambda)$. [*Hint:* Use Exercise 69 in Section 4.2.]

40. Let $\lambda_1, \lambda_2, \ldots, \lambda_n$ be a complete set of eigenvalues (repetitions included) of the $n \times n$ matrix A. Prove that

$$\det(A) = \lambda_1\lambda_2\cdots\lambda_n \quad \text{and}$$
$$\operatorname{tr}(A) = \lambda_1 + \lambda_2 + \cdots + \lambda_n$$

[*Hint:* The characteristic polynomial of A factors as

$$\det(A - \lambda I) = (-1)^n(\lambda - \lambda_1)(\lambda - \lambda_2)\cdots(\lambda - \lambda_n)$$

Find the constant term and the coefficient of λ^{n-1} on the left and right sides of this equation.]

41. Let A and B be $n \times n$ matrices. Prove that the sum of *all* the eigenvalues of $A + B$ is the sum of all the eigenvalues of A and B individually. Prove that the product of *all* the eigenvalues of AB is the product of all the eigenvalues of A and B individually. (Compare this exercise with Exercise 24.)

42. Prove Theorem 4.19.

Writing Project

The History of Eigenvalues

Like much of linear algebra, the way the topic of eigenvalues is taught today does not correspond to its historical development. Eigenvalues arose out of problems in systems of differential equations before the concept of a matrix was even formulated.

Write a report on the historical development of eigenvalues. Describe the types of mathematical problems in which they originally arose. Who were some of the key mathematicians involved with these problems? How did the terminology for eigenvalues change over time?

1. Thomas Hawkins, Cauchy and the Spectral Theory of Matrices, *Historia Mathematica 2* (1975), pp. 1–29.
2. Victor J. Katz, *A History of Mathematics: An Introduction* (Third Edition) (Reading, MA: Addison Wesley Longman, 2008).
3. Morris Kline, *Mathematical Thought from Ancient to Modern Times* (Oxford: Oxford University Press, 1972).

4.4 Similarity and Diagonalization

As you saw in the last section, triangular and diagonal matrices are nice in the sense that their eigenvalues are transparently displayed. It would be pleasant if we could relate a given square matrix to a triangular or diagonal one in such a way that they had exactly the same eigenvalues. Of course, we already know one procedure for converting a square matrix into triangular form—namely, Gaussian elimination. Unfortunately, this process does not preserve the eigenvalues of the matrix. In this section, we consider a different sort of transformation of a matrix that does behave well with respect to eigenvalues.

Similar Matrices

Definition Let A and B be $n \times n$ matrices. We say that A ***is similar to*** B if there is an invertible $n \times n$ matrix P such that $P^{-1}AP = B$. If A is similar to B, we write $A \sim B$.

Remarks

- If $A \sim B$, we can write, equivalently, that $A = PBP^{-1}$ or $AP = PB$.
- Similarity is a *relation* on square matrices in the same sense that "less than or equal to" is a relation on the integers. Note that there is a *direction* (or *order*) implicit in the definition. Just as $a \leq b$ does not necessarily imply $b \leq a$, we should not assume that $A \sim B$ implies $B \sim A$. (In fact, this *is* true, as we will prove in the next theorem, but it does not follow immediately from the definition.)
- The matrix P depends on A and B. It is not unique for a given pair of similar matrices A and B. To see this, simply take $A = B = I$, in which case $I \sim I$, since $P^{-1}IP = I$ for *any* invertible matrix P.

Example 4.22

Let $A = \begin{bmatrix} 1 & 2 \\ 0 & -1 \end{bmatrix}$ and $B = \begin{bmatrix} 1 & 0 \\ -2 & -1 \end{bmatrix}$. Then $A \sim B$, since

$$\begin{bmatrix} 1 & 2 \\ 0 & -1 \end{bmatrix}\begin{bmatrix} 1 & -1 \\ 1 & 1 \end{bmatrix} = \begin{bmatrix} 3 & 1 \\ -1 & -1 \end{bmatrix} = \begin{bmatrix} 1 & -1 \\ 1 & 1 \end{bmatrix}\begin{bmatrix} 1 & 0 \\ -2 & -1 \end{bmatrix}$$

Thus, $AP = PB$ with $P = \begin{bmatrix} 1 & -1 \\ 1 & 1 \end{bmatrix}$. (Note that it is not necessary to compute P^{-1}. See the first Remark before Example 4.22.)

Theorem 4.21 Let A, B, and C be $n \times n$ matrices.

a. $A \sim A$
b. If $A \sim B$, then $B \sim A$.
c. If $A \sim B$ and $B \sim C$, then $A \sim C$.

Proof (a) This property follows from the fact that $I^{-1}AI = A$.

(b) If $A \sim B$, then $P^{-1}AP = B$ for some invertible matrix P. As noted in the first Remark on the previous page, this is equivalent to $PBP^{-1} = A$. Setting $Q = P^{-1}$, we have $Q^{-1}BQ = (P^{-1})^{-1}BP^{-1} = PBP^{-1} = A$. Therefore, by definition, $B \sim A$.

(c) You are asked to prove property (c) in Exercise 30.

Remark Any relation satisfying the three properties of Theorem 4.21 is called an ***equivalence relation***. Equivalence relations arise frequently in mathematics, and objects that are related via an equivalence relation usually share important properties. We are about to see that this is true of similar matrices.

Theorem 4.22 Let A and B be $n \times n$ matrices with $A \sim B$. Then

a. $\det A = \det B$
b. A is invertible if and only if B is invertible.
c. A and B have the same rank.
d. A and B have the same characteristic polynomial.
e. A and B have the same eigenvalues.
f. $A^m \sim B^m$ for all integers $m \geq 0$.
g. If A is invertible, then $A^m \sim B^m$ for all integers m.

Proof We prove (a) and (d) and leave the remaining properties as exercises. If $A \sim B$, then $P^{-1}AP = B$ for some invertible matrix P.

(a) Taking determinants of both sides, we have

$$\begin{aligned} \det B &= \det(P^{-1}AP) = (\det P^{-1})(\det A)(\det P) \\ &= \left(\frac{1}{\det P}\right)(\det A)(\det P) = \det A \end{aligned}$$

(d) The characteristic polynomial of B is

$$\begin{aligned} \det(B - \lambda I) &= \det(P^{-1}AP - \lambda I) \\ &= \det(P^{-1}AP - \lambda P^{-1}IP) \end{aligned}$$

$$= \det(P^{-1}AP - P^{-1}(\lambda I)P)$$
$$= \det(P^{-1}(A - \lambda I)P) = \det(A - \lambda I)$$

with the last step following as in (a). Thus, $\det(B - \lambda I) = \det(A - \lambda I)$; that is, the characteristic polynomials of B and A are the same.

Remark Two matrices may have properties (a) through (e) (and more) in common and yet still not be similar. For example, $A = \begin{bmatrix} 1 & 0 \\ 0 & 1 \end{bmatrix}$ and $B = \begin{bmatrix} 1 & 1 \\ 0 & 1 \end{bmatrix}$ both have determinant 1 and rank 2, are invertible, and have characteristic polynomial $(1 - \lambda)^2$ and eigenvalues $\lambda_1 = \lambda_2 = 1$. But A is not similar to B, since $P^{-1}AP = P^{-1}IP = I \neq B$ for any invertible matrix P.

Theorem 4.22 is most useful in showing that two matrices are *not* similar, since A and B cannot be similar if any of properties (a) through (e) fails.

Example 4.23

(a) $A = \begin{bmatrix} 1 & 2 \\ 2 & 1 \end{bmatrix}$ and $B = \begin{bmatrix} 2 & 1 \\ 1 & 2 \end{bmatrix}$ are not similar, since $\det A = -3$ but $\det B = 3$.

(b) $A = \begin{bmatrix} 1 & 3 \\ 2 & 2 \end{bmatrix}$ and $B = \begin{bmatrix} 1 & 1 \\ 3 & -1 \end{bmatrix}$ are not similar, since the characteristic polynomial of A is $\lambda^2 - 3\lambda - 4$ while that of B is $\lambda^2 - 4$. (Check this.) Note that A and B do have the same determinant and rank, however.

Diagonalization

The best possible situation is when a square matrix is similar to a diagonal matrix. As you are about to see, whether a matrix is diagonalizable is closely related to the eigenvalues and eigenvectors of the matrix.

Definition An $n \times n$ matrix A is ***diagonalizable*** if there is a diagonal matrix D such that A is similar to D—that is, if there is an invertible $n \times n$ matrix P such that $P^{-1}AP = D$.

Example 4.24

$A = \begin{bmatrix} 1 & 3 \\ 2 & 2 \end{bmatrix}$ is diagonalizable since, if $P = \begin{bmatrix} 1 & 3 \\ 1 & -2 \end{bmatrix}$ and $D = \begin{bmatrix} 4 & 0 \\ 0 & -1 \end{bmatrix}$, then $P^{-1}AP = D$, as can be easily checked. (Actually, it is faster to check the equivalent statement $AP = PD$, since it does not require finding P^{-1}.)

Example 4.24 begs the question of where matrices P and D came from. Observe that the diagonal entries 4 and -1 of D are the eigenvalues of A, since they are the roots of its characteristic polynomial, which we found in Example 4.23(b). The origin of matrix P is less obvious, but, as we are about to demonstrate, its entries are obtained from the eigenvectors of A. Theorem 4.23 makes this connection precise.

Theorem 4.23 Let A be an $n \times n$ matrix. Then A is diagonalizable if and only if A has n linearly independent eigenvectors.

More precisely, there exist an invertible matrix P and a diagonal matrix D such that $P^{-1}AP = D$ if and only if the columns of P are n linearly independent eigenvectors of A and the diagonal entries of D are the eigenvalues of A corresponding to the eigenvectors in P in the same order.

Proof Suppose first that A is similar to the diagonal matrix D via $P^{-1}AP = D$ or, equivalently, $AP = PD$. Let the columns of P be $\mathbf{p}_1, \mathbf{p}_2, \ldots, \mathbf{p}_n$ and let the diagonal entries of D be $\lambda_1, \lambda_2, \ldots, \lambda_n$. Then

$$A[\mathbf{p}_1 \quad \mathbf{p}_2 \quad \cdots \quad \mathbf{p}_n] = [\mathbf{p}_1 \quad \mathbf{p}_2 \quad \cdots \quad \mathbf{p}_n]\begin{bmatrix} \lambda_1 & 0 & \cdots & 0 \\ 0 & \lambda_2 & \cdots & 0 \\ \vdots & \vdots & \ddots & \vdots \\ 0 & 0 & \cdots & \lambda_n \end{bmatrix} \tag{1}$$

or
$$[A\mathbf{p}_1 \quad A\mathbf{p}_2 \quad \cdots \quad A\mathbf{p}_n] = [\lambda_1\mathbf{p}_1 \quad \lambda_2\mathbf{p}_2 \quad \cdots \quad \lambda_n\mathbf{p}_n] \tag{2}$$

where the right-hand side is just the column-row representation of the product PD. Equating columns, we have

$$A\mathbf{p}_1 = \lambda_1\mathbf{p}_1, A\mathbf{p}_2 = \lambda_2\mathbf{p}_2, \ldots, A\mathbf{p}_n = \lambda_n\mathbf{p}_n$$

which proves that the column vectors of P are eigenvectors of A whose corresponding eigenvalues are the diagonal entries of D in the same order. Since P is invertible, its columns are linearly independent, by the Fundamental Theorem of Invertible Matrices.

Conversely, if A has n linearly independent eigenvectors $\mathbf{p}_1, \mathbf{p}_2, \ldots, \mathbf{p}_n$ with corresponding eigenvalues $\lambda_1, \lambda_2, \ldots, \lambda_n$, respectively, then

$$A\mathbf{p}_1 = \lambda_1\mathbf{p}_1, A\mathbf{p}_2 = \lambda_2\mathbf{p}_2, \ldots, A\mathbf{p}_n = \lambda_n\mathbf{p}_n$$

This implies Equation (2) above, which is equivalent to Equation (1). Consequently, if we take P to be the $n \times n$ matrix with columns $\mathbf{p}_1, \mathbf{p}_2, \ldots, \mathbf{p}_n$, then Equation (1) becomes $AP = PD$. Since the columns of P are linearly independent, the Fundamental Theorem of Invertible Matrices implies that P is invertible, so $P^{-1}AP = D$; that is, A is diagonalizable.

Example 4.25 If possible, find a matrix P that diagonalizes

$$A = \begin{bmatrix} 0 & 1 & 0 \\ 0 & 0 & 1 \\ 2 & -5 & 4 \end{bmatrix}$$

Solution We studied this matrix in Example 4.18, where we discovered that it has eigenvalues $\lambda_1 = \lambda_2 = 1$ and $\lambda_3 = 2$. The eigenspaces have the following bases:

$$\text{For } \lambda_1 = \lambda_2 = 1, E_1 \text{ has basis } \begin{bmatrix} 1 \\ 1 \\ 1 \end{bmatrix}.$$

$$\text{For } \lambda_3 = 2, E_2 \text{ has basis } \begin{bmatrix} 1 \\ 2 \\ 4 \end{bmatrix}.$$

Since all other eigenvectors are just multiples of one of these two basis vectors, there cannot be three linearly independent eigenvectors. By Theorem 4.23, therefore, A is not diagonalizable.

Example 4.26

If possible, find a matrix P that diagonalizes

$$A = \begin{bmatrix} -1 & 0 & 1 \\ 3 & 0 & -3 \\ 1 & 0 & -1 \end{bmatrix}$$

Solution This is the matrix of Example 4.19. There, we found that the eigenvalues of A are $\lambda_1 = \lambda_2 = 0$ and $\lambda_3 = -2$, with the following bases for the eigenspaces:

$$\text{For } \lambda_1 = \lambda_2 = 0, E_0 \text{ has basis } \mathbf{p}_1 = \begin{bmatrix} 0 \\ 1 \\ 0 \end{bmatrix} \text{ and } \mathbf{p}_2 = \begin{bmatrix} 1 \\ 0 \\ 1 \end{bmatrix}.$$

$$\text{For } \lambda_3 = -2, E_{-2} \text{ has basis } \mathbf{p}_3 = \begin{bmatrix} -1 \\ 3 \\ 1 \end{bmatrix}.$$

It is straightforward to check that these three vectors are linearly independent. Thus, if we take

$$P = [\mathbf{p}_1 \quad \mathbf{p}_2 \quad \mathbf{p}_3] = \begin{bmatrix} 0 & 1 & -1 \\ 1 & 0 & 3 \\ 0 & 1 & 1 \end{bmatrix}$$

then P is invertible. Furthermore,

$$P^{-1}AP = \begin{bmatrix} 0 & 0 & 0 \\ 0 & 0 & 0 \\ 0 & 0 & -2 \end{bmatrix} = D$$

as can be easily checked. (If you are checking by hand, it is much easier to check the equivalent equation $AP = PD$.)

Remarks

- When there are enough eigenvectors, they can be placed into the columns of P in any order. However, the eigenvalues will come up on the diagonal of D in the same order as their corresponding eigenvectors in P. For example, if we had chosen

$$P = [\mathbf{p}_1 \quad \mathbf{p}_3 \quad \mathbf{p}_2] = \begin{bmatrix} 0 & -1 & 1 \\ 1 & 3 & 0 \\ 0 & 1 & 1 \end{bmatrix}$$

then we would have found

$$P^{-1}AP = \begin{bmatrix} 0 & 0 & 0 \\ 0 & -2 & 0 \\ 0 & 0 & 0 \end{bmatrix}$$

- In Example 4.26, you were asked to check that the eigenvectors $\mathbf{p}_1$, $\mathbf{p}_2$, and $\mathbf{p}_3$ were linearly independent. Was it necessary to check this? We knew that $\{\mathbf{p}_1, \mathbf{p}_2\}$ was linearly independent, since it was a basis for the eigenspace E_0. We also knew that the sets $\{\mathbf{p}_1, \mathbf{p}_3\}$ and $\{\mathbf{p}_2, \mathbf{p}_3\}$ were linearly independent, by Theorem 4.20. But we could *not* conclude from this information that $\{\mathbf{p}_1, \mathbf{p}_2, \mathbf{p}_3\}$ was linearly independent. The next theorem, however, guarantees that linear independence is preserved when the bases of different eigenspaces are combined.

Theorem 4.24 Let A be an $n \times n$ matrix and let $\lambda_1, \lambda_2, \ldots, \lambda_k$ be distinct eigenvalues of A. If $\mathcal{B}_i$ is a basis for the eigenspace E_{λ_i}, then $\mathcal{B} = \mathcal{B}_1 \cup \mathcal{B}_2 \cup \cdots \cup \mathcal{B}_k$ (i.e., the total collection of basis vectors for all of the eigenspaces) is linearly independent.

Proof Let $\mathcal{B}_i = \{\mathbf{v}_{i1}, \mathbf{v}_{i2}, \ldots, \mathbf{v}_{in_i}\}$ for $i = 1, \ldots, k$. We have to show that

$$\mathcal{B} = \{\mathbf{v}_{11}, \mathbf{v}_{12}, \ldots, \mathbf{v}_{1n_1}, \mathbf{v}_{21}, \mathbf{v}_{22}, \ldots, \mathbf{v}_{2n_2}, \ldots, \mathbf{v}_{k1}, \mathbf{v}_{k2}, \ldots, \mathbf{v}_{kn_k}\}$$

is linearly independent. Suppose some nontrivial linear combination of these vectors is the zero vector—say,

$$(c_{11}\mathbf{v}_{11} + \cdots + c_{1n_1}\mathbf{v}_{1n_1}) + (c_{21}\mathbf{v}_{21} + \cdots + c_{2n_2}\mathbf{v}_{2n_2}) + \cdots + (c_{k1}\mathbf{v}_{k1} + \cdots + c_{kn_k}\mathbf{v}_{kn_k}) = \mathbf{0} \quad (3)$$

Denoting the sums in parentheses by $\mathbf{x}_1, \mathbf{x}_2, \ldots \mathbf{x}_k$, we can write Equation (3) as

$$\mathbf{x}_1 + \mathbf{x}_2 + \cdots + \mathbf{x}_k = \mathbf{0} \quad (4)$$

Now each $\mathbf{x}_i$ is in E_{λ_i} (why?) and so either is an eigenvector corresponding to λ_i or is $\mathbf{0}$. But, since the eigenvalues λ_i are distinct, if *any* of the factors $\mathbf{x}_i$ is an eigenvector, they are linearly independent, by Theorem 4.20. Yet Equation (4) is a linear *dependence* relationship; this is a contradiction. We conclude that Equation (3) must be trivial; that is, all of its coefficients are zero. Hence, $\mathcal{B}$ is linearly independent.

There is one case in which diagonalizability is automatic: an $n \times n$ matrix with n distinct eigenvalues.

Theorem 4.25 If A is an $n \times n$ matrix with n distinct eigenvalues, then A is diagonalizable.

Proof Let $\mathbf{v}_1, \mathbf{v}_2, \ldots, \mathbf{v}_n$ be eigenvectors corresponding to the n distinct eigenvalues of A. (Why could there not be *more* than n such eigenvectors?) By Theorem 4.20, $\mathbf{v}_1, \mathbf{v}_2, \ldots, \mathbf{v}_n$ are linearly independent, so, by Theorem 4.23, A is diagonalizable.

Example 4.27 The matrix

$$A = \begin{bmatrix} 2 & -3 & 7 \\ 0 & 5 & 1 \\ 0 & 0 & -1 \end{bmatrix}$$

has eigenvalues $\lambda_1 = 2$, $\lambda_2 = 5$, and $\lambda_3 = -1$, by Theorem 4.15. Since these are three distinct eigenvalues for a 3×3 matrix, A is diagonalizable, by Theorem 4.25. (If we actually require a matrix P such that $P^{-1}AP$ is diagonal, we must still compute bases for the eigenspaces, as in Example 4.19 and Example 4.26 above.)

The final theorem of this section is an important result that characterizes diagonalizable matrices in terms of the two notions of multiplicity that were introduced following Example 4.18. It gives precise conditions under which an $n \times n$ matrix can be diagonalized, even when it has fewer than n eigenvalues, as in Example 4.26. We first prove a lemma that holds whether or not a matrix is diagonalizable.

Lemma 4.26 If A is an $n \times n$ matrix, then the geometric multiplicity of each eigenvalue is less than or equal to its algebraic multiplicity.

Proof Suppose λ_1 is an eigenvalue of A with geometric multiplicity p; that is, $\dim E_{\lambda_1} = p$. Specifically, let E_{λ_1} have basis $\mathcal{B}_1 = \{\mathbf{v}_1, \mathbf{v}_2, \ldots, \mathbf{v}_p\}$. Let Q be any invertible $n \times n$ matrix having $\mathbf{v}_1, \mathbf{v}_2, \ldots, \mathbf{v}_p$ as its first p columns—say,

$$Q = [\mathbf{v}_1 \quad \cdots \quad \mathbf{v}_p \quad \mathbf{v}_{p+1} \quad \cdots \quad \mathbf{v}_n]$$

or, as a partitioned matrix,

$$Q = [U \mid V]$$

Let

$$Q^{-1} = \left[\begin{array}{c} C \\ \hline D \end{array}\right]$$

where C is $p \times n$.

Since the columns of U are eigenvectors corresponding to λ_1, $AU = \lambda_1 U$. We also have

$$\left[\begin{array}{c|c} I_p & O \\ \hline O & I_{n-p} \end{array}\right] = I_n = Q^{-1}Q = \left[\begin{array}{c} C \\ \hline D \end{array}\right][U \mid V] = \left[\begin{array}{c|c} CU & CV \\ \hline DU & DV \end{array}\right]$$

from which we obtain $CU = I_p$, $CV = O$, $DU = O$, and $DV = I_{n-p}$. Therefore,

$$Q^{-1}AQ = \left[\begin{array}{c} C \\ \hline D \end{array}\right]A[U \mid V] = \left[\begin{array}{c|c} CAU & CAV \\ \hline DAU & DAV \end{array}\right] = \left[\begin{array}{c|c} \lambda_1 CU & CAV \\ \hline \lambda_1 DU & DAV \end{array}\right] = \left[\begin{array}{c|c} \lambda_1 I_p & CAV \\ \hline O & DAV \end{array}\right]$$

By Exercise 69 in Section 4.2, it follows that

$$\det(Q^{-1}AQ - \lambda I) = (\lambda_1 - \lambda)^p \det(DAV - \lambda I) \tag{5}$$

But $\det(Q^{-1}AQ - \lambda I)$ is the characteristic polynomial of $Q^{-1}AQ$, which is the same as the characteristic polynomial of A, by Theorem 4.22(d). Thus, Equation (5) implies that the algebraic multiplicity of λ_1 is at least p, its geometric multiplicity.

Theorem 4.27 **The Diagonalization Theorem**

Let A be an $n \times n$ matrix whose distinct eigenvalues are $\lambda_1, \lambda_2, \ldots, \lambda_k$. The following statements are equivalent:

a. A is diagonalizable.
b. The union $\mathcal{B}$ of the bases of the eigenspaces of A (as in Theorem 4.24) contains n vectors.
c. The algebraic multiplicity of each eigenvalue equals its geometric multiplicity.

Proof (a) ⇒ (b) If A is diagonalizable, then it has n linearly independent eigenvectors, by Theorem 4.23. If n_i of these eigenvectors correspond to the eigenvalue λ_i, then $\mathcal{B}_i$ contains at least n_i vectors. (We already know that these n_i vectors are linearly independent; the only thing that might prevent them from being a basis for E_{λ_i} is that they might not span it.) Thus, $\mathcal{B}$ contains at least n vectors. But, by Theorem 4.24, $\mathcal{B}$ is a linearly independent set in $\mathbb{R}^n$; hence, it contains exactly n vectors.

(b) ⇒ (c) Let the geometric multiplicity of λ_i be $d_i = \dim E_{\lambda_i}$ and let the algebraic multiplicity of λ_i be m_i. By Lemma 4.26, $d_i \le m_i$ for $i = 1, \ldots, k$. Now assume that property (b) holds. Then we also have

$$n = d_1 + d_2 + \cdots + d_k \le m_1 + m_2 + \cdots + m_k$$

But $m_1 + m_2 + \cdots + m_k = n$, since the sum of the algebraic multiplicities of the eigenvalues of A is just the degree of the characteristic polynomial of A—namely, n.

It follows that $d_1 + d_2 + \cdots + d_k = m_1 + m_2 + \cdots + m_k$, which implies that

$$(m_1 - d_1) + (m_2 - d_2) + \cdots + (m_k - d_k) = 0 \qquad (6)$$

Using Lemma 4.26 again, we know that $m_i - d_i \ge 0$ for $i = 1, \ldots, k$, from which we can deduce that each summand in Equation (6) is zero; that is, $m_i = d_i$ for $i = 1, \ldots, k$.

(c) ⇒ (a) If the algebraic multiplicity m_i and the geometric multiplicity d_i are equal for each eigenvalue λ_i of A, then $\mathcal{B}$ has $d_1 + d_2 + \cdots + d_k = m_1 + m_2 + \cdots + m_k = n$ vectors, which are linearly independent, by Theorem 4.24. Thus, these are n linearly independent eigenvectors of A, and A is diagonalizable, by Theorem 4.23.

Example 4.28

(a) The matrix $A = \begin{bmatrix} 0 & 1 & 0 \\ 0 & 0 & 1 \\ 2 & -5 & 4 \end{bmatrix}$ from Example 4.18 has two distinct eigenvalues, $\lambda_1 = \lambda_2 = 1$ and $\lambda_3 = 2$. Since the eigenvalue $\lambda_1 = \lambda_2 = 1$ has algebraic multiplicity 2 but geometric multiplicity 1, A is not diagonalizable, by the Diagonalization Theorem. (See also Example 4.25.)

(b) The matrix $A = \begin{bmatrix} -1 & 0 & 1 \\ 3 & 0 & -3 \\ 1 & 0 & -1 \end{bmatrix}$ from Example 4.19 also has two distinct eigenvalues, $\lambda_1 = \lambda_2 = 0$ and $\lambda_3 = -2$. The eigenvalue 0 has algebraic and geometric multiplicity 2, and the eigenvalue -2 has algebraic and geometric multiplicity 1. Thus, this matrix *is* diagonalizable, by the Diagonalization Theorem. (This agrees with our findings in Example 4.26.)

We conclude this section with an application of diagonalization to the computation of the powers of a matrix.

Example 4.29

Compute A^{10} if $A = \begin{bmatrix} 0 & 1 \\ 2 & 1 \end{bmatrix}$.

Solution In Example 4.21, we found that this matrix has eigenvalues $\lambda_1 = -1$ and $\lambda_2 = 2$, with corresponding eigenvectors $\mathbf{v}_1 = \begin{bmatrix} 1 \\ -1 \end{bmatrix}$ and $\mathbf{v}_2 = \begin{bmatrix} 1 \\ 2 \end{bmatrix}$. It follows (from any one of a number of theorems in this section) that A is diagonalizable and $P^{-1}AP = D$, where

$$P = [\mathbf{v}_1 \quad \mathbf{v}_2] = \begin{bmatrix} 1 & 1 \\ -1 & 2 \end{bmatrix} \quad \text{and} \quad D = \begin{bmatrix} -1 & 0 \\ 0 & 2 \end{bmatrix}$$

Solving for A, we have $A = PDP^{-1}$ and, by Theorem 4.22(f), $A^n = PD^nP^{-1}$ for all $n \geq 1$.

Since

$$D^n = \begin{bmatrix} -1 & 0 \\ 0 & 2 \end{bmatrix}^n = \begin{bmatrix} (-1)^n & 0 \\ 0 & 2^n \end{bmatrix}$$

we have

$$\begin{aligned} A^n = PD^nP^{-1} &= \begin{bmatrix} 1 & 1 \\ -1 & 2 \end{bmatrix} \begin{bmatrix} (-1)^n & 0 \\ 0 & 2^n \end{bmatrix} \begin{bmatrix} 1 & 1 \\ -1 & 2 \end{bmatrix}^{-1} \\ &= \begin{bmatrix} 1 & 1 \\ -1 & 2 \end{bmatrix} \begin{bmatrix} (-1)^n & 0 \\ 0 & 2^n \end{bmatrix} \begin{bmatrix} \frac{2}{3} & -\frac{1}{3} \\ \frac{1}{3} & \frac{1}{3} \end{bmatrix} \\ &= \begin{bmatrix} \dfrac{2(-1)^n + 2^n}{3} & \dfrac{(-1)^{n+1} + 2^n}{3} \\ \dfrac{2(-1)^{n+1} + 2^{n+1}}{3} & \dfrac{(-1)^{n+2} + 2^{n+1}}{3} \end{bmatrix} \end{aligned}$$

Since we were only asked for A^{10}, this is more than we needed. But now we can simply set $n = 10$ to find

$$A^{10} = \begin{bmatrix} \dfrac{2(-1)^{10} + 2^{10}}{3} & \dfrac{(-1)^{11} + 2^{10}}{3} \\ \dfrac{2(-1)^{11} + 2^{11}}{3} & \dfrac{(-1)^{12} + 2^{11}}{3} \end{bmatrix} = \begin{bmatrix} 342 & 341 \\ 682 & 683 \end{bmatrix}$$

Exercises 4.4

In Exercises 1–4, show that A and B are not similar matrices.

1. $A = \begin{bmatrix} 4 & 1 \\ 3 & 1 \end{bmatrix}, B = \begin{bmatrix} 1 & 0 \\ 0 & 1 \end{bmatrix}$

2. $A = \begin{bmatrix} 2 & 1 \\ -4 & 6 \end{bmatrix}, B = \begin{bmatrix} 3 & -1 \\ -5 & 7 \end{bmatrix}$

3. $A = \begin{bmatrix} 2 & 1 & 4 \\ 0 & 2 & 3 \\ 0 & 0 & 4 \end{bmatrix}, B = \begin{bmatrix} 1 & 0 & 0 \\ -1 & 4 & 0 \\ 2 & 3 & 4 \end{bmatrix}$

4. $A = \begin{bmatrix} 1 & 2 & 0 \\ 0 & 1 & -1 \\ 0 & -1 & 1 \end{bmatrix}, B = \begin{bmatrix} 2 & 1 & 1 \\ 0 & 1 & 0 \\ 2 & 0 & 1 \end{bmatrix}$

In Exercises 5–7, a diagonalization of the matrix A is given in the form $P^{-1}AP = D$. List the eigenvalues of A and bases for the corresponding eigenspaces.

5. $\begin{bmatrix} 2 & -1 \\ -1 & 1 \end{bmatrix} \begin{bmatrix} 5 & -1 \\ 2 & 2 \end{bmatrix} \begin{bmatrix} 1 & 1 \\ 1 & 2 \end{bmatrix} = \begin{bmatrix} 4 & 0 \\ 0 & 3 \end{bmatrix}$

6. $\begin{bmatrix} \frac{1}{6} & \frac{1}{6} & \frac{1}{6} \\ \frac{1}{2} & -\frac{1}{2} & -\frac{1}{2} \\ \frac{1}{3} & \frac{1}{3} & -\frac{2}{3} \end{bmatrix} \begin{bmatrix} 1 & 1 & 1 \\ 0 & 0 & 1 \\ 1 & 1 & 0 \end{bmatrix} \begin{bmatrix} 3 & 1 & 0 \\ 1 & -1 & 1 \\ 2 & 0 & -1 \end{bmatrix}$

$= \begin{bmatrix} 2 & 0 & 0 \\ 0 & 0 & 0 \\ 0 & 0 & -1 \end{bmatrix}$

7. $\begin{bmatrix} \frac{1}{8} & \frac{1}{8} & \frac{1}{8} \\ -\frac{1}{4} & \frac{3}{4} & -\frac{1}{4} \\ \frac{5}{8} & -\frac{3}{8} & -\frac{3}{8} \end{bmatrix}\begin{bmatrix} 1 & 3 & 3 \\ 2 & 0 & 2 \\ 3 & 3 & 1 \end{bmatrix}\begin{bmatrix} 3 & 0 & 1 \\ 2 & 1 & 0 \\ 3 & -1 & -1 \end{bmatrix} = \begin{bmatrix} 6 & 0 & 0 \\ 0 & -2 & 0 \\ 0 & 0 & -2 \end{bmatrix}$

In Exercises 8–15, determine whether A is diagonalizable and, if so, find an invertible matrix P and a diagonal matrix D such that $P^{-1}AP = D$.

8. $A = \begin{bmatrix} 5 & 2 \\ 2 & 5 \end{bmatrix}$

9. $A = \begin{bmatrix} -3 & 4 \\ -1 & 1 \end{bmatrix}$

10. $A = \begin{bmatrix} 3 & 1 & 0 \\ 0 & 3 & 1 \\ 0 & 0 & 3 \end{bmatrix}$

11. $A = \begin{bmatrix} 1 & 0 & 1 \\ 0 & 1 & 1 \\ 1 & 1 & 0 \end{bmatrix}$

12. $A = \begin{bmatrix} 1 & 0 & 0 \\ 2 & 2 & 1 \\ 3 & 0 & 1 \end{bmatrix}$

13. $A = \begin{bmatrix} 1 & 2 & 1 \\ -1 & 0 & 1 \\ 1 & 1 & 0 \end{bmatrix}$

14. $A = \begin{bmatrix} 2 & 0 & 0 & 2 \\ 0 & 3 & 2 & 1 \\ 0 & 0 & 3 & 0 \\ 0 & 0 & 0 & 1 \end{bmatrix}$

15. $A = \begin{bmatrix} 2 & 0 & 0 & 4 \\ 0 & 2 & 0 & 0 \\ 0 & 0 & -2 & 0 \\ 0 & 0 & 0 & -2 \end{bmatrix}$

In Exercises 16–23, use the method of Example 4.29 to compute the indicated power of the matrix.

16. $\begin{bmatrix} -4 & 6 \\ -3 & 5 \end{bmatrix}^9$

17. $\begin{bmatrix} -1 & 6 \\ 1 & 0 \end{bmatrix}^{10}$

18. $\begin{bmatrix} 4 & -3 \\ -1 & 2 \end{bmatrix}^{-6}$

19. $\begin{bmatrix} 0 & 3 \\ 1 & 2 \end{bmatrix}^k$

20. $\begin{bmatrix} 2 & 1 & 2 \\ 2 & 1 & 2 \\ 2 & 1 & 2 \end{bmatrix}^8$

21. $\begin{bmatrix} 1 & 1 & 1 \\ 0 & -1 & 0 \\ 0 & 0 & -1 \end{bmatrix}^{2015}$

22. $\begin{bmatrix} 2 & 0 & 1 \\ 1 & 1 & 1 \\ 1 & 0 & 2 \end{bmatrix}^k$

23. $\begin{bmatrix} 1 & 1 & 0 \\ 2 & -2 & 2 \\ 0 & 1 & 1 \end{bmatrix}^k$

In Exercises 24–29, find all (real) values of k for which A is diagonalizable.

24. $A = \begin{bmatrix} 1 & 1 \\ 0 & k \end{bmatrix}$

25. $A = \begin{bmatrix} 1 & k \\ 0 & 1 \end{bmatrix}$

26. $A = \begin{bmatrix} k & 1 \\ 1 & 0 \end{bmatrix}$

27. $A = \begin{bmatrix} 1 & 0 & k \\ 0 & 1 & 0 \\ 0 & 0 & 1 \end{bmatrix}$

28. $A = \begin{bmatrix} 1 & k & 0 \\ 0 & 2 & 0 \\ 0 & 0 & 1 \end{bmatrix}$

29. $A = \begin{bmatrix} 1 & 1 & k \\ 1 & 1 & k \\ 1 & 1 & k \end{bmatrix}$

30. Prove Theorem 4.21(c).

31. Prove Theorem 4.22(b).

32. Prove Theorem 4.22(c).

33. Prove Theorem 4.22(e).

34. Prove Theorem 4.22(f).

35. Prove Theorem 4.22(g).

36. If A and B are invertible matrices, show that AB and BA are similar.

37. Prove that if A and B are similar matrices, then $\operatorname{tr}(A) = \operatorname{tr}(B)$. [*Hint:* Find a way to use Exercise 45 from Section 3.2.]

In general, it is difficult to show that two matrices are similar. However, if two similar matrices are diagonalizable, the task becomes easier. In Exercises 38–41, show that A and B are similar by showing that they are similar to the same diagonal matrix. Then find an invertible matrix P such that $P^{-1}AP = B$.

38. $A = \begin{bmatrix} 3 & 1 \\ 0 & -1 \end{bmatrix}, B = \begin{bmatrix} 1 & 2 \\ 2 & 1 \end{bmatrix}$

39. $A = \begin{bmatrix} 5 & -3 \\ 4 & -2 \end{bmatrix}, B = \begin{bmatrix} -1 & 1 \\ -6 & 4 \end{bmatrix}$

40. $A = \begin{bmatrix} 2 & 1 & 0 \\ 0 & -2 & 1 \\ 0 & 0 & 1 \end{bmatrix}, B = \begin{bmatrix} 3 & 2 & -5 \\ 1 & 2 & -1 \\ 2 & 2 & -4 \end{bmatrix}$

41. $A = \begin{bmatrix} 1 & 0 & 2 \\ 1 & -1 & 1 \\ 2 & 0 & 1 \end{bmatrix}, B = \begin{bmatrix} -3 & -2 & 0 \\ 6 & 5 & 0 \\ 4 & 4 & -1 \end{bmatrix}$

42. Prove that if A is similar to B, then A^T is similar to B^T.

43. Prove that if A is diagonalizable, so is A^T.

44. Let A be an invertible matrix. Prove that if A is diagonalizable, so is A^{-1}.

45. Prove that if A is a diagonalizable matrix with only one eigenvalue λ, then A is of the form $A = \lambda I$. (Such a matrix is called a ***scalar matrix***.)

46. Let A and B be $n \times n$ matrices, each with n distinct eigenvalues. Prove that A and B have the same eigenvectors if and only if $AB = BA$.

47. Let A and B be similar matrices. Prove that the algebraic multiplicities of the eigenvalues of A and B are the same.

48. Let A and B be similar matrices. Prove that the geometric multiplicities of the eigenvalues of A and B are the same. [*Hint:* Show that, if $B = P^{-1}AP$, then every eigenvector of B is of the form $P^{-1}\mathbf{v}$ for some eigenvector $\mathbf{v}$ of A.]

49. Prove that if A is a diagonalizable matrix such that every eigenvalue of A is either 0 or 1, then A is idempotent (that is, $A^2 = A$).

50. Let A be a nilpotent matrix (that is, $A^m = O$ for some $m > 1$). Prove that if A is diagonalizable, then A must be the zero matrix.

51. Suppose that A is a 6×6 matrix with characteristic polynomial $c_A(\lambda) = (1 + \lambda)(1 - \lambda)^2(2 - \lambda)^3$.

(a) Prove that it is not possible to find three linearly independent vectors $\mathbf{v}_1, \mathbf{v}_2, \mathbf{v}_3$ in $\mathbb{R}^6$ such that $A\mathbf{v}_1 = \mathbf{v}_1, A\mathbf{v}_2 = \mathbf{v}_2$, and $A\mathbf{v}_3 = \mathbf{v}_3$.

(b) If A is diagonalizable, what are the dimensions of the eigenspaces E_{-1}, E_1, and E_2?

52. Let $A = \begin{bmatrix} a & b \\ c & d \end{bmatrix}$.

(a) Prove that A is diagonalizable if $(a - d)^2 + 4bc > 0$ and is not diagonalizable if $(a - d)^2 + 4bc < 0$.

(b) Find two examples to demonstrate that if $(a - d)^2 + 4bc = 0$, then A may or may not be diagonalizable.

4.5 Iterative Methods for Computing Eigenvalues

At this point, the only method we have for computing the eigenvalues of a matrix is to solve the characteristic equation. However, there are several problems with this method that render it impractical in all but small examples. The first problem is that it depends on the computation of a determinant, which is a very time-consuming process for large matrices. The second problem is that the characteristic equation is a polynomial equation, and there are no formulas for solving polynomial equations of degree higher than 4 (polynomials of degrees 2, 3, and 4 can be solved using the quadratic formula and its analogues). Thus, we are forced to *approximate* eigenvalues in most practical problems. Unfortunately, methods for approximating the roots of a polynomial are quite sensitive to roundoff error and are therefore unreliable.

Instead, we bypass the characteristic polynomial altogether and take a different approach, approximating an eigenvector first and then using this eigenvector to find the corresponding eigenvalue. In this section, we will explore several variations on one such method that is based on a simple iterative technique.

In 1824, the Norwegian mathematician Niels Henrik Abel (1802–1829) proved that a general fifth-degree (quintic) polynomial equation is not *solvable by radicals;* that is, there is no formula for its roots in terms of its coefficients that uses only the operations of addition, subtraction, multiplication, division, and taking nth roots. In a paper written in 1830 and published posthumously in 1846, the French mathematician Evariste Galois (1811–1832) gave a more complete theory that established conditions under which an arbitrary polynomial equation can be solved by radicals. Galois's work was instrumental in establishing the branch of algebra called *group theory;* his approach to polynomial equations is now known as *Galois theory*.

The Power Method

The power method applies to an $n \times n$ matrix that has a ***dominant eigenvalue*** λ_1—that is, an eigenvalue that is larger in absolute value than all of the other eigenvalues. For example, if a matrix has eigenvalues -4, -3, 1, and 3, then -4 is the dominant eigenvalue, since $4 = |-4| > |-3| \geq |3| \geq |1|$. On the other hand, a matrix with eigenvalues -4, -3, 3, and 4 has no dominant eigenvalue.

The power method proceeds iteratively to produce a sequence of scalars that converges to λ_1 and a sequence of vectors that converges to the corresponding eigenvector $\mathbf{v}_1$, the ***dominant eigenvector***. For simplicity, we will assume that the matrix A is diagonalizable. The following theorem is the basis for the power method.

Theorem 4.28 Let A be an $n \times n$ diagonalizable matrix with dominant eigenvalue λ_1. Then there exists a nonzero vector $\mathbf{x}_0$ such that the sequence of vectors $\mathbf{x}_k$ defined by

$$\mathbf{x}_1 = A\mathbf{x}_0, \mathbf{x}_2 = A\mathbf{x}_1, \mathbf{x}_3 = A\mathbf{x}_2, \ldots, \mathbf{x}_k = A\mathbf{x}_{k-1}, \ldots$$

approaches a dominant eigenvector of A.

Proof We may assume that the eigenvalues of A have been labeled so that

$$|\lambda_1| > |\lambda_2| \geq |\lambda_3| \geq \cdots \geq |\lambda_n|$$

Let $\mathbf{v}_1, \mathbf{v}_2, \ldots, \mathbf{v}_n$ be the corresponding eigenvectors. Since $\mathbf{v}_1, \mathbf{v}_2, \ldots, \mathbf{v}_n$ are linearly independent (why?), they form a basis for $\mathbb{R}^n$. Consequently, we can write $\mathbf{x}_0$ as a linear combination of these eigenvectors—say,

$$\mathbf{x}_0 = c_1\mathbf{v}_1 + c_2\mathbf{v}_2 + \cdots + c_n\mathbf{v}_n$$

Now $\mathbf{x}_1 = A\mathbf{x}_0, \mathbf{x}_2 = A\mathbf{x}_1 = A(A\mathbf{x}_0) = A^2\mathbf{x}_0, \mathbf{x}_3 = A\mathbf{x}_2 = A(A^2\mathbf{x}_0) = A^3\mathbf{x}_0$, and, generally,

$$\mathbf{x}_k = A^k\mathbf{x}_0 \quad \text{for } k \geq 1$$

As we saw in Example 4.21,

$$\begin{aligned} A^k\mathbf{x}_0 &= c_1\lambda_1^k\mathbf{v}_1 + c_2\lambda_2^k\mathbf{v}_2 + \cdots + c_n\lambda_n^k\mathbf{v}_n \\ &= \lambda_1^k\left(c_1\mathbf{v}_1 + c_2\left(\frac{\lambda_2}{\lambda_1}\right)^k\mathbf{v}_2 + \cdots + c_n\left(\frac{\lambda_n}{\lambda_1}\right)^k\mathbf{v}_n\right) \end{aligned} \tag{1}$$

where we have used the fact that $\lambda_1 \neq 0$.

The fact that λ_1 is the dominant eigenvalue means that each of the fractions $\lambda_2/\lambda_1, \lambda_3/\lambda_1, \ldots, \lambda_n/\lambda_1$, is less than 1 in absolute value. Thus,

$$\left(\frac{\lambda_2}{\lambda_1}\right)^k, \left(\frac{\lambda_3}{\lambda_1}\right)^k, \ldots, \left(\frac{\lambda_n}{\lambda_1}\right)^k$$

all go to zero as $k \to \infty$. It follows that

$$\mathbf{x}_k = A^k\mathbf{x}_0 \to \lambda_1^k c_1\mathbf{v}_1 \quad \text{as } k \to \infty \tag{2}$$

Now, since $\lambda_1 \neq 0$ and $\mathbf{v}_1 \neq \mathbf{0}$, $\mathbf{x}_k$ is approaching a *nonzero* multiple of $\mathbf{v}_1$ (that is, an eigenvector corresponding to λ_1) *provided* $c_1 \neq 0$. (This is the required condition on the initial vector $\mathbf{x}_0$: It must have a nonzero component c_1 in the direction of the dominant eigenvector $\mathbf{v}_1$.)

Example 4.30

Approximate the dominant eigenvector of $A = \begin{bmatrix} 1 & 1 \\ 2 & 0 \end{bmatrix}$ using the method of Theorem 4.28.

Solution We will take $\mathbf{x}_0 = \begin{bmatrix} 1 \\ 0 \end{bmatrix}$ as the initial vector. Then

$$\mathbf{x}_1 = A\mathbf{x}_0 = \begin{bmatrix} 1 & 1 \\ 2 & 0 \end{bmatrix}\begin{bmatrix} 1 \\ 0 \end{bmatrix} = \begin{bmatrix} 1 \\ 2 \end{bmatrix}$$

$$\mathbf{x}_2 = A\mathbf{x}_1 = \begin{bmatrix} 1 & 1 \\ 2 & 0 \end{bmatrix}\begin{bmatrix} 1 \\ 2 \end{bmatrix} = \begin{bmatrix} 3 \\ 2 \end{bmatrix}$$

We continue in this fashion to obtain the values of $\mathbf{x}_k$ in Table 4.1.

Table 4.1

k	0	1	2	3	4	5	6	7	8
$\mathbf{x}_k$	$\begin{bmatrix}1\\0\end{bmatrix}$	$\begin{bmatrix}1\\2\end{bmatrix}$	$\begin{bmatrix}3\\2\end{bmatrix}$	$\begin{bmatrix}5\\6\end{bmatrix}$	$\begin{bmatrix}11\\10\end{bmatrix}$	$\begin{bmatrix}21\\22\end{bmatrix}$	$\begin{bmatrix}43\\42\end{bmatrix}$	$\begin{bmatrix}85\\86\end{bmatrix}$	$\begin{bmatrix}171\\170\end{bmatrix}$
r_k	—	0.50	1.50	0.83	1.10	0.95	1.02	0.99	1.01
l_k	—	1.00	3.00	1.67	2.20	1.91	2.05	1.98	2.01

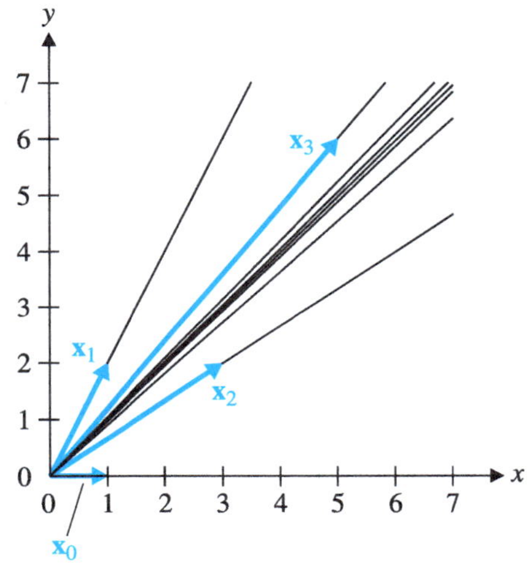

Figure 4.14

Figure 4.14 shows what is happening geometrically. We know that the eigenspace for the dominant eigenvector will have dimension 1. (Why? See Exercise 46.) Therefore, it is a line through the origin in $\mathbb{R}^2$. The first few iterates $\mathbf{x}_k$ are shown along with the directions they determine. It appears as though the iterates are converging on the line whose direction vector is $\begin{bmatrix}1\\1\end{bmatrix}$. To confirm that this is the dominant eigenvector we seek, we need only observe that the ratio r_k of the first to the second component of $\mathbf{x}_k$ gets very close to 1 as k increases. The second line in the body of Table 4.1 gives these values, and you can see clearly that r_k is indeed approaching 1. We deduce that a dominant eigenvector of A is $\begin{bmatrix}1\\1\end{bmatrix}$.

Once we have found a dominant eigenvector, how can we find the corresponding dominant eigenvalue? One approach is to observe that if an $\mathbf{x}_k$ is approximately a dominant eigenvector of A for the dominant eigenvalue λ_1, then

$$\mathbf{x}_{k+1} = A\mathbf{x}_k \approx \lambda_1\mathbf{x}_k$$

It follows that the ratio l_k of the first component of $\mathbf{x}_{k+1}$ to that of $\mathbf{x}_k$ will approach λ_1 as k increases. Table 4.1 gives the values of l_k, and you can see that they are approaching 2, which is the dominant eigenvalue.

There is a drawback to the method of Example 4.30: The components of the iterates $\mathbf{x}_k$ get very large very quickly and can cause significant roundoff errors. To avoid this drawback, we can multiply each iterate by some scalar that reduces the magnitude of its components. Since scalar multiples of the iterates $\mathbf{x}_k$ will still converge to a dominant eigenvector, this approach is acceptable. There are various ways to accomplish it. One is to normalize each $\mathbf{x}_k$ by dividing it by $\|\mathbf{x}_k\|$ (i.e., to make each iterate a *unit* vector). An easier method—and the one we will use—is to divide each $\mathbf{x}_k$ by the component with the maximum absolute value, so that the largest component is now 1. This method is called ***scaling.*** Thus, if m_k denotes the component of $\mathbf{x}_k$ with the maximum absolute value, we will replace $\mathbf{x}_k$ by $\mathbf{y}_k = (1/m_k)\mathbf{x}_k$.

We illustrate this approach with the calculations from Example 4.30. For $\mathbf{x}_0$, there is nothing to do, since $m_0 = 1$. Hence,

$$\mathbf{y}_0 = \mathbf{x}_0 = \begin{bmatrix} 1 \\ 0 \end{bmatrix}$$

We then compute $\mathbf{x}_1 = \begin{bmatrix} 1 \\ 2 \end{bmatrix}$ as before, but now we scale with $m_1 = 2$ to get

$$\mathbf{y}_1 = \left(\frac{1}{2}\right)\mathbf{x}_1 = \left(\frac{1}{2}\right)\begin{bmatrix} 1 \\ 2 \end{bmatrix} = \begin{bmatrix} 0.5 \\ 1 \end{bmatrix}$$

Now the calculations change. We take

$$\mathbf{x}_2 = A\mathbf{y}_1 = \begin{bmatrix} 1 & 1 \\ 2 & 0 \end{bmatrix}\begin{bmatrix} 0.5 \\ 1 \end{bmatrix} = \begin{bmatrix} 1.5 \\ 1 \end{bmatrix}$$

and scale to get

$$\mathbf{y}_2 = \left(\frac{1}{1.5}\right)\begin{bmatrix} 1.5 \\ 1 \end{bmatrix} = \begin{bmatrix} 1 \\ 0.67 \end{bmatrix}$$

The next few calculations are summarized in Table 4.2.

You can now see clearly that the sequence of vectors $\mathbf{y}_k$ is converging to $\begin{bmatrix} 1 \\ 1 \end{bmatrix}$, a dominant eigenvector. Moreover, the sequence of scalars m_k converges to the corresponding dominant eigenvalue $\lambda_1 = 2$.

Table 4.2

k	0	1	2	3	4	5	6	7	8
$\mathbf{x}_k$	$\begin{bmatrix} 1 \\ 0 \end{bmatrix}$	$\begin{bmatrix} 1 \\ 2 \end{bmatrix}$	$\begin{bmatrix} 1.5 \\ 1 \end{bmatrix}$	$\begin{bmatrix} 1.67 \\ 2 \end{bmatrix}$	$\begin{bmatrix} 1.83 \\ 1.67 \end{bmatrix}$	$\begin{bmatrix} 1.91 \\ 2 \end{bmatrix}$	$\begin{bmatrix} 1.95 \\ 1.91 \end{bmatrix}$	$\begin{bmatrix} 1.98 \\ 2 \end{bmatrix}$	$\begin{bmatrix} 1.99 \\ 1.98 \end{bmatrix}$
$\mathbf{y}_k$	$\begin{bmatrix} 1 \\ 0 \end{bmatrix}$	$\begin{bmatrix} 0.5 \\ 1 \end{bmatrix}$	$\begin{bmatrix} 1 \\ 0.67 \end{bmatrix}$	$\begin{bmatrix} 0.83 \\ 1 \end{bmatrix}$	$\begin{bmatrix} 1 \\ 0.91 \end{bmatrix}$	$\begin{bmatrix} 0.95 \\ 1 \end{bmatrix}$	$\begin{bmatrix} 1 \\ 0.98 \end{bmatrix}$	$\begin{bmatrix} 0.99 \\ 1 \end{bmatrix}$	$\begin{bmatrix} 1 \\ 0.99 \end{bmatrix}$
m_k	1	2	1.5	2	1.83	2	1.95	2	1.99

This method, called the ***power method,*** is summarized below.

The Power Method

Let A be a diagonalizable $n \times n$ matrix with a corresponding dominant eigenvalue λ_1.

1. Let $\mathbf{x}_0 = \mathbf{y}_0$ be any initial vector in $\mathbb{R}^n$ whose largest component is 1.
2. Repeat the following steps for $k = 1, 2, \ldots$:
   (a) Compute $\mathbf{x}_k = A\mathbf{y}_{k-1}$.
   (b) Let m_k be the component of $\mathbf{x}_k$ with the largest absolute value.
   (c) Set $\mathbf{y}_k = (1/m_k)\mathbf{x}_k$.

For most choices of $\mathbf{x}_0$, m_k converges to the dominant eigenvalue λ_1 and $\mathbf{y}_k$ converges to a dominant eigenvector.

Example 4.31

Use the power method to approximate the dominant eigenvalue and a dominant eigenvector of

$$A = \begin{bmatrix} 0 & 5 & -6 \\ -4 & 12 & -12 \\ -2 & -2 & 10 \end{bmatrix}$$

Solution Taking as our initial vector

$$\mathbf{x}_0 = \begin{bmatrix} 1 \\ 1 \\ 1 \end{bmatrix}$$

we compute the entries in Table 4.3.

You can see that the vectors $\mathbf{y}_k$ are approaching $\begin{bmatrix} 0.50 \\ 1 \\ -0.50 \end{bmatrix}$ and the scalars m_k are approaching 16. This suggests that they are, respectively, a dominant eigenvector and the dominant eigenvalue of A.

Remarks

- If the initial vector $\mathbf{x}_0$ has a zero component in the direction of the dominant eigenvector $\mathbf{v}_1$ (i.e., if $c = 0$ in the proof of Theorem 4.28), then the power method

Table 4.3

k	0	1	2	3	4	5	6	7
$\mathbf{x}_k$	$\begin{bmatrix} 1 \\ 1 \\ 1 \end{bmatrix}$	$\begin{bmatrix} -1 \\ -4 \\ 6 \end{bmatrix}$	$\begin{bmatrix} -9.33 \\ -19.33 \\ 11.67 \end{bmatrix}$	$\begin{bmatrix} 8.62 \\ 17.31 \\ -9.00 \end{bmatrix}$	$\begin{bmatrix} 8.12 \\ 16.25 \\ -8.20 \end{bmatrix}$	$\begin{bmatrix} 8.03 \\ 16.05 \\ -8.04 \end{bmatrix}$	$\begin{bmatrix} 8.01 \\ 16.01 \\ -8.01 \end{bmatrix}$	$\begin{bmatrix} 8.00 \\ 16.00 \\ -8.00 \end{bmatrix}$
$\mathbf{y}_k$	$\begin{bmatrix} 1 \\ 1 \\ 1 \end{bmatrix}$	$\begin{bmatrix} -0.17 \\ -0.67 \\ 1 \end{bmatrix}$	$\begin{bmatrix} 0.48 \\ 1 \\ -0.60 \end{bmatrix}$	$\begin{bmatrix} 0.50 \\ 1 \\ -0.52 \end{bmatrix}$	$\begin{bmatrix} 0.50 \\ 1 \\ -0.50 \end{bmatrix}$	$\begin{bmatrix} 0.50 \\ 1 \\ -0.50 \end{bmatrix}$	$\begin{bmatrix} 0.50 \\ 1 \\ -0.50 \end{bmatrix}$	$\begin{bmatrix} 0.50 \\ 1 \\ -0.50 \end{bmatrix}$
m_k	1	6	−19.33	17.31	16.25	16.05	16.01	16.00

will not converge to a dominant eigenvector. However, it is quite likely that during the calculation of the subsequent iterates, at some point roundoff error will produce an $\mathbf{x}_k$ with a nonzero component in the direction of $\mathbf{v}_1$. The power method will then start to converge to a multiple of $\mathbf{v}_1$. (This is one instance where roundoff errors actually help!)

- The power method still works when there is a *repeated* dominant eigenvalue, or even when the matrix is not diagonalizable, under certain conditions. Details may be found in most modern textbooks on numerical analysis. (See Exercises 21–24.)

- For some matrices the power method converges rapidly to a dominant eigenvector, while for others the convergence may be quite slow. A careful look at the proof of Theorem 4.28 reveals why. Since $|\lambda_2/\lambda_1| \geq |\lambda_3/\lambda_1| \geq \cdots \geq |\lambda_n/\lambda_1|$, if $|\lambda_2/\lambda_1|$ is close to zero, then $(\lambda_2/\lambda_1)^k, \ldots, (\lambda_n/\lambda_1)^k$ will all approach zero rapidly. Equation (2) then shows that $\mathbf{x}_k = A^k\mathbf{x}_0$ will approach $\lambda_1^k c_1\mathbf{v}_1$ rapidly too.

As an illustration, consider Example 4.31. The eigenvalues are 16, 4, and 2, so $\lambda_2/\lambda_1 = 4/16 = 0.25$. Since $0.25^7 \approx 0.00006$, by the seventh iteration we should have close to four-decimal-place accuracy. This is exactly what we saw.

- There is an alternative way to estimate the dominant eigenvalue λ_1 of a matrix A in conjunction with the power method. First, observe that if $A\mathbf{x} = \lambda_1\mathbf{x}$, then

$$\frac{(A\mathbf{x})\cdot\mathbf{x}}{\mathbf{x}\cdot\mathbf{x}} = \frac{(\lambda_1\mathbf{x})\cdot\mathbf{x}}{\mathbf{x}\cdot\mathbf{x}} = \frac{\lambda_1(\mathbf{x}\cdot\mathbf{x})}{\mathbf{x}\cdot\mathbf{x}} = \lambda_1$$

The expression $R(\mathbf{x}) = ((A\mathbf{x})\cdot\mathbf{x})/(\mathbf{x}\cdot\mathbf{x})$ is called a ***Rayleigh quotient.*** As we compute the iterates $\mathbf{x}_k$, the successive Rayleigh quotients $R(\mathbf{x}_k)$ should approach λ_1. In fact, for symmetric matrices, the Rayleigh quotient method is about twice as fast as the scaling factor method. (See Exercises 17–20.)

John William Strutt (1842–1919), Baron Rayleigh, was a British physicist who made major contributions to the fields of acoustics and optics. In 1871, he gave the first correct explanation of why the sky is blue, and in 1895, he discovered the inert gas argon, for which discovery he received the Nobel Prize in 1904. Rayleigh was president of the Royal Society from 1905 to 1908 and became chancellor of Cambridge University in 1908. He used Rayleigh quotients in an 1873 paper on vibrating systems and later in his book *The Theory of Sound.*

The Shifted Power Method and the Inverse Power Method

The power method can help us approximate the *dominant* eigenvalue of a matrix, but what should we do if we want the other eigenvalues? Fortunately, there are several variations of the power method that can be applied.

The ***shifted power method*** uses the observation that, if λ is an eigenvalue of A, then $\lambda - \alpha$ is an eigenvalue of $A - \alpha I$ for any scalar α (Exercise 22 in Section 4.3). Thus, if λ_1 is the dominant eigenvalue of A, the eigenvalues of $A - \lambda_1 I$ will be 0, $\lambda_2 - \lambda_1, \lambda_3 - \lambda_1, \ldots, \lambda_n - \lambda_1$. We can then apply the power method to compute $\lambda_2 - \lambda_1$, and from this value we can find λ_2. Repeating this process will allow us to compute all of the eigenvalues.

Example 4.32

Use the shifted power method to compute the second eigenvalue of the matrix $A = \begin{bmatrix} 1 & 1 \\ 2 & 0 \end{bmatrix}$ from Example 4.30.

Solution In Example 4.30, we found that $\lambda_1 = 2$. To find λ_2, we apply the power method to

$$A - 2I = \begin{bmatrix} -1 & 1 \\ 2 & -2 \end{bmatrix}$$

We take $\mathbf{x}_0 = \begin{bmatrix} 1 \\ 0 \end{bmatrix}$, but other choices will also work. The calculations are summarized in Table 4.4.

Table 4.4

k	0	1	2	3	4
$\mathbf{x}_k$	$\begin{bmatrix}1\\0\end{bmatrix}$	$\begin{bmatrix}-1\\2\end{bmatrix}$	$\begin{bmatrix}1.5\\-3\end{bmatrix}$	$\begin{bmatrix}1.5\\-3\end{bmatrix}$	$\begin{bmatrix}1.5\\-3\end{bmatrix}$
$\mathbf{y}_k$	$\begin{bmatrix}1\\0\end{bmatrix}$	$\begin{bmatrix}-0.5\\1\end{bmatrix}$	$\begin{bmatrix}-0.5\\1\end{bmatrix}$	$\begin{bmatrix}-0.5\\1\end{bmatrix}$	$\begin{bmatrix}-0.5\\1\end{bmatrix}$
m_k	1	2	-3	-3	-3

Our choice of $\mathbf{x}_0$ has produced the eigenvalue -3 after only two iterations. Therefore, $\lambda_2 - \lambda_1 = -3$, so $\lambda_2 = \lambda_1 - 3 = 2 - 3 = -1$ is the second eigenvalue of A.

Recall from property (b) of Theorem 4.18 that if A is invertible with eigenvalue λ, then A^{-1} has eigenvalue $1/\lambda$. Therefore, if we apply the power method to A^{-1}, its dominant eigenvalue will be the *reciprocal of the smallest* (in magnitude) eigenvalue of A. To use this ***inverse power method,*** we follow the same steps as in the power method, except that in step 2(a) we compute $\mathbf{x}_k = A^{-1}\mathbf{y}_{k-1}$. (In practice, we don't actually compute A^{-1} explicitly; instead, we solve the equivalent equation $A\mathbf{x}_k = \mathbf{y}_{k-1}$ for $\mathbf{x}_k$ using Gaussian elimination. This turns out to be faster.)

Example 4.33

Use the inverse power method to compute the second eigenvalue of the matrix $A = \begin{bmatrix}1 & 1\\2 & 0\end{bmatrix}$ from Example 4.30.

Solution We start, as in Example 4.30, with $\mathbf{x}_0 = \mathbf{y}_0 = \begin{bmatrix}1\\0\end{bmatrix}$. To solve $A\mathbf{x}_1 = \mathbf{y}_0$, we use row reduction:

$$[A \mid \mathbf{y}_0] = \left[\begin{array}{cc|c}1 & 1 & 1\\2 & 0 & 0\end{array}\right] \longrightarrow \left[\begin{array}{cc|c}1 & 0 & 0\\0 & 1 & 1\end{array}\right]$$

Thus, $\mathbf{x}_1 = \begin{bmatrix}0\\1\end{bmatrix}$, so $\mathbf{y}_1 = \begin{bmatrix}0\\1\end{bmatrix}$. Then we get $\mathbf{x}_2$ from $A\mathbf{x}_2 = \mathbf{y}_1$:

$$[A \mid \mathbf{y}_1] = \left[\begin{array}{cc|c}1 & 1 & 0\\2 & 0 & 1\end{array}\right] \longrightarrow \left[\begin{array}{cc|c}1 & 0 & 0.5\\0 & 1 & -0.5\end{array}\right]$$

Hence, $\mathbf{x}_2 = \begin{bmatrix}0.5\\-0.5\end{bmatrix}$, and, by scaling, we get $\mathbf{y}_2 = \begin{bmatrix}1\\-1\end{bmatrix}$. Continuing, we get the values shown in Table 4.5, where the values m_k are converging to -1. Thus, the smallest eigenvalue of A is the reciprocal of -1 (which is also -1). This agrees with our previous finding in Example 4.32.

Table 4.5

k	0	1	2	3	4	5	6	7	8	9
$\mathbf{x}_k$	$\begin{bmatrix}1\\0\end{bmatrix}$	$\begin{bmatrix}0\\1\end{bmatrix}$	$\begin{bmatrix}0.5\\-0.5\end{bmatrix}$	$\begin{bmatrix}-0.5\\1.5\end{bmatrix}$	$\begin{bmatrix}0.5\\-0.83\end{bmatrix}$	$\begin{bmatrix}0.5\\-1.1\end{bmatrix}$	$\begin{bmatrix}0.5\\-0.95\end{bmatrix}$	$\begin{bmatrix}0.5\\-1.02\end{bmatrix}$	$\begin{bmatrix}0.5\\-0.99\end{bmatrix}$	$\begin{bmatrix}0.5\\-1.01\end{bmatrix}$
$\mathbf{y}_k$	$\begin{bmatrix}0\\1\end{bmatrix}$	$\begin{bmatrix}1\\0\end{bmatrix}$	$\begin{bmatrix}1\\-1\end{bmatrix}$	$\begin{bmatrix}-0.33\\1\end{bmatrix}$	$\begin{bmatrix}-0.6\\1\end{bmatrix}$	$\begin{bmatrix}-0.45\\1\end{bmatrix}$	$\begin{bmatrix}-0.52\\1\end{bmatrix}$	$\begin{bmatrix}-0.49\\1\end{bmatrix}$	$\begin{bmatrix}-0.51\\1\end{bmatrix}$	$\begin{bmatrix}-0.50\\1\end{bmatrix}$
m_k	1	1	0.5	1.5	−0.83	−1.1	−0.95	−1.02	−0.99	−1.01

The Shifted Inverse Power Method

The most versatile of the variants of the power method is one that combines the two just mentioned. It can be used to find an approximation for *any* eigenvalue, provided we have a close approximation to that eigenvalue. In other words, if a scalar α is given, the ***shifted inverse power method*** will find the eigenvalue λ of A that is closest to α.

If λ is an eigenvalue of A and $\alpha \neq \lambda$, then $A - \alpha I$ is invertible if α is not an eigenvalue of A and $1/(\lambda - \alpha)$ is an eigenvalue of $(A - \alpha I)^{-1}$. (See Exercise 45.) If α is close to λ, then $1/(\lambda - \alpha)$ will be a dominant eigenvalue of $(A - \alpha I)^{-1}$. In fact, if α is *very* close to λ, then $1/(\lambda - \alpha)$ will be *much* bigger in magnitude than the next eigenvalue, so (as noted in the third Remark following Example 4.31) the convergence will be very rapid.

Example 4.34 Use the shifted inverse power method to approximate the eigenvalue of

$$A = \begin{bmatrix} 0 & 5 & -6 \\ -4 & 12 & -12 \\ -2 & -2 & 10 \end{bmatrix}$$

that is closest to 5.

Solution Shifting, we have

$$A - 5I = \begin{bmatrix} -5 & 5 & -6 \\ -4 & 7 & -12 \\ -2 & -2 & 5 \end{bmatrix}$$

Now we apply the inverse power method with

$$\mathbf{x}_0 = \mathbf{y}_0 = \begin{bmatrix} 1 \\ 1 \\ 1 \end{bmatrix}$$

We solve $(A - 5I)\mathbf{x}_1 = \mathbf{y}_0$ for $\mathbf{x}_1$:

$$[A - 5I \mid \mathbf{y}_0] = \left[\begin{array}{ccc|c} -5 & 5 & -6 & 1 \\ -4 & 7 & -12 & 1 \\ -2 & -2 & 5 & 1 \end{array}\right] \longrightarrow \left[\begin{array}{ccc|c} 1 & 0 & 0 & -0.61 \\ 0 & 1 & 0 & -0.88 \\ 0 & 0 & 1 & -0.39 \end{array}\right]$$

Table 4.6

k	0	1	2	3	4	5	6	7
$\mathbf{x}_k$	$\begin{bmatrix}1\\1\\1\end{bmatrix}$	$\begin{bmatrix}-0.61\\-0.88\\-0.39\end{bmatrix}$	$\begin{bmatrix}-0.41\\-0.69\\-0.35\end{bmatrix}$	$\begin{bmatrix}-0.47\\-0.89\\-0.44\end{bmatrix}$	$\begin{bmatrix}-0.49\\-0.95\\-0.48\end{bmatrix}$	$\begin{bmatrix}-0.50\\-0.98\\-0.49\end{bmatrix}$	$\begin{bmatrix}-0.50\\-0.99\\-0.50\end{bmatrix}$	$\begin{bmatrix}-0.50\\-1.00\\-0.50\end{bmatrix}$
$\mathbf{y}_k$	$\begin{bmatrix}1\\1\\1\end{bmatrix}$	$\begin{bmatrix}0.69\\1.00\\0.45\end{bmatrix}$	$\begin{bmatrix}0.59\\1.00\\0.51\end{bmatrix}$	$\begin{bmatrix}0.53\\1.00\\0.50\end{bmatrix}$	$\begin{bmatrix}0.51\\1.00\\0.50\end{bmatrix}$	$\begin{bmatrix}0.50\\1.00\\0.50\end{bmatrix}$	$\begin{bmatrix}0.50\\1.00\\0.50\end{bmatrix}$	$\begin{bmatrix}0.50\\1.00\\0.50\end{bmatrix}$
m_k	1	−0.88	−0.69	−0.89	−0.95	−0.98	−0.99	−1.00

This gives

$$\mathbf{x}_1 = \begin{bmatrix}-0.61\\-0.88\\-0.39\end{bmatrix}, \quad m_1 = -0.88, \quad \text{and} \quad \mathbf{y}_1 = \frac{1}{m_1}\mathbf{x}_1 = -\frac{1}{0.88}\begin{bmatrix}-0.61\\-0.88\\-0.39\end{bmatrix} = \begin{bmatrix}0.69\\1\\0.45\end{bmatrix}$$

We continue in this fashion to obtain the values in Table 4.6, from which we deduce that the eigenvalue of A closest to 5 is approximately $5 + 1/m_7 \approx 5 + 1/(-1) = 4$, which, in fact, is exact.

The power method and its variants represent only one approach to the computation of eigenvalues. In Chapter 5, we will discuss another method based on the QR factorization of a matrix. For a more complete treatment of this topic, you can consult almost any textbook on numerical methods.

a + bi

Gerschgorin's Theorem

We owe this theorem to the Russian mathematician S. Gerschgorin (1901–1933), who stated it in 1931. It did not receive much attention until 1949, when it was resurrected by Olga Taussky-Todd in a note she published in the *American Mathematical Monthly.*

In this section, we have discussed several variations on the power method for approximating the eigenvalues of a matrix. All of these methods are iterative, and the speed with which they converge depends on the choice of initial vector. If only we had some "inside information" about the location of the eigenvalues of a given matrix, then we could make a judicious choice of the initial vector and perhaps speed up the convergence of the iterative process.

Fortunately, there is a way to estimate the location of the eigenvalues of any matrix. ***Gerschgorin's Disk Theorem*** states that the eigenvalues of a (real or complex) $n \times n$ matrix all lie inside the union of n circular disks in the complex plane.

Definition Let $A = [a_{ij}]$ be a (real or complex) $n \times n$ matrix, and let r_i denote the sum of the absolute values of the off-diagonal entries in the ith row of A; that is, $r_i = \sum_{j \neq i} |a_{ij}|$. The ***ith Gerschgorin disk*** is the circular disk D_i in the complex plane with center a_{ii} and radius r_i. That is,

$$D_i = \{z \text{ in } \mathbb{C} : |z - a_{ii}| \le r_i\}$$

Archives, California Institute of Technology

Olga Taussky-Todd (1906–1995) was born in Olmütz in the Austro-Hungarian Empire (now Olmuac in the Czech Republic). She received her doctorate in number theory from the University of Vienna in 1930. During World War II, she worked for the National Physical Laboratory in London, where she investigated the problem of flutter in the wings of supersonic aircraft. Although the problem involved differential equations, the stability of an aircraft depended on the eigenvalues of a related matrix. Taussky-Todd remembered Gerschgorin's Theorem from her graduate studies in Vienna and was able to use it to simplify the otherwise laborious computations needed to determine the eigenvalues relevant to the flutter problem.

Taussky-Todd moved to the United States in 1947, and ten years later she became the first woman appointed to the California Institute of Technology. In her career, she produced over 200 publications and received numerous awards. She was instrumental in the development of the branch of mathematics now known as matrix theory.

Example 4.35

Sketch the Gerschgorin disks and the eigenvalues for the following matrices:

(a) $A = \begin{bmatrix} 2 & 1 \\ 2 & -3 \end{bmatrix}$ (b) $A = \begin{bmatrix} 1 & -3 \\ 2 & 3 \end{bmatrix}$

Solution (a) The two Gerschgorin disks are centered at 2 and -3 with radii 1 and 2, respectively. The characteristic polynomial of A is $\lambda^2 + \lambda - 8$, so the eigenvalues are

$$\lambda = (-1 \pm \sqrt{1^2 - 4(-8)})/2 \approx 2.37, -3.37$$

Figure 4.15 shows that the eigenvalues are contained within the two Gerschgorin disks.

(b) The two Gerschgorin disks are centered at 1 and 3 with radii $|-3| = 3$ and 2, respectively. The characteristic polynomial of A is $\lambda^2 - 4\lambda + 9$, so the eigenvalues are

$$\lambda = (4 \pm \sqrt{(-4)^2 - 4(9)})/2 = 2 \pm i\sqrt{5} \approx 2 + 2.23i,\ 2 - 2.23i$$

Figure 4.16 plots the location of the eigenvalues relative to the Gerschgorin disks.

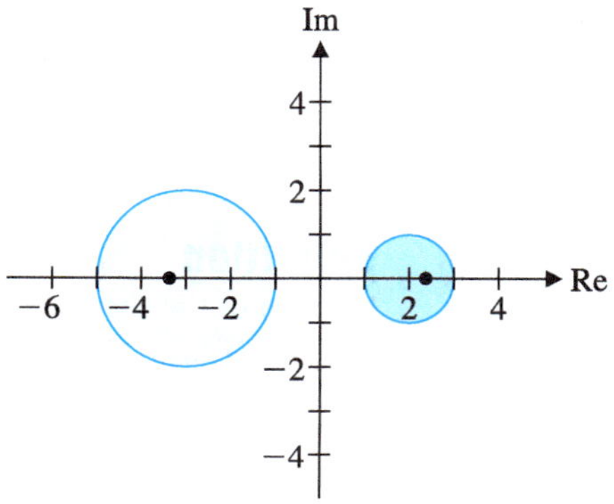

Figure 4.15

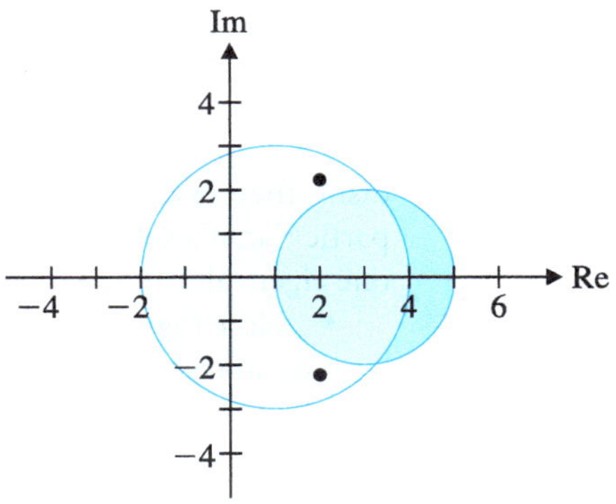

Figure 4.16

As Example 4.35 suggests, the eigenvalues of a matrix are contained within its Gerschgorin disks. The next theorem verifies that this is so.

Theorem 4.29 **Gerschgorin's Disk Theorem**

Let A be an $n \times n$ (real or complex) matrix. Then every eigenvalue of A is contained within a Gerschgorin disk.

Proof Let λ be an eigenvalue of A with corresponding eigenvector $\mathbf{x}$. Let x_i be the entry of $\mathbf{x}$ with the largest absolute value—and hence nonzero. (Why?) Then $A\mathbf{x} = \lambda\mathbf{x}$, the ith row of which is

$$[a_{i1} \quad a_{i2} \cdots a_{in}]\begin{bmatrix} x_1 \\ x_2 \\ \vdots \\ x_n \end{bmatrix} = \lambda x_i \quad \text{or} \quad \sum_{j=1}^{n} a_{ij}x_j = \lambda x_i$$

Rearranging, we have

$$(\lambda - a_{ii})x_i = \sum_{j \neq i} a_{ij}x_j \quad \text{or} \quad \lambda - a_{ii} = \frac{\sum\limits_{j \neq i} a_{ij}x_j}{x_i}$$

because $x_i \neq 0$. Taking absolute values and using properties of absolute value (see Appendix C), we obtain

$$|\lambda - a_{ii}| = \left| \frac{\sum\limits_{j \neq i} a_{ij}x_j}{x_i} \right| = \frac{\left|\sum\limits_{j \neq i} a_{ij}x_j\right|}{|x_i|} \leq \frac{\sum\limits_{j \neq i} |a_{ij}x_j|}{|x_i|} = \frac{\sum\limits_{j \neq i} |a_{ij}||x_j|}{|x_i|} \leq \sum_{j \neq i} |a_{ij}| = r_i$$

because $|x_j| \leq |x_i|$ for $j \neq i$.

This establishes that the eigenvalue λ is contained within the Gerschgorin disk centered at a_{ii} with radius r_i.

Remarks

- There is a corresponding version of the preceding theorem for Gerschgorin disks whose radii are the sum of the off-diagonal entries in the ith *column* of A.
- It can be shown that if k of the Gerschgorin disks are disjoint from the other disks, then exactly k eigenvalues are contained within the union of these k disks. In particular, if a single disk is disjoint from the other disks, then it must contain exactly one eigenvalue of the matrix. Example 4.35(a) illustrates this.
- Note that in Example 4.35(a), 0 is not contained in a Gerschgorin disk; that is, 0 is not an eigenvalue of A. Hence, without any further computation, we can deduce that the matrix A is invertible by Theorem 4.16. This observation is particularly useful when applied to larger matrices, because the Gerschgorin disks can be determined directly from the entries of the matrix.

Example 4.36

Consider the matrix $A = \begin{bmatrix} 2 & 1 & 0 \\ \frac{1}{2} & 6 & \frac{1}{2} \\ 2 & 0 & 8 \end{bmatrix}$. Gerschgorin's Theorem tells us that the eigenvalues of A are contained within three disks centered at 2, 6, and 8 with radii 1, 1, and 2, respectively. See Figure 4.17(a). Because the first disk is disjoint from the other two, it must contain exactly one eigenvalue, by the second Remark after Theorem 4.29. Because the characteristic polynomial of A has real coefficients, if it has complex roots (i.e., eigenvalues of A), they must occur in conjugate pairs. (See Appendix D.) Hence there is a unique real eigenvalue between 1 and 3, and the union of the other two disks contains two (possibly complex) eigenvalues whose real parts lie between 5 and 10.

On the other hand, the first Remark after Theorem 4.29 tells us that the same three eigenvalues of A are contained in disks centered at 2, 6, and 8 with radii $\frac{5}{2}$, 1, and $\frac{1}{2}$, respectively. See Figure 4.17(b). These disks are mutually disjoint, so each contains a single (and hence real) eigenvalue. Combining these results, we deduce that A has three real eigenvalues, one in each of the intervals [1, 3], [5, 7], and [7.5, 8.5]. (Compute the actual eigenvalues of A to verify this.)

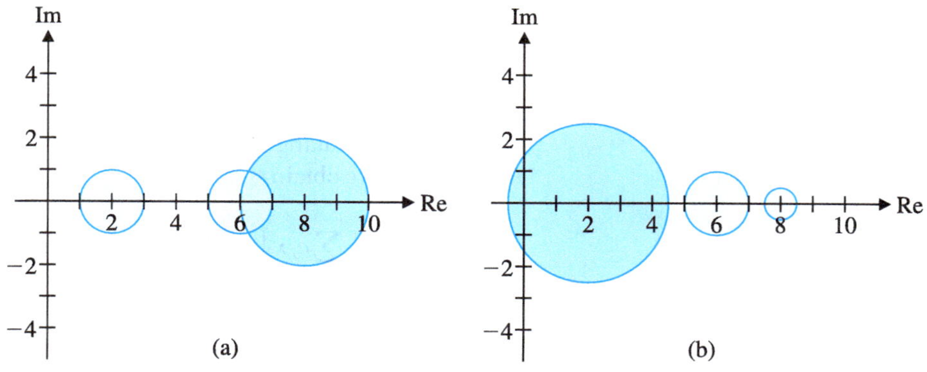

Figure 4.17

CAS

Exercises 4.5

In Exercises 1–4, a matrix A is given along with an iterate $\mathbf{x}_5$*, produced as in Example 4.30.*
(a) Use these data to approximate a dominant eigenvector whose first component is 1 and a corresponding dominant eigenvalue. (Use three-decimal-place accuracy.)
(b) Compare your approximate eigenvalue in part (a) with the actual dominant eigenvalue.

1. $A = \begin{bmatrix} 1 & 2 \\ 5 & 4 \end{bmatrix}, \mathbf{x}_5 = \begin{bmatrix} 4443 \\ 11109 \end{bmatrix}$

2. $A = \begin{bmatrix} 7 & 4 \\ -3 & -1 \end{bmatrix}, \mathbf{x}_5 = \begin{bmatrix} 7811 \\ -3904 \end{bmatrix}$

3. $A = \begin{bmatrix} 2 & 1 \\ 1 & 1 \end{bmatrix}, \mathbf{x}_5 = \begin{bmatrix} 144 \\ 89 \end{bmatrix}$

4. $A = \begin{bmatrix} 1.5 & 0.5 \\ 2.0 & 3.0 \end{bmatrix}, \mathbf{x}_5 = \begin{bmatrix} 60.625 \\ 239.500 \end{bmatrix}$

In Exercises 5–8, a matrix A is given along with an iterate $\mathbf{x}_k$*, produced using the power method, as in Example 4.31. (a) Approximate the dominant eigenvalue and eigenvector by computing the corresponding* m_k *and* $\mathbf{y}_k$*. (b) Verify that you have approximated an eigenvalue and an eigenvector of A by comparing* $A\mathbf{y}_k$ *with* $m_k\mathbf{y}_k$*.*

5. $A = \begin{bmatrix} 2 & -3 \\ -3 & 10 \end{bmatrix}, \mathbf{x}_5 = \begin{bmatrix} -3.667 \\ 11.001 \end{bmatrix}$

6. $A = \begin{bmatrix} 5 & 2 \\ 2 & -2 \end{bmatrix}, \mathbf{x}_{10} = \begin{bmatrix} 5.530 \\ 1.470 \end{bmatrix}$

7. $A = \begin{bmatrix} 4 & 0 & 6 \\ -1 & 3 & 1 \\ 6 & 0 & 4 \end{bmatrix}, \mathbf{x}_8 = \begin{bmatrix} 10.000 \\ 0.001 \\ 10.000 \end{bmatrix}$

8. $A = \begin{bmatrix} 1 & 2 & -2 \\ 1 & 1 & -3 \\ 0 & -1 & 1 \end{bmatrix}, \mathbf{x}_{10} = \begin{bmatrix} 3.415 \\ 2.914 \\ -1.207 \end{bmatrix}$

In Exercises 9–14, use the power method to approximate the dominant eigenvalue and eigenvector of A. Use the given initial vector $\mathbf{x}_0$*, the specified number of iterations k, and three-decimal-place accuracy.*

9. $A = \begin{bmatrix} 14 & 12 \\ 5 & 3 \end{bmatrix}, \mathbf{x}_0 = \begin{bmatrix} 1 \\ 1 \end{bmatrix}, k = 5$

10. $A = \begin{bmatrix} -6 & 4 \\ 8 & -2 \end{bmatrix}, \mathbf{x}_0 = \begin{bmatrix} 1 \\ 0 \end{bmatrix}, k = 6$

11. $A = \begin{bmatrix} 7 & 2 \\ 2 & 3 \end{bmatrix}, \mathbf{x}_0 = \begin{bmatrix} 1 \\ 0 \end{bmatrix}, k = 6$

12. $A = \begin{bmatrix} 3.5 & 1.5 \\ 1.5 & -0.5 \end{bmatrix}, \mathbf{x}_0 = \begin{bmatrix} 1 \\ 0 \end{bmatrix}, k = 6$

13. $A = \begin{bmatrix} 9 & 4 & 8 \\ 4 & 15 & -4 \\ 8 & -4 & 9 \end{bmatrix}, \mathbf{x}_0 = \begin{bmatrix} 1 \\ 1 \\ 1 \end{bmatrix}, k = 5$

14. $A = \begin{bmatrix} 3 & 1 & 0 \\ 1 & 3 & 1 \\ 0 & 1 & 3 \end{bmatrix}, \mathbf{x}_0 = \begin{bmatrix} 1 \\ 1 \\ 1 \end{bmatrix}, k = 6$

In Exercises 15 and 16, use the power method to approximate the dominant eigenvalue and eigenvector of A to two-decimal-place accuracy. Choose any initial vector you like (but keep the first Remark after Example 4.31 in mind!) and apply the method until the digit in the second decimal place of the iterates stops changing.

15. $A = \begin{bmatrix} 4 & 1 & 3 \\ 0 & 2 & 0 \\ 1 & 1 & 2 \end{bmatrix}$

16. $A = \begin{bmatrix} 12 & 6 & -6 \\ 2 & 0 & -2 \\ -6 & 6 & 12 \end{bmatrix}$

Rayleigh quotients are described on page 316. In Exercises 17–20, to see how the Rayleigh quotient method approximates the dominant eigenvalue more rapidly than the ordinary power method, compute the successive Rayleigh quotients $R(\mathbf{x}_i)$ *for* $i = 1, \ldots, k$ *for the matrix A in the given exercise.*

17. Exercise 11

18. Exercise 12

19. Exercise 13

20. Exercise 14

The matrices in Exercises 21–24 either are not diagonalizable or do not have a dominant eigenvalue (or both). Apply the power method anyway with the given initial vector $\mathbf{x}_0$*, performing eight iterations in each case. Compute the exact eigenvalues and eigenvectors and explain what is happening.*

21. $A = \begin{bmatrix} 4 & 1 \\ 0 & 4 \end{bmatrix}, \mathbf{x}_0 = \begin{bmatrix} 1 \\ 1 \end{bmatrix}$

22. $A = \begin{bmatrix} 3 & 1 \\ -1 & 1 \end{bmatrix}, \mathbf{x}_0 = \begin{bmatrix} 1 \\ 1 \end{bmatrix}$

23. $A = \begin{bmatrix} 4 & 0 & 1 \\ 0 & 4 & 0 \\ 0 & 0 & 1 \end{bmatrix}, \mathbf{x}_0 = \begin{bmatrix} 1 \\ 1 \\ 1 \end{bmatrix}$

24. $A = \begin{bmatrix} 0 & 0 & 0 \\ 0 & 5 & 1 \\ 0 & 0 & 5 \end{bmatrix}, \mathbf{x}_0 = \begin{bmatrix} 1 \\ 1 \\ 1 \end{bmatrix}$

In Exercises 25–28, the power method does not converge to the dominant eigenvalue and eigenvector. Verify this, using the given initial vector $\mathbf{x}_0$*. Compute the exact eigenvalues and eigenvectors and explain what is happening.*

25. $A = \begin{bmatrix} -1 & 2 \\ -1 & 1 \end{bmatrix}, \mathbf{x}_0 = \begin{bmatrix} 1 \\ 1 \end{bmatrix}$

26. $A = \begin{bmatrix} 2 & 1 \\ -2 & 5 \end{bmatrix}, \mathbf{x}_0 = \begin{bmatrix} 1 \\ 1 \end{bmatrix}$

27. $A = \begin{bmatrix} -5 & 1 & 7 \\ 0 & 4 & 0 \\ 7 & 1 & -5 \end{bmatrix}, \mathbf{x}_0 = \begin{bmatrix} 1 \\ 1 \\ 1 \end{bmatrix}$

28. $A = \begin{bmatrix} 1 & -1 & 0 \\ 1 & 1 & 0 \\ 1 & -1 & 1 \end{bmatrix}, \mathbf{x}_0 = \begin{bmatrix} 1 \\ 1 \\ 1 \end{bmatrix}$

In Exercises 29–32, apply the shifted power method to approximate the second eigenvalue of the matrix A in the given exercise. Use the given initial vector $\mathbf{x}_0$*, k iterations, and three-decimal-place accuracy.*

29. Exercise 9 **30.** Exercise 10

31. Exercise 13 **32.** Exercise 14

In Exercises 33–36, apply the inverse power method to approximate, for the matrix A in the given exercise, the eigenvalue that is smallest in magnitude. Use the given initial vector $\mathbf{x}_0$*, k iterations, and three-decimal-place accuracy.*

33. Exercise 9 **34.** Exercise 10

35. Exercise 7, $\mathbf{x}_0 = \begin{bmatrix} 1 \\ 1 \\ -1 \end{bmatrix}, k = 5$

36. Exercise 14

In Exercises 37–40, use the shifted inverse power method to approximate, for the matrix A in the given exercise, the eigenvalue closest to α*.*

37. Exercise 9, $\alpha = 0$ **38.** Exercise 12, $\alpha = 0$

39. Exercise 7, $\alpha = 5$ **40.** Exercise 13, $\alpha = -2$

Exercise 32 in Section 4.3 demonstrates that every polynomial is (plus or minus) the characteristic polynomial of its own companion matrix. Therefore, the roots of a polynomial p are the eigenvalues of $C(p)$*. Hence, we can use the methods of this section to approximate the roots of any polynomial when exact results are not readily available. In Exercises 41–44, apply the shifted inverse power method to the companion matrix* $C(p)$ *of p to approximate the root of p closest to* α *to three decimal places.*

41. $p(x) = x^2 + 2x - 2, \alpha = 0$

42. $p(x) = x^2 - x - 3, \alpha = 2$

43. $p(x) = x^3 - 2x^2 + 1, \alpha = 0$

44. $p(x) = x^3 - 5x^2 + x + 1, \alpha = 5$

45. Let λ be an eigenvalue of A with corresponding eigenvector $\mathbf{x}$. If $\alpha \neq \lambda$ and α is not an eigenvalue of A, show that $1/(\lambda - \alpha)$ is an eigenvalue of $(A - \alpha I)^{-1}$ with corresponding eigenvector $\mathbf{x}$. (Why must $A - \alpha I$ be invertible?)

46. If A has a dominant eigenvalue λ_1, prove that the eigenspace E_{λ_1} is one-dimensional.

a + bi *In Exercises 47–50, draw the Gerschgorin disks for the given matrix.*

47. $\begin{bmatrix} 1 & 1 & 0 \\ \frac{1}{2} & 4 & \frac{1}{2} \\ 1 & 0 & 5 \end{bmatrix}$

48. $\begin{bmatrix} 2 & -i & 0 \\ 1 & 2i & 1+i \\ 0 & 1 & -2i \end{bmatrix}$

49. $\begin{bmatrix} 4-3i & i & 2 & -2 \\ i & -1+i & 0 & 0 \\ 1+i & -i & 5+6i & 2i \\ 1 & -2i & 2i & -5-5i \end{bmatrix}$

50. $\begin{bmatrix} 2 & \frac{1}{2} & 0 & 0 \\ \frac{1}{4} & 4 & \frac{1}{4} & 0 \\ 0 & \frac{1}{6} & 6 & \frac{1}{6} \\ 0 & 0 & \frac{1}{8} & 8 \end{bmatrix}$

51. A square matrix is ***strictly diagonally dominant*** if the absolute value of each diagonal entry is greater than the sum of the absolute values of the remaining entries in that row. (See Section 2.5.) Use Gerschgorin's Disk Theorem to prove that a strictly diagonally dominant matrix must be invertible. [*Hint:* See the third Remark after Theorem 4.29.]

52. If A is an $n \times n$ matrix, let $\|A\|$ denote the maximum of the sums of the absolute values of the rows of A; that is, $\|A\| = \max\limits_{1 \le i \le n} \left(\sum_{j=1}^{n} |a_{ij}| \right)$. (See Section 7.2.) Prove that if λ is an eigenvalue of A, then $|\lambda| \le \|A\|$.

53. Let λ be an eigenvalue of a stochastic matrix A (see Section 3.7). Prove that $|\lambda| \le 1$. [*Hint:* Apply Exercise 52 to A^T.]

54. Prove that the eigenvalues of $A = \begin{bmatrix} 0 & 1 & 0 & 0 \\ 2 & 5 & 0 & 0 \\ \frac{1}{2} & 0 & 3 & \frac{1}{2} \\ 0 & 0 & \frac{3}{4} & 7 \end{bmatrix}$ are all real, and locate each of these eigenvalues within a closed interval on the real line.

4.6 Applications and the Perron-Frobenius Theorem

In this section, we will explore several applications of eigenvalues and eigenvectors. We begin by revisiting some applications from previous chapters.

Markov Chains

Section 3.7 introduced Markov chains and made several observations about the transition (stochastic) matrices associated with them. In particular, we observed that if P is the transition matrix of a Markov chain, then P has a steady state vector $\mathbf{x}$. That is, there is a vector $\mathbf{x}$ such that $P\mathbf{x} = \mathbf{x}$. This is equivalent to saying that P has 1 as an eigenvalue. We are now in a position to prove this fact.

Theorem 4.30 If P is the $n \times n$ transition matrix of a Markov chain, then 1 is an eigenvalue of P.

Proof Recall that every transition matrix is stochastic; hence, each of its columns sums to 1. Therefore, if $\mathbf{j}$ is a row vector consisting of n 1s, then $\mathbf{j}P = \mathbf{j}$. (See Exercise 13 in Section 3.7.) Taking transposes, we have

$$P^T\mathbf{j}^T = (\mathbf{j}P)^T = \mathbf{j}^T$$

which implies that $\mathbf{j}^T$ is an eigenvector of P^T with corresponding eigenvalue 1. By Exercise 19 in Section 4.3, P and P^T have the same eigenvalues, so 1 is also an eigenvalue of P.

In fact, much more is true. For most transition matrices, *every* eigenvalue λ satisfies $|\lambda| \le 1$ and the eigenvalue 1 is *dominant;* that is, if $\lambda \ne 1$, then $|\lambda| < 1$. We need the following two definitions: A matrix is called ***positive*** if all of its entries are positive, and a square matrix is called ***regular*** if some power of it is positive. For example, $A = \begin{bmatrix} 3 & 1 \\ 2 & 2 \end{bmatrix}$ is positive but $B = \begin{bmatrix} 3 & 1 \\ 2 & 0 \end{bmatrix}$ is not. However, B is regular, since $B^2 = \begin{bmatrix} 11 & 3 \\ 6 & 2 \end{bmatrix}$ is positive.

Theorem 4.31 Let P be an $n \times n$ transition matrix with eigenvalue λ.

a. $|\lambda| \le 1$
b. If P is regular and $\lambda \ne 1$, then $|\lambda| < 1$.

Proof As in Theorem 4.30, the trick to proving this theorem is to use the fact that P^T has the same eigenvalues as P.

(a) Let $\mathbf{x}$ be an eigenvector of P^T corresponding to λ and let x_k be the component of $\mathbf{x}$ with the largest absolute value m. Then $|x_i| \le |x_k| = m$ for $i = 1, 2, \ldots, n$. Comparing the kth components of the equation $P^T\mathbf{x} = \lambda\mathbf{x}$, we have

$$p_{1k}x_1 + p_{2k}x_2 + \cdots + p_{nk}x_n = \lambda x_k$$

(Remember that the rows of P^T are the columns of P.) Taking absolute values, we obtain

$$\begin{aligned} |\lambda|m = |\lambda||x_k| = |\lambda x_k| &= |p_{1k}x_1 + p_{2k}x_2 + \cdots + p_{nk}x_n| \\ &\leq |p_{1k}x_1| + |p_{2k}x_2| + \cdots + |p_{nk}x_n| \\ &= p_{1k}|x_1| + p_{2k}|x_2| + \cdots + p_{nk}|x_n| \qquad (1) \\ &\leq p_{1k}m + p_{2k}m + \cdots + p_{nk}m \\ &= (p_{1k} + p_{2k} + \cdots + p_{nk})m = m \end{aligned}$$

The first inequality follows from the Triangle Inequality in $\mathbb{R}$, and the last equality comes from the fact that the rows of P^T sum to 1. Thus, $|\lambda|m \leq m$. After dividing by m, we have $|\lambda| \leq 1$, as desired.

(b) We will prove the equivalent implication: If $|\lambda| = 1$, then $\lambda = 1$. First, we show that it is true when P (and therefore P^T) is a positive matrix. If $|\lambda| = 1$, then all of the inequalities in Equations (1) are actually equalities. In particular,

$$p_{1k}|x_1| + p_{2k}|x_2| + \cdots + p_{nk}|x_n| = p_{1k}m + p_{2k}m + \cdots + p_{nk}m$$

Equivalently,

$$p_{1k}(m - |x_1|) + p_{2k}(m - |x_2|) + \cdots + p_{nk}(m - |x_n|) = 0 \qquad (2)$$

Now, since P is positive, $p_{ik} > 0$ for $i = 1, 2, \ldots, n$. Also, $m - |x_i| \geq 0$ for $i = 1, 2, \ldots, n$. Therefore, each summand in Equation (2) must be zero, and this can happen only if $|x_i| = m$ for $i = 1, 2, \ldots, n$. Furthermore, we get equality in the Triangle Inequality in $\mathbb{R}$ if and only if all of the summands are positive or all are negative; in other words, the $p_{ik}x_i$'s all have the same sign. This implies that

$$\mathbf{x} = \begin{bmatrix} m \\ m \\ \vdots \\ m \end{bmatrix} = m\mathbf{j}^T \quad \text{or} \quad \mathbf{x} = \begin{bmatrix} -m \\ -m \\ \vdots \\ -m \end{bmatrix} = -m\mathbf{j}^T$$

where $\mathbf{j}$ is a row vector of n 1s, as in Theorem 4.30. Thus, in either case, the eigenspace of P^T corresponding to λ is $E_\lambda = \text{span}(\mathbf{j}^T)$.

But, using the proof of Theorem 4.30, we see that $\mathbf{j}^T = P^T\mathbf{j}^T = \lambda\mathbf{j}^T$, and, comparing components, we find that $\lambda = 1$. This handles the case where P is positive.

If P is regular, then some power of P is positive—say, P^k. It follows that P^{k+1} must also be positive. (Why?) Since λ^k and λ^{k+1} are eigenvalues of P^k and P^{k+1}, respectively, by Theorem 4.18, we have just proved that $\lambda^k = \lambda^{k+1} = 1$. Therefore, $\lambda^k(\lambda - 1) = 0$, which implies that $\lambda = 1$, since $\lambda = 0$ is impossible if $|\lambda| = 1$.

We can now explain some of the behavior of Markov chains that we observed in Chapter 3. In Example 3.64, we saw that for the transition matrix

$$P = \begin{bmatrix} 0.7 & 0.2 \\ 0.3 & 0.8 \end{bmatrix}$$

and initial state vector $\mathbf{x}_0 = \begin{bmatrix} 0.6 \\ 0.4 \end{bmatrix}$, the state vectors $\mathbf{x}_k$ converge to the vector $\mathbf{x} = \begin{bmatrix} 0.4 \\ 0.6 \end{bmatrix}$, a steady state vector for P (i.e., $P\mathbf{x} = \mathbf{x}$). We are going to prove that for regular

Markov chains, this always happens. Indeed, we will prove much more. Recall that the state vectors $\mathbf{x}_k$ satisfy $\mathbf{x}_k = P^k\mathbf{x}_0$. Let's investigate what happens to the powers P^k as P becomes large.

Example 4.37

The transition matrix $P = \begin{bmatrix} 0.7 & 0.2 \\ 0.3 & 0.8 \end{bmatrix}$ has characteristic equation

$$0 = \det(P - \lambda I) = \begin{vmatrix} 0.7 - \lambda & 0.2 \\ 0.3 & 0.8 - \lambda \end{vmatrix} = \lambda^2 - 1.5\lambda + 0.5 = (\lambda - 1)(\lambda - 0.5)$$

so its eigenvalues are $\lambda_1 = 1$ and $\lambda_2 = 0.5$. (Note that, thanks to Theorems 4.30 and 4.31, we knew in advance that 1 would be an eigenvalue and the other eigenvalue would be less than 1 in absolute value. However, we still needed to compute λ_2.) The eigenspaces are

$$E_1 = \text{span}\left(\begin{bmatrix} 2 \\ 3 \end{bmatrix}\right) \quad \text{and} \quad E_{0.5} = \text{span}\left(\begin{bmatrix} 1 \\ -1 \end{bmatrix}\right)$$

So, taking $Q = \begin{bmatrix} 2 & 1 \\ 3 & -1 \end{bmatrix}$, we know that $Q^{-1}PQ = \begin{bmatrix} 1 & 0 \\ 0 & 0.5 \end{bmatrix} = D$. From the method used in Example 4.29 in Section 4.4, we have

$$P^k = QD^kQ^{-1} = \begin{bmatrix} 2 & 1 \\ 3 & -1 \end{bmatrix}\begin{bmatrix} 1^k & 0 \\ 0 & (0.5)^k \end{bmatrix}\begin{bmatrix} 2 & 1 \\ 3 & -1 \end{bmatrix}^{-1}$$

Now, as $k \to \infty$, $(0.5)^k \to 0$, so

$$D^k \to \begin{bmatrix} 1 & 0 \\ 0 & 0 \end{bmatrix} \quad \text{and} \quad P^k \to \begin{bmatrix} 2 & 1 \\ 3 & -1 \end{bmatrix}\begin{bmatrix} 1 & 0 \\ 0 & 0 \end{bmatrix}\begin{bmatrix} 2 & 1 \\ 3 & -1 \end{bmatrix}^{-1} = \begin{bmatrix} 0.4 & 0.4 \\ 0.6 & 0.6 \end{bmatrix}$$

(Observe that the columns of this "limit matrix" are identical and each is a steady state vector for P.) Now let $\mathbf{x}_0 = \begin{bmatrix} a \\ b \end{bmatrix}$ be any initial probability vector (i.e., $a + b = 1$). Then

$$\mathbf{x}_k = P^k\mathbf{x}_0 \to \begin{bmatrix} 0.4 & 0.4 \\ 0.6 & 0.6 \end{bmatrix}\begin{bmatrix} a \\ b \end{bmatrix} = \begin{bmatrix} 0.4a + 0.4b \\ 0.6a + 0.6b \end{bmatrix} = \begin{bmatrix} 0.4 \\ 0.6 \end{bmatrix}$$

Not only does this explain what we saw in Example 3.64, it also tells us that the state vectors $\mathbf{x}_k$ will converge to the steady state vector $\mathbf{x} = \begin{bmatrix} 0.4 \\ 0.6 \end{bmatrix}$ for *any* choice of $\mathbf{x}_0$!

There is nothing special about Example 4.37. The next theorem shows that this type of behavior *always* occurs with regular transition matrices. Before we can present the theorem, we need the following lemma.

Lemma 4.32 Let P be a regular $n \times n$ transition matrix. If P is diagonalizable, then the dominant eigenvalue $\lambda_1 = 1$ has algebraic multiplicity 1.

Proof The eigenvalues of P and P^T are the same. From the proof of Theorem 4.31(b), $\lambda_1 = 1$ has geometric multiplicity 1 as an eigenvalue of P^T. Since P is diagonalizable, so is P^T, by Exercise 41 in Section 4.4. Therefore, the eigenvalue $\lambda_1 = 1$ has algebraic multiplicity 1, by the Diagonalization Theorem.

Theorem 4.33 Let P be a regular $n \times n$ transition matrix. Then as $k \to \infty$, P^k approaches an $n \times n$ matrix L whose columns are identical, each equal to the same vector $\mathbf{x}$. This vector $\mathbf{x}$ is a steady state probability vector for P.

Proof To simplify the proof, we will consider only the case where P is diagonalizable. The theorem is true, however, without this assumption.

See *Finite Markov Chains* by J. G. Kemeny and J. L. Snell (New York: Springer-Verlag, 1976).

We diagonalize P as $Q^{-1}PQ = D$ or, equivalently, $P = QDQ^{-1}$, where

$$D = \begin{bmatrix} \lambda_1 & 0 & \cdots & 0 \\ 0 & \lambda_2 & \cdots & 0 \\ \vdots & \vdots & \ddots & \vdots \\ 0 & 0 & \cdots & \lambda_n \end{bmatrix}$$

From Theorems 4.30 and 4.31, we know that each eigenvalue λ_i either is 1 or satisfies $|\lambda_i| < 1$. Hence, as $k \to \infty$, λ_i^k approaches 1 or 0 for $i = 1, \ldots, n$. It follows that D^k approaches a diagonal matrix—say, D^*—each of whose diagonal entries is 1 or 0. Thus, $P^k = QD^kQ^{-1}$ approaches $L = QD^*Q^{-1}$. We write

$$\lim_{k\to\infty} P^k = L$$

Observe that

$$PL = P \lim_{k\to\infty} P^k = \lim_{k\to\infty} PP^k = \lim_{k\to\infty} P^{k+1} = L$$

We are taking some liberties with the notion of a limit. Nevertheless, these steps should be intuitively clear. Rigorous proofs follow from the properties of limits, which you may have encountered in a calculus course. Rather than get sidetracked with a discussion of matrix limits, we will omit the proofs.

Therefore, each column of L is an eigenvector of P corresponding to $\lambda_1 = 1$. To see that each of these columns is a *probability* vector (i.e., L is a stochastic matrix), we need only observe that, if $\mathbf{j}$ is the row vector with n 1s, then

$$\mathbf{j}L = \mathbf{j} \lim_{k\to\infty} P^k = \lim_{k\to\infty} \mathbf{j}P^k = \lim_{k\to\infty} \mathbf{j} = \mathbf{j}$$

since P^k is a stochastic matrix, by Exercise 14 in Section 3.7. Exercise 13 in Section 3.7 now implies that L is stochastic.

We need only show that the columns of L are identical. The ith column of L is just $L\mathbf{e}_i$, where $\mathbf{e}_i$ is the ith standard basis vector. Let $\mathbf{v}_1, \mathbf{v}_2, \ldots, \mathbf{v}_n$ be eigenvectors of P forming a basis of $\mathbb{R}^n$, with $\mathbf{v}_1$ corresponding to $\lambda_1 = 1$. Write

$$\mathbf{e}_i = c_1\mathbf{v}_1 + c_2\mathbf{v}_2 + \cdots + c_n\mathbf{v}_n$$

for scalars $c_1, c_2, \ldots, c_n$. Then, by Theorem 4.19,

$$P^k\mathbf{e}_i = c_1 1^k\mathbf{v}_1 + c_2\lambda_2^k\mathbf{v}_2 + \cdots + c_n\lambda_n^k\mathbf{v}_n$$

By Lemma 4.32, $\lambda_j \neq 1$ for $j \neq 1$, so, by Theorem 4.31(b), $|\lambda_j| < 1$ for $j \neq 1$. Hence, $\lambda_j^k \to 0$ as $k \to \infty$, for $j \neq 1$. It follows that

$$L\mathbf{e}_i = \lim_{k\to\infty} P^k\mathbf{e}_i = c_1\mathbf{v}_1$$

In other words, column i of L is an eigenvector corresponding to $\lambda_1 = 1$. But we have shown that the columns of L are probability vectors, so $L\mathbf{e}_i$ is the *unique* multiple $\mathbf{x}$ of $\mathbf{v}_1$ whose components sum to 1. Since this is true for each column of L, it implies that all of the columns of L are identical, each equal to this vector $\mathbf{x}$.

Remark Since L is a stochastic matrix, we can interpret it as the ***long range transition matrix*** of the Markov chain. That is, L_{ij} represents the probability of being in state i, having started from state j, *if the transitions were to continue indefinitely*. The fact that the columns of L are identical says that the *starting state does not matter*, as the next example illustrates.

Example 4.38

Recall the rat in a box from Example 3.65. The transition matrix was

$$P = \begin{bmatrix} 0 & \frac{1}{3} & \frac{1}{3} \\ \frac{1}{2} & 0 & \frac{2}{3} \\ \frac{1}{2} & \frac{2}{3} & 0 \end{bmatrix}$$

We determined that the steady state probability vector was

$$\mathbf{x} = \begin{bmatrix} \frac{1}{4} \\ \frac{3}{8} \\ \frac{3}{8} \end{bmatrix}$$

Hence, the powers of P approach

$$L = \begin{bmatrix} \frac{1}{4} & \frac{1}{4} & \frac{1}{4} \\ \frac{3}{8} & \frac{3}{8} & \frac{3}{8} \\ \frac{3}{8} & \frac{3}{8} & \frac{3}{8} \end{bmatrix} = \begin{bmatrix} 0.250 & 0.250 & 0.250 \\ 0.375 & 0.375 & 0.375 \\ 0.375 & 0.375 & 0.375 \end{bmatrix}$$

from which we can see that the rat will *eventually* spend 25% of its time in compartment 1 and 37.5% of its time in each of the other two compartments.

We conclude our discussion of regular Markov chains by proving that the steady state vector $\mathbf{x}$ is independent of the initial state. The proof is easily adapted to cover the case of state vectors whose components sum to an arbitrary constant—say, s. In the exercises, you are asked to prove some other properties of regular Markov chains.

Theorem 4.34 Let P be a regular $n \times n$ transition matrix, with $\mathbf{x}$ the steady state probability vector for P, as in Theorem 4.33. Then, for any initial probability vector $\mathbf{x}_0$, the sequence of iterates $\mathbf{x}_k$ approaches $\mathbf{x}$.

Proof Let

$$\mathbf{x}_0 = \begin{bmatrix} x_1 \\ x_2 \\ \vdots \\ x_n \end{bmatrix}$$

where $x_1 + x_2 + \cdots + x_n = 1$. Since $\mathbf{x}_k = P^k\mathbf{x}_0$, we must show that $\lim_{k\to\infty} P^k\mathbf{x}_0 = \mathbf{x}$. Now, by Theorem 4.33, the long range transition matrix is $L = [\mathbf{x}\ \ \mathbf{x}\ \ \cdots\ \ \mathbf{x}]$ and $\lim_{k\to\infty} P^k = L$. Therefore,

$$\begin{aligned}\lim_{k\to\infty} P^k\mathbf{x}_0 &= (\lim_{k\to\infty} P^k)\mathbf{x}_0 = L\mathbf{x}_0 \\ &= [\mathbf{x}\ \ \mathbf{x}\ \ \cdots\ \ \mathbf{x}]\begin{bmatrix} x_1 \\ x_2 \\ \vdots \\ x_n \end{bmatrix} \\ &= x_1\mathbf{x} + x_2\mathbf{x} + \cdots + x_n\mathbf{x} \\ &= (x_1 + x_2 + \cdots + x_n)\mathbf{x} = \mathbf{x}\end{aligned}$$

Population Growth

We return to the Leslie model of population growth, which we first explored in Section 3.7. In Example 3.67 in that section, we saw that for the Leslie matrix

$$L = \begin{bmatrix} 0 & 4 & 3 \\ 0.5 & 0 & 0 \\ 0 & 0.25 & 0 \end{bmatrix}$$

iterates of the population vectors began to approach a multiple of the vector

$$\mathbf{x} = \begin{bmatrix} 18 \\ 6 \\ 1 \end{bmatrix}$$

In other words, the three age classes of this population eventually ended up in the ratio 18:6:1. Moreover, once this state is reached, it is stable, since the ratios for the following year are given by

$$L\mathbf{x} = \begin{bmatrix} 0 & 4 & 3 \\ 0.5 & 0 & 0 \\ 0 & 0.25 & 0 \end{bmatrix}\begin{bmatrix} 18 \\ 6 \\ 1 \end{bmatrix} = \begin{bmatrix} 27 \\ 9 \\ 1.5 \end{bmatrix} = 1.5\mathbf{x}$$

and the components are still in the ratio 27:9:1.5 = 18:6:1. Observe that 1.5 represents the *growth rate* of this population when it has reached its steady state.

We can now recognize that $\mathbf{x}$ is an eigenvector of L corresponding to the eigenvalue $\lambda = 1.5$. Thus, the steady state growth rate is a *positive* eigenvalue of L, and an eigenvector corresponding to this eigenvalue represents the *relative* sizes of the age classes when the steady state has been reached. We can compute these directly, without having to iterate as we did before.

Example 4.39

Find the steady state growth rate and the corresponding ratios between the age classes for the Leslie matrix L above.

Solution We need to find all positive eigenvalues and corresponding eigenvectors of L. The characteristic polynomial of L is

$$\det(L - \lambda I) = \begin{vmatrix} -\lambda & 4 & 3 \\ 0.5 & -\lambda & 0 \\ 0 & 0.25 & -\lambda \end{vmatrix} = -\lambda^3 + 2\lambda + 0.375$$

so we must solve $-\lambda^3 + 2\lambda + 0.375 = 0$ or, equivalently, $8\lambda^3 - 16\lambda - 3 = 0$. Factoring, we have

$$(2\lambda - 3)(4\lambda^2 + 6\lambda + 1) = 0$$

(See Appendix D.) Since the second factor has only the roots $(-3 + \sqrt{5})/4 \approx -0.19$ and $(-3 - \sqrt{5})/4 \approx -1.31$, the only positive root of this equation is $\lambda = \frac{3}{2} = 1.5$. The corresponding eigenvectors are in the null space of $L - 1.5I$, which we find by row reduction:

$$[L - 1.5I|0] = \left[\begin{array}{ccc|c} -1.5 & 4 & 3 & 0 \\ 0.5 & -1.5 & 0 & 0 \\ 0 & 0.25 & -1.5 & 0 \end{array}\right] \longrightarrow \left[\begin{array}{ccc|c} 1 & 0 & -18 & 0 \\ 0 & 1 & -6 & 0 \\ 0 & 0 & 0 & 0 \end{array}\right]$$

Thus, if $\mathbf{x} = \begin{bmatrix} x_1 \\ x_2 \\ x_3 \end{bmatrix}$ is an eigenvector corresponding to $\lambda = 1.5$, it satisfies $x_1 = 18x_3$ and $x_2 = 6x_3$. That is,

$$E_{1.5} = \left\{ \begin{bmatrix} 18x_3 \\ 6x_3 \\ x_3 \end{bmatrix} \right\} = \text{span}\left(\begin{bmatrix} 18 \\ 6 \\ 1 \end{bmatrix} \right)$$

Hence, the steady state growth rate is 1.5, and when this rate has been reached, the age classes are in the ratio 18:6:1, as we saw before.

In Example 4.39, there was only one candidate for the steady state growth rate: the unique positive eigenvalue of L. But what would we have done if L had had more than one positive eigenvalue or none? We were also apparently fortunate that there was a corresponding eigenvector all of whose components were positive, which allowed us to relate these components to the size of the population. We can prove that this situation is not accidental; that is, *every* Leslie matrix has exactly one positive eigenvalue and a corresponding eigenvector with positive components.

Recall that the form of a Leslie matrix is

$$L = \begin{bmatrix} b_1 & b_2 & b_3 & \cdots & b_{n-1} & b_n \\ s_1 & 0 & 0 & \cdots & 0 & 0 \\ 0 & s_2 & 0 & \cdots & 0 & 0 \\ 0 & 0 & s_3 & \cdots & 0 & 0 \\ \vdots & \vdots & \vdots & \ddots & \vdots & \vdots \\ 0 & 0 & 0 & \cdots & s_{n-1} & 0 \end{bmatrix} \tag{3}$$

Since the entries s_j represent survival probabilities, we will assume that they are all nonzero (otherwise, the population would rapidly die out). We will also assume that at least one of the birth parameters b_i is nonzero (otherwise, there would be no births and, again, the population would die out). With these standing assumptions, we can now prove the assertion we made above as a theorem.

Theorem 4.35 Every Leslie matrix has a unique positive eigenvalue and a corresponding eigenvector with positive components.

Proof Let L be as in Equation (3). The characteristic polynomial of L is

$$\begin{aligned} c_L(\lambda) &= \det(L - \lambda I) \\ &= (-1)^n(\lambda^n - b_1\lambda^{n-1} - b_2s_1\lambda^{n-2} - b_3s_1s_2\lambda^{n-3} - \cdots - b_ns_1s_2\cdots s_{n-1}) \\ &= (-1)^n f(\lambda) \end{aligned}$$

(You are asked to prove this in Exercise 16.) The eigenvalues of L are therefore the roots of $f(\lambda)$. Since at least one of the birth parameters b_i is positive and all of the survival probabilities s_j are positive, the coefficients of $f(\lambda)$ change sign exactly once. By Descartes's Rule of Signs (Appendix D), therefore, $f(\lambda)$ has exactly one positive root. Let us call it λ_1.

By direct calculation, we can check that an eigenvector corresponding to λ_1 is

$$\mathbf{x}_1 = \begin{bmatrix} 1 \\ s_1/\lambda_1 \\ s_1s_2/\lambda_1^2 \\ s_1s_2s_3/\lambda_1^3 \\ \vdots \\ s_1s_2s_3\cdots s_{n-1}/\lambda_1^{n-1} \end{bmatrix}$$

(You are asked to prove this in Exercise 18.) Clearly, all of the components of $\mathbf{x}_1$ are positive.

In fact, more is true. With the additional requirement that *two consecutive* birth parameters b_i and b_{i+1} are positive, it turns out that the unique positive eigenvalue λ_1 of L is *dominant;* that is, every other (real or complex) eigenvalue λ of L satisfies $|\lambda| < \lambda_1$. (It is beyond the scope of this book to prove this result, but a partial proof is outlined in Exercise 27 for readers who are familiar with the algebra of complex numbers.) This explains why we get convergence to a steady state vector when we iterate the population vectors: It is just the power method working for us!

The Perron-Frobenius Theorem

In the previous two applications, Markov chains and Leslie matrices, we saw that the eigenvalue of interest was positive and dominant. Moreover, there was a corresponding eigenvector with positive components. It turns out that a remarkable theorem guarantees that this will be the case for a large class of matrices, including many of the ones we have been considering. The first version of this theorem is for positive matrices.

Oskar Perron (1880–1975) was a German mathematician who did work in many fields of mathematics, including analysis, differential equations, algebra, geometry, and number theory. Perron's Theorem was published in 1907 in a paper on continued fractions.

First, we need some terminology and notation. Let's agree to refer to a vector as *positive* if all of its components are positive. For two $m \times n$ matrices $A = [a_{ij}]$ and $B = [b_{ij}]$, we will write $A \geq B$ if $a_{ij} \geq b_{ij}$ for all i and j. (Similar definitions will apply for $A > B$, $A \leq B$, and so on.) Thus, a positive vector $\mathbf{x}$ satisfies $\mathbf{x} > \mathbf{0}$. Let us define $|A| = [|a_{ij}|]$ to be the matrix of the absolute values of the entries of A.

Theorem 4.36 **Perron's Theorem**

Let A be a positive $n \times n$ matrix. Then A has a real eigenvalue λ_1 with the following properties:

a. $\lambda_1 > 0$
b. λ_1 has a corresponding positive eigenvector.
c. If λ is any other eigenvalue of A, then $|\lambda| \leq \lambda_1$.

Intuitively, we can see why the first two statements should be true. Consider the case of a 2×2 positive matrix A. The corresponding matrix transformation maps the first quadrant of the plane properly into itself, since all components are positive. If we repeatedly allow A to act on the images we get, they necessarily converge toward some ray in the first quadrant (Figure 4.18). A direction vector for this ray will be a positive vector $\mathbf{x}$, which must be mapped into some positive multiple of itself (say, λ_1), since A leaves the ray fixed. In other words, $A\mathbf{x} = \lambda_1\mathbf{x}$, with $\mathbf{x}$ and λ_1 both positive.

Proof For some nonzero vectors $\mathbf{x}$, $A\mathbf{x} \geq \lambda\mathbf{x}$ for some scalar λ. When this happens, then $A(k\mathbf{x}) \geq \lambda(k\mathbf{x})$ for all $k > 0$; thus, we need only consider *unit* vectors $\mathbf{x}$. In Chapter 7, we will see that A maps the set of all unit vectors in $\mathbb{R}^n$ (the *unit sphere*) into a "generalized ellipsoid." So, as $\mathbf{x}$ ranges over the nonnegative vectors on this unit sphere, there will be a maximum value of λ such that $A\mathbf{x} \geq \lambda\mathbf{x}$. (See Figure 4.19.) Denote this number by λ_1 and the corresponding unit vector by $\mathbf{x}_1$.

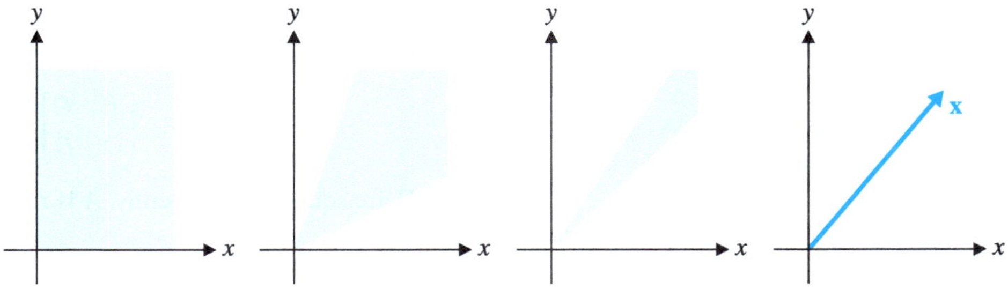

Figure 4.18

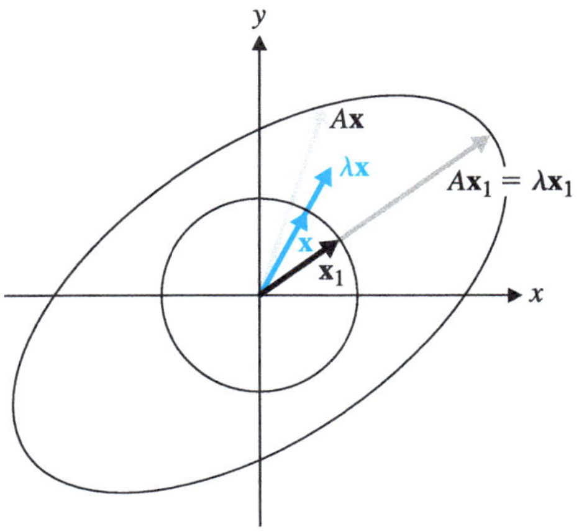

Figure 4.19

We now show that $A\mathbf{x}_1 = \lambda_1\mathbf{x}_1$. If not, then $A\mathbf{x}_1 > \lambda_1\mathbf{x}_1$, and, applying A again, we obtain

$$A(A\mathbf{x}_1) > A(\lambda_1\mathbf{x}_1) = \lambda_1(A\mathbf{x}_1)$$

where the inequality is preserved, since A is positive. (See Exercise 40 and Section 3.7 Exercise 36.) But then $\mathbf{y} = (1/\|A\mathbf{x}_1\|)A\mathbf{x}_1$ is a unit vector that satisfies $A\mathbf{y} > \lambda_1\mathbf{y}$, so there will be some $\lambda_2 > \lambda_1$ such that $A\mathbf{y} \geq \lambda_2\mathbf{y}$. This contradicts the fact that λ_1 was the maximum value with this property. Consequently, it must be the case that $A\mathbf{x}_1 = \lambda_1\mathbf{x}_1$; that is, λ_1 is an eigenvalue of A.

Now A is positive and $\mathbf{x}_1$ is positive, so $\lambda_1\mathbf{x}_1 = A\mathbf{x}_1 > \mathbf{0}$. This means that $\lambda_1 > 0$ and $\mathbf{x}_1 > \mathbf{0}$, which completes the proof of (a) and (b).

To prove (c), suppose λ is any other (real or complex) eigenvalue of A with corresponding eigenvector $\mathbf{z}$. Then $A\mathbf{z} = \lambda\mathbf{z}$, and, taking absolute values, we have

$$A|\mathbf{z}| = |A||\mathbf{z}| \geq |A\mathbf{z}| = |\lambda\mathbf{z}| = |\lambda||\mathbf{z}| \quad (4)$$

where the middle inequality follows from the Triangle Inequality. (See Exercise 40.) Since $|\mathbf{z}| > \mathbf{0}$, the unit vector $\mathbf{u}$ in the direction of $|\mathbf{z}|$ is also positive and satisfies $A\mathbf{u} \geq |\lambda|\mathbf{u}$. By the maximality of λ_1 from the first part of this proof, we must have $|\lambda| \leq \lambda_1$.

In fact, more is true. It turns out that λ_1 is dominant, so $|\lambda| < \lambda_1$ for any eigenvalue $\lambda \neq \lambda_1$. It is also the case that λ_1 has algebraic, and hence geometric, multiplicity 1. We will not prove these facts.

Perron's Theorem can be generalized from positive to certain nonnegative matrices. Frobenius did so in 1912. The result requires a technical condition on the matrix. A square matrix A is called ***reducible*** if, subject to some permutation of the rows and the same permutation of the columns, A can be written in block form as

$$\begin{bmatrix} B & C \\ O & D \end{bmatrix}$$

where B and D are square. Equivalently, A is reducible if there is some permutation matrix P such that

$$PAP^T = \begin{bmatrix} B & C \\ O & D \end{bmatrix}$$

(See page 187.) For example, the matrix

$$A = \begin{bmatrix} 2 & 0 & 0 & 1 & 3 \\ 4 & 2 & 1 & 5 & 5 \\ 1 & 2 & 7 & 3 & 0 \\ 6 & 0 & 0 & 2 & 1 \\ 1 & 0 & 0 & 7 & 2 \end{bmatrix}$$

is reducible, since interchanging rows 1 and 3 and then columns 1 and 3 produces

$$\left[\begin{array}{cc|ccc} 7 & 2 & 1 & 3 & 0 \\ 1 & 2 & 4 & 5 & 5 \\ \hline 0 & 0 & 2 & 1 & 3 \\ 0 & 0 & 6 & 2 & 1 \\ 0 & 0 & 1 & 7 & 2 \end{array}\right]$$

(This is just PAP^T, where

$$P = \begin{bmatrix} 0 & 0 & 1 & 0 & 0 \\ 0 & 1 & 0 & 0 & 0 \\ 1 & 0 & 0 & 0 & 0 \\ 0 & 0 & 0 & 1 & 0 \\ 0 & 0 & 0 & 0 & 1 \end{bmatrix}$$

Check this!)

A square matrix A that is not reducible is called ***irreducible.*** If $A^k > O$ for some k, then A is called ***primitive.*** For example, every regular Markov chain has a primitive transition matrix, by definition. It is not hard to show that every primitive matrix is irreducible. (Do you see why? Try showing the contrapositive of this.)

Theorem 4.37 **The Perron-Frobenius Theorem**

Let A be an irreducible nonnegative $n \times n$ matrix. Then A has a real eigenvalue λ_1 with the following properties:

a. $\lambda_1 > 0$
b. λ_1 has a corresponding positive eigenvector.
c. If λ is any other eigenvalue of A, then $|\lambda| \leq \lambda_1$. If A is primitive, then this inequality is strict.
d. If λ is an eigenvalue of A such that $|\lambda| = \lambda_1$, then λ is a (complex) root of the equation $\lambda^n - \lambda_1^n = 0$.
e. λ_1 has algebraic multiplicity 1.

See *Matrix Analysis* by R. A. Horn and C. R. Johnson (Cambridge, England: Cambridge University Press, 1985).

The interested reader can find a proof of the Perron-Frobenius Theorem in many texts on nonnegative matrices or matrix analysis. The eigenvalue λ_1 is often called the ***Perron root*** of A, and a corresponding *probability* eigenvector (which is necessarily unique) is called the ***Perron eigenvector*** of A.

Linear Recurrence Relations

The ***Fibonacci numbers*** are the numbers in the sequence 0, 1, 1, 2, 3, 5, 8, 13, 21, . . . , where, after the first two terms, each new term is obtained by summing the two terms preceding it. If we denote the nth Fibonacci number by f_n, then this sequence is completely defined by the equations $f_0 = 0, f_1 = 1$, and, for $n \geq 2$,

$$f_n = f_{n-1} + f_{n-2}$$

This last equation is an example of a linear recurrence relation. We will return to the Fibonacci numbers, but first we will consider linear recurrence relations somewhat more generally.

Bettmann/CORBIS

Leonardo of Pisa (1170–1250), pictured left, is better known by his nickname, Fibonacci, which means "son of Bonaccio." He wrote a number of important books, many of which have survived, including *Liber abaci* and *Liber quadratorum*. The Fibonacci sequence appears as the solution to a problem in *Liber abaci:* "A certain man put a pair of rabbits in a place surrounded on all sides by a wall. How many pairs of rabbits can be produced from that pair in a year if it is supposed that every month each pair begets a new pair which from the second month on becomes productive?" The name *Fibonacci numbers* was given to the terms of this sequence by the French mathematician Edouard Lucas (1842–1891).

Definition Let $(x_n) = (x_0, x_1, x_2, \ldots)$ be a sequence of numbers that is defined as follows:

1. $x_0 = a_0, x_1 = a_1, \ldots, x_{k-1} = a_{k-1}$, where $a_0, a_1, \ldots, a_{k-1}$ are scalars.
2. For all $n \geq k$, $x_n = c_1x_{n-1} + c_2x_{n-2} + \cdots + c_kx_{n-k}$, where $c_1, c_2, \ldots, c_k$ are scalars.

If $c_k \neq 0$, the equation in (2) is called a ***linear recurrence relation of order k.*** The equations in (1) are referred to as the ***initial conditions*** of the recurrence.

Thus, the Fibonacci numbers satisfy a linear recurrence relation of order 2.

Remarks

- If, in order to define the nth term in a recurrence relation, we require the $(n - k)$th term but no term before it, then the recurrence relation has order k.
- The number of initial conditions is the order of the recurrence relation.
- It is not necessary that the first term of the sequence be called x_0. We could start at x_1 or anywhere else.
- It is possible to have even more general linear recurrence relations by allowing the coefficients c_i to be *functions* rather than scalars and by allowing an extra, isolated coefficient, which may also be a function. An example would be the recurrence

$$x_n = 2x_{n-1} - n^2x_{n-2} + \frac{1}{n}x_{n-3} + n$$

We will not consider such recurrences here.

Example 4.40

Consider the sequence (x_n) defined by the initial conditions $x_1 = 1$, $x_2 = 5$ and the recurrence relation $x_n = 5x_{n-1} - 6x_{n-2}$ for $n \geq 2$. Write out the first five terms of this sequence.

Solution We are given the first two terms. We use the recurrence relation to calculate the next three terms. We have

$$\begin{aligned} x_3 &= 5x_2 - 6x_1 = 5 \cdot 5 - 6 \cdot 1 = 19 \\ x_4 &= 5x_3 - 6x_2 = 5 \cdot 19 - 6 \cdot 5 = 65 \\ x_5 &= 5x_4 - 6x_3 = 5 \cdot 65 - 6 \cdot 19 = 211 \end{aligned}$$

so the sequence begins 1, 5, 19, 65, 211, ...

Clearly, if we were interested in, say, the 100th term of the sequence in Example 4.40, then the approach used there would be rather tedious, since we would have to apply the recurrence relation 98 times. It would be nice if we could find an *explicit* formula for x_n as a function of n. We refer to finding such a formula as ***solving*** the recurrence relation. We will illustrate the process with the sequence from Example 4.40.

To begin, we rewrite the recurrence relation as a matrix equation. Let

$$A = \begin{bmatrix} 5 & -6 \\ 1 & 0 \end{bmatrix}$$

and introduce vectors $\mathbf{x}_n = \begin{bmatrix} x_n \\ x_{n-1} \end{bmatrix}$ for $n \geq 2$. Thus, $\mathbf{x}_2 = \begin{bmatrix} x_2 \\ x_1 \end{bmatrix} = \begin{bmatrix} 5 \\ 1 \end{bmatrix}$, $\mathbf{x}_3 = \begin{bmatrix} x_3 \\ x_2 \end{bmatrix} = \begin{bmatrix} 19 \\ 5 \end{bmatrix}$, $\mathbf{x}_4 = \begin{bmatrix} x_4 \\ x_3 \end{bmatrix} = \begin{bmatrix} 65 \\ 19 \end{bmatrix}$, and so on. Now observe that, for $n \geq 2$, we have

$$A\mathbf{x}_{n-1} = \begin{bmatrix} 5 & -6 \\ 1 & 0 \end{bmatrix}\begin{bmatrix} x_{n-1} \\ x_{n-2} \end{bmatrix} = \begin{bmatrix} 5x_{n-1} - 6x_{n-2} \\ x_{n-1} \end{bmatrix} = \begin{bmatrix} x_n \\ x_{n-1} \end{bmatrix} = \mathbf{x}_n$$

Notice that this is the same type of equation we encountered with Markov chains and Leslie matrices. As in those cases, we can write

$$\mathbf{x}_n = A\mathbf{x}_{n-1} = A^2\mathbf{x}_{n-2} = \cdots = A^{n-2}\mathbf{x}_2$$

We now use the technique of Example 4.29 to compute the powers of A.

The characteristic equation of A is

$$\lambda^2 - 5\lambda + 6 = 0$$

from which we find that the eigenvalues are $\lambda_1 = 3$ and $\lambda_2 = 2$. (Notice that the form of the characteristic equation follows that of the recurrence relation. If we write the recurrence as $x_n - 5x_{n-1} + 6x_{n-2} = 0$, it is apparent that the coefficients are exactly the same!) The corresponding eigenspaces are

$$E_3 = \text{span}\left(\begin{bmatrix} 3 \\ 1 \end{bmatrix}\right) \quad \text{and} \quad E_2 = \text{span}\left(\begin{bmatrix} 2 \\ 1 \end{bmatrix}\right)$$

Setting $P = \begin{bmatrix} 3 & 2 \\ 1 & 1 \end{bmatrix}$, we know that $P^{-1}AP = D = \begin{bmatrix} 3 & 0 \\ 0 & 2 \end{bmatrix}$. Then $A = PDP^{-1}$ and

$$\begin{aligned} A^k = PD^kP^{-1} &= \begin{bmatrix} 3 & 2 \\ 1 & 1 \end{bmatrix}\begin{bmatrix} 3^k & 0 \\ 0 & 2^k \end{bmatrix}\begin{bmatrix} 3 & 2 \\ 1 & 1 \end{bmatrix}^{-1} \\ &= \begin{bmatrix} 3 & 2 \\ 1 & 1 \end{bmatrix}\begin{bmatrix} 3^k & 0 \\ 0 & 2^k \end{bmatrix}\begin{bmatrix} 1 & -2 \\ -1 & 3 \end{bmatrix} \\ &= \begin{bmatrix} 3^{k+1} - 2^{k+1} & -2(3^{k+1}) + 3(2^{k+1}) \\ 3^k - 2^k & -2(3^k) + 3(2^k) \end{bmatrix} \end{aligned}$$

It now follows that

$$\begin{bmatrix} x_n \\ x_{n-1} \end{bmatrix} = \mathbf{x}_n = A^{n-2}\mathbf{x}_2 = \begin{bmatrix} 3^{n-1} - 2^{n-1} & -2(3^{n-1}) + 3(2^{n-1}) \\ 3^{n-2} - 2^{n-2} & -2(3^{n-2}) + 3(2^{n-2}) \end{bmatrix}\begin{bmatrix} 5 \\ 1 \end{bmatrix} = \begin{bmatrix} 3^n - 2^n \\ 3^{n-1} - 2^{n-1} \end{bmatrix}$$

from which we read off the solution $x_n = 3^n - 2^n$. (To check our work, we could plug in $n = 1, 2, \ldots, 5$ to verify that this formula gives the same terms that we calculated using the recurrence relation. Try it!)

Observe that x_n is a linear combination of powers of the eigenvalues. This is necessarily the case as long as the eigenvalues are distinct [as Theorem 4.38(a) will make explicit]. Using this observation, we can save ourselves some work. Once we have computed the eigenvalues $\lambda_1 = 3$ and $\lambda_2 = 2$, we can immediately write

$$x_n = c_1 3^n + c_2 2^n$$

where c_1 and c_2 are to be determined. Using the initial conditions, we have

$$1 = x_1 = c_1 3^1 + c_2 2^1 = 3c_1 + 2c_2$$

when $n = 1$ and

$$5 = x_2 = c_1 3^2 + c_2 2^2 = 9c_1 + 4c_2$$

when $n = 2$. We now solve the system

$$\begin{aligned} 3c_1 + 2c_2 &= 1 \\ 9c_1 + 4c_2 &= 5 \end{aligned}$$

for c_1 and c_2 to obtain $c_1 = 1$ and $c_2 = -1$. Thus, $x_n = 3^n - 2^n$, as before.

This is the method we will use in practice. We now illustrate its use to find an explicit formula for the Fibonacci numbers.

Example 4.41

Solve the Fibonacci recurrence $f_0 = 0, f_1 = 1$, and $f_n = f_{n-1} + f_{n-2}$ for $n \geq 2$.

Solution Writing the recurrence as $f_n - f_{n-1} - f_{n-2} = 0$, we see that the characteristic equation is $\lambda^2 - \lambda - 1 = 0$, so the eigenvalues are

$$\lambda_1 = \frac{1 + \sqrt{5}}{2} \quad \text{and} \quad \lambda_2 = \frac{1 - \sqrt{5}}{2}$$

It follows from the discussion above that the solution to the recurrence relation has the form

$$f_n = c_1\lambda_1^n + c_2\lambda_2^n = c_1\left(\frac{1 + \sqrt{5}}{2}\right)^n + c_2\left(\frac{1 - \sqrt{5}}{2}\right)^n$$

for some scalars c_1 and c_2.

Using the initial conditions, we find

$$0 = f_0 = c_1\lambda_1^0 + c_2\lambda_2^0 = c_1 + c_2$$

and

$$1 = f_1 = c_1\lambda_1^1 + c_2\lambda_2^1 = c_1\left(\frac{1 + \sqrt{5}}{2}\right) + c_2\left(\frac{1 - \sqrt{5}}{2}\right)$$

Solving for c_1 and c_2, we obtain $c_1 = 1/\sqrt{5}$ and $c_2 = -1/\sqrt{5}$. Hence, an explicit formula for the nth Fibonacci number is

$$f_n = \frac{1}{\sqrt{5}}\left(\frac{1 + \sqrt{5}}{2}\right)^n - \frac{1}{\sqrt{5}}\left(\frac{1 - \sqrt{5}}{2}\right)^n \tag{5}$$

Jacques Binet (1786–1856) made contributions to matrix theory, number theory, physics, and astronomy. He discovered the rule for matrix multiplication in 1812. Binet's formula for the Fibonacci numbers is actually due to Euler, who published it in 1765; however, it was forgotten until Binet published his version in 1843. Like Cauchy, Binet was a royalist, and he lost his university position when Charles X abdicated in 1830. He received many honors for his work, including his election, in 1843, to the Académie des Sciences.

Formula (5) is a remarkable formula, because it is defined in terms of the *irrational* number $\sqrt{5}$ yet the Fibonacci numbers are all integers! Try plugging in a few values for n to see how the $\sqrt{5}$ terms cancel out to leave the integer values f_n. Formula (5) is known as ***Binet's formula.***

The method we have just outlined works for any second order linear recurrence relation whose associated eigenvalues are all distinct. When there is a repeated eigenvalue, the technique must be modified, since the diagonalization method we used may no longer work. The next theorem summarizes both situations.

Theorem 4.38 Let $x_n = ax_{n-1} + bx_{n-2}$ be a recurrence relation that is satisfied by a sequence (x_n). Let λ_1 and λ_2 be the eigenvalues of the associated characteristic equation $\lambda^2 - a\lambda - b = 0$.

a. If $\lambda_1 \neq \lambda_2$, then $x_n = c_1\lambda_1^n + c_2\lambda_2^n$ for some scalars c_1 and c_2.
b. If $\lambda_1 = \lambda_2 = \lambda$, then $x_n = c_1\lambda^n + c_2 n\lambda^n$ for some scalars c_1 and c_2.

In either case, c_1 and c_2 can be determined using the initial conditions.

Proof (a) Generalizing our discussion above, we can write the recurrence as $\mathbf{x}_n = A\mathbf{x}_{n-1}$, where

$$\mathbf{x}_n = \begin{bmatrix} x_n \\ x_{n-1} \end{bmatrix} \quad \text{and} \quad A = \begin{bmatrix} a & b \\ 1 & 0 \end{bmatrix}$$

Since A has distinct eigenvalues, it can be diagonalized. The rest of the details are left for Exercise 53.

(b) We will show that $x_n = c_1\lambda^n + c_2 n\lambda^n$ satisfies the recurrence relation $x_n = ax_{n-1} + bx_{n-2}$ or, equivalently,

$$x_n - ax_{n-1} - bx_{n-2} = 0 \tag{6}$$

if $\lambda^2 - a\lambda - b = 0$. Since

$$x_{n-1} = c_1\lambda^{n-1} + c_2(n-1)\lambda^{n-1} \quad \text{and} \quad x_{n-2} = c_1\lambda^{n-2} + c_2(n-2)\lambda^{n-2}$$

substitution into Equation (6) yields

$$\begin{aligned} x_n - ax_{n-1} - bx_{n-2} &= (c_1\lambda^n + c_2 n\lambda^n) - a(c_1\lambda^{n-1} + c_2(n-1)\lambda^{n-1}) \\ &\quad - b(c_1\lambda^{n-2} + c_2(n-2)\lambda^{n-2}) \\ &= c_1(\lambda^n - a\lambda^{n-1} - b\lambda^{n-2}) + c_2(n\lambda^n - a(n-1)\lambda^{n-1} \\ &\quad - b(n-2)\lambda^{n-2}) \\ &= c_1\lambda^{n-2}(\lambda^2 - a\lambda - b) + c_2 n\lambda^{n-2}(\lambda^2 - a\lambda - b) + c_2\lambda^{n-2}(a\lambda + 2b) \\ &= c_1\lambda^{n-2}(0) + c_2 n\lambda^{n-2}(0) + c_2\lambda^{n-2}(a\lambda + 2b) \\ &= c_2\lambda^{n-2}(a\lambda + 2b) \end{aligned}$$

But, since λ is a double root of $\lambda^2 - a\lambda - b = 0$, we must have $a^2 + 4b = 0$ and $\lambda = a/2$, using the quadratic formula. Consequently, $a\lambda + 2b = a^2/2 + 2b = -4b/2 + 2b = 0$, so

$$x_n - ax_{n-1} - bx_{n-2} = c_2\lambda^{n-2}(a\lambda + 2b) = c_2\lambda^{n-2}(0) = 0$$

Suppose the initial conditions are $x_0 = r$ and $x_1 = s$. Then, in either (a) or (b) there is a unique solution for c_1 and c_2. (See Exercise 54.)

Example 4.42 Solve the recurrence relation $x_0 = 1$, $x_1 = 6$, and $x_n = 6x_{n-1} - 9x_{n-2}$ for $n \geq 2$.

Solution The characteristic equation is $\lambda^2 - 6\lambda + 9 = 0$, which has $\lambda = 3$ as a double root. By Theorem 4.38(b), we must have $x_n = c_1 3^n + c_2 n3^n = (c_1 + c_2 n)3^n$. Since $1 = x_0 = c_1$ and $6 = x_1 = (c_1 + c_2)3$, we find that $c_2 = 1$, so

$$x_n = (1 + n)3^n$$

The techniques outlined in Theorem 4.38 can be extended to higher order recurrence relations. We state, without proof, the general result.

Theorem 4.39 Let $x_n = a_{m-1}x_{n-1} + a_{m-2}x_{n-2} + \cdots + a_0 x_{n-m}$ be a recurrence relation of order m that is satisfied by a sequence (x_n). Suppose the associated characteristic polynomial

$$\lambda^m - a_{m-1}\lambda^{m-1} - a_{m-2}\lambda^{m-2} - \cdots - a_0$$

factors as $(\lambda - \lambda_1)^{m_1}(\lambda - \lambda_2)^{m_2}\cdots(\lambda - \lambda_k)^{m_k}$, where $m_1 + m_2 + \cdots + m_k = m$. Then x_n has the form

$$\begin{aligned} x_n = {} & (c_{11}\lambda_1^n + c_{12}n\lambda_1^n + c_{13}n^2\lambda_1^n + \cdots + c_{1m_1}n^{m_1-1}\lambda_1^n) + \cdots \\ & + (c_{k1}\lambda_k^n + c_{k2}n\lambda_k^n + c_{k3}n^2\lambda_k^n + \cdots + c_{km_k}n^{m_k-1}\lambda_k^n) \end{aligned}$$

Systems of Linear Differential Equations

In calculus, you learn that if $x = x(t)$ is a differentiable function satisfying a differential equation of the form $x' = kx$, where k is a constant, then the general solution is $x = Ce^{kt}$, where C is a constant. If an initial condition $x(0) = x_0$ is specified, then, by substituting $t = 0$ in the general solution, we find that $C = x_0$. Hence, the unique solution to the differential equation that satisfies the initial condition is

$$x = x_0 e^{kt}$$

Suppose we have n differentiable functions of t—say, $x_1, x_2, \ldots, x_n$—that satisfy a ***system of differential equations***

$$\begin{aligned} x_1' &= a_{11}x_1 + a_{12}x_2 + \cdots + a_{1n}x_n \\ x_2' &= a_{21}x_1 + a_{22}x_2 + \cdots + a_{2n}x_n \\ &\vdots \\ x_n' &= a_{n1}x_1 + a_{n2}x_2 + \cdots + a_{nn}x_n \end{aligned}$$

We can write this system in matrix form as $\mathbf{x}' = A\mathbf{x}$, where

$$\mathbf{x}(t) = \begin{bmatrix} x_1(t) \\ x_2(t) \\ \vdots \\ x_n(t) \end{bmatrix}, \quad \mathbf{x}'(t) = \begin{bmatrix} x_1'(t) \\ x_2'(t) \\ \vdots \\ x_n'(t) \end{bmatrix}, \quad \text{and} \quad A = \begin{bmatrix} a_{11} & a_{12} & \cdots & a_{1n} \\ a_{21} & a_{22} & \cdots & a_{2n} \\ \vdots & \vdots & \ddots & \vdots \\ a_{n1} & a_{n2} & \cdots & a_{nn} \end{bmatrix}$$

Now we can use matrix methods to help us find the solution.

First, we make a useful observation. Suppose we want to solve the following system of differential equations:

$$\begin{aligned} x_1' &= 2x_1 \\ x_2' &= 5x_2 \end{aligned}$$

Each equation can be solved separately, as above, to give

$$\begin{aligned} x_1 &= C_1 e^{2t} \\ x_2 &= C_2 e^{5t} \end{aligned}$$

where C_1 and C_2 are constants. Notice that, in matrix form, our equation $\mathbf{x}' = A\mathbf{x}$ has a *diagonal* coefficient matrix

$$A = \begin{bmatrix} 2 & 0 \\ 0 & 5 \end{bmatrix}$$

and the eigenvalues 2 and 5 occur in the exponentials e^{2t} and e^{5t} of the solution. This suggests that, for an arbitrary system, we should start by diagonalizing the coefficient matrix, if possible.

Example 4.43 Solve the following system of differential equations:

$$\begin{aligned} x_1' &= x_1 + 2x_2 \\ x_2' &= 3x_1 + 2x_2 \end{aligned}$$

Solution Here the coefficient matrix is $A = \begin{bmatrix} 1 & 2 \\ 3 & 2 \end{bmatrix}$, and we find that the eigenvalues are $\lambda_1 = 4$ and $\lambda_2 = -1$, with corresponding eigenvectors $\mathbf{v}_1 = \begin{bmatrix} 2 \\ 3 \end{bmatrix}$ and $\mathbf{v}_2 = \begin{bmatrix} -1 \\ 1 \end{bmatrix}$, respectively. Therefore, A is diagonalizable, and the matrix P that does the job is

$$P = [\mathbf{v}_1 \;\; \mathbf{v}_2] = \begin{bmatrix} 2 & -1 \\ 3 & 1 \end{bmatrix}$$

We know that

$$P^{-1}AP = \begin{bmatrix} 4 & 0 \\ 0 & -1 \end{bmatrix} = D$$

Let $\mathbf{x} = P\mathbf{y}$ (so that $\mathbf{x}' = P\mathbf{y}'$) and substitute these results into the original equation $\mathbf{x}' = A\mathbf{x}$ to get $P\mathbf{y}' = AP\mathbf{y}$ or, equivalently,

$$\mathbf{y}' = P^{-1}AP\mathbf{y} = D\mathbf{y}$$

This is just the system

$$\begin{aligned} y_1' &= 4y_1 \\ y_2' &= -y_2 \end{aligned}$$

whose general solution is

$$\begin{aligned} y_1 &= C_1 e^{4t} \\ y_2 &= C_2 e^{-t} \end{aligned} \quad \text{or} \quad \mathbf{y} = \begin{bmatrix} C_1 e^{4t} \\ C_2 e^{-t} \end{bmatrix}$$

To find **x**, we just compute

$$\mathbf{x} = P\mathbf{y} = \begin{bmatrix} 2 & -1 \\ 3 & 1 \end{bmatrix}\begin{bmatrix} C_1e^{4t} \\ C_2e^{-t} \end{bmatrix} = \begin{bmatrix} 2C_1e^{4t} - C_2e^{-t} \\ 3C_1e^{4t} + C_2e^{-t} \end{bmatrix}$$

so $x_1 = 2C_1e^{4t} - C_2e^{-t}$ and $x_2 = 3C_1e^{4t} + C_2e^{-t}$. (Check that these values satisfy the given system.)

Remark Observe that we could also express the solution in Example 4.43 as

$$\mathbf{x} = C_1e^{4t}\begin{bmatrix} 2 \\ 3 \end{bmatrix} + C_2e^{-t}\begin{bmatrix} -1 \\ 1 \end{bmatrix} = C_1e^{4t}\mathbf{v}_1 + C_2e^{-t}\mathbf{v}_2$$

This technique generalizes easily to $n \times n$ systems where the coefficient matrix is diagonalizable. The next theorem, whose proof is left as an exercise, summarizes the situation.

Theorem 4.40 Let A be an $n \times n$ diagonalizable matrix and let $P = [\mathbf{v}_1 \;\; \mathbf{v}_2 \;\; \cdots \;\; \mathbf{v}_n]$ be such that

$$P^{-1}AP = \begin{bmatrix} \lambda_1 & 0 & \cdots & 0 \\ 0 & \lambda_2 & \cdots & 0 \\ \vdots & \vdots & \ddots & \vdots \\ 0 & 0 & \cdots & \lambda_n \end{bmatrix}$$

Then the general solution to the system $\mathbf{x}' = A\mathbf{x}$ is

$$\mathbf{x} = C_1e^{\lambda_1 t}\mathbf{v}_1 + C_2e^{\lambda_2 t}\mathbf{v}_2 + \cdots + C_ne^{\lambda_n t}\mathbf{v}_n$$

The next example involves a biological model in which two species live in the same ecosystem. It is reasonable to assume that the growth rate of each species depends on the sizes of *both* populations. (Of course, there are other factors that govern growth, but we will keep our model simple by ignoring these.)

If $x_1(t)$ and $x_2(t)$ denote the sizes of the two populations at time t, then $x_1'(t)$ and $x_2'(t)$ are their rates of growth at time t. Our model is of the form

$$\begin{aligned} x_1'(t) &= ax_1(t) + bx_2(t) \\ x_2'(t) &= cx_1(t) + dx_2(t) \end{aligned}$$

where the coefficients a, b, c, and d depend on the conditions.

Example 4.44 Raccoons and squirrels inhabit the same ecosystem and compete with each other for food, water, and space. Let the raccoon and squirrel populations at time t years be given by $r(t)$ and $s(t)$, respectively. In the absence of squirrels, the raccoon growth rate is $r'(t) = 2.5r(t)$, but when squirrels are present, the competition slows the raccoon growth rate to $r'(t) = 2.5r(t) - s(t)$. The squirrel population is similarly affected by the raccoons. In the absence of raccoons, the growth rate of the squirrel population is $s'(t) = 2.5s(t)$, and the population growth rate for squirrels when they are sharing the ecosystem with raccoons is $s'(t) = -0.25r(t) + 2.5s(t)$. Suppose that initially there are 60 raccoons and 60 squirrels in the ecosystem. Determine what happens to these two populations.

Solution Our system is $\mathbf{x}' = A\mathbf{x}$, where

$$\mathbf{x} = \mathbf{x}(t) = \begin{bmatrix} r(t) \\ s(t) \end{bmatrix} \quad \text{and} \quad A = \begin{bmatrix} 2.5 & -1.0 \\ -0.25 & 2.5 \end{bmatrix}$$

The eigenvalues of A are $\lambda_1 = 3$ and $\lambda_2 = 2$, with corresponding eigenvectors $\mathbf{v}_1 = \begin{bmatrix} -2 \\ 1 \end{bmatrix}$ and $\mathbf{v}_2 = \begin{bmatrix} 2 \\ 1 \end{bmatrix}$. By Theorem 4.40, the general solution to our system is

$$\mathbf{x}(t) = C_1 e^{3t}\mathbf{v}_1 + C_2 e^{2t}\mathbf{v}_2 = C_1 e^{3t}\begin{bmatrix} -2 \\ 1 \end{bmatrix} + C_2 e^{2t}\begin{bmatrix} 2 \\ 1 \end{bmatrix} \tag{7}$$

The initial population vector is $\mathbf{x}(0) = \begin{bmatrix} r(0) \\ s(0) \end{bmatrix} = \begin{bmatrix} 60 \\ 60 \end{bmatrix}$, so, setting $t = 0$ in Equation (7), we have

$$C_1\begin{bmatrix} -2 \\ 1 \end{bmatrix} + C_2\begin{bmatrix} 2 \\ 1 \end{bmatrix} = \begin{bmatrix} 60 \\ 60 \end{bmatrix}$$

Solving this equation, we find $C_1 = 15$ and $C_2 = 45$. Hence,

$$\mathbf{x}(t) = 15e^{3t}\begin{bmatrix} -2 \\ 1 \end{bmatrix} + 45e^{2t}\begin{bmatrix} 2 \\ 1 \end{bmatrix}$$

from which we find $r(t) = -30e^{3t} + 90e^{2t}$ and $s(t) = 15e^{3t} + 45e^{2t}$. Figure 4.20 shows the graphs of these two functions, and you can see clearly that the raccoon population dies out after a little more than 1 year. (Can you determine *exactly* when it dies out?)

We now consider a similar example, in which one species is a source of food for the other. Such a model is called a ***predator-prey model.*** Once again, our model will be drastically oversimplified in order to illustrate its main features.

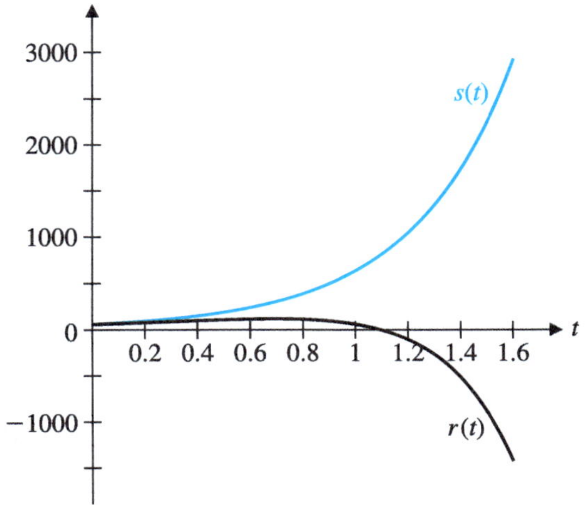

Figure 4.20
Raccoon and squirrel populations

a + bi

Example 4.45 Robins and worms cohabit an ecosystem. The robins eat the worms, which are their only source of food. The robin and worm populations at time t years are denoted by $r(t)$ and $w(t)$, respectively, and the equations governing the growth of the two populations are

$$\begin{aligned} r'(t) &= w(t) - 12 \\ w'(t) &= -r(t) + 10 \end{aligned} \tag{8}$$

If initially 6 robins and 20 worms occupy the ecosystem, determine the behavior of the two populations over time.

Solution The first thing we notice about this example is the presence of the extra constants, -12 and 10, in the two equations. Fortunately, we can get rid of them with a simple change of variables. If we let $r(t) = x(t) + 10$ and $w(t) = y(t) + 12$, then $r'(t) = x'(t)$ and $w'(t) = y'(t)$. Substituting into Equations (8), we have

$$\begin{aligned} x'(t) &= y(t) \\ y'(t) &= -x(t) \end{aligned} \tag{9}$$

which is easier to work with. Equations (9) have the form $\mathbf{x}' = A\mathbf{x}$, where $A = \begin{bmatrix} 0 & 1 \\ -1 & 0 \end{bmatrix}$. Our new initial conditions are

$$x(0) = r(0) - 10 = 6 - 10 = -4 \quad \text{and} \quad y(0) = w(0) - 12 = 20 - 12 = 8$$

so $\mathbf{x}(0) = \begin{bmatrix} -4 \\ 8 \end{bmatrix}$.

Proceeding as in the last example, we find the eigenvalues and eigenvectors of A. The characteristic polynomial is $\lambda^2 + 1$, which has no real roots. What should we do? We have no choice but to use the complex roots, which are $\lambda_1 = i$ and $\lambda_2 = -i$. The corresponding eigenvectors are also complex—namely, $\mathbf{v}_1 = \begin{bmatrix} 1 \\ i \end{bmatrix}$ and $\mathbf{v}_2 = \begin{bmatrix} 1 \\ -i \end{bmatrix}$. By Theorem 4.40, our solution has the form

$$\mathbf{x}(t) = C_1 e^{it}\mathbf{v}_1 + C_2 e^{-it}\mathbf{v}_2 = C_1 e^{it}\begin{bmatrix} 1 \\ i \end{bmatrix} + C_2 e^{-it}\begin{bmatrix} 1 \\ -i \end{bmatrix}$$

From $\mathbf{x}(0) = \begin{bmatrix} -4 \\ 8 \end{bmatrix}$, we get

$$C_1\begin{bmatrix} 1 \\ i \end{bmatrix} + C_2\begin{bmatrix} 1 \\ -i \end{bmatrix} = \begin{bmatrix} -4 \\ 8 \end{bmatrix}$$

whose solution is $C_1 = -2 - 4i$ and $C_2 = -2 + 4i$. So the solution to system (9) is

$$\mathbf{x}(t) = (-2 - 4i)e^{it}\begin{bmatrix} 1 \\ i \end{bmatrix} + (-2 + 4i)e^{-it}\begin{bmatrix} 1 \\ -i \end{bmatrix}$$

What are we to make of this solution? Robins and worms inhabit a real world—yet our solution involves complex numbers! Fearlessly proceeding, we apply Euler's formula

$$e^{it} = \cos t + i \sin t$$

(Appendix C) to get $e^{-it} = \cos(-t) + i\sin(-t) = \cos t - i\sin t$. Substituting, we have

$$\begin{aligned}\mathbf{x}(t) &= (-2-4i)(\cos t + i\sin t)\begin{bmatrix}1\\ i\end{bmatrix} + (-2+4i)(\cos t - i\sin t)\begin{bmatrix}1\\ -i\end{bmatrix}\\ &= \begin{bmatrix}(-2\cos t + 4\sin t) + i(-4\cos t - 2\sin t)\\ (4\cos t + 2\sin t) + i(-2\cos t + 4\sin t)\end{bmatrix}\\ &\quad + \begin{bmatrix}(-2\cos t + 4\sin t) + i(4\cos t + 2\sin t)\\ (4\cos t + 2\sin t) + i(2\cos t - 4\sin t)\end{bmatrix}\\ &= \begin{bmatrix}-4\cos t + 8\sin t\\ 8\cos t + 4\sin t\end{bmatrix}\end{aligned}$$

This gives $x(t) = -4\cos t + 8\sin t$ and $y(t) = 8\cos t + 4\sin t$. Putting everything in terms of our original variables, we conclude that

$$r(t) = x(t) + 10 = -4\cos t + 8\sin t + 10$$

and $$w(t) = y(t) + 12 = 8\cos t + 4\sin t + 12$$

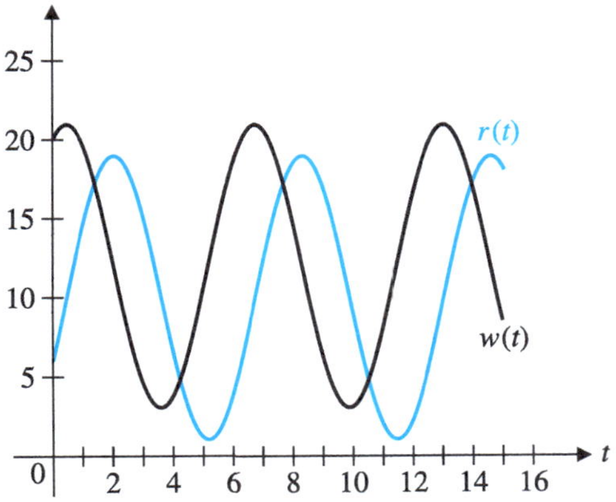

Figure 4.21

Robin and worm populations

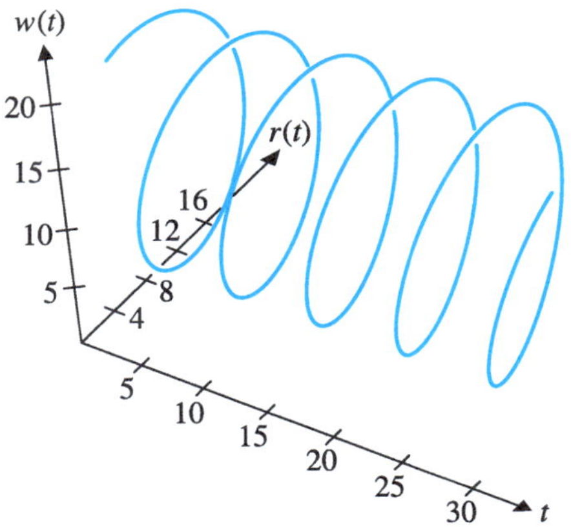

Figure 4.22

So our solution is real after all! The graphs of $r(t)$ and $w(t)$ in Figure 4.21 show that the two populations oscillate periodically. As the robin population increases, the worm population starts to decrease, but as the robins' only food source diminishes, their numbers start to decline as well. As the predators disappear, the worm population begins to recover. As its food supply increases, so does the robin population, and the cycle repeats itself. This oscillation is typical of examples in which the eigenvalues are complex.

Plotting robins, worms, and time on separate axes, as in Figure 4.22, clearly reveals the cyclic nature of the two populations.

We conclude this section by looking at what we have done from a different point of view. If $x = x(t)$ is a differentiable function of t, then the general solution of the ordinary differential equation $x' = ax$ is $x = ce^{at}$, where c is a scalar. The systems of linear differential equations we have been considering have the form $\mathbf{x}' = A\mathbf{x}$, so if we simply plowed ahead without thinking, we might be tempted to deduce that the solution would be $\mathbf{x} = \mathbf{c}e^{At}$, where $\mathbf{c}$ is a vector. But what on earth could this mean? On the right-hand side, we have the *number* e raised to the power of a *matrix*. This appears to be nonsense, yet you will see that there is a way to make sense of it.

Let's start by considering the expression e^A. In calculus, you learn that the function e^x has a power series expansion

$$e^x = 1 + x + \frac{x^2}{2!} + \frac{x^3}{3!} + \cdots$$

that converges for every real number x. By analogy, let us define

$$e^A = I + A + \frac{A^2}{2!} + \frac{A^3}{3!} + \cdots$$

The right-hand side is just defined in terms of powers of A, and it can be shown that it converges for any real matrix A. So now e^A is a matrix, called the ***exponential*** of A. But how can we compute e^A or e^{At}? For diagonal matrices, it is easy.

Example 4.46

Compute e^{Dt} for $D = \begin{bmatrix} 4 & 0 \\ 0 & -1 \end{bmatrix}$.

Solution From the definition, we have

$$\begin{aligned} e^{Dt} &= I + Dt + \frac{(Dt)^2}{2!} + \frac{(Dt)^3}{3!} + \cdots \\ &= \begin{bmatrix} 1 & 0 \\ 0 & 1 \end{bmatrix} + \begin{bmatrix} 4t & 0 \\ 0 & -t \end{bmatrix} + \frac{1}{2!}\begin{bmatrix} (4t)^2 & 0 \\ 0 & (-t)^2 \end{bmatrix} + \frac{1}{3!}\begin{bmatrix} (4t)^3 & 0 \\ 0 & (-t)^3 \end{bmatrix} + \cdots \\ &= \begin{bmatrix} 1 + (4t) + \frac{1}{2!}(4t)^2 + \frac{1}{3!}(4t)^3 + \cdots & 0 \\ 0 & 1 + (-t) + \frac{1}{2!}(-t)^2 + \frac{1}{3!}(-t)^3 + \cdots \end{bmatrix} \\ &= \begin{bmatrix} e^{4t} & 0 \\ 0 & e^{-t} \end{bmatrix} \end{aligned}$$

The matrix exponential is also nice if A is diagonalizable.

Example 4.47

Compute e^A for $A = \begin{bmatrix} 1 & 2 \\ 3 & 2 \end{bmatrix}$.

Solution In Example 4.43, we found the eigenvalues of A to be $\lambda_1 = 4$ and $\lambda_2 = -1$, with corresponding eigenvectors $\mathbf{v}_1 = \begin{bmatrix} 2 \\ 3 \end{bmatrix}$ and $\mathbf{v}_2 = \begin{bmatrix} -1 \\ 1 \end{bmatrix}$, respectively. Hence, with $P = [\mathbf{v}_1 \quad \mathbf{v}_2] = \begin{bmatrix} 2 & -1 \\ 3 & 1 \end{bmatrix}$, we have $P^{-1}AP = D = \begin{bmatrix} 4 & 0 \\ 0 & -1 \end{bmatrix}$. Since $A = PDP^{-1}$, we have $A^k = PD^kP^{-1}$, so

$$\begin{aligned} e^A &= I + A + \frac{A^2}{2!} + \frac{A^3}{3!} + \cdots \\ &= PIP^{-1} + PDP^{-1} + \frac{1}{2!}PD^2P^{-1} + \frac{1}{3!}PD^3P^{-1} + \cdots \\ &= P\left(I + D + \frac{D^2}{2!} + \frac{D^3}{3!} + \cdots\right)P^{-1} \\ &= Pe^DP^{-1} \\ &= \begin{bmatrix} 2 & -1 \\ 3 & 1 \end{bmatrix}\begin{bmatrix} e^4 & 0 \\ 0 & e^{-1} \end{bmatrix}\begin{bmatrix} 2 & -1 \\ 3 & 1 \end{bmatrix}^{-1} \\ &= \frac{1}{5}\begin{bmatrix} 2e^4 + 3e^{-1} & 2e^4 - 2e^{-1} \\ 3e^4 - 3e^{-1} & 3e^4 + 2e^{-1} \end{bmatrix} \end{aligned}$$

We are now in a position to show that our bold (and seemingly foolish) guess at an "exponential" solution of $\mathbf{x}' = A\mathbf{x}$ was not so far off after all!

Theorem 4.41

Let A be an $n \times n$ diagonalizable matrix with eigenvalues $\lambda_1, \lambda_2, \ldots, \lambda_n$. Then the general solution to the system $\mathbf{x}' = A\mathbf{x}$ is $\mathbf{x} = e^{At}\mathbf{c}$, where $\mathbf{c}$ is an arbitrary constant vector. If an initial condition $\mathbf{x}(0)$ is specified, then $\mathbf{c} = \mathbf{x}(0)$.

Proof Let P diagonalize A. Then $A = PDP^{-1}$, and, as in Example 4.47,

$$e^{At} = Pe^{Dt}P^{-1}$$

Hence, we need to check that $\mathbf{x}' = A\mathbf{x}$ is satisfied by $\mathbf{x} = Pe^{Dt}P^{-1}\mathbf{c}$. Now, everything is constant except for e^{Dt}, so

$$\mathbf{x}' = \frac{d\mathbf{x}}{dt} = \frac{d}{dt}(Pe^{Dt}P^{-1}\mathbf{c}) = P\frac{d}{dt}(e^{Dt})P^{-1}\mathbf{c} \qquad (10)$$

If
$$D = \begin{bmatrix} \lambda_1 & 0 & \cdots & 0 \\ 0 & \lambda_2 & \cdots & 0 \\ \vdots & \vdots & \ddots & \vdots \\ 0 & 0 & \cdots & \lambda_n \end{bmatrix}$$

then $$e^{Dt} = \begin{bmatrix} e^{\lambda_1 t} & 0 & \cdots & 0 \\ 0 & e^{\lambda_2 t} & \cdots & 0 \\ \vdots & \vdots & \ddots & \vdots \\ 0 & 0 & \cdots & e^{\lambda_n t} \end{bmatrix}$$

Taking derivatives, we have

$$\begin{aligned} \frac{d}{dt}(e^{Dt}) &= \begin{bmatrix} \frac{d}{dt}(e^{\lambda_1 t}) & 0 & \cdots & 0 \\ 0 & \frac{d}{dt}(e^{\lambda_2 t}) & \cdots & 0 \\ \vdots & \vdots & \ddots & \vdots \\ 0 & 0 & \cdots & \frac{d}{dt}(e^{\lambda_n t}) \end{bmatrix} \\ &= \begin{bmatrix} \lambda_1 e^{\lambda_1 t} & 0 & \cdots & 0 \\ 0 & \lambda_2 e^{\lambda_2 t} & \cdots & 0 \\ \vdots & \vdots & \ddots & \vdots \\ 0 & 0 & \cdots & \lambda_n e^{\lambda_n t} \end{bmatrix} \\ &= \begin{bmatrix} \lambda_1 & 0 & \cdots & 0 \\ 0 & \lambda_2 & \cdots & 0 \\ \vdots & \vdots & \ddots & \vdots \\ 0 & 0 & \cdots & \lambda_n \end{bmatrix} \begin{bmatrix} e^{\lambda_1 t} & 0 & \cdots & 0 \\ 0 & e^{\lambda_2 t} & \cdots & 0 \\ \vdots & \vdots & \ddots & \vdots \\ 0 & 0 & \cdots & e^{\lambda_n t} \end{bmatrix} \\ &= De^{Dt} \end{aligned}$$

Substituting this result into Equation (10), we obtain

$$\mathbf{x}' = PDe^{Dt}P^{-1}\mathbf{c} = PDP^{-1}Pe^{Dt}P^{-1}\mathbf{c} = (PDP^{-1})(Pe^{Dt}P^{-1})\mathbf{c} = Ae^{At}\mathbf{c} = A\mathbf{x}$$

as required.

The last statement follows easily from the fact that if $\mathbf{x} = \mathbf{x}(t) = e^{At}\mathbf{c}$, then

$$\mathbf{x}(0) = e^{A \cdot 0}\mathbf{c} = e^{O}\mathbf{c} = I\mathbf{c} = \mathbf{c}$$

since $e^{O} = I$. (Why?)

In fact, Theorem 4.41 is true even if A is not diagonalizable, but we will not prove this. Computation of matrix exponentials for nondiagonalizable matrices requires the *Jordan normal form* of a matrix, a topic that may be found in more advanced linear algebra texts.

For example, see *Linear Algebra* by S. H. Friedberg, A. J. Insel, and L. E. Spence (Englewood Cliffs, NJ: Prentice-Hall, 1979).

Ideally, this short digression has served to illustrate the power of mathematics to generalize and the value of creative thinking. Matrix exponentials turn out to be very important tools in many applications of linear algebra, both theoretical and applied.

a + bi

Discrete Linear Dynamical Systems

We conclude this chapter as we began it—by looking at dynamical systems. Markov chains and the Leslie model of population growth are examples of ***discrete linear dynamical systems.*** Each can be described by a matrix equation of the form

$$\mathbf{x}_{k+1} = A\mathbf{x}_k$$

where the vector $\mathbf{x}_k$ records the state of the system at "time" k and A is a square matrix. As we have seen, the long-term behavior of these systems is related to the eigenvalues and eigenvectors of the matrix A. The power method exploits the iterative nature of such dynamical systems to approximate eigenvalues and eigenvectors, and the Perron-Frobenius Theorem gives specialized information about the long-term behavior of a discrete linear dynamical system whose coefficient matrix A is nonnegative.

When A is a 2×2 matrix, we can describe the evolution of a dynamical system geometrically. The equation $\mathbf{x}_{k+1} = A\mathbf{x}_k$ is really an infinite collection of equations. Beginning with an initial vector $\mathbf{x}_0$, we have:

$$\begin{aligned} \mathbf{x}_1 &= A\mathbf{x}_0 \\ \mathbf{x}_2 &= A\mathbf{x}_1 \\ \mathbf{x}_3 &= A\mathbf{x}_2 \\ &\vdots \end{aligned}$$

The set $\{\mathbf{x}_0, \mathbf{x}_1, \mathbf{x}_2, \ldots\}$ is called a ***trajectory*** of the system. (For graphical purposes, we will identify each vector in a trajectory with its head so that we can plot it as a point.) Note that $\mathbf{x}_k = A^k\mathbf{x}_0$.

Example 4.48

Let $A = \begin{bmatrix} 0.5 & 0 \\ 0 & 0.8 \end{bmatrix}$. For the dynamical system $\mathbf{x}_{k+1} = A\mathbf{x}_k$, plot the first five points in the trajectories with the following initial vectors:

(a) $\mathbf{x}_0 = \begin{bmatrix} 5 \\ 0 \end{bmatrix}$ (b) $\mathbf{x}_0 = \begin{bmatrix} 0 \\ -5 \end{bmatrix}$ (c) $\mathbf{x}_0 = \begin{bmatrix} 4 \\ 4 \end{bmatrix}$ (d) $\mathbf{x}_0 = \begin{bmatrix} -2 \\ 4 \end{bmatrix}$

Solution (a) We compute $\mathbf{x}_1 = A\mathbf{x}_0 = \begin{bmatrix} 2.5 \\ 0 \end{bmatrix}$, $\mathbf{x}_2 = A\mathbf{x}_1 = \begin{bmatrix} 1.25 \\ 0 \end{bmatrix}$, $\mathbf{x}_3 = A\mathbf{x}_2 = \begin{bmatrix} 0.625 \\ 0 \end{bmatrix}$, $\mathbf{x}_4 = A\mathbf{x}_3 = \begin{bmatrix} 0.3125 \\ 0 \end{bmatrix}$. These are plotted in Figure 4.23, and the points are connected to highlight the trajectory. Similar calculations produce the trajectories marked (b), (c), and (d) in Figure 4.23.

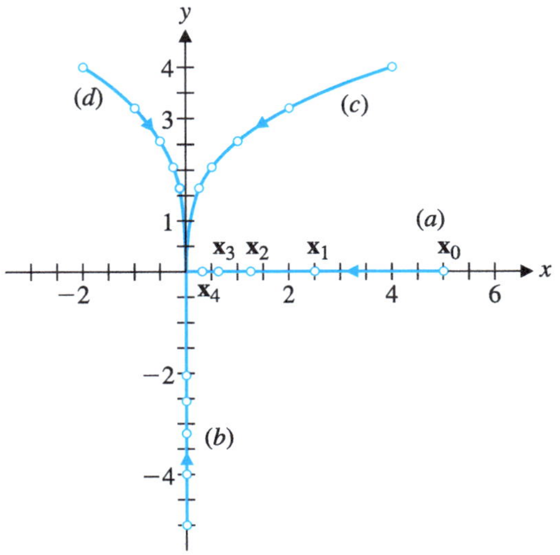

Figure 4.23

In Example 4.48, every trajectory converges to **0**. The origin is called an ***attractor*** in this case. We can understand why this is so from Theorem 4.19. The matrix A in Example 4.48 has eigenvectors $\begin{bmatrix}1\\0\end{bmatrix}$ and $\begin{bmatrix}0\\1\end{bmatrix}$ corresponding to its eigenvalues 0.5 and 0.8, respectively. (Check this.) Accordingly, for any initial vector

$$\mathbf{x}_0 = \begin{bmatrix}c_1\\c_2\end{bmatrix} = c_1\begin{bmatrix}1\\0\end{bmatrix} + c_2\begin{bmatrix}0\\1\end{bmatrix}$$

we have

$$\mathbf{x}_k = A^k\mathbf{x}_0 = c_1(0.5)^k\begin{bmatrix}1\\0\end{bmatrix} + c_2(0.8)^k\begin{bmatrix}0\\1\end{bmatrix}$$

Because both $(0.5)^k$ and $(0.8)^k$ approach zero as k gets large, $\mathbf{x}_k$ approaches **0** for any choice of $\mathbf{x}_0$. In addition, we know from Theorem 4.28 that because 0.8 is the dominant eigenvalue of A, $\mathbf{x}_k$ will approach a multiple of the corresponding eigenvector $\begin{bmatrix}0\\1\end{bmatrix}$ as long as $c_2 \neq 0$ (the coefficient of $\mathbf{x}_0$ corresponding to $\begin{bmatrix}0\\1\end{bmatrix}$). In other words, all trajectories except those that begin on the x-axis (where $c_2 = 0$) will approach the y-axis, as Figure 4.23 shows.

Example 4.49

Discuss the behavior of the dynamical system $\mathbf{x}_{k+1} = A\mathbf{x}_k$ corresponding to the matrix $A = \begin{bmatrix}0.65 & -0.15\\-0.15 & 0.65\end{bmatrix}$.

Solution The eigenvalues of A are 0.5 and 0.8 with corresponding eigenvectors $\begin{bmatrix}1\\1\end{bmatrix}$ and $\begin{bmatrix}-1\\1\end{bmatrix}$, respectively. (Check this.) Hence for an initial vector $\mathbf{x}_0 = c_1\begin{bmatrix}1\\1\end{bmatrix} + c_2\begin{bmatrix}-1\\1\end{bmatrix}$, we have

$$\mathbf{x}_k = A^k\mathbf{x}_0 = c_1(0.5)^k\begin{bmatrix}1\\1\end{bmatrix} + c_2(0.8)^k\begin{bmatrix}-1\\1\end{bmatrix}$$

Once again the origin is an attractor, because $\mathbf{x}_k$ approaches **0** for any choice of $\mathbf{x}_0$. If $c_2 \neq 0$, the trajectory will approach the line through the origin with direction vector $\begin{bmatrix}-1\\1\end{bmatrix}$. Several such trajectories are shown in Figure 4.24. The vectors $\mathbf{x}_0$ where $c_2 = 0$ are on the line through the origin with direction vector $\begin{bmatrix}1\\1\end{bmatrix}$, and the corresponding trajectory in this case follows this line into the origin.

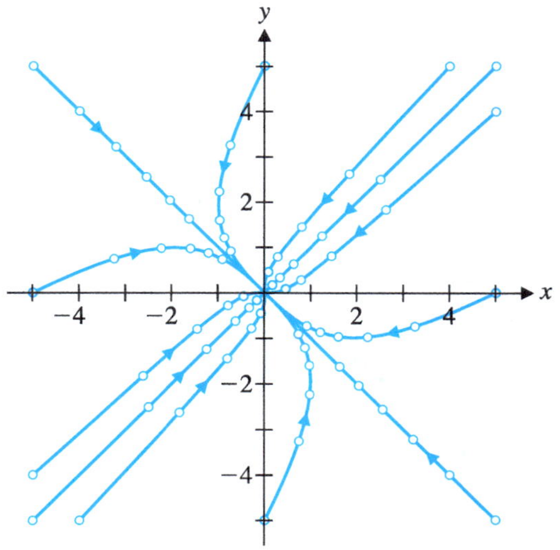

Figure 4.24

Example 4.50

Discuss the behavior of the dynamical systems $\mathbf{x}_{k+1} = A\mathbf{x}_k$ corresponding to the following matrices:

(a) $A = \begin{bmatrix} 4 & 1 \\ 1 & 4 \end{bmatrix}$ (b) $A = \begin{bmatrix} 1 & 0.5 \\ 0.5 & 1 \end{bmatrix}$

Solution (a) The eigenvalues of A are 5 and 3 with corresponding eigenvectors $\begin{bmatrix} 1 \\ 1 \end{bmatrix}$ and $\begin{bmatrix} -1 \\ 1 \end{bmatrix}$, respectively. Hence for an initial vector $\mathbf{x}_0 = c_1\begin{bmatrix} 1 \\ 1 \end{bmatrix} + c_2\begin{bmatrix} -1 \\ 1 \end{bmatrix}$, we have

$$\mathbf{x}_k = A^k\mathbf{x}_0 = c_1 5^k\begin{bmatrix} 1 \\ 1 \end{bmatrix} + c_2 3^k\begin{bmatrix} -1 \\ 1 \end{bmatrix}$$

As k becomes large, so do both 5^k and 3^k. Hence, $\mathbf{x}_k$ tends away from the origin. Because the dominant eigenvalue of 5 has corresponding eigenvector $\begin{bmatrix} 1 \\ 1 \end{bmatrix}$, all trajectories for which $c_1 \neq 0$ will eventually end up in the first or the third quadrant. Trajectories with $c_1 = 0$ start and stay on the line $y = -x$ whose direction vector is $\begin{bmatrix} -1 \\ 1 \end{bmatrix}$. See Figure 4.25(a).

(b) In this example, the eigenvalues are 1.5 and 0.5 with corresponding eigenvectors $\begin{bmatrix} 1 \\ 1 \end{bmatrix}$ and $\begin{bmatrix} -1 \\ 1 \end{bmatrix}$, respectively. Hence,

$$\mathbf{x}_k = c_1(1.5)^k\begin{bmatrix} 1 \\ 1 \end{bmatrix} + c_2(0.5)^k\begin{bmatrix} -1 \\ 1 \end{bmatrix} \quad \text{if} \quad \mathbf{x}_0 = c_1\begin{bmatrix} 1 \\ 1 \end{bmatrix} + c_2\begin{bmatrix} -1 \\ 1 \end{bmatrix}$$

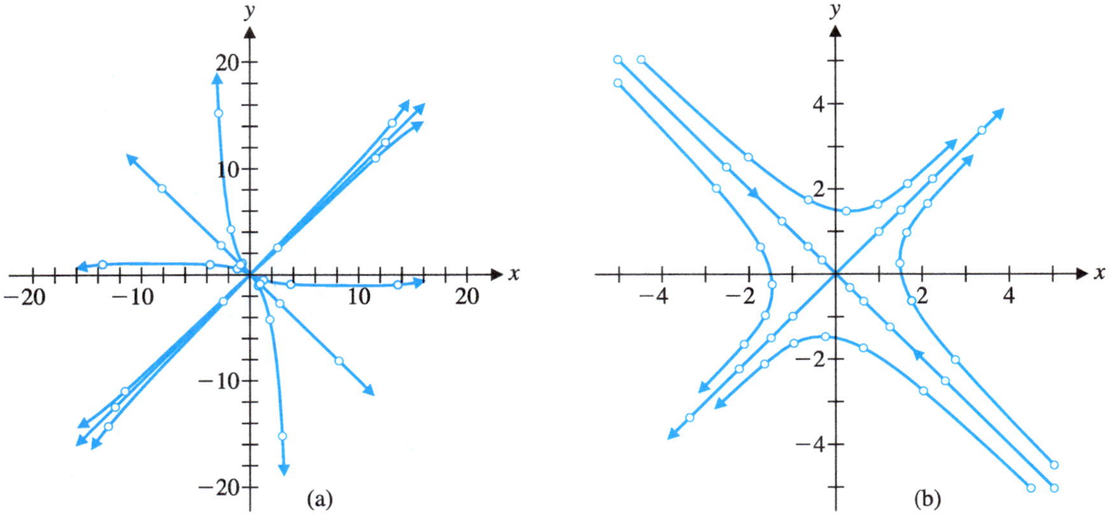

Figure 4.25

If $c_1 = 0$, then $\mathbf{x}_k = c_2(0.5)^k\begin{bmatrix}-1\\1\end{bmatrix} \to \begin{bmatrix}0\\0\end{bmatrix}$ as $k \to \infty$. But if $c_1 \neq 0$, then

$$\mathbf{x}_k = c_1(1.5)^k\begin{bmatrix}1\\1\end{bmatrix} + c_2(0.5)^k\begin{bmatrix}-1\\1\end{bmatrix} \approx c_1(1.5)^k\begin{bmatrix}1\\1\end{bmatrix} \quad \text{as} \quad k \to \infty$$

and such trajectories asymptotically approach the line $y = x$. See Figure 4.25(b).

In Example 4.50(a), all points that start out near the origin become increasingly large in magnitude because $|\lambda| > 1$ for both eigenvalues; $\mathbf{0}$ is called a ***repeller.*** In Example 4.50(b), $\mathbf{0}$ is called a ***saddle point*** because the origin attracts points in some directions and repels points in other directions. In this case, $|\lambda_1| < 1$ and $|\lambda_2| > 1$.

The next example shows what can happen when the eigenvalues of a real 2×2 matrix are complex (and hence conjugates of one another).

Example 4.51

Plot the trajectory beginning with $\mathbf{x}_0 = \begin{bmatrix}4\\4\end{bmatrix}$ for the dynamical systems $\mathbf{x}_{k+1} = A\mathbf{x}_k$ corresponding to the following matrices:

(a) $A = \begin{bmatrix}0.5 & -0.5\\0.5 & 0.5\end{bmatrix}$ (b) $A = \begin{bmatrix}0.2 & -1.2\\0.6 & 1.4\end{bmatrix}$

Solution The trajectories are shown in Figure 4.26(a) and (b), respectively. Note that (a) is a trajectory spiraling into the origin, whereas (b) appears to follow an elliptical orbit.

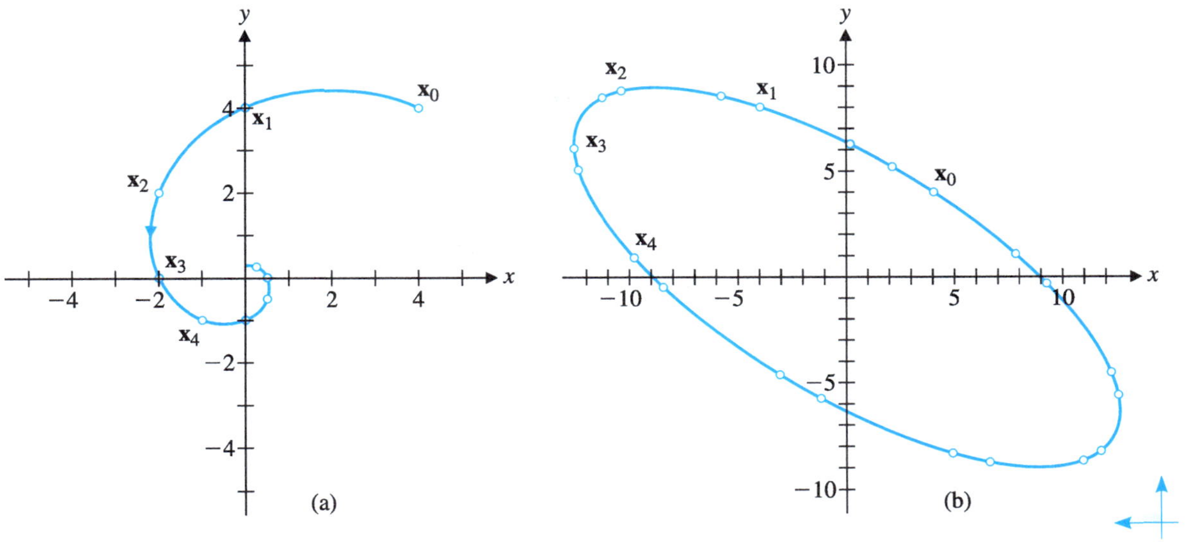

Figure 4.26

The following theorem explains the spiral behavior of the trajectory in Example 4.51(a).

Theorem 4.42 Let $A = \begin{bmatrix} a & -b \\ b & a \end{bmatrix}$. The eigenvalues of A are $\lambda = a \pm bi$, and if a and b are not both zero, then A can be factored as

$$A = \begin{bmatrix} a & -b \\ b & a \end{bmatrix} = \begin{bmatrix} r & 0 \\ 0 & r \end{bmatrix} \begin{bmatrix} \cos\theta & -\sin\theta \\ \sin\theta & \cos\theta \end{bmatrix}$$

where $r = |\lambda| = \sqrt{a^2 + b^2}$ and θ is the principal argument of $a + bi$.

Proof The eigenvalues of A are

$$\lambda = \tfrac{1}{2}(2a \pm \sqrt{4(-b^2)}) = \tfrac{1}{2}(2a \pm 2\sqrt{b^2}\sqrt{-1}) = a \pm |b|i = a \pm bi$$

by Exercise 35(b) in Section 4.1. Figure 4.27 displays $a + bi$, r, and θ. It follows that

$$A = \begin{bmatrix} a & -b \\ b & a \end{bmatrix} = r\begin{bmatrix} a/r & -b/r \\ b/r & a/r \end{bmatrix} = \begin{bmatrix} r & 0 \\ 0 & r \end{bmatrix} \begin{bmatrix} \cos\theta & -\sin\theta \\ \sin\theta & \cos\theta \end{bmatrix}$$

Remark Geometrically, Theorem 4.42 implies that when $A = \begin{bmatrix} a & -b \\ b & a \end{bmatrix} \neq O$, the linear transformation $T(\mathbf{x}) = A\mathbf{x}$ is the composition of a rotation $R = \begin{bmatrix} \cos\theta & -\sin\theta \\ \sin\theta & \cos\theta \end{bmatrix}$ through the angle θ followed by a scaling $S = \begin{bmatrix} r & 0 \\ 0 & r \end{bmatrix}$ with factor r (Figure 4.28). In Example 4.51(a), the eigenvalues are $\lambda = 0.5 \pm 0.5i$ so $r = |\lambda| = \sqrt{2}/2 \approx 0.707 < 1$, and hence the trajectories all spiral inward toward $\mathbf{0}$.

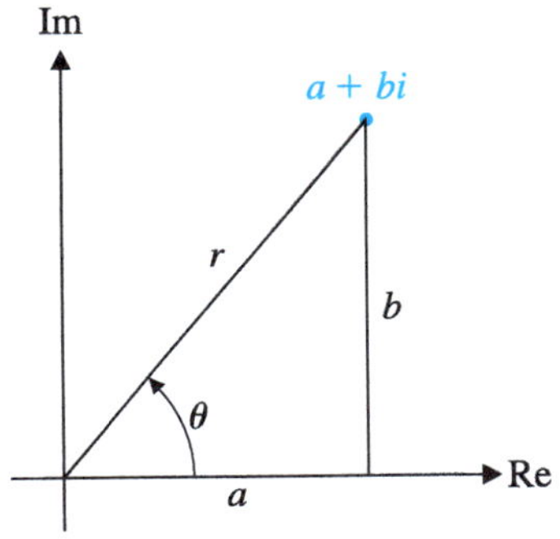

Figure 4.27

The next theorem shows that, in general, when a real 2×2 matrix has complex eigenvalues, it is similar to a matrix of the form $\begin{bmatrix} a & -b \\ b & a \end{bmatrix}$. For a complex vector

$$\mathbf{x} = \begin{bmatrix} z \\ w \end{bmatrix} = \begin{bmatrix} a + bi \\ c + di \end{bmatrix} = \begin{bmatrix} a \\ b \end{bmatrix} + \begin{bmatrix} c \\ d \end{bmatrix} i$$

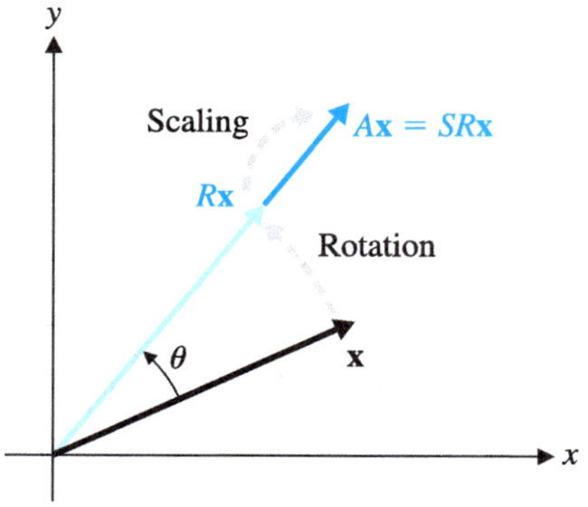

Figure 4.28
A rotation followed by a scaling

we define the real part, Re **x**, and the imaginary part, Im **x**, of **x** to be

$$\text{Re}\,\mathbf{x} = \begin{bmatrix} a \\ b \end{bmatrix} = \begin{bmatrix} \text{Re}\,z \\ \text{Re}\,w \end{bmatrix} \quad \text{and} \quad \text{Im}\,\mathbf{x} = \begin{bmatrix} c \\ d \end{bmatrix} = \begin{bmatrix} \text{Im}\,z \\ \text{Im}\,w \end{bmatrix}$$

Theorem 4.43 Let A be a real 2×2 matrix with a complex eigenvalue $\lambda = a - bi$ (where $b \neq 0$) and corresponding eigenvector $\mathbf{x}$. Then the matrix $P = [\text{Re}\,\mathbf{x} \quad \text{Im}\,\mathbf{x}]$ is invertible and

$$A = P\begin{bmatrix} a & -b \\ b & a \end{bmatrix}P^{-1}$$

Proof Let $\mathbf{x} = \mathbf{u} + \mathbf{v}i$ so that $\text{Re}\,\mathbf{x} = \mathbf{u}$ and $\text{Im}\,\mathbf{x} = \mathbf{v}$. From $A\mathbf{x} = \lambda\mathbf{x}$, we have

$$\begin{aligned} A\mathbf{u} + A\mathbf{v}i = A\mathbf{x} = \lambda\mathbf{x} &= (a - bi)(\mathbf{u} + \mathbf{v}i) \\ &= a\mathbf{u} + a\mathbf{v}i - b\mathbf{u}i + b\mathbf{v} = (a\mathbf{u} + b\mathbf{v}) + (-b\mathbf{u} + a\mathbf{v})i \end{aligned}$$

Equating real and imaginary parts, we obtain

$$A\mathbf{u} = a\mathbf{u} + b\mathbf{v} \quad \text{and} \quad A\mathbf{v} = -b\mathbf{u} + a\mathbf{v}$$

Now $P = [\mathbf{u} \mid \mathbf{v}]$, so

$$\begin{aligned} P\begin{bmatrix} a & -b \\ b & a \end{bmatrix} = [\mathbf{u} \mid \mathbf{v}]\begin{bmatrix} a & -b \\ b & a \end{bmatrix} &= [a\mathbf{u} + b\mathbf{v} \mid -b\mathbf{u} + a\mathbf{v}] = [A\mathbf{u} \mid A\mathbf{v}] = A[\mathbf{u} \mid \mathbf{v}] \\ &= AP \end{aligned}$$

To show that P is invertible, it is enough to show that $\mathbf{u}$ and $\mathbf{v}$ are linearly independent. If $\mathbf{u}$ and $\mathbf{v}$ were not linearly independent, then it would follow that $\mathbf{v} = k\mathbf{u}$ for some (nonzero complex) scalar k, because neither $\mathbf{u}$ nor $\mathbf{v}$ is $\mathbf{0}$. Thus,

$$\mathbf{x} = \mathbf{u} + \mathbf{v}i = \mathbf{u} + k\mathbf{u}i = (1 + ki)\mathbf{u}$$

Now, because A is real, $A\mathbf{x} = \lambda\mathbf{x}$ implies that

$$A\overline{\mathbf{x}} = \overline{A\mathbf{x}} = \overline{A\mathbf{x}} = \overline{\lambda\mathbf{x}} = \overline{\lambda}\overline{\mathbf{x}}$$

so $\overline{\mathbf{x}} = \mathbf{u} - \mathbf{v}i$ is an eigenvector corresponding to the other eigenvalue $\overline{\lambda} = a + bi$. But

$$\overline{\mathbf{x}} = \overline{(1 + ki)\mathbf{u}} = (1 - \overline{k}i)\mathbf{u}$$

because **u** is a real vector. Hence, the eigenvectors **x** and $\bar{\mathbf{x}}$ of A are both nonzero multiples of **u** and therefore are multiples of one another. This is impossible because eigenvectors corresponding to distinct eigenvalues must be linearly independent by Theorem 4.20. (This theorem is valid over the complex numbers as well as the real numbers.)

This contradiction implies that **u** and **v** are linearly independent and hence P is invertible. It now follows that

$$A = P\begin{bmatrix} a & -b \\ b & a \end{bmatrix}P^{-1}$$

Theorem 4.43 serves to explain Example 4.51(b). The eigenvalues of $A = \begin{bmatrix} 0.2 & -1.2 \\ 0.6 & 1.4 \end{bmatrix}$ are $0.8 \pm 0.6i$. For $\lambda = 0.8 - 0.6i$, a corresponding eigenvector is

$$\mathbf{x} = \begin{bmatrix} -1-i \\ 1 \end{bmatrix} = \begin{bmatrix} -1 \\ 1 \end{bmatrix} + \begin{bmatrix} -1 \\ 0 \end{bmatrix}i$$

From Theorem 4.43, it follows that for $P = \begin{bmatrix} -1 & -1 \\ 1 & 0 \end{bmatrix}$ and $C = \begin{bmatrix} 0.8 & -0.6 \\ 0.6 & 0.8 \end{bmatrix}$, we have

$$A = PCP^{-1} \quad \text{and} \quad P^{-1}AP = C$$

For the given dynamical system $\mathbf{x}_{k+1} = A\mathbf{x}_k$, we perform a change of variable. Let

$$\mathbf{x}_k = P\mathbf{y}_k \quad \text{(or, equivalently, } \mathbf{y}_k = P^{-1}\mathbf{x}_k\text{)}$$

Then

$$P\mathbf{y}_{k+1} = \mathbf{x}_{k+1} = A\mathbf{x}_k = AP\mathbf{y}_k$$

so

$$\mathbf{y}_{k+1} = \mathbf{x}_{k+1} = A\mathbf{x}_k = P^{-1}AP\mathbf{y}_k = C\mathbf{y}_k$$

Now C has the same eigenvalues as A (why?) and $|0.8 \pm 0.6i| = 1$. Thus, the dynamical system $\mathbf{y}_{k+1} = C\mathbf{y}_k$ simply rotates the points in every trajectory in a circle about the origin by Theorem 4.42.

To determine a trajectory of the dynamical system in Example 4.51(b), we iteratively apply the linear transformation $T(\mathbf{x}) = A\mathbf{x} = PCP^{-1}\mathbf{x}$. The transformation can be thought of as the composition of a change of variable (**x** to **y**), followed by the rotation determined by C, followed by the reverse change of variable (**y** back to **x**). We will encounter this idea again in the application to graphing quadratic equations in Section 5.5 and, more generally, as "change of basis" in Section 6.3. In Exercise 74 of Section 5.5, you will show that the trajectory in Example 4.51(b) is indeed an ellipse, as it appears to be from Figure 4.26(b).

To summarize then: If a real 2×2 matrix A has complex eigenvalues $\lambda = a \pm bi$, then the trajectories of the dynamical system $\mathbf{x}_{k+1} = A\mathbf{x}_k$ spiral inward if $|\lambda| < 1$ (**0** is a ***spiral attractor***), spiral outward if $|\lambda| > 1$ (**0** is a ***spiral repeller***), and lie on a closed orbit if $|\lambda| = 1$ (**0** is an ***orbital center***).

Vignette

Ranking Sports Teams and Searching the Internet

In any competitive sports league, it is not necessarily a straightforward process to rank the players or teams. Counting wins and losses alone overlooks the possibility that one team may accumulate a large number of victories against weak teams, while another team may have fewer victories but all of them against strong teams. Which of these teams is better? How should we compare two teams that never play one another? Should points scored be taken into account? Points against?

Despite these complexities, the ranking of athletes and sports teams has become a commonplace and much-anticipated feature in the media. For example, there are various annual rankings of U.S. college football and basketball teams, and golfers and tennis players are also ranked internationally. There are many copyrighted schemes used to produce such rankings, but we can gain some insight into how to approach the problem by using the ideas from this chapter.

To establish the basic idea, let's revisit Example 3.69. Five tennis players play one another in a round-robin tournament. Wins and losses are recorded in the form of a digraph in which a directed edge from i to j indicates that player i defeats player j. The corresponding adjacency matrix A therefore has $a_{ij} = 1$ if player i defeats player j and has $a_{ij} = 0$ otherwise.

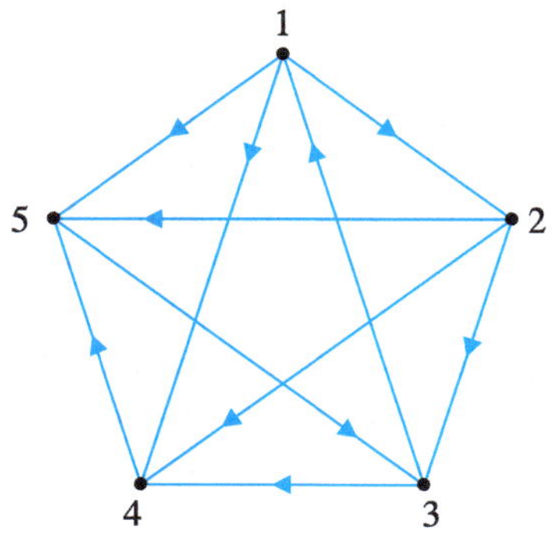

$$A = \begin{bmatrix} 0 & 1 & 0 & 1 & 1 \\ 0 & 0 & 1 & 1 & 1 \\ 1 & 0 & 0 & 1 & 0 \\ 0 & 0 & 0 & 0 & 1 \\ 0 & 0 & 1 & 0 & 0 \end{bmatrix}$$

We would like to associate a ranking r_i with player i in such a way that $r_i > r_j$ indicates that player i is ranked more highly than player j. For this purpose, let's require that the r_i's be probabilities (that is, $0 \le r_i \le 1$ for all i, and $r_1 + r_2 + r_3 + r_4 + r_5 = 1$) and then organize the rankings in a *ranking vector*

$$\mathbf{r} = \begin{bmatrix} r_1 \\ r_2 \\ r_3 \\ r_4 \\ r_5 \end{bmatrix}$$

Furthermore, let's insist that player i's ranking should be proportional to the sum of the rankings of the players defeated by player i. For example, player 1 defeated players 2, 4, and 5, so we want

$$r_1 = \alpha(r_2 + r_4 + r_5)$$

where α is the constant of proportionality. Writing out similar equations for the other players produces the following system:

$$\begin{aligned} r_1 &= \alpha(r_2 + r_4 + r_5) \\ r_2 &= \alpha(r_3 + r_4 + r_5) \\ r_3 &= \alpha(r_1 + r_4) \\ r_4 &= \alpha r_5 \\ r_5 &= \alpha r_3 \end{aligned}$$

Observe that we can write this system in matrix form as

$$\begin{bmatrix} r_1 \\ r_2 \\ r_3 \\ r_4 \\ r_5 \end{bmatrix} = \alpha \begin{bmatrix} 0 & 1 & 0 & 1 & 1 \\ 0 & 0 & 1 & 1 & 1 \\ 1 & 0 & 0 & 1 & 0 \\ 0 & 0 & 0 & 0 & 1 \\ 0 & 0 & 1 & 0 & 0 \end{bmatrix} \begin{bmatrix} r_1 \\ r_2 \\ r_3 \\ r_4 \\ r_5 \end{bmatrix} \quad \text{or} \quad \mathbf{r} = \alpha A\mathbf{r}$$

Equivalently, we see that the ranking vector $\mathbf{r}$ must satisfy $A\mathbf{r} = \frac{1}{\alpha}\mathbf{r}$. In other words, $\mathbf{r}$ is an eigenvector corresponding to the matrix A!

Furthermore, A is a primitive nonnegative matrix, so the Perron-Frobenius Theorem guarantees that there is a *unique* ranking vector $\mathbf{r}$. In this example, the ranking vector turns out to be

$$\mathbf{r} = \begin{bmatrix} 0.29 \\ 0.27 \\ 0.22 \\ 0.08 \\ 0.14 \end{bmatrix}$$

so we would rank the players in the order 1, 2, 3, 5, 4.

By modifying the matrix A, it is possible to take into account many of the complexities mentioned in the opening paragraph. However, this simple example has served to indicate one useful approach to the problem of ranking teams.

The same idea can be used to understand how an Internet search engine such as Google works. Older search engines used to return the results of a search *unordered*. Useful sites would often be buried among irrelevant ones. Much scrolling was often needed to uncover what you were looking for. By contrast, Google returns search results *ordered* according to their likely relevance. Thus, a method for ranking websites is needed.

Instead of teams playing one another, we now have websites linking to one another. We can once again use a digraph to model the situation, only now an edge from i to j indicates that website i links to (or refers to) website j. So whereas for the sports team digraph, incoming directed edges are bad (they indicate losses), for the Internet digraph, incoming directed edges are good (they indicate links from other sites). In this setting, we want the ranking of website i to be proportional to sum of the rankings of all the websites that link to i.

Using the digraph on page 356 to represent just five websites, we have

$$r_4 = \alpha(r_1 + r_2 + r_3)$$

for example. It is easy to see that we now want to use the *transpose* of the adjacency matrix of the digraph. Therefore, the ranking vector $\mathbf{r}$ must satisfy $A^T\mathbf{r} = \frac{1}{a}\mathbf{r}$ and will thus be the Perron eigenvector of A^T. In this example, we obtain

$$A^T = \begin{bmatrix} 0 & 0 & 1 & 0 & 0 \\ 1 & 0 & 0 & 0 & 0 \\ 0 & 1 & 0 & 0 & 1 \\ 1 & 1 & 1 & 0 & 0 \\ 1 & 1 & 0 & 1 & 0 \end{bmatrix} \quad \text{and} \quad \mathbf{r} = \begin{bmatrix} 0.14 \\ 0.08 \\ 0.22 \\ 0.27 \\ 0.29 \end{bmatrix}$$

so a search that turns up these five sites would list them in the order 5, 4, 3, 1, 2.

Google actually uses a variant of the method described here and computes the ranking vector via an iterative method very similar to the power method (Section 4.5).

Exercises 4.6

Markov Chains

Which of the stochastic matrices in Exercises 1–6 are regular?

1. $\begin{bmatrix} 0 & 1 \\ 1 & 0 \end{bmatrix}$

2. $\begin{bmatrix} 1 & \frac{1}{2} \\ 0 & \frac{1}{2} \end{bmatrix}$

3. $\begin{bmatrix} \frac{1}{3} & 1 \\ \frac{2}{3} & 0 \end{bmatrix}$

4. $\begin{bmatrix} \frac{1}{2} & 0 & 1 \\ \frac{1}{2} & 0 & 0 \\ 0 & 1 & 0 \end{bmatrix}$

5. $\begin{bmatrix} 0.1 & 0 & 0.5 \\ 0.5 & 1 & 0 \\ 0.4 & 0 & 0.5 \end{bmatrix}$

6. $\begin{bmatrix} 0.5 & 1 & 0 \\ 0.5 & 0 & 1 \\ 0 & 0 & 0 \end{bmatrix}$

In Exercises 7–9, P is the transition matrix of a regular Markov chain. Find the long range transition matrix L of P.

7. $P = \begin{bmatrix} \frac{1}{3} & \frac{1}{6} \\ \frac{2}{3} & \frac{5}{6} \end{bmatrix}$

8. $P = \begin{bmatrix} \frac{1}{2} & \frac{1}{3} & \frac{1}{6} \\ \frac{1}{2} & \frac{1}{2} & \frac{1}{3} \\ 0 & \frac{1}{6} & \frac{1}{2} \end{bmatrix}$

9. $P = \begin{bmatrix} 0.2 & 0.3 & 0.4 \\ 0.6 & 0.1 & 0.4 \\ 0.2 & 0.6 & 0.2 \end{bmatrix}$

10. Prove that the steady state probability vector of a regular Markov chain is unique. [*Hint:* Use Theorem 4.33 or Theorem 4.34.]

Population Growth

In Exercises 11–14, calculate the positive eigenvalue and a corresponding positive eigenvector of the Leslie matrix L.

11. $L = \begin{bmatrix} 0 & 2 \\ 0.5 & 0 \end{bmatrix}$

12. $L = \begin{bmatrix} 1 & 1.5 \\ 0.5 & 0 \end{bmatrix}$

13. $L = \begin{bmatrix} 0 & 7 & 4 \\ 0.5 & 0 & 0 \\ 0 & 0.5 & 0 \end{bmatrix}$

14. $L = \begin{bmatrix} 1 & 5 & 3 \\ \frac{1}{3} & 0 & 0 \\ 0 & \frac{2}{3} & 0 \end{bmatrix}$

15. If a Leslie matrix has a unique positive eigenvalue λ_1, what is the significance for the population if $\lambda_1 > 1$? $\lambda_1 < 1$? $\lambda_1 = 1$?

16. Verify that the characteristic polynomial of the Leslie matrix L in Equation (3) is

$$c_L(\lambda) = (-1)^n(\lambda^n - b_1\lambda^{n-1} - b_2s_1\lambda^{n-2} - b_3s_1s_2\lambda^{n-3} - \cdots - b_ns_1s_2\cdots s_{n-1})$$

[*Hint:* Use mathematical induction and expand $\det(L - \lambda I)$ along the last column.]

17. If all of the survival rates s_i are nonzero, let

$$P = \begin{bmatrix} 1 & 0 & 0 & \cdots & 0 \\ 0 & s_1 & 0 & \cdots & 0 \\ 0 & 0 & s_1s_2 & \cdots & 0 \\ \vdots & \vdots & \vdots & \ddots & \vdots \\ 0 & 0 & 0 & \cdots & s_1s_2\cdots s_{n-1} \end{bmatrix}$$

Compute $P^{-1}LP$ and use it to find the characteristic polynomial of L. [*Hint:* Refer to Exercise 32 in Section 4.3.]

18. Verify that an eigenvector of L corresponding to λ_1 is

$$\mathbf{x}_1 = \begin{bmatrix} 1 \\ s_1/\lambda_1 \\ s_1s_2/\lambda_1^2 \\ s_1s_2s_3/\lambda_1^3 \\ \vdots \\ s_1s_2s_3\cdots s_{n-1}/\lambda_1^{n-1} \end{bmatrix}$$

[*Hint:* Combine Exercise 17 above with Exercise 32 in Section 4.3 and Exercise 46 in Section 4.4.]

CAS *In Exercises 19–21, compute the steady state growth rate of the population with the Leslie matrix L from the given exercise. Then use Exercise 18 to help find the corresponding distribution of the age classes.*

19. Exercise 39 in Section 3.7

20. Exercise 40 in Section 3.7

21. Exercise 44 in Section 3.7

CAS **22.** Many species of seal have suffered from commercial hunting. They have been killed for their skin, blubber, and meat. The fur trade, in particular, reduced some seal populations to the point of extinction. Today, the greatest threats to seal populations are decline of fish stocks due to overfishing, pollution, disturbance of habitat, entanglement in marine debris, and culling by fishery owners. Some seals have been declared endangered species; other species are carefully managed. Table 4.7 gives the birth and survival rates for the northern fur seal, divided into 2-year age classes. [The data are based on A. E. York and J. R. Hartley, "Pup Production Following Harvest of Female Northern Fur Seals," *Canadian Journal of Fisheries and Aquatic Science*, 38 (1981), pp. 84–90.]

Table 4.7

Age (years)	Birth Rate	Survival Rate
0–2	0.00	0.91
2–4	0.02	0.88
4–6	0.70	0.85
6–8	1.53	0.80
8–10	1.67	0.74
10–12	1.65	0.67
12–14	1.56	0.59
14–16	1.45	0.49
16–18	1.22	0.38
18–20	0.91	0.27
20–22	0.70	0.17
22–24	0.22	0.15
24–26	0.00	0.00

(a) Construct the Leslie matrix L for these data and compute the positive eigenvalue and a corresponding positive eigenvector.

(b) In the long run, what percentage of seals will be in each age class and what will the growth rate be?

Exercise 23 shows that the long-run behavior of a population can be determined directly from the entries of its Leslie matrix.

23. The *net reproduction rate* of a population is defined as

$$r = b_1 + b_2s_1 + b_3s_1s_2 + \cdots + b_ns_1s_2\cdots s_{n-1}$$

where the b_i are the birth rates and the s_j are the survival rates for the population.

(a) Explain why r can be interpreted as the average number of daughters born to a single female over her lifetime.

(b) Show that $r = 1$ if and only if $\lambda_1 = 1$. (This represents *zero population growth.*) [*Hint:* Let

$$g(\lambda) = \frac{b_1}{\lambda} + \frac{b_2s_1}{\lambda^2} + \frac{b_3s_1s_2}{\lambda^3} + \cdots + \frac{b_ns_1s_2\cdots s_{n-1}}{\lambda^n}$$

Show that λ is an eigenvalue of L if and only if $g(\lambda) = 1$.]

(c) Assuming that there is a unique positive eigenvalue λ_1, show that $r < 1$ if and only if the population is decreasing and $r > 1$ if and only if the population is increasing.

A sustainable harvesting policy is a procedure that allows a certain fraction of a population (represented by a population distribution vector **x***) to be harvested so that the population returns to* **x** *after one time interval (where a time interval is the length of one age class). If h is the fraction of each age class that is harvested, then we can express the harvesting procedure mathematically as follows: If we start with a population vector* **x***, after one time interval we have L***x***; harvesting removes hL***x***, leaving*

$$L\mathbf{x} - hL\mathbf{x} = (1 - h)L\mathbf{x}$$

Sustainability requires that

$$(1 - h)L\mathbf{x} = \mathbf{x}$$

24. If λ_1 is the unique positive eigenvalue of a Leslie matrix L and h is the sustainable harvest ratio, prove that $h = 1 - 1/\lambda_1$.

CAS **25. (a)** Find the sustainable harvest ratio for the woodland caribou in Exercise 44 in Section 3.7.

(b) Using the data in Exercise 44 in Section 3.7, reduce the caribou herd according to your answer to part (a). Verify that the population returns to its original level after one time interval.

26. Find the sustainable harvest ratio for the seal in Exercise 22. (Conservationists have had to harvest seal populations when overfishing has reduced the available food supply to the point where the seals are in danger of starvation.)

a + bi **27.** Let L be a Leslie matrix with a unique positive eigenvalue λ_1. Show that if λ is any other (real or complex) eigenvalue of L, then $|\lambda| \le \lambda_1$. [*Hint:* Write $\lambda = r(\cos\theta + i\sin\theta)$ and substitute it into the equation $g(\lambda) = 1$, as in part (b) of Exercise 23. Use De Moivre's Theorem and then take the real part of both sides. The Triangle Inequality should prove useful.]

The Perron-Frobenius Theorem

In Exercises 28–31, find the Perron root and the corresponding Perron eigenvector of A.

28. $A = \begin{bmatrix} 2 & 0 \\ 1 & 1 \end{bmatrix}$ **29.** $A = \begin{bmatrix} 1 & 3 \\ 2 & 0 \end{bmatrix}$

30. $A = \begin{bmatrix} 0 & 1 & 1 \\ 1 & 0 & 1 \\ 1 & 1 & 0 \end{bmatrix}$ **31.** $A = \begin{bmatrix} 2 & 1 & 1 \\ 1 & 1 & 0 \\ 1 & 0 & 1 \end{bmatrix}$

It can be shown that a nonnegative $n \times n$ matrix is irreducible if and only if $(I + A)^{n-1} > O$. In Exercises 32–35, use this criterion to determine whether the matrix A is irreducible. If A is reducible, find a permutation of its rows and columns that puts A into the block form

$$\left[\begin{array}{c|c} B & C \\ \hline O & D \end{array}\right]$$

32. $A = \begin{bmatrix} 0 & 0 & 1 & 0 \\ 0 & 0 & 0 & 1 \\ 0 & 1 & 0 & 0 \\ 1 & 0 & 0 & 0 \end{bmatrix}$ **33.** $A = \begin{bmatrix} 0 & 0 & 1 & 0 \\ 0 & 0 & 1 & 1 \\ 1 & 0 & 0 & 0 \\ 1 & 1 & 0 & 0 \end{bmatrix}$

34. $A = \begin{bmatrix} 0 & 1 & 0 & 0 & 0 \\ 0 & 0 & 1 & 0 & 1 \\ 1 & 0 & 1 & 0 & 1 \\ 0 & 0 & 1 & 1 & 0 \\ 1 & 0 & 0 & 0 & 0 \end{bmatrix}$ **35.** $A = \begin{bmatrix} 0 & 1 & 0 & 0 & 0 \\ 0 & 0 & 0 & 0 & 1 \\ 1 & 0 & 0 & 0 & 1 \\ 0 & 0 & 1 & 0 & 0 \\ 0 & 0 & 0 & 1 & 1 \end{bmatrix}$

36. **(a)** If A is the adjacency matrix of a graph G, show that A is irreducible if and only if G is connected. (A graph is ***connected*** if there is a path between every pair of vertices.)

(b) Which of the graphs in Section 4.0 have an irreducible adjacency matrix? Which have a primitive adjacency matrix?

37. Let G be a bipartite graph with adjacency matrix A.

(a) Show that A is not primitive.

(b) Show that if λ is an eigenvalue of A, so is $-\lambda$. [*Hint:* Use Exercise 80 in Section 3.7 and partition an eigenvector for λ so that it is compatible with this partitioning of A. Use this partitioning to find an eigenvector for $-\lambda$.]

38. A graph is called ***k-regular*** if k edges meet at each vertex. Let G be a k-regular graph.

(a) Show that the adjacency matrix A of G has $\lambda = k$ as an eigenvalue. [*Hint:* Adapt Theorem 4.30.]

(b) Show that if A is primitive, then the other eigenvalues are all less than k in absolute value. [*Hint:* Adapt Theorem 4.31.]

39. Explain the results of your exploration in Section 4.0 in light of Exercises 36–38 and Section 4.5.

In Exercise 40, the absolute value of a matrix $A = [a_{ij}]$ is defined to be the matrix $|A| = [|a_{ij}|]$.

40. Let A and B be $n \times n$ matrices, $\mathbf{x}$ a vector in $\mathbb{R}^n$, and c a scalar. Prove the following matrix inequalities:

(a) $|cA| = |c|\,|A|$ **(b)** $|A + B| \leq |A| + |B|$
(c) $|A\mathbf{x}| \leq |A|\,|\mathbf{x}|$ **(d)** $|AB| \leq |A|\,|B|$

41. Prove that a 2×2 matrix $A = \begin{bmatrix} a_{11} & a_{12} \\ a_{21} & a_{22} \end{bmatrix}$ is reducible if and only if $a_{12} = 0$ or $a_{21} = 0$.

42. Let A be a nonnegative, irreducible matrix such that $I - A$ is invertible and $(I - A)^{-1} \geq O$. Let λ_1 and $\mathbf{v}_1$ be the Perron root and Perron eigenvector of A.

(a) Prove that $0 < \lambda_1 < 1$. [*Hint:* Apply Exercise 22 in Section 4.3 and Theorem 4.18(b).]

(b) Deduce from (a) that $\mathbf{v}_1 > A\mathbf{v}_1$.

Linear Recurrence Relations

In Exercises 43–46, write out the first six terms of the sequence defined by the recurrence relation with the given initial conditions.

43. $x_0 = 1, x_n = 2x_{n-1}$ for $n \geq 1$

44. $a_1 = 128, a_n = a_{n-1}/2$ for $n \geq 2$

45. $y_0 = 0, y_1 = 1, y_n = y_{n-1} - y_{n-2}$ for $n \geq 2$

46. $b_0 = 1, b_1 = 1, b_n = 2b_{n-1} + b_{n-2}$ for $n \geq 2$

In Exercises 47–52, solve the recurrence relation with the given initial conditions.

47. $x_0 = 0, x_1 = 5, x_n = 3x_{n-1} + 4x_{n-2}$ for $n \geq 2$

48. $x_0 = 0, x_1 = 1, x_n = 4x_{n-1} - 3x_{n-2}$ for $n \geq 2$

49. $y_1 = 1, y_2 = 6, y_n = 4y_{n-1} - 4y_{n-2}$ for $n \geq 3$

50. $a_0 = 4, a_1 = 1, a_n = a_{n-1} - a_{n-2}/4$ for $n \geq 2$

51. $b_0 = 0, b_1 = 1, b_n = 2b_{n-1} + 2b_{n-2}$ for $n \geq 2$

52. The recurrence relation in Exercise 45. Show that your solution agrees with the answer to Exercise 45.

53. Complete the proof of Theorem 4.38(a) by showing that if the recurrence relation $x_n = ax_{n-1} + bx_{n-2}$ has distinct eigenvalues $\lambda_1 \neq \lambda_2$, then the solution will be of the form

$$x_n = c_1\lambda_1^n + c_2\lambda_2^n$$

[*Hint:* Show that the method of Example 4.40 works in general.]

54. **(a)** Show that for any choice of initial conditions $x_0 = r$ and $x_1 = s$, the scalars c_1 and c_2 can be found, as stated in Theorem 4.38(a) and (b).

(b) If the eigenvalues λ_1 and λ_2 are distinct and the initial conditions are $\mathbf{x}_0 = 0$, $\mathbf{x}_1 = 1$, show that

$$\mathbf{x}_n = \left(\frac{1}{\lambda_1 - \lambda_2}\right)(\lambda_1^n - \lambda_2^n).$$

55. The Fibonacci recurrence $f_n = f_{n-1} + f_{n-2}$ has the associated matrix equation $\mathbf{x}_n = A\mathbf{x}_{n-1}$, where

$$\mathbf{x}_n = \begin{bmatrix} f_n \\ f_{n-1} \end{bmatrix} \text{ and } A = \begin{bmatrix} 1 & 1 \\ 1 & 0 \end{bmatrix}$$

(a) With $f_0 = 0$ and $f_1 = 1$, use mathematical induction to prove that

$$A^n = \begin{bmatrix} f_{n+1} & f_n \\ f_n & f_{n-1} \end{bmatrix}$$

for all $n \geq 1$.

(b) Using part (a), prove that

$$f_{n+1}f_{n-1} - f_n^2 = (-1)^n$$

for all $n \geq 1$. [This is called ***Cassini's Identity***, after the astronomer Giovanni Domenico Cassini (1625–1712). Cassini was born in Italy but, on the invitation of Louis XIV, moved in 1669 to France, where he became director of the Paris Observatory. He became a French citizen and adopted the French version of his name: Jean-Dominique Cassini. Mathematics was one of his many interests other than astronomy. Cassini's Identity was published in 1680 in a paper submitted to the Royal Academy of Sciences in Paris.]

(c) An 8×8 checkerboard can be dissected as shown in Figure 4.29(a) and the pieces reassembled to form the 5×13 rectangle in Figure 4.29(b).

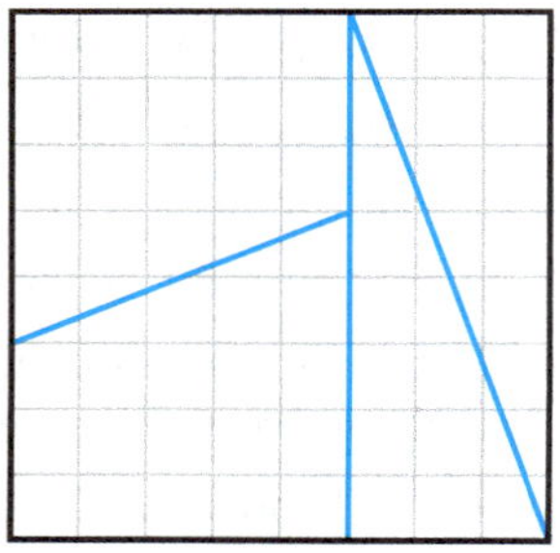
(a)

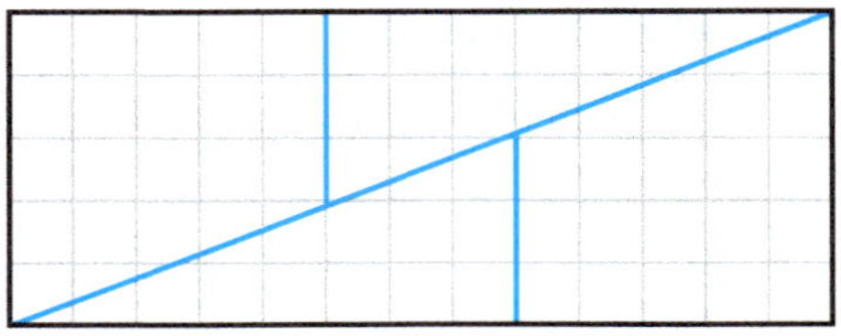
(b)

Figure 4.29

The area of the square is 64 square units, but the rectangle's area is 65 square units! Where did the extra square come from? [*Hint:* What does this have to do with the Fibonacci sequence?]

56. You have a supply of three kinds of tiles: two kinds of 1×2 tiles and one kind of 1×1 tile, as shown in Figure 4.30.

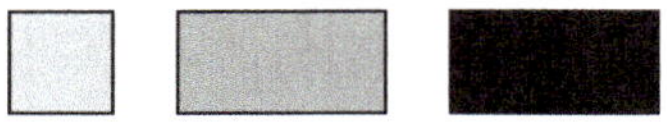

Figure 4.30

Let t_n be the number of different ways to cover a $1 \times n$ rectangle with these tiles. For example, Figure 4.31 shows that $t_3 = 5$.

(a) Find $t_1, \ldots, t_5$.
(Does t_0 make any sense? If so, what is it?)

(b) Set up a second order recurrence relation for t_n.

(c) Using t_1 and t_2 as the initial conditions, solve the recurrence relation in part (b). Check your answer against the data in part (a).

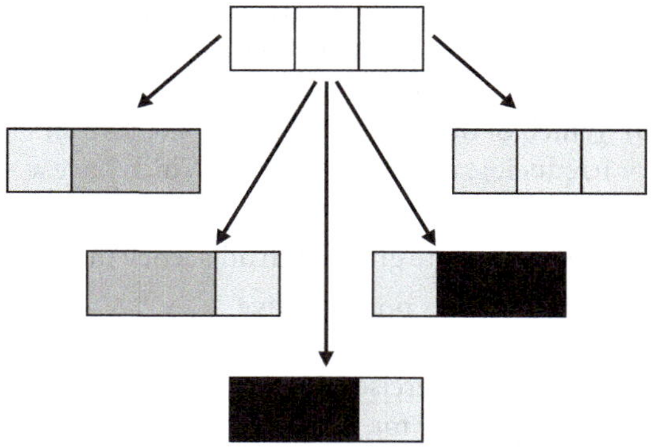

Figure 4.31
The five ways to tile a 1×3 rectangle

57. You have a supply of 1×2 dominoes with which to cover a $2 \times n$ rectangle. Let d_n be the number of different ways to cover the rectangle. For example, Figure 4.32 shows that $d_3 = 3$.

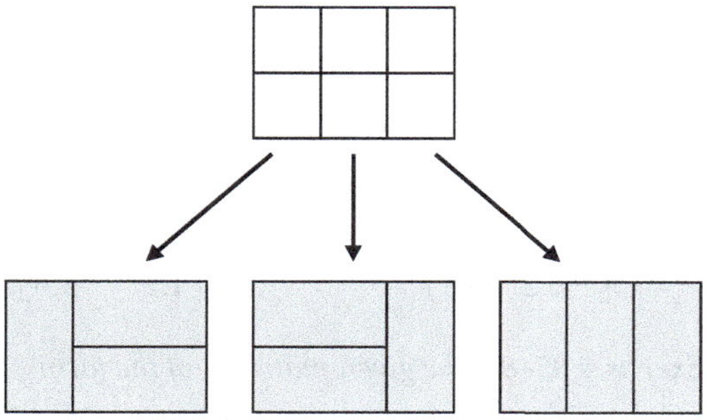

Figure 4.32

The three ways to cover a 2×3 rectangle with 1×2 dominoes

(a) Find $d_1, \ldots, d_5$.
(Does d_0 make any sense? If so, what is it?)

(b) Set up a second order recurrence relation for d_n.

(c) Using d_1 and d_2 as the initial conditions, solve the recurrence relation in part (b). Check your answer against the data in part (a).

58. In Example 4.41, find eigenvectors $\mathbf{v}_1$ and $\mathbf{v}_2$ corresponding to $\lambda_1 = \dfrac{1+\sqrt{5}}{2}$ and $\lambda_2 = \dfrac{1-\sqrt{5}}{2}$. With $\mathbf{x}_k = \begin{bmatrix} f_k \\ f_{k-1} \end{bmatrix}$, verify formula (2) in Section 4.5. That is, show that, for some scalar c_1,

$$\lim_{k\to\infty} \frac{\mathbf{x}_k}{\lambda_1^k} = c_1\mathbf{v}_1$$

Systems of Linear Differential Equations

In Exercises 59–64, find the general solution to the given system of differential equations. Then find the specific solution that satisfies the initial conditions. (Consider all functions to be functions of t.)

59. $\begin{aligned} x' &= x + 3y, & x(0) &= 0 \\ y' &= 2x + 2y, & y(0) &= 5 \end{aligned}$

60. $\begin{aligned} x' &= 2x - y, & x(0) &= 1 \\ y' &= -x + 2y, & y(0) &= 1 \end{aligned}$

61. $\begin{aligned} x_1' &= x_1 + x_2, & x_1(0) &= 1 \\ x_2' &= x_1 - x_2, & x_2(0) &= 0 \end{aligned}$

62. $\begin{aligned} y_1' &= y_1 - y_2, & y_1(0) &= 1 \\ y_2' &= y_1 + y_2, & y_2(0) &= 1 \end{aligned}$

63. $\begin{aligned} x' &= y - z, & x(0) &= 1 \\ y' &= x + z, & y(0) &= 0 \\ z' &= x + y, & z(0) &= -1 \end{aligned}$

64. $\begin{aligned} x' &= x + 3z, & x(0) &= 2 \\ y' &= x - 2y + z, & y(0) &= 3 \\ z' &= 3x + z, & z(0) &= 4 \end{aligned}$

65. A scientist places two strains of bacteria, X and Y, in a petri dish. Initially, there are 400 of X and 500 of Y. The two bacteria compete for food and space but do not feed on each other. If $x = x(t)$ and $y = y(t)$ are the numbers of the strains at time t days, the growth rates of the two populations are given by the system

$$\begin{aligned} x' &= 1.2x - 0.2y \\ y' &= -0.2x + 1.5y \end{aligned}$$

(a) Determine what happens to these two populations by solving the system of differential equations.

(b) Explore the effect of changing the initial populations by letting $x(0) = a$ and $y(0) = b$. Describe what happens to the populations in terms of a and b.

66. Two species, X and Y, live in a *symbiotic* relationship. That is, neither species can survive on its own and each depends on the other for its survival. Initially, there are 15 of X and 10 of Y. If $x = x(t)$ and $y = y(t)$ are the sizes of the populations at time t months, the growth rates of the two populations are given by the system

$$\begin{aligned} x' &= -0.8x + 0.4y \\ y' &= 0.4x - 0.2y \end{aligned}$$

Determine what happens to these two populations.

In Exercises 67 and 68, species X preys on species Y. The sizes of the populations are represented by $x = x(t)$ and $y = y(t)$. The growth rate of each population is governed by the system of differential equations $\mathbf{x}' = A\mathbf{x} + \mathbf{b}$, where $\mathbf{x} = \begin{bmatrix} x \\ y \end{bmatrix}$ and $\mathbf{b}$ is a constant vector. Determine what happens to the two populations for the given A and $\mathbf{b}$ and initial conditions $\mathbf{x}(0)$. (First show that there are constants a and b such that the substitutions $x = u + a$ and $y = v + b$ convert the system into an equivalent one with no constant terms.)

67. $A = \begin{bmatrix} 1 & 1 \\ -1 & 1 \end{bmatrix}, \mathbf{b} = \begin{bmatrix} -30 \\ -10 \end{bmatrix}, \mathbf{x}(0) = \begin{bmatrix} 20 \\ 30 \end{bmatrix}$

68. $A = \begin{bmatrix} -1 & 1 \\ -1 & -1 \end{bmatrix}, \mathbf{b} = \begin{bmatrix} 0 \\ 40 \end{bmatrix}, \mathbf{x}(0) = \begin{bmatrix} 10 \\ 30 \end{bmatrix}$

69. Let $x = x(t)$ be a twice-differentiable function and consider the *second order differential equation*

$$x'' + ax' + bx = 0 \qquad (11)$$

(a) Show that the change of variables $y = x'$ and $z = x$ allows Equation (11) to be written as a system of two linear differential equations in y and z.

(b) Show that the characteristic equation of the system in part (a) is $\lambda^2 + a\lambda + b = 0$.

70. Show that there is a change of variables that converts the nth *order differential equation*

$$x^{(n)} + a_{n-1}x^{(n-1)} + \cdots + a_1x' + a_0 = 0$$

into a system of n linear differential equations whose coefficient matrix is the companion matrix $C(p)$ of the polynomial $p(\lambda) = \lambda^n + a_{n-1}\lambda^{n-1} + \cdots + a_1\lambda + a_0$. [The notation $x^{(k)}$ denotes the kth derivative of x. See Exercises 26–32 in Section 4.3 for the definition of a companion matrix.]

In Exercises 71 and 72, use Exercise 69 to find the general solution of the given equation.

71. $x'' - 5x' + 6x = 0$

72. $x'' + 4x' + 3x = 0$

In Exercises 73–76, solve the system of differential equations in the given exercise using Theorem 4.41.

73. Exercise 59

74. Exercise 60

75. Exercise 63

76. Exercise 64

Discrete Linear Dynamical Systems

In Exercises 77–84, consider the dynamical system $\mathbf{x}_{k+1} = A\mathbf{x}_k$.

(a) Compute and plot $\mathbf{x}_0, \mathbf{x}_1, \mathbf{x}_2, \mathbf{x}_3$ for $\mathbf{x}_0 = \begin{bmatrix} 1 \\ 1 \end{bmatrix}$.

(b) Compute and plot $\mathbf{x}_0, \mathbf{x}_1, \mathbf{x}_2, \mathbf{x}_3$ for $\mathbf{x}_0 = \begin{bmatrix} 1 \\ 0 \end{bmatrix}$.

(c) Using eigenvalues and eigenvectors, classify the origin as an attractor, repeller, saddle point, or none of these.

(d) Sketch several typical trajectories of the system.

77. $A = \begin{bmatrix} 2 & 1 \\ 0 & 3 \end{bmatrix}$

78. $A = \begin{bmatrix} 0.5 & -0.5 \\ 0 & 0.5 \end{bmatrix}$

79. $A = \begin{bmatrix} 2 & -1 \\ -1 & 2 \end{bmatrix}$

80. $A = \begin{bmatrix} -4 & 2 \\ 1 & -3 \end{bmatrix}$

81. $A = \begin{bmatrix} 1.5 & -1 \\ -1 & 0 \end{bmatrix}$

82. $A = \begin{bmatrix} 0.1 & 0.9 \\ 0.5 & 0.5 \end{bmatrix}$

83. $A = \begin{bmatrix} 0.2 & 0.4 \\ -0.2 & 0.8 \end{bmatrix}$

84. $A = \begin{bmatrix} 0 & -1.5 \\ 1.2 & 3.6 \end{bmatrix}$

In Exercises 85–88, the given matrix is of the form $A = \begin{bmatrix} a & -b \\ b & a \end{bmatrix}$. In each case, A can be factored as the product of a scaling matrix and a rotation matrix. Find the scaling factor r and the angle θ of rotation. Sketch the first four points of the trajectory for the dynamical system $\mathbf{x}_{k+1} = A\mathbf{x}_k$ with $\mathbf{x}_0 = \begin{bmatrix} 1 \\ 1 \end{bmatrix}$ and classify the origin as a spiral attractor, spiral repeller, or orbital center.

85. $A = \begin{bmatrix} 1 & -1 \\ 1 & 1 \end{bmatrix}$

86. $A = \begin{bmatrix} 0 & 0.5 \\ -0.5 & 0 \end{bmatrix}$

87. $A = \begin{bmatrix} 1 & \sqrt{3} \\ -\sqrt{3} & 1 \end{bmatrix}$

88. $A = \begin{bmatrix} -\sqrt{3}/2 & -1/2 \\ 1/2 & -\sqrt{3}/2 \end{bmatrix}$

In Exercises 89–92, find an invertible matrix P and a matrix C of the form $C = \begin{bmatrix} a & -b \\ b & a \end{bmatrix}$ such that $A = PCP^{-1}$. Sketch the first six points of the trajectory for the dynamical system $\mathbf{x}_{k+1} = A\mathbf{x}_k$ with $\mathbf{x}_0 = \begin{bmatrix} 1 \\ 1 \end{bmatrix}$ and classify the origin as a spiral attractor, spiral repeller, or orbital center.

89. $A = \begin{bmatrix} 0.1 & -0.2 \\ 0.1 & 0.3 \end{bmatrix}$

90. $A = \begin{bmatrix} 2 & 1 \\ -2 & 0 \end{bmatrix}$

91. $A = \begin{bmatrix} 1 & -1 \\ 1 & 0 \end{bmatrix}$

92. $A = \begin{bmatrix} 0 & -1 \\ 1 & \sqrt{3} \end{bmatrix}$

Chapter Review

Key Definitions and Concepts

adjoint of a matrix, 276
algebraic multiplicity of an eigenvalue, 294
characteristic equation, 292
characteristic polynomial, 292
cofactor expansion, 266
Cramer's Rule, 274–275
determinant, 263–265
diagonalizable matrix, 303
eigenvalue, 254
eigenvector, 254
eigenspace, 256
Fundamental Theorem of Invertible Matrices, 296
geometric multiplicity of an eigenvalue, 294
Gerschgorin disk, 319
Gerschgorin's Disk Theorem, 321
Laplace Expansion Theorem, 266
power method (and its variants), 311–319
properties of determinants, 269–274
similar matrices, 301

Review Questions

1. Mark each of the following statements true or false:
   (a) For all square matrices A, $\det(-A) = -\det A$.
   (b) If A and B are $n \times n$ matrices, then $\det(AB) = \det(BA)$.
   (c) If A and B are $n \times n$ matrices whose columns are the same but in different orders, then $\det B = -\det A$.
   (d) If A is invertible, then $\det(A^{-1}) = \det A^T$.
   (e) If 0 is the only eigenvalue of a square matrix A, then A is the zero matrix.
   (f) Two eigenvectors corresponding to the same eigenvalue must be linearly dependent.
   (g) If an $n \times n$ matrix has n distinct eigenvalues, then it must be diagonalizable.
   (h) If an $n \times n$ matrix is diagonalizable, then it must have n distinct eigenvalues.
   (i) Similar matrices have the same eigenvectors.
   (j) If A and B are two $n \times n$ matrices with the same reduced row echelon form, then A is similar to B.

2. Let $A = \begin{bmatrix} 1 & 3 & 5 \\ 3 & 5 & 7 \\ 7 & 9 & 11 \end{bmatrix}$.
   (a) Compute $\det A$ by cofactor expansion along any row or column.
   (b) Compute $\det A$ by first reducing A to triangular form.

3. If $\begin{vmatrix} a & b & c \\ d & e & f \\ g & h & i \end{vmatrix} = 3$, find $\begin{vmatrix} 3d & 2e - 4f & f \\ 3a & 2b - 4c & c \\ 3g & 2h - 4i & i \end{vmatrix}$.

4. Let A and B be 4×4 matrices with $\det A = 2$ and $\det B = -\frac{1}{4}$. Find $\det C$ for the indicated matrix C:
   (a) $C = (AB)^{-1}$ (b) $C = A^2B(3A^T)$

5. If A is a skew-symmetric $n \times n$ matrix and n is odd, prove that $\det A = 0$.

6. Find all values of k for which $\begin{vmatrix} 1 & -1 & 2 \\ 1 & 1 & k \\ 2 & 4 & k^2 \end{vmatrix} = 0$.

In Questions 7 and 8, show that $\mathbf{x}$ *is an eigenvector of A and find the corresponding eigenvalue.*

7. $\mathbf{x} = \begin{bmatrix} 1 \\ 2 \end{bmatrix}$, $A = \begin{bmatrix} 3 & 1 \\ 4 & 3 \end{bmatrix}$

8. $\mathbf{x} = \begin{bmatrix} 3 \\ -1 \\ 2 \end{bmatrix}$, $A = \begin{bmatrix} 13 & -60 & -45 \\ -5 & 18 & 15 \\ 10 & -40 & -32 \end{bmatrix}$

9. Let $A = \begin{bmatrix} -5 & -6 & 3 \\ 3 & 4 & -3 \\ 0 & 0 & -2 \end{bmatrix}$.
   (a) Find the characteristic polynomial of A.
   (b) Find all of the eigenvalues of A.
   (c) Find a basis for each of the eigenspaces of A.
   (d) Determine whether A is diagonalizable. If A is not diagonalizable, explain why not. If A is diagonalizable, find an invertible matrix P and a diagonal matrix D such that $P^{-1}AP = D$.

10. If A is a 3×3 diagonalizable matrix with eigenvalues -2, 3, and 4, find $\det A$.

11. If A is a 2×2 matrix with eigenvalues $\lambda_1 = \frac{1}{2}$, $\lambda_2 = -1$, and corresponding eigenvectors $\mathbf{v}_1 = \begin{bmatrix} 1 \\ 1 \end{bmatrix}$, $\mathbf{v}_2 = \begin{bmatrix} 1 \\ -1 \end{bmatrix}$, find $A^{-5}\begin{bmatrix} 3 \\ 7 \end{bmatrix}$.

12. If A is a diagonalizable matrix and all of its eigenvalues satisfy $|\lambda| < 1$, prove that A^n approaches the zero matrix as n gets large.

In Questions 13–15, determine, with reasons, whether A is similar to B. If $A \sim B$*, give an invertible matrix P such that* $P^{-1}AP = B$.

13. $A = \begin{bmatrix} 4 & 2 \\ 3 & 1 \end{bmatrix}$, $B = \begin{bmatrix} 2 & 2 \\ 3 & 2 \end{bmatrix}$

14. $A = \begin{bmatrix} 2 & 0 \\ 0 & 3 \end{bmatrix}$, $B = \begin{bmatrix} 3 & 0 \\ 0 & 2 \end{bmatrix}$

15. $A = \begin{bmatrix} 1 & 1 & 0 \\ 0 & 1 & 1 \\ 0 & 0 & 1 \end{bmatrix}$, $B = \begin{bmatrix} 1 & 1 & 0 \\ 0 & 1 & 0 \\ 0 & 0 & 1 \end{bmatrix}$

16. Let $A = \begin{bmatrix} 2 & k \\ 1 & 0 \end{bmatrix}$. Find all values of k for which:
    (a) A has eigenvalues 3 and -1.
    (b) A has an eigenvalue with algebraic multiplicity 2.
    (c) A has no real eigenvalues.

17. If $A^3 = A$, what are the possible eigenvalues of A?

18. If a square matrix A has two equal rows, why must A have 0 as one of its eigenvalues?

19. If $\mathbf{x}$ is an eigenvector of A with eigenvalue $\lambda = 3$, show that $\mathbf{x}$ is also an eigenvector of $A^2 - 5A + 2I$. What is the corresponding eigenvalue?

20. If A is similar to B with $P^{-1}AP = B$ and $\mathbf{x}$ is an eigenvector of A, show that $P^{-1}\mathbf{x}$ is an eigenvector of B.

5 Orthogonality

... that sprightly Scot of Scots, Douglas, that runs a-horseback up a hill perpendicular—

—William Shakespeare
Henry IV, Part I
Act II, Scene IV

5.0 Introduction: Shadows on a Wall

In this chapter, we will extend the notion of orthogonal projection that we encountered first in Chapter 1 and then again in Chapter 3. Until now, we have discussed only projection onto a single vector (or, equivalently, the one-dimensional subspace spanned by that vector). In this section, we will see if we can find the analogous formulas for projection onto a plane in $\mathbb{R}^3$. Figure 5.1 shows what happens, for example, when parallel light rays create a shadow on a wall. A similar process occurs when a three-dimensional object is displayed on a two-dimensional screen, such as a computer monitor. Later in this chapter, we will consider these ideas in full generality.

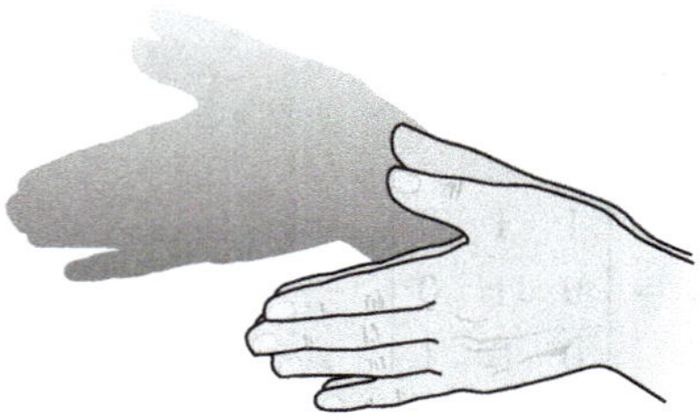

Figure 5.1
Shadows on a wall are projections

To begin, let's take another look at what we already know about projections. In Section 3.6, we showed that, in $\mathbb{R}^2$, the standard matrix of a projection onto the line through the origin with direction vector $\mathbf{d} = \begin{bmatrix} d_1 \\ d_2 \end{bmatrix}$ is

$$P = \frac{1}{d_1^2 + d_2^2}\begin{bmatrix} d_1^2 & d_1d_2 \\ d_1d_2 & d_2^2 \end{bmatrix} = \begin{bmatrix} d_1^2/(d_1^2 + d_2^2) & d_1d_2/(d_1^2 + d_2^2) \\ d_1d_2/(d_1^2 + d_2^2) & d_2^2/(d_1^2 + d_2^2) \end{bmatrix}$$

Hence, the projection of the vector $\mathbf{v}$ onto this line is just $P\mathbf{v}$.

Problem 1 Show that P can be written in the equivalent form

$$P = \begin{bmatrix} \cos^2\theta & \cos\theta\,\sin\theta \\ \cos\theta\,\sin\theta & \sin^2\theta \end{bmatrix}$$

(What does θ represent here?)

Problem 2 Show that P can also be written in the form $P = \mathbf{u}\mathbf{u}^T$, where $\mathbf{u}$ is a *unit* vector in the direction of $\mathbf{d}$.

Problem 3 Using Problem 2, find P and then find the projection of $\mathbf{v} = \begin{bmatrix} 3 \\ -4 \end{bmatrix}$ onto the lines with the following unit direction vectors:

$$\text{(a) } \mathbf{u} = \begin{bmatrix} 1/\sqrt{2} \\ -1/\sqrt{2} \end{bmatrix} \quad \text{(b) } \mathbf{u} = \begin{bmatrix} \frac{4}{5} \\ \frac{3}{5} \end{bmatrix} \quad \text{(c) } \mathbf{u} = \begin{bmatrix} -\frac{3}{5} \\ \frac{4}{5} \end{bmatrix}$$

Problem 4 Using the form $P = \mathbf{u}\mathbf{u}^T$, show that (a) $P^T = P$ (i.e., P is symmetric) and (b) $P^2 = P$ (i.e., P is idempotent).

Problem 5 Explain why, if P is a 2×2 projection matrix, the line onto which it projects vectors is the column space of P.

Now we will move into $\mathbb{R}^3$ and consider projections onto planes through the origin. We will explore several approaches.

Figure 5.2 shows one way to proceed. If $\mathcal{P}$ is a plane through the origin in $\mathbb{R}^3$ with normal vector $\mathbf{n}$ and if $\mathbf{v}$ is a vector in $\mathbb{R}^3$, then $\mathbf{p} = \text{proj}_{\mathcal{P}}(\mathbf{v})$ is a vector in $\mathcal{P}$ such that $\mathbf{v} - c\mathbf{n} = \mathbf{p}$ for some scalar c.

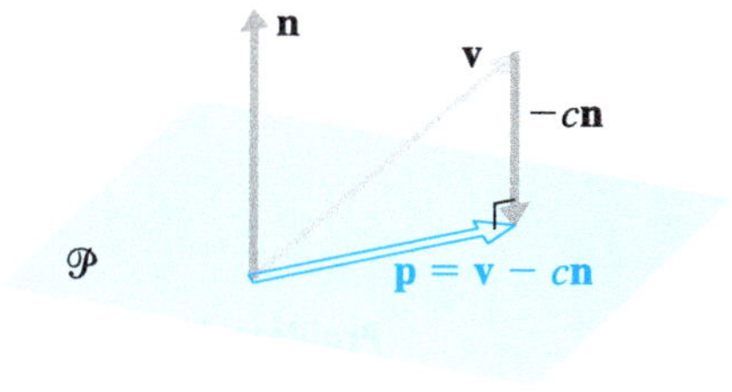

Figure 5.2
Projection onto a plane

Problem 6 Using the fact that $\mathbf{n}$ is orthogonal to every vector in $\mathcal{P}$, solve $\mathbf{v} - c\mathbf{n} = \mathbf{p}$ for c to find an expression for $\mathbf{p}$ in terms of $\mathbf{v}$ and $\mathbf{n}$.

Problem 7 Use the method of Problem 6 to find the projection of

$$\mathbf{v} = \begin{bmatrix} 1 \\ 0 \\ -2 \end{bmatrix}$$

onto the planes with the following equations:

$$\text{(a) } x + y + z = 0 \quad \text{(b) } x - 2z = 0 \quad \text{(c) } 2x - 3y + z = 0$$

Another approach to the problem of finding the projection of a vector onto a plane is suggested by Figure 5.3. We can decompose the projection of $\mathbf{v}$ onto $\mathcal{P}$ into the *sum* of its projections onto the direction vectors for $\mathcal{P}$. This works only if the direction vectors are orthogonal unit vectors. Accordingly, let $\mathbf{u}_1$ and $\mathbf{u}_2$ be direction vectors for $\mathcal{P}$ with the property that

$$\|\mathbf{u}_1\| = \|\mathbf{u}_2\| = 1 \quad \text{and} \quad \mathbf{u}_1 \cdot \mathbf{u}_2 = 0$$

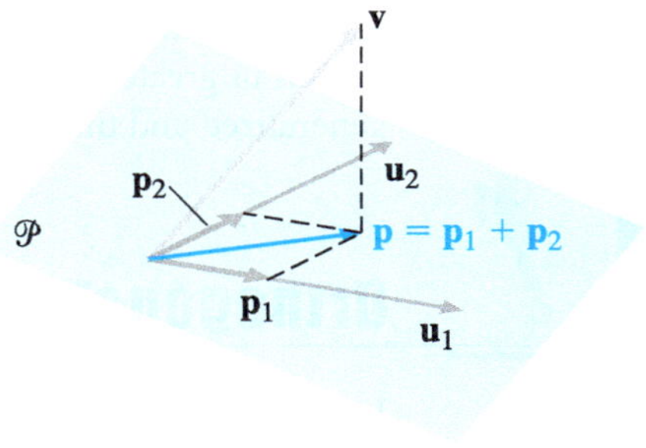

Figure 5.3

By Problem 2, the projections of $\mathbf{v}$ onto $\mathbf{u}_1$ and $\mathbf{u}_2$ are

$$\mathbf{p}_1 = \mathbf{u}_1\mathbf{u}_1^T\mathbf{v} \quad \text{and} \quad \mathbf{p}_2 = \mathbf{u}_2\mathbf{u}_2^T\mathbf{v}$$

respectively. To show that $\mathbf{p}_1 + \mathbf{p}_2$ gives the projection of $\mathbf{v}$ onto $\mathcal{P}$, we need to show that $\mathbf{v} - (\mathbf{p}_1 + \mathbf{p}_2)$ is orthogonal to $\mathcal{P}$. It is enough to show that $\mathbf{v} - (\mathbf{p}_1 + \mathbf{p}_2)$ is orthogonal to both $\mathbf{u}_1$ and $\mathbf{u}_2$. (Why?)

Problem 8 Show that $\mathbf{u}_1 \cdot (\mathbf{v} - (\mathbf{p}_1 + \mathbf{p}_2)) = 0$ and $\mathbf{u}_2 \cdot (\mathbf{v} - (\mathbf{p}_1 + \mathbf{p}_2)) = 0$. [*Hint:* Use the alternative form of the dot product, $\mathbf{x}^T\mathbf{y} = \mathbf{x} \cdot \mathbf{y}$, together with the fact that $\mathbf{u}_1$ and $\mathbf{u}_2$ are orthogonal unit vectors.]

It follows from Problem 8 and the comments preceding it that the matrix of the projection onto the subspace $\mathcal{P}$ of $\mathbb{R}^3$ spanned by orthogonal unit vectors $\mathbf{u}_1$ and $\mathbf{u}_2$ is

$$P = \mathbf{u}_1\mathbf{u}_1^T + \mathbf{u}_2\mathbf{u}_2^T \tag{1}$$

Problem 9 Repeat Problem 7, using the formula for P given by Equation (1). Use the same $\mathbf{v}$ and use $\mathbf{u}_1$ and $\mathbf{u}_2$, as indicated below. (First, verify that $\mathbf{u}_1$ and $\mathbf{u}_2$ are orthogonal unit vectors in the given plane.)

(a) $x + y + z = 0$ with $\mathbf{u}_1 = \begin{bmatrix} -2/\sqrt{6} \\ 1/\sqrt{6} \\ 1/\sqrt{6} \end{bmatrix}$ and $\mathbf{u}_2 = \begin{bmatrix} 0 \\ 1/\sqrt{2} \\ -1/\sqrt{2} \end{bmatrix}$

(b) $x - 2z = 0$ with $\mathbf{u}_1 = \begin{bmatrix} 2/\sqrt{5} \\ 0 \\ 1/\sqrt{5} \end{bmatrix}$ and $\mathbf{u}_2 = \begin{bmatrix} 0 \\ 1 \\ 0 \end{bmatrix}$

(c) $2x - 3y + z = 0$ with $\mathbf{u}_1 = \begin{bmatrix} 1/\sqrt{3} \\ -1/\sqrt{3} \\ 1/\sqrt{3} \end{bmatrix}$ and $\mathbf{u}_2 = \begin{bmatrix} 2/\sqrt{6} \\ 1/\sqrt{6} \\ -1/\sqrt{6} \end{bmatrix}$

Problem 10 Show that a projection matrix given by Equation (1) satisfies properties (a) and (b) of Problem 4.

Problem 11 Show that the matrix P of a projection onto a plane in $\mathbb{R}^3$ can be expressed as

$$P = AA^T$$

for some 3×2 matrix A. [*Hint:* Show that Equation (1) is an outer product expansion.]

Problem 12 Show that if P is the matrix of a projection onto a plane in $\mathbb{R}^3$, then $\operatorname{rank}(P) = 2$.

In this chapter, we will look at the concepts of orthogonality and orthogonal projection in greater detail. We will see that the ideas introduced in this section can be generalized and that they have many important applications.

5.1 Orthogonality in $\mathbb{R}^n$

In this section, we will generalize the notion of orthogonality of vectors in $\mathbb{R}^n$ from two vectors to sets of vectors. In doing so, we will see that two properties make the standard basis $\{\mathbf{e}_1, \mathbf{e}_2, \ldots, \mathbf{e}_n\}$ of $\mathbb{R}^n$ easy to work with: First, any two distinct vectors in the set are orthogonal. Second, each vector in the set is a unit vector. These two properties lead us to the notion of orthogonal bases and orthonormal bases—concepts that we will be able to fruitfully apply to a variety of applications.

Orthogonal and Orthonormal Sets of Vectors

Definition A set of vectors $\{\mathbf{v}_1, \mathbf{v}_2, \ldots, \mathbf{v}_k\}$ in $\mathbb{R}^n$ is called an ***orthogonal set*** if all pairs of distinct vectors in the set are orthogonal—that is, if

$$\mathbf{v}_i \cdot \mathbf{v}_j = 0 \quad \text{whenever} \quad i \neq j \quad \text{for } i, j = 1, 2, \ldots, k$$

The standard basis $\{\mathbf{e}_1, \mathbf{e}_2, \ldots, \mathbf{e}_n\}$ of $\mathbb{R}^n$ is an orthogonal set, as is any subset of it. As the first example illustrates, there are many other possibilities.

Example 5.1 Show that $\{\mathbf{v}_1, \mathbf{v}_2, \mathbf{v}_3\}$ is an orthogonal set in $\mathbb{R}^3$ if

$$\mathbf{v}_1 = \begin{bmatrix} 2 \\ 1 \\ -1 \end{bmatrix}, \quad \mathbf{v}_2 = \begin{bmatrix} 0 \\ 1 \\ 1 \end{bmatrix}, \quad \mathbf{v}_3 = \begin{bmatrix} 1 \\ -1 \\ 1 \end{bmatrix}$$

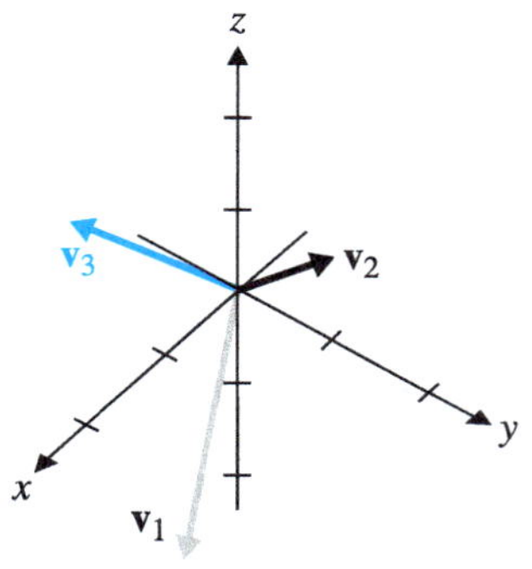

Figure 5.4
An orthogonal set of vectors

Solution We must show that every pair of vectors from this set is orthogonal. This is true, since

$$\mathbf{v}_1 \cdot \mathbf{v}_2 = 2(0) + 1(1) + (-1)(1) = 0$$
$$\mathbf{v}_2 \cdot \mathbf{v}_3 = 0(1) + 1(-1) + (1)(1) = 0$$
$$\mathbf{v}_1 \cdot \mathbf{v}_3 = 2(1) + 1(-1) + (-1)(1) = 0$$

Geometrically, the vectors in Example 5.1 are mutually perpendicular, as Figure 5.4 shows.

One of the main advantages of working with orthogonal sets of vectors is that they are necessarily linearly independent, as Theorem 5.1 shows.

Theorem 5.1 If $\{\mathbf{v}_1, \mathbf{v}_2, \ldots, \mathbf{v}_k\}$ is an orthogonal set of nonzero vectors in $\mathbb{R}^n$, then these vectors are linearly independent.

Proof If $c_1, \ldots, c_k$ are scalars such that $c_1\mathbf{v}_1 + \cdots + c_k\mathbf{v}_k = \mathbf{0}$, then

$$(c_1\mathbf{v}_1 + \cdots + c_k\mathbf{v}_k) \cdot \mathbf{v}_i = 0 \cdot \mathbf{v}_i = 0$$

or, equivalently,

$$c_1(\mathbf{v}_1 \cdot \mathbf{v}_i) + \cdots + c_i(\mathbf{v}_i \cdot \mathbf{v}_i) + \cdots + c_k(\mathbf{v}_k \cdot \mathbf{v}_i) = 0 \qquad (1)$$

Since $\{\mathbf{v}_1, \mathbf{v}_2, \ldots, \mathbf{v}_k\}$ is an orthogonal set, all of the dot products in Equation (1) are zero, except $\mathbf{v}_i \cdot \mathbf{v}_i$. Thus, Equation (1) reduces to

$$c_i(\mathbf{v}_i \cdot \mathbf{v}_i) = 0$$

Now, $\mathbf{v}_i \cdot \mathbf{v}_i \neq 0$ because $\mathbf{v}_i \neq \mathbf{0}$ by hypothesis. So we must have $c_i = 0$. The fact that this is true for all $i = 1, \ldots, k$ implies that $\{\mathbf{v}_1, \mathbf{v}_2, \ldots, \mathbf{v}_k\}$ is a linearly independent set.

Remark Thanks to Theorem 5.1, we know that if a set of vectors is orthogonal, it is automatically linearly independent. For example, we can immediately deduce that the three vectors in Example 5.1 are linearly independent. Contrast this approach with the work needed to establish their linear independence directly!

Definition An ***orthogonal basis*** for a subspace W of $\mathbb{R}^n$ is a basis of W that is an orthogonal set.

Example 5.2

The vectors

$$\mathbf{v}_1 = \begin{bmatrix} 2 \\ 1 \\ -1 \end{bmatrix}, \quad \mathbf{v}_2 = \begin{bmatrix} 0 \\ 1 \\ 1 \end{bmatrix}, \quad \mathbf{v}_3 = \begin{bmatrix} 1 \\ -1 \\ 1 \end{bmatrix}$$

from Example 5.1 are orthogonal and, hence, linearly independent. Since any three linearly independent vectors in $\mathbb{R}^3$ form a basis for $\mathbb{R}^3$, by the Fundamental Theorem of Invertible Matrices, it follows that $\{\mathbf{v}_1, \mathbf{v}_2, \mathbf{v}_3\}$ is an orthogonal basis for $\mathbb{R}^3$.

Remark In Example 5.2, suppose only the orthogonal vectors $\mathbf{v}_1$ and $\mathbf{v}_2$ were given and you were asked to find a third vector $\mathbf{v}_3$ to make $\{\mathbf{v}_1, \mathbf{v}_2, \mathbf{v}_3\}$ an orthogonal basis for $\mathbb{R}^3$. One way to do this is to remember that in $\mathbb{R}^3$, the cross product of two vectors $\mathbf{v}_1$ and $\mathbf{v}_2$ is orthogonal to each of them. (See Exploration: The Cross Product in Chapter 1.) Hence we may take

$$\mathbf{v}_3 = \mathbf{v}_1 \times \mathbf{v}_2 = \begin{bmatrix} 2 \\ 1 \\ -1 \end{bmatrix} \times \begin{bmatrix} 0 \\ 1 \\ 1 \end{bmatrix} = \begin{bmatrix} 2 \\ -2 \\ 2 \end{bmatrix}$$

Note that the resulting vector is a multiple of the vector $\mathbf{v}_3$ in Example 5.2, as it must be.

Example 5.3

Find an orthogonal basis for the subspace W of $\mathbb{R}^3$ given by

$$W = \left\{ \begin{bmatrix} x \\ y \\ z \end{bmatrix} : x - y + 2z = 0 \right\}$$

Solution Section 5.3 gives a general procedure for problems of this sort. For now, we will find the orthogonal basis by brute force. The subspace W is a plane through the origin in $\mathbb{R}^3$. From the equation of the plane, we have $x = y - 2z$, so W consists of vectors of the form

$$\begin{bmatrix} y - 2z \\ y \\ z \end{bmatrix} = y\begin{bmatrix} 1 \\ 1 \\ 0 \end{bmatrix} + z\begin{bmatrix} -2 \\ 0 \\ 1 \end{bmatrix}$$

It follows that $\mathbf{u} = \begin{bmatrix} 1 \\ 1 \\ 0 \end{bmatrix}$ and $\mathbf{v} = \begin{bmatrix} -2 \\ 0 \\ 1 \end{bmatrix}$ are a basis for W, but they are *not* orthogonal. It suffices to find another nonzero vector in W that is orthogonal to either one of these.

Suppose $\mathbf{w} = \begin{bmatrix} x \\ y \\ z \end{bmatrix}$ is a vector in W that is orthogonal to $\mathbf{u}$. Then $x - y + 2z = 0$, since $\mathbf{w}$ is in the plane W. Since $\mathbf{u} \cdot \mathbf{w} = 0$, we also have $x + y = 0$. Solving the linear system

$$\begin{aligned} x - y + 2z &= 0 \\ x + y &= 0 \end{aligned}$$

we find that $x = -z$ and $y = z$. (Check this.) Thus, any nonzero vector $\mathbf{w}$ of the form

$$\mathbf{w} = \begin{bmatrix} -z \\ z \\ z \end{bmatrix}$$

will do. To be specific, we could take $\mathbf{w} = \begin{bmatrix} -1 \\ 1 \\ 1 \end{bmatrix}$. It is easy to check that $\{\mathbf{u}, \mathbf{w}\}$ is an orthogonal set in W and, hence, an orthogonal basis for W, since $\dim W = 2$.

Another advantage of working with an orthogonal basis is that the coordinates of a vector with respect to such a basis are easy to compute. Indeed, there is a formula for these coordinates, as the following theorem establishes.

Theorem 5.2 Let $\{\mathbf{v}_1, \mathbf{v}_2, \ldots, \mathbf{v}_k\}$ be an orthogonal basis for a subspace W of $\mathbb{R}^n$ and let $\mathbf{w}$ be any vector in W. Then the unique scalars $c_1, \ldots, c_k$ such that

$$\mathbf{w} = c_1\mathbf{v}_1 + \cdots + c_k\mathbf{v}_k$$

are given by

$$c_i = \frac{\mathbf{w} \cdot \mathbf{v}_i}{\mathbf{v}_i \cdot \mathbf{v}_i} \quad \text{for } i = 1, \ldots, k$$

Proof Since $\{\mathbf{v}_1, \mathbf{v}_2, \ldots, \mathbf{v}_k\}$ is a basis for W, we know that there are unique scalars $c_1, \ldots, c_k$ such that $\mathbf{w} = c_1\mathbf{v}_1 + \cdots + c_k\mathbf{v}_k$ (from Theorem 3.29). To establish the formula for c_i, we take the dot product of this linear combination with $\mathbf{v}_i$ to obtain

$$\begin{aligned} \mathbf{w} \cdot \mathbf{v}_i &= (c_1\mathbf{v}_1 + \cdots + c_k\mathbf{v}_k) \cdot \mathbf{v}_i \\ &= c_1(\mathbf{v}_1 \cdot \mathbf{v}_i) + \cdots + c_i(\mathbf{v}_i \cdot \mathbf{v}_i) + \cdots + c_k(\mathbf{v}_k \cdot \mathbf{v}_i) \\ &= c_i(\mathbf{v}_i \cdot \mathbf{v}_i) \end{aligned}$$

since $\mathbf{v}_j \cdot \mathbf{v}_i = 0$ for $j \neq i$. Since $\mathbf{v}_i \neq \mathbf{0}$, $\mathbf{v}_i \cdot \mathbf{v}_i \neq 0$. Dividing by $\mathbf{v}_i \cdot \mathbf{v}_i$, we obtain the desired result.

Example 5.4

Find the coordinates of $\mathbf{w} = \begin{bmatrix} 1 \\ 2 \\ 3 \end{bmatrix}$ with respect to the orthogonal basis $\mathcal{B} = \{\mathbf{v}_1, \mathbf{v}_2, \mathbf{v}_3\}$ of Examples 5.1 and 5.2.

Solution Using Theorem 5.2, we compute

$$c_1 = \frac{\mathbf{w} \cdot \mathbf{v}_1}{\mathbf{v}_1 \cdot \mathbf{v}_1} = \frac{2 + 2 - 3}{4 + 1 + 1} = \frac{1}{6}$$

$$c_2 = \frac{\mathbf{w} \cdot \mathbf{v}_2}{\mathbf{v}_2 \cdot \mathbf{v}_2} = \frac{0 + 2 + 3}{0 + 1 + 1} = \frac{5}{2}$$

$$c_3 = \frac{\mathbf{w} \cdot \mathbf{v}_3}{\mathbf{v}_3 \cdot \mathbf{v}_3} = \frac{1 - 2 + 3}{1 + 1 + 1} = \frac{2}{3}$$

Thus,

$$\mathbf{w} = c_1\mathbf{v}_1 + c_2\mathbf{v}_2 + c_3\mathbf{v}_3 = \tfrac{1}{6}\mathbf{v}_1 + \tfrac{5}{2}\mathbf{v}_2 + \tfrac{2}{3}\mathbf{v}_3$$

(Check this.) With the notation introduced in Section 3.5, we can also write the above equation as

$$[\mathbf{w}]_{\mathcal{B}} = \begin{bmatrix} \frac{1}{6} \\ \frac{5}{2} \\ \frac{2}{3} \end{bmatrix}$$

Compare the procedure in Example 5.4 with the work required to find these coordinates directly and you should start to appreciate the value of orthogonal bases.

As noted at the beginning of this section, the other property of the standard basis in $\mathbb{R}^n$ is that each standard basis vector is a unit vector. Combining this property with orthogonality, we have the following definition.

Definition A set of vectors in $\mathbb{R}^n$ is an ***orthonormal set*** if it is an orthogonal set of unit vectors. An ***orthonormal basis*** for a subspace W of $\mathbb{R}^n$ is a basis of W that is an orthonormal set.

Remark If $S = \{\mathbf{q}_1, \ldots, \mathbf{q}_k\}$ is an orthonormal set of vectors, then $\mathbf{q}_i \cdot \mathbf{q}_j = 0$ for $i \neq j$ and $\|\mathbf{q}_i\| = 1$. The fact that each $\mathbf{q}_i$ is a unit vector is equivalent to $\mathbf{q}_i \cdot \mathbf{q}_i = 1$. It follows that we can summarize the statement that S is orthonormal as

$$\mathbf{q}_i \cdot \mathbf{q}_j = \begin{cases} 0 & \text{if } i \neq j \\ 1 & \text{if } i = j \end{cases}$$

Example 5.5

Show that $S = \{\mathbf{q}_1, \mathbf{q}_2\}$ is an orthonormal set in $\mathbb{R}^3$ if

$$\mathbf{q}_1 = \begin{bmatrix} 1/\sqrt{3} \\ -1/\sqrt{3} \\ 1/\sqrt{3} \end{bmatrix} \quad \text{and} \quad \mathbf{q}_2 = \begin{bmatrix} 1/\sqrt{6} \\ 2/\sqrt{6} \\ 1/\sqrt{6} \end{bmatrix}$$

Solution We check that

$$\mathbf{q}_1 \cdot \mathbf{q}_2 = 1/\sqrt{18} - 2/\sqrt{18} + 1/\sqrt{18} = 0$$
$$\mathbf{q}_1 \cdot \mathbf{q}_1 = 1/3 + 1/3 + 1/3 = 1$$
$$\mathbf{q}_2 \cdot \mathbf{q}_2 = 1/6 + 4/6 + 1/6 = 1$$

If we have an orthogonal set, we can easily obtain an orthonormal set from it: We simply normalize each vector.

Example 5.6

Construct an orthonormal basis for $\mathbb{R}^3$ from the vectors in Example 5.1.

Solution Since we already know that $\mathbf{v}_1$, $\mathbf{v}_2$, and $\mathbf{v}_3$ are an orthogonal basis, we normalize them to get

$$\mathbf{q}_1 = \frac{1}{\|\mathbf{v}_1\|}\mathbf{v}_1 = \frac{1}{\sqrt{6}}\begin{bmatrix} 2 \\ 1 \\ -1 \end{bmatrix} = \begin{bmatrix} 2/\sqrt{6} \\ 1/\sqrt{6} \\ -1/\sqrt{6} \end{bmatrix}$$

$$\mathbf{q}_2 = \frac{1}{\|\mathbf{v}_2\|}\mathbf{v}_2 = \frac{1}{\sqrt{2}}\begin{bmatrix} 0 \\ 1 \\ 1 \end{bmatrix} = \begin{bmatrix} 0 \\ 1/\sqrt{2} \\ 1/\sqrt{2} \end{bmatrix}$$

$$\mathbf{q}_3 = \frac{1}{\|\mathbf{v}_3\|}\mathbf{v}_3 = \frac{1}{\sqrt{3}}\begin{bmatrix} 1 \\ -1 \\ 1 \end{bmatrix} = \begin{bmatrix} 1/\sqrt{3} \\ -1/\sqrt{3} \\ 1/\sqrt{3} \end{bmatrix}$$

Then $\{\mathbf{q}_1, \mathbf{q}_2, \mathbf{q}_3\}$ is an orthonormal basis for $\mathbb{R}^3$.

Since any orthonormal set of vectors is, in particular, orthogonal, it is linearly independent, by Theorem 5.1. If we have an orthonormal basis, Theorem 5.2 becomes even simpler.

Theorem 5.3 Let $\{\mathbf{q}_1, \mathbf{q}_2, \ldots, \mathbf{q}_k\}$ be an orthonormal basis for a subspace W of $\mathbb{R}^n$ and let $\mathbf{w}$ be any vector in W. Then

$$\mathbf{w} = (\mathbf{w} \cdot \mathbf{q}_1)\mathbf{q}_1 + (\mathbf{w} \cdot \mathbf{q}_2)\mathbf{q}_2 + \cdots + (\mathbf{w} \cdot \mathbf{q}_k)\mathbf{q}_k$$

and this representation is unique.

Proof Apply Theorem 5.2 and use the fact that $\mathbf{q}_i \cdot \mathbf{q}_i = 1$ for $i = 1, \ldots, k$.

Orthogonal Matrices

Matrices whose columns form an orthonormal set arise frequently in applications, as you will see in Section 5.5. Such matrices have several attractive properties, which we now examine.

Theorem 5.4 The columns of an $m \times n$ matrix Q form an orthonormal set if and only if $Q^TQ = I_n$.

Proof We need to show that

$$(Q^TQ)_{ij} = \begin{cases} 0 & \text{if } i \neq j \\ 1 & \text{if } i = j \end{cases}$$

Let $\mathbf{q}_i$ denote the ith column of Q (and, hence, the ith row of Q^T). Since the (i, j) entry of Q^TQ is the dot product of the ith row of Q^T and the jth column of Q, it follows that

$$(Q^TQ)_{ij} = \mathbf{q}_i \cdot \mathbf{q}_j \qquad (2)$$

by the definition of matrix multiplication.

Now the columns Q form an orthonormal set if and only if

$$\mathbf{q}_i \cdot \mathbf{q}_j = \begin{cases} 0 & \text{if } i \neq j \\ 1 & \text{if } i = j \end{cases}$$

which, by Equation (2), holds if and only if

$$(Q^TQ)_{ij} = \begin{cases} 0 & \text{if } i \neq j \\ 1 & \text{if } i = j \end{cases}$$

This completes the proof.

Orthogonal matrix is an unfortunate bit of terminology. "Orthonormal matrix" would clearly be a better term, but it is not standard. Moreover, there is no term for a nonsquare matrix with orthonormal columns.

If the matrix Q in Theorem 5.4 is a *square* matrix, it has a special name.

Definition An $n \times n$ matrix Q whose columns form an orthonormal set is called an ***orthogonal matrix.***

The most important fact about orthogonal matrices is given by the next theorem.

Theorem 5.5 A square matrix Q is orthogonal if and only if $Q^{-1} = Q^T$.

Proof By Theorem 5.4, Q is orthogonal if and only if $Q^TQ = I$. This is true if and only if Q is invertible and $Q^{-1} = Q^T$, by Theorem 3.13.

Example 5.7 Show that the following matrices are orthogonal and find their inverses:

$$A = \begin{bmatrix} 0 & 1 & 0 \\ 0 & 0 & 1 \\ 1 & 0 & 0 \end{bmatrix} \quad \text{and} \quad B = \begin{bmatrix} \cos\theta & -\sin\theta \\ \sin\theta & \cos\theta \end{bmatrix}$$

Solution The columns of A are just the standard basis vectors for $\mathbb{R}^3$, which are clearly orthonormal. Hence, A is orthogonal and

$$A^{-1} = A^T = \begin{bmatrix} 0 & 0 & 1 \\ 1 & 0 & 0 \\ 0 & 1 & 0 \end{bmatrix}$$

For B, we check directly that

$$B^TB = \begin{bmatrix} \cos\theta & \sin\theta \\ -\sin\theta & \cos\theta \end{bmatrix}\begin{bmatrix} \cos\theta & -\sin\theta \\ \sin\theta & \cos\theta \end{bmatrix}$$
$$= \begin{bmatrix} \cos^2\theta + \sin^2\theta & -\cos\theta\sin\theta + \sin\theta\cos\theta \\ -\sin\theta\cos\theta + \cos\theta\sin\theta & \sin^2\theta + \cos^2\theta \end{bmatrix} = \begin{bmatrix} 1 & 0 \\ 0 & 1 \end{bmatrix} = I$$

Therefore, B is orthogonal, by Theorem 5.5, and

$$B^{-1} = B^T = \begin{bmatrix} \cos\theta & \sin\theta \\ -\sin\theta & \cos\theta \end{bmatrix}$$

Remark Matrix A in Example 5.7 is an example of a permutation matrix, a matrix obtained by permuting the columns of an identity matrix. In general, any $n \times n$ permutation matrix is orthogonal (see Exercise 25). Matrix B is the matrix of a rotation through the angle θ in $\mathbb{R}^2$. Any rotation has the property that it is a *length-preserving* transformation (known as an ***isometry*** in geometry). The next theorem shows that every orthogonal matrix transformation is an isometry. Orthogonal matrices also preserve dot products. In fact, orthogonal matrices are characterized by either one of these properties.

The word *isometry* literally means "length preserving," since it is derived from the Greek roots *isos* ("equal") and *metron* ("measure").

Theorem 5.6 Let Q be an $n \times n$ matrix. The following statements are equivalent:

a. Q is orthogonal.
b. $\|Q\mathbf{x}\| = \|\mathbf{x}\|$ for every $\mathbf{x}$ in $\mathbb{R}^n$.
c. $Q\mathbf{x}\cdot Q\mathbf{y} = \mathbf{x}\cdot\mathbf{y}$ for every $\mathbf{x}$ and $\mathbf{y}$ in $\mathbb{R}^n$.

Proof We will prove that (a) $\Rightarrow$ (c) $\Rightarrow$ (b) $\Rightarrow$ (a). To do so, we will need to make use of the fact that if $\mathbf{x}$ and $\mathbf{y}$ are (column) vectors in $\mathbb{R}^n$, then $\mathbf{x}\cdot\mathbf{y} = \mathbf{x}^T\mathbf{y}$.

(a) $\Rightarrow$ (c) Assume that Q is orthogonal. Then $Q^TQ = I$, and we have

$$Q\mathbf{x}\cdot Q\mathbf{y} = (Q\mathbf{x})^TQ\mathbf{y} = \mathbf{x}^TQ^TQ\mathbf{y} = \mathbf{x}^TI\mathbf{y} = \mathbf{x}^T\mathbf{y} = \mathbf{x}\cdot\mathbf{y}$$

(c) $\Rightarrow$ (b) Assume that $Q\mathbf{x}\cdot Q\mathbf{y} = \mathbf{x}\cdot\mathbf{y}$ for every $\mathbf{x}$ and $\mathbf{y}$ in $\mathbb{R}^n$. Then, taking $\mathbf{y} = \mathbf{x}$, we have $Q\mathbf{x}\cdot Q\mathbf{x} = \mathbf{x}\cdot\mathbf{x}$, so $\|Q\mathbf{x}\| = \sqrt{Q\mathbf{x}\cdot Q\mathbf{x}} = \sqrt{\mathbf{x}\cdot\mathbf{x}} = \|\mathbf{x}\|$.

(b) $\Rightarrow$ (a) Assume that property (b) holds and let $\mathbf{q}_i$ denote the ith column of Q. Using Exercise 63 in Section 1.2 and property (b), we have

$$\begin{aligned} \mathbf{x}\cdot\mathbf{y} &= \tfrac{1}{4}(\|\mathbf{x}+\mathbf{y}\|^2 - \|\mathbf{x}-\mathbf{y}\|^2) \\ &= \tfrac{1}{4}(\|Q(\mathbf{x}+\mathbf{y})\|^2 - \|Q(\mathbf{x}-\mathbf{y})\|^2) \\ &= \tfrac{1}{4}(\|Q\mathbf{x}+Q\mathbf{y}\|^2 - \|Q\mathbf{x}-Q\mathbf{y}\|^2) \\ &= Q\mathbf{x}\cdot Q\mathbf{y} \end{aligned}$$

for all $\mathbf{x}$ and $\mathbf{y}$ in $\mathbb{R}^n$. [This shows that (b) $\Rightarrow$ (c).]

Now if $\mathbf{e}_i$ is the ith standard basis vector, then $\mathbf{q}_i = Q\mathbf{e}_i$. Consequently,

$$\mathbf{q}_i\cdot\mathbf{q}_j = Q\mathbf{e}_i\cdot Q\mathbf{e}_j = \mathbf{e}_i\cdot\mathbf{e}_j = \begin{cases} 0 & \text{if } i \neq j \\ 1 & \text{if } i = j \end{cases}$$

Thus, the columns of Q form an orthonormal set, so Q is an orthogonal matrix.

Looking at the orthogonal matrices A and B in Example 5.7, you may notice that not only do their columns form orthonormal sets—so do their *rows*. In fact, every orthogonal matrix has this property, as the next theorem shows.

Theorem 5.7 If Q is an orthogonal matrix, then its rows form an orthonormal set.

Proof From Theorem 5.5, we know that $Q^{-1} = Q^T$. Therefore,

$$(Q^T)^{-1} = (Q^{-1})^{-1} = Q = (Q^T)^T$$

so Q^T is an orthogonal matrix. Thus, the columns of Q^T—which are just the rows of Q—form an orthonormal set.

The final theorem in this section lists some other properties of orthogonal matrices.

Theorem 5.8 Let Q be an orthogonal matrix.

a. Q^{-1} is orthogonal.
b. $\det Q = \pm 1$
c. If λ is an eigenvalue of Q, then $|\lambda| = 1$.
d. If Q_1 and Q_2 are orthogonal $n \times n$ matrices, then so is Q_1Q_2.

Proof We will prove property (c) and leave the proofs of the remaining properties as exercises.

(c) Let λ be an eigenvalue of Q with corresponding eigenvector $\mathbf{v}$. Then $Q\mathbf{v} = \lambda\mathbf{v}$, and, using Theorem 5.6(b), we have

$$\|\mathbf{v}\| = \|Q\mathbf{v}\| = \|\lambda\mathbf{v}\| = |\lambda|\,\|\mathbf{v}\|$$

Since $\|\mathbf{v}\| \neq 0$, this implies that $|\lambda| = 1$.

a + bi

Remark Property (c) holds even for complex eigenvalues. The matrix $\begin{bmatrix} 0 & -1 \\ 1 & 0 \end{bmatrix}$ is orthogonal with eigenvalues i and $-i$, both of which have absolute value 1.

Exercises 5.1

In Exercises 1–6, determine which sets of vectors are orthogonal.

1. $\begin{bmatrix} -3 \\ 1 \\ 2 \end{bmatrix}, \begin{bmatrix} 2 \\ 4 \\ 1 \end{bmatrix}, \begin{bmatrix} 1 \\ -1 \\ 2 \end{bmatrix}$

2. $\begin{bmatrix} 4 \\ 2 \\ -5 \end{bmatrix}, \begin{bmatrix} -1 \\ 2 \\ 0 \end{bmatrix}, \begin{bmatrix} 2 \\ 1 \\ 2 \end{bmatrix}$

3. $\begin{bmatrix} 3 \\ 1 \\ -1 \end{bmatrix}, \begin{bmatrix} -1 \\ 2 \\ 1 \end{bmatrix}, \begin{bmatrix} 2 \\ -2 \\ 4 \end{bmatrix}$

4. $\begin{bmatrix} 5 \\ 3 \\ 1 \end{bmatrix}, \begin{bmatrix} 1 \\ -2 \\ 1 \end{bmatrix}, \begin{bmatrix} 3 \\ 1 \\ -1 \end{bmatrix}$

5. $\begin{bmatrix} 2 \\ 3 \\ -1 \\ 4 \end{bmatrix}, \begin{bmatrix} -2 \\ 1 \\ -1 \\ 0 \end{bmatrix}, \begin{bmatrix} -4 \\ -6 \\ 2 \\ 7 \end{bmatrix}$

6. $\begin{bmatrix} 1 \\ 0 \\ -1 \\ 1 \end{bmatrix}, \begin{bmatrix} -1 \\ 0 \\ 1 \\ 2 \end{bmatrix}, \begin{bmatrix} 1 \\ 1 \\ 1 \\ 0 \end{bmatrix}, \begin{bmatrix} 0 \\ -1 \\ 1 \\ 1 \end{bmatrix}$

In Exercises 7–10, show that the given vectors form an orthogonal basis for $\mathbb{R}^2$ or $\mathbb{R}^3$. Then use Theorem 5.2 to express $\mathbf{w}$ *as a linear combination of these basis vectors. Give the coordinate vector* $[\mathbf{w}]_{\mathcal{B}}$ *of* $\mathbf{w}$ *with respect to the basis* $\mathcal{B} = \{\mathbf{v}_1, \mathbf{v}_2\}$ *of* $\mathbb{R}^2$ *or* $\mathcal{B} = \mathbf{v}_1, \mathbf{v}_2, \mathbf{v}_3$ *of* $\mathbb{R}^3$.

7. $\mathbf{v}_1 = \begin{bmatrix} 4 \\ -2 \end{bmatrix}, \mathbf{v}_2 = \begin{bmatrix} 1 \\ 2 \end{bmatrix}; \mathbf{w} = \begin{bmatrix} 1 \\ -3 \end{bmatrix}$

8. $\mathbf{v}_1 = \begin{bmatrix} 1 \\ 3 \end{bmatrix}, \mathbf{v}_2 = \begin{bmatrix} -6 \\ 2 \end{bmatrix}; \mathbf{w} = \begin{bmatrix} 1 \\ 1 \end{bmatrix}$

9. $\mathbf{v}_1 = \begin{bmatrix} 1 \\ 0 \\ -1 \end{bmatrix}, \mathbf{v}_2 = \begin{bmatrix} 1 \\ 2 \\ 1 \end{bmatrix}, \mathbf{v}_3 = \begin{bmatrix} 1 \\ -1 \\ 1 \end{bmatrix}; \mathbf{w} = \begin{bmatrix} 1 \\ 1 \\ 1 \end{bmatrix}$

10. $\mathbf{v}_1 = \begin{bmatrix} 1 \\ 1 \\ 1 \end{bmatrix}, \mathbf{v}_2 = \begin{bmatrix} 1 \\ -1 \\ 0 \end{bmatrix}, \mathbf{v}_3 = \begin{bmatrix} 1 \\ 1 \\ -2 \end{bmatrix}; \mathbf{w} = \begin{bmatrix} 1 \\ 2 \\ 3 \end{bmatrix}$

In Exercises 11–15, determine whether the given orthogonal set of vectors is orthonormal. If it is not, normalize the vectors to form an orthonormal set.

11. $\begin{bmatrix} \frac{3}{5} \\ \frac{4}{5} \end{bmatrix}, \begin{bmatrix} -\frac{4}{5} \\ \frac{3}{5} \end{bmatrix}$

12. $\begin{bmatrix} \frac{1}{2} \\ \frac{1}{2} \end{bmatrix}, \begin{bmatrix} \frac{1}{2} \\ -\frac{1}{2} \end{bmatrix}$

13. $\begin{bmatrix} \frac{1}{3} \\ \frac{2}{3} \\ \frac{2}{3} \end{bmatrix}, \begin{bmatrix} \frac{2}{3} \\ -\frac{1}{3} \\ 0 \end{bmatrix}, \begin{bmatrix} 1 \\ 2 \\ -\frac{5}{2} \end{bmatrix}$

14. $\begin{bmatrix} \frac{1}{2} \\ \frac{1}{2} \\ -\frac{1}{2} \\ \frac{1}{2} \end{bmatrix}, \begin{bmatrix} 0 \\ \frac{1}{3} \\ \frac{2}{3} \\ \frac{1}{3} \end{bmatrix}, \begin{bmatrix} \frac{1}{2} \\ -\frac{1}{6} \\ \frac{1}{6} \\ -\frac{1}{6} \end{bmatrix}$

15. $\begin{bmatrix} 1/2 \\ 1/2 \\ -1/2 \\ 1/2 \end{bmatrix}, \begin{bmatrix} 0 \\ \sqrt{6}/3 \\ 1/\sqrt{6} \\ -1/\sqrt{6} \end{bmatrix}, \begin{bmatrix} \sqrt{3}/2 \\ -\sqrt{3}/6 \\ \sqrt{3}/6 \\ -\sqrt{3}/6 \end{bmatrix}, \begin{bmatrix} 0 \\ 0 \\ 1/\sqrt{2} \\ 1/\sqrt{2} \end{bmatrix}$

In Exercises 16–21, determine whether the given matrix is orthogonal. If it is, find its inverse.

16. $\begin{bmatrix} 0 & -1 \\ 1 & 0 \end{bmatrix}$

17. $\begin{bmatrix} 1/\sqrt{2} & 1/\sqrt{2} \\ -1/\sqrt{2} & 1/\sqrt{2} \end{bmatrix}$

18. $\begin{bmatrix} \frac{1}{3} & \frac{1}{2} & \frac{1}{5} \\ \frac{1}{3} & -\frac{1}{2} & \frac{1}{5} \\ -\frac{1}{3} & 0 & \frac{2}{5} \end{bmatrix}$

19. $\begin{bmatrix} \cos\theta \sin\theta & -\cos\theta & -\sin^2\theta \\ \cos^2\theta & \sin\theta & -\cos\theta\sin\theta \\ \sin\theta & 0 & \cos\theta \end{bmatrix}$

20. $\begin{bmatrix} \frac{1}{2} & -\frac{1}{2} & \frac{1}{2} & \frac{1}{2} \\ \frac{1}{2} & \frac{1}{2} & \frac{1}{2} & -\frac{1}{2} \\ -\frac{1}{2} & \frac{1}{2} & \frac{1}{2} & \frac{1}{2} \\ \frac{1}{2} & \frac{1}{2} & -\frac{1}{2} & \frac{1}{2} \end{bmatrix}$

21. $\begin{bmatrix} 1 & 0 & 0 & 1/\sqrt{6} \\ 0 & 2/3 & 1/\sqrt{2} & 1/\sqrt{6} \\ 0 & -2/3 & 1/\sqrt{2} & -1/\sqrt{6} \\ 0 & 1/3 & 0 & 1/\sqrt{2} \end{bmatrix}$

22. Prove Theorem 5.8(a).

23. Prove Theorem 5.8(b).

24. Prove Theorem 5.8(d).

25. Prove that every permutation matrix is orthogonal.

26. If Q is an orthogonal matrix, prove that any matrix obtained by rearranging the rows of Q is also orthogonal.

27. Let Q be an orthogonal 2×2 matrix and let $\mathbf{x}$ and $\mathbf{y}$ be vectors in $\mathbb{R}^2$. If θ is the angle between $\mathbf{x}$ and $\mathbf{y}$, prove that the angle between $Q\mathbf{x}$ and $Q\mathbf{y}$ is also θ. (This proves that the linear transformations defined by orthogonal matrices are *angle-preserving* in $\mathbb{R}^2$, a fact that is true in general.)

28. **(a)** Prove that an orthogonal 2×2 matrix must have the form

$$\begin{bmatrix} a & -b \\ b & a \end{bmatrix} \quad \text{or} \quad \begin{bmatrix} a & b \\ b & -a \end{bmatrix}$$

where $\begin{bmatrix} a \\ b \end{bmatrix}$ is a unit vector.

(b) Using part (a), show that every orthogonal 2×2 matrix is of the form

$$\begin{bmatrix} \cos\theta & -\sin\theta \\ \sin\theta & \cos\theta \end{bmatrix} \quad \text{or} \quad \begin{bmatrix} \cos\theta & \sin\theta \\ \sin\theta & -\cos\theta \end{bmatrix}$$

where $0 \le \theta < 2\pi$.

(c) Show that every orthogonal 2×2 matrix corresponds to either a rotation or a reflection in $\mathbb{R}^2$.

(d) Show that an orthogonal 2×2 matrix Q corresponds to a rotation in $\mathbb{R}^2$ if $\det Q = 1$ and a reflection in $\mathbb{R}^2$ if $\det Q = -1$.

In Exercises 29–32, use Exercise 28 to determine whether the given orthogonal matrix represents a rotation or a reflection. If it is a rotation, give the angle of rotation; if it is a reflection, give the line of reflection.

29. $\begin{bmatrix} 1/\sqrt{2} & -1/\sqrt{2} \\ 1/\sqrt{2} & 1/\sqrt{2} \end{bmatrix}$

30. $\begin{bmatrix} -1/2 & \sqrt{3}/2 \\ -\sqrt{3}/2 & -1/2 \end{bmatrix}$

31. $\begin{bmatrix} -1/2 & \sqrt{3}/2 \\ \sqrt{3}/2 & 1/2 \end{bmatrix}$

32. $\begin{bmatrix} -\frac{3}{5} & -\frac{4}{5} \\ -\frac{4}{5} & \frac{3}{5} \end{bmatrix}$

33. Let A and B be $n \times n$ orthogonal matrices.

(a) Prove that $A(A^T + B^T)B = A + B$.

(b) Use part (a) to prove that, if $\det A + \det B = 0$, then $A + B$ is not invertible.

34. Let $\mathbf{x}$ be a unit vector in $\mathbb{R}^n$. Partition $\mathbf{x}$ as

$$\mathbf{x} = \begin{bmatrix} x_1 \\ \hline x_2 \\ \vdots \\ x_n \end{bmatrix} = \begin{bmatrix} x_1 \\ \mathbf{y} \end{bmatrix}$$

Let

$$Q = \left[\begin{array}{c|c} x_1 & \mathbf{y}^T \\ \hline \mathbf{y} & I - \left(\dfrac{1}{1 - x_1}\right)\mathbf{y}\mathbf{y}^T \end{array}\right]$$

Prove that Q is orthogonal. (This procedure gives a quick method for finding an orthonormal basis for $\mathbb{R}^n$ with a prescribed first vector $\mathbf{x}$, a construction that is frequently useful in applications.)

35. Prove that if an upper triangular matrix is orthogonal, then it must be a diagonal matrix.

36. Prove that if $n > m$, then there is no $m \times n$ matrix A such that $\|A\mathbf{x}\| = \|\mathbf{x}\|$ for all $\mathbf{x}$ in $\mathbb{R}^n$.

37. Let $\mathcal{B} = \{\mathbf{v}_1, \ldots, \mathbf{v}_n\}$ be an orthonormal basis for $\mathbb{R}^n$.

(a) Prove that, for any $\mathbf{x}$ and $\mathbf{y}$ in $\mathbb{R}^n$,

$$\mathbf{x}\cdot\mathbf{y} = (\mathbf{x}\cdot\mathbf{v}_1)(\mathbf{y}\cdot\mathbf{v}_1) + (\mathbf{x}\cdot\mathbf{v}_2)(\mathbf{y}\cdot\mathbf{v}_2) + \cdots + (\mathbf{x}\cdot\mathbf{v}_n)(\mathbf{y}\cdot\mathbf{v}_n)$$

(This identity is called ***Parseval's Identity.***)

(b) What does Parseval's Identity imply about the relationship between the dot products $\mathbf{x}\cdot\mathbf{y}$ and $[\mathbf{x}]_{\mathcal{B}} \cdot [\mathbf{y}]_{\mathcal{B}}$?

5.2 Orthogonal Complements and Orthogonal Projections

In this section, we generalize two concepts that we encountered in Chapter 1. The notion of a normal vector to a plane will be extended to orthogonal complements, and the projection of one vector onto another will give rise to the concept of orthogonal projection onto a subspace.

Orthogonal Complements

$W^\perp$ is pronounced "W perp."

A normal vector $\mathbf{n}$ to a plane is orthogonal to every vector in that plane. If the plane passes through the origin, then it is a subspace W of $\mathbb{R}^3$, as is span($\mathbf{n}$). Hence, we have two subspaces of $\mathbb{R}^3$ with the property that every vector of one is orthogonal to every vector of the other. This is the idea behind the following definition.

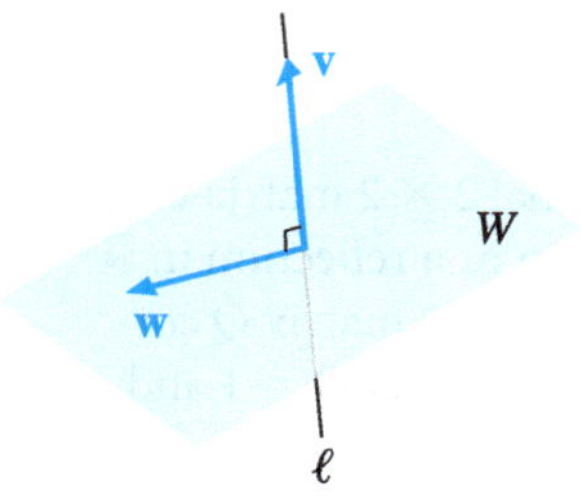

Figure 5.5
$\ell = W^\perp$ and $W = \ell^\perp$

Definition Let W be a subspace of $\mathbb{R}^n$. We say that a vector $\mathbf{v}$ in $\mathbb{R}^n$ is ***orthogonal to*** W if $\mathbf{v}$ is orthogonal to every vector in W. The set of all vectors that are orthogonal to W is called the ***orthogonal complement of*** W, denoted $W^\perp$. That is,

$$W^\perp = \{\mathbf{v} \text{ in } \mathbb{R}^n : \mathbf{v}\cdot\mathbf{w} = 0 \quad \text{for all } \mathbf{w} \text{ in } W\}$$

Example 5.8

If W is a plane through the origin in $\mathbb{R}^3$ and ℓ is the line through the origin perpendicular to W (i.e., parallel to the normal vector to W), then every vector $\mathbf{v}$ on ℓ is orthogonal to every vector $\mathbf{w}$ in W; hence, $\ell = W^\perp$. Moreover, W consists *precisely* of those vectors $\mathbf{w}$ that are orthogonal to every $\mathbf{v}$ on ℓ; hence, we also have $W = \ell^\perp$. Figure 5.5 illustrates this situation.

In Example 5.8, the orthogonal complement of a subspace turned out to be another subspace. Also, the complement of the complement of a subspace was the original subspace. These properties are true in general and are proved as properties (a) and (b) of Theorem 5.9. Properties (c) and (d) will also be useful. (Recall that the *intersection* $A \cap B$ of sets A and B consists of their common elements. See Appendix A.)

Theorem 5.9 Let W be a subspace of $\mathbb{R}^n$.

a. $W^\perp$ is a subspace of $\mathbb{R}^n$.
b. $(W^\perp)^\perp = W$
c. $W \cap W^\perp = \{\mathbf{0}\}$
d. If $W = \text{span}(\mathbf{w}_1, \ldots, \mathbf{w}_k)$, then $\mathbf{v}$ is in $W^\perp$ if and only if $\mathbf{v} \cdot \mathbf{w}_i = 0$ for all $i = 1, \ldots, k$.

Proof (a) Since $\mathbf{0} \cdot \mathbf{w} = 0$ for all $\mathbf{w}$ in W, $\mathbf{0}$ is in $W^\perp$. Let $\mathbf{u}$ and $\mathbf{v}$ be in $W^\perp$ and let c be a scalar. Then

$$\mathbf{u} \cdot \mathbf{w} = \mathbf{v} \cdot \mathbf{w} = 0 \quad \text{for all } \mathbf{w} \text{ in } W$$

Therefore,

$$(\mathbf{u} + \mathbf{v}) \cdot \mathbf{w} = \mathbf{u} \cdot \mathbf{w} + \mathbf{v} \cdot \mathbf{w} = 0 + 0 = 0$$

so $\mathbf{u} + \mathbf{v}$ is in $W^\perp$.

We also have

$$(c\mathbf{u}) \cdot \mathbf{w} = c(\mathbf{u} \cdot \mathbf{w}) = c(0) = 0$$

from which we see that $c\mathbf{u}$ is in $W^\perp$. It follows that $W^\perp$ is a subspace of $\mathbb{R}^n$.

(b) We will prove this property as Corollary 5.12.

(c) You are asked to prove this property in Exercise 23.

(d) You are asked to prove this property in Exercise 24.

We can now express some fundamental relationships involving the subspaces associated with an $m \times n$ matrix.

Theorem 5.10 Let A be an $m \times n$ matrix. Then the orthogonal complement of the row space of A is the null space of A, and the orthogonal complement of the column space of A is the null space of A^T:

$$(\text{row}(A))^\perp = \text{null}(A) \quad \text{and} \quad (\text{col}(A))^\perp = \text{null}(A^T)$$

Proof If $\mathbf{x}$ is a vector in $\mathbb{R}^n$, then $\mathbf{x}$ is in $(\text{row}(A))^\perp$ if and only if $\mathbf{x}$ is orthogonal to every row of A. But this is true if and only if $A\mathbf{x} = \mathbf{0}$, which is equivalent to $\mathbf{x}$ being in $\text{null}(A)$, so we have established the first identity. To prove the second identity, we simply replace A by A^T and use the fact that $\text{row}(A^T) = \text{col}(A)$.

Thus, an $m \times n$ matrix has four subspaces: $\text{row}(A)$, $\text{null}(A)$, $\text{col}(A)$, and $\text{null}(A^T)$. The first two are orthogonal complements in $\mathbb{R}^n$, and the last two are orthogonal

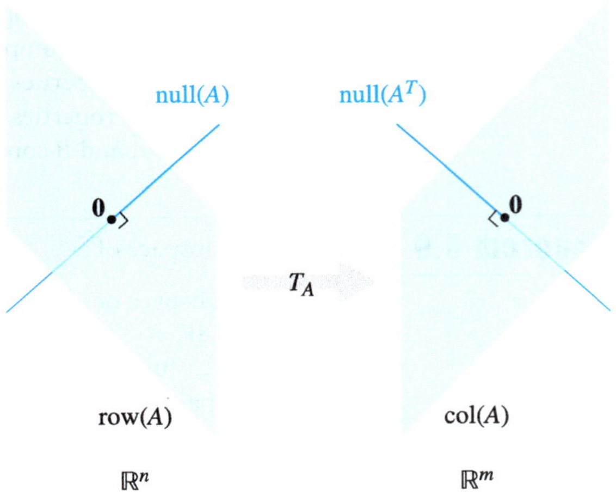

Figure 5.6
The four fundamental subspaces

complements in $\mathbb{R}^m$. The $m \times n$ matrix A defines a linear transformation from $\mathbb{R}^n$ into $\mathbb{R}^m$ whose range is col(A). Moreover, this transformation sends null(A) to $\mathbf{0}$ in $\mathbb{R}^m$. Figure 5.6 illustrates these ideas schematically. These four subspaces are called the ***fundamental subspaces*** of the $m \times n$ matrix A.

Example 5.9

Find bases for the four fundamental subspaces of

$$A = \begin{bmatrix} 1 & 1 & 3 & 1 & 6 \\ 2 & -1 & 0 & 1 & -1 \\ -3 & 2 & 1 & -2 & 1 \\ 4 & 1 & 6 & 1 & 3 \end{bmatrix}$$

and verify Theorem 5.10.

Solution In Examples 3.45, 3.47, and 3.48, we computed bases for the row space, column space, and null space of A. We found that row(A) = span($\mathbf{u}_1, \mathbf{u}_2, \mathbf{u}_3$), where

$$\mathbf{u}_1 = [1 \quad 0 \quad 1 \quad 0 \quad -1], \quad \mathbf{u}_2 = [0 \quad 1 \quad 2 \quad 0 \quad 3], \quad \mathbf{u}_3 = [0 \quad 0 \quad 0 \quad 1 \quad 4]$$

Also, null(A) = span($\mathbf{x}_1, \mathbf{x}_2$), where

$$\mathbf{x}_1 = \begin{bmatrix} -1 \\ -2 \\ 1 \\ 0 \\ 0 \end{bmatrix}, \quad \mathbf{x}_2 = \begin{bmatrix} 1 \\ -3 \\ 0 \\ -4 \\ 1 \end{bmatrix}$$

To show that $(\text{row}(A))^\perp = \text{null}(A)$, it is enough to show that every $\mathbf{u}_i$ is orthogonal to each $\mathbf{x}_j$, which is an easy exercise. (Why is this sufficient?)

The column space of A is $\text{col}(A) = \text{span}(\mathbf{a}_1, \mathbf{a}_2, \mathbf{a}_3)$, where

$$\mathbf{a}_1 = \begin{bmatrix} 1 \\ 2 \\ -3 \\ 4 \end{bmatrix}, \quad \mathbf{a}_2 = \begin{bmatrix} 1 \\ -1 \\ 2 \\ 1 \end{bmatrix}, \quad \mathbf{a}_3 = \begin{bmatrix} 1 \\ 1 \\ -2 \\ 1 \end{bmatrix}$$

We still need to compute the null space of A^T. Row reduction produces

$$[A^T \mid \mathbf{0}] = \left[\begin{array}{rrrr|r} 1 & 2 & -3 & 4 & 0 \\ 1 & -1 & 2 & 1 & 0 \\ 3 & 0 & 1 & 6 & 0 \\ 1 & 1 & -2 & 1 & 0 \\ 6 & -1 & 1 & 3 & 0 \end{array}\right] \longrightarrow \left[\begin{array}{rrrr|r} 1 & 0 & 0 & 1 & 0 \\ 0 & 1 & 0 & 6 & 0 \\ 0 & 0 & 1 & 3 & 0 \\ 0 & 0 & 0 & 0 & 0 \\ 0 & 0 & 0 & 0 & 0 \end{array}\right]$$

So, if $\mathbf{y}$ is in the null space of A^T, then $y_1 = -y_4$, $y_2 = -6y_4$, and $y_3 = -3y_4$. It follows that

$$\text{null}(A^T) = \left\{ \begin{bmatrix} -y_4 \\ -6y_4 \\ -3y_4 \\ y_4 \end{bmatrix} \right\} = \text{span}\left(\begin{bmatrix} -1 \\ -6 \\ -3 \\ 1 \end{bmatrix} \right)$$

and it is easy to check that this vector is orthogonal to $\mathbf{a}_1$, $\mathbf{a}_2$, and $\mathbf{a}_3$.

The method of Example 5.9 is easily adapted to other situations.

Example 5.10

Let W be the subspace of $\mathbb{R}^5$ spanned by

$$\mathbf{w}_1 = \begin{bmatrix} 1 \\ -3 \\ 5 \\ 0 \\ 5 \end{bmatrix}, \quad \mathbf{w}_2 = \begin{bmatrix} -1 \\ 1 \\ 2 \\ -2 \\ 3 \end{bmatrix}, \quad \mathbf{w}_3 = \begin{bmatrix} 0 \\ -1 \\ 4 \\ -1 \\ 5 \end{bmatrix}$$

Find a basis for $W^\perp$.

Solution The subspace W spanned by $\mathbf{w}_1$, $\mathbf{w}_2$, and $\mathbf{w}_3$ is the same as the column space of

$$A = \begin{bmatrix} 1 & -1 & 0 \\ -3 & 1 & -1 \\ 5 & 2 & 4 \\ 0 & -2 & -1 \\ 5 & 3 & 5 \end{bmatrix}$$

Therefore, by Theorem 5.10, $W^{\perp} = (\text{col}(A))^{\perp} = \text{null}(A^T)$, and we may proceed as in the previous example. We compute

$$[A^T \mid \mathbf{0}] = \left[\begin{array}{rrrrr|r} 1 & -3 & 5 & 0 & 5 & 0 \\ -1 & 1 & 2 & -2 & 3 & 0 \\ 0 & -1 & 4 & -1 & 5 & 0 \end{array}\right] \longrightarrow \left[\begin{array}{rrrrr|r} 1 & 0 & 0 & 3 & 4 & 0 \\ 0 & 1 & 0 & 1 & 3 & 0 \\ 0 & 0 & 1 & 0 & 2 & 0 \end{array}\right]$$

Hence, $\mathbf{y}$ is in $W^{\perp}$ if and only if $y_1 = -3y_4 - 4y_5$, $y_2 = -y_4 - 3y_5$, and $y_3 = -2y_5$. It follows that

$$W^{\perp} = \left\{ \begin{bmatrix} -3y_4 - 4y_5 \\ -y_4 - 3y_5 \\ -2y_5 \\ y_4 \\ y_5 \end{bmatrix} \right\} = \text{span}\left(\begin{bmatrix} -3 \\ -1 \\ 0 \\ 1 \\ 0 \end{bmatrix}, \begin{bmatrix} -4 \\ -3 \\ -2 \\ 0 \\ 1 \end{bmatrix} \right)$$

and these two vectors form a basis for $W^{\perp}$.

Orthogonal Projections

Recall that, in $\mathbb{R}^2$, the projection of a vector $\mathbf{v}$ onto a nonzero vector $\mathbf{u}$ is given by

$$\text{proj}_{\mathbf{u}}(\mathbf{v}) = \left(\frac{\mathbf{u} \cdot \mathbf{v}}{\mathbf{u} \cdot \mathbf{u}}\right)\mathbf{u}$$

Furthermore, the vector $\text{perp}_{\mathbf{u}}(\mathbf{v}) = \mathbf{v} - \text{proj}_{\mathbf{u}}(\mathbf{v})$ is orthogonal to $\text{proj}_{\mathbf{u}}(\mathbf{v})$, and we can decompose $\mathbf{v}$ as

$$\mathbf{v} = \text{proj}_{\mathbf{u}}(\mathbf{v}) + \text{perp}_{\mathbf{u}}(\mathbf{v})$$

as shown in Figure 5.7.

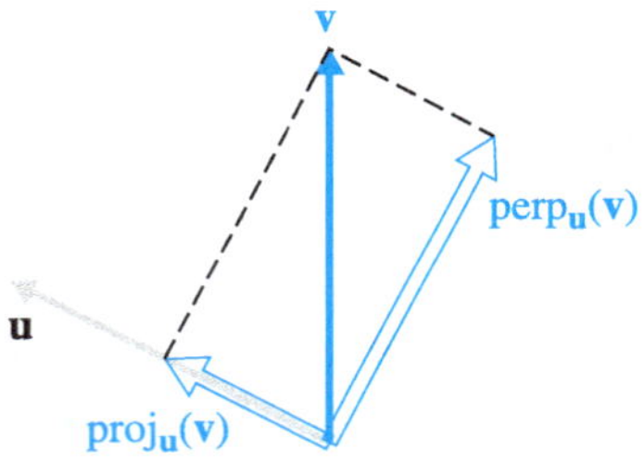

Figure 5.7
$\mathbf{v} = \text{proj}_{\mathbf{u}}(\mathbf{v}) + \text{perp}_{\mathbf{u}}(\mathbf{v})$

If we let $W = \text{span}(\mathbf{u})$, then $\mathbf{w} = \text{proj}_{\mathbf{u}}(\mathbf{v})$ is in W and $\mathbf{w}^{\perp} = \text{perp}_{\mathbf{u}}(\mathbf{v})$ is in $W^{\perp}$. We therefore have a way of "decomposing" $\mathbf{v}$ into the sum of two vectors, one from W and the other orthogonal to W—namely, $\mathbf{v} = \mathbf{w} + \mathbf{w}^{\perp}$. We now generalize this idea to $\mathbb{R}^n$.

Definition Let W be a subspace of $\mathbb{R}^n$ and let $\{\mathbf{u}_1, \ldots, \mathbf{u}_k\}$ be an orthogonal basis for W. For any vector $\mathbf{v}$ in $\mathbb{R}^n$, the ***orthogonal projection of v onto W*** is defined as

$$\text{proj}_W(\mathbf{v}) = \left(\frac{\mathbf{u}_1 \cdot \mathbf{v}}{\mathbf{u}_1 \cdot \mathbf{u}_1}\right)\mathbf{u}_1 + \cdots + \left(\frac{\mathbf{u}_k \cdot \mathbf{v}}{\mathbf{u}_k \cdot \mathbf{u}_k}\right)\mathbf{u}_k$$

The ***component of v orthogonal to W*** is the vector

$$\text{perp}_W(\mathbf{v}) = \mathbf{v} - \text{proj}_W(\mathbf{v})$$

Each summand in the definition of $\text{proj}_W(\mathbf{v})$ is also a projection onto a single vector (or, equivalently, the one-dimensional subspace spanned by it—in our previous sense). Therefore, with the notation of the preceding definition, we can write

$$\text{proj}_W(\mathbf{v}) = \text{proj}_{\mathbf{u}_1}(\mathbf{v}) + \cdots + \text{proj}_{\mathbf{u}_k}(\mathbf{v})$$

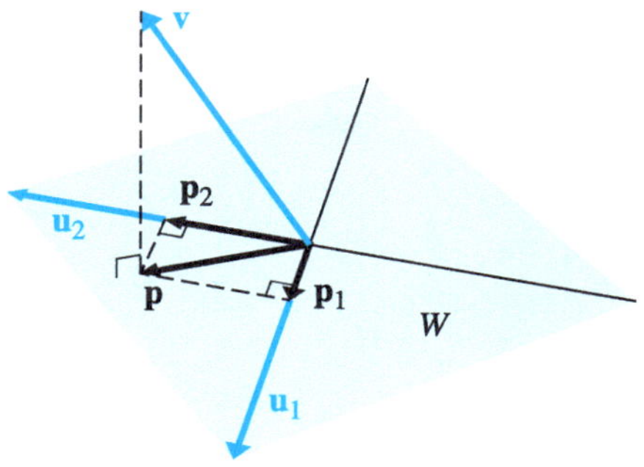

Figure 5.8
$\mathbf{p} = \mathbf{p}_1 + \mathbf{p}_2$

Since the vectors $\mathbf{u}_i$ are orthogonal, the orthogonal projection of $\mathbf{v}$ onto W is the sum of its projections onto one-dimensional subspaces that are mutually orthogonal. Figure 5.8 illustrates this situation with $W = \text{span}(\mathbf{u}_1, \mathbf{u}_2)$, $\mathbf{p} = \text{proj}_W(\mathbf{v})$, $\mathbf{p}_1 = \text{proj}_{\mathbf{u}_1}(\mathbf{v})$, and $\mathbf{p}_2 = \text{proj}_{\mathbf{u}_2}(\mathbf{v})$.

As a special case of the definition of $\text{proj}_W(\mathbf{v})$, we now also have a nice geometric interpretation of Theorem 5.2. In terms of our present notation and terminology, that theorem states that if $\mathbf{w}$ is in the subspace W of $\mathbb{R}^n$, which has orthogonal basis $\{\mathbf{v}_1, \mathbf{v}_2, \ldots, \mathbf{v}_k\}$, then

$$\begin{aligned}\mathbf{w} &= \left(\frac{\mathbf{w}\cdot\mathbf{v}_1}{\mathbf{v}_1\cdot\mathbf{v}_1}\right)\mathbf{v}_1 + \cdots + \left(\frac{\mathbf{w}\cdot\mathbf{v}_k}{\mathbf{v}_k\cdot\mathbf{v}_k}\right)\mathbf{v}_k \\ &= \text{proj}_{\mathbf{v}_1}(\mathbf{w}) + \cdots + \text{proj}_{\mathbf{v}_k}(\mathbf{w})\end{aligned}$$

Thus, $\mathbf{w}$ is decomposed into a sum of orthogonal projections onto mutually orthogonal one-dimensional subspaces of W.

The definition above seems to depend on the choice of orthogonal basis; that is, a different basis $\{\mathbf{u}'_1, \ldots, \mathbf{u}'_k\}$ for W would appear to give a "different" $\text{proj}_W(\mathbf{v})$ and $\text{perp}_W(\mathbf{v})$. Fortunately, this is not the case, as we will soon prove. For now, let's be content with an example.

Example 5.11

Let W be the plane in $\mathbb{R}^3$ with equation $x - y + 2z = 0$, and let $\mathbf{v} = \begin{bmatrix} 3 \\ -1 \\ 2 \end{bmatrix}$. Find the orthogonal projection of $\mathbf{v}$ onto W and the component of $\mathbf{v}$ orthogonal to W.

Solution In Example 5.3, we found an orthogonal basis for W. Taking

$$\mathbf{u}_1 = \begin{bmatrix} 1 \\ 1 \\ 0 \end{bmatrix} \quad \text{and} \quad \mathbf{u}_2 = \begin{bmatrix} -1 \\ 1 \\ 1 \end{bmatrix}$$

we have

$$\begin{aligned} \mathbf{u}_1\cdot\mathbf{v} &= 2 \quad & \mathbf{u}_2\cdot\mathbf{v} &= -2 \\ \mathbf{u}_1\cdot\mathbf{u}_1 &= 2 \quad & \mathbf{u}_2\cdot\mathbf{u}_2 &= 3 \end{aligned}$$

Therefore,

$$\begin{aligned}\text{proj}_W(\mathbf{v}) &= \left(\frac{\mathbf{u}_1 \cdot \mathbf{v}}{\mathbf{u}_1 \cdot \mathbf{u}_1}\right)\mathbf{u}_1 + \left(\frac{\mathbf{u}_2 \cdot \mathbf{v}}{\mathbf{u}_2 \cdot \mathbf{u}_2}\right)\mathbf{u}_2 \\ &= \tfrac{2}{2}\begin{bmatrix}1\\1\\0\end{bmatrix} - \tfrac{2}{3}\begin{bmatrix}-1\\1\\1\end{bmatrix} = \begin{bmatrix}\frac{5}{3}\\ \frac{1}{3}\\ -\frac{2}{3}\end{bmatrix}\end{aligned}$$

and $$\text{perp}_W(\mathbf{v}) = \mathbf{v} - \text{proj}_W(\mathbf{v}) = \begin{bmatrix}3\\-1\\2\end{bmatrix} - \begin{bmatrix}\frac{5}{3}\\ \frac{1}{3}\\ -\frac{2}{3}\end{bmatrix} = \begin{bmatrix}\frac{4}{3}\\ -\frac{4}{3}\\ \frac{8}{3}\end{bmatrix}$$

It is easy to see that $\text{proj}_W(\mathbf{v})$ is in W, since it satisfies the equation of the plane. It is equally easy to see that $\text{perp}_W(\mathbf{v})$ is orthogonal to W, since it is a scalar multiple of the normal vector $\begin{bmatrix}1\\-1\\2\end{bmatrix}$ to W. (See Figure 5.9.)

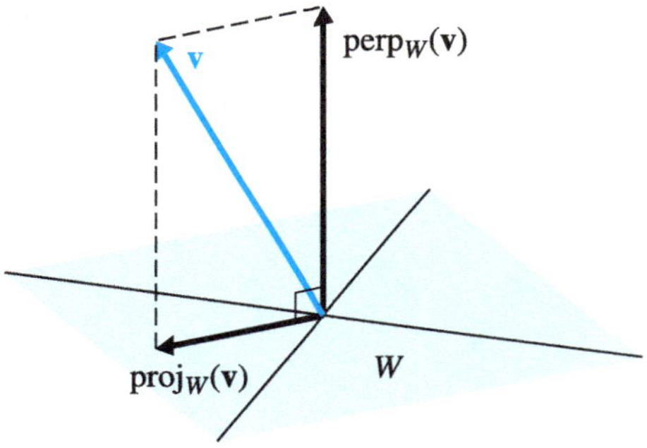

Figure 5.9

$\mathbf{v} = \text{proj}_W(\mathbf{v}) + \text{perp}_W(\mathbf{v})$

The next theorem shows that we can always find a decomposition of a vector with respect to a subspace and its orthogonal complement.

Theorem 5.11 **The Orthogonal Decomposition Theorem**

Let W be a subspace of $\mathbb{R}^n$ and let $\mathbf{v}$ be a vector in $\mathbb{R}^n$. Then there are unique vectors $\mathbf{w}$ in W and $\mathbf{w}^\perp$ in $W^\perp$ such that

$$\mathbf{v} = \mathbf{w} + \mathbf{w}^\perp$$

Proof We need to show two things: that such a decomposition *exists* and that it is *unique*.

To show existence, we choose an orthogonal basis $\{\mathbf{u}_1, \ldots, \mathbf{u}_k\}$ for W. Let $\mathbf{w} = \text{proj}_W(\mathbf{v})$ and let $\mathbf{w}^\perp = \text{perp}_W(\mathbf{v})$. Then

$$\mathbf{w} + \mathbf{w}^\perp = \text{proj}_W(\mathbf{v}) + \text{perp}_W(\mathbf{v}) = \text{proj}_W(\mathbf{v}) + (\mathbf{v} - \text{proj}_W(\mathbf{v})) = \mathbf{v}$$

Clearly, $\mathbf{w} = \text{proj}_W(\mathbf{v})$ is in W, since it is a linear combination of the basis vectors $\mathbf{u}_1, \ldots, \mathbf{u}_k$. To show that $\mathbf{w}^\perp$ is in $W^\perp$, it is enough to show that $\mathbf{w}^\perp$ is orthogonal to each of the basis vectors $\mathbf{u}_i$, by Theorem 5.9(d). We compute

$$\begin{aligned}
\mathbf{u}_i \cdot \mathbf{w}^\perp &= u_i \cdot \text{perp}_W(\mathbf{v}) \\
&= \mathbf{u}_i \cdot (\mathbf{v} - \text{proj}_W(\mathbf{v})) \\
&= \mathbf{u}_i \cdot \left(\mathbf{v} - \left(\frac{\mathbf{u}_1 \cdot \mathbf{v}}{\mathbf{u}_1 \cdot \mathbf{u}_1}\right)\mathbf{u}_1 - \cdots - \left(\frac{\mathbf{u}_k \cdot \mathbf{v}}{\mathbf{u}_k \cdot \mathbf{u}_k}\right)\mathbf{u}_k\right) \\
&= \mathbf{u}_i \cdot \mathbf{v} - \left(\frac{\mathbf{u}_1 \cdot \mathbf{v}}{\mathbf{u}_1 \cdot \mathbf{u}_1}\right)(\mathbf{u}_i \cdot \mathbf{u}_1) - \cdots - \left(\frac{\mathbf{u}_i \cdot \mathbf{v}}{\mathbf{u}_i \cdot \mathbf{u}_i}\right)(\mathbf{u}_i \cdot \mathbf{u}_i) - \cdots \\
&\quad - \left(\frac{\mathbf{u}_k \cdot \mathbf{v}}{\mathbf{u}_k \cdot \mathbf{u}_k}\right)(\mathbf{u}_i \cdot \mathbf{u}_k) \\
&= \mathbf{u}_i \cdot \mathbf{v} - 0 - \cdots - \left(\frac{\mathbf{u}_i \cdot \mathbf{v}}{\mathbf{u}_i \cdot \mathbf{u}_i}\right)(\mathbf{u}_i \cdot \mathbf{u}_i) - \cdots - 0 \\
&= \mathbf{u}_i \cdot \mathbf{v} - \mathbf{u}_i \cdot \mathbf{v} = 0
\end{aligned}$$

since $\mathbf{u}_i \cdot \mathbf{u}_j = 0$ for $j \neq i$. This proves that $\mathbf{w}^\perp$ is in $W^\perp$ and completes the existence part of the proof.

To show the uniqueness of this decomposition, let's suppose we have another decomposition $\mathbf{v} = \mathbf{w}_1 + \mathbf{w}_1^\perp$, where $\mathbf{w}_1$ is in W and $\mathbf{w}_1^\perp$ is in $W^\perp$. Then $\mathbf{w} + \mathbf{w}^\perp = \mathbf{w}_1 + \mathbf{w}_1^\perp$, so

$$\mathbf{w} - \mathbf{w}_1 = \mathbf{w}_1^\perp - \mathbf{w}^\perp$$

But since $\mathbf{w} - \mathbf{w}_1$ is in W and $\mathbf{w}_1^\perp - \mathbf{w}^\perp$ is in $W^\perp$ (because these are subspaces), we know that this common vector is in $W \cap W^\perp = \{\mathbf{0}\}$ [using Theorem 5.9(c)]. Thus,

$$\mathbf{w} - \mathbf{w}_1 = \mathbf{w}_1^\perp - \mathbf{w}^\perp = 0$$

so $\mathbf{w}_1 = \mathbf{w}$ and $\mathbf{w}_1^\perp = \mathbf{w}_1$.

Example 5.11 illustrated the Orthogonal Decomposition Theorem. When W is the subspace of $\mathbb{R}^3$ given by the plane with equation $x - y + 2z = 0$, the orthogonal decomposition of $\mathbf{v} = \begin{bmatrix} 3 \\ -1 \\ 2 \end{bmatrix}$ with respect to W is $\mathbf{v} = \mathbf{w} + \mathbf{w}^\perp$, where

$$\mathbf{w} = \text{proj}_W(\mathbf{v}) = \begin{bmatrix} \frac{5}{3} \\ \frac{1}{3} \\ -\frac{2}{3} \end{bmatrix} \quad \text{and} \quad \mathbf{w}^\perp = \text{perp}_W(\mathbf{v}) = \begin{bmatrix} \frac{4}{3} \\ -\frac{4}{3} \\ \frac{8}{3} \end{bmatrix}$$

The uniqueness of the orthogonal decomposition guarantees that the definitions of $\text{proj}_W(\mathbf{v})$ and $\text{perp}_W(\mathbf{v})$ do not depend on the choice of orthogonal basis. The Orthogonal Decomposition Theorem also allows us to prove property (b) of Theorem 5.9. We state that property here as a corollary to the Orthogonal Decomposition Theorem.

Corollary 5.12 If W is a subspace of $\mathbb{R}^n$, then

$$(W^\perp)^\perp = W$$

Proof If $\mathbf{w}$ is in W and $\mathbf{x}$ is in $W^\perp$, then $\mathbf{w}\cdot\mathbf{x} = 0$. But this now implies that $\mathbf{w}$ is in $(W^\perp)^\perp$. Hence, $W \subseteq (W^\perp)^\perp$. Now let $\mathbf{v}$ be in $(W^\perp)^\perp$. By Theorem 5.11, we can write $\mathbf{v} = \mathbf{w} + \mathbf{w}^\perp$ for (unique) vectors $\mathbf{w}$ in W and $\mathbf{w}^\perp$ in $W^\perp$. But now

$$0 = \mathbf{v}\cdot\mathbf{w}^\perp = (\mathbf{w} + \mathbf{w}^\perp)\cdot\mathbf{w}^\perp = \mathbf{w}\cdot\mathbf{w}^\perp + \mathbf{w}^\perp\cdot\mathbf{w}^\perp = 0 + \mathbf{w}^\perp\cdot\mathbf{w}^\perp = \mathbf{w}^\perp\cdot\mathbf{w}^\perp$$

so $\mathbf{w}^\perp = \mathbf{0}$. Therefore, $\mathbf{v} = \mathbf{w} + \mathbf{w}^\perp = \mathbf{w}$, and thus $\mathbf{v}$ is in W. This shows that $(W^\perp)^\perp \subseteq W$ and, since the reverse inclusion is also true, we conclude that $(W^\perp)^\perp = W$, as required.

There is also a nice relationship between the dimensions of W and $W^\perp$, expressed in Theorem 5.13.

Theorem 5.13 If W is a subspace of $\mathbb{R}^n$, then

$$\dim W + \dim W^\perp = n$$

Proof Let $\{\mathbf{u}_1, \ldots, \mathbf{u}_k\}$ be an orthogonal basis for W and let $\{\mathbf{v}_1, \ldots, \mathbf{v}_l\}$ be an orthogonal basis for $W^\perp$. Then $\dim W = k$ and $\dim W^\perp = l$. Let $\mathcal{B} = \{\mathbf{u}_1, \ldots, \mathbf{u}_k, \mathbf{v}_1, \ldots, \mathbf{v}_l\}$. We claim that $\mathcal{B}$ is an orthogonal basis for $\mathbb{R}^n$.

We first note that, since each $\mathbf{u}_i$ is in W and each $\mathbf{v}_j$ is in $W^\perp$,

$$\mathbf{u}_i\cdot\mathbf{v}_j = 0 \quad \text{for } i = 1, \ldots, k \text{ and } j = 1, \ldots, l$$

Thus, $\mathcal{B}$ is an orthogonal set and, hence, is linearly independent, by Theorem 5.1. Next, if $\mathbf{v}$ is a vector in $\mathbb{R}^n$, the Orthogonal Decomposition Theorem tells us that $\mathbf{v} = \mathbf{w} + \mathbf{w}^\perp$ for some $\mathbf{w}$ in W and $\mathbf{w}^\perp$ in $W^\perp$. Since $\mathbf{w}$ can be written as a linear combination of the vectors $\mathbf{u}_i$ and $\mathbf{w}^\perp$ can be written as a linear combination of the vectors $\mathbf{v}_j$, $\mathbf{v}$ can be written as a linear combination of the vectors in $\mathcal{B}$. Therefore, $\mathcal{B}$ spans $\mathbb{R}^n$ also and so is a basis for $\mathbb{R}^n$. It follows that $k + l = \dim \mathbb{R}^n$, or

$$\dim W + \dim W^\perp = n$$

As a lovely bonus, when we apply this result to the fundamental subspaces of a matrix, we get a quick proof of the Rank Theorem (Theorem 3.26), restated here as Corollary 5.14.

Corollary 5.14 The Rank Theorem

If A is an $m \times n$ matrix, then

$$\operatorname{rank}(A) + \operatorname{nullity}(A) = n$$

Proof In Theorem 5.13, take $W = \operatorname{row}(A)$. Then $W^\perp = \operatorname{null}(A)$, by Theorem 5.10, so $\dim W = \operatorname{rank}(A)$ and $\dim W^\perp = \operatorname{nullity}(A)$. The result follows.

Note that we get a counterpart identity by taking $W = \operatorname{col}(A)$ [and therefore $W^\perp = \operatorname{null}(A^T)$]:

$$\operatorname{rank}(A) + \operatorname{nullity}(A^T) = m$$

Sections 5.1 and 5.2 have illustrated some of the advantages of working with orthogonal bases. However, we have not established that every subspace *has* an orthogonal basis, nor have we given a method for constructing such a basis (except in particular examples, such as Example 5.3). These issues are the subject of the next section.

Exercises 5.2

In Exercises 1–6, find the orthogonal complement $W^\perp$ of W and give a basis for $W^\perp$.

1. $W = \left\{ \begin{bmatrix} x \\ y \end{bmatrix} : 2x - y = 0 \right\}$

2. $W = \left\{ \begin{bmatrix} x \\ y \end{bmatrix} : 3x + 4y = 0 \right\}$

3. $W = \left\{ \begin{bmatrix} x \\ y \\ z \end{bmatrix} : x + y - z = 0 \right\}$

4. $W = \left\{ \begin{bmatrix} x \\ y \\ z \end{bmatrix} : 2x - y + 3z = 0 \right\}$

5. $W = \left\{ \begin{bmatrix} x \\ y \\ z \end{bmatrix} : x = t, y = -t, z = 3t \right\}$

6. $W = \left\{ \begin{bmatrix} x \\ y \\ z \end{bmatrix} : x = \frac{1}{2}t, y = -\frac{1}{2}t, z = 2t \right\}$

In Exercises 7 and 8, find bases for the row space and null space of A. Verify that every vector in row(A) is orthogonal to every vector in null(A).

7. $A = \begin{bmatrix} 1 & -1 & 3 \\ 5 & 2 & 1 \\ 0 & 1 & -2 \\ -1 & -1 & 1 \end{bmatrix}$

8. $A = \begin{bmatrix} 1 & 1 & -1 & 0 & 2 \\ -2 & 0 & 2 & 4 & 4 \\ 2 & 2 & -2 & 0 & 1 \\ -3 & -1 & 3 & 4 & 5 \end{bmatrix}$

In Exercises 9 and 10, find bases for the column space of A and the null space of A^T for the given exercise. Verify that every vector in col(A) is orthogonal to every vector in null(A^T).

9. Exercise 7 **10.** Exercise 8

In Exercises 11–14, let W be the subspace spanned by the given vectors. Find a basis for $W^\perp$.

11. $\mathbf{w}_1 = \begin{bmatrix} 2 \\ 1 \\ -2 \end{bmatrix}, \mathbf{w}_2 = \begin{bmatrix} 4 \\ 0 \\ 1 \end{bmatrix}$

12. $\mathbf{w}_1 = \begin{bmatrix} 1 \\ -1 \\ 3 \\ -2 \end{bmatrix}, \mathbf{w}_2 = \begin{bmatrix} 0 \\ 1 \\ -2 \\ 1 \end{bmatrix}$

13. $\mathbf{w}_1 = \begin{bmatrix} 2 \\ -1 \\ 6 \\ 3 \end{bmatrix}, \mathbf{w}_2 = \begin{bmatrix} -1 \\ 2 \\ -3 \\ -2 \end{bmatrix}, \mathbf{w}_3 = \begin{bmatrix} 2 \\ 5 \\ 6 \\ 1 \end{bmatrix}$

14. $\mathbf{w}_1 = \begin{bmatrix} 4 \\ 6 \\ -1 \\ 1 \\ -1 \end{bmatrix}, \mathbf{w}_2 = \begin{bmatrix} 1 \\ 2 \\ 0 \\ 1 \\ -3 \end{bmatrix}, \mathbf{w}_3 = \begin{bmatrix} 2 \\ 2 \\ 2 \\ -1 \\ 2 \end{bmatrix}$

In Exercises 15–18, find the orthogonal projection of **v** *onto the subspace W spanned by the vectors $\mathbf{u}_i$. (You may assume that the vectors $\mathbf{u}_i$ are orthogonal.)*

15. $\mathbf{v} = \begin{bmatrix} 7 \\ -4 \end{bmatrix}, \mathbf{u}_1 = \begin{bmatrix} 1 \\ 1 \end{bmatrix}$

16. $\mathbf{v} = \begin{bmatrix} 3 \\ 1 \\ -2 \end{bmatrix}, \mathbf{u}_1 = \begin{bmatrix} 1 \\ 1 \\ 1 \end{bmatrix}, \mathbf{u}_2 = \begin{bmatrix} 1 \\ -1 \\ 0 \end{bmatrix}$

17. $\mathbf{v} = \begin{bmatrix} 1 \\ 2 \\ 3 \end{bmatrix}, \mathbf{u}_1 = \begin{bmatrix} 2 \\ -2 \\ 1 \end{bmatrix}, \mathbf{u}_2 = \begin{bmatrix} -1 \\ 1 \\ 4 \end{bmatrix}$

18. $\mathbf{v} = \begin{bmatrix} 3 \\ -2 \\ 4 \\ -3 \end{bmatrix}, \mathbf{u}_1 = \begin{bmatrix} 1 \\ 1 \\ 0 \\ 0 \end{bmatrix}, \mathbf{u}_2 = \begin{bmatrix} 1 \\ -1 \\ -1 \\ 1 \end{bmatrix}, \mathbf{u}_3 = \begin{bmatrix} 0 \\ 0 \\ 1 \\ 1 \end{bmatrix}$

In Exercises 19–22, find the orthogonal decomposition of $\mathbf{v}$ *with respect to W.*

19. $\mathbf{v} = \begin{bmatrix} 2 \\ -2 \end{bmatrix}, W = \text{span}\left(\begin{bmatrix} 1 \\ 3 \end{bmatrix}\right)$

20. $\mathbf{v} = \begin{bmatrix} 3 \\ 2 \\ -1 \end{bmatrix}, W = \text{span}\left(\begin{bmatrix} 1 \\ 1 \\ 1 \end{bmatrix}\right)$

21. $\mathbf{v} = \begin{bmatrix} 4 \\ -2 \\ 3 \end{bmatrix}, W = \text{span}\left(\begin{bmatrix} 1 \\ 2 \\ 1 \end{bmatrix}, \begin{bmatrix} 1 \\ -1 \\ 1 \end{bmatrix}\right)$

22. $\mathbf{v} = \begin{bmatrix} 2 \\ -1 \\ 5 \\ 6 \end{bmatrix}, W = \text{span}\left(\begin{bmatrix} 1 \\ 1 \\ 1 \\ 0 \end{bmatrix}, \begin{bmatrix} 1 \\ 0 \\ -1 \\ 1 \end{bmatrix}\right)$

23. Prove Theorem 5.9(c).

24. Prove Theorem 5.9(d).

25. Let W be a subspace of $\mathbb{R}^n$ and $\mathbf{v}$ a vector in $\mathbb{R}^n$. Suppose that $\mathbf{w}$ and $\mathbf{w}'$ are orthogonal vectors with $\mathbf{w}$ in W and that $\mathbf{v} = \mathbf{w} + \mathbf{w}'$. Is it necessarily true that $\mathbf{w}'$ is in $W^\perp$? Either prove that it is true or find a counterexample.

26. Let $\{\mathbf{v}_1, \ldots, \mathbf{v}_n\}$ be an orthogonal basis for $\mathbb{R}^n$ and let $W = \text{span}(\mathbf{v}_1, \ldots, \mathbf{v}_k)$. Is it necessarily true that $W^\perp = \text{span}(\mathbf{v}_{k+1}, \ldots, \mathbf{v}_n)$? Either prove that it is true or find a counterexample.

In Exercises 27–29, let W be a subspace of $\mathbb{R}^n$, *and let* $\mathbf{x}$ *be a vector in* $\mathbb{R}^n$.

27. Prove that $\mathbf{x}$ is in W if and only if $\text{proj}_W(\mathbf{x}) = \mathbf{x}$.

28. Prove that $\mathbf{x}$ is orthogonal to W if and only if $\text{proj}_W(\mathbf{x}) = \mathbf{0}$.

29. Prove that $\text{proj}_W(\text{proj}_W(\mathbf{x})) = \text{proj}_W(\mathbf{x})$.

30. Let $S = \{\mathbf{v}_1, \ldots, \mathbf{v}_k\}$ be an orthonormal set in $\mathbb{R}^n$, and let $\mathbf{x}$ be a vector in $\mathbb{R}^n$.

(a) Prove that

$$\|\mathbf{x}\|^2 \geq |\mathbf{x}\cdot\mathbf{v}_1|^2 + |\mathbf{x}\cdot\mathbf{v}_2|^2 + \cdots + |\mathbf{x}\cdot\mathbf{v}_k|^2$$

(This inequality is called ***Bessel's Inequality.***)

(b) Prove that Bessel's Inequality is an equality if and only if $\mathbf{x}$ is in $\text{span}(S)$.

5.3 The Gram-Schmidt Process and the *QR* Factorization

In this section, we present a simple method for constructing an orthogonal (or orthonormal) basis for any subspace of $\mathbb{R}^n$. This method will then lead us to one of the most useful of all matrix factorizations.

The Gram-Schmidt Process

We would like to be able to find an orthogonal basis for a subspace W of $\mathbb{R}^n$. The idea is to begin with an arbitrary basis $\{\mathbf{x}_1, \ldots, \mathbf{x}_k\}$ for W and to "orthogonalize" it one vector at a time. We will illustrate the basic construction with the subspace W from Example 5.3.

Example 5.12

Let $W = \text{span}(\mathbf{x}_1, \mathbf{x}_2)$, where

$$\mathbf{x}_1 = \begin{bmatrix} 1 \\ 1 \\ 0 \end{bmatrix} \quad \text{and} \quad \mathbf{x}_2 = \begin{bmatrix} -2 \\ 0 \\ 1 \end{bmatrix}$$

Construct an orthogonal basis for W.

Solution Starting with $\mathbf{x}_1$, we get a second vector that is orthogonal to it by taking the component of $\mathbf{x}_2$ orthogonal to $\mathbf{x}_1$ (Figure 5.10).

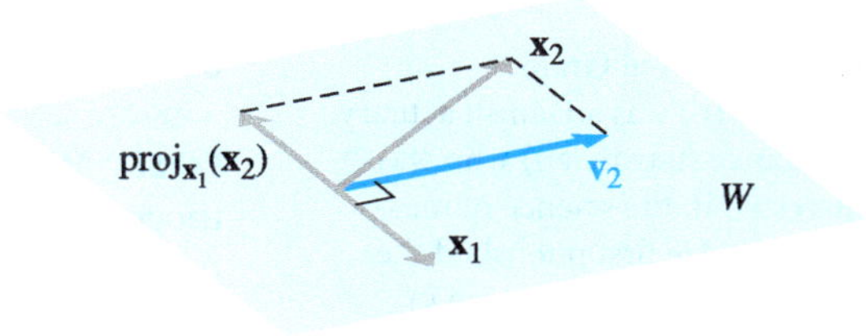

Figure 5.10
Constructing $\mathbf{v}_2$ orthogonal to $\mathbf{x}_1$

Algebraically, we set $\mathbf{v}_1 = \mathbf{x}_1$, so

$$\begin{aligned}
\mathbf{v}_2 &= \text{perp}_{\mathbf{x}_1}(\mathbf{x}_2) = \mathbf{x}_2 - \text{proj}_{\mathbf{x}_1}(\mathbf{x}_2) \\
&= \mathbf{x}_2 - \left(\frac{\mathbf{x}_1 \cdot \mathbf{x}_2}{\mathbf{x}_1 \cdot \mathbf{x}_1}\right)\mathbf{x}_1 \\
&= \begin{bmatrix} -2 \\ 0 \\ 1 \end{bmatrix} - \left(\frac{-2}{2}\right)\begin{bmatrix} 1 \\ 1 \\ 0 \end{bmatrix} = \begin{bmatrix} -1 \\ 1 \\ 1 \end{bmatrix}
\end{aligned}$$

Then $\{\mathbf{v}_1, \mathbf{v}_2\}$ is an orthogonal set of vectors in W. Hence, $\{\mathbf{v}_1, \mathbf{v}_2\}$ is a linearly independent set and therefore a basis for W, since $\dim W = 2$.

Remark Observe that this method depends on the *order* of the original basis vectors. In Example 5.12, if we had taken $\mathbf{x}_1 = \begin{bmatrix} -2 \\ 0 \\ 1 \end{bmatrix}$ and $\mathbf{x}_2 = \begin{bmatrix} 1 \\ 1 \\ 0 \end{bmatrix}$, we would have obtained a different orthogonal basis for W. (Verify this.)

The generalization of this method to more than two vectors begins as in Example 5.12. Then the process is to iteratively construct the components of subsequent vectors orthogonal to all of the vectors that have already been constructed. The method is known as the ***Gram-Schmidt Process.***

Theorem 5.15 The Gram-Schmidt Process

Let $\{\mathbf{x}_1, \ldots, \mathbf{x}_k\}$ be a basis for a subspace W of $\mathbb{R}^n$ and define the following:

$$\begin{aligned}
\mathbf{v}_1 &= \mathbf{x}_1, & W_1 &= \text{span}(\mathbf{x}_1) \\
\mathbf{v}_2 &= \mathbf{x}_2 - \left(\frac{\mathbf{v}_1 \cdot \mathbf{x}_2}{\mathbf{v}_1 \cdot \mathbf{v}_1}\right)\mathbf{v}_1, & W_2 &= \text{span}(\mathbf{x}_1, \mathbf{x}_2) \\
\mathbf{v}_3 &= \mathbf{x}_3 - \left(\frac{\mathbf{v}_1 \cdot \mathbf{x}_3}{\mathbf{v}_1 \cdot \mathbf{v}_1}\right)\mathbf{v}_1 - \left(\frac{\mathbf{v}_2 \cdot \mathbf{x}_3}{\mathbf{v}_2 \cdot \mathbf{v}_2}\right)\mathbf{v}_2, & W_3 &= \text{span}(\mathbf{x}_1, \mathbf{x}_2, \mathbf{x}_3) \\
&\vdots \\
\mathbf{v}_k &= \mathbf{x}_k - \left(\frac{\mathbf{v}_1 \cdot \mathbf{x}_k}{\mathbf{v}_1 \cdot \mathbf{v}_1}\right)\mathbf{v}_1 - \left(\frac{\mathbf{v}_2 \cdot \mathbf{x}_k}{\mathbf{v}_2 \cdot \mathbf{v}_2}\right)\mathbf{v}_2 - \cdots \\
&\quad - \left(\frac{\mathbf{v}_{k-1} \cdot \mathbf{x}_k}{\mathbf{v}_{k-1} \cdot \mathbf{v}_{k-1}}\right)\mathbf{v}_{k-1}, & W_k &= \text{span}(\mathbf{x}_1, \ldots, \mathbf{x}_k)
\end{aligned}$$

Then for each $i = 1, \ldots, k$, $\{\mathbf{v}_1, \ldots, \mathbf{v}_i\}$ is an orthogonal basis for W_i. In particular, $\{\mathbf{v}_1, \ldots, \mathbf{v}_k\}$ is an orthogonal basis for W.

Jörgen Pedersen Gram (1850–1916) was a Danish actuary (insurance statistician) who was interested in the science of measurement. He first published the process that bears his name in an 1883 paper on least squares. Erhard Schmidt (1876–1959) was a German mathematician who studied under the great David Hilbert and is considered one of the founders of the branch of mathematics known as functional analysis. His contribution to the Gram-Schmidt Process came in a 1907 paper on integral equations, in which he wrote out the details of the method more explicitly than Gram had done.

Stated succinctly, Theorem 5.15 says that every subspace of $\mathbb{R}^n$ has an orthogonal basis, and it gives an algorithm for constructing such a basis.

Proof We will prove by induction that, for each $i = 1, \ldots, k$, $\{\mathbf{v}_1, \ldots, \mathbf{v}_i\}$ is an orthogonal basis for W_i.

Since $\mathbf{v}_1 = \mathbf{x}_1$, clearly $\{\mathbf{v}_1\}$ is an (orthogonal) basis for $W_1 = \text{span}(\mathbf{x}_1)$. Now assume that, for some $i < k$, $\{\mathbf{v}_1, \ldots, \mathbf{v}_i\}$ is an orthogonal basis for W_i. Then

$$\mathbf{v}_{i+1} = \mathbf{x}_{i+1} - \left(\frac{\mathbf{v}_1 \cdot \mathbf{x}_{i+1}}{\mathbf{v}_1 \cdot \mathbf{v}_1}\right)\mathbf{v}_1 - \left(\frac{\mathbf{v}_2 \cdot \mathbf{x}_{i+1}}{\mathbf{v}_2 \cdot \mathbf{v}_2}\right)\mathbf{v}_2 - \cdots - \left(\frac{\mathbf{v}_i \cdot \mathbf{x}_{i+1}}{\mathbf{v}_i \cdot \mathbf{v}_i}\right)\mathbf{v}_i$$

By the induction hypothesis, $\{\mathbf{v}_1, \ldots, \mathbf{v}_i\}$ is an orthogonal basis for $\text{span}(\mathbf{x}_1, \ldots, \mathbf{x}_i) = W_i$. Hence,

$$\mathbf{v}_{i+1} = \mathbf{x}_{i+1} - \text{proj}_{W_i}(\mathbf{x}_{i+1}) = \text{perp}_{W_i}(\mathbf{x}_{i+1})$$

So, by the Orthogonal Decomposition Theorem, $\mathbf{v}_{i+1}$ is orthogonal to W_i. By definition, $\mathbf{v}_1, \ldots, \mathbf{v}_i$ are linear combinations of $\mathbf{x}_1, \ldots, \mathbf{x}_i$ and, hence, are in W_i. Therefore, $\{\mathbf{v}_1, \ldots, \mathbf{v}_{i+1}\}$ is an orthogonal set of vectors in W_{i+1}.

Moreover, $\mathbf{v}_{i+1} \neq \mathbf{0}$, since otherwise $\mathbf{x}_{i+1} = \text{proj}_{W_i}(\mathbf{x}_{i+1})$, which in turn implies that $\mathbf{x}_{i+1}$ is in W_i. But this is impossible, since $W_i = \text{span}(\mathbf{x}_1, \ldots, \mathbf{x}_i)$ and $\{\mathbf{x}_1, \ldots, \mathbf{x}_{i+1}\}$ is linearly independent. (Why?) We conclude that $\{\mathbf{v}_1, \ldots, \mathbf{v}_{i+1}\}$ is a set of $i + 1$ linearly independent vectors in W_{i+1}. Consequently, $\{\mathbf{v}_1, \ldots, \mathbf{v}_{i+1}\}$ is a basis for W_{i+1}, since $\dim W_{i+1} = i + 1$. This completes the proof.

If we require an orthonormal basis for W, we simply need to normalize the orthogonal vectors produced by the Gram-Schmidt Process. That is, for each i, we replace $\mathbf{v}_i$ by the unit vector $\mathbf{q}_i = (1/\|\mathbf{v}_i\|)\mathbf{v}_i$.

Example 5.13

Apply the Gram-Schmidt Process to construct an orthonormal basis for the subspace $W = \text{span}(\mathbf{x}_1, \mathbf{x}_2, \mathbf{x}_3)$ of $\mathbb{R}^4$, where

$$\mathbf{x}_1 = \begin{bmatrix} 1 \\ -1 \\ -1 \\ 1 \end{bmatrix}, \quad \mathbf{x}_2 = \begin{bmatrix} 2 \\ 1 \\ 0 \\ 1 \end{bmatrix}, \quad \mathbf{x}_3 = \begin{bmatrix} 2 \\ 2 \\ 1 \\ 2 \end{bmatrix}$$

Solution First we note that $\{\mathbf{x}_1, \mathbf{x}_2, \mathbf{x}_3\}$ is a linearly independent set, so it forms a basis for W. We begin by setting $\mathbf{v}_1 = \mathbf{x}_1$. Next, we compute the component of $\mathbf{x}_2$ orthogonal to $W_1 = \text{span}(\mathbf{v}_1)$:

$$\begin{aligned}
\mathbf{v}_2 = \text{perp}_{W_1}(\mathbf{x}_2) &= \mathbf{x}_2 - \left(\frac{\mathbf{v}_1 \cdot \mathbf{x}_2}{\mathbf{v}_1 \cdot \mathbf{v}_1}\right)\mathbf{v}_1 \\
&= \begin{bmatrix} 2 \\ 1 \\ 0 \\ 1 \end{bmatrix} - \left(\tfrac{2}{4}\right)\begin{bmatrix} 1 \\ -1 \\ -1 \\ 1 \end{bmatrix} \\
&= \begin{bmatrix} \frac{3}{2} \\ \frac{3}{2} \\ \frac{1}{2} \\ \frac{1}{2} \end{bmatrix}
\end{aligned}$$

For hand calculations, it is a good idea to "scale" $\mathbf{v}_2$ at this point to eliminate fractions. When we are finished, we can rescale the orthogonal set we are constructing to obtain an orthonormal set; thus, we can replace each $\mathbf{v}_i$ by any convenient scalar multiple without affecting the final result. Accordingly, we replace $\mathbf{v}_2$ by

$$\mathbf{v}_2' = 2\mathbf{v}_2 = \begin{bmatrix} 3 \\ 3 \\ 1 \\ 1 \end{bmatrix}$$

We now find the component of $\mathbf{x}_3$ orthogonal to

$$W_2 = \text{span}(\mathbf{x}_1, \mathbf{x}_2) = \text{span}(\mathbf{v}_1, \mathbf{v}_2) = \text{span}(\mathbf{v}_1, \mathbf{v}_2')$$

using the orthogonal basis $\{\mathbf{v}_1, \mathbf{v}_2'\}$:

$$\begin{aligned} \mathbf{v}_3 = \text{perp}_{W_2}(\mathbf{x}_3) &= \mathbf{x}_3 - \left(\frac{\mathbf{v}_1 \cdot \mathbf{x}_3}{\mathbf{v}_1 \cdot \mathbf{v}_1}\right)\mathbf{v}_1 - \left(\frac{\mathbf{v}_2' \cdot \mathbf{x}_3}{\mathbf{v}_2' \cdot \mathbf{v}_2'}\right)\mathbf{v}_2' \\ &= \begin{bmatrix} 2 \\ 2 \\ 1 \\ 2 \end{bmatrix} - \left(\tfrac{1}{4}\right)\begin{bmatrix} 1 \\ -1 \\ -1 \\ 1 \end{bmatrix} - \left(\tfrac{15}{20}\right)\begin{bmatrix} 3 \\ 3 \\ 1 \\ 1 \end{bmatrix} \\ &= \begin{bmatrix} -\frac{1}{2} \\ 0 \\ \frac{1}{2} \\ 1 \end{bmatrix} \end{aligned}$$

Again, we rescale and use $\mathbf{v}_3' = 2\mathbf{v}_3 = \begin{bmatrix} -1 \\ 0 \\ 1 \\ 2 \end{bmatrix}$.

We now have an orthogonal basis $\{\mathbf{v}_1, \mathbf{v}_2', \mathbf{v}_3'\}$ for W. (Check to make sure that these vectors are orthogonal.) To obtain an orthonormal basis, we normalize each vector:

$$\mathbf{q}_1 = \left(\frac{1}{\|\mathbf{v}_1\|}\right)\mathbf{v}_1 = \left(\frac{1}{2}\right)\begin{bmatrix} 1 \\ -1 \\ -1 \\ 1 \end{bmatrix} = \begin{bmatrix} 1/2 \\ -1/2 \\ -1/2 \\ 1/2 \end{bmatrix}$$

$$\mathbf{q}_2 = \left(\frac{1}{\|\mathbf{v}_2'\|}\right)\mathbf{v}_2' = \left(\frac{1}{2\sqrt{5}}\right)\begin{bmatrix} 3 \\ 3 \\ 1 \\ 1 \end{bmatrix} = \begin{bmatrix} 3/2\sqrt{5} \\ 3/2\sqrt{5} \\ 1/2\sqrt{5} \\ 1/2\sqrt{5} \end{bmatrix} = \begin{bmatrix} 3\sqrt{5}/10 \\ 3\sqrt{5}/10 \\ \sqrt{5}/10 \\ \sqrt{5}/10 \end{bmatrix}$$

$$\mathbf{q}_3 = \left(\frac{1}{\|\mathbf{v}_3'\|}\right)\mathbf{v}_3' = \left(\frac{1}{\sqrt{6}}\right)\begin{bmatrix} -1 \\ 0 \\ 1 \\ 2 \end{bmatrix} = \begin{bmatrix} -1/\sqrt{6} \\ 0 \\ 1/\sqrt{6} \\ 2/\sqrt{6} \end{bmatrix} = \begin{bmatrix} -\sqrt{6}/6 \\ 0 \\ \sqrt{6}/6 \\ \sqrt{6}/3 \end{bmatrix}$$

Then $\{\mathbf{q}_1, \mathbf{q}_2, \mathbf{q}_3\}$ is an orthonormal basis for W.

One of the important uses of the Gram-Schmidt Process is to construct an orthogonal basis that contains a specified vector. The next example illustrates this application.

Example 5.14 Find an orthogonal basis for $\mathbb{R}^3$ that contains the vector

$$\mathbf{v}_1 = \begin{bmatrix} 1 \\ 2 \\ 3 \end{bmatrix}$$

Solution We first find *any* basis for $\mathbb{R}^3$ containing $\mathbf{v}_1$. If we take

$$\mathbf{x}_2 = \begin{bmatrix} 0 \\ 1 \\ 0 \end{bmatrix} \quad \text{and} \quad \mathbf{x}_3 = \begin{bmatrix} 0 \\ 0 \\ 1 \end{bmatrix}$$

then $\{\mathbf{v}_1, \mathbf{x}_2, \mathbf{x}_3\}$ is clearly a basis for $\mathbb{R}^3$. (Why?) We now apply the Gram-Schmidt Process to this basis to obtain

$$\mathbf{v}_2 = \mathbf{x}_2 - \left(\frac{\mathbf{v}_1 \cdot \mathbf{x}_2}{\mathbf{v}_1 \cdot \mathbf{v}_1}\right)\mathbf{v}_1 = \begin{bmatrix} 0 \\ 1 \\ 0 \end{bmatrix} - \left(\tfrac{2}{14}\right)\begin{bmatrix} 1 \\ 2 \\ 3 \end{bmatrix} = \begin{bmatrix} -\frac{1}{7} \\ \frac{5}{7} \\ -\frac{3}{7} \end{bmatrix}, \quad \mathbf{v}_2' = \begin{bmatrix} -1 \\ 5 \\ -3 \end{bmatrix}$$

and finally

$$\mathbf{v}_3 = \mathbf{x}_3 - \left(\frac{\mathbf{v}_1 \cdot \mathbf{x}_3}{\mathbf{v}_1 \cdot \mathbf{v}_1}\right)\mathbf{v}_1 - \left(\frac{\mathbf{v}_2' \cdot \mathbf{x}_3}{\mathbf{v}_2' \cdot \mathbf{v}_2'}\right)\mathbf{v}_2' = \begin{bmatrix} 0 \\ 0 \\ 1 \end{bmatrix} - \left(\tfrac{3}{14}\right)\begin{bmatrix} 1 \\ 2 \\ 3 \end{bmatrix} - \left(\tfrac{-3}{35}\right)\begin{bmatrix} -1 \\ 5 \\ -3 \end{bmatrix} = \begin{bmatrix} \frac{-3}{10} \\ 0 \\ \frac{1}{10} \end{bmatrix},$$

$$\mathbf{v}_3' = \begin{bmatrix} -3 \\ 0 \\ 1 \end{bmatrix}$$

Then $\{\mathbf{v}_1, \mathbf{v}_2', \mathbf{v}_3'\}$ is an orthogonal basis for $\mathbb{R}^3$ that contains $\mathbf{v}_1$.

Similarly, given a unit vector, we can find an orthonormal basis that contains it by using the preceding method and then normalizing the resulting orthogonal vectors.

Remark When the Gram-Schmidt Process is implemented on a computer, there is almost always some roundoff error, leading to a loss of orthogonality in the vectors $\mathbf{q}_i$. To avoid this loss of orthogonality, some modifications are usually made. The vectors $\mathbf{v}_i$ are normalized as soon as they are computed, rather than at the end, to give the vectors $\mathbf{q}_i$, and as each $\mathbf{q}_i$ is computed, the remaining vectors $\mathbf{x}_j$ are modified to be orthogonal to $\mathbf{q}_i$. This procedure is known as the ***Modified Gram-Schmidt Process.*** In practice, however, a version of the *QR* factorization is used to compute orthonormal bases.

The *QR* Factorization

If A is an $m \times n$ matrix with linearly independent columns (requiring that $m \geq n$), then applying the Gram-Schmidt Process to these columns yields a very useful factorization of A into the product of a matrix Q with orthonormal columns and an upper triangular matrix R. This is the ***QR factorization,*** and it has applications to the numerical approximation of eigenvalues, which we explore at the end of this section, and to the problem of least squares approximation, which we discuss in Chapter 7.

To see how the QR factorization arises, let $\mathbf{a}_1, \ldots, \mathbf{a}_n$ be the (linearly independent) columns of A and let $\mathbf{q}_1, \ldots, \mathbf{q}_n$ be the orthonormal vectors obtained by applying the Gram-Schmidt Process to A with normalizations. From Theorem 5.15, we know that, for each $i = 1, \ldots, n$,

$$W_i = \text{span}(\mathbf{a}_1, \ldots, \mathbf{a}_i) = \text{span}(\mathbf{q}_1, \ldots, \mathbf{q}_i)$$

Therefore, there are scalars $r_{1i}, r_{2i}, \ldots, r_{ii}$ such that

$$\mathbf{a}_i = r_{1i}\mathbf{q}_1 + r_{2i}\mathbf{q}_2 + \cdots + r_{ii}\mathbf{q}_i \quad \text{for } i = 1, \ldots, n$$

That is,

$$\begin{aligned}
\mathbf{a}_1 &= r_{11}\mathbf{q}_1 \\
\mathbf{a}_2 &= r_{12}\mathbf{q}_1 + r_{22}\mathbf{q}_2 \\
&\vdots \\
\mathbf{a}_n &= r_{1n}\mathbf{q}_1 + r_{2n}\mathbf{q}_2 + \cdots + r_{nn}\mathbf{q}_n
\end{aligned}$$

which can be written in matrix form as

$$A = [\mathbf{a}_1 \quad \mathbf{a}_2 \quad \cdots \quad \mathbf{a}_n] = [\mathbf{q}_1 \quad \mathbf{q}_2 \quad \cdots \quad \mathbf{q}_n]\begin{bmatrix} r_{11} & r_{12} & \cdots & r_{1n} \\ 0 & r_{22} & \cdots & r_{2n} \\ \vdots & \vdots & \ddots & \vdots \\ 0 & 0 & \cdots & r_{nn} \end{bmatrix} = QR$$

Clearly, the matrix Q has orthonormal columns. It is also the case that the diagonal entries of R are all nonzero. To see this, observe that if $r_{ii} = 0$, then $\mathbf{a}_i$ is a linear combination of $\mathbf{q}_1, \ldots, \mathbf{q}_{i-1}$ and, hence, is in W_{i-1}. But then $\mathbf{a}_i$ would be a linear combination of $\mathbf{a}_1, \ldots, \mathbf{a}_{i-1}$, which is impossible, since $\mathbf{a}_1, \ldots, \mathbf{a}_i$ are linearly independent. We conclude that $r_{ii} \neq 0$ for $i = 1, \ldots, n$. Since R is upper triangular, it follows that it must be invertible. (See Exercise 23.)

We have proved the following theorem.

Theorem 5.16 **The QR Factorization**

Let A be an $m \times n$ matrix with linearly independent columns. Then A can be factored as $A = QR$, where Q is an $m \times n$ matrix with orthonormal columns and R is an invertible upper triangular matrix.

Remarks

- We can also arrange for the diagonal entries of R to be *positive*. If any $r_{ii} < 0$, simply replace $\mathbf{q}_i$ by $-\mathbf{q}_i$ and r_{ii} by $-r_{ii}$.
- The requirement that A have linearly independent columns is a necessary one. To prove this, suppose that A is an $m \times n$ matrix that has a QR factorization, as in Theorem 5.16. Then, since R is invertible, we have $Q = AR^{-1}$. Hence, rank(Q) = rank(A), by Exercise 61 in Section 3.5. But rank$(Q) = n$, since its columns are orthonormal and, therefore, linearly independent. So rank$(A) = n$ too, and consequently the columns of A are linearly independent, by the Fundamental Theorem.

- The QR factorization can be extended to arbitrary matrices in a slightly modified form. If A is $m \times n$, it is possible to find a sequence of orthogonal matrices $Q_1, \ldots, Q_{m-1}$ such that $Q_{m-1} \cdots Q_2 Q_1 A$ is an upper triangular $m \times n$ matrix R. Then $A = QR$, where $Q = (Q_{m-1} \cdots Q_2 Q_1)^{-1}$ is an orthogonal matrix. We will examine this approach in Exploration: The Modified QR Factorization.

Example 5.15

Find a QR factorization of

$$A = \begin{bmatrix} 1 & 2 & 2 \\ -1 & 1 & 2 \\ -1 & 0 & 1 \\ 1 & 1 & 2 \end{bmatrix}$$

Solution The columns of A are just the vectors from Example 5.13. The orthonormal basis for col(A) produced by the Gram-Schmidt Process was

$$\mathbf{q}_1 = \begin{bmatrix} 1/2 \\ -1/2 \\ -1/2 \\ 1/2 \end{bmatrix}, \quad \mathbf{q}_2 = \begin{bmatrix} 3\sqrt{5}/10 \\ 3\sqrt{5}/10 \\ \sqrt{5}/10 \\ \sqrt{5}/10 \end{bmatrix}, \quad \mathbf{q}_3 = \begin{bmatrix} -\sqrt{6}/6 \\ 0 \\ \sqrt{6}/6 \\ \sqrt{6}/3 \end{bmatrix}$$

so

$$Q = [\mathbf{q}_1 \quad \mathbf{q}_2 \quad \mathbf{q}_3] = \begin{bmatrix} 1/2 & 3\sqrt{5}/10 & -\sqrt{6}/6 \\ -1/2 & 3\sqrt{5}/10 & 0 \\ -1/2 & \sqrt{5}/10 & \sqrt{6}/6 \\ 1/2 & \sqrt{5}/10 & \sqrt{6}/3 \end{bmatrix}$$

From Theorem 5.16, $A = QR$ for some upper triangular matrix R. To find R, we use the fact that Q has orthonormal columns and, hence, $Q^TQ = I$. Therefore,

$$Q^TA = Q^TQR = IR = R$$

We compute

$$R = Q^TA = \begin{bmatrix} 1/2 & -1/2 & -1/2 & 1/2 \\ 3\sqrt{5}/10 & 3\sqrt{5}/10 & \sqrt{5}/10 & \sqrt{5}/10 \\ -\sqrt{6}/6 & 0 & \sqrt{6}/6 & \sqrt{6}/3 \end{bmatrix} \begin{bmatrix} 1 & 2 & 2 \\ -1 & 1 & 2 \\ -1 & 0 & 1 \\ 1 & 1 & 2 \end{bmatrix}$$

$$= \begin{bmatrix} 2 & 1 & 1/2 \\ 0 & \sqrt{5} & 3\sqrt{5}/2 \\ 0 & 0 & \sqrt{6}/2 \end{bmatrix}$$

Exercises 5.3

In Exercises 1–4, the given vectors form a basis for $\mathbb{R}^2$ or $\mathbb{R}^3$. Apply the Gram-Schmidt Process to obtain an orthogonal basis. Then normalize this basis to obtain an orthonormal basis.

1. $\mathbf{x}_1 = \begin{bmatrix} 1 \\ 1 \end{bmatrix}, \mathbf{x}_2 = \begin{bmatrix} 1 \\ 2 \end{bmatrix}$

2. $\mathbf{x}_1 = \begin{bmatrix} 3 \\ -3 \end{bmatrix}, \mathbf{x}_2 = \begin{bmatrix} 3 \\ 1 \end{bmatrix}$

3. $\mathbf{x}_1 = \begin{bmatrix} 1 \\ -1 \\ -1 \end{bmatrix}, \mathbf{x}_2 = \begin{bmatrix} 0 \\ 3 \\ 3 \end{bmatrix}, \mathbf{x}_3 = \begin{bmatrix} 3 \\ 2 \\ 4 \end{bmatrix}$

4. $\mathbf{x}_1 = \begin{bmatrix} 1 \\ 1 \\ 1 \end{bmatrix}, \mathbf{x}_2 = \begin{bmatrix} 1 \\ 1 \\ 0 \end{bmatrix}, \mathbf{x}_3 = \begin{bmatrix} 1 \\ 0 \\ 0 \end{bmatrix}$

In Exercises 5 and 6, the given vectors form a basis for a subspace W of $\mathbb{R}^3$ or $\mathbb{R}^4$. Apply the Gram-Schmidt Process to obtain an orthogonal basis for W.

5. $\mathbf{x}_1 = \begin{bmatrix} 1 \\ 1 \\ 0 \end{bmatrix}, \mathbf{x}_2 = \begin{bmatrix} 3 \\ 4 \\ 2 \end{bmatrix}$

6. $\mathbf{x}_1 = \begin{bmatrix} 2 \\ -1 \\ 1 \\ 2 \end{bmatrix}, \mathbf{x}_2 = \begin{bmatrix} 3 \\ -1 \\ 0 \\ 4 \end{bmatrix}, \mathbf{x}_3 = \begin{bmatrix} 1 \\ 1 \\ 1 \\ 1 \end{bmatrix}$

In Exercises 7 and 8, find the orthogonal decomposition of $\mathbf{v}$ with respect to the subspace W.

7. $\mathbf{v} = \begin{bmatrix} 4 \\ -4 \\ 3 \end{bmatrix}$, W as in Exercise 5

8. $\mathbf{v} = \begin{bmatrix} 1 \\ 4 \\ 0 \\ 2 \end{bmatrix}$, W as in Exercise 6

Use the Gram-Schmidt Process to find an orthogonal basis for the column spaces of the matrices in Exercises 9 and 10.

9. $\begin{bmatrix} 0 & 1 & 1 \\ 1 & 0 & 1 \\ 1 & 1 & 0 \end{bmatrix}$

10. $\begin{bmatrix} 1 & 1 & 1 \\ 1 & -1 & 2 \\ -1 & 1 & 0 \\ 1 & 5 & 1 \end{bmatrix}$

11. Find an orthogonal basis for $\mathbb{R}^3$ that contains the vector $\begin{bmatrix} 3 \\ 1 \\ 5 \end{bmatrix}$.

12. Find an orthogonal basis for $\mathbb{R}^4$ that contains the vectors

$$\begin{bmatrix} 1 \\ 2 \\ -1 \\ 0 \end{bmatrix} \text{ and } \begin{bmatrix} 1 \\ 0 \\ 1 \\ 3 \end{bmatrix}$$

In Exercises 13 and 14, fill in the missing entries of Q to make Q an orthogonal matrix.

13. $Q = \begin{bmatrix} 1/\sqrt{2} & 1/\sqrt{3} & * \\ 0 & 1/\sqrt{3} & * \\ -1/\sqrt{2} & 1/\sqrt{3} & * \end{bmatrix}$

14. $Q = \begin{bmatrix} 1/2 & 2/\sqrt{14} & * & * \\ 1/2 & 1/\sqrt{14} & * & * \\ 1/2 & 0 & * & * \\ 1/2 & -3/\sqrt{14} & * & * \end{bmatrix}$

In Exercises 15 and 16, find a QR factorization of the matrix in the given exercise.

15. Exercise 9

16. Exercise 10

In Exercises 17 and 18, the columns of Q were obtained by applying the Gram-Schmidt Process to the columns of A. Find the upper triangular matrix R such that $A = QR$.

17. $A = \begin{bmatrix} 2 & 8 & 2 \\ 1 & 7 & -1 \\ -2 & -2 & 1 \end{bmatrix}, Q = \begin{bmatrix} \frac{2}{3} & \frac{1}{3} & \frac{2}{3} \\ \frac{1}{3} & \frac{2}{3} & -\frac{2}{3} \\ -\frac{2}{3} & \frac{2}{3} & \frac{1}{3} \end{bmatrix}$

18. $A = \begin{bmatrix} 1 & 3 \\ 2 & 4 \\ -1 & -1 \\ 0 & 1 \end{bmatrix}, Q = \begin{bmatrix} 1/\sqrt{6} & 1/\sqrt{3} \\ 2/\sqrt{6} & 0 \\ -1/\sqrt{6} & 1/\sqrt{3} \\ 0 & 1/\sqrt{3} \end{bmatrix}$

19. If A is an orthogonal matrix, find a QR factorization of A.

20. Prove that A is invertible if and only if $A = QR$, where Q is orthogonal and R is upper triangular with nonzero entries on its diagonal.

In Exercises 21 and 22, use the method suggested by Exercise 20 to compute A^{-1} for the matrix A in the given exercise.

21. Exercise 9

22. Exercise 15

23. Let A be an $m \times n$ matrix with linearly independent columns. Give an alternative proof that the upper triangular matrix R in a QR factorization of A must be invertible, using property (c) of the Fundamental Theorem.

24. Let A be an $m \times n$ matrix with linearly independent columns and let $A = QR$ be a QR factorization of A. Show that A and Q have the same column space.

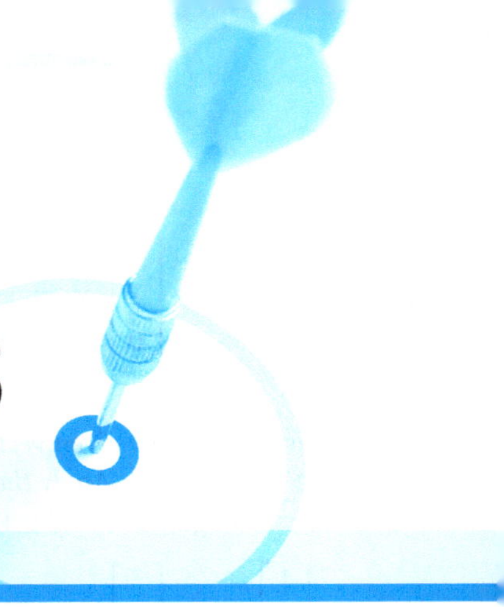

Explorations

The Modified QR Factorization

When the matrix A does not have linearly independent columns, the Gram-Schmidt Process as we have stated it does not work and so cannot be used to develop a generalized QR factorization of A. There is a modification of the Gram-Schmidt Process that can be used, but instead we will explore a method that converts A into upper triangular form one column at a time, using a sequence of orthogonal matrices. The method is analogous to that of LU factorization, in which the matrix L is formed using a sequence of elementary matrices.

The first thing we need is the "orthogonal analogue" of an elementary matrix; that is, we need to know how to construct an orthogonal matrix Q that will transform a given column of A—call it $\mathbf{x}$—into the corresponding column of R—call it $\mathbf{y}$. By Theorem 5.6, it will be necessary that $\|\mathbf{x}\| = \|Q\mathbf{x}\| = \|\mathbf{y}\|$. Figure 5.11 suggests a way to proceed: We can reflect $\mathbf{x}$ in a line perpendicular to $\mathbf{x} - \mathbf{y}$. If

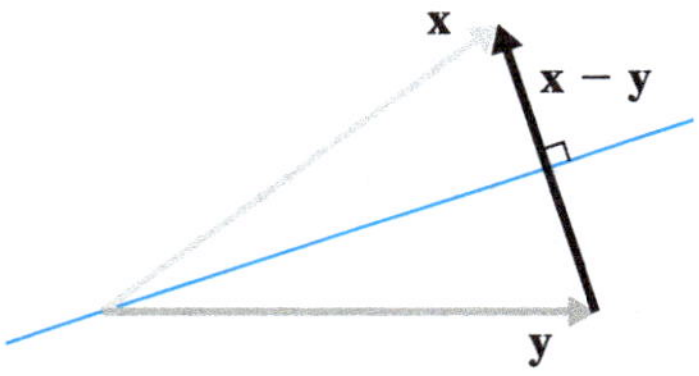

Figure 5.11

$$\mathbf{u} = \left(\frac{1}{\|\mathbf{x} - \mathbf{y}\|}\right)(\mathbf{x} - \mathbf{y}) = \begin{bmatrix} d_1 \\ d_2 \end{bmatrix}$$

is the unit vector in the direction of $\mathbf{x} - \mathbf{y}$, then $\mathbf{u}^\perp = \begin{bmatrix} -d_2 \\ d_1 \end{bmatrix}$ is orthogonal to $\mathbf{u}$, and we can use Exercise 26 in Section 3.6 to find the standard matrix Q of the reflection in the line through the origin in the direction of $\mathbf{u}^\perp$.

1. Show that $Q = \begin{bmatrix} 1 - 2d_1^2 & -2d_1d_2 \\ -2d_1d_2 & 1 - 2d_2^2 \end{bmatrix} = I - 2\mathbf{u}\mathbf{u}^T$.

2. Compute Q for

$$\text{(a) } \mathbf{u} = \begin{bmatrix} \frac{3}{5} \\ \frac{4}{5} \end{bmatrix} \quad \text{(b) } \mathbf{x} = \begin{bmatrix} 5 \\ 5 \end{bmatrix}, \mathbf{y} = \begin{bmatrix} 1 \\ 7 \end{bmatrix}$$

We can generalize the definition of Q as follows. If $\mathbf{u}$ is any unit vector in $\mathbb{R}^n$, we define an $n \times n$ matrix Q as

$$Q = I - 2\mathbf{u}\mathbf{u}^T$$

Courtesy Oak Ridge National Lab

Alston Householder (1904–1993) was one of the pioneers in the field of numerical linear algebra. He was the first to present a systematic treatment of algorithms for solving problems involving linear systems. In addition to introducing the widely used Householder transformations that bear his name, he was one of the first to advocate the systematic use of norms in linear algebra. His 1964 book *The Theory of Matrices in Numerical Analysis* is considered a classic.

Such a matrix is called a ***Householder matrix*** (or an ***elementary reflector***).

3. Prove that every Householder matrix Q satisfies the following properties:

   (a) Q is symmetric. (b) Q is orthogonal. (c) $Q^2 = I$

4. Prove that if Q is a Householder matrix corresponding to the unit vector $\mathbf{u}$, then

$$Q\mathbf{v} = \begin{cases} -\mathbf{v} & \text{if } \mathbf{v} \text{ is in span}(\mathbf{u}) \\ \mathbf{v} & \text{if } \mathbf{v} \cdot \mathbf{u} = \mathbf{0} \end{cases}$$

5. Compute Q for $\mathbf{u} = \begin{bmatrix} 1 \\ -1 \\ 2 \end{bmatrix}$ and verify Problems 3 and 4.

6. Let $\mathbf{x} \neq \mathbf{y}$ with $\|\mathbf{x}\| = \|\mathbf{y}\|$ and set $\mathbf{u} = (1/\|\mathbf{x} - \mathbf{y}\|)(\mathbf{x} - \mathbf{y})$. Prove that the corresponding Householder matrix Q satisfies $Q\mathbf{x} = \mathbf{y}$. [*Hint:* Apply Exercise 57 in Section 1.2 to the result in Problem 4.]

7. Find Q and verify Problem 6 for

$$\mathbf{x} = \begin{bmatrix} 1 \\ 2 \\ 2 \end{bmatrix} \quad \text{and} \quad \mathbf{y} = \begin{bmatrix} 3 \\ 0 \\ 0 \end{bmatrix}$$

We are now ready to perform the triangularization of an $m \times n$ matrix A, column by column.

8. Let $\mathbf{x}$ be the first column of A and let

$$\mathbf{y} = \begin{bmatrix} \|\mathbf{x}\| \\ 0 \\ \vdots \\ 0 \end{bmatrix}$$

Show that if Q_1 is the Householder matrix given by Problem 6, then Q_1A is a matrix with the block form

$$Q_1A = \begin{bmatrix} * & * \\ \mathbf{0} & A_1 \end{bmatrix}$$

where A_1 is $(m-1)\times(n-1)$.

If we repeat Problem 8 on the matrix A_1, we use a Householder matrix P_2 such that

$$P_2A_1 = \begin{bmatrix} * & * \\ \mathbf{0} & A_2 \end{bmatrix}$$

where A_2 is $(m-2)\times(n-2)$.

9. Set $Q_2 = \begin{bmatrix} 1 & \mathbf{0} \\ \mathbf{0} & P_2 \end{bmatrix}$. Show that Q_2 is an orthogonal matrix and that

$$Q_2Q_1A = \begin{bmatrix} * & * & * \\ 0 & * & * \\ \mathbf{0} & \mathbf{0} & A_2 \end{bmatrix}$$

10. Show that we can continue in this fashion to find a sequence of orthogonal matrices $Q_1, \ldots, Q_{m-1}$ such that $Q_{m-1} \cdots Q_2 Q_1 A = R$ is an upper triangular $m \times n$ matrix (i.e., $r_{ij} = 0$ if $i > j$).

11. Deduce that $A = QR$ with $Q = Q_1 Q_2 \cdots Q_{m-1}$ orthogonal.

12. Use the method of this exploration to find a QR factorization of

$$\text{(a) } A = \begin{bmatrix} 3 & 9 & 1 \\ -4 & 3 & 2 \end{bmatrix} \quad \text{(b) } A = \begin{bmatrix} 1 & 3 & 3 & 2 \\ 2 & -4 & 1 & 1 \\ 2 & -5 & -1 & -2 \end{bmatrix}$$

Approximating Eigenvalues with the QR Algorithm

One of the best (and most widely used) methods for numerically approximating the eigenvalues of a matrix makes use of the QR factorization. The purpose of this exploration is to introduce this method, the ***QR algorithm,*** and to show it at work in a few examples. For a more complete treatment of this topic, consult any good text on numerical linear algebra. (You will find it helpful to use a CAS to perform the calculations in the problems below.)

See G. H. Golub and C. F. Van Loan, *Matrix Computations* (Baltimore: Johns Hopkins University Press, 1983).

Given a square matrix A, the first step is to factor it as $A = QR$ (using whichever method is appropriate). Then we define $A_1 = RQ$.

1. First prove that A_1 is similar to A. Then prove that A_1 has the same eigenvalues as A.

2. If $A = \begin{bmatrix} 1 & 0 \\ 1 & 3 \end{bmatrix}$, find A_1 and verify that it has the same eigenvalues as A.

Continuing the algorithm, we factor A_1 as $A_1 = Q_1 R_1$ and set $A_2 = R_1 Q_1$. Then we factor $A_2 = Q_2 R_2$ and set $A_3 = R_2 Q_2$, and so on. That is, for $k \geq 1$, we compute $A_k = Q_k R_k$ and then set $A_{k+1} = R_k Q_k$.

3. Prove that A_k is similar to A for all $k \geq 1$.

4. Continuing Problem 2, compute A_2, A_3, A_4, and A_5, using two-decimal-place accuracy. What do you notice?

It can be shown that if the eigenvalues of A are all real and have distinct absolute values, then the matrices A_k approach an upper triangular matrix U.

5. What will be true of the diagonal entries of this matrix U?

6. Approximate the eigenvalues of the following matrices by applying the QR algorithm. Use two-decimal-place accuracy and perform at least five iterations.

$$\text{(a) } \begin{bmatrix} 2 & 3 \\ 2 & 1 \end{bmatrix} \qquad \text{(b) } \begin{bmatrix} 1 & 1 \\ 2 & 1 \end{bmatrix}$$

$$\text{(c) } \begin{bmatrix} 1 & 0 & -1 \\ 1 & 2 & 1 \\ -4 & 0 & 1 \end{bmatrix} \quad \text{(d) } \begin{bmatrix} 1 & 1 & -1 \\ 0 & 2 & 0 \\ -2 & 4 & 2 \end{bmatrix}$$

7. Apply the QR algorithm to the matrix $A = \begin{bmatrix} 2 & 3 \\ -1 & -2 \end{bmatrix}$. What happens? Why?

8. Shift the eigenvalues of the matrix in Problem 7 by replacing A with $B = A + 0.9I$. Apply the QR algorithm to B and then shift back by subtracting 0.9 from the (approximate) eigenvalues of B. Verify that this method approximates the eigenvalues of A.

9. Let $Q_0 = Q$ and $R_0 = R$. First show that

$$Q_0Q_1 \cdots Q_{k-1}A_k = AQ_0Q_1 \cdots Q_{k-1}$$

for all $k \geq 1$. Then show that

$$(Q_0Q_1 \cdots Q_k)(R_k \cdots R_1R_0) = A(Q_0Q_1 \cdots Q_{k-1})(R_{k-1} \cdots R_1R_0)$$

[*Hint:* Repeatedly use the same approach used for the first equation, working from the "inside out."] Finally, deduce that $(Q_0Q_1 \cdots Q_k)(R_k \cdots R_1R_0)$ is the QR factorization of A^{k+1}.

5.4 Orthogonal Diagonalization of Symmetric Matrices

We saw in Chapter 4 that a square matrix with real entries will not necessarily have real eigenvalues. Indeed, the matrix $\begin{bmatrix} 0 & -1 \\ 1 & 0 \end{bmatrix}$ has complex eigenvalues i and $-i$. We also discovered that not all square matrices are diagonalizable. The situation changes dramatically if we restrict our attention to real *symmetric* matrices. As we will show in this section, all of the eigenvalues of a real symmetric matrix are real, and such a matrix is always diagonalizable.

Recall that a symmetric matrix is one that equals its own transpose. Let's begin by studying the diagonalization process for a symmetric 2×2 matrix.

Example 5.16

If possible, diagonalize the matrix $A = \begin{bmatrix} 1 & 2 \\ 2 & -2 \end{bmatrix}$.

Solution The characteristic polynomial of A is $\lambda^2 + \lambda - 6 = (\lambda + 3)(\lambda - 2)$, from which we see that A has eigenvalues $\lambda_1 = -3$ and $\lambda_2 = 2$. Solving for the corresponding eigenvectors, we find

$$\mathbf{v}_1 = \begin{bmatrix} 1 \\ -2 \end{bmatrix} \quad \text{and} \quad \mathbf{v}_2 = \begin{bmatrix} 2 \\ 1 \end{bmatrix}$$

respectively. So A is diagonalizable, and if we set $P = [\mathbf{v}_1 \ \ \mathbf{v}_2]$, then we know that $P^{-1}AP = \begin{bmatrix} -3 & 0 \\ 0 & 2 \end{bmatrix} = D$.

However, we can do better. Observe that $\mathbf{v}_1$ and $\mathbf{v}_2$ are orthogonal. So, if we normalize them to get the unit eigenvectors

$$\mathbf{u}_1 = \begin{bmatrix} 1/\sqrt{5} \\ -2/\sqrt{5} \end{bmatrix} \quad \text{and} \quad \mathbf{u}_2 = \begin{bmatrix} 2/\sqrt{5} \\ 1/\sqrt{5} \end{bmatrix}$$

and then take

$$Q = [\mathbf{u}_1 \ \ \mathbf{u}_2] = \begin{bmatrix} 1/\sqrt{5} & 2/\sqrt{5} \\ -2/\sqrt{5} & 1/\sqrt{5} \end{bmatrix}$$

we have $Q^{-1}AQ = D$ also. But now Q is an *orthogonal* matrix, since $\{\mathbf{u}_1, \mathbf{u}_2\}$ is an orthonormal set of vectors. Therefore, $Q^{-1} = Q^T$, and we have $Q^TAQ = D$. (Note that checking is easy, since computing Q^{-1} only involves taking a transpose!)

The situation in Example 5.16 is the one that interests us. It is important enough to warrant a new definition.

Definition A square matrix A is ***orthogonally diagonalizable*** if there exists an orthogonal matrix Q and a diagonal matrix D such that $Q^TAQ = D$.

We are interested in finding conditions under which a matrix is orthogonally diagonalizable. Theorem 5.17 shows us where to look.

Theorem 5.17 If A is orthogonally diagonalizable, then A is symmetric.

Proof If A is orthogonally diagonalizable, then there exists an orthogonal matrix Q and a diagonal matrix D such that $Q^TAQ = D$. Since $Q^{-1} = Q^T$, we have $Q^TQ = I = QQ^T$, so

$$QDQ^T = QQ^TAQQ^T = IAI = A$$

But then

$$A^T = (QDQ^T)^T = (Q^T)^TD^TQ^T = QDQ^T = A$$

since every diagonal matrix is symmetric. Hence, A is symmetric.

Remark Theorem 5.17 shows that the orthogonally diagonalizable matrices are all to be found *among* the symmetric matrices. It does *not* say that every symmetric matrix must be orthogonally diagonalizable. However, it is a remarkable fact that this indeed is true! Finding a proof for this amazing result will occupy us for much of the rest of this section.

a + bi

We next prove that we don't need to worry about *complex* eigenvalues when working with symmetric matrices with *real* entries.

Theorem 5.18 If A is a real symmetric matrix, then the eigenvalues of A are real.

Recall that the *complex conjugate* of a complex number $z = a + bi$ is the number $\overline{z} = a - bi$ (see Appendix C). To show that z is real, we need to show that $b = 0$. One way to do this is to show that $z = \overline{z}$, for then $bi = -bi$ (or $2bi = 0$), from which it follows that $b = 0$.

We can also extend the notion of complex conjugate to vectors and matrices by, for example, defining $\overline{A}$ to be the matrix whose entries are the complex conjugates of the entries of A; that is, if $A = [a_{ij}]$, then $\overline{A} = [\overline{a}_{ij}]$. The rules for complex conjugation extend easily to matrices; in particular, we have $\overline{AB} = \overline{A}\,\overline{B}$ for compatible matrices A and B.

Proof Suppose that λ is an eigenvalue of A with corresponding eigenvector $\mathbf{v}$. Then $A\mathbf{v} = \lambda\mathbf{v}$, and, taking complex conjugates, we have $\overline{A\mathbf{v}} = \overline{\lambda\mathbf{v}}$. But then

$$A\overline{\mathbf{v}} = \overline{A}\,\overline{\mathbf{v}} = \overline{A\mathbf{v}} = \overline{\lambda\mathbf{v}} = \overline{\lambda}\,\overline{\mathbf{v}}$$

since A is real. Taking transposes and using the fact that A is symmetric, we have

$$\overline{\mathbf{v}}^TA = \overline{\mathbf{v}}^TA^T = (A\overline{\mathbf{v}})^T = (\overline{\lambda}\,\overline{\mathbf{v}})^T = \overline{\lambda}\,\overline{\mathbf{v}}^T$$

Therefore,

$$\lambda(\overline{\mathbf{v}}^T\mathbf{v}) = \overline{\mathbf{v}}^T(\lambda\mathbf{v}) = \overline{\mathbf{v}}^T(A\mathbf{v}) = (\overline{\mathbf{v}}^TA)\mathbf{v} = (\overline{\lambda}\,\overline{\mathbf{v}}^T)\mathbf{v} = \overline{\lambda}(\overline{\mathbf{v}}^T\mathbf{v})$$

or $(\lambda - \overline{\lambda})(\overline{\mathbf{v}}^T\mathbf{v}) = 0$.

Now if $\mathbf{v} = \begin{bmatrix} a_1 + b_1 i \\ \vdots \\ a_n + b_n i \end{bmatrix}$, then $\overline{\mathbf{v}} = \begin{bmatrix} a_1 - b_1 i \\ \vdots \\ a_n - b_n i \end{bmatrix}$, so

$$\overline{\mathbf{v}}^T\mathbf{v} = (a_1^2 + b_1^2) + \cdots + (a_n^2 + b_n^2) \neq 0$$

since $\mathbf{v} \neq \mathbf{0}$ (because it is an eigenvector). We conclude that $\lambda - \bar{\lambda} = 0$, or $\lambda = \bar{\lambda}$. Hence, λ is real.

Theorem 4.20 showed that, for any square matrix, eigenvectors corresponding to distinct eigenvalues are linearly independent. For symmetric matrices, something stronger is true: Such eigenvectors are *orthogonal.*

Theorem 5.19 If A is a symmetric matrix, then any two eigenvectors corresponding to distinct eigenvalues of A are orthogonal.

Proof Let $\mathbf{v}_1$ and $\mathbf{v}_2$ be eigenvectors corresponding to the distinct eigenvalues $\lambda_1 \neq \lambda_2$ so that $A\mathbf{v}_1 = \lambda_1\mathbf{v}_1$ and $A\mathbf{v}_2 = \lambda_2\mathbf{v}_2$. Using $A^T = A$ and the fact that $\mathbf{x}\cdot\mathbf{y} = \mathbf{x}^T\mathbf{y}$ for any two vectors $\mathbf{x}$ and $\mathbf{y}$ in $\mathbb{R}^n$, we have

$$\begin{aligned}\lambda_1(\mathbf{v}_1\cdot\mathbf{v}_2) &= (\lambda_1\mathbf{v}_1)\cdot\mathbf{v}_2 = A\mathbf{v}_1\cdot\mathbf{v}_2 = (A\mathbf{v}_1)^T\mathbf{v}_2\\ &= (\mathbf{v}_1^TA^T)\mathbf{v}_2 = (\mathbf{v}_1^TA)\mathbf{v}_2 = \mathbf{v}_1^T(A\mathbf{v}_2)\\ &= \mathbf{v}_1^T(\lambda_2\mathbf{v}_2) = \lambda_2(\mathbf{v}_1^T\mathbf{v}_2) = \lambda_2(\mathbf{v}_1\cdot\mathbf{v}_2)\end{aligned}$$

Hence, $(\lambda_1 - \lambda_2)(\mathbf{v}_1\cdot\mathbf{v}_2) = 0$. But $\lambda_1 - \lambda_2 \neq 0$, so $\mathbf{v}_1\cdot\mathbf{v}_2 = 0$, as we wished to show.

Example 5.17 Verify the result of Theorem 5.19 for

$$A = \begin{bmatrix} 2 & 1 & 1 \\ 1 & 2 & 1 \\ 1 & 1 & 2 \end{bmatrix}$$

Solution The characteristic polynomial of A is $-\lambda^3 + 6\lambda^2 - 9\lambda + 4 = -(\lambda - 4)\cdot(\lambda - 1)^2$, from which it follows that the eigenvalues of A are $\lambda_1 = 4$ and $\lambda_2 = 1$. The corresponding eigenspaces are

$$E_4 = \text{span}\left(\begin{bmatrix} 1 \\ 1 \\ 1 \end{bmatrix}\right) \quad \text{and} \quad E_1 = \text{span}\left(\begin{bmatrix} -1 \\ 0 \\ 1 \end{bmatrix}, \begin{bmatrix} -1 \\ 1 \\ 0 \end{bmatrix}\right)$$

(Check this.) We easily verify that

$$\begin{bmatrix} 1 \\ 1 \\ 1 \end{bmatrix}\cdot\begin{bmatrix} -1 \\ 0 \\ 1 \end{bmatrix} = 0 \quad \text{and} \quad \begin{bmatrix} 1 \\ 1 \\ 1 \end{bmatrix}\cdot\begin{bmatrix} -1 \\ 1 \\ 0 \end{bmatrix} = 0$$

from which it follows that every vector in E_4 is orthogonal to every vector in E_1. (Why?)

Remark Note that $\begin{bmatrix} -1 \\ 0 \\ 1 \end{bmatrix}\cdot\begin{bmatrix} -1 \\ 1 \\ 0 \end{bmatrix} = 1$. Thus, eigenvectors corresponding to the *same* eigenvalue need not be orthogonal.

We can now prove the main result of this section. It is called the Spectral Theorem, since the set of eigenvalues of a matrix is sometimes called the ***spectrum*** of the matrix. (Technically, we should call Theorem 5.20 the *Real* Spectral Theorem, since there is a corresponding result for matrices with complex entries.)

Theorem 5.20 **The Spectral Theorem**

Let A be an $n \times n$ real matrix. Then A is symmetric if and only if it is orthogonally diagonalizable.

Spectrum is a Latin word meaning "image." When atoms vibrate, they emit light. And when light passes through a prism, it spreads out into a spectrum—a band of rainbow colors. Vibration frequencies correspond to the eigenvalues of a certain operator and are visible as bright lines in the spectrum of light that is emitted from a prism. Thus, we can literally see the eigenvalues of the atom in its spectrum, and for this reason, it is appropriate that the word *spectrum* has come to be applied to the set of all eigenvalues of a matrix (or operator).

Proof We have already proved the "if" part as Theorem 5.17. To prove the "only if" implication, we proceed by induction on n. For $n = 1$, there is nothing to do, since a 1×1 matrix is already in diagonal form. Now assume that every $k \times k$ real symmetric matrix with real eigenvalues is orthogonally diagonalizable. Let $n = k + 1$ and let A be an $n \times n$ real symmetric matrix with real eigenvalues.

Let λ_1 be one of the eigenvalues of A and let $\mathbf{v}_1$ be a corresponding eigenvector. Then $\mathbf{v}_1$ is a *real* vector (why?) and we can assume that $\mathbf{v}_1$ is a unit vector, since otherwise we can normalize it and we will still have an eigenvector corresponding to λ_1. Using the Gram-Schmidt Process, we can extend $\mathbf{v}_1$ to an orthonormal basis $\{\mathbf{v}_1, \mathbf{v}_2, \ldots, \mathbf{v}_n\}$ of $\mathbb{R}^n$. Now we form the matrix

$$Q_1 = [\mathbf{v}_1 \quad \mathbf{v}_2 \cdots \mathbf{v}_n]$$

Then Q_1 is orthogonal, and

$$\begin{aligned} Q_1^T A Q_1 &= \begin{bmatrix} \mathbf{v}_1^T \\ \mathbf{v}_2^T \\ \vdots \\ \mathbf{v}_n^T \end{bmatrix} A [\mathbf{v}_1 \quad \mathbf{v}_2 \cdots \mathbf{v}_n] = \begin{bmatrix} \mathbf{v}_1^T \\ \mathbf{v}_2^T \\ \vdots \\ \mathbf{v}_n^T \end{bmatrix} [A\mathbf{v}_1 \quad A\mathbf{v}_2 \cdots A\mathbf{v}_n] \\ &= \begin{bmatrix} \mathbf{v}_1^T \\ \mathbf{v}_2^T \\ \vdots \\ \mathbf{v}_n^T \end{bmatrix} [\lambda_1\mathbf{v}_1 \quad A\mathbf{v}_2 \cdots A\mathbf{v}_n] \\ &= \left[\begin{array}{c|c} \lambda_1 & * \\ \hline \mathbf{0} & A_1 \end{array}\right] = B \end{aligned}$$

Baldwin H. Ward & Kathryn C. Ward/CORBIS

In a lecture he delivered at the University of Göttingen in 1905, the German mathematician David Hilbert (1862–1943) considered linear operators acting on certain infinite-dimensional vector spaces. Out of this lecture arose the notion of a quadratic form in infinitely many variables, and it was in this context that Hilbert first used the term *spectrum* to mean a complete set of eigenvalues. The spaces in question are now called *Hilbert spaces*.

Hilbert made major contributions to many areas of mathematics, among them integral equations, number theory, geometry, and the foundations of mathematics. In 1900, at the Second International Congress of Mathematicians in Paris, Hilbert gave an address entitled "The Problems of Mathematics." In it, he challenged mathematicians to solve 23 problems of fundamental importance during the coming century. Many of the problems have been solved—some were proved true, others false—and some may never be solved. Nevertheless, Hilbert's speech energized the mathematical community and is often regarded as the most influential speech ever given about mathematics.

since $\mathbf{v}_1^T(\lambda_1\mathbf{v}_1) = \lambda_1(\mathbf{v}_1^T\mathbf{v}_1) = \lambda_1(\mathbf{v}_1 \cdot \mathbf{v}_1) = \lambda_1$ and $\mathbf{v}_i^T(\lambda_1\mathbf{v}_1) = \lambda_1(\mathbf{v}_i^T\mathbf{v}_1) = \lambda_1(\mathbf{v}_i \cdot \mathbf{v}_1) = 0$ for $i \neq 1$, because $\{\mathbf{v}_1, \mathbf{v}_2, \ldots, \mathbf{v}_n\}$ is an orthonormal set.

But

$$B^T = (Q_1^TAQ_1)^T = Q_1^TA^T(Q_1^T)^T = Q_1^TAQ_1 = B$$

so B is symmetric. Therefore, B has the block form

$$B = \left[\begin{array}{c|c} \lambda_1 & \mathbf{0} \\ \hline \mathbf{0} & A_1 \end{array}\right]$$

and A_1 is symmetric. Furthermore, B is similar to A (why?), so the characteristic polynomial of B is equal to the characteristic polynomial of A, by Theorem 4.22. By Exercise 39 in Section 4.3, the characteristic polynomial of A_1 divides the characteristic polynomial of A. It follows that the eigenvalues of A_1 are also eigenvalues of A and, hence, are real. We also see that A_1 has real entries. (Why?) Thus, A_1 is a $k \times k$ real symmetric matrix with real eigenvalues, so the induction hypothesis applies to it. Hence, there is an orthogonal matrix P_2 such that $P_2^TA_1P_2$ is a diagonal matrix—say, D_1. Now let

$$Q_2 = \left[\begin{array}{c|c} 1 & \mathbf{0} \\ \hline \mathbf{0} & P_2 \end{array}\right]$$

Then Q_2 is an orthogonal $(k + 1)\times(k + 1)$ matrix, and therefore so is $Q = Q_1Q_2$. Consequently,

$$\begin{aligned} Q^TAQ &= (Q_1Q_2)^TA(Q_1Q_2) = (Q_2^TQ_1^T)A(Q_1Q_2) = Q_2^T(Q_1^TAQ_1)Q_2 = Q_2^TBQ_2 \\ &= \left[\begin{array}{c|c} 1 & \mathbf{0} \\ \hline \mathbf{0} & P_2^T \end{array}\right]\left[\begin{array}{c|c} \lambda_1 & \mathbf{0} \\ \hline \mathbf{0} & A_1 \end{array}\right]\left[\begin{array}{c|c} 1 & \mathbf{0} \\ \hline \mathbf{0} & P_2 \end{array}\right] \\ &= \left[\begin{array}{c|c} \lambda_1 & \mathbf{0} \\ \hline \mathbf{0} & P_2^TA_1P_2 \end{array}\right] \\ &= \left[\begin{array}{c|c} \lambda_1 & \mathbf{0} \\ \hline \mathbf{0} & D_1 \end{array}\right] \end{aligned}$$

which is a diagonal matrix. This completes the induction step, and we conclude that, for all $n \geq 1$, an $n \times n$ real symmetric matrix with real eigenvalues is orthogonally diagonalizable.

Example 5.18

Orthogonally diagonalize the matrix

$$A = \begin{bmatrix} 2 & 1 & 1 \\ 1 & 2 & 1 \\ 1 & 1 & 2 \end{bmatrix}$$

Solution This is the matrix from Example 5.17. We have already found that the eigenspaces of A are

$$E_4 = \text{span}\left(\begin{bmatrix} 1 \\ 1 \\ 1 \end{bmatrix}\right) \quad \text{and} \quad E_1 = \text{span}\left(\begin{bmatrix} -1 \\ 0 \\ 1 \end{bmatrix}, \begin{bmatrix} -1 \\ 1 \\ 0 \end{bmatrix}\right)$$

We need three orthonormal eigenvectors. First, we apply the Gram-Schmidt Process to

$$\begin{bmatrix} -1 \\ 0 \\ 1 \end{bmatrix} \quad \text{and} \quad \begin{bmatrix} -1 \\ 1 \\ 0 \end{bmatrix}$$

to obtain

$$\begin{bmatrix} -1 \\ 0 \\ 1 \end{bmatrix} \quad \text{and} \quad \begin{bmatrix} -\frac{1}{2} \\ 1 \\ -\frac{1}{2} \end{bmatrix}$$

The new vector, which has been constructed to be orthogonal to $\begin{bmatrix} -1 \\ 0 \\ 1 \end{bmatrix}$, is still in E_1 (why?) and so is orthogonal to $\begin{bmatrix} 1 \\ 1 \\ 1 \end{bmatrix}$. Thus, we have three mutually orthogonal vectors, and all we need to do is normalize them and construct a matrix Q with these vectors as its columns. We find that

$$Q = \begin{bmatrix} 1/\sqrt{3} & -1/\sqrt{2} & -1/\sqrt{6} \\ 1/\sqrt{3} & 0 & 2/\sqrt{6} \\ 1/\sqrt{3} & 1/\sqrt{2} & -1/\sqrt{6} \end{bmatrix}$$

and it is straightforward to verify that

$$Q^TAQ = \begin{bmatrix} 4 & 0 & 0 \\ 0 & 1 & 0 \\ 0 & 0 & 1 \end{bmatrix}$$

The Spectral Theorem allows us to write a real symmetric matrix A in the form $A = QDQ^T$, where Q is orthogonal and D is diagonal. The diagonal entries of D are just the eigenvalues of A, and if the columns of Q are the orthonormal vectors $\mathbf{q}_1, \ldots, \mathbf{q}_n$, then, using the column-row representation of the product, we have

$$\begin{aligned} A = QDQ^T &= [\mathbf{q}_1 \ \cdots \ \mathbf{q}_n] \begin{bmatrix} \lambda_1 & \cdots & 0 \\ \vdots & \ddots & \vdots \\ 0 & \cdots & \lambda_n \end{bmatrix} \begin{bmatrix} \mathbf{q}_1^T \\ \vdots \\ \mathbf{q}_n^T \end{bmatrix} \\ &= [\lambda_1\mathbf{q}_1 \ \cdots \ \lambda_n\mathbf{q}_n] \begin{bmatrix} \mathbf{q}_1^T \\ \vdots \\ \mathbf{q}_n^T \end{bmatrix} \\ &= \lambda_1\mathbf{q}_1\mathbf{q}_1^T + \lambda_2\mathbf{q}_2\mathbf{q}_2^T + \cdots + \lambda_n\mathbf{q}_n\mathbf{q}_n^T \end{aligned}$$

This is called the ***spectral decomposition*** of A. Each of the terms $\lambda_i\mathbf{q}_i\mathbf{q}_i^T$ is a rank 1 matrix, by Exercise 62 in Section 3.5, and $\mathbf{q}_i\mathbf{q}_i^T$ is actually the matrix of the projection onto the subspace spanned by $\mathbf{q}_i$. (See Exercise 25.) For this reason, the spectral decomposition

$$A = \lambda_1\mathbf{q}_1\mathbf{q}_1^T + \lambda_2\mathbf{q}_2\mathbf{q}_2^T + \cdots + \lambda_n\mathbf{q}_n\mathbf{q}_n^T$$

is sometimes referred to as the ***projection form of the Spectral Theorem.***

Example 5.19

Find the spectral decomposition of the matrix A from Example 5.18.

Solution From Example 5.18, we have:

$$\lambda_1 = 4, \qquad \lambda_2 = 1, \qquad \lambda_3 = 1$$

$$\mathbf{q}_1 = \begin{bmatrix} 1/\sqrt{3} \\ 1/\sqrt{3} \\ 1/\sqrt{3} \end{bmatrix}, \quad \mathbf{q}_2 = \begin{bmatrix} -1/\sqrt{2} \\ 0 \\ 1/\sqrt{2} \end{bmatrix}, \quad \mathbf{q}_3 = \begin{bmatrix} -1/\sqrt{6} \\ 2/\sqrt{6} \\ -1/\sqrt{6} \end{bmatrix}$$

Therefore,

$$\mathbf{q}_1\mathbf{q}_1^T = \begin{bmatrix} 1/\sqrt{3} \\ 1/\sqrt{3} \\ 1/\sqrt{3} \end{bmatrix}[1/\sqrt{3} \quad 1/\sqrt{3} \quad 1/\sqrt{3}] = \begin{bmatrix} 1/3 & 1/3 & 1/3 \\ 1/3 & 1/3 & 1/3 \\ 1/3 & 1/3 & 1/3 \end{bmatrix}$$

$$\mathbf{q}_2\mathbf{q}_2^T = \begin{bmatrix} -1/\sqrt{2} \\ 0 \\ 1/\sqrt{2} \end{bmatrix}[-1/\sqrt{2} \quad 0 \quad 1/\sqrt{2}] = \begin{bmatrix} 1/2 & 0 & -1/2 \\ 0 & 0 & 0 \\ -1/2 & 0 & 1/2 \end{bmatrix}$$

$$\mathbf{q}_3\mathbf{q}_3^T = \begin{bmatrix} -1/\sqrt{6} \\ 2/\sqrt{6} \\ -1/\sqrt{6} \end{bmatrix}[-1/\sqrt{6} \quad 2/\sqrt{6} \quad -1/\sqrt{6}] = \begin{bmatrix} 1/6 & -1/3 & 1/6 \\ -1/3 & 2/3 & -1/3 \\ 1/6 & -1/3 & 1/6 \end{bmatrix}$$

so

$$\begin{aligned} A &= \lambda_1\mathbf{q}_1\mathbf{q}_1^T + \lambda_2\mathbf{q}_2\mathbf{q}_2^T + \lambda_3\mathbf{q}_3\mathbf{q}_3^T \\ &= 4\begin{bmatrix} \frac{1}{3} & \frac{1}{3} & \frac{1}{3} \\ \frac{1}{3} & \frac{1}{3} & \frac{1}{3} \\ \frac{1}{3} & \frac{1}{3} & \frac{1}{3} \end{bmatrix} + \begin{bmatrix} \frac{1}{2} & 0 & -\frac{1}{2} \\ 0 & 0 & 0 \\ -\frac{1}{2} & 0 & \frac{1}{2} \end{bmatrix} + \begin{bmatrix} \frac{1}{6} & -\frac{1}{3} & \frac{1}{6} \\ -\frac{1}{3} & \frac{2}{3} & -\frac{1}{3} \\ \frac{1}{6} & -\frac{1}{3} & \frac{1}{6} \end{bmatrix} \end{aligned}$$

which can be easily verified.

In this example, $\lambda_2 = \lambda_3$, so we could combine the last two terms $\lambda_2\mathbf{q}_2\mathbf{q}_2^T + \lambda_3\mathbf{q}_3\mathbf{q}_3^T$ to get

$$\begin{bmatrix} \frac{2}{3} & -\frac{1}{3} & -\frac{1}{3} \\ -\frac{1}{3} & \frac{2}{3} & -\frac{1}{3} \\ -\frac{1}{3} & -\frac{1}{3} & \frac{2}{3} \end{bmatrix}$$

The rank 2 matrix $\mathbf{q}_2\mathbf{q}_2^T + \mathbf{q}_3\mathbf{q}_3^T$ is the matrix of a projection onto the two-dimensional subspace (i.e., the plane) spanned by $\mathbf{q}_2$ and $\mathbf{q}_3$. (See Exercise 26.)

Observe that the spectral decomposition expresses a symmetric matrix A explicitly in terms of its eigenvalues and eigenvectors. This gives us a way of constructing a matrix with given eigenvalues and (orthonormal) eigenvectors.

Example 5.20

Find a 2×2 matrix with eigenvalues $\lambda_1 = 3$ and $\lambda_2 = -2$ and corresponding eigenvectors

$$\mathbf{v}_1 = \begin{bmatrix} 3 \\ 4 \end{bmatrix} \quad \text{and} \quad \mathbf{v}_2 = \begin{bmatrix} -4 \\ 3 \end{bmatrix}$$

Solution We begin by normalizing the vectors to obtain an orthonormal basis $\{\mathbf{q}_1, \mathbf{q}_2\}$, with

$$\mathbf{q}_1 = \begin{bmatrix} \frac{3}{5} \\ \frac{4}{5} \end{bmatrix} \quad \text{and} \quad \mathbf{q}_2 = \begin{bmatrix} -\frac{4}{5} \\ \frac{3}{5} \end{bmatrix}$$

Now, we compute the matrix A whose spectral decomposition is

$$\begin{aligned} A &= \lambda_1 \mathbf{q}_1 \mathbf{q}_1^T + \lambda_2 \mathbf{q}_2 \mathbf{q}_2^T \\ &= 3\begin{bmatrix} \frac{3}{5} \\ \frac{4}{5} \end{bmatrix}\begin{bmatrix} \frac{3}{5} & \frac{4}{5} \end{bmatrix} - 2\begin{bmatrix} -\frac{4}{5} \\ \frac{3}{5} \end{bmatrix}\begin{bmatrix} -\frac{4}{5} & \frac{3}{5} \end{bmatrix} \\ &= 3\begin{bmatrix} \frac{9}{25} & \frac{12}{25} \\ \frac{12}{25} & \frac{16}{25} \end{bmatrix} - 2\begin{bmatrix} \frac{16}{25} & -\frac{12}{25} \\ -\frac{12}{25} & \frac{9}{25} \end{bmatrix} \\ &= \begin{bmatrix} -\frac{1}{5} & \frac{12}{5} \\ \frac{12}{5} & \frac{6}{5} \end{bmatrix} \end{aligned}$$

It is easy to check that A has the desired properties. (Do this.)

Exercises 5.4

Orthogonally diagonalize the matrices in Exercises 1–10 by finding an orthogonal matrix Q and a diagonal matrix D such that $Q^TAQ = D$.

1. $A = \begin{bmatrix} 4 & 1 \\ 1 & 4 \end{bmatrix}$

2. $A = \begin{bmatrix} -1 & 3 \\ 3 & -1 \end{bmatrix}$

3. $A = \begin{bmatrix} 1 & \sqrt{2} \\ \sqrt{2} & 0 \end{bmatrix}$

4. $A = \begin{bmatrix} 9 & -2 \\ -2 & 6 \end{bmatrix}$

5. $A = \begin{bmatrix} 5 & 0 & 0 \\ 0 & 1 & 3 \\ 0 & 3 & 1 \end{bmatrix}$

6. $A = \begin{bmatrix} 2 & 3 & 0 \\ 3 & 2 & 4 \\ 0 & 4 & 2 \end{bmatrix}$

7. $A = \begin{bmatrix} 1 & 0 & -1 \\ 0 & 1 & 0 \\ -1 & 0 & 1 \end{bmatrix}$

8. $A = \begin{bmatrix} 1 & 2 & 2 \\ 2 & 1 & 2 \\ 2 & 2 & 1 \end{bmatrix}$

9. $A = \begin{bmatrix} 1 & 1 & 0 & 0 \\ 1 & 1 & 0 & 0 \\ 0 & 0 & 1 & 1 \\ 0 & 0 & 1 & 1 \end{bmatrix}$

10. $A = \begin{bmatrix} 2 & 0 & 0 & 1 \\ 0 & 1 & 0 & 0 \\ 0 & 0 & 1 & 0 \\ 1 & 0 & 0 & 2 \end{bmatrix}$

11. If $b \neq 0$, orthogonally diagonalize $A = \begin{bmatrix} a & b \\ b & a \end{bmatrix}$.

12. If $b \neq 0$, orthogonally diagonalize $A = \begin{bmatrix} a & 0 & b \\ 0 & a & 0 \\ b & 0 & a \end{bmatrix}$.

13. Let A and B be orthogonally diagonalizable $n \times n$ matrices and let c be a scalar. Use the Spectral Theorem to prove that the following matrices are orthogonally diagonalizable:

(a) $A + B$ **(b)** cA **(c)** A^2

14. If A is an invertible matrix that is orthogonally diagonalizable, show that A^{-1} is orthogonally diagonalizable.

15. If A and B are orthogonally diagonalizable and $AB = BA$, show that AB is orthogonally diagonalizable.

16. If A is a symmetric matrix, show that every eigenvalue of A is nonnegative if and only if $A = B^2$ for some symmetric matrix B.

In Exercises 17–20, find a spectral decomposition of the matrix in the given exercise.

17. Exercise 1

18. Exercise 2

19. Exercise 5

20. Exercise 8

In Exercises 21 and 22, find a symmetric 2 × 2 matrix with eigenvalues λ_1 and λ_2 and corresponding orthogonal eigenvectors $\mathbf{v}_1$ and $\mathbf{v}_2$.

21. $\lambda_1 = -1, \lambda_2 = 2, \mathbf{v}_1 = \begin{bmatrix} 1 \\ 1 \end{bmatrix}, \mathbf{v}_2 = \begin{bmatrix} 1 \\ -1 \end{bmatrix}$

22. $\lambda_1 = 3, \lambda_2 = -3, \mathbf{v}_1 = \begin{bmatrix} 1 \\ 2 \end{bmatrix}, \mathbf{v}_2 = \begin{bmatrix} -2 \\ 1 \end{bmatrix}$

In Exercises 23 and 24, find a symmetric 3 × 3 matrix with eigenvalues λ_1, λ_2, and λ_3 and corresponding orthogonal eigenvectors $\mathbf{v}_1$, $\mathbf{v}_2$, and $\mathbf{v}_3$.

23. $\lambda_1 = 1, \lambda_2 = 2, \lambda_3 = 3, \mathbf{v}_1 = \begin{bmatrix} 1 \\ 1 \\ 0 \end{bmatrix}, \mathbf{v}_2 = \begin{bmatrix} 1 \\ -1 \\ 1 \end{bmatrix},$
$\mathbf{v}_3 = \begin{bmatrix} -1 \\ 1 \\ 2 \end{bmatrix}$

24. $\lambda_1 = 1, \lambda_2 = -4, \lambda_3 = -4, \mathbf{v}_1 = \begin{bmatrix} 4 \\ 5 \\ -1 \end{bmatrix}, \mathbf{v}_2 = \begin{bmatrix} -1 \\ 1 \\ 1 \end{bmatrix},$
$\mathbf{v}_3 = \begin{bmatrix} 2 \\ -1 \\ 3 \end{bmatrix}$

25. Let $\mathbf{q}$ be a unit vector in $\mathbb{R}^n$ and let W be the subspace spanned by $\mathbf{q}$. Show that the orthogonal projection of a vector $\mathbf{v}$ onto W (as defined in Sections 1.2 and 5.2) is given by

$$\text{proj}_W(\mathbf{v}) = (\mathbf{q}\mathbf{q}^T)\mathbf{v}$$

and that the matrix of this projection is thus $\mathbf{q}\mathbf{q}^T$. [*Hint:* Remember that, for $\mathbf{x}$ and $\mathbf{y}$ in $\mathbb{R}^n$, $\mathbf{x} \cdot \mathbf{y} = \mathbf{x}^T\mathbf{y}$.]

26. Let $\{\mathbf{q}_1, \ldots, \mathbf{q}_k\}$ be an orthonormal set of vectors in $\mathbb{R}^n$ and let W be the subspace spanned by this set.

(a) Show that the matrix of the orthogonal projection onto W is given by

$$P = \mathbf{q}_1\mathbf{q}_1^T + \cdots + \mathbf{q}_k\mathbf{q}_k^T$$

(b) Show that the projection matrix P in part (a) is symmetric and satisfies $P^2 = P$.

(c) Let $Q = [\mathbf{q}_1 \ \cdots \ \mathbf{q}_k]$ be the $n \times k$ matrix whose columns are the orthonormal basis vectors of W. Show that $P = QQ^T$ and deduce that $\text{rank}(P) = k$.

27. Let A be an $n \times n$ real matrix, all of whose eigenvalues are real. Prove that there exist an orthogonal matrix Q and an upper triangular matrix T such that $Q^TAQ = T$. This very useful result is known as ***Schur's Triangularization Theorem.*** [*Hint:* Adapt the proof of the Spectral Theorem.]

28. Let A be a nilpotent matrix (see Exercise 56 in Section 4.2). Prove that there is an orthogonal matrix Q such that Q^TAQ is upper triangular with zeros on its diagonal. [*Hint:* Use Exercise 27.]

5.5 Applications

Quadratic Forms

An expression of the form

$$ax^2 + by^2 + cxy$$

is called a ***quadratic form*** in x and y. Similarly,

$$ax^2 + by^2 + cz^2 + dxy + exz + fyz$$

is a quadratic form in x, y, and z. In words, a quadratic form is a sum of terms, each of which has total degree *two* in the variables. Therefore, $5x^2 - 3y^2 + 2xy$ is a quadratic form, but $x^2 + y^2 + x$ is not.

We can represent quadratic forms using matrices as follows:

$$ax^2 + by^2 + cxy = [x \quad y]\begin{bmatrix} a & c/2 \\ c/2 & b \end{bmatrix}\begin{bmatrix} x \\ y \end{bmatrix}$$

and

$$ax^2 + by^2 + cz^2 + dxy + exz + fyz = [x \quad y \quad z]\begin{bmatrix} a & d/2 & e/2 \\ d/2 & b & f/2 \\ e/2 & f/2 & c \end{bmatrix}\begin{bmatrix} x \\ y \\ z \end{bmatrix}$$

(Verify these.) Each has the form $\mathbf{x}^T A\mathbf{x}$, where the matrix A is symmetric. This observation leads us to the following general definition.

Definition A ***quadratic form*** in n variables is a function $f: \mathbb{R}^n \to \mathbb{R}$ of the form

$$f(\mathbf{x}) = \mathbf{x}^T A\mathbf{x}$$

where A is a symmetric $n \times n$ matrix and $\mathbf{x}$ is in $\mathbb{R}^n$. We refer to A as the ***matrix associated with f***.

Example 5.21

What is the quadratic form with associated matrix $A = \begin{bmatrix} 2 & -3 \\ -3 & 5 \end{bmatrix}$?

Solution If $\mathbf{x} = \begin{bmatrix} x_1 \\ x_2 \end{bmatrix}$, then

$$f(\mathbf{x}) = \mathbf{x}^T A\mathbf{x} = [x_1 \quad x_2]\begin{bmatrix} 2 & -3 \\ -3 & 5 \end{bmatrix}\begin{bmatrix} x_1 \\ x_2 \end{bmatrix} = 2x_1^2 + 5x_2^2 - 6x_1x_2$$

Observe that the *off-diagonal* entries $a_{12} = a_{21} = -3$ of A are *combined* to give the coefficient -6 of x_1x_2. This is true generally. We can expand a quadratic form in n variables $\mathbf{x}^T A\mathbf{x}$ as follows:

$$\mathbf{x}^T A\mathbf{x} = a_{11}x_1^2 + a_{22}x_2^2 + \cdots + a_{nn}x_n^2 + \sum_{i<j} 2a_{ij}x_ix_j$$

Thus, if $i \neq j$, the coefficient of x_ix_j is $2a_{ij}$.

Example 5.22

Find the matrix associated with the quadratic form

$$f(x_1, x_2, x_3) = 2x_1^2 - x_2^2 + 5x_3^2 + 6x_1x_2 - 3x_1x_3$$

Solution The coefficients of the squared terms x_i^2 go on the diagonal as a_{ii}, and the coefficients of the cross-product terms x_ix_j are split between a_{ij} and a_{ji}. This gives

$$A = \begin{bmatrix} 2 & 3 & -\frac{3}{2} \\ 3 & -1 & 0 \\ -\frac{3}{2} & 0 & 5 \end{bmatrix}$$

so
$$f(x_1, x_2, x_3) = [x_1 \quad x_2 \quad x_3]\begin{bmatrix} 2 & 3 & -\frac{3}{2} \\ 3 & -1 & 0 \\ -\frac{3}{2} & 0 & 5 \end{bmatrix}\begin{bmatrix} x_1 \\ x_2 \\ x_3 \end{bmatrix}$$

as you can easily check.

In the case of a quadratic form $f(x, y)$ in two variables, the graph of $z = f(x, y)$ is a surface in $\mathbb{R}^3$. Some examples are shown in Figure 5.12.

Observe that the effect of holding x or y constant is to take a cross section of the graph parallel to the yz or xz planes, respectively. For the graphs in Figure 5.12, all of these cross sections are easy to identify. For example, in Figure 5.12(a), the cross sections we get by holding x or y constant are all parabolas opening upward, so $f(x, y) \geq 0$ for all values of x and y. In Figure 5.12(c), holding x constant gives parabolas opening downward and holding y constant gives parabolas opening upward, producing a *saddle point*.

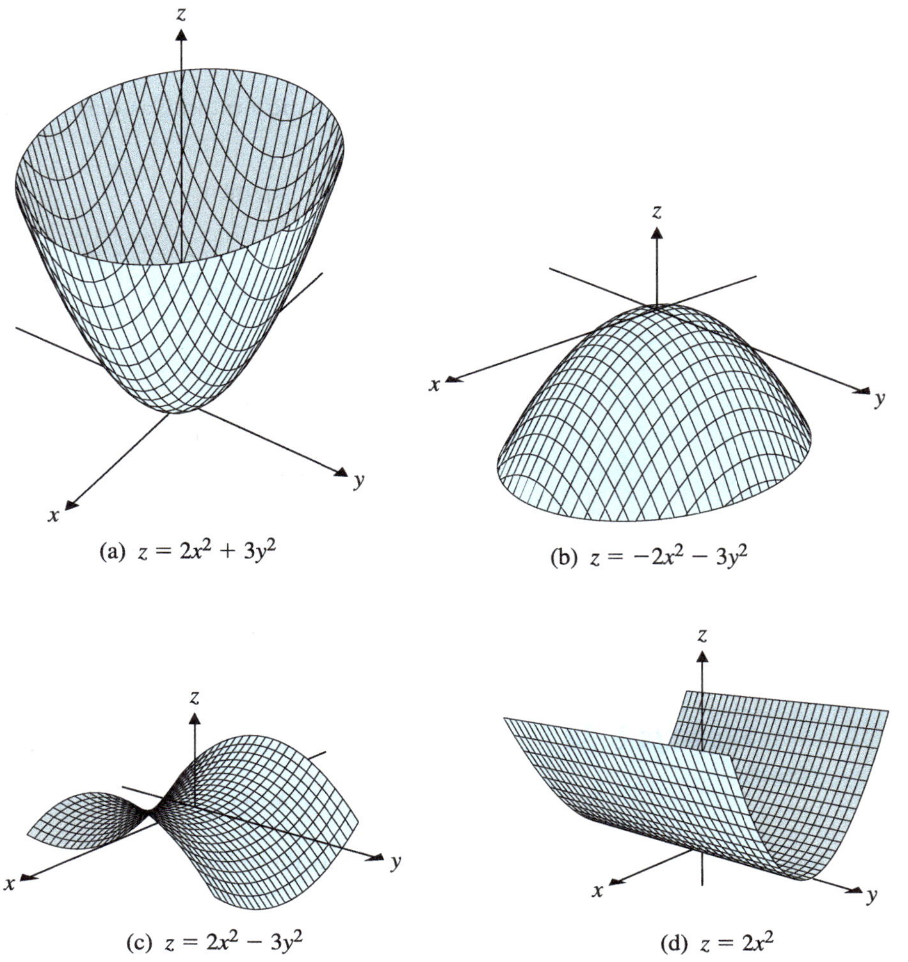

(a) $z = 2x^2 + 3y^2$

(b) $z = -2x^2 - 3y^2$

(c) $z = 2x^2 - 3y^2$

(d) $z = 2x^2$

Figure 5.12

Graphs of quadratic forms $f(x, y)$

What makes this type of analysis quite easy is the fact that these quadratic forms have no cross-product terms. The matrix associated with such a quadratic form is a diagonal matrix. For example,

$$2x^2 - 3y^2 = [x \quad y]\begin{bmatrix} 2 & 0 \\ 0 & -3 \end{bmatrix}\begin{bmatrix} x \\ y \end{bmatrix}$$

In general, the matrix of a quadratic form is a symmetric matrix, and we saw in Section 5.4 that such matrices can always be diagonalized. We will now use this fact to show that, for *every* quadratic form, we can eliminate the cross-product terms by means of a suitable change of variable.

Let $f(\mathbf{x}) = \mathbf{x}^T A\mathbf{x}$ be a quadratic form in n variables, with A a symmetric $n \times n$ matrix. By the Spectral Theorem, there is an orthogonal matrix Q that diagonalizes A; that is, $Q^T AQ = D$, where D is a diagonal matrix displaying the eigenvalues of A. We now set

$$\mathbf{x} = Q\mathbf{y} \quad \text{or, equivalently,} \quad \mathbf{y} = Q^{-1}\mathbf{x} = Q^T\mathbf{x}$$

Substitution into the quadratic form yields

$$\begin{aligned} \mathbf{x}^T A\mathbf{x} &= (Q\mathbf{y})^T A(Q\mathbf{y}) \\ &= \mathbf{y}^T Q^T AQ\mathbf{y} \\ &= \mathbf{y}^T D\mathbf{y} \end{aligned}$$

which is a quadratic form without cross-product terms, since D is diagonal. Furthermore, if the eigenvalues of A are $\lambda_1, \ldots, \lambda_n$, then Q can be chosen so that

$$D = \begin{bmatrix} \lambda_1 & \cdots & 0 \\ \vdots & \ddots & \vdots \\ 0 & \cdots & \lambda_n \end{bmatrix}$$

If $\mathbf{y} = [y_1 \quad \cdots \quad y_n]^T$, then, with respect to these new variables, the quadratic form becomes

$$\mathbf{y}^T D\mathbf{y} = \lambda_1 y_1^2 + \cdots + \lambda_n y_n^2$$

This process is called ***diagonalizing a quadratic form.*** We have just proved the following theorem, known as the ***Principal Axes Theorem.*** (The reason for this name will become clear in the next subsection.)

Theorem 5.21 **The Principal Axes Theorem**

Every quadratic form can be diagonalized. Specifically, if A is the $n \times n$ symmetric matrix associated with the quadratic form $\mathbf{x}^T A\mathbf{x}$ and if Q is an orthogonal matrix such that $Q^T AQ = D$ is a diagonal matrix, then the change of variable $\mathbf{x} = Q\mathbf{y}$ transforms the quadratic form $\mathbf{x}^T A\mathbf{x}$ into the quadratic form $\mathbf{y}^T D\mathbf{y}$, which has no cross-product terms. If the eigenvalues of A are $\lambda_1, \ldots, \lambda_n$ and $\mathbf{y} = [y_1 \quad \cdots \quad y_n]^T$, then

$$\mathbf{x}^T A\mathbf{x} = \mathbf{y}^T D\mathbf{y} = \lambda_1 y_1^2 + \cdots + \lambda_n y_n^2$$

Example 5.23

Find a change of variable that transforms the quadratic form

$$f(x_1, x_2) = 5x_1^2 + 4x_1x_2 + 2x_2^2$$

into one with no cross-product terms.

Solution The matrix of f is

$$A = \begin{bmatrix} 5 & 2 \\ 2 & 2 \end{bmatrix}$$

with eigenvalues $\lambda_1 = 6$ and $\lambda_2 = 1$. Corresponding unit eigenvectors are

$$\mathbf{q}_1 = \begin{bmatrix} 2/\sqrt{5} \\ 1/\sqrt{5} \end{bmatrix} \quad \text{and} \quad \mathbf{q}_2 = \begin{bmatrix} 1/\sqrt{5} \\ -2/\sqrt{5} \end{bmatrix}$$

(Check this.) If we set

$$Q = \begin{bmatrix} 2/\sqrt{5} & 1/\sqrt{5} \\ 1/\sqrt{5} & -2/\sqrt{5} \end{bmatrix} \quad \text{and} \quad D = \begin{bmatrix} 6 & 0 \\ 0 & 1 \end{bmatrix}$$

then $Q^TAQ = D$. The change of variable $\mathbf{x} = Q\mathbf{y}$, where

$$\mathbf{x} = \begin{bmatrix} x_1 \\ x_2 \end{bmatrix} \quad \text{and} \quad \mathbf{y} = \begin{bmatrix} y_1 \\ y_2 \end{bmatrix}$$

converts f into

$$f(\mathbf{y}) = f(y_1, y_2) = [y_1 \quad y_2]\begin{bmatrix} 6 & 0 \\ 0 & 1 \end{bmatrix}\begin{bmatrix} y_1 \\ y_2 \end{bmatrix} = 6y_1^2 + y_2^2$$

The original quadratic form $\mathbf{x}^TA\mathbf{x}$ and the new one $\mathbf{y}^TD\mathbf{y}$ (referred to in the Principal Axes Theorem) are *equal* in the following sense. In Example 5.23, suppose we want to evaluate $f(\mathbf{x}) = \mathbf{x}^TA\mathbf{x}$ at $\mathbf{x} = \begin{bmatrix} -1 \\ 3 \end{bmatrix}$. We have

$$f(-1, 3) = 5(-1)^2 + 4(-1)(3) + 2(3)^2 = 11$$

In terms of the new variables,

$$\begin{bmatrix} y_1 \\ y_2 \end{bmatrix} = \mathbf{y} = Q^T\mathbf{x} = \begin{bmatrix} 2/\sqrt{5} & 1/\sqrt{5} \\ 1/\sqrt{5} & -2/\sqrt{5} \end{bmatrix}\begin{bmatrix} -1 \\ 3 \end{bmatrix} = \begin{bmatrix} 1/\sqrt{5} \\ -7/\sqrt{5} \end{bmatrix}$$

so

$$f(y_1, y_2) = 6y_1^2 + y_2^2 = 6(1/\sqrt{5})^2 + (-7/\sqrt{5})^2 = 55/5 = 11$$

exactly as before.

The Principal Axes Theorem has some interesting and important consequences. We will consider two of these. The first relates to the possible *values* that a quadratic form can take on.

Definition A quadratic form $f(\mathbf{x}) = \mathbf{x}^TA\mathbf{x}$ is classified as one of the following:

1. ***positive definite*** if $f(\mathbf{x}) > 0$ for all $\mathbf{x} \neq \mathbf{0}$
2. ***positive semidefinite*** if $f(\mathbf{x}) \geq 0$ for all $\mathbf{x}$
3. ***negative definite*** if $f(\mathbf{x}) < 0$ for all $\mathbf{x} \neq \mathbf{0}$
4. ***negative semidefinite*** if $f(\mathbf{x}) \leq 0$ for all $\mathbf{x}$
5. ***indefinite*** if $f(\mathbf{x})$ takes on both positive and negative values

A symmetric matrix A is called ***positive definite, positive semidefinite, negative definite, negative semidefinite,*** or ***indefinite*** if the associated quadratic form $f(\mathbf{x}) = \mathbf{x}^T A\mathbf{x}$ has the corresponding property.

The quadratic forms in parts (a), (b), (c), and (d) of Figure 5.12 are positive definite, negative definite, indefinite, and positive semidefinite, respectively. The Principal Axes Theorem makes it easy to tell if a quadratic form has one of these properties.

Theorem 5.22

Let A be an $n \times n$ symmetric matrix. The quadratic form $f(\mathbf{x}) = \mathbf{x}^T A\mathbf{x}$ is

a. positive definite if and only if all of the eigenvalues of A are positive.
b. positive semidefinite if and only if all of the eigenvalues of A are nonnegative.
c. negative definite if and only if all of the eigenvalues of A are negative.
d. negative semidefinite if and only if all of the eigenvalues of A are nonpositive.
e. indefinite if and only if A has both positive and negative eigenvalues.

You are asked to prove Theorem 5.22 in Exercise 27.

Example 5.24

Classify $f(x, y, z) = 3x^2 + 3y^2 + 3z^2 - 2xy - 2xz - 2yz$ as positive definite, negative definite, indefinite, or none of these.

Solution The matrix associated with f is

$$\begin{bmatrix} 3 & -1 & -1 \\ -1 & 3 & -1 \\ -1 & -1 & 3 \end{bmatrix}$$

which has eigenvalues 1, 4, and 4. (Verify this.) Since all of these eigenvalues are positive, f is a positive definite quadratic form.

If a quadratic form $f(\mathbf{x}) = \mathbf{x}^T A\mathbf{x}$ is positive definite, then, since $f(\mathbf{0}) = 0$, the *minimum* value of $f(\mathbf{x})$ is 0 and it occurs at the origin. Similarly, a negative definite quadratic form has a maximum at the origin. Thus, Theorem 5.22 allows us to solve certain types of maxima/minima problems easily, without resorting to calculus. A type of problem that falls into this category is the ***constrained optimization problem.***

It is often important to know the maximum or minimum values of a quadratic form subject to certain constraints. (Such problems arise not only in mathematics but also in statistics, physics, engineering, and economics.) We will be interested in finding the extreme values of $f(\mathbf{x}) = \mathbf{x}^T A\mathbf{x}$ subject to the constraint that $\|\mathbf{x}\| = 1$. In the case of a quadratic form in two variables, we can visualize what the problem means. The graph of $z = f(x, y)$ is a surface in $\mathbb{R}^3$, and the constraint $\|\mathbf{x}\| = 1$ restricts the point (x, y) to the unit circle in the xy-plane. Thus, we are considering those points that lie simultaneously on the surface and on the unit cylinder perpendicular to the xy plane. These points form a curve lying on the surface, and we want the highest and lowest points on this curve. Figure 5.13 shows this situation for the quadratic form and corresponding surface in Figure 5.12(c).

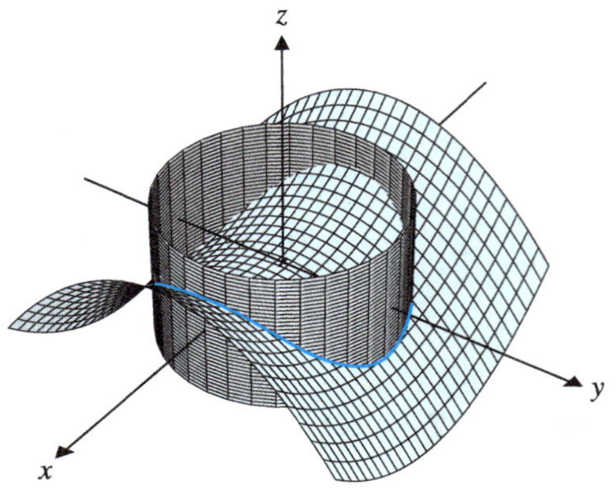

Figure 5.13

The intersection of $z = 2x^2 - 3y^2$ with the cylinder $x^2 + y^2 = 1$

In this case, the maximum and minimum values of $f(x, y) = 2x^2 - 3y^2$ (the highest and lowest points on the curve of intersection) are 2 and -3, respectively, which are just the eigenvalues of the associated matrix. Theorem 5.23 shows that this is always the case.

Theorem 5.23 Let $f(\mathbf{x}) = \mathbf{x}^T A\mathbf{x}$ be a quadratic form with associated $n \times n$ symmetric matrix A. Let the eigenvalues of A be $\lambda_1 \geq \lambda_2 \geq \cdots \geq \lambda_n$. Then the following are true, subject to the constraint $\|\mathbf{x}\| = 1$:

a. $\lambda_1 \geq f(\mathbf{x}) \geq \lambda_n$
b. The maximum value of $f(\mathbf{x})$ is λ_1, and it occurs when $\mathbf{x}$ is a unit eigenvector corresponding to λ_1.
c. The minimum value of $f(\mathbf{x})$ is λ_n, and it occurs when $\mathbf{x}$ is a unit eigenvector corresponding to λ_n.

Proof As usual, we begin by orthogonally diagonalizing A. Accordingly, let Q be an orthogonal matrix such that $Q^T AQ$ is the diagonal matrix

$$D = \begin{bmatrix} \lambda_1 & \cdots & 0 \\ \vdots & \ddots & \vdots \\ 0 & \cdots & \lambda_n \end{bmatrix}$$

Then, by the Principal Axes Theorem, the change of variable $\mathbf{x} = Q\mathbf{y}$ gives $\mathbf{x}^T A\mathbf{x} = \mathbf{y}^T D\mathbf{y}$. Now note that $\mathbf{y} = Q^T\mathbf{x}$ implies that

$$\mathbf{y}^T\mathbf{y} = (Q^T\mathbf{x})^T(Q^T\mathbf{x}) = \mathbf{x}^T(Q^T)^T Q^T\mathbf{x} = \mathbf{x}^T QQ^T\mathbf{x} = \mathbf{x}^T\mathbf{x}$$

since $Q^T = Q^{-1}$. Hence, using $\mathbf{x} \cdot \mathbf{x} = \mathbf{x}^T\mathbf{x}$, we see that $\|\mathbf{y}\| = \sqrt{\mathbf{y}^T\mathbf{y}} = \sqrt{\mathbf{x}^T\mathbf{x}} = \|\mathbf{x}\| = 1$. Thus, if $\mathbf{x}$ is a unit vector, so is the corresponding $\mathbf{y}$, and the values of $\mathbf{x}^T A\mathbf{x}$ and $\mathbf{y}^T D\mathbf{y}$ are the same.

(a) To prove property (a), we observe that if $\mathbf{y} = [y_1 \quad \cdots \quad y_n]^T$, then

$$\begin{aligned} f(\mathbf{x}) &= \mathbf{x}^T A \mathbf{x} = \mathbf{y}^T D \mathbf{y} \\ &= \lambda_1 y_1^2 + \lambda_2 y_2^2 + \cdots + \lambda_n y_n^2 \\ &\le \lambda_1 y_1^2 + \lambda_1 y_2^2 + \cdots + \lambda_1 y_n^2 \\ &= \lambda_1 (y_1^2 + y_2^2 + \cdots + y_n^2) \\ &= \lambda_1 \|\mathbf{y}\|^2 \\ &= \lambda_1 \end{aligned}$$

Thus, $f(\mathbf{x}) \le \lambda_1$ for all $\mathbf{x}$ such that $\|\mathbf{x}\| = 1$. The proof that $f(\mathbf{x}) \ge \lambda_n$ is similar. (See Exercise 37.)

(b) If $\mathbf{q}_1$ is a unit eigenvector corresponding to λ_1, then $A\mathbf{q}_1 = \lambda_1 \mathbf{q}_1$ and

$$f(\mathbf{q}_1) = \mathbf{q}_1^T A \mathbf{q}_1 = \mathbf{q}_1^T \lambda_1 \mathbf{q}_1 = \lambda_1(\mathbf{q}_1^T \mathbf{q}_1) = \lambda_1$$

This shows that the quadratic form actually takes on the value λ_1, and so, by property (a), it is the maximum value of $f(\mathbf{x})$ and it occurs when $\mathbf{x} = \mathbf{q}_1$.

(c) You are asked to prove this property in Exercise 38.

Example 5.25

Find the maximum and minimum values of the quadratic form $f(x_1, x_2) = 5x_1^2 + 4x_1x_2 + 2x_2^2$ subject to the constraint $x_1^2 + x_2^2 = 1$, and determine values of x_1 and x_2 for which each of these occurs.

Solution In Example 5.23, we found that f has the associated eigenvalues $\lambda_1 = 6$ and $\lambda_2 = 1$, with corresponding unit eigenvectors

$$\mathbf{q}_1 = \begin{bmatrix} 2/\sqrt{5} \\ 1/\sqrt{5} \end{bmatrix} \quad \text{and} \quad \mathbf{q}_2 = \begin{bmatrix} 1/\sqrt{5} \\ -2/\sqrt{5} \end{bmatrix}$$

Therefore, the maximum value of f is 6 when $x_1 = 2/\sqrt{5}$ and $x_2 = 1/\sqrt{5}$. The minimum value of f is 1 when $x_1 = 1/\sqrt{5}$ and $x_2 = -2/\sqrt{5}$. (Observe that these extreme values occur twice—in opposite directions—since $-\mathbf{q}_1$ and $-\mathbf{q}_2$ are also unit eigenvectors for λ_1 and λ_2, respectively.)

Graphing Quadratic Equations

The general form of a quadratic equation in two variables x and y is

$$ax^2 + by^2 + cxy + dx + ey + f = 0$$

where at least one of a, b, and c is nonzero. The graphs of such quadratic equations are called ***conic sections*** (or ***conics***), since they can be obtained by taking cross sections of a (double) cone (i.e., slicing it with a plane). The most important of the conic sections are the ellipses (with circles as a special case), hyperbolas, and parabolas. These are called the ***nondegenerate*** conics. Figure 5.14 shows how they arise.

It is also possible for a cross section of a cone to result in a single point, a straight line, or a pair of lines. These are called ***degenerate*** conics. (See Exercises 59–64.)

The graph of a nondegenerate conic is said to be in ***standard position*** relative to the coordinate axes if its equation can be expressed in one of the forms in Figure 5.15.

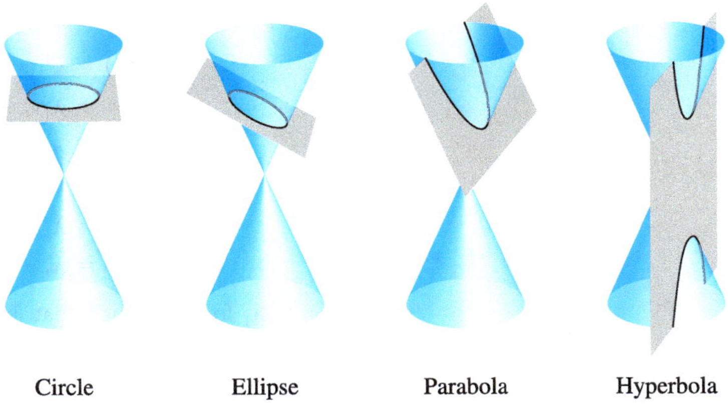

Figure 5.14
The nondegenerate conics

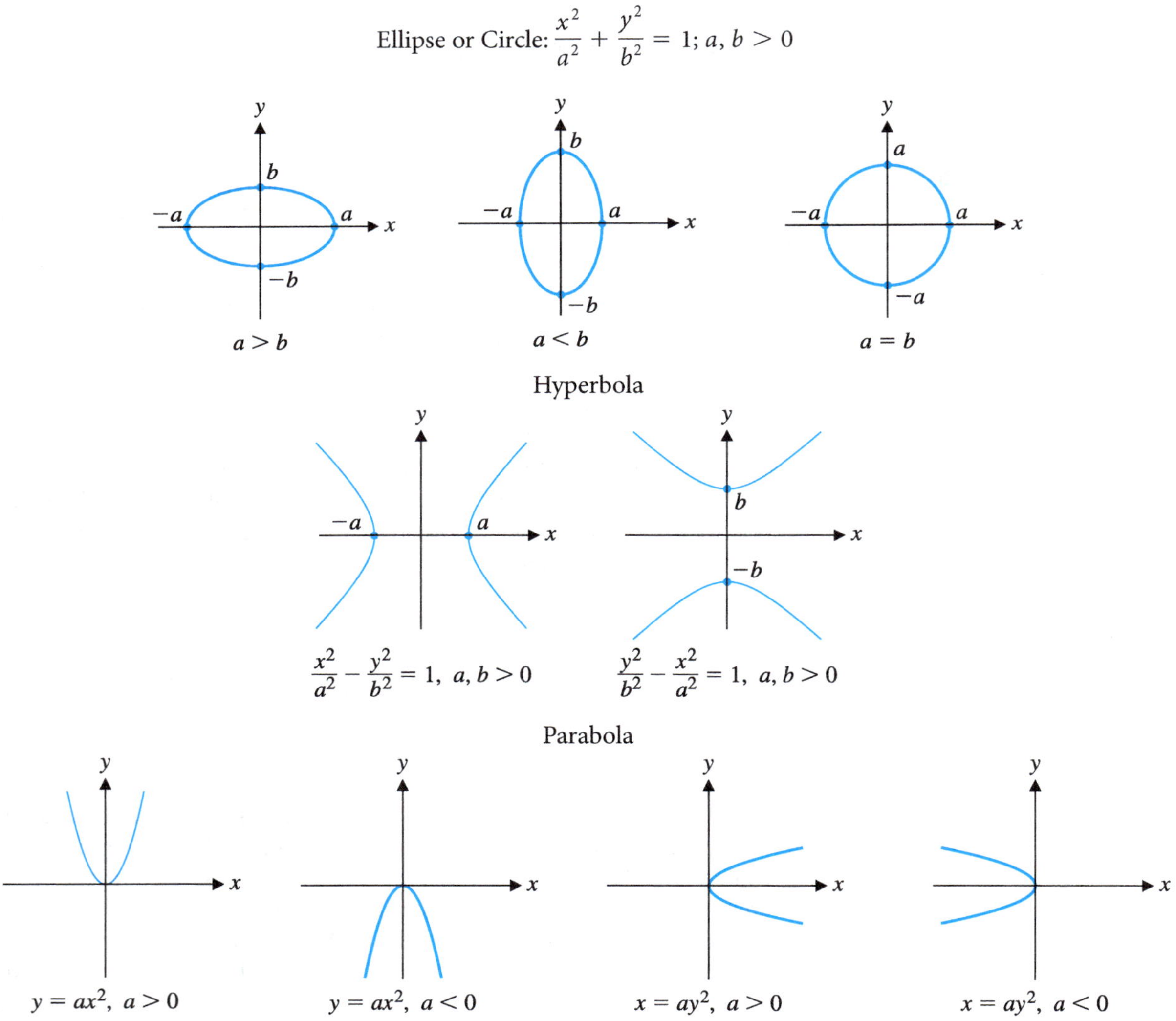

Figure 5.15
Nondegenerate conics in standard position

Example 5.26

If possible, write each of the following quadratic equations in the form of a conic in standard position and identify the resulting graph.

(a) $4x^2 + 9y^2 = 36$ (b) $4x^2 - 9y^2 + 1 = 0$ (c) $4x^2 - 9y = 0$

Solution (a) The equation $4x^2 + 9y^2 = 36$ can be written in the form

$$\frac{x^2}{9} + \frac{y^2}{4} = 1$$

so its graph is an ellipse intersecting the x-axis at $(\pm3, 0)$ and the y-axis at $(0, \pm2)$.

(b) The equation $4x^2 - 9y^2 + 1 = 0$ can be written in the form

$$\frac{y^2}{\frac{1}{9}} - \frac{x^2}{\frac{1}{4}} = 1$$

so its graph is a hyperbola, opening up and down, intersecting the y-axis at $(0, \pm\frac{1}{3})$.

(c) The equation $4x^2 - 9y = 0$ can be written in the form

$$y = \frac{4}{9}x^2$$

so its graph is a parabola opening upward.

If a quadratic equation contains too many terms to be written in one of the forms in Figure 5.15, then its graph is not in standard position. When there are additional terms but no xy term, the graph of the conic has been *translated* out of standard position.

Example 5.27

Identify and graph the conic whose equation is

$$x^2 + 2y^2 - 6x + 8y + 9 = 0$$

Solution We begin by grouping the x and y terms separately to get

$$(x^2 - 6x) + (2y^2 + 8y) = -9$$

or

$$(x^2 - 6x) + 2(y^2 + 4y) = -9$$

Next, we complete the squares on the two expressions in parentheses to obtain

$$(x^2 - 6x + 9) + 2(y^2 + 4y + 4) = -9 + 9 + 8$$

or

$$(x - 3)^2 + 2(y + 2)^2 = 8$$

We now make the substitutions $x' = x - 3$ and $y' = y + 2$, turning the above equation into

$$(x')^2 + 2(y')^2 = 8 \quad \text{or} \quad \frac{(x')^2}{8} + \frac{(y')^2}{4} = 1$$

This is the equation of an ellipse in standard position in the $x'y'$ coordinate system, intersecting the x'-axis at $(\pm 2\sqrt{2}, 0)$ and the y'-axis at $(0, \pm 2)$. The origin in the $x'y'$ coordinate system is at $x = 3$, $y = -2$, so the ellipse has been translated out of standard position 3 units to the right and 2 units down. Its graph is shown in Figure 5.16.

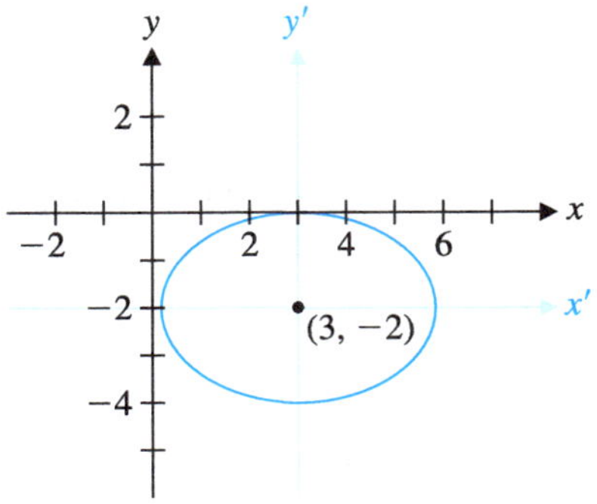

Figure 5.16
A translated ellipse

If a quadratic equation contains a cross-product term, then it represents a conic that has been *rotated*.

Example 5.28

Identify and graph the conic whose equation is

$$5x^2 + 4xy + 2y^2 = 6$$

Solution The left-hand side of the equation is a quadratic form, so we can write it in matrix form as $\mathbf{x}^T A\mathbf{x} = 6$, where

$$A = \begin{bmatrix} 5 & 2 \\ 2 & 2 \end{bmatrix}$$

In Example 5.23, we found that the eigenvalues of A are 6 and 1, and a matrix Q that orthogonally diagonalizes A is

$$Q = \begin{bmatrix} 2/\sqrt{5} & 1/\sqrt{5} \\ 1/\sqrt{5} & -2/\sqrt{5} \end{bmatrix}$$

Observe that $\det Q = -1$. In this example, we will interchange the columns of this matrix to make the determinant equal to $+1$. Then Q will be the matrix of a *rotation*, by Exercise 28 in Section 5.1. It is always possible to rearrange the columns of an orthogonal matrix Q to make its determinant equal to $+1$. (Why?) We set

$$Q = \begin{bmatrix} 1/\sqrt{5} & 2/\sqrt{5} \\ -2/\sqrt{5} & 1/\sqrt{5} \end{bmatrix}$$

instead, so that

$$Q^TAQ = \begin{bmatrix} 1 & 0 \\ 0 & 6 \end{bmatrix} = D$$

The change of variable $\mathbf{x} = Q\mathbf{x}'$ converts the given equation into the form $(\mathbf{x}')^T D\mathbf{x}' = 6$ by means of a rotation. If $\mathbf{x}' = \begin{bmatrix} x' \\ y' \end{bmatrix}$, then this equation is just

$$(x')^2 + 6(y')^2 = 6 \quad \text{or} \quad \frac{(x')^2}{6} + (y')^2 = 1$$

which represents an ellipse in the $x'y'$ coordinate system.

To graph this ellipse, we need to know which vectors play the roles of $\mathbf{e}_1' = \begin{bmatrix} 1 \\ 0 \end{bmatrix}$ and $\mathbf{e}_2' = \begin{bmatrix} 0 \\ 1 \end{bmatrix}$ in the new coordinate system. (These two vectors locate the positions of the x' and y' axes.) But, from $\mathbf{x} = Q\mathbf{x}'$, we have

$$Q\mathbf{e}_1' = \begin{bmatrix} 1/\sqrt{5} & 2/\sqrt{5} \\ -2/\sqrt{5} & 1/\sqrt{5} \end{bmatrix} \begin{bmatrix} 1 \\ 0 \end{bmatrix} = \begin{bmatrix} 1/\sqrt{5} \\ -2/\sqrt{5} \end{bmatrix}$$

and

$$Q\mathbf{e}_2' = \begin{bmatrix} 1/\sqrt{5} & 2/\sqrt{5} \\ -2/\sqrt{5} & 1/\sqrt{5} \end{bmatrix} \begin{bmatrix} 0 \\ 1 \end{bmatrix} = \begin{bmatrix} 2/\sqrt{5} \\ 1/\sqrt{5} \end{bmatrix}$$

These are just the columns $\mathbf{q}_1$ and $\mathbf{q}_2$ of Q, which are the eigenvectors of A! The fact that these are orthonormal vectors agrees perfectly with the fact that the change of variable is just a rotation. The graph is shown in Figure 5.17.

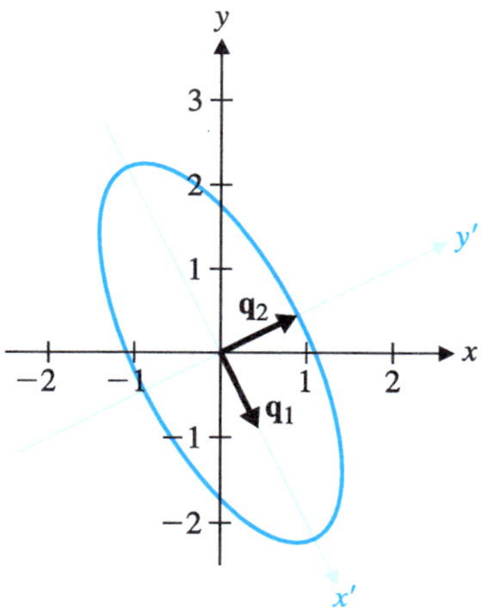

Figure 5.17
A rotated ellipse

You can now see why the Principal Axes Theorem is so named. If a real symmetric matrix A arises as the coefficient matrix of a quadratic equation, the eigenvectors of A give the directions of the principal axes of the corresponding graph.

It is possible for the graph of a conic to be both rotated and translated out of standard position, as illustrated in Example 5.29.

Example 5.29

Identify and graph the conic whose equation is

$$5x^2 + 4xy + 2y^2 - \frac{28}{\sqrt{5}}x - \frac{4}{\sqrt{5}}y + 4 = 0$$

Solution The strategy is to eliminate the cross-product term first. In matrix form, the equation is $\mathbf{x}^T A\mathbf{x} + B\mathbf{x} + 4 = 0$, where

$$A = \begin{bmatrix} 5 & 2 \\ 2 & 2 \end{bmatrix} \quad \text{and} \quad B = \begin{bmatrix} -\frac{28}{\sqrt{5}} & -\frac{4}{\sqrt{5}} \end{bmatrix}$$

The cross-product term comes from the quadratic form $\mathbf{x}^T A\mathbf{x}$, which we diagonalize as in Example 5.28 by setting $\mathbf{x} = Q\mathbf{x}'$, where

$$Q = \begin{bmatrix} 1/\sqrt{5} & 2/\sqrt{5} \\ -2/\sqrt{5} & 1/\sqrt{5} \end{bmatrix}$$

Then, as in Example 5.28,

$$\mathbf{x}^T A\mathbf{x} = (\mathbf{x}')^T D\mathbf{x}' = (x')^2 + 6(y')^2$$

But now we also have

$$B\mathbf{x} = BQ\mathbf{x}' = \begin{bmatrix} -\frac{28}{\sqrt{5}} & -\frac{4}{\sqrt{5}} \end{bmatrix} \begin{bmatrix} 1/\sqrt{5} & 2/\sqrt{5} \\ -2/\sqrt{5} & 1/\sqrt{5} \end{bmatrix} \begin{bmatrix} x' \\ y' \end{bmatrix} = -4x' - 12y'$$

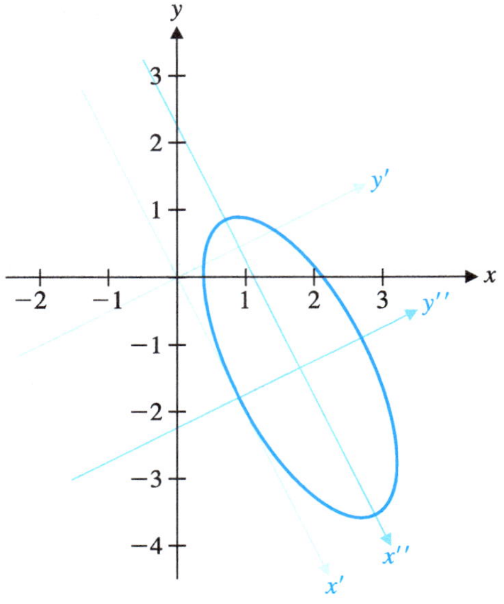

Figure 5.18

Thus, in terms of x' and y', the given equation becomes

$$(x')^2 + 6(y')^2 - 4x' - 12y' + 4 = 0$$

To bring the conic represented by this equation into standard position, we need to *translate* the $x'y'$ axes. We do so by completing the squares, as in Example 5.27. We have

$$((x')^2 - 4x' + 4) + 6((y')^2 - 2y' + 1) = -4 + 4 + 6 = 6$$

or

$$(x' - 2)^2 + 6(y' - 1)^2 = 6$$

This gives us the translation equations

$$x'' = x' - 2 \quad \text{and} \quad y'' = y' - 1$$

In the $x''y''$ coordinate system, the equation is simply

$$(x'')^2 + 6(y'')^2 = 6$$

which is the equation of an ellipse (as in Example 5.28). We can sketch this ellipse by first rotating and then translating. The resulting graph is shown in Figure 5.18.

The general form of a quadratic equation in three variables x, y, and z is

$$ax^2 + by^2 + cz^2 + dxy + exz + fyz + gx + hy + iz + j = 0$$

where at least one of $a, b, \ldots, f$ is nonzero. The graph of such a quadratic equation is called a ***quadric surface*** (or ***quadric***). Once again, to recognize a quadric we need

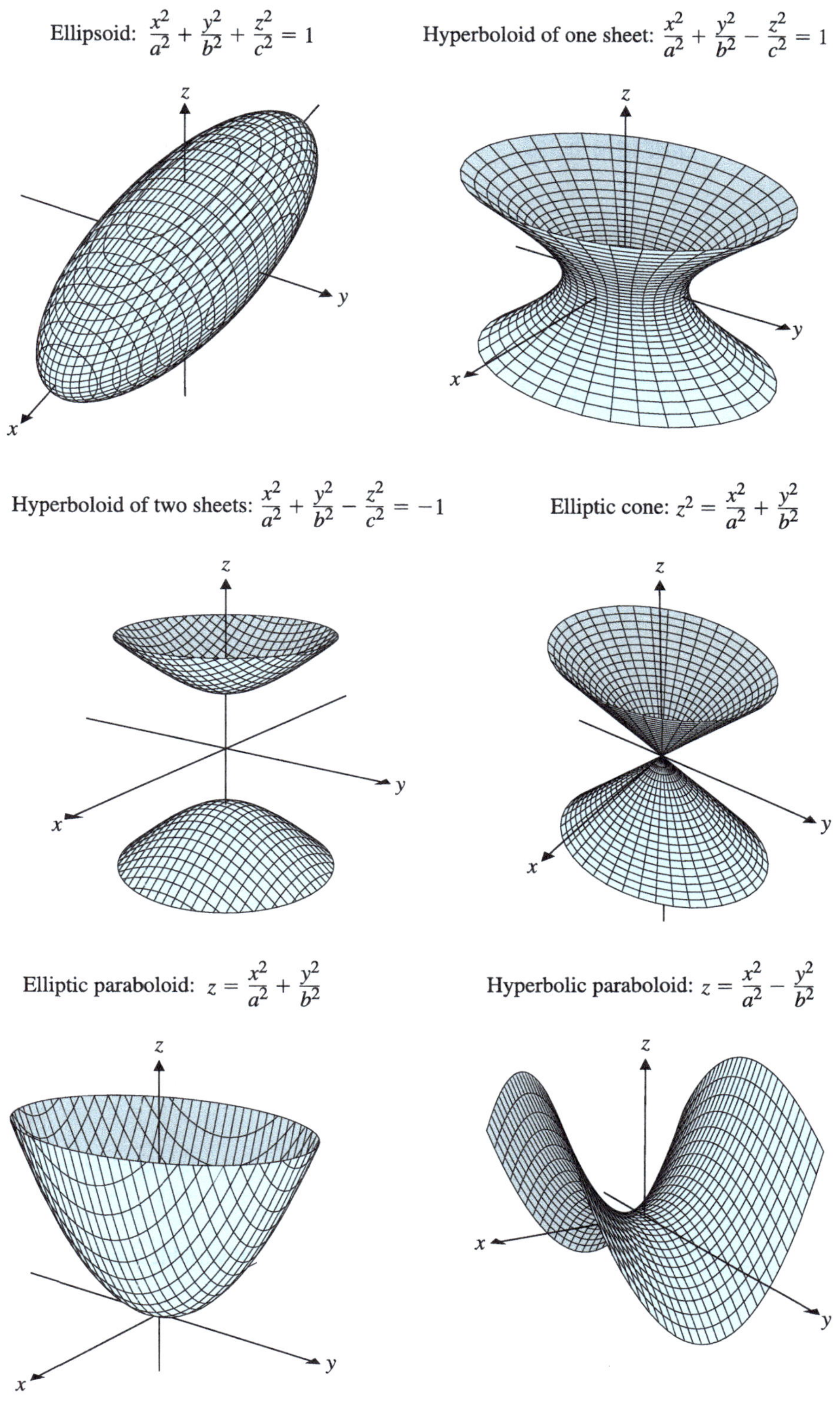

Figure 5.19
Quadric surfaces

to put it into standard position. Some quadrics in standard position are shown in Figure 5.19; others are obtained by permuting the variables.

Example 5.30

Identify the quadric surface whose equation is

$$5x^2 + 11y^2 + 2z^2 + 16xy + 20xz - 4yz = 36$$

Solution The equation can be written in matrix form as $\mathbf{x}^T A\mathbf{x} = 36$, where

$$A = \begin{bmatrix} 5 & 8 & 10 \\ 8 & 11 & -2 \\ 10 & -2 & 2 \end{bmatrix}$$

We find the eigenvalues of A to be 18, 9, and -9, with corresponding orthogonal eigenvectors

$$\begin{bmatrix} 2 \\ 2 \\ 1 \end{bmatrix}, \quad \begin{bmatrix} 1 \\ -2 \\ 2 \end{bmatrix}, \quad \text{and} \quad \begin{bmatrix} 2 \\ -1 \\ -2 \end{bmatrix}$$

respectively. We normalize them to obtain

$$\mathbf{q}_1 = \begin{bmatrix} \frac{2}{3} \\ \frac{2}{3} \\ \frac{1}{3} \end{bmatrix}, \quad \mathbf{q}_2 = \begin{bmatrix} \frac{1}{3} \\ -\frac{2}{3} \\ \frac{2}{3} \end{bmatrix}, \quad \text{and} \quad \mathbf{q}_3 = \begin{bmatrix} \frac{2}{3} \\ -\frac{1}{3} \\ -\frac{2}{3} \end{bmatrix}$$

and form the orthogonal matrix

$$Q = [\mathbf{q}_1 \ \ \mathbf{q}_2 \ \ \mathbf{q}_3] = \begin{bmatrix} \frac{2}{3} & \frac{1}{3} & \frac{2}{3} \\ \frac{2}{3} & -\frac{2}{3} & -\frac{1}{3} \\ \frac{1}{3} & \frac{2}{3} & -\frac{2}{3} \end{bmatrix}$$

Note that in order for Q to be the matrix of a rotation, we require $\det Q = 1$, which is true in this case. (Otherwise, $\det Q = -1$, and swapping two columns changes the sign of the determinant.) Therefore,

$$Q^T A Q = D = \begin{bmatrix} 18 & 0 & 0 \\ 0 & 9 & 0 \\ 0 & 0 & -9 \end{bmatrix}$$

and, with the change of variable $\mathbf{x} = Q\mathbf{x}'$, we get $\mathbf{x}^T A\mathbf{x} = (\mathbf{x}')D\mathbf{x}' = 36$, so

$$18(x')^2 + 9(y')^2 - 9(z')^2 = 36 \quad \text{or} \quad \frac{(x')^2}{2} + \frac{(y')^2}{4} - \frac{(z')^2}{4} = 1$$

From Figure 5.19, we recognize this equation as the equation of a hyperboloid of one sheet. The x', y', and z' axes are in the directions of the eigenvectors $\mathbf{q}_1$, $\mathbf{q}_2$, and $\mathbf{q}_3$, respectively. The graph is shown in Figure 5.20.

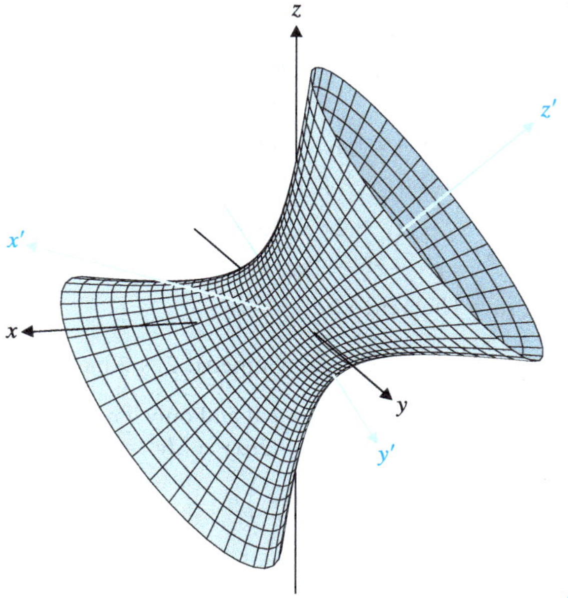

Figure 5.20
A hyperboloid of one sheet in nonstandard position

We can also identify and graph quadrics that have been translated out of standard position using the "complete-the-squares method" of Examples 5.27 and 5.29. You will be asked to do so in the exercises.

Exercises 5.5

Quadratic Forms

In Exercises 1–6, evaluate the quadratic form $f(\mathbf{x}) = \mathbf{x}^T A\mathbf{x}$ *for the given* A *and* $\mathbf{x}$.

1. $A = \begin{bmatrix} 2 & 3 \\ 3 & 4 \end{bmatrix}, \mathbf{x} = \begin{bmatrix} x \\ y \end{bmatrix}$

2. $A = \begin{bmatrix} 5 & 1 \\ 1 & -1 \end{bmatrix}, \mathbf{x} = \begin{bmatrix} x_1 \\ x_2 \end{bmatrix}$

3. $A = \begin{bmatrix} 3 & -2 \\ -2 & 4 \end{bmatrix}, \mathbf{x} = \begin{bmatrix} 1 \\ 6 \end{bmatrix}$

4. $A = \begin{bmatrix} 1 & 0 & -3 \\ 0 & 2 & 1 \\ -3 & 1 & 3 \end{bmatrix}, \mathbf{x} = \begin{bmatrix} x \\ y \\ z \end{bmatrix}$

5. $A = \begin{bmatrix} 1 & 0 & -3 \\ 0 & 2 & 1 \\ -3 & 1 & 3 \end{bmatrix}, \mathbf{x} = \begin{bmatrix} 2 \\ -1 \\ 1 \end{bmatrix}$

6. $A = \begin{bmatrix} 2 & 2 & 0 \\ 2 & 0 & 1 \\ 0 & 1 & 1 \end{bmatrix}, \mathbf{x} = \begin{bmatrix} 1 \\ 2 \\ 3 \end{bmatrix}$

In Exercises 7–12, find the symmetric matrix A associated with the given quadratic form.

7. $x_1^2 + 2x_2^2 + 6x_1x_2$

8. x_1x_2

9. $3x^2 - 3xy - y^2$

10. $x_1^2 - x_3^2 + 8x_1x_2 - 6x_2x_3$

11. $5x_1^2 - x_2^2 + 2x_3^2 + 2x_1x_2 - 4x_1x_3 + 4x_2x_3$

12. $2x^2 - 3y^2 + z^2 - 4xz$

Diagonalize the quadratic forms in Exercises 13–18 by finding an orthogonal matrix Q such that the change of variable $\mathbf{x} = Q\mathbf{y}$ *transforms the given form into one with no cross-product terms. Give Q and the new quadratic form.*

13. $2x_1^2 + 5x_2^2 - 4x_1x_2$

14. $x^2 + 8xy + y^2$

15. $7x_1^2 + x_2^2 + x_3^2 + 8x_1x_2 + 8x_1x_3 - 16x_2x_3$

16. $x_1^2 + x_2^2 + 3x_3^2 - 4x_1x_2$

17. $x^2 + z^2 - 2xy + 2yz$

18. $2xy + 2xz + 2yz$

Classify each of the quadratic forms in Exercises 19–26 as positive definite, positive semidefinite, negative definite, negative semidefinite, or indefinite.

19. $x_1^2 + 2x_2^2$

20. $x_1^2 + x_2^2 - 2x_1x_2$

21. $-2x^2 - 2y^2 + 2xy$

22. $x^2 + y^2 + 4xy$

23. $2x_1^2 + 2x_2^2 + 2x_3^2 + 2x_1x_2 + 2x_1x_3 + 2x_2x_3$

24. $x_1^2 + x_2^2 + x_3^2 + 2x_1x_3$

25. $x_1^2 + x_2^2 - x_3^2 + 4x_1x_2$

26. $-x^2 - y^2 - z^2 - 2xy - 2xz - 2yz$

27. Prove Theorem 5.22.

28. Let $A = \begin{bmatrix} a & b \\ b & d \end{bmatrix}$ be a symmetric 2×2 matrix. Prove that A is positive definite if and only if $a > 0$ and $\det A > 0$. [*Hint:* $ax^2 + 2bxy + dy^2 = a\left(x + \frac{b}{a}y\right)^2 + \left(d - \frac{b^2}{a}\right)y^2$.]

29. Let B be an invertible matrix. Show that $A = B^TB$ is positive definite.

30. Let A be a positive definite symmetric matrix. Show that there exists an invertible matrix B such that $A = B^TB$. [*Hint:* Use the Spectral Theorem to write $A = QDQ^T$. Then show that D can be factored as C^TC for some invertible matrix C.]

31. Let A and B be positive definite symmetric $n \times n$ matrices and let c be a positive scalar. Show that the following matrices are positive definite.

(a) cA **(b)** A^2 **(c)** $A + B$

(d) A^{-1} (First show that A is necessarily invertible.)

32. Let A be a positive definite symmetric matrix. Show that there is a positive definite symmetric matrix B such that $A = B^2$. (Such a matrix B is called a ***square root*** of A.)

In Exercises 33–36, find the maximum and minimum values of the quadratic form $f(\mathbf{x})$ in the given exercise, subject to the constraint $\|\mathbf{x}\| = 1$, and determine the values of $\mathbf{x}$ for which these occur.

33. Exercise 20

34. Exercise 22

35. Exercise 23

36. Exercise 24

37. Finish the proof of Theorem 5.23(a).

38. Prove Theorem 5.23(c).

Graphing Quadratic Equations

In Exercises 39–44, identify the graph of the given equation.

39. $x^2 + 5y^2 = 25$

40. $x^2 - y^2 - 4 = 0$

41. $x^2 - y - 1 = 0$

42. $2x^2 + y^2 - 8 = 0$

43. $3x^2 = y^2 - 1$

44. $x = -2y^2$

In Exercises 45–50, use a translation of axes to put the conic in standard position. Identify the graph, give its equation in the translated coordinate system, and sketch the curve.

45. $x^2 + y^2 - 4x - 4y + 4 = 0$

46. $4x^2 + 2y^2 - 8x + 12y + 6 = 0$

47. $9x^2 - 4y^2 - 4y = 37$

48. $x^2 + 10x - 3y = -13$

49. $2y^2 + 4x + 8y = 0$

50. $2y^2 - 3x^2 - 18x - 20y + 11 = 0$

In Exercises 51–54, use a rotation of axes to put the conic in standard position. Identify the graph, give its equation in the rotated coordinate system, and sketch the curve.

51. $x^2 + xy + y^2 = 6$

52. $4x^2 + 10xy + 4y^2 = 9$

53. $4x^2 + 6xy - 4y^2 = 5$

54. $3x^2 - 2xy + 3y^2 = 8$

In Exercises 55–58, identify the conic with the given equation and give its equation in standard form.

55. $3x^2 - 4xy + 3y^2 - 28\sqrt{2}x + 22\sqrt{2}y + 84 = 0$

56. $6x^2 - 4xy + 9y^2 - 20x - 10y - 5 = 0$

57. $2xy + 2\sqrt{2}x - 1 = 0$

58. $x^2 - 2xy + y^2 + 4\sqrt{2}x - 4 = 0$

Sometimes the graph of a quadratic equation is a straight line, a pair of straight lines, or a single point. We refer to such a graph as a ***degenerate conic.*** *It is also possible that the equation is not satisfied for any values of the variables, in which case there is no graph at all and we refer to the conic as an* ***imaginary conic.*** *In Exercises 59–64, identify the conic with the given equation as either degenerate or imaginary and, where possible, sketch the graph.*

59. $x^2 - y^2 = 0$

60. $x^2 + 2y^2 + 2 = 0$

61. $3x^2 + y^2 = 0$

62. $x^2 + 2xy + y^2 = 0$

63. $x^2 - 2xy + y^2 + 2\sqrt{2}x - 2\sqrt{2}y = 0$

64. $2x^2 + 2xy + 2y^2 + 2\sqrt{2}x - 2\sqrt{2}y + 6 = 0$

65. Let A be a symmetric 2×2 matrix and let k be a scalar. Prove that the graph of the quadratic equation $\mathbf{x}^TA\mathbf{x} = k$ is

(a) a hyperbola if $k \neq 0$ and $\det A < 0$

(b) an ellipse, circle, or imaginary conic if $k \neq 0$ and $\det A > 0$

(c) a pair of straight lines or an imaginary conic if $k \neq 0$ and $\det A = 0$

(d) a pair of straight lines or a single point if $k = 0$ and $\det A \neq 0$

(e) a straight line if $k = 0$ and $\det A = 0$

[*Hint:* Use the Principal Axes Theorem.]

In Exercises 66–73, identify the quadric with the given equation and give its equation in standard form.

66. $4x^2 + 4y^2 + 4z^2 + 4xy + 4xz + 4yz = 8$

67. $x^2 + y^2 + z^2 - 4yz = 1$

68. $-x^2 - y^2 - z^2 + 4xy + 4xz + 4yz = 12$

69. $2xy + z = 0$

70. $16x^2 + 100y^2 + 9z^2 - 24xz - 60x - 80z = 0$

71. $x^2 + y^2 - 2z^2 + 4xy - 2xz + 2yz - x + y + z = 0$

72. $10x^2 + 25y^2 + 10z^2 - 40xz + 20\sqrt{2}x + 50y + 20\sqrt{2}z = 15$

73. $11x^2 + 11y^2 + 14z^2 + 2xy + 8xz - 8yz - 12x + 12y + 12z = 6$

74. Let A be a real 2×2 matrix with complex eigenvalues $\lambda = a \pm bi$ such that $b \neq 0$ and $|\lambda| = 1$. Prove that every trajectory of the dynamical system $\mathbf{x}_{k+1} = A\mathbf{x}_k$ lies on an ellipse. [*Hint:* Theorem 4.43 shows that if $\mathbf{v}$ is an eigenvector corresponding to $\lambda = a - bi$, then the matrix $P = [\text{Re}\,\mathbf{v} \quad \text{Im}\,\mathbf{v}]$ is invertible and $A = P\begin{bmatrix} a & -b \\ b & a \end{bmatrix}P^{-1}$. Set $B = (PP^T)^{-1}$. Show that the quadratic $\mathbf{x}^T B\mathbf{x} = k$ defines an ellipse for all $k > 0$, and prove that if $\mathbf{x}$ lies on this ellipse, so does $A\mathbf{x}$.]

Chapter Review

Key Definitions and Concepts

fundamental subspaces of a matrix, 380
Gram-Schmidt Process, 389
orthogonal basis, 370
orthogonal complement of a subspace, 378
orthogonal matrix, 374
orthogonal projection, 382
orthogonal set of vectors, 369
Orthogonal Decomposition Theorem, 384
orthogonally diagonalizable matrix, 400
orthonormal basis, 372
orthonormal set of vectors, 372
properties of orthogonal matrices, 374–376
QR factorization, 393
Rank Theorem, 386
spectral decomposition, 405
Spectral Theorem, 403

Review Questions

1. Mark each of the following statements true or false:

(a) Every orthonormal set of vectors is linearly independent.

(b) Every nonzero subspace of $\mathbb{R}^n$ has an orthogonal basis.

(c) If A is a square matrix with orthonormal rows, then A is an orthogonal matrix.

(d) Every orthogonal matrix is invertible.

(e) If A is a matrix with $\det A = 1$, then A is an orthogonal matrix.

(f) If A is an $m \times n$ matrix such that $(\text{row}(A))^\perp = \mathbb{R}^n$, then A must be the zero matrix.

(g) If W is a subspace of $\mathbb{R}^n$ and $\mathbf{v}$ is a vector in $\mathbb{R}^n$ such that $\text{proj}_W(\mathbf{v}) = \mathbf{0}$, then $\mathbf{v}$ must be the zero vector.

(h) If A is a symmetric, orthogonal matrix, then $A^2 = I$.

(i) Every orthogonally diagonalizable matrix is invertible.

(j) Given any n real numbers $\lambda_1, \ldots, \lambda_n$, there exists a symmetric $n \times n$ matrix with $\lambda_1, \ldots, \lambda_n$ as its eigenvalues.

2. Find all values of a and b such that

$$\left\{\begin{bmatrix} 1 \\ 2 \\ 3 \end{bmatrix}, \begin{bmatrix} 4 \\ 1 \\ -2 \end{bmatrix}, \begin{bmatrix} a \\ b \\ 3 \end{bmatrix}\right\} \text{ is an orthogonal set of vectors.}$$

3. Find the coordinate vector $[\mathbf{v}]_{\mathcal{B}}$ of $\mathbf{v} = \begin{bmatrix} 7 \\ -3 \\ 2 \end{bmatrix}$ with respect to the orthogonal basis

$$\mathcal{B} = \left\{\begin{bmatrix} 1 \\ 0 \\ 1 \end{bmatrix}, \begin{bmatrix} 1 \\ 1 \\ -1 \end{bmatrix}, \begin{bmatrix} -1 \\ 2 \\ 1 \end{bmatrix}\right\} \text{ of } \mathbb{R}^3$$

4. The coordinate vector of a vector $\mathbf{v}$ with respect to an orthonormal basis $\mathcal{B} = \{\mathbf{v}_1, \mathbf{v}_2\}$ of $\mathbb{R}^2$ is $[\mathbf{v}]_{\mathcal{B}} = \begin{bmatrix} -3 \\ 1/2 \end{bmatrix}$. If $\mathbf{v}_1 = \begin{bmatrix} 3/5 \\ 4/5 \end{bmatrix}$, find all possible vectors $\mathbf{v}$.

5. Show that $\begin{bmatrix} 6/7 & 2/7 & 3/7 \\ -1/\sqrt{5} & 0 & 2/\sqrt{5} \\ 4/7\sqrt{5} & -15/7\sqrt{5} & 2/7\sqrt{5} \end{bmatrix}$ is an orthogonal matrix.

6. If $\begin{bmatrix} 1/2 & a \\ b & c \end{bmatrix}$ is an orthogonal matrix, find all possible values of a, b, and c.

7. If Q is an orthogonal $n \times n$ matrix and $\{\mathbf{v}_1, \dots, \mathbf{v}_k\}$ is an orthonormal set in $\mathbb{R}^n$, prove that $\{Q\mathbf{v}_1, \dots, Q\mathbf{v}_k\}$ is an orthonormal set.

8. If Q is an $n \times n$ matrix such that the angles $\angle(Q\mathbf{x}, Q\mathbf{y})$ and $\angle(\mathbf{x}, \mathbf{y})$ are equal for all vectors $\mathbf{x}$ and $\mathbf{y}$ in $\mathbb{R}^n$, prove that Q is an orthogonal matrix.

In Questions 9–12, find a basis for $W^{\perp}$.

9. W is the line in $\mathbb{R}^2$ with general equation $2x - 5y = 0$

10. W is the line in $\mathbb{R}^3$ with parametric equations

$$x = t$$
$$y = 2t$$
$$z = -t$$

11. $W = \text{span}\left\{ \begin{bmatrix} 1 \\ -1 \\ 4 \end{bmatrix}, \begin{bmatrix} 0 \\ 1 \\ -3 \end{bmatrix} \right\}$

12. $W = \text{span}\left\{ \begin{bmatrix} 1 \\ 1 \\ 1 \\ 1 \end{bmatrix}, \begin{bmatrix} 1 \\ -1 \\ 1 \\ 2 \end{bmatrix} \right\}$

13. Find bases for each of the four fundamental subspaces of

$$A = \begin{bmatrix} 1 & -1 & 2 & 1 & 3 \\ -1 & 2 & -2 & 1 & -2 \\ 2 & 1 & 4 & 8 & 9 \\ 3 & -5 & 6 & -1 & 7 \end{bmatrix}$$

14. Find the orthogonal decomposition of

$$\mathbf{v} = \begin{bmatrix} 1 \\ 0 \\ -1 \\ 2 \end{bmatrix}$$

with respect to

$$W = \text{span}\left\{ \begin{bmatrix} 0 \\ 1 \\ 1 \\ 1 \end{bmatrix}, \begin{bmatrix} 1 \\ 0 \\ 1 \\ -1 \end{bmatrix}, \begin{bmatrix} 3 \\ 1 \\ -2 \\ 1 \end{bmatrix} \right\}$$

15. (a) Apply the Gram-Schmidt Process to

$$\mathbf{x}_1 = \begin{bmatrix} 1 \\ 1 \\ 1 \\ 1 \end{bmatrix}, \mathbf{x}_2 = \begin{bmatrix} 1 \\ 1 \\ 1 \\ 0 \end{bmatrix}, \mathbf{x}_3 = \begin{bmatrix} 0 \\ 1 \\ 1 \\ 1 \end{bmatrix}$$

to find an orthogonal basis for $W = \text{span}\{\mathbf{x}_1, \mathbf{x}_2, \mathbf{x}_3\}$.

(b) Use the result of part (a) to find a QR factorization of $A = \begin{bmatrix} 1 & 1 & 0 \\ 1 & 1 & 1 \\ 1 & 1 & 1 \\ 1 & 0 & 1 \end{bmatrix}$.

16. Find an orthogonal basis for $\mathbb{R}^4$ that contains the vectors $\begin{bmatrix} 1 \\ 0 \\ 2 \\ 2 \end{bmatrix}$ and $\begin{bmatrix} 0 \\ 1 \\ 1 \\ -1 \end{bmatrix}$.

17. Find an orthogonal basis for the subspace

$$W = \left\{ \begin{bmatrix} x_1 \\ x_2 \\ x_3 \\ x_4 \end{bmatrix} : x_1 + x_2 + x_3 + x_4 = 0 \right\} \text{ of } \mathbb{R}^4$$

18. Let $A = \begin{bmatrix} 2 & 1 & -1 \\ 1 & 2 & 1 \\ -1 & 1 & 2 \end{bmatrix}$.

(a) Orthogonally diagonalize A.
(b) Give the spectral decomposition of A.

19. Find a symmetric matrix with eigenvalues $\lambda_1 = \lambda_2 = 1$, $\lambda_3 = -2$ and eigenspaces

$$E_1 = \text{span}\left(\begin{bmatrix} 1 \\ 1 \\ 0 \end{bmatrix}, \begin{bmatrix} 1 \\ 1 \\ 1 \end{bmatrix} \right), E_{-2} = \text{span}\left(\begin{bmatrix} 1 \\ -1 \\ 0 \end{bmatrix} \right)$$

20. If $\{\mathbf{v}_1, \mathbf{v}_2, \dots, \mathbf{v}_n\}$ is an orthonormal basis for $\mathbb{R}^n$ and

$$A = c_1\mathbf{v}_1\mathbf{v}_1^T + c_2\mathbf{v}_2\mathbf{v}_2^T + \cdots + c_n\mathbf{v}_n\mathbf{v}_n^T$$

prove that A is a symmetric matrix with eigenvalues $c_1, c_2, \dots, c_n$ and corresponding eigenvectors $\mathbf{v}_1, \mathbf{v}_2, \dots, \mathbf{v}_n$.

6 Vector Spaces

6.0 Introduction: Fibonacci in (Vector) Space

Algebra is generous; she often gives more than is asked of her.
—Jean le Rond d'Alembert (1717–1783)
In Carl B. Boyer
A History of Mathematics
Wiley, 1968, p. 481

The Fibonacci sequence was introduced in Section 4.6. It is the sequence

$$0, 1, 1, 2, 3, 5, 8, 13, \ldots$$

of nonnegative integers with the property that after the first two terms, each term is the sum of the two terms preceding it. Thus $0 + 1 = 1$, $1 + 1 = 2$, $1 + 2 = 3$, $2 + 3 = 5$, and so on.

If we denote the terms of the Fibonacci sequence by $f_0, f_1, f_2, \ldots$, then the entire sequence is completely determined by specifying that

$$f_0 = 0, f_1 = 1 \quad \text{and} \quad f_n = f_{n-1} + f_{n-2} \quad \text{for } n \geq 2$$

By analogy with vector notation, let's write a sequence $x_0, x_1, x_2, x_3, \ldots$ as

$$\mathbf{x} = [x_0, x_1, x_2, x_3, \ldots)$$

The Fibonacci sequence then becomes

$$\mathbf{f} = [f_0, f_1, f_2, f_3, \ldots) = [0, 1, 1, 2, \ldots)$$

We now generalize this notion.

Definition A ***Fibonacci-type sequence*** is *any* sequence $\mathbf{x} = [x_0, x_1, x_2, x_3, \ldots)$ such that x_0 and x_1 are real numbers and $x_n = x_{n-1} + x_{n-2}$ for $n \geq 2$.

For example, $[1, \sqrt{2}, 1 + \sqrt{2}, 1 + 2\sqrt{2}, 2 + 3\sqrt{2}, \ldots)$ is a Fibonacci-type sequence.

Problem 1 Write down the first five terms of three more Fibonacci-type sequences.

By analogy with vectors again, let's define the *sum* of two sequences $\mathbf{x} = [x_0, x_1, x_2, \ldots)$ and $\mathbf{y} = [y_0, y_1, y_2, \ldots)$ to be the sequence

$$\mathbf{x} + \mathbf{y} = [x_0 + y_0, x_1 + y_1, x_2 + y_2, \ldots)$$

If c is a scalar, we can likewise define the scalar multiple of a sequence by

$$c\mathbf{x} = [cx_0, cx_1, cx_2, \ldots)$$

Problem 2 (a) Using your examples from Problem 1 or other examples, compute the sums of various pairs of Fibonacci-type sequences. Do the resulting sequences appear to be Fibonacci-type?

(b) Compute various scalar multiples of your Fibonacci-type sequences from Problem 1. Do the resulting sequences appear to be Fibonacci-type?

Problem 3 (a) Prove that if $\mathbf{x}$ and $\mathbf{y}$ are Fibonacci-type sequences, then so is $\mathbf{x} + \mathbf{y}$.

(b) Prove that if $\mathbf{x}$ is a Fibonacci-type sequence and c is a scalar, then $c\mathbf{x}$ is also a Fibonacci-type sequence.

Let's denote the set of all Fibonacci-type sequences by *Fib*. Problem 3 shows that, like $\mathbb{R}^n$, *Fib* is closed under addition and scalar multiplication. The next exercises show that *Fib* has much more in common with $\mathbb{R}^n$.

Problem 4 Review the algebraic properties of vectors in Theorem 1.1. Does *Fib* satisfy all of these properties? What Fibonacci-type sequence plays the role of $\mathbf{0}$? For a Fibonacci-type sequence $\mathbf{x}$, what is $-\mathbf{x}$? Is $-\mathbf{x}$ also a Fibonacci-type sequence?

Problem 5 In $\mathbb{R}^n$, we have the standard basis vectors $\mathbf{e}_1, \mathbf{e}_2, \ldots, \mathbf{e}_n$. The Fibonacci sequence $\mathbf{f} = [0, 1, 1, 2, \ldots)$ can be thought of as the analogue of $\mathbf{e}_2$ because its first two terms are 0 and 1. What sequence $\mathbf{e}$ in *Fib* plays the role of $\mathbf{e}_1$?

What about $\mathbf{e}_3, \mathbf{e}_4, \ldots$? Do these vectors have analogues in *Fib*?

Problem 6 Let $\mathbf{x} = [x_0, x_1, x_2, \ldots)$ be a Fibonacci-type sequence. Show that $\mathbf{x}$ is a linear combination of $\mathbf{e}$ and $\mathbf{f}$.

Problem 7 Show that $\mathbf{e}$ and $\mathbf{f}$ are linearly independent. (That is, show that if $c\mathbf{e} + d\mathbf{f} = \mathbf{0}$, then $c = d = 0$.)

Problem 8 Given your answers to Problems 6 and 7, what would be a sensible value to assign to the "dimension" of *Fib*? Why?

Problem 9 Are there any geometric sequences in *Fib*? That is, if

$$[1, r, r^2, r^3, \ldots)$$

is a Fibonacci-type sequence, what are the possible values of r?

Problem 10 Find a "basis" for *Fib* consisting of geometric Fibonacci-type sequences.

Problem 11 Using your answer to Problem 10, give an alternative derivation of *Binet's formula* [formula (5) in Section 4.6]:

$$f_n = \frac{1}{\sqrt{5}}\left(\frac{1+\sqrt{5}}{2}\right)^n - \frac{1}{\sqrt{5}}\left(\frac{1-\sqrt{5}}{2}\right)^n$$

for the terms of the Fibonacci sequence $\mathbf{f} = [f_0, f_1, f_2, \ldots)$. [*Hint:* Express $\mathbf{f}$ in terms of the basis from Problem 10.]

The Lucas sequence is named after Edouard Lucas (see page 336).

The ***Lucas sequence*** is the Fibonacci-type sequence

$$\mathbf{l} = [l_0, l_1, l_2, l_3, \ldots) = [2, 1, 3, 4, \ldots)$$

Problem 12 Use the basis from Problem 10 to find an analogue of Binet's formula for the nth term l_n of the Lucas sequence.

Problem 13 Prove that the Fibonacci and Lucas sequences are related by the identity

$$f_{n-1} + f_{n+1} = l_n \quad \text{for } n \geq 1$$

[*Hint:* The Fibonacci-type sequences $\mathbf{f}^- = [1, 1, 2, 3, \ldots)$ and $\mathbf{f}^+ = [1, 0, 1, 1, \ldots)$ form a basis for *Fib*. (Why?)]

In this Introduction, we have seen that the collection *Fib* of all Fibonacci-type sequences behaves in many respects like $\mathbb{R}^2$, even though the "vectors" are actually infinite sequences. This useful analogy leads to the general notion of a *vector space* that is the subject of this chapter.

6.1 Vector Spaces and Subspaces

In Chapters 1 and 3, we saw that the algebra of vectors and the algebra of matrices are similar in many respects. In particular, we can add both vectors and matrices, and we can multiply both by scalars. The properties that result from these two operations (Theorem 1.1 and Theorem 3.2) are identical in both settings. In this section, we use these properties to define generalized "vectors" that arise in a wide variety of examples. By proving general theorems about these "vectors," we will therefore simultaneously be proving results about all of these examples. This is the real power of algebra: its ability to take properties from a concrete setting, like $\mathbb{R}^n$, and *abstract* them into a general setting.

Definition Let V be a set on which two operations, called *addition* and *scalar multiplication,* have been defined. If $\mathbf{u}$ and $\mathbf{v}$ are in V, the *sum* of $\mathbf{u}$ and $\mathbf{v}$ is denoted by $\mathbf{u} + \mathbf{v}$, and if c is a scalar, the *scalar multiple* of $\mathbf{u}$ by c is denoted by $c\mathbf{u}$. If the following axioms hold for all $\mathbf{u}$, $\mathbf{v}$, and $\mathbf{w}$ in V and for all scalars c and d, then V is called a ***vector space*** and its elements are called ***vectors***.

1. $\mathbf{u} + \mathbf{v}$ is in V. Closure under addition
2. $\mathbf{u} + \mathbf{v} = \mathbf{v} + \mathbf{u}$ Commutativity
3. $(\mathbf{u} + \mathbf{v}) + \mathbf{w} = \mathbf{u} + (\mathbf{v} + \mathbf{w})$ Associativity
4. There exists an element $\mathbf{0}$ in V, called a ***zero vector,*** such that $\mathbf{u} + \mathbf{0} = \mathbf{u}$.
5. For each $\mathbf{u}$ in V, there is an element $-\mathbf{u}$ in V such that $\mathbf{u} + (-\mathbf{u}) = \mathbf{0}$.
6. $c\mathbf{u}$ is in V. Closure under scalar multiplication
7. $c(\mathbf{u} + \mathbf{v}) = c\mathbf{u} + c\mathbf{v}$ Distributivity
8. $(c + d)\mathbf{u} = c\mathbf{u} + d\mathbf{u}$ Distributivity
9. $c(d\mathbf{u}) = (cd)\mathbf{u}$
10. $1\mathbf{u} = \mathbf{u}$

The German mathematician Hermann Grassmann (1809–1877) is generally credited with first introducing the idea of a vector space (although he did not call it that) in 1844. Unfortunately, his work was very difficult to read and did not receive the attention it deserved. One person who did study it was the Italian mathematician Giuseppe Peano (1858–1932). In his 1888 book *Calcolo Geometrico,* Peano clarified Grassmann's earlier work and laid down the axioms for a vector space as we know them today. Peano's book is also remarkable for introducing operations on sets. His notations $\cup$, $\cap$, and $\in$ (for "union," "intersection," and "is an element of") are the ones we still use, although they were not immediately accepted by other mathematicians. Peano's axiomatic definition of a vector space also had very little influence for many years. Acceptance came in 1918, after Hermann Weyl (1885–1955) repeated it in his book *Space, Time, Matter,* an introduction to Einstein's general theory of relativity.

Remarks

- By "scalars" we will usually mean the real numbers. Accordingly, we should refer to V as a *real vector space* (or a *vector space over the real numbers*). It is also possible for scalars to be complex numbers or to belong to $\mathbb{Z}_p$, where p is prime. In these cases, V is called a *complex vector space* or a *vector space over* $\mathbb{Z}_p$, respectively. Most of our examples will be real vector spaces, so we will usually omit the adjective "real." If something is referred to as a "vector space," assume that we are working over the real number system.

  In fact, the scalars can be chosen from any number system in which, roughly speaking, we can add, subtract, multiply, and divide according to the usual laws of arithmetic. In abstract algebra, such a number system is called a ***field***.
- The definition of a vector space does not specify what the set V consists of. Neither does it specify what the operations called "addition" and "scalar multiplication" look like. Often, they will be familiar, but they need not be. See Example 6.6 and Exercises 5–7.

We will now look at several examples of vector spaces. In each case, we need to specify the set V and the operations of addition and scalar multiplication and to verify axioms 1 through 10. We need to pay particular attention to axioms 1 and 6 (closure), axiom 4 (the existence of a zero vector in V), and axiom 5 (each vector in V must have a negative in V).

Example 6.1

For any $n \geq 1$, $\mathbb{R}^n$ is a vector space with the usual operations of addition and scalar multiplication. Axioms 1 and 6 follow from the definitions of these operations, and the remaining axioms follow from Theorem 1.1.

Example 6.2

The set of all 2×3 matrices is a vector space with the usual operations of matrix addition and matrix scalar multiplication. Here the "vectors" are actually matrices. We know that the sum of two 2×3 matrices is also a 2×3 matrix and that multiplying a 2×3 matrix by a scalar gives another 2×3 matrix; hence, we have closure. The remaining axioms follow from Theorem 3.2. In particular, the zero vector $\mathbf{0}$ is the 2×3 zero matrix, and the negative of a 2×3 matrix A is just the 2×3 matrix $-A$.

There is nothing special about 2×3 matrices. For any positive integers m and n, the set of all $m \times n$ matrices forms a vector space with the usual operations of matrix addition and matrix scalar multiplication. This vector space is denoted M_{mn}.

Example 6.3

Let $\mathscr{P}_2$ denote the set of all polynomials of degree 2 or less with real coefficients. Define addition and scalar multiplication in the usual way. (See Appendix D.) If

$$p(x) = a_0 + a_1x + a_2x^2 \quad \text{and} \quad q(x) = b_0 + b_1x + b_2x^2$$

are in $\mathscr{P}_2$, then

$$p(x) + q(x) = (a_0 + b_0) + (a_1 + b_1)x + (a_2 + b_2)x^2$$

has degree at most 2 and so is in $\mathscr{P}_2$. If c is a scalar, then

$$cp(x) = ca_0 + ca_1x + ca_2x^2$$

is also in $\mathscr{P}_2$. This verifies axioms 1 and 6.

The zero vector $\mathbf{0}$ is the zero polynomial—that is, the polynomial all of whose coefficients are zero. The negative of a polynomial $p(x) = a_0 + a_1x + a_2x^2$ is the polynomial $-p(x) = -a_0 - a_1x - a_2x^2$. It is now easy to verify the remaining axioms. We will check axiom 2 and leave the others for Exercise 12. With $p(x)$ and $q(x)$ as above, we have

$$\begin{aligned} p(x) + q(x) &= (a_0 + a_1x + a_2x^2) + (b_0 + b_1x + b_2x^2) \\ &= (a_0 + b_0) + (a_1 + b_1)x + (a_2 + b_2)x^2 \\ &= (b_0 + a_0) + (b_1 + a_1)x + (b_2 + a_2)x^2 \\ &= (b_0 + b_1x + b_2x^2) + (a_0 + a_1x + a_2x^2) \\ &= q(x) + p(x) \end{aligned}$$

where the third equality follows from the fact that addition of real numbers is commutative.

In general, for any fixed $n \geq 0$, the set $\mathscr{P}_n$ of all polynomials of degree less than or equal to n is a vector space, as is the set $\mathscr{P}$ of *all* polynomials.

Example 6.4

Let $\mathscr{F}$ denote the set of all real-valued functions defined on the real line. If f and g are two such functions and c is a scalar, then $f + g$ and cf are defined by

$$(f + g)(x) = f(x) + g(x) \quad \text{and} \quad (cf)(x) = cf(x)$$

In other words, the *value* of $f + g$ at x is obtained by adding together the values of f and g at x [Figure 6.1(a)]. Similarly, the value of cf at x is just the value of f at x multiplied by the scalar c [Figure 6.1(b)]. The zero vector in $\mathscr{F}$ is the constant function f_0 that is identically zero; that is, $f_0(x) = 0$ for all x. The negative of a function f is the function $-f$ defined by $(-f)(x) = -f(x)$ [Figure 6.1(c)].

Axioms 1 and 6 are obviously true. Verification of the remaining axioms is left as Exercise 13. Thus, $\mathscr{F}$ is a vector space.

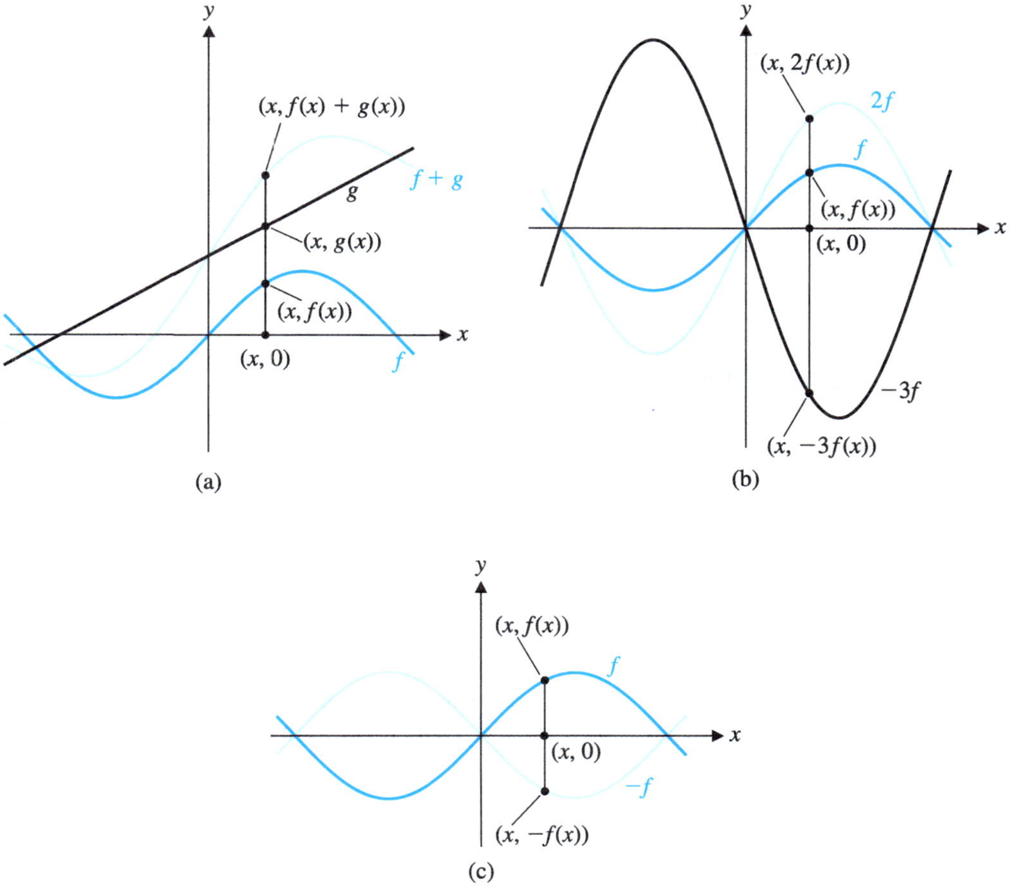

Figure 6.1
The graphs of (a) f, g, and $f + g$, (b) f, $2f$, and $-3f$, and (c) f and $-f$

In Example 6.4, we could also have considered only those functions defined on some *closed interval* $[a, b]$ of the real line. This approach also produces a vector space, denoted by $\mathscr{F}[a, b]$.

Example 6.5

The set $\mathbb{Z}$ of integers with the usual operations is *not* a vector space. To demonstrate this, it is enough to find that *one* of the ten axioms fails and to give a specific instance in which it fails (a *counterexample*). In this case, we find that we do not have closure under scalar multiplication. For example, the multiple of the integer 2 by the scalar $\frac{1}{3}$ is $(\frac{1}{3})(2) = \frac{2}{3}$, which is not an integer. Thus, it is not true that cx is in $\mathbb{Z}$ for *every* x in $\mathbb{Z}$ and *every* scalar c (i.e., axiom 6 fails).

Example 6.6

Let $V = \mathbb{R}^2$ with the usual definition of addition but the following definition of scalar multiplication:

$$c\begin{bmatrix} x \\ y \end{bmatrix} = \begin{bmatrix} cx \\ 0 \end{bmatrix}$$

Then, for example,
$$1\begin{bmatrix} 2 \\ 3 \end{bmatrix} = \begin{bmatrix} 2 \\ 0 \end{bmatrix} \neq \begin{bmatrix} 2 \\ 3 \end{bmatrix}$$

so axiom 10 fails. [In fact, the other nine axioms are all true (check this), but we do not need to look into them, because V has already failed to be a vector space. This example shows the value of looking ahead, rather than working through the list of axioms in the order in which they have been given.]

a + bi

Example 6.7

Let $\mathbb{C}^2$ denote the set of all ordered pairs of complex numbers. Define addition and scalar multiplication as in $\mathbb{R}^2$, except here the scalars are complex numbers. For example,

$$\begin{bmatrix} 1+i \\ 2-3i \end{bmatrix} + \begin{bmatrix} -3+2i \\ 4 \end{bmatrix} = \begin{bmatrix} -2+3i \\ 6-3i \end{bmatrix}$$

and
$$(1-i)\begin{bmatrix} 1+i \\ 2-3i \end{bmatrix} = \begin{bmatrix} (1-i)(1+i) \\ (1-i)(2-3i) \end{bmatrix} = \begin{bmatrix} 2 \\ -1-5i \end{bmatrix}$$

Using properties of the complex numbers, it is straightforward to check that all ten axioms hold. Therefore, $\mathbb{C}^2$ is a complex vector space.

In general, $\mathbb{C}^n$ is a complex vector space for all $n \geq 1$.

Example 6.8

If p is prime, the set $\mathbb{Z}_p^n$ (with the usual definitions of addition and multiplication by scalars from $\mathbb{Z}_p$) is a vector space over $\mathbb{Z}_p$ for all $n \geq 1$.

Before we consider further examples, we state a theorem that contains some useful properties of vector spaces. It is important to note that, by proving this theorem for vector spaces in *general*, we are actually proving it for *every specific* vector space.

Theorem 6.1 Let V be a vector space, $\mathbf{u}$ a vector in V, and c a scalar.

a. $0\mathbf{u} = \mathbf{0}$
b. $c\mathbf{0} = \mathbf{0}$
c. $(-1)\mathbf{u} = -\mathbf{u}$
d. If $c\mathbf{u} = \mathbf{0}$, then $c = 0$ or $\mathbf{u} = \mathbf{0}$.

Proof We prove properties (b) and (d) and leave the proofs of the remaining properties as exercises.

(b) We have

$$c\mathbf{0} = c(\mathbf{0} + \mathbf{0}) = c\mathbf{0} + c\mathbf{0}$$

by vector space axioms 4 and 7. Adding the negative of $c\mathbf{0}$ to both sides produces

$$c\mathbf{0} + (-c\mathbf{0}) = (c\mathbf{0} + c\mathbf{0}) + (-c\mathbf{0})$$

which implies

$$\begin{aligned} \mathbf{0} &= c\mathbf{0} + (c\mathbf{0} + (-c\mathbf{0})) && \text{by axioms 5 and 3} \\ &= c\mathbf{0} + \mathbf{0} && \text{by axiom 5} \\ &= c\mathbf{0} && \text{by axiom 4} \end{aligned}$$

(d) Suppose $c\mathbf{u} = \mathbf{0}$. To show that either $c = 0$ or $\mathbf{u} = \mathbf{0}$, let's assume that $c \neq 0$. (If $c = 0$, there is nothing to prove.) Then, since $c \neq 0$, its reciprocal $1/c$ is defined, and

$$\begin{aligned} \mathbf{u} &= 1\mathbf{u} && \text{by axiom 10} \\ &= \left(\frac{1}{c}c\right)\mathbf{u} \\ &= \frac{1}{c}(c\mathbf{u}) && \text{by axiom 9} \\ &= \frac{1}{c}\mathbf{0} \\ &= \mathbf{0} && \text{by property (b)} \end{aligned}$$

We will write $\mathbf{u} - \mathbf{v}$ for $\mathbf{u} + (-\mathbf{v})$, thereby defining ***subtraction*** of vectors. We will also exploit the associativity property of addition to unambiguously write $\mathbf{u} + \mathbf{v} + \mathbf{w}$ for the sum of three vectors and, more generally,

$$c_1\mathbf{u}_1 + c_2\mathbf{u}_2 + \cdots + c_n\mathbf{u}_n$$

for a ***linear combination*** of vectors.

Subspaces

We have seen that, in $\mathbb{R}^n$, it is possible for one vector space to sit inside another one, giving rise to the notion of a subspace. For example, a plane through the origin is a subspace of $\mathbb{R}^3$. We now extend this concept to general vector spaces.

Definition A subset W of a vector space V is called a ***subspace*** of V if W is itself a vector space with the same scalars, addition, and scalar multiplication as V.

As in $\mathbb{R}^n$, checking to see whether a subset W of a vector space V is a subspace of V involves testing only two of the ten vector space axioms. We prove this observation as a theorem.

Theorem 6.2 Let V be a vector space and let W be a nonempty subset of V. Then W is a subspace of V if and only if the following conditions hold:

a. If $\mathbf{u}$ and $\mathbf{v}$ are in W, then $\mathbf{u} + \mathbf{v}$ is in W.
b. If $\mathbf{u}$ is in W and c is a scalar, then $c\mathbf{u}$ is in W.

Proof Assume that W is a subspace of V. Then W satisfies vector space axioms 1 to 10. In particular, axiom 1 is condition (a) and axiom 6 is condition (b).

Conversely, assume that W is a subset of a vector space V, satisfying conditions (a) and (b). By hypothesis, axioms 1 and 6 hold. Axioms 2, 3, 7, 8, 9, and 10 hold in W because they are true for *all* vectors in V and thus are true in particular for those vectors in W. (We say that W *inherits* these properties from V.) This leaves axioms 4 and 5 to be checked.

Since W is nonempty, it contains at least one vector $\mathbf{u}$. Then condition (b) and Theorem 6.1(a) imply that $0\mathbf{u} = \mathbf{0}$ is also in W. This is axiom 4.

If $\mathbf{u}$ is in V, then, by taking $c = -1$ in condition (b), we have that $-\mathbf{u} = (-1)\mathbf{u}$ is also in W, using Theorem 6.1(c).

Remark Since Theorem 6.2 generalizes the notion of a subspace from the context of $\mathbb{R}^n$ to general vector spaces, all of the subspaces of $\mathbb{R}^n$ that we encountered in Chapter 3 are subspaces of $\mathbb{R}^n$ in the current context. In particular, lines and planes through the origin are subspaces of $\mathbb{R}^3$.

Example 6.9 We have already shown that the set $\mathscr{P}_n$ of all polynomials with degree at most n is a vector space. Hence, $\mathscr{P}_n$ is a subspace of the vector space $\mathscr{P}$ of *all* polynomials.

Example 6.10 Let W be the set of symmetric $n \times n$ matrices. Show that W is a subspace of M_{nn}.

Solution Clearly, W is nonempty, so we need only check conditions (a) and (b) in Theorem 6.2. Let A and B be in W and let c be a scalar. Then $A^T = A$ and $B^T = B$, from which it follows that

$$(A + B)^T = A^T + B^T = A + B$$

Therefore, $A + B$ is symmetric and, hence, is in W. Similarly,

$$(cA)^T = cA^T = cA$$

so cA is symmetric and, thus, is in W. We have shown that W is closed under addition and scalar multiplication. Therefore, it is a subspace of M_{nn}, by Theorem 6.2.

Example 6.11

Let $\mathscr{C}$ be the set of all continuous real-valued functions defined on $\mathbb{R}$ and let $\mathscr{D}$ be the set of all differentiable real-valued functions defined on $\mathbb{R}$. Show that $\mathscr{C}$ and $\mathscr{D}$ are subspaces of $\mathscr{F}$, the vector space of all real-valued functions defined on $\mathbb{R}$.

Solution From calculus, we know that if f and g are continuous functions and c is a scalar, then $f + g$ and cf are also continuous. Hence, $\mathscr{C}$ is closed under addition and scalar multiplication and so is a subspace of $\mathscr{F}$. If f and g are differentiable, then so are $f + g$ and cf. Indeed,

$$(f + g)' = f' + g' \quad \text{and} \quad (cf)' = c(f')$$

So $\mathscr{D}$ is also closed under addition and scalar multiplication, making it a subspace of $\mathscr{F}$.

It is a theorem of calculus that every differentiable function is continuous. Consequently, $\mathscr{D}$ is contained in $\mathscr{C}$ (denoted by $\mathscr{D} \subset \mathscr{C}$), making $\mathscr{D}$ a subspace of $\mathscr{C}$. It is also the case that every polynomial function is differentiable, so $\mathscr{P} \subset \mathscr{D}$, and thus $\mathscr{P}$ is a subspace of $\mathscr{D}$. We therefore have a *hierarchy* of subspaces of $\mathscr{F}$, one inside the other:

$$\mathscr{P} \subset \mathscr{D} \subset \mathscr{C} \subset \mathscr{F}$$

This hierarchy is depicted in Figure 6.2.

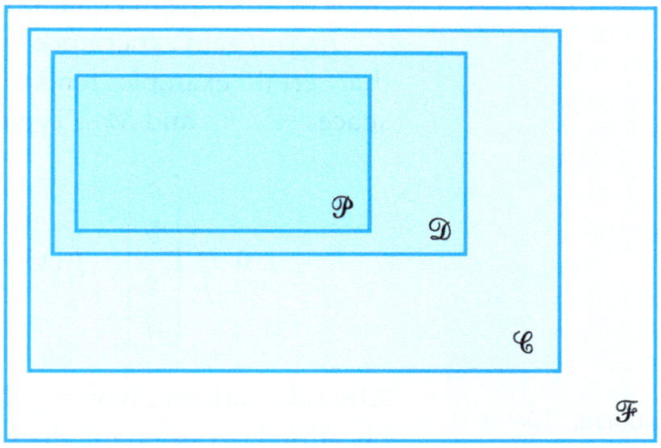

Figure 6.2
The hierarchy of subspaces of $\mathscr{F}$

There are other subspaces of $\mathscr{F}$ that can be placed into this hierarchy. Some of these are explored in the exercises.

In the preceding discussion, we could have restricted our attention to functions defined on a closed interval $[a, b]$. Then the corresponding subspaces of $\mathscr{F}[a, b]$ would be $\mathscr{C}[a, b]$, $\mathscr{D}[a, b]$, and $\mathscr{P}[a, b]$.

Example 6.12

Let S be the set of all functions that satisfy the differential equation

$$f'' + f = 0 \tag{1}$$

That is, S is the solution set of Equation (1). Show that S is a subspace of $\mathscr{F}$.

Solution S is nonempty, since the zero function clearly satisfies Equation (1). Let f and g be in S and let c be a scalar. Then

$$\begin{aligned}(f+g)'' + (f+g) &= (f'' + g'') + (f+g) \\ &= (f'' + f) + (g'' + g) \\ &= 0 + 0 \\ &= 0\end{aligned}$$

which shows that $f + g$ is in S. Similarly,

$$\begin{aligned}(cf)'' + cf &= cf'' + cf \\ &= c(f'' + f) \\ &= c0 \\ &= 0\end{aligned}$$

so cf is also in S.

Therefore, S is closed under addition and scalar multiplication and is a subspace of $\mathscr{F}$.

The differential Equation (1) is an example of a ***homogeneous linear differential equation.*** The solution sets of such equations are always subspaces of $\mathscr{F}$. Note that in Example 6.12 we did not actually solve Equation (1) (i.e., we did not find any specific solutions, other than the zero function). We will discuss techniques for finding solutions to this type of equation in Section 6.7.

As you gain experience working with vector spaces and subspaces, you will notice that certain examples tend to resemble one another. For example, consider the vector spaces $\mathbb{R}^4$, $\mathscr{P}_3$, and M_{22}. Typical elements of these vector spaces are, respectively,

$$\mathbf{u} = \begin{bmatrix} a \\ b \\ c \\ d \end{bmatrix}, \quad p(x) = a + bx + cx^2 + dx^3, \quad \text{and} \quad A = \begin{bmatrix} a & b \\ c & d \end{bmatrix}$$

In the words of Yogi Berra, "It's déjà vu all over again."

Any calculations involving the vector space operations of addition and scalar multiplication are essentially the same in all three settings. To highlight the similarities, in the next example we will perform the necessary steps in the three vector spaces side by side.

Example 6.13

(a) Show that the set W of all vectors of the form

$$\begin{bmatrix} a \\ b \\ -b \\ a \end{bmatrix}$$

is a subspace of $\mathbb{R}^4$.

(b) Show that the set W of all polynomials of the form $a + bx - bx^2 + ax^3$ is a subspace of $\mathscr{P}_3$.

(c) Show that the set W of all matrices of the form $\begin{bmatrix} a & b \\ -b & a \end{bmatrix}$ is a subspace of M_{22}.

Solution

(a) W is nonempty because it contains the zero vector $\mathbf{0}$. (Take $a = b = 0$.) Let $\mathbf{u}$ and $\mathbf{v}$ be in W—say,

$$\mathbf{u} = \begin{bmatrix} a \\ b \\ -b \\ a \end{bmatrix} \quad \text{and} \quad \mathbf{v} = \begin{bmatrix} c \\ d \\ -d \\ c \end{bmatrix}$$

Then

$$\mathbf{u} + \mathbf{v} = \begin{bmatrix} a + c \\ b + d \\ -b - d \\ a + c \end{bmatrix} = \begin{bmatrix} a + c \\ b + d \\ -(b + d) \\ a + c \end{bmatrix}$$

so $\mathbf{u} + \mathbf{v}$ is also in W (because it has the right *form*).

Similarly, if k is a scalar, then

$$k\mathbf{u} = \begin{bmatrix} ka \\ kb \\ -kb \\ ka \end{bmatrix}$$

so $k\mathbf{u}$ is in W.

Thus, W is a nonempty subset of $\mathbb{R}^4$ that is closed under addition and scalar multiplication. Therefore, W is a subspace of $\mathbb{R}^4$, by Theorem 6.2.

(b) W is nonempty because it contains the zero polynomial. (Take $a = b = 0$.) Let $p(x)$ and $q(x)$ be in W—say,

$$p(x) = a + bx - bx^2 + ax^3$$

and

$$q(x) = c + dx - dx^2 + cx^3$$

Then

$$p(x) + q(x) = (a + c) + (b + d)x - (b + d)x^2 + (a + c)x^3$$

so $p(x) + q(x)$ is also in W (because it has the right *form*).

Similarly, if k is a scalar, then

$$kp(x) = ka + kbx - kbx^2 + kax^3$$

so $kp(x)$ is in W.

Thus, W is a nonempty subset of $\mathscr{P}_3$ that is closed under addition and scalar multiplication. Therefore, W is a subspace of $\mathscr{P}_3$ by Theorem 6.2.

(c) W is nonempty because it contains the zero matrix O. (Take $a = b = 0$.) Let A and B be in W—say,

$$A = \begin{bmatrix} a & b \\ -b & a \end{bmatrix}$$

and

$$B = \begin{bmatrix} c & d \\ -d & c \end{bmatrix}$$

Then

$$A + B = \begin{bmatrix} a + c & b + d \\ -(b + d) & a + c \end{bmatrix}$$

so $A + B$ is also in W (because it has the right *form*).

Similarly, if k is a scalar, then

$$kA = \begin{bmatrix} ka & kb \\ -kb & ka \end{bmatrix}$$

so kA is in W.

Thus, W is a nonempty subset of M_{22} that is closed under addition and scalar multiplication. Therefore, W is a subspace of M_{22}, by Theorem 6.2.

Example 6.13 shows that it is often possible to relate examples that, on the surface, appear to have nothing in common. Consequently, we can apply our knowledge of $\mathbb{R}^n$ to polynomials, matrices, and other examples. We will encounter this idea several times in this chapter and will make it precise in Section 6.5.

Example 6.14

If V is a vector space, then V is clearly a subspace of itself. The set $\{\mathbf{0}\}$, consisting of only the zero vector, is also a subspace of V, called the ***zero subspace.*** To show this, we simply note that the two closure conditions of Theorem 6.2 are satisfied:

$$\mathbf{0} + \mathbf{0} = \mathbf{0} \quad \text{and} \quad c\mathbf{0} = \mathbf{0} \quad \text{for any scalar } c$$

The subspaces $\{\mathbf{0}\}$ and V are called the ***trivial subspaces*** of V.

An examination of the proof of Theorem 6.2 reveals the following useful fact:

If W is a subspace of a vector space V, then W contains the zero vector $\mathbf{0}$ of V.

This fact is consistent with, and analogous to, the fact that lines and planes are subspaces of $\mathbb{R}^3$ if and only if they contain the origin. The requirement that every subspace must contain $\mathbf{0}$ is sometimes useful in showing that a set is *not* a subspace.

Example 6.15

Let W be the set of all 2×2 matrices of the form

$$\begin{bmatrix} a & a+1 \\ 0 & b \end{bmatrix}$$

Is W a subspace of M_{22}?

Solution Each matrix in W has the property that its (1, 2) entry is one more than its (1, 1) entry. Since the zero matrix

$$O = \begin{bmatrix} 0 & 0 \\ 0 & 0 \end{bmatrix}$$

does not have this property, it is not in W. Hence, W is not a subspace of M_{22}.

Example 6.16

Let W be the set of all 2×2 matrices with determinant equal to 0. Is W a subspace of M_{22}? (Since $\det O = 0$, the zero matrix is in W, so the method of Example 6.15 is of no use to us.)

Solution Let

$$A = \begin{bmatrix} 1 & 0 \\ 0 & 0 \end{bmatrix} \quad \text{and} \quad B = \begin{bmatrix} 0 & 0 \\ 0 & 1 \end{bmatrix}$$

Then $\det A = \det B = 0$, so A and B are in W. But

$$A + B = \begin{bmatrix} 1 & 0 \\ 0 & 1 \end{bmatrix}$$

so $\det(A + B) = 1 \neq 0$, and therefore $A + B$ is not in W. Thus, W is not closed under addition and so is not a subspace of M_{22}.

Spanning Sets

The notion of a spanning set of vectors carries over easily from $\mathbb{R}^n$ to general vector spaces.

Definition If $S = \{\mathbf{v}_1, \mathbf{v}_2, \ldots, \mathbf{v}_k\}$ is a set of vectors in a vector space V, then the set of all linear combinations of $\mathbf{v}_1, \mathbf{v}_2, \ldots, \mathbf{v}_k$ is called the ***span*** of $\mathbf{v}_1, \mathbf{v}_2, \ldots, \mathbf{v}_k$ and is denoted by $\text{span}(\mathbf{v}_1, \mathbf{v}_2, \ldots, \mathbf{v}_k)$ or $\text{span}(S)$. If $V = \text{span}(S)$, then S is called a ***spanning set*** for V and V is said to be ***spanned*** by S.

Example 6.17

Show that the polynomials 1, x, and x^2 span $\mathscr{P}_2$.

Solution By its very definition, a polynomial $p(x) = a + bx + cx^2$ is a linear combination of 1, x, and x^2. Therefore, $\mathscr{P}_2 = \text{span}(1, x, x^2)$.

Example 6.17 can clearly be generalized to show that $\mathscr{P}_n = \text{span}(1, x, x^2, \ldots, x^n)$. However, no finite set of polynomials can possibly span $\mathscr{P}$, the vector space of all polynomials. (See Exercise 44 in Section 6.2.) But, if we allow a spanning set to be infinite, then clearly the set of *all* nonnegative powers of x will do. That is, $\mathscr{P} = \text{span}(1, x, x^2, \ldots)$.

Example 6.18

Show that $M_{23} = \text{span}(E_{11}, E_{12}, E_{13}, E_{21}, E_{22}, E_{23})$, where

$$E_{11} = \begin{bmatrix} 1 & 0 & 0 \\ 0 & 0 & 0 \end{bmatrix} \quad E_{12} = \begin{bmatrix} 0 & 1 & 0 \\ 0 & 0 & 0 \end{bmatrix} \quad E_{13} = \begin{bmatrix} 0 & 0 & 1 \\ 0 & 0 & 0 \end{bmatrix}$$

$$E_{21} = \begin{bmatrix} 0 & 0 & 0 \\ 1 & 0 & 0 \end{bmatrix} \quad E_{22} = \begin{bmatrix} 0 & 0 & 0 \\ 0 & 1 & 0 \end{bmatrix} \quad E_{23} = \begin{bmatrix} 0 & 0 & 0 \\ 0 & 0 & 1 \end{bmatrix}$$

(That is, E_{ij} is the matrix with a 1 in row i, column j and zeros elsewhere.)

Solution We need only observe that

$$\begin{bmatrix} a_{11} & a_{12} & a_{13} \\ a_{21} & a_{22} & a_{23} \end{bmatrix} = a_{11}E_{11} + a_{12}E_{12} + a_{13}E_{13} + a_{21}E_{21} + a_{22}E_{22} + a_{23}E_{23}$$

Extending this example, we see that, in general, M_{mn} is spanned by the mn matrices E_{ij}, where $i = 1, \ldots, m$ and $j = 1, \ldots, n$.

Example 6.19

In $\mathscr{P}_2$, determine whether $r(x) = 1 - 4x + 6x^2$ is in span$(p(x), q(x))$, where

$$p(x) = 1 - x + x^2 \quad \text{and} \quad q(x) = 2 + x - 3x^2$$

Solution We are looking for scalars c and d such that $cp(x) + dq(x) = r(x)$. This means that

$$c(1 - x + x^2) + d(2 + x - 3x^2) = 1 - 4x + 6x^2$$

Regrouping according powers of x, we have

$$(c + 2d) + (-c + d)x + (c - 3d)x^2 = 1 - 4x + 6x^2$$

Equating the coefficients of like powers of x gives

$$\begin{aligned} c + 2d &= 1 \\ -c + d &= -4 \\ c - 3d &= 6 \end{aligned}$$

which is easily solved to give $c = 3$ and $d = -1$. Therefore, $r(x) = 3p(x) - q(x)$, so $r(x)$ is in span$(p(x), q(x))$. (Check this.)

Example 6.20

In $\mathscr{F}$, determine whether $\sin 2x$ is in span$(\sin x, \cos x)$.

Solution We set $c \sin x + d \cos x = \sin 2x$ and try to determine c and d so that this equation is true. Since these are functions, the equation must be true for *all* values of x. Setting $x = 0$, we have

$$c \sin 0 + d \cos 0 = \sin 0 \quad \text{or} \quad c(0) + d(1) = 0$$

from which we see that $d = 0$. Setting $x = \pi/2$, we get

$$c \sin(\pi/2) + d \cos(\pi/2) = \sin(\pi) \quad \text{or} \quad c(1) + d(0) = 0$$

giving $c = 0$. But this implies that $\sin 2x = 0(\sin x) + 0(\cos x) = 0$ for all x, which is absurd, since $\sin 2x$ is not the zero function. We conclude that $\sin 2x$ is not in span$(\sin x, \cos x)$.

Remark It *is* true that $\sin 2x$ can be written in terms of $\sin x$ and $\cos x$. For example, we have the double angle formula $\sin 2x = 2 \sin x \cos x$. However, this is not a *linear* combination.

Example 6.21

In M_{22}, describe the span of $A = \begin{bmatrix} 1 & 1 \\ 1 & 0 \end{bmatrix}$, $B = \begin{bmatrix} 1 & 0 \\ 0 & 1 \end{bmatrix}$, and $C = \begin{bmatrix} 0 & 1 \\ 1 & 0 \end{bmatrix}$.

Solution Every linear combination of A, B, and C is of the form

$$\begin{aligned} cA + dB + eC &= c\begin{bmatrix} 1 & 1 \\ 1 & 0 \end{bmatrix} + d\begin{bmatrix} 1 & 0 \\ 0 & 1 \end{bmatrix} + e\begin{bmatrix} 0 & 1 \\ 1 & 0 \end{bmatrix} \\ &= \begin{bmatrix} c + d & c + e \\ c + e & d \end{bmatrix} \end{aligned}$$

This matrix is symmetric, so span(A, B, C) is contained within the subspace of symmetric 2×2 matrices. In fact, we have equality; that is, *every* symmetric 2×2 matrix is in span(A, B, C). To show this, we let $\begin{bmatrix} x & y \\ y & z \end{bmatrix}$ be a symmetric 2×2 matrix. Setting

$$\begin{bmatrix} x & y \\ y & z \end{bmatrix} = \begin{bmatrix} c + d & c + e \\ c + e & d \end{bmatrix}$$

and solving for c and d, we find that $c = x - z$, $d = z$, and $e = -x + y + z$. Therefore,

$$\begin{bmatrix} x & y \\ y & z \end{bmatrix} = (x - z)\begin{bmatrix} 1 & 1 \\ 1 & 0 \end{bmatrix} + z\begin{bmatrix} 1 & 0 \\ 0 & 1 \end{bmatrix} + (-x + y + z)\begin{bmatrix} 0 & 1 \\ 1 & 0 \end{bmatrix}$$

(Check this.) It follows that span(A, B, C) is the subspace of symmetric 2×2 matrices.

As was the case in $\mathbb{R}^n$, the span of a set of vectors is always a subspace of the vector space that contains them. The next theorem makes this result precise. It generalizes Theorem 3.19.

Theorem 6.3 Let $\mathbf{v}_1, \mathbf{v}_2, \ldots, \mathbf{v}_k$ be vectors in a vector space V.

a. $\text{span}(\mathbf{v}_1, \mathbf{v}_2, \ldots, \mathbf{v}_k)$ is a subspace of V.
b. $\text{span}(\mathbf{v}_1, \mathbf{v}_2, \ldots, \mathbf{v}_k)$ is the smallest subspace of V that contains $\mathbf{v}_1, \mathbf{v}_2, \ldots, \mathbf{v}_k$.

Proof (a) The proof of property (a) is identical to the proof of Theorem 3.19, with $\mathbb{R}^n$ replaced by V.

(b) To establish property (b), we need to show that any subspace of V that contains $\mathbf{v}_1, \mathbf{v}_2, \ldots, \mathbf{v}_k$ also contains $\text{span}(\mathbf{v}_1, \mathbf{v}_2, \ldots, \mathbf{v}_k)$. Accordingly, let W be a subspace of V that contains $\mathbf{v}_1, \mathbf{v}_2, \ldots, \mathbf{v}_k$. Then, since W is closed under addition and scalar multiplication, it contains every linear combination $c_1\mathbf{v}_1 + c_2\mathbf{v}_2 + \cdots + c_k\mathbf{v}_k$ of $\mathbf{v}_1, \mathbf{v}_2, \ldots, \mathbf{v}_k$. Therefore, $\text{span}(\mathbf{v}_1, \mathbf{v}_2, \ldots, \mathbf{v}_k)$ is contained in W.

Exercises 6.1

In Exercises 1–11, determine whether the given set, together with the specified operations of addition and scalar multiplication, is a vector space. If it is not, list all of the axioms that fail to hold.

1. The set of all vectors in $\mathbb{R}^2$ of the form $\begin{bmatrix} x \\ x \end{bmatrix}$, with the usual vector addition and scalar multiplication

2. The set of all vectors $\begin{bmatrix} x \\ y \end{bmatrix}$ in $\mathbb{R}^2$ with $x \geq 0, y \geq 0$ (i.e., the first quadrant), with the usual vector addition and scalar multiplication

3. The set of all vectors $\begin{bmatrix} x \\ y \end{bmatrix}$ in $\mathbb{R}^2$ with $xy \geq 0$ (i.e., the union of the first and third quadrants), with the usual vector addition and scalar multiplication

4. The set of all vectors $\begin{bmatrix} x \\ y \end{bmatrix}$ in $\mathbb{R}^2$ with $x \geq y$, with the usual vector addition and scalar multiplication

5. $\mathbb{R}^2$, with the usual addition but scalar multiplication defined by

$$c\begin{bmatrix} x \\ y \end{bmatrix} = \begin{bmatrix} cx \\ y \end{bmatrix}$$

6. $\mathbb{R}^2$, with the usual scalar multiplication but addition defined by

$$\begin{bmatrix} x_1 \\ y_1 \end{bmatrix} + \begin{bmatrix} x_2 \\ y_2 \end{bmatrix} = \begin{bmatrix} x_1 + x_2 + 1 \\ y_1 + y_2 + 1 \end{bmatrix}$$

7. The set of all positive real numbers, with addition $\oplus$ defined by $x \oplus y = xy$ and scalar multiplication $\odot$ defined by $c \odot x = x^c$

8. The set of all rational numbers, with the usual addition and multiplication

9. The set of all upper triangular 2×2 matrices, with the usual matrix addition and scalar multiplication

10. The set of all 2×2 matrices of the form $\begin{bmatrix} a & b \\ c & d \end{bmatrix}$, where $ad = 0$, with the usual matrix addition and scalar multiplication

11. The set of all skew-symmetric $n \times n$ matrices, with the usual matrix addition and scalar multiplication (see page 162).

12. Finish verifying that $\mathscr{P}_2$ is a vector space (see Example 6.3).

13. Finish verifying that $\mathscr{F}$ is a vector space (see Example 6.4).

In Exercises 14–17, determine whether the given set, together with the specified operations of addition and scalar multiplication, is a complex vector space. If it is not, list all of the axioms that fail to hold.

14. The set of all vectors in $\mathbb{C}^2$ of the form $\begin{bmatrix} z \\ \bar{z} \end{bmatrix}$, with the usual vector addition and scalar multiplication

15. The set $M_{mn}(\mathbb{C})$ of all $m \times n$ complex matrices, with the usual matrix addition and scalar multiplication

16. The set $\mathbb{C}^2$, with the usual vector addition but scalar multiplication defined by $c\begin{bmatrix} z_1 \\ z_2 \end{bmatrix} = \begin{bmatrix} \bar{c}z_1 \\ \bar{c}z_2 \end{bmatrix}$

17. $\mathbb{R}^n$, with the usual vector addition and scalar multiplication

In Exercises 18–21, determine whether the given set, together with the specified operations of addition and scalar multiplication, is a vector space over the indicated $\mathbb{Z}_p$. If it is not, list all of the axioms that fail to hold.

18. The set of all vectors in $\mathbb{Z}_2^n$ with an even number of 1s, over $\mathbb{Z}_2$ with the usual vector addition and scalar multiplication

19. The set of all vectors in $\mathbb{Z}_2^n$ with an odd number of 1s, over $\mathbb{Z}_2$ with the usual vector addition and scalar multiplication

20. The set $M_{mn}(\mathbb{Z}_p)$ of all $m \times n$ matrices with entries from $\mathbb{Z}_p$, over $\mathbb{Z}_p$ with the usual matrix addition and scalar multiplication

21. $\mathbb{Z}_6$, over $\mathbb{Z}_3$ with the usual addition and multiplication (Think this one through carefully!)

22. Prove Theorem 6.1(a).

23. Prove Theorem 6.1(c).

In Exercises 24–45, use Theorem 6.2 to determine whether W is a subspace of V.

24. $V = \mathbb{R}^3$, $W = \left\{ \begin{bmatrix} a \\ 0 \\ a \end{bmatrix} \right\}$

25. $V = \mathbb{R}^3$, $W = \left\{ \begin{bmatrix} a \\ -a \\ 2a \end{bmatrix} \right\}$

26. $V = \mathbb{R}^3$, $W = \left\{ \begin{bmatrix} a \\ b \\ a+b+1 \end{bmatrix} \right\}$

27. $V = \mathbb{R}^3$, $W = \left\{ \begin{bmatrix} a \\ b \\ |a| \end{bmatrix} \right\}$

28. $V = M_{22}$, $W = \left\{ \begin{bmatrix} a & b \\ b & 2a \end{bmatrix} \right\}$

29. $V = M_{22}$, $W = \left\{ \begin{bmatrix} a & b \\ c & d \end{bmatrix} : ad \geq bc \right\}$

30. $V = M_{nn}$, $W = \{A \text{ in } M_{nn} : \det A = 1\}$

31. $V = M_{nn}$, W is the set of diagonal $n \times n$ matrices

32. $V = M_{nn}$, W is the set of idempotent $n \times n$ matrices

33. $V = M_{nn}$, $W = \{A \text{ in } M_{nn} : AB = BA\}$, where B is a given (fixed) matrix

34. $V = \mathscr{P}_2$, $W = \{bx + cx^2\}$

35. $V = \mathscr{P}_2$, $W = \{a + bx + cx^2 : a + b + c = 0\}$

36. $V = \mathscr{P}_2$, $W = \{a + bx + cx^2 : abc = 0\}$

37. $V = \mathscr{P}$, W is the set of all polynomials of degree 3

38. $V = \mathscr{F}$, $W = \{f \text{ in } \mathscr{F} : f(-x) = f(x)\}$

39. $V = \mathscr{F}$, $W = \{f \text{ in } \mathscr{F} : f(-x) = -f(x)\}$

40. $V = \mathscr{F}$, $W = \{f \text{ in } \mathscr{F} : f(0) = 1\}$

41. $V = \mathscr{F}$, $W = \{f \text{ in } \mathscr{F} : f(0) = 0\}$

42. $V = \mathscr{F}$, W is the set of all integrable functions

43. $V = \mathscr{D}$, $W = \{f \text{ in } \mathscr{D} : f'(x) \geq 0 \text{ for all } x\}$

44. $V = \mathscr{F}$, $W = \mathscr{C}^{(2)}$, the set of all functions with continuous second derivatives

45. $V = \mathscr{F}$, $W = \{f \text{ in } \mathscr{F} : = \lim_{x \to 0} f(x) = \infty\}$

46. Let V be a vector space with subspaces U and W. Prove that $U \cap W$ is a subspace of V.

47. Let V be a vector space with subspaces U and W. Give an example with $V = \mathbb{R}^2$ to show that $U \cup W$ need not be a subspace of V.

48. Let V be a vector space with subspaces U and W. Define the ***sum of U and W*** to be

$$U + W = \{\mathbf{u} + \mathbf{w} : \mathbf{u} \text{ is in } U, \mathbf{w} \text{ is in } W\}$$

(a) If $V = \mathbb{R}^3$, U is the x-axis, and W is the y-axis, what is $U + W$?

(b) If U and W are subspaces of a vector space V, prove that $U + W$ is a subspace of V.

49. If U and V are vector spaces, define the ***Cartesian product*** of U and V to be

$$U \times V = \{(\mathbf{u}, \mathbf{v}) : \mathbf{u} \text{ is in } U \text{ and } \mathbf{v} \text{ is in } V\}$$

Prove that $U \times V$ is a vector space.

50. Let W be a subspace of a vector space V. Prove that $\Delta = \{(\mathbf{w}, \mathbf{w}) : \mathbf{w} \text{ is in } W\}$ is a subspace of $V \times V$.

In Exercises 51 and 52, let $A = \begin{bmatrix} 1 & 1 \\ -1 & 1 \end{bmatrix}$ and $B = \begin{bmatrix} 1 & -1 \\ 1 & 0 \end{bmatrix}$. Determine whether C is in span(A, B).

51. $C = \begin{bmatrix} 1 & 2 \\ 3 & 4 \end{bmatrix}$

52. $C = \begin{bmatrix} 3 & -5 \\ 5 & -1 \end{bmatrix}$

In Exercises 53 and 54, let $p(x) = 1 - 2x$, $q(x) = x - x^2$, and $r(x) = -2 + 3x + x^2$. Determine whether $s(x)$ is in span$(p(x), q(x), r(x))$.

53. $s(x) = 3 - 5x - x^2$ **54.** $s(x) = 1 + x + x^2$

In Exercises 55–58, let $f(x) = \sin^2 x$ and $g(x) = \cos^2 x$. Determine whether $h(x)$ is in span$(f(x), g(x))$.

55. $h(x) = 1$ **56.** $h(x) = \cos 2x$

57. $h(x) = \sin 2x$ **58.** $h(x) = \sin x$

59. Is M_{22} spanned by $\begin{bmatrix} 1 & 1 \\ 0 & 1 \end{bmatrix}, \begin{bmatrix} 0 & 1 \\ 1 & 0 \end{bmatrix}, \begin{bmatrix} 1 & 0 \\ 1 & 1 \end{bmatrix}, \begin{bmatrix} 0 & -1 \\ 1 & 0 \end{bmatrix}$?

60. Is M_{22} spanned by $\begin{bmatrix} 1 & 0 \\ 1 & 0 \end{bmatrix}, \begin{bmatrix} 1 & 1 \\ 1 & 0 \end{bmatrix}, \begin{bmatrix} 1 & 1 \\ 1 & 1 \end{bmatrix}, \begin{bmatrix} 0 & -1 \\ 1 & 0 \end{bmatrix}$?

61. Is $\mathscr{P}_2$ spanned by $1 + x, x + x^2, 1 + x^2$?

62. Is $\mathscr{P}_2$ spanned by $1 + x + 2x^2, 2 + x + 2x^2, -1 + x + 2x^2$?

63. Prove that every vector space has a unique zero vector.

64. Prove that for every vector $\mathbf{v}$ in a vector space V, there is a unique $\mathbf{v}'$ in V such that $\mathbf{v} + \mathbf{v}' = \mathbf{0}$.

Writing Project

The Rise of Vector Spaces

As noted in the sidebar on page 429, in the late 19th century, the mathematicians Hermann Grassmann and Giuseppe Peano were instrumental in introducing the idea of a vector space and the vector space axioms that we use today. Grassmann's work had its origins in barycentric coordinates, a technique invented in 1827 by August Ferdinand Möbius (of Möbius strip fame). However, widespread acceptance of the vector space concept did not come until the early 20th century.

Write a report on the history of vector spaces. Discuss the origins of the notion of a vector space and the contributions of Grassmann and Peano. Why was the mathematical community slow to adopt these ideas, and how did acceptance come about?

1. Carl B. Boyer and Uta C. Merzbach, *A History of Mathematics* (Third Edition) (Hoboken, NJ: Wiley, 2011).
2. Jean-Luc, Dorier (1995), A General Outline of the Genesis of Vector Space Theory, *Historia Mathematica* 22 (1995), pp. 227–261.
3. Victor J. Katz, *A History of Mathematics: An Introduction* (Third Edition) (Reading, MA: Addison Wesley Longman, 2008).

6.2 Linear Independence, Basis, and Dimension

In this section, we extend the notions of linear independence, basis, and dimension to general vector spaces, generalizing the results of Sections 2.3 and 3.5. In most cases, the proofs of the theorems carry over; we simply replace $\mathbb{R}^n$ by the vector space V.

Linear Independence

Definition A set of vectors $\{\mathbf{v}_1, \mathbf{v}_2, \ldots, \mathbf{v}_k\}$ in a vector space V is ***linearly dependent*** if there are scalars $c_1, c_2, \ldots, c_k$, *at least one of which is not zero,* such that

$$c_1\mathbf{v}_1 + c_2\mathbf{v}_2 + \cdots + c_k\mathbf{v}_k = \mathbf{0}$$

A set of vectors that is not linearly dependent is said to be ***linearly independent.***

As in $\mathbb{R}^n$, $\{\mathbf{v}_1, \mathbf{v}_2, \ldots, \mathbf{v}_k\}$ is linearly independent in a vector space V if and only if

$$c_1\mathbf{v}_1 + c_2\mathbf{v}_2 + \cdots + c_k\mathbf{v}_k = \mathbf{0} \quad \text{implies} \quad c_1 = 0, c_2 = 0, \ldots, c_k = 0$$

We also have the following useful alternative formulation of linear dependence.

Theorem 6.4 A set of vectors $\{\mathbf{v}_1, \mathbf{v}_2, \ldots, \mathbf{v}_k\}$ in a vector space V is linearly dependent if and only if at least one of the vectors can be expressed as a linear combination of the others.

Proof The proof is identical to that of Theorem 2.5.

As a special case of Theorem 6.4, note that a set of *two* vectors is linearly dependent if and only if one is a scalar multiple of the other.

Example 6.22

In $\mathscr{P}_2$, the set $\{1 + x + x^2, 1 - x + 3x^2, 1 + 3x - x^2\}$ is linearly dependent, since

$$2(1 + x + x^2) - (1 - x + 3x^2) = 1 + 3x - x^2$$

Example 6.23

In M_{22}, let

$$A = \begin{bmatrix} 1 & 1 \\ 0 & 1 \end{bmatrix}, \quad B = \begin{bmatrix} 1 & -1 \\ 1 & 0 \end{bmatrix}, \quad C = \begin{bmatrix} 2 & 0 \\ 1 & 1 \end{bmatrix}$$

Then $A + B = C$, so the set $\{A, B, C\}$ is linearly dependent.

Example 6.24

In $\mathscr{F}$, the set $\{\sin^2 x, \cos^2 x, \cos 2x\}$ is linearly dependent, since

$$\cos 2x = \cos^2 x - \sin^2 x$$

Example 6.25

Show that the set $\{1, x, x^2, \ldots, x^n\}$ is linearly independent in $\mathscr{P}_n$.

Solution 1 Suppose that $c_0, c_1, \ldots, c_n$ are scalars such that

$$c_0 \cdot 1 + c_1 x + c_2 x^2 + \cdots + c_n x^n = 0$$

Then the polynomial $p(x) = c_0 + c_1 x + c_2 x^2 + \cdots + c_n x^n$ is zero for all values of x. But a polynomial of degree at most n cannot have more than n zeros (see Appendix D). So $p(x)$ must be the zero polynomial, meaning that $c_0 = c_1 = c_2 = \cdots = c_n = 0$. Therefore, $\{1, x, x^2, \ldots, x^n\}$ is linearly independent.

Solution 2 We begin, as in the first solution, by assuming that

$$p(x) = c_0 + c_1 x + c_2 x^2 + \cdots + c_n x^n = 0$$

Since this is true for all x, we can substitute $x = 0$ to obtain $c_0 = 0$. This leaves

$$c_1 x + c_2 x^2 + \cdots + c_n x^n = 0$$

Taking derivatives, we obtain

$$c_1 + 2c_2x + 3c_3x^2 + \cdots + nc_nx^{n-1} = 0$$

and setting $x = 0$, we see that $c_1 = 0$. Differentiating $2c_2x + 3c_3x^2 + \cdots + nc_nx^{n-1} = 0$ and setting $x = 0$, we find that $2c_2 = 0$, so $c_2 = 0$. Continuing in this fashion, we find that $k!c_k = 0$ for $k = 0, \ldots, n$. Therefore, $c_0 = c_1 = c_2 = \cdots = c_n = 0$, and $\{1, x, x^2, \ldots, x^n\}$ is linearly independent.

Example 6.26

In $\mathscr{P}_2$, determine whether the set $\{1 + x, x + x^2, 1 + x^2\}$ is linearly independent.

Solution Let c_1, c_2, and c_3 be scalars such that

$$c_1(1 + x) + c_2(x + x^2) + c_3(1 + x^2) = 0$$

Then

$$(c_1 + c_3) + (c_1 + c_2)x + (c_2 + c_3)x^2 = 0$$

This implies that

$$\begin{aligned} c_1 + c_3 &= 0 \\ c_1 + c_2 &= 0 \\ c_2 + c_3 &= 0 \end{aligned}$$

the solution to which is $c_1 = c_2 = c_3 = 0$. It follows that $\{1 + x, x + x^2, 1 + x^2\}$ is linearly independent.

Remark Compare Example 6.26 with Example 2.23(b). The system of equations that arises is exactly the same. This is because of the correspondence between $\mathscr{P}_2$ and $\mathbb{R}^3$ that relates

$$1 + x \leftrightarrow \begin{bmatrix} 1 \\ 1 \\ 0 \end{bmatrix}, \quad x + x^2 \leftrightarrow \begin{bmatrix} 0 \\ 1 \\ 1 \end{bmatrix}, \quad 1 + x^2 \leftrightarrow \begin{bmatrix} 1 \\ 0 \\ 1 \end{bmatrix}$$

and produces the columns of the coefficient matrix of the linear system that we have to solve. Thus, showing that $\{1 + x, x + x^2, 1 + x^2\}$ is linearly independent is equivalent to showing that

$$\left\{ \begin{bmatrix} 1 \\ 1 \\ 0 \end{bmatrix}, \begin{bmatrix} 0 \\ 1 \\ 1 \end{bmatrix}, \begin{bmatrix} 1 \\ 0 \\ 1 \end{bmatrix} \right\}$$

is linearly independent. This can be done simply by establishing that the matrix

$$\begin{bmatrix} 1 & 0 & 1 \\ 1 & 1 & 0 \\ 0 & 1 & 1 \end{bmatrix}$$

has rank 3, by the Fundamental Theorem of Invertible Matrices.

Example 6.27

In $\mathcal{F}$, determine whether the set $\{\sin x, \cos x\}$ is linearly independent.

Solution The functions $f(x) = \sin x$ and $g(x) = \cos x$ are linearly *dependent* if and only if one of them is a scalar multiple of the other. But it is clear from their graphs that this is not the case, since, for example, any nonzero multiple of $f(x) = \sin x$ has the same zeros, none of which are zeros of $g(x) = \cos x$.

This approach may not always be appropriate to use, so we offer the following direct, more computational method. Suppose c and d are scalars such that

$$c \sin x + d \cos x = 0$$

Setting $x = 0$, we obtain $d = 0$, and setting $x = \pi/2$, we obtain $c = 0$. Therefore, the set $\{\sin x, \cos x\}$ is linearly independent.

Although the definitions of linear dependence and independence are phrased in terms of *finite* sets of vectors, we can extend the concepts to *infinite* sets as follows:

A set S of vectors in a vector space V is ***linearly dependent*** if it contains finitely many linearly dependent vectors. A set of vectors that is not linearly dependent is said to be ***linearly independent.***

Note that for finite sets of vectors, this is just the original definition. Following is an example of an infinite set of linearly independent vectors.

Example 6.28

In $\mathcal{P}$, show that $S = \{1, x, x^2, \ldots\}$ is linearly independent.

Solution Suppose there is a finite subset T of S that is linearly dependent. Let x^m be the highest power of x in T and let x^n be the lowest power of x in T. Then there are scalars $c_n, c_{n+1}, \ldots, c_m$, not all zero, such that

$$c_n x^n + c_{n+1} x^{n+1} + \cdots + c_m x^m = 0$$

But, by an argument similar to that used in Example 6.25, this implies that $c_n = c_{n+1} = \cdots = c_m = 0$, which is a contradiction. Hence, S cannot contain finitely many linearly dependent vectors, so it is linearly independent.

Bases

The important concept of a basis now can be extended easily to arbitrary vector spaces.

Definition A subset $\mathcal{B}$ of a vector space V is a ***basis*** for V if

1. $\mathcal{B}$ spans V and
2. $\mathcal{B}$ is linearly independent.

Example 6.29

If $\mathbf{e}_i$ is the ith column of the $n \times n$ identity matrix, then $\{\mathbf{e}_1, \mathbf{e}_2, \ldots, \mathbf{e}_n\}$ is a basis for $\mathbb{R}^n$, called the ***standard basis*** for $\mathbb{R}^n$.

Example 6.30

$\{1, x, x^2, \ldots, x^n\}$ is a basis for $\mathscr{P}_n$, called the ***standard basis*** for $\mathscr{P}_n$.

Example 6.31

The set $\mathcal{E} = \{E_{11}, \ldots, E_{1n}, E_{21}, \ldots, E_{2n}, E_{m1}, \ldots, E_{mn}\}$ is a basis for M_{mn}, where the matrices E_{ij} are as defined in Example 6.18. $\mathcal{E}$ is called the ***standard basis*** for M_{mn}.

We have already seen that $\mathcal{E}$ spans M_{mn}. It is easy to show that $\mathcal{E}$ is linearly independent. (Verify this!) Hence, $\mathcal{E}$ is a basis for M_{mn}.

Example 6.32

Show that $\mathcal{B} = \{1 + x, x + x^2, 1 + x^2\}$ is a basis for $\mathscr{P}_2$.

Solution We have already shown that $\mathcal{B}$ is linearly independent, in Example 6.26. To show that $\mathcal{B}$ spans $\mathscr{P}_2$, let $a + bx + cx^2$ be an arbitrary polynomial in $\mathscr{P}_2$. We must show that there are scalars c_1, c_2, and c_3 such that

$$c_1(1 + x) + c_2(x + x^2) + c_3(1 + x^2) = a + bx + cx^2$$

or, equivalently,

$$(c_1 + c_3) + (c_1 + c_2)x + (c_2 + c_3)x^2 = a + bx + cx^2$$

Equating coefficients of like powers of x, we obtain the linear system

$$\begin{aligned} c_1 + \phantom{c_2 + {}} c_3 &= a \\ c_1 + c_2 \phantom{{} + c_3} &= b \\ c_2 + c_3 &= c \end{aligned}$$

which has a solution, since the coefficient matrix $\begin{bmatrix} 1 & 0 & 1 \\ 1 & 1 & 0 \\ 0 & 1 & 1 \end{bmatrix}$ has rank 3 and, hence, is invertible. (We do not need to know *what* the solution is; we only need to know that it exists.) Therefore, $\mathcal{B}$ is a basis for $\mathscr{P}_2$.

Remark Observe that the matrix $\begin{bmatrix} 1 & 0 & 1 \\ 1 & 1 & 0 \\ 0 & 1 & 1 \end{bmatrix}$ is the key to Example 6.32. We can immediately obtain it using the correspondence between $\mathscr{P}_2$ and $\mathbb{R}^3$, as indicated in the Remark following Example 6.26.

Example 6.33 Show that $\mathcal{B} = \{1, x, x^2, \ldots\}$ is a basis for $\mathcal{P}$.

Solution In Example 6.28, we saw that $\mathcal{B}$ is linearly independent. It also spans $\mathcal{P}$, since clearly every polynomial is a linear combination of (finitely many) powers of x.

Example 6.34 Find bases for the three vector spaces in Example 6.13:

(a) $W_1 = \left\{\begin{bmatrix} a \\ b \\ -b \\ a \end{bmatrix}\right\}$ (b) $W_2 = \{a + bx - bx^2 + ax^3\}$ (c) $W_3 = \left\{\begin{bmatrix} a & b \\ -b & a \end{bmatrix}\right\}$

Solution Once again, we will work the three examples side by side to highlight the similarities among them. In a strong sense, they are all the *same* example, but it will take us until Section 6.5 to make this idea perfectly precise.

(a) Since

$$\begin{bmatrix} a \\ b \\ -b \\ a \end{bmatrix} = a\begin{bmatrix} 1 \\ 0 \\ 0 \\ 1 \end{bmatrix} + b\begin{bmatrix} 0 \\ 1 \\ -1 \\ 0 \end{bmatrix}$$

we have $W_1 = \text{span}(\mathbf{u}, \mathbf{v})$, where

$$\mathbf{u} = \begin{bmatrix} 1 \\ 0 \\ 0 \\ 1 \end{bmatrix} \quad \text{and} \quad \mathbf{v} = \begin{bmatrix} 0 \\ 1 \\ -1 \\ 0 \end{bmatrix}$$

Since $\{\mathbf{u}, \mathbf{v}\}$ is clearly linearly independent, it is also a basis for W_1.

(b) Since

$$a + bx - bx^2 + ax^3 = a(1 + x^3) + b(x - x^2)$$

we have $W_2 = \text{span}(u(x), v(x))$, where

$$u(x) = 1 + x^3$$

and

$$v(x) = x - x^2$$

Since $\{u(x), v(x)\}$ is clearly linearly independent, it is also a basis for W_2.

(c) Since

$$\begin{bmatrix} a & b \\ -b & a \end{bmatrix} = a\begin{bmatrix} 1 & 0 \\ 0 & 1 \end{bmatrix} + b\begin{bmatrix} 0 & 1 \\ -1 & 0 \end{bmatrix}$$

we have $W_3 = \text{span}(U, V)$, where

$$U = \begin{bmatrix} 1 & 0 \\ 0 & 1 \end{bmatrix} \quad \text{and} \quad V = \begin{bmatrix} 0 & 1 \\ -1 & 0 \end{bmatrix}$$

Since $\{U, V\}$ is clearly linearly independent, it is also a basis for W_3.

Coordinates

Section 3.5 introduced the idea of the coordinates of a vector with respect to a basis for subspaces of $\mathbb{R}^n$. We now extend this concept to arbitrary vector spaces.

Theorem 6.5 Let V be a vector space and let $\mathcal{B}$ be a basis for V. For every vector $\mathbf{v}$ in V, there is exactly one way to write $\mathbf{v}$ as a linear combination of the basis vectors in $\mathcal{B}$.

Proof The proof is the same as the proof of Theorem 3.29. It works even if the basis $\mathcal{B}$ is infinite, since linear combinations are, by definition, finite.

The converse of Theorem 6.5 is also true. That is, if $\mathcal{B}$ is a set of vectors in a vector space V with the property that every vector in V can be written uniquely as a linear combination of the vectors in $\mathcal{B}$, then $\mathcal{B}$ is a basis for V (see Exercise 30). In this sense, the *unique representation property* characterizes a basis.

Since representation of a vector with respect to a basis is unique, the next definition makes sense.

Definition Let $\mathcal{B} = \{\mathbf{v}_1, \mathbf{v}_2, \ldots, \mathbf{v}_n\}$ be a basis for a vector space V. Let $\mathbf{v}$ be a vector in V, and write $\mathbf{v} = c_1\mathbf{v}_1 + c_2\mathbf{v}_2 + \cdots + c_n\mathbf{v}_n$. Then $c_1, c_2, \ldots, c_n$ are called the ***coordinates of* v *with respect to* $\mathcal{B}$**, and the column vector

$$[\mathbf{v}]_{\mathcal{B}} = \begin{bmatrix} c_1 \\ c_2 \\ \vdots \\ c_n \end{bmatrix}$$

is called the ***coordinate vector of* v *with respect to* $\mathcal{B}$**.

Observe that if the basis $\mathcal{B}$ of V has n vectors, then $[\mathbf{v}]_{\mathcal{B}}$ is a (column) vector in $\mathbb{R}^n$.

Example 6.35

Find the coordinate vector $[p(x)]_{\mathcal{B}}$ of $p(x) = 2 - 3x + 5x^2$ with respect to the standard basis $\mathcal{B} = \{1, x, x^2\}$ of $\mathscr{P}_2$.

Solution The polynomial $p(x)$ is already a linear combination of 1, x, and x^2, so

$$[p(x)]_{\mathcal{B}} = \begin{bmatrix} 2 \\ -3 \\ 5 \end{bmatrix}$$

This is the correspondence between $\mathscr{P}_2$ and $\mathbb{R}^3$ that we remarked on after Example 6.26, and it can easily be generalized to show that the coordinate vector of a polynomial

$$p(x) = a_0 + a_1x + a_2x^2 + \cdots + a_nx^n \quad \text{in } \mathscr{P}_n$$

with respect to the standard basis $\mathcal{B} = \{1, x, x^2, \ldots, x^n\}$ is just the vector

$$[p(x)]_{\mathcal{B}} = \begin{bmatrix} a_0 \\ a_1 \\ a_2 \\ \vdots \\ a_n \end{bmatrix} \quad \text{in } \mathbb{R}^{n+1}$$

Remark The *order* in which the basis vectors appear in $\mathcal{B}$ affects the order of the entries in a coordinate vector. For example, in Example 6.35, assume that the standard basis vectors are ordered as $\mathcal{B}' = \{x^2, x, 1\}$. Then the coordinate vector of $p(x) = 2 - 3x + 5x^2$ with respect to $\mathcal{B}'$ is

$$[p(x)]_{\mathcal{B}'} = \begin{bmatrix} 5 \\ -3 \\ 2 \end{bmatrix}$$

Example 6.36

Find the coordinate vector $[A]_{\mathcal{B}}$ of $A = \begin{bmatrix} 2 & -1 \\ 4 & 3 \end{bmatrix}$ with respect to the standard basis $\mathcal{B} = \{E_{11}, E_{12}, E_{21}, E_{22}\}$ of M_{22}.

Solution Since

$$\begin{aligned} A = \begin{bmatrix} 2 & -1 \\ 4 & 3 \end{bmatrix} &= 2\begin{bmatrix} 1 & 0 \\ 0 & 0 \end{bmatrix} - \begin{bmatrix} 0 & 1 \\ 0 & 0 \end{bmatrix} + 4\begin{bmatrix} 0 & 0 \\ 1 & 0 \end{bmatrix} + 3\begin{bmatrix} 0 & 0 \\ 0 & 1 \end{bmatrix} \\ &= 2E_{11} - E_{12} + 4E_{21} + 3E_{22} \end{aligned}$$

we have

$$[A]_{\mathcal{B}} = \begin{bmatrix} 2 \\ -1 \\ 4 \\ 3 \end{bmatrix}$$

This is the correspondence between M_{22} and $\mathbb{R}^4$ that we noted before the introduction to Example 6.13. It too can easily be generalized to give a correspondence between M_{mn} and $\mathbb{R}^{mn}$.

Example 6.37

Find the coordinate vector $[p(x)]_{\mathcal{B}}$ of $p(x) = 1 + 2x - x^2$ with respect to the basis $\mathcal{C} = \{1 + x, x + x^2, 1 + x^2\}$ of $\mathscr{P}_2$.

Solution We need to find c_1, c_2, and c_3 such that

$$c_1(1 + x) + c_2(x + x^2) + c_3(1 + x^2) = 1 + 2x - x^2$$

or, equivalently,

$$(c_1 + c_3) + (c_1 + c_2)x + (c_2 + c_3)x^2 = 1 + 2x - x^2$$

As in Example 6.32, this means we need to solve the system

$$\begin{aligned} c_1 + c_3 &= 1 \\ c_1 + c_2 &= 2 \\ c_2 + c_3 &= -1 \end{aligned}$$

whose solution is found to be $c_1 = 2$, $c_2 = 0$, $c_3 = -1$. Therefore,

$$[p(x)]_{\mathcal{C}} = \begin{bmatrix} 2 \\ 0 \\ -1 \end{bmatrix}$$

[Since this result says that $p(x) = 2(1 + x) - (1 + x^2)$, it is easy to check that it is correct.]

The next theorem shows that the process of forming coordinate vectors is compatible with the vector space operations of addition and scalar multiplication.

Theorem 6.6 Let $\mathcal{B} = \{\mathbf{v}_1, \mathbf{v}_2, \ldots, \mathbf{v}_n\}$ be a basis for a vector space V. Let $\mathbf{u}$ and $\mathbf{v}$ be vectors in V and let c be a scalar. Then

a. $[\mathbf{u} + \mathbf{v}]_{\mathcal{B}} = [\mathbf{u}]_{\mathcal{B}} + [\mathbf{v}]_{\mathcal{B}}$
b. $[c\mathbf{u}]_{\mathcal{B}} = c[\mathbf{u}]_{\mathcal{B}}$

Proof We begin by writing $\mathbf{u}$ and $\mathbf{v}$ in terms of the basis vectors—say, as

$$\mathbf{u} = c_1\mathbf{v}_1 + c_2\mathbf{v}_2 + \cdots + c_n\mathbf{v}_n \quad \text{and} \quad \mathbf{v} = d_1\mathbf{v}_1 + d_2\mathbf{v}_2 + \cdots + d_n\mathbf{v}_n$$

Then, using vector space properties, we have

$$\mathbf{u} + \mathbf{v} = (c_1 + d_1)\mathbf{v}_1 + (c_2 + d_2)\mathbf{v}_2 + \cdots + (c_n + d_n)\mathbf{v}_n$$

and

$$c\mathbf{u} = (cc_1)\mathbf{v}_1 + (cc_2)\mathbf{v}_2 + \cdots + (cc_n)\mathbf{v}_n$$

so

$$[\mathbf{u} + \mathbf{v}]_{\mathcal{B}} = \begin{bmatrix} c_1 + d_1 \\ c_2 + d_2 \\ \vdots \\ c_n + d_n \end{bmatrix} = \begin{bmatrix} c_1 \\ c_2 \\ \vdots \\ c_n \end{bmatrix} + \begin{bmatrix} d_1 \\ d_2 \\ \vdots \\ d_n \end{bmatrix} = [\mathbf{u}]_{\mathcal{B}} + [\mathbf{v}]_{\mathcal{B}}$$

and

$$[c\mathbf{u}]_{\mathcal{B}} = \begin{bmatrix} cc_1 \\ cc_2 \\ \vdots \\ cc_n \end{bmatrix} = c\begin{bmatrix} c_1 \\ c_2 \\ \vdots \\ c_n \end{bmatrix} = c[\mathbf{u}]_{\mathcal{B}}$$

An easy corollary to Theorem 6.6 states that coordinate vectors preserve linear combinations:

$$[c_1\mathbf{u}_1 + \cdots + c_k\mathbf{u}_k]_{\mathcal{B}} = c_1[\mathbf{u}_1]_{\mathcal{B}} + \cdots + c_k[\mathbf{u}_k]_{\mathcal{B}} \tag{1}$$

You are asked to prove this corollary in Exercise 31.

The most useful aspect of coordinate vectors is that they allow us to transfer information from a general vector space to $\mathbb{R}^n$, where we have the tools of Chapters 1 to 3 at our disposal. We will explore this idea in some detail in Sections 6.3 and 6.6. For now, we have the following useful theorem.

Theorem 6.7 Let $\mathcal{B} = \{\mathbf{v}_1, \mathbf{v}_2, \ldots, \mathbf{v}_n\}$ be a basis for a vector space V and let $\mathbf{u}_1, \ldots, \mathbf{u}_k$ be vectors in V. Then $\{\mathbf{u}_1, \ldots, \mathbf{u}_k\}$ is linearly independent in V if and only if $\{[\mathbf{u}_1]_{\mathcal{B}}, \ldots, [\mathbf{u}_k]_{\mathcal{B}}\}$ is linearly independent in $\mathbb{R}^n$.

Proof Assume that $\{\mathbf{u}_1, \ldots, \mathbf{u}_k\}$ is linearly independent in V and let

$$c_1[\mathbf{u}_1]_{\mathcal{B}} + \cdots + c_k[\mathbf{u}_k]_{\mathcal{B}} = \mathbf{0}$$

in $\mathbb{R}^n$. But then we have

$$[c_1\mathbf{u}_1 + \cdots + c_k\mathbf{u}_k]_{\mathcal{B}} = \mathbf{0}$$

using Equation (1), so the coordinates of the vector $c_1\mathbf{u}_1 + \cdots + c_k\mathbf{u}_k$ with respect to $\mathcal{B}$ are all zero. That is,

$$c_1\mathbf{u}_1 + \cdots + c_k\mathbf{u}_k = 0\mathbf{v}_1 + 0\mathbf{v}_2 + \cdots + 0\mathbf{v}_n = \mathbf{0}$$

The linear independence of $\{\mathbf{u}_1, \ldots, \mathbf{u}_k\}$ now forces $c_1 = c_2 = \cdots = c_k = 0$, so $\{[\mathbf{u}_1]_{\mathcal{B}}, \ldots, [\mathbf{u}_k]_{\mathcal{B}}\}$ is linearly independent.

The converse implication, which uses similar ideas, is left as Exercise 32.

Observe that, in the special case where $\mathbf{u}_i = \mathbf{v}_i$, we have

$$\mathbf{v}_i = 0 \cdot \mathbf{v}_1 + \cdots + 1 \cdot \mathbf{v}_i + \cdots + 0 \cdot \mathbf{v}_n$$

so $[\mathbf{v}_i]_{\mathcal{B}} = \mathbf{e}_i$ and $\{[\mathbf{v}_1]_{\mathcal{B}}, \ldots, [\mathbf{v}_n]_{\mathcal{B}}\} = \{\mathbf{e}_1, \ldots, \mathbf{e}_n\}$ is the standard basis in $\mathbb{R}^n$.

Dimension

The definition of dimension is the same for a vector space as for a subspace of $\mathbb{R}^n$—the number of vectors in a basis for the space. Since a vector space can have more than one basis, we need to show that this definition makes sense; that is, we need to establish that different bases for the same vector space contain the same number of vectors.

Part (a) of the next theorem generalizes Theorem 2.8.

Theorem 6.8 Let $\mathcal{B} = \{\mathbf{v}_1, \mathbf{v}_2, \ldots, \mathbf{v}_n\}$ be a basis for a vector space V.

a. Any set of more than n vectors in V must be linearly dependent.
b. Any set of fewer than n vectors in V cannot span V.

Proof (a) Let $\{\mathbf{u}_1, \ldots, \mathbf{u}_m\}$ be a set of vectors in V, with $m > n$. Then $\{[\mathbf{u}_1]_{\mathcal{B}}, \ldots, [\mathbf{u}_m]_{\mathcal{B}}\}$ is a set of more than n vectors in $\mathbb{R}^n$ and, hence, is linearly dependent, by Theorem 2.8. This means that $\{\mathbf{u}_1, \ldots, \mathbf{u}_m\}$ is linearly dependent as well, by Theorem 6.7.

(b) Let $\{\mathbf{u}_1, \ldots, \mathbf{u}_m\}$ be a set of vectors in V, with $m < n$. Then $S = \{[\mathbf{u}_1]_{\mathcal{B}}, \ldots, [\mathbf{u}_m]_{\mathcal{B}}\}$ is a set of fewer than n vectors in $\mathbb{R}^n$. Now $\text{span}(\mathbf{u}_1, \ldots, \mathbf{u}_m) = V$ if and only if $\text{span}(S) = \mathbb{R}^n$ (see Exercise 33). But $\text{span}(S)$ is just the column space of the $n \times m$ matrix

$$A = [\,[\mathbf{u}_1]_{\mathcal{B}} \cdots [\mathbf{u}_m]_{\mathcal{B}}\,]$$

so $\dim(\text{span}(S)) = \dim(\text{col}(A)) \le m < n$. Hence, S cannot span $\mathbb{R}^n$, so $\{\mathbf{u}_1, \ldots, \mathbf{u}_m\}$ does not span V.

Now we extend Theorem 3.23.

Theorem 6.9 **The Basis Theorem**

If a vector space V has a basis with n vectors, then every basis for V has exactly n vectors.

The proof of Theorem 3.23 also works here, virtually word for word. However, it is easier to make use of Theorem 6.8.

Proof Let $\mathcal{B}$ be a basis for V with n vectors and let $\mathcal{B}'$ be another basis for V with m vectors. By Theorem 6.8, $m \leq n$; otherwise, $\mathcal{B}'$ would be linearly dependent.

Now use Theorem 6.8 with the roles of $\mathcal{B}$ and $\mathcal{B}'$ interchanged. Since $\mathcal{B}'$ is a basis of V with m vectors, Theorem 6.8 implies that any set of more than m vectors in V is linearly dependent. Hence, $n \leq m$, since $\mathcal{B}$ is a basis and is, therefore, linearly independent.

Since $n \leq m$ and $m \leq n$, we must have $n = m$, as required.

The following definition now makes sense, since the number of vectors in a (finite) basis does not depend on the choice of basis.

Definition A vector space V is called ***finite-dimensional*** if it has a basis consisting of finitely many vectors. The ***dimension*** of V, denoted by dim V, is the number of vectors in a basis for V. The dimension of the zero vector space $\{\mathbf{0}\}$ is defined to be zero. A vector space that has no finite basis is called ***infinite-dimensional.***

Example 6.38

Since the standard basis for $\mathbb{R}^n$ has n vectors, dim $\mathbb{R}^n = n$. In the case of $\mathbb{R}^3$, a one-dimensional subspace is just the span of a single nonzero vector and thus is a line through the origin. A two-dimensional subspace is spanned by its basis of two linearly independent (i.e., nonparallel) vectors and therefore is a plane through the origin. Any three linearly independent vectors must span $\mathbb{R}^3$, by the Fundamental Theorem. The subspaces of $\mathbb{R}^3$ are now completely classified according to dimension, as shown in Table 6.1.

Table 6.1

dim V	V
3	$\mathbb{R}^3$
2	Plane through the origin
1	Line through the origin
0	$\{\mathbf{0}\}$

Example 6.39

The standard basis for $\mathscr{P}_n$ contains $n + 1$ vectors (see Example 6.30), so dim $\mathscr{P}_n = n + 1$.

Example 6.40 The standard basis for M_{mn} contains mn vectors (see Example 6.31), so $\dim M_{mn} = mn$.

Example 6.41 Both $\mathscr{P}$ and $\mathscr{F}$ are infinite-dimensional, since they each contain the infinite linearly independent set $\{1, x, x^2, \ldots\}$ (see Exercise 44).

Example 6.42 Find the dimension of the vector space W of symmetric 2×2 matrices (see Example 6.10).

Solution A symmetric 2×2 matrix is of the form

$$\begin{bmatrix} a & b \\ b & c \end{bmatrix} = a\begin{bmatrix} 1 & 0 \\ 0 & 0 \end{bmatrix} + b\begin{bmatrix} 0 & 1 \\ 1 & 0 \end{bmatrix} + c\begin{bmatrix} 0 & 0 \\ 0 & 1 \end{bmatrix}$$

so W is spanned by the set

$$S = \left\{ \begin{bmatrix} 1 & 0 \\ 0 & 0 \end{bmatrix}, \begin{bmatrix} 0 & 1 \\ 1 & 0 \end{bmatrix}, \begin{bmatrix} 0 & 0 \\ 0 & 1 \end{bmatrix} \right\}$$

If S is linearly independent, then it will be a basis for W. Setting

$$a\begin{bmatrix} 1 & 0 \\ 0 & 0 \end{bmatrix} + b\begin{bmatrix} 0 & 1 \\ 1 & 0 \end{bmatrix} + c\begin{bmatrix} 0 & 0 \\ 0 & 1 \end{bmatrix} = \begin{bmatrix} 0 & 0 \\ 0 & 0 \end{bmatrix}$$

we obtain

$$\begin{bmatrix} a & b \\ b & c \end{bmatrix} = \begin{bmatrix} 0 & 0 \\ 0 & 0 \end{bmatrix}$$

from which it immediately follows that $a = b = c = 0$. Hence, S is linearly independent and is, therefore, a basis for W. We conclude that $\dim W = 3$.

The dimension of a vector space is its "magic number." Knowing the dimension of a vector space V provides us with much information about V and can greatly simplify the work needed in certain types of calculations, as the next few theorems and examples illustrate.

Theorem 6.10 Let V be a vector space with $\dim V = n$. Then:

a. Any linearly independent set in V contains at most n vectors.
b. Any spanning set for V contains at least n vectors.
c. Any linearly independent set of exactly n vectors in V is a basis for V.
d. Any spanning set for V consisting of exactly n vectors is a basis for V.
e. Any linearly independent set in V can be extended to a basis for V.
f. Any spanning set for V can be reduced to a basis for V.

Proof The proofs of properties (a) and (b) follow from parts (a) and (b) of Theorem 6.8, respectively.

(c) Let S be a linearly independent set of exactly n vectors in V. If S does not span V, then there is some vector $\mathbf{v}$ in V that is not a linear combination of the vectors in S. Inserting $\mathbf{v}$ into S produces a set S' with $n + 1$ vectors that is still linearly independent (see Exercise 54). But this is impossible, by Theorem 6.8(a). We conclude that S must span V and therefore be a basis for V.

(d) Let S be a spanning set for V consisting of exactly n vectors. If S is linearly dependent, then some vector $\mathbf{v}$ in S is a linear combination of the others. Throwing $\mathbf{v}$ away leaves a set S' with $n - 1$ vectors that still spans V (see Exercise 55). But this is impossible, by Theorem 6.8(b). We conclude that S must be linearly independent and therefore be a basis for V.

(e) Let S be a linearly independent set of vectors in V. If S spans V, it is a basis for V and so consists of exactly n vectors, by the Basis Theorem. If S does not span V, then, as in the proof of property (c), there is some vector $\mathbf{v}$ in V that is not a linear combination of the vectors in S. Inserting $\mathbf{v}$ into S produces a set S' that is still linearly independent. If S' still does not span V, we can repeat the process and expand it into a larger, linearly independent set. Eventually, this process must stop, since no linearly independent set in V can contain more than n vectors, by Theorem 6.8(a). When the process stops, we have a linearly independent set S^* that contains S and also spans V. Therefore, S^* is a basis for V that extends S.

(f) You are asked to prove this property in Exercise 56.

You should view Theorem 6.10 as, in part, a labor-saving device. In many instances, it can dramatically decrease the amount of work needed to check that a set of vectors is linearly independent, a spanning set, or a basis.

Example 6.43

In each case, determine whether S is a basis for V.

(a) $V = \mathscr{P}_2$, $S = \{1 + x, 2 - x + x^2, 3x - 2x^2, -1 + 3x + x^2\}$

(b) $V = M_{22}$, $S = \left\{\begin{bmatrix} 1 & 0 \\ 1 & 1 \end{bmatrix}, \begin{bmatrix} 0 & -1 \\ 1 & 0 \end{bmatrix}, \begin{bmatrix} 1 & 1 \\ 0 & -1 \end{bmatrix}\right\}$

(c) $V = \mathscr{P}_2$, $S = \{1 + x, x + x^2, 1 + x^2\}$

Solution (a) Since $\dim(\mathscr{P}_2) = 3$ and S contains four vectors, S is linearly dependent, by Theorem 6.10(a). Hence, S is not a basis for $\mathscr{P}_2$.

(b) Since $\dim(M_{22}) = 4$ and S contains three vectors, S cannot span M_{22}, by Theorem 6.10(b). Hence, S is not a basis for M_{22}.

(c) Since $\dim(\mathscr{P}_2) = 3$ and S contains three vectors, S will be a basis for $\mathscr{P}_2$ if it is linearly independent or if it spans $\mathscr{P}_2$, by Theorem 6.10(c) or (d). It is easier to show that S is linearly independent; we did this in Example 6.26. Therefore, S is a basis for $\mathscr{P}_2$. (This is the same problem as in Example 6.32—but see how much easier it becomes using Theorem 6.10!)

Example 6.44

Extend $\{1 + x, 1 - x\}$ to a basis for $\mathscr{P}_2$.

Solution First note that $\{1 + x, 1 - x\}$ is linearly independent. (Why?) Since $\dim(\mathscr{P}_2) = 3$, we need a third vector—one that is not linearly dependent on the first two.

We could proceed, as in the proof of Theorem 6.10(e), to find such a vector using trial and error. However, it is easier in practice to proceed in a different way.

We enlarge the given set of vectors by throwing in the *entire* standard basis for $\mathcal{P}_2$. This gives

$$S = \{1 + x, 1 - x, 1, x, x^2\}$$

Now S is linearly dependent, by Theorem 6.10(a), so we need to throw away some vectors—in this case, two. Which ones? We use Theorem 6.10(f), starting with the first vector that was added, 1. Since $1 = \frac{1}{2}(1 + x) + \frac{1}{2}(1 - x)$, the set $\{1 + x, 1 - x, 1\}$ is linearly dependent, so we throw away 1. Similarly, $x = \frac{1}{2}(1 + x) - \frac{1}{2}(1 - x)$, so $\{1 + x, 1 - x, x\}$ is linearly dependent also. Finally, we check that $\{1 + x, 1 - x, x^2\}$ is linearly independent. (Can you see a quick way to tell this?) Therefore, $\{1 + x, 1 - x, x^2\}$ is a basis for $\mathcal{P}_2$ that extends $\{1 + x, 1 - x\}$.

In Example 6.42, the vector space W of symmetric 2×2 matrices is a subspace of the vector space M_{22} of all 2×2 matrices. As we showed, $\dim W = 3 \leq 4 = \dim M_{22}$. This is an example of a general result, as the final theorem of this section shows.

Theorem 6.11 Let W be a subspace of a finite-dimensional vector space V. Then:

a. W is finite-dimensional and $\dim W \leq \dim V$.
b. $\dim W = \dim V$ if and only if $W = V$.

Proof (a) Let $\dim V = n$. If $W = \{\mathbf{0}\}$, then $\dim(W) = 0 \leq n = \dim V$. If W is nonzero, then any basis $\mathcal{B}$ for V (containing n vectors) certainly spans W, since W is contained in V. But $\mathcal{B}$ can be reduced to a basis $\mathcal{B}'$ for W (containing at most n vectors), by Theorem 6.10(f). Hence, W is finite-dimensional and $\dim(W) \leq n = \dim V$.

(b) If $W = V$, then certainly $\dim W = \dim V$. On the other hand, if $\dim W = \dim V = n$, then any basis $\mathcal{B}$ for W consists of exactly n vectors. But these are then n linearly independent vectors in V and, hence, a basis for V, by Theorem 6.10(c). Therefore, $V = \text{span}(\mathcal{B}) = W$.

Exercises 6.2

In Exercises 1–4, test the sets of matrices for linear independence in M_{22}. For those that are linearly dependent, express one of the matrices as a linear combination of the others.

1. $\left\{\begin{bmatrix} 1 & 1 \\ 0 & -1 \end{bmatrix}, \begin{bmatrix} 1 & -1 \\ 1 & 0 \end{bmatrix}, \begin{bmatrix} 1 & 0 \\ 3 & 2 \end{bmatrix}\right\}$

2. $\left\{\begin{bmatrix} 2 & -3 \\ 4 & 2 \end{bmatrix}, \begin{bmatrix} 1 & -1 \\ 3 & 3 \end{bmatrix}, \begin{bmatrix} -1 & 3 \\ 1 & 5 \end{bmatrix}\right\}$

3. $\left\{\begin{bmatrix} -1 & 1 \\ -2 & 2 \end{bmatrix}, \begin{bmatrix} 3 & 0 \\ 1 & 1 \end{bmatrix}, \begin{bmatrix} 0 & 2 \\ -3 & 1 \end{bmatrix}, \begin{bmatrix} -1 & 0 \\ -1 & 7 \end{bmatrix}\right\}$

4. $\left\{\begin{bmatrix} 1 & 1 \\ 0 & 1 \end{bmatrix}, \begin{bmatrix} 1 & 0 \\ 1 & 1 \end{bmatrix}, \begin{bmatrix} 0 & 1 \\ 1 & 1 \end{bmatrix}, \begin{bmatrix} 1 & 1 \\ 1 & 0 \end{bmatrix}\right\}$

In Exercises 5–9, test the sets of polynomials for linear independence. For those that are linearly dependent, express one of the polynomials as a linear combination of the others.

5. $\{x, 1 + x\}$ in $\mathcal{P}_1$

6. $\{1 + x, 1 + x^2, 1 - x + x^2\}$ in $\mathcal{P}_2$

7. $\{x, 2x - x^2, 3x + 2x^2\}$ in $\mathcal{P}_2$

8. $\{2x, x - x^2, 1 + x^3, 2 - x^2 + x^3\}$ in $\mathscr{P}_3$

9. $\{1 - 2x, 3x + x^2 - x^3, 1 + x^2 + 2x^3, 3 + 2x + 3x^3\}$ in $\mathscr{P}_3$

In Exercises 10–14, test the sets of functions for linear independence in $\mathscr{F}$. For those that are linearly dependent, express one of the functions as a linear combination of the others.

10. $\{1, \sin x, \cos x\}$

11. $\{1, \sin^2 x, \cos^2 x\}$

12. $\{e^x, e^{-x}\}$

13. $\{1, \ln(2x), \ln(x^2)\}$

14. $\{\sin x, \sin 2x, \sin 3x\}$

15. If f and g are in $\mathscr{C}^{(1)}$, the vector space of all functions with continuous derivatives, then the determinant

$$W(x) = \begin{vmatrix} f(x) & g(x) \\ f'(x) & g'(x) \end{vmatrix}$$

is called the ***Wronskian*** of f and g [named after the Polish-French mathematician Jósef Maria Hoëné-Wronski (1776–1853), who worked on the theory of determinants and the philosophy of mathematics]. Show that f and g are linearly independent if their Wronskian is not identically zero (that is, if there is some x such that $W(x) \neq 0$).

16. In general, the Wronskian of $f_1, \ldots, f_n$ in $\mathscr{C}^{(n-1)}$ is the determinant

$$W(x) = \begin{vmatrix} f_1(x) & f_2(x) & \cdots & f_n(x) \\ f_1'(x) & f_2'(x) & \cdots & f_n'(x) \\ \vdots & \vdots & \ddots & \vdots \\ f_1^{(n-1)}(x) & f_2^{(n-1)}(x) & \cdots & f_n^{(n-1)}(x) \end{vmatrix}$$

and $f_1, \ldots, f_n$ are linearly independent, provided $W(x)$ is not identically zero. Repeat Exercises 10–14 using the Wronskian test.

17. Let $\{\mathbf{u}, \mathbf{v}, \mathbf{w}\}$ be a linearly independent set of vectors in a vector space V.

(a) Is $\{\mathbf{u} + \mathbf{v}, \mathbf{v} + \mathbf{w}, \mathbf{u} + \mathbf{w}\}$ linearly independent? Either prove that it is or give a counterexample to show that it is not.

(b) Is $\{\mathbf{u} - \mathbf{v}, \mathbf{v} - \mathbf{w}, \mathbf{u} - \mathbf{w}\}$ linearly independent? Either prove that it is or give a counterexample to show that it is not.

In Exercises 18–25, determine whether the set $\mathcal{B}$ is a basis for the vector space V.

18. $V = M_{22}$, $\mathcal{B} = \left\{ \begin{bmatrix} 1 & 0 \\ 0 & 1 \end{bmatrix}, \begin{bmatrix} 1 & 1 \\ 1 & 0 \end{bmatrix}, \begin{bmatrix} 1 & -1 \\ -1 & 1 \end{bmatrix} \right\}$

19. $V = M_{22}$, $\mathcal{B} = \left\{ \begin{bmatrix} 1 & 0 \\ 0 & 1 \end{bmatrix}, \begin{bmatrix} 0 & -1 \\ 1 & 0 \end{bmatrix}, \begin{bmatrix} 1 & 1 \\ 1 & 1 \end{bmatrix}, \begin{bmatrix} 1 & 1 \\ 1 & -1 \end{bmatrix} \right\}$

20. $V = M_{22}$,
$\mathcal{B} = \left\{ \begin{bmatrix} 1 & 0 \\ 0 & 1 \end{bmatrix}, \begin{bmatrix} 0 & 1 \\ 1 & 0 \end{bmatrix}, \begin{bmatrix} 1 & 1 \\ 0 & 1 \end{bmatrix}, \begin{bmatrix} 1 & 0 \\ 1 & 1 \end{bmatrix} \right\}$

21. $V = M_{22}$,
$\mathcal{B} = \left\{ \begin{bmatrix} 1 & 2 \\ 2 & 1 \end{bmatrix}, \begin{bmatrix} 2 & 1 \\ 1 & 2 \end{bmatrix}, \begin{bmatrix} 1 & 3 \\ -3 & 1 \end{bmatrix}, \begin{bmatrix} 2 & 3 \\ 3 & 1 \end{bmatrix}, \begin{bmatrix} 1 & 2 \\ 3 & 2 \end{bmatrix} \right\}$

22. $V = \mathscr{P}_2$, $\mathcal{B} = \{x, 1 + x, x - x^2\}$

23. $V = \mathscr{P}_2$, $\mathcal{B} = \{1 - x, 1 - x^2, x - x^2\}$

24. $V = \mathscr{P}_2$, $\mathcal{B} = \{1, 1 + 2x + 3x^2\}$

25. $V = \mathscr{P}_2$, $\mathcal{B} = \{1, 2 - x, 3 - x^2, x + 2x^2\}$

26. Find the coordinate vector of $A = \begin{bmatrix} 1 & 2 \\ 3 & 4 \end{bmatrix}$ with respect to the basis $\mathcal{B} = \{E_{22}, E_{21}, E_{12}, E_{11}\}$ of M_{22}.

27. Find the coordinate vector of $A = \begin{bmatrix} 1 & 2 \\ 3 & 4 \end{bmatrix}$ with respect to the basis $\mathcal{B} = \left\{ \begin{bmatrix} 1 & 0 \\ 0 & 0 \end{bmatrix}, \begin{bmatrix} 1 & 1 \\ 0 & 0 \end{bmatrix}, \begin{bmatrix} 1 & 1 \\ 1 & 0 \end{bmatrix}, \begin{bmatrix} 1 & 1 \\ 1 & 1 \end{bmatrix} \right\}$ of M_{22}.

28. Find the coordinate vector of $p(x) = 1 + 2x + 3x^2$ with respect to the basis $\mathcal{B} = \{1 + x, 1 - x, x^2\}$ of $\mathscr{P}_2$.

29. Find the coordinate vector of $p(x) = 2 - x + 3x^2$ with respect to the basis $\mathcal{B} = \{1, 1 + x, -1 + x^2\}$ of $\mathscr{P}_2$.

30. Let $\mathcal{B}$ be a set of vectors in a vector space V with the property that every vector in V can be written uniquely as a linear combination of the vectors in $\mathcal{B}$. Prove that $\mathcal{B}$ is a basis for V.

31. Let $\mathcal{B}$ be a basis for a vector space V, let $\mathbf{u}_1, \ldots, \mathbf{u}_k$ be vectors in V, and let $c_1, \ldots, c_k$ be scalars. Show that $[c_1\mathbf{u}_1 + \cdots + c_k\mathbf{u}_k]_{\mathcal{B}} = c_1[\mathbf{u}_1]_{\mathcal{B}} + \cdots + c_k[\mathbf{u}_k]_{\mathcal{B}}$.

32. Finish the proof of Theorem 6.7 by showing that if $\{[\mathbf{u}_1]_{\mathcal{B}}, \ldots, [\mathbf{u}_k]_{\mathcal{B}}\}$ is linearly independent in $\mathbb{R}^n$ then $\{\mathbf{u}_1, \ldots, \mathbf{u}_k\}$ is linearly independent in V.

33. Let $\{\mathbf{u}_1, \ldots, \mathbf{u}_m\}$ be a set of vectors in an n-dimensional vector space V and let $\mathcal{B}$ be a basis for V. Let $S = \{[\mathbf{u}_1]_{\mathcal{B}}, \ldots, [\mathbf{u}_m]_{\mathcal{B}}\}$ be the set of coordinate vectors of $\{\mathbf{u}_1, \ldots, \mathbf{u}_m\}$ with respect to $\mathcal{B}$. Prove that $\text{span}(\mathbf{u}_1, \ldots, \mathbf{u}_m) = V$ if and only if $\text{span}(S) = \mathbb{R}^n$.

In Exercises 34–39, find the dimension of the vector space V and give a basis for V.

34. $V = \{p(x) \text{ in } \mathscr{P}_2 : p(0) = 0\}$

35. $V = \{p(x) \text{ in } \mathscr{P}_2 : p(1) = 0\}$

36. $V = \{p(x) \text{ in } \mathscr{P}_2 : xp'(x) = p(x)\}$

37. $V = \{A \text{ in } M_{22} : A \text{ is upper triangular}\}$

38. $V = \{A \text{ in } M_{22} : A \text{ is skew-symmetric}\}$

39. $V = \{A \text{ in } M_{22} : AB = BA\}$, where $B = \begin{bmatrix} 1 & 1 \\ 0 & 1 \end{bmatrix}$

40. Find a formula for the dimension of the vector space of symmetric $n \times n$ matrices.

41. Find a formula for the dimension of the vector space of skew-symmetric $n \times n$ matrices.

42. Let U and W be subspaces of a finite-dimensional vector space V. Prove ***Grassmann's Identity:***

$$\dim(U + W) = \dim U + \dim W - \dim(U \cap W)$$

[*Hint:* The subspace $U + W$ is defined in Exercise 48 of Section 6.1. Let $\mathcal{B} = \{\mathbf{v}_1, \ldots, \mathbf{v}_k\}$ be a basis for $U \cap W$. Extend $\mathcal{B}$ to a basis $\mathcal{C}$ of U and a basis $\mathcal{D}$ of W. Prove that $\mathcal{C} \cup \mathcal{D}$ is a basis for $U + W$.]

43. Let U and V be finite-dimensional vector spaces.

(a) Find a formula for $\dim(U \times V)$ in terms of $\dim U$ and $\dim V$. (See Exercise 49 in Section 6.1.)

(b) If W is a subspace of V, show that $\dim \Delta = \dim W$, where $\Delta = \{(\mathbf{w}, \mathbf{w}) : \mathbf{w} \text{ is in } W\}$.

44. Prove that the vector space $\mathcal{P}$ is infinite-dimensional. [*Hint:* Suppose it has a finite basis. Show that there is some polynomial that is not a linear combination of this basis.]

45. Extend $\{1 + x, 1 + x + x^2\}$ to a basis for $\mathcal{P}_2$.

46. Extend $\left\{\begin{bmatrix} 0 & 1 \\ 0 & 1 \end{bmatrix}, \begin{bmatrix} 1 & 1 \\ 0 & 1 \end{bmatrix}\right\}$ to a basis for M_{22}.

47. Extend $\left\{\begin{bmatrix} 1 & 0 \\ 0 & 1 \end{bmatrix}, \begin{bmatrix} 0 & 1 \\ 1 & 0 \end{bmatrix}, \begin{bmatrix} 0 & -1 \\ 1 & 0 \end{bmatrix}\right\}$ to a basis for M_{22}.

48. Extend $\left\{\begin{bmatrix} 1 & 0 \\ 0 & 1 \end{bmatrix}, \begin{bmatrix} 0 & 1 \\ 1 & 0 \end{bmatrix}\right\}$ to a basis for the vector space of symmetric 2×2 matrices.

49. Find a basis for $\text{span}(1, 1 + x, 2x)$ in $\mathcal{P}_1$.

50. Find a basis for $\text{span}(1 - 2x, 2x - x^2, 1 - x^2, 1 + x^2)$ in $\mathcal{P}_2$.

51. Find a basis for $\text{span}(1 - x, x - x^2, 1 - x^2, 1 - 2x + x^2)$ in $\mathcal{P}_2$.

52. Find a basis for $\text{span}\left(\begin{bmatrix} 1 & 0 \\ 0 & 1 \end{bmatrix}, \begin{bmatrix} 0 & 1 \\ 1 & 0 \end{bmatrix}, \begin{bmatrix} -1 & 1 \\ 1 & -1 \end{bmatrix}, \begin{bmatrix} 1 & -1 \\ -1 & 1 \end{bmatrix}\right)$ in M_{22}.

53. Find a basis for $\text{span}(\sin^2 x, \cos^2 x, \cos 2x)$ in $\mathcal{F}$.

54. Let $S = \{\mathbf{v}_1, \ldots, \mathbf{v}_n\}$ be a linearly independent set in a vector space V. Show that if $\mathbf{v}$ is a vector in V that is not in $\text{span}(S)$, then $S' = \{\mathbf{v}_1, \ldots, \mathbf{v}_n, \mathbf{v}\}$ is still linearly independent.

55. Let $S = \{\mathbf{v}_1, \ldots, \mathbf{v}_n\}$ be a spanning set for a vector space V. Show that if $\mathbf{v}_n$ is in $\text{span}(\mathbf{v}_1, \ldots, \mathbf{v}_{n-1})$, then $S' = \{\mathbf{v}_1, \ldots, \mathbf{v}_{n-1}\}$ is still a spanning set for V.

56. Prove Theorem 6.10(f).

57. Let $\{\mathbf{v}_1, \ldots, \mathbf{v}_n\}$ be a basis for a vector space V and let $c_1, \ldots, c_n$ be nonzero scalars. Prove that $\{c_1\mathbf{v}_1, \ldots, c_n\mathbf{v}_n\}$ is also a basis for V.

58. Let $\{\mathbf{v}_1, \ldots, \mathbf{v}_n\}$ be a basis for a vector space V. Prove that

$$\{\mathbf{v}_1, \mathbf{v}_1 + \mathbf{v}_2, \mathbf{v}_1 + \mathbf{v}_2 + \mathbf{v}_3, \ldots, \mathbf{v}_1 + \cdots + \mathbf{v}_n\}$$

is also a basis for V.

Let $a_0, a_1, \ldots, a_n$ be $n + 1$ distinct real numbers. Define polynomials $p_0(x), p_1(x), \ldots, p_n(x)$ by

$$p_i(x) = \frac{(x - a_0)\cdots(x - a_{i-1})(x - a_{i+1})\cdots(x - a_n)}{(a_i - a_0)\cdots(a_i - a_{i-1})(a_i - a_{i+1})\cdots(a_i - a_n)}$$

*These are called the **Lagrange polynomials** associated with $a_0, a_1, \ldots, a_n$. [Joseph-Louis Lagrange (1736–1813) was born in Italy but spent most of his life in Germany and France. He made important contributions to such fields as number theory, algebra, astronomy, mechanics, and the calculus of variations. In 1773, Lagrange was the first to give the volume interpretation of a determinant (see Chapter 4).]*

59. (a) Compute the Lagrange polynomials associated with $a_0 = 1, a_1 = 2, a_2 = 3$.

(b) Show, in general, that

$$p_i(a_j) = \begin{cases} 0 & \text{if } i \neq j \\ 1 & \text{if } i = j \end{cases}$$

60. (a) Prove that the set $\mathcal{B} = \{p_0(x), p_1(x), \ldots, p_n(x)\}$ of Lagrange polynomials is linearly independent in $\mathcal{P}_n$. [*Hint:* Set $c_0p_0(x) + \cdots + c_np_n(x) = 0$ and use Exercise 59(b).]

(b) Deduce that $\mathcal{B}$ is a basis for $\mathcal{P}_n$.

61. If $q(x)$ is an arbitrary polynomial in $\mathcal{P}_n$, it follows from Exercise 60(b) that

$$q(x) = c_0p_0(x) + \cdots + c_np_n(x) \qquad (1)$$

for some scalars $c_0, \ldots, c_n$.

(a) Show that $c_i = q(a_i)$ for $i = 0, \ldots, n$, and deduce that $q(x) = q(a_0)p_0(x) + \cdots + q(a_n)p_n(x)$ is the unique representation of $q(x)$ with respect to the basis $\mathcal{B}$.

(b) Show that for any $n + 1$ points $(a_0, c_0), (a_1, c_1), \ldots, (a_n, c_n)$ with distinct first components, the function $q(x)$ defined by Equation (1) is the unique polynomial of degree at most n that passes through all of the points. This formula is known as the ***Lagrange interpolation formula.*** (Compare this formula with Problem 19 in Exploration: Geometric Applications of Determinants in Chapter 4.)

(c) Use the Lagrange interpolation formula to find the polynomial of degree at most 2 that passes through the points

(i) $(1, 6)$, $(2, -1)$, and $(3, -2)$

(ii) $(-1, 10)$, $(0, 5)$, and $(3, 2)$

62. Use the Lagrange interpolation formula to show that if a polynomial in $\mathscr{P}_n$ has $n + 1$ zeros, then it must be the zero polynomial.

63. Find a formula for the number of invertible matrices in $M_{nn}(\mathbb{Z}_p)$. [*Hint:* This is the same as determining the number of different bases for $\mathbb{Z}_p^n$. (Why?) Count the number of ways to construct a basis for $\mathbb{Z}_p^n$, one vector at a time.]

Exploration

Magic Squares

The engraving shown on page 461 is Albrecht Dürer's *Melancholia I* (1514). Among the many mathematical artifacts in this engraving is the chart of numbers that hangs on the wall in the upper right-hand corner. (It is enlarged in the detail shown.) Such an array of numbers is known as a *magic square*. We can think of it as a 4×4 matrix

$$\begin{bmatrix} 16 & 3 & 2 & 13 \\ 5 & 10 & 11 & 8 \\ 9 & 6 & 7 & 12 \\ 4 & 15 & 14 & 1 \end{bmatrix}$$

Observe that the numbers in each row, in each column, and in both diagonals have the same sum: 34. Observe further that the entries are the integers $1, 2, \ldots, 16$. (Note that Dürer cleverly placed the 15 and 14 adjacent to each other in the last row, giving the date of the engraving.) These observations lead to the following definition.

Definition An $n \times n$ matrix M is called a ***magic square*** if the sum of the entries is the same in each row, each column, and both diagonals. This common sum is called the ***weight*** of M, denoted $\text{wt}(M)$. If M is an $n \times n$ magic square that contains each of the entries $1, 2, \ldots, n^2$ exactly once, then M is called a ***classical magic square.***

1. If M is a classical $n \times n$ magic square, show that

$$\text{wt}(M) = \frac{n(n^2 + 1)}{2}$$

[*Hint:* Use Exercise 51 in Section 2.4.]

2. Find a classical 3×3 magic square. Find a different one. Are your two examples related in any way?

Burstein Collection/CORBIS

Burstein Collection/CORBIS

3. Clearly, the 3×3 matrix with all entries equal to $\frac{1}{3}$ is a magic square with weight 1. Using your answer to Problem 2, find a 3×3 magic square with weight 1, *all of whose entries are different.* Describe a method for constructing a 3×3 magic square with distinct entries and weight w for any real number w.

Let Mag_n denote the set of all $n \times n$ magic squares, and let Mag_n^0 denote the set of all $n \times n$ magic squares of weight 0.

4. (a) Prove that Mag_3 is a subspace of M_{33}.
   (b) Prove that Mag_3^0 is a subspace of Mag_3.

5. Use Problems 3 and 4 to show that if M is a 3×3 magic square with weight w, then we can write M as

$$M = M_0 + kJ$$

where M_0 is a 3×3 magic square of weight 0, J is the 3×3 matrix consisting entirely of ones, and k is a scalar. What must k be? [*Hint:* Show that $M - kJ$ is in Mag_3^0 for an appropriate value of k.]

Let's try to find a way of describing *all* 3×3 magic squares. Let

$$M = \begin{bmatrix} a & b & c \\ d & e & f \\ g & h & i \end{bmatrix}$$

be a magic square with weight 0. The conditions on the rows, columns, and diagonals give rise to a system of eight homogeneous linear equations in the variables $a, b, \ldots, i$.

6. Write out this system of equations and solve it. [*Note:* Using a CAS will facilitate the calculations.]

7. Find the dimension of Mag_3^0. *Hint:* By doing a substitution, if necessary, use your solution to Problem 6 to show that M can be written in the form

$$M = \begin{bmatrix} s & -s-t & t \\ -s+t & 0 & s-t \\ -t & s+t & -s \end{bmatrix}$$

8. Find the dimension of Mag_3. [*Hint:* Combine the results of Problems 5 and 7.]

9. Can you find a direct way of showing that the (2, 2) entry of a 3×3 magic square with weight w must be $w/3$? [*Hint:* Add and subtract certain rows, columns, and diagonals to leave a multiple of the central entry.]

10. Let M be a 3×3 magic square of weight 0, obtained from a classical 3×3 magic square as in Problem 5. If M has the form given in Problem 7, write out an equation for the sum of the squares of the entries of M. Show that this is the equation of a circle in the variables s and t, and carefully plot it. Show that there are exactly eight points (s, t) on this circle with both s and t integers. Using Problem 8, show that these eight points give rise to eight classical 3×3 magic squares. How are these magic squares related to one another?

6.3 Change of Basis

In many applications, a problem described using one coordinate system may be solved more easily by switching to a new coordinate system. This switch is usually accomplished by performing a change of variables, a process that you have probably encountered in other mathematics courses. In linear algebra, a basis provides us with a coordinate system for a vector space, via the notion of coordinate vectors. Choosing the right basis will often greatly simplify a particular problem. For example, consider the molecular structure of zinc, shown in Figure 6.3(a). A scientist studying zinc might wish to measure the lengths of the bonds between the atoms, the angles between these bonds, and so on. Such an analysis will be greatly facilitated by introducing coordinates and making use of the tools of linear algebra. The standard basis and the associated standard *xyz* coordinate axes are not always the best choice. As Figure 6.3(b) shows, in this case $\{\mathbf{u}, \mathbf{v}, \mathbf{w}\}$ is probably a better choice of basis for $\mathbb{R}^3$ than the standard basis, since these vectors align nicely with the bonds between the atoms of zinc.

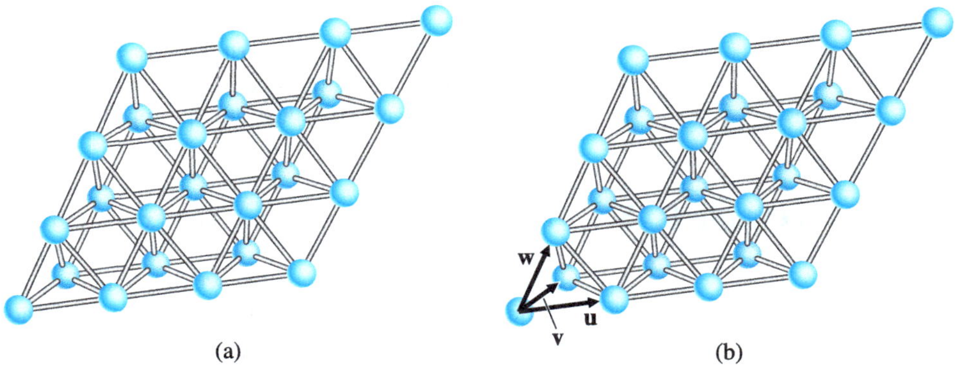

(a) (b)

Figure 6.3

Change-of-Basis Matrices

Figure 6.4 shows two different coordinate systems for $\mathbb{R}^2$, each arising from a different basis. Figure 6.4(a) shows the coordinate system related to the basis $\mathcal{B} = \{\mathbf{u}_1, \mathbf{u}_2\}$, while Figure 6.4(b) arises from the basis $\mathcal{C} = \{\mathbf{v}_1, \mathbf{v}_2\}$, where

$$\mathbf{u}_1 = \begin{bmatrix} -1 \\ 2 \end{bmatrix}, \quad \mathbf{u}_2 = \begin{bmatrix} 2 \\ -1 \end{bmatrix}, \quad \mathbf{v}_1 = \begin{bmatrix} 1 \\ 0 \end{bmatrix}, \quad \mathbf{v}_2 = \begin{bmatrix} 1 \\ 1 \end{bmatrix}$$

The same vector $\mathbf{x}$ is shown relative to each coordinate system. It is clear from the diagrams that the coordinate vectors of $\mathbf{x}$ with respect to $\mathcal{B}$ and $\mathcal{C}$ are

$$[\mathbf{x}]_{\mathcal{B}} = \begin{bmatrix} 1 \\ 3 \end{bmatrix} \quad \text{and} \quad [\mathbf{x}]_{\mathcal{C}} = \begin{bmatrix} 6 \\ -1 \end{bmatrix}$$

respectively. It turns out that there is a direct connection between the two coordinate vectors. One way to find the relationship is to use $[\mathbf{x}]_{\mathcal{B}}$ to calculate

$$\mathbf{x} = \mathbf{u}_1 + 3\mathbf{u}_2 = \begin{bmatrix} -1 \\ 2 \end{bmatrix} + 3\begin{bmatrix} 2 \\ -1 \end{bmatrix} = \begin{bmatrix} 5 \\ -1 \end{bmatrix}$$

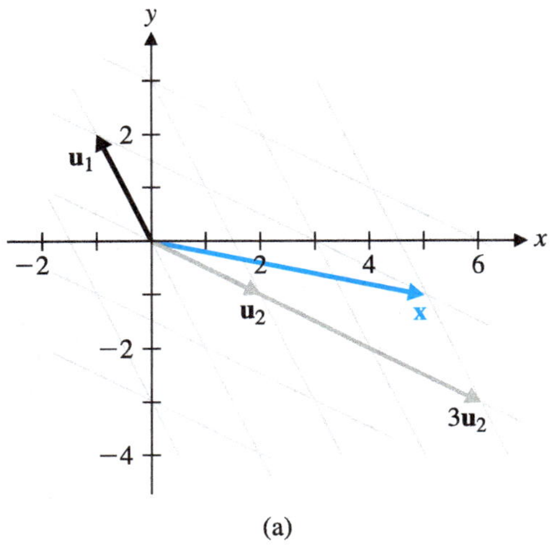

Figure 6.4

Then we can find $[\mathbf{x}]_{\mathcal{C}}$ by writing $\mathbf{x}$ as a linear combination of $\mathbf{v}_1$ and $\mathbf{v}_2$. However, there is a better way to proceed—one that will provide us with a general mechanism for such problems. We illustrate this approach in the next example.

Example 6.45 Using the bases $\mathcal{B}$ and $\mathcal{C}$ above, find $[\mathbf{x}]_{\mathcal{C}}$, given that $[\mathbf{x}]_{\mathcal{B}} = \begin{bmatrix} 1 \\ 3 \end{bmatrix}$.

Solution Since $\mathbf{x} = \mathbf{u}_1 + 3\mathbf{u}_2$, writing $\mathbf{u}_1$ and $\mathbf{u}_2$ in terms of $\mathbf{v}_1$ and $\mathbf{v}_2$ will give us the required coordinates of $\mathbf{x}$ with respect to $\mathcal{C}$. We find that

$$\mathbf{u}_1 = \begin{bmatrix} -1 \\ 2 \end{bmatrix} = -3\begin{bmatrix} 1 \\ 0 \end{bmatrix} + 2\begin{bmatrix} 1 \\ 1 \end{bmatrix} = -3\mathbf{v}_1 + 2\mathbf{v}_2$$

and

$$\mathbf{u}_2 = \begin{bmatrix} 2 \\ -1 \end{bmatrix} = 3\begin{bmatrix} 1 \\ 0 \end{bmatrix} - \begin{bmatrix} 1 \\ 1 \end{bmatrix} = 3\mathbf{v}_1 - \mathbf{v}_2$$

so

$$\begin{aligned} \mathbf{x} &= \mathbf{u}_1 + 3\mathbf{u}_2 \\ &= (-3\mathbf{v}_1 + 2\mathbf{v}_2) + 3(3\mathbf{v}_1 - \mathbf{v}_2) \\ &= 6\mathbf{v}_1 - \mathbf{v}_2 \end{aligned}$$

This gives

$$[\mathbf{x}]_{\mathcal{C}} = \begin{bmatrix} 6 \\ -1 \end{bmatrix}$$

in agreement with Figure 6.4(b).

This method may not look any easier than the one suggested prior to Example 6.45, but it has one big advantage: We can now find $[\mathbf{y}]_{\mathcal{C}}$ from $[\mathbf{y}]_{\mathcal{B}}$ for *any* vector $\mathbf{y}$ in $\mathbb{R}^2$

with very little additional work. Let's look at the calculations in Example 6.45 from a different point of view. From $\mathbf{x} = \mathbf{u}_1 + 3\mathbf{u}_2$, we have

$$[\mathbf{x}]_{\mathcal{C}} = [\mathbf{u}_1 + 3\mathbf{u}_2]_{\mathcal{C}} = [\mathbf{u}_1]_{\mathcal{C}} + 3[\mathbf{u}_2]_{\mathcal{C}}$$

by Theorem 6.6. Thus,

$$\begin{aligned}[\mathbf{x}]_{\mathcal{C}} &= [[\mathbf{u}_1]_{\mathcal{C}}[\mathbf{u}_2]_{\mathcal{C}}]\begin{bmatrix}1\\3\end{bmatrix}\\ &= \begin{bmatrix}-3 & 3\\ 2 & -1\end{bmatrix}\begin{bmatrix}1\\3\end{bmatrix}\\ &= P[\mathbf{x}]_{\mathcal{B}}\end{aligned}$$

where P is the matrix whose columns are $[\mathbf{u}_1]_{\mathcal{C}}$ and $[\mathbf{u}_2]_{\mathcal{C}}$. This procedure generalizes very nicely.

Definition Let $\mathcal{B} = \{\mathbf{u}_1, \ldots, \mathbf{u}_n\}$ and $\mathcal{C} = \{\mathbf{v}_1, \ldots, \mathbf{v}_n\}$ be bases for a vector space V. The $n \times n$ matrix whose columns are the coordinate vectors $[\mathbf{u}_1]_{\mathcal{C}}, \ldots, [\mathbf{u}_n]_{\mathcal{C}}$ of the vectors in $\mathcal{B}$ with respect to $\mathcal{C}$ is denoted by $P_{\mathcal{C}\leftarrow\mathcal{B}}$ and is called the ***change-of-basis matrix*** from $\mathcal{B}$ to $\mathcal{C}$. That is,

$$P_{\mathcal{C}\leftarrow\mathcal{B}} = [[\mathbf{u}_1]_{\mathcal{C}}[\mathbf{u}_2]_{\mathcal{C}} \cdots [\mathbf{u}_n]_{\mathcal{C}}]$$

Think of $\mathcal{B}$ as the "old" basis and $\mathcal{C}$ as the "new" basis. Then the columns of $P_{\mathcal{C}\leftarrow\mathcal{B}}$ are just the coordinate vectors obtained by writing the old basis vectors in terms of the new ones. Theorem 6.12 shows that Example 6.45 is a special case of a general result.

Theorem 6.12 Let $\mathcal{B} = \{\mathbf{u}_1, \ldots, \mathbf{u}_n\}$ and $\mathcal{C} = \{\mathbf{v}_1, \ldots, \mathbf{v}_n\}$ be bases for a vector space V and let $P_{\mathcal{C}\leftarrow\mathcal{B}}$ be the change-of-basis matrix from $\mathcal{B}$ to $\mathcal{C}$. Then

a. $P_{\mathcal{C}\leftarrow\mathcal{B}}[\mathbf{x}]_{\mathcal{B}} = [\mathbf{x}]_{\mathcal{C}}$ for all $\mathbf{x}$ in V.
b. $P_{\mathcal{C}\leftarrow\mathcal{B}}$ is the unique matrix P with the property that $P[\mathbf{x}]_{\mathcal{B}} = [\mathbf{x}]_{\mathcal{C}}$ for all $\mathbf{x}$ in V.
c. $P_{\mathcal{C}\leftarrow\mathcal{B}}$ is invertible and $(P_{\mathcal{C}\leftarrow\mathcal{B}})^{-1} = P_{\mathcal{B}\leftarrow\mathcal{C}}$.

Proof (a) Let $\mathbf{x}$ be in V and let

$$[\mathbf{x}]_{\mathcal{B}} = \begin{bmatrix}c_1\\ \vdots\\ c_n\end{bmatrix}$$

That is, $\mathbf{x} = c_1\mathbf{u}_1 + \cdots + c_n\mathbf{u}_n$. Then

$$\begin{aligned}[\mathbf{x}]_{\mathcal{C}} &= [c_1\mathbf{u}_1 + \cdots + c_n\mathbf{u}_n]_{\mathcal{C}}\\ &= c_1[\mathbf{u}_1]_{\mathcal{C}} + \cdots + c_n[\mathbf{u}_n]_{\mathcal{C}}\\ &= [[\mathbf{u}_1]_{\mathcal{C}} \cdots [\mathbf{u}_n]_{\mathcal{C}}]\begin{bmatrix}c_1\\ \vdots\\ c_n\end{bmatrix}\\ &= P_{\mathcal{C}\leftarrow\mathcal{B}}[\mathbf{x}]_{\mathcal{B}}\end{aligned}$$

(b) Suppose that P is an $n \times n$ matrix with the property that $P[\mathbf{x}]_{\mathcal{B}} = [\mathbf{x}]_{\mathcal{C}}$ for all $\mathbf{x}$ in V. Taking $\mathbf{x} = \mathbf{u}_i$, the ith basis vector in $\mathcal{B}$, we see that $[\mathbf{x}]_{\mathcal{B}} = [\mathbf{u}_i]_{\mathcal{B}} = \mathbf{e}_i$, so the ith column of P is

$$\mathbf{p}_i = P\mathbf{e}_i = P[\mathbf{u}_i]_{\mathcal{B}} = [\mathbf{u}_i]_{\mathcal{C}}$$

which is the ith column of $P_{\mathcal{C}\leftarrow\mathcal{B}}$, by definition. It follows that $P = P_{\mathcal{C}\leftarrow\mathcal{B}}$.

(c) Since $\{\mathbf{u}_1, \ldots, \mathbf{u}_n\}$ is linearly independent in V, the set $\{[\mathbf{u}_1]_{\mathcal{C}}, \ldots, [\mathbf{u}_n]_{\mathcal{C}}\}$ is linearly independent in $\mathbb{R}^n$, by Theorem 6.7. Hence, $P_{\mathcal{C}\leftarrow\mathcal{B}} = [[\mathbf{u}_1]_{\mathcal{C}} \ \cdots \ [\mathbf{u}]_{\mathcal{C}}$ is invertible, by the Fundamental Theorem.

For all $\mathbf{x}$ in V, we have $P_{\mathcal{C}\leftarrow\mathcal{B}}[\mathbf{x}]_{\mathcal{B}} = [\mathbf{x}]_{\mathcal{C}}$. Solving for $[\mathbf{x}]_{\mathcal{B}}$, we find that

$$[\mathbf{x}]_{\mathcal{B}} = (P_{\mathcal{C}\leftarrow\mathcal{B}})^{-1}[\mathbf{x}]_{\mathcal{C}}$$

for all $\mathbf{x}$ in V. Therefore, $(P_{\mathcal{C}\leftarrow\mathcal{B}})^{-1}$ is a matrix that changes bases from $\mathcal{C}$ to $\mathcal{B}$. Thus, by the uniqueness property (b), we must have $(P_{\mathcal{C}\leftarrow\mathcal{B}})^{-1} = \mathcal{P}_{\mathcal{B}\leftarrow\mathcal{C}}$.

Remarks

- You may find it helpful to think of change of basis as a transformation (indeed, it is a linear transformation) from $\mathbb{R}^n$ to itself that simply switches from one coordinate system to another. The transformation corresponding to $P_{\mathcal{C}\leftarrow\mathcal{B}}$ accepts $[\mathbf{x}]_{\mathcal{B}}$ as input and returns $[\mathbf{x}]_{\mathcal{C}}$ as output; $(P_{\mathcal{C}\leftarrow\mathcal{B}})^{-1} = \mathcal{P}_{\mathcal{B}\leftarrow\mathcal{C}}$ does just the opposite. Figure 6.5 gives a schematic representation of the process.

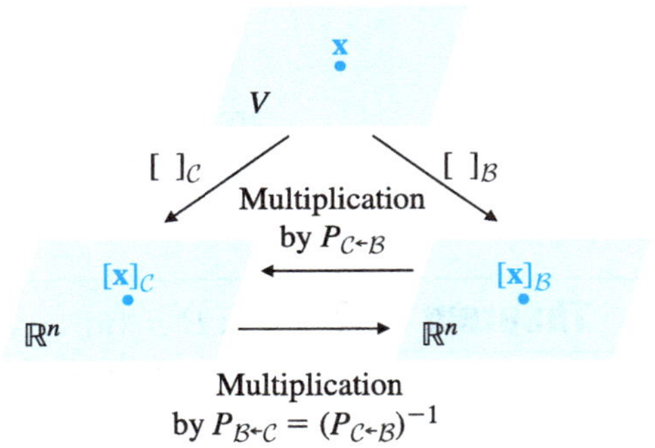

Figure 6.5
Change of basis

- The columns of $P_{\mathcal{C}\leftarrow\mathcal{B}}$ are the coordinate vectors of one basis with respect to the other basis. To remember which basis is which, think of the notation $\mathcal{C} \leftarrow \mathcal{B}$ as saying "$\mathcal{B}$ in terms of $\mathcal{C}$." It is also helpful to remember that $P_{\mathcal{C}\leftarrow\mathcal{B}}[\mathbf{x}]_{\mathcal{B}}$ is a linear combination of the columns of $P_{\mathcal{C}\leftarrow\mathcal{B}}$. But since the result of this combination is $[\mathbf{x}]_{\mathcal{C}}$, the columns of $P_{\mathcal{C}\leftarrow\mathcal{B}}$ must themselves be coordinate vectors with respect to $\mathcal{C}$.

Example 6.46

Find the change-of-basis matrices $P_{\mathcal{C}\leftarrow\mathcal{B}}$ and $P_{\mathcal{B}\leftarrow\mathcal{C}}$ for the bases $\mathcal{B} = \{1, x, x^2\}$ and $\mathcal{C} = \{1 + x, x + x^2, 1 + x^2\}$ of $\mathscr{P}_2$. Then find the coordinate vector of $p(x) = 1 + 2x - x^2$ with respect to $\mathcal{C}$.

Solution Changing *to* a standard basis is easy, so we find $P_{\mathcal{B}\leftarrow\mathcal{C}}$ first. Observe that the coordinate vectors for $\mathcal{C}$ in terms of $\mathcal{B}$ are

$$[1 + x]_{\mathcal{B}} = \begin{bmatrix} 1 \\ 1 \\ 0 \end{bmatrix}, \quad [x + x^2]_{\mathcal{B}} = \begin{bmatrix} 0 \\ 1 \\ 1 \end{bmatrix}, \quad [1 + x^2]_{\mathcal{B}} = \begin{bmatrix} 1 \\ 0 \\ 1 \end{bmatrix}$$

(Look back at the Remark following Example 6.26.) It follows that

$$P_{\mathcal{B}\leftarrow\mathcal{C}} = \begin{bmatrix} 1 & 0 & 1 \\ 1 & 1 & 0 \\ 0 & 1 & 1 \end{bmatrix}$$

To find $P_{\mathcal{C}\leftarrow\mathcal{B}}$, we could express each vector in $\mathcal{B}$ as a linear combination of the vectors in $\mathcal{C}$ (do this), but it is much easier to use the fact that $P_{\mathcal{C}\leftarrow\mathcal{B}} = (P_{\mathcal{B}\leftarrow\mathcal{C}})^{-1}$, by Theorem 6.12(c). We find that

$$P_{\mathcal{C}\leftarrow\mathcal{B}} = (P_{\mathcal{B}\leftarrow\mathcal{C}})^{-1} = \begin{bmatrix} \frac{1}{2} & \frac{1}{2} & -\frac{1}{2} \\ -\frac{1}{2} & \frac{1}{2} & \frac{1}{2} \\ \frac{1}{2} & -\frac{1}{2} & \frac{1}{2} \end{bmatrix}$$

It now follows that

$$\begin{aligned} [p(x)]_{\mathcal{C}} &= P_{\mathcal{C}\leftarrow\mathcal{B}}[p(x)]_{\mathcal{B}} \\ &= \begin{bmatrix} \frac{1}{2} & \frac{1}{2} & -\frac{1}{2} \\ -\frac{1}{2} & \frac{1}{2} & \frac{1}{2} \\ \frac{1}{2} & -\frac{1}{2} & \frac{1}{2} \end{bmatrix}\begin{bmatrix} 1 \\ 2 \\ -1 \end{bmatrix} \\ &= \begin{bmatrix} 2 \\ 0 \\ -1 \end{bmatrix} \end{aligned}$$

which agrees with Example 6.37.

Remark If we do not need $P_{\mathcal{C}\leftarrow\mathcal{B}}$ explicitly, we can find $[p(x)]_{\mathcal{C}}$ from $[p(x)]_{\mathcal{B}}$ and $P_{\mathcal{B}\leftarrow\mathcal{C}}$ using Gaussian elimination. Row reduction produces

$$[P_{\mathcal{B}\leftarrow\mathcal{C}} \mid [p(x)]_{\mathcal{B}}] \longrightarrow [I \mid (P_{\mathcal{B}\leftarrow\mathcal{C}})^{-1}[p(x)]_{\mathcal{B}}] = [I \mid P_{\mathcal{C}\leftarrow\mathcal{B}}[p(x)]_{\mathcal{B}}] = [I \mid [p(x)]_{\mathcal{C}}]$$

(See the next section on using Gauss-Jordan elimination.)

It is worth repeating the observation in Example 6.46: Changing *to* a standard basis is easy. If $\mathcal{E}$ is the standard basis for a vector space V and $\mathcal{B}$ is any other basis, then the columns of $P_{\mathcal{E}\leftarrow\mathcal{B}}$ are the coordinate vectors of $\mathcal{B}$ with respect to $\mathcal{E}$, and these are usually "visible." We make use of this observation again in the next example.

Example 6.47

In M_{22}, let $\mathcal{B}$ be the basis $\{E_{11}, E_{21}, E_{12}, E_{22}\}$ and let $\mathcal{C}$ be the basis $\{A, B, C, D\}$, where

$$A = \begin{bmatrix} 1 & 0 \\ 0 & 0 \end{bmatrix}, \quad B = \begin{bmatrix} 1 & 1 \\ 0 & 0 \end{bmatrix}, \quad C = \begin{bmatrix} 1 & 1 \\ 1 & 0 \end{bmatrix}, \quad D = \begin{bmatrix} 1 & 1 \\ 1 & 1 \end{bmatrix}$$

Find the change-of-basis matrix $P_{\mathcal{C}\leftarrow\mathcal{B}}$ and verify that $[X]_{\mathcal{C}} = P_{\mathcal{C}\leftarrow\mathcal{B}}[X]_{\mathcal{B}}$ for $X = \begin{bmatrix} 1 & 2 \\ 3 & 4 \end{bmatrix}$.

Solution 1 To solve this problem directly, we must find the coordinate vectors of $\mathcal{B}$ with respect to $\mathcal{C}$. This involves solving four linear combination problems of the form $X = aA + bB + cC + dD$, where X is in $\mathcal{B}$ and we must find a, b, c, and d. However, here we are lucky, since we can find the required coefficients by inspection.

Clearly, $E_{11} = A$, $E_{21} = -B + C$, $E_{12} = -A + B$, and $E_{22} = -C + D$. Thus,

$$[E_{11}]_{\mathcal{C}} = \begin{bmatrix} 1 \\ 0 \\ 0 \\ 0 \end{bmatrix}, \quad [E_{21}]_{\mathcal{C}} = \begin{bmatrix} 0 \\ -1 \\ 1 \\ 0 \end{bmatrix}, \quad [E_{12}]_{\mathcal{C}} = \begin{bmatrix} -1 \\ 1 \\ 0 \\ 0 \end{bmatrix}, \quad [E_{22}]_{\mathcal{C}} = \begin{bmatrix} 0 \\ 0 \\ -1 \\ 1 \end{bmatrix}$$

so $$P_{\mathcal{C}\leftarrow\mathcal{B}} = [[E_{11}]_{\mathcal{C}} \; [E_{21}]_{\mathcal{C}} \; [E_{12}]_{\mathcal{C}} \; [E_{22}]_{\mathcal{C}}] = \begin{bmatrix} 1 & 0 & -1 & 0 \\ 0 & -1 & 1 & 0 \\ 0 & 1 & 0 & -1 \\ 0 & 0 & 0 & 1 \end{bmatrix}$$

If $X = \begin{bmatrix} 1 & 2 \\ 3 & 4 \end{bmatrix}$, then

$$[X]_{\mathcal{B}} = \begin{bmatrix} 1 \\ 3 \\ 2 \\ 4 \end{bmatrix}$$

and $$P_{\mathcal{C}\leftarrow\mathcal{B}}[X]_{\mathcal{B}} = \begin{bmatrix} 1 & 0 & -1 & 0 \\ 0 & -1 & 1 & 0 \\ 0 & 1 & 0 & -1 \\ 0 & 0 & 0 & 1 \end{bmatrix}\begin{bmatrix} 1 \\ 3 \\ 2 \\ 4 \end{bmatrix} = \begin{bmatrix} -1 \\ -1 \\ -1 \\ 4 \end{bmatrix}$$

This is the coordinate vector with respect to $\mathcal{C}$ of the matrix

$$\begin{aligned} -A - B - C + 4D &= -\begin{bmatrix} 1 & 0 \\ 0 & 0 \end{bmatrix} - \begin{bmatrix} 1 & 1 \\ 0 & 0 \end{bmatrix} - \begin{bmatrix} 1 & 1 \\ 1 & 0 \end{bmatrix} + 4\begin{bmatrix} 1 & 1 \\ 1 & 1 \end{bmatrix} \\ &= \begin{bmatrix} 1 & 2 \\ 3 & 4 \end{bmatrix} = X \end{aligned}$$

as it should be.

Solution 2 We can compute $P_{\mathcal{C}\leftarrow\mathcal{B}}$ in a different way, as follows. As you will be asked to prove in Exercise 21, if $\mathcal{E}$ is another basis for M_{22}, then $P_{\mathcal{C}\leftarrow\mathcal{B}} = P_{\mathcal{C}\leftarrow\mathcal{E}}P_{\mathcal{E}\leftarrow\mathcal{B}} = (P_{\mathcal{E}\leftarrow\mathcal{C}})^{-1}P_{\mathcal{E}\leftarrow\mathcal{B}}$. If $\mathcal{E}$ is the standard basis, then $P_{\mathcal{E}\leftarrow\mathcal{B}}$ and $P_{\mathcal{E}\leftarrow\mathcal{C}}$ can be found by inspection. We have

$$P_{\mathcal{E}\leftarrow\mathcal{B}} = \begin{bmatrix} 1 & 0 & 0 & 0 \\ 0 & 0 & 1 & 0 \\ 0 & 1 & 0 & 0 \\ 0 & 0 & 0 & 1 \end{bmatrix} \quad \text{and} \quad P_{\mathcal{E}\leftarrow\mathcal{C}} = \begin{bmatrix} 1 & 1 & 1 & 1 \\ 0 & 1 & 1 & 1 \\ 0 & 0 & 1 & 1 \\ 0 & 0 & 0 & 1 \end{bmatrix}$$

(Do you see why?) Therefore,

$$P_{\mathcal{C}\leftarrow\mathcal{B}} = (P_{\mathcal{E}\leftarrow\mathcal{C}})^{-1}P_{\mathcal{E}\leftarrow\mathcal{B}}$$

$$= \begin{bmatrix} 1 & 1 & 1 & 1 \\ 0 & 1 & 1 & 1 \\ 0 & 0 & 1 & 1 \\ 0 & 0 & 0 & 1 \end{bmatrix}^{-1} \begin{bmatrix} 1 & 0 & 0 & 0 \\ 0 & 0 & 1 & 0 \\ 0 & 1 & 0 & 0 \\ 0 & 0 & 0 & 1 \end{bmatrix}$$

$$= \begin{bmatrix} 1 & -1 & 0 & 0 \\ 0 & 1 & -1 & 0 \\ 0 & 0 & 1 & -1 \\ 0 & 0 & 0 & 1 \end{bmatrix} \begin{bmatrix} 1 & 0 & 0 & 0 \\ 0 & 0 & 1 & 0 \\ 0 & 1 & 0 & 0 \\ 0 & 0 & 0 & 1 \end{bmatrix}$$

$$= \begin{bmatrix} 1 & 0 & -1 & 0 \\ 0 & -1 & 1 & 0 \\ 0 & 1 & 0 & -1 \\ 0 & 0 & 0 & 1 \end{bmatrix}$$

which agrees with the first solution.

Remark The second method has the advantage of not requiring the computation of any linear combinations. It has the disadvantage of requiring that we find a matrix inverse. However, using a CAS will facilitate finding a matrix inverse, so in general the second method is preferable to the first. For certain problems, though, the first method may be just as easy to use. In any event, we are about to describe yet a third approach, which you may find best of all.

The Gauss-Jordan Method for Computing a Change-of-Basis Matrix

Finding the change-of-basis matrix to a standard basis is easy and can be done by inspection. Finding the change-of-basis matrix from a standard basis is almost as easy, but requires the calculation of a matrix inverse, as in Example 6.46. If we do it by hand, then (except for the 2×2 case) we will usually find the necessary inverse by Gauss-Jordan elimination. We now look at a modification of the Gauss-Jordan method that can be used to find the change-of-basis matrix between two nonstandard bases, as in Example 6.47.

Suppose $\mathcal{B} = \{\mathbf{u}_1, \ldots, \mathbf{u}_n\}$ and $\mathcal{C} = \{\mathbf{v}_1, \ldots, \mathbf{v}_n\}$ are bases for a vector space V and $P_{\mathcal{C}\leftarrow\mathcal{B}}$ is the change-of-basis matrix from $\mathcal{B}$ to $\mathcal{C}$. The ith column of P is

$$[\mathbf{u}_i]_{\mathcal{C}} = \begin{bmatrix} p_{1i} \\ \vdots \\ p_{ni} \end{bmatrix}$$

so $\mathbf{u}_i = p_{1i}\mathbf{v}_1 + \cdots + p_{ni}\mathbf{v}_n$. If $\mathcal{E}$ is any basis for V, then

$$[\mathbf{u}_i]_{\mathcal{E}} = [p_{1i}\mathbf{v}_1 + \cdots + p_{ni}\mathbf{v}_n]_{\mathcal{E}} = p_{1i}[\mathbf{v}_1]_{\mathcal{E}} + \cdots + p_{ni}[\mathbf{v}_n]_{\mathcal{E}}$$

This can be rewritten in matrix form as

$$[\,[\mathbf{v}_1]_{\mathcal{E}} \;\cdots\; [\mathbf{v}_n]_{\mathcal{E}}] \begin{bmatrix} p_{1i} \\ \vdots \\ p_{ni} \end{bmatrix} = [\mathbf{u}_i]_{\mathcal{E}}$$

which we can solve by applying Gauss-Jordan elimination to the augmented matrix

$$[[\mathbf{v}_1]_\mathcal{E} \;\cdots\; [\mathbf{v}_n]_\mathcal{E} \,|\, [\mathbf{u}_i]_\mathcal{E}]$$

There are n such systems of equations to be solved, one for each column of $P_{\mathcal{C}\leftarrow\mathcal{B}}$, but *the coefficient matrix* $[[\mathbf{v}_1]_\mathcal{E} \;\cdots\; [\mathbf{v}_n]_\mathcal{E}]$ *is the same in each case*. Hence, we can solve all the systems simultaneously by row reducing the $n \times 2n$ augmented matrix

$$[[\mathbf{v}_1]_\mathcal{E} \;\cdots\; [\mathbf{v}_n]_\mathcal{E} \,|\, [\mathbf{u}_1]_\mathcal{E} \;\cdots\; [\mathbf{u}_n]_\mathcal{E}] = [C \,|\, B]$$

Since $\{\mathbf{v}_1, \ldots, \mathbf{v}_n\}$ is linearly independent, so is $\{[\mathbf{v}_1]_\mathcal{E}, \ldots, [\mathbf{v}_n]_\mathcal{E}\}$, by Theorem 6.7. Therefore, the matrix C whose columns are $[\mathbf{v}_1]_\mathcal{E}, \ldots, [\mathbf{v}_n]_\mathcal{E}$ has the $n \times n$ identity matrix I for its reduced row echelon form, by the Fundamental Theorem. It follows that Gauss-Jordan elimination will necessarily produce

$$[C \,|\, B] \to [I \,|\, P]$$

where $P = P_{\mathcal{C}\leftarrow\mathcal{B}}$.

We have proved the following theorem.

Theorem 6.13 Let $\mathcal{B} = \{\mathbf{u}_1, \ldots, \mathbf{u}_n\}$ and $\mathcal{C} = \{\mathbf{v}_1, \ldots, \mathbf{v}_n\}$ be bases for a vector space V. Let $B = [[\mathbf{u}_1]_\mathcal{E} \;\cdots\; [\mathbf{u}_n]_\mathcal{E}]$ and $C = [[\mathbf{v}_1]_\mathcal{E} \ldots [\mathbf{v}_n]_\mathcal{E}]$, where $\mathcal{E}$ is any basis for V. Then row reduction applied to the $n \times 2n$ augmented matrix $[C \,|\, B]$ produces

$$[C \,|\, B] \to [I \,|\, P_{\mathcal{C}\leftarrow\mathcal{B}}]$$

If $\mathcal{E}$ is a standard basis, this method is particularly easy to use, since in that case $B = P_{\mathcal{E}\leftarrow\mathcal{B}}$ and $C = P_{\mathcal{E}\leftarrow\mathcal{C}}$. We illustrate this method by reworking the problem in Example 6.47.

Example 6.48 Rework Example 6.47 using the Gauss-Jordan method.

Solution Taking $\mathcal{E}$ to be the standard basis for M_{22}, we see that

$$B = P_{\mathcal{E}\leftarrow\mathcal{B}} = \begin{bmatrix} 1 & 0 & 0 & 0 \\ 0 & 0 & 1 & 0 \\ 0 & 1 & 0 & 0 \\ 0 & 0 & 0 & 1 \end{bmatrix} \quad \text{and} \quad C = P_{\mathcal{E}\leftarrow\mathcal{C}} = \begin{bmatrix} 1 & 1 & 1 & 1 \\ 0 & 1 & 1 & 1 \\ 0 & 0 & 1 & 1 \\ 0 & 0 & 0 & 1 \end{bmatrix}$$

Row reduction produces

$$[C \,|\, B] = \left[\begin{array}{cccc|cccc} 1 & 1 & 1 & 1 & 1 & 0 & 0 & 0 \\ 0 & 1 & 1 & 1 & 0 & 0 & 1 & 0 \\ 0 & 0 & 1 & 1 & 0 & 1 & 0 & 0 \\ 0 & 0 & 0 & 1 & 0 & 0 & 0 & 1 \end{array}\right] \longrightarrow \left[\begin{array}{cccc|cccc} 1 & 0 & 0 & 0 & 1 & 0 & -1 & 0 \\ 0 & 1 & 0 & 0 & 0 & -1 & 1 & 0 \\ 0 & 0 & 1 & 0 & 0 & 1 & 0 & -1 \\ 0 & 0 & 0 & 1 & 0 & 0 & 0 & 1 \end{array}\right]$$

(Verify this row reduction.) It follows that

$$P_{\mathcal{C}\leftarrow\mathcal{B}} = \begin{bmatrix} 1 & 0 & -1 & 0 \\ 0 & -1 & 1 & 0 \\ 0 & 1 & 0 & -1 \\ 0 & 0 & 0 & 1 \end{bmatrix}$$

as we found before.

Exercises 6.3

In Exercises 1–4:
(a) Find the coordinate vectors $[\mathbf{x}]_{\mathcal{B}}$ *and* $[\mathbf{x}]_{\mathcal{C}}$ *of* $\mathbf{x}$ *with respect to the bases* $\mathcal{B}$ *and* $\mathcal{C}$*, respectively.*
(b) Find the change-of-basis matrix $P_{\mathcal{C}\leftarrow\mathcal{B}}$ *from* $\mathcal{B}$ *to* $\mathcal{C}$*.*
(c) Use your answer to part (b) to compute $[\mathbf{x}]_{\mathcal{C}}$*, and compare your answer with the one found in part (a).*
(d) Find the change-of-basis matrix $P_{\mathcal{B}\leftarrow\mathcal{C}}$ *from* $\mathcal{C}$ *to* $\mathcal{B}$*.*
(e) Use your answers to parts (c) and (d) to compute $[\mathbf{x}]_{\mathcal{B}}$*, and compare your answer with the one found in part (a).*

1. $\mathbf{x} = \begin{bmatrix} 2 \\ 3 \end{bmatrix}, \mathcal{B} = \left\{ \begin{bmatrix} 1 \\ 0 \end{bmatrix}, \begin{bmatrix} 0 \\ 1 \end{bmatrix} \right\},$
$\mathcal{C} = \left\{ \begin{bmatrix} 1 \\ 1 \end{bmatrix}, \begin{bmatrix} 1 \\ -1 \end{bmatrix} \right\}$ in $\mathbb{R}^2$

2. $\mathbf{x} = \begin{bmatrix} 4 \\ -1 \end{bmatrix}, \mathcal{B} = \left\{ \begin{bmatrix} 1 \\ 0 \end{bmatrix}, \begin{bmatrix} 1 \\ 1 \end{bmatrix} \right\},$
$\mathcal{C} = \left\{ \begin{bmatrix} 0 \\ 1 \end{bmatrix}, \begin{bmatrix} 2 \\ 3 \end{bmatrix} \right\}$ in $\mathbb{R}^2$

3. $\mathbf{x} = \begin{bmatrix} 1 \\ 0 \\ -1 \end{bmatrix}, \mathcal{B} = \left\{ \begin{bmatrix} 1 \\ 0 \\ 0 \end{bmatrix}, \begin{bmatrix} 0 \\ 1 \\ 0 \end{bmatrix}, \begin{bmatrix} 0 \\ 0 \\ 1 \end{bmatrix} \right\},$
$\mathcal{C} = \left\{ \begin{bmatrix} 1 \\ 1 \\ 1 \end{bmatrix}, \begin{bmatrix} 0 \\ 1 \\ 1 \end{bmatrix}, \begin{bmatrix} 0 \\ 0 \\ 1 \end{bmatrix} \right\}$ in $\mathbb{R}^3$

4. $\mathbf{x} = \begin{bmatrix} 3 \\ 1 \\ 5 \end{bmatrix}, \mathcal{B} = \left\{ \begin{bmatrix} 0 \\ 1 \\ 0 \end{bmatrix}, \begin{bmatrix} 0 \\ 0 \\ 1 \end{bmatrix}, \begin{bmatrix} 1 \\ 0 \\ 0 \end{bmatrix} \right\},$
$\mathcal{C} = \left\{ \begin{bmatrix} 1 \\ 1 \\ 0 \end{bmatrix}, \begin{bmatrix} 0 \\ 1 \\ 1 \end{bmatrix}, \begin{bmatrix} 1 \\ 0 \\ 1 \end{bmatrix} \right\}$ in $\mathbb{R}^3$

In Exercises 5–8, follow the instructions for Exercises 1–4 using $p(x)$ *instead of* $\mathbf{x}$*.*

5. $p(x) = 2 - x, \mathcal{B} = \{1, x\}, \mathcal{C} = \{x, 1 + x\}$ in $\mathcal{P}_1$

6. $p(x) = 1 + 3x, \mathcal{B} = \{1 + x, 1 - x\},$
$\mathcal{C} = \{2x, 4\}$ in $\mathcal{P}_1$

7. $p(x) = 1 + x^2, \mathcal{B} = \{1 + x + x^2, x + x^2, x^2\},$
$\mathcal{C} = \{1, x, x^2\}$ in $\mathcal{P}_2$

8. $p(x) = 4 - 2x - x^2, \mathcal{B} = \{x, 1 + x^2, x + x^2\},$
$\mathcal{C} = \{1, 1 + x, x^2\}$ in $\mathcal{P}_2$

In Exercises 9 and 10, follow the instructions for Exercises 1–4 using A instead of $\mathbf{x}$*.*

9. $A = \begin{bmatrix} 4 & 2 \\ 0 & -1 \end{bmatrix}$, $\mathcal{B}$ = the standard basis,
$\mathcal{C} = \left\{ \begin{bmatrix} 1 & 2 \\ 0 & -1 \end{bmatrix}, \begin{bmatrix} 2 & 1 \\ 1 & 0 \end{bmatrix}, \begin{bmatrix} 1 & 1 \\ 0 & 1 \end{bmatrix}, \begin{bmatrix} 1 & 0 \\ 0 & 1 \end{bmatrix} \right\}$ in M_{22}

10. $A = \begin{bmatrix} 1 & 1 \\ 1 & 1 \end{bmatrix},$
$\mathcal{B} = \left\{ \begin{bmatrix} 1 & 0 \\ 0 & 1 \end{bmatrix}, \begin{bmatrix} 0 & 1 \\ 1 & 0 \end{bmatrix}, \begin{bmatrix} 1 & 1 \\ 0 & 0 \end{bmatrix}, \begin{bmatrix} 1 & 0 \\ 1 & 0 \end{bmatrix} \right\},$
$\mathcal{C} = \left\{ \begin{bmatrix} 1 & 1 \\ 0 & 1 \end{bmatrix}, \begin{bmatrix} 1 & 1 \\ 1 & 0 \end{bmatrix}, \begin{bmatrix} 1 & 0 \\ 1 & 1 \end{bmatrix}, \begin{bmatrix} 0 & 1 \\ 1 & 1 \end{bmatrix} \right\}$ in M_{22}

In Exercises 11 and 12, follow the instructions for Exercises 1–4 using $f(x)$ *instead of* $\mathbf{x}$*.*

11. $f(x) = 2 \sin x - 3 \cos x, \mathcal{B} = \{\sin x + \cos x, \cos x\},$
$\mathcal{C} = \{\sin x + \cos x, \sin x - \cos x\}$ in $\operatorname{span}(\sin x, \cos x)$

12. $f(x) = \sin x, \mathcal{B} = \{\sin x + \cos x, \cos x\},$
$\mathcal{C} = \{\cos x - \sin x, \sin x + \cos x\}$ in $\operatorname{span}(\sin x, \cos x)$

13. Rotate the xy-axes in the plane counterclockwise through an angle $\theta = 60°$ to obtain new $x'y'$-axes. Use the methods of this section to find (a) the $x'y'$-coordinates of the point whose xy-coordinates are $(3, 2)$ and (b) the xy-coordinates of the point whose $x'y'$-coordinates are $(4, -4)$.

14. Repeat Exercise 13 with $\theta = 135°$.

15. Let $\mathcal{B}$ and $\mathcal{C}$ be bases for $\mathbb{R}^2$. If $\mathcal{C} = \left\{ \begin{bmatrix} 1 \\ 2 \end{bmatrix}, \begin{bmatrix} 2 \\ 3 \end{bmatrix} \right\}$ and the change-of-basis matrix from $\mathcal{B}$ to $\mathcal{C}$ is

$$P_{\mathcal{C}\leftarrow\mathcal{B}} = \begin{bmatrix} 1 & -1 \\ -1 & 2 \end{bmatrix}$$

find $\mathcal{B}$.

16. Let $\mathcal{B}$ and $\mathcal{C}$ be bases for $\mathcal{P}_2$. If $\mathcal{B} = \{x, 1 + x, 1 - x + x^2\}$ and the change-of-basis matrix from $\mathcal{B}$ to $\mathcal{C}$ is

$$P_{\mathcal{C}\leftarrow\mathcal{B}} = \begin{bmatrix} 1 & 0 & 0 \\ 0 & 2 & 1 \\ -1 & 1 & 1 \end{bmatrix}$$

find $\mathcal{C}$.

In calculus, you learn that a ***Taylor polynomial of degree n about a*** *is a polynomial of the form*

$$p(x) = a_0 + a_1(x - a) + a_2(x - a)^2 + \cdots + a_n(x - a)^n$$

where $a_n \neq 0$. In other words, it is a polynomial that has been expanded in terms of powers of $x - a$ instead of powers of x. Taylor polynomials are very useful for approximating functions that are "well behaved" near $x = a$.

The set $\mathcal{B} = \{1, x - a, (x - a)^2, \ldots, (x - a)^n\}$ is a basis for $\mathcal{P}_n$ for any real number a. (Do you see a quick way to show this? Try using Theorem 6.7.) This fact allows us to use the techniques of this section to rewrite a polynomial as a Taylor polynomial about a given a.

17. Express $p(x) = 1 + 2x - 5x^2$ as a Taylor polynomial about $a = 1$.

18. Express $p(x) = 1 + 2x - 5x^2$ as a Taylor polynomial about $a = -2$.

19. Express $p(x) = x^3$ as a Taylor polynomial about $a = -1$.

20. Express $p(x) = x^3$ as a Taylor polynomial about $a = \frac{1}{2}$.

21. Let $\mathcal{B}$, $\mathcal{C}$, and $\mathcal{D}$ be bases for a finite-dimensional vector space V. Prove that

$$P_{\mathcal{D}\leftarrow\mathcal{C}}P_{\mathcal{C}\leftarrow\mathcal{B}} = P_{\mathcal{D}\leftarrow\mathcal{B}}$$

22. Let V be an n-dimensional vector space with basis $\mathcal{B} = \{\mathbf{v}_1, \ldots, \mathbf{v}_n\}$. Let P be an invertible $n \times n$ matrix and set

$$\mathbf{u}_i = p_{1i}\mathbf{v}_1 + \cdots + p_{ni}\mathbf{v}_n$$

for $i = 1, \ldots, n$. Prove that $\mathcal{C} = \{\mathbf{u}_1, \ldots, \mathbf{u}_n\}$ is a basis for V and show that $P = P_{\mathcal{B}\leftarrow\mathcal{C}}$.

6.4 Linear Transformations

We encountered linear transformations in Section 3.6 in the context of matrix transformations from $\mathbb{R}^n$ to $\mathbb{R}^m$. In this section, we extend this concept to linear transformations between arbitrary vector spaces.

Definition A ***linear transformation*** from a vector space V to a vector space W is a mapping $T: V \to W$ such that, for all $\mathbf{u}$ and $\mathbf{v}$ in V and for all scalars c,

1. $T(\mathbf{u} + \mathbf{v}) = T(\mathbf{u}) + T(\mathbf{v})$
2. $T(c\mathbf{u}) = cT(\mathbf{u})$

It is straightforward to show that this definition is equivalent to the requirement that T preserve all linear combinations. That is,

$T: V \to W$ is a linear transformation if and only if

$$T(c_1\mathbf{v}_1 + c_2\mathbf{v}_2 + \cdots + c_k\mathbf{v}_k) = c_1T(\mathbf{v}_1) + c_2T(\mathbf{v}_2) + \cdots + c_kT(\mathbf{v}_k)$$

for all $\mathbf{v}_1, \ldots, \mathbf{v}_k$ in V and scalars $c_1, \ldots, c_k$.

Example 6.49

Every matrix transformation is a linear transformation. That is, if A is an $m \times n$ matrix, then the transformation $T_A: \mathbb{R}^n \to \mathbb{R}^m$ defined by

$$T_A(\mathbf{x}) = A\mathbf{x} \quad \text{for } \mathbf{x} \text{ in } \mathbb{R}^n$$

is a linear transformation. This is a restatement of Theorem 3.30.

Example 6.50

Define $T: M_{nn} \to M_{nn}$ by $T(A) = A^T$. Show that T is a linear transformation.

Solution We check that, for A and B in M_{nn} and scalars c,

$$T(A + B) = (A + B)^T = A^T + B^T = T(A) + T(B)$$

and

$$T(cA) = (cA)^T = cA^T = cT(A)$$

Therefore, T is a linear transformation.

Example 6.51

Let D be the ***differential operator*** $D: \mathscr{D} \to \mathscr{F}$ defined by $D(f) = f'$. Show that D is a linear transformation.

Solution Let f and g be differentiable functions and let c be a scalar. Then, from calculus, we know that

$$D(f + g) = (f + g)' = f' + g' = D(f) + D(g)$$

and

$$D(cf) = (cf)' = cf' = cD(f)$$

Hence, D is a linear transformation.

In calculus, you learn that every continuous function on $[a, b]$ is integrable. The next example shows that integration is a linear transformation.

Example 6.52

Define $S: \mathscr{C}[a, b] \to \mathbb{R}$ by $S(f) = \int_a^b f(x)\,dx$. Show that S is a linear transformation.

Solution Let f and g be in $\mathscr{C}[a, b]$. Then

$$\begin{aligned} S(f + g) &= \int_a^b (f + g)(x)\,dx \\ &= \int_a^b (f(x) + g(x))\,dx \\ &= \int_a^b f(x)\,dx + \int_a^b g(x)\,dx \\ &= S(f) + S(g) \end{aligned}$$

and

$$\begin{aligned} S(cf) &= \int_a^b (cf)(x)\,dx \\ &= \int_a^b cf(x)\,dx \\ &= c\int_a^b f(x)\,dx \\ &= cS(f) \end{aligned}$$

It follows that S is linear.

Example 6.53

Show that none of the following transformations is linear:

(a) $T: M_{22} \to \mathbb{R}$ defined by $T(A) = \det A$

(b) $T: \mathbb{R} \to \mathbb{R}$ defined by $T(x) = 2^x$

(c) $T: \mathbb{R} \to \mathbb{R}$ defined by $T(x) = x + 1$

Solution In each case, we give a specific counterexample to show that one of the properties of a linear transformation fails to hold.

(a) Let $A = \begin{bmatrix} 1 & 0 \\ 0 & 0 \end{bmatrix}$ and $B = \begin{bmatrix} 0 & 0 \\ 0 & 1 \end{bmatrix}$. Then $A + B = \begin{bmatrix} 1 & 0 \\ 0 & 1 \end{bmatrix}$, so

$$T(A + B) = \det(A + B) = \begin{vmatrix} 1 & 0 \\ 0 & 1 \end{vmatrix} = 1$$

But

$$T(A) + T(B) = \det A + \det B = \begin{vmatrix} 1 & 0 \\ 0 & 0 \end{vmatrix} + \begin{vmatrix} 0 & 0 \\ 0 & 1 \end{vmatrix} = 0 + 0 = 0$$

so $T(A + B) \neq T(A) + T(B)$ and T is not linear.

(b) Let $x = 1$ and $y = 2$. Then

$$T(x + y) = T(3) = 2^3 = 8 \neq 6 = 2^1 + 2^2 = T(x) + T(y)$$

so T is not linear.

(c) Let $x = 1$ and $y = 2$. Then

$$T(x + y) = T(3) = 3 + 1 = 4 \neq 5 = (1 + 1) + (2 + 1) = T(x) + T(y)$$

Therefore, T is not linear.

Remark Example 6.53(c) shows that you need to be careful when you encounter the word "linear." As a *function*, $T(x) = x + 1$ is linear, since its graph is a straight line. However, it is not a *linear transformation* from the vector space $\mathbb{R}$ to itself, since it fails to satisfy the definition. (Which linear functions from $\mathbb{R}$ to $\mathbb{R}$ will also be linear transformations?)

There are two special linear transformations that deserve to be singled out.

Example 6.54

(a) For any vector spaces V and W, the transformation $T_0 : V \to W$ that maps every vector in V to the zero vector in W is called the ***zero transformation.*** That is,

$$T_0(\mathbf{v}) = \mathbf{0} \quad \text{for all } \mathbf{v} \text{ in } V$$

(b) For any vector space V, the transformation $I: V \to V$ that maps every vector in V to itself is called the ***identity transformation.*** That is,

$$I(\mathbf{v}) = \mathbf{v} \quad \text{for all } \mathbf{v} \text{ in } V$$

(If it is important to identify the vector space V, we may write I_V for clarity.) The proofs that the zero and identity transformations are linear are left as easy exercises.

Properties of Linear Transformations

In Chapter 3, all linear transformations were matrix transformations, and their properties were directly related to properties of the matrices involved. The following theorem is easy to prove for matrix transformations. (Do it!) The full proof for linear transformations in general takes a bit more care, but it is still straightforward.

Theorem 6.14 Let $T: V \to W$ be a linear transformation. Then:

a. $T(\mathbf{0}) = \mathbf{0}$
b. $T(-\mathbf{v}) = -T(\mathbf{v})$ for all $\mathbf{v}$ in V.
c. $T(\mathbf{u} - \mathbf{v}) = T(\mathbf{u}) - T(\mathbf{v})$ for all $\mathbf{u}$ and $\mathbf{v}$ in V.

Proof We prove properties (a) and (c) and leave the proof of property (b) for Exercise 21.

(a) Let $\mathbf{v}$ be any vector in V. Then $T(\mathbf{0}) = T(0\mathbf{v}) = 0T(\mathbf{v}) = \mathbf{0}$, as required. (Can you give a reason for each step?)

(c) $T(\mathbf{u} - \mathbf{v}) = T(\mathbf{u} + (-1)\mathbf{v}) = T(\mathbf{u}) + (-1)T(\mathbf{v}) = T(\mathbf{u}) - T(\mathbf{v})$

Remark Property (a) can be useful in showing that certain transformations are *not* linear. As an illustration, consider Example 6.53(b). If $T(x) = 2^x$, then $T(0) = 2^0 = 1 \neq 0$, so T is not linear, by Theorem 6.14(a). Be warned, however, that there are lots of transformations that *do* map the zero vector to the zero vector but that are still *not* linear. Example 6.53(a) is a case in point: The zero vector is the 2×2 zero matrix O, so $T(O) = \det O = 0$, but we have seen that $T(A) = \det A$ is not linear.

The most important property of a linear transformation $T: V \to W$ is that T is completely determined by its effect on a basis for V. The next example shows what this means.

Example 6.55 Suppose T is a linear transformation from $\mathbb{R}^2$ to $\mathscr{P}_2$ such that

$$T\begin{bmatrix}1\\1\end{bmatrix} = 2 - 3x + x^2 \quad \text{and} \quad T\begin{bmatrix}2\\3\end{bmatrix} = 1 - x^2$$

Find $T\begin{bmatrix}-1\\2\end{bmatrix}$ and $T\begin{bmatrix}a\\b\end{bmatrix}$.

Solution Since $\mathcal{B} = \left\{\begin{bmatrix}1\\1\end{bmatrix}, \begin{bmatrix}2\\3\end{bmatrix}\right\}$ is a basis for $\mathbb{R}^2$ (why?), every vector in $\mathbb{R}^2$ is in span($\mathcal{B}$). Solving

$$c_1\begin{bmatrix}1\\1\end{bmatrix} + c_2\begin{bmatrix}2\\3\end{bmatrix} = \begin{bmatrix}-1\\2\end{bmatrix}$$

we find that $c_1 = -7$ and $c_2 = 3$. Therefore,

$$\begin{aligned} T\begin{bmatrix} -1 \\ 2 \end{bmatrix} &= T\left(-7\begin{bmatrix} 1 \\ 1 \end{bmatrix} + 3\begin{bmatrix} 2 \\ 3 \end{bmatrix}\right) \\ &= -7T\begin{bmatrix} 1 \\ 1 \end{bmatrix} + 3T\begin{bmatrix} 2 \\ 3 \end{bmatrix} \\ &= -7(2 - 3x + x^2) + 3(1 - x^2) \\ &= -11 + 21x - 10x^2 \end{aligned}$$

Similarly, we discover that

$$\begin{bmatrix} a \\ b \end{bmatrix} = (3a - 2b)\begin{bmatrix} 1 \\ 1 \end{bmatrix} + (b - a)\begin{bmatrix} 2 \\ 3 \end{bmatrix}$$

so

$$\begin{aligned} T\begin{bmatrix} a \\ b \end{bmatrix} &= T\left((3a - 2b)\begin{bmatrix} 1 \\ 1 \end{bmatrix} + (b - a)\begin{bmatrix} 2 \\ 3 \end{bmatrix}\right) \\ &= (3a - 2b)T\begin{bmatrix} 1 \\ 1 \end{bmatrix} + (b - a)T\begin{bmatrix} 2 \\ 3 \end{bmatrix} \\ &= (3a - 2b)(2 - 3x + x^2) + (b - a)(1 - x^2) \\ &= (5a - 3b) + (-9a + 6b)x + (4a - 3b)x^2 \end{aligned}$$

(Note that by setting $a = -1$ and $b = 2$, we recover the solution $T\begin{bmatrix} -1 \\ 2 \end{bmatrix} = -11 + 21x - 10x^2$.)

The proof of the general theorem is quite straightforward.

Theorem 6.15 Let $T: V \to W$ be a linear transformation and let $\mathcal{B} = \{\mathbf{v}_1, \ldots, \mathbf{v}_n\}$ be a spanning set for V. Then $T(\mathcal{B}) = \{T(\mathbf{v}_1), \ldots, T(\mathbf{v}_n)\}$ spans the range of T.

Proof The range of T is the set of all vectors in W that are of the form $T(\mathbf{v})$, where $\mathbf{v}$ is in V. Let $T(\mathbf{v})$ be in the range of T. Since $\mathcal{B}$ spans V, there are scalars $c_1, \ldots, c_n$ such that

$$\mathbf{v} = c_1\mathbf{v}_1 + \cdots + c_n\mathbf{v}_n$$

Applying T and using the fact that it is a linear transformation, we see that

$$T(\mathbf{v}) = T(c_1\mathbf{v}_1 + \cdots + c_n\mathbf{v}_n) = c_1T(\mathbf{v}_1) + \cdots + c_nT(\mathbf{v}_n)$$

In other words, $T(\mathbf{v})$ is in $\text{span}(T(\mathcal{B}))$, as required.

Theorem 6.15 applies, in particular, when $\mathcal{B}$ is a basis for V. You might guess that, in this case, $T(\mathcal{B})$ would then be a basis for the range of T. Unfortunately, this is not always the case. We will address this issue in Section 6.5.

Composition of Linear Transformations

In Section 3.6, we defined the composition of matrix transformations. The definition extends to general linear transformations in an obvious way.

$S \circ T$ is read "S of T."

Definition If $T : U \to V$ and $S : V \to W$ are linear transformations, then the ***composition of S with T*** is the mapping $S \circ T$, defined by

$$(S \circ T)(\mathbf{u}) = S(T(\mathbf{u}))$$

where $\mathbf{u}$ is in U.

Observe that $S \circ T$ is a mapping from U to W (see Figure 6.6). Notice also that for the definition to make sense, the range of T must be contained in the domain of S.

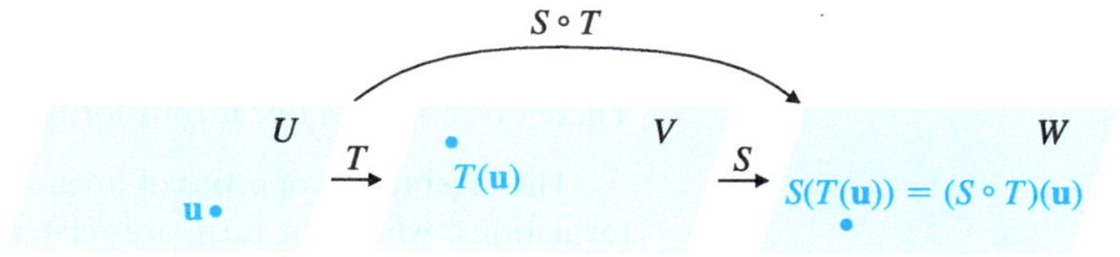

Figure 6.6
Composition of linear transformations

Example 6.56

Let $T : \mathbb{R}^2 \to \mathscr{P}_1$ and $S : \mathscr{P}_1 \to \mathscr{P}_2$ be the linear transformations defined by

$$T\begin{bmatrix} a \\ b \end{bmatrix} = a + (a + b)x \quad \text{and} \quad S(p(x)) = xp(x)$$

Find $(S \circ T)\begin{bmatrix} 3 \\ -2 \end{bmatrix}$ and $(S \circ T)\begin{bmatrix} a \\ b \end{bmatrix}$.

Solution We compute

$$(S \circ T)\begin{bmatrix} 3 \\ -2 \end{bmatrix} = S\left(T\begin{bmatrix} 3 \\ -2 \end{bmatrix}\right) = S(3 + (3 - 2)x) = S(3 + x) = x(3 + x)$$
$$= 3x + x^2$$

and

$$(S \circ T)\begin{bmatrix} a \\ b \end{bmatrix} = S\left(T\begin{bmatrix} a \\ b \end{bmatrix}\right) = S(a + (a + b)x) = x(a + (a + b)x)$$
$$= ax + (a + b)x^2$$

Chapter 3 showed that the composition of two matrix transformations was another matrix transformation. In general, we have the following theorem.

Theorem 6.16 If $T : U \to V$ and $S : V \to W$ are linear transformations, then $S \circ T : U \to W$ is a linear transformation.

Proof Let $\mathbf{u}$ and $\mathbf{v}$ be in U and let c be a scalar. Then

$$\begin{aligned}(S \circ T)(\mathbf{u} + \mathbf{v}) &= S(T(\mathbf{u} + \mathbf{v})) \\ &= S(T(\mathbf{u}) + T(\mathbf{v})) && \text{since } T \text{ is linear} \\ &= S(T(\mathbf{u})) + S(T(\mathbf{v})) && \text{since } S \text{ is linear} \\ &= (S \circ T)(\mathbf{u}) + (S \circ T)(\mathbf{v})\end{aligned}$$

and

$$\begin{aligned}(S \circ T)(c\mathbf{u}) &= S(T(c\mathbf{u})) \\ &= S(cT(\mathbf{u})) && \text{since } T \text{ is linear} \\ &= cS(T(\mathbf{u})) && \text{since } S \text{ is linear} \\ &= c(S \circ T)(\mathbf{u})\end{aligned}$$

Therefore, $S \circ T$ is a linear transformation.

The algebraic properties of linear transformations mirror those of matrix transformations, which, in turn, are related to the algebraic properties of matrices. For example, composition of linear transformations is associative. That is, if R, S, and T are linear transformations, then

$$R \circ (S \circ T) = (R \circ S) \circ T$$

provided these compositions make sense. The proof of this property is identical to that given in Section 3.6.

The next example gives another useful (but not surprising) property of linear transformations.

Example 6.57

Let $S: U \to V$ and $T: V \to W$ be linear transformations and let $I: V \to V$ be the identity transformation. Then for every $\mathbf{v}$ in V, we have

$$(T \circ I)(\mathbf{v}) = T(I(\mathbf{v})) = T(\mathbf{v})$$

Since $T \circ I$ and T have the same value at every $\mathbf{v}$ in their domain, it follows that $T \circ I = T$. Similarly, $I \circ S = S$.

Remark The method of Example 6.57 is worth noting. Suppose we want to show that two linear transformations T_1 and T_2 (both from V to W) are equal. It suffices to show that $T_1(\mathbf{v}) = T_2(\mathbf{v})$ for every $\mathbf{v}$ in V.

Further properties of linear transformations are explored in the exercises.

Inverses of Linear Transformations

Definition A linear transformation $T: V \to W$ is ***invertible*** if there is a linear transformation $T': W \to V$ such that

$$T' \circ T = I_V \quad \text{and} \quad T \circ T' = I_W$$

In this case, T' is called an ***inverse*** for T.

Remarks

- The domain V and codomain W of T do not have to be the same, as they do in the case of invertible matrix transformations. However, we will see in the next section that V and W must be very closely related.
- The requirement that T' be linear could have been omitted from this definition. For, as we will see in Theorem 6.24, if T' is *any* mapping from W to V such that $T' \circ T = I_V$ and $T \circ T' = I_W$, then T' is forced to be linear as well.
- If T' is an inverse for T, then the definition implies that T is an inverse for T'. Hence, T' is invertible too.

Example 6.58

Verify that the mappings $T: \mathbb{R}^2 \to \mathcal{P}_1$ and $T': \mathcal{P}_1 \to \mathbb{R}^2$ defined by

$$T\begin{bmatrix} a \\ b \end{bmatrix} = a + (a + b)x \quad \text{and} \quad T'(c + dx) = \begin{bmatrix} c \\ d - c \end{bmatrix}$$

are inverses.

Solution We compute

$$(T' \circ T)\begin{bmatrix} a \\ b \end{bmatrix} = T'\left(T\begin{bmatrix} a \\ b \end{bmatrix}\right) = T'(a + (a + b)x) = \begin{bmatrix} a \\ (a + b) - a \end{bmatrix} = \begin{bmatrix} a \\ b \end{bmatrix}$$

and

$$(T \circ T')(c + dx) = T(T'(c + dx)) = T\begin{bmatrix} c \\ d - c \end{bmatrix} = c + (c + (d - c))x = c + dx$$

Hence, $T' \circ T = I_{\mathbb{R}^2}$ and $T \circ T' = I_{\mathcal{P}_1}$. Therefore, T and T' are inverses of each other.

As was the case for invertible matrices, inverses of linear transformations are unique if they exist. The following theorem is the analogue of Theorem 3.6.

Theorem 6.17 If T is an invertible linear transformation, then its inverse is unique.

Proof The proof is the same as that of Theorem 3.6, with products of matrices replaced by compositions of linear transformations. (You are asked to complete this proof in Exercise 31.)

Thanks to Theorem 6.17, if T is invertible, we can refer to *the* inverse of T. It will be denoted by T^{-1} (pronounced "T inverse"). In the next two sections, we will address the issue of determining when a given linear transformation is invertible and finding its inverse when it exists.

Exercises 6.4

In Exercises 1–12, determine whether T is a linear transformation.

1. $T: M_{22} \to M_{22}$ defined by

$$T\begin{bmatrix} a & b \\ c & d \end{bmatrix} = \begin{bmatrix} a+b & 0 \\ 0 & c+d \end{bmatrix}$$

2. $T: M_{22} \to M_{22}$ defined by

$$T\begin{bmatrix} w & x \\ y & z \end{bmatrix} = \begin{bmatrix} 1 & w-z \\ x-y & 1 \end{bmatrix}$$

3. $T: M_{nn} \to M_{nn}$ defined by $T(A) = AB$, where B is a fixed $n \times n$ matrix

4. $T: M_{nn} \to M_{nn}$ defined by $T(A) = AB - BA$, where B is a fixed $n \times n$ matrix

5. $T: M_{nn} \to \mathbb{R}$ defined by $T(A) = \operatorname{tr}(A)$

6. $T: M_{nn} \to \mathbb{R}$ defined by $T(A) = a_{11}a_{22}\cdots a_{nn}$

7. $T: M_{nn} \to \mathbb{R}$ defined by $T(A) = \operatorname{rank}(A)$

8. $T: \mathcal{P}_2 \to \mathcal{P}_2$ defined by $T(a + bx + cx^2) = (a+1) + (b+1)x + (c+1)x^2$

9. $T: \mathcal{P}_2 \to \mathcal{P}_2$ defined by $T(a + bx + cx^2) = a + b(x+1) + b(x+1)^2$

10. $T: \mathcal{F} \to \mathcal{F}$ defined by $T(f) = f(x^2)$

11. $T: \mathcal{F} \to \mathcal{F}$ defined by $T(f) = (f(x))^2$

12. $T: \mathcal{F} \to \mathbb{R}$ defined by $T(f) = f(c)$, where c is a fixed scalar

13. Show that the transformations S and T in Example 6.56 are both linear.

14. Let $T: \mathbb{R}^2 \to \mathbb{R}^3$ be a linear transformation for which

$$T\begin{bmatrix} 1 \\ 0 \end{bmatrix} = \begin{bmatrix} 1 \\ 2 \\ -1 \end{bmatrix} \quad \text{and} \quad T\begin{bmatrix} 0 \\ 1 \end{bmatrix} = \begin{bmatrix} 3 \\ 0 \\ 4 \end{bmatrix}$$

Find $T\begin{bmatrix} 5 \\ 2 \end{bmatrix}$ and $T\begin{bmatrix} a \\ b \end{bmatrix}$.

15. Let $T: \mathbb{R}^2 \to \mathcal{P}_2$ be a linear transformation for which

$$T\begin{bmatrix} 1 \\ 1 \end{bmatrix} = 1 - 2x \quad \text{and} \quad T\begin{bmatrix} 3 \\ -1 \end{bmatrix} = x + 2x^2$$

Find $T\begin{bmatrix} -7 \\ 9 \end{bmatrix}$ and $T\begin{bmatrix} a \\ b \end{bmatrix}$.

16. Let $T: \mathcal{P}_2 \to \mathcal{P}_2$ be a linear transformation for which

$$T(1) = 3 - 2x, \quad T(x) = 4x - x^2, \quad \text{and} \quad T(x^2) = 2 + 2x^2$$

Find $T(6 + x - 4x^2)$ and $T(a + bx + cx^2)$.

17. Let $T: \mathcal{P}_2 \to \mathcal{P}_2$ be a linear transformation for which

$$T(1 + x) = 1 + x^2, \quad T(x + x^2) = x - x^2,$$
$$T(1 + x^2) = 1 + x + x^2$$

Find $T(4 - x + 3x^2)$ and $T(a + bx + cx^2)$.

18. Let $T: M_{22} \to \mathbb{R}$ be a linear transformation for which

$$T\begin{bmatrix} 1 & 0 \\ 0 & 0 \end{bmatrix} = 1, \quad T\begin{bmatrix} 1 & 1 \\ 0 & 0 \end{bmatrix} = 2,$$
$$T\begin{bmatrix} 1 & 1 \\ 1 & 0 \end{bmatrix} = 3, \quad T\begin{bmatrix} 1 & 1 \\ 1 & 1 \end{bmatrix} = 4$$

Find $T\begin{bmatrix} 1 & 3 \\ 4 & 2 \end{bmatrix}$ and $T\begin{bmatrix} a & b \\ c & d \end{bmatrix}$.

19. Let $T: M_{22} \to \mathbb{R}$ be a linear transformation. Show that there are scalars a, b, c, and d such that

$$T\begin{bmatrix} w & x \\ y & z \end{bmatrix} = aw + bx + cy + dz$$

for all $\begin{bmatrix} w & x \\ y & z \end{bmatrix}$ in M_{22}.

20. Show that there is no linear transformation $T: \mathbb{R}^3 \to \mathcal{P}_2$ such that

$$T\begin{bmatrix} 2 \\ 1 \\ 0 \end{bmatrix} = 1 + x, \quad T\begin{bmatrix} 3 \\ 0 \\ 2 \end{bmatrix} = 2 - x + x^2,$$
$$T\begin{bmatrix} 0 \\ 6 \\ -8 \end{bmatrix} = -2 + 2x^2$$

21. Prove Theorem 6.14(b).

22. Let $\{\mathbf{v}_1, \ldots, \mathbf{v}_n\}$ be a basis for a vector space V and let $T: V \to V$ be a linear transformation. Prove that if $T(\mathbf{v}_1) = \mathbf{v}_1, T(\mathbf{v}_2) = \mathbf{v}_2 \ldots, T(\mathbf{v}_n) = \mathbf{v}_n$, then T is the identity transformation on V.

23. Let $T: \mathcal{P}_n \to \mathcal{P}_n$ be a linear transformation such that $T(x^k) = kx^{k-1}$ for $k = 0, 1, \ldots, n$. Show that T must be the differential operator D.

24. Let $\mathbf{v}_1, \ldots, \mathbf{v}_n$ be vectors in a vector space V and let $T: V \to W$ be a linear transformation.

(a) If $\{T(\mathbf{v}_1), \ldots, T(\mathbf{v}_n)\}$ is linearly independent in W, show that $\{\mathbf{v}_1, \ldots, \mathbf{v}_n\}$ is linearly independent in V.

(b) Show that the converse of part (a) is false. That is, it is not necessarily true that if $\{\mathbf{v}_1, \ldots, \mathbf{v}_n\}$ is linearly independent in V, then $\{T(\mathbf{v}_1), \ldots, T(\mathbf{v}_n)\}$ is linearly independent in W. Illustrate this with an example $T: \mathbb{R}^2 \to \mathbb{R}^2$.

25. Define linear transformations $S: \mathbb{R}^2 \to M_{22}$ and $T: \mathbb{R}^2 \to \mathbb{R}^2$ by

$$S\begin{bmatrix} a \\ b \end{bmatrix} = \begin{bmatrix} a+b & b \\ 0 & a-b \end{bmatrix} \quad \text{and} \quad T\begin{bmatrix} c \\ d \end{bmatrix} = \begin{bmatrix} 2c+d \\ -d \end{bmatrix}$$

Compute $(S \circ T)\begin{bmatrix} 2 \\ 1 \end{bmatrix}$ and $(S \circ T)\begin{bmatrix} x \\ y \end{bmatrix}$. Can you compute $(T \circ S)\begin{bmatrix} x \\ y \end{bmatrix}$? If so, compute it.

26. Define linear transformations $S: \mathscr{P}_1 \to \mathscr{P}_2$ and $T: \mathscr{P}_2 \to \mathscr{P}_1$ by

$$S(a+bx) = a + (a+b)x + 2bx^2$$

and $$T(a+bx+cx^2) = b + 2cx$$

Compute $(S \circ T)(3 + 2x - x^2)$ and $(S \circ T)(a + bx + cx^2)$. Can you compute $(T \circ S)(a+bx)$? If so, compute it.

27. Define linear transformations $S: \mathscr{P}_n \to \mathscr{P}_n$ and $T: \mathscr{P}_n \to \mathscr{P}_n$ by

$$S(p(x)) = p(x+1) \quad \text{and} \quad T(p(x)) = p'(x)$$

Find $(S \circ T)(p(x))$ and $(T \circ S)(p(x))$. [*Hint:* Remember the Chain Rule.]

28. Define linear transformations $S: \mathscr{P}_n \to \mathscr{P}_n$ and $T: \mathscr{P}_n \to \mathscr{P}_n$ by

$$S(p(x)) = p(x+1) \quad \text{and} \quad T(p(x)) = xp'(x)$$

Find $(S \circ T)(p(x))$ and $(T \circ S)(p(x))$.

In Exercises 29 and 30, verify that S and T are inverses.

29. $S: \mathbb{R}^2 \to \mathbb{R}^2$ defined by $S\begin{bmatrix} x \\ y \end{bmatrix} = \begin{bmatrix} 4x+y \\ 3x+y \end{bmatrix}$ and $T: \mathbb{R}^2 \to \mathbb{R}^2$ defined by $T\begin{bmatrix} x \\ y \end{bmatrix} = \begin{bmatrix} x-y \\ -3x+4y \end{bmatrix}$

30. $S: \mathscr{P}_1 \to \mathscr{P}_1$ defined by $S(a+bx) = (-4a+b) + 2ax$ and $T: \mathscr{P}_1 \to \mathscr{P}_1$ defined by

$$T(a+bx) = b/2 + (a+2b)x$$

31. Prove Theorem 6.17.

32. Let $T: V \to V$ be a linear transformation such that $T \circ T = I$.

(a) Show that $\{\mathbf{v}, T(\mathbf{v})\}$ is linearly dependent if and only if $T(\mathbf{v}) = \pm\mathbf{v}$.

(b) Give an example of such a linear transformation with $V = \mathbb{R}^2$.

33. Let $T: V \to V$ be a linear transformation such that $T \circ T = T$.

(a) Show that $\{\mathbf{v}, T(\mathbf{v})\}$ is linearly dependent if and only if $T(\mathbf{v}) = \mathbf{v}$ or $T(\mathbf{v}) = \mathbf{0}$.

(b) Give an example of such a linear transformation with $V = \mathbb{R}^2$.

The set of all linear transformations from a vector space V to a vector space W is denoted by $\mathscr{L}(V, W)$. If S and T are in $\mathscr{L}(V, W)$, we can define the **sum** *$S + T$ of S and T by*

$$(S+T)(\mathbf{v}) = S(\mathbf{v}) + T(\mathbf{v})$$

for all $\mathbf{v}$ *in V. If c is a scalar, we define the* ***scalar multiple*** *cT of T by c to be*

$$(cT)(\mathbf{v}) = cT(\mathbf{v})$$

for all $\mathbf{v}$ *in V. Then $S + T$ and cT are both transformations from V to W.*

34. Prove that $S + T$ and cT are linear transformations.

35. Prove that $\mathscr{L}(V, W)$ is a vector space with this addition and scalar multiplication.

36. Let R, S, and T be linear transformations such that the following operations make sense. Prove that:

(a) $R \circ (S + T) = R \circ S + R \circ T$

(b) $c(R \circ S) = (cR) \circ S = R \circ (cS)$ for any scalar c

6.5 The Kernel and Range of a Linear Transformation

The null space and column space are two of the fundamental subspaces associated with a matrix. In this section, we extend these notions to the kernel and range of a linear transformation.

The word *kernel* is derived from the Old English word *cyrnel,* a form of the word *corn,* meaning "seed" or "grain." Like a kernel of corn, the kernel of a linear transformation is its "core" or "seed" in the sense that it carries information about many of the important properties of the transformation.

Definition Let $T: V \to W$ be a linear transformation. The ***kernel*** of T, denoted $\ker(T)$, is the set of all vectors in V that are mapped by T to $\mathbf{0}$ in W. That is,

$$\ker(T) = \{\mathbf{v} \text{ in } V: T(\mathbf{v}) = \mathbf{0}\}$$

The ***range*** of T, denoted $\text{range}(T)$, is the set of all vectors in W that are images of vectors in V under T. That is,

$$\begin{aligned}\text{range}(T) &= \{T(\mathbf{v}) : \mathbf{v} \text{ in } V\} \\ &= \{\mathbf{w} \text{ in } W : \mathbf{w} = T(\mathbf{v}) \text{ for some } \mathbf{v} \text{ in } V\}\end{aligned}$$

Example 6.59

Let A be an $m \times n$ matrix and let $T = T_A$ be the corresponding matrix transformation from $\mathbb{R}^n$ to $\mathbb{R}^m$ defined by $T(\mathbf{v}) = A\mathbf{v}$. Then, as we saw in Chapter 3, the range of T is the column space of A.

The kernel of T is

$$\begin{aligned}\ker(T) &= \{\mathbf{v} \text{ in } \mathbb{R}^n : T(\mathbf{v}) = \mathbf{0}\} \\ &= \{\mathbf{v} \text{ in } \mathbb{R}^n : A\mathbf{v} = \mathbf{0}\} \\ &= \text{null}(A)\end{aligned}$$

In words, the kernel of a matrix transformation is just the null space of the corresponding matrix.

Example 6.60

Find the kernel and range of the differential operator $D : \mathscr{P}_3 \to \mathscr{P}_2$ defined by $D(p(x)) = p'(x)$.

Solution Since $D(a + bx + cx^2 + dx^3) = b + 2cx + 3dx^2$, we have

$$\begin{aligned}\ker(D) &= \{a + bx + cx^2 + dx^3 : D(a + bx + cx^2 + dx^3) = 0\} \\ &= \{a + bx + cx^2 + dx^3 : b + 2cx + 3dx^2 = 0\}\end{aligned}$$

But $b + 2cx + 3dx^2 = 0$ if and only if $b = 2c = 3d = 0$, which implies that $b = c = d = 0$. Therefore,

$$\begin{aligned}\ker(D) &= \{a + bx + cx^2 + dx^3 : b = c = d = 0\} \\ &= \{a : a \text{ in } \mathbb{R}\}\end{aligned}$$

In other words, the kernel of D is the set of constant polynomials.

The range of D is all of $\mathscr{P}_2$, since *every* polynomial in $\mathscr{P}_2$ is the image under D (i.e., the derivative) of *some* polynomial in $\mathscr{P}_3$. To be specific, if $a + bx + cx^2$ is in $\mathscr{P}_2$, then

$$a + bx + cx^2 = D\left(ax + \left(\frac{b}{2}\right)x^2 + \left(\frac{c}{3}\right)x^3\right)$$

Example 6.61

Let $S : \mathscr{P}_1 \to \mathbb{R}$ be the linear transformation defined by

$$S(p(x)) = \int_0^1 p(x)\,dx$$

Find the kernel and range of S.

Solution In detail, we have

$$\begin{aligned} S(a + bx) &= \int_0^1 (a + bx)\,dx \\ &= \left[ax + \frac{b}{2}x^2\right]_0^1 \\ &= \left(a + \frac{b}{2}\right) - 0 = a + \frac{b}{2} \end{aligned}$$

Therefore,

$$\begin{aligned} \ker(S) &= \{a + bx : S(a + bx) = 0\} \\ &= \left\{a + bx : a + \frac{b}{2} = 0\right\} \\ &= \left\{a + bx : a = -\frac{b}{2}\right\} \\ &= \left\{-\frac{b}{2} + bx\right\} \end{aligned}$$

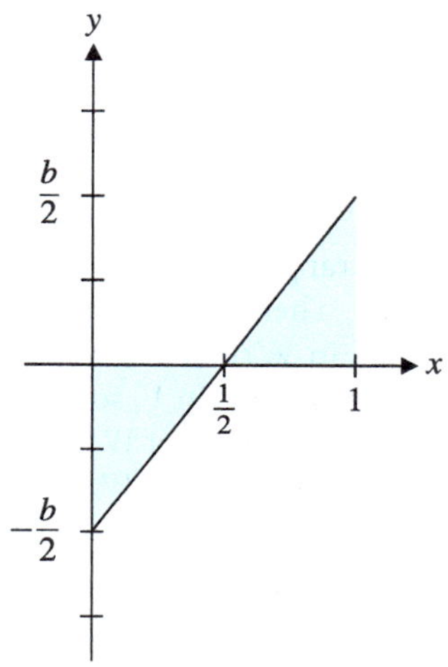

Figure 6.7

If $y = -\dfrac{b}{2} + bx$,

then $\displaystyle\int_0^1 y\,dx = 0$

Geometrically, $\ker(S)$ consists of all those linear polynomials whose graphs have the property that the area between the line and the x-axis is equally distributed above and below the axis on the interval $[0, 1]$ (see Figure 6.7).

The range of S is $\mathbb{R}$, since every real number can be obtained as the image under S of some polynomial in $\mathscr{P}_1$. For example, if a is an arbitrary real number, then

$$\int_0^1 a\,dx = [ax]_0^1 = a - 0 = a$$

so $a = S(a)$.

Example 6.62

Let $T : M_{22} \to M_{22}$ be the linear transformation defined by taking transposes: $T(A) = A^T$. Find the kernel and range of T.

Solution We see that

$$\begin{aligned} \ker(T) &= \{A \text{ in } M_{22} : T(A) = O\} \\ &= \{A \text{ in } M_{22} : A^T = O\} \end{aligned}$$

But if $A^T = O$, then $A = (A^T)^T = O^T = O$. It follows that $\ker(T) = \{O\}$.

Since, for any matrix A in M_{22}, we have $A = (A^T)^T = T(A^T)$ (and A^T is in M_{22}), we deduce that $\text{range}(T) = M_{22}$.

In all of these examples, the kernel and range of a linear transformation are subspaces of the domain and codomain, respectively, of the transformation. Since we are generalizing the null space and column space of a matrix, this is perhaps not surprising. Nevertheless, we should not take anything for granted, so we need to prove that it is not a coincidence.

Theorem 6.18 Let $T : V \to W$ be a linear transformation. Then:

a. The kernel of T is a subspace of V.
b. The range of T is a subspace of W.

Proof (a) Since $T(\mathbf{0}) = \mathbf{0}$, the zero vector of V is in $\ker(T)$, so $\ker(T)$ is nonempty. Let $\mathbf{u}$ and $\mathbf{v}$ be in $\ker(T)$ and let c be a scalar. Then $T(\mathbf{u}) = T(\mathbf{v}) = \mathbf{0}$, so

$$T(\mathbf{u} + \mathbf{v}) = T(\mathbf{u}) + T(\mathbf{v}) = \mathbf{0} + \mathbf{0} = \mathbf{0}$$

and
$$T(c\mathbf{u}) = cT(\mathbf{u}) = c\mathbf{0} = \mathbf{0}$$

Therefore, $\mathbf{u} + \mathbf{v}$ and $c\mathbf{u}$ are in $\ker(T)$, and $\ker(T)$ is a subspace of V.

(b) Since $\mathbf{0} = T(\mathbf{0})$, the zero vector of W is in range(T), so range(T) is nonempty. Let $T(\mathbf{u})$ and $T(\mathbf{v})$ be in the range of T and let c be a scalar. Then $T(\mathbf{u}) + T(\mathbf{v}) = T(\mathbf{u} + \mathbf{v})$ is the image of the vector $\mathbf{u} + \mathbf{v}$. Since $\mathbf{u}$ and $\mathbf{v}$ are in V, so is $\mathbf{u} + \mathbf{v}$, and hence $T(\mathbf{u}) + T(\mathbf{v})$ is in range (T). Similarly, $cT(\mathbf{u}) = T(c\mathbf{u})$. Since $\mathbf{u}$ is in V, so is $c\mathbf{u}$, and hence $cT(\mathbf{u})$ is in range(T). Therefore, range(T) is a nonempty subset of W that is closed under addition and scalar multiplication, and thus it is a subspace of W.

Figure 6.8 gives a schematic representation of the kernel and range of a linear transformation.

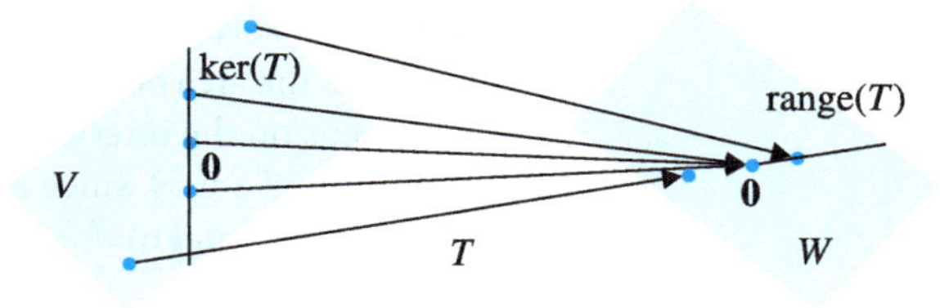

Figure 6.8
The kernel and range of $T : V \to W$

In Chapter 3, we defined the rank of a matrix to be the dimension of its column space and the nullity of a matrix to be the dimension of its null space. We now extend these definitions to linear transformations.

Definition Let $T : V \to W$ be a linear transformation. The ***rank*** of T is the dimension of the range of T and is denoted by rank(T). The ***nullity*** of T is the dimension of the kernel of T and is denoted by nullity(T).

Example 6.63 If A is a matrix and $T = T_A$ is the matrix transformation defined by $T(\mathbf{v}) = A\mathbf{v}$, then the range and kernel of T are the column space and the null space of A, respectively, by Example 6.59. Hence, from Section 3.5, we have

$$\text{rank}(T) = \text{rank}(A) \quad \text{and} \quad \text{nullity}(T) = \text{nullity}(A)$$

Example 6.64

Find the rank and the nullity of the linear transformation $D : \mathscr{P}_3 \to \mathscr{P}_2$ defined by $D(p(x)) = p'(x)$.

Solution In Example 6.60, we computed range $(D) = \mathscr{P}_2$, so

$$\operatorname{rank}(D) = \dim \mathscr{P}_2 = 3$$

The kernel of D is the set of all constant polynomials: $\ker(D) = \{a : a \text{ in } \mathbb{R}\} = \{a \cdot 1 : a \text{ in } \mathbb{R}\}$. Hence, $\{1\}$ is a basis for $\ker(D)$, so

$$\operatorname{nullity}(D) = \dim(\ker(D)) = 1$$

Example 6.65

Find the rank and the nullity of the linear transformation $S : \mathscr{P}_1 \to \mathbb{R}$ defined by

$$S(p(x)) = \int_0^1 p(x)\,dx$$

Solution From Example 6.61, $\operatorname{range}(S) = \mathbb{R}$ and $\operatorname{rank}(S) = \dim \mathbb{R} = 1$. Also,

$$\begin{aligned}\ker(S) &= \left\{-\frac{b}{2} + bx : b \text{ in } \mathbb{R}\right\} \\ &= \{b(-\tfrac{1}{2} + x) : b \text{ in } \mathbb{R}\} \\ &= \operatorname{span}(-\tfrac{1}{2} + x)\end{aligned}$$

so $\{-\frac{1}{2} + x\}$ is a basis for $\ker(S)$. Therefore, $\operatorname{nullity}(S) = \dim(\ker(S)) = 1$.

Example 6.66

Find the rank and the nullity of the linear transformation $T : M_{22} \to M_{22}$ defined by $T(A) = A^T$.

Solution In Example 6.62, we found that $\operatorname{range}(T) = M_{22}$ and $\ker(T) = \{O\}$. Hence,

$$\operatorname{rank}(T) = \dim M_{22} = 4 \quad \text{and} \quad \operatorname{nullity}(T) = \dim\{O\} = 0$$

In Chapter 3, we saw that the rank and nullity of an $m \times n$ matrix A are related by the formula $\operatorname{rank}(A) + \operatorname{nullity}(A) = n$. This is the Rank Theorem (Theorem 3.26). Since the matrix transformation $T = T_A$ has $\mathbb{R}^n$ as its domain, we could rewrite the relationship as

$$\operatorname{rank}(A) + \operatorname{nullity}(A) = \dim \mathbb{R}^n$$

This version of the Rank Theorem extends very nicely to general linear transformations, as you can see from the last three examples:

$$\operatorname{rank}(D) + \operatorname{nullity}(D) = 3 + 1 = 4 = \dim \mathscr{P}_3 \qquad \text{Example 6.64}$$

$$\operatorname{rank}(S) + \operatorname{nullity}(S) = 1 + 1 = 2 = \dim \mathscr{P}_1 \qquad \text{Example 6.65}$$

$$\operatorname{rank}(T) + \operatorname{nullity}(T) = 4 + 0 = 4 = \dim M_{22} \qquad \text{Example 6.66}$$

Theorem 6.19 **The Rank Theorem**

Let $T: V \to W$ be a linear transformation from a finite-dimensional vector space V into a vector space W. Then

$$\text{rank}(T) + \text{nullity}(T) = \dim V$$

In the next section, you will see how to adapt the proof of Theorem 3.26 to prove this version of the result. For now, we give an alternative proof that does not use matrices.

Proof Let $\dim V = n$ and let $\{\mathbf{v}_1, \ldots, \mathbf{v}_k\}$ be a basis for $\ker(T)$[so that $\text{nullity}(T) = \dim(\ker(T)) = k$]. Since $\{\mathbf{v}_1, \ldots, \mathbf{v}_k\}$ is a linearly independent set, it can be extended to a basis for V, by Theorem 6.28. Let $\mathcal{B} = \{\mathbf{v}_1, \ldots, \mathbf{v}_k, \mathbf{v}_{k+1}, \ldots, \mathbf{v}_n\}$ be such a basis. If we can show that the set $\mathcal{C} = \{T(\mathbf{v}_{k+1}), \ldots, T(\mathbf{v}_n)\}$ is a basis for $\text{range}(T)$, then we will have $\text{rank}(T) = \dim(\text{range}(T)) = n - k$ and thus

$$\text{rank}(T) + \text{nullity}(T) = k + (n - k) = n = \dim V$$

as required.

Certainly $\mathcal{C}$ is contained in the range of T. To show that $\mathcal{C}$ spans the range of T, let $T(\mathbf{v})$ be a vector in the range of T. Then $\mathbf{v}$ is in V, and since $\mathcal{B}$ is a basis for V, we can find scalars $c_1, \ldots, c_n$ such that

$$\mathbf{v} = c_1\mathbf{v}_1 + \cdots + c_k\mathbf{v}_k + c_{k+1}\mathbf{v}_{k+1} + \cdots + c_n\mathbf{v}_n$$

Since $\mathbf{v}_1, \ldots, \mathbf{v}_k$ are in the kernel of T, we have $T(\mathbf{v}_1) = \cdots = T(\mathbf{v}_k) = \mathbf{0}$, so

$$\begin{aligned} T(\mathbf{v}) &= T(c_1\mathbf{v}_1 + \cdots + c_k\mathbf{v}_k + c_{k+1}\mathbf{v}_{k+1} + \cdots + c_n\mathbf{v}_n) \\ &= c_1T(\mathbf{v}_1) + \cdots + c_kT(\mathbf{v}_k) + c_{k+1}T(\mathbf{v}_{k+1}) + \cdots + c_nT(\mathbf{v}_n) \\ &= c_{k+1}T(\mathbf{v}_{k+1}) + \cdots + c_nT(\mathbf{v}_n) \end{aligned}$$

This shows that the range of T is spanned by $\mathcal{C}$.

To show that $\mathcal{C}$ is linearly independent, suppose that there are scalars $c_{k+1}, \ldots, c_n$ such that

$$c_{k+1}T(\mathbf{v}_{k+1}) + \cdots + c_nT(\mathbf{v}_n) = \mathbf{0}$$

Then $T(c_{k+1}\mathbf{v}_{k+1} + \cdots + c_n\mathbf{v}_n) = \mathbf{0}$, which means that $c_{k+1}\mathbf{v}_{k+1} + \cdots + c_n\mathbf{v}_n$ is in the kernel of T and is, hence, expressible as a linear combination of the basis vectors $\mathbf{v}_1, \ldots, \mathbf{v}_k$ of $\ker(T)$—say,

$$c_{k+1}\mathbf{v}_{k+1} + \cdots + c_n\mathbf{v}_n = c_1\mathbf{v}_1 + \cdots + c_k\mathbf{v}_k$$

But now
$$c_1\mathbf{v}_1 + \cdots + c_k\mathbf{v}_k - c_{k+1}\mathbf{v}_{k+1} - \cdots - c_n\mathbf{v}_n = \mathbf{0}$$

and the linear independence of $\mathcal{B}$ forces $c_1 = \cdots = c_n = 0$. In particular, $c_{k+1} = \cdots = c_n = 0$, which means $\mathcal{C}$ is linearly independent.

We have shown that $\mathcal{C}$ is a basis for the range of T, so, by our comments above, the proof is complete.

We have verified the Rank Theorem for Examples 6.64, 6.65, and 6.66. In practice, this theorem allows us to find the rank and nullity of a linear transformation with only half the work. The following examples illustrate the process.

Example 6.67

Find the rank and nullity of the linear transformation $T: \mathscr{P}_2 \to \mathscr{P}_3$ defined by $T(p(x)) = xp(x)$. (Check that T really is linear.)

Solution In detail, we have

$$T(a + bx + cx^2) = ax + bx^2 + cx^3$$

It follows that

$$\begin{aligned}\ker(T) &= \{a + bx + cx^2 : T(a + bx + cx^2) = 0\}\\ &= \{a + bx + cx^2 : ax + bx^2 + cx^3 = 0\}\\ &= \{a + bx + cx^2 : a = b = c = 0\}\\ &= \{0\}\end{aligned}$$

so we have $\text{nullity}(T) = \dim(\ker(T)) = 0$. The Rank Theorem implies that

$$\text{rank}(T) = \dim \mathscr{P}_2 - \text{nullity}(T) = 3 - 0 = 3$$

Remark In Example 6.67, it would be just as easy to find the rank of T first, since $\{x, x^2, x^3\}$ is easily seen to be a basis for the range of T. Usually, though, one of the two (the rank or the nullity of a linear transformation) will be easier to compute; the Rank Theorem can then be used to find the other. With practice, you will become better at knowing which way to proceed.

Example 6.68

Let W be the vector space of all symmetric 2×2 matrices. Define a linear transformation $T: W \to \mathscr{P}_2$ by

$$T\begin{bmatrix} a & b \\ b & c \end{bmatrix} = (a - b) + (b - c)x + (c - a)x^2$$

(Check that T is linear.) Find the rank and nullity of T.

Solution The nullity of T is easier to compute directly than the rank, so we proceed as follows:

$$\begin{aligned}\ker(T) &= \left\{\begin{bmatrix} a & b \\ b & c \end{bmatrix} : T\begin{bmatrix} a & b \\ b & c \end{bmatrix} = 0\right\}\\ &= \left\{\begin{bmatrix} a & b \\ b & c \end{bmatrix} : (a - b) + (b - c)x + (c - a)x^2 = 0\right\}\\ &= \left\{\begin{bmatrix} a & b \\ b & c \end{bmatrix} : (a - b) = (b - c) = (c - a) = 0\right\}\\ &= \left\{\begin{bmatrix} a & b \\ b & c \end{bmatrix} : a = b = c\right\}\\ &= \left\{\begin{bmatrix} c & c \\ c & c \end{bmatrix}\right\} = \text{span}\left(\begin{bmatrix} 1 & 1 \\ 1 & 1 \end{bmatrix}\right)\end{aligned}$$

Therefore, $\left\{\begin{bmatrix} 1 & 1 \\ 1 & 1 \end{bmatrix}\right\}$ is a basis for the kernel of T, so $\text{nullity}(T) = \dim(\ker(T)) = 1$. The Rank Theorem and Example 6.42 tell us that $\text{rank}(T) = \dim W - \text{nullity}(T) = 3 - 1 = 2$.

One-to-One and Onto Linear Transformations

We now investigate criteria for a linear transformation to be invertible. The keys to the discussion are the very important properties one-to-one and onto.

Definition A linear transformation $T: V \to W$ is called ***one-to-one*** if T maps distinct vectors in V to distinct vectors in W. If range$(T) = W$, then T is called ***onto.***

Remarks

- The definition of one-to-one may be written more formally as follows:

$T: V \to W$ is one-to-one if, for all $\mathbf{u}$ and $\mathbf{v}$ in V,

$$\mathbf{u} \neq \mathbf{v} \text{ implies that } T(\mathbf{u}) \neq T(\mathbf{v})$$

The above statement is equivalent to the following:

$T: V \to W$ is one-to-one if, for all $\mathbf{u}$ and $\mathbf{v}$ in V,

$$T(\mathbf{u}) = T(\mathbf{v}) \text{ implies that } \mathbf{u} = \mathbf{v}$$

Figure 6.9 illustrates these two statements.

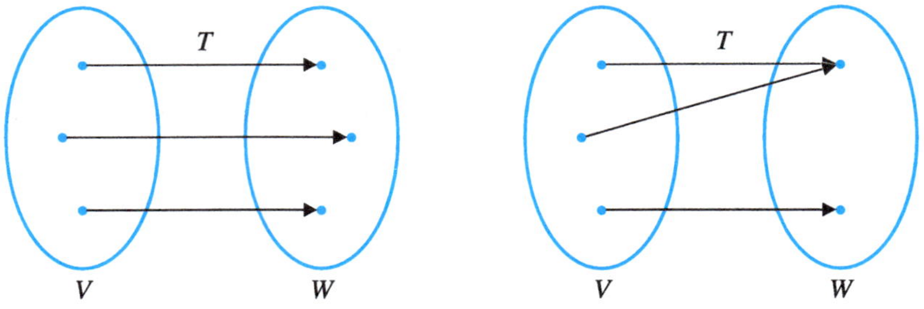

(a) T is one-to-one (b) T is not one-to-one

Figure 6.9

- Another way to write the definition of onto is as follows:

$T: V \to W$ is onto if, for all $\mathbf{w}$ in W, there is at least one $\mathbf{v}$ in V such that

$$\mathbf{w} = T(\mathbf{v})$$

In other words, *given* $\mathbf{w}$ in W, does there exist some $\mathbf{v}$ in V such that $\mathbf{w} = T(\mathbf{v})$? If, for an arbitrary $\mathbf{w}$, we can solve this equation for $\mathbf{v}$, then T is onto (see Figure 6.10).

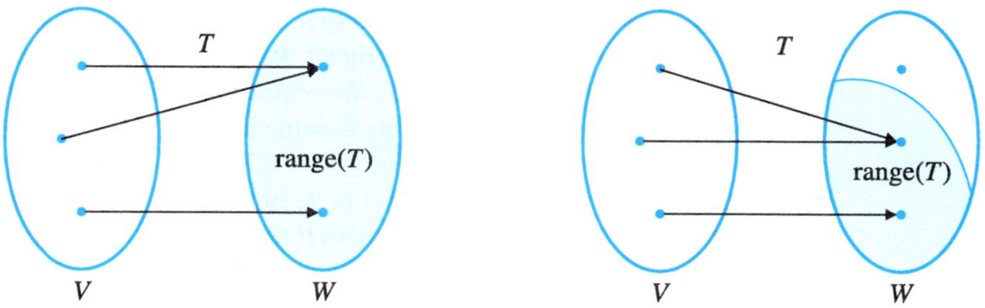

(a) T is onto (b) T is not onto

Figure 6.10

Example 6.69

Which of the following linear transformations are one-to-one? onto?

(a) $T: \mathbb{R}^2 \to \mathbb{R}^3$ defined by $T\begin{bmatrix} x \\ y \end{bmatrix} = \begin{bmatrix} 2x \\ x - y \\ 0 \end{bmatrix}$

(b) $D: \mathscr{P}_3 \to \mathscr{P}_2$ defined by $D(p(x)) = p'(x)$

(c) $T: M_{22} \to M_{22}$ defined by $T(A) = A^T$

Solution (a) Let $T\begin{bmatrix} x_1 \\ y_1 \end{bmatrix} = T\begin{bmatrix} x_2 \\ y_2 \end{bmatrix}$. Then

$$\begin{bmatrix} 2x_1 \\ x_1 - y_1 \\ 0 \end{bmatrix} = \begin{bmatrix} 2x_2 \\ x_2 - y_2 \\ 0 \end{bmatrix}$$

so $2x_1 = 2x_2$ and $x_1 - y_1 = x_2 - y_2$. Solving these equations, we see that $x_1 = x_2$ and $y_1 = y_2$. Hence, $\begin{bmatrix} x_1 \\ y_1 \end{bmatrix} = \begin{bmatrix} x_2 \\ y_2 \end{bmatrix}$, so T is one-to-one.

T is not onto, since its range is not all of $\mathbb{R}^3$. To be specific, there is no vector $\begin{bmatrix} x \\ y \end{bmatrix}$ in $\mathbb{R}^2$ such that $T\begin{bmatrix} x \\ y \end{bmatrix} = \begin{bmatrix} 0 \\ 0 \\ 1 \end{bmatrix}$. (Why not?)

(b) In Example 6.60, we showed that range$(D) = \mathscr{P}_2$, so D is onto. D is not one-to-one, since distinct polynomials in $\mathscr{P}_3$ can have the same derivative. For example, $x^3 \neq x^3 + 1$, but $D(x^3) = 3x^2 = D(x^3 + 1)$.

(c) Let A and B be in M_{22}, with $T(A) = T(B)$. Then $A^T = B^T$, so $A = (A^T)^T = (B^T)^T = B$. Hence, T is one-to-one. In Example 6.62, we showed that range$(T) = M_{22}$. Hence, T is onto.

It turns out that there is a very simple criterion for determining whether a linear transformation is one-to-one.

Theorem 6.20 A linear transformation $T: V \to W$ is one-to-one if and only if $\ker(T) = \{\mathbf{0}\}$.

Proof Assume that T is one-to-one. If $\mathbf{v}$ is in the kernel of T, then $T(\mathbf{v}) = \mathbf{0}$. But we also know that $T(\mathbf{0}) = \mathbf{0}$, so $T(\mathbf{v}) = T(\mathbf{0})$. Since T is one-to-one, this implies that $\mathbf{v} = \mathbf{0}$, so the only vector in the kernel of T is the zero vector.

Conversely, assume that $\ker(T) = \{\mathbf{0}\}$. To show that T is one-to-one, let $\mathbf{u}$ and $\mathbf{v}$ be in V with $T(\mathbf{u}) = T(\mathbf{v})$. Then $T(\mathbf{u} - \mathbf{v}) = T(\mathbf{u}) - T(\mathbf{v}) = \mathbf{0}$, which implies that $\mathbf{u} - \mathbf{v}$ is in the kernel of T. But $\ker(T) = \{\mathbf{0}\}$, so we must have $\mathbf{u} - \mathbf{v} = \mathbf{0}$ or, equivalently, $\mathbf{u} = \mathbf{v}$. This proves that T is one-to-one.

Example 6.70

Show that the linear transformation $T: \mathbb{R}^2 \to \mathscr{P}_1$ defined by

$$T\begin{bmatrix} a \\ b \end{bmatrix} = a + (a + b)x$$

is one-to-one and onto.

Solution If $\begin{bmatrix} a \\ b \end{bmatrix}$ is in the kernel of T, then

$$0 = T\begin{bmatrix} a \\ b \end{bmatrix} = a + (a + b)x$$

It follows that $a = 0$ and $a + b = 0$. Hence, $b = 0$, and therefore $\begin{bmatrix} a \\ b \end{bmatrix} = \begin{bmatrix} 0 \\ 0 \end{bmatrix}$. Consequently, $\ker(T) = \left\{\begin{bmatrix} 0 \\ 0 \end{bmatrix}\right\}$, and T is one-to-one, by Theorem 6.20.

By the Rank Theorem,

$$\text{rank}(T) = \dim \mathbb{R}^2 - \text{nullity}(T) = 2 - 0 = 2$$

Therefore, the range of T is a two-dimensional subspace of $\mathbb{R}^2$, and hence $\text{range}(T) = \mathbb{R}^2$. It follows that T is onto.

For linear transformations between two n-dimensional vector spaces, the properties of one-to-one and onto are closely related. Observe first that for a linear transformation $T: V \to W$, $\ker(T) = \{\mathbf{0}\}$ if and only if $\text{nullity}(T) = 0$, and T is onto if and only if $\text{rank}(T) = \dim W$. (Why?) The proof of the next theorem essentially uses the method of Example 6.70.

Theorem 6.21

Let $\dim V = \dim W = n$. Then a linear transformation $T: V \to W$ is one-to-one if and only if it is onto.

Proof Assume that T is one-to-one. Then $\text{nullity}(T) = 0$ by Theorem 6.20 and the remark preceding Theorem 6.21. The Rank Theorem implies that

$$\text{rank}(T) = \dim V - \text{nullity}(T) = n - 0 = n$$

Therefore, T is onto.

Conversely, assume that T is onto. Then $\text{rank}(T) = \dim W = n$. By the Rank Theorem,

$$\text{nullity}(T) = \dim V - \text{rank}(T) = n - n = 0$$

Hence, $\ker(T) = \{\mathbf{0}\}$, and T is one-to-one.

In Section 6.4, we pointed out that if $T: V \to W$ is a linear transformation, then the image of a basis for V under T need not be a basis for the range of T. We can now give a condition that ensures that a basis for V will be mapped by T to a basis for W.

Theorem 6.22 Let $T: V \to W$ be a one-to-one linear transformation. If $S = \{\mathbf{v}_1, \ldots, \mathbf{v}_k\}$ is a linearly independent set in V, then $T(S) = \{T(\mathbf{v}_1), \ldots, T(\mathbf{v}_k)\}$ is a linearly independent set in W.

Proof Let $c_1, \ldots, c_k$ be scalars such that

$$c_1T(\mathbf{v}_1) + \cdots + c_kT(\mathbf{v}_k) = \mathbf{0}$$

Then $T(c_1\mathbf{v}_1 + \cdots + c_k\mathbf{v}_k) = \mathbf{0}$, which implies that $c_1\mathbf{v}_1 + \cdots + c_k\mathbf{v}_k$ is in the kernel of T. But, since T is one-to-one, $\ker(T) = \{\mathbf{0}\}$, by Theorem 6.20. Hence,

$$c_1\mathbf{v}_1 + \cdots + c_k\mathbf{v}_k = \mathbf{0}$$

But, since $\{\mathbf{v}_1, \ldots, \mathbf{v}_k\}$ is linearly independent, all of the scalars c_i must be 0. Therefore, $\{T(\mathbf{v}_1), \ldots, T(\mathbf{v}_k)\}$ is linearly independent.

Corollary 6.23 Let $\dim V = \dim W = n$. Then a one-to-one linear transformation $T: V \to W$ maps a basis for V to a basis for W.

Proof Let $\mathcal{B} = \{\mathbf{v}_1, \ldots, \mathbf{v}_n\}$ be a basis for V. By Theorem 6.22, $T(\mathcal{B}) = \{T(\mathbf{v}_1), \ldots, T(\mathbf{v}_n)\}$ is a linearly independent set in W, so we need only show that $T(\mathcal{B})$ spans W. But, by Theorem 6.15, $T(\mathcal{B})$ spans the range of T. Moreover, T is onto, by Theorem 6.21, so $\text{range}(T) = W$. Therefore, $T(\mathcal{B})$ spans W, which completes the proof.

Example 6.71 Let $T: \mathbb{R}^2 \to \mathscr{P}_1$ be the linear transformation from Example 6.70, defined by

$$T\begin{bmatrix} a \\ b \end{bmatrix} = a + (a + b)x$$

Then, by Corollary 6.23, the standard basis $\mathcal{E} = \{\mathbf{e}_1, \mathbf{e}_2\}$ for $\mathbb{R}^2$ is mapped to a basis $T(\mathcal{E}) = \{T(\mathbf{e}_1), T(\mathbf{e}_2)\}$ of $\mathscr{P}_1$. We find that

$$T(\mathbf{e}_1) = T\begin{bmatrix} 1 \\ 0 \end{bmatrix} = 1 + x \quad \text{and} \quad T(\mathbf{e}_2) = T\begin{bmatrix} 0 \\ 1 \end{bmatrix} = x$$

It follows that $\{1 + x, x\}$ is a basis for $\mathscr{P}_1$.

We can now determine which linear transformations $T: V \to W$ are invertible.

Theorem 6.24 A linear transformation $T: V \to W$ is invertible if and only if it is one-to-one and onto.

Proof Assume that T is invertible. Then there exists a linear transformation $T^{-1}: W \to V$ such that

$$T^{-1} \circ T = I_V \quad \text{and} \quad T \circ T^{-1} = I_W$$

To show that T is one-to-one, let $\mathbf{v}$ be in the kernel of T. Then $T(\mathbf{v}) = \mathbf{0}$. Therefore,

$$\begin{aligned} T^{-1}(T(\mathbf{v})) = T^{-1}(\mathbf{0}) &\Rightarrow (T^{-1} \circ T)(\mathbf{v}) = \mathbf{0} \\ &\Rightarrow I(\mathbf{v}) = \mathbf{0} \\ &\Rightarrow \mathbf{v} = \mathbf{0} \end{aligned}$$

which establishes that $\ker(T) = \{\mathbf{0}\}$. Therefore, T is one-to-one, by Theorem 6.20.

To show that T is onto, let $\mathbf{w}$ be in W and let $\mathbf{v} = T^{-1}(\mathbf{w})$. Then

$$\begin{aligned} T(\mathbf{v}) &= T(T^{-1}(\mathbf{w})) \\ &= (T \circ T^{-1})(\mathbf{w}) \\ &= I(\mathbf{w}) \\ &= \mathbf{w} \end{aligned}$$

which shows that $\mathbf{w}$ is the image of $\mathbf{v}$ under T. Since $\mathbf{v}$ is in V, this shows that T is onto.

Conversely, assume that T is one-to-one and onto. This means that $\text{nullity}(T) = 0$ and $\text{rank}(T) = \dim W$. We need to show that there exists a linear transformation $T': W \to V$ such that $T' \circ T = I_V$ and $T \circ T' = I_W$.

Let $\mathbf{w}$ be in W. Since T is onto, there exists some vector $\mathbf{v}$ in V such that $T(\mathbf{v}) = \mathbf{w}$. There is only one such vector $\mathbf{v}$, since, if $\mathbf{v}'$ is another vector in V such that $T(\mathbf{v}') = \mathbf{w}$, then $T(\mathbf{v}) = T(\mathbf{v}')$; the fact that T is one-to-one then implies that $\mathbf{v} = \mathbf{v}'$. It therefore makes sense to define a mapping $T': W \to V$ by setting $T'(\mathbf{w}) = \mathbf{v}$.

It follows that

$$(T' \circ T)(\mathbf{v}) = T'(T(\mathbf{v})) = T'(\mathbf{w}) = \mathbf{v}$$

and

$$(T \circ T')(\mathbf{w}) = T(T'(\mathbf{w})) = T(\mathbf{v}) = \mathbf{w}$$

It then follows that $T' \circ T = I_V$ and $T \circ T' = I_W$. Now we must show that T' is a *linear* transformation.

To this end, let $\mathbf{w}_1$ and $\mathbf{w}_2$ be in W and let c_1 and c_2 be scalars. As above, let $T(\mathbf{v}_1) = \mathbf{w}_1$ and $T(\mathbf{v}_2) = \mathbf{w}_2$. Then $\mathbf{v}_1 = T'(\mathbf{w}_1)$ and $\mathbf{v}_2 = T'(\mathbf{w}_2)$ and

$$\begin{aligned} T'(c_1\mathbf{w}_1 + c_2\mathbf{w}_2) &= T'(c_1T(\mathbf{v}_1) + c_2T(\mathbf{v}_2)) \\ &= T'(T(c_1\mathbf{v}_1 + c_2\mathbf{v}_2)) \\ &= I(c_1\mathbf{v}_1 + c_2\mathbf{v}_2) \\ &= c_1\mathbf{v}_1 + c_2\mathbf{v}_2 \\ &= c_1T'(\mathbf{w}_1) + c_2T'(\mathbf{w}_2) \end{aligned}$$

Consequently, T' is linear, so, by Theorem 6.17, $T' = T^{-1}$.

Isomorphisms of Vector Spaces

The words *isomorphism* and *isomorphic* are derived from the Greek words *isos,* meaning "equal," and *morph,* meaning "shape." Thus, figuratively speaking, isomorphic vector spaces have "equal shapes."

We now are in a position to describe, in concrete terms, what it means for two vector spaces to be "essentially the same."

Definition A linear transformation $T: V \to W$ is called an ***isomorphism*** if it is one-to-one and onto. If V and W are two vector spaces such that there is an isomorphism from V to W, then we say that V is ***isomorphic*** to W and write $V \cong W$.

Example 6.72

Show that $\mathscr{P}_{n-1}$ and $\mathbb{R}^n$ are isomorphic.

Solution The process of forming the coordinate vector of a polynomial provides us with one possible isomorphism (as we observed already in Section 6.2, although we did not use the term *isomorphism* there). Specifically, define $T: \mathscr{P}_{n-1} \to \mathbb{R}^n$ by $T(p(x)) = [p(x)]_{\mathcal{E}}$, where $\mathcal{E} = \{1, x, \ldots, x^{n-1}\}$ is the standard basis for $\mathscr{P}_{n-1}$. That is,

$$T(a_0 + a_1x + \cdots + a_{n-1}x^{n-1}) = \begin{bmatrix} a_0 \\ a_1 \\ \vdots \\ a_{n-1} \end{bmatrix}$$

Theorem 6.6 shows that T is a linear transformation. If $p(x) = a_0 + a_1x + \cdots + a_{n-1}x^{n-1}$ is in the kernel of T, then

$$\begin{bmatrix} a_0 \\ \vdots \\ a_{n-1} \end{bmatrix} = T(a_0 + a_1x + \cdots + a_{n-1}x^{n-1}) = \begin{bmatrix} 0 \\ \vdots \\ 0 \end{bmatrix}$$

Hence, $a_0 = a_1 = \cdots = a_{n-1} = 0$, so $p(x) = 0$. Therefore, $\ker(T) = \{\mathbf{0}\}$, and T is one-to-one. Since $\dim \mathscr{P}_{n-1} = \dim \mathbb{R}^n = n$, T is also onto, by Theorem 6.21. Thus, T is an isomorphism, and $\mathscr{P}_{n-1} \cong \mathbb{R}^n$.

Example 6.73

Show that M_{mn} and $\mathbb{R}^{mn}$ are isomorphic.

Solution Once again, the coordinate mapping from M_{mn} to $\mathbb{R}^{mn}$ (as in Example 6.36) is an isomorphism. The details of the proof are left as an exercise.

In fact, the easiest way to tell if two vector spaces are isomorphic is simply to check their dimensions, as the next theorem shows.

Theorem 6.25 Let V and W be two finite-dimensional vector spaces (over the same field of scalars). Then V is isomorphic to W if and only if $\dim V = \dim W$.

Proof Let $n = \dim V$. If V is isomorphic to W, then there is an isomorphism $T: V \to W$. Since T is one-to-one, $\text{nullity}(T) = 0$. The Rank Theorem then implies that

$$\text{rank}(T) = \dim V - \text{nullity}(T) = n - 0 = n$$

Therefore, the range of T is an n-dimensional subspace of W. But, since T is onto, $W = \text{range}(T)$, so $\dim W = n$, as we wished to show.

Conversely, assume that V and W have the same dimension, n. Let $\mathcal{B} = \{\mathbf{v}_1, \ldots, \mathbf{v}_n\}$ be a basis for V and let $\mathcal{C} = \{\mathbf{w}_1, \ldots, \mathbf{w}_n\}$ be a basis for W. We will define a linear transformation $T: V \to W$ and then show that T is one-to-one and onto. An arbitrary vector $\mathbf{v}$ in V can be written uniquely as a linear combination of the vectors in the basis $\mathcal{B}$—say,

$$\mathbf{v} = c_1\mathbf{v}_1 + \cdots + c_n\mathbf{v}_n$$

We define T by

$$T(\mathbf{v}) = c_1\mathbf{w}_1 + \cdots + c_n\mathbf{w}_n$$

It is straightforward to check that T is linear. (Do so.) To see that T is one-to-one, suppose $\mathbf{v}$ is in the kernel of T. Then

$$c_1\mathbf{w}_1 + \cdots + c_n\mathbf{w}_n = T(\mathbf{v}) = \mathbf{0}$$

and the linear independence of $\mathcal{C}$ forces $c_1 = \cdots = c_n = 0$. But then

$$\mathbf{v} = c_1\mathbf{v}_1 + \cdots + c_n\mathbf{v}_n = \mathbf{0}$$

so $\ker(T) = \{\mathbf{0}\}$, meaning that T is one-to-one. Since $\dim V = \dim W$, T is also onto, by Theorem 6.21. Therefore, T is an isomorphism, and $V \cong W$.

Example 6.74 Show that $\mathbb{R}^n$ and $\mathscr{P}_n$ are not isomorphic.

Solution Since $\dim \mathbb{R}^n = n \neq n + 1 = \dim \mathscr{P}_n$, $\mathbb{R}^n$ and $\mathscr{P}_n$ are not isomorphic, by Theorem 6.25.

Example 6.75 Let W be the vector space of all symmetric 2×2 matrices. Show that W is isomorphic to $\mathbb{R}^3$.

Solution In Example 6.42, we showed that $\dim W = 3$. Hence, $\dim W = \dim \mathbb{R}^3$, so $W \cong \mathbb{R}^3$, by Theorem 6.25. (There is an obvious candidate for an isomorphism $T: W \to \mathbb{R}^3$. What is it?)

Remark Our examples have all been *real* vector spaces, but the theorems we have proved are true for vector spaces over the complex numbers $\mathbb{C}$ or $\mathbb{Z}_p$, where p is prime. For example, the vector space $M_{22}(\mathbb{Z}_2)$ of all 2×2 matrices with entries from $\mathbb{Z}_2$ has dimension 4 as a vector space over $\mathbb{Z}_2$, and hence $M_{22}(\mathbb{Z}_2) \cong \mathbb{Z}_2^4$.

Exercises 6.5

1. Let $T: M_{22} \to M_{22}$ be the linear transformation defined by

$$T\begin{bmatrix} a & b \\ c & d \end{bmatrix} = \begin{bmatrix} a & 0 \\ 0 & d \end{bmatrix}$$

(a) Which, if any, of the following matrices are in $\ker(T)$?

(i) $\begin{bmatrix} 1 & 2 \\ -1 & 3 \end{bmatrix}$ **(ii)** $\begin{bmatrix} 0 & 4 \\ 2 & 0 \end{bmatrix}$ **(iii)** $\begin{bmatrix} 3 & 0 \\ 0 & -3 \end{bmatrix}$

(b) Which, if any, of the matrices in part (a) are in $\text{range}(T)$?

(c) Describe $\ker(T)$ and $\text{range}(T)$.

2. Let $T: M_{22} \to \mathbb{R}$ be the linear transformation defined by $T(A) = \text{tr}(A)$.

(a) Which, if any, of the following matrices are in $\ker(T)$?

(i) $\begin{bmatrix} 1 & 2 \\ -1 & 3 \end{bmatrix}$ **(ii)** $\begin{bmatrix} 0 & 4 \\ 2 & 0 \end{bmatrix}$ **(iii)** $\begin{bmatrix} 1 & 3 \\ 0 & -1 \end{bmatrix}$

(b) Which, if any, of the following scalars are in $\text{range}(T)$?

(i) 0 **(ii)** 2 **(iii)** $\sqrt{2}/2$

(c) Describe $\ker(T)$ and $\text{range}(T)$.

3. Let $T: \mathscr{P}_2 \to \mathbb{R}^2$ be the linear transformation defined by

$$T(a + bx + cx^2) = \begin{bmatrix} a - b \\ b + c \end{bmatrix}$$

(a) Which, if any, of the following polynomials are in $\ker(T)$?

(i) $1 + x$ **(ii)** $x - x^2$ **(iii)** $1 + x - x^2$

(b) Which, if any, of the following vectors are in $\text{range}(T)$?

(i) $\begin{bmatrix} 0 \\ 0 \end{bmatrix}$ **(ii)** $\begin{bmatrix} 1 \\ 0 \end{bmatrix}$ **(iii)** $\begin{bmatrix} 0 \\ 1 \end{bmatrix}$

(c) Describe $\ker(T)$ and $\text{range}(T)$.

4. Let $T: \mathscr{P}_2 \to \mathscr{P}_2$ be the linear transformation defined by $T(p(x)) = xp'(x)$.

(a) Which, if any, of the following polynomials are in $\ker(T)$?

(i) 1 **(ii)** x **(iii)** x^2

(b) Which, if any, of the polynomials in part (a) are in $\text{range}(T)$?

(c) Describe $\ker(T)$ and $\text{range}(T)$.

In Exercises 5–8, find bases for the kernel and range of the linear transformations T in the indicated exercises. In each case, state the nullity and rank of T and verify the Rank Theorem.

5. Exercise 1

6. Exercise 2

7. Exercise 3

8. Exercise 4

In Exercises 9–14, find either the nullity or the rank of T and then use the Rank Theorem to find the other.

9. $T: M_{22} \to \mathbb{R}^2$ defined by $T\begin{bmatrix} a & b \\ c & d \end{bmatrix} = \begin{bmatrix} a - b \\ c - d \end{bmatrix}$

10. $T: \mathscr{P}_2 \to \mathbb{R}^2$ defined by $T(p(x)) = \begin{bmatrix} p(0) \\ p(1) \end{bmatrix}$

11. $T: M_{22} \to M_{22}$ defined by $T(A) = AB$, where $B = \begin{bmatrix} 1 & -1 \\ -1 & 1 \end{bmatrix}$

12. $T: M_{22} \to M_{22}$ defined by $T(A) = AB - BA$, where $B = \begin{bmatrix} 1 & 1 \\ 0 & 1 \end{bmatrix}$

13. $T: \mathscr{P}_2 \to \mathbb{R}$ defined by $T(p(x)) = p'(0)$

14. $T: M_{33} \to M_{33}$ defined by $T(A) = A - A^T$

In Exercises 15–20, determine whether the linear transformation T is (a) one-to-one and (b) onto.

15. $T: \mathbb{R}^2 \to \mathbb{R}^2$ defined by $T\begin{bmatrix} x \\ y \end{bmatrix} = \begin{bmatrix} 2x - y \\ x + 2y \end{bmatrix}$

16. $T: \mathbb{R}^2 \to \mathcal{P}_2$ defined by

$$T\begin{bmatrix} a \\ b \end{bmatrix} = (a - 2b) + (3a + b)x + (a + b)x^2$$

17. $T: \mathcal{P}_2 \to \mathbb{R}^3$ defined by

$$T(a + bx + cx^2) = \begin{bmatrix} 2a - b \\ a + b - 3c \\ c - a \end{bmatrix}$$

18. $T: \mathcal{P}_2 \to \mathbb{R}^2$ defined by $T(p(x)) = \begin{bmatrix} p(0) \\ p(1) \end{bmatrix}$

19. $T: \mathbb{R}^3 \to M_{22}$ defined by $T\begin{bmatrix} a \\ b \\ c \end{bmatrix} = \begin{bmatrix} a - b & b - c \\ a + b & b + c \end{bmatrix}$

20. $T: \mathbb{R}^3 \to W$ defined by $T\begin{bmatrix} a \\ b \\ c \end{bmatrix} =$ $\begin{bmatrix} a + b + c & b - 2c \\ b - 2c & a - c \end{bmatrix}$, where W is the vector space of all symmetric 2×2 matrices

In Exercises 21–26, determine whether V and W are isomorphic. If they are, give an explicit isomorphism $T: V \to W$.

21. $V = D_3$ (diagonal 3×3 matrices), $W = \mathbb{R}^3$

22. $V = S_3$ (symmetric 3×3 matrices), $W = U_3$ (upper triangular 3×3 matrices)

23. $V = S_3$ (symmetric 3×3 matrices), $W = S_3'$ (skew-symmetric 3×3 matrices)

24. $V = \mathcal{P}_2$, $W = \{p(x) \text{ in } \mathcal{P}_3 : p(0) = 0\}$

25. $V = \mathbb{C}$, $W = \mathbb{R}^2$

26. $V = \{A \text{ in } M_{22} : \text{tr}(A) = 0\}$, $W = \mathbb{R}^2$

27. Show that $T: \mathcal{P}_n \to \mathcal{P}_n$ defined by $T(p(x)) = p(x) + p'(x)$ is an isomorphism.

28. Show that $T: \mathcal{P}_n \to \mathcal{P}_n$ defined by $T(p(x)) = p(x - 2)$ is an isomorphism.

29. Show that $T: \mathcal{P}_n \to \mathcal{P}_n$ defined by $T(p(x)) = x^n p\left(\frac{1}{x}\right)$ is an isomorphism.

30. **(a)** Show that $\mathscr{C}[0, 1] \cong \mathscr{C}[2, 3]$. [*Hint:* Define $T: \mathscr{C}[0, 1] \to \mathscr{C}[2, 3]$ by letting $T(f)$ be the function whose value at x is $(T(f))(x) = f(x - 2)$ for x in $[2, 3]$.]
(b) Show that $\mathscr{C}[0, 1] \cong \mathscr{C}[a, a + 1]$ for all a.

31. Show that $\mathscr{C}[0, 1] \cong \mathscr{C}[0, 2]$.

32. Show that $\mathscr{C}[a, b] \cong \mathscr{C}[c, d]$ for all $a < b$ and $c < d$.

33. Let $S: V \to W$ and $T: U \to V$ be linear transformations.
(a) Prove that if S and T are both one-to-one, so is $S \circ T$.
(b) Prove that if S and T are both onto, so is $S \circ T$.

34. Let $S: V \to W$ and $T: U \to V$ be linear transformations.
(a) Prove that if $S \circ T$ is one-to-one, so is T.
(b) Prove that if $S \circ T$ is onto, so is S.

35. Let $T: V \to W$ be a linear transformation between two finite-dimensional vector spaces.
(a) Prove that if $\dim V < \dim W$, then T cannot be onto.
(b) Prove that if $\dim V > \dim W$, then T cannot be one-to-one.

36. Let $a_0, a_1, \ldots, a_n$ be $n + 1$ distinct real numbers. Define $T: \mathcal{P}_n \to \mathbb{R}^{n+1}$ by

$$T(p(x)) = \begin{bmatrix} p(a_0) \\ p(a_1) \\ \vdots \\ p(a_n) \end{bmatrix}$$

Prove that T is an isomorphism.

37. If V is a finite-dimensional vector space and $T: V \to V$ is a linear transformation such that $\text{rank}(T) = \text{rank}(T^2)$, prove that $\text{range}(T) \cap \ker(T) = \{\mathbf{0}\}$. [*Hint:* T^2 denotes $T \circ T$. Use the Rank Theorem to help show that the kernels of T and T^2 are the same.]

38. Let U and W be subspaces of a finite-dimensional vector space V. Define $T: U \times W \to V$ by $T(\mathbf{u}, \mathbf{w}) = \mathbf{u} - \mathbf{w}$.
(a) Prove that T is a linear transformation.
(b) Show that $\text{range}(T) = U + W$.
(c) Show that $\ker(T) \cong U \cap W$. [*Hint:* See Exercise 50 in Section 6.1.]
(d) Prove ***Grassmann's Identity:***

$$\dim(U + W) = \dim U + \dim W - \dim(U \cap W)$$

[*Hint:* Apply the Rank Theorem, using results (a) and (b) and Exercise 43(b) in Section 6.2.]

6.6 The Matrix of a Linear Transformation

Theorem 6.15 showed that a linear transformation $T: V \to W$ is completely determined by its effect on a spanning set for V. In particular, if we know how T acts on a basis for V, then we can compute $T(\mathbf{v})$ for any vector $\mathbf{v}$ in V. Example 6.55 illustrated the process. We implicitly used this important property of linear transformations in Theorem 3.31 to help us compute the standard matrix of a linear transformation $T: \mathbb{R}^n \to \mathbb{R}^m$. In this section, we will show that every linear transformation between finite-dimensional vector spaces can be represented as a matrix transformation.

Suppose that V is an n-dimensional vector space, W is an m-dimensional vector space, and $T: V \to W$ is a linear transformation. Let $\mathcal{B}$ and $\mathcal{C}$ be bases for V and W, respectively. Then the coordinate vector mapping $R(\mathbf{v}) = [\mathbf{v}]_{\mathcal{B}}$ defines an isomorphism $R: V \to \mathbb{R}^n$. At the same time, we have an isomorphism $S: W \to \mathbb{R}^m$ given by $S(\mathbf{w}) = [\mathbf{w}]_{\mathcal{C}}$, which allows us to associate the image $T(\mathbf{v})$ with the vector $[T(\mathbf{v})]_{\mathcal{C}}$ in $\mathbb{R}^m$. Figure 6.11 illustrates the relationships.

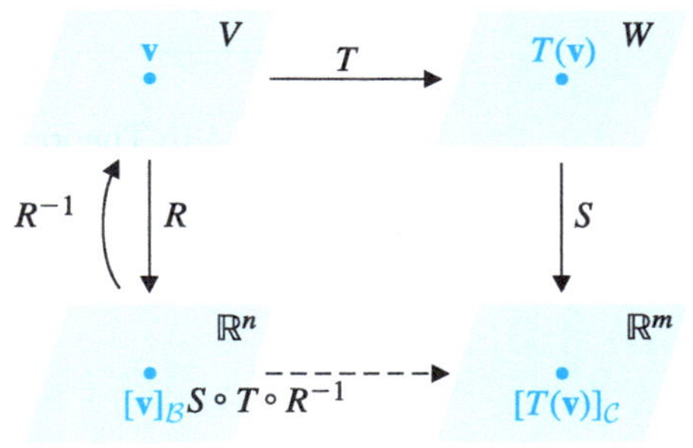

Figure 6.11

Since R is an isomorphism, it is invertible, so we may form the composite mapping

$$S \circ T \circ R^{-1}: \mathbb{R}^n \to \mathbb{R}^m$$

which maps $[\mathbf{v}]_{\mathcal{B}}$ to $[T(\mathbf{v})]_{\mathcal{C}}$. Since this mapping goes from $\mathbb{R}^n$ to $\mathbb{R}^m$, we know from Chapter 3 that it is a matrix transformation. What, then, is the standard matrix of $S \circ T \circ R^{-1}$? We would like to find the $m \times n$ matrix A such that $A[\mathbf{v}]_{\mathcal{B}} = (S \circ T \circ R^{-1})([\mathbf{v}]_{\mathcal{B}})$. Or, since $(S \circ T \circ R^{-1})([\mathbf{v}]_{\mathcal{B}}) = [T(\mathbf{v})]_{\mathcal{C}}$, we require

$$A[\mathbf{v}]_{\mathcal{B}} = [T(\mathbf{v})]_{\mathcal{C}}$$

It turns out to be surprisingly easy to find. The basic idea is that of Theorem 3.31. The columns of A are the images of the standard basis vectors for $\mathbb{R}^n$ under $S \circ T \circ R^{-1}$. But, if $\mathcal{B} = \{\mathbf{v}_1, \ldots, \mathbf{v}_n\}$ is a basis for V, then

$$\begin{aligned} R(\mathbf{v}_i) &= [\mathbf{v}_i]_{\mathcal{B}} \\ &= \begin{bmatrix} 0 \\ \vdots \\ 1 \\ \vdots \\ 0 \end{bmatrix} \quad \leftarrow i\text{th entry} \\ &= \mathbf{e}_i \end{aligned}$$

so $R^{-1}(\mathbf{e}_i) = \mathbf{v}_i$. Therefore, the ith column of the matrix A we seek is given by

$$\begin{aligned}(S \circ T \circ R^{-1})(\mathbf{e}_i) &= S(T(R^{-1}(\mathbf{e}_i))) \\ &= S(T(\mathbf{v}_i)) \\ &= [T(\mathbf{v}_i)]_{\mathcal{C}}\end{aligned}$$

which is the coordinate vector of $T(\mathbf{v}_i)$ with respect to the basis $\mathcal{C}$ of W.

We summarize this discussion as a theorem.

Theorem 6.26 Let V and W be two finite-dimensional vector spaces with bases $\mathcal{B}$ and $\mathcal{C}$, respectively, where $\mathcal{B} = \{\mathbf{v}_1, \ldots, \mathbf{v}_n\}$. If $T: V \to W$ is a linear transformation, then the $m \times n$ matrix A defined by

$$A = \left[\, [T(\mathbf{v}_1)]_{\mathcal{C}} \mid [T(\mathbf{v}_2)]_{\mathcal{C}} \mid \cdots \mid [T(\mathbf{v}_n)]_{\mathcal{C}} \,\right]$$

satisfies

$$A[\mathbf{v}]_{\mathcal{B}} = [T(\mathbf{v})]_{\mathcal{C}}$$

for every vector $\mathbf{v}$ in V.

The matrix A in Theorem 6.26 is called the ***matrix of T with respect to the bases $\mathcal{B}$ and $\mathcal{C}$.*** The relationship is illustrated below. (Recall that T_A denotes multiplication by A.)

$$\begin{array}{ccc} \mathbf{v} & \xrightarrow{T} & T(\mathbf{v}) \\ \downarrow & & \downarrow \\ [\mathbf{v}]_{\mathcal{B}} & \xrightarrow{T_A} & A[\mathbf{v}]_{\mathcal{B}} = [T(\mathbf{v})]_{\mathcal{C}} \end{array}$$

Remarks

- The matrix of a linear transformation T with respect to bases $\mathcal{B}$ and $\mathcal{C}$ is sometimes denoted by $[T]_{\mathcal{C} \leftarrow \mathcal{B}}$. Note the direction of the arrow: right-to-left (not left-to-right, as for $T: V \to W$). With this notation, the final equation in Theorem 6.26 becomes

$$[T]_{\mathcal{C} \leftarrow \mathcal{B}}[\mathbf{v}]_{\mathcal{B}} = [T(\mathbf{v})]_{\mathcal{C}}$$

Observe that the $\mathcal{B}$s in the subscripts appear side by side and appear to "cancel" each other. In words, this equation says, "The matrix for T times the coordinate vector for $\mathbf{v}$ gives the coordinate vector for $T(\mathbf{v})$."

In the special case where $V = W$ and $\mathcal{B} = \mathcal{C}$, we write $[T]_{\mathcal{B}}$ (instead of $[T]_{\mathcal{B} \leftarrow \mathcal{B}}$). Theorem 6.26 then states that

$$[T]_{\mathcal{B}}[\mathbf{v}]_{\mathcal{B}} = [T(\mathbf{v})]_{\mathcal{B}}$$

- The matrix of a linear transformation with respect to given bases is unique. That is, for every vector $\mathbf{v}$ in V, there is only *one* matrix A with the property specified by Theorem 6.26—namely,

$$A[\mathbf{v}]_{\mathcal{B}} = [T(\mathbf{v})]_{\mathcal{C}}$$

(You are asked to prove this in Exercise 39.)

- The diagram that follows Theorem 6.26 is sometimes called a *commutative diagram* because we can start in the upper left-hand corner with the vector $\mathbf{v}$ and get to $[T(\mathbf{v})]_{\mathcal{C}}$ in the lower right-hand corner in two different, but equivalent, ways. If, as before, we denote the coordinate mappings that map $\mathbf{v}$ to $[\mathbf{v}]_{\mathcal{B}}$ and $\mathbf{w}$ to $[\mathbf{w}]_{\mathcal{C}}$ by R and S, respectively, then we can summarize this "commutativity" by

$$S \circ T = T_A \circ R$$

The reason for the term *commutative* becomes clearer when $V = W$ and $\mathcal{B} = \mathcal{C}$, for then $R = S$ too, and we have

$$R \circ T = T_A \circ R$$

suggesting that the coordinate mapping R commutes with the linear transformation T (provided we use the matrix version of T—namely, $T_A = T_{[T]_{\mathcal{B}}}$—where it is required).

- The matrix $[T]_{\mathcal{C} \leftarrow \mathcal{B}}$ depends on the *order* of the vectors in the bases $\mathcal{B}$ and $\mathcal{C}$. Rearranging the vectors within either basis will affect the matrix $[T]_{\mathcal{C} \leftarrow \mathcal{B}}$. [See Example 6.77(b).]

Example 6.76

Let $T : \mathbb{R}^3 \to \mathbb{R}^2$ be the linear transformation defined by

$$T\begin{bmatrix} x \\ y \\ z \end{bmatrix} = \begin{bmatrix} x - 2y \\ x + y - 3z \end{bmatrix}$$

and let $\mathcal{B} = \{\mathbf{e}_1, \mathbf{e}_2, \mathbf{e}_3\}$ and $\mathcal{C} = \{\mathbf{e}_2, \mathbf{e}_1\}$ be bases for $\mathbb{R}^3$ and $\mathbb{R}^2$, respectively. Find the matrix of T with respect to $\mathcal{B}$ and $\mathcal{C}$ and verify Theorem 6.26 for $\mathbf{v} = \begin{bmatrix} 1 \\ 3 \\ -2 \end{bmatrix}$.

Solution First, we compute

$$T(\mathbf{e}_1) = \begin{bmatrix} 1 \\ 1 \end{bmatrix}, \quad T(\mathbf{e}_2) = \begin{bmatrix} -2 \\ 1 \end{bmatrix}, \quad T(\mathbf{e}_3) = \begin{bmatrix} 0 \\ -3 \end{bmatrix}$$

Next, we need their coordinate vectors with respect to $\mathcal{C}$. Since

$$\begin{bmatrix} 1 \\ 1 \end{bmatrix} = \mathbf{e}_2 + \mathbf{e}_1, \quad \begin{bmatrix} -2 \\ 1 \end{bmatrix} = \mathbf{e}_2 - 2\mathbf{e}_1, \quad \begin{bmatrix} 0 \\ -3 \end{bmatrix} = -3\mathbf{e}_2 + 0\mathbf{e}_1$$

we have

$$[T(\mathbf{e}_1)]_{\mathcal{C}} = \begin{bmatrix} 1 \\ 1 \end{bmatrix}, \quad [T(\mathbf{e}_2)]_{\mathcal{C}} = \begin{bmatrix} 1 \\ -2 \end{bmatrix}, \quad [T(\mathbf{e}_3)]_{\mathcal{C}} = \begin{bmatrix} -3 \\ 0 \end{bmatrix}$$

Therefore, the matrix of T with respect to $\mathcal{B}$ and $\mathcal{C}$ is

$$\begin{aligned} A = [T]_{\mathcal{C} \leftarrow \mathcal{B}} &= [\,[T(\mathbf{e}_1)]_{\mathcal{C}} \;\; [T(\mathbf{e}_2)]_{\mathcal{C}} \;\; [T(\mathbf{e}_3)]_{\mathcal{C}}\,] \\ &= \begin{bmatrix} 1 & 1 & -3 \\ 1 & -2 & 0 \end{bmatrix} \end{aligned}$$

To verify Theorem 6.26 for $\mathbf{v}$, we first compute

$$T(\mathbf{v}) = T\begin{bmatrix} 1 \\ 3 \\ -2 \end{bmatrix} = \begin{bmatrix} -5 \\ 10 \end{bmatrix}$$

Then

$$[\mathbf{v}]_{\mathcal{B}} = \begin{bmatrix} 1 \\ 3 \\ -2 \end{bmatrix}$$

and

$$[T(\mathbf{v})]_{\mathcal{C}} = \begin{bmatrix} -5 \\ 10 \end{bmatrix}_{\mathcal{C}} = \begin{bmatrix} 10 \\ -5 \end{bmatrix}$$

(Check these.)

Using all of these facts, we confirm that

$$A[\mathbf{v}]_{\mathcal{B}} = \begin{bmatrix} 1 & 1 & -3 \\ 1 & -2 & 0 \end{bmatrix}\begin{bmatrix} 1 \\ 3 \\ -2 \end{bmatrix} = \begin{bmatrix} 10 \\ -5 \end{bmatrix} = [T(\mathbf{v})]_{\mathcal{C}}$$

Example 6.77

Let $D : \mathscr{P}_3 \to \mathscr{P}_2$ be the differential operator $D(p(x)) = p'(x)$. Let $\mathcal{B} = \{1, x, x^2, x^3\}$ and $\mathcal{C} = \{1, x, x^2\}$ be bases for $\mathscr{P}_3$ and $\mathscr{P}_2$, respectively.

(a) Find the matrix A of D with respect to $\mathcal{B}$ and $\mathcal{C}$.

(b) Find the matrix A' of D with respect to $\mathcal{B}'$ and $\mathcal{C}$, where $\mathcal{B}' = \{x^3, x^2, x, 1\}$.

(c) Using part (a), compute $D(5 - x + 2x^3)$ and $D(a + bx + cx^2 + dx^3)$ to verify Theorem 6.26.

Solution First note that $D(a + bx + cx^2 + dx^3) = b + 2cx + 3dx^2$. (See Example 6.60.)

(a) Since the images of the basis $\mathcal{B}$ under D are $D(1) = 0$, $D(x) = 1$, $D(x^2) = 2x$, and $D(x^3) = 3x^2$, their coordinate vectors with respect to $\mathcal{C}$ are

$$[D(1)]_{\mathcal{C}} = \begin{bmatrix} 0 \\ 0 \\ 0 \end{bmatrix}, \quad [D(x)]_{\mathcal{C}} = \begin{bmatrix} 1 \\ 0 \\ 0 \end{bmatrix}, \quad [D(x^2)]_{\mathcal{C}} = \begin{bmatrix} 0 \\ 2 \\ 0 \end{bmatrix}, \quad [D(x^3)]_{\mathcal{C}} = \begin{bmatrix} 0 \\ 0 \\ 3 \end{bmatrix}$$

Consequently,

$$\begin{aligned} A = [D]_{\mathcal{C}\leftarrow\mathcal{B}} &= [\,[D(1)]_{\mathcal{C}} \mid [D(x)]_{\mathcal{C}} \mid [D(x^2)]_{\mathcal{C}} \mid [D(x^3)]_{\mathcal{C}}\,] \\ &= \begin{bmatrix} 0 & 1 & 0 & 0 \\ 0 & 0 & 2 & 0 \\ 0 & 0 & 0 & 3 \end{bmatrix} \end{aligned}$$

(b) Since the basis $\mathcal{B}'$ is just $\mathcal{B}$ in the *reverse* order, we see that

$$\begin{aligned} A' = [D]_{\mathcal{C}\leftarrow\mathcal{B}'} &= [\,[D(x^3)]_{\mathcal{C}} \mid [D(x^2)]_{\mathcal{C}} \mid [D(x)]_{\mathcal{C}} \mid [D(1)]_{\mathcal{C}}\,] \\ &= \begin{bmatrix} 0 & 0 & 1 & 0 \\ 0 & 2 & 0 & 0 \\ 3 & 0 & 0 & 0 \end{bmatrix} \end{aligned}$$

(This shows that the *order* of the vectors in the bases $\mathcal{B}$ and $\mathcal{C}$ affects the matrix of a transformation with respect to these bases.)

(c) First we compute $D(5 - x + 2x^3) = -1 + 6x^2$ directly, getting the coordinate vector

$$[D(5 - x + 2x^3)]_{\mathcal{C}} = [-1 + 6x^2]_{\mathcal{C}} = \begin{bmatrix} -1 \\ 0 \\ 6 \end{bmatrix}$$

On the other hand,

$$[5 - x + 2x^3]_{\mathcal{B}} = \begin{bmatrix} 5 \\ -1 \\ 0 \\ 2 \end{bmatrix}$$

so

$$A[5 - x + 2x^3]_{\mathcal{B}} = \begin{bmatrix} 0 & 1 & 0 & 0 \\ 0 & 0 & 2 & 0 \\ 0 & 0 & 0 & 3 \end{bmatrix}\begin{bmatrix} 5 \\ -1 \\ 0 \\ 2 \end{bmatrix} = \begin{bmatrix} -1 \\ 0 \\ 6 \end{bmatrix} = [D(5 - x + 2x^3)]_{\mathcal{C}}$$

which agrees with Theorem 6.26. We leave proof of the general case as an exercise.

Since the linear transformation in Example 6.77 is easy to use directly, there is really no advantage to using the matrix of this transformation to do calculations. However, in other examples—especially large ones—the matrix approach may be simpler, as it is very well-suited to computer implementation. Example 6.78 illustrates the basic idea behind this indirect approach.

Example 6.78

Let $T: \mathscr{P}_2 \to \mathscr{P}_2$ be the linear transformation defined by

$$T(p(x)) = p(2x - 1)$$

(a) Find the matrix of T with respect to $\mathcal{E} = \{1, x, x^2\}$.

(b) Compute $T(3 + 2x - x^2)$ indirectly, using part (a).

Solution (a) We see that

$$T(1) = 1, \quad T(x) = 2x - 1, \quad T(x^2) = (2x - 1)^2 = 1 - 4x + 4x^2$$

so the coordinate vectors are

$$[T(1)]_{\mathcal{E}} = \begin{bmatrix} 1 \\ 0 \\ 0 \end{bmatrix}, \quad [T(x)]_{\mathcal{E}} = \begin{bmatrix} -1 \\ 2 \\ 0 \end{bmatrix}, \quad [T(x^2)]_{\mathcal{E}} = \begin{bmatrix} 1 \\ -4 \\ 4 \end{bmatrix}$$

Therefore,

$$[T]_{\mathcal{E}} = [[T(1)]_{\mathcal{E}} \mid [T(x)]_{\mathcal{E}} \mid [T(x^2)]_{\mathcal{E}}] = \begin{bmatrix} 1 & -1 & 1 \\ 0 & 2 & -4 \\ 0 & 0 & 4 \end{bmatrix}$$

(b) We apply Theorem 6.26 as follows: The coordinate vector of $p(x) = 3 + 2x - x^2$ with respect to $\mathcal{E}$ is

$$[p(x)]_{\mathcal{E}} = \begin{bmatrix} 3 \\ 2 \\ -1 \end{bmatrix}$$

Therefore, by Theorem 6.26,

$$\begin{aligned} [T(3 + 2x - x^2)]_{\mathcal{E}} &= [T(p(x))]_{\mathcal{E}} \\ &= [T]_{\mathcal{E}}[p(x)]_{\mathcal{E}} \\ &= \begin{bmatrix} 1 & -1 & 1 \\ 0 & 2 & -4 \\ 0 & 0 & 4 \end{bmatrix}\begin{bmatrix} 3 \\ 2 \\ -1 \end{bmatrix} = \begin{bmatrix} 0 \\ 8 \\ -4 \end{bmatrix} \end{aligned}$$

It follows that $T(3 + 2x - x^2) = 0 \cdot 1 + 8 \cdot x - 4 \cdot x^2 = 8x - 4x^2$. [Verify this by computing $T(3 + 2x - x^2) = 3 + 2(2x - 1) - (2x - 1)^2$ directly.]

The matrix of a linear transformation can sometimes be used in surprising ways. Example 6.79 shows its application to a traditional calculus problem.

Example 6.79

Let $\mathcal{D}$ be the vector space of all differentiable functions. Consider the subspace W of $\mathcal{D}$ given by $W = \text{span}(e^{3x}, xe^{3x}, x^2e^{3x})$. Since the set $\mathcal{B} = \{e^{3x}, xe^{3x}, x^2e^{3x}\}$ is linearly independent (why?), it is a basis for W.

(a) Show that the differential operator D maps W into itself.

(b) Find the matrix of D with respect to $\mathcal{B}$.

(c) Compute the derivative of $5e^{3x} + 2xe^{3x} - x^2e^{3x}$ indirectly, using Theorem 6.26, and verify it using part (a).

Solution (a) Applying D to a general element of W, we see that

$$D(ae^{3x} + bxe^{3x} + cx^2e^{3x}) = (3a + b)e^{3x} + (3b + 2c)xe^{3x} + 3cx^2e^{3x}$$

(check this), which is again in W.

(b) Using the formula in part (a), we see that

$$D(e^{3x}) = 3e^{3x}, \quad D(xe^{3x}) = e^{3x} + 3xe^{3x}, \quad D(x^2e^{3x}) = 2xe^{3x} + 3x^2e^{3x}$$

so

$$[D(e^{3x})]_{\mathcal{B}} = \begin{bmatrix} 3 \\ 0 \\ 0 \end{bmatrix}, \quad [D(xe^{3x})]_{\mathcal{B}} = \begin{bmatrix} 1 \\ 3 \\ 0 \end{bmatrix}, \quad [D(x^2e^{3x})]_{\mathcal{B}} = \begin{bmatrix} 0 \\ 2 \\ 3 \end{bmatrix}$$

It follows that

$$[D]_{\mathcal{B}} = [[D(e^{3x})]_{\mathcal{B}} \mid [D(xe^{3x})]_{\mathcal{B}} \mid [D(x^2e^{3x})]_{\mathcal{B}}] = \begin{bmatrix} 3 & 1 & 0 \\ 0 & 3 & 2 \\ 0 & 0 & 3 \end{bmatrix}$$

(c) For $f(x) = 5e^{3x} + 2xe^{3x} - x^2e^{3x}$, we see by inspection that

$$[f(x)]_{\mathcal{B}} = \begin{bmatrix} 5 \\ 2 \\ -1 \end{bmatrix}$$

Hence, by Theorem 6.26, we have

$$[D(f(x))]_{\mathcal{B}} = [D]_{\mathcal{B}}[f(x)]_{\mathcal{B}} = \begin{bmatrix} 3 & 1 & 0 \\ 0 & 3 & 2 \\ 0 & 0 & 3 \end{bmatrix}\begin{bmatrix} 5 \\ 2 \\ -1 \end{bmatrix} = \begin{bmatrix} 17 \\ 4 \\ -3 \end{bmatrix}$$

which, in turn, implies that $f'(x) = D(f(x)) = 17e^{3x} + 4xe^{3x} - 3x^2e^{3x}$, in agreement with the formula in part (a).

Remark The point of Example 6.79 is not that this method is easier than direct differentiation. Indeed, once the formula in part (a) has been established, there is little to do. What is significant is that matrix methods can be used at all in what appears, on the surface, to be a calculus problem. We will explore this idea further in Example 6.83.

Example 6.80

Let V be an n-dimensional vector space and let I be the identity transformation on V. What is the matrix of I with respect to bases $\mathcal{B}$ and $\mathcal{C}$ of V if $\mathcal{B} = \mathcal{C}$ (including the order of the basis vectors)? What if $\mathcal{B} \neq \mathcal{C}$?

Solution Let $\mathcal{B} = \{\mathbf{v}_1, \ldots, \mathbf{v}_n\}$. Then $I(\mathbf{v}_1) = \mathbf{v}_1, \ldots, I(\mathbf{v}_n) = \mathbf{v}_n$, so

$$[I(\mathbf{v}_1)]_{\mathcal{B}} = \begin{bmatrix} 1 \\ 0 \\ \vdots \\ 0 \end{bmatrix} = \mathbf{e}_1, \quad [I(\mathbf{v}_2)]_{\mathcal{B}} = \begin{bmatrix} 0 \\ 1 \\ \vdots \\ 0 \end{bmatrix} = \mathbf{e}_2, \quad \ldots, \quad [I(\mathbf{v}_n)]_{\mathcal{B}} = \begin{bmatrix} 0 \\ 0 \\ \vdots \\ 1 \end{bmatrix} = \mathbf{e}_n$$

and, if $\mathcal{B} = \mathcal{C}$,

$$\begin{aligned} [I]_{\mathcal{B}} &= [\,[I(\mathbf{v}_1)]_{\mathcal{B}} \mid [I(\mathbf{v}_2)]_{\mathcal{B}} \mid \cdots \mid [I(\mathbf{v}_n)]_{\mathcal{B}}\,] \\ &= [\,\mathbf{e}_1 \mid \mathbf{e}_2 \mid \cdots \mid \mathbf{e}_n\,] \\ &= I_n \end{aligned}$$

the $n \times n$ identity matrix. (This is what you expected, isn't it?)

In the case $\mathcal{B} \neq \mathcal{C}$, we have

$$[I(\mathbf{v}_1)]_{\mathcal{C}} = [\mathbf{v}_1]_{\mathcal{C}}, \quad \ldots, \quad [I(\mathbf{v}_n)]_{\mathcal{C}} = [\mathbf{v}_n]_{\mathcal{C}}$$

so

$$\begin{aligned} [I]_{\mathcal{C}\leftarrow\mathcal{B}} &= [\,[\mathbf{v}_1]_{\mathcal{C}} \mid \cdots \mid [\mathbf{v}_n]_{\mathcal{C}}\,] \\ &= P_{\mathcal{C}\leftarrow\mathcal{B}} \end{aligned}$$

the change-of-basis matrix from $\mathcal{B}$ to $\mathcal{C}$.

Matrices of Composite and Inverse Linear Transformations

We now generalize Theorems 3.32 and 3.33 to get a theorem that will allow us to easily find the inverse of a linear transformation between finite-dimensional vector spaces (if it exists).

Theorem 6.27 Let U, V, and W be finite-dimensional vector spaces with bases $\mathcal{B}$, $\mathcal{C}$, and $\mathcal{D}$, respectively. Let $T: U \to V$ and $S: V \to W$ be linear transformations. Then

$$[S \circ T]_{\mathcal{D} \leftarrow \mathcal{B}} = [S]_{\mathcal{D} \leftarrow \mathcal{C}}[T]_{\mathcal{C} \leftarrow \mathcal{B}}$$

Remarks

- In words, this theorem says, "The matrix of the composite is the product of the matrices."
- Notice how the "inner subscripts" $\mathcal{C}$ must match and appear to cancel each other out, leaving the "outer subscripts" in the form $\mathcal{D} \leftarrow \mathcal{B}$.

Proof We will show that corresponding columns of the matrices $[S \circ T]_{\mathcal{D} \leftarrow \mathcal{B}}$ and $[S]_{\mathcal{D} \leftarrow \mathcal{C}}[T]_{\mathcal{C} \leftarrow \mathcal{B}}$ are the same. Let $\mathbf{v}_i$ be the ith basis vector in $\mathcal{B}$. Then the ith column of $[S \circ T]_{\mathcal{D} \leftarrow \mathcal{B}}$ is

$$\begin{aligned} [(S \circ T)(\mathbf{v}_i)]_{\mathcal{D}} &= [S(T(\mathbf{v}_i)]_{\mathcal{D}} \\ &= [S]_{\mathcal{D} \leftarrow \mathcal{C}}[T(\mathbf{v}_i)]_{\mathcal{C}} \\ &= [S]_{\mathcal{D} \leftarrow \mathcal{C}}[T]_{\mathcal{C} \leftarrow \mathcal{B}}[\mathbf{v}_i]_{\mathcal{B}} \end{aligned}$$

by two applications of Theorem 6.26. But $[\mathbf{v}_i]_{\mathcal{B}} = \mathbf{e}_i$ (why?), so

$$[S]_{\mathcal{D} \leftarrow \mathcal{C}}[T]_{\mathcal{C} \leftarrow \mathcal{B}}[\mathbf{v}_i]_{\mathcal{B}} = [S]_{\mathcal{D} \leftarrow \mathcal{C}}[T]_{\mathcal{C} \leftarrow \mathcal{B}}\mathbf{e}_i$$

is the ith column of the matrix $[S]_{\mathcal{D} \leftarrow \mathcal{C}}[T]_{\mathcal{C} \leftarrow \mathcal{B}}$. Therefore, the ith columns of $[S \circ T]_{\mathcal{D} \leftarrow \mathcal{B}}$ and $[S]_{\mathcal{D} \leftarrow \mathcal{C}}[T]_{\mathcal{C} \leftarrow \mathcal{B}}$ are the same, as we wished to prove.

Example 6.81 Use matrix methods to compute $(S \circ T)\begin{bmatrix} a \\ b \end{bmatrix}$ for the linear transformations S and T of Example 6.56.

Solution Recall that $T: \mathbb{R}^2 \to \mathscr{P}_1$ and $S: \mathscr{P}_1 \to \mathscr{P}_2$ are defined by

$$T\begin{bmatrix} a \\ b \end{bmatrix} = a + (a + b)x \quad \text{and} \quad S(a + bx) = ax + bx^2$$

Choosing the standard bases $\mathcal{E}$, $\mathcal{E}'$, and $\mathcal{E}''$ for $\mathbb{R}^2$, $\mathscr{P}_1$, and $\mathscr{P}_2$, respectively, we see that

$$[T]_{\mathcal{E}' \leftarrow \mathcal{E}} = \begin{bmatrix} 1 & 0 \\ 1 & 1 \end{bmatrix} \quad \text{and} \quad [S]_{\mathcal{E}'' \leftarrow \mathcal{E}'} = \begin{bmatrix} 0 & 0 \\ 1 & 0 \\ 0 & 1 \end{bmatrix}$$

(Verify these.) By Theorem 6.27, the matrix of $S \circ T$ with respect to $\mathcal{E}$ and $\mathcal{E}''$ is

$$\begin{aligned} [(S \circ T)]_{\mathcal{E}'' \leftarrow \mathcal{E}} &= [S]_{\mathcal{E}'' \leftarrow \mathcal{E}'}[T]_{\mathcal{E}' \leftarrow \mathcal{E}} \\ &= \begin{bmatrix} 0 & 0 \\ 1 & 0 \\ 0 & 1 \end{bmatrix}\begin{bmatrix} 1 & 0 \\ 1 & 1 \end{bmatrix} = \begin{bmatrix} 0 & 0 \\ 1 & 0 \\ 1 & 1 \end{bmatrix} \end{aligned}$$

Thus, by Theorem 6.26,

$$\left[(S \circ T)\begin{bmatrix} a \\ b \end{bmatrix}\right]_{\mathcal{E}''} = [(S \circ T)]_{\mathcal{E}'' \leftarrow \mathcal{E}}\begin{bmatrix} a \\ b \end{bmatrix}_{\mathcal{E}}$$

$$= \begin{bmatrix} 0 & 0 \\ 1 & 0 \\ 1 & 1 \end{bmatrix}\begin{bmatrix} a \\ b \end{bmatrix} = \begin{bmatrix} 0 \\ a \\ a + b \end{bmatrix}$$

Consequently, $(S \circ T)\begin{bmatrix} a \\ b \end{bmatrix} = ax + (a + b)x^2$, which agrees with the solution to Example 6.56.

In Theorem 6.24, we proved that a linear transformation is invertible if and only if it is one-to-one and onto (i.e., if it is an isomorphism). When the vector spaces involved are finite-dimensional, we can use the matrix methods we have developed to find the inverse of such a linear transformation.

Theorem 6.28

Let $T: V \to W$ be a linear transformation between n-dimensional vector spaces V and W and let $\mathcal{B}$ and $\mathcal{C}$ be bases for V and W, respectively. Then T is invertible if and only if the matrix $[T]_{\mathcal{C} \leftarrow \mathcal{B}}$ is invertible. In this case,

$$([T]_{\mathcal{C} \leftarrow \mathcal{B}})^{-1} = [T^{-1}]_{\mathcal{B} \leftarrow \mathcal{C}}$$

Proof Observe that the matrices of T and T^{-1} (if T is invertible) are $n \times n$. If T is invertible, then $T^{-1} \circ T = I_V$. Applying Theorem 6.27, we have

$$\begin{aligned} I_n = [I_V]_{\mathcal{B}} &= [T^{-1} \circ T]_{\mathcal{B}} \\ &= [T^{-1}]_{\mathcal{B} \leftarrow \mathcal{C}}[T]_{\mathcal{C} \leftarrow \mathcal{B}} \end{aligned}$$

This shows that $[T]_{\mathcal{C} \leftarrow \mathcal{B}}$ is invertible and that $([T]_{\mathcal{C} \leftarrow \mathcal{B}})^{-1} = [T^{-1}]_{\mathcal{B} \leftarrow \mathcal{C}}$.

Conversely, assume that $A = [T]_{\mathcal{C} \leftarrow \mathcal{B}}$ is invertible. To show that T is invertible, it is enough to show that $\ker(T) = \{\mathbf{0}\}$. (Why?) To this end, let $\mathbf{v}$ be in the kernel of T. Then $T(\mathbf{v}) = \mathbf{0}$, so

$$A[\mathbf{v}]_{\mathcal{B}} = [T]_{\mathcal{C} \leftarrow \mathcal{B}}[\mathbf{v}]_{\mathcal{B}} = [T(\mathbf{v})]_{\mathcal{C}} = [\mathbf{0}]_{\mathcal{C}} = \mathbf{0}$$

which means that $[\mathbf{v}]_{\mathcal{B}}$ is in the null space of the invertible matrix A. By the Fundamental Theorem, this implies that $[\mathbf{v}]_{\mathcal{B}} = \mathbf{0}$, which, in turn, implies that $\mathbf{v} = \mathbf{0}$, as required.

Example 6.82

In Example 6.70, the linear transformation $T: \mathbb{R}^2 \to \mathscr{P}_1$ defined by

$$T\begin{bmatrix} a \\ b \end{bmatrix} = a + (a + b)x$$

was shown to be one-to-one and onto and hence invertible. Find T^{-1}.

Solution In Example 6.81, we found the matrix of T with respect to the standard bases $\mathcal{E}$ and $\mathcal{E}'$ for $\mathbb{R}^2$ and $\mathcal{P}_1$, respectively, to be

$$[T]_{\mathcal{E}'\leftarrow\mathcal{E}} = \begin{bmatrix} 1 & 0 \\ 1 & 1 \end{bmatrix}$$

By Theorem 6.28, it follows that the matrix of T^{-1} with respect to $\mathcal{E}'$ and $\mathcal{E}$ is

$$[T^{-1}]_{\mathcal{E}\leftarrow\mathcal{E}'} = ([T]_{\mathcal{E}'\leftarrow\mathcal{E}})^{-1} = \begin{bmatrix} 1 & 0 \\ 1 & 1 \end{bmatrix}^{-1} = \begin{bmatrix} 1 & 0 \\ -1 & 1 \end{bmatrix}$$

By Theorem 6.26,

$$\begin{aligned} [T^{-1}(a + bx)]_{\mathcal{E}} &= [T^{-1}]_{\mathcal{E}\leftarrow\mathcal{E}'}[a + bx]_{\mathcal{E}'} \\ &= \begin{bmatrix} 1 & 0 \\ -1 & 1 \end{bmatrix}\begin{bmatrix} a \\ b \end{bmatrix} \\ &= \begin{bmatrix} a \\ b - a \end{bmatrix} \end{aligned}$$

This means that

$$T^{-1}(a + bx) = a\mathbf{e}_1 + (b - a)\mathbf{e}_2 = \begin{bmatrix} a \\ b - a \end{bmatrix}$$

(Note that the choice of the standard basis makes this last calculation virtually irrelevant.)

The next example, a continuation of Example 6.79, shows that matrices can be used in certain integration problems in calculus. The specific integral we consider is usually evaluated in a calculus course by means of two applications of integration by parts. Contrast this approach with our method.

Example 6.83 Show that the differential operator, restricted to the subspace $W = \text{span}(e^{3x}, xe^{3x}, x^2e^{3x})$ of $\mathcal{D}$, is invertible, and use this fact to find the integral

$$\int x^2e^{3x}\,dx$$

Solution In Example 6.79, we found the matrix of D with respect to the basis $\mathcal{B} = \{e^{3x}, xe^{3x}, x^2e^{3x}\}$ of W to be

$$[D]_{\mathcal{B}} = \begin{bmatrix} 3 & 1 & 0 \\ 0 & 3 & 2 \\ 0 & 0 & 3 \end{bmatrix}$$

By Theorem 6.28, therefore, D is invertible on W, and the matrix of D^{-1} is

$$[D^{-1}]_{\mathcal{B}} = ([D]_{\mathcal{B}})^{-1} = \begin{bmatrix} 3 & 1 & 0 \\ 0 & 3 & 2 \\ 0 & 0 & 3 \end{bmatrix}^{-1} = \begin{bmatrix} \frac{1}{3} & -\frac{1}{9} & \frac{2}{27} \\ 0 & \frac{1}{3} & -\frac{2}{9} \\ 0 & 0 & \frac{1}{3} \end{bmatrix}$$

Since integration is *antidifferentiation*, this is the matrix corresponding to integration on W. We want to integrate the function x^2e^{3x} whose coordinate vector is

$$[x^2e^{3x}]_{\mathcal{B}} = \begin{bmatrix} 0 \\ 0 \\ 1 \end{bmatrix}$$

Consequently, by Theorem 6.26,

$$\begin{aligned}\left[\int x^2e^{3x}\,dx\right]_{\mathcal{B}} &= [D^{-1}(x^2e^{3x})]_{\mathcal{B}} \\ &= [D^{-1}]_{\mathcal{B}}[x^2e^{3x}]_{\mathcal{B}} \\ &= \begin{bmatrix} \frac{1}{3} & -\frac{1}{9} & \frac{2}{27} \\ 0 & \frac{1}{3} & -\frac{2}{9} \\ 0 & 0 & \frac{1}{3} \end{bmatrix}\begin{bmatrix} 0 \\ 0 \\ 1 \end{bmatrix} = \begin{bmatrix} \frac{2}{27} \\ -\frac{2}{9} \\ \frac{1}{3} \end{bmatrix}\end{aligned}$$

It follows that

$$\int x^2e^{3x}\,dx = \tfrac{2}{27}e^{3x} - \tfrac{2}{9}xe^{3x} + \tfrac{1}{3}x^2e^{3x}$$

(To be fully correct, we need to add a constant of integration. It does not show up here because we are working with *linear* transformations, which must send zero vectors to zero vectors, forcing the constant of integration to be zero as well.)

Warning In general, differentiation is *not* an invertible transformation. (See Exercise 22.) What the preceding example shows is that, suitably restricted, it sometimes is. Exercises 27–30 explore this idea further.

Change of Basis and Similarity

Suppose $T: V \to V$ is a linear transformation and $\mathcal{B}$ and $\mathcal{C}$ are two different bases for V. It is natural to wonder how, if at all, the matrices $[T]_{\mathcal{B}}$ and $[T]_{\mathcal{C}}$ are related. It turns out that the answer to this question is quite satisfying and relates to some questions we first considered in Chapter 4.

Figure 6.12 suggests one way to address this problem. Chasing the arrows around the diagram from the upper left-hand corner to the lower right-hand corner in two different, but equivalent, ways shows that $I \circ T = T \circ I$, something we already knew, since both are equal to T. However, if the "upper" version of T is with respect to the

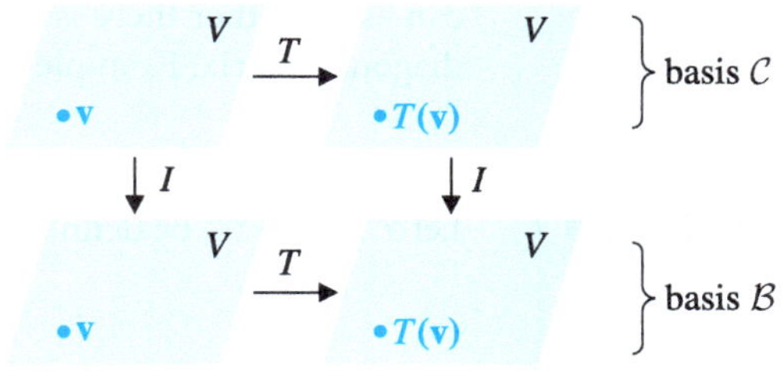

Figure 6.12
$I \circ T = T \circ I$

basis $\mathcal{C}$ and the "lower" version is with respect to $\mathcal{B}$, then $T = I \circ T = T \circ I$ is with respect to $\mathcal{C}$ in its domain and with respect to $\mathcal{B}$ in its codomain. Thus, the matrix of T in this case is $[T]_{\mathcal{B}\leftarrow\mathcal{C}}$. But

$$[T]_{\mathcal{B}\leftarrow\mathcal{C}} = [I \circ T]_{\mathcal{B}\leftarrow\mathcal{C}} = [I]_{\mathcal{B}\leftarrow\mathcal{C}}[T]_{\mathcal{C}\leftarrow\mathcal{C}}$$

and

$$[T]_{\mathcal{B}\leftarrow\mathcal{C}} = [T \circ I]_{\mathcal{B}\leftarrow\mathcal{C}} = [T]_{\mathcal{B}\leftarrow\mathcal{B}}[I]_{\mathcal{B}\leftarrow\mathcal{C}}$$

Therefore, $[I]_{\mathcal{B}\leftarrow\mathcal{C}}[T]_{\mathcal{C}\leftarrow\mathcal{C}} = [T]_{\mathcal{B}\leftarrow\mathcal{B}}[I]_{\mathcal{B}\leftarrow\mathcal{C}}$.

From Example 6.80, we know that $[I]_{\mathcal{B}\leftarrow\mathcal{C}} = P_{\mathcal{B}\leftarrow\mathcal{C}}$, the (invertible) change-of-basis matrix from $\mathcal{C}$ to $\mathcal{B}$. If we denote this matrix by P, then we also have

$$P^{-1} = (P_{\mathcal{B}\leftarrow\mathcal{C}})^{-1} = P_{\mathcal{C}\leftarrow\mathcal{B}}$$

With this notation,

$$P[T]_{\mathcal{C}\leftarrow\mathcal{C}} = [T]_{\mathcal{B}\leftarrow\mathcal{B}}P$$

so

$$[T]_{\mathcal{C}\leftarrow\mathcal{C}} = P^{-1}[T]_{\mathcal{B}\leftarrow\mathcal{B}}P \quad \text{or} \quad [T]_{\mathcal{C}} = P^{-1}[T]_{\mathcal{B}}P$$

Thus, the matrices $[T]_{\mathcal{B}}$ and $[T]_{\mathcal{C}}$ are similar, in the terminology of Section 4.4.

We summarize the foregoing discussion as a theorem.

Theorem 6.29 Let V be a finite-dimensional vector space with bases $\mathcal{B}$ and $\mathcal{C}$ and let $T: V \to V$ be a linear transformation. Then

$$[T]_{\mathcal{C}} = P^{-1}[T]_{\mathcal{B}}P$$

where P is the change-of-basis matrix from $\mathcal{C}$ to $\mathcal{B}$.

Remark As an aid in remembering that P must be the change-of-basis matrix from $\mathcal{C}$ to $\mathcal{B}$, and not $\mathcal{B}$ to $\mathcal{C}$, it is instructive to look at what Theorem 6.29 says when written in full detail. As shown below, the "inner subscripts" must be the same (all $\mathcal{B}$s) and must appear to cancel, leaving the "outer subscripts," which are both $\mathcal{C}$s.

$$[T]_{\mathcal{C}\leftarrow\mathcal{C}} = P_{\mathcal{C}\leftarrow\mathcal{B}}[T]_{\mathcal{B}\leftarrow\mathcal{B}}P_{\mathcal{B}\leftarrow\mathcal{C}}$$

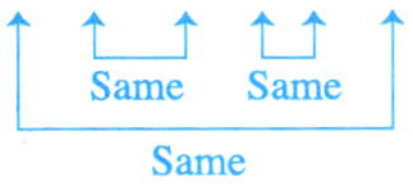

Theorem 6.29 is often used when we are trying to find a basis with respect to which the matrix of a linear transformation is particularly simple. For example, we can ask whether there is a basis $\mathcal{C}$ of V such that the matrix $[T]_{\mathcal{C}}$ of $T: V \to V$ is a diagonal matrix. Example 6.84 illustrates this application.

Example 6.84 Let $T: \mathbb{R}^2 \to \mathbb{R}^2$ be defined by

$$T\begin{bmatrix} x \\ y \end{bmatrix} = \begin{bmatrix} x + 3y \\ 2x + 2y \end{bmatrix}$$

If possible, find a basis $\mathcal{C}$ for $\mathbb{R}^2$ such that the matrix of T with respect to $\mathcal{C}$ is diagonal.

Solution The matrix of T with respect to the standard basis $\mathcal{E}$ is

$$[T]_{\mathcal{E}} = \begin{bmatrix} 1 & 3 \\ 2 & 2 \end{bmatrix}$$

This matrix is diagonalizable, as we saw in Example 4.24. Indeed, if

$$P = \begin{bmatrix} 1 & 3 \\ 1 & -2 \end{bmatrix} \quad \text{and} \quad D = \begin{bmatrix} 4 & 0 \\ 0 & -1 \end{bmatrix}$$

then $P^{-1}[T]_{\mathcal{E}}P = D$. If we let $\mathcal{C}$ be the basis of $\mathbb{R}^2$ consisting of the columns of P, then P is the change-of-basis matrix $P_{\mathcal{E}\leftarrow\mathcal{C}}$ from $\mathcal{C}$ to $\mathcal{E}$. By Theorem 6.29,

$$[T]_{\mathcal{C}} = P^{-1}[T]_{\mathcal{E}}P = D$$

so the matrix of T with respect to the basis $\mathcal{C} = \left\{ \begin{bmatrix} 1 \\ 1 \end{bmatrix}, \begin{bmatrix} 3 \\ -2 \end{bmatrix} \right\}$ is diagonal.

Remarks

- It is easy to check that the solution above is correct by computing $[T]_{\mathcal{C}}$ directly. We find that

$$T\begin{bmatrix} 1 \\ 1 \end{bmatrix} = \begin{bmatrix} 4 \\ 4 \end{bmatrix} = 4\begin{bmatrix} 1 \\ 1 \end{bmatrix} + 0\begin{bmatrix} 3 \\ -2 \end{bmatrix} \quad \text{and} \quad T\begin{bmatrix} 3 \\ -2 \end{bmatrix} = \begin{bmatrix} -3 \\ 2 \end{bmatrix} = 0\begin{bmatrix} 1 \\ 1 \end{bmatrix} - \begin{bmatrix} 3 \\ -2 \end{bmatrix}$$

Thus, the coordinate vectors that form the columns of $[T]_{\mathcal{C}}$ are

$$\left[T\begin{bmatrix} 1 \\ 1 \end{bmatrix}\right]_{\mathcal{C}} = \begin{bmatrix} 4 \\ 0 \end{bmatrix} \quad \text{and} \quad \left[T\begin{bmatrix} 3 \\ -2 \end{bmatrix}\right]_{\mathcal{C}} = \begin{bmatrix} 0 \\ -1 \end{bmatrix}$$

in agreement with our solution above.

- The general procedure for a problem like Example 6.84 is to take the standard matrix $[T]_{\mathcal{E}}$ and determine whether it is diagonalizable by finding bases for its eigenspaces, as in Chapter 4. The solution then proceeds exactly as in the preceding example.

Example 6.84 motivates the following definition.

Definition Let V be a finite-dimensional vector space and let $T: V \to V$ be a linear transformation. Then T is called ***diagonalizable*** if there is a basis $\mathcal{C}$ for V such that the matrix $[T]_{\mathcal{C}}$ is a diagonal matrix.

It is not hard to show that if $\mathcal{B}$ is *any* basis for V, then T is diagonalizable if and only if the matrix $[T]_{\mathcal{B}}$ is diagonalizable. This is essentially what we did, for a special case, in the last example. You are asked to prove this result in general in Exercise 42.

Sometimes it is easiest to write down the matrix of a linear transformation with respect to a "nonstandard" basis. We can then reverse the process of Example 6.84 to find the standard matrix. We illustrate this idea by revisiting Example 3.59.

Example 6.85 Let ℓ be the line through the origin in $\mathbb{R}^2$ with direction vector $\mathbf{d} = \begin{bmatrix} d_1 \\ d_2 \end{bmatrix}$. Find the standard matrix of the projection onto ℓ.

Solution Let T denote the projection. There is no harm in assuming that $\mathbf{d}$ is a unit vector (i.e., $d_1^2 + d_2^2 = 1$), since any nonzero multiple of $\mathbf{d}$ can serve as a direction vector for ℓ. Let $\mathbf{d}' = \begin{bmatrix} -d_2 \\ d_1 \end{bmatrix}$ so that $\mathbf{d}$ and $\mathbf{d}'$ are orthogonal. Since $\mathbf{d}'$ is also a unit vector, the set $\mathcal{D} = \{\mathbf{d}, \mathbf{d}'\}$ is an orthonormal basis for $\mathbb{R}^2$. As Figure 6.13 shows, $T(\mathbf{d}) = \mathbf{d}$ and $T(\mathbf{d}') = \mathbf{0}$. Therefore,

$$[T(\mathbf{d})]_{\mathcal{D}} = \begin{bmatrix} 1 \\ 0 \end{bmatrix} \quad \text{and} \quad [T(\mathbf{d}')]_{\mathcal{D}} = \begin{bmatrix} 0 \\ 0 \end{bmatrix}$$

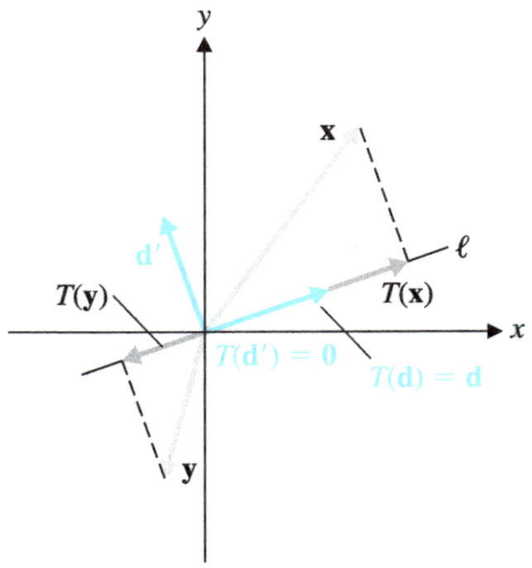

Figure 6.13
Projection onto ℓ

so
$$[T]_{\mathcal{D}} = \begin{bmatrix} 1 & 0 \\ 0 & 0 \end{bmatrix}$$

The change-of-basis matrix from $\mathcal{D}$ to the standard basis $\mathcal{E}$ is

$$P_{\mathcal{E}\leftarrow\mathcal{D}} = \begin{bmatrix} d_1 & -d_2 \\ d_2 & d_1 \end{bmatrix}$$

so the change-of-basis matrix from $\mathcal{E}$ to $\mathcal{D}$ is

$$P_{\mathcal{D}\leftarrow\mathcal{E}} = (P_{\mathcal{E}\leftarrow\mathcal{D}})^{-1} = \begin{bmatrix} d_1 & -d_2 \\ d_2 & d_1 \end{bmatrix}^{-1} = \begin{bmatrix} d_1 & d_2 \\ -d_2 & d_1 \end{bmatrix}$$

By Theorem 6.29, then, the standard matrix of T is

$$\begin{aligned} [T]_{\mathcal{E}} &= P_{\mathcal{E}\leftarrow\mathcal{D}}[T]_{\mathcal{D}}P_{\mathcal{D}\leftarrow\mathcal{E}} \\ &= \begin{bmatrix} d_1 & -d_2 \\ d_2 & d_1 \end{bmatrix}\begin{bmatrix} 1 & 0 \\ 0 & 0 \end{bmatrix}\begin{bmatrix} d_1 & d_2 \\ -d_2 & d_1 \end{bmatrix} \\ &= \begin{bmatrix} d_1^2 & d_1d_2 \\ d_1d_2 & d_2^2 \end{bmatrix} \end{aligned}$$

which agrees with part (b) of Example 3.59.

Example 6.86

Let $T : \mathscr{P}_2 \to \mathscr{P}_2$ be the linear transformation defined by

$$T(p(x)) = p(2x - 1)$$

(a) Find the matrix of T with respect to the basis $\mathcal{B} = \{1 + x, 1 - x, x^2\}$ of $\mathscr{P}_2$.

(b) Show that T is diagonalizable and find a basis $\mathcal{C}$ for $\mathscr{P}_2$ such that $[T]_{\mathcal{C}}$ is a diagonal matrix.

Solution (a) In Example 6.78, we found that the matrix of T with respect to the standard basis $\mathcal{E} = \{1, x, x^2\}$ is

$$[T]_{\mathcal{E}} = \begin{bmatrix} 1 & -1 & 1 \\ 0 & 2 & -4 \\ 0 & 0 & 4 \end{bmatrix}$$

The change-of-basis matrix from $\mathcal{B}$ to $\mathcal{E}$ is

$$P = P_{\mathcal{E} \leftarrow \mathcal{B}} = \begin{bmatrix} 1 & 1 & 0 \\ 1 & -1 & 0 \\ 0 & 0 & 1 \end{bmatrix}$$

It follows that the matrix of T with respect to $\mathcal{B}$ is

$$\begin{aligned} [T]_{\mathcal{B}} &= P^{-1}[T]_{\mathcal{E}}P \\ &= \begin{bmatrix} \frac{1}{2} & \frac{1}{2} & 0 \\ \frac{1}{2} & -\frac{1}{2} & 0 \\ 0 & 0 & 1 \end{bmatrix} \begin{bmatrix} 1 & -1 & 1 \\ 0 & 2 & -4 \\ 0 & 0 & 4 \end{bmatrix} \begin{bmatrix} 1 & 1 & 0 \\ 1 & -1 & 0 \\ 0 & 0 & 1 \end{bmatrix} \\ &= \begin{bmatrix} 1 & 0 & -\frac{3}{2} \\ -1 & 2 & \frac{5}{2} \\ 0 & 0 & 4 \end{bmatrix} \end{aligned}$$

(Check this.)

(b) The eigenvalues of $[T]_{\mathcal{E}}$ are 1, 2, and 4 (why?), so we know from Theorem 4.25 that $[T]_{\mathcal{E}}$ is diagonalizable. Eigenvectors corresponding to these eigenvalues are

$$\begin{bmatrix} 1 \\ 0 \\ 0 \end{bmatrix}, \begin{bmatrix} -1 \\ 1 \\ 0 \end{bmatrix}, \begin{bmatrix} 1 \\ -2 \\ 1 \end{bmatrix}$$

respectively. Therefore, setting

$$P = \begin{bmatrix} 1 & -1 & 1 \\ 0 & 1 & -2 \\ 0 & 0 & 1 \end{bmatrix} \quad \text{and} \quad D = \begin{bmatrix} 1 & 0 & 0 \\ 0 & 2 & 0 \\ 0 & 0 & 4 \end{bmatrix}$$

we have $P^{-1}[T]_{\mathcal{E}}P = D$. Furthermore, P is the change-of-basis matrix from a basis $\mathcal{C}$ to $\mathcal{E}$, and the columns of P are thus the coordinate vectors of $\mathcal{C}$ in terms of $\mathcal{E}$. It follows that

$$\mathcal{C} = \{1, -1 + x, 1 - 2x + x^2\}$$

and $[T]_{\mathcal{C}} = D$.

The preceding ideas can be generalized to relate the matrices $[T]_{\mathcal{C}\leftarrow\mathcal{B}}$ and $[T]_{\mathcal{C}'\leftarrow\mathcal{B}'}$ of a linear transformation $T: V \to W$, where $\mathcal{B}$ and $\mathcal{B}'$ are bases for V and $\mathcal{C}$ and $\mathcal{C}'$ are bases for W. (See Exercise 44.)

We conclude this section by revisiting the Fundamental Theorem of Invertible Matrices and incorporating some results from this chapter.

Theorem 6.30 **The Fundamental Theorem of Invertible Matrices: Version 4**

Let A be an $n \times n$ matrix and let $T: V \to W$ be a linear transformation whose matrix $[T]_{\mathcal{C}\leftarrow\mathcal{B}}$ with respect to bases $\mathcal{B}$ and $\mathcal{C}$ of V and W, respectively, is A. The following statements are equivalent:

a. A is invertible.
b. $A\mathbf{x} = \mathbf{b}$ has a unique solution for every $\mathbf{b}$ in $\mathbb{R}^n$.
c. $A\mathbf{x} = \mathbf{0}$ has only the trivial solution.
d. The reduced row echelon form of A is I_n.
e. A is a product of elementary matrices.
f. $\text{rank}(A) = n$
g. $\text{nullity}(A) = 0$
h. The column vectors of A are linearly independent.
i. The column vectors of A span $\mathbb{R}^n$.
j. The column vectors of A form a basis for $\mathbb{R}^n$.
k. The row vectors of A are linearly independent.
l. The row vectors of A span $\mathbb{R}^n$.
m. The row vectors of A form a basis for $\mathbb{R}^n$.
n. $\det A \neq 0$
o. 0 is not an eigenvalue of A.
p. T is invertible.
q. T is one-to-one.
r. T is onto.
s. $\ker(T) = \{\mathbf{0}\}$
t. $\text{range}(T) = W$

Proof The equivalence (q) $\Leftrightarrow$ (s) is Theorem 6.20, and (r) $\Leftrightarrow$ (t) is the definition of onto. Since A is $n \times n$, we must have $\dim V = \dim W = n$. From Theorems 6.21 and 6.24, we get (p) $\Leftrightarrow$ (q) $\Leftrightarrow$ (r). Finally, we connect the last five statements to the others by Theorem 6.28, which implies that (a) $\Leftrightarrow$ (p).

Exercises 6.6

In Exercises 1–12, find the matrix $[T]_{\mathcal{C}\leftarrow\mathcal{B}}$ of the linear transformation $T: V \to W$ with respect to the bases $\mathcal{B}$ and $\mathcal{C}$ of V and W, respectively. Verify Theorem 6.26 for the vector $\mathbf{v}$ *by computing $T(\mathbf{v})$ directly and using the theorem.*

1. $T: \mathscr{P}_1 \to \mathscr{P}_1$ defined by $T(a + bx) = b - ax$, $\mathcal{B} = \mathcal{C} = \{1, x\}$, $\mathbf{v} = p(x) = 4 + 2x$

2. $T: \mathscr{P}_1 \to \mathscr{P}_1$ defined by $T(a + bx) = b - ax$, $\mathcal{B} = \{1 + x, 1 - x\}$, $\mathcal{C} = \{1, x\}$, $\mathbf{v} = p(x) = 4 + 2x$

3. $T: \mathscr{P}_2 \to \mathscr{P}_2$ defined by $T(p(x)) = p(x + 2)$, $\mathcal{B} = \{1, x, x^2\}$, $\mathcal{C} = \{1, x + 2, (x + 2)^2\}$, $\mathbf{v} = p(x) = a + bx + cx^2$

4. $T: \mathscr{P}_2 \to \mathscr{P}_2$ defined by $T(p(x)) = p(x + 2)$, $\mathcal{B} = \{1, x + 2, (x + 2)^2\}, \mathcal{C} = \{1, x, x^2\}$, $\mathbf{v} = p(x) = a + bx + cx^2$

5. $T: \mathscr{P}_2 \to \mathbb{R}^2$ defined by $T(p(x)) = \begin{bmatrix} p(0) \\ p(1) \end{bmatrix}$, $\mathcal{B} = \{1, x, x^2\}, \mathcal{C} = \{\mathbf{e}_1, \mathbf{e}_2\}$, $\mathbf{v} = p(x) = a + bx + cx^2$

6. $T: \mathscr{P}_2 \to \mathbb{R}^2$ defined by $T(p(x)) = \begin{bmatrix} p(0) \\ p(1) \end{bmatrix}$, $\mathcal{B} = \{x^2, x, 1\}, \mathcal{C} = \left\{\begin{bmatrix} 1 \\ 0 \end{bmatrix}, \begin{bmatrix} 1 \\ 1 \end{bmatrix}\right\}$, $\mathbf{v} = p(x) = a + bx + cx^2$

7. $T: \mathbb{R}^2 \to \mathbb{R}^3$ defined by

$$T\begin{bmatrix} a \\ b \end{bmatrix} = \begin{bmatrix} a + 2b \\ -a \\ b \end{bmatrix}, \quad \mathcal{B} = \left\{\begin{bmatrix} 1 \\ 2 \end{bmatrix}, \begin{bmatrix} 3 \\ -1 \end{bmatrix}\right\},$$

$$\mathcal{C} = \left\{\begin{bmatrix} 1 \\ 0 \\ 0 \end{bmatrix}, \begin{bmatrix} 1 \\ 1 \\ 0 \end{bmatrix}, \begin{bmatrix} 1 \\ 1 \\ 1 \end{bmatrix}\right\}, \quad \mathbf{v} = \begin{bmatrix} -7 \\ 7 \end{bmatrix}$$

8. Repeat Exercise 7 with $\mathbf{v} = \begin{bmatrix} a \\ b \end{bmatrix}$.

9. $T: M_{22} \to M_{22}$ defined by $T(A) = A^T, \mathcal{B} = \mathcal{C} = \{E_{11}, E_{12}, E_{21}, E_{22}\}, \mathbf{v} = A = \begin{bmatrix} a & b \\ c & d \end{bmatrix}$

10. Repeat Exercise 9 with $\mathcal{B} = \{E_{22}, E_{21}, E_{12}, E_{11}\}$ and $\mathcal{C} = \{E_{12}, E_{21}, E_{22}, E_{11}\}$.

11. $T: M_{22} \to M_{22}$ defined by $T(A) = AB - BA$, where $B = \begin{bmatrix} 1 & -1 \\ -1 & 1 \end{bmatrix}, \mathcal{B} = \mathcal{C} = \{E_{11}, E_{12}, E_{21}, E_{22}\}$, $\mathbf{v} = A = \begin{bmatrix} a & b \\ c & d \end{bmatrix}$

12. $T: M_{22} \to M_{22}$ defined by $T(A) = A - A^T, \mathcal{B} = \mathcal{C} = \{E_{11}, E_{12}, E_{21}, E_{22}\}, \mathbf{v} = A = \begin{bmatrix} a & b \\ c & d \end{bmatrix}$

13. Consider the subspace W of $\mathscr{D}$, given by $W = \text{span}(\sin x, \cos x)$.

(a) Show that the differential operator D maps W into itself.

(b) Find the matrix of D with respect to $\mathcal{B} = \{\sin x, \cos x\}$.

(c) Compute the derivative of $f(x) = 3 \sin x - 5 \cos x$ indirectly, using Theorem 6.26, and verify that it agrees with $f'(x)$ as computed directly.

14. Consider the subspace W of $\mathscr{D}$, given by $W = \text{span}(e^{2x}, e^{-2x})$.

(a) Show that the differential operator D maps W into itself.

(b) Find the matrix of D with respect to $\mathcal{B} = \{e^{2x}, e^{-2x}\}$.

(c) Compute the derivative of $f(x) = e^{2x} - 3e^{-2x}$ indirectly, using Theorem 6.26, and verify that it agrees with $f'(x)$ as computed directly.

15. Consider the subspace W of $\mathscr{D}$, given by $W = \text{span}(e^{2x}, e^{2x} \cos x, e^{2x} \sin x)$.

(a) Find the matrix of D with respect to $\mathcal{B} = \{e^{2x}, e^{2x} \cos x, e^{2x} \sin x\}$.

(b) Compute the derivative of $f(x) = 3e^{2x} - e^{2x} \cos x + 2e^{2x} \sin x$ indirectly, using Theorem 6.26, and verify that it agrees with $f'(x)$ as computed directly.

16. Consider the subspace W of $\mathscr{D}$, given by $W = \text{span}(\cos x, \sin x, x \cos x, x \sin x)$.

(a) Find the matrix of D with respect to $\mathcal{B} = \{\cos x, \sin x, x \cos x, x \sin x\}$.

(b) Compute the derivative of $f(x) = \cos x + 2x \cos x$ indirectly, using Theorem 6.26, and verify that it agrees with $f'(x)$ as computed directly.

In Exercises 17 and 18, $T: U \to V$ and $S: V \to W$ are linear transformations and $\mathcal{B}$, $\mathcal{C}$, and $\mathcal{D}$ are bases for U, V, and W, respectively. Compute $[S \circ T]_{\mathcal{D} \leftarrow \mathcal{B}}$ in two ways: (a) by finding $S \circ T$ directly and then computing its matrix and (b) by finding the matrices of S and T separately and using Theorem 6.27.

17. $T: \mathscr{P}_1 \to \mathbb{R}^2$ defined by $T(p(x)) = \begin{bmatrix} p(0) \\ p(1) \end{bmatrix}$, $S: \mathbb{R}^2 \to \mathbb{R}^2$ defined by $S\begin{bmatrix} a \\ b \end{bmatrix} = \begin{bmatrix} a - 2b \\ 2a - b \end{bmatrix}$, $\mathcal{B} = \{1, x\}$, $\mathcal{C} = \mathcal{D} = \{\mathbf{e}_1, \mathbf{e}_2\}$

18. $T: \mathscr{P}_1 \to \mathscr{P}_2$ defined by $T(p(x)) = p(x + 1)$, $S: \mathscr{P}_2 \to \mathscr{P}_2$ defined by $S(p(x)) = p(x + 1)$, $\mathcal{B} = \{1, x\}, \mathcal{C} = \mathcal{D} = \{1, x, x^2\}$

In Exercises 19–26, determine whether the linear transformation T is invertible by considering its matrix with respect to the standard bases. If T is invertible, use Theorem 6.28 and the method of Example 6.82 to find T^{-1}.

19. T in Exercise 1

20. T in Exercise 5

21. T in Exercise 3

22. $T: \mathscr{P}_2 \to \mathscr{P}_2$ defined by $T(p(x)) = p'(x)$

23. $T: \mathscr{P}_2 \to \mathscr{P}_2$ defined by $T(p(x)) = p(x) + p'(x)$

24. $T: M_{22} \to M_{22}$ defined by $T(A) = AB$, where

$$B = \begin{bmatrix} 3 & 2 \\ 2 & 1 \end{bmatrix}$$

25. T in Exercise 11

26. T in Exercise 12

In Exercises 27–30, use the method of Example 6.83 to evaluate the given integral.

27. $\int (\sin x - 3\cos x)\, dx$. (See Exercise 13.)

28. $\int 5e^{-2x}\, dx$. (See Exercise 14.)

29. $\int (e^{2x}\cos x - 2e^{2x}\sin x)\, dx$. (See Exercise 15.)

30. $\int (x\cos x + x\sin x)\, dx$. (See Exercise 16.)

In Exercises 31–36, a linear transformation $T: V \to V$ is given. If possible, find a basis $\mathcal{C}$ for V such that the matrix $[T]_{\mathcal{C}}$ of T with respect to $\mathcal{C}$ is diagonal.

31. $T: \mathbb{R}^2 \to \mathbb{R}^2$ defined by $T\begin{bmatrix} a \\ b \end{bmatrix} = \begin{bmatrix} -4b \\ a + 5b \end{bmatrix}$

32. $T: \mathbb{R}^2 \to \mathbb{R}^2$ defined by $T\begin{bmatrix} a \\ b \end{bmatrix} = \begin{bmatrix} a - b \\ a + b \end{bmatrix}$

33. $T: \mathscr{P}_1 \to \mathscr{P}_1$ defined by $T(a + bx) = (4a + 2b) + (a + 3b)x$

34. $T: \mathscr{P}_2 \to \mathscr{P}_2$ defined by $T(p(x)) = p(x + 1)$

35. $T: \mathscr{P}_1 \to \mathscr{P}_1$ defined by $T(p(x)) = p(x) + xp'(x)$

36. $T: \mathscr{P}_2 \to \mathscr{P}_2$ defined by $T(p(x)) = p(3x + 2)$

37. Let ℓ be the line through the origin in $\mathbb{R}^2$ with direction vector $\mathbf{d} = \begin{bmatrix} d_1 \\ d_2 \end{bmatrix}$. Use the method of Example 6.85 to find the standard matrix of a reflection in ℓ.

38. Let W be the plane in $\mathbb{R}^3$ with equation $x - y + 2z = 0$. Use the method of Example 6.85 to find the standard matrix of an orthogonal projection onto W. Verify that your answer is correct by using it to compute the orthogonal projection of $\mathbf{v}$ onto W, where

$$\mathbf{v} = \begin{bmatrix} 3 \\ -1 \\ 2 \end{bmatrix}$$

Compare your answer with Example 5.11. [*Hint:* Find an orthogonal decomposition of $\mathbb{R}^3$ as $\mathbb{R}^3 = W + W^{\perp}$ using an orthogonal basis for W. See Example 5.3.]

39. Let $T: V \to W$ be a linear transformation between finite-dimensional vector spaces and let $\mathcal{B}$ and $\mathcal{C}$ be bases for V and W, respectively. Show that the matrix of T with respect to $\mathcal{B}$ and $\mathcal{C}$ is unique. That is, if A is a matrix such that $A[\mathbf{v}]_{\mathcal{B}} = [T(\mathbf{v})]_{\mathcal{C}}$ for all $\mathbf{v}$ in V, then $A = [T]_{\mathcal{C}\leftarrow\mathcal{B}}$. [*Hint:* Find values of $\mathbf{v}$ that will show this, one column at a time.]

In Exercises 40–45, let $T: V \to W$ be a linear transformation between finite-dimensional vector spaces V and W. Let $\mathcal{B}$ and $\mathcal{C}$ be bases for V and W, respectively, and let $A = [T]_{\mathcal{C}\leftarrow\mathcal{B}}$.

40. Show that nullity(T) = nullity(A).

41. Show that rank(T) = rank(A).

42. If $V = W$ and $\mathcal{B} = \mathcal{C}$, show that T is diagonalizable if and only if A is diagonalizable.

43. Use the results of this section to give a matrix-based proof of the Rank Theorem (Theorem 6.19).

44. If $\mathcal{B}'$ and $\mathcal{C}'$ are also bases for V and W, respectively, what is the relationship between $[T]_{\mathcal{C}\leftarrow\mathcal{B}}$ and $[T]_{\mathcal{C}'\leftarrow\mathcal{B}'}$? Prove your assertion.

45. If dim $V = n$ and dim $W = m$, prove that $\mathscr{L}(V, W) \cong M_{mn}$. (See the exercises for Section 6.4.) [*Hint:* Let $\mathcal{B}$ and $\mathcal{C}$ be bases for V and W, respectively. Show that the mapping $\varphi(T) = [T]_{\mathcal{C}\leftarrow\mathcal{B}}$, for T in $\mathscr{L}(V, W)$, defines a linear transformation $\varphi: \mathscr{L}(V, W) \to M_{mn}$ that is an isomorphism.]

46. If V is a vector space, then the ***dual space*** of V is the vector space $V^* = \mathscr{L}(V, \mathbb{R})$. Prove that if V is finite-dimensional, then $V^* \cong V$.

Exploration

Tilings, Lattices, and the Crystallographic Restriction

Repeating patterns are frequently found in nature and in art. The molecular structure of crystals often exhibits repetition, as do the tilings and mosaics found in the artwork of many cultures. *Tiling* (or *tessellation*) is covering of a plane by shapes that do not overlap and leave no gaps. The Dutch artist M. C. Escher (1898–1972) produced many works in which he explored the possibility of tiling a plane using fanciful shapes (Figure 6.14).

Figure 6.14

M. C. Escher's "Symmetry Drawing E103"

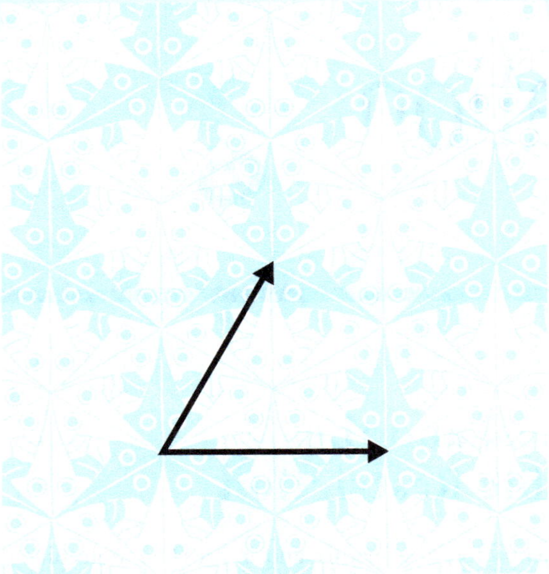

Figure 6.15
Invariance under translation
M. C. Escher's "Symmetry Drawing E103"

Figure 6.17
Rotational symmetry
M. C. Escher's "Symmetry Drawing E103"

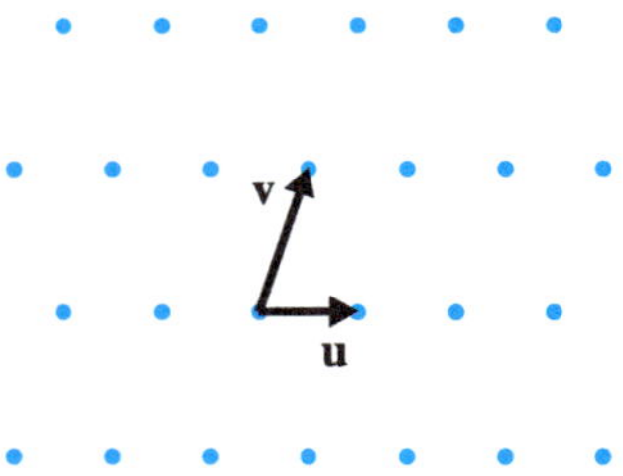

Figure 6.16
A lattice

In this exploration, we will be interested in patterns such as those in Figure 6.14, which we assume to be infinite and repeating in all directions of the plane. Such a pattern has the property that it can be shifted (or *translated*) in at least two directions (corresponding to two linearly independent vectors) so that it appears not to have been moved at all. We say that the pattern is *invariant* under translations and has ***translational symmetry*** in these directions. For example, the pattern in Figure 6.14 has translational symmetry in the directions shown in Figure 6.15.

If a pattern has translational symmetry in two directions, it has translational symmetry in infinitely many directions.

1. Let the two vectors shown in Figure 6.15 be denoted by $\mathbf{u}$ and $\mathbf{v}$. Show that the pattern in Figure 6.14 is invariant under translation by any *integer* linear combination of $\mathbf{u}$ and $\mathbf{v}$—that is, by any vector of the form $a\mathbf{u} + b\mathbf{v}$, where a and b are integers.

For any two linearly independent vectors $\mathbf{u}$ and $\mathbf{v}$ in $\mathbb{R}^2$, the set of points determined by all integer linear combinations of $\mathbf{u}$ and $\mathbf{v}$ is called a ***lattice.*** Figure 6.16 shows an example of a lattice.

2. Draw the lattice corresponding to the vectors $\mathbf{u}$ and $\mathbf{v}$ of Figure 6.15.

Figure 6.14 also exhibits ***rotational symmetry.*** That is, it is possible to rotate the entire pattern about some point and have it appear unchanged. We say that it is *invariant* under such a rotation. For example, the pattern of Figure 6.14 is invariant under a rotation of 120° about the point O, as shown in Figure 6.17. We call O a ***center*** of rotational symmetry (or a ***rotation center***).

Note that if a pattern is based on an underlying lattice, then any symmetries of the pattern must also be possessed by the lattice.

3. Explain why, if a point O is a rotation center through an angle θ, then it is a rotation center through every integer multiple of θ. Deduce that if $0 < \theta \leq 360°$, then $360/\theta$ must be an integer. (If $360/\theta = n$, we say the pattern or lattice has ***n-fold*** rotational symmetry.)

4. What is the smallest positive angle of rotational symmetry for the lattice in Problem 2? Does the pattern in Figure 6.14 also have rotational symmetry through this angle?

5. Take various values of θ such that $0 < \theta \leq 360°$ and $360/\theta$ is an integer. Try to draw a lattice that has rotational symmetry through the angle θ. In particular, can you draw a lattice with eight-fold rotational symmetry?

We will show that values of θ that are possible angles of rotational symmetry for a lattice are severely restricted. The technique we will use is to consider rotation transformations in terms of different bases. Accordingly, let R_θ denote a rotation about the origin through an angle θ and let $\mathcal{E}$ be the standard basis for $\mathbb{R}^2$. Then the standard matrix of R_θ is

$$[R_\theta]_{\mathcal{E}} = \begin{bmatrix} \cos\theta & -\sin\theta \\ \sin\theta & \cos\theta \end{bmatrix}$$

6. Referring to Problems 2 and 4, take the origin to be at the tails of **u** and **v**.
   (a) What is the actual (i.e., numerical) value of $[R_\theta]_{\mathcal{E}}$ in this case?
   (b) Let $\mathcal{B}$ be the basis $\{\mathbf{u}, \mathbf{v}\}$. Compute the matrix $[R_\theta]_{\mathcal{B}}$.

7. In general, let **u** and **v** be any two linearly independent vectors in $\mathbb{R}^2$ and suppose that the lattice determined by **u** and **v** is invariant under a rotation through an angle θ. If $\mathcal{B} = \{\mathbf{u}, \mathbf{v}\}$, show that the matrix of R_θ with respect to $\mathcal{B}$ must have the form

$$[R_\theta]_{\mathcal{B}} = \begin{bmatrix} a & b \\ c & d \end{bmatrix}$$

where a, b, c, and d are *integers*.

8. In the terminology and notation of Problem 7, show that $2\cos\theta$ must be an integer. [*Hint:* Use Exercise 35 in Section 4.4 and Theorem 6.29.]

9. Using Problem 8, make a list of all possible values of θ, with $0 < \theta \leq 360°$, that can be angles of rotational symmetry of a lattice. Record the corresponding values of n, where $n = 360/\theta$, to show that a lattice can have n-fold rotational symmetry if and only if $n = 1, 2, 3, 4,$ or 6. This result, known as the ***crystallographic restriction,*** was first proved by W. Barlow in 1894.

10. In the library or on the Internet, see whether you can find an Escher tiling for each of the five possible types of rotational symmetry—that is, where the *smallest* angle of rotational symmetry of the pattern is one of those specified by the crystallographic restriction.

6.7 Applications

Homogeneous Linear Differential Equations

In Exercises 69–72 in Section 4.6, we showed that if $y = y(t)$ is a twice-differentiable function that satisfies the differential equation

$$y'' + ay' + by = 0 \tag{1}$$

then y is of the form

$$y = c_1e^{\lambda_1 t} + c_2e^{\lambda_2 t}$$

if λ_1 and λ_2 are *distinct* roots of the associated characteristic equation $\lambda^2 + a\lambda + b = 0$. (The case where $\lambda_1 = \lambda_2$ was left unresolved.) Example 6.12 and Exercise 20 in this section show that the set of solutions to Equation (1) forms a subspace of $\mathscr{F}$, the vector space of functions. In this section, we pursue these ideas further, paying particular attention to the role played by vector spaces, bases, and dimension.

To set the stage, we consider a simpler class of examples. A differential equation of the form

$$y' + ay = 0 \tag{2}$$

is called a ***first-order, homogeneous, linear differential equation.*** ("First-order" refers to the fact that the highest derivative that is involved is a first derivative, and "homogeneous" means that the right-hand side is zero. Do you see why the equation is "linear"?) A ***solution*** to Equation (2) is a differentiable function $y = y(t)$ that satisfies Equation (2) for all values of t.

It is easy to check that one solution to Equation (2) is $y = e^{-at}$. (Do it.) However, we would like to describe *all* solutions—and this is where vector spaces come in. We have the following theorem.

Theorem 6.31 The set S of all solutions to $y' + ay = 0$ is a subspace of $\mathscr{F}$.

Proof Since the zero function certainly satisfies Equation (2), S is nonempty. Let x and y be two differentiable functions of t that are in S and let c be a scalar. Then

$$x' + ax = 0 \quad \text{and} \quad y' + ay = 0$$

so, using rules for differentiation, we have

$$(x + y)' + a(x + y) = x' + y' + ax + ay = (x' + ax) + (y' + ay) = 0 + 0 = 0$$

and

$$(cy)' + a(cy) = cy' + c(ay) = c(y' + ay) = c \cdot 0 = 0$$

Hence, $x + y$ and cy are also in S, so S is a subspace of $\mathscr{F}$.

Now we will show that S is a one-dimensional subspace of $\mathscr{F}$ and that $\{e^{-at}\}$ is a basis. To this end, let $x = x(t)$ be in S. Then, for all t,

$$x'(t) + ax(t) = 0 \quad \text{or} \quad x'(t) = -ax(t)$$

Define a new function $z(t) = x(t)e^{at}$. Then, by the Chain Rule for differentiation,

$$\begin{aligned} z'(t) &= x(t)ae^{at} + x'(t)e^{at} \\ &= ax(t)e^{at} - ax(t)e^{at} \\ &= 0 \end{aligned}$$

Since z' is identically zero, z must be a constant function—say, $z(t) = k$. But this means that

$$x(t)e^{at} = z(t) = k \quad \text{for all } t$$

so $x(t) = ke^{-at}$. Therefore, all solutions to Equation (2) are scalar multiples of the single solution $y = e^{-at}$. We have proved the following theorem.

Theorem 6.32 If S is the solution space of $y' + ay = 0$, then $\dim S = 1$ and $\{e^{-at}\}$ is a basis for S.

One model for population growth assumes that the growth rate of the population is proportional to the size of the population. This model works well if there are few restrictions (such as limited space, food, or the like) on growth. If the size of the population at time t is $p(t)$, then the growth rate, or rate of change of the population, is its derivative $p'(t)$. Our assumption that the growth rate of the population is proportional to its size can be written as

$$p'(t) = kp(t)$$

where k is the proportionality constant. Thus, p satisfies the differential equation $p' - kp = 0$, so, by Theorem 6.32,

$$p(t) = ce^{kt}$$

for some scalar c. The constants c and k are determined using experimental data.

Example 6.87 The bacterium *Escherichia coli* (or *E. coli,* for short) is commonly found in the intestines of humans and other mammals. It poses severe health risks if it escapes into the environment. Under laboratory conditions, each cell of the bacterium divides into two every 20 minutes. If we start with a single *E. coli* cell, how many will there be after 1 day?

E. coli is mentioned in Michael Crichton's novel *The Andromeda Strain* (New York: Dell, 1969), although the "villain" in that novel was supposedly an alien virus. In real life, *E. coli* contaminated the town water supply of Walkerton, Ontario, in 2000, resulting in seven deaths and causing hundreds of people to become seriously ill.

Solution We do not need to use differential equations to solve this problem, but we will, in order to illustrate the basic method.

To determine c and k, we use the data given in the statement of the problem. If we take 1 unit of time to be 20 minutes, then we are given that $p(0) = 1$ and $p(1) = 2$. Therefore,

$$c = c \cdot 1 = ce^{k \cdot 0} = 1 \quad \text{and} \quad 2 = ce^{k \cdot 1} = e^k$$

It follows that $k = \ln 2$, so

$$p(t) = e^{t \ln 2} = e^{\ln 2^t} = 2^t$$

After 1 day, $t = 72$, so the number of bacteria cells will be $p(72) = 2^{72} \approx 4.72 \times 10^{21}$ (see Figure 6.18).

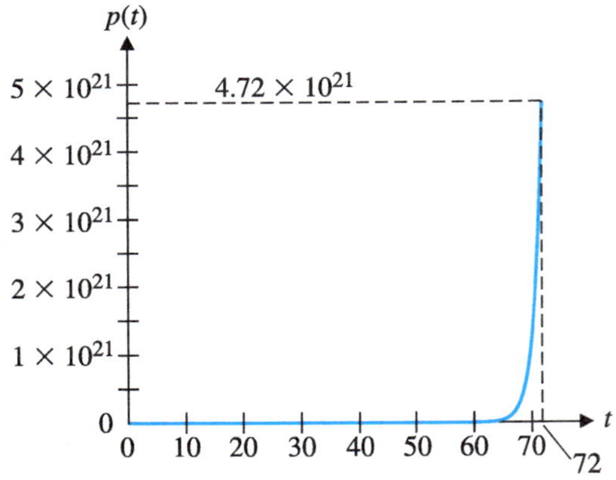

Figure 6.18
Exponential growth

Radioactive substances decay by emitting radiation. If $m(t)$ denotes the mass of the substance at time t, then the rate of decay is $m'(t)$. Physicists have found that the rate of decay of a substance is proportional to its mass; that is,

$$m'(t) = km(t) \quad \text{or} \quad m' - km = 0$$

where k is a negative constant. Applying Theorem 6.32, we have

$$m(t) = ce^{kt}$$

for some constant c. The time required for half of a radioactive substance to decay is called its ***half-life.***

Example 6.88 After 5.5 days, a 100 mg sample of radon-222 decayed to 37 mg.

(a) Find a formula for $m(t)$, the mass remaining after t days.

(b) What is the half-life of radon-222?

(c) When will only 10 mg remain?

Solution (a) From $m(t) = ce^{kt}$, we have

$$100 = m(0) = ce^{k \cdot 0} = c \cdot 1 = c$$

so

$$m(t) = 100e^{kt}$$

With time measured in days, we are given that $m(5.5) = 37$. Therefore,

$$100e^{5.5k} = 37$$

so

$$e^{5.5k} = 0.37$$

Solving for k, we find

$$5.5k = \ln(0.37)$$

so

$$k = \frac{\ln(0.37)}{5.5} \approx -0.18$$

Therefore, $m(t) = 100e^{-0.18t}$.

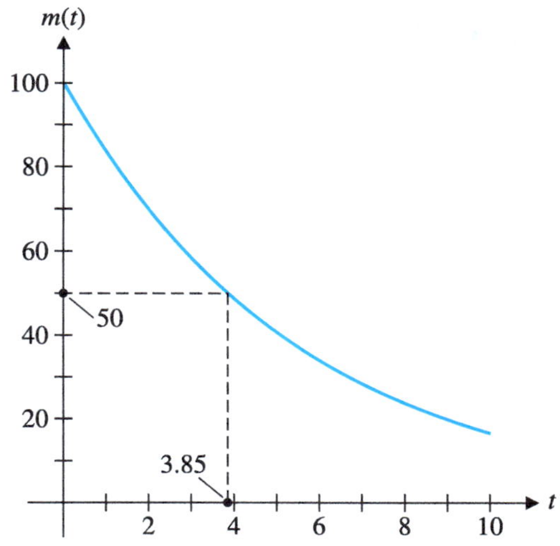

Figure 6.19
Radioactive decay

(b) To find the half-life of radon-222, we need the value of t for which $m(t) = 50$. Solving this equation, we find

$$100e^{-0.18t} = 50$$

so

$$e^{-0.18t} = 0.50$$

Hence,

$$-0.18t = \ln(\tfrac{1}{2}) = -\ln 2$$

and

$$t = \frac{\ln 2}{0.18} \approx 3.85$$

Thus, radon-222 has a half-life of approximately 3.85 days. (See Figure 6.19.)

(c) We need to determine the value of t such that $m(t) = 10$. That is, we must solve the equation

$$100e^{-0.18t} = 10 \quad \text{or} \quad e^{-0.18t} = 0.1$$

Taking the natural logarithm of both sides yields $-0.18t = \ln 0.1$. Thus,

$$t = \frac{\ln 0.1}{-0.18} \approx 12.79$$

so 10 mg of the sample will remain after approximately 12.79 days.

The solution set S of the second-order differential equation $y'' + ay' + by = 0$ is also a subspace of $\mathscr{F}$ (Exercise 20), and it turns out that the dimension of S is 2. Part (a) of Theorem 6.33, which extends Theorem 6.32, is implied by Theorem 4.40. Our approach here is to use the power of vector spaces; doing so allows us to obtain part (b) of Theorem 6.33 as well, a result that we could not obtain with our previous methods.

See *Linear Algebra* by S. H. Friedberg, A. J. Insel, and L. E. Spence (Englewood Cliffs, NJ: Prentice-Hall, 1979).

Theorem 6.33 Let S be the solution space of

$$y'' + ay' + by = 0$$

and let λ_1 and λ_2 be the roots of the characteristic equation $\lambda^2 + a\lambda + b = 0$.

a. If $\lambda_1 \neq \lambda_2$, then $\{e^{\lambda_1 t}, e^{\lambda_2 t}\}$ is a basis for S.
b. If $\lambda_1 = \lambda_2$, then $\{e^{\lambda_1 t}, te^{\lambda_1 t}\}$ is a basis for S.

Remarks

- Observe that what the theorem says, in other words, is that the solutions of $y'' + ay' + by = 0$ are of the form

$$y = c_1 e^{\lambda_1 t} + c_2 e^{\lambda_2 t}$$

in the first case and

$$y = c_1 e^{\lambda_1 t} + c_2 t e^{\lambda_1 t}$$

in the second case.

- Compare Theorem 6.33 with Theorem 4.38. Linear differential equations and linear recurrence relations have much in common. Although the former belong to *continuous* mathematics and the latter to *discrete* mathematics, there are many parallels.

Proof (a) We first show that $\{e^{\lambda_1 t}, e^{\lambda_2 t}\}$ is contained in S. Let λ be any root of the characteristic equation and let $f(t) = e^{\lambda t}$. Then

$$f'(t) = \lambda e^{\lambda t} \quad \text{and} \quad f''(t) = \lambda^2 e^{\lambda t}$$

from which it follows that

$$\begin{aligned} f'' + af' + bf &= \lambda^2 e^{\lambda t} + a\lambda e^{\lambda t} + be^{\lambda t} \\ &= (\lambda^2 + a\lambda + b)e^{\lambda t} \\ &= 0 \cdot e^{\lambda t} = 0 \end{aligned}$$

Therefore, f is in S. But, since λ_1 and λ_2 are roots of the characteristic equation, this means that $e^{\lambda_1 t}$ and $e^{\lambda_2 t}$ are in S.

The set $\{e^{\lambda_1 t}, e^{\lambda_2 t}\}$ is also linearly independent, since if

$$c_1 e^{\lambda_1 t} + c_2 e^{\lambda_2 t} = 0$$

then, setting $t = 0$, we have

$$c_1 + c_2 = 0 \quad \text{or} \quad c_2 = -c_1$$

Next, we set $t = 1$ to obtain

$$c_1 e^{\lambda_1} - c_1 e^{\lambda_2} = 0 \quad \text{or} \quad c_1(e^{\lambda_1} - e^{\lambda_2}) = 0$$

But $e^{\lambda_1} - e^{\lambda_2} \neq 0$, since $e^{\lambda_1} - e^{\lambda_2} = 0$ implies that $e^{\lambda_1} = e^{\lambda_2}$, which is clearly impossible if $\lambda_1 \neq \lambda_2$. (See Figure 6.20.) We deduce that $c_1 = 0$ and, hence, $c_2 = 0$, so $\{e^{\lambda_1 t}, e^{\lambda_2 t}\}$ is linearly independent.

Since $\dim S = 2$, $\{e^{\lambda_1 t}, e^{\lambda_2 t}\}$ must be a basis for S.

(b) You are asked to prove this property in Exercise 21.

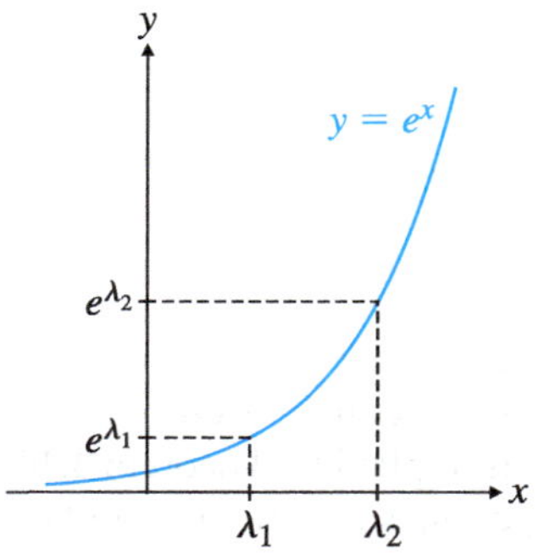

Figure 6.20

Example 6.89 Find all solutions of $y'' - 5y' + 6y = 0$.

Solution The characteristic equation is $\lambda^2 - 5\lambda + 6 = (\lambda - 2)(\lambda - 3) = 0$. Thus, the roots are 2 and 3, so $\{e^{2t}, e^{3t}\}$ is a basis for the solution space. It follows that the solutions to the given equation are of the form

$$y = c_1e^{2t} + c_2e^{3t}$$

The constants c_1 and c_2 can be determined if additional equations, called ***boundary conditions,*** are specified.

Example 6.90 Find the solution of $y'' + 6y' + 9y = 0$ that satisfies $y(0) = 1, y'(0) = 0$.

Solution The characteristic equation is $\lambda^2 + 6\lambda + 9 = (\lambda + 3)^2 = 0$, so -3 is a repeated root. Therefore, $\{e^{-3t}, te^{-3t}\}$ is a basis for the solution space, and the general solution is of the form

$$y = c_1e^{-3t} + c_2te^{-3t}$$

The first boundary condition gives

$$1 = y(0) = c_1e^{-3\cdot 0} + 0 = c_1$$

so $y = e^{-3t} + c_2te^{-3t}$. Differentiating, we have

$$y' = -3e^{-3t} + c_2(-3te^{-3t} + e^{-3t})$$

so the second boundary condition gives

$$0 = y'(0) = -3e^{-3\cdot 0} + c_2(0 + e^{-3\cdot 0}) = -3 + c_2$$

or

$$c_2 = 3$$

Therefore, the required solution is

$$y = e^{-3t} + 3te^{-3t} = (1 + 3t)e^{-3t}$$

a + bi

Theorem 6.33 includes the case in which the roots of the characteristic equation are complex. If $\lambda = p + qi$ is a complex root of the equation $\lambda^2 + a\lambda + b = 0$, then so is its conjugate $\bar{\lambda} = p - qi$. (See Appendices C and D.) By Theorem 6.33(a), the solution space S of the differential equation $y'' + ay' + by = 0$ has $\{e^{\lambda t}, e^{\bar{\lambda} t}\}$ as a basis. Now

$$e^{\lambda t} = e^{(p+qi)t} = e^{pt}e^{i(qt)} = e^{pt}(\cos qt + i\sin qt)$$

and

$$e^{\bar{\lambda} t} = e^{(p-qi)t} = e^{pt}e^{i(-qt)} = e^{pt}(\cos qt - i\sin qt)$$

so

$$e^{pt}\cos qt = \frac{e^{\lambda t} + e^{\bar{\lambda} t}}{2} \quad \text{and} \quad e^{pt}\sin qt = \frac{e^{\lambda t} - e^{\bar{\lambda} t}}{2i}$$

It follows that $\{e^{pt}\cos qt, e^{pt}\sin qt\}$ is contained in $\operatorname{span}(e^{\lambda t}, e^{\bar{\lambda} t}) = S$. Since $e^{pt}\cos qt$ and $e^{pt}\sin qt$ are linearly independent (see Exercise 22) and $\dim S = 2$, $\{e^{pt}\cos qt, e^{pt}\sin qt\}$ is also a basis for S. Thus, when its characteristic equation has a complex root $p + qi$, the differential equation $y'' + ay' + by = 0$ has solutions of the form

$$y = c_1e^{pt}\cos qt + c_2e^{pt}\sin qt$$

a + bi

Example 6.91

Find all solutions of $y'' - 2y' + 4 = 0$.

Solution The characteristic equation is $\lambda^2 - 2\lambda + 4 = 0$ with roots $1 \pm i\sqrt{3}$. The foregoing discussion tells us that the general solution to the given differential equation is

$$y = c_1 e^t \cos \sqrt{3}t + c_2 e^t \sin \sqrt{3}t$$

a + bi

Example 6.92

A mass is attached to the end of a vertical spring (Figure 6.21). If the mass is pulled downward and released, it will oscillate up and down. Two laws of physics govern this situation. The first, ***Hooke's law,*** states that if the spring is stretched (or compressed) x units, the force F needed to restore it to its original position is proportional to x:

$$F = -kx$$

where k is a positive constant (called the spring constant). ***Newton's Second Law of Motion*** states that force equals mass times acceleration. Since $x = x(t)$ represents distance, or displacement, of the spring at time t, x' gives its velocity and x'' its acceleration. Thus, we have

$$mx'' = -kx \quad \text{or} \quad x'' + \left(\frac{k}{m}\right)x = 0$$

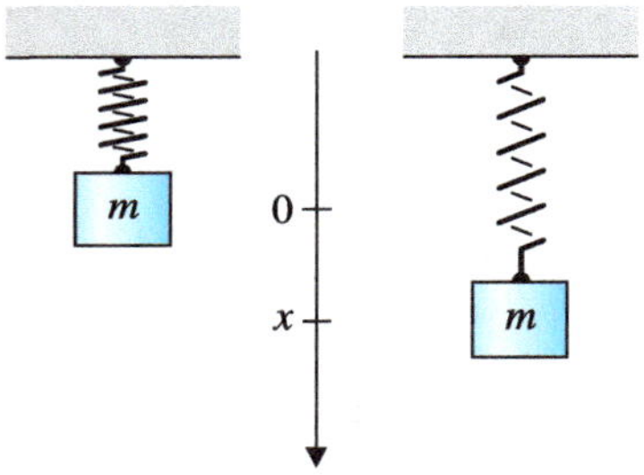

Figure 6.21

Since both k and m are positive, so is $K = k/m$, and our differential equation has the form $x'' + Kx = 0$, where K is positive.

The characteristic equation is $\lambda^2 + K = 0$ with roots $\pm i\sqrt{K}$. Therefore, the general solution to the differential equation of the oscillating spring is

$$x = c_1 \cos \sqrt{K}t + c_2 \sin \sqrt{K}t$$

Suppose the spring is at rest ($x = 0$) at time $t = 0$ seconds and is stretched as far as possible, to a length of 20 cm, before it is released. Then

$$0 = x(0) = c_1 \cos 0 + c_2 \sin 0 = c_1$$

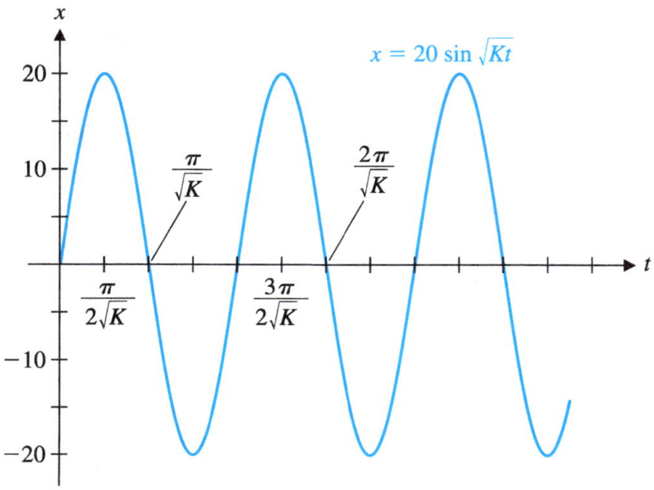

Figure 6.22

so $x = c_2 \sin \sqrt{K}t$. Since the maximum value of the sine function is 1, we must have $c_2 = 20$ (occurring for the first time when $t = \pi/2\sqrt{K}$), giving us the solution

$$x = 20 \sin \sqrt{K}t$$

(See Figure 6.22.)

Of course, this is an idealized solution, since it neglects any form of resistance and predicts that the spring will oscillate forever. It is possible to take damping effects (such as friction) into account, but this simple model has served to introduce an important application of differential equations and the techniques we have developed.

Exercises 6.7

Homogeneous Linear Differential Equations

In Exercises 1–12, find the solution of the differential equation that satisfies the given boundary condition(s).

1. $y' - 3y = 0, y(1) = 2$

2. $x' + x = 0, x(1) = 1$

3. $y'' - 7y' + 12y = 0, y(0) = y(1) = 1$

4. $x'' + x' - 12x = 0, x(0) = 0, x'(0) = 1$

5. $f'' - f' - f = 0, f(0) = 0, f(1) = 1$

6. $g'' - 2g = 0, g(0) = 1, g(1) = 0$

7. $y'' - 2y' + y = 0, y(0) = y(1) = 1$

8. $x'' + 4x' + 4x = 0, x(0) = 1, x'(0) = 1$

9. $y'' - k^2y = 0, k \neq 0, y(0) = y'(0) = 1$

10. $y'' - 2ky' + k^2y = 0, k \neq 0, y(0) = 1, y(1) = 0$

11. $f'' - 2f' + 5f = 0, f(0) = 1, f(\pi/4) = 0$

12. $h'' - 4h' + 5h = 0, h(0) = 0, h'(0) = -1$

13. A strain of bacteria has a growth rate that is proportional to the size of the population. Initially, there are 100 bacteria; after 3 hours, there are 1600.

(a) If $p(t)$ denotes the number of bacteria after t hours, find a formula for $p(t)$.

(b) How long does it take for the population to double?

(c) When will the population reach one million?

CAS **14.** Table 6.2 gives the population of the United States at 10-year intervals for the years 1900–2000.

(a) Assuming an exponential growth model, use the data for 1900 and 1910 to find a formula for $p(t)$, the population in year t. [*Hint:* Let $t = 0$ be 1900 and let $t = 1$ be 1910.] How accurately does your formula calculate the U.S. population in 2000?

(b) Repeat part (a), but use the data for the years 1970 and 1980 to solve for $p(t)$. Does this approach give a better approximation for the year 2000?

(c) What can you conclude about U.S. population growth?

Table 6.2

Year	**Population (in millions)**
1900	76
1910	92
1920	106
1930	123
1940	131
1950	150
1960	179
1970	203
1980	227
1990	250
2000	281

Source: U.S. Bureau of the Census

15. The half-life of radium-226 is 1590 years. Suppose we start with a sample of radium-226 whose mass is 50 mg.

(a) Find a formula for the mass $m(t)$ remaining after t years and use this formula to predict the mass remaining after 1000 years.

(b) When will only 10 mg remain?

16. *Radiocarbon dating* is a method used by scientists to estimate the age of ancient objects that were once living matter, such as bone, leather, wood, or paper.

All of these contain carbon, a proportion of which is carbon-14, a radioactive isotope that is continuously being formed in the upper atmosphere. Since living organisms take up radioactive carbon along with other carbon atoms, the ratio between the two forms remains constant. However, when an organism dies, the carbon-14 in its cells decays and is not replaced. Carbon-14 has a known half-life of 5730 years, so by measuring the concentration of carbon-14 in an object, scientists can determine its approximate age.

One of the most successful applications of radiocarbon dating has been to determine the age of the Stonehenge monument in England (Figure 6.23). Samples taken from the remains of wooden posts were found to have a concentration of carbon-14 that was 45% of that found in living material. What is the estimated age of these posts?

Figure 6.23
Stonehenge

17. A mass is attached to a spring, as in Example 6.92. At time $t = 0$ second, the spring is stretched to a length of 10 cm below its position at rest. The spring is released, and its length 10 seconds later is observed to be 5 cm. Find a formula for the length of the spring at time t seconds.

18. A 50 g mass is attached to a spring, as in Example 6.92. If the period of oscillation is 10 seconds, find the spring constant.

19. A pendulum consists of a mass, called a *bob*, that is affixed to the end of a string of length L (see Figure 6.24). When the bob is moved from its rest position and released, it swings back and forth. The time it takes the pendulum to swing from its farthest right position to its farthest left position and back to its next farthest right position is called the *period* of the pendulum.

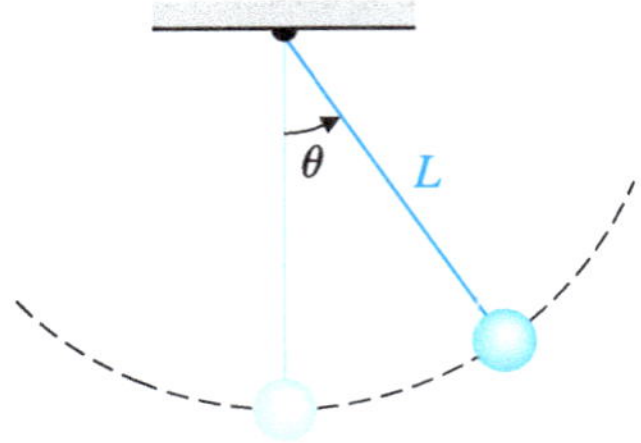

Figure 6.24

Let $\theta = \theta(t)$ be the angle of the pendulum from the vertical. It can be shown that if there is no resistance, then when θ is small it satisfies the differential equation

$$\theta'' + \frac{g}{L}\theta = 0$$

where g is the constant of acceleration due to gravity, approximately 9.7 m/s^2. Suppose that $L = 1$ m and that the pendulum is at rest (i.e., $\theta = 0$) at time $t = 0$ second. The bob is then drawn to the right at an angle of θ_1 radians and released.

(a) Find the period of the pendulum.

(b) Does the period depend on the angle θ_1 at which the pendulum is released? This question was posed and answered by Galileo in 1638. [Galileo Galilei (1564–1642) studied medicine as a student at the University of Pisa, but his real interest was always mathematics. In 1592, Galileo was appointed professor of mathematics at the University of Padua in Venice, where he taught primarily geometry and astronomy. He was the first to use a telescope to look at the stars and planets, and in so doing, he produced experimental data in support of the Copernican view that the planets revolve around the sun and not the earth. For this, Galileo was summoned before the Inquisition, placed under house arrest, and forbidden to publish his results. While under house arrest, he was able to write up his research on falling objects and pendulums. His notes were smuggled out of Italy and published as *Discourses on Two New Sciences* in 1638.]

20. Show that the solution set S of the second-order differential equation $y'' + ay' + by = 0$ is a subspace of $\mathcal{F}$.

21. Prove Theorem 6.33(b).

22. Show that $e^{pt}\cos qt$ and $e^{pt}\sin qt$ are linearly independent.

Chapter Review

Key Definitions and Concepts

Review Questions

1. Mark each of the following statements true or false:

(a) If $V = \text{span}(\mathbf{v}_1, \ldots, \mathbf{v}_n)$, then every spanning set for V contains at least n vectors.

(b) If $\{\mathbf{u}, \mathbf{v}, \mathbf{w}\}$ is a linearly independent set of vectors, then so is $\{\mathbf{u} + \mathbf{v}, \mathbf{v} + \mathbf{w}, \mathbf{u} + \mathbf{w}\}$.

(c) M_{22} has a basis consisting of invertible matrices.

(d) M_{22} has a basis consisting of matrices whose trace is zero.

(e) The transformation $T: \mathbb{R}^n \to \mathbb{R}$ defined by $T(\mathbf{x}) = \|\mathbf{x}\|$ is a linear transformation.

(f) If $T: V \to W$ is a linear transformation and dim $V \neq$ dim W, then T cannot be both one-to-one and onto.

(g) If $T: V \to W$ is a linear transformation and $\ker(T) = V$, then $W = \{\mathbf{0}\}$.

(h) If $T: M_{33} \to \mathscr{P}_4$ is a linear transformation and $\text{nullity}(T) = 4$, then T is onto.

(i) The vector space $V = \{p(x) \text{ in } \mathscr{P}_4 : p(1) = 0\}$ is isomorphic to $\mathscr{P}_3$.

(j) If $I: V \to V$ is the identity transformation, then the matrix $[I]_{\mathcal{C} \leftarrow \mathcal{B}}$ is the identity matrix for any bases $\mathcal{B}$ and $\mathcal{C}$ of V.

In Questions 2–5, determine whether W is a subspace of V.

2. $V = \mathbb{R}^2$, $W = \left\{ \begin{bmatrix} x \\ y \end{bmatrix} : x^2 + 3y^2 = 0 \right\}$

3. $V = M_{22}$, $W = \left\{ \begin{bmatrix} a & b \\ c & d \end{bmatrix} : a + b = c + d = a + c = b + d \right\}$

4. $V = \mathscr{P}_3$, $W = \{p(x) \text{ in } \mathscr{P}_3 : x^3 p(1/x) = p(x)\}$

5. $V = \mathscr{F}$, $W = \{f \text{ in } \mathscr{F} : f(x + \pi) = f(x) \text{ for all } x\}$

6. Determine whether $\{1, \cos 2x, 3\sin^2 x\}$ is linearly dependent or independent.

7. Let A and B be nonzero $n \times n$ matrices such that A is symmetric and B is skew-symmetric. Prove that $\{A, B\}$ is linearly independent.

In Questions 8 and 9, find a basis for W and state the dimension of W.

8. $W = \left\{ \begin{bmatrix} a & b \\ c & d \end{bmatrix} : a + d = b + c \right\}$

9. $W = \{p(x) \text{ in } \mathscr{P}_5 : p(-x) = p(x)\}$

10. Find the change-of-basis matrices $P_{\mathcal{C} \leftarrow \mathcal{B}}$ and $P_{\mathcal{B} \leftarrow \mathcal{C}}$ with respect to the bases $\mathcal{B} = \{1, 1 + x, 1 + x + x^2\}$ and $\mathcal{C} = \{1 + x, x + x^2, 1 + x^2\}$ of $\mathscr{P}_2$.

In Questions 11–13, determine whether T is a linear transformation.

11. $T: \mathbb{R}^2 \to \mathbb{R}^2$ defined by $T(\mathbf{x}) = \mathbf{y}\mathbf{x}^T\mathbf{y}$, where $\mathbf{y} = \begin{bmatrix} 1 \\ 2 \end{bmatrix}$

12. $T: M_{nn} \to M_{nn}$ defined by $T(A) = A^T A$

13. $T: \mathscr{P}_n \to \mathscr{P}_n$ defined by $T(p(x)) = p(2x - 1)$

14. If $T: \mathscr{P}_2 \to M_{22}$ is a linear transformation such that

$$T(1) = \begin{bmatrix} 1 & 0 \\ 0 & 1 \end{bmatrix}, T(1 + x) = \begin{bmatrix} 1 & 1 \\ 0 & 1 \end{bmatrix} \text{ and}$$

$$T(1 + x + x^2) = \begin{bmatrix} 0 & -1 \\ 1 & 0 \end{bmatrix}, \text{ find } T(5 - 3x + 2x^2).$$

15. Find the nullity of the linear transformation $T: M_{nn} \to \mathbb{R}$ defined by $T(A) = \text{tr}(A)$.

16. Let W be the vector space of upper triangular 2×2 matrices.

    (a) Find a linear transformation $T: M_{22} \to M_{22}$ such that $\ker(T) = W$.

    (b) Find a linear transformation $T: M_{22} \to M_{22}$ such that $\text{range}(T) = W$.

17. Find the matrix $[T]_{\mathcal{C} \leftarrow \mathcal{B}}$ of the linear transformation T in Question 14 with respect to the standard bases $\mathcal{B} = \{1, x, x^2\}$ of $\mathscr{P}_2$ and $\mathcal{C} = \{E_{11}, E_{12}, E_{21}, E_{22}\}$ of M_{22}.

18. Let $S = \{\mathbf{v}_1, \ldots, \mathbf{v}_n\}$ be a set of vectors in a vector space V with the property that every vector in V can be written as a linear combination of $\mathbf{v}_1, \ldots, \mathbf{v}_n$ in exactly one way. Prove that S is a basis for V.

19. If $T: U \to V$ and $S: V \to W$ are linear transformations such that $\text{range}(T) \subseteq \ker(S)$, what can be deduced about $S \circ T$?

20. Let $T: V \to V$ be a linear transformation, and let $\{\mathbf{v}_1, \ldots, \mathbf{v}_n\}$ be a basis for V such that $\{T(\mathbf{v}_1), \ldots, T(\mathbf{v}_n)\}$ is also a basis for V. Prove that T is invertible.

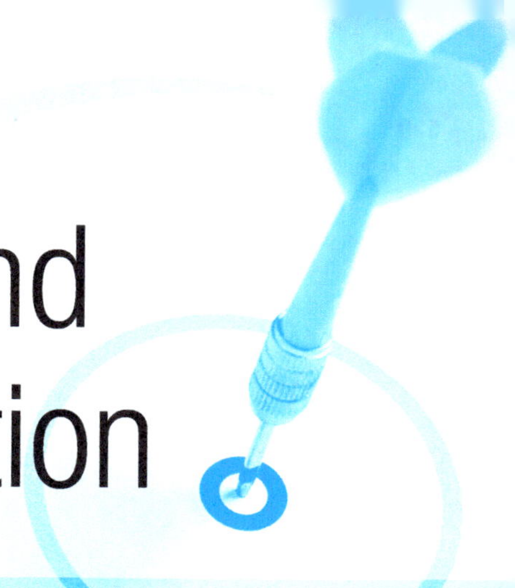

7 Distance and Approximation

A straight line may be the shortest distance between two points, but it is by no means the most interesting.
—Doctor Who
In "The Time Monster"
By Robert Sloman
BBC, 1972

Although this may seem a paradox, all exact science is dominated by the idea of approximation.
—Bertrand Russell
In W. H. Auden and
L. Kronenberger, eds.
The Viking Book of Aphorisms
Viking, 1962, p. 263

7.0 Introduction: Taxicab Geometry

We live in a three-dimensional Euclidean world, and, therefore, concepts from Euclidean geometry govern our way of looking at the world. In particular, imagine stopping people on the street and asking them to fill in the blank in the following sentence: "The shortest distance between two points is a ______." They will almost certainly respond with "straight line." There are, however, other equally sensible and intuitive notions of distance. By allowing ourselves to think of "distance" in a more flexible way, we will open the door to the possibility of having a "distance" between polynomials, functions, matrices, and many other objects that arise in linear algebra.

In this section, you will discover a type of "distance" that is every bit as real as the straight-line distance you are used to from Euclidean geometry (the one that is a consequence of Pythagoras' Theorem). As you'll see, this new type of "distance" still behaves in some familiar ways.

Suppose you are standing at an intersection in a city, trying to get to a restaurant at another intersection. If you ask someone how far it is to the restaurant, that person is unlikely to measure distance "as the crow flies" (i.e., using the Euclidean version of distance). Instead, the response will be something like "It's five blocks away." Since this is the way taxicab drivers measure distance, we will refer to this notion of "distance" as ***taxicab distance.***

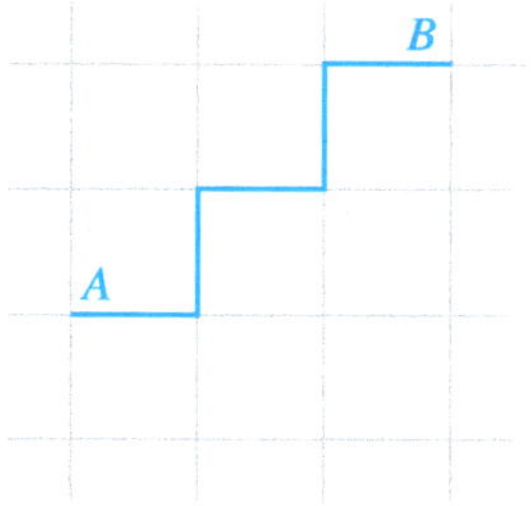

Figure 7.1
Taxicab distance

Figure 7.1 shows an example of taxicab distance. The shortest path from A to B requires traversing the sides of five city blocks. Notice that although there is more than one route from A to B, all shortest routes require three horizontal moves and two vertical moves, where a "move" corresponds to the side of one city block. (How many shortest routes are there from A to B?) Therefore, the taxicab distance from A to B is 5.

Idealizing this situation, we will assume that all blocks are unit squares, and we will use the notation $\mathrm{d}_t(A, B)$ for the taxicab distance from A to B.

Problem 1 Find the taxicab distance between the following pairs of points:

(a) $(1, 2)$ and $(5, 5)$ (b) $(2, 4)$ and $(3, -2)$

(c) $(0, 0)$ and $(-4, -3)$ (d) $(-2, 3)$ and $(1, 3)$

(e) $(1, \frac{1}{2})$ and $(-\frac{3}{2}, \frac{3}{2})$ (f) $(2.5, 4.6)$ and $(3.1, 1.5)$

Problem 2 Which of the following is the correct formula for the taxicab distance $d_t(A, B)$ between $A = (a_1, a_2)$ and $B = (b_1, b_2)$?

(a) $d_t(A, B) = (a_1 - b_1) + (a_2 - b_2)$
(b) $d_t(A, B) = (|a_1| - |b_1|) + (|a_2| - |b_2|)$
(c) $d_t(A, B) = |a_1 - b_1| + |a_2 - b_2|$

We can define the ***taxicab norm*** of a vector $\mathbf{v}$ as

$$\|\mathbf{v}\|_t = d_t(\mathbf{v}, \mathbf{0})$$

Problem 3 Find $\|\mathbf{v}\|_t$ for the following vectors:

(a) $\mathbf{v} = \begin{bmatrix} 3 \\ -2 \end{bmatrix}$ (b) $\mathbf{v} = \begin{bmatrix} 6 \\ -4 \end{bmatrix}$

(c) $\mathbf{v} = \begin{bmatrix} -3 \\ -6 \end{bmatrix}$ (d) $\mathbf{v} = \begin{bmatrix} 1 \\ 2 \end{bmatrix}$

Problem 4 Show that Theorem 1.3 is true for the taxicab norm.

Problem 5 Verify the Triangle Inequality (Theorem 1.5), using the taxicab norm and the following pairs of vectors:

(a) $\mathbf{u} = \begin{bmatrix} 3 \\ 1 \end{bmatrix}, \mathbf{v} = \begin{bmatrix} 1 \\ 2 \end{bmatrix}$ (b) $\mathbf{u} = \begin{bmatrix} 1 \\ -1 \end{bmatrix}, \mathbf{v} = \begin{bmatrix} -2 \\ 3 \end{bmatrix}$

Problem 6 Show that the Triangle Inequality is true, in general, for the taxicab norm.

In Euclidean geometry, we can define a circle of radius r, centered at the origin, as the set of all $\mathbf{x}$ such that $\|\mathbf{x}\| = r$. Analogously, we can define a ***taxicab circle*** of radius r, centered at the origin, as the set of all $\mathbf{x}$ such that $\|\mathbf{x}\|_t = r$.

Problem 7 Draw taxicab circles centered at the origin with the following radii:

(a) $r = 3$ (b) $r = 4$ (c) $r = 1$

Problem 8 In Euclidean geometry, the value of π is half the circumference of a unit circle (a circle of radius 1). Let's define ***taxicab pi*** to be the number π_t that is half the circumference of a taxicab unit circle. What is the value of π_t?

In Euclidean geometry, the perpendicular bisector of a line segment $\overline{AB}$ can be defined as the set of all points that are equidistant from A and B. If we use taxicab distance instead of Euclidean distance, it is reasonable to ask what the perpendicular bisector of a line segment now looks like. To be precise, the ***taxicab perpendicular bisector*** of $\overline{AB}$ is the set of all points X such that

$$d_t(X, A) = d_t(X, B)$$

Problem 9 Draw the taxicab perpendicular bisector of $\overline{AB}$ for the following pairs of points:

(a) $A = (2, 1), B = (4, 1)$ (b) $A = (-1, 3), B = (-1, -2)$
(c) $A = (1, 1), B = (5, 3)$ (d) $A = (1, 1), B = (5, 5)$

As these problems illustrate, taxicab geometry shares some properties with Euclidean geometry, but it also differs in some striking ways. In this chapter, we will encounter several other types of distances and norms, each of which is useful in its own way. We will try to discover what they have in common and use these common properties to our advantage. We will also explore a variety of approximation problems in which the notion of "distance" plays an important role.

7.1 Inner Product Spaces

In Chapter 1, we defined the dot product $\mathbf{u} \cdot \mathbf{v}$ of vectors $\mathbf{u}$ and $\mathbf{v}$ in $\mathbb{R}^n$, and we have made repeated use of this operation throughout this book. In this section, we will use the properties of the dot product as a means of defining the general notion of an *inner product*. In the next section, we will show that inner products can be used to define analogues of "length" and "distance" in vector spaces other than $\mathbb{R}^n$.

The following definition is our starting point; it is based on the properties of the dot product proved in Theorem 1.2.

Definition An ***inner product*** on a vector space V is an operation that assigns to every pair of vectors $\mathbf{u}$ and $\mathbf{v}$ in V a real number $\langle \mathbf{u}, \mathbf{v} \rangle$ such that the following properties hold for all vectors $\mathbf{u}$, $\mathbf{v}$, and $\mathbf{w}$ in V and all scalars c:

1. $\langle \mathbf{u}, \mathbf{v} \rangle = \langle \mathbf{v}, \mathbf{u} \rangle$
2. $\langle \mathbf{u}, \mathbf{v} + \mathbf{w} \rangle = \langle \mathbf{u}, \mathbf{v} \rangle + \langle \mathbf{u}, \mathbf{w} \rangle$
3. $\langle c\mathbf{u}, \mathbf{v} \rangle = c\langle \mathbf{u}, \mathbf{v} \rangle$
4. $\langle \mathbf{u}, \mathbf{u} \rangle \geq 0$ and $\langle \mathbf{u}, \mathbf{u} \rangle = 0$ if and only if $\mathbf{u} = \mathbf{0}$

A vector space with an inner product is called an ***inner product space.***

Remark Technically, this definition defines a *real* inner product space, since it assumes that V is a real vector space and since the inner product of two vectors is a real number. There are *complex* inner product spaces too, but their definition is somewhat different. (See Exploration: Vectors and Matrices with Complex Entries at the end of this section.)

Example 7.1

$\mathbb{R}^n$ is an inner product space with $\langle \mathbf{u}, \mathbf{v} \rangle = \mathbf{u} \cdot \mathbf{v}$. Properties (1) through (4) were verified as Theorem 1.2.

The dot product is not the only inner product that can be defined on $\mathbb{R}^n$.

Example 7.2

Let $\mathbf{u} = \begin{bmatrix} u_1 \\ u_2 \end{bmatrix}$ and $\mathbf{v} = \begin{bmatrix} v_1 \\ v_2 \end{bmatrix}$ be two vectors in $\mathbb{R}^2$. Show that

$$\langle \mathbf{u}, \mathbf{v} \rangle = 2u_1v_1 + 3u_2v_2$$

defines an inner product.

Solution We must verify properties (1) through (4). Property (1) holds because

$$\langle \mathbf{u}, \mathbf{v}\rangle = 2u_1v_1 + 3u_2v_2 = 2v_1u_1 + 3v_2u_2 = \langle \mathbf{v}, \mathbf{u}\rangle$$

Next, let $\mathbf{w} = \begin{bmatrix} w_1 \\ w_2 \end{bmatrix}$. We check that

$$\begin{aligned}\langle \mathbf{u}, \mathbf{v} + \mathbf{w}\rangle &= 2u_1(v_1 + w_1) + 3u_2(v_2 + w_2) \\ &= 2u_1v_1 + 2u_1w_1 + 3u_2v_2 + 3u_2w_2 \\ &= (2u_1v_1 + 3u_2v_2) + (2u_1w_1 + 3u_2w_2) \\ &= \langle \mathbf{u}, \mathbf{v}\rangle + \langle \mathbf{u}, \mathbf{w}\rangle\end{aligned}$$

which proves property (2).

If c is a scalar, then

$$\begin{aligned}\langle c\mathbf{u}, \mathbf{v}\rangle &= 2(cu_1)v_1 + 3(cu_2)v_2 \\ &= c(2u_1v_1 + 3u_2v_2) \\ &= c\langle \mathbf{u}, \mathbf{v}\rangle\end{aligned}$$

which verifies property (3).

Finally,

$$\langle \mathbf{u}, \mathbf{u}\rangle = 2u_1u_1 + 3u_2u_2 = 2u_1^2 + 3u_2^2 \geq 0$$

and it is clear that $\langle \mathbf{u}, \mathbf{u}\rangle = 2u_1^2 + 3u_2^2 = 0$ if and only if $u_1 = u_2 = 0$ (that is, if and only if $\mathbf{u} = \mathbf{0}$). This verifies property (4), completing the proof that $\langle \mathbf{u}, \mathbf{v}\rangle$, as defined, is an inner product.

Example 7.2 can be generalized to show that if $w_1, \ldots, w_n$ are *positive* scalars and

$$\mathbf{u} = \begin{bmatrix} u_1 \\ \vdots \\ u_n \end{bmatrix} \quad \text{and} \quad \mathbf{v} = \begin{bmatrix} v_1 \\ \vdots \\ v_n \end{bmatrix}$$

are vectors in $\mathbb{R}^n$, then

$$\langle \mathbf{u}, \mathbf{v}\rangle = w_1u_1v_1 + \cdots + w_nu_nv_n \tag{1}$$

defines an inner product on $\mathbb{R}^n$, called a ***weighted dot product.*** If any of the weights w_i is negative or zero, then Equation (1) does not define an inner product. (See Exercises 13 and 14.)

Recall that the dot product can be expressed as $\mathbf{u} \cdot \mathbf{v} = \mathbf{u}^T\mathbf{v}$. Observe that we can write the weighted dot product in Equation (1) as

$$\langle \mathbf{u}, \mathbf{v}\rangle = \mathbf{u}^T W\mathbf{v}$$

where W is the $n \times n$ diagonal matrix

$$W = \begin{bmatrix} w_1 & \cdots & 0 \\ \vdots & \ddots & \vdots \\ 0 & \cdots & w_n \end{bmatrix}$$

The next example further generalizes this type of inner product.

Example 7.3

Let A be a symmetric, positive definite $n \times n$ matrix (see Section 5.5) and let $\mathbf{u}$ and $\mathbf{v}$ be vectors in $\mathbb{R}^n$. Show that

$$\langle \mathbf{u}, \mathbf{v} \rangle = \mathbf{u}^T A\mathbf{v}$$

defines an inner product.

Solution We check that

$$\begin{aligned} \langle \mathbf{u}, \mathbf{v} \rangle &= u^T A\mathbf{v} = \mathbf{u} \cdot A\mathbf{v} = A\mathbf{v} \cdot \mathbf{u} \\ &= A^T\mathbf{v} \cdot \mathbf{u} = (\mathbf{v}^T A)^T \cdot \mathbf{u} = \mathbf{v}^T A\mathbf{u} = \langle \mathbf{v}, \mathbf{u} \rangle \end{aligned}$$

Also,

$$\langle \mathbf{u}, \mathbf{v} + \mathbf{w} \rangle = \mathbf{u}^T A(\mathbf{v} + \mathbf{w}) = \mathbf{u}^T A\mathbf{v} + \mathbf{u}^T A\mathbf{w} = \langle \mathbf{u}, \mathbf{v} \rangle + \langle \mathbf{u}, \mathbf{w} \rangle$$

and

$$\langle c\mathbf{u}, \mathbf{v} \rangle = (c\mathbf{u})^T A\mathbf{v} = c(\mathbf{u}^T A\mathbf{v}) = c\langle \mathbf{u}, \mathbf{v} \rangle$$

Finally, since A is positive definite, $\langle \mathbf{u}, \mathbf{u} \rangle = \mathbf{u}^T A\mathbf{u} > 0$ for all $\mathbf{u} \neq \mathbf{0}$, so $\langle \mathbf{u}, \mathbf{u} \rangle = \mathbf{u}^T A\mathbf{u} = \mathbf{0}$ if and only if $\mathbf{u} = \mathbf{0}$. This establishes the last property.

To illustrate Example 7.3, let $A = \begin{bmatrix} 4 & -2 \\ -2 & 7 \end{bmatrix}$. Then

$$\langle \mathbf{u}, \mathbf{v} \rangle = \mathbf{u}^T A\mathbf{v} = [u_1 \;\; u_2] \begin{bmatrix} 4 & -2 \\ -2 & 7 \end{bmatrix} \begin{bmatrix} v_1 \\ v_2 \end{bmatrix} = 4u_1v_1 - 2u_1v_2 - 2u_2v_1 + 7u_2v_2$$

The matrix A is positive definite, by Theorem 5.24, since its eigenvalues are 3 and 8. Hence, $\langle \mathbf{u}, \mathbf{v} \rangle$ defines an inner product on $\mathbb{R}^2$.

We now define some inner products on vector spaces other than $\mathbb{R}^n$.

Example 7.4

In $\mathscr{P}_2$, let $p(x) = a_0 + a_1x + a_2x^2$ and $q(x) = b_0 + b_1x + b_2x^2$. Show that

$$\langle p(x), q(x) \rangle = a_0b_0 + a_1b_1 + a_2b_2$$

defines an inner product on $\mathscr{P}_2$. (For example, if $p(x) = 1 - 5x + 3x^2$ and $q(x) = 6 + 2x - x^2$, then $\langle p(x), q(x) \rangle = 1 \cdot 6 + (-5) \cdot 2 + 3 \cdot (-1) = -7$.)

Solution Since $\mathscr{P}_2$ is isomorphic to $\mathbb{R}^3$, we need only show that the dot product in $\mathbb{R}^3$ is an inner product, which we have already established.

Example 7.5 Let f and g be in $\mathscr{C}[a, b]$, the vector space of all continuous functions on the closed interval $[a, b]$. Show that

$$\langle f, g\rangle = \int_a^b f(x)g(x)\,dx$$

defines an inner product on $\mathscr{C}[a, b]$.

Solution We have

$$\langle f, g\rangle = \int_a^b f(x)g(x)\,dx = \int_a^b g(x)f(x)\,dx = \langle g, f\rangle$$

Also, if h is in $\mathscr{C}[a, b]$, then

$$\begin{aligned}\langle f, g + h\rangle &= \int_a^b f(x)(g(x) + h(x))\,dx \\ &= \int_a^b (f(x)g(x) + f(x)h(x))\,dx \\ &= \int_a^b f(x)g(x)\,dx + \int_a^b f(x)h(x)\,dx \\ &= \langle f, g\rangle + \langle f, h\rangle\end{aligned}$$

If c is a scalar, then

$$\begin{aligned}\langle cf, g\rangle &= \int_a^b cf(x)g(x)\,dx \\ &= c\int_a^b f(x)g(x)\,dx \\ &= c\langle f, g\rangle\end{aligned}$$

Finally, $\langle f, f\rangle = \int_a^b (f(x))^2\,dx \geq 0$, and it follows from a theorem of calculus that, since f is continuous, $\langle f, f\rangle = \int_a^b (f(x))^2\,dx = 0$ if and only if f is the zero function. Therefore, $\langle f, g\rangle$ is an inner product on $\mathscr{C}[a, b]$.

Example 7.5 also defines an inner product on any *subspace* of $\mathscr{C}[a, b]$. For example, we could restrict our attention to polynomials defined on the interval $[a, b]$. Suppose we consider $\mathscr{P}[0, 1]$, the vector space of all polynomials on the interval $[0, 1]$. Then, using the inner product of Example 7.5, we have

$$\begin{aligned}\langle x^2, 1 + x\rangle &= \int_0^1 x^2(1 + x)\,dx = \int_0^1 (x^2 + x^3)\,dx \\ &= \left[\frac{x^3}{3} + \frac{x^4}{4}\right]_0^1 = \frac{1}{3} + \frac{1}{4} = \frac{7}{12}\end{aligned}$$

Properties of Inner Products

The following theorem summarizes some additional properties that follow from the definition of inner product.

Theorem 7.1 Let **u**, **v**, and **w** be vectors in an inner product space V and let c be a scalar.

a. $\langle \mathbf{u} + \mathbf{v}, \mathbf{w}\rangle = \langle \mathbf{u}, \mathbf{w}\rangle + \langle \mathbf{v}, \mathbf{w}\rangle$
b. $\langle \mathbf{u}, c\mathbf{v}\rangle = c\langle \mathbf{u}, \mathbf{v}\rangle$
c. $\langle \mathbf{u}, \mathbf{0}\rangle = \langle \mathbf{0}, \mathbf{v}\rangle = 0$

Proof We prove property (a), leaving the proofs of properties (b) and (c) as Exercises 23 and 24. Referring to the definition of inner product, we have

$$\begin{aligned} \langle \mathbf{u} + \mathbf{v}, \mathbf{w}\rangle &= \langle \mathbf{w}, \mathbf{u} + \mathbf{v}\rangle && \text{by (1)} \\ &= \langle \mathbf{w}, \mathbf{u}\rangle + \langle \mathbf{w}, \mathbf{v}\rangle && \text{by (2)} \\ &= \langle \mathbf{u}, \mathbf{w}\rangle + \langle \mathbf{v}, \mathbf{w}\rangle && \text{by (1)} \end{aligned}$$

Length, Distance, and Orthogonality

In an inner product space, we can define the length of a vector, distance between vectors, and orthogonal vectors, just as we did in Section 1.2. We simply have to replace every use of the dot product $\mathbf{u} \cdot \mathbf{v}$ by the more general inner product $\langle \mathbf{u}, \mathbf{v}\rangle$.

Definition Let **u** and **v** be vectors in an inner product space V.

1. The ***length*** (or ***norm***) of **v** is $\|\mathbf{v}\| = \sqrt{\langle \mathbf{v}, \mathbf{v}\rangle}$.
2. The ***distance*** between **u** and **v** is $d(\mathbf{u}, \mathbf{v}) = \|\mathbf{u} - \mathbf{v}\|$.
3. **u** and **v** are ***orthogonal*** if $\langle \mathbf{u}, \mathbf{v}\rangle = 0$.

Note that $\|\mathbf{v}\|$ is always defined, since $\langle \mathbf{v}, \mathbf{v}\rangle \geq 0$ by the definition of inner product, so we can take the square root of this nonnegative quantity. As in $\mathbb{R}^n$, a vector of length 1 is called a ***unit vector.*** The ***unit sphere*** in V is the set S of all unit vectors in V.

Example 7.6 Consider the inner product on $\mathscr{C}[0, 1]$ given in Example 7.5. If $f(x) = x$ and $g(x) = 3x - 2$, find

(a) $\|f\|$ (b) $d(f, g)$ (c) $\langle f, g\rangle$

Solution (a) We find that

$$\langle f, f\rangle = \int_0^1 f^2(x)\,dx = \int_0^1 x^2\,dx = \frac{x^3}{3}\bigg]_0^1 = \frac{1}{3}$$

so $\|f\| = \sqrt{\langle f, f\rangle} = 1/\sqrt{3}$.

(b) Since $d(f, g) = \|f - g\| = \sqrt{\langle f - g, f - g\rangle}$ and

$$f(x) - g(x) = x - (3x - 2) = 2 - 2x = 2(1 - x)$$

we have

$$\langle f - g, f - g\rangle = \int_0^1 (f(x) - g(x))^2\, dx = \int_0^1 4(1 - 2x + x^2)\, dx$$

$$= 4\left[x - x^2 + \frac{x^3}{3}\right]_0^1 = \frac{4}{3}$$

Combining these facts, we see that $d(f, g) = \sqrt{4/3} = 2/\sqrt{3}$.

(c) We compute

$$\langle f, g\rangle = \int_0^1 f(x)g(x)\, dx = \int_0^1 x(3x - 2)\, dx = \int_0^1 (3x^2 - 2x)\, dx = [x^3 - x^2]_0^1 = 0$$

Thus, f and g are orthogonal.

It is important to remember that the "distance" between f and g in Example 7.6 does *not* refer to any measurement related to the graphs of these functions. Neither does the fact that f and g are orthogonal mean that their graphs intersect at right angles. We are simply applying the definition of a particular inner product. However, in doing so, we should be guided by the corresponding notions in $\mathbb{R}^2$ and $\mathbb{R}^3$, where the inner product is the dot product. The geometry of Euclidean space can still guide us here, even though we cannot visualize things in the same way.

Example 7.7

Using the inner product on $\mathbb{R}^2$ defined in Example 7.2, draw a sketch of the unit sphere (circle).

Solution If $\mathbf{x} = \begin{bmatrix} x \\ y \end{bmatrix}$, then $\langle \mathbf{x}, \mathbf{x}\rangle = 2x^2 + 3y^2$. Since the unit sphere (circle) consists of all $\mathbf{x}$ such that $\|\mathbf{x}\| = 1$, we have

$$1 = \|\mathbf{x}\| = \sqrt{\langle \mathbf{x}, \mathbf{x}\rangle} = \sqrt{2x^2 + 3y^2} \quad \text{or} \quad 2x^2 + 3y^2 = 1$$

This is the equation of an ellipse, and its graph is shown in Figure 7.2.

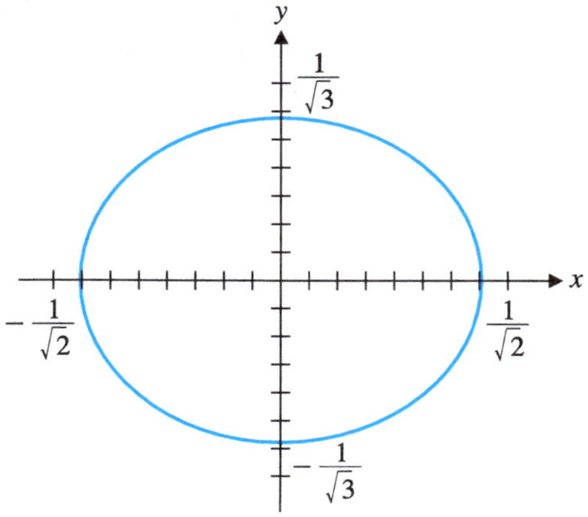

Figure 7.2
A unit circle that is an ellipse

We will discuss properties of length, distance, and orthogonality in the next section and in the exercises. One result that we will need in this section is the generalized version of Pythagoras' Theorem, which extends Theorem 1.6.

Theorem 7.2 **Pythagoras' Theorem**

Let $\mathbf{u}$ and $\mathbf{v}$ be vectors in an inner product space V. Then $\mathbf{u}$ and $\mathbf{v}$ are orthogonal if and only if

$$\|\mathbf{u}+\mathbf{v}\|^2 = \|\mathbf{u}\|^2 + \|\mathbf{v}\|^2$$

Proof As you will be asked to prove in Exercise 32, we have

$$\|\mathbf{u}+\mathbf{v}\|^2 = \langle \mathbf{u}+\mathbf{v}, \mathbf{u}+\mathbf{v}\rangle = \|\mathbf{u}\|^2 + 2\langle \mathbf{u}, \mathbf{v}\rangle + \|\mathbf{v}\|^2$$

It follows immediately that $\|\mathbf{u}+\mathbf{v}\|^2 = \|\mathbf{u}\|^2 + \|\mathbf{v}\|^2$ if and only if $\langle \mathbf{u}, \mathbf{v}\rangle = 0$.

Orthogonal Projections and the Gram-Schmidt Process

In Chapter 5, we discussed orthogonality in $\mathbb{R}^n$. Most of this material generalizes nicely to general inner product spaces. For example, an ***orthogonal set*** of vectors in an inner product space V is a set $\{\mathbf{v}_1, \ldots, \mathbf{v}_k\}$ of vectors from V such that $\langle \mathbf{v}_i, \mathbf{v}_j\rangle = 0$ whenever $\mathbf{v}_i \neq \mathbf{v}_j$. An ***orthonormal set*** of vectors is then an orthogonal set of *unit* vectors. An ***orthogonal basis*** for a subspace W of V is just a basis for W that is an orthogonal set; similarly, an ***orthonormal basis*** for a subspace W of V is a basis for W that is an orthonormal set.

In $\mathbb{R}^n$, the Gram-Schmidt Process (Theorem 5.15) shows that every subspace has an orthogonal basis. We can mimic the construction of the Gram-Schmidt Process to show that every finite-dimensional subspace of an inner product space has an orthogonal basis—all we need to do is replace the dot product by the more general inner product. We illustrate this approach with an example. (Compare the steps here with those in Example 5.13.)

Example 7.8

Construct an orthogonal basis for $\mathcal{P}_2$ with respect to the inner product

$$\langle f, g\rangle = \int_{-1}^{1} f(x)g(x)\,dx$$

by applying the Gram-Schmidt Process to the basis $\{1, x, x^2\}$.

Solution Let $\mathbf{x}_1 = 1$, $\mathbf{x}_2 = x$, and $\mathbf{x}_3 = x^2$. We begin by setting $\mathbf{v}_1 = \mathbf{x}_1 = 1$. Next we compute

$$\langle \mathbf{v}_1, \mathbf{v}_1\rangle = \int_{-1}^{1} dx = x\Big]_{-1}^{1} = 2 \quad \text{and} \quad \langle \mathbf{v}_1, \mathbf{x}_2\rangle = \int_{-1}^{1} x\,dx = \frac{x^2}{2}\Big]_{-1}^{1} = 0$$

Science Source/Photo Researchers

Adrien Marie Legendre (1752–1833) was a French mathematician who worked in astronomy, number theory, and elliptic functions. He was involved in several heated disputes with Gauss. Legendre gave the first published statement of the law of quadratic reciprocity in number theory in 1765. Gauss, however, gave the first rigorous proof of this result in 1801 and claimed credit for the result, prompting understandable outrage from Legendre. Then in 1806, Legendre gave the first published application of the method of least squares in a book on the orbits of comets. Gauss published on the same topic in 1809 but claimed he had been using the method since 1795, once again infuriating Legendre.

Therefore,

$$\mathbf{v}_2 = \mathbf{x}_2 - \frac{\langle \mathbf{v}_1, \mathbf{x}_2 \rangle}{\langle \mathbf{v}_1, \mathbf{v}_1 \rangle}\mathbf{v}_1 = x - \frac{0}{2}(1) = x$$

To find $\mathbf{v}_3$, we first compute

$$\langle \mathbf{v}_1, \mathbf{x}_3 \rangle = \int_{-1}^{1} x^2\, dx = \frac{x^3}{3}\Big]_{-1}^{1} = \frac{2}{3}, \quad \langle \mathbf{v}_2, \mathbf{x}_3 \rangle = \int_{-1}^{1} x^3\, dx = \frac{x^4}{4}\Big]_{-1}^{1} = 0,$$

$$\langle \mathbf{v}_2, \mathbf{v}_2 \rangle = \int_{-1}^{1} x^2\, dx = \frac{2}{3}$$

Then

$$\mathbf{v}_3 = \mathbf{x}_3 - \frac{\langle \mathbf{v}_1, \mathbf{x}_3 \rangle}{\langle \mathbf{v}_1, \mathbf{v}_1 \rangle}\mathbf{v}_1 - \frac{\langle \mathbf{v}_2, \mathbf{x}_3 \rangle}{\langle \mathbf{v}_2, \mathbf{v}_2 \rangle}\mathbf{v}_2 = x^2 - \frac{\frac{2}{3}}{2}(1) - \frac{0}{\frac{2}{3}}x = x^2 - \frac{1}{3}$$

It follows that $\{\mathbf{v}_1, \mathbf{v}_2, \mathbf{v}_3\}$ is an orthogonal basis for $\mathscr{P}_2$ on the interval $[-1, 1]$. The polynomials

$$1, \quad x, \quad x^2 - \tfrac{1}{3}$$

are the first three ***Legendre polynomials.*** If we divide each of these polynomials by its length relative to the same inner product, we obtain ***normalized Legendre polynomials*** (see Exercise 41).

Just as we did in Section 5.2, we can define the ***orthogonal projection*** $\text{proj}_W(\mathbf{v})$ of a vector $\mathbf{v}$ onto a subspace W of an inner product space. If $\{\mathbf{u}_1, \ldots, \mathbf{u}_k\}$ is an orthogonal basis for W, then

$$\text{proj}_W(\mathbf{v}) = \frac{\langle \mathbf{u}_1, \mathbf{v} \rangle}{\langle \mathbf{u}_1, \mathbf{u}_1 \rangle}\mathbf{u}_1 + \cdots + \frac{\langle \mathbf{u}_k, \mathbf{v} \rangle}{\langle \mathbf{u}_k, \mathbf{u}_k \rangle}\mathbf{u}_k$$

Then the ***component of* v *orthogonal to W*** is the vector

$$\text{perp}_W(\mathbf{v}) = \mathbf{v} - \text{proj}_W(\mathbf{v})$$

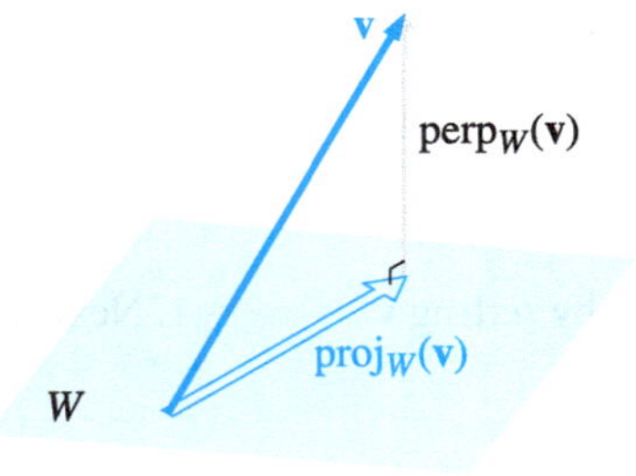

Figure 7.3

As in the Orthogonal Decomposition Theorem (Theorem 5.11), $\text{proj}_W(\mathbf{v})$ and $\text{perp}_W(\mathbf{v})$ are orthogonal (see Exercise 43), and so, schematically, we have the situation illustrated in Figure 7.3.

We will make use of these formulas in Sections 7.3 and 7.5 when we consider approximation problems—in particular, the problem of how best to approximate a given function by "nice" functions. Consequently, we will defer any examples until then, when they will make more sense. Our immediate use of orthogonal projection will be to prove an inequality that we first encountered in Chapter 1.

The Cauchy-Schwarz and Triangle Inequalities

The proofs of identities and inequalities involving the dot product in $\mathbb{R}^n$ are easily adapted to give corresponding results in general inner product spaces. Some of these are given in Exercises 31–36. In Section 1.2, we first encountered the Cauchy-Schwarz Inequality, which is important in many branches of mathematics. We now give a proof of this result for inner product spaces.

Theorem 7.3 **The Cauchy-Schwarz Inequality**

Let $\mathbf{u}$ and $\mathbf{v}$ be vectors in an inner product space V. Then

$$|\langle \mathbf{u}, \mathbf{v} \rangle| \le \|\mathbf{u}\| \, \|\mathbf{v}\|$$

with equality holding if and only if $\mathbf{u}$ and $\mathbf{v}$ are scalar multiples of each other.

This inequality was discovered by several different mathematicians, in several different contexts. It is no surprise that the name of the prolific Cauchy is attached to it. The second name associated with this result is that of Karl Herman Amandus Schwarz (1843–1921), a German mathematician who taught at the University of Berlin. His version of the inequality that bears his name was published in 1885 in a paper that used integral equations to study surfaces of minimal area. A third name also associated with this important result is that of the Russian mathematician Viktor Yakovlevitch Bunyakovsky (1804–1889). Bunyakovsky published the inequality in 1859, a full quarter-century before Schwarz's work on the same subject. Hence, it is more proper to refer to the result as the Cauchy-Bunyakovsky-Schwarz Inequality.

Proof If $\mathbf{u} = \mathbf{0}$, then the inequality is actually an equality, since

$$|\langle \mathbf{0}, \mathbf{v} \rangle| = 0 = \|\mathbf{0}\| \, \|\mathbf{v}\|$$

If $\mathbf{u} \neq \mathbf{0}$, then let W be the subspace of V spanned by $\mathbf{u}$. Since $\text{proj}_W(\mathbf{v}) = \dfrac{\langle \mathbf{u}, \mathbf{v} \rangle}{\langle \mathbf{u}, \mathbf{u} \rangle}\mathbf{u}$ and $\text{perp}_W \mathbf{v} = \mathbf{v} - \text{proj}_W(\mathbf{v})$ are orthogonal, we can apply Pythagoras' Theorem to obtain

$$\begin{aligned} \|\mathbf{v}\|^2 &= \|\text{proj}_W(\mathbf{v}) + (\mathbf{v} - \text{proj}_W(\mathbf{v}))\|^2 = \|\text{proj}_W(\mathbf{v}) + \text{perp}_W(\mathbf{v})\|^2 \\ &= \|\text{proj}_W(\mathbf{v})\|^2 + \|\text{perp}_W(\mathbf{v})\|^2 \end{aligned} \tag{2}$$

It follows that $\|\text{proj}_W(\mathbf{v})\|^2 \le \|\mathbf{v}\|^2$. Now

$$\|\text{proj}_W(\mathbf{v})\|^2 = \left\langle \frac{\langle \mathbf{u}, \mathbf{v} \rangle}{\langle \mathbf{u}, \mathbf{u} \rangle}\mathbf{u}, \frac{\langle \mathbf{u}, \mathbf{v} \rangle}{\langle \mathbf{u}, \mathbf{u} \rangle}\mathbf{u} \right\rangle = \left(\frac{\langle \mathbf{u}, \mathbf{v} \rangle}{\langle \mathbf{u}, \mathbf{u} \rangle} \right)^2 \langle \mathbf{u}, \mathbf{u} \rangle = \frac{\langle \mathbf{u}, \mathbf{v} \rangle^2}{\langle \mathbf{u}, \mathbf{u} \rangle} = \frac{\langle \mathbf{u}, \mathbf{v} \rangle^2}{\|\mathbf{u}\|^2}$$

so we have

$$\frac{\langle \mathbf{u}, \mathbf{v} \rangle^2}{\|\mathbf{u}\|^2} \le \|\mathbf{v}\|^2 \quad \text{or, equivalently,} \quad \langle \mathbf{u}, \mathbf{v} \rangle^2 \le \|\mathbf{u}\|^2 \|\mathbf{v}\|^2$$

Taking square roots, we obtain $|\langle \mathbf{u}, \mathbf{v} \rangle| \le \|\mathbf{u}\| \, \|\mathbf{v}\|$.

Clearly this last inequality is an equality if and only if $\|\text{proj}_W(\mathbf{v})\|^2 = \|\mathbf{v}\|^2$. By Equation (2) this is true if and only if $\text{perp}_W(\mathbf{v}) = \mathbf{0}$ or, equivalently,

$$\mathbf{v} = \text{proj}_W(\mathbf{v}) = \frac{\langle \mathbf{u}, \mathbf{v} \rangle}{\langle \mathbf{u}, \mathbf{u} \rangle}\mathbf{u}$$

If this is so, then $\mathbf{v}$ is a scalar multiple of $\mathbf{u}$. Conversely, if $\mathbf{v} = c\mathbf{u}$, then

$$\text{perp}_W(\mathbf{v}) = \mathbf{v} - \text{proj}_W(\mathbf{v}) = c\mathbf{u} - \frac{\langle \mathbf{u}, c\mathbf{u}\rangle}{\langle \mathbf{u}, \mathbf{u}\rangle}\mathbf{u} = c\mathbf{u} - \frac{c\langle \mathbf{u}, \mathbf{u}\rangle}{\langle \mathbf{u}, \mathbf{u}\rangle}\mathbf{u} = \mathbf{0}$$

so equality holds in the Cauchy-Schwarz Inequality.

For an alternative proof of this inequality, see Exercise 44. We will investigate some interesting consequences of the Cauchy-Schwarz Inequality and related inequalities in Exploration: Geometric Inequalities and Optimization Problems, which follows this section. For the moment, we use it to prove a generalized version of the Triangle Inequality (Theorem 1.5).

Theorem 7.4 **The Triangle Inequality**

Let $\mathbf{u}$ and $\mathbf{v}$ be vectors in an inner product space V. Then

$$\|\mathbf{u} + \mathbf{v}\| \le \|\mathbf{u}\| + \|\mathbf{v}\|$$

Proof Starting with the equality you will be asked to prove in Exercise 32, we have

$$\begin{aligned} \|\mathbf{u} + \mathbf{v}\|^2 &= \|\mathbf{u}\|^2 + 2\langle \mathbf{u}, \mathbf{v}\rangle + \|\mathbf{v}\|^2 \\ &\le \|\mathbf{u}\|^2 + 2|\langle \mathbf{u}, \mathbf{v}\rangle| + \|\mathbf{v}\|^2 \\ &\le \|\mathbf{u}\|^2 + 2\|\mathbf{u}\|\|\mathbf{v}\| + \|\mathbf{v}\|^2 \qquad \text{by Cauchy-Schwarz} \\ &= (\|\mathbf{u}\| + \|\mathbf{v}\|)^2 \end{aligned}$$

Taking square roots yields the result.

Exercises 7.1

In Exercises 1–4, let $\mathbf{u} = \begin{bmatrix} 1 \\ -2 \end{bmatrix}$ and $\mathbf{v} = \begin{bmatrix} 4 \\ 3 \end{bmatrix}$.

1. $\langle \mathbf{u}, \mathbf{v}\rangle$ is the inner product of Example 7.2. Compute

(a) $\langle \mathbf{u}, \mathbf{v}\rangle$ **(b)** $\|\mathbf{u}\|$ **(c)** $d(\mathbf{u}, \mathbf{v})$

2. $\langle \mathbf{u}, \mathbf{v}\rangle$ is the inner product of Example 7.3 with $A = \begin{bmatrix} 6 & 2 \\ 2 & 3 \end{bmatrix}$. Compute

(a) $\langle \mathbf{u}, \mathbf{v}\rangle$ **(b)** $\|\mathbf{u}\|$ **(c)** $d(\mathbf{u}, \mathbf{v})$

3. In Exercise 1, find a nonzero vector orthogonal to $\mathbf{u}$.

4. In Exercise 2, find a nonzero vector orthogonal to $\mathbf{u}$.

In Exercises 5–8, let $p(x) = 3 - 2x$ and $q(x) = 1 + x + x^2$.

5. $\langle p(x), q(x)\rangle$ is the inner product of Example 7.4. Compute

(a) $\langle p(x), q(x)\rangle$ **(b)** $\|p(x)\|$ **(c)** $d(p(x), q(x))$

6. $\langle p(x), q(x)\rangle$ is the inner product of Example 7.5 on the vector space $\mathscr{P}_2[0, 1]$. Compute

(a) $\langle p(x), q(x)\rangle$ **(b)** $\|p(x)\|$ **(c)** $d(p(x), q(x))$

7. In Exercise 5, find a nonzero vector orthogonal to $p(x)$.

8. In Exercise 6, find a nonzero vector orthogonal to $p(x)$.

In Exercises 9 and 10, let $f(x) = \sin x$ *and* $g(x) = \sin x + \cos x$ *in the vector space* $\mathscr{C}[0, 2\pi]$ *with the inner product defined by Example 7.5.*

9. Compute

(a) $\langle f, g\rangle$ **(b)** $\|f\|$ **(c)** $d(f, g)$

10. Find a nonzero vector orthogonal to f.

11. Let a, b, and c be distinct real numbers. Show that

$$\langle p(x), q(x)\rangle = p(a)q(a) + p(b)q(b) + p(c)q(c)$$

defines an inner product on $\mathscr{P}_2$. [*Hint:* You will need the fact that a polynomial of degree n has at most n zeros. See Appendix D.]

12. Repeat Exercise 5 using the inner product of Exercise 11 with $a = 0, b = 1, c = 2$.

In Exercises 13–18, determine which of the four inner product axioms do not hold. Give a specific example in each case.

13. Let $\mathbf{u} = \begin{bmatrix} u_1 \\ u_2 \end{bmatrix}$ and $\mathbf{v} = \begin{bmatrix} v_1 \\ v_2 \end{bmatrix}$ in $\mathbb{R}^2$. Define $\langle \mathbf{u}, \mathbf{v} \rangle = u_1 v_1$.

14. Let $\mathbf{u} = \begin{bmatrix} u_1 \\ u_2 \end{bmatrix}$ and $\mathbf{v} = \begin{bmatrix} v_1 \\ v_2 \end{bmatrix}$ in $\mathbb{R}^2$. Define $\langle \mathbf{u}, \mathbf{v} \rangle = u_1 v_1 - u_2 v_2$.

15. Let $\mathbf{u} = \begin{bmatrix} u_1 \\ u_2 \end{bmatrix}$ and $\mathbf{v} = \begin{bmatrix} v_1 \\ v_2 \end{bmatrix}$ in $\mathbb{R}^2$. Define $\langle \mathbf{u}, \mathbf{v} \rangle = u_1 v_2 + u_2 v_1$.

16. In $\mathscr{P}_2$, define $\langle p(x), q(x) \rangle = p(0)q(0)$.

17. In $\mathscr{P}_2$, define $\langle p(x), q(x) \rangle = p(1)q(1)$.

18. In M_{22}, define $\langle A, B \rangle = \det(AB)$.

In Exercises 19 and 20, $\langle \mathbf{u}, \mathbf{v} \rangle$ defines an inner product on $\mathbb{R}^2$, where $\mathbf{u} = \begin{bmatrix} u_1 \\ u_2 \end{bmatrix}$ and $\mathbf{v} = \begin{bmatrix} v_1 \\ v_2 \end{bmatrix}$. Find a symmetric matrix A such that $\langle \mathbf{u}, \mathbf{v} \rangle = \mathbf{u}^T A \mathbf{v}$.

19. $\langle \mathbf{u}, \mathbf{v} \rangle = 4u_1 v_1 + u_1 v_2 + u_2 v_1 + 4u_2 v_2$

20. $\langle \mathbf{u}, \mathbf{v} \rangle = u_1 v_1 + 2u_1 v_2 + 2u_2 v_1 + 5u_2 v_2$

In Exercises 21 and 22, sketch the unit circle in $\mathbb{R}^2$ for the given inner product, where $\mathbf{u} = \begin{bmatrix} u_1 \\ u_2 \end{bmatrix}$ and $\mathbf{v} = \begin{bmatrix} v_1 \\ v_2 \end{bmatrix}$.

21. $\langle \mathbf{u}, \mathbf{v} \rangle = u_1 v_1 + \frac{1}{4} u_2 v_2$

22. $\langle \mathbf{u}, \mathbf{v} \rangle = 4u_1 v_1 + u_1 v_2 + u_2 v_1 + 4u_2 v_2$

23. Prove Theorem 7.1(b).

24. Prove Theorem 7.1(c).

In Exercises 25–29, suppose that **u**, **v**, *and* **w** *are vectors in an inner product space such that*

$$\langle \mathbf{u}, \mathbf{v} \rangle = 1, \quad \langle \mathbf{u}, \mathbf{w} \rangle = 5, \quad \langle \mathbf{v}, \mathbf{w} \rangle = 0$$
$$\|\mathbf{u}\| = 1, \quad \|\mathbf{v}\| = \sqrt{3}, \quad \|\mathbf{w}\| = 2$$

Evaluate the expressions in Exercises 25–28.

25. $\langle \mathbf{u} + \mathbf{w}, \mathbf{v} - \mathbf{w} \rangle$

26. $\langle 2\mathbf{v} - \mathbf{w}, 3\mathbf{u} + 2\mathbf{w} \rangle$

27. $\|\mathbf{u} + \mathbf{v}\|$

28. $\|2\mathbf{u} - 3\mathbf{v} + \mathbf{w}\|$

29. Show that $\mathbf{u} + \mathbf{v} = \mathbf{w}$. [*Hint:* How can you use the properties of inner product to verify that $\mathbf{u} + \mathbf{v} - \mathbf{w} = \mathbf{0}$?]

30. Show that, in an inner product space, there cannot be unit vectors **u** and **v** with $\langle \mathbf{u}, \mathbf{v} \rangle < -1$.

In Exercises 31–36, $\langle \mathbf{u}, \mathbf{v} \rangle$ is an inner product. In Exercises 31–34, prove that the given statement is an identity.

31. $\langle \mathbf{u} + \mathbf{v}, \mathbf{u} - \mathbf{v} \rangle = \|\mathbf{u}\|^2 - \|\mathbf{v}\|^2$

32. $\|\mathbf{u} + \mathbf{v}\|^2 = \|\mathbf{u}\|^2 + 2\langle \mathbf{u}, \mathbf{v} \rangle + \|\mathbf{v}\|^2$

33. $\|\mathbf{u}\|^2 + \|\mathbf{v}\|^2 = \frac{1}{2}\|\mathbf{u} + \mathbf{v}\|^2 + \frac{1}{2}\|\mathbf{u} - \mathbf{v}\|^2$

34. $\langle \mathbf{u}, \mathbf{v} \rangle = \frac{1}{4}\|\mathbf{u} + \mathbf{v}\|^2 - \frac{1}{4}\|\mathbf{u} - \mathbf{v}\|^2$

35. Prove that $\|\mathbf{u} + \mathbf{v}\| = \|\mathbf{u} - \mathbf{v}\|$ if and only if **u** and **v** are orthogonal.

36. Prove that $d(\mathbf{u}, \mathbf{v}) = \sqrt{\|\mathbf{u}\|^2 + \|\mathbf{v}\|^2}$ if and only if **u** and **v** are orthogonal.

In Exercises 37–40, apply the Gram-Schmidt Process to the basis $\mathcal{B}$ to obtain an orthogonal basis for the inner product space V relative to the given inner product.

37. $V = \mathbb{R}^2$, $\mathcal{B} = \left\{ \begin{bmatrix} 1 \\ 0 \end{bmatrix}, \begin{bmatrix} 1 \\ 1 \end{bmatrix} \right\}$, with the inner product in Example 7.2

38. $V = \mathbb{R}^2$, $\mathcal{B} = \left\{ \begin{bmatrix} 1 \\ 0 \end{bmatrix}, \begin{bmatrix} 1 \\ 1 \end{bmatrix} \right\}$, with the inner product immediately following Example 7.3

39. $V = \mathscr{P}_2$, $\mathcal{B} = \{1, 1 + x, 1 + x + x^2\}$, with the inner product in Example 7.4

40. $V = \mathscr{P}_2[0, 1]$, $\mathcal{B} = \{1, 1 + x, 1 + x + x^2\}$, with the inner product in Example 7.5

41. (a) Compute the first three normalized Legendre polynomials. (See Example 7.8.)

(b) Use the Gram-Schmidt Process to find the fourth normalized Legendre polynomial.

42. If we multiply the Legendre polynomial of degree n by an appropriate scalar we can obtain a polynomial $L_n(x)$ such that $L_n(1) = 1$.

(a) Find $L_0(x)$, $L_1(x)$, $L_2(x)$, and $L_3(x)$.

(b) It can be shown that $L_n(x)$ satisfies the recurrence relation

$$L_n(x) = \frac{2n - 1}{n} x L_{n-1}(x) - \frac{n - 1}{n} L_{n-2}(x)$$

for all $n \geq 2$. Verify this recurrence for $L_2(x)$ and $L_3(x)$. Then use it to compute $L_4(x)$ and $L_5(x)$.

43. Verify that if W is a subspace of an inner product space V and $\mathbf{v}$ is in V, then $\text{perp}_W(\mathbf{v})$ is orthogonal to all $\mathbf{w}$ in W.

44. Let $\mathbf{u}$ and $\mathbf{v}$ be vectors in an inner product space V. Prove the Cauchy-Schwarz Inequality for $\mathbf{u} \neq \mathbf{0}$ as follows:

(a) Let t be a real scalar. Then $\langle t\mathbf{u} + \mathbf{v}, t\mathbf{u} + \mathbf{v} \rangle \geq 0$ for all values of t. Expand this inequality to obtain a quadratic inequality of the form

$$at^2 + bt + c \geq 0$$

What are a, b, and c in terms of $\mathbf{u}$ and $\mathbf{v}$?

(b) Use your knowledge of quadratic equations and their graphs to obtain a condition on a, b, and c for which the inequality in part (a) is true.

(c) Show that, in terms of $\mathbf{u}$ and $\mathbf{v}$, your condition in part (b) is equivalent to the Cauchy-Schwarz Inequality.

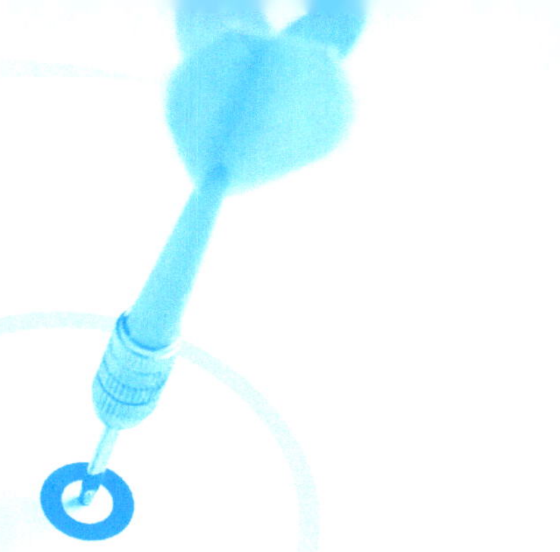

Explorations

Vectors and Matrices with Complex Entries

In this book, we have developed the theory and applications of real vector spaces, the most basic example of which is $\mathbb{R}^n$. We have also explored the finite vector spaces $\mathbb{Z}_p^n$ and their applications. The set $\mathbb{C}^n$ of n-tuples of complex numbers is also a vector space, with the complex numbers $\mathbb{C}$ as scalars. The vector space axioms (Section 6.1) all hold for $\mathbb{C}^n$, and concepts such as linear independence, basis, and dimension carry over from $\mathbb{R}^n$ without difficulty.

The first notable difference between $\mathbb{R}^n$ and $\mathbb{C}^n$ is in the definition of dot product. If we define the dot product in $\mathbb{C}^n$ as in $\mathbb{R}^n$, then for the nonzero vector $\mathbf{v} = \begin{bmatrix} i \\ 1 \end{bmatrix}$ we have

$$\|\mathbf{v}\| = \sqrt{\mathbf{v}\cdot\mathbf{v}} = \sqrt{i^2 + 1^2} = \sqrt{-1+1} = \sqrt{0} = 0$$

This is clearly an undesirable situation (a nonzero vector whose length is zero) and violates Theorems 1.2(d) and 1.3. We now generalize the real dot product to $\mathbb{C}^n$ in a way that avoids this type of difficulty.

Definition If $\mathbf{u} = \begin{bmatrix} u_1 \\ \vdots \\ u_n \end{bmatrix}$ and $\mathbf{v} = \begin{bmatrix} v_1 \\ \vdots \\ v_n \end{bmatrix}$ are vectors in $\mathbb{C}^n$, then the ***complex dot product*** of $\mathbf{u}$ and $\mathbf{v}$ is defined by

$$\mathbf{u}\cdot\mathbf{v} = \overline{u}_1 v_1 + \cdots + \overline{u}_n v_n$$

The norm (or length) of a complex vector $\mathbf{v}$ is defined as in the real case: $\|\mathbf{v}\| = \sqrt{\mathbf{v}\cdot\mathbf{v}}$. Likewise, the distance between two complex vectors $\mathbf{u}$ and $\mathbf{v}$ is still defined as $d(\mathbf{u}, \mathbf{v}) = \|\mathbf{u}-\mathbf{v}\|$.

1. Show that, for $\mathbf{v} = \begin{bmatrix} v_1 \\ \vdots \\ v_n \end{bmatrix}$ in $\mathbb{C}^n$, $\|\mathbf{v}\| = \sqrt{|v_1^2| + |v_2^2| + \cdots + |v_n^2|}$.

2. Let $\mathbf{u} = \begin{bmatrix} i \\ 1 \end{bmatrix}$ and $\mathbf{v} = \begin{bmatrix} 2 - 3i \\ 1 + 5i \end{bmatrix}$. Find:

(a) $\mathbf{u} \cdot \mathbf{v}$ (b) $\|\mathbf{u}\|$ (c) $\|\mathbf{v}\|$ (d) $d(\mathbf{u}, \mathbf{v})$ (e) a nonzero vector orthogonal to $\mathbf{u}$
(f) a nonzero vector orthogonal to $\mathbf{v}$

The complex dot product is an example of the more general notion of a complex inner product, which satisfies the same conditions as a real inner product with two exceptions. Problem 3 provides a summary.

3. Prove that the complex dot product satisfies the following properties for all vectors $\mathbf{u}$, $\mathbf{v}$, and $\mathbf{w}$ in $\mathbb{C}^n$ and all complex scalars.

(a) $\mathbf{u} \cdot \mathbf{v} = \overline{\mathbf{v} \cdot \mathbf{u}}$
(b) $\mathbf{u} \cdot (\mathbf{v} + \mathbf{w}) = \mathbf{u} \cdot \mathbf{v} + \mathbf{u} \cdot \mathbf{w}$
(c) $(c\mathbf{u}) \cdot \mathbf{v} = \bar{c}\,(\mathbf{u} \cdot \mathbf{v})$ and $\mathbf{u} \cdot (c\mathbf{v}) = c(\mathbf{u} \cdot \mathbf{v})$
(d) $\mathbf{u} \cdot \mathbf{u} \geqslant 0$ and $\mathbf{u} \cdot \mathbf{u} = 0$ if and only if $\mathbf{u} = \mathbf{0}$.

For matrices with complex entries, addition, multiplication by complex scalars, transpose, and matrix multiplication are all defined exactly as we did for real matrices in Section 3.1, and the algebraic properties of these operations still hold. (See Section 3.2.) Likewise, we have the notion of the inverse and determinant of a square complex matrix just as in the real case, and the techniques and properties all carry over to the complex case. (See Sections 3.3 and 4.2.)

The notion of transpose is, however, less useful in the complex case than in the real case. The following definition provides an alternative.

Definition If A is a complex matrix, then the ***conjugate transpose*** of A is the matrix A^* defined by

$$A^* = \overline{A}^T$$

In the preceding definition, $\overline{A}$ refers to the matrix whose entries are the complex conjugates of the corresponding entries of A; that is, if $A = [a_{ij}]$, then $\overline{A} = [\overline{a}_{ij}]$.

4. Find the conjugate transpose A^* of the given matrix:

(a) $A = \begin{bmatrix} i & 2i \\ -i & 3 \end{bmatrix}$ (b) $A = \begin{bmatrix} 2 & 5 - 2i \\ 5 + 2i & -1 \end{bmatrix}$

(c) $A = \begin{bmatrix} 2 - i & 1 + 3i & -2 \\ 4 & 0 & 3 - 4i \end{bmatrix}$ (d) $A = \begin{bmatrix} 3i & 0 & 1 + i \\ 1 - i & 4 & i \\ 1 + i & 0 & -i \end{bmatrix}$

Properties of the complex conjugate (Appendix C) extend to matrices, as the next problem shows.

5. Let A and B be complex matrices, and let c be a complex scalar. Prove the following properties:

(a) $\overline{\overline{A}} = A$ (b) $\overline{A + B} = \overline{A} + \overline{B}$
(c) $\overline{cA} = \bar{c}\,\overline{A}$ (d) $\overline{AB} = \overline{A}\,\overline{B}$
(e) $(\overline{A})^T = (\overline{A^T})$

The properties in Problem 5 can be used to establish the following properties of the conjugate transpose, which are analogous to the properties of the transpose for real matrices (Theorem 3.4).

6. Let A and B be complex matrices, and let c be a complex scalar. Prove the following properties:

(a) $(A^*)^* = A$ (b) $(A + B)^* = A^* + B^*$
(c) $(cA)^* = \bar{c}A^*$ (d) $(AB)^* = B^*A^*$

7. Show that for vectors $\mathbf{u}$ and $\mathbf{v}$ in $\mathbb{C}^n$, the complex dot product satisfies $\mathbf{u} \cdot \mathbf{v} = \mathbf{u}^*\mathbf{v}$. (This result is why we defined the complex dot product as we did. It gives us the analogue of the formula $\mathbf{u} \cdot \mathbf{v} = \mathbf{u}^T\mathbf{v}$ for vectors in $\mathbb{R}^n$.)

For real matrices, we have seen the importance of symmetric matrices, especially in our study of diagonalization. Recall that a real matrix A is symmetric if $A^T = A$. For complex matrices, the following definition is the correct generalization.

Hermitian matrices are named after the French mathematician Charles Hermite (1822–1901). Hermite is best known for his proof that the number e is transcendental, but he also was the first to use the term *orthogonal matrices,* and he proved that symmetric (and Hermitian) matrices have real eigenvalues.

Definition A square complex matrix A is called ***Hermitian*** if $A^* = A$—that is, if it is equal to its own conjugate transpose.

8. Prove that the diagonal entries of a Hermitian matrix must be real.
9. Which of the following matrices are Hermitian?

(a) $A = \begin{bmatrix} 2 & 1 + i \\ 1 - i & i \end{bmatrix}$ (b) $A = \begin{bmatrix} -1 & 2 - 3i \\ 2 - 3i & 5 \end{bmatrix}$

(c) $A = \begin{bmatrix} -3 & -1 + 5i \\ 1 - 5i & 3 \end{bmatrix}$ (d) $A = \begin{bmatrix} 1 & 1 + 4i & 3 - i \\ 1 - 4i & 2 & i \\ 3 + i & -i & 0 \end{bmatrix}$

(e) $A = \begin{bmatrix} 0 & 3 & 2 \\ -3 & 0 & -1 \\ -2 & 1 & 0 \end{bmatrix}$ (f) $A = \begin{bmatrix} 3 & 0 & -2 \\ 0 & 2 & 1 \\ -2 & 1 & 5 \end{bmatrix}$

10. Prove that the eigenvalues of a Hermitian matrix are real numbers. [*Hint:* The proof of Theorem 5.18 can be adapted by making use of the conjugate transpose operation.]

11. Prove that if A is a Hermitian matrix, then eigenvectors corresponding to distinct eigenvalues of A are orthogonal. [*Hint:* Adapt the proof of Theorem 5.19 using $\mathbf{u} \cdot \mathbf{v} = \mathbf{u}^*\mathbf{v}$ instead of $\mathbf{u} \cdot \mathbf{v} = \mathbf{u}^T\mathbf{v}$.]

Recall that a square real matrix Q is orthogonal if $Q^{-1} = Q^T$. The next definition provides the complex analogue.

Definition A square complex matrix U is called ***unitary*** if $U^{-1} = U^*$.

Just as for orthogonal matrices, in practice it is not necessary to compute U^{-1} directly. You need only show that $U^*U = I$ to verify that U is unitary.

12. Which of the following matrices are unitary? For those that are unitary, give their inverses.

(a) $\begin{bmatrix} i/\sqrt{2} & -i/\sqrt{2} \\ i/\sqrt{2} & i/\sqrt{2} \end{bmatrix}$ (b) $\begin{bmatrix} 1+i & 1+i \\ 1-i & -1+i \end{bmatrix}$

(c) $\begin{bmatrix} 3/5 & -4/5 \\ 4i/5 & 3i/5 \end{bmatrix}$ (d) $\begin{bmatrix} (1+i)/\sqrt{6} & 0 & 2/\sqrt{6} \\ 0 & 1 & 0 \\ (-1-i)/\sqrt{3} & 0 & 1/\sqrt{3} \end{bmatrix}$

Unitary matrices behave in most respects like orthogonal matrices. The following problem gives some alternative characterizations of unitary matrices.

13. Prove that the following statements are equivalent for a square complex matrix U:

(a) U is unitary.
(b) The columns of U form an orthonormal set in $\mathbb{C}^n$ with respect to the complex dot product.
(c) The rows of U form an orthonormal set in $\mathbb{C}^n$ with respect to the complex dot product.
(d) $\|U\mathbf{x}\| = \|\mathbf{x}\|$ for every $\mathbf{x}$ in $\mathbb{C}^n$.
(e) $U\mathbf{x} \cdot U\mathbf{y} = \mathbf{x} \cdot \mathbf{y}$ for every $\mathbf{x}$ and $\mathbf{y}$ in $\mathbb{C}^n$.

[*Hint:* Adapt the proofs of Theorems 5.4–5.7.]

14. Repeat Problem 12, this time by applying the criterion in part (b) or part (c) of Problem 13.

The next definition is the natural generalization of orthogonal diagonalizability to complex matrices.

Definition A square complex matrix A is called ***unitarily diagonalizable*** if there exists a unitary matrix U and a diagonal matrix D such that

$$U^*AU = D$$

The process for diagonalizing a unitarily diagonalizable $n \times n$ matrix A mimics the real case. The columns of U must form an orthonormal basis for $\mathbb{C}^n$ consisting of eigenvectors of A. Therefore, we (1) compute the eigenvalues of A, (2) find a basis for each eigenspace, (3) ensure that each eigenspace basis consists of orthonormal vectors (using the Gram-Schmidt Process, with the complex dot product, if necessary), (4) form the matrix U whose columns are the orthonormal eigenvectors just found. Then U^*AU will be a diagonal matrix D whose diagonal entries are the eigenvalues of A, arranged in the same order as the corresponding eigenvectors in the columns of U.

15. In each of the following, find a unitary matrix U and a diagonal matrix D such that $U^*AU = D$.

(a) $A = \begin{bmatrix} 2 & i \\ -i & 2 \end{bmatrix}$ (b) $A = \begin{bmatrix} 0 & -1 \\ 1 & 0 \end{bmatrix}$

(c) $A = \begin{bmatrix} -1 & 1+i \\ 1-i & 0 \end{bmatrix}$ (d) $A = \begin{bmatrix} 1 & 0 & 0 \\ 0 & 2 & 1-i \\ 0 & 1+i & 3 \end{bmatrix}$

The matrices in (a), (c), and (d) of the preceding problem are all Hermitian. It turns out that every Hermitian matrix is unitarily diagonalizable. (This is the *Complex Spectral Theorem,* which can be proved by adapting the proof of Theorem 5.20.) At this point you probably suspect that the converse of this result must also be true—namely, that every unitarily diagonalizable matrix must be Hermitian. But unfortunately this is *false*! (Can you see where the complex analogue of the proof of Theorem 5.17 breaks down?)

For a specific counterexample, take the matrix in part (b) of Problem 15. It is not Hermitian, but it *is* unitarily diagonalizable.

See *Linear Algebra with Applications* by S. J. Leon (Upper Saddle River, NJ: Prentice-Hall, 2002).

It turns out that the correct characterization of unitary diagonalizability is the following theorem, the proof of which can be found in more advanced textbooks.

A square complex matrix A is ***unitarily diagonalizable*** if and only if

$$A^*A = AA^*$$

A matrix A for which $A^*A = AA^*$ is called ***normal.***

16. Show that every Hermitian matrix, every unitary matrix, and every *skew-Hermitian* matrix ($A^* = -A$) is normal. (Note that in the real case, this result refers to symmetric, orthogonal, and skew-symmetric matrices, respectively.)

17. Prove that if a square complex matrix is unitarily diagonalizable, then it must be normal.

Geometric Inequalities and Optimization Problems

This exploration will introduce some powerful (and perhaps surprising) applications of various inequalities, such as the Cauchy-Schwarz Inequality. As you will see, certain maximization/minimization problems (*optimization problems*) that typically arise in a calculus course can be solved without using calculus at all!

Recall that the Cauchy-Schwarz Inequality in $\mathbb{R}^n$ states that for all vectors $\mathbf{u}$ and $\mathbf{v}$,

$$|\mathbf{u} \cdot \mathbf{v}| \leq \|\mathbf{u}\| \|\mathbf{v}\|$$

with equality if and only if $\mathbf{u}$ and $\mathbf{v}$ are scalar multiples of each other. If $\mathbf{u} = [x_1 \ \cdots \ x_n]^T$ and $\mathbf{v} = [y_1 \ \cdots \ y_n]^T$, the above inequality is equivalent to

$$|x_1y_1 + \cdots + x_ny_n| \leq \sqrt{x_1^2 + \cdots + x_n^2}\sqrt{y_1^2 + \cdots + y_n^2}$$

Squaring both sides and using summation notation, we have

$$\left(\sum_{i=1}^{n} x_iy_i\right)^2 \leq \left(\sum_{i=1}^{n} x_i^2\right)\left(\sum_{i=1}^{n} y_i^2\right)$$

Equality holds if and only if there is some scalar k such that $y_i = kx_i$ for $i = 1, \ldots, n$.

Let's begin by using Cauchy-Schwarz to derive a special case of one of the most useful of all inequalities.

1. Let x and y be nonnegative real numbers. Apply the Cauchy-Schwarz Inequality to $\mathbf{u} = \begin{bmatrix} \sqrt{x} \\ \sqrt{y} \end{bmatrix}$ and $\mathbf{v} = \begin{bmatrix} \sqrt{y} \\ \sqrt{x} \end{bmatrix}$ to show that

$$\sqrt{xy} \leq \frac{x+y}{2} \tag{1}$$

with equality if and only if $x = y$.

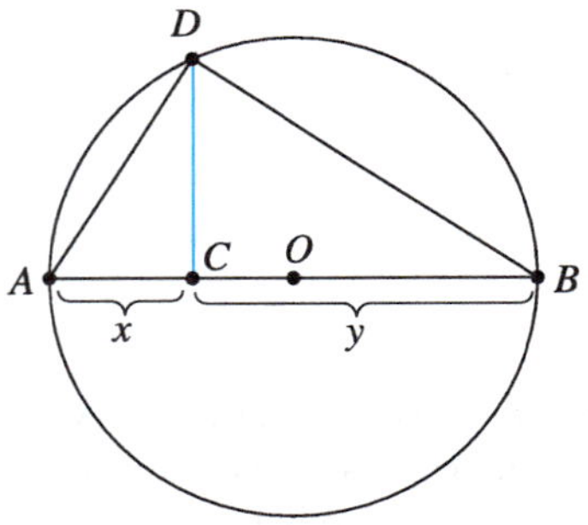

Figure 7.4

2. (a) Prove inequality (1) directly. [*Hint:* Square both sides.] (b) Figure 7.4 shows a circle with center O and diameter $AB = AC + CB = x + y$. The segment $\overline{CD}$ is perpendicular to $\overline{AB}$. Prove that $CD = \sqrt{xy}$ and use this result to deduce inequality (1). [*Hint:* Use similar triangles.]

The right-hand side of inequality (1) is the familiar ***arithmetic mean*** (or *average*) of the numbers x and y. The left-hand side shows the less familiar ***geometric mean*** of x and y. Accordingly, inequality (1) is known as the ***Arithmetic Mean–Geometric Mean Inequality (AMGM)***. It holds more generally; for n nonnegative variables $x_1, \ldots, x_n$, it states

$$\sqrt[n]{x_1 x_2 \cdots x_n} \leq \frac{x_1 + x_2 + \cdots + x_n}{n}$$

with equality if and only if $x_1 = x_2 = \cdots = x_n$.

In words, the AMGM Inequality says that the geometric mean of a set of nonnegative numbers is always less than or equal to their arithmetic mean, and the two are the same precisely when all of the numbers are the same. (For the general proof, see Appendix B.)

We now explore how such an inequality can be applied to optimization problems. Here is a typical calculus problem.

Example 7.9

Prove that among all rectangles whose perimeter is 100 units, the square has the largest area.

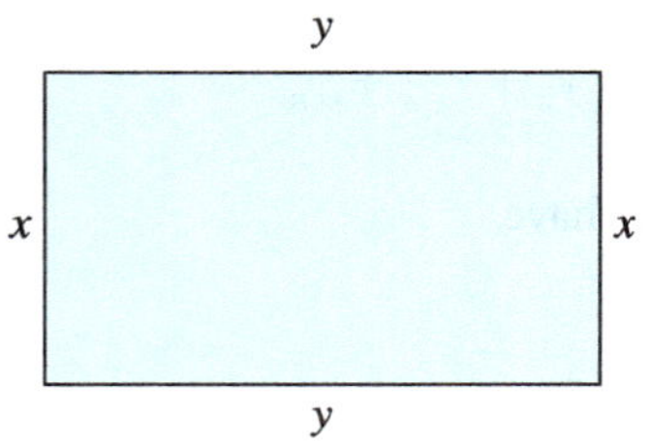

Figure 7.5

Solution If we let x and y be the dimensions of the rectangle (see Figure 7.5), then the area we want to maximize is given by

$$A = xy$$

We are given that the perimeter satisfies

$$2x + 2y = 100$$

which is the same as $x + y = 50$. We can relate xy and $x + y$ using the AMGM Inequality:

$$\sqrt{xy} \le \frac{x+y}{2} \quad \text{or, equivalently,} \quad xy \le \tfrac{1}{4}(x+y)^2$$

Since $x + y = 50$ is a *constant* (and this is the key), we see that the maximum value of $A = xy$ is $50^2/4 = 625$ and it occurs when $x = y = 25$.

Not a derivative in sight! Isn't that impressive? Notice that in this maximization problem, the crucial step was showing that the right-hand side of the AMGM Inequality was *constant*. In a similar fashion, we may be able to apply the inequality to a *minimization* problem if we can arrange for the left-hand side to be constant.

Example 7.10 Prove that among all rectangular prisms with volume 8 m^3, the cube has the minimum surface area.

Solution As shown in Figure 7.6, if the dimensions of such a prism are x, y, and z, then its volume is given by

$$V = xyz$$

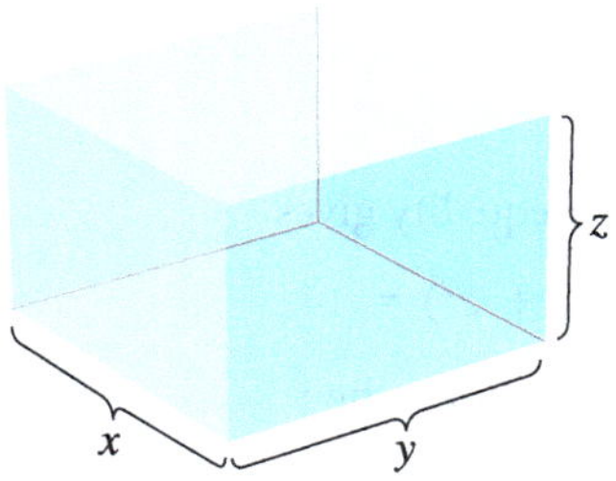

Figure 7.6

Thus, we are given that $xyz = 8$. The surface area to be minimized is

$$S = 2xy + 2yz + 2zx$$

Since this is a three-variable problem, the obvious thing to try is the version of the AMGM Inequality for $n = 3$—namely,

$$\sqrt[3]{xyz} \le \frac{x+y+z}{3}$$

Unfortunately, the expression for S does not appear here. However, the AMGM Inequality also implies that

$$\begin{aligned} \frac{S}{3} &= \frac{2xy + 2yz + 2zx}{3} \\ &\ge \sqrt[3]{(2xy)(2yz)(2zx)} \\ &= 2\sqrt[3]{(xyz)^2} \\ &= 2\sqrt[3]{64} = 8 \end{aligned}$$

which is equivalent to $S \ge 24$. Therefore, the minimum value of S is 24, and it occurs when

$$2xy = 2yz = 2zx$$

(Why?) This implies that $x = y = z = 2$ (i.e., the rectangular prism is a cube).

3. Prove that among all rectangles with area 100 square units, the square has the smallest perimeter.

4. What is the minimum value of $f(x) = x + \dfrac{1}{x}$ for $x > 0$?

5. A cardboard box with a square base and an open top is to be constructed from a square of cardboard 10 cm on a side by cutting out four squares at the corners and folding up the sides. What should the dimensions of the box be in order to make the enclosed volume as large as possible?

6. Find the minimum value of $f(x, y, z) = (x + y)(y + z)(z + x)$ if x, y, and z are positive real numbers such that $xyz = 1$.

7. For $x > y > 0$, find the minimum value of $x + \dfrac{8}{y(x - y)}$. [*Hint:* A substitution might help.]

The Cauchy-Schwarz Inequality itself can be applied to similar problems, as the next example illustrates.

Example 7.11

Find the maximum value of the function $f(x, y, z) = 3x + y + 2z$ subject to the constraint $x^2 + y^2 + z^2 = 1$. Where does the maximum value occur?

Solution This sort of problem is usually handled by techniques covered in a multivariable calculus course. Here's how to use the Cauchy-Schwarz Inequality. The function $3x + y + 2z$ has the form of a dot product, so we let

$$\mathbf{u} = \begin{bmatrix} 3 \\ 1 \\ 2 \end{bmatrix} \quad \text{and} \quad \mathbf{v} = \begin{bmatrix} x \\ y \\ z \end{bmatrix}$$

Then the componentwise form of the Cauchy-Schwarz Inequality gives

$$(3x + y + 2z)^2 \le (3^2 + 1^2 + 2^2)(x^2 + y^2 + z^2) = 14$$

Thus, the maximum value of our function is $\sqrt{14}$, and it occurs when

$$\begin{bmatrix} x \\ y \\ z \end{bmatrix} = k\begin{bmatrix} 3 \\ 1 \\ 2 \end{bmatrix}$$

Therefore, $x = 3k$, $y = k$, and $z = 2k$, so $3(3k) + k + 2(2k) = \sqrt{14}$. It follows that $k = 1/\sqrt{14}$, and hence

$$\begin{bmatrix} x \\ y \\ z \end{bmatrix} = \begin{bmatrix} 3/\sqrt{14} \\ 1/\sqrt{14} \\ 2/\sqrt{14} \end{bmatrix}$$

8. Find the maximum value of $f(x, y, z) = x + 2y + 4z$ subject to $x^2 + 2y^2 + z^2 = 1$.

9. Find the minimum value of $f(x, y, z) = x^2 + y^2 + \dfrac{z^2}{2}$ subject to $x + y + z = 10$.

10. Find the maximum value of $\sin\theta + \cos\theta$.

11. Find the point on the line $x + 2y = 5$ that is closest to the origin.

There are many other inequalities that can be used to solve optimization problems. The ***quadratic mean*** of the numbers $x_1, \ldots, x_n$ is defined as

$$\sqrt{\frac{x_1^2 + \cdots + x_n^2}{n}}$$

If $x_1, \ldots, x_n$ are nonzero, their ***harmonic mean*** is given by

$$\frac{n}{1/x_1 + 1/x_2 + \cdots + 1/x_n}$$

It turns out that the quadratic, arithmetic, geometric, and harmonic means are all related.

12. Let x and y be positive real numbers. Show that

$$\sqrt{\frac{x^2 + y^2}{2}} \geq \frac{x + y}{2} \geq \sqrt{xy} \geq \frac{2}{1/x + 1/y}$$

with equality if and only if $x = y$. (The middle inequality is just AMGM, so you need only establish the first and third inequalities.)

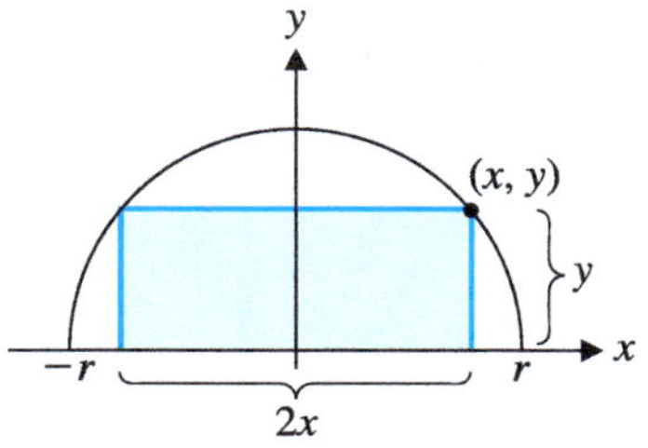

Figure 7.7

13. Find the area of the largest rectangle that can be inscribed in a semicircle of radius r (Figure 7.7).

14. Find the minimum value of the function

$$f(x, y) = \frac{(x + y)^2}{xy}$$

for $x, y > 0$. [*Hint:* $(x + y)^2/xy = (x + y)(1/x + 1/y)$.]

15. Let x and y be positive real numbers with $x + y = 1$. Show that the minimum value of

$$f(x, y) = \left(x + \frac{1}{x}\right)^2 + \left(y + \frac{1}{y}\right)^2$$

is $\frac{25}{2}$, and determine the values of x and y for which it occurs.

7.2 Norms and Distance Functions

In the last section, you saw that it is possible to define length and distance in an inner product space. As you will see shortly, there are also some versions of these two concepts that are not defined in terms of an inner product.

To begin, we need to specify the properties that we want a "length function" to have. The following definition does this, using as its basis Theorem 1.3 and the Triangle Inequality.

Definition A ***norm*** on a vector space V is a mapping that associates with each vector $\mathbf{v}$ a real number $\|\mathbf{v}\|$, called the ***norm*** of $\mathbf{v}$, such that the following properties are satisfied for all vectors $\mathbf{u}$ and $\mathbf{v}$ and all scalars c:

1. $\|\mathbf{v}\| \geq 0$, and $\|\mathbf{v}\| = 0$ if and only if $\mathbf{v} = \mathbf{0}$.
2. $\|c\mathbf{v}\| = |c|\|\mathbf{v}\|$
3. $\|\mathbf{u} + \mathbf{v}\| \leq \|\mathbf{u}\| + \|\mathbf{v}\|$

A vector space with a norm is called a ***normed linear space.***

Example 7.12 Show that in an inner product space, $\|\mathbf{v}\| = \sqrt{\langle \mathbf{v}, \mathbf{v} \rangle}$ defines a norm.

Solution Clearly, $\sqrt{\langle \mathbf{v}, \mathbf{v} \rangle} \geq 0$. Moreover,

$$\sqrt{\langle \mathbf{v}, \mathbf{v} \rangle} = 0 \Leftrightarrow \langle \mathbf{v}, \mathbf{v} \rangle = 0 \Leftrightarrow \mathbf{v} = \mathbf{0}$$

by the definition of inner product. This proves property (1).

For property (2), we only need to note that

$$\|c\mathbf{v}\| = \sqrt{\langle c\mathbf{v}, c\mathbf{v} \rangle} = \sqrt{c^2 \langle \mathbf{v}, \mathbf{v} \rangle} = \sqrt{c^2}\sqrt{\langle \mathbf{v}, \mathbf{v} \rangle} = |c|\|\mathbf{v}\|$$

Property (3) is just the Triangle Inequality, which we verified in Theorem 7.4.

We now look at some examples of norms that are not defined in terms of an inner product. Example 7.13 is the mathematical generalization to $\mathbb{R}^n$ of the taxicab norm that we explored in the Introduction to this chapter.

Example 7.13 The ***sum norm*** $\|\mathbf{v}\|_s$ of a vector $\mathbf{v}$ in $\mathbb{R}^n$ is the sum of the absolute values of its components. That is, if $\mathbf{v} = [v_1 \ \cdots \ v_n]^T$, then

$$\|\mathbf{v}\|_s = |v_1| + \cdots + |v_n|$$

Show that the sum norm is a norm.

Solution Clearly, $\|\mathbf{v}\|_s = |v_1| + \cdots + |v_n| \geq 0$, and the only way to achieve equality is if $|v_1| = \cdots = |v_n| = 0$. But this is so if and only if $v_1 = \cdots = v_n = 0$ or, equivalently, $\mathbf{v} = \mathbf{0}$, proving property (1). For property (2), we see that $c\mathbf{v} = [cv_1 \ \cdots \ cv_n]^T$, so

$$\|c\mathbf{v}\|_s = |cv_1| + \cdots + |cv_n| = |c|(|v_1| + \cdots + |v_n|) = |c|\|\mathbf{v}\|_s$$

Finally, the Triangle Inequality holds, because if $\mathbf{u} = [u_1 \quad \cdots \quad u_n]^T$, then

$$\begin{aligned}\|\mathbf{u} + \mathbf{v}\|_s &= |u_1 + v_1| + \cdots + |u_n + v_n| \\ &\leq (|u_1| + |v_1|) + \cdots + (|u_n| + |v_n|) \\ &= (|u_1| + \cdots + |u_n|) + (|v_1| + \cdots + |v_n|) = \|\mathbf{u}\|_s + \|\mathbf{v}\|_s\end{aligned}$$

The sum norm is also known as the **1-*norm*** and is often denoted by $\|\mathbf{v}\|_1$. On $\mathbb{R}^2$, it is the same as the taxicab norm. As Example 7.13 shows, it is possible to have several norms on the same vector space. Example 7.14 illustrates another norm on $\mathbb{R}^n$.

Example 7.14

The ***max norm*** $\|\mathbf{v}\|_m$ of a vector $\mathbf{v}$ in $\mathbb{R}^n$ is the largest number among the absolute values of its components. That is, if $\mathbf{v} = [v_1 \quad \cdots \quad v_n]^T$, then

$$\|\mathbf{v}\|_m = \max\{|v_1|, \ldots, |v_n|\}$$

Show that the max norm is a norm.

Solution Again, it is clear that $\|\mathbf{v}\|_m \geq 0$. If $\|\mathbf{v}\|_m = 0$, then the largest of $|v_1|, \ldots, |v_n|$ is zero, and so they all are. Hence, $v_1 = \cdots = v_n = 0$, so $\mathbf{v} = \mathbf{0}$. This verifies property (1). Next, we observe that for any scalar c,

$$\|c\mathbf{v}\|_m = \max\{|cv_1|, \ldots, |cv_n|\} = |c|\max\{|v_1|, \ldots, |v_n|\} = |c|\,\|\mathbf{v}\|_m$$

Finally, for $\mathbf{u} = [u_1 \quad \cdots \quad u_n]^T$, we have

$$\begin{aligned}\|\mathbf{u} + \mathbf{v}\|_m &= \max\{|u_1 + v_1|, \ldots, |u_n + v_n|\} \\ &\leq \max\{|u_1| + |v_1|, \ldots, |u_n| + |v_n|\} \\ &\leq \max\{|u_1|, \ldots, |u_n|\} + \max\{|v_1|, \ldots, |v_n|\} = \|\mathbf{u}\|_m + \|\mathbf{v}\|_m\end{aligned}$$

(Why is the second inequality true?) This verifies the Triangle Inequality.

The max norm is also known as the **∞-*norm*** or ***uniform norm*** and is often denoted by $\|\mathbf{v}\|_\infty$. In general, it is possible to define a norm $\|\mathbf{v}\|_p$ on $\mathbb{R}^n$ by

$$\|\mathbf{v}\|_p = (|v_1|^p + \cdots + |v_n|^p)^{1/p}$$

for any real number $p \geq 1$. For $p = 1$, $\|\mathbf{v}\|_1 = \|\mathbf{v}\|_s$, justifying the term 1-norm. For $p = 2$,

$$\|\mathbf{v}\|_2 = (|v_1|^2 + \cdots + |v_n|^2)^{1/2} = \sqrt{v_1^2 + \cdots + v_n^2}$$

which is just the familiar norm on $\mathbb{R}^n$ obtained from the dot product. Called the **2-*norm*** or ***Euclidean norm,*** it is often denoted by $\|\mathbf{v}\|_E$. As p gets large, it can be shown using calculus that $\|\mathbf{v}\|_p$ approaches the max norm $\|\mathbf{v}\|_m$. This justifies the use of the alternative notation $\|\mathbf{v}\|_\infty$ for this norm.

Example 7.15

For a vector $\mathbf{v}$ in $\mathbb{Z}_2^n$, define $\|\mathbf{v}\|_H$ to be $w(\mathbf{v})$, the weight of $\mathbf{v}$. Show that it defines a norm.

Solution Certainly, $\|\mathbf{v}\|_H = w(\mathbf{v}) \geq 0$, and the only vector whose weight is zero is the zero vector. Therefore, property (1) is true. Since the only candidates for a scalar c are 0 and 1, property (2) is immediate.

To verify the Triangle Inequality, first observe that if $\mathbf{u}$ and $\mathbf{v}$ are vectors in $\mathbb{Z}_2^n$, then $w(\mathbf{u} + \mathbf{v})$ counts the number of places in which $\mathbf{u}$ and $\mathbf{v}$ differ. [For example, if

$$\mathbf{u} = [1 \quad 1 \quad 0 \quad 1 \quad 0]^T \quad \text{and} \quad \mathbf{v} = [0 \quad 1 \quad 1 \quad 1 \quad 1]^T$$

then $\mathbf{u} + \mathbf{v} = [1 \quad 0 \quad 1 \quad 0 \quad 1]^T$, so $w(\mathbf{u} + \mathbf{v}) = 3$, in agreement with the fact that $\mathbf{u}$ and $\mathbf{v}$ differ in exactly three positions.] Suppose that both $\mathbf{u}$ and $\mathbf{v}$ have zeros in n_0 positions and 1s in n_1 positions, $\mathbf{u}$ has a 0 and $\mathbf{v}$ has a 1 in n_{01} positions, and $\mathbf{u}$ has a 1 and $\mathbf{v}$ has a 0 in n_{10} positions. (In the example above, $n_0 = 0$, $n_1 = 2$, $n_{01} = 2$, and $n_{10} = 1$.) Now

$$w(\mathbf{u}) = n_1 + n_{10}, \; w(\mathbf{v}) = n_1 + n_{01}, \quad \text{and} \quad w(\mathbf{u} + \mathbf{v}) = n_{10} + n_{01}$$

Therefore,

$$\begin{aligned} \|\mathbf{u} + \mathbf{v}\|_H = w(\mathbf{u} + \mathbf{v}) &= n_{10} + n_{01} \\ &= (n_1 + n_{10}) + (n_1 + n_{01}) - 2n_1 \\ &\leq (n_1 + n_{10}) + (n_1 + n_{01}) \\ &= w(\mathbf{u}) + w(\mathbf{v}) = \|\mathbf{u}\|_H + \|\mathbf{v}\|_H \end{aligned}$$

The norm $\|\mathbf{v}\|_H$ is called the ***Hamming norm***.

Distance Functions

For any norm, we can define a distance function just as we did in the last section—namely,

$$\mathrm{d}(\mathbf{u}, \mathbf{v}) = \|\mathbf{u} - \mathbf{v}\|$$

Example 7.16

Let $\mathbf{u} = \begin{bmatrix} 3 \\ -2 \end{bmatrix}$ and $\mathbf{v} = \begin{bmatrix} -1 \\ 1 \end{bmatrix}$. Compute $\mathrm{d}(\mathbf{u}, \mathbf{v})$ relative to (a) the Euclidean norm, (b) the sum norm, and (c) the max norm.

Solution Each calculation requires knowing that $\mathbf{u} - \mathbf{v} = \begin{bmatrix} 4 \\ -3 \end{bmatrix}$.

(a) As is by now quite familiar,

$$\mathrm{d}_E(\mathbf{u}, \mathbf{v}) = \|\mathbf{u} - \mathbf{v}\|_E = \sqrt{4^2 + (-3)^2} = \sqrt{25} = 5$$

(b) $\mathrm{d}_s(\mathbf{u}, \mathbf{v}) = \|\mathbf{u} - \mathbf{v}\|_s = |4| + |-3| = 7$

(c) $\mathrm{d}_m(\mathbf{u}, \mathbf{v}) = \|\mathbf{u} - \mathbf{v}\|_m = \max\{|4|, |-3|\} = 4$

The distance function on $\mathbb{Z}_2^n$ determined by the Hamming norm is called the ***Hamming distance.*** We will explore its use in error-correcting codes in Section 8.5. Example 7.17 provides an illustration of the Hamming distance.

Example 7.17

Find the Hamming distance between

$$\mathbf{u} = [1 \quad 1 \quad 0 \quad 1 \quad 0]^T \quad \text{and} \quad \mathbf{v} = [0 \quad 1 \quad 1 \quad 1 \quad 1]^T$$

Solution Since we are working over $\mathbb{Z}_2$, $\mathbf{u} - \mathbf{v} = \mathbf{u} + \mathbf{v}$. But

$$d_H(\mathbf{u}, \mathbf{v}) = \|\mathbf{u} + \mathbf{v}\|_H = w(\mathbf{u} + \mathbf{v})$$

As we noted in Example 7.15, this is just the number of positions in which $\mathbf{u}$ and $\mathbf{v}$ differ. The given vectors are the same ones used in that example; the calculation is therefore exactly the same. Hence, $d_H(\mathbf{u}, \mathbf{v}) = 3$.

Theorem 7.5 summarizes the most important properties of a distance function.

Theorem 7.5

Let d be a distance function defined on a normed linear space V. The following properties hold for all vectors $\mathbf{u}$, $\mathbf{v}$, and $\mathbf{w}$ in V:

a. $d(\mathbf{u}, \mathbf{v}) \geq 0$, and $d(\mathbf{u}, \mathbf{v}) = 0$ if and only if $\mathbf{u} = \mathbf{v}$.
b. $d(\mathbf{u}, \mathbf{v}) = d(\mathbf{v}, \mathbf{u})$
c. $d(\mathbf{u}, \mathbf{w}) \leq d(\mathbf{u}, \mathbf{v}) + d(\mathbf{v}, \mathbf{w})$

Proof (a) Using property (1) from the definition of a norm, it is easy to check that $d(\mathbf{u}, \mathbf{v}) = \|\mathbf{u} - \mathbf{v}\| \geq 0$, with equality holding if and only if $\mathbf{u} - \mathbf{v} = \mathbf{0}$ or, equivalently, $\mathbf{u} = \mathbf{v}$.

(b) You are asked to prove property (b) in Exercise 19.

(c) We apply the Triangle Inequality to obtain

$$\begin{aligned} d(\mathbf{u}, \mathbf{v}) + d(\mathbf{v}, \mathbf{w}) &= \|\mathbf{u} - \mathbf{v}\| + \|\mathbf{v} - \mathbf{w}\| \\ &\geq \|(\mathbf{u} - \mathbf{v}) + (\mathbf{v} - \mathbf{w})\| \\ &= \|\mathbf{u} - \mathbf{w}\| = d(\mathbf{u}, \mathbf{w}) \end{aligned}$$

A function d satisfying the three properties of Theorem 7.5 is also called a ***metric***, and a vector space that possesses such a function is called a ***metric space***. These are very important in many branches of mathematics and are studied in detail in more advanced courses.

Matrix Norms

We can define norms for matrices exactly as we defined norms for vectors in $\mathbb{R}^n$. After all, the vector space M_{mn} of all $m \times n$ matrices is isomorphic to $\mathbb{R}^{mn}$, so this is not difficult to do. Of course, properties (1), (2), and (3) of a norm will also hold in the setting of matrices. It turns out that, for matrices, the norms that are most useful satisfy an additional property. (We will restrict our attention to square matrices, but it is possible to generalize everything to arbitrary matrices.)

Definition A ***matrix norm*** on M_{nn} is a mapping that associates with each $n \times n$ matrix A a real number $\|A\|$, called the ***norm*** of A, such that the following properties are satisfied for all $n \times n$ matrices A and B and all scalars c.

1. $\|A\| \geq 0$ and $\|A\| = 0$ if and only if $A = O$.
2. $\|cA\| = |c|\|A\|$
3. $\|A + B\| \leq \|A\| + \|B\|$
4. $\|AB\| \leq \|A\|\|B\|$

A matrix norm on M_{nn} is said to be ***compatible*** with a vector norm $\|\mathbf{x}\|$ on $\mathbb{R}^n$ if, for all $n \times n$ matrices A and all vectors $\mathbf{x}$ in $\mathbb{R}^n$, we have

$$\|A\mathbf{x}\| \leq \|A\|\|\mathbf{x}\|$$

Example 7.18

The ***Frobenius norm*** $\|A\|_F$ of a matrix A is obtained by stringing out the entries of the matrix into a vector and then taking the Euclidean norm. In other words, $\|A\|_F$ is just the square root of the sum of the squares of the entries of A. So, if $A = [a_{ij}]$, then

$$\|A\|_F = \sqrt{\sum_{i,j=1}^{n} a_{ij}^2}$$

(a) Find the Frobenius norm of

$$A = \begin{bmatrix} 3 & -1 \\ 2 & 4 \end{bmatrix}$$

(b) Show that the Frobenius norm is compatible with the Euclidean norm.

(c) Show that the Frobenius norm is a matrix norm.

Solution (a) $\|A\|_F = \sqrt{3^2 + (-1)^2 + 2^2 + 4^2} = \sqrt{30}$

Before we continue, observe that if $\mathbf{A}_1 = [3 \quad -1]$ and $\mathbf{A}_2 = [2 \quad 4]$ are the row vectors of A, then $\|\mathbf{A}_1\|_E = \sqrt{3^2 + (-1)^2}$ and $\|\mathbf{A}_2\|_E = \sqrt{2^2 + 4^2}$. Thus,

$$\|A\|_F = \sqrt{\|\mathbf{A}_1\|_E^2 + \|\mathbf{A}_2\|_E^2}$$

Similarly, if $\mathbf{a}_1 = \begin{bmatrix} 3 \\ 2 \end{bmatrix}$ and $\mathbf{a}_2 = \begin{bmatrix} -1 \\ 4 \end{bmatrix}$ are the column vectors of A, then

$$\|A\|_F = \sqrt{\|\mathbf{a}_1\|_E^2 + \|\mathbf{a}_2\|_E^2}$$

It is easy to see that these facts extend to $n \times n$ matrices in general. We will use these observations to solve parts (b) and (c).

(b) Write
$$A = \begin{bmatrix} \mathbf{A}_1 \\ \vdots \\ \mathbf{A}_n \end{bmatrix}$$

Then

$$\begin{aligned}\|A\mathbf{x}\|_E &= \left\| \begin{bmatrix} A_1\mathbf{x} \\ \vdots \\ A_n\mathbf{x} \end{bmatrix} \right\|_E \\ &= \sqrt{\|\mathbf{A}_1\mathbf{x}\|_E^2 + \cdots + \|\mathbf{A}_n\mathbf{x}\|_E^2} \\ &\le \sqrt{\|\mathbf{A}_1\|_E^2\|\mathbf{x}\|_E^2 + \cdots + \mathbf{A}_n\|_E^2\|\mathbf{x}\|_E^2} \\ &= (\sqrt{\|\mathbf{A}_1\|_E^2 + \cdots + \|\mathbf{A}_n\|_E^2})\|\mathbf{x}\|_E \\ &= \|\mathbf{A}\|_F\|\mathbf{x}\|_E\end{aligned}$$

where the inequality arises from the Cauchy-Schwarz Inequality applied to the dot products of the row vectors $\mathbf{A}_i$ with the column vector $\mathbf{x}$. (Do you see how Cauchy-Schwarz has been applied?) Hence, the Frobenius norm is compatible with the Euclidean norm.

(c) Let $\mathbf{b}_i$ denote the ith column of B. Using the matrix-column representation of the product AB, we have

$$\begin{aligned}\|AB\|_F &= \|[A\mathbf{b}_1 \cdots A\mathbf{b}_n]\|_F \\ &= \sqrt{\|A\mathbf{b}_1\|_E^2 + \cdots + \|A\mathbf{b}_n\|_E^2} \\ &\le \sqrt{\|A\|_F^2\|\mathbf{b}_1\|_E^2 + \cdots + \|A\|_F^2\|\mathbf{b}_n\|_E^2} && \text{by part (b)} \\ &= \|A\|_F\sqrt{\|\mathbf{b}_1\|_E^2 + \cdots + \|\mathbf{b}_n\|_E^2} \\ &= \|A\|_F\|B\|_F\end{aligned}$$

which proves property (4) of the definition of a matrix norm. Properties (1) through (3) are true, since the Frobenius norm is derived from the Euclidean norm, which satisfies these properties. Therefore, the Frobenius norm is a matrix norm.

For many applications, the Frobenius matrix norm is not the best (or the easiest) one to use. The most useful types of matrix norms arise from considering the effect of the matrix transformation corresponding to the square matrix A. This transformation maps a vector $\mathbf{x}$ into $A\mathbf{x}$. One way to measure the "size" of A is to compare $\|\mathbf{x}\|$ and $\|A\mathbf{x}\|$ using any convenient (vector) norm. Let's think ahead. Whatever definition of $\|A\|$ we arrive at, we know we are going to want it to be compatible with the vector norm we are using; that is, we will need

$$\|A\mathbf{x}\| \le \|A\|\,\|\mathbf{x}\| \quad \text{or} \quad \frac{\|A\mathbf{x}\|}{\|\mathbf{x}\|} \le \|A\| \quad \text{for } \mathbf{x} \ne \mathbf{0}$$

The expression $\dfrac{\|A\mathbf{x}\|}{\|\mathbf{x}\|}$ measures the "stretching capability" of A. If we normalize each nonzero vector $\mathbf{x}$ by dividing it by its norm, we get unit vectors $\hat{\mathbf{x}} = \dfrac{1}{\|\mathbf{x}\|}\mathbf{x}$ and thus

$$\frac{\|A\mathbf{x}\|}{\|\mathbf{x}\|} = \frac{1}{\|\mathbf{x}\|}\|A\mathbf{x}\| = \left\|\frac{1}{\|\mathbf{x}\|}(A\mathbf{x})\right\| = \left\|A\left(\frac{1}{\|\mathbf{x}\|}\mathbf{x}\right)\right\| = \|A\hat{\mathbf{x}}\|$$

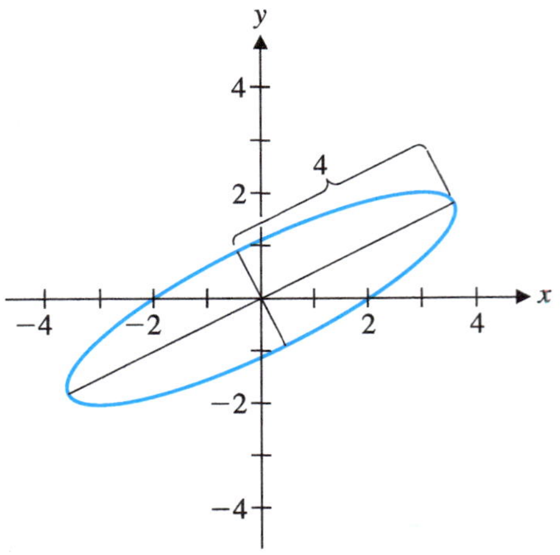

Figure 7.8

If $\mathbf{x}$ ranges over all nonzero vectors in $\mathbb{R}^n$, then $\hat{\mathbf{x}}$ ranges over all *unit* vectors (i.e., the unit sphere) and the set of all vectors $A\hat{\mathbf{x}}$ determines some curve in $\mathbb{R}^n$. For example, Figure 7.8 shows how the matrix $A = \begin{bmatrix} 3 & 2 \\ 2 & 0 \end{bmatrix}$ affects the unit circle in $\mathbb{R}^2$—it maps it into an ellipse. With the Euclidean norm, the maximum value of $\|A\hat{\mathbf{x}}\|$ is clearly just half the length of the principal axis—in this case, 4 units. We express this by writing $\max_{\|\hat{\mathbf{x}}\|=1} \|A\hat{\mathbf{x}}\| = 4$.

In Section 7.4, we will see that this is not an isolated phenomenon. That is,

$$\max_{\mathbf{x}\neq\mathbf{0}} \frac{\|A\mathbf{x}\|}{\|\mathbf{x}\|} = \max_{\|\hat{\mathbf{x}}\|=1} \|A\hat{\mathbf{x}}\|$$

always exists, and there is a particular unit vector $\mathbf{y}$ for which $\|A\mathbf{y}\|$ *is* maximum. Now we prove that $\|A\| = \max_{\|\mathbf{x}\|=1} \|A\mathbf{x}\|$ defines a matrix norm.

Theorem 7.6 If $\|\mathbf{x}\|$ is a vector norm on $\mathbb{R}^n$, then $\|A\| = \max_{\|\mathbf{x}\|=1} \|A\mathbf{x}\|$ defines a matrix norm on M_{nn} that is compatible with the vector norm that induces it.

Proof (1) Certainly, $\|A\mathbf{x}\| \geq 0$ for all vectors $\mathbf{x}$, so, in particular, this inequality is true if $\|\mathbf{x}\| = 1$. Hence, $\|A\| = \max_{\|\mathbf{x}\|=1} \|A\mathbf{x}\| \geq 0$ also. If $\|A\| = 0$, then we must have $\|A\mathbf{x}\| = 0$—and, hence, $A\mathbf{x} = \mathbf{0}$—for all $\mathbf{x}$ with $\|\mathbf{x}\| = 1$. In particular, $A\mathbf{e}_i = \mathbf{0}$ for each of the standard basis vectors $\mathbf{e}_i$ in $\mathbb{R}^n$. But $A\mathbf{e}_i$ is just the ith column of A, so we must have $A = O$. Conversely, if $A = O$, it is clear that $\|A\| = 0$. (Why?)

(2) Let c be a scalar. Then

$$\|cA\| = \max_{\|\mathbf{x}\|=1} \|cA\mathbf{x}\| = \max_{\|\mathbf{x}\|=1} |c|\,\|A\mathbf{x}\| = |c| \max_{\|\mathbf{x}\|=1} \|A\mathbf{x}\| = |c|\,\|A\|$$

(3) Let B be an $n \times n$ matrix and let $\mathbf{y}$ be a unit vector for which

$$\|A + B\| = \max_{\|\mathbf{x}\|=1} \|(A + B)\mathbf{x}\| = \|(A + B)\mathbf{y}\|$$

Then

$$\begin{aligned} \|A + B\| &= \|(A + B)\mathbf{y}\| \\ &= \|A\mathbf{y} + B\mathbf{y}\| \\ &\le \|A\mathbf{y}\| + \|B\mathbf{y}\| \\ &\le \|A\| + \|B\| \end{aligned}$$

(Where does the second inequality come from?) Next, we show that our definition is compatible with the vector norm [property (5)] and then use this fact to complete the proof that we have a matrix norm.

(5) If $\mathbf{x} = \mathbf{0}$, then the inequality $\|A\mathbf{x}\| \le \|A\|\,\|\mathbf{x}\|$ is true, since both sides are zero. If $\mathbf{x} \ne \mathbf{0}$, then from the comments preceding this theorem,

$$\frac{\|A\mathbf{x}\|}{\|\mathbf{x}\|} \le \max_{\mathbf{x}\ne\mathbf{0}} \frac{\|A\mathbf{x}\|}{\|\mathbf{x}\|} = \|A\|$$

Hence, $\|A\mathbf{x}\| \le \|A\|\,\|\mathbf{x}\|$.

(4) Let $\mathbf{z}$ be a unit vector such that $\|AB\| = \max_{\|\mathbf{x}\|=1} \|(AB)\mathbf{x}\| = \|AB\mathbf{z}\|$. Then

$$\begin{aligned} \|AB\| &= \|AB\mathbf{z}\| \\ &= \|A(B\mathbf{z})\| \\ &\le \|A\|\,\|B\mathbf{z}\| && \text{by property (5)} \\ &\le \|A\|\,\|B\|\,\|\mathbf{z}\| && \text{by property (5)} \\ &= \|A\|\,\|B\| \end{aligned}$$

This completes the proof that $\|A\| = \max_{\|\mathbf{x}\|=1} \|A\mathbf{x}\|$ defines a matrix norm on M_{nn} that is compatible with the vector norm that induces it.

Definition The matrix norm $\|A\|$ in Theorem 7.6 is called the ***operator norm*** induced by the vector norm $\|\mathbf{x}\|$.

The term *operator norm* reflects the fact that a matrix transformation arising from a square matrix is also called a *linear operator*. This norm is therefore a measure of the stretching capability of a linear operator.

The three most commonly used operator norms are those induced by the sum norm, the Euclidean norm, and the max norm—namely,

$$\|A\|_1 = \max_{\|\mathbf{x}\|_\sigma=1} \|A\mathbf{x}\|_s, \quad \|A\|_2 = \max_{\|\mathbf{x}\|_E=1} \|A\mathbf{x}\|_E, \quad \|A\|_\infty = \max_{\|\mathbf{x}\|_\mu=1} \|A\mathbf{x}\|_m$$

respectively. The first and last of these turn out to have especially nice formulas that make them very easy to compute.

Theorem 7.7 Let A be an $n \times n$ matrix with column vectors $\mathbf{a}_i$ and row vectors $\mathbf{A}_i$ for $i = 1, \ldots, n$.

a. $\|A\|_1 = \max_{j=1,\ldots,n} \{\|\mathbf{a}_j\|_s\} = \max_{j=1,\ldots,n} \left\{ \sum_{i=1}^{n} |a_{ij}| \right\}$

b. $\|A\|_\infty = \max_{i=1,\ldots,n} \{\|\mathbf{A}_i\|_s\} = \max_{i=1,\ldots,n} \left\{ \sum_{j=1}^{n} |a_{ij}| \right\}$

In other words, $\|A\|_1$ is the largest absolute column sum, and $\|A\|_\infty$ is the largest absolute row sum. Before we prove the theorem, let's look at an example to see how easy it is to use.

Example 7.19 Let

$$A = \begin{bmatrix} 1 & -3 & 2 \\ 4 & -1 & -2 \\ -5 & 1 & 3 \end{bmatrix}$$

Find $\|A\|_1$ and $\|A\|_\infty$.

Solution Clearly, the largest absolute column sum is in the first column, so

$$\|A\|_1 = \|\mathbf{a}_1\|_s = |1| + |4| + |-5| = 10$$

The third row has the largest absolute row sum, so

$$\|A\|_\infty = \|\mathbf{A}_3\|_s = |-5| + |1| + |3| = 9$$

With reference to the definition $\|A\|_1 = \max_{\|\mathbf{x}\|_s=1} \|A\mathbf{x}\|_s$, we see that the maximum value of 10 is actually achieved when we take $\mathbf{x} = \mathbf{e}_1$, for then

$$\|A\mathbf{e}_1\|_s = \|\mathbf{a}_1\|_s = 10 = \|A\|_1$$

For $\|A\|_\infty = \max_{\|\mathbf{x}\|_m=1} \|A\mathbf{x}\|_m$, if we take

$$\mathbf{x} = \begin{bmatrix} -1 \\ 1 \\ 1 \end{bmatrix}$$

we obtain

$$\|A\mathbf{x}\|_m = \left\| \begin{bmatrix} 1 & -3 & 2 \\ 4 & -1 & -2 \\ -5 & 1 & 3 \end{bmatrix} \begin{bmatrix} -1 \\ 1 \\ 1 \end{bmatrix} \right\|_m = \left\| \begin{bmatrix} -2 \\ -7 \\ 9 \end{bmatrix} \right\|_m$$
$$= \max\{|-2|, |-7|, |9|\} = 9 = \|A\|_\infty$$

We will use these observations in proving Theorem 7.7.

Proof of Theorem 7.7 The strategy is the same in the case of both the column sum and the row sum. If M represents the maximum value, we show that $\|A\mathbf{x}\| \le M$ for all unit vectors $\mathbf{x}$. Then we find a specific unit vector $\mathbf{x}$ for which equality occurs. It is important to remember that for property (a) the vector norm is the sum norm whereas for property (b) it is the max norm.

(a) To prove (a), let $M = \max_{j=1,\ldots,n} \{\|\mathbf{a}_j\|_s\}$, the maximum absolute column sum, and let $\|\mathbf{x}\|_s = 1$. Then $|x_1| + \cdots + |x_n| = 1$, so

$$\begin{aligned} \|A\mathbf{x}\|_s &= \|x_1\mathbf{a}_1 + \cdots + x_n\mathbf{a}_n\|_s \\ &\le |x_1|\,\|\mathbf{a}_1\|_s + \cdots + |x_n|\,\|\mathbf{a}_n\|_s \\ &\le |x_1|M + \cdots + |x_n|M \\ &= (|x_1| + \cdots + |x_n|)M = 1 \cdot M = M \end{aligned}$$

If the maximum absolute column sum occurs in column k, then with $\mathbf{x} = \mathbf{e}_k$ we obtain

$$\|A\mathbf{e}_k\|_s = \|\mathbf{a}_k\|_s = M$$

Therefore, $\|A\|_1 = \max_{\|\mathbf{x}\|_s=1} \|A\mathbf{x}\|_s = M = \max_{j=1,\ldots,n} \{\|\mathbf{a}_j\|_s\}$, as required.

(b) The proof of property (b) is left as Exercise 32.

In Section 7.4, we will discover a formula for the operator norm $\|A\|_2$, although it is not as computationally feasible as the formula for $\|A\|_1$ or $\|A\|_\infty$.

The Condition Number of a Matrix

In Exploration: Lies My Computer Told Me in Chapter 2, we encountered the notion of an *ill-conditioned* system of linear equations. Here is the definition as it applies to matrices.

Definition A matrix A is ***ill-conditioned*** if small changes in its entries can produce large changes in the solutions to $A\mathbf{x} = \mathbf{b}$. If small changes in the entries of A produce only small changes in the solutions to $A\mathbf{x} = \mathbf{b}$, then A is called ***well-conditioned.***

Although the definition applies to arbitrary matrices, we will restrict our attention to square matrices.

Example 7.20

Show that $A = \begin{bmatrix} 1 & 1 \\ 1 & 1.0005 \end{bmatrix}$ is ill-conditioned.

Solution If we take $\mathbf{b} = \begin{bmatrix} 3 \\ 3.0010 \end{bmatrix}$, then the solution to $A\mathbf{x} = \mathbf{b}$ is $\mathbf{x} = \begin{bmatrix} 1 \\ 2 \end{bmatrix}$. However, if A changes to

$$A' = \begin{bmatrix} 1 & 1 \\ 1 & 1.0010 \end{bmatrix}$$

then the solution changes to $\mathbf{x}' = \begin{bmatrix} 2 \\ 1 \end{bmatrix}$. (Check these assertions.) Therefore, a relative change of $0.0005/1.0005 \approx 0.0005$, or about 0.05%, causes a change of $(2-1)/1 = 1$, or 100%, in x_1 and $(1-2)/2 = -0.5$, or -50%, in x_2. Hence, A is ill-conditioned.

We can use matrix norms to give a more precise way of determining when a matrix is ill-conditioned. Think of the change from A to A' as an error ΔA that, in turn, introduces an error $\Delta\mathbf{x}$ in the solution $\mathbf{x}$ to $A\mathbf{x} = \mathbf{b}$. Then $A' = A + \Delta A$ and $\mathbf{x}' = \mathbf{x} + \Delta\mathbf{x}$. In Example 7.20,

$$\Delta A = \begin{bmatrix} 0 & 0 \\ 0 & 0.0005 \end{bmatrix} \quad \text{and} \quad \Delta\mathbf{x} = \begin{bmatrix} 1 \\ -1 \end{bmatrix}$$

Then, since $A\mathbf{x} = \mathbf{b}$ and $A'\mathbf{x}' = \mathbf{b}$, we have $(A + \Delta A)(\mathbf{x} + \Delta\mathbf{x}) = \mathbf{b}$. Expanding and canceling off $A\mathbf{x} = \mathbf{b}$, we obtain

$$A(\Delta\mathbf{x}) + (\Delta A)\mathbf{x} + (\Delta A)(\Delta\mathbf{x}) = 0 \quad \text{or} \quad A(\Delta\mathbf{x}) = -\Delta A(\mathbf{x} + \Delta\mathbf{x})$$

Since we are assuming that $A\mathbf{x} = \mathbf{b}$ has a solution, A must be invertible. Therefore, we can rewrite the last equation as

$$\Delta\mathbf{x} = -A^{-1}(\Delta A)(\mathbf{x} + \Delta\mathbf{x}) = -A^{-1}(\Delta A)\mathbf{x}'$$

Taking norms of both sides (using a matrix norm that is compatible with a vector norm), we have

$$\begin{aligned} \|\Delta\mathbf{x}\| &= \|-A^{-1}(\Delta A)\mathbf{x}'\| = \|A^{-1}(\Delta A)\mathbf{x}'\| \\ &\le \|A^{-1}(\Delta A)\|\,\|\mathbf{x}'\| \\ &\le \|A^{-1}\|\,\|\Delta A\|\,\|\mathbf{x}'\| \end{aligned}$$

(What is the justification for each step?) Therefore,

$$\frac{\|\Delta\mathbf{x}\|}{\|\mathbf{x}'\|} \le \|A^{-1}\|\,\|\Delta A\| = (\|A^{-1}\|\,\|A\|)\frac{\|\Delta A\|}{\|A\|}$$

The expression $\|A^{-1}\|\,\|A\|$ is called the ***condition number*** of A and is denoted by $\operatorname{cond}(A)$. If A is not invertible, we define $\operatorname{cond}(A) = \infty$.

What are we to make of the inequality just above? The ratio $\|\Delta A\|/\|A\|$ is a measure of the *relative change* in the matrix A, which we are assuming to be small. Similarly, $\|\Delta\mathbf{x}\|/\|\mathbf{x}'\|$ is a measure of the relative error created in the solution to $A\mathbf{x} = \mathbf{b}$ (although, in this case, the error is measured relative to the *new* solution, $\mathbf{x}'$, not the original one, $\mathbf{x}$). Thus, the inequality

$$\frac{\|\Delta\mathbf{x}\|}{\|\mathbf{x}'\|} \le \operatorname{cond}(A)\frac{\|\Delta A\|}{\|A\|} \tag{1}$$

gives an upper bound on how large the relative error in the solution can be in terms of the relative error in the coefficient matrix. The larger the condition number, the more ill-conditioned the matrix, since there is more "room" for the error to be large relative to the solution.

Remarks

- The condition number of a matrix depends on the choice of norm. The most commonly used norms are the operator norms $\|A\|_1$ and $\|A\|_\infty$.
- For any norm, $\text{cond}(A) \geq 1$. (See Exercise 45.)

Example 7.21

Find the condition number of $A = \begin{bmatrix} 1 & 1 \\ 1 & 1.0005 \end{bmatrix}$ relative to the ∞-norm.

Solution We first compute

$$A^{-1} = \begin{bmatrix} 2001 & -2000 \\ -2000 & 2000 \end{bmatrix}$$

Therefore, in the ∞-norm (maximum absolute row sum),

$$\|A\|_\infty = 1 + 1.0005 = 2.0005 \quad \text{and} \quad \|A^{-1}\|_\infty = 2001 + |-2000| = 4001$$

so $\text{cond}_\infty(A) = \|A^{-1}\|_\infty \|A\|_\infty = 4001(2.0005) \approx 8004$.

It turns out that if the condition number is large relative to one compatible matrix norm, it will be large relative to *any* compatible matrix norm. For example, it can be shown that for matrix A in Examples 7.20 and 7.21, $\text{cond}_1(A) \approx 8004$, $\text{cond}_2(A) \approx 8002$ (relative to the 2-norm), and $\text{cond}_F(A) \approx 8002$ (relative to the Frobenius norm).

The Convergence of Iterative Methods

In Section 2.5, we explored two iterative methods for solving a system of linear equations: Jacobi's method and the Gauss-Seidel method. In Theorem 2.9, we stated without proof that if A is a strictly diagonally dominant $n \times n$ matrix, then both of these methods converge to the solution of $A\mathbf{x} = \mathbf{b}$. We are now in a position to prove this theorem. Indeed, one of the important uses of matrix norms is to establish the convergence properties of various iterative methods.

We will deal only with Jacobi's method here. (The Gauss-Seidel method can be handled using similar techniques, but it requires a bit more care.) The key is to rewrite the iterative process in terms of matrices. Let's revisit Example 2.37 with this in mind. The system of equations is

$$\begin{aligned} 7x_1 - x_2 &= 5 \\ 3x_1 - 5x_2 &= -7 \end{aligned} \qquad (2)$$

so

$$A = \begin{bmatrix} 7 & -1 \\ 3 & -5 \end{bmatrix} \quad \text{and} \quad \mathbf{b} = \begin{bmatrix} 5 \\ -7 \end{bmatrix}$$

We rewrote Equation (2) as

$$\begin{aligned} x_1 &= \frac{5 + x_2}{7} \\ x_2 &= \frac{7 + 3x_1}{5} \end{aligned} \tag{3}$$

which is equivalent to

$$\begin{aligned} 7x_1 &= \; x_2 + 5 \\ -5x_2 &= -3x_1 - 7 \end{aligned} \tag{4}$$

or, in terms of matrices,

$$\begin{bmatrix} 7 & 0 \\ 0 & -5 \end{bmatrix}\begin{bmatrix} x_1 \\ x_2 \end{bmatrix} = \begin{bmatrix} 0 & 1 \\ -3 & 0 \end{bmatrix}\begin{bmatrix} x_1 \\ x_2 \end{bmatrix} + \begin{bmatrix} 5 \\ -7 \end{bmatrix} \tag{5}$$

Study Equation (5) carefully: The matrix on the left-hand side contains the diagonal entries of A, while on the right-hand side we see the *negative* of the off-diagonal entries of A and the vector $\mathbf{b}$. So, if we decompose A as

$$A = \begin{bmatrix} 7 & -1 \\ 3 & -5 \end{bmatrix} = \begin{bmatrix} 0 & 0 \\ 3 & 0 \end{bmatrix} + \begin{bmatrix} 7 & 0 \\ 0 & -5 \end{bmatrix} + \begin{bmatrix} 0 & -1 \\ 0 & 0 \end{bmatrix} = L + D + U$$

then Equation (5) can be written as

$$D\mathbf{x} = -(L + U)\mathbf{x} + \mathbf{b}$$

or, equivalently,

$$\mathbf{x} = -D^{-1}(L + U)\mathbf{x} + D^{-1}\mathbf{b} \tag{6}$$

since the matrix D is invertible. Equation (6) is the matrix version of Equation (3). It is easy to see that we can do this in general: An $n \times n$ matrix A can be written as $A = L + D + U$, where D is the diagonal part of A and L and U are, respectively, the portions of A below and above the diagonal. The system $A\mathbf{x} = \mathbf{b}$ can then be written in the form of Equation (6), provided D is invertible—which it is if A is strictly diagonally dominant. (Why?) To simplify the notation, let's let $M = -D^{-1}(L + U)$ and $\mathbf{c} = D^{-1}\mathbf{b}$ so that Equation (6) becomes

$$\mathbf{x} = M\mathbf{x} + \mathbf{c} \tag{7}$$

Recall how we use this equation in Jacobi's method. We start with an initial vector $\mathbf{x}_0$ and plug it into the right-hand side of Equation (7) to get the first iterate $\mathbf{x}_1$—that is, $\mathbf{x}_1 = M\mathbf{x}_0 + \mathbf{c}$. Then we plug $\mathbf{x}_1$ into the right-hand side of Equation (7) to get the second iterate $\mathbf{x}_2 = M\mathbf{x}_1 + \mathbf{c}$. In general, we have

$$\mathbf{x}_{k+1} = M\mathbf{x}_k + \mathbf{c} \tag{8}$$

for $k \geq 0$. For Example 2.37, we have

$$M = -D^{-1}(L + U) = -\begin{bmatrix} 7 & 0 \\ 0 & -5 \end{bmatrix}^{-1}\begin{bmatrix} 0 & -1 \\ 3 & 0 \end{bmatrix} = \begin{bmatrix} 0 & \frac{1}{7} \\ \frac{3}{5} & 0 \end{bmatrix}$$

and $$\mathbf{c} = D^{-1}\mathbf{b} = \begin{bmatrix} 7 & 0 \\ 0 & -5 \end{bmatrix}^{-1}\begin{bmatrix} 5 \\ -7 \end{bmatrix} = \begin{bmatrix} \frac{5}{7} \\ \frac{7}{5} \end{bmatrix}$$

so
$$\mathbf{x}_1 = \begin{bmatrix} 0 & \frac{1}{7} \\ \frac{3}{5} & 0 \end{bmatrix}\begin{bmatrix} 0 \\ 0 \end{bmatrix} + \begin{bmatrix} \frac{5}{7} \\ \frac{7}{5} \end{bmatrix} = \begin{bmatrix} \frac{5}{7} \\ \frac{7}{5} \end{bmatrix} \approx \begin{bmatrix} 0.714 \\ 1.400 \end{bmatrix}$$
$$\mathbf{x}_2 = \begin{bmatrix} 0 & \frac{1}{7} \\ \frac{3}{5} & 0 \end{bmatrix}\begin{bmatrix} 0.714 \\ 1.400 \end{bmatrix} + \begin{bmatrix} \frac{5}{7} \\ \frac{7}{5} \end{bmatrix} \approx \begin{bmatrix} 0.914 \\ 1.829 \end{bmatrix}$$

and so on. (These are exactly the same calculations we did in Example 2.37, but written in matrix form.)

To show that Jacobi's method will converge, we need to show that the iterates $\mathbf{x}_k$ approach the actual solution $\mathbf{x}$ of $A\mathbf{x} = \mathbf{b}$. It is enough to show that the ***error vectors*** $\mathbf{x}_k - \mathbf{x}$ approach the zero vector. From our calculations above, $A\mathbf{x} = \mathbf{b}$ is equivalent to $\mathbf{x} = M\mathbf{x} + \mathbf{c}$. Using Equation (8), we then have

$$\begin{aligned}\mathbf{x}_{k+1} - \mathbf{x} &= M\mathbf{x}_k + \mathbf{c} - (M\mathbf{x} + \mathbf{c}) \\ &= M(\mathbf{x}_k - \mathbf{x})\end{aligned}$$

Now we take the norm of both sides of this equation. (At this point, it is not important which norm we use as long as we choose a matrix norm that is compatible with a vector norm.) We have

$$\|\mathbf{x}_{k+1} - \mathbf{x}\| = \|M(\mathbf{x}_k - \mathbf{x})\| \le \|M\|\,\|\mathbf{x}_k - \mathbf{x}\| \tag{9}$$

If we can show that $\|M\| < 1$, then we will have $\|\mathbf{x}_{k+1} - \mathbf{x}\| < \|\mathbf{x}_k - \mathbf{x}\|$ for all $k \ge 0$, and it follows that $\|\mathbf{x}_k - \mathbf{x}\|$ approaches zero, so the error vectors $\mathbf{x}_k - \mathbf{x}$ approach the zero vector.

The fact that strict diagonal dominance is defined in terms of the absolute values of the entries in the *rows* of a matrix suggests that the ∞-norm of a matrix (the operator norm induced by the max norm) is the one to choose. If $A = [a_{ij}]$, then

$$M = \begin{bmatrix} 0 & -a_{12}/a_{11} & \cdots & -a_{1n}/a_{11} \\ -a_{21}/a_{22} & 0 & \cdots & -a_{2n}/a_{22} \\ \vdots & \vdots & \ddots & \vdots \\ -a_{n1}/a_{nn} & -a_{n2}/a_{nn} & \cdots & 0 \end{bmatrix}$$

(verify this), so, by Theorem 7.7, $\|M\|_\infty$ is the maximum absolute row sum of M. Suppose it occurs in the kth row. Then

$$\begin{aligned}\|M\|_\infty &= \left|\frac{-a_{k1}}{a_{kk}}\right| + \cdots + \left|\frac{-a_{k,k-1}}{a_{kk}}\right| + \left|\frac{-a_{k,k+1}}{a_{kk}}\right| + \cdots + \left|\frac{-a_{kn}}{a_{kk}}\right| \\ &= \frac{|a_{k1}| + \cdots + |a_{k,k-1}| + |a_{k,k+1}| + \cdots + |a_{kn}|}{|a_{kk}|} < 1\end{aligned}$$

since A is strictly diagonally dominant. Thus, $\|M\|_\infty < 1$, so $\|\mathbf{x}_k - \mathbf{x}\| \to 0$, as we wished to show.

Example 7.22

Compute $\|M\|_\infty$ in Example 2.37 and use this value to find the number of iterations required to approximate the solution to three-decimal-place accuracy (after rounding) if the initial vector is $\mathbf{x}_0 = \mathbf{0}$.

Solution We have already computed $M = \begin{bmatrix} 0 & \frac{1}{7} \\ \frac{3}{5} & 0 \end{bmatrix}$, so $\|M\|_\infty = \frac{3}{5} = 0.6 < 1$ (implying that Jacobi's method converges in Example 2.37, as we saw). The approximate solution $\mathbf{x}_k$ will be accurate to three decimal places if the error vector $\mathbf{x}_k - \mathbf{x}$ has the property that each of its components is less than 0.0005 in absolute value. (Why?) Thus, we need only guarantee that the *maximum* absolute component of $\mathbf{x}_k - \mathbf{x}$ is less than 0.0005. In other words, we need to find the smallest value of k such that

$$\|\mathbf{x}_k - \mathbf{x}\|_m < 0.0005$$

Using Equation (9) above, we see that

$$\|\mathbf{x}_k - \mathbf{x}\|_m \le \|M\|_\infty \|\mathbf{x}_{k-1} - \mathbf{x}\|_m \le \|M\|_\infty^2 \|\mathbf{x}_{k-2} - \mathbf{x}\|_m \le \cdots \le \|M\|_\infty^k \|\mathbf{x}_0 - \mathbf{x}\|_m$$

Now $\|M\|_\infty = 0.6$ and $\|\mathbf{x}_0 - \mathbf{x}\|_m \approx \|\mathbf{x}_0 - \mathbf{x}_1\|_m = \|\mathbf{x}_1\|_m = \left\| \begin{bmatrix} 0.714 \\ 1.400 \end{bmatrix} \right\|_m = 1.4$, so

$$\|M\|_\infty^k \|\mathbf{x}_0 - \mathbf{x}\|_m \approx (0.6)^k(1.4)$$

(If we knew the exact solution in advance, we could use it instead of $\mathbf{x}_1$. In practice, this is not the case, so we use an approximation to the solution, as we have done here.) Therefore, we need to find k such that

$$(0.6)^k(1.4) < 0.0005$$

We can solve this inequality by taking logarithms (base 10) of both sides. We have

$$\begin{aligned} \log_{10}((0.6)^k(1.4)) < \log_{10}(5 \times 10^{-4}) &\Rightarrow k\log_{10}(0.6) + \log_{10}(1.4) < \log_{10}5 - 4 \\ &\Rightarrow -0.222k + 0.146 < -3.301 \\ &\Rightarrow k > 15.5 \end{aligned}$$

Since k must be an integer, we can therefore conclude that $k = 16$ will work and that 16 iterations of Jacobi's method will give us three-decimal-place accuracy in this example. (In fact, it appears from our calculations in Example 2.37 that we get this degree of accuracy sooner, but our goal here was only to come up with an estimate.)

Exercises 7.2

In Exercises 1–3, let $\mathbf{u} = \begin{bmatrix} -1 \\ 4 \\ -5 \end{bmatrix}$ *and* $\mathbf{v} = \begin{bmatrix} 2 \\ -2 \\ 0 \end{bmatrix}$.

1. Compute the Euclidean norm, the sum norm, and the max norm of $\mathbf{u}$.

2. Compute the Euclidean norm, the sum norm, and the max norm of $\mathbf{v}$.

3. Compute $d(\mathbf{u}, \mathbf{v})$ relative to the Euclidean norm, the sum norm, and the max norm.

4. **(a)** What does $d_s(\mathbf{u}, \mathbf{v})$ measure?
(b) What does $d_m(\mathbf{u}, \mathbf{v})$ measure?

In Exercises 5 and 6, let $\mathbf{u} = [1 \quad 0 \quad 1 \quad 1 \quad 0 \quad 0 \quad 1]^T$ *and* $\mathbf{v} = [0 \quad 1 \quad 1 \quad 0 \quad 1 \quad 1 \quad 1]^T$.

5. Compute the Hamming norms of $\mathbf{u}$ and $\mathbf{v}$.

6. Compute the Hamming distance between $\mathbf{u}$ and $\mathbf{v}$.

7. **(a)** For which vectors $\mathbf{v}$ is $\|\mathbf{v}\|_E = \|\mathbf{v}\|_m$? Explain your answer.

(b) For which vectors $\mathbf{v}$ is $\|\mathbf{v}\|_s = \|\mathbf{v}\|_m$? Explain your answer.

(c) For which vectors $\mathbf{v}$ is $\|\mathbf{v}\|_s = \|\mathbf{v}\|_m = \|\mathbf{v}\|_E$? Explain your answer.

8. (a) Under what conditions on $\mathbf{u}$ and $\mathbf{v}$ is $\|\mathbf{u} + \mathbf{v}\|_E = \|\mathbf{u}\|_E + \|\mathbf{v}\|_E$? Explain your answer.

(b) Under what conditions on $\mathbf{u}$ and $\mathbf{v}$ is $\|\mathbf{u} + \mathbf{v}\|_s = \|\mathbf{u}\|_s + \|\mathbf{v}\|_s$? Explain your answer.

(c) Under what conditions on $\mathbf{u}$ and $\mathbf{v}$ is $\|\mathbf{u} + \mathbf{v}\|_m = \|\mathbf{u}\|_m + \|\mathbf{v}\|_m$? Explain your answer.

9. Show that for all $\mathbf{v}$ in $\mathbb{R}^n$, $\|\mathbf{v}\|_m \le \|\mathbf{v}\|_E$.

10. Show that for all $\mathbf{v}$ in $\mathbb{R}^n$, $\|\mathbf{v}\|_E \le \|\mathbf{v}\|_s$.

11. Show that for all $\mathbf{v}$ in $\mathbb{R}^n$, $\|\mathbf{v}\|_s \le n\|\mathbf{v}\|_m$.

12. Show that for all $\mathbf{v}$ in $\mathbb{R}^n$, $\|\mathbf{v}\|_E \le \sqrt{n}\|\mathbf{v}\|_m$.

13. Draw the unit circles in $\mathbb{R}^2$ relative to the sum norm and the max norm.

14. By showing that the identity of Exercise 33 in Section 7.1 fails, show that the sum norm does not arise from any inner product.

In Exercises 15–18, prove that $\|\ \|$ defines a norm on the vector space V.

15. $V = \mathbb{R}^2$, $\left\|\begin{bmatrix} a \\ b \end{bmatrix}\right\| = \max\{|2a|, |3b|\}$

16. $V = M_{mn}$, $\|A\| = \max\limits_{i,j}\{|a_{ij}|\}$

17. $V = \mathscr{C}[0, 1]$, $\|f\| = \int_0^1 |f(x)|\,dx$

18. $\|f\| = \max\limits_{0 \le x \le 1} |f(x)|$

19. Prove Theorem 7.5(b).

In Exercises 20–25, compute $\|A\|_F$, $\|A\|_1$, and $\|A\|_\infty$.

20. $A = \begin{bmatrix} 2 & 3 \\ 4 & 1 \end{bmatrix}$ **21.** $A = \begin{bmatrix} 0 & -1 \\ -3 & 3 \end{bmatrix}$

22. $A = \begin{bmatrix} 1 & 5 \\ -2 & -1 \end{bmatrix}$ **23.** $A = \begin{bmatrix} 2 & 1 & 1 \\ 1 & 3 & 2 \\ 1 & 1 & 3 \end{bmatrix}$

24. $A = \begin{bmatrix} 0 & -5 & 2 \\ 3 & 1 & -3 \\ -4 & -4 & 3 \end{bmatrix}$ **25.** $A = \begin{bmatrix} 4 & -2 & -1 \\ 0 & -1 & 2 \\ 3 & -3 & 0 \end{bmatrix}$

In Exercises 26–31, find vectors $\mathbf{x}$ and $\mathbf{y}$ with $\|\mathbf{x}\|_s = 1$ and $\|\mathbf{y}\|_m = 1$ such that $\|A\|_1 = \|A\mathbf{x}\|_s$ and $\|A\|_\infty = \|A\mathbf{y}\|_m$, where A is the matrix in the given exercise.

26. Exercise 20 **27.** Exercise 21 **28.** Exercise 22

29. Exercise 23 **30.** Exercise 24 **31.** Exercise 25

32. Prove Theorem 7.7(b).

33. (a) If $\|A\|$ is an operator norm, prove that $\|I\| = 1$, where I is an identity matrix.

(b) Is there a vector norm that induces the Frobenius norm as an operator norm? Why or why not?

34. Let $\|A\|$ be a matrix norm that is compatible with a vector norm $\|\mathbf{x}\|$. Prove that $\|A\| \ge |\lambda|$ for every eigenvalue λ of A.

In Exercises 35–40, find $\text{cond}_1(A)$ and $\text{cond}_\infty(A)$. State whether the given matrix is ill-conditioned.

35. $A = \begin{bmatrix} 3 & 1 \\ 4 & 2 \end{bmatrix}$ **36.** $A = \begin{bmatrix} 1 & -2 \\ -3 & 6 \end{bmatrix}$

37. $A = \begin{bmatrix} 1 & 0.99 \\ 1 & 1 \end{bmatrix}$ **38.** $A = \begin{bmatrix} 150 & 200 \\ 3001 & 4002 \end{bmatrix}$

39. $A = \begin{bmatrix} 1 & 1 & 1 \\ 5 & 5 & 6 \\ 1 & 0 & 0 \end{bmatrix}$ **40.** $A = \begin{bmatrix} 1 & \frac{1}{2} & \frac{1}{3} \\ \frac{1}{2} & \frac{1}{3} & \frac{1}{4} \\ \frac{1}{3} & \frac{1}{4} & \frac{1}{5} \end{bmatrix}$

41. Let $A = \begin{bmatrix} 1 & k \\ 1 & 1 \end{bmatrix}$.

(a) Find a formula for $\text{cond}_\infty(A)$ in terms of k.

(b) What happens to $\text{cond}_\infty(A)$ as k approaches 1?

42. Consider the linear system $A\mathbf{x} = \mathbf{b}$, where A is invertible. Suppose an error $\Delta\mathbf{b}$ changes $\mathbf{b}$ to $\mathbf{b}' = \mathbf{b} + \Delta\mathbf{b}$. Let $\mathbf{x}'$ be the solution to the new system; that is, $A\mathbf{x}' = \mathbf{b}'$. Let $\mathbf{x}' = \mathbf{x} + \Delta\mathbf{x}$ so that $\Delta\mathbf{x}$ represents the resulting error in the solution of the system. Show that

$$\frac{\|\Delta\mathbf{x}\|}{\|\mathbf{x}\|} \le \text{cond}(A)\frac{\|\Delta\mathbf{b}\|}{\|\mathbf{b}\|}$$

for any compatible matrix norm.

43. Let $A = \begin{bmatrix} 10 & 10 \\ 10 & 9 \end{bmatrix}$ and $\mathbf{b} = \begin{bmatrix} 100 \\ 99 \end{bmatrix}$.

(a) Compute $\text{cond}_\infty(A)$.

(b) Suppose A is changed to $A' = \begin{bmatrix} 10 & 10 \\ 10 & 11 \end{bmatrix}$. How large a relative change can this change produce in the solution to $A\mathbf{x} = \mathbf{b}$? [*Hint:* Use inequality (1) from this section.]

(c) Solve the systems using A and A' and determine the actual relative error.

(d) Suppose $\mathbf{b}$ is changed to $\mathbf{b}' = \begin{bmatrix} 100 \\ 101 \end{bmatrix}$. How large a relative change can this change produce in the solution to $A\mathbf{x} = \mathbf{b}$? [*Hint:* Use Exercise 42.]

(e) Solve the systems using $\mathbf{b}$ and $\mathbf{b}'$ and determine the actual relative error.

44. Let $A = \begin{bmatrix} 1 & 1 & 1 \\ 2 & 5 & 0 \\ 1 & -1 & 2 \end{bmatrix}$ and $\mathbf{b} = \begin{bmatrix} 1 \\ 2 \\ 3 \end{bmatrix}$.

(a) Compute $\text{cond}_1(A)$.

(b) Suppose A is changed to $A' = \begin{bmatrix} 1 & 1 & 1 \\ 1 & 5 & 0 \\ 1 & -1 & 2 \end{bmatrix}$. How large a relative change can this change produce in the solution to $A\mathbf{x} = \mathbf{b}$? [*Hint:* Use inequality (1) from this section.]

(c) Solve the systems using A and A' and determine the actual relative error.

(d) Suppose $\mathbf{b}$ is changed to $\mathbf{b}' = \begin{bmatrix} 1 \\ 1 \\ 3 \end{bmatrix}$. How large a relative change can this change produce in the solution to $A\mathbf{x} = \mathbf{b}$? [*Hint:* Use Exercise 42.]

(e) Solve the systems using $\mathbf{b}$ and $\mathbf{b}'$ and determine the actual relative error.

45. Show that if A is an invertible matrix, then $\text{cond}(A) \geq 1$ with respect to any matrix norm.

46. Show that if A and B are invertible matrices, then $\text{cond}(AB) \leq \text{cond}(A)\,\text{cond}(B)$ with respect to any matrix norm.

47. Let A be an invertible matrix and let λ_1 and λ_n be the eigenvalues with the largest and smallest absolute values, respectively. Show that

$$\text{cond}\,(A) \geq \frac{|\lambda_1|}{|\lambda_n|}$$

[*Hint:* See Exercise 34 and Theorem 4.18(b) in Section 4.3.]

CAS *In Exercises 48–51, write the given system in the form of Equation (7). Then use the method of Example 7.22 to estimate the number of iterations of Jacobi's method that will be needed to approximate the solution to three-decimal-place accuracy. (Use $\mathbf{x}_0 = \mathbf{0}$.) Compare your answer with the solution computed in the given exercise from Section 2.5.*

48. Exercise 1, Section 2.5

49. Exercise 3, Section 2.5

50. Exercise 4, Section 2.5

51. Exercise 5, Section 2.5

Exercise 52(c) refers to the Leontief model of an open economy, as discussed in Sections 2.4 and 3.7.

52. Let A be an $n \times n$ matrix such that $\|A\| < 1$, where the norm is either the sum norm or the max norm.

(a) Prove that $A^n \to O$ as $n \to \infty$.

(b) Deduce from (a) that $I - A$ is invertible and

$$(I - A)^{-1} = I + A + A^2 + A^3 + \cdots$$

[*Hint:* See the proof of Theorem 3.34.]

(c) Show that (b) can be used to prove Corollaries 3.35 and 3.36.

7.3 Least Squares Approximation

In many branches of science, experimental data are used to infer a mathematical relationship among the variables being measured. For example, we might measure the height of a tree at various points in time and try to deduce a function that expresses the tree's height h in terms of time t. Or, we might measure the size p of a population over time and try to find a rule that relates p to t. Relationships between variables are also of interest in business; for example, a company producing widgets may be interested in knowing the relationship between its total costs c and the number n of widgets produced.

In each of these examples, the data come in the form of two measurements: one for the independent variable and one for the (supposedly) dependent variable. Thus, we have a set of *data points* (x_i, y_i), and we are looking for a function that best approximates the relationship between the independent variable x and the dependent variable y. Figure 7.9 shows examples in which experimental data points are plotted, along with a curve that approximately "fits" the data.

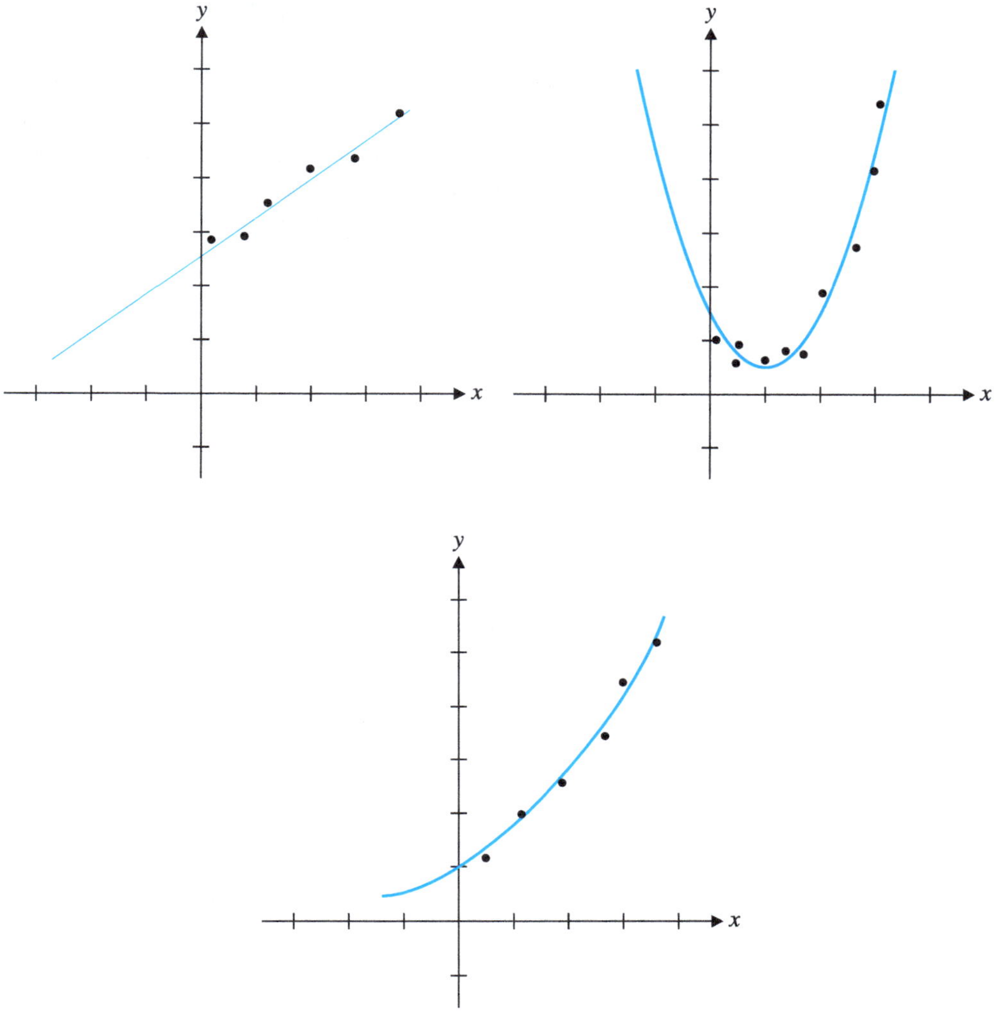

Figure 7.9
Curves of "best fit"

Roger Cotes (1682–1716) was an English mathematician who, while a fellow at Cambridge, edited the second edition of Newton's *Principia*. Although he published little, he made important discoveries in the theory of logarithms, integral calculus, and numerical methods.

The method of least squares, which we are about to consider, is attributed to Gauss. A new asteroid, Ceres, was discovered on New Year's Day, 1801, but it disappeared behind the sun shortly after it was observed. Astronomers predicted when and where Ceres would reappear, but their calculations differed greatly from those done, independently, by Gauss. Ceres reappeared on December 7, 1801, almost exactly where Gauss had predicted it would be. Although he did not disclose his methods at the time, Gauss had used his least squares approximation method, which he described in a paper in 1809. The same method was actually known earlier; Cotes anticipated the method in the early 18th century, and Legendre published a paper on it in 1806. Nevertheless, Gauss is generally given credit for the method of least squares approximation.

We begin our exploration of approximation with a more general result.

The Best Approximation Theorem

In the sciences, there are many problems that can be phrased generally as "What is the best approximation to X of type Y?" X might be a set of data points, a function, a vector, or many other things, while Y might be a particular type of function, a vector belonging to a certain vector space, etc. A typical example of such a problem is finding the vector $\mathbf{w}$ in a subspace W of a vector space V that best approximates (i.e., is closest to) a given vector $\mathbf{v}$ in V. This problem gives rise to the following definition.

Definition If W is a subspace of a normed linear space V and if $\mathbf{v}$ is a vector in V, then the ***best approximation to* v *in* W** is the vector $\overline{\mathbf{v}}$ in W such that

$$\|\mathbf{v} - \overline{\mathbf{v}}\| < \|\mathbf{v} - \mathbf{w}\|$$

for every vector $\mathbf{w}$ in W different from $\overline{\mathbf{v}}$.

In $\mathbb{R}^2$ or $\mathbb{R}^3$, we are used to thinking of "shortest distance" as corresponding to "perpendicular distance." In algebraic terminology, "shortest distance" relates to the notion of orthogonal projection: If W is a subspace of $\mathbb{R}^n$ and $\mathbf{v}$ is a vector in $\mathbb{R}^n$, then we expect $\text{proj}_W(\mathbf{v})$ to be the vector in W that is closest to $\mathbf{v}$ (Figure 7.10).

Since orthogonal projection can be defined in any inner product space, we have the following theorem.

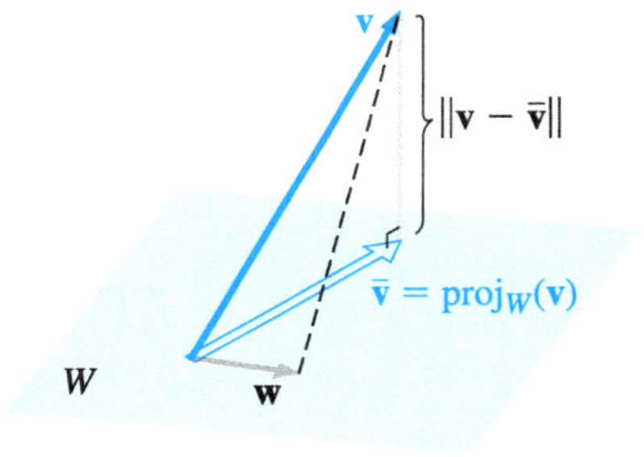

Figure 7.10
If $\overline{\mathbf{v}} = \text{proj}_W(\mathbf{v})$, then
$\|\mathbf{v} - \overline{\mathbf{v}}\| < \|\mathbf{v} - \mathbf{w}\|$ for all $\mathbf{w} \neq \overline{\mathbf{v}}$

Theorem 7.8 **The Best Approximation Theorem**

If W is a finite-dimensional subspace of an inner product space V and if $\mathbf{v}$ is a vector in V, then $\text{proj}_W(\mathbf{v})$ is the best approximation to $\mathbf{v}$ in W.

Proof Let $\mathbf{w}$ be a vector in W different from $\text{proj}_W(\mathbf{v})$. Then $\text{proj}_W(\mathbf{v}) - \mathbf{w}$ is also in W, so $\mathbf{v} - \text{proj}_W(\mathbf{v}) = \text{perp}_W(\mathbf{v})$ is orthogonal to $\text{proj}_W(\mathbf{v}) - \mathbf{w}$, by Exercise 43 in Section 7.1. Pythagoras' Theorem now implies that

$$\begin{aligned}\|\mathbf{v} - \text{proj}_W(\mathbf{v})\|^2 + \|\text{proj}_W(\mathbf{v}) - \mathbf{w}\|^2 &= \|(\mathbf{v} - \text{proj}_W(\mathbf{v})) + (\text{proj}_W(\mathbf{v}) - \mathbf{w})\|^2 \\ &= \|\mathbf{v} - \mathbf{w}\|^2\end{aligned}$$

as Figure 7.10 illustrates. However, $\|\text{proj}_W(\mathbf{v}) - \mathbf{w}\|^2 > 0$, since $\mathbf{w} \neq \text{proj}_W(\mathbf{v})$, so

$$\|\mathbf{v} - \text{proj}_W(\mathbf{v})\|^2 < \|\mathbf{v} - \text{proj}_W(\mathbf{v})\|^2 + \|\text{proj}_W(\mathbf{v}) - \mathbf{w}\|^2 = \|\mathbf{v} - \mathbf{w}\|^2$$

or, equivalently,

$$\|\mathbf{v} - \text{proj}_W(\mathbf{v})\| < \|\mathbf{v} - \mathbf{w}\|$$

Example 7.23

Let $\mathbf{u}_1 = \begin{bmatrix} 1 \\ 2 \\ -1 \end{bmatrix}$, $\mathbf{u}_2 = \begin{bmatrix} 5 \\ -2 \\ 1 \end{bmatrix}$, and $\mathbf{v} = \begin{bmatrix} 3 \\ 2 \\ 5 \end{bmatrix}$. Find the best approximation to $\mathbf{v}$ in the plane $W = \text{span}(\mathbf{u}_1, \mathbf{u}_2)$ and find the Euclidean distance from $\mathbf{v}$ to W.

Solution The vector in W that best approximates $\mathbf{v}$ is $\text{proj}_W(\mathbf{v})$. Since $\mathbf{u}_1$ and $\mathbf{u}_2$ are orthogonal,

$$\begin{aligned} \text{proj}_W(\mathbf{v}) &= \left(\frac{\mathbf{u}_1 \cdot \mathbf{v}}{\mathbf{u}_1 \cdot \mathbf{u}_1}\right)\mathbf{u}_1 + \left(\frac{\mathbf{u}_2 \cdot \mathbf{v}}{\mathbf{u}_2 \cdot \mathbf{u}_2}\right)\mathbf{u}_2 \\ &= \tfrac{2}{6}\begin{bmatrix} 1 \\ 2 \\ -1 \end{bmatrix} + \tfrac{16}{30}\begin{bmatrix} 5 \\ -2 \\ 1 \end{bmatrix} = \begin{bmatrix} 3 \\ -\frac{2}{5} \\ \frac{1}{5} \end{bmatrix} \end{aligned}$$

The distance from $\mathbf{v}$ to W is the distance from $\mathbf{v}$ to the point in W closest to $\mathbf{v}$. But this distance is just $\|\text{perp}_W(\mathbf{v})\| = \|\mathbf{v} - \text{proj}_W(\mathbf{v})\|$. We compute

$$\mathbf{v} - \text{proj}_W(\mathbf{v}) = \begin{bmatrix} 3 \\ 2 \\ 5 \end{bmatrix} - \begin{bmatrix} 3 \\ -\frac{2}{5} \\ \frac{1}{5} \end{bmatrix} = \begin{bmatrix} 0 \\ \frac{12}{5} \\ \frac{24}{5} \end{bmatrix}$$

so

$$\|\mathbf{v} - \text{proj}_W(\mathbf{v})\| = \sqrt{0^2 + (\tfrac{12}{5})^2 + (\tfrac{24}{5})^2} = \sqrt{\tfrac{720}{25}} = 12\sqrt{5}/5$$

which is the distance from $\mathbf{v}$ to W.

In Section 7.5, we will look at other examples of the Best Approximation Theorem when we explore the problem of approximating functions.

Remark The orthogonal projection of a vector $\mathbf{v}$ onto a subspace W is defined in terms of an orthogonal basis for W. The Best Approximation Theorem gives us an alternative proof that $\text{proj}_W(\mathbf{v})$ does not depend on the choice of this basis, since there can be only one vector in W that is closest to $\mathbf{v}$—namely, $\text{proj}_W(\mathbf{v})$.

Least Squares Approximation

We now turn to the problem of finding a curve that "best fits" a set of data points. Before we can proceed, however, we need to define what we mean by "best fit." Suppose the data points (1, 2), (2, 2), and (3, 4) have arisen from measurements taken during some experiment. Also suppose we have reason to believe that the x and y values are related by a linear function; that is, we expect the points to lie on some line with equation $y = a + bx$. If our measurements were accurate, all three points would satisfy this equation and we would have

$$2 = a + b \cdot 1 \qquad 2 = a + b \cdot 2 \qquad 4 = a + b \cdot 3$$

This is a system of three linear equations in two variables:

$$\begin{aligned} a + b &= 2 \\ a + 2b &= 2 \\ a + 3b &= 4 \end{aligned} \quad \text{or} \quad \begin{bmatrix} 1 & 1 \\ 1 & 2 \\ 1 & 3 \end{bmatrix}\begin{bmatrix} a \\ b \end{bmatrix} = \begin{bmatrix} 2 \\ 2 \\ 4 \end{bmatrix}$$

Unfortunately, this system is inconsistent (since the three points do not lie on a straight line). So we will settle for a line that comes "as close as possible" to passing through our points. For any line, we will measure the vertical distance from each data point to the line (representing the *errors* in the y-direction), and then we will try to choose the line that minimizes the *total error*. Figure 7.11 illustrates the situation.

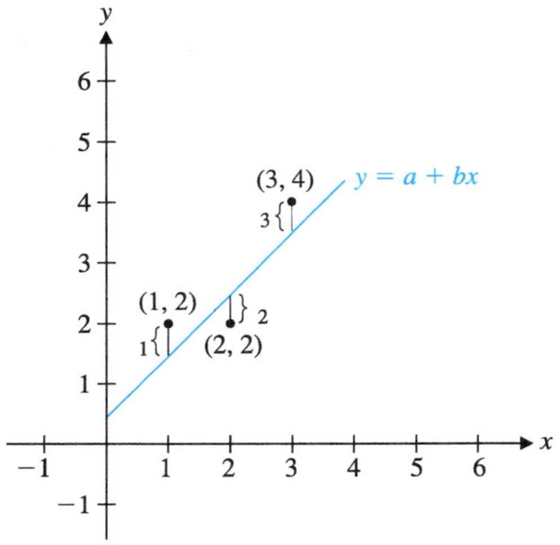

Figure 7.11
Finding the line that minimizes $\varepsilon_1^2 + \varepsilon_2^2 + \varepsilon_3^2$

If the errors are denoted by ε_1, ε_2, and ε_3, then we can form the ***error vector***

$$\mathbf{e} = \begin{bmatrix} \varepsilon_1 \\ \varepsilon_2 \\ \varepsilon_3 \end{bmatrix}$$

We want $\mathbf{e}$ to be as small as possible, so $\|\mathbf{e}\|$ must be as close to zero as possible. Which norm should we use? It turns out that the familiar Euclidean norm is the best choice. (The sum norm would also be a sensible choice, since $\|\mathbf{e}\|_s = |\varepsilon_1| + |\varepsilon_2| + |\varepsilon_3|$ is the actual sum of the errors in Figure 7.11. However, the absolute value signs are hard to work with, and, as you will soon see, the choice of the Euclidean norm leads to some very nice formulas.) So we are going to minimize

$$\|\mathbf{e}\| = \sqrt{\varepsilon_1^2 + \varepsilon_2^2 + \varepsilon_3^2} \quad \text{or, equivalently,} \quad \|\mathbf{e}\|^2 = \varepsilon_1^2 + \varepsilon_2^2 + \varepsilon_3^2$$

This is where the term "least squares" comes from: We need to find the smallest sum of squares, in the sense of the foregoing equation. The number $\|\mathbf{e}\|$ is called the ***least squares error*** of the approximation.

From Figure 7.11, we also obtain the following formulas for ε_1, ε_2, and ε_3 in our example:

$$\varepsilon_1 = 2 - (a + b \cdot 1) \quad \varepsilon_2 = 2 - (a + b \cdot 2) \quad \varepsilon_3 = 4 - (a + b \cdot 3)$$

Example 7.24

Which of the following lines gives the smallest least squares error for the data points (1, 2), (2, 2), and (3, 4)?

(a) $y = 1 + x$

(b) $y = -2 + 2x$

(c) $y = \frac{2}{3} + x$

Solution Table 7.1 shows the necessary calculations.

Table 7.1

	$y = 1 + x$	$y = -2 + 2x$	$y = \frac{2}{3} + x$
ε_1	$2 - (1 + 1) = 0$	$2 - (-2 + 2) = 2$	$2 - (\frac{2}{3} + 1) = \frac{1}{3}$
ε_2	$2 - (1 + 2) = -1$	$2 - (-2 + 4) = 0$	$2 - (\frac{2}{3} + 2) = -\frac{2}{3}$
ε_3	$4 - (1 + 3) = 0$	$4 - (-2 + 6) = 0$	$4 - (\frac{2}{3} + 3) = \frac{1}{3}$
$\varepsilon_1^2 + \varepsilon_2^2 + \varepsilon_3^2$	$0^2 + (-1)^2 + 0^2 = 1$	$2^2 + 0^2 + 0^2 = 4$	$(\frac{1}{3})^2 + (-\frac{2}{3})^2 + (\frac{1}{3})^2 = \frac{2}{3}$
$\|\mathbf{e}\|$	1	2	$\sqrt{\frac{2}{3}} \approx 0.816$

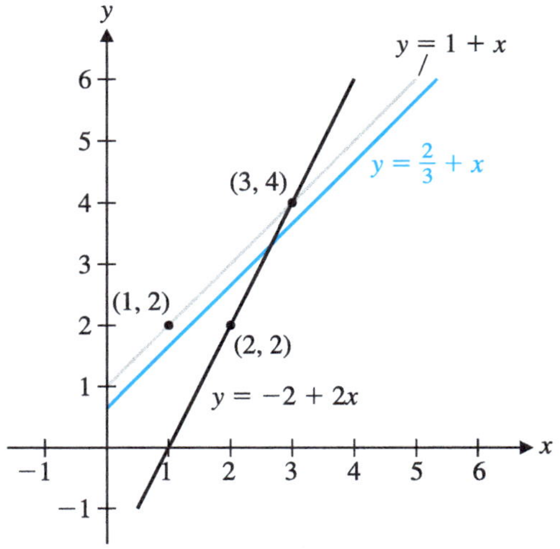

Figure 7.12

We see that the line $y = \frac{2}{3} + x$ produces the smallest least squares error among these three lines. Figure 7.12 shows the data points and all three lines.

It turns out that the line $y = \frac{2}{3} + x$ in Example 7.24 gives the smallest least squares error of *any* line, even though it passes through *none* of the given points. The rest of this section is devoted to illustrating why this is so.

In general, suppose we have n data points $(x_1, y_1), \ldots, (x_n, y_n)$ and a line $y = a + bx$. Our error vector is

$$\mathbf{e} = \begin{bmatrix} \varepsilon_1 \\ \vdots \\ \varepsilon_n \end{bmatrix}$$

where $\varepsilon_i = y_i - (a + bx_i)$. The line $y = a + bx$ that minimizes $\varepsilon_1^2 + \cdots + \varepsilon_n^2$ is called the ***least squares approximating line*** (or the ***line of best fit***) for the points $(x_1, y_1), \ldots, (x_n, y_n)$. As noted prior to Example 7.24, we can express this problem in matrix form. If the given points were actually on the line $y = a + bx$, then the n linear equations

$$\begin{aligned} a + bx_1 &= y_1 \\ &\vdots \\ a + bx_n &= y_n \end{aligned}$$

would all be true (i.e., the system would be consistent). Our interest is in the case where the points are *not* collinear, in which case the system is *inconsistent*. In matrix form, we have

$$\begin{bmatrix} 1 & x_1 \\ 1 & x_2 \\ \vdots & \vdots \\ 1 & x_n \end{bmatrix} \begin{bmatrix} a \\ b \end{bmatrix} = \begin{bmatrix} y_1 \\ y_2 \\ \vdots \\ y_n \end{bmatrix}$$

which is of the form $A\mathbf{x} = \mathbf{b}$, where

$$A = \begin{bmatrix} 1 & x_1 \\ 1 & x_2 \\ \vdots & \vdots \\ 1 & x_n \end{bmatrix}, \quad \mathbf{x} = \begin{bmatrix} a \\ b \end{bmatrix}, \quad \mathbf{b} = \begin{bmatrix} y_1 \\ y_2 \\ \vdots \\ y_n \end{bmatrix}$$

The error vector $\mathbf{e}$ is just $\mathbf{b} - A\mathbf{x}$ (check this), and we want to minimize $\|\mathbf{e}\|^2$ or, equivalently, $\|\mathbf{e}\|$. We can therefore rephrase our problem in terms of matrices as follows.

Definition If A is an $m \times n$ matrix and $\mathbf{b}$ is in $\mathbb{R}^m$, a ***least squares solution*** of $A\mathbf{x} = \mathbf{b}$ is a vector $\overline{\mathbf{x}}$ in $\mathbb{R}^n$ such that

$$\|\mathbf{b} - A\overline{\mathbf{x}}\| \leq \|\mathbf{b} - A\mathbf{x}\|$$

for all $\mathbf{x}$ in $\mathbb{R}^n$.

Solution of the Least Squares Problem

Any vector of the form $A\mathbf{x}$ is in the column space of A, and as $\mathbf{x}$ varies over all vectors in $\mathbb{R}^n$, $A\mathbf{x}$ varies over all vectors in col(A). A least squares solution of $A\mathbf{x} = \mathbf{b}$ is therefore equivalent to a vector $\overline{\mathbf{y}}$ in col(A) such that

$$\|\mathbf{b} - \overline{\mathbf{y}}\| \leq \|\mathbf{b} - \mathbf{y}\|$$

for all $\mathbf{y}$ in col(A). In other words, we need the closest vector in col(A) to $\mathbf{b}$. By the Best Approximation Theorem, the vector we want is the orthogonal projection of $\mathbf{b}$ onto col(A). Thus, if $\overline{\mathbf{x}}$ is a least squares solution of $A\mathbf{x} = \mathbf{b}$, we have

$$A\overline{\mathbf{x}} = \text{proj}_{\text{col}(A)}(\mathbf{b}) \tag{1}$$

In order to find $\overline{\mathbf{x}}$, it would appear that we need to first compute $\text{proj}_{\text{col}(A)}(\mathbf{b})$ and then solve the system (1). However, there is a better way to proceed.

We know that

$$\mathbf{b} - A\overline{\mathbf{x}} = \mathbf{b} - \text{proj}_{\text{col}(A)}(\mathbf{b}) = \text{perp}_{\text{col}(A)}(\mathbf{b})$$

is orthogonal to col(A). So $\mathbf{b} - A\overline{\mathbf{x}}$ is in $(\text{col}(A))^\perp = \text{null}(A^T)$. Therefore $A^T(\mathbf{b} - A\overline{\mathbf{x}}) = \mathbf{0}$, which, in turn, is equivalent to $A^T\mathbf{b} - A^TA\overline{\mathbf{x}} = \mathbf{0}$ or

$$A^TA\overline{\mathbf{x}} = A^T\mathbf{b}$$

This represents a system of equations known as the ***normal equations*** for $\overline{\mathbf{x}}$.

We have just established that the solutions of the normal equations for $\overline{\mathbf{x}}$ are precisely the least squares solutions of $A\mathbf{x} = \mathbf{b}$. This proves the first part of the following theorem.

Theorem 7.9 **The Least Squares Theorem**

Let A be an $m \times n$ matrix and let $\mathbf{b}$ be in $\mathbb{R}^m$. Then $A\mathbf{x} = \mathbf{b}$ always has at least one least squares solution $\overline{\mathbf{x}}$. Moreover:

a. $\overline{\mathbf{x}}$ is a least squares solution of $A\mathbf{x} = \mathbf{b}$ if and only if $\overline{\mathbf{x}}$ is a solution of the normal equations $A^TA\overline{\mathbf{x}} = A^T\mathbf{b}$.

b. A has linearly independent columns if and only if A^TA is invertible. In this case, the least squares solution of $A\mathbf{x} = \mathbf{b}$ is unique and is given by

$$\overline{\mathbf{x}} = (A^TA)^{-1}A^T\mathbf{b}$$

Proof We have already established property (a). For property (b), we note that the n columns of A are linearly independent if and only if rank(A) $= n$. But this is true if and only if A^TA is invertible, by Theorem 3.28. If A^TA is invertible, then the unique solution of $A^TA\overline{\mathbf{x}} = A^T\mathbf{b}$ is clearly $\overline{\mathbf{x}} = (A^TA)^{-1}A^T\mathbf{b}$.

Example 7.25 Find a least squares solution to the inconsistent system $A\mathbf{x} = \mathbf{b}$, where

$$A = \begin{bmatrix} 1 & 5 \\ 2 & -2 \\ -1 & 1 \end{bmatrix} \quad \text{and} \quad \mathbf{b} = \begin{bmatrix} 3 \\ 2 \\ 5 \end{bmatrix}$$

Solution We compute

$$A^TA = \begin{bmatrix} 1 & 2 & -1 \\ 5 & -2 & 1 \end{bmatrix}\begin{bmatrix} 1 & 5 \\ 2 & -2 \\ -1 & 1 \end{bmatrix} = \begin{bmatrix} 6 & 0 \\ 0 & 30 \end{bmatrix}$$

and

$$A^T\mathbf{b} = \begin{bmatrix} 1 & 2 & -1 \\ 5 & -2 & 1 \end{bmatrix}\begin{bmatrix} 3 \\ 2 \\ 5 \end{bmatrix} = \begin{bmatrix} 2 \\ 16 \end{bmatrix}$$

The normal equations $A^TA\overline{\mathbf{x}} = A^T\mathbf{b}$ are just

$$\begin{bmatrix} 6 & 0 \\ 0 & 30 \end{bmatrix}\overline{\mathbf{x}} = \begin{bmatrix} 2 \\ 16 \end{bmatrix}$$

which yield $\overline{\mathbf{x}} = \begin{bmatrix} \frac{1}{3} \\ \frac{8}{15} \end{bmatrix}$. The fact that this solution is unique was guaranteed by Theorem 7.9(b), since the columns of A are clearly linearly independent.

Remark We could have phrased Example 7.25 as follows: Find the best approximation to $\mathbf{b}$ in the column space of A. The resulting equations give the system $A\mathbf{x} = \mathbf{b}$ whose least squares solution we just found. (Verify this.) In this case, the components of $\overline{\mathbf{x}}$ are the *coefficients* of that linear combination of the columns of A that produces the best approximation to $\mathbf{b}$—namely,

$$\frac{1}{3}\begin{bmatrix} 1 \\ 2 \\ -1 \end{bmatrix} + \frac{8}{15}\begin{bmatrix} 5 \\ -2 \\ 1 \end{bmatrix} = \begin{bmatrix} 3 \\ -\frac{2}{5} \\ \frac{1}{5} \end{bmatrix}$$

This is exactly the result of Example 7.23. Compare the two approaches.

Example 7.26 Find the least squares approximating line for the data points (1, 2), (2, 2), and (3, 4) from Example 7.24.

Solution We have already seen that the corresponding system $A\mathbf{x} = \mathbf{b}$ is

$$\begin{bmatrix} 1 & 1 \\ 1 & 2 \\ 1 & 3 \end{bmatrix}\begin{bmatrix} a \\ b \end{bmatrix} = \begin{bmatrix} 2 \\ 2 \\ 4 \end{bmatrix}$$

where $y = a + bx$ is the line we seek. Since the columns of A are clearly linearly independent, there will be a unique least squares solution, by part (b) of the Least Squares Theorem. We compute

$$A^TA = \begin{bmatrix} 1 & 1 & 1 \\ 1 & 2 & 3 \end{bmatrix}\begin{bmatrix} 1 & 1 \\ 1 & 2 \\ 1 & 3 \end{bmatrix} = \begin{bmatrix} 3 & 6 \\ 6 & 14 \end{bmatrix} \quad \text{and} \quad A^T\mathbf{b} = \begin{bmatrix} 1 & 1 & 1 \\ 1 & 2 & 3 \end{bmatrix}\begin{bmatrix} 2 \\ 2 \\ 4 \end{bmatrix} = \begin{bmatrix} 8 \\ 18 \end{bmatrix}$$

Hence, we can solve the normal equations $A^TA\overline{\mathbf{x}} = A^T\mathbf{b}$, using Gaussian elimination to obtain

$$[A^TA \mid A^T\mathbf{b}] = \left[\begin{array}{cc|c} 3 & 6 & 8 \\ 6 & 14 & 18 \end{array}\right] \longrightarrow \left[\begin{array}{cc|c} 1 & 0 & \frac{2}{3} \\ 0 & 1 & 1 \end{array}\right]$$

So $\overline{\mathbf{x}} = \begin{bmatrix} \frac{2}{3} \\ 1 \end{bmatrix}$, from which we see that $a = \frac{2}{3}$, $b = 1$ are the coefficients of the least squares approximating line: $y = \frac{2}{3} + x$.

The line we just found is the line in Example 7.24(c), so we have justified our claim that this line produces the smallest least squares error for the data points (1, 2), (2, 2), and (3, 4). Notice that if $\overline{\mathbf{x}}$ is a least squares solution of $A\mathbf{x} = \mathbf{b}$, we may compute the least squares error as

$$\|\mathbf{e}\| = \|\mathbf{b} - A\overline{\mathbf{x}}\|$$

Since $A\overline{\mathbf{x}} = \text{proj}_{\text{col}(A)}(\mathbf{b})$, this is just the length of $\text{perp}_{\text{col}(A)}(\mathbf{b})$—that is, the distance from $\mathbf{b}$ to the column space of A. In Example 7.26, we had

$$\mathbf{e} = \mathbf{b} - A\overline{\mathbf{x}} = \begin{bmatrix} 2 \\ 2 \\ 4 \end{bmatrix} - \begin{bmatrix} 1 & 1 \\ 1 & 2 \\ 1 & 3 \end{bmatrix}\begin{bmatrix} \frac{2}{3} \\ 1 \end{bmatrix} = \begin{bmatrix} \frac{1}{3} \\ -\frac{2}{3} \\ \frac{1}{3} \end{bmatrix}$$

so, as in Example 7.24(c), we have a least squares error of $\|\mathbf{e}\| = \sqrt{\frac{2}{3}} \approx 0.816$.

Remark Note that the columns of A in Example 7.26 are linearly independent, so $(A^TA)^{-1}$ exists, and we could calculate $\overline{\mathbf{x}}$ as $\overline{\mathbf{x}} = (A^TA)^{-1}A^T\mathbf{b}$. However, it is almost always easier to solve the normal equations using Gaussian elimination (or to let your CAS do it for you!).

It is interesting to look at Example 7.26 from two different geometric points of view. On the one hand, we have the least squares approximating line $y = \frac{2}{3} + x$, with corresponding errors $\varepsilon_1 = \frac{1}{3}$, $\varepsilon_2 = -\frac{2}{3}$, and $\varepsilon_3 = \frac{1}{3}$, as shown in Figure 7.13(a). Equivalently, we have the projection of $\mathbf{b}$ onto the column space of A, as shown in Figure 7.13(b). Here,

$$\mathbf{p} = \text{proj}_{\text{col}(A)}(\mathbf{b}) = A\overline{\mathbf{x}} = \begin{bmatrix} 1 & 1 \\ 1 & 2 \\ 1 & 3 \end{bmatrix}\begin{bmatrix} \frac{2}{3} \\ 1 \end{bmatrix} = \begin{bmatrix} \frac{5}{3} \\ \frac{8}{3} \\ \frac{11}{3} \end{bmatrix}$$

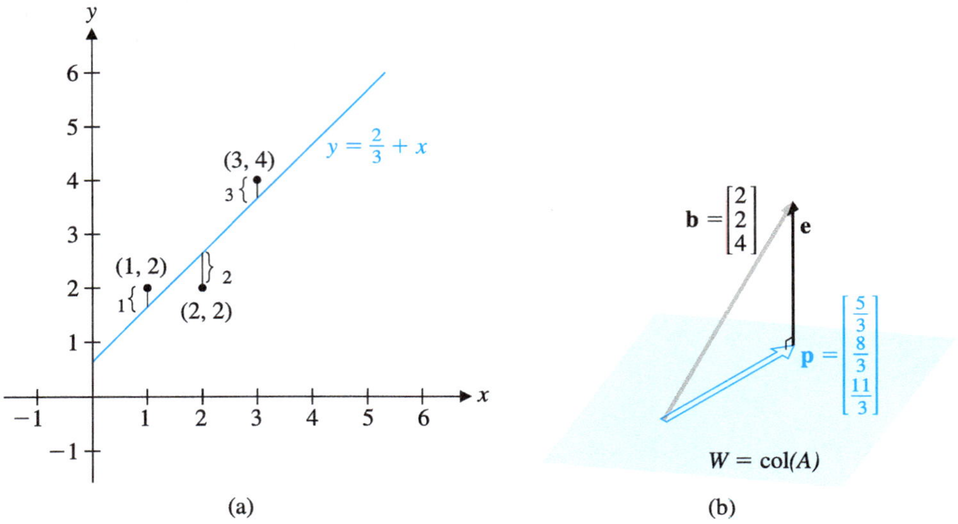

Figure 7.13

and the least squares error vector is $\mathbf{e} = \begin{bmatrix} \varepsilon_1 \\ \varepsilon_2 \\ \varepsilon_3 \end{bmatrix}$. [What would Figure 7.13(b) look like if the data points *were* collinear?]

Example 7.27

Find the least squares approximating line and the least squares error for the points (1, 1), (2, 2), (3, 2), and (4, 3).

Solution Let $y = a + bx$ be the equation of the line we seek. Then, substituting the four points into this equation, we obtain

$$\begin{aligned} a + b &= 1 \\ a + 2b &= 2 \\ a + 3b &= 2 \\ a + 4b &= 3 \end{aligned} \quad \text{or} \quad \begin{bmatrix} 1 & 1 \\ 1 & 2 \\ 1 & 3 \\ 1 & 4 \end{bmatrix} \begin{bmatrix} a \\ b \end{bmatrix} = \begin{bmatrix} 1 \\ 2 \\ 2 \\ 3 \end{bmatrix}$$

So we want the least squares solution of $A\mathbf{x} = \mathbf{b}$, where

$$A = \begin{bmatrix} 1 & 1 \\ 1 & 2 \\ 1 & 3 \\ 1 & 4 \end{bmatrix} \quad \text{and} \quad \mathbf{b} = \begin{bmatrix} 1 \\ 2 \\ 2 \\ 3 \end{bmatrix}$$

Since the columns of A are linearly independent, the solution we want is

$$\overline{\mathbf{x}} = (A^TA)^{-1}A^T\mathbf{b} = \begin{bmatrix} \frac{1}{2} \\ \frac{3}{5} \end{bmatrix}$$

(Check this calculation.) Therefore, we take $a = \frac{1}{2}$ and $b = \frac{3}{5}$, producing the least squares approximating line $y = \frac{1}{2} + \frac{3}{5}x$, as shown in Figure 7.14.

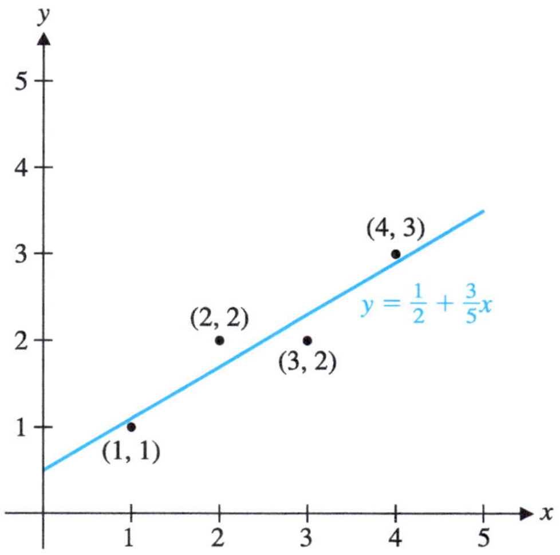

Figure 7.14

Since

$$\mathbf{e} = \mathbf{b} - A\overline{\mathbf{x}} = \begin{bmatrix} 1 \\ 2 \\ 2 \\ 3 \end{bmatrix} - \begin{bmatrix} 1 & 1 \\ 1 & 2 \\ 1 & 3 \\ 1 & 4 \end{bmatrix} \begin{bmatrix} \frac{1}{2} \\ \frac{3}{5} \end{bmatrix} = \begin{bmatrix} -\frac{1}{10} \\ \frac{3}{10} \\ -\frac{3}{10} \\ \frac{1}{10} \end{bmatrix}$$

the least squares error is $\|\mathbf{e}\| = \sqrt{5}/5 \approx 0.447$.

We can use the method of least squares to approximate data points by curves other than straight lines.

Example 7.28

Find the parabola that gives the best least squares approximation to the points $(-1, 1)$, $(0, -1)$, $(1, 0)$, and $(2, 2)$.

Solution The equation of a parabola is a quadratic $y = a + bx + cx^2$. Substituting the given points into this quadratic, we obtain the linear system

$$\begin{aligned} a - b + c &= 1 \\ a &= -1 \\ a + b + c &= 0 \\ a + 2b + 4c &= 2 \end{aligned} \quad \text{or} \quad \begin{bmatrix} 1 & -1 & 1 \\ 1 & 0 & 0 \\ 1 & 1 & 1 \\ 1 & 2 & 4 \end{bmatrix} \begin{bmatrix} a \\ b \\ c \end{bmatrix} = \begin{bmatrix} 1 \\ -1 \\ 0 \\ 2 \end{bmatrix}$$

Thus, we want the least squares approximation of $A\mathbf{x} = \mathbf{b}$, where

$$A = \begin{bmatrix} 1 & -1 & 1 \\ 1 & 0 & 0 \\ 1 & 1 & 1 \\ 1 & 2 & 4 \end{bmatrix} \quad \text{and} \quad \mathbf{b} = \begin{bmatrix} 1 \\ -1 \\ 0 \\ 2 \end{bmatrix}$$

We compute

$$A^TA = \begin{bmatrix} 4 & 2 & 6 \\ 2 & 6 & 8 \\ 6 & 8 & 18 \end{bmatrix} \quad \text{and} \quad A^T\mathbf{b} = \begin{bmatrix} 2 \\ 3 \\ 9 \end{bmatrix}$$

so the normal equations are given by

$$\begin{bmatrix} 4 & 2 & 6 \\ 2 & 6 & 8 \\ 6 & 8 & 18 \end{bmatrix} \overline{\mathbf{x}} = \begin{bmatrix} 2 \\ 3 \\ 9 \end{bmatrix}$$

whose solution is

$$\overline{\mathbf{x}} = \begin{bmatrix} -\frac{7}{10} \\ -\frac{3}{5} \\ 1 \end{bmatrix}$$

Thus, the least squares approximating parabola has the equation

$$y = -\tfrac{7}{10} - \tfrac{3}{5}x + x^2$$

as shown in Figure 7.15.

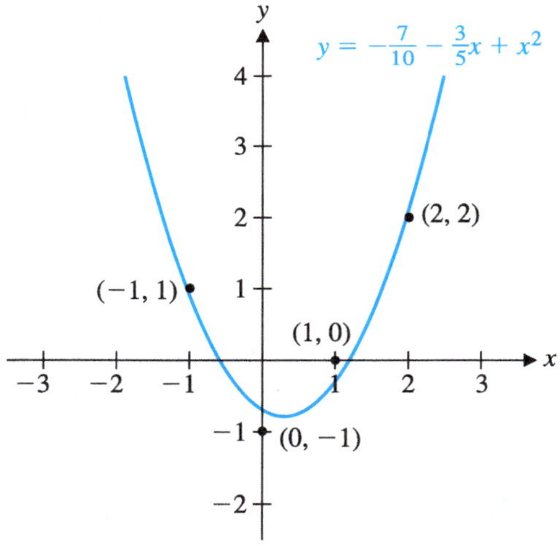

Figure 7.15
A least squares approximating parabola

One of the important uses of least squares approximation is to estimate constants associated with various processes. The next example illustrates this application in the context of population growth. Recall from Section 6.7 that a population that is growing (or decaying) exponentially satisfies an equation of the form $p(t) = ce^{kt}$, where $p(t)$ is the size of the population at time t and c and k are constants. Clearly, $c = p(0)$, but k is not so easy to determine. It is easy to see that

$$k = \frac{p'(t)}{p(t)}$$

which explains why k is sometimes referred to as the *relative growth rate* of the population: It is the ratio of the growth rate $p'(t)$ to the size of the population $p(t)$.

CAS **Example 7.29**

Table 7.2 gives the population of the world at 10-year intervals for the second half of the 20th century. Assuming an exponential growth model, find the relative growth rate and predict the world's population in 2010.

Table 7.2

Year	Population (in billions)
1950	2.56
1960	3.04
1970	3.71
1980	4.46
1990	5.28
2000	6.08

Source: U.S. Bureau of the Census, International Data Base

Solution Let's agree to measure time t in 10-year intervals so that $t = 0$ is 1950, $t = 1$ is 1960, and so on. Since $c = p(0) = 2.56$, the equation for the growth rate of the population is

$$p = 2.56e^{kt}$$

How can we use the method of least squares on this equation? If we take the natural logarithm of both sides, we convert the equation into a linear one:

$$\begin{aligned} \ln p &= \ln(2.56e^{kt}) \\ &= \ln 2.56 + \ln(e^{kt}) \\ &\approx 0.94 + kt \end{aligned}$$

Plugging in the values of t and p from Table 7.2 yields the following system (where we have rounded calculations to three decimal places):

$$\begin{aligned} 0.94 &= 0.94 \\ k &= 0.172 \\ 2k &= 0.371 \\ 3k &= 0.555 \\ 4k &= 0.724 \\ 5k &= 0.865 \end{aligned}$$

We can ignore the first equation (it just corresponds to the initial condition $c = p(0) = 2.56$). The remaining equations correspond to a system $A\mathbf{x} = \mathbf{b}$, with

$$A = \begin{bmatrix} 1 \\ 2 \\ 3 \\ 4 \\ 5 \end{bmatrix} \quad \text{and} \quad \mathbf{b} = \begin{bmatrix} 0.172 \\ 0.371 \\ 0.555 \\ 0.724 \\ 0.865 \end{bmatrix}$$

Since $A^TA = 55$ and $A^T\mathbf{b} = 9.80$, the corresponding normal equations are just the single equation

$$55\overline{\mathbf{x}} = 9.80$$

Therefore, $k = \overline{\mathbf{x}} = 9.80/55 \approx 0.178$. Consequently, the least squares solution has the form $p = 2.56e^{0.178t}$ (see Figure 7.16).

The world's population in 2010 corresponds to $t = 6$, from which we obtain

$$p(6) = 2.56e^{0.178(6)} \approx 7.448$$

Thus, if our model is accurate, there will be approximately 7.45 billion people on Earth in the year 2010. (The U.S. Census Bureau estimates that the global population will be "only" 6.82 billion in 2010. Why do you think our estimate is higher?)

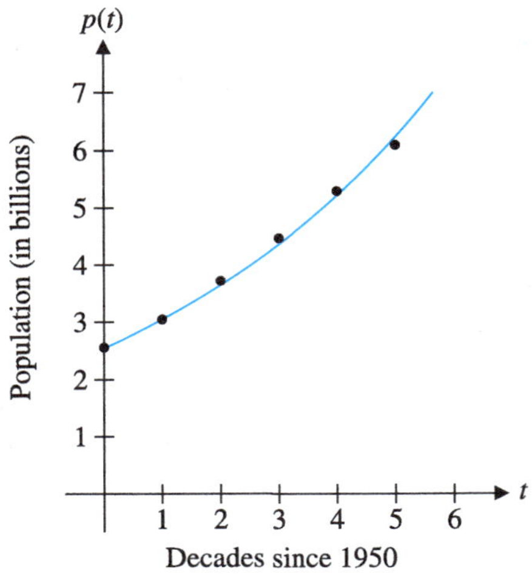

Figure 7.16

Least Squares via the QR Factorization

It is often the case that the normal equations for a least squares problem are ill-conditioned. Therefore, a small numerical error in performing Gaussian elimination will result in a large error in the least squares solution. Consequently, in practice, other methods are usually used to compute least squares approximations.

It turns out that the QR factorization of A yields a more reliable way of computing the least squares approximation of $A\mathbf{x} = \mathbf{b}$.

Theorem 7.10 Let A be an $m \times n$ matrix with linearly independent columns and let $\mathbf{b}$ be in $\mathbb{R}^m$. If $A = QR$ is a QR factorization of A, then the unique least squares solution $\overline{\mathbf{x}}$ of $A\mathbf{x} = \mathbf{b}$ is

$$\overline{\mathbf{x}} = R^{-1}Q^T\mathbf{b}$$

Proof Recall from Theorem 5.16 that the QR factorization $A = QR$ involves an $m \times n$ matrix Q with orthonormal columns and an invertible upper triangular matrix R. From the Least Squares Theorem, we have

$$\begin{aligned} A^TA\overline{\mathbf{x}} &= A^T\mathbf{b} \\ \Rightarrow (QR)^TQR\overline{\mathbf{x}} &= (QR)^T\mathbf{b} \\ \Rightarrow R^TQ^TQR\overline{\mathbf{x}} &= R^TQ^T\mathbf{b} \\ \Rightarrow R^TR\overline{\mathbf{x}} &= R^TQ^T\mathbf{b} \end{aligned}$$

since $Q^TQ = I$. (Why?)

Since R is invertible, so is R^T, and hence we have

$$R\overline{\mathbf{x}} = Q^T\mathbf{b} \quad \text{or, equivalently,} \quad \overline{\mathbf{x}} = R^{-1}Q^T\mathbf{b}$$

Remark Since R is upper triangular, in practice it is easier to solve $R\overline{\mathbf{x}} = Q^T\mathbf{b}$ directly than to invert R and compute $R^{-1}Q^T\mathbf{b}$.

Example 7.30

Use the QR factorization to find a least squares solution of $A\mathbf{x} = \mathbf{b}$, where

$$A = \begin{bmatrix} 1 & 2 & 2 \\ -1 & 1 & 2 \\ -1 & 0 & 1 \\ 1 & 1 & 2 \end{bmatrix} \quad \text{and} \quad \mathbf{b} = \begin{bmatrix} 2 \\ -3 \\ -2 \\ 0 \end{bmatrix}$$

Solution From Example 5.15,

$$A = QR = \begin{bmatrix} 1/2 & 3\sqrt{5}/10 & -\sqrt{6}/6 \\ -1/2 & 3\sqrt{5}/10 & 0 \\ -1/2 & \sqrt{5}/10 & \sqrt{6}/6 \\ 1/2 & \sqrt{5}/10 & \sqrt{6}/3 \end{bmatrix} \begin{bmatrix} 2 & 1 & 1/2 \\ 0 & \sqrt{5} & 3\sqrt{5}/2 \\ 0 & 0 & \sqrt{6}/2 \end{bmatrix}$$

We have

$$Q^T\mathbf{b} = \begin{bmatrix} 1/2 & -1/2 & -1/2 & 1/2 \\ 3\sqrt{5}/10 & 3\sqrt{5}/10 & \sqrt{5}/10 & \sqrt{5}/10 \\ -\sqrt{6}/6 & 0 & \sqrt{6}/6 & \sqrt{6}/3 \end{bmatrix} \begin{bmatrix} 2 \\ -3 \\ -2 \\ 0 \end{bmatrix} = \begin{bmatrix} 7/2 \\ -\sqrt{5}/2 \\ -2\sqrt{6}/3 \end{bmatrix}$$

so we require the solution to $R\overline{\mathbf{x}} = Q^T\mathbf{b}$, or

$$\begin{bmatrix} 2 & 1 & 1/2 \\ 0 & \sqrt{5} & 3\sqrt{5}/2 \\ 0 & 0 & \sqrt{6}/2 \end{bmatrix} \overline{\mathbf{x}} = \begin{bmatrix} 7/2 \\ -\sqrt{5}/2 \\ -2\sqrt{6}/3 \end{bmatrix}$$

Back substitution quickly yields

$$\overline{\mathbf{x}} = \begin{bmatrix} 4/3 \\ 3/2 \\ -4/3 \end{bmatrix}$$

Orthogonal Projection Revisited

One of the nice byproducts of the least squares method is a new formula for the orthogonal projection of a vector onto a subspace of $\mathbb{R}^m$.

Theorem 7.11

Let W be a subspace of $\mathbb{R}^m$ and let A be an $m \times n$ matrix whose columns form a basis for W. If $\mathbf{v}$ is any vector in $\mathbb{R}^m$, then the orthogonal projection of $\mathbf{v}$ onto W is the vector

$$\text{proj}_W(\mathbf{v}) = A(A^TA)^{-1}A^T\mathbf{v}$$

The linear transformation $P : \mathbb{R}^m \to \mathbb{R}^m$ that projects $\mathbb{R}^m$ onto W has $A(A^TA)^{-1}A^T$ as its standard matrix.

Proof Given the way we have constructed A, its column space is W. Since the columns of A are linearly independent, the Least Squares Theorem guarantees that there is a unique least squares solution to $A\mathbf{x} = \mathbf{v}$ given by

$$\overline{\mathbf{x}} = (A^TA)^{-1}A^T\mathbf{v}$$

By Equation (1),

$$A\bar{\mathbf{x}} = \text{proj}_{\text{col}(A)}(\mathbf{v}) = \text{proj}_W(\mathbf{v})$$

Therefore, $$\text{proj}_W(\mathbf{v}) = A((A^TA)^{-1}A^T\mathbf{v}) = (A(A^TA)^{-1}A^T)\mathbf{v}$$

as required. Since this equation holds for all $\mathbf{v}$ in $\mathbb{R}^m$, the last statement of the theorem follows immediately.

We will illustrate Theorem 7.11 by revisiting Example 5.11.

Example 7.31

Find the orthogonal projection of $\mathbf{v} = \begin{bmatrix} 3 \\ -1 \\ 2 \end{bmatrix}$ onto the plane W in $\mathbb{R}^3$ with equation $x - y + 2z = 0$, and give the standard matrix of the orthogonal projection transformation onto W.

Solution As in Example 5.11, we will take as a basis for W the set

$$\left\{ \begin{bmatrix} 1 \\ 1 \\ 0 \end{bmatrix}, \begin{bmatrix} -1 \\ 1 \\ 1 \end{bmatrix} \right\}$$

We form the matrix

$$A = \begin{bmatrix} 1 & -1 \\ 1 & 1 \\ 0 & 1 \end{bmatrix}$$

with these basis vectors as its columns. Then

$$A^TA = \begin{bmatrix} 1 & 1 & 0 \\ -1 & 1 & 1 \end{bmatrix}\begin{bmatrix} 1 & -1 \\ 1 & 1 \\ 0 & 1 \end{bmatrix} = \begin{bmatrix} 2 & 0 \\ 0 & 3 \end{bmatrix}$$

so $$(A^TA)^{-1} = \begin{bmatrix} \frac{1}{2} & 0 \\ 0 & \frac{1}{3} \end{bmatrix}$$

By Theorem 7.11, the standard matrix of the orthogonal projection transformation onto W is

$$A(A^TA)^{-1}A^T = A = \begin{bmatrix} 1 & -1 \\ 1 & 1 \\ 0 & 1 \end{bmatrix}\begin{bmatrix} \frac{1}{2} & 0 \\ 0 & \frac{1}{3} \end{bmatrix}\begin{bmatrix} 1 & 1 & 0 \\ -1 & 1 & 1 \end{bmatrix} = \begin{bmatrix} \frac{5}{6} & \frac{1}{6} & -\frac{1}{3} \\ \frac{1}{6} & \frac{5}{6} & \frac{1}{3} \\ -\frac{1}{3} & \frac{1}{3} & \frac{1}{3} \end{bmatrix}$$

so the orthogonal projection of $\mathbf{v}$ onto W is

$$\text{proj}_W(\mathbf{v}) = A(A^TA)^{-1}A^T\mathbf{v} = \begin{bmatrix} \frac{5}{6} & \frac{1}{6} & -\frac{1}{3} \\ \frac{1}{6} & \frac{5}{6} & \frac{1}{3} \\ -\frac{1}{3} & \frac{1}{3} & \frac{1}{3} \end{bmatrix}\begin{bmatrix} 3 \\ -1 \\ 2 \end{bmatrix} = \begin{bmatrix} \frac{5}{3} \\ \frac{1}{3} \\ -\frac{2}{3} \end{bmatrix}$$

which agrees with our solution to Example 5.11.

Remark Since the projection of a vector onto a subspace W is unique, the standard matrix of this linear transformation (as given by Theorem 7.11) cannot depend on the choice of basis for W. In other words, with a different basis for W, we have a different matrix A, but the matrix $A(A^TA)^{-1}A^T$ will be the same! (You are asked to verify this in Exercise 43.)

The Pseudoinverse of a Matrix

If A is an $n \times n$ matrix with linearly independent columns, then it is invertible, and the unique solution to $A\mathbf{x} = \mathbf{b}$ is $\mathbf{x} = A^{-1}\mathbf{b}$. If $m > n$ and A is $m \times n$ with linearly independent columns, then $A\mathbf{x} = \mathbf{b}$ has no exact solution, but the best approximation is given by the unique least squares solution $\overline{\mathbf{x}} = (A^TA)^{-1}A^T\mathbf{b}$. The matrix $(A^TA)^{-1}A^T$ therefore plays the role of an "inverse of A" in this situation.

Definition If A is a matrix with linearly independent columns, then the ***pseudoinverse*** of A is the matrix A^+ defined by

$$A^+ = (A^TA)^{-1}A^T$$

Observe that if A is $m \times n$, then A^+ is $n \times m$.

Example 7.32

Find the pseudoinverse of $A = \begin{bmatrix} 1 & 1 \\ 1 & 2 \\ 1 & 3 \end{bmatrix}$.

Solution We have already done most of the calculations in Example 7.26. Using our previous work, we have

$$A^+ = (A^TA)^{-1}A^T = \begin{bmatrix} \frac{7}{3} & -1 \\ -1 & \frac{1}{2} \end{bmatrix}\begin{bmatrix} 1 & 1 & 1 \\ 1 & 2 & 3 \end{bmatrix} = \begin{bmatrix} \frac{4}{3} & \frac{1}{3} & -\frac{2}{3} \\ -\frac{1}{2} & 0 & \frac{1}{2} \end{bmatrix}$$

The pseudoinverse is a convenient shorthand notation for some of the concepts we have been exploring. For example, if A is $m \times n$ with linearly independent columns, the least squares solution of $A\mathbf{x} = \mathbf{b}$ is given by

$$\overline{\mathbf{x}} = A^+\mathbf{b}$$

and the standard matrix of the orthogonal projection P from $\mathbb{R}^m$ onto col(A) is

$$[P] = AA^+$$

If A is actually a square matrix, then it is easy to show that $A^+ = A^{-1}$ (see Exercise 53). In this case, the least squares solution of $A\mathbf{x} = \mathbf{b}$ is the *exact* solution, since

$$\overline{\mathbf{x}} = A^+\mathbf{b} = A^{-1}\mathbf{b} = \mathbf{x}$$

The projection matrix becomes $[P] = AA^+ = AA^{-1} = I$. (What is the geometric interpretation of this equality?)

Theorem 7.12 summarizes the key properties of the pseudoinverse of a matrix. (Before reading the proof of this theorem, verify these properties for the matrix in Example 7.32.)

Theorem 7.12 Let A be a matrix with linearly independent columns. Then the pseudoinverse A^+ of A satisfies the following properties, called the ***Penrose conditions*** for A:

a. $AA^+A = A$
b. $A^+AA^+ = A^+$
c. AA^+ and A^+A are symmetric.

Proof We prove condition (a) and half of condition (c) and leave the proofs of the remaining conditions as Exercises 54 and 55.

(a) We compute

$$\begin{aligned} AA^+A &= A((A^TA)^{-1}A^T)A \\ &= A(A^TA)^{-1}(A^TA) \\ &= AI = A \end{aligned}$$

(c) By Theorem 3.4, A^TA is symmetric. Therefore, $(A^TA)^{-1}$ is also symmetric, by Exercise 46 in Section 3.3. Taking the transpose of AA^+, we have

$$\begin{aligned} (AA^+)^T &= (A(A^TA)^{-1}A^T)^T \\ &= (A^T)^T((A^TA)^{-1})^TA^T \\ &= A(A^TA)^{-1}A^T \\ &= AA^+ \end{aligned}$$

Exercise 56 explores further properties of the pseudoinverse. In the next section, we will see how to extend the definition of A^+ to handle *all* matrices, whether or not the columns of A are linearly independent.

Exercises 7.3

CAS

In Exercises 1–3, consider the data points (1, 0), (2, 1), *and* (3, 5). *Compute the least squares error for the given line. In each case, plot the points and the line.*

1. $y = -2 + 2x$ **2.** $y = x$ **3.** $y = -3 + \frac{5}{2}x$

In Exercises 4–6, consider the data points (−5, 3), (0, 3), (5, 2), *and* (10, 0). *Compute the least squares error for the given line. In each case, plot the points and the line.*

4. $y = 3 - \frac{1}{3}x$ **5.** $y = \frac{5}{2}$ **6.** $y = 2 - \frac{1}{5}x$

In Exercises 7–14, find the least squares approximating line for the given points and compute the corresponding least squares error.

7. (1, 0), (2, 1), (3, 5)

8. (1, 6), (2, 3), (3, 1)

9. (0, 4), (1, 1), (2, 0)

10. (0, 3), (1, 3), (2, 5)

11. (−5, −1), (0, 1), (5, 2), (10, 4)

12. $(-5, 3), (0, 3), (5, 2), (10, 0)$

13. $(1, 1), (2, 3), (3, 4), (4, 5), (5, 7)$

14. $(1, 10), (2, 8), (3, 5), (4, 3), (5, 0)$

In Exercises 15–18, find the least squares approximating parabola for the given points.

15. $(1, 1), (2, -2), (3, 3), (4, 4)$

16. $(1, 6), (2, 0), (3, 0), (4, 2)$

17. $(-2, 4), (-1, 7), (0, 3), (1, 0), (2, -1)$

18. $(-2, 0), (-1, -11), (0, -10), (1, -9), (2, 8)$

In Exercises 19–22, find a least squares solution of $A\mathbf{x} = \mathbf{b}$ *by constructing and solving the normal equations.*

19. $A = \begin{bmatrix} 3 & 1 \\ 1 & 1 \\ 1 & 2 \end{bmatrix}, \mathbf{b} = \begin{bmatrix} 1 \\ 1 \\ 1 \end{bmatrix}$

20. $A = \begin{bmatrix} 1 & -2 \\ 3 & -2 \\ 2 & 1 \end{bmatrix}, \mathbf{b} = \begin{bmatrix} 1 \\ 1 \\ 1 \end{bmatrix}$

21. $A = \begin{bmatrix} 1 & -2 \\ 0 & -3 \\ 2 & 5 \\ 3 & 0 \end{bmatrix}, \mathbf{b} = \begin{bmatrix} 4 \\ 1 \\ -2 \\ 4 \end{bmatrix}$

22. $A = \begin{bmatrix} 1 & 0 \\ 2 & -1 \\ -1 & 1 \\ 0 & 2 \end{bmatrix}, \mathbf{b} = \begin{bmatrix} 1 \\ 5 \\ -1 \\ 2 \end{bmatrix}$

In Exercises 23 and 24, show that the least squares solution of $A\mathbf{x} = \mathbf{b}$ *is not unique and solve the normal equations to find all the least squares solutions.*

23. $A = \begin{bmatrix} 1 & 1 & 0 & 0 \\ 1 & 0 & 1 & 1 \\ 0 & -1 & 1 & 1 \\ 1 & -1 & 1 & 0 \end{bmatrix}, \mathbf{b} = \begin{bmatrix} 1 \\ -3 \\ 2 \\ 4 \end{bmatrix}$

24. $A = \begin{bmatrix} 0 & 1 & 1 & 0 \\ 1 & -1 & 1 & -1 \\ 1 & 0 & 1 & 0 \\ 1 & 1 & 1 & 1 \end{bmatrix}, \mathbf{b} = \begin{bmatrix} 5 \\ 3 \\ -1 \\ 1 \end{bmatrix}$

In Exercises 25 and 26, find the best approximation to a solution of the given system of equations.

25.
$$\begin{aligned} x + y - z &= 2 \\ -y + 2z &= 6 \\ 3x + 2y - z &= 11 \\ -x + z &= 0 \end{aligned}$$

26.
$$\begin{aligned} 2x + 3y + z &= 21 \\ x + y + z &= 7 \\ -x + y - z &= 14 \\ 2y + z &= 0 \end{aligned}$$

In Exercises 27 and 28, a QR factorization of A is given. Use it to find a least squares solution of $A\mathbf{x} = \mathbf{b}$.

27. $A = \begin{bmatrix} 2 & 1 \\ 2 & 0 \\ 1 & 1 \end{bmatrix}, Q = \begin{bmatrix} \frac{2}{3} & \frac{1}{3} \\ \frac{2}{3} & -\frac{2}{3} \\ \frac{1}{3} & \frac{2}{3} \end{bmatrix}, R = \begin{bmatrix} 3 & 1 \\ 0 & 1 \end{bmatrix}, \mathbf{b} = \begin{bmatrix} 2 \\ 3 \\ -1 \end{bmatrix}$

28. $A = \begin{bmatrix} 1 & 0 \\ 2 & -1 \\ -1 & 1 \end{bmatrix}, Q = \begin{bmatrix} 1/\sqrt{6} & 1/\sqrt{2} \\ 2/\sqrt{6} & 0 \\ -1/\sqrt{6} & 1/\sqrt{2} \end{bmatrix},$

$R = \begin{bmatrix} \sqrt{6} & -\sqrt{6}/2 \\ 0 & 1/\sqrt{2} \end{bmatrix}, \mathbf{b} = \begin{bmatrix} 1 \\ 1 \\ 1 \end{bmatrix}$

29. A tennis ball is dropped from various heights, and the height of the ball on the first bounce is measured. Use the data in Table 7.3 to find the least squares approximating line for bounce height b as a linear function of initial height h.

Table 7.3

h **(cm)**	20	40	48	60	80	100
b **(cm)**	14.5	31	36	45.5	59	73.5

30. Hooke's Law states that the length L of a spring is a linear function of the force F applied to it. (See Figure 7.17 and Example 6.92.) Accordingly, there are constants a and b such that

$$L = a + bF$$

Table 7.4 shows the results of attaching various weights to a spring.

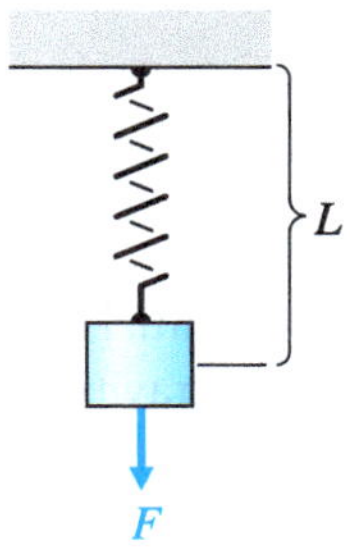

Figure 7.17

Table 7.4

F **(oz)**	2	4	6	8
L **(in.)**	7.4	9.6	11.5	13.6

Table 7.5

Year of Birth	1920	1930	1940	1950	1960	1970	1980	1990
Life Expectancy (years)	54.1	59.7	62.9	68.2	69.7	70.8	73.7	75.4

Source: World Almanac and Book of Facts. New York: World Almanac Books, 1999

(a) Determine the constants a and b by finding the least squares approximating line for these data. What does a represent?

(b) Estimate the length of the spring when a weight of 5 ounces is attached.

31. Table 7.5 gives life expectancies for people born in the United States in the given years.

(a) Determine the least squares approximating line for these data and use it to predict the life expectancy of someone born in 2000.

(b) How good is this model? Explain.

32. When an object is thrown straight up into the air, Newton's Second Law of Motion states that its height $s(t)$ at time t is given by

$$s(t) = s_0 + v_0t + \tfrac{1}{2}gt^2$$

where v_0 is its initial velocity and g is the constant of acceleration due to gravity. Suppose we take the measurements shown in Table 7.6.

Table 7.6

Time (s)	0.5	1	1.5	2	3
Height (m)	11	17	21	23	18

(a) Find the least squares approximating quadratic for these data.

(b) Estimate the height at which the object was released (in m), its initial velocity (in m/s), and its acceleration due to gravity (in m/s^2).

(c) Approximately when will the object hit the ground?

33. Table 7.7 gives the population of the United States at 10-year intervals for the years 1950–2000.

(a) Assuming an exponential growth model of the form $p(t) = ce^{kt}$, where $p(t)$ is the population at time t, use least squares to find the equation for the growth rate of the population. [*Hint:* Let $t = 0$ be 1950.]

(b) Use the equation to estimate the U.S. population in 2010.

Table 7.7

Year	Population (in millions)
1950	150
1960	179
1970	203
1980	227
1990	250
2000	281

Source: U.S. Bureau of the Census

34. Table 7.8 shows average major league baseball salaries for the years 1970–2005.

(a) Find the least squares approximating quadratic for these data.

(b) Find the least squares approximating exponential for these data.

(c) Which equation gives the better approximation? Why?

(d) What do you estimate the average major league baseball salary will be in 2010 and 2015?

Table 7.8

Year	Average Salary (thousands of dollars)
1970	29.3
1975	44.7
1980	143.8
1985	371.6
1990	597.5
1995	1110.8
2000	1895.6
2005	2476.6

Source: Major League Baseball Players Association

35. A 200 mg sample of radioactive polonium-210 is observed as it decays. Table 7.9 shows the mass remaining at various times.

Assuming an exponential decay model, use least squares to find the half-life of polonium-210. (See Section 6.7.)

Table 7.9

Time (days)	0	30	60	90
Mass (mg)	200	172	148	128

36. Find the plane $z = a + bx + cy$ that best fits the data points $(0, -4, 0)$, $(5, 0, 0)$, $(4, -1, 1)$, $(1, -3, 1)$, and $(-1, -5, -2)$.

In Exercises 37–42, find the standard matrix of the orthogonal projection onto the subspace W. Then use this matrix to find the orthogonal projection of **v** *onto W.*

37. $W = \text{span}\left(\begin{bmatrix} 1 \\ 1 \end{bmatrix}\right), \mathbf{v} = \begin{bmatrix} 3 \\ 4 \end{bmatrix}$

38. $W = \text{span}\left(\begin{bmatrix} 1 \\ -2 \end{bmatrix}\right), \mathbf{v} = \begin{bmatrix} 1 \\ 1 \end{bmatrix}$

39. $W = \text{span}\left(\begin{bmatrix} 1 \\ 1 \\ 1 \end{bmatrix}\right), \mathbf{v} = \begin{bmatrix} 1 \\ 2 \\ 3 \end{bmatrix}$

40. $W = \text{span}\left(\begin{bmatrix} 2 \\ 2 \\ -1 \end{bmatrix}\right), \mathbf{v} = \begin{bmatrix} 1 \\ 0 \\ 0 \end{bmatrix}$

41. $W = \text{span}\left(\begin{bmatrix} 1 \\ 0 \\ -1 \end{bmatrix}, \begin{bmatrix} 1 \\ 1 \\ 1 \end{bmatrix}\right), \mathbf{v} = \begin{bmatrix} 1 \\ 0 \\ 0 \end{bmatrix}$

42. $W = \text{span}\left(\begin{bmatrix} 1 \\ -2 \\ 1 \end{bmatrix}, \begin{bmatrix} 1 \\ 0 \\ -1 \end{bmatrix}\right), \mathbf{v} = \begin{bmatrix} 1 \\ 2 \\ 3 \end{bmatrix}$

43. Verify that the standard matrix of the projection onto W in Example 7.31 (as constructed by Theorem 7.11) does not depend on the choice of basis. Take

$$\left\{\begin{bmatrix} 1 \\ 1 \\ 0 \end{bmatrix}, \begin{bmatrix} 1 \\ 3 \\ 1 \end{bmatrix}\right\}$$

as a basis for W and repeat the calculations to show that the resulting projection matrix is the same.

44. Let A be a matrix with linearly independent columns and let $P = A(A^TA)^{-1}A^T$ be the matrix of orthogonal projection onto col(A).

(a) Show that P is symmetric.
(b) Show that P is idempotent.

In Exercises 45–52, compute the pseudoinverse of A.

45. $A = \begin{bmatrix} 1 \\ 2 \end{bmatrix}$

46. $A = \begin{bmatrix} 1 \\ -1 \\ 2 \end{bmatrix}$

47. $A = \begin{bmatrix} 1 & 3 \\ -1 & 1 \\ 0 & 2 \end{bmatrix}$

48. $A = \begin{bmatrix} 1 & 3 \\ 3 & 1 \\ 2 & 2 \end{bmatrix}$

49. $A = \begin{bmatrix} 1 & 1 \\ 0 & 1 \end{bmatrix}$

50. $A = \begin{bmatrix} 1 & 2 \\ 3 & 4 \end{bmatrix}$

51. $A = \begin{bmatrix} 1 & 0 & 0 \\ 1 & 0 & 1 \\ 0 & 1 & 1 \\ 1 & 1 & 1 \end{bmatrix}$

52. $A = \begin{bmatrix} 1 & 2 & 0 \\ 0 & 1 & -1 \\ 1 & 1 & -2 \\ 0 & 0 & 2 \end{bmatrix}$

53. (a) Show that if A is a square matrix with linearly independent columns, then $A^+ = A^{-1}$.
(b) If A is an $m \times n$ matrix with orthonormal columns, what is A^+?

54. Prove Theorem 7.12(b).

55. Prove the remaining part of Theorem 7.12(c).

56. Let A be a matrix with linearly independent columns. Prove the following:

(a) $(cA)^+ = (1/c)A^+$ for all scalars $c \neq 0$.
(b) $(A^+)^+ = A$ if A is a square matrix.
(c) $(A^T)^+ = (A^+)^T$ if A is a square matrix.

57. Let n data points $(x_1, y_1), \ldots, (x_n, y_n)$ be given. Show that if the points do not all lie on the same vertical line, then they have a unique least squares approximating line.

58. Let n data points $(x_1, y_1), \ldots, (x_n, y_n)$ be given. Generalize Exercise 57 to show that if at least $k + 1$ of $x_1, \ldots, x_n$ are distinct, then the given points have a unique least squares approximating polynomial of degree at most k.

7.4 The Singular Value Decomposition

In Chapter 5, we saw that every symmetric matrix A can be factored as $A = PDP^T$, where P is an orthogonal matrix and D is a diagonal matrix displaying the eigenvalues of A. If A is not symmetric, such a factorization is not possible, but as we learned in Chapter 4, we may still be able to factor a square matrix A as $A = PDP^{-1}$, where D is as before but P is now simply an invertible matrix. However, not every matrix is diagonalizable, so it may surprise you that we will now show that *every* matrix (symmetric or not, square or not) has a factorization of the form $A = PDQ^T$, where P and Q are orthogonal and D is a diagonal matrix! This remarkable result is the *singular value decomposition* (SVD), and it is one of the most important of all matrix factorizations.

In this section, we will show how to compute the SVD of a matrix and then consider some of its many applications. Along the way, we will tie up some loose ends by answering a few questions that were left open in previous sections.

The Singular Values of a Matrix

For any $m \times n$ matrix A, the $n \times n$ matrix A^TA is symmetric and hence can be orthogonally diagonalized, by the Spectral Theorem. Not only are the eigenvalues of A^TA all real (Theorem 5.18), they are all *nonnegative*. To show this, let λ be an eigenvalue of A^TA with corresponding unit eigenvector $\mathbf{v}$. Then

$$\begin{aligned} 0 \le \|A\mathbf{v}\|^2 &= (A\mathbf{v})\cdot(A\mathbf{v}) = (A\mathbf{v})^TA\mathbf{v} = \mathbf{v}^TA^TA\mathbf{v} \\ &= \mathbf{v}^T\lambda\mathbf{v} = \lambda(\mathbf{v}\cdot\mathbf{v}) = \lambda\|\mathbf{v}\|^2 = \lambda \end{aligned}$$

It therefore makes sense to take (positive) square roots of these eigenvalues.

Definition If A is an $m \times n$ matrix, the ***singular values*** of A are the square roots of the eigenvalues of A^TA and are denoted by $\sigma_1, \ldots, \sigma_n$. It is conventional to arrange the singular values so that $\sigma_1 \ge \sigma_2 \ge \cdots \ge \sigma_n$.

Example 7.33 Find the singular values of

$$A = \begin{bmatrix} 1 & 1 \\ 1 & 0 \\ 0 & 1 \end{bmatrix}$$

Solution The matrix

$$A^TA = \begin{bmatrix} 1 & 1 & 0 \\ 1 & 0 & 1 \end{bmatrix}\begin{bmatrix} 1 & 1 \\ 1 & 0 \\ 0 & 1 \end{bmatrix} = \begin{bmatrix} 2 & 1 \\ 1 & 2 \end{bmatrix}$$

has eigenvalues $\lambda_1 = 3$ and $\lambda_2 = 1$. Consequently, the singular values of A are $\sigma_1 = \sqrt{\lambda_1} = \sqrt{3}$ and $\sigma_2 = \sqrt{\lambda_2} = 1$.

To understand the significance of the singular values of an $m \times n$ matrix A, consider the eigenvectors of A^TA. Since A^TA is symmetric, we know that there is an *orthonormal* basis for $\mathbb{R}^n$ that consists of eigenvectors of A^TA. Let $\{\mathbf{v}_1, \ldots, \mathbf{v}_n\}$ be such a basis corresponding to the eigenvalues of A^TA, ordered so that $\lambda_1 \geq \lambda_2 \geq \cdots \geq \lambda_n$. From our calculations just before the definition,

$$\lambda_i = \|A\mathbf{v}_i\|^2$$

Therefore,

$$\sigma_i = \sqrt{\lambda_i} = \|A\mathbf{v}_i\|$$

In other words, the singular values of A are the lengths of the vectors $A\mathbf{v}_1, \ldots, A\mathbf{v}_n$.

Geometrically, this result has a nice interpretation. Consider Example 7.33 again. If $\mathbf{x}$ lies on the unit circle in $\mathbb{R}^2$ (i.e., $\|\mathbf{x}\| = 1$), then

$$\|A\mathbf{x}\|^2 = (A\mathbf{x})\cdot(A\mathbf{x}) = (A\mathbf{x})^T(A\mathbf{x}) = \mathbf{x}^TA^TA\mathbf{x}$$

$$= [x_1 \quad x_2]\begin{bmatrix} 2 & 1 \\ 1 & 2 \end{bmatrix}\begin{bmatrix} x_1 \\ x_2 \end{bmatrix} = 2x_1^2 + 2x_1x_2 + 2x_2^2$$

which we recognize is a quadratic form. By Theorem 5.25, the maximum and minimum values of this quadratic form, subject to the constraint $\|\mathbf{x}\| = 1$, are $\lambda_1 = 3$ and $\lambda_2 = 1$, respectively, and they occur at the corresponding eigenvectors of A^TA—that is, when $\mathbf{x} = \mathbf{v}_1 = \begin{bmatrix} 1/\sqrt{2} \\ 1/\sqrt{2} \end{bmatrix}$ and $\mathbf{x} = \mathbf{v}_2 = \begin{bmatrix} -1/\sqrt{2} \\ 1/\sqrt{2} \end{bmatrix}$, respectively. Since

$$\|A\mathbf{v}_i\|^2 = \mathbf{v}_i^TA^TA\mathbf{v}_i = \lambda_i$$

for $i = 1, 2$, we see that $\sigma_1 = \|A\mathbf{v}_1\| = \sqrt{3}$ and $\sigma_2 = \|A\mathbf{v}_2\| = 1$ are the maximum and minimum values of the lengths $\|A\mathbf{x}\|$ as $\mathbf{x}$ traverses the unit circle in $\mathbb{R}^2$.

Now, the linear transformation corresponding to A maps $\mathbb{R}^2$ onto the plane in $\mathbb{R}^3$ with equation $x - y - z = 0$ (verify this), and the image of the unit circle under this transformation is an ellipse that lies in this plane. (We will verify this fact in general shortly; see Figure 7.18.) So σ_1 and σ_2 are the lengths of half of the major and minor axes of this ellipse, as shown in Figure 7.19.

We can now describe the singular value decomposition of a matrix.

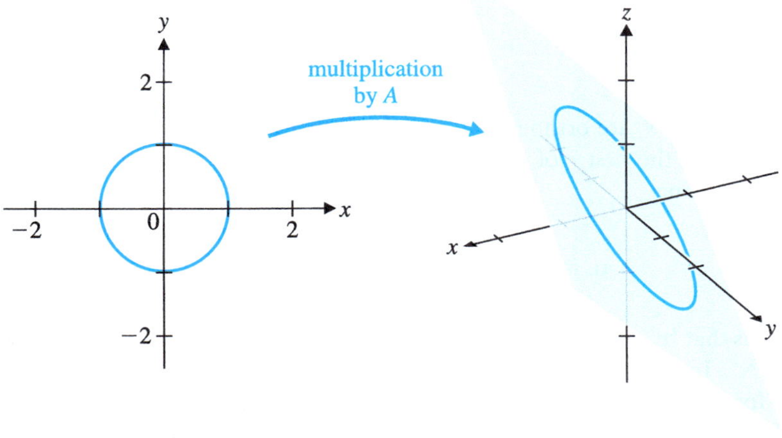

Figure 7.18
The matrix A transforms the unit circle in $\mathbb{R}^2$ into an ellipse in $\mathbb{R}^3$

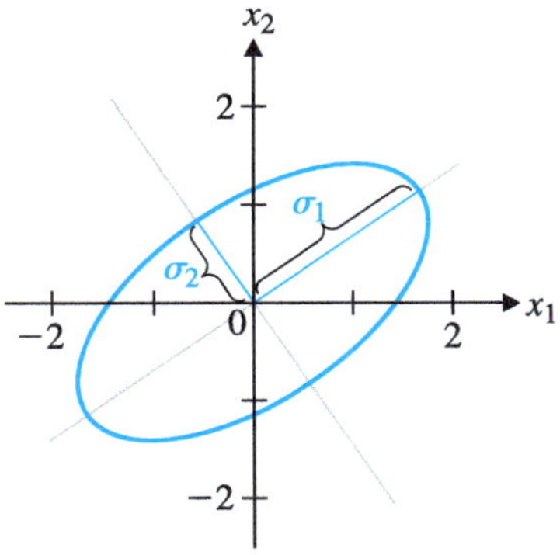

Figure 7.19

The Singular Value Decomposition

We want to show that an $m \times n$ matrix A can be factored as

$$A = U\Sigma V^T$$

where U is an $m \times m$ orthogonal matrix, V is an $n \times n$ orthogonal matrix, and Σ is an $m \times n$ "diagonal" matrix. If the *nonzero* singular values of A are

$$\sigma_1 \geq \sigma_2 \geq \cdots \geq \sigma_r > 0$$

and $\sigma_{r+1} = \sigma_{r+2} = \cdots = \sigma_n = 0$, then Σ will have the block form

$$\Sigma = \overset{\quad r \quad n-r}{\left[\begin{array}{c|c} D & O \\ \hline O & O \end{array}\right]} \begin{array}{l} \}\, r \\ \}\, m-r \end{array}, \quad \text{where} \quad D = \begin{bmatrix} \sigma_1 & \cdots & 0 \\ \vdots & \ddots & \vdots \\ 0 & \cdots & \sigma_r \end{bmatrix} \tag{1}$$

and each matrix O is a zero matrix of the appropriate size. (If $r = m$ or $r = n$, some of these will not appear.) Some examples of such a matrix Σ with $r = 2$ are

$$\Sigma = \begin{bmatrix} 4 & 0 & 0 \\ 0 & 3 & 0 \end{bmatrix}, \quad \Sigma = \begin{bmatrix} 2 & 0 \\ 0 & 2 \\ 0 & 0 \end{bmatrix}, \quad \Sigma = \begin{bmatrix} 8 & 0 & 0 \\ 0 & 3 & 0 \\ 0 & 0 & 0 \end{bmatrix}, \quad \Sigma = \begin{bmatrix} 5 & 0 & 0 \\ 0 & 2 & 0 \\ 0 & 0 & 0 \\ 0 & 0 & 0 \end{bmatrix}$$

(What is D in each case?)

To construct the orthogonal matrix V, we first find an orthonormal basis $\{\mathbf{v}_1, \ldots, \mathbf{v}_n\}$ for $\mathbb{R}^n$ consisting of eigenvectors of the $n \times n$ symmetric matrix A^TA. Then

$$V = [\mathbf{v}_1 \quad \cdots \quad \mathbf{v}_n]$$

is an orthogonal $n \times n$ matrix.

For the orthogonal matrix U, we first note that $\{A\mathbf{v}_1, \ldots, A\mathbf{v}_n\}$ is an orthogonal set of vectors in $\mathbb{R}^m$. To see this, suppose that $\mathbf{v}_i$ is the eigenvector of A^TA corresponding to the eigenvalue λ_i. Then, for $i \neq j$, we have

$$\begin{aligned}(A\mathbf{v}_i) \cdot (A\mathbf{v}_j) &= (A\mathbf{v}_i)^T A\mathbf{v}_j \\ &= \mathbf{v}_i^T A^T A\mathbf{v}_j \\ &= \mathbf{v}_i^T \lambda_j \mathbf{v}_j \\ &= \lambda_j(\mathbf{v}_i \cdot \mathbf{v}_j) = 0\end{aligned}$$

since the eigenvectors $\mathbf{v}_i$ are orthogonal. Now recall that the singular values satisfy $\sigma_i = \|A\mathbf{v}_i\|$ and that the first r of these are nonzero. Therefore, we can normalize $A\mathbf{v}_1, \ldots, A\mathbf{v}_r$ by setting

$$\mathbf{u}_i = \frac{1}{\sigma_i} A\mathbf{v}_i \quad \text{for } i = 1, \ldots, r$$

This guarantees that $\{\mathbf{u}_1, \ldots, \mathbf{u}_r\}$ is an orthonormal set in $\mathbb{R}^m$, but if $r < m$ it will not be a basis for $\mathbb{R}^m$. In this case, we extend the set $\{\mathbf{u}_1, \ldots, \mathbf{u}_r\}$ to an orthonormal basis $\{\mathbf{u}_1, \ldots, \mathbf{u}_m\}$ for $\mathbb{R}^m$. (This is the only tricky part of the construction; we will describe techniques for carrying it out in the examples below and in the exercises.) Then we set

$$U = [\mathbf{u}_1 \quad \cdots \quad \mathbf{u}_m]$$

All that remains to be shown is that this works; that is, we need to verify that with U, V, and Σ as described, we have $A = U\Sigma V^T$. Since $V^T = V^{-1}$, this is equivalent to showing that

$$AV = U\Sigma$$

We know that $$A\mathbf{v}_i = \sigma_i \mathbf{u}_i \quad \text{for } i = 1, \ldots, r$$

and $\|A\mathbf{v}_i\| = \sigma_i = 0$ for $i = r + 1, \ldots, n$. Hence,

$$A\mathbf{v}_i = \mathbf{0} \quad \text{for } i = r + 1, \ldots, n$$

Therefore,
$$\begin{aligned} AV &= A[\mathbf{v}_1 \ \cdots \ \mathbf{v}_n] \\ &= [A\mathbf{v}_1 \ \cdots \ A\mathbf{v}_n] \\ &= [A\mathbf{v}_1 \ \cdots \ A\mathbf{v}_r \ \ \mathbf{0} \ \cdots \ \mathbf{0}] \\ &= [\sigma_1\mathbf{u}_1 \ \cdots \ \sigma_r\mathbf{u}_r \ \ \mathbf{0} \ \cdots \ \mathbf{0}] \\ &= [\mathbf{u}_1 \ \cdots \ \mathbf{u}_m] \left[\begin{array}{ccc|c} \sigma_1 & \cdots & 0 & \\ \vdots & \ddots & \vdots & O \\ 0 & \cdots & \sigma_r & \\ \hline & O & & O \end{array}\right] \\ &= U\Sigma \end{aligned}$$

as required.

We have just proved the following extremely important theorem.

Theorem 7.13 **The Singular Value Decomposition**

Let A be an $m \times n$ matrix with singular values $\sigma_1 \geq \sigma_2 \geq \cdots \geq \sigma_r > 0$ and $\sigma_{r+1} = \sigma_{r+2} = \cdots = \sigma_n = 0$. Then there exist an $m \times m$ orthogonal matrix U, an $n \times n$ orthogonal matrix V, and an $m \times n$ matrix Σ of the form shown in Equation (1) such that

$$A = U\Sigma V^T$$

A factorization of A as in Theorem 7.13 is called a ***singular value decomposition (SVD)*** of A. The columns of U are called ***left singular vectors*** of A, and the columns of V are called ***right singular vectors*** of A. The matrices U and V are not uniquely determined by A, but Σ *must* contain the singular values of A, as in Equation (1). (See Exercise 25.)

Example 7.34

Find a singular value decomposition for the following matrices:

(a) $A = \begin{bmatrix} 1 & 1 & 0 \\ 0 & 0 & 1 \end{bmatrix}$ (b) $A = \begin{bmatrix} 1 & 1 \\ 1 & 0 \\ 0 & 1 \end{bmatrix}$

Solution (a) We compute

$$A^TA = \begin{bmatrix} 1 & 1 & 0 \\ 1 & 1 & 0 \\ 0 & 0 & 1 \end{bmatrix}$$

and find that its eigenvalues are $\lambda_1 = 2$, $\lambda_2 = 1$, and $\lambda_3 = 0$, with corresponding eigenvectors

$$\begin{bmatrix} 1 \\ 1 \\ 0 \end{bmatrix}, \begin{bmatrix} 0 \\ 0 \\ 1 \end{bmatrix}, \begin{bmatrix} -1 \\ 1 \\ 0 \end{bmatrix}$$

(Verify this.) These vectors are orthogonal, so we normalize them to obtain

$$\mathbf{v}_1 = \begin{bmatrix} 1/\sqrt{2} \\ 1/\sqrt{2} \\ 0 \end{bmatrix}, \quad \mathbf{v}_2 = \begin{bmatrix} 0 \\ 0 \\ 1 \end{bmatrix}, \quad \mathbf{v}_3 = \begin{bmatrix} -1/\sqrt{2} \\ 1/\sqrt{2} \\ 0 \end{bmatrix}$$

The singular values of A are $\sigma_1 = \sqrt{2}$, $\sigma_2 = \sqrt{1} = 1$, and $\sigma_3 = \sqrt{0} = 0$. Thus,

$$V = \begin{bmatrix} 1/\sqrt{2} & 0 & -1/\sqrt{2} \\ 1/\sqrt{2} & 0 & 1/\sqrt{2} \\ 0 & 1 & 0 \end{bmatrix} \quad \text{and} \quad \Sigma = \begin{bmatrix} \sqrt{2} & 0 & 0 \\ 0 & 1 & 0 \end{bmatrix}$$

To find U, we compute

$$\mathbf{u}_1 = \frac{1}{\sigma_1}A\mathbf{v}_1 = \frac{1}{\sqrt{2}}\begin{bmatrix} 1 & 1 & 0 \\ 0 & 0 & 1 \end{bmatrix}\begin{bmatrix} 1/\sqrt{2} \\ 1/\sqrt{2} \\ 0 \end{bmatrix} = \begin{bmatrix} 1 \\ 0 \end{bmatrix}$$

and

$$\mathbf{u}_2 = \frac{1}{\sigma_2}A\mathbf{v}_2 = \frac{1}{1}\begin{bmatrix} 1 & 1 & 0 \\ 0 & 0 & 1 \end{bmatrix}\begin{bmatrix} 0 \\ 0 \\ 1 \end{bmatrix} = \begin{bmatrix} 0 \\ 1 \end{bmatrix}$$

These vectors already form an orthonormal basis (the standard basis) for $\mathbb{R}^2$, so we have

$$U = \begin{bmatrix} 1 & 0 \\ 0 & 1 \end{bmatrix}$$

This yields the SVD

$$A = \begin{bmatrix} 1 & 1 & 0 \\ 0 & 0 & 1 \end{bmatrix} = \begin{bmatrix} 1 & 0 \\ 0 & 1 \end{bmatrix}\begin{bmatrix} \sqrt{2} & 0 & 0 \\ 0 & 1 & 0 \end{bmatrix}\begin{bmatrix} 1/\sqrt{2} & 1/\sqrt{2} & 0 \\ 0 & 0 & 1 \\ -1/\sqrt{2} & 1/\sqrt{2} & 0 \end{bmatrix} = U\Sigma V^T$$

which can be easily checked. (Note that V had to be transposed. Also note that the singular value σ_3 does not appear in Σ.)

(b) This is the matrix in Example 7.33, so we already know that the singular values are $\sigma_1 = \sqrt{3}$ and $\sigma_2 = 1$, corresponding to $\mathbf{v}_1 = \begin{bmatrix} 1/\sqrt{2} \\ 1/\sqrt{2} \end{bmatrix}$ and $\mathbf{v}_2 = \begin{bmatrix} -1/\sqrt{2} \\ 1/\sqrt{2} \end{bmatrix}$. So

$$\Sigma = \begin{bmatrix} \sqrt{3} & 0 \\ 0 & 1 \\ 0 & 0 \end{bmatrix} \quad \text{and} \quad V = \begin{bmatrix} 1/\sqrt{2} & -1/\sqrt{2} \\ 1/\sqrt{2} & 1/\sqrt{2} \end{bmatrix}$$

For U, we compute

$$\mathbf{u}_1 = \frac{1}{\sigma_1}A\mathbf{v}_1 = \frac{1}{\sqrt{3}}\begin{bmatrix} 1 & 1 \\ 1 & 0 \\ 0 & 1 \end{bmatrix}\begin{bmatrix} 1/\sqrt{2} \\ 1/\sqrt{2} \end{bmatrix} = \begin{bmatrix} 2/\sqrt{6} \\ 1/\sqrt{6} \\ 1/\sqrt{6} \end{bmatrix}$$

and
$$\mathbf{u}_2 = \frac{1}{\sigma_2}A\mathbf{v}_2 = \frac{1}{1}\begin{bmatrix} 1 & 1 \\ 1 & 0 \\ 0 & 1 \end{bmatrix}\begin{bmatrix} -1/\sqrt{2} \\ 1/\sqrt{2} \end{bmatrix} = \begin{bmatrix} 0 \\ -1/\sqrt{2} \\ 1/\sqrt{2} \end{bmatrix}$$

This time, we need to extend $\{\mathbf{u}_1, \mathbf{u}_2\}$ to an orthonormal basis for $\mathbb{R}^3$. There are several ways to proceed; one method is to use the Gram-Schmidt Process, as in Example 5.14. We first need to find a linearly independent set of three vectors that contains $\mathbf{u}_1$ and $\mathbf{u}_2$. If $\mathbf{e}_3$ is the third standard basis vector in $\mathbb{R}^3$, it is clear that $\{\mathbf{u}_1, \mathbf{u}_2, \mathbf{e}_3\}$ is linearly independent. (Here, you should be able to determine this by inspection, but a reliable method to use in general is to row reduce the matrix with these vectors as its columns and use the Fundamental Theorem.) Applying Gram-Schmidt (with normalization) to $\{\mathbf{u}_1, \mathbf{u}_2, \mathbf{e}_3\}$ (only the last step is needed), we find

$$\mathbf{u}_3 = \begin{bmatrix} -1/\sqrt{3} \\ 1/\sqrt{3} \\ 1/\sqrt{3} \end{bmatrix}$$

so
$$U = \begin{bmatrix} 2/\sqrt{6} & 0 & -1/\sqrt{3} \\ 1/\sqrt{6} & -1/\sqrt{2} & 1/\sqrt{3} \\ 1/\sqrt{6} & 1/\sqrt{2} & 1/\sqrt{3} \end{bmatrix}$$

and we have the SVD

$$A = \begin{bmatrix} 1 & 1 \\ 1 & 0 \\ 0 & 1 \end{bmatrix} = \begin{bmatrix} 2/\sqrt{6} & 0 & -1/\sqrt{3} \\ 1/\sqrt{6} & -1/\sqrt{2} & 1/\sqrt{3} \\ 1/\sqrt{6} & 1/\sqrt{2} & 1/\sqrt{3} \end{bmatrix}\begin{bmatrix} \sqrt{3} & 0 \\ 0 & 1 \\ 0 & 0 \end{bmatrix}\begin{bmatrix} 1/\sqrt{2} & 1/\sqrt{2} \\ -1/\sqrt{2} & 1/\sqrt{2} \end{bmatrix} = U\Sigma V^T$$

There is another form of the singular value decomposition, analogous to the spectral decomposition of a symmetric matrix. It is obtained from the SVD by an outer product expansion and is very useful in applications. We can obtain this version of the SVD by imitating what we did to obtain the spectral decomposition.

Accordingly, we have

$$
\begin{aligned}
A = U\Sigma V^T &= [\mathbf{u}_1 \ \cdots \ \mathbf{u}_m] \left[\begin{array}{ccc|c} \sigma_1 & \cdots & 0 & \\ \vdots & \ddots & \vdots & O \\ 0 & \cdots & \sigma_r & \\ \hline & O & & O \end{array}\right] \begin{bmatrix} \mathbf{v}_1^T \\ \vdots \\ \mathbf{v}_n^T \end{bmatrix} \\
&= [\mathbf{u}_1 \ \cdots \ \mathbf{u}_r \,|\, \mathbf{u}_{r+1} \ \cdots \ \mathbf{u}_m] \left[\begin{array}{ccc|c} \sigma_1 & \cdots & 0 & \\ \vdots & \ddots & \vdots & O \\ 0 & \cdots & \sigma_r & \\ \hline & O & & O \end{array}\right] \left[\begin{array}{c} \mathbf{v}_1^T \\ \vdots \\ \mathbf{v}_r^T \\ \hline \mathbf{v}_{r+1}^T \\ \vdots \\ \mathbf{v}_n^T \end{array}\right] \\
&= [\mathbf{u}_1 \ \cdots \ \mathbf{u}_r] \begin{bmatrix} \sigma_1 & \cdots & 0 \\ \vdots & \ddots & \vdots \\ 0 & \cdots & \sigma_r \end{bmatrix} \begin{bmatrix} \mathbf{v}_1^T \\ \vdots \\ \mathbf{v}_r^T \end{bmatrix} + [\mathbf{u}_{r+1} \ \cdots \ \mathbf{u}_m][O] \begin{bmatrix} \mathbf{v}_{r+1}^T \\ \vdots \\ \mathbf{v}_n^T \end{bmatrix} \\
&= [\mathbf{u}_1 \ \cdots \ \mathbf{u}_r] \begin{bmatrix} \sigma_1 & \cdots & 0 \\ \vdots & \ddots & \vdots \\ 0 & \cdots & \sigma_r \end{bmatrix} \begin{bmatrix} \mathbf{v}_1^T \\ \vdots \\ \mathbf{v}_r^T \end{bmatrix} \\
&= [\sigma_1\mathbf{u}_1 \ \cdots \ \sigma_r\mathbf{u}_r] \begin{bmatrix} \mathbf{v}_1^T \\ \vdots \\ \mathbf{v}_r^T \end{bmatrix} \\
&= \sigma_1\mathbf{u}_1\mathbf{v}_1^T + \cdots + \sigma_r\mathbf{u}_r\mathbf{v}_r^T
\end{aligned}
$$

using block multiplication and the column-row representation of the product. The following theorem summarizes the process for obtaining this ***outer product form of the SVD.***

Theorem 7.14 **The Outer Product Form of the SVD**

Let A be an $m \times n$ matrix with singular values $\sigma_1 \geq \sigma_2 \geq \cdots \geq \sigma_r > 0$ and $\sigma_{r+1} = \sigma_{r+2} = \cdots = \sigma_n = 0$. Let $\mathbf{u}_1, \ldots, \mathbf{u}_r$ be left singular vectors and let $\mathbf{v}_1, \ldots, \mathbf{v}_r$ be right singular vectors of A corresponding to these singular values. Then

$$A = \sigma_1\mathbf{u}_1\mathbf{v}_1^T + \cdots + \sigma_r\mathbf{u}_r\mathbf{v}_r^T$$

Remark If A is a positive definite, symmetric matrix, then Theorems 7.13 and 7.14 both reduce to results that we already know. In this case, it is not hard to show that the SVD generalizes the Spectral Theorem and that Theorem 7.14 generalizes the spectral decomposition. (See Exercise 27.)

The SVD of a matrix A contains much important information about A, as outlined in the crucial Theorem 7.15.

Theorem 7.15 Let $A = U\Sigma V^T$ be a singular value decomposition of an $m \times n$ matrix A. Let $\sigma_1, \ldots, \sigma_r$ be all the nonzero singular values of A. Then:

a. The rank of A is r.

b. $\{\mathbf{u}_1, \ldots, \mathbf{u}_r\}$ is an orthonormal basis for $\operatorname{col}(A)$.

c. $\{\mathbf{u}_{r+1}, \ldots, \mathbf{u}_m\}$ is an orthonormal basis for $\operatorname{null}(A^T)$.

d. $\{\mathbf{v}_1, \ldots, \mathbf{v}_r\}$ is an orthonormal basis for $\operatorname{row}(A)$.

e. $\{\mathbf{v}_{r+1}, \ldots, \mathbf{v}_n\}$ is an orthonormal basis for $\operatorname{null}(A)$.

Proof (a) By Exercise 61 in Section 3.5, we have

$$\begin{aligned}\operatorname{rank}(A) &= \operatorname{rank}(U\Sigma V^T)\\ &= \operatorname{rank}(\Sigma V^T)\\ &= \operatorname{rank}(\Sigma) = r\end{aligned}$$

(b) We already know that $\{\mathbf{u}_1, \ldots, \mathbf{u}_r\}$ is an orthonormal set. Therefore, it is linearly independent, by Theorem 5.1. Since $\mathbf{u}_i = (1/\sigma_i)A\mathbf{v}_i$ for $i = 1, \ldots, r$, each $\mathbf{u}_i$ is in the column space of A. (Why?) Furthermore,

$$r = \operatorname{rank}(A) = \dim(\operatorname{col}(A))$$

Therefore, $\{\mathbf{u}_1, \ldots, \mathbf{u}_r\}$ is an orthonormal basis for $\operatorname{col}(A)$, by Theorem 6.10(c).

(c) Since $\{\mathbf{u}_1, \ldots, \mathbf{u}_m\}$ is an orthonormal basis for $\mathbb{R}^m$ and $\{\mathbf{u}_1, \ldots, \mathbf{u}_r\}$ is a basis for $\operatorname{col}(A)$, by property (b), it follows that $\{\mathbf{u}_{r+1}, \ldots, \mathbf{u}_m\}$ is an orthonormal basis for the orthogonal complement of $\operatorname{col}(A)$. But $(\operatorname{col}(A)) = \operatorname{null}(A^T)$, by Theorem 5.10.

(e) Since

$$A\mathbf{v}_{r+1} = \cdots = A\mathbf{v}_n = \mathbf{0}$$

the set $\{\mathbf{v}_{r+1}, \ldots, \mathbf{v}_n\}$ is an orthonormal set contained in the null space of A. Therefore, $\{\mathbf{v}_{r+1}, \ldots, \mathbf{v}_n\}$ is a linearly independent set of $n - r$ vectors in $\operatorname{null}(A)$. But

$$\dim(\operatorname{null}(A)) = n - r$$

by the Rank Theorem, so $\{\mathbf{v}_{r+1}, \ldots, \mathbf{v}_n\}$ is an orthonormal basis for $\operatorname{null}(A)$, by Theorem 6.10(c).

(d) Property (d) follows from property (e) and Theorem 5.10. (You are asked to prove this in Exercise 32.)

The SVD provides new geometric insight into the effect of matrix transformations. We have noted several times (without proof) that an $m \times n$ matrix transforms the unit sphere in $\mathbb{R}^n$ into an ellipsoid in $\mathbb{R}^m$. This point arose, for example, in our discussions of Perron's Theorem and of operator norms, as well as in the introduction to singular values in this section. We now prove this result.

Theorem 7.16 Let A be an $m \times n$ matrix with rank r. Then the image of the unit sphere in $\mathbb{R}^n$ under the matrix transformation that maps $\mathbf{x}$ to $A\mathbf{x}$ is

a. the surface of an ellipsoid in $\mathbb{R}^m$ if $r = n$.
b. a solid ellipsoid in $\mathbb{R}^m$ if $r < n$.

Proof Let $A = U\Sigma V^T$ be a singular value decomposition of the $m \times n$ matrix A. Let the left and right singular vectors of A be $\mathbf{u}_1, \ldots, \mathbf{u}_m$ and $\mathbf{v}_1, \ldots, \mathbf{v}_n$, respectively. Since $\operatorname{rank}(A) = r$, the singular values of A satisfy

$$\sigma_1 \geq \sigma_2 \geq \cdots \geq \sigma_r > 0 \quad \text{and} \quad \sigma_{r+1} = \sigma_{r+2} = \cdots = \sigma_n = 0$$

by Theorem 7.15(a). Let $\mathbf{x} = \begin{bmatrix} x_1 \\ \vdots \\ x_n \end{bmatrix}$ be a unit vector in $\mathbb{R}^n$. Now, since V is an orthogonal matrix, so is V^T, and hence $V^T\mathbf{x}$ is a unit vector, by Theorem 5.6. Now

$$V^T\mathbf{x} = \begin{bmatrix} \mathbf{v}_1^T \\ \vdots \\ \mathbf{v}_n^T \end{bmatrix}\mathbf{x} = \begin{bmatrix} \mathbf{v}_1^T\mathbf{x} \\ \vdots \\ \mathbf{v}_n^T\mathbf{x} \end{bmatrix}$$

so $(\mathbf{v}_1^T\mathbf{x})^2 + \cdots + (\mathbf{v}_n^T\mathbf{x})^2 = 1$.

By the outer product form of the SVD, we have $A = \sigma_1\mathbf{u}_1\mathbf{v}_1^T + \cdots + \sigma_r\mathbf{u}_r\mathbf{v}_r^T$. Therefore,

$$\begin{aligned} A\mathbf{x} &= \sigma_1\mathbf{u}_1\mathbf{v}_1^T\mathbf{x} + \cdots + \sigma_r\mathbf{u}_r\mathbf{v}_r^T\mathbf{x} \\ &= (\sigma_1\mathbf{v}_1^T\mathbf{x})\mathbf{u}_1 + \cdots + (\sigma_r\mathbf{v}_r^T\mathbf{x})\mathbf{u}_r \\ &= y_1\mathbf{u}_1 + \cdots + y_r\mathbf{u}_r \end{aligned}$$

where we are letting y_i denote the scalar $\sigma_i\mathbf{v}_i^T\mathbf{x}$.

(a) If $r = n$, then we must have $n \leq m$ and

$$\begin{aligned} A\mathbf{x} &= y_1\mathbf{u}_1 + \cdots + y_n\mathbf{u}_n \\ &= U\mathbf{y} \end{aligned}$$

where $\mathbf{y} = \begin{bmatrix} y_1 \\ \vdots \\ y_n \end{bmatrix}$. Therefore, again by Theorem 5.6, $\|A\mathbf{x}\| = \|U\mathbf{y}\| = \|\mathbf{y}\|$, since U is orthogonal. But

$$\left(\frac{y_1}{\sigma_1}\right)^2 + \cdots + \left(\frac{y_n}{\sigma_n}\right)^2 = (\mathbf{v}_1^T\mathbf{x})^2 + \cdots + (\mathbf{v}_n^T\mathbf{x})^2 = 1$$

which shows that the vectors $A\mathbf{x}$ form the surface of an ellipsoid in $\mathbb{R}^m$. (Why?)

(b) If $r < n$, the only difference in the above steps is that the equation becomes

$$\left(\frac{y_1}{\sigma_1}\right)^2 + \cdots + \left(\frac{y_r}{\sigma_r}\right)^2 \leq 1$$

since we are missing some terms. This inequality corresponds to a solid ellipsoid in $\mathbb{R}^m$.

Example 7.35

Describe the image of the unit sphere in $\mathbb{R}^3$ under the action of the matrix

$$A = \begin{bmatrix} 1 & 1 & 0 \\ 0 & 0 & 1 \end{bmatrix}$$

Solution In Example 7.34(a), we found the following SVD of A:

$$\begin{bmatrix} 1 & 1 & 0 \\ 0 & 0 & 1 \end{bmatrix} = \begin{bmatrix} 1 & 0 \\ 0 & 1 \end{bmatrix}\begin{bmatrix} \sqrt{2} & 0 & 0 \\ 0 & 1 & 0 \end{bmatrix}\begin{bmatrix} 1/\sqrt{2} & 1/\sqrt{2} & 0 \\ 0 & 0 & 1 \\ -1/\sqrt{2} & 1/\sqrt{2} & 0 \end{bmatrix}$$

Since $r = \text{rank}(A) = 2 < 3 = n$, the second part of Theorem 7.16 applies. The image of the unit sphere will satisfy the inequality

$$\left(\frac{y_1}{\sqrt{2}}\right)^2 + \left(\frac{y_2}{1}\right)^2 \leq 1 \quad \text{or} \quad \frac{y_1^2}{2} + y_2^2 \leq 1$$

relative to y_1y_2 coordinate axes in $\mathbb{R}^2$ (corresponding to the left singular vectors $\mathbf{u}_1$ and $\mathbf{u}_2$). Since $\mathbf{u}_1 = \mathbf{e}_1$ and $\mathbf{u}_2 = \mathbf{e}_2$, the image is as shown in Figure 7.20.

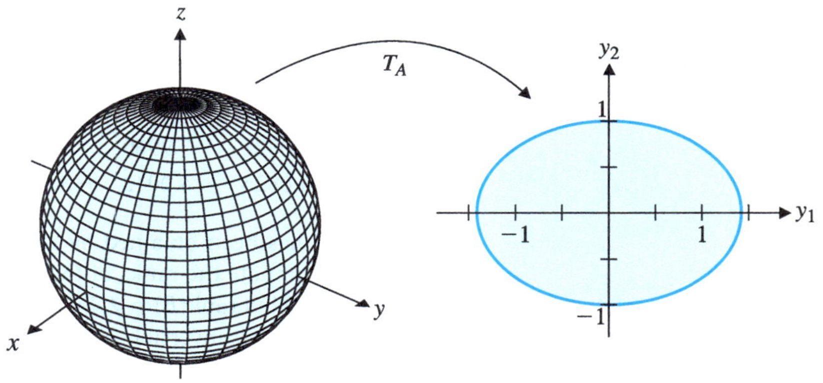

Figure 7.20

In general, we can describe the effect of an $m \times n$ matrix A on the unit sphere in $\mathbb{R}^n$ in terms of the effect of each factor in its SVD, $A = U\Sigma V^T$, from right to left. Since V^T is an orthogonal matrix, it maps the unit sphere to itself. The $m \times n$ matrix Σ does two things: The diagonal entries $\sigma_{r+1} = \sigma_{r+2} = \cdots = \sigma_n = 0$ collapse $n - r$ of the dimensions of the unit sphere, leaving an r-dimensional unit sphere, which the nonzero diagonal entries $\sigma_1, \ldots, \sigma_r$ then distort into an ellipsoid. The orthogonal matrix U then aligns the axes of this ellipsoid with the orthonormal basis vectors $\mathbf{u}_1, \ldots, \mathbf{u}_r$ in $\mathbb{R}^m$. (See Figure 7.21.)

Applications of the SVD

The singular value decomposition is an extremely useful tool, both practically and theoretically. We will look at just a few of its many applications.

Figure 7.21

Rank Until now, we have not worried about calculating the rank of a matrix from a computational point of view. We compute the rank of a matrix by row reducing it to echelon form and counting the number of nonzero rows. However, as we have seen, roundoff errors can affect this process, especially if the matrix is ill-conditioned. Entries that should be zero may end up as very small nonzero numbers, affecting our ability to accurately determine the rank and other quantities associated with the matrix. In practice, the SVD is often used to find the rank of a matrix, since it is much more reliable when roundoff errors are present. The basic idea behind this approach is that the orthogonal matrices U and V in the SVD preserve lengths and thus do not introduce additional errors; any errors that occur will tend to show up in the matrix Σ.

CAS **Example 7.36**

Let

$$A = \begin{bmatrix} 8.1650 & -0.0041 & -0.0041 \\ 4.0825 & -3.9960 & 4.0042 \\ 4.0825 & 4.0042 & -3.9960 \end{bmatrix} \quad \text{and} \quad B = \begin{bmatrix} 8.17 & 0 & 0 \\ 4.08 & -4 & 4 \\ 4.08 & 4 & -4 \end{bmatrix}$$

The matrix B has been obtained by rounding off the entries in A to two decimal places. If we compute the ranks of these two approximately equal matrices, we find that $\text{rank}(A) = 3$ but $\text{rank}(B) = 2$. By the Fundamental Theorem, this implies, among other things, that A is invertible but B is not.

The explanation for this critical difference between two matrices that are approximately equal lies in their SVDs. The singular values of A are 10, 8, and 0.01, so A has rank 3. The singular values of B are 10, 8, and 0, so B has rank 2.

In practical applications, it is often assumed that if a singular value is computed to be close to zero, then roundoff error has crept in and the actual value should be zero. In this way, "noise" can be filtered out. In this example, if we compute $A = U\Sigma V^T$ and replace

$$\Sigma = \begin{bmatrix} 10 & 0 & 0 \\ 0 & 8 & 0 \\ 0 & 0 & 0.01 \end{bmatrix} \quad \text{by} \quad \Sigma' = \begin{bmatrix} 10 & 0 & 0 \\ 0 & 8 & 0 \\ 0 & 0 & 0 \end{bmatrix}$$

then $U\Sigma' V^T = B$. (Try it!)

Matrix Norms and the Condition Number The SVD can provide simple formulas for certain expressions involving matrix norms. Consider, for example, the Frobenius norm of a matrix. The following theorem shows that it is completely determined by the singular values of the matrix.

Theorem 7.17 Let A be an $m \times n$ matrix and let $\sigma_1, \ldots, \sigma_r$ be all the nonzero singular values of A. Then

$$\|A\|_F = \sqrt{\sigma_1^2 + \cdots + \sigma_r^2}$$

The proof of this result depends on the following analogue of Theorem 5.6:

If A is an $m \times n$ matrix and Q is an $m \times m$ orthogonal matrix, then

$$\|QA\|_F = \|A\|_F \tag{2}$$

To show that this is true, we compute

$$\begin{aligned}\|QA\|_F^2 &= \|[Q\mathbf{a}_1 \quad \cdots \quad Q\mathbf{a}_n]\|_F^2 \\ &= \|Q\mathbf{a}_1\|_E^2 + \cdots + \|Q\mathbf{a}_n\|_E^2 \\ &= \|\mathbf{a}_1\|_E^2 + \cdots + \|\mathbf{a}_n\|_E^2 \\ &= \|A\|_F^2\end{aligned}$$

Proof of Theorem 7.17 Let $A = U\Sigma V^T$ be a singular value decomposition of A. Then, using Equation (2) twice, we have

$$\begin{aligned}\|A\|_F^2 &= \|U\Sigma V^T\|_F^2 \\ &= \|\Sigma V^T\|_F^2 = \|(\Sigma V^T)^T\|_F^2 \\ &= \|V\Sigma^T\|_F^2 = \|\Sigma^T\|_F^2 = \sigma_1^2 + \cdots + \sigma_r^2\end{aligned}$$

which establishes the result.

CAS **Example 7.37** Verify Theorem 7.17 for the matrix A in Example 7.18.

Solution The matrix $A = \begin{bmatrix} 3 & -1 \\ 2 & 4 \end{bmatrix}$ has singular values 4.5150 and 3.1008. We check that

$$\sqrt{4.5150^2 + 3.1008^2} = \sqrt{30} = \|A\|_F$$

which agrees with Example 7.18.

In Section 7.2, we commented that there is no easy formula for the operator 2-norm of a matrix A. Although that is true, the SVD of A provides us with a very nice expression for $\|A\|_2$. Recall that

$$\|A\|_2 = \max_{\|\mathbf{x}\|=1} \|A\mathbf{x}\|$$

where the vector norm is the ordinary Euclidean norm. By Theorem 7.16, for $\|\mathbf{x}\| = 1$, the set of vectors $\|A\mathbf{x}\|$ lies on or inside an ellipsoid whose semi-axes have lengths equal to the nonzero singular values of A. It follows immediately that the largest of these is σ_1, so

$$\|A\|_2 = \sigma_1$$

This provides us with a neat way to express the condition number of a (square) matrix with respect to the operator 2-norm. Recall that the condition number (with respect to the operator 2-norm) of an invertible matrix A is defined as

$$\text{cond}_2(A) = \|A^{-1}\|_2\|A\|_2$$

As you will be asked to show in Exercise 28, if $A = U\Sigma V^T$, then $A^{-1} = V\Sigma^{-1}U^T$. Therefore, the singular values of A^{-1} are $1/\sigma_1, \ldots, 1/\sigma_n$ (why?), and

$$1/\sigma_n \geq \cdots \geq 1/\sigma_1$$

It follows that $\|A^{-1}\|_2 = 1/\sigma_n$, so

$$\text{cond}_2(A) = \frac{\sigma_1}{\sigma_n}$$

Example 7.38

Find the 2-condition number of the matrix A in Example 7.36.

Solution Since $\sigma_1 = 10$ and $\sigma_3 = 0.01$,

$$\text{cond}_2(A) = \frac{\sigma_1}{\sigma_3} = \frac{10}{0.01} = 1000$$

This value is large enough to suggest that A may be ill-conditioned and we should be wary of the effect of roundoff errors.

The Pseudoinverse and Least Squares Approximation In Section 7.3, we produced the formula $A^+ = (A^TA)^{-1}A^T$ for the pseudoinverse of a matrix A. Clearly, this formula is valid only if A^TA is invertible, as we noted at the time. Equipped with the SVD, we can now define the pseudoinverse of *any* matrix, generalizing our previous formula.

E. H. Moore (1862–1932) was an American mathematician who worked in group theory, number theory, and geometry. He was the first head of the mathematics department at the University of Chicago when it opened in 1892. In 1920, he introduced a generalized matrix inverse that included rectangular matrices. His work did not receive much attention because of his obscure writing style.

Definition Let $A = U\Sigma V^T$ be an SVD for an $m \times n$ matrix A, where $\Sigma = \begin{bmatrix} D & O \\ O & O \end{bmatrix}$ and D is an $r \times r$ diagonal matrix containing the nonzero singular values $\sigma_1 \geq \sigma_2 \geq \cdots \geq \sigma_r > 0$ of A. The ***pseudoinverse*** (or ***Moore-Penrose inverse***) of A is the $n \times m$ matrix A^+ defined by

$$A^+ = V\Sigma^+U^T$$

where Σ^+ is the $n \times m$ matrix

$$\Sigma^+ = \begin{bmatrix} D^{-1} & O \\ O & O \end{bmatrix}$$

Example 7.39 Find the pseudoinverses of the matrices in Example 7.34.

Solution (a) From the SVD

$$A = \begin{bmatrix} 1 & 1 & 0 \\ 0 & 0 & 1 \end{bmatrix} = \begin{bmatrix} 1 & 0 \\ 0 & 1 \end{bmatrix}\begin{bmatrix} \sqrt{2} & 0 & 0 \\ 0 & 1 & 0 \end{bmatrix}\begin{bmatrix} 1/\sqrt{2} & 1/\sqrt{2} & 0 \\ 0 & 0 & 1 \\ -1/\sqrt{2} & 1/\sqrt{2} & 0 \end{bmatrix} = U\Sigma V^T$$

we form

$$\Sigma^+ = \begin{bmatrix} 1/\sqrt{2} & 0 \\ 0 & 1 \\ 0 & 0 \end{bmatrix}$$

Then

$$A^+ = V\Sigma^+U^T = \begin{bmatrix} 1/\sqrt{2} & 0 & -1/\sqrt{2} \\ 1/\sqrt{2} & 0 & 1/\sqrt{2} \\ 0 & 1 & 0 \end{bmatrix}\begin{bmatrix} 1/\sqrt{2} & 0 \\ 0 & 1 \\ 0 & 0 \end{bmatrix}\begin{bmatrix} 1 & 0 \\ 0 & 1 \end{bmatrix} = \begin{bmatrix} 1/2 & 0 \\ 1/2 & 0 \\ 0 & 1 \end{bmatrix}$$

(b) We have the SVD

$$A = \begin{bmatrix} 1 & 1 \\ 1 & 0 \\ 0 & 1 \end{bmatrix} = \begin{bmatrix} 2/\sqrt{6} & 0 & -1/\sqrt{3} \\ 1/\sqrt{6} & -1/\sqrt{2} & 1/\sqrt{3} \\ 1/\sqrt{6} & 1/\sqrt{2} & 1/\sqrt{3} \end{bmatrix}\begin{bmatrix} \sqrt{3} & 0 \\ 0 & 1 \\ 0 & 0 \end{bmatrix}\begin{bmatrix} 1/\sqrt{2} & 1/\sqrt{2} \\ -1/\sqrt{2} & 1/\sqrt{2} \end{bmatrix} = U\Sigma V^T$$

so

$$\Sigma^+ = \begin{bmatrix} 1/\sqrt{3} & 0 & 0 \\ 0 & 1 & 0 \end{bmatrix}$$

and

$$A^+ = V\Sigma^+U^T = \begin{bmatrix} 1/\sqrt{2} & -1/\sqrt{2} \\ 1/\sqrt{2} & 1/\sqrt{2} \end{bmatrix}\begin{bmatrix} 1/\sqrt{3} & 0 & 0 \\ 0 & 1 & 0 \end{bmatrix}\begin{bmatrix} 2/\sqrt{6} & 1/\sqrt{6} & 1/\sqrt{6} \\ 0 & -1/\sqrt{2} & 1/\sqrt{2} \\ -1/\sqrt{3} & 1/\sqrt{3} & 1/\sqrt{3} \end{bmatrix}$$

$$= \begin{bmatrix} 1/3 & 2/3 & -1/3 \\ 1/3 & -1/3 & 2/3 \end{bmatrix}$$

Jerry Bauer

One of those who was unaware of Moore's work on matrix inverses was Roger Penrose (b.1931), who introduced his own notion of a generalized matrix inverse in 1955. Penrose has made many contributions to geometry and theoretical physics. He is also the inventor of a type of *nonperiodic tiling* that covers the plane with only two different shapes of tile, yet has no repeating pattern. He has received many awards, including the 1988 Wolf Prize in Physics, which he shared with Stephen Hawking. In 1994, he was knighted for services to science. Sir Roger Penrose is currently the Emeritus Rouse Ball Professor of Mathematics at the University of Oxford.

It is straightforward to check that this new definition of the pseudoinverse generalizes the old one, for if the $m \times n$ matrix $A = U\Sigma V^T$ has linearly independent columns, then direct substitution shows that $(A^TA)^{-1}A^T = V\Sigma^+U^T$. (You are asked to verify this in Exercise 50.) Other properties of the pseudoinverse are explored in the exercises.

We have seen that when A has linearly independent columns, there is a unique least squares solution $\overline{\mathbf{x}}$ to $A\mathbf{x} = \mathbf{b}$; that is, the normal equations $A^TA\mathbf{x} = A^T\mathbf{b}$ have the unique solution

$$\overline{\mathbf{x}} = (A^TA)^{-1}A^T\mathbf{b} = A^+\mathbf{b}$$

When the columns of A are linearly dependent, then A^TA is not invertible, so the normal equations have infinitely many solutions. In this case, we will ask for the solution $\overline{\mathbf{x}}$ of *minimum length* (i.e., the one closest to the origin). It turns out that this time we simply use the general version of the pseudoinverse.

Theorem 7.18 The least squares problem $A\mathbf{x} = \mathbf{b}$ has a unique least squares solution $\bar{\mathbf{x}}$ of minimal length that is given by

$$\bar{\mathbf{x}} = A^+\mathbf{b}$$

Proof Let A be an $m \times n$ matrix of rank r with SVD $A = U\Sigma V^T$ (so that $A^+ = V\Sigma^+U^T$). Let $\mathbf{y} = V^T\mathbf{x}$ and let $\mathbf{c} = U^T\mathbf{b}$. Write $\mathbf{y}$ and $\mathbf{c}$ in block form as

$$\mathbf{y} = \begin{bmatrix} \mathbf{y}_1 \\ \mathbf{y}_2 \end{bmatrix} \quad \text{and} \quad \mathbf{c} = \begin{bmatrix} \mathbf{c}_1 \\ \mathbf{c}_2 \end{bmatrix}$$

where $\mathbf{y}_1$ and $\mathbf{c}_1$ are in $\mathbb{R}^r$.

We wish to minimize $\|\mathbf{b} - A\mathbf{x}\|$ or, equivalently, $\|\mathbf{b} - A\mathbf{x}\|^2$. Using Theorem 5.6 and the fact that U^T is orthogonal (because U is), we have

$$\begin{aligned} \|\mathbf{b} - A\mathbf{x}\|^2 &= \|U^T(\mathbf{b} - A\mathbf{x})\|^2 = \|U^T(\mathbf{b} - U\Sigma V^T\mathbf{x})\|^2 = \|U^T\mathbf{b} - U^TU\Sigma V^T\mathbf{x}\|^2 \\ &= \|\mathbf{c} - \Sigma\mathbf{y}\|^2 = \left\| \begin{bmatrix} \mathbf{c}_1 \\ \mathbf{c}_2 \end{bmatrix} - \begin{bmatrix} D & O \\ O & O \end{bmatrix} \begin{bmatrix} \mathbf{y}_1 \\ \mathbf{y}_2 \end{bmatrix} \right\|^2 = \left\| \begin{bmatrix} \mathbf{c}_1 - D\mathbf{y}_1 \\ \mathbf{c}_2 \end{bmatrix} \right\|^2 \end{aligned}$$

The only part of this expression that we have any control over is $\mathbf{y}_1$, so the minimum value occurs when $\mathbf{c}_1 - D\mathbf{y}_1 = \mathbf{0}$ or, equivalently, when $\mathbf{y}_1 = D^{-1}\mathbf{c}_1$. So all least squares solutions $\mathbf{x}$ are of the form

$$\mathbf{x} = V\mathbf{y} = V\begin{bmatrix} D^{-1}\mathbf{c}_1 \\ \mathbf{y}_2 \end{bmatrix}$$

Set

$$\bar{\mathbf{x}} = V\bar{\mathbf{y}} = V\begin{bmatrix} D^{-1}\mathbf{c}_1 \\ \mathbf{0} \end{bmatrix}$$

We claim that this $\bar{\mathbf{x}}$ is the least squares solution of minimal length. To show this, let's suppose that

$$\mathbf{x}' = V\mathbf{y}' = V\begin{bmatrix} D^{-1}\mathbf{c}_1 \\ \mathbf{y}_2 \end{bmatrix}$$

is a different least squares solution (hence, $\mathbf{y}_2 \neq \mathbf{0}$). Then

$$\|\bar{\mathbf{x}}\| = \|V\bar{\mathbf{y}}\| = \|\bar{\mathbf{y}}\| < \|\mathbf{y}'\| = \|V\mathbf{y}'\| = \|\mathbf{x}'\|$$

as claimed.

We still must show that $\bar{\mathbf{x}}$ is equal to $A^+\mathbf{b}$. To do so, we simply compute

$$\begin{aligned} \bar{\mathbf{x}} = V\bar{\mathbf{y}} &= V\begin{bmatrix} D^{-1}\mathbf{c}_1 \\ \mathbf{0} \end{bmatrix} = V\begin{bmatrix} D^{-1} & O \\ O & O \end{bmatrix}\begin{bmatrix} \mathbf{c}_1 \\ \mathbf{c}_2 \end{bmatrix} \\ &= V\Sigma^+\mathbf{c} = V\Sigma^+U^T\mathbf{b} = A^+\mathbf{b} \end{aligned}$$

Example 7.40 Find the minimum length least squares solution of $A\mathbf{x} = \mathbf{b}$, where

$$A = \begin{bmatrix} 1 & 1 \\ 1 & 1 \end{bmatrix} \quad \text{and} \quad \mathbf{b} = \begin{bmatrix} 0 \\ 1 \end{bmatrix}$$

Solution The corresponding equations

$$x + y = 0$$
$$x + y = 1$$

are clearly inconsistent, so a least squares solution is our only hope. Moreover, the columns of A are linearly dependent, so there will be infinitely many least squares solutions—among which we want the one with minimal length.

An SVD of A is given by

$$A = \begin{bmatrix} 1 & 1 \\ 1 & 1 \end{bmatrix} = \begin{bmatrix} 1/\sqrt{2} & 1/\sqrt{2} \\ 1/\sqrt{2} & -1/\sqrt{2} \end{bmatrix} \begin{bmatrix} 2 & 0 \\ 0 & 0 \end{bmatrix} \begin{bmatrix} 1/\sqrt{2} & 1/\sqrt{2} \\ 1/\sqrt{2} & -1/\sqrt{2} \end{bmatrix} = U\Sigma V^T$$

(Verify this.) It follows that

$$A^+ = V\Sigma^+U^T = \begin{bmatrix} 1/\sqrt{2} & 1/\sqrt{2} \\ 1/\sqrt{2} & -1/\sqrt{2} \end{bmatrix} \begin{bmatrix} 1/2 & 0 \\ 0 & 0 \end{bmatrix} \begin{bmatrix} 1/\sqrt{2} & 1/\sqrt{2} \\ 1/\sqrt{2} & -1/\sqrt{2} \end{bmatrix} = \begin{bmatrix} 1/4 & 1/4 \\ 1/4 & 1/4 \end{bmatrix}$$

so
$$\overline{\mathbf{x}} = A^+\mathbf{b} = \begin{bmatrix} \frac{1}{4} & \frac{1}{4} \\ \frac{1}{4} & \frac{1}{4} \end{bmatrix} \begin{bmatrix} 0 \\ 1 \end{bmatrix} = \begin{bmatrix} \frac{1}{4} \\ \frac{1}{4} \end{bmatrix}$$

You can see that the minimum least squares solution in Example 7.40 satisfies $x + y = \frac{1}{2}$. In a sense, this is a compromise between the two equations we started with. In Exercise 49, you are asked to solve the normal equations for this problem directly and to verify that this solution really is the one closest to the origin.

The Fundamental Theorem of Invertible Matrices It is appropriate to conclude by revisiting the Fundamental Theorem of Invertible Matrices one more time. Not surprisingly, the singular values of a square matrix tell us when the matrix is invertible.

Theorem 7.19 **The Fundamental Theorem of Invertible Matrices: Final Version**

Let A be an $n \times n$ matrix and let $T : V \to W$ be a linear transformation whose matrix $[T]_{\mathcal{C} \leftarrow \mathcal{B}}$ with respect to bases $\mathcal{B}$ and $\mathcal{C}$ of V and W, respectively, is A. The following statements are equivalent:

a. A is invertible.
b. $A\mathbf{x} = \mathbf{b}$ has a unique solution for every $\mathbf{b}$ in $\mathbb{R}^n$.
c. $A\mathbf{x} = \mathbf{0}$ has only the trivial solution.
d. The reduced row echelon form of A is I_n.
e. A is a product of elementary matrices.
f. $\text{rank}(A) = n$
g. $\text{nullity}(A) = 0$
h. The column vectors of A are linearly independent.
i. The column vectors of A span $\mathbb{R}^n$.
j. The column vectors of A form a basis for $\mathbb{R}^n$.
k. The row vectors of A are linearly independent.
l. The row vectors of A span $\mathbb{R}^n$.

m. The row vectors of A form a basis for $\mathbb{R}^n$.
n. $\det A \neq 0$
o. 0 is not an eigenvalue of A.
p. T is invertible.
q. T is one-to-one.
r. T is onto.
s. $\ker(T) = \{\mathbf{0}\}$
t. $\text{range}(T) = W$
u. 0 is not a singular value of A.

Proof First note that, by the definition of singular values, 0 is a singular value of A if and only if 0 is an eigenvalue of A^TA.

(a) $\Rightarrow$ (u) If A is invertible, so is A^T, and hence A^TA is as well. Therefore, property (o) implies that 0 is not an eigenvalue of A^TA, so 0 is not a singular value of A.

(u) $\Rightarrow$ (a) If 0 is not a singular value of A, then 0 is not an eigenvalue of A^TA. Therefore, A^TA is invertible, by the equivalence of properties (a) and (o). But then $\text{rank}(A) = n$, by Theorem 3.28, so A is invertible, by the equivalence of properties (a) and (f).

Vignette

Digital Image Compression

Among the many applications of the SVD, one of the most impressive is its use in compressing digital images so that they can be efficiently transmitted electronically (by satellite, fax, Internet, or the like). We have already discussed the problem of detecting and correcting errors in such transmissions. The problem we now wish to consider has to do with reducing the amount of information that has to be transmitted, without losing any essential information.

In the case of digital images, let's suppose we have a grayscale picture that is 340×280 pixels in size. Each pixel is one of 256 shades of gray, which we can represent by a number between 0 and 255. We can store this information in a 340×280 matrix A, but transmitting and manipulating these 95,200 numbers is very expensive. The idea behind image compression is that some parts of the picture are less interesting than others. For example, in a photograph of someone standing outside, there may be a lot of sky in the background, while the person's face contains a lot of detail. We can probably get away with transmitting every second or third pixel in the background, but we would like to keep all the pixels in the region of the face.

It turns out that the small singular values in the SVD of the matrix A come from the "boring" parts of the image, and we can ignore many of them. Suppose, then, that we have the SVD of A in outer product form

$$A = \sigma_1 \mathbf{u}_1 \mathbf{v}_1^T + \cdots + \sigma_r \mathbf{u}_r \mathbf{v}_r^T$$

Let $k \leq r$ and define $$A_k = \sigma_1 \mathbf{u}_1 \mathbf{v}_1^T + \cdots + \sigma_k \mathbf{u}_k \mathbf{v}_k^T$$

Then A_k is an approximation to A that corresponds to keeping only the first k singular values and the corresponding singular vectors. For our 340×280 example, we may discover that it is enough to transmit only the data corresponding to the first 20 singular values. Then, instead of transmitting 95,200 numbers, we need only send 20 singular values plus the 20 vectors $\mathbf{u}_1, \ldots, \mathbf{u}_{20}$ in $\mathbb{R}^{340}$ and the 20 vectors $\mathbf{v}_1, \ldots, \mathbf{v}_{20}$ in $\mathbb{R}^{280}$, for a total of

$$20 + 20 \cdot 340 + 20 \cdot 280 = 12{,}420$$

numbers. This represents a substantial saving!

Bettmann/Corbis

Figure 7.22

The picture of the mathematician Gauss in Figure 7.22 is a 340×280 pixel image. It has 256 shades of gray, so the corresponding matrix A is 340×280, with entries between 0 and 255.

It turns out that the matrix A has rank 280. If we approximate A by A_k, as described above, we get an image that corresponds to the first k singular values of A. Figure 7.23 shows several of these images for values of k from 2 to 256. At first, the image is very blurry, but fairly quickly it takes shape. Notice that A_{32} already gives a pretty good approximation to the actual image (which comes from $A = A_{280}$, as shown in the upper left-hand corner of Figure 7.23).

Some of the singular values of A are $\sigma_1 = 49{,}096$, $\sigma_{16} = 22{,}589$, $\sigma_{32} = 10{,}187$, $\sigma_{64} = 484$, $\sigma_{128} = 182$, $\sigma_{256} = 5$, and $\sigma_{280} = 0.5$. The smaller singular values contribute very little to the image, which is why the approximations quickly look so close to the original.

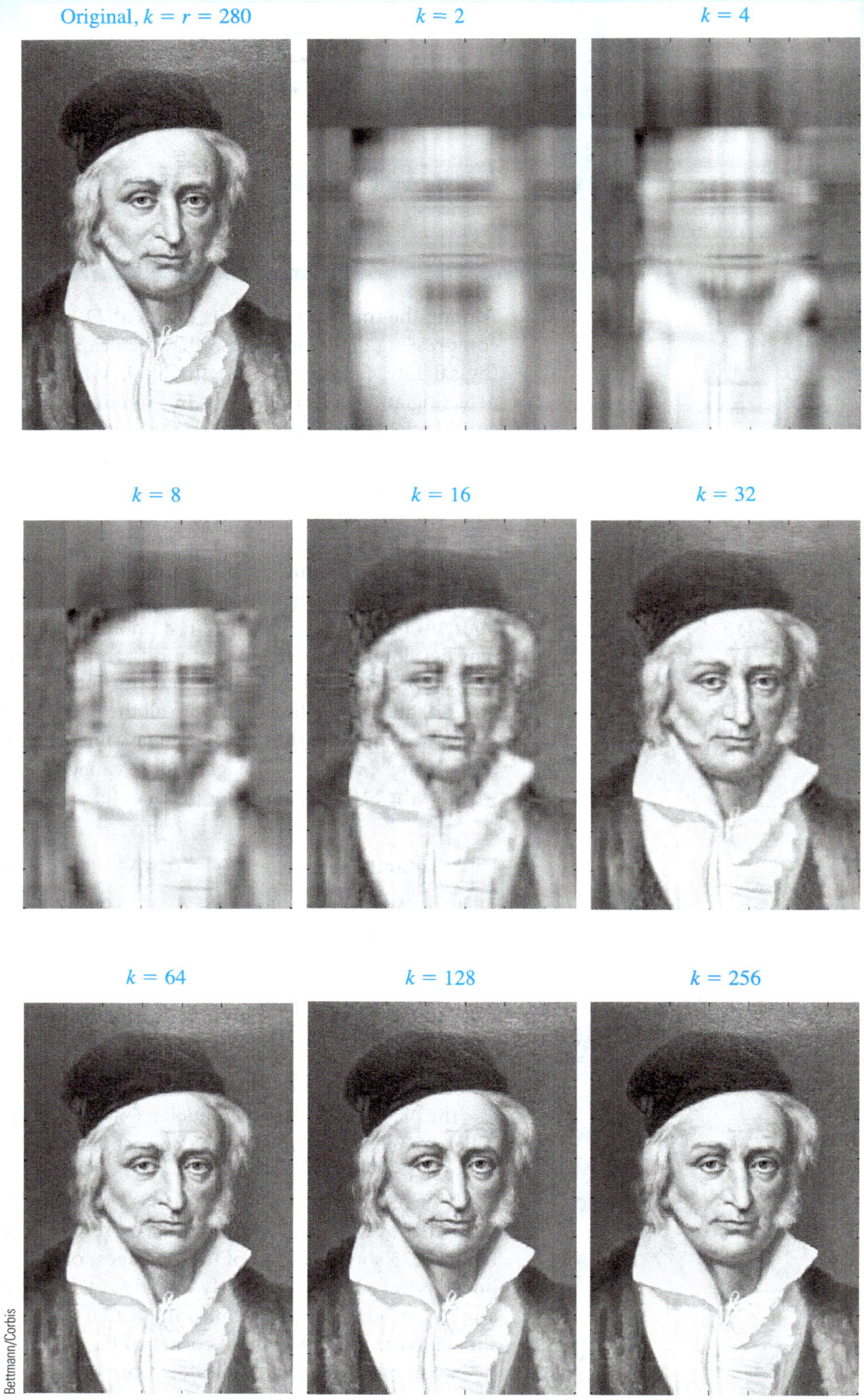
Original, $k = r = 280$
$k = 2$
$k = 4$
$k = 8$
$k = 16$
$k = 32$
$k = 64$
$k = 128$
$k = 256$
Bettmann/Corbis

Figure 7.23

Exercises 7.4

In Exercises 1–10, find the singular values of the given matrix.

1. $A = \begin{bmatrix} 2 & 0 \\ 0 & 3 \end{bmatrix}$ **2.** $A = \begin{bmatrix} 2 & 1 \\ 1 & 2 \end{bmatrix}$

3. $A = \begin{bmatrix} 1 & 1 \\ 0 & 0 \end{bmatrix}$ **4.** $A = \begin{bmatrix} \sqrt{2} & 1 \\ 0 & \sqrt{2} \end{bmatrix}$

5. $A = \begin{bmatrix} 3 \\ 4 \end{bmatrix}$ **6.** $A = \begin{bmatrix} 3 & 4 \end{bmatrix}$

7. $A = \begin{bmatrix} 0 & 0 \\ 0 & 3 \\ -2 & 0 \end{bmatrix}$ **8.** $A = \begin{bmatrix} 0 & 1 \\ 1 & 0 \\ 2 & -2 \end{bmatrix}$

9. $A = \begin{bmatrix} 2 & 0 & 1 \\ 0 & 2 & 0 \end{bmatrix}$ **10.** $A = \begin{bmatrix} 1 & 0 & 1 \\ 0 & -3 & 0 \\ 1 & 0 & 1 \end{bmatrix}$

In Exercises 11–20, find an SVD of the indicated matrix.

11. A in Exercise 3 **12.** $A = \begin{bmatrix} 1 & 1 \\ 1 & 1 \end{bmatrix}$

13. $A = \begin{bmatrix} 0 & -2 \\ -3 & 0 \end{bmatrix}$ **14.** $A = \begin{bmatrix} 1 & -1 \\ 1 & 1 \end{bmatrix}$

15. A in Exercise 5 **16.** A in Exercise 6

17. A in Exercise 7 **18.** A in Exercise 8

19. A in Exercise 9 **20.** $A = \begin{bmatrix} 1 & 1 & 1 \\ 1 & 1 & 1 \end{bmatrix}$

In Exercises 21–24, find the outer product form of the SVD for the matrix in the given exercises.

21. Exercises 3 and 11 **22.** Exercise 14

23. Exercises 7 and 17 **24.** Exercises 9 and 19

25. Show that the matrices U and V in the SVD are not uniquely determined. [*Hint:* Find an example in which it would be possible to make different choices in the construction of these matrices.]

26. Let A be a symmetric matrix. Show that the singular values of A are:

(a) the absolute values of the eigenvalues of A.

(b) the eigenvalues of A if A is positive definite.

27. **(a)** Show that, for a positive definite, symmetric matrix A, Theorem 7.13 gives the orthogonal diagonalization of A, as guaranteed by the Spectral Theorem.

(b) Show that, for a positive definite, symmetric matrix A, Theorem 7.14 gives the spectral decomposition of A.

28. If A is an invertible matrix with SVD $A = U\Sigma V^T$, show that Σ is invertible and that $A^{-1} = V\Sigma^{-1}U^T$ is an SVD of A^{-1}.

29. Show that if $A = U\Sigma V^T$ is an SVD of A, then the left singular vectors are eigenvectors of AA^T.

30. Show that A and A^T have the same singular values.

31. Let Q be an orthogonal matrix such that QA makes sense. Show that A and QA have the same singular values.

32. Prove Theorem 7.15(d).

33. What is the image of the unit circle in $\mathbb{R}^2$ under the action of the matrix in Exercise 3?

34. What is the image of the unit circle in $\mathbb{R}^2$ under the action of the matrix in Exercise 7?

35. What is the image of the unit sphere in $\mathbb{R}^3$ under the action of the matrix in Exercise 9?

36. What is the image of the unit sphere in $\mathbb{R}^3$ under the action of the matrix in Exercise 10?

In Exercises 37–40, compute (a) $\|A\|_2$ and (b) $\text{cond}_2(A)$ for the indicated matrix.

37. A in Exercise 3 **38.** A in Exercise 8

39. $A = \begin{bmatrix} 1 & 0.9 \\ 1 & 1 \end{bmatrix}$ **40.** $A = \begin{bmatrix} 10 & 10 & 0 \\ 100 & 100 & 1 \end{bmatrix}$

In Exercises 41–44, compute the pseudoinverse A^+ of A in the given exercise.

41. Exercise 3 **42.** Exercise 8

43. Exercise 9 **44.** Exercise 10

In Exercises 45–48, find A^+ and use it to compute the minimal length least squares solution to $A\mathbf{x} = \mathbf{b}$.

45. $A = \begin{bmatrix} 1 & 2 \\ 2 & 4 \end{bmatrix}, \mathbf{b} = \begin{bmatrix} 3 \\ 5 \end{bmatrix}$

46. $A = \begin{bmatrix} 3 & 0 & 0 \\ 0 & 0 & 2 \end{bmatrix}, \mathbf{b} = \begin{bmatrix} 3 \\ 0 \end{bmatrix}$

47. $A = \begin{bmatrix} 1 & 1 \\ 1 & 1 \\ 1 & 1 \end{bmatrix}, \mathbf{b} = \begin{bmatrix} 1 \\ 2 \\ 3 \end{bmatrix}$

48. $A = \begin{bmatrix} 1 & 0 & 1 \\ 0 & 1 & 0 \\ 1 & 0 & 1 \end{bmatrix}, \mathbf{b} = \begin{bmatrix} 1 \\ 1 \\ 1 \end{bmatrix}$

49. (a) Set up and solve the normal equations for the system of equations in Example 7.40.

(b) Find a parametric expression for the length of a solution vector in part (a).

(c) Find the solution vector of minimal length and verify that it is the one produced by the method of Example 7.40. [*Hint:* Recall how to find the coordinates of the vertex of a parabola.]

50. Verify that when A has linearly independent columns, the definitions of pseudoinverse in this section and in Section 7.3 are the same.

51. Verify that the pseudoinverse (as defined in this section) satisfies the Penrose conditions for A (Theorem 7.12 in Section 7.3).

52. Show that A^+ is the *only* matrix that satisfies the Penrose conditions for A. To do this, assume that A' is a matrix satisfying the Penrose conditions: (a) $AA'A = A$, (b) $A'AA' = A'$, and (c) AA' and $A'A$ are symmetric. Prove that $A' = A^+$. [*Hint:* Use the Penrose conditions for A^+ and A' to show that $A^+ = A'AA^+$ and $A' = A'AA^+$. It is helpful to note that condition (c) can be written as $AA' = (A')^TA^T$ and $A'A = A^T(A')^T$, with similar versions for A^+.]

53. Show that $(A^+)^+ = A$. [*Hint:* Show that A satisfies the Penrose conditions for A^+. By Exercise 52, A must therefore be $(A^+)^+$.]

54. Show that $(A^+)^T = (A^T)^+$. [*Hint:* Show that $(A^+)^T$ satisfies the Penrose conditions for A^T. By Exercise 52, $(A^+)^T$ must therefore be $(A^T)^+$.]

55. Show that if A is a symmetric, idempotent matrix, then $A^+ = A$.

56. Let Q be an orthogonal matrix such that QA makes sense. Show that $(QA)^+ = A^+Q^T$.

57. Prove that if A is a positive definite matrix with SVD $A = U\Sigma V^T$, then $U = V$.

58. Prove that for a diagonal matrix, the 1-, 2-, and ∞-norms are the same.

59. Prove that for any square matrix A, $\|A\|_2^2 \le \|A\|_1\|A\|_\infty$. [*Hint:* $\|A\|_2^2$ is the square of the largest singular value of A and hence is equal to the largest eigenvalue of A^TA. Now use Exercise 34 in Section 7.2.]

Every complex number can be written in polar form as $z = re^{i\theta}$, where $r = |z|$ is a nonnegative real number and θ is its argument, with $|e^{i\theta}| = 1$. (See Appendix C.) Thus, z has been decomposed into a stretching factor r and a rotation factor $e^{i\theta}$. There is an analogous decomposition $A = RQ$ for square matrices, called the ***polar decomposition.***

60. Show that every square matrix A can be factored as $A = RQ$, where R is symmetric, positive semidefinite and Q is orthogonal. [*Hint:* Show that the SVD can be rewritten to give

$$A = U\Sigma V^T = U\Sigma(U^TU)V^T = (U\Sigma U^T)(UV^T)$$

Then show that $R = U\Sigma U^T$ and $Q = UV^T$ have the right properties.]

Find a polar decomposition of the matrices in Exercises 61–64.

61. A in Exercise 3

62. A in Exercise 14

63. $A = \begin{bmatrix} 1 & 2 \\ -3 & -1 \end{bmatrix}$

64. $A = \begin{bmatrix} 4 & 2 & -3 \\ -2 & 2 & 6 \\ 4 & -1 & 6 \end{bmatrix}$

7.5 Applications

Approximation of Functions

In many applications, it is necessary to approximate a given function by a "nicer" function. For example, we might want to approximate $f(x) = e^x$ by a linear function $g(x) = c + dx$ on some interval $[a, b]$. In this case, we have a continuous function f, and we want to approximate it as closely as possible on the interval $[a, b]$

by a function g in the subspace $\mathscr{P}_1$. The general problem can be phrased as follows:

> Given a continuous function f on an interval $[a, b]$ and a subspace W of $\mathscr{C}[a, b]$, find the function "closest" to f in W.

The problem is analogous to the least squares fitting of data points, except now we have infinitely many data points—namely, the points on the graph of the function f. What should "approximate" mean in this context? Once again, the Best Approximation Theorem holds the answer.

The given function f lives in the vector space $\mathscr{C}[a, b]$ of continuous functions on the interval $[a, b]$. This is an inner product space, with inner product

$$\langle f, g\rangle = \int_a^b f(x)g(x)\,dx$$

If W is a finite-dimensional subspace of $\mathscr{C}[a, b]$, then the best approximation to f in W is given by the projection of f onto W, by Theorem 7.8. Furthermore, if $\{\mathbf{u}_1, \ldots, \mathbf{u}_k\}$ is an orthogonal basis for W, then

$$\text{proj}_W(f) = \frac{\langle \mathbf{u}_1, f\rangle}{\langle \mathbf{u}_1, \mathbf{u}_1\rangle}\mathbf{u}_1 + \cdots + \frac{\langle \mathbf{u}_k, f\rangle}{\langle \mathbf{u}_k, \mathbf{u}_k\rangle}\mathbf{u}_k$$

Example 7.41

Find the best linear approximation to $f(x) = e^x$ on the interval $[-1, 1]$.

Solution Linear functions are polynomials of degree 1, so we use the subspace $W = \mathscr{P}_1[-1, 1]$ of $\mathscr{C}[-1, 1]$ with the inner product

$$\langle f, g\rangle = \int_{-1}^1 f(x)g(x)\,dx$$

A basis for $\mathscr{P}_1[-1, 1]$ is given by $\{1, x\}$. Since

$$\langle 1, x\rangle = \int_{-1}^1 x\,dx = 0$$

this is an orthogonal basis, so the best approximation to f in W is

$$\begin{aligned} g(x) = \text{proj}_W(e^x) &= \frac{\langle 1, e^x\rangle}{\langle 1, 1\rangle}1 + \frac{\langle x, e^x\rangle}{\langle x, x\rangle}x \\ &= \frac{\int_{-1}^1 (1\cdot e^x)\,dx}{\int_{-1}^1 (1\cdot 1)\,dx} + \frac{\int_{-1}^1 xe^x\,dx}{\int_{-1}^1 x^2\,dx}x \\ &= \frac{e - e^{-1}}{2} + \frac{2e^{-1}}{\frac{2}{3}}x \\ &= \tfrac{1}{2}(e - e^{-1}) + 3e^{-1}x \approx 1.18 + 1.10x \end{aligned}$$

where we have used integration by parts to evaluate $\int_{-1}^{1} xe^x\,dx$. (Check these calculations.) See Figure 7.24.

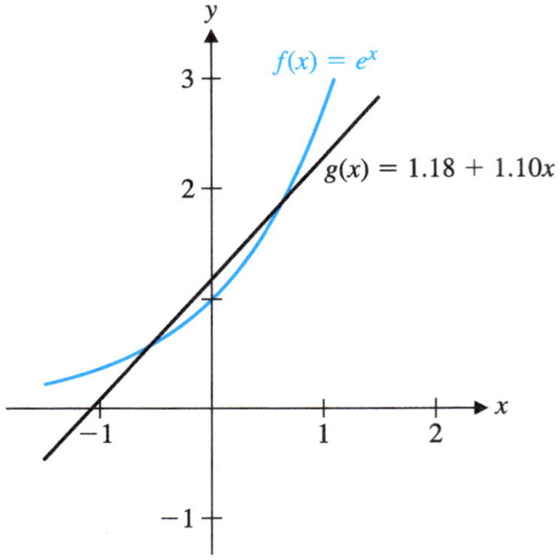

Figure 7.24

The error in approximating f by g is the one specified by the Best Approximation Theorem: the distance $\|f - g\|$ between f and g relative to the inner product on $\mathscr{C}[-1, 1]$. This error is just

$$\|f - g\| = \sqrt{\int_{-1}^{1} (f(x) - g(x))^2\,dx}$$

and is often called the ***root mean square error***. With the aid of a CAS, we find that the root mean square error in Example 7.41 is

$$\|e^x - (\tfrac{1}{2}(e - e^{-1}) + 3e^{-1}x)\| = \sqrt{\int_{-1}^{1} (e^x - \tfrac{1}{2}(e - e^{-1}) - 3e^{-1}x)^2\,dx} \approx 0.23$$

Remark The root mean square error can be thought of as analogous to the area between the graphs of f and g on the specified interval. Recall that the area between the graphs of f and g on the interval $[a, b]$ is given by

$$\int_a^b |f(x) - g(x)|\,dx$$

(See Figure 7.25.)

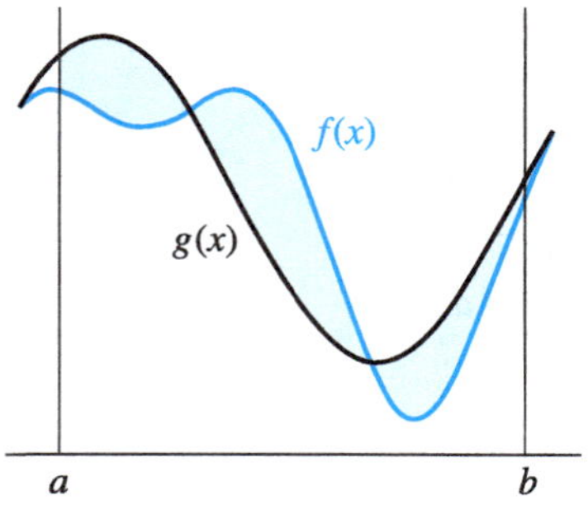

Figure 7.25

Although the equation in the above Remark is a sensible measure of the "error" between f and g, the absolute value sign makes it hard to work with. The root mean square error is easier to use and therefore preferable. The square root is necessary to "compensate" for the squaring and to keep the unit of measurement the same as it would be for the area between the curves. For comparison purposes, the area between the graphs of f and g in Example 7.41 is

$$\int_{-1}^{1} |e^x - \tfrac{1}{2}(e - e^{-1}) - 3e^{-1}x|\,dx \approx 0.28$$

Example 4.30 Find the best quadratic approximation to $f(x) = e^x$ on the interval $[-1, 1]$.

Solution A quadratic function is a polynomial of the form $g(x) = a + bx + cx^2$ in $W = \mathscr{P}_2[-1, 1]$. This time, the standard basis $\{1, x, x^2\}$ is not orthogonal. However, we can construct an orthogonal basis using the Gram-Schmidt Process, as we did in Example 7.8. The result is the set of Legendre polynomials

$$\{1, x, x^2 - \tfrac{1}{3}\}$$

Using this set as our basis, we compute the best approximation to f in W as $g(x) = \text{proj}_W(e^x)$. The linear terms in this calculation are exactly as in Example 7.41, so we only require the additional calculations

$$\langle x^2 - \tfrac{1}{3}, e^x\rangle = \int_{-1}^{1} (x^2 - \tfrac{1}{3})e^x\,dx = \int_{-1}^{1} x^2e^x\,dx - \tfrac{1}{3}\int_{-1}^{1} e^x\,dx = \tfrac{2}{3}(e - 7e^{-1})$$

and $$\langle x^2 - \tfrac{1}{3}, x^2 - \tfrac{1}{3}\rangle = \int_{-1}^{1} (x^2 - \tfrac{1}{3})^2\,dx = \int_{-1}^{1} (x^4 - \tfrac{2}{3}x^2 + \tfrac{1}{9})\,dx = \tfrac{8}{45}$$

Then the best quadratic approximation to $f(x) = e^x$ on the interval $[-1, 1]$ is

$$\begin{aligned} g(x) = \text{proj}_W(e^x) &= \frac{\langle 1, e^x\rangle}{\langle 1, 1\rangle}1 + \frac{\langle x, e^x\rangle}{\langle x, x\rangle}x + \frac{\langle x^2 - \frac{1}{3}, e^x\rangle}{\langle x^2 - \frac{1}{3}, x^2 - \frac{1}{3}\rangle}(x^2 - \tfrac{1}{3}) \\ &= \tfrac{1}{2}(e - e^{-1}) + 3e^{-1}x + \frac{\frac{2}{3}(e - 7e^{-1})}{\frac{8}{45}}(x^2 - \tfrac{1}{3}) \\ &= \frac{3(11e^{-1} - e)}{4} + 3e^{-1}x + \frac{15(e - 7e^{-1})}{4}x^2 \approx 1.00 + 1.10x + 0.54x^2 \end{aligned}$$

(See Figure 7.26.)

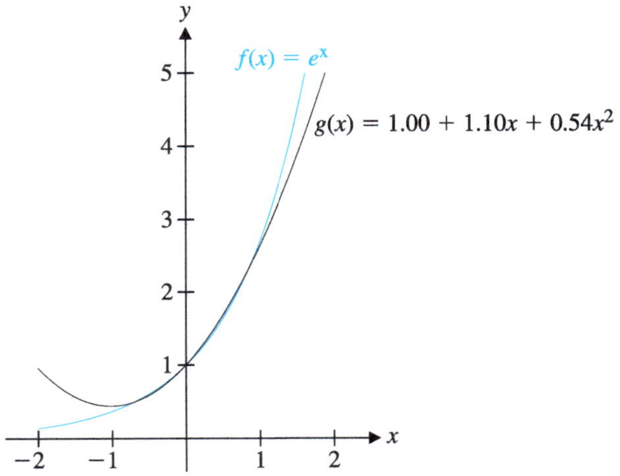

Figure 7.26

Notice how much better the quadratic approximation in Example 7.42 is than the linear approximation in Example 7.41. It turns out that, in the quadratic case, the root mean square error is

$$\|e^x - g(x)\| = \sqrt{\int_{-1}^{1} (e^x - g(x))^2\, dx} \approx 0.04$$

In general, the higher the degree of the approximating polynomial, the smaller the error and the better the approximation.

In many applications, functions are approximated by combinations of sine and cosine functions. This method is particularly useful if the function being approximated displays periodic or almost periodic behavior (such as that of a sound wave, an electrical impulse, or the motion of a vibrating system). A function of the form

$$\begin{aligned} p(x) = a_0 + a_1 \cos x + a_2 \cos 2x + \cdots + a_n \cos nx + b_1 \sin x \\ + b_2 \sin 2x + \cdots + b_n \sin nx \end{aligned} \tag{1}$$

is called a ***trigonometric polynomial;*** if a_n and b_n are not both zero, then $p(x)$ is said to have ***order n.*** For example,

$$p(x) = 3 - \cos x + \sin 2x + 4 \sin 3x$$

is a trigonometric polynomial of order 3.

Let's restrict our attention to the vector space $\mathscr{C}[-\pi, \pi]$ with the inner product

$$\langle f, g \rangle = \int_{-\pi}^{\pi} f(x)g(x)\, dx$$

The trigonometric polynomials of the form in Equation (1) are linear combinations of the set

$$\mathcal{B} = \{1, \cos x, \ldots, \cos nx, \sin x, \ldots, \sin nx\}$$

The best approximation to a function f in $\mathscr{C}[-\pi, \pi]$ by a trigonometric polynomial of order n will therefore be $\text{proj}_W(f)$, where $W = \text{span}(\mathcal{B})$. It turns out that $\mathcal{B}$ is an orthogonal set and, hence, a basis for W. Verification of this fact involves showing that any two distinct functions in $\mathcal{B}$ are orthogonal with respect to the given inner product. Example 7.43 presents some of the necessary calculations; you are asked to provide the remaining ones in Exercises 17–19.

Example 7.43

Show that $\sin jx$ is orthogonal to $\cos kx$ in $\mathscr{C}[-\pi, \pi]$ for $j, k \geq 1$.

Solution Using a trigonometric identity, we compute as follows: If $j \neq k$, then

$$\begin{aligned} \int_{-\pi}^{\pi} \sin jx \cos kx\, dx &= \tfrac{1}{2}\int_{-\pi}^{\pi} [\sin(j + k)x + \sin(j - k)x]\, dx \\ &= -\tfrac{1}{2}\left[\frac{\cos(j + k)x}{j + k} + \frac{\cos(j - k)x}{j - k}\right]_{-\pi}^{\pi} \\ &= 0 \end{aligned}$$

since the cosine function is periodic with period 2π.

If $j = k$, then

$$\int_{-\pi}^{\pi} \sin kx \cos kx \, dx = \frac{1}{2k}[\sin^2 kx]_{-\pi}^{\pi} = 0$$

since $\sin k\pi = 0$ for any integer k.

In order to find the orthogonal projection of a function f in $\mathscr{C}[-\pi, \pi]$ onto the subspace W spanned by the orthogonal basis $\mathcal{B}$, we need to know the squares of the norms of the basis vectors. For example, using a half-angle formula, we have

$$\begin{aligned} \langle \sin kx, \sin kx \rangle &= \int_{-\pi}^{\pi} \sin^2 kx \, dx \\ &= \tfrac{1}{2}\int_{-\pi}^{\pi} (1 - \cos 2kx) \, dx \\ &= \tfrac{1}{2}\left[x - \frac{\sin 2kx}{2k} \right]_{-\pi}^{\pi} \\ &= \pi \end{aligned}$$

In Exercise 20, you are asked to show that $\langle \cos kx, \cos kx \rangle = \pi$ and $\langle 1, 1 \rangle = 2\pi$.

We now have

$$\text{proj}_W(f) = a_0 + a_1 \cos x + \cdots + a_n \cos nx + b_1 \sin x + \cdots + b_n \sin nx \qquad (2)$$

where

$$\begin{aligned} a_0 &= \frac{\langle 1, f \rangle}{\langle 1, 1 \rangle} = \frac{1}{2\pi}\int_{-\pi}^{\pi} f(x) \, dx \\ a_k &= \frac{\langle \cos kx, f \rangle}{\langle \cos kx, \cos kx \rangle} = \frac{1}{\pi}\int_{-\pi}^{\pi} f(x) \cos kx \, dx \\ b_k &= \frac{\langle \sin kx, f \rangle}{\langle \sin kx, \sin kx \rangle} = \frac{1}{\pi}\int_{-\pi}^{\pi} f(x) \sin kx \, dx \end{aligned} \qquad (3)$$

for $k \geq 1$. The approximation to f given by Equations (2) and (3) is called the ***nth-order Fourier approximation*** to f on $[-\pi, \pi]$. The coefficients $a_0, a_1, \ldots, a_n, b_1, \ldots, b_n$ are called the ***Fourier coefficients*** of f.

Example 7.44

Find the fourth-order Fourier approximation to $f(x) = x$ on $[-\pi, \pi]$.

Solution Using formulas (3), we obtain

$$a_0 = \frac{1}{2\pi}\int_{-\pi}^{\pi} x \, dx = \frac{1}{2\pi}\left[\frac{x^2}{2}\right]_{-\pi}^{\pi} = 0$$

and for $k \geq 1$, integration by parts yields

$$a_k = \frac{1}{\pi}\int_{-\pi}^{\pi} x \cos kx \, dx = \frac{1}{\pi}\left[\frac{x}{k}\sin kx + \frac{1}{k^2}\cos kx\right]_{-\pi}^{\pi} = 0$$

Jean-Baptiste Joseph Fourier (1768–1830) was a French mathematician and physicist who gained prominence through his investigation into the theory of heat. In his landmark solution of the so-called heat equation, he introduced techniques related to what are now known as Fourier series, a tool widely used in many branches of mathematics, physics, and engineering. Fourier was a political activist during the French revolution and became a favorite of Napoleon, accompanying him on his Egyptian campaign in 1798. Later Napoleon appointed Fourier Prefect of Isère, where he oversaw many important engineering projects. In 1808, Fourier was made a baron. He is commemorated by a plaque on the Eiffel Tower.

and

$$\begin{aligned} b_k &= \frac{1}{\pi}\int_{-\pi}^{\pi} x \sin kx\, dx = \frac{1}{\pi}\left[-\frac{x}{k}\cos kx + \frac{1}{k^2}\sin kx\right]_{-\pi}^{\pi} \\ &= \frac{1}{\pi}\left[\frac{-\pi\cos k\pi - \pi\cos(-k\pi)}{k}\right] \\ &= \begin{cases} -\dfrac{2}{k} & \text{if } k \text{ is even} \\ \dfrac{2}{k} & \text{if } k \text{ is odd} \end{cases} \\ &= \frac{2(-1)^{k+1}}{k} \end{aligned}$$

It follows that the fourth-order Fourier approximation to $f(x) = x$ on $[-\pi, \pi]$ is

$$2(\sin x - \tfrac{1}{2}\sin 2x + \tfrac{1}{3}\sin 3x - \tfrac{1}{4}\sin 4x)$$

Figure 7.27 shows the first four Fourier approximations to $f(x) = x$ on $[-\pi, \pi]$.

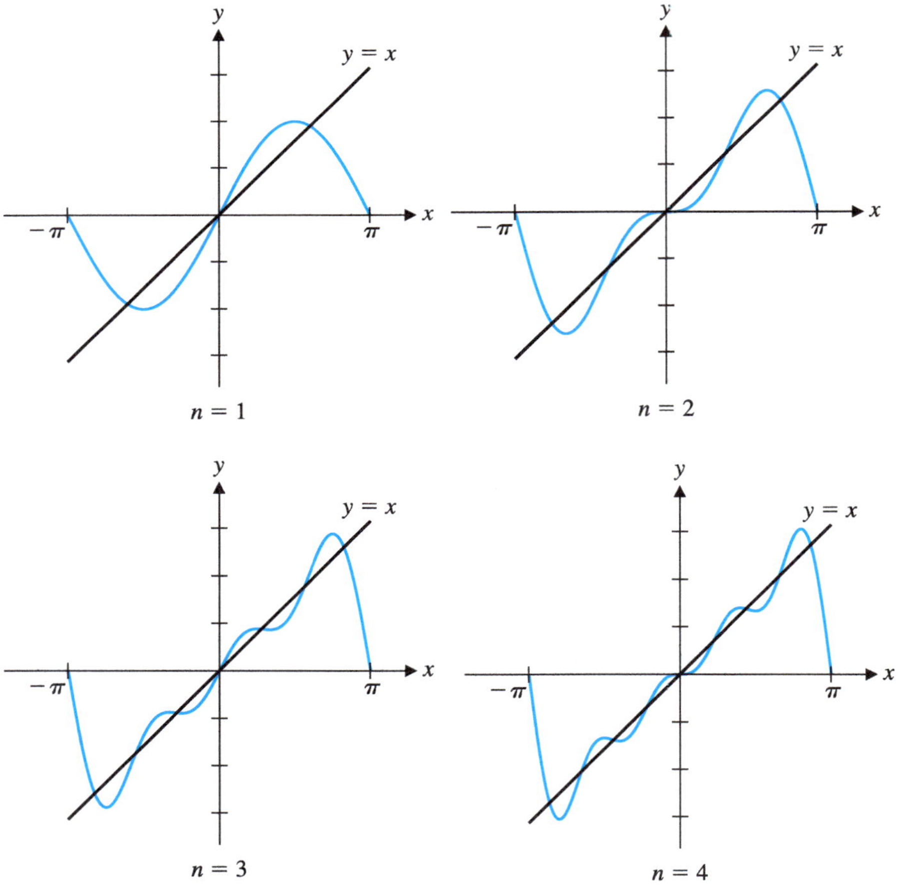

Figure 7.27

You can clearly see the approximations in Figure 7.27 improving, a fact that can be confirmed by computing the root mean square error in each case. As the order of the Fourier approximation increases, it can be shown that this error approaches zero. The trigonometric polynomial then becomes an *infinite series,* and we write

$$f(x) = a_0 + \sum_{k=1}^{\infty}(a_k \cos kx + b_k \sin kx)$$

This is called the ***Fourier series*** of f on $[-\pi, \pi]$.

Mariner 9 used the Reed-Muller code R_5, whose minimum distance is $2^4 = 16$. By Theorem 1, this code can correct k errors, where $2k + 1 \leq 16$. The largest value of k for which this inequality is true is $k = 7$. Thus, R_5 not only contains exactly the right number of code vectors for transmitting 64 shades of gray but also is capable of correcting up to 7 errors, making it quite reliable. This explains why the images transmitted by *Mariner 9* were so sharp!

Exercises 7.5

Approximation of Functions

In Exercises 1–4, find the best linear approximation to f on the interval $[-1, 1]$.

1. $f(x) = x^2$
2. $f(x) = x^2 + 2x$
3. $f(x) = x^3$
4. $f(x) = \sin(\pi x/2)$

In Exercises 5 and 6, find the best quadratic approximation to f on the interval $[-1, 1]$.

5. $f(x) = |x|$
6. $f(x) = \cos(\pi x/2)$
7. Apply the Gram-Schmidt Process to the basis $\{1, x\}$ to construct an orthogonal basis for $\mathscr{P}_1[0, 1]$.
8. Apply the Gram-Schmidt Process to the basis $\{1, x, x^2\}$ to construct an orthogonal basis for $\mathscr{P}_2[0, 1]$.

In Exercises 9–12, find the best linear approximation to f on the interval $[0, 1]$.

9. $f(x) = x^2$
10. $f(x) = \sqrt{x}$
11. $f(x) = e^x$
12. $f(x) = \sin(\pi x/2)$

In Exercises 13–16, find the best quadratic approximation to f on the interval $[0, 1]$.

13. $f(x) = x^3$
14. $f(x) = \sqrt{x}$
15. $f(x) = e^x$
16. $f(x) = \sin(\pi x/2)$
17. Show that 1 is orthogonal to $\cos kx$ and $\sin kx$ in $\mathscr{C}[-\pi, \pi]$ for $k \geq 1$.
18. Show that $\cos jx$ is orthogonal to $\cos kx$ in $\mathscr{C}[-\pi, \pi]$ for $j \neq k, j, k \geq 1$.
19. Show that $\sin jx$ is orthogonal to $\sin kx$ in $\mathscr{C}[-\pi, \pi]$ for $j \neq k, j, k \geq 1$.
20. Show that $\|1\|^2 = 2\pi$ and $\|\cos kx\|^2 = \pi$ in $\mathscr{C}[-\pi, \pi]$.

In Exercises 21 and 22, find the third-order Fourier approximation to f on $[-\pi, \pi]$.

21. $f(x) = |x|$
22. $f(x) = x^2$

In Exercises 23–26, find the Fourier coefficients a_0, a_k, *and* b_k *of f on* $[-\pi, \pi]$.

23. $f(x) = \begin{cases} 0 & \text{if } -\pi \leq x < 0 \\ 1 & \text{if } 0 \leq x \leq \pi \end{cases}$
24. $f(x) = \begin{cases} -1 & \text{if } -\pi \leq x < 0 \\ 1 & \text{if } 0 \leq x \leq \pi \end{cases}$
25. $f(x) = \pi - x$
26. $f(x) = |x|$

*Recall that a function f is an **even function** if* $f(-x) = f(x)$ *for all x; f is called an **odd function** if* $f(-x) = -f(x)$ *for all x.*

27. (a) Prove that $\int_{-\pi}^{\pi} f(x)\,dx = 0$ if f is an odd function.
    (b) Prove that the Fourier coefficients a_k are all zero if f is odd.
28. (a) Prove that $\int_{-\pi}^{\pi} f(x)\,dx = 2\int_{0}^{\pi} f(x)\,dx$ if f is an even function.
    (b) Prove that the Fourier coefficients b_k are all zero if f is even.

Chapter Review

Key Definitions and Concepts

Review Questions

1. Mark each of the following statements true or false:

(a) If $\mathbf{u} = \begin{bmatrix} u_1 \\ u_2 \end{bmatrix}$ and $\mathbf{v} = \begin{bmatrix} v_1 \\ v_2 \end{bmatrix}$, then $\langle \mathbf{u}, \mathbf{v} \rangle = u_1v_1 + \pi u_2v_2$ defines an inner product on $\mathbb{R}^2$.

(b) If $\mathbf{u} = \begin{bmatrix} u_1 \\ u_2 \end{bmatrix}$ and $\mathbf{v} = \begin{bmatrix} v_1 \\ v_2 \end{bmatrix}$, then $\langle \mathbf{u}, \mathbf{v} \rangle = 4u_1v_1 - 2u_1v_2 - 2u_2v_1 + 4u_2v_2$ defines an inner product on $\mathbb{R}^2$.

(c) $\langle A, B \rangle = \text{tr}(A) + \text{tr}(B)$ defines an inner product on M_{22}.

(d) If $\mathbf{u}$ and $\mathbf{v}$ are vectors in an inner product space with $\|\mathbf{u}\| = 4$, $\|\mathbf{v}\| = \sqrt{5}$, and $\langle \mathbf{u}, \mathbf{v} \rangle = 2$, then $\|\mathbf{u} + \mathbf{v}\| = 5$.

(e) The sum norm, max norm, and Euclidean norm on $\mathbb{R}^n$ are all equal to the absolute value function when $n = 1$.

(f) If a matrix A is well-conditioned, then $\text{cond}(A)$ is small.

(g) If $\text{cond}(A)$ is small, then the matrix A is well-conditioned.

(h) Every linear system has a unique least squares solution.

(i) If A is a matrix with orthonormal columns, then the standard matrix of an orthogonal projection onto the column space of A is $P = AA^T$.

(j) If A is a symmetric matrix, then the singular values of A are the same as the eigenvalues of A.

In Questions 2–4, determine whether the definition gives an inner product.

2. $\langle p(x), q(x) \rangle = p(0)q(1) + p(1)q(0)$ for $p(x)$, $q(x)$ in $\mathscr{P}_1$

3. $\langle A, B \rangle = \text{tr}(A^TB)$ for A, B in M_{22}

4. $\langle f, g \rangle = (\max_{0 \le x \le 1} f(x))(\max_{0 \le x \le 1} g(x))$ for f, g in $\mathscr{C}[0, 1]$

In Questions 5 and 6, compute the indicated quantity using the specified inner product.

5. $\|1 + x + x^2\|$ if $\langle a_0 + a_1x + a_2x^2, b_0 + b_1x + b_2x^2 \rangle = a_0b_0 + a_1b_1 + a_2b_2$

6. $d(x, x^2)$ if $\langle p(x), q(x) \rangle = \int_0^1 p(x)q(x)\,dx$

In Questions 7 and 8, construct an orthogonal set of vectors by applying the Gram-Schmidt Process to the given set of vectors using the specified inner product.

7. $\left\{ \begin{bmatrix} 1 \\ 1 \end{bmatrix}, \begin{bmatrix} 1 \\ 2 \end{bmatrix} \right\}$ if $\langle \mathbf{u}, \mathbf{v} \rangle = \mathbf{u}^T A\mathbf{v}$, where $A = \begin{bmatrix} 6 & 4 \\ 4 & 6 \end{bmatrix}$

8. $\{1, x, x^2\}$ if $\langle p(x), q(x) \rangle = \int_0^1 p(x)q(x)\,dx$

In Questions 9 and 10, determine whether the definition gives a norm.

9. $\|\mathbf{v}\| = \mathbf{v}^T\mathbf{v}$ for $\mathbf{v}$ in $\mathbb{R}^n$

10. $\|p(x)\| = |p(0)| + |p(1) - p(0)|$ for $p(x)$ in $\mathscr{P}_1$

11. Show that the matrix $A = \begin{bmatrix} 1 & 0.1 & 0.11 \\ 0.1 & 0.11 & 0.111 \\ 0.11 & 0.111 & 0.1111 \end{bmatrix}$ is ill-conditioned.

12. Prove that if Q is an orthogonal $n \times n$ matrix, then its Frobenius norm is $\|Q\|_F = \sqrt{n}$.

13. Find the line of best fit through the points (1, 2), (2, 3), (3, 5), and (4, 7).

14. Find the least squares solution of

$$\begin{bmatrix} 1 & 2 \\ 1 & 0 \\ 2 & -1 \\ 0 & 5 \end{bmatrix} \begin{bmatrix} x_1 \\ x_2 \end{bmatrix} = \begin{bmatrix} 1 \\ 0 \\ -1 \\ 3 \end{bmatrix}.$$

15. Find the orthogonal projection of $\mathbf{x} = \begin{bmatrix} 1 \\ 2 \\ 3 \end{bmatrix}$ onto the column space of $A = \begin{bmatrix} 1 & 1 \\ 0 & 1 \\ 1 & 0 \end{bmatrix}$.

16. If $\mathbf{u}$ and $\mathbf{v}$ are orthonormal vectors, show that $P = \mathbf{u}\mathbf{u}^T + \mathbf{v}\mathbf{v}^T$ is the standard matrix of an orthogonal projection onto span $(\mathbf{u}, \mathbf{v})$. [*Hint:* Show that $P = A(A^TA)^{-1}A^T$ for some matrix A.]

In Questions 17 and 18, find (a) the singular values, (b) a singular value decomposition, and (c) the pseudoinverse of the matrix A.

17. $A = \begin{bmatrix} 1 & 1 \\ 0 & 0 \\ 1 & -1 \end{bmatrix}$

18. $A = \begin{bmatrix} 1 & 1 & -1 \\ 1 & 1 & -1 \end{bmatrix}$

19. If P and Q are orthogonal matrices for which PAQ is defined, prove that PAQ has the same singular values as A.

20. If A is a square matrix for which $A^2 = O$, prove that $(A^+)^2 = O$.

Appendix A*

Mathematical Notation and Methods of Proof

Please, sir, I want some more.
—Oliver
Charles Dickens, *Oliver Twist*

Anyone who understands algebraic notation reads at a glance in an equation results reached arithmetically only with great labour and pains.
—Augustin Cournot
Researches into the Mathematical Principles of the Theory of Wealth
Translated by Nathaniel T. Bacon
Macmillan, 1897, p. 4

In this book, an effort has been made to use "mathematical English" as much as possible, keeping mathematical notation to a minimum. However, mathematical notation is a convenient shorthand that can greatly simplify the amount of writing we have to do. Moreover, it is commonly used in every branch of mathematics, so the ability to read and write mathematical notation is an essential ingredient of mathematical understanding. Finally, there are some theorems whose proofs become "obvious" if the right notation is used.

Proving theorems in mathematics is as much an art as a science. For the beginner, it is often hard to know what approach to use in proving a theorem; there are many approaches, any one of which might turn out to be the best. To become proficient at proofs, it is important to study as many examples as possible and to get plenty of practice.

This appendix summarizes basic mathematical notation applied to sets. Summation notation, a useful shorthand for dealing with sums, is also discussed. Finally, some approaches to proofs are illustrated with generic examples.

Set Notation

A ***set*** is a collection of objects, called the ***elements*** (or ***members***) of the set. Examples of sets include the set of all words in this text, the set of all books in your college library, the set of positive integers, and the set of all vectors in the plane whose equation is $2x + 3y - z = 0$.

It is often possible to list the elements of a set, in which case it is conventional to enclose the list within braces. For example, we have

$$\{1, 2, 3\}, \quad \{a, t, x, z\}, \quad \{2, 4, 6, \ldots, 100\}, \quad \left\{\frac{\pi}{4}, \frac{2\pi}{5}, \frac{\pi}{2}, \frac{4\pi}{7}, \ldots, \frac{5\pi}{6}\right\}$$

Note that ellipses (. . .) denote elements omitted when a pattern is present. (What is the pattern in the last two examples?) Infinite sets are often expressed using ellipses. For example, the set of positive integers is usually denoted by $\mathbb{N}$ or $\mathbb{Z}^+$, so

$$\mathbb{N} = \mathbb{Z}^+ = \{1, 2, 3, \ldots\}$$

The set of all integers is denoted by $\mathbb{Z}$, so

$$\mathbb{Z} = \{\ldots, -2, -1, 0, 1, 2, \ldots\}$$

Two sets are considered to be ***equal*** if they contain exactly the same elements. The *order* in which elements are listed does not matter, and repetitions are not counted. Thus,

*Exercises and selected odd-numbered answers for this appendix can be found on the student companion website.

$$\{1, 2, 3\} = \{2, 1, 3\} = \{1, 3, 2, 1\}$$

The symbol $\in$ means "is an element of" or "is in," and the symbol $\notin$ denotes the negation—that is, "is not an element of" or "is not in." For example,

$$5 \in \mathbb{Z}^+ \quad \text{but} \quad 0 \notin \mathbb{Z}^+$$

It is often more convenient to describe a set in terms of a rule satisfied by all of its elements. In such cases, ***set builder notation*** is appropriate. The format is

$$\{x : x \text{ satisfies } P\}$$

where P represents a property or a collection of properties that the element x must satisfy. The colon is pronounced "such that." For example,

$$\{n : n \in \mathbb{Z}, n > 0\}$$

is read as "the set of all n such that n is an integer and n is greater than zero." This is just another way of describing the positive integers $\mathbb{Z}^+$. (We could also write $\mathbb{Z}^+ = \{n \in \mathbb{Z} : n > 0\}$.)

The ***empty set*** is the set with no elements. It is denoted by either $\varnothing$ or $\{\ \}$.

Example A.1

Describe in words the following sets:

(a) $A = \{n : n = 2k, k \in \mathbb{Z}\}$
(b) $B = \{m/n : m, n \in \mathbb{Z}, n \neq 0\}$
(c) $C = \{x \in \mathbb{R} : 4x^2 - 4x - 3 = 0\}$
(d) $D = \{x \in \mathbb{Z} : 4x^2 - 4x - 3 = 0\}$

Solution (a) A is the set of numbers n that are integer multiples of 2. Therefore, A is the set of all even integers.

(b) B is the set of all expressions of the form m/n, where m and n are integers and n is nonzero. This is the set of *rational numbers*, usually denoted by $\mathbb{Q}$. (Note that this way of describing $\mathbb{Q}$ produces many repetitions; however, our convention, as noted above, is that we include only one occurrence of each element. Thus, this expression precisely describes the set of all rational numbers.)

(c) C is the set of all real solutions of the equation $4x^2 - 4x - 3 = 0$. By factoring or using the quadratic formula, we find that the roots of this equation are $-\frac{1}{2}$ and $\frac{3}{2}$. (Verify this.) Therefore,

$$C = \{-\tfrac{1}{2}, \tfrac{3}{2}\}$$

(d) From the solution to (c) we see that there are *no* solutions to $4x^2 - 4x - 3 = 0$ in $\mathbb{R}$ that are integers. Therefore, D is the empty set, which we can express by writing $D = \varnothing$.

John Venn (1834–1923) was an English mathematician who studied at Cambridge University and later lectured there. He worked primarily in mathematical logic and is best known for inventing Venn diagrams.

If every element of a set A is also an element of a set B, then A is called a ***subset*** of B, denoted $A \subseteq B$. We can represent this situation schematically using a ***Venn diagram***, as shown in Figure A.1. (The rectangle represents the *universal set*, a set large enough to contain all of the other sets in question—in this case, A and B.)

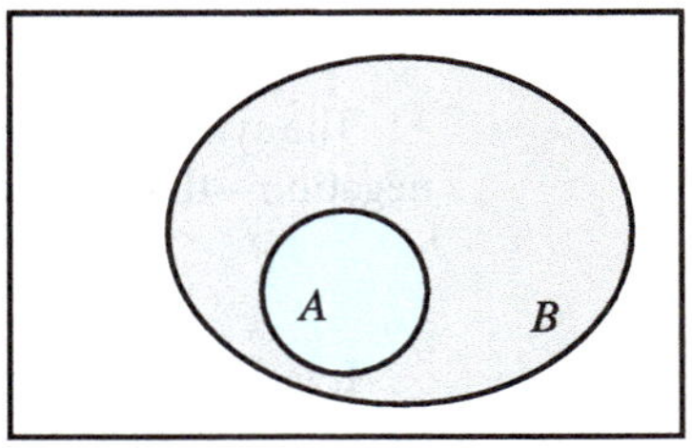

Figure A.1
$A \subseteq B$

Example A.2

(a) $\{1, 2, 3\} \subseteq \{1, 2, 3, 4, 5\}$

(b) $\mathbb{Z}^+ \subseteq \mathbb{Z} \subseteq \mathbb{R}$

(c) Let A be the set of all positive integers whose last two digits are 24 and let B be the set of all positive integers that are evenly divisible by 4. Then if n is in A, it is of the form

$$n = 100k + 24$$

for some integer k. (For example, $36{,}524 = 100 \cdot 365 + 24$.) But then

$$n = 100k + 24 = 4(25k + 6)$$

so $n/4 = 25k + 6$, which is an integer. Hence, n is evenly divisible by 4, so it is in B. Therefore, $A \subseteq B$.

We can show that two sets A and B are equal by showing that each is a subset of the other. This strategy is particularly useful if the sets are defined abstractly or if it is not easy to list and compare their elements.

Example A.3

Let A be the set of all positive integers whose last two digits form a number that is evenly divisible by 4. In the case of a one-digit number, we take its tens digit to be 0. Let B be the set of all positive integers that are evenly divisible by 4. Show that $A = B$.

Solution As in Example A.2(c), it is easy to see that $A \subseteq B$. If n is in A, then we can split off the number m formed by its last two digits by writing

$$n = 100k + m$$

for some integer k. But, since m is divisible by 4, we have $m = 4r$ for some integer r. Therefore,

$$n = 100k + m = 100k + 4r = 4(25k + r)$$

so n is also evenly divisible by 4. Hence, $A \subseteq B$.

To show that $B \subseteq A$, let n be in B. That is, n is evenly divisible by 4. Let's say that $n = 4s$, where s is an integer. If m is the number formed by the last two digits of n, then, as above, $n = 100k + m$ for some integer k. But now

$$m = n - 100k = 4s - 100k = 4(s - 25k)$$

which implies that m is evenly divisible by 4, since $s - 25k$ is an integer. Therefore, n is in A, and we have shown that $B \subseteq A$.

Since $A \subseteq B$ and $B \subseteq A$, we must have $A = B$.

The ***intersection*** of sets A and B is denoted by $A \cap B$ and consists of the elements that A and B have in common. That is,

$$A \cap B = \{x : x \in A \quad \text{and} \quad x \in B\}$$

Figure A.2 shows a Venn diagram of this case. The ***union*** of A and B is denoted by $A \cup B$ and consists of the elements that are in either A or B (or both). That is,

$$A \cup B = \{x : x \in A \quad \text{or} \quad x \in B\}$$

See Figure A.3.

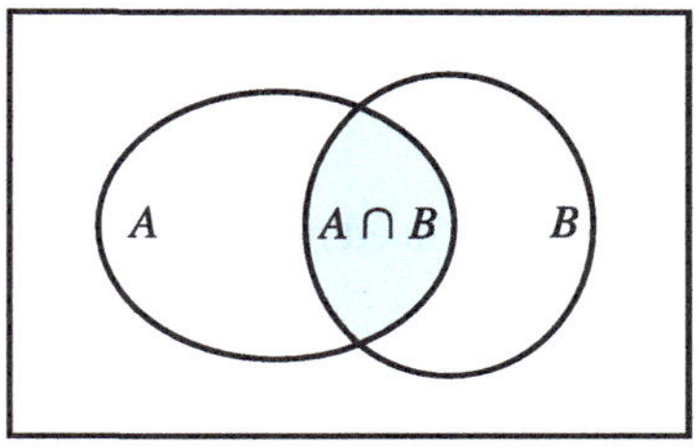

Figure A.2
$A \cap B$

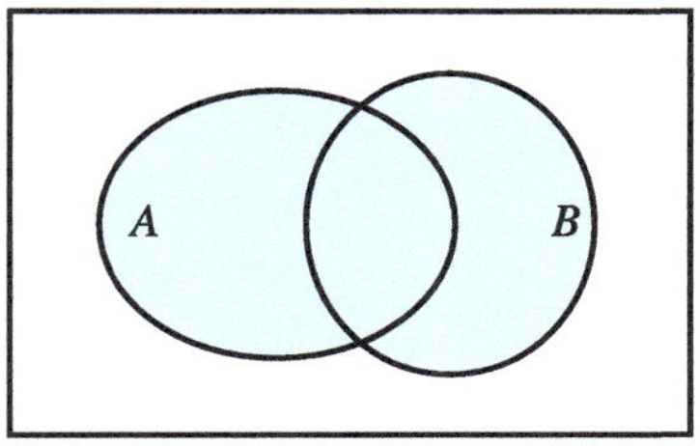

Figure A.3
$A \cup B$

Example A.4

Let $A = \{n^2 : n \in \mathbb{Z}^+, 1 \leq n \leq 4\}$ and let $B = \{n \in \mathbb{Z}^+ : n \leq 10 \text{ and } n \text{ is odd}\}$. Find $A \cap B$ and $A \cup B$.

Solution We see that

$$A = \{1^2, 2^2, 3^2, 4^2\} = \{1, 4, 9, 16\} \quad \text{and} \quad B = \{1, 3, 5, 7, 9\}$$

Therefore, $A \cap B = \{1, 9\}$ and $A \cup B = \{1, 3, 4, 5, 7, 9, 16\}$.

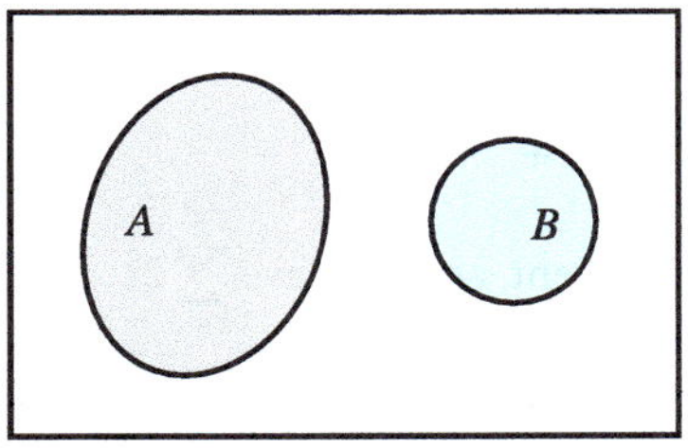

Figure A.4
Disjoint sets

If $A \cap B = \varnothing$, then A and B are called ***disjoint sets.*** (See Figure A.4.) For example, the set of even integers and the set of odd integers are disjoint.

Summation Notation

Summation notation is a convenient shorthand to use to write out a sum such as

$$1 + 2 + 3 + \cdots + 100$$

where we want to leave out all but a few terms. As in set notation, ellipses (. . .) convey that we have established a pattern and have simply left out some intermediate terms. In the above example, readers are expected to recognize that we are summing all of the positive integers from 1 to 100. However, ellipses can be ambiguous. For example, what would one make of the following sum?

$$1 + 2 + \cdots + 64$$

Is this the sum of all positive integers from 1 to 64 or just the powers of two, $1 + 2 + 4 + 8 + 16 + 32 + 64$? It is often clearer (and shorter) to use ***summation notation*** (or ***sigma notation***).

Σ is the capital Greek letter *sigma*, corresponding to S (for "sum"). Summation notation was introduced by Fourier in 1820 and was quickly adopted by the mathematical community.

We can abbreviate a sum of the form

$$a_1 + a_2 + \cdots + a_n \tag{1}$$

as

$$\sum_{k=1}^{n} a_k \tag{2}$$

which tells us to sum the terms a_k over all integers k ranging from 1 to n. An alternative version of this expression is

$$\sum_{1 \le k \le n} a_k$$

The subscript k is called the ***index of summation.*** It is a "dummy variable" in the sense that it does not appear in the actual sum in expression (1). Therefore, we can use any letter we like as the index of summation (as long as it doesn't already appear somewhere else in the expressions we are summing). Thus, expression (2) can also be written as

$$\sum_{i=1}^{n} a_i$$

The index of summation need not start at 1. The sum $a_3 + a_4 + \cdots + a_{99}$ becomes

$$\sum_{k=3}^{99} a_k$$

although we can arrange for the index to begin at 1 by rewriting the expression as

$$\sum_{k=1}^{97} a_{k+2}.$$

The key to using summation notation effectively is being able to recognize patterns.

Example A.5

Write the following sums using summation notation.

(a) $1 + 2 + 4 + \cdots + 64$ (b) $1 + 3 + 5 + \cdots + 99$ (c) $3 + 8 + 15 + \cdots + 99$

Solution (a) We recognize this expression as a sum of powers of 2:

$$1 + 2 + 4 + \cdots + 64 = 2^0 + 2^1 + 2^2 + \cdots + 2^6$$

Therefore, the index of summation appears as the exponent, and we have $\sum_{k=0}^{6} 2^k$.

(b) This expression is the sum of all the odd integers from 1 to 99. Every odd integer is of the form $2k + 1$, so the sum is

$$\begin{aligned} &1 + 3 + 5 + \cdots + 99 \\ &\quad = (2 \cdot 0 + 1) + (2 \cdot 1 + 1) + (2 \cdot 2 + 1) + \cdots + (2 \cdot 49 + 1) \\ &\quad = \sum_{k=0}^{49} (2k + 1) \end{aligned}$$

(c) The pattern here is less clear, but a little reflection reveals that each term is 1 less than a perfect square:

$$\begin{aligned} &3 + 8 + 15 + \cdots + 99 \\ &\quad = (2^2 - 1) + (3^2 - 1) + (4^2 - 1) + \cdots + (10^2 - 1) \\ &\quad = \sum_{k=2}^{10} (k^2 - 1) \end{aligned}$$

Example A.6

Rewrite each of the sums in Example A.5 so that the index of summation starts at 1.

Solution (a) If we use the change of variable $i = k + 1$, then, as k goes from 0 to 6, i goes from 1 to 7. Since $k = i - 1$, we obtain

$$\sum_{k=0}^{6} 2^k = \sum_{i=1}^{7} 2^{i-1}$$

(b) Using the same substitution as in part (a), we get

$$\sum_{k=0}^{49} (2k + 1) = \sum_{i=1}^{50} (2(i - 1) + 1) = \sum_{i=1}^{50} (2i - 1)$$

(c) The substitution $i = k - 2$ will work (try it), but it is easier just to add a term corresponding to $k = 1$, since $1^2 - 1 = 0$. Therefore,

$$\sum_{k=2}^{10} (k^2 - 1) = \sum_{k=1}^{10} (k^2 - 1)$$

Multiple summations arise when there is more than one index of summation, as there is with a matrix. The notation

$$\sum_{i,j=1}^{n} a_{ij} \tag{3}$$

means to sum the terms a_{ij} as i and j each range independently from 1 to n. The sum in expression (3) is equivalent to either

$$\sum_{i=1}^{n} \sum_{j=1}^{n} a_{ij}$$

where we sum first over j and then over i (we always work from the inside out), or

$$\sum_{j=1}^{n} \sum_{i=1}^{n} a_{ij}$$

where the order of summation is reversed.

Example A.7

Write out $\sum_{i,j=1}^{3} i^j$ using both possible orders of summation.

Solution

$$\begin{aligned}\sum_{i=1}^{3} \sum_{j=1}^{3} i^j &= \sum_{i=1}^{n} (i^1 + i^2 + i^3) \\ &= (1^1 + 1^2 + 1^3) + (2^1 + 2^2 + 2^3) + (3^1 + 3^2 + 3^3) \\ &= (1 + 1 + 1) + (2 + 4 + 8) + (3 + 9 + 27) = 56\end{aligned}$$

and

$$\sum_{j=1}^{3}\sum_{i=1}^{3} i^j = \sum_{j=1}^{n}(1^j + 2^j + 3^j)$$
$$= (1^1 + 2^1 + 3^1) + (1^2 + 2^2 + 3^2) + (1^3 + 2^3 + 3^3)$$
$$= (1 + 2 + 3) + (1 + 4 + 9) + (1 + 8 + 27) = 56$$

Remark Of course, the value of the sum in Example A.7 is the same no matter which order of summation we choose, because the sum is *finite*. It is also possible to consider *infinite sums* (known as *infinite series* in calculus), but such sums do not always have a value and great care must be taken when rearranging or manipulating their terms. For example, suppose we let

$$S = \sum_{k=0}^{\infty} 2^k$$

Then

$$S = 1 + 2 + 4 + 8 + \cdots$$
$$= 1 + 2(1 + 2 + 4 + \cdots)$$
$$= 1 + 2S$$

from which it follows that $S = -1$. This is clearly nonsense, since S is a sum of *non-negative* terms! (Where is the error?)

How to Solve It is the title of a book by the mathematician George Pólya (1887–1985). Since its publication in 1945, *How to Solve It* has sold over a million copies and has been translated into 17 languages. Pólya was born in Hungary, but because of the political situation in Europe, he moved to the United States in 1940. He subsequently taught at Brown and Stanford Universities, where he did mathematical research and developed a well-deserved reputation as an outstanding teacher. The Pólya Prize is awarded annually by the Society for Industrial and Applied Mathematics for major contributions to areas of mathematics close to those on which Pólya worked. The Mathematical Association of America annually awards Pólya Lectureships to mathematicians demonstrating the high-quality exposition for which Pólya was known.

Methods of Proof

The notion of proof is at the very heart of mathematics. It is one thing to know *what* is true; it is quite another to know *why* it is true and to be able to demonstrate its truth by means of a logically connected sequence of statements. The intention here is not to try to teach you how to do proofs; you will become better at doing proofs by studying examples and by practicing—something you should do often as you work through this text. The intention of this brief section is simply to provide a few elementary examples of some types of proofs. The proofs of theorems in the text will provide further illustrations of "how to solve it."

Roughly speaking, mathematical proofs fall into two categories: ***direct proofs*** and ***indirect proofs.*** Many theorems have the structure "if P, then Q," where P (the *hypothesis*, or *premise*) and Q (the *conclusion*) are statements that are either true or false. We denote such an implication by $P \Rightarrow Q$. A direct proof proceeds by establishing a chain of implications

$$P \Rightarrow P_1 \Rightarrow P_2 \Rightarrow \cdots \Rightarrow P_n \Rightarrow Q$$

leading directly from P to Q.

Example A.8

Prove that any two consecutive perfect squares differ by an odd number. This instruction can be rephrased as "Prove that if a and b are consecutive perfect squares, then $a - b$ is odd." Hence, it has the form $P \Rightarrow Q$, with P being "a and b are consecutive perfect squares" and Q being "$a - b$ is odd."

Solution Assume that a and b are consecutive perfect squares, with $a > b$. Then

$$a = (n + 1)^2 \quad \text{and} \quad b = n^2$$

for some integer n. But now

$$\begin{aligned} a - b &= (n + 1)^2 - n^2 \\ &= n^2 + 2n + 1 - n^2 \\ &= 2n + 1 \end{aligned}$$

so $a - b$ is odd.

There are two types of indirect proofs that can be used to establish a conditional statement of the form $P \Rightarrow Q$. A ***proof by contradiction*** assumes that the hypothesis P is true, just as in a direct proof, but then supposes that the conclusion Q is *false*. The strategy then is to show that this is not possible (i.e., to rule out the possibility that the conclusion is false) by finding a contradiction to the truth of P. It then follows that Q must be true.

Example A.9

Let n be a positive integer. Prove that if n^2 is even, so is n. (Take a few minutes to try to find a direct proof of this assertion; it will help you to appreciate the indirect proof that follows.)

Solution Assume that n is a positive integer such that n^2 is even. Now suppose that n is not even. Then n is odd, so

$$n = 2k + 1$$

for some integer k. But if so, we have

$$n^2 = (2k + 1)^2 = 4k^2 + 4k + 1$$

so n^2 is odd, since it is 1 more than the even number $4k^2 + 4k$. This contradicts our hypothesis that n^2 is even. We conclude that our supposition that n was *not* even must have been false; in other words, n must be even.

Closely related to the method of proof by contradiction is ***proof by contrapositive***. The *negative* of a statement P is the statement "it is not the case that P," abbreviated symbolically as $\neg P$ and pronounced "not P." For example, if P is "n is even," then $\neg P$ is "it is not the case that n is even"—in other words, "n is odd."

The *contrapositive* of the statement $P \Rightarrow Q$ is the statement $\neg Q \Rightarrow \neg P$. A conditional statement $P \Rightarrow Q$ and its contrapositive $\neg Q \Rightarrow \neg P$ are logically equivalent in the sense that they are either both true or both false. (For example, if $P \Rightarrow Q$ is a theorem, then so is $\neg Q \Rightarrow \neg P$. To see this, note that if the hypothesis $\neg Q$ is true, then Q is false. The conclusion $\neg P$ cannot be false, for if it were, then P would be true and our known theorem $P \Rightarrow Q$ would imply the truth of Q, giving us a contradiction. It follows that $\neg P$ is true and we have proved $\neg Q \Rightarrow \neg P$.) Here is a contrapositive proof of the assertion in Example A.9.

Example A.10

Let n be a positive integer. Prove that if n^2 is even, so is n.

Solution The contrapositive of the given statement is

"If n is not even, then n^2 is not even" or "If n is odd, so is n^2"

To prove this contrapositive, assume that n is odd. Then $n = 2k + 1$ for some integer k. As before, this means that $n^2 = (2k + 1)^2 = 4k^2 + 4k + 1$ is odd, which completes the proof of the contrapositive. Since the contrapositive is true, so is the original statement.

Although we do not require a new method of proof to handle it, we will briefly consider how to prove an "if and only if" theorem. A statement of the form "P if and only if Q" signals a *double implication,* which we denote by $P \Leftrightarrow Q$. To prove such a statement, we must prove $P \Rightarrow Q$ and $Q \Rightarrow P$. To do so, we can use the techniques described earlier, where appropriate. It is important to notice that the "*if*" part of $P \Leftrightarrow Q$ is "P if Q," which is $Q \Rightarrow P$; the "only if" part of $P \Leftrightarrow Q$ is "P only if Q," meaning $P \Rightarrow Q$. The implication $P \Rightarrow Q$ is sometimes read as "P is sufficient for Q" or "Q is necessary for P"; $Q \Rightarrow P$ is read "Q is sufficient for P" or "P is necessary for Q." Taken together, they are $P \Leftrightarrow Q$, or "P is necessary and sufficient for Q" and vice versa.

Example A.11

A pawn is placed on a chessboard and is allowed to move one square at a time, either horizontally or vertically. A *pawn's tour* of a chessboard is a path taken by a pawn, moving as described, that visits each square exactly once, starting and ending on the same square. Prove that there is a pawn's tour of an $n \times n$ chessboard if and only if n is even.

Solution [$\Leftarrow$] ("if") Assume that n is even. It is easy to see that the strategy illustrated in Figure A.5 for a 6×6 chessboard will always give a pawn's tour.

[$\Rightarrow$] ("only if") Suppose that there is a pawn's tour of an $n \times n$ chessboard. We will give a proof by contradiction that n must be even. To this end, let's assume that n is odd. At each move, the pawn moves to a square of a different color. The total number of moves in its tour is n^2, which is also an odd number, according to the proof in Example A.10. Therefore, the pawn must end up on a square of the opposite color from that of the square on which it started. (Why?) This is impossible, since the pawn ends where it started, so we have a contradiction. It follows that n cannot be odd; hence, n is even and the proof is complete.

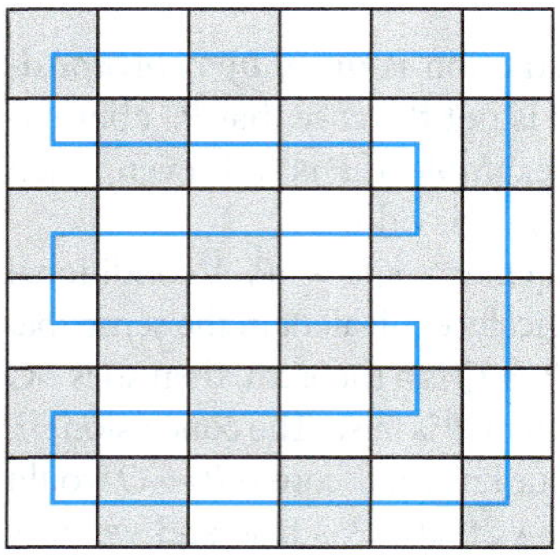

Figure A.5

Some theorems assert that several statements are *equivalent*. This means that each is true if and only if all of the others are true. Showing that n statements are equivalent requires $\binom{n}{2} = \frac{n!}{2!(n-2)!} = \frac{n^2 - n}{2}$ "if and only if" proofs. In practice, however, it is often easier to establish a "ring" of n implications that links all of the statements. The proof of the Fundamental Theorem of Invertible Matrices provides an excellent example of this approach.

Appendix B*

Mathematical Induction

The ability to spot patterns is one of the keys to success in mathematical problem solving. Consider the following pattern:

$$
\begin{aligned}
1 &= 1 \\
1 + 3 &= 4 \\
1 + 3 + 5 &= 9 \\
1 + 3 + 5 + 7 &= 16 \\
1 + 3 + 5 + 7 + 9 &= 25
\end{aligned}
$$

Great fleas have little fleas
upon their backs to bite 'em,
And little fleas have lesser fleas,
and so ad infinitum.
—Augustus De Morgan
A Budget of Paradoxes
Longmans, Green, and Company, 1872, p. 377

The sums are all perfect squares: $1^2, 2^2, 3^2, 4^2, 5^2$. It seems reasonable to conjecture that this pattern will continue to hold; that is, the sum of consecutive odd numbers, starting at 1, will always be a perfect square. Let's try to be more precise. If the sum is n^2, then the last odd number in the sum is $2n - 1$. (Check this in the five cases above.) In symbols, our conjecture becomes

$$1 + 3 + 5 + \cdots + (2n - 1) = n^2 \quad \text{for all } n \geq 1 \tag{1}$$

Notice that Equation (1) is really an *infinite* collection of statements, one for each value of $n \geq 1$. Although our conjecture seems reasonable, we cannot assume that the pattern continues—we need to prove it. This is where ***mathematical induction*** comes in.

First Principle of Mathematical Induction

Let $S(n)$ be a statement about the positive integer n. If

1. $S(1)$ is true and
2. for all $k \geq 1$, the truth of $S(k)$ implies the truth of $S(k + 1)$

then $S(n)$ is true for all $n \geq 1$.

Verifying that $S(1)$ is true is called the ***basis step.*** The assumption that $S(k)$ is true for some $k \geq 1$ is called the ***induction hypothesis.*** Using the induction hypothesis to prove that $S(k + 1)$ is then true is called the ***induction step.*** Mathematical induction has been referred to as the *domino principle* because it is analogous to showing that a line of dominoes will fall down if (1) the first domino can be knocked down (the basis step) and (2) knocking down any domino (the induction hypothesis) will knock over the next domino (the induction step). See Figure B.1.

We now use the principle of mathematical induction to prove Equation (1).

*Exercises and selected odd-numbered answers for this appendix can be found on the student companion website.

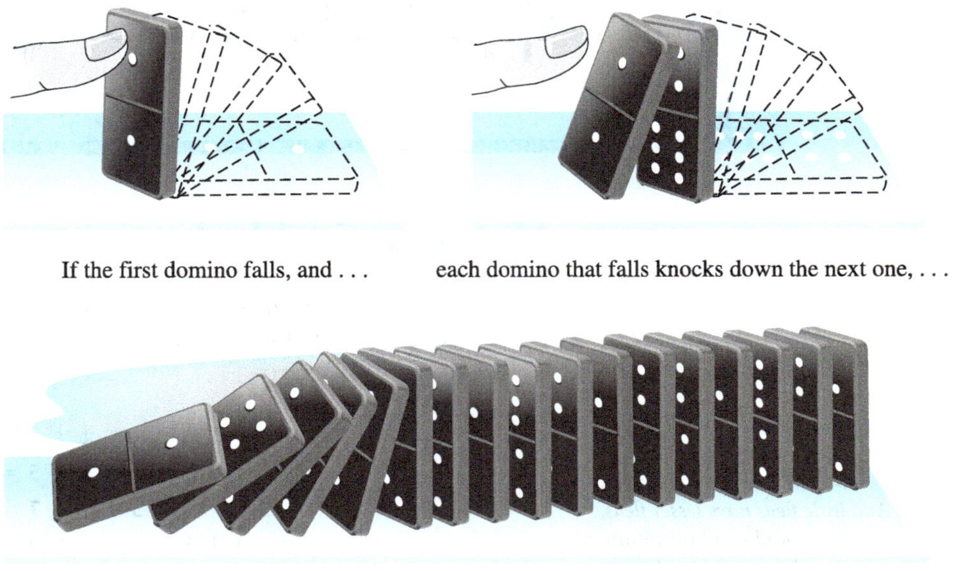

If the first domino falls, and . . . each domino that falls knocks down the next one, . . .

then all the dominoes can be made to fall by pushing over the first one.

Figure B.1

Example B.1 Use mathematical induction to prove that

$$1 + 3 + 5 + \cdots + (2n - 1) = n^2$$

for all $n \geq 1$.

Solution For $n = 1$, the sum on the left-hand side is just 1, while the right-hand side is 1^2. Since $1 = 1^2$, this completes the basis step.

Now assume that the formula is true for some integer $k \geq 1$. That is, assume that

$$1 + 3 + 5 + \cdots + (2k - 1) = k^2$$

(This is the induction hypothesis.) The induction step consists of proving that the formula is true when $n = k + 1$. We see that when $n = k + 1$, the left-hand side of formula (1) is

$$\begin{aligned} 1 + 3 + 5 + \cdots + (2(k+1) - 1) &= 1 + 3 + 5 + \cdots + (2k + 1) \\ &= \underbrace{1 + 3 + 5 + \cdots + (2k - 1)}_{} + (2k + 1) \\ &= \qquad k^2 \qquad + 2k + 1 \quad \leftarrow \text{by the induction hypothesis} \\ &= (k + 1)^2 \end{aligned}$$

which is the right-hand side of Equation (1) when $n = k + 1$.

This completes the induction step, and we conclude that Equation (1) is true for all $n \geq 1$, by the principle of mathematical induction.

The next example gives a proof of a useful formula for the sum of the first n positive integers. The formula appears several times in the text; for example, see the solution to Exercise 51 in Section 2.4.

Example B.2 Prove that

$$1 + 2 + \cdots + n = \frac{n(n + 1)}{2}$$

for all $n \geq 1$.

Solution The formula is true for $n = 1$, since

$$1 = \frac{1(1 + 1)}{2}$$

Assume that the formula is true for $n = k$; that is,

$$1 + 2 + \cdots + k = \frac{k(k + 1)}{2}$$

We need to show that the formula is true when $n = k + 1$; that is, we must prove that

$$1 + 2 + \cdots + (k + 1) = \frac{(k + 1)[(k + 1) + 1]}{2}$$

But we see that

$$\begin{aligned} 1 + 2 + \cdots + (k + 1) &= (1 + 2 + \cdots + k) + (k + 1) \\ &= \frac{k(k + 1)}{2} + (k + 1) \qquad \text{by the induction hypothesis} \\ &= \frac{k(k + 1) + 2(k + 1)}{2} \\ &= \frac{k^2 + 3k + 2}{2} \\ &= \frac{(k + 1)(k + 2)}{2} \\ &= \frac{(k + 1)[(k + 1) + 1]}{2} \end{aligned}$$

which is what we needed to show.

This completes the induction step, and we conclude that the formula is true for all $n \geq 1$, by the principle of mathematical induction.

In a similar vein, we can prove that the sum of the squares of the first n positive integers satisfies the formula

$$1^2 + 2^2 + 3^2 + \cdots + n^2 = \frac{n(n + 1)(2n + 1)}{6}$$

for all $n \geq 1$. (Verify this for yourself.)

The basis step need not be for $n = 1$, as the next two examples illustrate.

Example B.3

Prove that $n! > 2^n$ for all integers $n \geq 4$.

Solution The basis step here is when $n = 4$. The inequality is clearly true in this case, since

$$4! = 24 > 16 = 2^4$$

Assume that $k! > 2^k$ for some integer $k \geq 4$. Then

$$\begin{aligned} (k+1)! &= (k+1)k! \\ &> (k+1)2^k && \text{by the induction hypothesis} \\ &\geq 5 \cdot 2^k && \text{since } k \geq 4 \\ &> 2 \cdot 2^k = 2^{k+1} \end{aligned}$$

which verifies the inequality for $n = k + 1$ and completes the induction step.

We conclude that $n! > 2^n$ for all integers $n \geq 4$, by the principle of mathematical induction.

If a is a nonzero real number and $n \geq 0$ is an integer, we can give a recursive definition of the power a^n that is compatible with mathematical induction. We define $a^0 = 1$ and, for $n \geq 0$,

$$a^{n+1} = a^n a$$

(This form avoids the ellipses used in the version $a^n = \overbrace{aa \cdots a}^{n \text{ times}}$.) We can now use mathematical induction to verify a familiar property of exponents.

Example B.4

Let a be a nonzero real number. Prove that $a^m a^n = a^{m+n}$ for all integers $m, n \geq 0$.

Solution At first glance, it is not clear how to proceed, since there are *two* variables, m and n. But we simply need to keep one of them fixed and perform our induction using the other. So, let $m \geq 0$ be a fixed integer. When $n = 0$, we have

$$a^m a^0 = a^m \cdot 1 = a^m = a^{m+0}$$

using the definition $a^0 = 1$. Hence, the basis step is true.

Now assume that the formula holds when $n = k$, where $k \geq 0$. Then $a^m a^k = a^{m+k}$. For $n = k + 1$, using our recursive definition and the fact that addition and multiplication are associative, we see that

$$\begin{aligned} a^m a^{k+1} &= a^m(a^k a) && \text{by definition} \\ &= (a^m a^k)a \\ &= a^{m+k}a && \text{by the induction hypothesis} \\ &= a^{(m+k)+1} && \text{by definition} \\ &= a^{m+(k+1)} \end{aligned}$$

Therefore, the formula is true for $n = k + 1$, and the induction step is complete.

We conclude that $a^m a^n = a^{m+n}$ for all integers $m, n \geq 0$, by the principle of mathematical induction.

In Examples B.1 through B.4, the use of the induction hypothesis during the induction step is relatively straightforward. However, this is not always the case. An alternative version of the principle of mathematical induction is often more useful.

Second Principle of Mathematical Induction

Let $S(n)$ be a statement about the positive integer n. If

1. $S(1)$ is true and
2. the truth of $S(1), S(2), \ldots, S(k)$ implies the truth of $S(k + 1)$

then $S(n)$ is true for all $n \geq 1$.

The only difference between the two principles of mathematical induction is in the induction hypothesis: The first version assumes that $S(k)$ is true, whereas the second version assumes that all of $S(1), S(2), \ldots, S(k)$ are true. This makes the second principle seem weaker than the first, since we need to assume more in order to prove $S(k+1)$ (although, paradoxically, the second principle is sometimes called *strong* induction). In fact, however, the two principles are logically equivalent: Each one implies the other. (Can you see why?)

The next example presents an instance in which the second principle of mathematical induction is easier to use than the first. Recall that a prime number is a positive integer whose only positive integer factors are 1 and itself.

Example B.5

Prove that every positive integer $n \geq 2$ either is prime or can be factored into a product of primes.

Solution The result is clearly true when $n = 2$, since 2 is prime. Now assume that for all integers n between 2 and k, n either is prime or can be factored into a product of primes. Let $n = k + 1$. If $k + 1$ is prime, we are done. Otherwise, it must factor into a product of two smaller integers—say,

$$k + 1 = ab$$

Since $2 \leq a, b \leq k$ (why?), the induction hypothesis applies to a and b. Therefore,

$$a = p_1 \cdots p_r \quad \text{and} \quad b = q_1 \cdots q_s$$

where the p's and q's are all prime. Then

$$ab = p_1 \cdots p_r q_1 \cdots q_s$$

gives a factorization of ab into primes, completing the induction step.

We conclude that the result is true for all integers $n \geq 2$, by the second principle of mathematical induction.

Do you see why the first principle of mathematical induction would have been difficult to use here?

We conclude with a highly nontrivial example that involves a combination of induction and *backward* induction. The result is the Arithmetic Mean–Geometric Mean Inequality, discussed in Chapter 7 in Exploration: Geometric Inequalities and Optimization Problems. The clever proof in Example B.6 is due to Cauchy.

Example B.6 Let $x_1, \ldots, x_n$ be nonnegative real numbers. Prove that

$$\sqrt[n]{x_1x_2\cdots x_n} \leq \frac{x_1 + x_2 + \cdots + x_n}{n}$$

for all integers $n \geq 2$.

Solution For $n = 2$, the inequality becomes $\sqrt{xy} \leq (x + y)/2$. You are asked to verify this in Problems 1 and 2 of the Exploration mentioned above.

If $S(n)$ is the stated inequality, we will prove that $S(k)$ implies $S(2k)$. Assume that $S(k)$ is true; that is,

$$\sqrt[k]{x_1x_2\cdots x_k} \leq \frac{x_1 + x_2 + \cdots + x_k}{k}$$

for all nonnegative real numbers $x_1, \ldots, x_k$. Let

$$x_1 = \frac{y_1 + y_2}{2}, \quad x_2 = \frac{y_3 + y_4}{2}, \quad \ldots, \quad x_k = \frac{y_{2k-1} + y_{2k}}{2}$$

Then

$$\begin{aligned}
\sqrt[2k]{y_1\cdots y_{2k}} &= \sqrt[k]{\sqrt{y_1\cdots y_{2k}}} = \sqrt[k]{\sqrt{y_1y_2}\cdots\sqrt{y_{2k-1}y_{2k}}} \\
&\leq \sqrt[k]{\left(\frac{y_1 + y_2}{2}\right)\cdots\left(\frac{y_{2k-1} + y_{2k}}{2}\right)} && \text{by } S(2) \\
&= \sqrt[k]{x_1\cdots x_k} \\
&\leq \frac{x_1 + x_2 + \cdots + x_k}{k} && \text{by } S(k) \\
&= \frac{\left(\dfrac{y_1 + y_2}{2}\right) + \cdots + \left(\dfrac{y_{2k-1} + y_{2k}}{2}\right)}{k} \\
&= \frac{y_1 + \cdots + y_{2k}}{2k}
\end{aligned}$$

which verifies $S(2k)$.

Thus, the Arithmetic Mean–Geometric Mean Inequality is true for $n = 2, 4, 8, \ldots$—the powers of 2. In order to complete the proof, we need to "fill in the gaps." We will use backward induction to prove that $S(k)$ implies $S(k - 1)$. Assuming $S(k)$ is true, let

$$x_k = \frac{x_1 + x_2 + \cdots + x_{k-1}}{k - 1}$$

Then

$$\sqrt[k]{x_1x_2\cdots x_{k-1}\left(\frac{x_1 + x_2 + \cdots + x_{k-1}}{k-1}\right)} \le \frac{x_1 + x_2 + \cdots + \left(\frac{x_1 + x_2 + \cdots + x_{k-1}}{k-1}\right)}{k}$$

$$= \frac{kx_1 + kx_2 + \cdots + kx_{k-1}}{k(k-1)}$$

$$= \frac{x_1 + x_2 + \cdots + x_{k-1}}{k-1}$$

Equivalently,

$$x_1x_2\cdots x_{k-1}\left(\frac{x_1 + x_2 + \cdots + x_{k-1}}{k-1}\right) \le \left(\frac{x_1 + x_2 + \cdots + x_{k-1}}{k-1}\right)^k$$

or

$$x_1x_2\cdots x_{k-1} \le \left(\frac{x_1 + x_2 + \cdots + x_{k-1}}{k-1}\right)^{k-1}$$

Taking the $(k-1)$th root of both sides yields $S(k-1)$.

The two inductions, taken together, show that the Arithmetic Mean–Geometric Mean Inequality is true for all $n \ge 2$.

Remark Although mathematical induction is a powerful and indispensable tool, it cannot work miracles. That is, it cannot prove that a pattern or formula holds if it does not. Consider the diagrams in Figure B.2, which show the maximum number of regions $R(n)$ into which a circle can be subdivided by n straight lines.

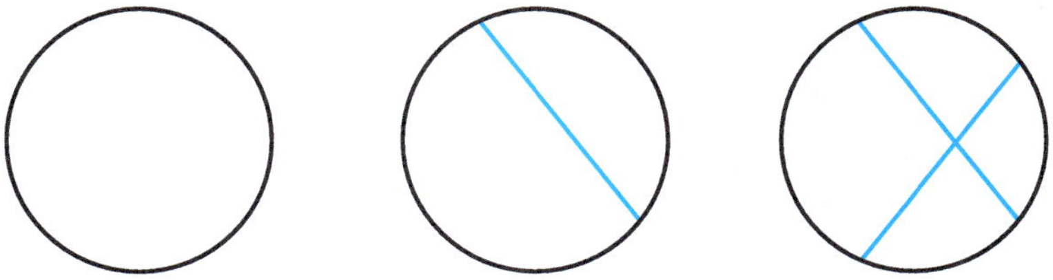

$R(0) = 1 = 2^0$ $R(1) = 2 = 2^1$ $R(2) = 4 = 2^2$

Figure B.2

Based on the evidence in Figure B.2, we might conjecture that $R(n) = 2^n$ for $n \ge 0$ and try to prove this conjecture using mathematical induction. We would not succeed, since this formula is not correct! If we had considered one more case, we would have discovered that $R(3) = 7 \ne 8 = 2^3$, thereby demolishing our conjecture. In fact, the correct formula turns out to be

$$R(n) = \frac{n^2 + n + 2}{2}$$

which *can* be verified by induction. (Can you do it?)

For other examples in which a pattern appears to be true, only to disappear when enough cases are considered, see Richard K. Guy's delightful article "The Strong Law of Small Numbers" in the *American Mathematical Monthly*, Vol. 95 (1988), pp. 697–712.

Appendix C*

Complex Numbers

[The] extension of the number concept to include the irrational, and we will at once add, the imaginary, is the greatest forward step which pure mathematics has ever taken.
—Hermann Hankel
Theorie der Complexen Zahlensysteme
Leipzig, 1867, p. 60

A ***complex number*** is a number of the form $a + bi$, where a and b are real numbers and i is a symbol with the property that $i^2 = -1$. The real number a is considered to be a special type of complex number, since $a = a + 0i$. If $z = a + bi$ is a complex number, then the ***real part*** of z, denoted by Re z, is a, and the ***imaginary part*** of z, denoted by Im z, is b. Two complex numbers $a + bi$ and $c + di$ are ***equal*** if their real parts are equal and their imaginary parts are equal—that is, if $a = c$ and $b = d$. A complex number $a + bi$ can be identified with the point (a, b) and plotted in the plane (called the ***complex plane,*** or the ***Argand plane***), as shown in Figure C.1. In the complex plane, the horizontal axis is called the ***real axis*** and the vertical axis is called the ***imaginary axis.***

There is nothing "imaginary" about complex numbers—they are just as "real" as the real numbers. The term *imaginary* arose from the study of polynomial equations such as $x^2 + 1 = 0$, whose solutions are not "real" (i.e., real numbers). It is worth remembering that at one time negative numbers were thought of as "imaginary" too.

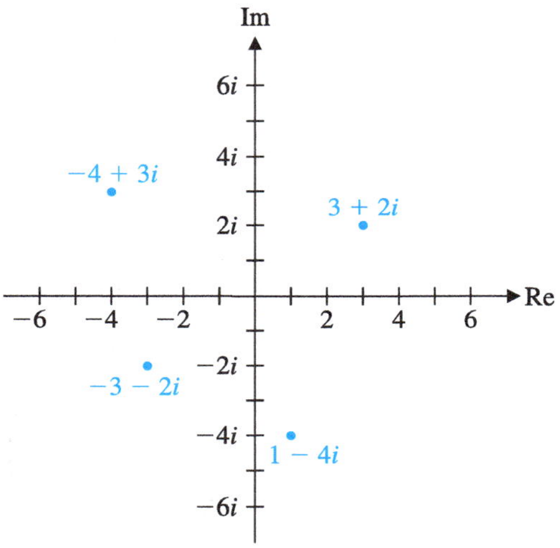

Figure C.1
The complex plane

Jean-Robert Argand (1768–1822) was a French accountant and amateur mathematician. His geometric interpretation of complex numbers appeared in 1806 in a book that he published privately. He was not, however, the first to give such an interpretation. The Norwegian-Danish surveyor Caspar Wessel (1745–1818) gave the same version of the complex plane in 1787, but his paper was not noticed by the mathematical community until after his death.

Operations on Complex Numbers

The ***sum*** of the complex numbers $a + bi$ and $c + di$ is defined as

$$(a + bi) + (c + di) = (a + c) + (b + d)i$$

Notice that, with the identification of $a + bi$ with (a, b), $c + di$ with (c, d), and $(a + c) + (b + d)i$ with $(a + c, b + d)$, addition of complex numbers is the same as vector addition. The ***product*** of $a + bi$ and $c + di$ is

$$\begin{aligned}(a + bi)(c + di) &= a(c + di) + bi(c + di)\\ &= ac + adi + bci + bdi^2\end{aligned}$$

*Exercises and selected odd-numbered answers for this appendix can be found on the student companion website.

Since $i^2 = -1$, this expression simplifies to $(ac - bd) + (ad + bc)i$. Thus, we have

$$(a + bi)(c + di) = (ac - bd) + (ad + bc)i$$

Observe that, as a special case, $a(c + di) = ac + adi$, so the ***negative*** of $c + di$ is $-(c + di) = (-1)(c + di) = -c - di$. This fact allows us to compute the ***difference*** of $a + bi$ and $c + di$ as

$$\begin{aligned}(a + bi) - (c + di) &= (a + bi) + (-1)(c + di)\\ &= (a + (-c)) + (b + (-d))i\\ &= (a - c) + (b - d)i\end{aligned}$$

Example C.1

Find the sum, difference, and product of $3 - 4i$ and $-1 + 2i$.

Solution The sum is

$$(3 - 4i) + (-1 + 2i) = (3 - 1) + (-4 + 2)i = 2 - 2i$$

The difference is

$$(3 - 4i) - (-1 + 2i) = (3 - (-1)) + (-4 - 2)i = 4 - 6i$$

The product is

$$\begin{aligned}(3 - 4i)(-1 + 2i) &= -3 + 6i + 4i - 8i^2\\ &= -3 + 10i - 8(-1) = 5 + 10i\end{aligned}$$

The ***conjugate*** of $z = a + bi$ is the complex number

$$\bar{z} = a - bi$$

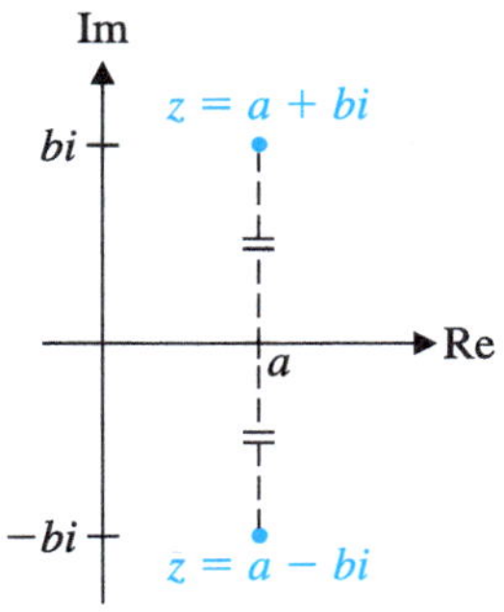

Figure C.2
Complex conjugates

($\bar{z}$ is pronounced "z bar.") Figure C.2 gives the geometric interpretation of the conjugate.

To find the quotient of two complex numbers, we multiply the numerator and the denominator by the conjugate of the denominator.

Example C.2

Express $\dfrac{-1 + 2i}{3 + 4i}$ in the form $a + bi$.

Solution We multiply the numerator and denominator by $\overline{3 + 4i} = 3 - 4i$. Using Example C.1, we obtain

$$\frac{-1 + 2i}{3 + 4i} = \frac{-1 + 2i}{3 + 4i} \cdot \frac{3 - 4i}{3 - 4i} = \frac{5 + 10i}{3^2 + 4^2} = \frac{5 + 10i}{25} = \frac{1}{5} + \frac{2}{5}i$$

On the following page is a summary of some of the properties of conjugates. The proofs follow from the definition of conjugate; you should verify them for yourself.

1. $\overline{\overline{z}} = z$
2. $\overline{z + w} = \overline{z} + \overline{w}$
3. $\overline{zw} = \overline{z}\,\overline{w}$
4. If $z \neq 0$, then $\overline{(w/z)} = \overline{w}/\overline{z}$.
5. z is real if and only if $\overline{z} = z$.

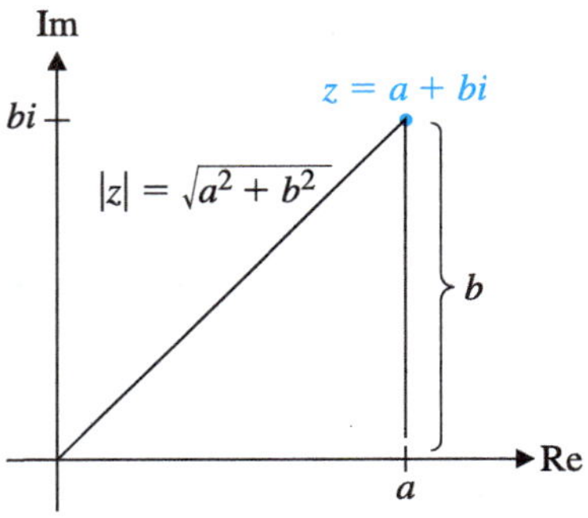

Figure C.3

The ***absolute value*** (or ***modulus***) $|z|$ of a complex number $z = a + bi$ is its distance from the origin. As Figure C.3 shows, Pythagoras' Theorem gives

$$|z| = |a + bi| = \sqrt{a^2 + b^2}$$

Observe that

$$z\overline{z} = (a + bi)(a - bi) = a^2 - abi + bai - b^2i^2 = a^2 + b^2$$

Hence,

$$z\overline{z} = |z|^2$$

This gives us an alternative way of describing the division process for the quotient of two complex numbers. If w and $z \neq 0$ are two complex numbers, then

$$\frac{w}{z} = \frac{w}{z}\cdot\frac{\overline{z}}{\overline{z}} = \frac{w\overline{z}}{z\overline{z}} = \frac{w\overline{z}}{|z|^2}$$

Below is a summary of some of the properties of absolute value. (You should try to prove these using the definition of absolute value and other properties of complex numbers.)

1. $|z| = 0$ if and only if $z = 0$.
2. $|z| = |\overline{z}|$
3. $|zw| = |z||w|$
4. If $z \neq 0$, then $\left|\frac{1}{z}\right| = \frac{1}{|z|}$.
5. $|z + w| \leq |z| + |w|$

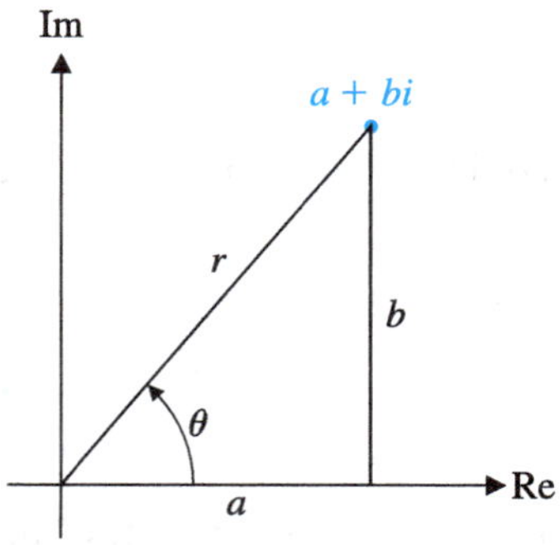

Figure C.4

Polar Form

As you have seen, the complex number $z = a + bi$ can be represented geometrically by the point (a, b). This point can also be expressed in terms of ***polar coordinates*** (r, θ), where $r \geq 0$, as shown in Figure C.4. We have

$$a = r\cos\theta \quad \text{and} \quad b = r\sin\theta$$

so

$$z = a + bi = r\cos\theta + (r\sin\theta)i$$

Thus, any complex number can be written in the ***polar form***

$$z = r(\cos\theta + i\sin\theta)$$

where $r = |z| = \sqrt{a^2 + b^2}$ and $\tan\theta = b/a$. The angle θ is called an ***argument*** of z and is denoted by arg z. Observe that arg z is not unique: Adding or subtracting any integer multiple of 2π gives another argument of z. However, there is only one argument θ that satisfies

$$-\pi < \theta \le \pi$$

This is called the ***principal argument*** of z and is denoted by Arg z.

Example C.3

Write the following complex numbers in polar form using their principal arguments:

(a) $z = 1 + i$ (b) $w = 1 - \sqrt{3}i$

Solution (a) We compute

$$r = |z| = \sqrt{1^2 + 1^2} = \sqrt{2} \quad \text{and} \quad \tan\theta = \frac{1}{1} = 1$$

Therefore, Arg $z = \theta = \dfrac{\pi}{4}$ (= 45°), and we have

$$z = \sqrt{2}\left(\cos\frac{\pi}{4} + i\sin\frac{\pi}{4}\right)$$

as shown in Figure C.5.

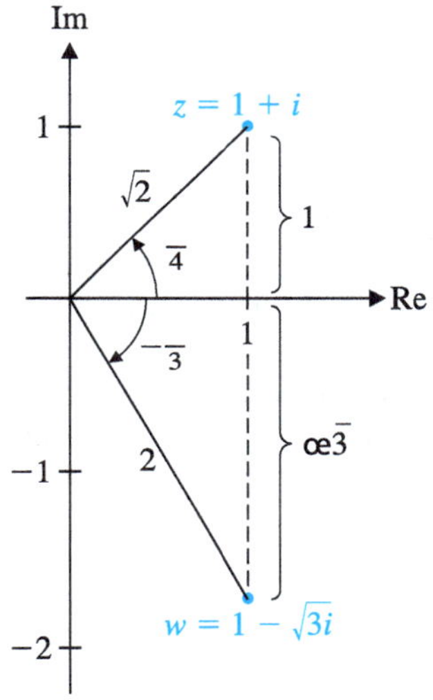

Figure C.5

(b) We have

$$r = |w| = \sqrt{1^2 + (-\sqrt{3})^2} = \sqrt{4} = 2 \quad \text{and} \quad \tan\theta = \frac{-\sqrt{3}}{1} = -\sqrt{3}$$

Since w lies in the fourth quadrant, we must have Arg $z = \theta = -\dfrac{\pi}{3}$ (= −60°). Therefore,

$$w = 2\left(\cos\left(-\frac{\pi}{3}\right) + i\sin\left(-\frac{\pi}{3}\right)\right)$$

See Figure C.5.

The polar form of complex numbers can be used to give geometric interpretations of multiplication and division. Let

$$z_1 = r_1(\cos\theta_1 + i\sin\theta_1) \quad \text{and} \quad z_2 = r_2(\cos\theta_2 + i\sin\theta_2)$$

Multiplying, we obtain

$$\begin{aligned} z_1z_2 &= r_1r_2(\cos\theta_1 + i\sin\theta_1)(\cos\theta_2 + i\sin\theta_2) \\ &= r_1r_2[(\cos\theta_1\cos\theta_2 - \sin\theta_1\sin\theta_2) + i(\sin\theta_1\cos\theta_2 + \cos\theta_1\sin\theta_2)] \end{aligned}$$

Using the trigonometric identities

$$\cos(\theta_1 + \theta_2) = \cos\theta_1\cos\theta_2 - \sin\theta_1\sin\theta_2$$
$$\sin(\theta_1 + \theta_2) = \sin\theta_1\cos\theta_2 + \cos\theta_1\sin\theta_2$$

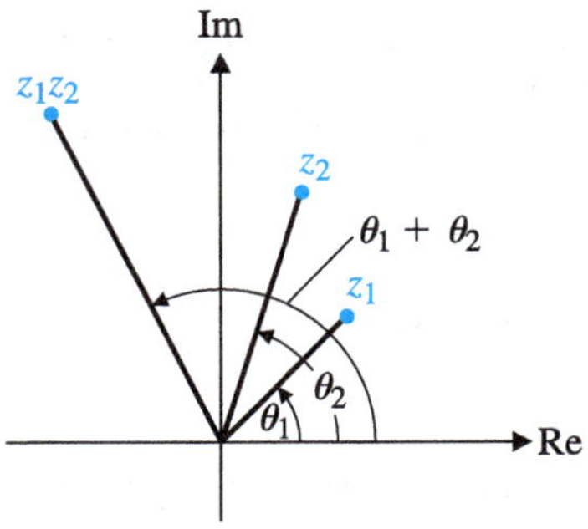

Figure C.6

we obtain

$$z_1z_2 = r_1r_2[\cos(\theta_1 + \theta_2) + i\sin(\theta_1 + \theta_2)] \quad (1)$$

which is the polar form of a complex number with absolute value r_1r_2 and argument $\theta_1 + \theta_2$. This shows that

$$|z_1z_2| = |z_1||z_2| \quad \text{and} \quad \arg(z_1z_2) = \arg z_1 + \arg z_2$$

Equation (1) says that *to multiply two complex numbers, we multiply their absolute values and add their arguments*. See Figure C.6.

Similarly, using the subtraction identities for sine and cosine, we can show that

$$\frac{z_1}{z_2} = \frac{r_1}{r_2}[\cos(\theta_1 - \theta_2) + i\sin(\theta_1 - \theta_2)] \quad \text{if } z \neq 0$$

(Verify this.) Therefore,

$$\left|\frac{z_1}{z_2}\right| = \frac{|z_1|}{|z_2|} \quad \text{and} \quad \arg\left(\frac{z_1}{z_2}\right) = \arg z_1 - \arg z_2$$

and we see that *to divide two complex numbers, we divide their absolute values and subtract their arguments*.

As a special case of the last result, we obtain a formula for the reciprocal of a complex number in polar form. Setting $z_1 = 1$ (and therefore $\theta_1 = 0$) and $z_2 = z$ (and therefore $\theta_2 = \theta$), we obtain the following:

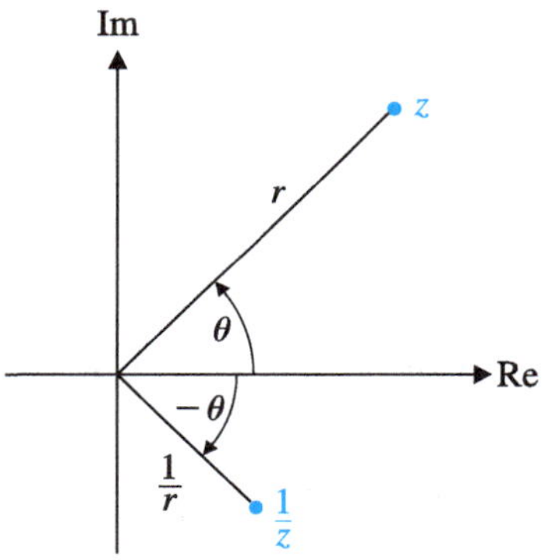

Figure C.7

If $z = r(\cos\theta + i\sin\theta)$ is nonzero, then

$$\frac{1}{z} = \frac{1}{r}(\cos\theta - i\sin\theta)$$

See Figure C.7.

Example C.4

Find the product of $1 + i$ and $1 - \sqrt{3}i$ in polar form.

Solution From Example C.3, we have

$$1 + i = \sqrt{2}\left(\cos\frac{\pi}{4} + i\sin\frac{\pi}{4}\right) \quad \text{and} \quad 1 - \sqrt{3}i = 2\left(\cos\left(-\frac{\pi}{3}\right) + i\sin\left(-\frac{\pi}{3}\right)\right)$$

Therefore,

$$\begin{aligned}(1 + i)(1 - \sqrt{3}i) &= 2\sqrt{2}\left[\cos\left(\frac{\pi}{4} - \frac{\pi}{3}\right) + i\sin\left(\frac{\pi}{4} - \frac{\pi}{3}\right)\right] \\ &= 2\sqrt{2}\left[\cos\left(-\frac{\pi}{12}\right) + i\sin\left(-\frac{\pi}{12}\right)\right]\end{aligned}$$

See Figure C.8.

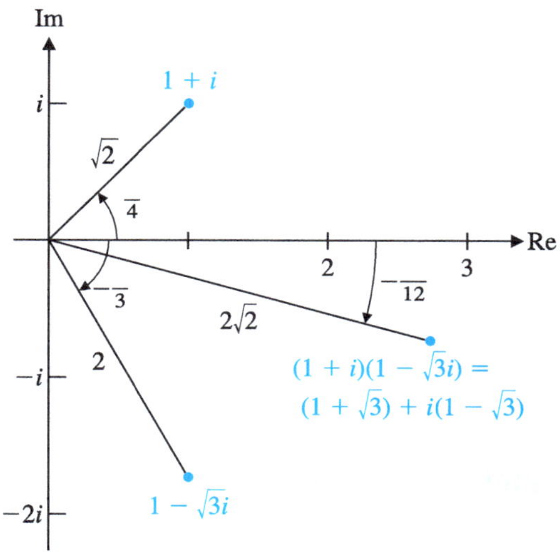

Figure C.8

Remark Since $(1 + i)(1 - \sqrt{3}i) = (1 + \sqrt{3}) + i(1 - \sqrt{3})$ (check this), we must have

$$1 + \sqrt{3} = 2\sqrt{2}\cos\left(-\frac{\pi}{12}\right) = -2\sqrt{2}\cos\left(\frac{\pi}{12}\right)$$

and

$$1 - \sqrt{3} = 2\sqrt{2}\sin\left(-\frac{\pi}{12}\right) = -2\sqrt{2}\sin\left(\frac{\pi}{12}\right)$$

(Why?) This implies that

$$\cos\left(\frac{\pi}{12}\right) = \frac{1 + \sqrt{3}}{2\sqrt{2}} \quad \text{and} \quad \sin\left(\frac{\pi}{12}\right) = \frac{\sqrt{3} - 1}{2\sqrt{2}}$$

We therefore have a method for finding the sine and cosine of an angle such as $\pi/12$ that is not a special angle but that can be obtained as a sum or difference of special angles.

De Moivre's Theorem

If n is a positive integer and $z = r(\cos\theta + i\sin\theta)$, then repeated use of Equation (1) yields formulas for the powers of z:

$$z^2 = r^2(\cos 2\theta + i\sin 2\theta)$$
$$z^3 = zz^2 = r^3(\cos 3\theta + i\sin 3\theta)$$
$$z^4 = zz^3 = r^4(\cos 4\theta + i\sin 4\theta)$$
$$\vdots$$

Abraham De Moivre (1667–1754) was a French mathematician who made important contributions to trigonometry, analytic geometry, probability, and statistics.

In general, we have the following result, known as ***De Moivre's Theorem.***

Theorem C.1 **De Moivre's Theorem**

If $z = r(\cos\theta + i\sin\theta)$ and n is a positive integer, then

$$z^n = r^n(\cos n\theta + i\sin n\theta)$$

Stated differently, we have

$$|z^n| = |z|^n \text{ and } \arg(z^n) = n \arg z$$

In words, De Moivre's Theorem says that *to take the* nth *power of a complex number, we take the* nth *power of its absolute value and multiply its argument by* n.

Example C.5

Find $(1 + i)^6$.

Solution From Example C.3(a), we have

$$1 + i = \sqrt{2}\left(\cos\frac{\pi}{4} + i\sin\frac{\pi}{4}\right)$$

Hence, De Moivre's Theorem gives

$$\begin{aligned}(1 + i)^6 &= (\sqrt{2})^6\left(\cos\frac{6\pi}{4} + i\sin\frac{6\pi}{4}\right) \\ &= 8\left(\cos\frac{3\pi}{2} + i\sin\frac{3\pi}{2}\right) \\ &= 8(0 + i(-1)) = -8i\end{aligned}$$

See Figure C.9, which shows $1 + i, (1 + i)^2, (1 + i)^3, \ldots, (1 + i)^6$.

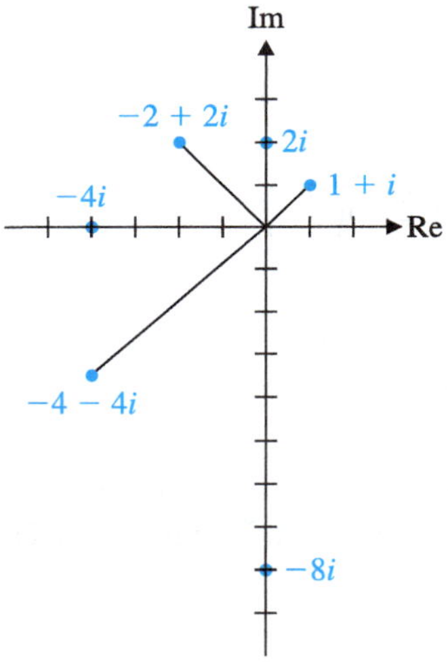

Figure C.9
Powers of $1 + i$

We can also use De Moivre's Theorem to find nth roots of complex numbers. An nth root of the complex number z is any complex number w such that

$$w^n = z$$

In polar form, we have

$$w = s(\cos\varphi + i\sin\varphi) \quad \text{and} \quad z = r(\cos\theta + i\sin\theta)$$

so, by De Moivre's Theorem,

$$s^n(\cos n\varphi + i\sin n\varphi) = r(\cos\theta + i\sin\theta)$$

Equating the absolute values, we see that

$$s^n = r \quad \text{or} \quad s = r^{1/n} = \sqrt[n]{r}$$

We must also have

$$\cos n\varphi = \cos\theta \quad \text{and} \quad \sin n\varphi = \sin\theta$$

(Why?) Since the sine and cosine functions each have period 2π, these equations imply that $n\varphi$ and θ differ by an integer multiple of 2π; that is,

$$n\varphi = \theta + 2k\pi \quad \text{or} \quad \varphi = \frac{\theta + 2k\pi}{n}$$

where k is an integer. Therefore,

$$w = r^{1/n}\left[\cos\left(\frac{\theta + 2k\pi}{n}\right) + i\sin\left(\frac{\theta + 2k\pi}{n}\right)\right]$$

describes the possible nth roots of z as k ranges over the integers. It is not hard to show that $k = 0, 1, 2, \ldots, n - 1$ produce distinct values of w, so there are exactly n different nth roots of $z = r(\cos\theta + i\sin\theta)$. We summarize this result as follows:

Let $z = r(\cos\theta + i\sin\theta)$ and let n be a positive integer. Then z has exactly n distinct nth roots given by

$$r^{1/n}\left[\cos\left(\frac{\theta + 2k\pi}{n}\right) + i\sin\left(\frac{\theta + 2k\pi}{n}\right)\right] \tag{2}$$

for $k = 0, 1, 2, \ldots, n - 1$.

Example C.6

Find the three cube roots of -27.

Solution In polar form, $-27 = 27(\cos\pi + i\sin\pi)$. It follows that the cube roots of -27 are given by

$$(-27)^{1/3} = 27^{1/3}\left[\cos\left(\frac{\pi + 2k\pi}{3}\right) + i\sin\left(\frac{\pi + 2k\pi}{3}\right)\right] \quad \text{for } k = 0, 1, 2$$

Using formula (2) with $n = 3$, we obtain

$$27^{1/3}\left[\cos\frac{\pi}{3} + i\sin\frac{\pi}{3}\right] = 3\left(\frac{1}{2} + \frac{\sqrt{3}}{2}i\right) = \frac{3}{2} + \frac{3\sqrt{3}}{2}i$$

$$27^{1/3}\left[\cos\left(\frac{\pi + 2\pi}{3}\right) + i\sin\left(\frac{\pi + 2\pi}{3}\right)\right] = 3(\cos\pi + i\sin\pi) = -3$$

$$27^{1/3}\left[\cos\left(\frac{\pi + 4\pi}{3}\right) + i\sin\left(\frac{\pi + 4\pi}{3}\right)\right] = 3\left(\cos\frac{5\pi}{3} + i\sin\frac{5\pi}{3}\right)$$

$$= 3\left(\frac{1}{2} - \frac{\sqrt{3}}{2}i\right) = \frac{3}{2} - \frac{3\sqrt{3}}{2}i$$

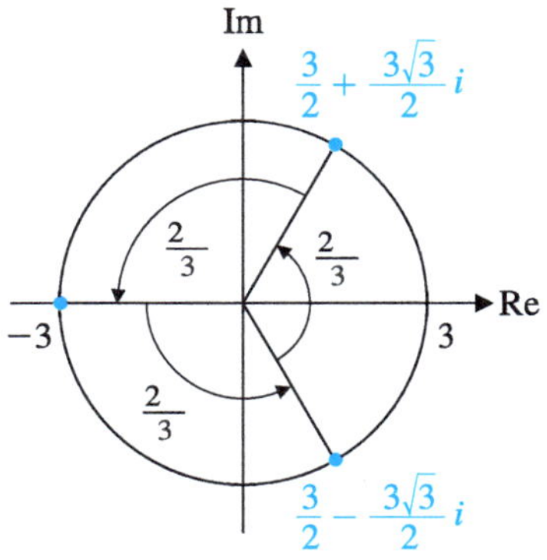

Figure C.10
The cube roots of -27

As Figure C.10 shows, the three cube roots of -27 are equally spaced $2\pi/3$ radians (120°) apart around a circle of radius 3 centered at the origin.

In general, formula (2) implies that the nth roots of $z = r(\cos\theta + i\sin\theta)$ will lie on a circle of radius $r^{1/n}$ centered at the origin. Moreover, they will be equally spaced $2\pi/n$ radians $(360/n°)$ apart. (Verify this.) Thus, if we can find one nth root of z, the remaining nth roots of z can be obtained by rotating the first root through successive increments of $2\pi/n$ radians. Had we known this in Example C.6, we could have used the fact that the real cube root of -27 is -3 and then rotated it twice through an angle of $2\pi/3$ radians (120°) to get the other two cube roots.

Leonhard Euler (1707–1783) was the most prolific mathematician of all time. He has over 900 publications to his name, and his collected works fill over 70 volumes. There are so many results attributed to him that "Euler's formula" or "Euler's Theorem" can mean many different things, depending on the context.

Euler worked in so many areas of mathematics, it is difficult to list them all. His contributions to calculus and analysis, differential equations, number theory, geometry, topology, mechanics, and other areas of applied mathematics continue to be influential. He also introduced much of the notation we currently use, including π, e, i, Σ for summation, Δ for difference, and $f(x)$ for a function, and was the first to treat sine and cosine as functions.

Euler was born in Switzerland but spent most of his mathematical life in Russia and Germany. In 1727, he joined the St. Petersburg Academy of Sciences, which had been founded by Catherine I, the wife of Peter the Great. He went to Berlin in 1741 at the invitation of Frederick the Great, but returned in 1766 to St. Petersburg, where he remained until his death. When he was young, he lost the vision in one eye as the result of an illness, and by 1776 he had lost the vision in the other eye and was totally blind. Remarkably, his mathematical output did not diminish, and he continued to be productive until the day he died.

Euler's Formula

In calculus, you learn that the function e^z has a power series expansion

$$e^z = 1 + z + \frac{z^2}{2!} + \frac{z^3}{3!} + \cdots$$

that converges for every real number z. It can be shown that this expansion also works when z is a complex number and that the complex exponential function e^z obeys the usual rules for exponents. The sine and cosine functions also have power series expansions:

$$\sin x = x - \frac{x^3}{3!} + \frac{x^5}{5!} - \frac{x^7}{7!} + - \cdots$$

$$\cos x = 1 - \frac{x^2}{2!} + \frac{x^4}{4!} - \frac{x^6}{6!} + - \cdots$$

If we let $z = ix$, where x is a real number, then we have

$$e^z = e^{ix} = 1 + ix + \frac{(ix)^2}{2!} + \frac{(ix)^3}{3!} + \cdots$$

Using the fact that $i^2 = -1$, $i^3 = -i$, $i^4 = 1$, $i^5 = i$, and so on, repeating in a cycle of length 4, we see that

$$\begin{aligned} e^{ix} &= 1 + ix - \frac{x^2}{2!} - \frac{ix^3}{3!} + \frac{x^4}{4!} + \frac{ix^5}{5!} - \frac{x^6}{6!} - \frac{ix^7}{7!} + + - - \cdots \\ &= \left(1 - \frac{x^2}{2!} + \frac{x^4}{4!} - \frac{x^6}{6!} + - \cdots\right) + i\left(x - \frac{x^3}{3!} + \frac{x^5}{5!} - \frac{x^7}{7!} + - \cdots\right) \\ &= \cos x + i \sin x \end{aligned}$$

This remarkable result is known as ***Euler's formula.***

Theorem C.2 **Euler's Formula**

For any real number x,

$$e^{ix} = \cos x + i \sin x$$

Using Euler's formula, we see that the polar form of a complex number can be written more compactly as

$$z = r(\cos\theta + i\sin\theta) = re^{i\theta}$$

For example, from Example C.3(a), we have

$$1 + i = \sqrt{2}\left(\cos\frac{\pi}{4} + i\sin\frac{\pi}{4}\right) = \sqrt{2}e^{i\pi/4}$$

We can also go in the other direction and convert a complex exponential back into polar or standard form.

Example C.7

Write the following in the form $a + bi$:

(a) $e^{i\pi}$ (b) $e^{2+i\pi/4}$

Solution (a) Using Euler's formula, we have

$$e^{i\pi} = \cos\pi + i\sin\pi = -1 + i\cdot 0 = -1$$

(If we write this equation as $e^{i\pi} + 1 = 0$, we obtain what is surely one of the most remarkable equations in mathematics. It contains the fundamental operations of addition, multiplication, and exponentiation; the additive identity 0 and the multiplicative identity 1; the two most important transcendental numbers, π and e; and the complex unit i—all in one equation!)

(b) Using rules for exponents together with Euler's formula, we obtain

$$e^{2+i\pi/4} = e^2e^{i\pi/4} = e^2\left(\cos\frac{\pi}{4} + i\sin\frac{\pi}{4}\right) = e^2\left(\frac{\sqrt{2}}{2} + i\frac{\sqrt{2}}{2}\right)$$

$$= \frac{e^2\sqrt{2}}{2} + \frac{e^2\sqrt{2}}{2}i$$

If $z = re^{i\theta} = r(\cos\theta + i\sin\theta)$, then

$$\bar{z} = r(\cos\theta - i\sin\theta) \tag{3}$$

The trigonometric identities

$$\cos(-\theta) = \cos\theta \text{ and } \sin(-\theta) = -\sin\theta$$

allow us to rewrite Equation (3) as

$$\bar{z} = r(\cos(-\theta) + i\sin(-\theta)) = re^{i(-\theta)}$$

This gives the following useful formula for the conjugate:

If $z = re^{i\theta}$, then

$$\bar{z} = re^{-i\theta}$$

Note Euler's formula gives a quick, one-line proof of De Moivre's Theorem:

$$[r(\cos\theta + i\sin\theta)]^n = (re^{i\theta})^n = r^n e^{in\theta} = r^n(\cos n\theta + i\sin n\theta)$$

Appendix D*

Polynomials

A ***polynomial*** is a function p of a single variable x that can be written in the form

$$p(x) = a_0 + a_1x + a_2x^2 + \cdots + a_nx^n \qquad (1)$$

Euler gave the most algebraic of the proofs of the existence of the roots of [a polynomial] equation. . . . I regard it as unjust to ascribe this proof exclusively to Gauss, who merely added the finishing touches.
—Georg Frobenius, 1907
Quoted on the MacTutor History of Mathematics archive, http://www-history.mcs.st-and.ac.uk/history/

where $a_0, a_1, \ldots, a_n$ are constants ($a_n \neq 0$), called the ***coefficients*** of p. With the convention that $x^0 = 1$, we can use summation notation to write p as

$$p(x) = \sum_{k=0}^{n} a_k x^k$$

The integer n is called the ***degree*** of p, which is denoted by writing $\deg p = n$. A polynomial of degree zero is called a ***constant polynomial.***

Example D.1

Which of the following are polynomials?

(a) $2 - \frac{1}{3}x + \sqrt{2}x^2$ (b) $2 - \dfrac{1}{3x^2}$ (c) $\sqrt{2x^2}$

(d) $\ln\left(\dfrac{2e^{5x^3}}{e^{3x}}\right)$ (e) $\dfrac{x^2 - 5x + 6}{x - 2}$ (f) $\sqrt{x}$

(g) $\cos(2\cos^{-1}x)$ (h) e^x

Solution (a) This is the only one that is obviously a polynomial.

(b) A polynomial of the form shown in Equation (1) cannot become infinite as x approaches a finite value $[\lim_{x \to c} p(x) \neq \pm\infty]$, whereas $2 - 1/3x^2$ approaches $-\infty$ as x approaches zero. Hence, it is not a polynomial.

(c) We have

$$\sqrt{2x^2} = \sqrt{2}\sqrt{x^2} = \sqrt{2}|x|$$

which is equal to $\sqrt{2}x$ when $x \geq 0$ and to $-\sqrt{2}x$ when $x < 0$. Therefore, this expression is formed by "splicing together" two polynomials (a *piecewise polynomial*), but it is not a polynomial itself.

*Exercises and selected odd-numbered answers for this appendix can be found on the student companion website.

(d) Using properties of exponents and logarithms, we have

$$\ln\left(\frac{2e^{5x^3}}{e^{3x}}\right) = \ln(2e^{5x^3-3x}) = \ln 2 + \ln(e^{5x^3-3x})$$
$$= \ln 2 + 5x^3 - 3x = \ln 2 - 3x + 5x^3$$

so this expression is a polynomial.

(e) The domain of this function consists of all real numbers $x \neq 2$. For these values of x, the function simplifies to

$$\frac{x^2 - 5x + 6}{x - 2} = \frac{(x-2)(x-3)}{x-2} = x - 3$$

so we can say that it is a polynomial *on its domain.*

(f) We see that this function cannot be a polynomial (even on its domain $x \geq 0$), since repeated differentiation of a polynomial of the form shown in Equation (1) eventually results in zero and $\sqrt{x}$ does not have this property. (Verify this.)

(g) The domain of this expression is $-1 \leq x \leq 1$. Let $\theta = \cos^{-1} x$ so that $\cos \theta = x$. Using a trigonometric identity, we see that

$$\cos(2\cos^{-1} x) = \cos 2\theta = 2\cos^2\theta - 1 = 2x^2 - 1$$

so this expression is a polynomial on its domain.

(h) Analyzing this expression as we did the one in (f), we conclude that it is not a polynomial.

Two polynomials are ***equal*** if the coefficients of corresponding powers of x are all equal. In particular, equal polynomials must have the same degree. The ***sum*** of two polynomials is obtained by adding together the coefficients of corresponding powers of x.

Example D.2 Find the sum of $2 - 4x + x^2$ and $1 + 2x - x^2 + 3x^3$.

Solution We compute

$$(2 - 4x + x^2) + (1 + 2x - x^2 + 3x^3) = (2 + 1) + (-4 + 2)x$$
$$+ (1 + (-1))x^2 + (0 + 3)x^3$$
$$= 3 - 2x + 3x^3$$

where we have "padded" the first polynomial by giving it an x^3 coefficient of zero.

We define the ***difference*** of two polynomials analogously, subtracting coefficients instead of adding them. The ***product*** of two polynomials is obtained by repeatedly using the distributive law and then gathering together corresponding powers of x.

Example D.3

Find the product of $2 - 4x + x^2$ and $1 + 2x - x^2 + 3x^3$.

Solution We obtain

$$\begin{aligned}
&(2 - 4x + x^2)(1 + 2x - x^2 + 3x^3) \\
&\quad = 2(1 + 2x - x^2 + 3x^3) - 4x(1 + 2x - x^2 + 3x^3) \\
&\qquad + x^2(1 + 2x - x^2 + 3x^3) \\
&\quad = (2 + 4x - 2x^2 + 6x^3) + (-4x - 8x^2 + 4x^3 - 12x^4) \\
&\qquad + (x^2 + 2x^3 - x^4 + 3x^5) \\
&\quad = 2 + (4x - 4x) + (-2x^2 - 8x^2 + x^2) + (6x^3 + 4x^3 + 2x^3) \\
&\qquad + (-12x^4 - x^4) + 3x^5 \\
&\quad = 2 - 9x^2 + 12x^3 - 13x^4 + 3x^5
\end{aligned}$$

Observe that for two polynomials p and q, we have

$$\deg(pq) = \deg p + \deg q$$

If p and q are polynomials with $\deg q \leq \deg p$, we can divide q into p, using long division to obtain the quotient p/q. The next example illustrates the procedure, which is the same as for long division of one integer into another. Just as the quotient of two integers is not, in general, an integer, the quotient of two polynomials is not, in general, another polynomial.

Example D.4

Compute $\dfrac{1 + 2x - x^2 + 3x^3}{2 - 4x + x^2}$.

Solution We will perform long division. It is helpful to write each polynomial with *decreasing* powers of x. Accordingly, we have

$$x^2 - 4x + 2\overline{)3x^3 - x^2 + 2x + 1}$$

We begin by dividing x^2 into $3x^3$ to obtain the partial quotient $3x$. We then multiply $3x$ by the divisor $x^2 - 4x + 2$, subtract the result, and bring down the next term from the dividend $(3x^3 - x^2 + 2x + 1)$:

$$\begin{array}{r@{}l}
 & 3x \\
x^2 - 4x + 2\big) & \overline{3x^3 - x^2 + 2x + 1} \\
 & \underline{3x^3 - 12x^2 + 6x\phantom{{}+1}} \\
 & 11x^2 - 4x + 1
\end{array}$$

Then we repeat the process with $11x^2$, multiplying 11 by $x^2 - 4x + 2$ and subtracting the result from $11x^2 - 4x + 1$. We obtain

$$\begin{array}{r@{}l}
 & 3x + 11 \\
x^2 - 4x + 2\big) & \overline{3x^3 - x^2 + 2x + 1} \\
 & \underline{3x^3 - 12x^2 + 6x\phantom{{}+22}} \\
 & 11x^2 - 4x + 1 \\
 & \underline{11x^2 - 44x + 22} \\
 & 40x - 21
\end{array}$$

We now have a remainder $40x - 21$. Its degree is less than that of the divisor $x^2 - 4x + 2$, so the process stops, and we have found that

$$3x^3 - x^2 + 2x + 1 = (x^2 - 4x + 2)(3x + 11) + (40x - 21)$$

or
$$\frac{3x^3 - x^2 + 2x + 1}{x^2 - 4x + 2} = 3x + 11 + \frac{40x - 21}{x^2 - 4x + 2}$$

Example D.4 can be generalized to give the following result, known as the ***division algorithm.***

Theorem D.1 **The Division Algorithm**

If f and g are polynomials with $\deg g \leq \deg f$, then there are polynomials q and r such that

$$f(x) = g(x)q(x) + r(x)$$

where either $r = 0$ or $\deg r < \deg g$.

In Example D.4,

$$f(x) = 3x^3 - x^2 + 2x + 1, \quad g(x) = x^2 - 4x + 2, \quad q(x) = 3x + 11,$$
$$\text{and} \quad r(x) = 40x - 21$$

In the division algorithm, if the remainder is zero, then

$$f(x) = g(x)q(x)$$

and we say that g is a ***factor*** of f. (Notice that q is also a factor of f.) There is a close connection between the factors of a polynomial and its zeros. A ***zero*** of a polynomial f is a number a such that $f(a) = 0$. [The number a is also called a ***root*** of the polynomial equation $f(x) = 0$.] The following result, known as the ***Factor Theorem,*** establishes the connection between factors of a polynomial and its zeros.

Theorem D.2 **The Factor Theorem**

Let f be a polynomial and let a be a constant. Then a is a zero of f if and only if $x - a$ is a factor of $f(x)$.

Proof By the division algorithm,

$$f(x) = (x - a)q(x) + r(x)$$

where either $r(x) = 0$ or $\deg r < \deg(x - a) = 1$. Thus, in either case, $r(x) = r$ is a constant. Now,

$$f(a) = (a - a)q(a) + r = r$$

so $f(a) = 0$ if and only if $r = 0$, which is equivalent to

$$f(x) = (x - a)q(x)$$

as we needed to prove.

There is no method that is guaranteed to find the zeros of a given polynomial. However, there are some guidelines that are useful in special cases. The case of a polynomial with *integer* coefficients is particularly interesting. The following result, known as the ***Rational Roots Theorem,*** gives criteria for a zero of such a polynomial to be a *rational* number.

Theorem D.3 The Rational Roots Theorem

Let

$$f(x) = a_0 + a_1x + \cdots + a_nx^n$$

be a polynomial with integer coefficients and let a/b be a rational number written in lowest terms. If a/b is a zero of f, then a_0 is a multiple of a and a_n is a multiple of b.

Proof If a/b is a zero of f, then

$$a_0 + a_1\left(\frac{a}{b}\right) + \cdots + a_{n-1}\left(\frac{a}{b}\right)^{n-1} + a_n\left(\frac{a}{b}\right)^n = 0$$

Multiplying through by b^n, we have

$$a_0b^n + a_1ab^{n-1} + \cdots + a_{n-1}a^{n-1}b + a_na = 0 \qquad (1)$$

which implies that

$$a_0b^n + a_1ab^{n-1} + \cdots + a_{n-1}a^{n-1}b = -a_na^n \qquad (2)$$

The left-hand side of Equation (2) is a multiple of b, so a_na^n must be a multiple of b also. Since a/b is in lowest terms, a and b have no common factors greater than 1. Therefore, a_n must be a multiple of b.

We can also write Equation (1) as

$$-a_0b^n = a_1ab^{n-1} + \cdots + a_{n-1}a^{n-1}b + a_na^n$$

and a similar argument shows that a_0 must be a multiple of a. (Show this.)

Example D.5

Find all the rational roots of the equation

$$6x^3 + 13x^2 - 4 = 0 \qquad (3)$$

Solution If a/b is a root of this equation, then 6 is a multiple of b and -4 is a multiple of a, by the Rational Roots Theorem. Therefore,

$$a \in \{\pm1, \pm2, \pm4\} \quad \text{and} \quad b \in \{\pm1, \pm2, \pm3, \pm6\}$$

Forming all possible rational numbers a/b with these choices of a and b, we see that the only possible rational roots of the given equation are

$$\pm 1, \pm 2, \pm 4, \pm \tfrac{1}{2}, \pm \tfrac{1}{3}, \pm \tfrac{2}{3}, \pm \tfrac{4}{3}, \pm \tfrac{1}{6}$$

Substituting these values into Equation (3) one at a time, we find that -2, $-\frac{2}{3}$, and $\frac{1}{2}$ are the only values from this list that are actually roots. (Check these.) As we will see shortly, a polynomial equation of degree 3 cannot have more than three roots, so these are not only all the *rational* roots of Equation (3) but also its *only* roots.

We can improve on the trial-and-error method of Example D.5 in various ways. For example, once we find one root a of a given polynomial equation $f(x) = 0$, we know that $x - a$ is a factor of $f(x)$—say, $f(x) = (x - a)g(x)$. We can therefore divide $f(x)$ by $x - a$ (using long division) to find $g(x)$. Since $\deg g < \deg f$, the roots of $g(x) = 0$ [which are also roots of $f(x) = 0$] may be easier to find. In particular, if $g(x)$ is a quadratic polynomial, we have access to the ***quadratic formula.***

Suppose

$$ax^2 + bx + c = 0$$

(We may assume that a is positive, since multiplying both sides by -1 would produce an equivalent equation otherwise.) Then, completing the square, we have

$$a\left(x^2 + \frac{b}{a}x + \frac{b^2}{4a^2}\right) = \frac{b^2}{4a} - c$$

(Verify this.) Equivalently,

$$a\left(x + \frac{b}{2a}\right)^2 = \frac{b^2}{4a} - c \quad \text{or} \quad \left(x + \frac{b}{2a}\right)^2 = \frac{b^2 - 4ac}{4a^2}$$

Therefore,

$$x + \frac{b}{2a} = \pm\sqrt{\frac{b^2 - 4ac}{4a^2}} = \frac{\pm\sqrt{b^2 - 4ac}}{2a}$$

or

$$x = \frac{-b \pm \sqrt{b^2 - 4ac}}{2a}$$

Let's revisit the equation from Example D.5 with the quadratic formula in mind.

Example D.6

Find the roots of $6x^3 + 13x^2 - 4 = 0$.

Solution Let's suppose we use the Rational Roots Theorem to discover that $x = -2$ is a rational root of $6x^3 + 13x^2 - 4 = 0$. Then $x + 2$ is a factor of $6x^3 + 13x^2 - 4$, and long division gives

$$6x^3 + 13x^2 - 4 = (x + 2)(6x^2 + x - 2)$$

(Check this.) We can now apply the quadratic formula to the second factor to find that its zeros are

$$\begin{aligned} x &= \frac{-1 \pm \sqrt{1^2 - 4(6)(-2)}}{2 \cdot 6} \\ &= \frac{-1 \pm \sqrt{49}}{12} = \frac{-1 \pm 7}{12} \\ &= \tfrac{6}{12}, -\tfrac{8}{12} \end{aligned}$$

or, in lowest terms, $\frac{1}{2}$ and $-\frac{2}{3}$. Thus, the three roots of Equation (3) are $-2, \frac{1}{2}$, and $-\frac{2}{3}$, as we determined in Example D.5.

Remark The Factor Theorem establishes a connection between the zeros of a polynomial and its *linear* factors. However, a polynomial without linear factors may still have factors of higher degree. Furthermore, when asked to factor a polynomial, we need to know the number system to which the coefficients of the factors are supposed to belong.

For example, consider the polynomial

$$p(x) = x^4 + 1$$

Over the *rational numbers* $\mathbb{Q}$, the only possible zeros of p are 1 and -1, by the Rational Roots Theorem. A quick check shows that neither of these actually works, so $p(x)$ has no *linear* factors with rational coefficients, by the Factor Theorem. However, $p(x)$ may still factor into a product of two *quadratics*. We will check for quadratic factors using the method of ***undetermined coefficients.***

Suppose that

$$x^4 + 1 = (x^2 + ax + b)(x^2 + cx + d)$$

Expanding the right-hand side and comparing coefficients, we obtain the equations

$$\begin{aligned} a + c &= 0 \\ b + ac + d &= 0 \\ bc + ad &= 0 \\ bd &= 1 \end{aligned}$$

If $a = 0$, then $c = 0$ and $d = -b$. This gives $-b^2 = 1$, which has no solutions in $\mathbb{Q}$. Hence, we may assume that $a \neq 0$. Then $c = -a$, and we obtain $d = b$. It now follows that $b^2 = 1$, so $b = 1$ or $b = -1$. This implies that $a^2 = 2$ or $a^2 = -2$, respectively, neither of which has solutions in $\mathbb{Q}$. It follows that $x^4 + 1$ cannot be factored over $\mathbb{Q}$. We say that it is ***irreducible*** over $\mathbb{Q}$.

However, over the *real numbers* $\mathbb{R}$, $x^4 + 1$ does factor. The calculations we have just done show that

$$x^4 + 1 = (x^2 + \sqrt{2}x + 1)(x^2 - \sqrt{2}x + 1)$$

(Why?) To see whether we can factor further, we apply the quadratic formula. We see that the first factor has zeros

$$x = \frac{-\sqrt{2} \pm \sqrt{(\sqrt{2})^2 - 4}}{2} = \frac{-\sqrt{2} \pm \sqrt{-2}}{2} = \frac{\sqrt{2}}{2}(-1 \pm i) = -\frac{1}{\sqrt{2}} \pm \frac{1}{\sqrt{2}}i$$

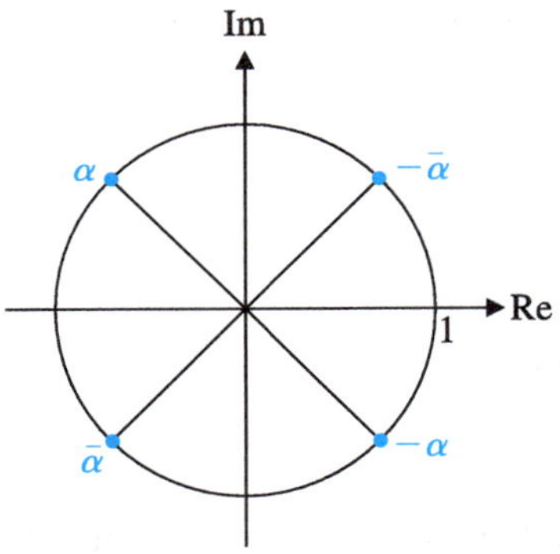

Figure D.1

which are in $\mathbb{C}$ but not in $\mathbb{R}$. Hence, $x^2 + \sqrt{2}x + 1$ cannot be factored into linear factors over $\mathbb{R}$. Similarly, $x^2 - \sqrt{2}x + 1$ cannot be factored into linear factors over $\mathbb{R}$.

Our calculations show that a complete factorization of $x^4 + 1$ is possible over the *complex numbers* $\mathbb{C}$. The four zeros of $x^4 + 1$ are

$$\alpha = -\frac{1}{\sqrt{2}} + \frac{1}{\sqrt{2}}i, \quad \bar{\alpha} = -\frac{1}{\sqrt{2}} - \frac{1}{\sqrt{2}}i, \quad -\bar{\alpha} = \frac{1}{\sqrt{2}} + \frac{1}{\sqrt{2}}i,$$
$$-\alpha = \frac{1}{\sqrt{2}} - \frac{1}{\sqrt{2}}i$$

which, as Figure D.1 shows, all lie on the unit circle in the complex plane. Thus, the factorization of $x^4 + 1$ is

$$x^4 + 1 = (x - \alpha)(x - \bar{\alpha})(x + \bar{\alpha})(x + \alpha)$$

The preceding Remark illustrates several important properties of polynomials. Notice that the polynomial $p(x) = x^4 + 1$ satisfies $\deg p = 4$ and has exactly four zeros in $\mathbb{C}$. Furthermore, its complex zeros occur in *conjugate pairs*; that is, its complex zeros can be paired up as

$$\{\alpha, \bar{\alpha}\} \quad \text{and} \quad \{-\alpha, -\bar{\alpha}\}$$

These last two facts are true in general. The first is an instance of the ***Fundamental Theorem of Algebra (FTA)***, a result that was first proved by Gauss in 1797.

Theorem D.4 **The Fundamental Theorem of Algebra**

Every polynomial of degree n with real or complex coefficients has exactly n zeros (counting multiplicities) in $\mathbb{C}$.

This important theorem is sometimes stated as

"Every polynomial with real or complex coefficients has a zero in $\mathbb{C}$."

Let's call this statement FTA′. Certainly, FTA implies FTA′. Conversely, if FTA′ is true, then if we have a polynomial p of degree n, it has a zero α in $\mathbb{C}$. The Factor Theorem then tells us that $x - \alpha$ is a factor of $p(x)$, so

$$p(x) = (x - \alpha)q(x)$$

where q is a polynomial of degree $n - 1$ (also with real or complex coefficients). We can now apply FTA′ to q to get another zero, and so on, making FTA true. This argument can be made into a nice induction proof. (Try it.)

It is not possible to give a formula (along the lines of the quadratic formula) for the zeros of polynomials of degree 5 or more. (The work of Abel and Galois confirmed this; see page 311.) Consequently, other methods must be used to prove FTA. The proof that Gauss gave uses topological methods and can be found in more advanced mathematics courses.

Now suppose that

$$p(x) = a_0 + a_1x + \cdots + a_nx^n$$

is a polynomial with real coefficients. Let α be a complex zero of p so that

$$a_0 + a_1\alpha + \cdots + a_n\alpha^n = p(\alpha) = 0$$

Then, using properties of conjugates, we have

$$\begin{aligned} p(\overline{\alpha}) &= a_0 + a_1\overline{\alpha} + \cdots + a_n\overline{\alpha}^n = \overline{a}_0 + \overline{a}_1\overline{\alpha} + \cdots + \overline{a}_n\overline{\alpha}^n \\ &= \overline{a_0 + a_1\alpha + \cdots + a_n\alpha^n} \\ &= \overline{p(\alpha)} = \overline{0} = 0 \end{aligned}$$

Thus, $\overline{\alpha}$ is also a zero of p. This proves the following result:

The complex zeros of a polynomial with real coefficients occur in conjugate pairs.

In some situations, we do not need to know *what* the zeros of a polynomial are—we only need to know *where* they are located. For example, we might only need to know whether the zeros are positive or negative (as in Theorem 4.35). One theorem that is useful in this regard is ***Descartes' Rule of Signs.*** It allows us to make certain predictions about the number of positive zeros of a polynomial with real coefficients based on the signs of these coefficients.

Descartes' stated this rule in his 1637 book *La Géometrie,* but did not give a proof. Several mathematicians later furnished a proof, and Gauss provided a somewhat sharper version of the theorem in 1828.

Given a polynomial $a_0 + a_1x + \cdots + a_nx^n$, write its nonzero coefficients in order. Replace each positive coefficient by a plus sign and each negative coefficient by a minus sign. We will say that the polynomial has k ***sign changes*** if there are k places where the coefficients change sign. For example, the polynomial $2 - 3x + 4x^3 + x^4 - 7x^5$ has the sign pattern

$$\underbrace{+\ -}\underbrace{\ +}\ \underbrace{+\ -}$$

so it has three sign changes, as indicated.

Theorem D.5 **Descartes' Rule of Signs**

Let p be a polynomial with real coefficients that has k sign changes. Then the number of positive zeros of p (counting multiplicities) is at most k.

In words, Descartes' Rule of Signs says that a real polynomial cannot have more positive zeros than it has sign changes.

Example D.7

Show that the polynomial $p(x) = 4 + 2x^2 - 7x^4$ has exactly one positive zero.

Solution The coefficients of p have the sign pattern $+ + -$, which has only one sign change. So, by Descartes' Rule of Signs, p has at most one positive zero. But $p(0) = 4$ and $p(1) = -1$, so there is a zero somewhere in the interval (0, 1). Hence, this is the only positive zero of p.

We can also use Descartes' Rule of Signs to give a bound on the number of *negative* zeros of a polynomial with real coefficients. Let

$$p(x) = a_0 + a_1x + a_2x^2 + \cdots + a_nx^n$$

and let b be a negative zero of p. Then $b = -c$ for $c > 0$, and we have

$$\begin{aligned} 0 = p(b) &= a_0 + a_1b + a_2b^2 + \cdots + a_nb^n \\ &= a_0 - a_1c + a_2c^2 - + \cdots + (-1)^n a_nc^n \end{aligned}$$

But $$p(-x) = a_0 - a_1x + a_2x^2 - + \cdots + (-1)^n a_nx^n$$

so c is a positive zero of $p(-x)$. Therefore, $p(x)$ has exactly as many negative zeros as $p(-x)$ has positive zeros. Combined with Descartes' Rule of Signs, this observation yields the following:

> Let p be a polynomial with real coefficients. Then the number of negative zeros of p is at most the number of sign changes of $p(-x)$.

Example D.8

Show that the zeros of $p(x) = 1 + 3x + 2x^2 + x^5$ cannot all be real.

Solution The coefficients of $p(x)$ have no sign changes, so p has no positive zeros. Since $p(-x) = 1 - 3x + 2x^2 - x^5$ has three sign changes among its coefficients, p has at most three negative zeros. We note that 0 is not a zero of p either, so p has at most three real zeros. Therefore, it has at least two complex (nonreal) zeros.

Answers to Selected Odd-Numbered Exercises

Answers are easy. It's asking the right questions [that's] hard.

—Doctor Who
"The Face of Evil,"
By Chris Boucher
BBC, 1977

Chapter 1

Exercises 1.1

1.

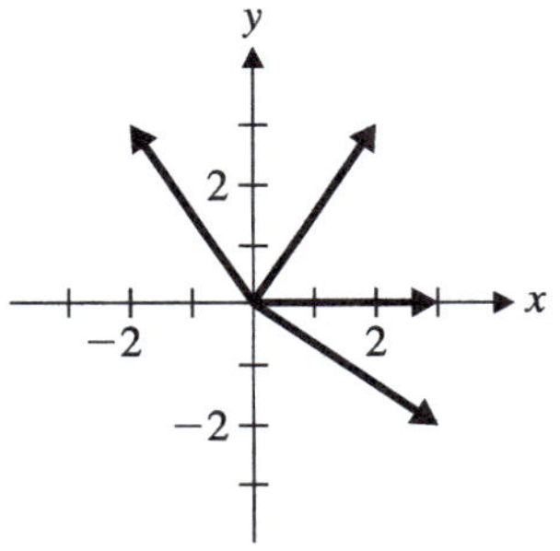

3. (a), (b)

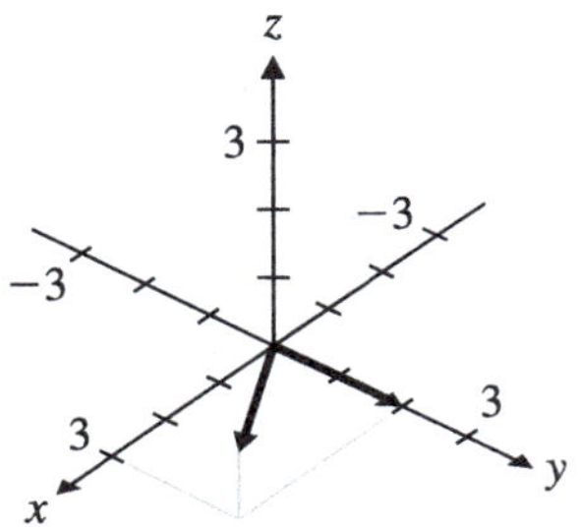

(c)

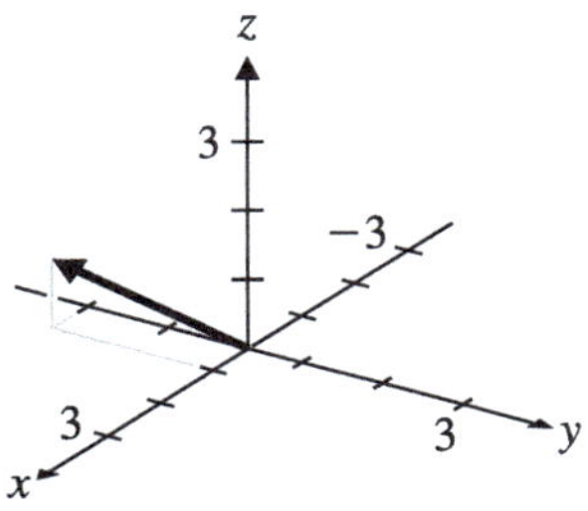

(d)

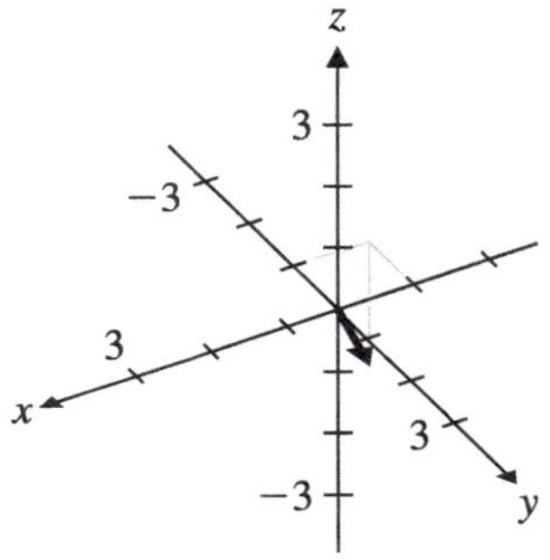

5. (a)

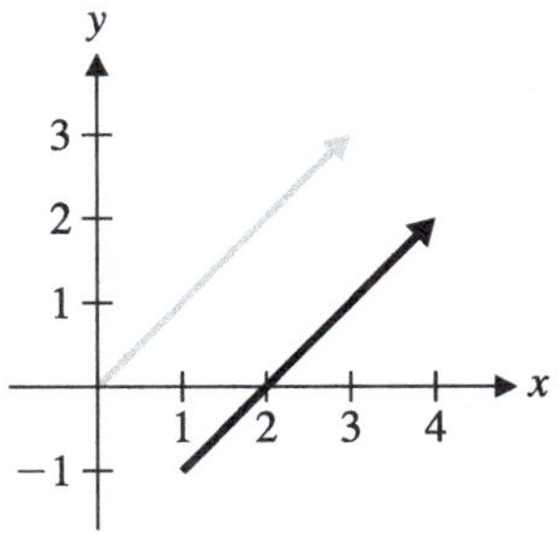

(b)

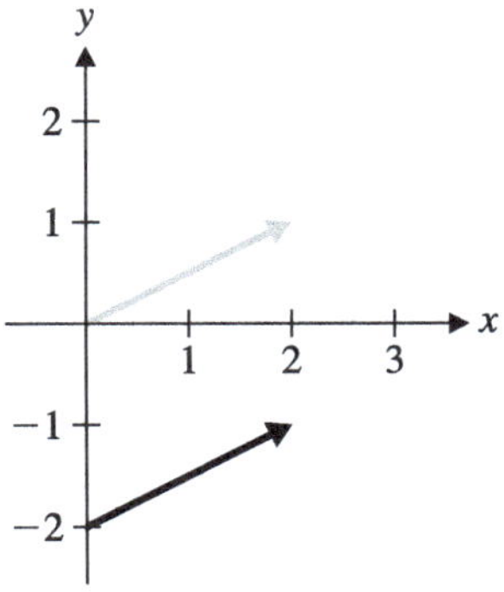

(c)

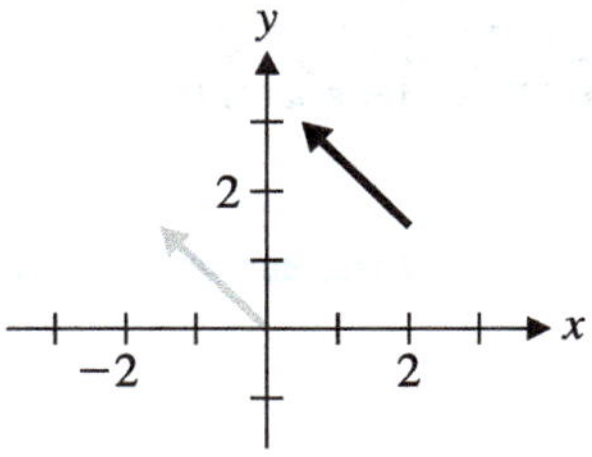

(d)

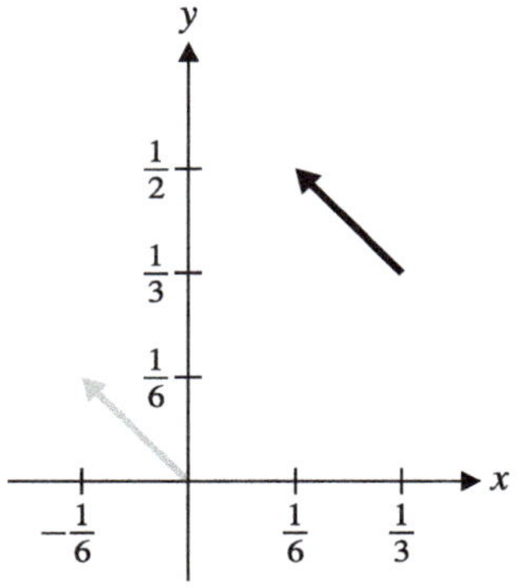

7. $\mathbf{a} + \mathbf{b} = [5, 3]$

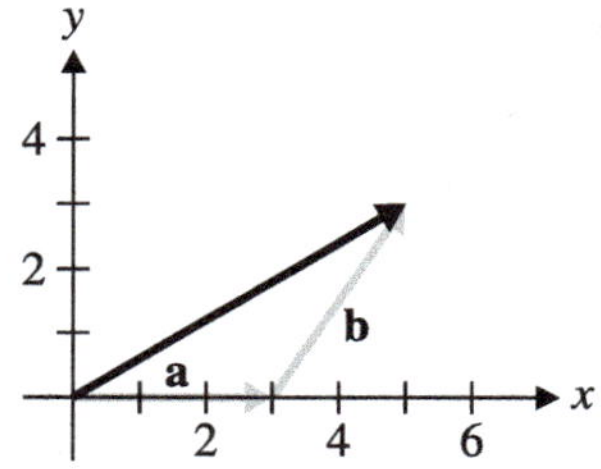

9. $\mathbf{d} - \mathbf{c} = [5, -5]$

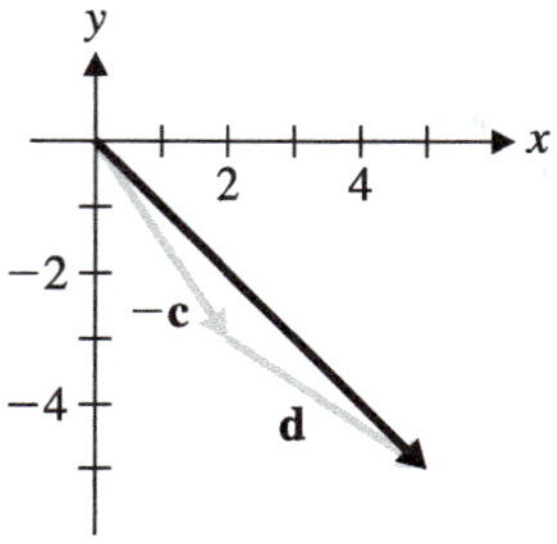

11. $[3, -2, 3]$

13. $\mathbf{u} = \begin{bmatrix} 1/2 \\ \sqrt{3}/2 \end{bmatrix}, \mathbf{v} = \begin{bmatrix} -\sqrt{3}/2 \\ -1/2 \end{bmatrix}, \mathbf{u} + \mathbf{v} = \begin{bmatrix} (1 - \sqrt{3})/2 \\ (\sqrt{3} - 1)/2 \end{bmatrix}, \mathbf{u} - \mathbf{v} = \begin{bmatrix} (1 + \sqrt{3})/2 \\ (1 + \sqrt{3})/2 \end{bmatrix}$

15. $5\mathbf{a}$

17. $\mathbf{x} = 3\mathbf{a}$

19.

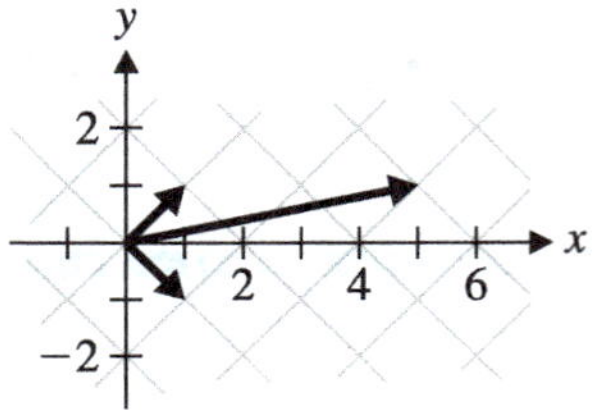

21. $\mathbf{w} = -2\mathbf{u} + 4\mathbf{v}$

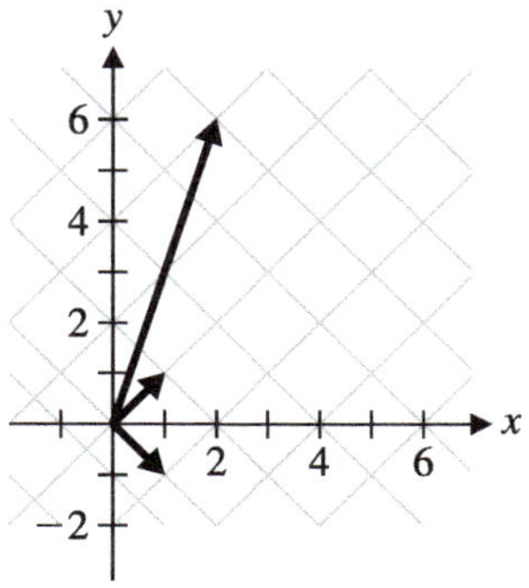

25. $\mathbf{u} + \mathbf{v} = \begin{bmatrix} 1 \\ 0 \end{bmatrix}$

27. $\mathbf{u} + \mathbf{v} = [0, 1, 0, 0]$

29.

+	0	1	2	3
0	0	1	2	3
1	1	2	3	0
2	2	3	0	1
3	3	0	1	2

·	0	1	2	3
0	0	0	0	0
1	0	1	2	3
2	0	2	0	2
3	0	3	2	1

31. 0

33. 1

35. 0

37. 2, 0, 3

39. 5

41. [1, 1, 0]

43. [0, 0, 2, 2], [2, 3, 1, 1]

45. $x = 2$

47. No solution

49. $x = 3$

51. No solution

53. $x = 2$

55. $x = 1$, or $x = 5$

57. **(a)** All $a \neq 0$ **(b)** $a = 1, 5$

(c) a and m can have no common factors other than 1 [i.e., the greatest common divisor (gcd) of a and m is 1].

Exercises 1.2

1. -1

3. 11

5. 2

7. $\sqrt{5}, \begin{bmatrix} -1/\sqrt{5} \\ 2/\sqrt{5} \end{bmatrix}$

9. $\sqrt{14}, \begin{bmatrix} 1/\sqrt{14} \\ 2/\sqrt{14} \\ 3/\sqrt{14} \end{bmatrix}$

11. $\sqrt{6}$, $[1/\sqrt{6}, 1/\sqrt{3}, 1/\sqrt{2}, 0]$

13. $\sqrt{17}$ **15.** $\sqrt{6}$

17. (a) $\mathbf{u} \cdot \mathbf{v}$ is a scalar, not a vector.

(c) $\mathbf{v} \cdot \mathbf{w}$ is a scalar and $\mathbf{u}$ is a vector.

19. Acute **21.** Acute **23.** Acute

25. 60° **27.** ≈88.10° **29.** ≈14.34°

31. Since $\overrightarrow{AB} \cdot \overrightarrow{AC} = \begin{bmatrix} -4 \\ 1 \\ -1 \end{bmatrix} \cdot \begin{bmatrix} 1 \\ 1 \\ -3 \end{bmatrix} = 0$, $\angle BAC$ is a right angle.

33. If we take the cube to be a unit cube (as in Figure 1.34), the four diagonals are given by the vectors

$$\mathbf{d}_1 = \begin{bmatrix} 1 \\ 1 \\ 1 \end{bmatrix}, \mathbf{d}_2 = \begin{bmatrix} 1 \\ 1 \\ -1 \end{bmatrix}, \mathbf{d}_3 = \begin{bmatrix} -1 \\ 1 \\ 1 \end{bmatrix}, \mathbf{d}_4 = \begin{bmatrix} 1 \\ -1 \\ 1 \end{bmatrix}$$

Since $\mathbf{d}_i \cdot \mathbf{d}_j \neq 0$ for all $i \neq j$ (six possibilities), no two diagonals are perpendicular.

35. $D = (-2, 1, 1)$

37. 5 mi/h at an angle of ≈ 53.13° to the bank

39. 60°

41. $\begin{bmatrix} -\frac{3}{5} \\ \frac{4}{5} \end{bmatrix}$ **43.** $\begin{bmatrix} \frac{3}{2} \\ -\frac{3}{2} \\ \frac{3}{2} \\ -\frac{3}{2} \end{bmatrix}$ **45.** $\begin{bmatrix} -0.301 \\ 0.033 \\ -0.252 \end{bmatrix}$

47. $\mathcal{A} = \sqrt{45}/2$ **49.** $k = -2, 3$

51. $\mathbf{v}$ is of the form $k\begin{bmatrix} b \\ -a \end{bmatrix}$, where k is a scalar.

53. The Cauchy-Schwarz Inequality would be violated.

Exercises 1.3

1. (a) $\begin{bmatrix} 3 \\ 2 \end{bmatrix} \cdot \begin{bmatrix} x \\ y \end{bmatrix} = 0$ **(b)** $3x + 2y = 0$

3. (a) $\begin{bmatrix} x \\ y \end{bmatrix} = \begin{bmatrix} 1 \\ 0 \end{bmatrix} + t\begin{bmatrix} -1 \\ 3 \end{bmatrix}$

(b) $x = 1 - t$
$y = 3t$

5. (a) $\begin{bmatrix} x \\ y \\ z \end{bmatrix} = t\begin{bmatrix} 1 \\ -1 \\ 4 \end{bmatrix}$ **(b)** $x = t$, $y = -t$, $z = 4t$

7. (a) $\begin{bmatrix} 3 \\ 2 \\ 1 \end{bmatrix} \cdot \begin{bmatrix} x \\ y \\ z \end{bmatrix} = 2$ **(b)** $3x + 2y + z = 2$

9. (a) $\begin{bmatrix} x \\ y \\ z \end{bmatrix} = s\begin{bmatrix} 2 \\ 1 \\ 2 \end{bmatrix} + t\begin{bmatrix} -3 \\ 2 \\ 1 \end{bmatrix}$

(b) $x = 2s - 3t$
$y = s + 2t$
$z = 2s + t$

11. $\begin{bmatrix} x \\ y \end{bmatrix} = \begin{bmatrix} 1 \\ -2 \end{bmatrix} + t\begin{bmatrix} 2 \\ 2 \end{bmatrix}$

13. $\begin{bmatrix} x \\ y \\ z \end{bmatrix} = \begin{bmatrix} 1 \\ 1 \\ 1 \end{bmatrix} + s\begin{bmatrix} 3 \\ -1 \\ 1 \end{bmatrix} + t\begin{bmatrix} -1 \\ 0 \\ -2 \end{bmatrix}$

15. (a) $x = t$, $y = -1 + 3t$ $\quad \begin{bmatrix} x \\ y \end{bmatrix} = \begin{bmatrix} 0 \\ -1 \end{bmatrix} + t\begin{bmatrix} 1 \\ 3 \end{bmatrix}$

17. Direction vectors for the two lines are given by $\mathbf{d}_1 = \begin{bmatrix} 1 \\ m_1 \end{bmatrix}$ and $\mathbf{d}_2 = \begin{bmatrix} 1 \\ m_2 \end{bmatrix}$. The lines are perpendicular if and only if $\mathbf{d}_1$ and $\mathbf{d}_2$ are orthogonal. But $\mathbf{d}_1 \cdot \mathbf{d}_2 = 0$ if and only if $1 + m_1 m_2 = 0$ or, equivalently, $m_1 m_2 = -1$.

19. (a) Perpendicular **(b)** Parallel
(c) Perpendicular **(d)** Perpendicular

21. $\begin{bmatrix} x \\ y \end{bmatrix} = \begin{bmatrix} 2 \\ -1 \end{bmatrix} + t\begin{bmatrix} 3 \\ 2 \end{bmatrix}$

23. $\begin{bmatrix} x \\ y \\ z \end{bmatrix} = \begin{bmatrix} -1 \\ 0 \\ 3 \end{bmatrix} + t\begin{bmatrix} -1 \\ 3 \\ -1 \end{bmatrix}$

25. (a) $x = 0, x = 1, y = 0, y = 1, z = 0, z = 1$
(b) $x - y = 0$ **(c)** $x + y - z = 0$

27. $3\sqrt{2}/2$ **29.** $2\sqrt{3}/3$ **31.** $(\frac{1}{2}, \frac{1}{2})$

33. $(\frac{4}{3}, \frac{4}{3}, \frac{8}{3})$ **35.** $18\sqrt{13}/13$ **37.** $\frac{5}{3}$

43. ≈78.9° **45.** ≈80.4°

Exercises 1.4

1. 13 N at approx N 67.38 E

3. $8\sqrt{3}$ N at an angle of 30° to $\mathbf{f}_1$

5. 4 N at an angle of 60° to $\mathbf{f}_2$

7. 5 N at an angle of 60° to the given force, $5\sqrt{3}$ N perpendicular to the 5 N force

9. $750\sqrt{2}$ N

11. 980 N

13. ≈ 117.6 N in the 15 cm wire, ≈ 88.2 N in the 20 cm wire

Review Questions

1. (a) T **(c)** F **(e)** T **(g)** F **(i)** T

3. $\mathbf{x} = \begin{bmatrix} 8 \\ 11 \end{bmatrix}$ **5.** 120° **7.** $\begin{bmatrix} -2/\sqrt{5} \\ 1/\sqrt{5} \\ 0 \end{bmatrix}$

9. $2x + 3y - z = 7$ **11.** $\sqrt{6}/2$

13. The Cauchy-Schwarz Inequality would be violated.

15. $2\sqrt{6}/3$ **17.** $x = 2$ **19.** 3

Chapter 2

Exercises 2.1

1. Linear **3.** Not linear because of the x^{-1} term

5. Not linear **7.** $2x + 4y = 7$

9. $x + y = 4 (x, y \neq 0)$

11. $\left\{\begin{bmatrix} 2t \\ t \end{bmatrix}\right\}$ **13.** $\left\{\begin{bmatrix} 4 - 2s - 3t \\ s \\ t \end{bmatrix}\right\}$

15. Unique solution, $x = 3, y = -3$

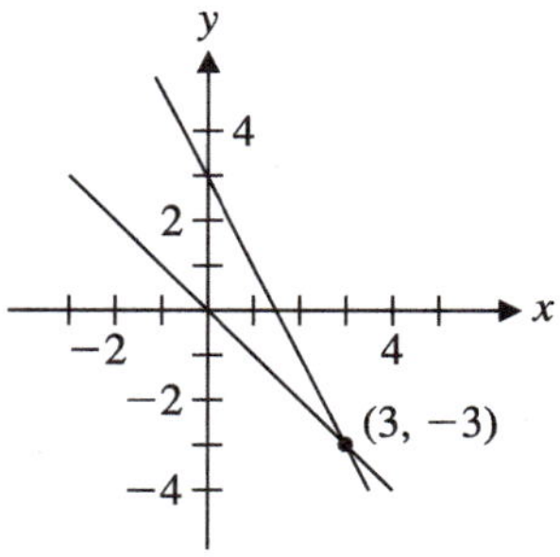

17. No solution

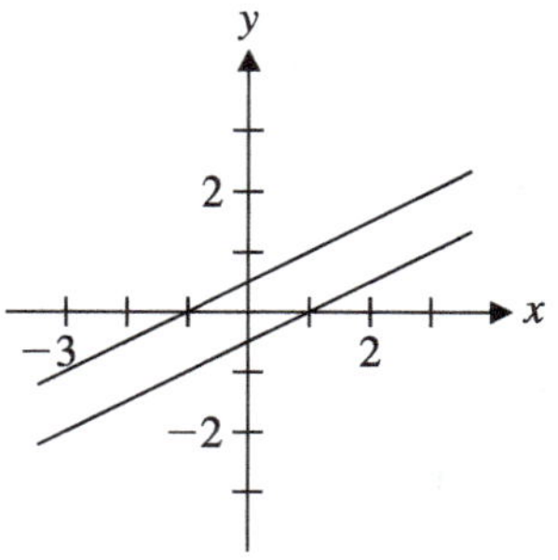

19. [7, 3] **21.** $[\frac{2}{3}, \frac{1}{3}, -\frac{1}{3}]$

23. [5, −2, 1, 1] **25.** [2, −7, −32]

27. $\left[\begin{array}{cc|c} 1 & -1 & 0 \\ 2 & 1 & 3 \end{array}\right]$ **29.** $\left[\begin{array}{cc|c} 1 & 5 & -1 \\ -1 & 1 & -5 \\ 2 & 4 & 4 \end{array}\right]$

31. $\begin{aligned} y + z &= 1 \\ x - y \quad &= 1 \\ 2x - y + z &= 1 \end{aligned}$ **33.** [1, 1]

35. [4, −1] **37.** No solution

39. (a) $\begin{aligned} 2x + y &= 3 \\ 4x + 2y &= 6 \end{aligned}$ **(b)** $\begin{aligned} x &= \frac{3}{2} - \frac{1}{2}s \\ y &= s \end{aligned}$

41. Let $u = \dfrac{1}{x}$ and $v = \dfrac{1}{y}$. The solution is $x = \frac{1}{3}, y = -\frac{1}{2}$.

43. Let $u = \tan x$, $v = \sin y$, $w = \cos z$. One solution is $x = \pi/4$, $y = -\pi/6$, $z = \pi/3$. (There are infinitely many solutions.)

Exercises 2.2

1. No **3.** Reduced row echelon form

5. No **7.** No

9. (a) $\begin{bmatrix} 1 & 1 & 1 \\ 0 & 1 & 1 \\ 0 & 0 & 1 \end{bmatrix}$ **11. (b)** $\begin{bmatrix} 1 & 0 \\ 0 & 1 \\ 0 & 0 \end{bmatrix}$

13. (b) $\begin{bmatrix} 1 & 0 & -1 \\ 0 & 1 & -1 \\ 0 & 0 & 0 \end{bmatrix}$

15. Perform elementary row operations in the order $R_4 + 29R_3$, $8R_3$, $R_4 - 3R_2$, $R_2 \leftrightarrow R_3$, $R_4 - R_1$, $R_3 + 2R_1$, and, finally, $R_2 + 2R_1$.

17. One possibility is to perform elementary row operations on A in the order $R_2 - 3R_1$, $\frac{1}{2}R_2$, $R_1 + 2R_2$, $R_2 + 3R_1$, $R_1 \leftrightarrow R_2$.

19. *Hint:* Pick a random 2×2 matrix and try this—carefully!

21. This is really two elementary row operations combined: $3R_2$ and $R_2 - 2R_1$.

23. Exercise 1: 3; Exercise 3: 2; Exercise 5: 2; Exercise 7: 3

25. $\begin{bmatrix} 2 \\ 5 \\ 1 \end{bmatrix}$ **27.** $t\begin{bmatrix} 1 \\ 1 \\ -1 \end{bmatrix}$ **29.** $\begin{bmatrix} 2 \\ -1 \end{bmatrix}$

31. $\begin{bmatrix} 24 \\ -10 \\ 0 \\ 0 \\ 0 \end{bmatrix} + r\begin{bmatrix} 6 \\ -2 \\ 1 \\ 0 \\ 0 \end{bmatrix} + s\begin{bmatrix} 0 \\ 6 \\ 0 \\ 1 \\ 0 \end{bmatrix} + t\begin{bmatrix} 12 \\ -6 \\ 0 \\ 0 \\ 1 \end{bmatrix}$

33. No solution

35. Unique solution

37. Infinitely many solutions

39. *Hint:* Show that if $ad - bc \neq 0$, the rank of $\begin{bmatrix} a & b \\ c & d \end{bmatrix}$ is 2. (There are two cases: $a = 0$ and $a \neq 0$.) Use the Rank Theorem to deduce that the given system must have a unique solution.

41. (a) No solution if $k = -1$
(b) A unique solution if $k \neq \pm 1$
(c) Infinitely many solutions if $k = 1$

43. (a) No solution if $k = 1$
(b) A unique solution if $k \neq -2, 1$
(c) Infinitely many solutions if $k = -2$

45. $\begin{bmatrix} x \\ y \\ z \end{bmatrix} = \begin{bmatrix} 0 \\ -1 \\ 1 \end{bmatrix} + t\begin{bmatrix} 9 \\ -10 \\ -7 \end{bmatrix}$

49. No intersection

51. The required vectors $\mathbf{x} = \begin{bmatrix} x_1 \\ x_2 \\ x_3 \end{bmatrix}$ are the solutions of the homogeneous system with augmented matrix

$$\left[\begin{array}{ccc|c} u_1 & u_2 & u_3 & 0 \\ v_1 & v_2 & v_3 & 0 \end{array}\right]$$

By Theorem 3, there are infinitely many solutions. If $u_1 \neq 0$ and $u_1v_2 - u_2v_1 \neq 0$, the solutions are given by

$$t\begin{bmatrix} u_2v_3 - u_3v_2 \\ u_3v_1 - u_1v_3 \\ u_1v_2 - u_2v_1 \end{bmatrix}$$

But a direct check shows that these are still solutions even if $u_1 = 0$ and/or $u_1v_2 - u_2v_1 = 0$.

53. $\begin{bmatrix} 0 \\ 2 \end{bmatrix}$ **55.** $\begin{bmatrix} 1 \\ 0 \\ 0 \end{bmatrix}$ **57.** $\begin{bmatrix} 3 \\ 3 \end{bmatrix}$

Exercises 2.3

1. Yes **3.** No **5.** Yes **7.** Yes

9. We need to show that the vector equation $x\begin{bmatrix} 1 \\ 1 \end{bmatrix} + y\begin{bmatrix} 1 \\ -1 \end{bmatrix} = \begin{bmatrix} a \\ b \end{bmatrix}$ has a solution for all values of a and b. This vector equation is equivalent to the linear system whose augmented matrix is $\left[\begin{array}{cc|c} 1 & 1 & a \\ 1 & -1 & b \end{array}\right]$. Row reduction yields $\left[\begin{array}{cc|c} 1 & 1 & a \\ 0 & -2 & b - a \end{array}\right]$, from which we can see that there is a (unique) solution. [Further row operations yield $x = (a + b)/2$, $y = (a - b)/2$.] Hence, $\mathbb{R}^2 = \text{span}\left(\begin{bmatrix} 1 \\ 1 \end{bmatrix}, \begin{bmatrix} 1 \\ -1 \end{bmatrix}\right)$.

11. We need to show that the vector equation $x\begin{bmatrix} 1 \\ 0 \\ 1 \end{bmatrix} + y\begin{bmatrix} 1 \\ 1 \\ 0 \end{bmatrix} + z\begin{bmatrix} 0 \\ 1 \\ 1 \end{bmatrix} = \begin{bmatrix} a \\ b \\ c \end{bmatrix}$ has a solution for all values of a, b, and c. This vector equation is equivalent to the linear system whose augmented matrix is $\left[\begin{array}{ccc|c} 1 & 1 & 0 & a \\ 0 & 1 & 1 & b \\ 1 & 0 & 1 & c \end{array}\right]$. Row reduction yields $\left[\begin{array}{ccc|c} 1 & 1 & 0 & a \\ 0 & 1 & 1 & b \\ 0 & 0 & 2 & b + c - a \end{array}\right]$, from which we can see that there is a (unique) solution. [Further row operations yield $x = (a - b + c)/2$, $y = (a + b - c)/2$, $z = (-a + b + c)/2$.]

Hence, $\mathbb{R}^3 = \text{span}\left(\begin{bmatrix} 1 \\ 0 \\ 1 \end{bmatrix}, \begin{bmatrix} 1 \\ 1 \\ 0 \end{bmatrix}, \begin{bmatrix} 0 \\ 1 \\ 1 \end{bmatrix}\right)$.

13. (a) The line through the origin with direction vector $\begin{bmatrix} -1 \\ 2 \end{bmatrix}$
(b) The line with general equation $2x + y = 0$

15. (a) The plane through the origin with direction vectors $\begin{bmatrix} 1 \\ 2 \\ 0 \end{bmatrix}, \begin{bmatrix} 3 \\ 2 \\ -1 \end{bmatrix}$
(b) The plane with general equation $2x - y + 4z = 0$

17. Substitution yields the linear system

$$\begin{aligned} a \qquad + 3c &= 0 \\ -a + b - 3c &= 0 \end{aligned}$$

whose solution is $t\begin{bmatrix} -3 \\ 0 \\ 1 \end{bmatrix}$. It follows that there are infinitely many solutions, the simplest perhaps being $a = -3, b = 0, c = 1$.

19. $\mathbf{u} = \mathbf{u} + 0(\mathbf{u} + \mathbf{v}) + 0(\mathbf{u} + \mathbf{v} + \mathbf{w})$
$\mathbf{v} = (-1)\mathbf{u} + (\mathbf{u} + \mathbf{v}) + 0(\mathbf{u} + \mathbf{v} + \mathbf{w})$
$\mathbf{w} = 0\mathbf{u} + (-1)(\mathbf{u} + \mathbf{v}) + (\mathbf{u} + \mathbf{v} + \mathbf{w})$

21. (c) We must show that span($\mathbf{e}_1, \mathbf{e}_2, \mathbf{e}_3$) = span($\mathbf{e}_1, \mathbf{e}_1 + \mathbf{e}_2, \mathbf{e}_1 + \mathbf{e}_2 + \mathbf{e}_3$). We know that span($\mathbf{e}_1, \mathbf{e}_1 + \mathbf{e}_2, \mathbf{e}_1 + \mathbf{e}_2 + \mathbf{e}_3$) $\subseteq \mathbb{R}^3$ = span($\mathbf{e}_1, \mathbf{e}_2, \mathbf{e}_3$). From Exercise 19, $\mathbf{e}_1$, $\mathbf{e}_2$, and $\mathbf{e}_3$ all belong to span($\mathbf{e}_1, \mathbf{e}_1 + \mathbf{e}_2, \mathbf{e}_1 + \mathbf{e}_2 + \mathbf{e}_3$). Therefore, by Exercise 21(b), span($\mathbf{e}_1, \mathbf{e}_2, \mathbf{e}_3$) = span($\mathbf{e}_1, \mathbf{e}_1 + \mathbf{e}_2, \mathbf{e}_1 + \mathbf{e}_2 + \mathbf{e}_3$).

23. Linearly independent

25. Linearly dependent, $-\begin{bmatrix}0\\1\\2\end{bmatrix} + \begin{bmatrix}2\\1\\3\end{bmatrix} = \begin{bmatrix}2\\0\\1\end{bmatrix}$

27. Linearly dependent, since the set contains the zero vector

29. Linearly independent

31. Linearly dependent, $\begin{bmatrix}3\\-1\\1\\-1\end{bmatrix} + \begin{bmatrix}-1\\3\\1\\-1\end{bmatrix} + \begin{bmatrix}-1\\-1\\1\\3\end{bmatrix} = \begin{bmatrix}1\\1\\3\\1\end{bmatrix}$

43. (a) Yes **(b)** No

Exercises 2.4

1. $x_1 = 160, x_2 = 120, x_3 = 160$

3. two small, three medium, four large

5. 65 bags of house blend, 30 bags of special blend, 45 bags of gourmet blend

7. $4FeS_2 + 11O_2 \longrightarrow 2Fe_2O_3 + 8SO_2$

9. $2C_4H_{10} + 13O_2 \longrightarrow 8CO_2 + 10H_2O$

11. $2C_5H_{11}OH + 15O_2 \longrightarrow 12H_2O + 10CO_2$

13. $Na_2CO_3 + 4C + N_2 \longrightarrow 2NaCN + 3CO$

15. (a) $\begin{aligned} f_1 &= 30 - t \\ f_2 &= -10 + t \\ f_3 &= t \end{aligned}$ **(b)** $f_1 = 15, f_3 = 15$

(c) $\begin{aligned} 0 \le f_1 \le 20 \\ 0 \le f_2 \le 20 \\ 10 \le f_3 \le 30 \end{aligned}$

(d) Negative flow would mean that water was flowing backward, against the direction of the arrow.

17. (a) $\begin{aligned} f_1 &= -200 + s + t \\ f_2 &= 300 - s - t \\ f_3 &= s \\ f_4 &= 150 - t \\ f_5 &= t \end{aligned}$ **(b)** $200 \le f_3 \le 300$

(c) If $f_3 = s = 0$, then $f_5 = t \ge 200$ (from the f_1 equation), but $f_5 = t \le 150$ (from the f_4 equation). This is a contradiction.

(d) $50 \le f_3 \le 300$

19. $I_1 = 3$ amps, $I_2 = 5$ amps, $I_3 = 2$ amps

21. (a) $I = 10$ amps, $I_1 = I_5 = 6$ amps, $I_2 = I_4 = 4$ amps, $I_3 = 2$ amps

(b) $R_{eff} = \frac{7}{5}$ ohms

(c) Yes; change it to 4 ohms.

23. Farming : Manufacturing = 2 : 3

25. The painter charges \$39/hr, the plumber \$42/hr, the electrician \$54/hr.

27. (a) Coal should produce \$100 million and steel \$160 million.

(b) Coal should reduce production by ≈ \$4.2 million and steel should increase production by ≈ \$5.7 million.

29. (a) Yes; push switches 1, 2, and 3 or switches 3, 4, and 5.

(b) No

31. The states that can be obtained are represented by those vectors

$$\begin{bmatrix}x_1\\x_2\\x_3\\x_4\\x_5\end{bmatrix}$$

in $\mathbb{Z}_2^5$ for which $x_1 + x_2 + x_4 + x_5 = 0$. (There are 16 such possibilities.)

33. If 0 = off, 1 = light blue, and 2 = dark blue, then the linear system that arises has augmented matrix

$$\left[\begin{array}{ccccc|c} 1 & 1 & 0 & 0 & 0 & 2 \\ 1 & 1 & 1 & 0 & 0 & 1 \\ 0 & 1 & 1 & 1 & 0 & 2 \\ 0 & 0 & 1 & 1 & 1 & 1 \\ 0 & 0 & 0 & 1 & 1 & 2 \end{array}\right]$$

which reduces over $\mathbb{Z}_3$ to

$$\left[\begin{array}{ccccc|c} 1 & 0 & 0 & 0 & 1 & 1 \\ 0 & 1 & 0 & 0 & 2 & 1 \\ 0 & 0 & 1 & 0 & 0 & 2 \\ 0 & 0 & 0 & 1 & 1 & 2 \\ 0 & 0 & 0 & 0 & 0 & 0 \end{array}\right]$$

This yields the solutions

$$\begin{bmatrix}x_1\\x_2\\x_3\\x_4\\x_5\end{bmatrix} = \begin{bmatrix}1\\1\\2\\2\\0\end{bmatrix} + t\begin{bmatrix}2\\1\\0\\2\\1\end{bmatrix}$$

where t is in $\mathbb{Z}_3$. Hence, there are exactly three solutions:

$$\begin{bmatrix}1\\1\\2\\2\\0\end{bmatrix}, \begin{bmatrix}0\\2\\2\\1\\1\end{bmatrix}, \begin{bmatrix}2\\0\\2\\0\\2\end{bmatrix}$$

where each entry indicates the number of times the corresponding switch should be pushed.

35. **(a)** Push squares 3 and 7.
(b) The 9×9 coefficient matrix A is row equivalent to $\mathbb{Z}_2$, so for any $\mathbf{b}$ in $\mathbb{Z}_2^9$, $A\mathbf{x} = \mathbf{b}$ has a unique solution.

37. Grace is 15, and Hans is 5.

39. 1200 and 600 square yards

41. **(a)** $a = 4 - d, b = 5 - d, c = -2 + d, d$ is arbitrary
(b) No solution

43. **(a)** No solution
(b) $[a, b, c, d, e, f] = [4, 5, 6, -3, -1, 0] + f[-1, -1, -1, 1, 1, 1]$

45. **(a)** $y = x^2 - 2x + 1$ **(b)** $y = x^2 + 6x + 10$

47. $A = 1, B = 2$

49. $A = -\frac{1}{5}, B = \frac{1}{3}, C = 0, D = -\frac{2}{15}, E = -\frac{1}{5}$

51. $a = \frac{1}{2}, b = \frac{1}{2}, c = 0$

Exercises 2.5

1.

n	0	1	2	3	4	5
x_1	0	0.8571	0.9714	0.9959	0.9991	0.9998
x_2	0	0.8000	0.9714	0.9943	0.9992	0.9998

Exact solution: $x_1 = 1, x_2 = 1$

3.

n	0	1	2	3	4	5	6
x_1	0	0.2222	0.2539	0.2610	0.2620	0.2622	0.2623
x_2	0	0.2857	0.3492	0.3582	0.3603	0.3606	0.3606

Exact solution (to four decimal places): $x_1 = 0.2623$, $x_2 = 0.3606$

5.

n	0	1	2	3	4	5	6	7	8
x_1	0	0.3333	0.2500	0.3055	0.2916	0.3009	0.2986	0.3001	0.2997
x_2	0	0.2500	0.0834	0.1250	0.0972	0.1042	0.0996	0.1008	0.1000
x_3	0	0.3333	0.2500	0.3055	0.2916	0.3009	0.2986	0.3001	0.2997

Exact solution: $x_1 = 0.3, x_2 = 0.1, x_3 = 0.3$

7.

n	0	1	2	3	4
x_1	0	0.8571	0.9959	0.9998	1.0000
x_2	0	0.9714	0.9992	1.0000	1.0000

After three iterations, the Gauss-Seidel method is within 0.001 of the exact solution. Jacobi's method took four iterations to reach the same accuracy.

9.

n	0	1	2	3	4
x_1	0	0.2222	0.2610	0.2622	0.2623
x_2	0	0.3492	0.3603	0.3606	0.3606

After three iterations, the Gauss-Seidel method is within 0.001 of the exact solution. Jacobi's method took four iterations to reach the same accuracy.

11.

n	0	1	2	3	4	5	6
x_1	0	0.3333	0.2777	0.2962	0.2993	0.2998	0.3000
x_2	0	0.1667	0.1112	0.1020	0.1004	0.1000	0.1000
x_3	0	0.2777	0.2962	0.2993	0.2998	0.3000	0.3000

After four iterations, the Gauss-Seidel method is within 0.001 of the exact solution. Jacobi's method took seven iterations to reach the same accuracy.

13.

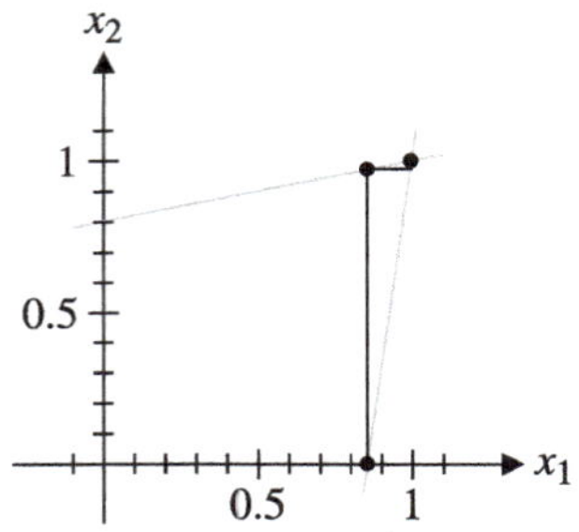

15.

n	0	1	2	3	4
x_1	0	3	−5	19	−53
x_2	0	−4	8	−28	80

If the equations are interchanged and the Gauss-Seidel method is applied to the equivalent system

$$\begin{aligned} 3x_1 + 2x_2 &= 1 \\ x_1 - 2x_2 &= 3 \end{aligned}$$

we obtain

n	0	1	2	3	4	5	6	7	8
x_1	0	0.3333	1.2222	0.9260	1.0247	0.9918	1.0027	0.9991	1.0003
x_2	0	−1.3333	−0.8889	−1.0370	−0.9876	−1.0041	−0.9986	−1.0004	−0.9998

After seven iterations, the process has converged to within 0.001 of the exact solution $x_1 = 1, x_2 = -1$.

17.

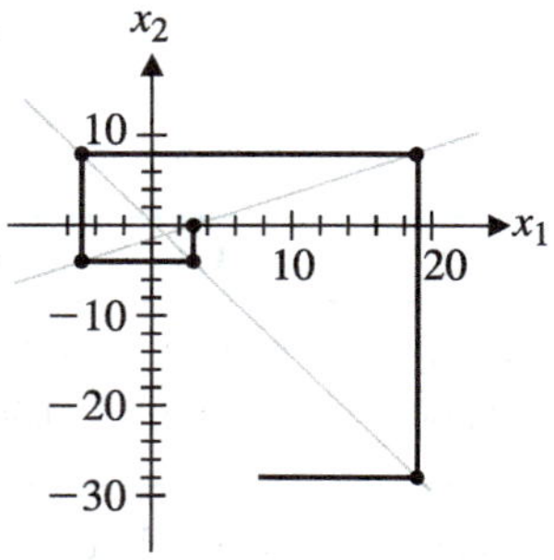

19.

n	0	1	2	3	4	5	6
x_1	0	−1.6	14.97	8.550	10.740	9.839	10.120
x_2	0	25.9	11.408	14.051	11.615	11.718	11.249
x_3	0	−10.35	−9.311	−11.200	−11.322	−11.721	−11.816

n	7	8	9	10	11	12
x_1	9.989	10.022	10.002	10.005	10.001	10.001
x_2	11.187	11.082	11.052	11.026	11.015	11.008
x_3	−11.912	−11.948	−11.973	−11.985	−11.992	−11.996

After 12 iterations, the Gauss-Seidel method has converged to within 0.01 of the exact solution $x_1 = 10, x_2 = 11, x_3 = -12$.

21.

n	13	14	15	16
x_1	10.0004	10.0003	10.0001	10.0001
x_2	11.0043	11.0023	11.0014	11.0007
x_3	−11.9976	−11.9986	−11.9993	−11.9996

23. The Gauss-Seidel method produces

n	0	1	2	3	4	5	6	7	8	9
x_1	0	0	12.5	21.875	24.219	24.805	24.951	24.988	24.997	24.999
x_2	0	0	18.75	21.438	24.609	24.902	24.976	24.994	24.998	24.999
x_3	0	50	68.75	73.438	74.609	74.902	74.976	74.994	74.998	74.999
x_4	0	62.5	71.875	74.219	74.805	74.951	74.988	74.997	74.999	75.000

The exact solution is $x_1 = 25$, $x_2 = 25$, $x_3 = 75$, $x_4 = 75$.

25. The Gauss-Seidel method produces the following iterates:

n	0	1	2	3	4	5	6
t_1	0	20	21.25	22.8125	23.3301	23.6596	23.7732
t_2	0	5	11.25	13.3203	14.6386	15.0926	15.2732
t_3	0	21.25	24.6094	26.9873	27.7303	27.9626	28.0352
t_4	0	2.5	5.8594	8.2373	8.9804	9.2126	9.2852
t_5	0	7.1875	14.6289	16.2829	16.7578	16.9036	16.9491
t_6	0	23.0469	24.9072	25.3207	25.4394	25.4759	25.4873

n	7	8	9	10	11	12
t_1	23.8093	23.8206	23.8242	23.8252	23.8256	23.8257
t_2	15.2824	15.2966	15.3010	15.3024	15.3029	15.3029
t_3	28.0579	28.0650	28.0671	28.0678	28.0681	28.0681
t_4	9.3079	9.3150	9.3172	9.3178	9.3181	9.3181
t_5	16.9633	16.9677	16.9690	16.9695	16.9696	16.9696
t_6	25.4908	25.4919	25.4922	25.4924	25.4924	25.4924

27. (a)

n	0	1	2	3	4	5	6
x_1	0	0	$\frac{1}{4}$	$\frac{1}{4}$	$\frac{5}{16}$	$\frac{5}{16}$	$\frac{21}{64}$
x_2	1	$\frac{1}{2}$	$\frac{1}{2}$	$\frac{3}{8}$	$\frac{3}{8}$	$\frac{11}{32}$	$\frac{11}{32}$

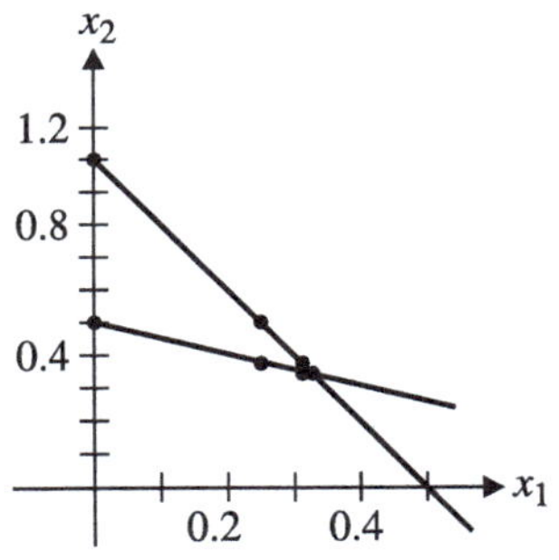

(b) $2x_1 + x_2 = 1$
$x_1 + 2x_2 = 1$

(c)

n	0	1	2	3	4	5	6	7
x_1	0	0	0.25	0.3125	0.3281	0.3320	0.3330	0.3332
x_2	1	0.5	0.375	0.3438	0.3360	0.3340	0.3335	0.3334

[Columns 1, 2, and 3 of this table are the *odd-numbered* columns 1, 3, and 5 from the table in part (a).] The iterates are converging to $x_1 = x_2 = 0.3333$.

(d) $x_1 = x_2 = \frac{1}{3}$

Review Questions

1. (a) F **(c)** F **(e)** T **(g)** T **(i)** F

3. $\begin{bmatrix} 0 \\ 2 \\ -1 \end{bmatrix}$ **5.** $\begin{bmatrix} 6 \\ 2 \end{bmatrix}$ **7.** $k = -1$ **9.** (0, 3, 1)

11. $x - 2y + z = 0$ **13. (a)** Yes **15.** 1 or 2

17. If $c_1(\mathbf{u} + \mathbf{v}) + c_2(\mathbf{u} - \mathbf{v}) = \mathbf{0}$, then $(c_1 + c_2)\mathbf{u} + (c_1 - c_2)\mathbf{v} = \mathbf{0}$. Linear independence of $\mathbf{u}$ and $\mathbf{v}$ implies $c_1 + c_2 = 0$ and $c_1 - c_2 = 0$. Solving this system, we get $c_1 = c_2 = 0$. Hence $\mathbf{u} + \mathbf{v}$ and $\mathbf{u} - \mathbf{v}$ are linearly independent.

19. Their ranks must be equal.

Chapter 3

Exercises 3.1

1. $\begin{bmatrix} 3 & -6 \\ -5 & 7 \end{bmatrix}$ **3.** Not possible

5. $\begin{bmatrix} 12 & -6 & 3 \\ -4 & 12 & 14 \end{bmatrix}$ **7.** $\begin{bmatrix} 3 & 3 \\ 19 & 27 \end{bmatrix}$

9. [10] **11.** $\begin{bmatrix} -4 & -2 \\ 8 & 4 \end{bmatrix}$

13. $\begin{bmatrix} 0 & 0 & 0 \\ 0 & 0 & 0 \\ 0 & 0 & 0 \end{bmatrix}$ **15.** $\begin{bmatrix} 27 & 0 \\ -49 & 125 \end{bmatrix}$

17. $\begin{bmatrix} 0 & 1 \\ 0 & 0 \end{bmatrix}$

19. $B = \begin{bmatrix} 1.50 & 1.00 & 2.00 \\ 1.75 & 1.50 & 1.00 \end{bmatrix}$, $BA = \begin{bmatrix} 650.00 & 462.50 \\ 675.00 & 406.25 \end{bmatrix}$

Column i corresponds to warehouse i, row 1 contains the costs of shipping by truck, and row 2 contains the costs of shipping by train.

21. $\begin{bmatrix} 1 & -2 & 3 \\ 2 & 1 & -5 \end{bmatrix}\begin{bmatrix} x_1 \\ x_2 \\ x_3 \end{bmatrix} = \begin{bmatrix} 0 \\ 4 \end{bmatrix}$

23. $AB = [2\mathbf{a}_1 + \mathbf{a}_2 - \mathbf{a}_3 \quad 3\mathbf{a}_1 - \mathbf{a}_2 + 6\mathbf{a}_3 \quad \mathbf{a}_2 + 4\mathbf{a}_3]$ (where $\mathbf{a}_i$ is the ith column of A)

25. $\begin{bmatrix} 2 & 3 & 0 \\ -6 & -9 & 0 \\ 4 & 6 & 0 \end{bmatrix} + \begin{bmatrix} 0 & 0 & 0 \\ 1 & -1 & 1 \\ 0 & 0 & 0 \end{bmatrix} + \begin{bmatrix} 2 & -12 & -8 \\ -1 & 6 & 4 \\ 1 & -6 & -4 \end{bmatrix}$

27. $BA = \begin{bmatrix} 2\mathbf{A}_1 + 3\mathbf{A}_2 \\ \mathbf{A}_1 - \mathbf{A}_2 + \mathbf{A}_3 \\ -\mathbf{A}_1 + 6\mathbf{A}_2 + 4\mathbf{A}_3 \end{bmatrix}$ (where $\mathbf{A}_i$ is the ith row of A)

29. If $\mathbf{b}_i$ is the ith column of B, then $A\mathbf{b}_i$ is the ith column of AB. If the columns of B are linearly dependent, then there are scalars $c_1, \ldots, c_n$ (not all zero) such that $c_1\mathbf{b}_1 + \cdots + c_n\mathbf{b}_n = \mathbf{0}$. But then $c_1(A\mathbf{b}_1) + \cdots + c_n(A\mathbf{b}_n) = A(c_1\mathbf{b}_1 + \cdots + c_n\mathbf{b}_n) = A\mathbf{0} = \mathbf{0}$, so the columns of AB are linearly dependent.

31. $\left[\begin{array}{cc|c} 3 & 2 & 0 \\ -1 & 1 & 0 \\ \hline 0 & 0 & 5 \end{array}\right]$ **33.** $\left[\begin{array}{cc|cc} 1 & 2 & 2 & 0 \\ 3 & 4 & 5 & 3 \\ \hline 1 & 0 & 1 & 2 \\ 0 & 1 & 0 & -1 \end{array}\right]$

35. (a) $A^2 = \begin{bmatrix} -1 & 1 \\ -1 & 0 \end{bmatrix}$, $A^3 = \begin{bmatrix} -1 & 0 \\ 0 & -1 \end{bmatrix}$,

$A^4 = \begin{bmatrix} 0 & -1 \\ 1 & -1 \end{bmatrix}$, $A^5 = \begin{bmatrix} 1 & -1 \\ 1 & 0 \end{bmatrix}$, $A^6 = \begin{bmatrix} 1 & 0 \\ 0 & 1 \end{bmatrix}$,

$A^7 = \begin{bmatrix} 0 & 1 \\ -1 & 1 \end{bmatrix}$

(b) $A^{2001} = \begin{bmatrix} -1 & 0 \\ 0 & -1 \end{bmatrix}$

37. $A^n = \begin{bmatrix} 1 & n \\ 0 & 1 \end{bmatrix}$

39. (a) $\begin{bmatrix} 1 & -1 & 1 & -1 \\ -1 & 1 & -1 & 1 \\ 1 & -1 & 1 & -1 \\ -1 & 1 & -1 & 1 \end{bmatrix}$ **(c)** $\begin{bmatrix} 0 & 0 & 0 & 0 \\ 1 & 1 & 1 & 1 \\ 2 & 4 & 8 & 16 \\ 3 & 9 & 27 & 81 \end{bmatrix}$

Exercises 3.2

1. $X = \begin{bmatrix} 5 & 4 \\ 3 & 5 \end{bmatrix}$ **3.** $X = \begin{bmatrix} -\frac{2}{3} & \frac{4}{3} \\ \frac{10}{3} & 4 \end{bmatrix}$

5. $B = 2A_1 + A_2$ **7.** Not possible

9. $\text{span}(A_1, A_2) = \left\{\begin{bmatrix} c_1 & 2c_1 + c_2 \\ -c_1 + 2c_2 & c_1 + c_2 \end{bmatrix}\right\} = \left\{\begin{bmatrix} w & x \\ 2x - 5w & x - w \end{bmatrix}\right\}$

11. $\text{span}(A_1, A_2, A_3) = \left\{\begin{bmatrix} c_1 - c_2 + c_3 & 2c_2 + c_3 & -c_1 + c_3 \\ 0 & c_1 + c_2 & 0 \end{bmatrix}\right\} = \left\{\begin{bmatrix} -3b + 4c + 5e & b & c \\ 0 & e & 0 \end{bmatrix}\right\}$

13. Linearly independent **15.** Linearly independent

23. $a = d, c = 0$ **25.** $3b = 2c, a = d - c$

27. $a = d, b = c = 0$

29. Let $A = [a_{ij}]$ and $B = [b_{ij}]$ be upper triangular $n \times n$ matrices and let $i > j$. Then, by the definition of an upper triangular matrix,

$a_{i1} = a_{i2} = \cdots = a_{i,i-1} = 0$ and
$b_{ij} = b_{i+1,j} = \cdots = b_{nj} = 0$

Now let $C = AB$. Then

$$\begin{aligned} c_{ij} &= a_{i1}b_{1j} + a_{i2}b_{2j} + \cdots a_{i,i-1}b_{i-1,j} + a_{ii}b_{ij} \\ &\quad + a_{i,i+1}b_{i+1,j} + \cdots + a_{in}b_{nj} \\ &= 0 \cdot b_{1j} + 0 \cdot b_{2j} + \cdots 0 \cdot b_{i-1,j} + a_{ii} \cdot 0 \\ &\quad + a_{i,i+1} \cdot 0 + \cdots + a_{in} \cdot 0 = 0 \end{aligned}$$

from which it follows that C is upper triangular.

35. (a) A, B symmetric $\Rightarrow (A + B)^T = A^T + B^T = A + B \Rightarrow A + B$ is symmetric

37. Matrices (b) and (c) are skew-symmetric.

41. Either A or B (or both) must be the zero matrix.

43. (b) $\begin{bmatrix} 1 & 2 & 3 \\ 4 & 5 & 6 \\ 7 & 8 & 9 \end{bmatrix} = \begin{bmatrix} 1 & 3 & 5 \\ 3 & 5 & 7 \\ 5 & 7 & 9 \end{bmatrix} + \begin{bmatrix} 0 & -1 & -2 \\ 1 & 0 & -1 \\ 2 & 1 & 0 \end{bmatrix}$

47. *Hint:* Use the trace.

Exercises 3.3

1. $\begin{bmatrix} 2 & -7 \\ -1 & 4 \end{bmatrix}$ **3.** Not invertible

5. Not invertible **7.** $\begin{bmatrix} -1.6 & -2.8 \\ 0.\overline{3} & 1 \end{bmatrix}$

9. $\begin{bmatrix} a/(a^2 + b^2) & b/(a^2 + b^2) \\ -b/(a^2 + b^2) & a/(a^2 + b^2) \end{bmatrix}$ **11.** $\begin{bmatrix} -5 \\ 9 \end{bmatrix}$

13. (a) $\mathbf{x}_1 = \begin{bmatrix} 4 \\ -\frac{1}{2} \end{bmatrix}, \mathbf{x}_2 = \begin{bmatrix} -5 \\ 2 \end{bmatrix}, \mathbf{x}_3 = \begin{bmatrix} 6 \\ -2 \end{bmatrix}$

(c) The method in part (b) uses fewer multiplications.

17. (b) $(AB)^{-1} = A^{-1}B^{-1}$ if and only if $AB = BA$

21. $X = A^{-1}(BA)^2B^{-1}$

23. $X = (AB)^{-1}BA + A$

25. $E = \begin{bmatrix} 0 & 0 & 1 \\ 0 & 1 & 0 \\ 1 & 0 & 0 \end{bmatrix}$ **27.** $E = \begin{bmatrix} 1 & 0 & 0 \\ 0 & 1 & 0 \\ -1 & 0 & 1 \end{bmatrix}$

29. $E = \begin{bmatrix} 1 & 0 & 0 \\ 0 & 1 & 2 \\ 0 & 0 & 1 \end{bmatrix}$ **31.** $\begin{bmatrix} \frac{1}{3} & 0 \\ 0 & 1 \end{bmatrix}$

33. $\begin{bmatrix} 0 & 1 \\ 1 & 0 \end{bmatrix}$ **35.** $\begin{bmatrix} 1 & 0 & 0 \\ 0 & 1 & 2 \\ 0 & 0 & 1 \end{bmatrix}$ **37.** $\begin{bmatrix} 1 & 0 & 0 \\ 0 & 1/c & 0 \\ 0 & 0 & 1 \end{bmatrix}$

39. $A = \begin{bmatrix} 1 & 0 \\ -1 & 1 \end{bmatrix}\begin{bmatrix} 1 & 0 \\ 0 & -2 \end{bmatrix}, A^{-1} = \begin{bmatrix} 1 & 0 \\ 0 & -\frac{1}{2} \end{bmatrix}\begin{bmatrix} 1 & 0 \\ 1 & 1 \end{bmatrix}$

43. (a) If A is invertible, then $BA = CA \Rightarrow (BA)A^{-1} = (CA)A^{-1} \Rightarrow B(AA^{-1}) = C(AA^{-1}) \Rightarrow BI = CI \Rightarrow B = C$.

45. *Hint:* Rewrite $A^2 - 2A + I = O$ as $A(2I - A) = I$.

47. If AB is invertible, then there exists a matrix X such that $(AB)X = I$. But then $A(BX) = I$ too, so A is invertible (with inverse BX).

49. $\begin{bmatrix} \frac{1}{10} & \frac{2}{5} \\ \frac{3}{10} & \frac{1}{5} \end{bmatrix}$ **51.** $\begin{bmatrix} 1/(a^2 + 1) & -a/(a^2 + 1) \\ a/(a^2 + 1) & 1/(a^2 + 1) \end{bmatrix}$

53. Not invertible

55. $\begin{bmatrix} 1/a & 0 & 0 \\ -1/a^2 & 1/a & 0 \\ 1/a^3 & -1/a^2 & 1/a \end{bmatrix}, a \neq 0$

57. $\begin{bmatrix} -11 & -2 & 5 & -4 \\ 4 & 1 & -2 & 2 \\ 5 & 1 & -2 & 2 \\ 9 & 2 & -4 & 3 \end{bmatrix}$

59. $\begin{bmatrix} 1 & 0 & 0 & 0 \\ 0 & 1 & 0 & 0 \\ 0 & 0 & 1 & 0 \\ -a/d & -b/d & -c/d & 1/d \end{bmatrix}, d \neq 0$

61. Not invertible **63.** $\begin{bmatrix} 4 & 6 & 4 \\ 5 & 3 & 2 \\ 0 & 6 & 5 \end{bmatrix}$

69. $\left[\begin{array}{cc|cc} 1 & 0 & 0 & 0 \\ 0 & 1 & 0 & 0 \\ \hline -2 & -3 & 1 & 0 \\ -1 & -2 & 0 & 1 \end{array}\right]$ **71.** $\left[\begin{array}{cc|cc} -1 & 0 & 1 & 1 \\ 0 & 1 & -1 & 0 \\ \hline 0 & 1 & 0 & 0 \\ 1 & -1 & 0 & 0 \end{array}\right]$

Exercises 3.4

1. $\begin{bmatrix} -2 \\ 1 \end{bmatrix}$ **3.** $\begin{bmatrix} -3/2 \\ -2 \\ -1 \end{bmatrix}$ **5.** $\begin{bmatrix} -7 \\ -15 \\ -2 \\ 2 \end{bmatrix}$

7. $\begin{bmatrix} 1 & 0 \\ -3 & 1 \end{bmatrix}\begin{bmatrix} 1 & 2 \\ 0 & 5 \end{bmatrix}$

9. $\begin{bmatrix} 1 & 0 & 0 \\ 4 & 1 & 0 \\ 8 & 3 & 1 \end{bmatrix}\begin{bmatrix} 1 & 2 & 3 \\ 0 & -3 & -6 \\ 0 & 0 & 3 \end{bmatrix}$

11. $\begin{bmatrix} 1 & 0 & 0 & 0 \\ 2 & 1 & 0 & 0 \\ 0 & 3 & 1 & 0 \\ -1 & 0 & -2 & 1 \end{bmatrix}\begin{bmatrix} 1 & 2 & 3 & -1 \\ 0 & 2 & -3 & 2 \\ 0 & 0 & 3 & 1 \\ 0 & 0 & 0 & 1 \end{bmatrix}$

13. $\begin{bmatrix} 1 & 0 & 0 \\ 0 & 1 & 0 \\ 0 & 0 & 1 \end{bmatrix}\begin{bmatrix} 1 & 0 & 1 & -2 \\ 0 & 3 & 3 & 1 \\ 0 & 0 & 0 & 5 \end{bmatrix}$

15. $L^{-1} = \begin{bmatrix} 1 & 0 \\ 1 & 1 \end{bmatrix}$, $U^{-1} = \begin{bmatrix} -\frac{1}{2} & \frac{1}{12} \\ 0 & \frac{1}{6} \end{bmatrix}$,

$A^{-1} = \begin{bmatrix} -5/12 & 1/12 \\ 1/6 & 1/6 \end{bmatrix}$

19. $\begin{bmatrix} 0 & 0 & 1 \\ 0 & 1 & 0 \\ 1 & 0 & 0 \end{bmatrix}\begin{bmatrix} 0 & 1 & 0 \\ 1 & 0 & 0 \\ 0 & 0 & 1 \end{bmatrix}$

21. $\begin{bmatrix} 0 & 1 & 0 & 0 \\ 1 & 0 & 0 & 0 \\ 0 & 0 & 1 & 0 \\ 0 & 0 & 0 & 1 \end{bmatrix}\begin{bmatrix} 0 & 0 & 0 & 1 \\ 0 & 1 & 0 & 0 \\ 0 & 0 & 1 & 0 \\ 1 & 0 & 0 & 0 \end{bmatrix}\begin{bmatrix} 0 & 0 & 1 & 0 \\ 0 & 1 & 0 & 0 \\ 1 & 0 & 0 & 0 \\ 0 & 0 & 0 & 1 \end{bmatrix}$

23. $\begin{bmatrix} 0 & 1 & 0 \\ 1 & 0 & 0 \\ 0 & 0 & 1 \end{bmatrix}\begin{bmatrix} 1 & 0 & 0 \\ 0 & 1 & 0 \\ -1 & 5 & 1 \end{bmatrix}\begin{bmatrix} -1 & 2 & 1 \\ 0 & 1 & 4 \\ 0 & 0 & -16 \end{bmatrix}$

25. $\begin{bmatrix} 0 & 1 & 0 & 0 \\ 1 & 0 & 0 & 0 \\ 0 & 0 & 0 & 1 \\ 0 & 0 & 1 & 0 \end{bmatrix}\begin{bmatrix} 1 & 0 & 0 & 0 \\ 0 & 1 & 0 & 0 \\ 0 & 0 & 1 & 0 \\ 0 & -1 & 0 & 1 \end{bmatrix}\begin{bmatrix} -1 & 1 & 1 & 2 \\ 0 & -1 & 1 & 3 \\ 0 & 0 & 1 & 1 \\ 0 & 0 & 0 & 4 \end{bmatrix}$

27. $\begin{bmatrix} 4 \\ -1 \\ -2 \end{bmatrix}$ **31.** $\begin{bmatrix} 1 & 0 \\ -1 & 1 \end{bmatrix}\begin{bmatrix} -2 & 0 \\ 0 & 6 \end{bmatrix}\begin{bmatrix} 1 & -\frac{1}{2} \\ 0 & 1 \end{bmatrix}$

Exercises 3.5

1. Subspace **3.** Subspace

5. Subspace **7.** Not a subspace

11. $\mathbf{b}$ is in col(A), $\mathbf{w}$ is not in row(A).

15. No

17. $\{[1 \;\; 0 \;\; -1], [0 \;\; 1 \;\; 2]\}$ is a basis for row(A); $\left\{\begin{bmatrix} 1 \\ 1 \end{bmatrix}, \begin{bmatrix} 0 \\ 1 \end{bmatrix}\right\}$ is a basis for col(A); $\left\{\begin{bmatrix} 1 \\ -2 \\ 1 \end{bmatrix}\right\}$ is a basis for null(A).

19. $\{[1 \;\; 0 \;\; 1 \;\; 0], [0 \;\; 1 \;\; -1 \;\; 0], [0 \;\; 0 \;\; 0 \;\; 1]\}$ is a basis for row(A); $\left\{\begin{bmatrix} 1 \\ 0 \\ 0 \end{bmatrix}, \begin{bmatrix} 1 \\ 1 \\ 1 \end{bmatrix}, \begin{bmatrix} 1 \\ 1 \\ -1 \end{bmatrix}\right\}$ is a basis for col(A); $\left\{\begin{bmatrix} -1 \\ 1 \\ 1 \\ 0 \end{bmatrix}\right\}$ is a basis for null(A).

21. $\{[1 \;\; 0 \;\; -1], [1 \;\; 1 \;\; 1]\}$ is a basis for row(A); $\left\{\begin{bmatrix} 1 \\ 0 \end{bmatrix}, \begin{bmatrix} 0 \\ 1 \end{bmatrix}\right\}$ is a basis for col(A)

23. $\{[1 \;\; 1 \;\; 0 \;\; 1], [0 \;\; 1 \;\; -1 \;\; 1], [0 \;\; 1 \;\; -1 \;\; -1]\}$ is a basis for row(A); $\left\{\begin{bmatrix} 1 \\ 0 \\ 0 \end{bmatrix}, \begin{bmatrix} 0 \\ 1 \\ 0 \end{bmatrix}, \begin{bmatrix} 0 \\ 0 \\ 1 \end{bmatrix}\right\}$ is a basis for col(A)

25. Both $\{[1 \;\; 0 \;\; -1], [0 \;\; 1 \;\; 2]\}$ and $\{[1 \;\; 0 \;\; -1], [1 \;\; 1 \;\; 1]\}$ are linearly independent spanning sets for row(A) $= \{[a \;\; b \;\; -a + 2b]\}$. Both $\left\{\begin{bmatrix} 1 \\ 1 \end{bmatrix}, \begin{bmatrix} 0 \\ 1 \end{bmatrix}\right\}$ and $\left\{\begin{bmatrix} 1 \\ 0 \end{bmatrix}, \begin{bmatrix} 0 \\ 1 \end{bmatrix}\right\}$ are linearly independent spanning sets for col(A) $= \mathbb{R}^2$.

27. $\left\{\begin{bmatrix} 1 \\ -1 \\ 0 \end{bmatrix}, \begin{bmatrix} -1 \\ 0 \\ 1 \end{bmatrix}\right\}$

29. $\{[1 \;\; 0 \;\; 0], [0 \;\; 1 \;\; 0], [0 \;\; 0 \;\; 1]\}$

31. $\{[2 \;\; -3 \;\; 1], [1 \;\; -1 \;\; 0], [4 \;\; -4 \;\; 1]\}$

35. rank(A) $= 2$, nullity(A) $= 1$

37. rank(A) $= 3$, nullity(A) $= 1$

39. If A is 3×5, then rank(A) ≤ 3, so there cannot be more than three linearly independent columns.

41. nullity(A) $= 2, 3, 4,$ or 5

43. If $a = -1$, then $\text{rank}(A) = 1$; if $a = 2$, then $\text{rank}(A) = 2$; for $a \neq -1, 2$, $\text{rank}(A) = 3$.

45. Yes **47.** Yes **49.** No

51. $\mathbf{w}$ is in span($\mathcal{B}$) if and only if the linear system with augmented matrix $[\mathcal{B} \mid \mathbf{w}]$ is consistent, which is true in this case, since

$$[\mathcal{B} \mid \mathbf{w}] = \left[\begin{array}{cc|c} 1 & 1 & 1 \\ 2 & 0 & 6 \\ 0 & -1 & 2 \end{array}\right] \longrightarrow \left[\begin{array}{cc|c} 1 & 0 & 3 \\ 0 & 1 & -2 \\ 0 & 0 & 0 \end{array}\right]$$

From this reduced row echelon form, it is also clear that $[\mathbf{w}]_{\mathcal{B}} = \begin{bmatrix} 3 \\ -2 \end{bmatrix}$.

53. $\text{rank}(A) = 2$, $\text{nullity}(A) = 1$

55. $\text{rank}(A) = 3$, $\text{nullity}(A) = 1$

57. Let $\mathbf{A}_1, \ldots, \mathbf{A}_m$ be the row vectors of A so that $\text{row}(A) = \text{span}(\mathbf{A}_1, \ldots, \mathbf{A}_m)$. If $\mathbf{x}$ is in null(A), then, since $A\mathbf{x} = \mathbf{0}$, we also have $\mathbf{A}_i \cdot \mathbf{x} = 0$ for $i = 1, \ldots, m$, by the row-column definition of matrix multiplication. If $\mathbf{r}$ is in row(A), then $\mathbf{r}$ is of the form $\mathbf{r} = c_1\mathbf{A}_1 + \cdots + c_m\mathbf{A}_m$. Therefore,

$$\begin{aligned} \mathbf{r} \cdot \mathbf{x} &= (c_1\mathbf{A}_1 + \cdots + c_m\mathbf{A}_m) \cdot \mathbf{x} \\ &= c_1(\mathbf{A}_1 \cdot \mathbf{x}) + \cdots + c_m(\mathbf{A}_m \cdot \mathbf{x}) = 0 \end{aligned}$$

59. (a) If a set of columns of AB is linearly independent, then the corresponding columns of B are linearly independent (by an argument similar to that needed to prove Exercise 29 in Section 3.1). It follows that the *maximum* number k of linearly independent columns of AB [i.e., $k = \text{rank}(AB)$] is not more than the *maximum* number r of linearly independent columns of B [i.e., $r = \text{rank}(B)$]. In other words, $\text{rank}(AB) \leq \text{rank}(B)$.

61. (a) From Exercise 59(a), $\text{rank}(UA) \leq \text{rank}(A)$ and $\text{rank}(A) = \text{rank}((U^{-1}U)A) = \text{rank}(U^{-1}(UA)) \leq \text{rank}(UA)$. Hence, $\text{rank}(UA) = \text{rank}(A)$.

Exercises 3.6

1. $T(\mathbf{u}) = \begin{bmatrix} 0 \\ 11 \end{bmatrix}$, $T(\mathbf{v}) = \begin{bmatrix} 8 \\ 1 \end{bmatrix}$

11. $\begin{bmatrix} 1 & 1 \\ 1 & -1 \end{bmatrix}$ **13.** $\begin{bmatrix} 1 & -1 & 1 \\ 2 & 1 & -3 \end{bmatrix}$

15. $[F] = \begin{bmatrix} -1 & 0 \\ 0 & 1 \end{bmatrix}$ **17.** $[D] = \begin{bmatrix} 2 & 0 \\ 0 & 3 \end{bmatrix}$

19. $\begin{bmatrix} k & 0 \\ 0 & 1 \end{bmatrix}$ stretches or contracts in the x-direction (combined with a reflection in the y-axis if $k < 0$); $\begin{bmatrix} 1 & 0 \\ 0 & k \end{bmatrix}$ stretches or contracts in the y-direction (combined with a reflection in the x-axis if $k < 0$); $\begin{bmatrix} 0 & 1 \\ 1 & 0 \end{bmatrix}$ is a reflection in the line $y = x$; $\begin{bmatrix} 1 & k \\ 0 & 1 \end{bmatrix}$ is a *shear* in the x-direction; $\begin{bmatrix} 1 & 0 \\ k & 1 \end{bmatrix}$ is a *shear* in the y-direction. For example,

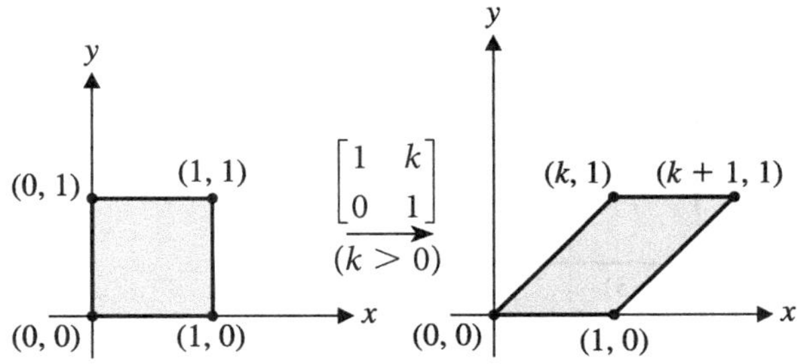

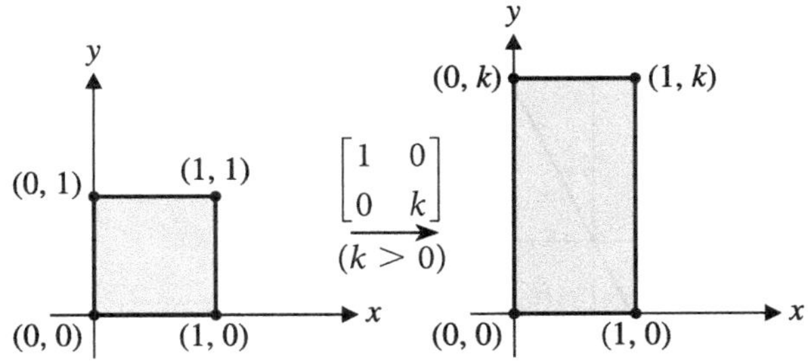

21. $\begin{bmatrix} \sqrt{3}/2 & 1/2 \\ -1/2 & \sqrt{3}/2 \end{bmatrix}$ **23.** $\begin{bmatrix} \frac{1}{2} & -\frac{1}{2} \\ -\frac{1}{2} & \frac{1}{2} \end{bmatrix}$

25. $\begin{bmatrix} 0 & -1 \\ -1 & 0 \end{bmatrix}$ **27.** $\begin{bmatrix} -\frac{3}{5} & \frac{4}{5} \\ \frac{4}{5} & \frac{3}{5} \end{bmatrix}$

31. $[S \circ T] = \begin{bmatrix} -8 & 5 \\ 4 & 1 \end{bmatrix}$

33. $[S \circ T] = \begin{bmatrix} 0 & 6 & -6 \\ 1 & -2 & 2 \end{bmatrix}$

35. $[S \circ T] = \begin{bmatrix} 1 & 0 & -1 \\ -1 & 1 & 0 \\ 0 & -1 & 1 \end{bmatrix}$

37. $\begin{bmatrix} -\sqrt{3}/2 & 1/2 \\ 1/2 & \sqrt{3}/2 \end{bmatrix}$ **39.** $\begin{bmatrix} -\sqrt{3}/2 & -1/2 \\ 1/2 & -\sqrt{3}/2 \end{bmatrix}$

45. In vector form, let the parallel lines be given by $\mathbf{x} = \mathbf{p} + t\mathbf{d}$ and $\mathbf{x}' = \mathbf{p}' + t\mathbf{d}$. Their images are $T(\mathbf{x}) = T(\mathbf{p} + t\mathbf{d}) = T(\mathbf{p}) + tT(\mathbf{d})$ and $T(\mathbf{x}') = T(\mathbf{p}' + t\mathbf{d}) = T(\mathbf{p}') + tT(\mathbf{d})$. Suppose $T(\mathbf{d}) \neq \mathbf{0}$. If $T(\mathbf{p}') - T(\mathbf{p})$ is parallel to $T(\mathbf{d})$, then the images represent the same line; otherwise the images represent distinct parallel lines. On the other hand, if $T(\mathbf{d}) = \mathbf{0}$, then the images represent two distinct points if $T(\mathbf{p}') \neq T(\mathbf{p})$ and single point otherwise.

47.

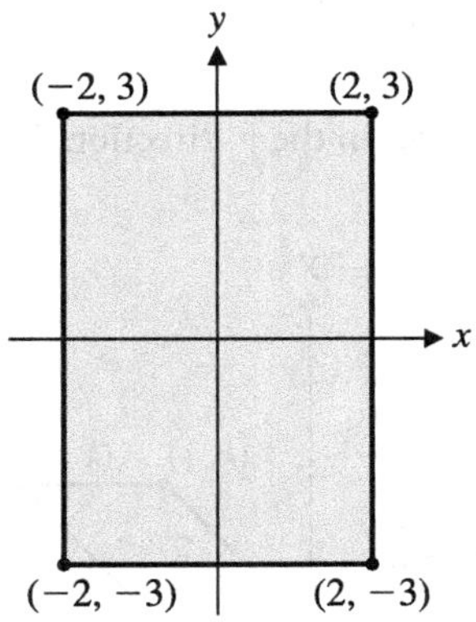

49.

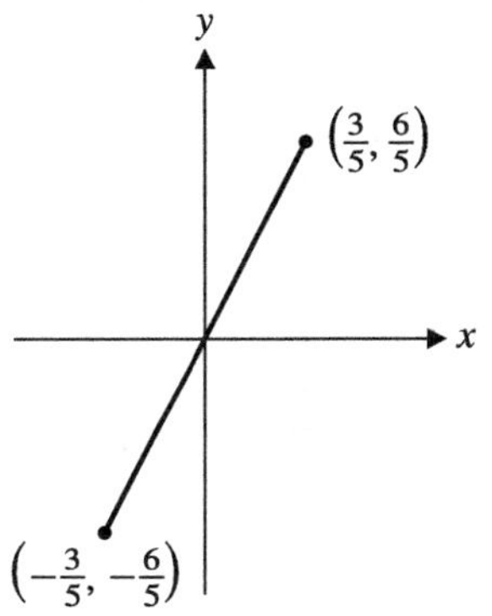

51.

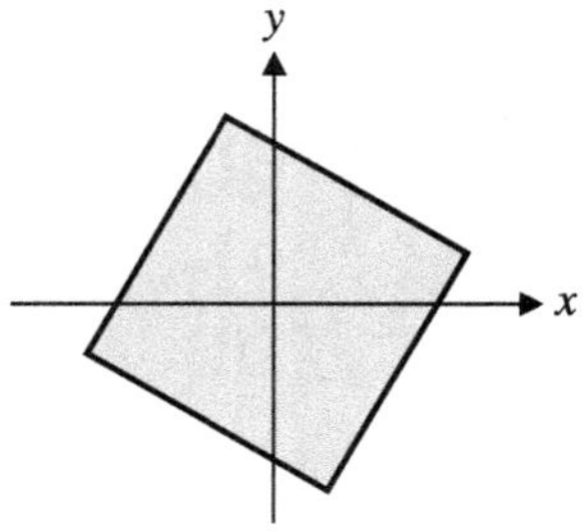

Exercises 3.7

1. $\mathbf{x}_1 = \begin{bmatrix} 0.4 \\ 0.6 \end{bmatrix}, \mathbf{x}_2 = \begin{bmatrix} 0.38 \\ 0.62 \end{bmatrix}$

3. 64%

5. $\mathbf{x}_1 = \begin{bmatrix} 150 \\ 120 \\ 120 \end{bmatrix}, \mathbf{x}_2 = \begin{bmatrix} 155 \\ 120 \\ 115 \end{bmatrix}$

7. $\frac{5}{18}$

9. (a) $P = \begin{bmatrix} 0.662 & 0.250 \\ 0.338 & 0.750 \end{bmatrix}$ **(b)** 0.353

(c) 42.5% wet, 57.5% dry

11. (a) $P = \begin{bmatrix} 0.08 & 0.09 & 0.11 \\ 0.07 & 0.11 & 0.05 \\ 0.85 & 0.80 & 0.84 \end{bmatrix}$

(b) 0.08, 0.1062, 0.1057, 0.1057, 0.1057

(c) 10.6% good, 5.5% fair, 83.9% poor

13. The entries of the vector $\mathbf{j}P$ are just the column sums of the matrix P. So P is stochastic if and only if $\mathbf{j}P = \mathbf{j}$.

15. 4

17. 9.375

19. Yes, $\mathbf{x} = \begin{bmatrix} 1 \\ 2 \end{bmatrix}$

21. No

23. No

25. Yes, $\mathbf{x} = \begin{bmatrix} 10 \\ 27 \\ 35 \end{bmatrix}$

27. Productive

29. Not productive

31. $\mathbf{x} = \begin{bmatrix} 10 \\ 16 \end{bmatrix}$

33. Yes, $\mathbf{x} = \begin{bmatrix} 10 \\ 6 \\ 8 \end{bmatrix}$

37. $\mathbf{x}_1 = \begin{bmatrix} 45 \\ 6 \end{bmatrix}, \mathbf{x}_2 = \begin{bmatrix} 120 \\ 27 \end{bmatrix}, \mathbf{x}_3 = \begin{bmatrix} 375 \\ 72 \end{bmatrix}$

39. $\mathbf{x}_1 = \begin{bmatrix} 500 \\ 70 \\ 50 \end{bmatrix}, \mathbf{x}_2 = \begin{bmatrix} 720 \\ 350 \\ 35 \end{bmatrix}, \mathbf{x}_3 = \begin{bmatrix} 1175 \\ 504 \\ 175 \end{bmatrix}$

41. (a) For L_1, we have $\mathbf{x}_1 = \begin{bmatrix} 50 \\ 8 \end{bmatrix}, \mathbf{x}_2 = \begin{bmatrix} 40 \\ 40 \end{bmatrix}$,

$\mathbf{x}_3 = \begin{bmatrix} 200 \\ 32 \end{bmatrix}, \mathbf{x}_4 = \begin{bmatrix} 160 \\ 160 \end{bmatrix}, \mathbf{x}_5 = \begin{bmatrix} 800 \\ 128 \end{bmatrix}, \mathbf{x}_6 = \begin{bmatrix} 640 \\ 640 \end{bmatrix}$,

$\mathbf{x}_7 = \begin{bmatrix} 3200 \\ 512 \end{bmatrix}, \mathbf{x}_8 = \begin{bmatrix} 2560 \\ 2560 \end{bmatrix}, \mathbf{x}_9 = \begin{bmatrix} 12800 \\ 2048 \end{bmatrix}$,

$\mathbf{x}_{10} = \begin{bmatrix} 10240 \\ 10240 \end{bmatrix}$.

(b) The first population oscillates between two states,while the second approaches a steady state.

43. The population oscillates through a cycle of three states (for the relative population): If $0.1 < s \leq 1$, the actual population is growing; if $s = 0.1$, the actual population goes through a cycle of length 3; and if $0 \leq s < 0.1$, the actual population is declining (and will eventually die out).

45. $A = \begin{bmatrix} 0 & 1 & 0 & 1 \\ 1 & 0 & 1 & 0 \\ 0 & 1 & 0 & 1 \\ 1 & 0 & 1 & 0 \end{bmatrix}$

47. $A = \begin{bmatrix} 0 & 1 & 1 & 1 & 1 \\ 1 & 0 & 1 & 0 & 0 \\ 1 & 1 & 0 & 1 & 0 \\ 1 & 0 & 1 & 0 & 1 \\ 1 & 0 & 0 & 1 & 0 \end{bmatrix}$

49.

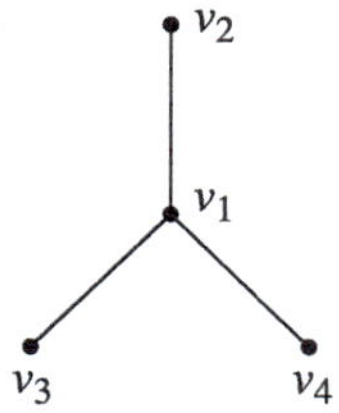

51.

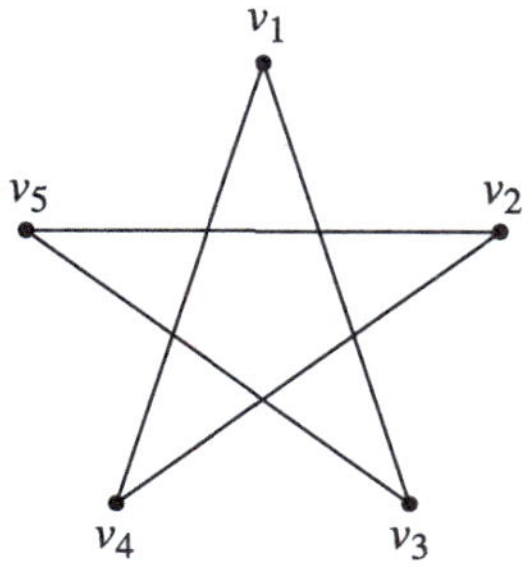

53. $A = \begin{bmatrix} 0 & 1 & 1 & 0 \\ 0 & 0 & 0 & 0 \\ 0 & 1 & 0 & 1 \\ 1 & 0 & 0 & 0 \end{bmatrix}$ **55.** $A = \begin{bmatrix} 0 & 1 & 0 & 1 & 0 \\ 1 & 0 & 0 & 1 & 0 \\ 1 & 1 & 0 & 0 & 0 \\ 1 & 0 & 0 & 0 & 1 \\ 1 & 0 & 0 & 0 & 0 \end{bmatrix}$

57.

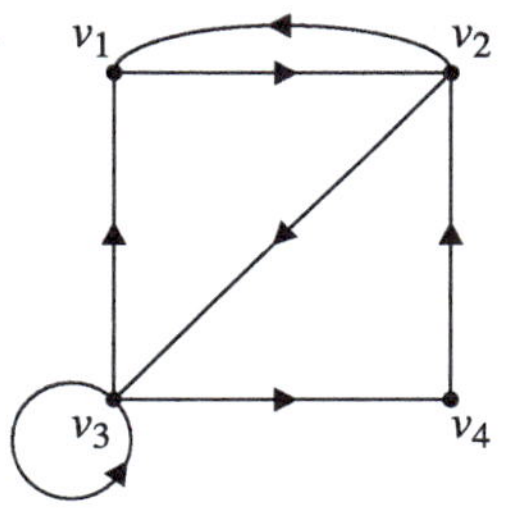

59.

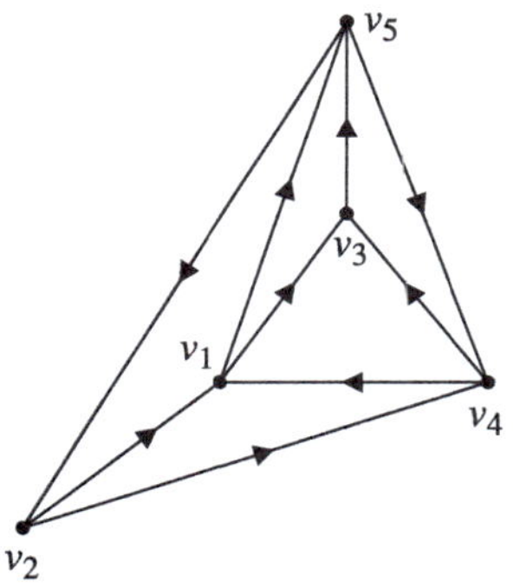

61. 2 **63.** 3 **65.** 0 **67.** 3

69. (a) Vertex i is not adjacent to any other vertices.

71. If we use direct wins only, P_2 is in first place; P_3, P_4, and P_6 tie for second place; and P_1 and P_5 tie for third place. If we combine direct and indirect wins, the players rank as follows: P_2 in first place, followed by P_6, P_4, P_3, P_5, and P_1.

73. (a)

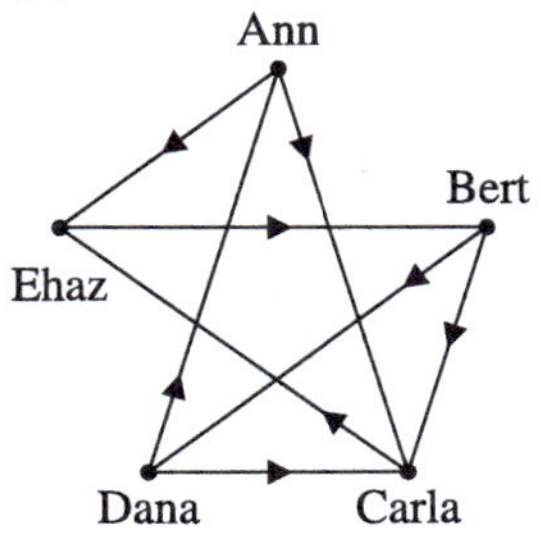

$A = \begin{bmatrix} 0 & 0 & 1 & 0 & 1 \\ 0 & 0 & 1 & 1 & 0 \\ 0 & 0 & 0 & 0 & 1 \\ 1 & 0 & 1 & 0 & 0 \\ 0 & 1 & 0 & 0 & 0 \end{bmatrix}$

(b) two steps; all of the off-diagonal entries of the second row of $A + A^2$ are nonzero.

(d) If the graph has n vertices, check the (i, j) entry of the powers A^k for $k = 1, \dots, n - 1$. Vertex i is connected to vertex j by a path of length k if and only if $(A^k)_{ij} \neq 0$.

75. $(AA^T)_{ij}$ counts the number of vertices adjacent to *both* vertex i and vertex j.

77. Bipartite **79.** Bipartite

Review Questions

1. (a) T **(c)** F **(e)** T **(g)** T **(i)** T

3. Impossible **5.** $\begin{bmatrix} \frac{17}{83} & -\frac{1}{83} \\ -\frac{1}{83} & \frac{5}{166} \end{bmatrix}$

7. $\begin{bmatrix} 1 & 3 \\ 3 & 9 \end{bmatrix} + \begin{bmatrix} 4 & 10 \\ 10 & 25 \end{bmatrix}$ **9.** $\begin{bmatrix} 0 & -9 \\ 2 & 4 \\ 1 & -6 \end{bmatrix}$

11. Because $(I - A)(I + A + A^2) = I - A^3 = I - O = I$, $(I - A)^{-1} = I + A + A^2$.

13. A basis for row(A) is $\{[1, -2, 0, -1, 0], [0, 0, 1, 2, 0], [0, 0, 0, 0, 1]\}$; a basis for col($A$) is $\left\{ \begin{bmatrix} 2 \\ 1 \\ 4 \end{bmatrix}, \begin{bmatrix} 5 \\ 2 \\ 3 \end{bmatrix}, \begin{bmatrix} 5 \\ 1 \\ 6 \end{bmatrix} \right\}$ (or the standard basis for $\mathbf{R}^3$); and a basis for null(A) is $\left\{ \begin{bmatrix} 2 \\ 1 \\ 0 \\ 0 \\ 0 \end{bmatrix}, \begin{bmatrix} 1 \\ 0 \\ -2 \\ 1 \\ 0 \end{bmatrix} \right\}$.

15. An invertible matrix has a trivial (zero) null space. If A is invertible, then so is A^T, and so both A and A^T have trivial null spaces. If A is not invertible, then A and A^T need not have the same null space. For example, take $A = \begin{bmatrix} 1 & 1 \\ 0 & 0 \end{bmatrix}$.

17. Because A has n linearly independent columns, rank(A) $= n$. Hence rank(A^TA) $= n$ by Theorem 3.28. Because A^TA is $n \times n$, this implies that A^TA is invertible, by the Fundamental Theorem of Invertible Matrices. AA^T need not be invertible. For example, take $A = \begin{bmatrix} 1 \\ 0 \end{bmatrix}$.

19. $\begin{bmatrix} -1/5\sqrt{2} & -3/5\sqrt{2} \\ 2/5\sqrt{2} & 6/5\sqrt{2} \end{bmatrix}$

Chapter 4

Exercises 4.1

1. $A\mathbf{v} = \begin{bmatrix} 3 \\ 3 \end{bmatrix} = 3\mathbf{v}$, $\lambda = 3$

3. $A\mathbf{v} = \begin{bmatrix} -3 \\ 6 \end{bmatrix} = -3\mathbf{v}, \lambda = -3$

5. $A\mathbf{v} = \begin{bmatrix} 6 \\ -3 \\ 3 \end{bmatrix} = 3\mathbf{v}, \lambda = 3$

7. $\begin{bmatrix} 2 \\ 1 \end{bmatrix}$ **9.** $\begin{bmatrix} 4 \\ 1 \end{bmatrix}$ **11.** $\begin{bmatrix} 1 \\ 1 \\ -1 \end{bmatrix}$

13. $\lambda = 1, E_1 = \text{span}\left(\begin{bmatrix} 0 \\ 1 \end{bmatrix}\right); \lambda = -1, E_{-1} = \text{span}\left(\begin{bmatrix} 1 \\ 0 \end{bmatrix}\right)$

15. $\lambda = 0, E_0 = \text{span}\left(\begin{bmatrix} 0 \\ 1 \end{bmatrix}\right); \lambda = 1, E_1 = \text{span}\left(\begin{bmatrix} 1 \\ 0 \end{bmatrix}\right)$

17. $\lambda = 2, E_2 = \text{span}\left(\begin{bmatrix} 1 \\ 0 \end{bmatrix}\right); \lambda = 3, E_3 = \text{span}\left(\begin{bmatrix} 0 \\ 1 \end{bmatrix}\right)$

19. $\mathbf{v} = \begin{bmatrix} 1 \\ 0 \end{bmatrix}, \lambda = 1; \mathbf{v} = \begin{bmatrix} 0 \\ 1 \end{bmatrix}, \lambda = 2$

21. $\mathbf{v} = \begin{bmatrix} 1/\sqrt{2} \\ 1/\sqrt{2} \end{bmatrix}, \lambda = 2; \mathbf{v} = \begin{bmatrix} -1/\sqrt{2} \\ 1/\sqrt{2} \end{bmatrix}, \lambda = 0$

23. $\lambda = 2, E_2 = \text{span}\left(\begin{bmatrix} 1 \\ 2 \end{bmatrix}\right); \lambda = 3, E_3 = \text{span}\left(\begin{bmatrix} 1 \\ 1 \end{bmatrix}\right)$

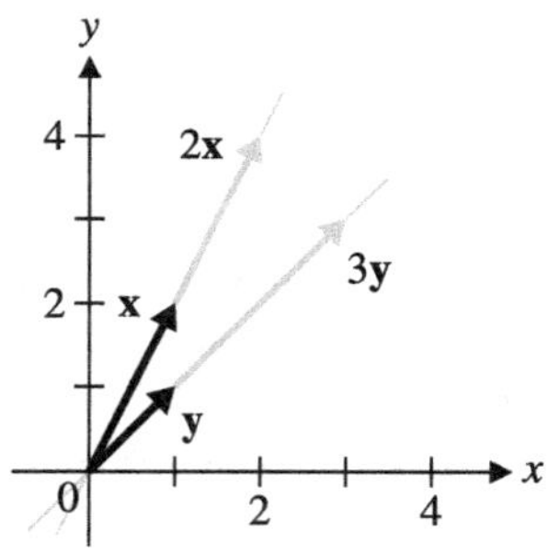

25. $\lambda = 2, E_2 = \text{span}\left(\begin{bmatrix} 1 \\ 0 \end{bmatrix}\right)$

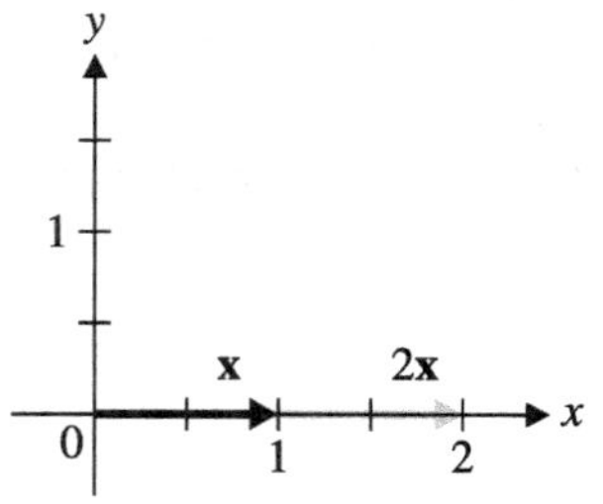

27. $\lambda = 1 + i, E_{1+i} = \text{span}\left(\begin{bmatrix} 1 \\ i \end{bmatrix}\right); \lambda = 1 - i, E_{1-i} = \text{span}\left(\begin{bmatrix} 1 \\ -i \end{bmatrix}\right)$

29. $\lambda = 1 + i, E_{1+i} = \text{span}\left(\begin{bmatrix} 1 \\ 1 \end{bmatrix}\right); \lambda = 1 - i, E_{1-i} = \text{span}\left(\begin{bmatrix} 1 \\ -1 \end{bmatrix}\right)$

31. $\lambda = 1, 2$ **33.** $\lambda = 4$

Exercises 4.2

1. 16 **3.** 0 **5.** −18 **7.** 6
9. −12 **11.** $a^2b + ab^2$ **13.** 4 **15.** $abdg$
17. 0 **25.** 2 **27.** −24 **29.** 0
31. 0 **33.** −24 **35.** 8 **37.** −4
39. −8 **45.** $k \neq 0, 2$ **47.** −6 **49.** $-\frac{3}{2}$
51. $(-2)3^n$
53. $\det(AB) = (\det A)(\det B) = (\det B)(\det A) = \det(BA)$
55. 0, 1 **57.** $x = \frac{3}{2}, y = -\frac{1}{2}$
59. $x = -1, y = 0, z = 1$ **61.** $\begin{bmatrix} \frac{1}{2} & \frac{1}{2} \\ \frac{1}{2} & -\frac{1}{2} \end{bmatrix}$

63. $\begin{bmatrix} \frac{1}{2} & -\frac{1}{2} & -1 \\ 0 & 1 & -1 \\ 0 & 0 & 1 \end{bmatrix}$

Exercises 4.3

1. (a) $\lambda^2 - 7\lambda + 12$ **(b)** $\lambda = 3, 4$
(c) $E_3 = \text{span}\left(\begin{bmatrix} 3 \\ 2 \end{bmatrix}\right); E_4 = \text{span}\left(\begin{bmatrix} 1 \\ 1 \end{bmatrix}\right)$
(d) The algebraic and geometric multiplicities are all 1.

3. (a) $-\lambda^3 + 2\lambda^2 + 5\lambda - 6$
(b) $\lambda = -2, 1, 3$
(c) $E_{-2} = \text{span}\left(\begin{bmatrix} 1 \\ -3 \\ 0 \end{bmatrix}\right); E_1 = \text{span}\left(\begin{bmatrix} 1 \\ 0 \\ 0 \end{bmatrix}\right);$
$E_3 = \text{span}\left(\begin{bmatrix} 1 \\ 2 \\ 10 \end{bmatrix}\right)$
(d) The algebraic and geometric multiplicities are all 1.

5. (a) $-\lambda^3 + \lambda^2$ **(b)** $\lambda = 0, 1$
(c) $E_0 = \text{span}\left(\begin{bmatrix} 2 \\ -1 \\ 1 \end{bmatrix}\right); E_1 = \text{span}\left(\begin{bmatrix} 1 \\ 0 \\ 1 \end{bmatrix}\right)$
(d) $\lambda = 0$ has algebraic multiplicity 2 and geometric multiplicity 1; $\lambda = 1$ has algebraic and geometric multiplicity 1.

7. (a) $-\lambda^3 + 9\lambda^2 - 27\lambda + 27$
(b) $\lambda = 3$

(c) $E_3 = \text{span}\left(\begin{bmatrix}-1\\0\\1\end{bmatrix}, \begin{bmatrix}0\\1\\0\end{bmatrix}\right)$

(d) $\lambda = 3$ has algebraic multiplicity 3 and geometric multiplicity 2.

9. (a) $\lambda^4 - 6\lambda^3 + 9\lambda^2 + 4\lambda - 12$

(b) $\lambda = -1, 2, 3$

(c) $E_{-1} = \text{span}\left(\begin{bmatrix}0\\0\\-2\\1\end{bmatrix}\right)$; $E_2 = \text{span}\left(\begin{bmatrix}1\\-1\\0\\0\end{bmatrix}\right)$;

$E_3 = \text{span}\left(\begin{bmatrix}0\\0\\2\\1\end{bmatrix}\right)$

(d) $\lambda = -1$ and $\lambda = 3$ have algebraic and geometric multiplicity 1; $\lambda = 2$ has algebraic multiplicity 2 and geometric multiplicity 1.

11. (a) $\lambda^4 - 4\lambda^3 + 2\lambda^2 + 4\lambda - 3$

(b) $\lambda = -1, 1, 3$

(c) $E_{-1} = \text{span}\left(\begin{bmatrix}0\\0\\0\\1\end{bmatrix}\right)$;

$E_1 = \text{span}\left(\begin{bmatrix}-2\\0\\1\\3\end{bmatrix}, \begin{bmatrix}-2\\2\\0\\3\end{bmatrix}\right)$;

$E_3 = \text{span}\left(\begin{bmatrix}0\\0\\2\\1\end{bmatrix}\right)$

(d) $\lambda = -1$ and $\lambda = 3$ have algebraic and geometric multiplicity 1; $\lambda = 1$ has algebraic and geometric multiplicity 2.

15. $\begin{bmatrix}2^{-9} + 3 \cdot 2^{10}\\ -2^{-9} + 3 \cdot 2^{10}\end{bmatrix}$ **17.** $\begin{bmatrix}2\\(2 \cdot 3^{20} - 1)/3^{20}\\2\end{bmatrix}$

23. (a) $\lambda = -2, E_{-2} = \text{span}\left(\begin{bmatrix}2\\-5\end{bmatrix}\right)$; $\lambda = 5, E_5 = \text{span}\left(\begin{bmatrix}1\\1\end{bmatrix}\right)$

(b) (i) $\lambda = -\frac{1}{2}, E_{-1/2} = \text{span}\left(\begin{bmatrix}2\\-5\end{bmatrix}\right)$; $\lambda = \frac{1}{5}, E_{1/5} = \text{span}\left(\begin{bmatrix}1\\1\end{bmatrix}\right)$

(iii) $\lambda = 0, E_0 = \text{span}\left(\begin{bmatrix}2\\-5\end{bmatrix}\right)$; $\lambda = 7$, $E_7 = \text{span}\left(\begin{bmatrix}1\\1\end{bmatrix}\right)$

27. $\begin{bmatrix}-3 & 4 & -12\\1 & 0 & 0\\0 & 1 & 0\end{bmatrix}$, $-\lambda^3 - 3\lambda^2 + 4\lambda - 12$

35. $A^2 = 4A - 5I, A^3 = 11A - 20I$
$A^4 = 24A - 55I$

37. $A^{-1} = -\frac{1}{5}A + \frac{4}{5}I, A^{-2} = -\frac{4}{25}A + \frac{11}{25}I$

Exercises 4.4

1. The characteristic polynomial of A is $\lambda^2 - 5\lambda + 1$, but that of B is $\lambda^2 - 2\lambda + 1$.

3. The eigenvalues of A are $\lambda = 2$ and $\lambda = 4$, but those of B are $\lambda = 1$ and $\lambda = 4$.

5. $\lambda_1 = 4, E_4 = \text{span}\left(\begin{bmatrix}1\\1\end{bmatrix}\right)$; $\lambda_2 = 3, E_3 = \text{span}\left(\begin{bmatrix}1\\2\end{bmatrix}\right)$

7. $\lambda_1 = 6, E_6 = \text{span}\left(\begin{bmatrix}3\\2\\3\end{bmatrix}\right)$; $\lambda_2 = -2, E_{-2} = \text{span}\left(\begin{bmatrix}0\\1\\-1\end{bmatrix}, \begin{bmatrix}1\\0\\-1\end{bmatrix}\right)$

9. Not diagonalizable

11. $P = \begin{bmatrix}1 & 1 & -1\\1 & 1 & 1\\1 & -2 & 0\end{bmatrix}, D = \begin{bmatrix}2 & 0 & 0\\0 & -1 & 0\\0 & 0 & 1\end{bmatrix}$

13. Not diagonalizable

15. $P = \begin{bmatrix}1 & 0 & 0 & -1\\0 & 1 & 0 & 0\\0 & 0 & 1 & 0\\0 & 0 & 0 & 1\end{bmatrix}, D = \begin{bmatrix}2 & 0 & 0 & 0\\0 & 2 & 0 & 0\\0 & 0 & -2 & 0\\0 & 0 & 0 & -2\end{bmatrix}$

17. $\begin{bmatrix}35839 & -69630\\-11605 & 24234\end{bmatrix}$

19. $\begin{bmatrix}(3^k + 3(-1)^k)/4 & (3^{k+1} - 3(-1)^k)/4\\(3^k - (-1)^k)/4 & (3^{k+1} + (-1)^k)/4\end{bmatrix}$

21. $\begin{bmatrix} 1 & 1 & 1 \\ 0 & -1 & 0 \\ 0 & 0 & -1 \end{bmatrix}$

23.

$$\begin{bmatrix} (5 + 2^{k+2} + (-3)^k)/10 & (2^k - (-3)^k)/5 & (-5 + 2^{k+2} + (-3)^k)/10 \\ (2^{k+1} - 2(-3)^k)/5 & (2^k + 4(-3)^k)/5 & (2^{k+1} - 2(-3)^k)/5 \\ (-5 + 2^{k+2} + (-3)^k)/10 & (2^k - (-3)^k)/5 & (5 + 2^{k+2} + (-3)^k)/10 \end{bmatrix}$$

25. $k = 0$ **27.** $k = 0$

29. All real values of k

37. If $A \sim B$, then there is an invertible matrix P such that $B = P^{-1}AP$. Therefore, we have

$$\begin{aligned} \operatorname{tr}(B) &= \operatorname{tr}(P^{-1}AP) = \operatorname{tr}(P^{-1}(AP)) = \operatorname{tr}((AP)P^{-1}) \\ &= \operatorname{tr}(APP^{-1}) = \operatorname{tr}(AI) = \operatorname{tr}(A) \end{aligned}$$

using Exercise 45 in Section 3.2.

39. $P = \begin{bmatrix} 7 & -2 \\ 10 & -3 \end{bmatrix}$

41. $P = \begin{bmatrix} \frac{1}{2} & -\frac{1}{2} & 0 \\ -\frac{3}{2} & -\frac{3}{2} & 1 \\ -\frac{5}{2} & -\frac{3}{2} & 0 \end{bmatrix}$

51. (b) $\dim E_{-1} = 1, \dim E_1 = 2, \dim E_2 = 3$

Exercises 4.5

1. (a) $\begin{bmatrix} 1 \\ 2.5 \end{bmatrix}$, 6.000

(b) $\lambda_1 = 6$

3. (a) $\begin{bmatrix} 1 \\ 0.618 \end{bmatrix}$, 2.618

(b) $\lambda_1 = (3 + \sqrt{5})/2 \approx 2.618$

5. (a) $m_5 = 11.001$, $\mathbf{y}_5 = \begin{bmatrix} -0.333 \\ 1.000 \end{bmatrix}$

7. (a) $m_8 = 10.000$, $\mathbf{y}_8 = \begin{bmatrix} 1 \\ 0 \\ 1 \end{bmatrix}$

9.

k	0	1	2	3	4	5
$\mathbf{x}_k$	$\begin{bmatrix} 1 \\ 1 \end{bmatrix}$	$\begin{bmatrix} 26 \\ 8 \end{bmatrix}$	$\begin{bmatrix} 17.692 \\ 5.923 \end{bmatrix}$	$\begin{bmatrix} 18.018 \\ 6.004 \end{bmatrix}$	$\begin{bmatrix} 17.999 \\ 6.000 \end{bmatrix}$	$\begin{bmatrix} 18.000 \\ 6.000 \end{bmatrix}$
$\mathbf{y}_k$	$\begin{bmatrix} 1 \\ 1 \end{bmatrix}$	$\begin{bmatrix} 1 \\ 0.308 \end{bmatrix}$	$\begin{bmatrix} 1 \\ 0.335 \end{bmatrix}$	$\begin{bmatrix} 1 \\ 0.333 \end{bmatrix}$	$\begin{bmatrix} 1 \\ 0.333 \end{bmatrix}$	$\begin{bmatrix} 1 \\ 0.333 \end{bmatrix}$
m_k	1	26	17.692	18.018	17.999	18.000

Therefore, $\lambda_1 \approx 18$, $\mathbf{v}_1 \approx \begin{bmatrix} 1 \\ 0.333 \end{bmatrix}$.

11.

k	0	1	2	3	4	5	6
$\mathbf{x}_k$	$\begin{bmatrix} 1 \\ 0 \end{bmatrix}$	$\begin{bmatrix} 7 \\ 2 \end{bmatrix}$	$\begin{bmatrix} 7.571 \\ 2.857 \end{bmatrix}$	$\begin{bmatrix} 7.755 \\ 3.132 \end{bmatrix}$	$\begin{bmatrix} 7.808 \\ 3.212 \end{bmatrix}$	$\begin{bmatrix} 7.823 \\ 3.234 \end{bmatrix}$	$\begin{bmatrix} 7.827 \\ 3.240 \end{bmatrix}$
$\mathbf{y}_k$	$\begin{bmatrix} 1 \\ 0 \end{bmatrix}$	$\begin{bmatrix} 1 \\ 0.286 \end{bmatrix}$	$\begin{bmatrix} 1 \\ 0.377 \end{bmatrix}$	$\begin{bmatrix} 1 \\ 0.404 \end{bmatrix}$	$\begin{bmatrix} 1 \\ 0.411 \end{bmatrix}$	$\begin{bmatrix} 1 \\ 0.413 \end{bmatrix}$	$\begin{bmatrix} 1 \\ 0.414 \end{bmatrix}$
m_k	1	7	7.571	7.755	7.808	7.823	7.827

Therefore, $\lambda_1 \approx 7.827$, $\mathbf{v}_1 \approx \begin{bmatrix} 1 \\ 0.414 \end{bmatrix}$.

13.

k	0	1	2	3	4	5
$\mathbf{x}_k$	$\begin{bmatrix}1\\1\\1\end{bmatrix}$	$\begin{bmatrix}21\\15\\13\end{bmatrix}$	$\begin{bmatrix}16.809\\12.238\\10.714\end{bmatrix}$	$\begin{bmatrix}17.011\\12.371\\10.824\end{bmatrix}$	$\begin{bmatrix}16.999\\12.363\\10.818\end{bmatrix}$	$\begin{bmatrix}17.000\\12.363\\10.818\end{bmatrix}$
$\mathbf{y}_k$	$\begin{bmatrix}1\\1\\1\end{bmatrix}$	$\begin{bmatrix}1\\0.714\\0.619\end{bmatrix}$	$\begin{bmatrix}1\\0.728\\0.637\end{bmatrix}$	$\begin{bmatrix}1\\0.727\\0.636\end{bmatrix}$	$\begin{bmatrix}1\\0.727\\0.636\end{bmatrix}$	$\begin{bmatrix}1\\0.727\\0.636\end{bmatrix}$
m_k	1	21	16.809	17.011	16.999	17.000

Therefore, $\lambda_1 \approx 17, \mathbf{v}_1 \approx \begin{bmatrix}1\\0.727\\0.636\end{bmatrix}$.

15. $\lambda_1 \approx 5, \mathbf{v}_1 \approx \begin{bmatrix}1\\0\\0.333\end{bmatrix}$

17.

k	0	1	2	3	4	5	6
$\mathbf{x}_k$	$\begin{bmatrix}1\\0\end{bmatrix}$	$\begin{bmatrix}7\\2\end{bmatrix}$	$\begin{bmatrix}7.571\\2.857\end{bmatrix}$	$\begin{bmatrix}7.755\\3.132\end{bmatrix}$	$\begin{bmatrix}7.808\\3.212\end{bmatrix}$	$\begin{bmatrix}7.823\\3.234\end{bmatrix}$	$\begin{bmatrix}7.827\\3.240\end{bmatrix}$
$R(\mathbf{x}_k)$	7	7.755	7.823	7.828	7.828	7.828	7.828
$\mathbf{y}_k$	$\begin{bmatrix}1\\0\end{bmatrix}$	$\begin{bmatrix}1\\0.286\end{bmatrix}$	$\begin{bmatrix}1\\0.377\end{bmatrix}$	$\begin{bmatrix}1\\0.404\end{bmatrix}$	$\begin{bmatrix}1\\0.411\end{bmatrix}$	$\begin{bmatrix}1\\0.413\end{bmatrix}$	$\begin{bmatrix}1\\0.414\end{bmatrix}$

19.

k	0	1	2	3	4	5
$\mathbf{x}_k$	$\begin{bmatrix}1\\1\\1\end{bmatrix}$	$\begin{bmatrix}21\\15\\13\end{bmatrix}$	$\begin{bmatrix}16.809\\12.238\\10.714\end{bmatrix}$	$\begin{bmatrix}17.011\\12.371\\10.824\end{bmatrix}$	$\begin{bmatrix}16.999\\12.363\\10.818\end{bmatrix}$	$\begin{bmatrix}17.000\\12.363\\10.818\end{bmatrix}$
$R(\mathbf{x}_k)$	16.333	16.998	17.000	17.000	17.000	17.000
$\mathbf{y}_k$	$\begin{bmatrix}1\\1\\1\end{bmatrix}$	$\begin{bmatrix}1\\0.714\\0.619\end{bmatrix}$	$\begin{bmatrix}1\\0.728\\0.637\end{bmatrix}$	$\begin{bmatrix}1\\0.727\\0.636\end{bmatrix}$	$\begin{bmatrix}1\\0.727\\0.636\end{bmatrix}$	$\begin{bmatrix}1\\0.727\\0.636\end{bmatrix}$

21.

k	0	1	2	3	4	5	6	7	8
$\mathbf{x}_k$	$\begin{bmatrix}1\\1\end{bmatrix}$	$\begin{bmatrix}5\\4\end{bmatrix}$	$\begin{bmatrix}4.8\\3.2\end{bmatrix}$	$\begin{bmatrix}4.667\\2.667\end{bmatrix}$	$\begin{bmatrix}4.571\\2.286\end{bmatrix}$	$\begin{bmatrix}4.500\\2.000\end{bmatrix}$	$\begin{bmatrix}4.444\\1.778\end{bmatrix}$	$\begin{bmatrix}4.400\\1.600\end{bmatrix}$	$\begin{bmatrix}4.364\\1.455\end{bmatrix}$
$\mathbf{y}_k$	$\begin{bmatrix}1\\1\end{bmatrix}$	$\begin{bmatrix}1\\0.8\end{bmatrix}$	$\begin{bmatrix}1\\0.667\end{bmatrix}$	$\begin{bmatrix}1\\0.571\end{bmatrix}$	$\begin{bmatrix}1\\0.500\end{bmatrix}$	$\begin{bmatrix}1\\0.444\end{bmatrix}$	$\begin{bmatrix}1\\0.400\end{bmatrix}$	$\begin{bmatrix}1\\0.364\end{bmatrix}$	$\begin{bmatrix}1\\0.333\end{bmatrix}$
m_k	1	5	4.8	4.667	4.571	4.500	4.444	4.400	4.364

Since $\lambda_1 = \lambda_2 = 4, \mathbf{v}_1 = \begin{bmatrix}1\\0\end{bmatrix}$, m_k is converging slowly to the exact answer.

23.

k	0	1	2	3	4	5	6	7	8
$\mathbf{x}_k$	$\begin{bmatrix}1\\1\\1\end{bmatrix}$	$\begin{bmatrix}5\\4\\1\end{bmatrix}$	$\begin{bmatrix}4.2\\3.2\\0.2\end{bmatrix}$	$\begin{bmatrix}4.048\\3.048\\0.048\end{bmatrix}$	$\begin{bmatrix}4.012\\3.012\\0.012\end{bmatrix}$	$\begin{bmatrix}4.003\\3.003\\0.003\end{bmatrix}$	$\begin{bmatrix}4.001\\3.001\\0.001\end{bmatrix}$	$\begin{bmatrix}4.000\\3.000\\0.000\end{bmatrix}$	$\begin{bmatrix}4.000\\3.000\\0.000\end{bmatrix}$
$\mathbf{y}_k$	$\begin{bmatrix}1\\1\\1\end{bmatrix}$	$\begin{bmatrix}1\\0.8\\0.2\end{bmatrix}$	$\begin{bmatrix}1\\0.762\\0.048\end{bmatrix}$	$\begin{bmatrix}1\\0.753\\0.012\end{bmatrix}$	$\begin{bmatrix}1\\0.751\\0.003\end{bmatrix}$	$\begin{bmatrix}1\\0.750\\0.001\end{bmatrix}$	$\begin{bmatrix}1\\0.750\\0\end{bmatrix}$	$\begin{bmatrix}1\\0.750\\0\end{bmatrix}$	$\begin{bmatrix}1\\0.750\\0\end{bmatrix}$
m_k	1	5	4.2	4.048	4.012	4.003	4.001	4.000	4.000

In this case, $\lambda_1 = \lambda_2 = 4$ and $E_4 = \text{span}\left(\begin{bmatrix}1\\0\\0\end{bmatrix}, \begin{bmatrix}0\\1\\0\end{bmatrix}\right)$.
Clearly, m_k is converging to 4 and $\mathbf{y}_k$ is converging to a vector in the eigenspace E_4—namely, $\begin{bmatrix}1\\0\\0\end{bmatrix} + 0.75\begin{bmatrix}0\\1\\0\end{bmatrix}$.

25.

k	0	1	2	3	4	5
$\mathbf{x}_k$	$\begin{bmatrix}1\\1\end{bmatrix}$	$\begin{bmatrix}1\\0\end{bmatrix}$	$\begin{bmatrix}-1\\-1\end{bmatrix}$	$\begin{bmatrix}1\\0\end{bmatrix}$	$\begin{bmatrix}-1\\-1\end{bmatrix}$	$\begin{bmatrix}1\\0\end{bmatrix}$
$\mathbf{y}_k$	$\begin{bmatrix}1\\1\end{bmatrix}$	$\begin{bmatrix}1\\0\end{bmatrix}$	$\begin{bmatrix}1\\1\end{bmatrix}$	$\begin{bmatrix}1\\0\end{bmatrix}$	$\begin{bmatrix}1\\1\end{bmatrix}$	$\begin{bmatrix}1\\0\end{bmatrix}$
m_k	1	1	-1	1	-1	1

The exact eigenvalues are complex (i and $-i$), so the power method cannot possibly converge to either the dominant eigenvalue or the dominant eigenvector if we start with a *real* initial iterate. Instead, the power method oscillates between two sets of real vectors.

27.

k	0	1	2	3	4	5
$\mathbf{x}_k$	$\begin{bmatrix}1\\1\\1\end{bmatrix}$	$\begin{bmatrix}3\\4\\3\end{bmatrix}$	$\begin{bmatrix}2.500\\4.000\\2.500\end{bmatrix}$	$\begin{bmatrix}2.250\\4.000\\2.250\end{bmatrix}$	$\begin{bmatrix}2.125\\4.000\\2.125\end{bmatrix}$	$\begin{bmatrix}2.063\\4.000\\2.063\end{bmatrix}$
$\mathbf{y}_k$	$\begin{bmatrix}1\\1\\1\end{bmatrix}$	$\begin{bmatrix}0.750\\1\\0.750\end{bmatrix}$	$\begin{bmatrix}0.625\\1\\0.625\end{bmatrix}$	$\begin{bmatrix}0.562\\1\\0.562\end{bmatrix}$	$\begin{bmatrix}0.531\\1\\0.531\end{bmatrix}$	$\begin{bmatrix}0.516\\1\\0.516\end{bmatrix}$
m_k	1	4	4	4	4	4

The eigenvalues are $\lambda_1 = -12$, $\lambda_2 = 4$, $\lambda_3 = 2$, with corresponding eigenvectors

$$\mathbf{v}_1 = \begin{bmatrix} 1 \\ 0 \\ -1 \end{bmatrix}, \mathbf{v}_2 = \begin{bmatrix} 1 \\ 2 \\ 1 \end{bmatrix}, \mathbf{v}_3 = \begin{bmatrix} 1 \\ 0 \\ 1 \end{bmatrix}.$$

Since $\mathbf{x}_0 = \frac{1}{2}\mathbf{v}_2 + \frac{1}{2}\mathbf{v}_3$, the initial vector $\mathbf{x}_0$ has a zero component in the direction of the dominant eigenvector, so the power method cannot converge to the dominant eigenvalue/eigenvector. Instead, it converges to a *second* eigenvalue/eigenvector pair, as the calculations show.

29. Apply the power method to $A - 18I = \begin{bmatrix} -4 & 12 \\ 5 & -15 \end{bmatrix}$.

k	0	1	2	3
$\mathbf{x}_k$	$\begin{bmatrix} 1 \\ 1 \end{bmatrix}$	$\begin{bmatrix} 8 \\ -10 \end{bmatrix}$	$\begin{bmatrix} 15.2 \\ -19 \end{bmatrix}$	$\begin{bmatrix} 15.2 \\ -19 \end{bmatrix}$
$\mathbf{y}_k$	$\begin{bmatrix} 1 \\ 1 \end{bmatrix}$	$\begin{bmatrix} -0.8 \\ 1 \end{bmatrix}$	$\begin{bmatrix} -0.8 \\ 1 \end{bmatrix}$	$\begin{bmatrix} -0.8 \\ 1 \end{bmatrix}$
m_k	1	-10	-19	-19

Thus, -19 is the dominant eigenvalue of $A - 18I$, and $\lambda_2 = -19 + 18 = -1$ is the second eigenvalue of A.

31. Apply the power method to $A - 17I = \begin{bmatrix} -8 & 4 & 8 \\ 4 & -2 & -4 \\ 8 & -4 & -8 \end{bmatrix}$.

k	0	1	2	3
$\mathbf{x}_k$	$\begin{bmatrix} 1 \\ 1 \\ 1 \end{bmatrix}$	$\begin{bmatrix} 4 \\ -2 \\ -4 \end{bmatrix}$	$\begin{bmatrix} -18 \\ 9 \\ 18 \end{bmatrix}$	$\begin{bmatrix} -18 \\ 9 \\ 18 \end{bmatrix}$
$\mathbf{y}_k$	$\begin{bmatrix} 1 \\ 1 \\ 1 \end{bmatrix}$	$\begin{bmatrix} 1 \\ -0.5 \\ -1 \end{bmatrix}$	$\begin{bmatrix} 1 \\ -0.5 \\ -1 \end{bmatrix}$	$\begin{bmatrix} 1 \\ -0.5 \\ -1 \end{bmatrix}$
m_k	1	4	-18	-18
$R(\mathbf{x}_k)$	-0.667	-18	-18	-18

In this case, there is no dominant eigenvalue. (We could choose either 18 or -18 for m_k, $k \geq 2$.) However, the Rayleigh quotient method (Exercises 17–20) converges to -18. Thus, -18 is the dominant eigenvalue of $A - 17I$, and $\lambda_2 = -18 + 17 = -1$ is the second eigenvalue of A.

33.

k	0	1	2	3	4	5
$\mathbf{x}_k$	$\begin{bmatrix} 1 \\ 1 \end{bmatrix}$	$\begin{bmatrix} 0.5 \\ -0.5 \end{bmatrix}$	$\begin{bmatrix} -0.833 \\ 1.056 \end{bmatrix}$	$\begin{bmatrix} 0.798 \\ -0.997 \end{bmatrix}$	$\begin{bmatrix} 0.800 \\ -1.000 \end{bmatrix}$	$\begin{bmatrix} 0.800 \\ -1.000 \end{bmatrix}$
$\mathbf{y}_k$	$\begin{bmatrix} 1 \\ 1 \end{bmatrix}$	$\begin{bmatrix} 1 \\ -1 \end{bmatrix}$	$\begin{bmatrix} -0.789 \\ 1 \end{bmatrix}$	$\begin{bmatrix} -0.801 \\ 1 \end{bmatrix}$	$\begin{bmatrix} -0.800 \\ 1 \end{bmatrix}$	$\begin{bmatrix} -0.800 \\ 1 \end{bmatrix}$
m_k	1	0.5	1.056	-0.997	-1.000	-1.000

Thus, the eigenvalue of A that is smallest in magnitude is $1/(-1) = -1$.

35.

k	0	1	2	3	4	5
$\mathbf{x}_k$	$\begin{bmatrix}1\\1\\-1\end{bmatrix}$	$\begin{bmatrix}-0.500\\0.000\\0.500\end{bmatrix}$	$\begin{bmatrix}0.500\\0.333\\-0.500\end{bmatrix}$	$\begin{bmatrix}0.500\\0.111\\-0.500\end{bmatrix}$	$\begin{bmatrix}0.500\\0.259\\-0.500\end{bmatrix}$	$\begin{bmatrix}0.500\\0.160\\-0.500\end{bmatrix}$
$\mathbf{y}_k$	$\begin{bmatrix}1\\1\\-1\end{bmatrix}$	$\begin{bmatrix}-1.000\\0.000\\1.000\end{bmatrix}$	$\begin{bmatrix}-1.000\\-0.667\\1.000\end{bmatrix}$	$\begin{bmatrix}-1.000\\-0.222\\1.000\end{bmatrix}$	$\begin{bmatrix}-1.000\\-0.518\\1.000\end{bmatrix}$	$\begin{bmatrix}-1.000\\-0.321\\1.000\end{bmatrix}$
m_k	1	-0.500	-0.500	-0.500	-0.500	-0.500

Clearly, m_k converges to -0.5, so the smallest eigenvalue of A is $1/(-0.5) = -2$.

37. The calculations are the same as for Exercise 33.

39. We apply the inverse power method to $A - 5I =$ $\begin{bmatrix}-1 & 0 & 6\\-1 & -2 & 1\\6 & 0 & -1\end{bmatrix}$. Taking $\mathbf{x}_0 = \begin{bmatrix}1\\1\\1\end{bmatrix}$, we have

k	0	1	2	3
$\mathbf{x}_k$	$\begin{bmatrix}1\\1\\1\end{bmatrix}$	$\begin{bmatrix}0.200\\-0.500\\0.200\end{bmatrix}$	$\begin{bmatrix}-0.080\\-0.500\\-0.080\end{bmatrix}$	$\begin{bmatrix}0.032\\-0.500\\0.032\end{bmatrix}$
$\mathbf{y}_k$	$\begin{bmatrix}1\\1\\1\end{bmatrix}$	$\begin{bmatrix}-0.400\\1\\-0.400\end{bmatrix}$	$\begin{bmatrix}0.160\\1\\0.160\end{bmatrix}$	$\begin{bmatrix}-0.064\\1\\-0.064\end{bmatrix}$
m_k	1	-0.500	-0.500	-0.500

Clearly, m_k converges to -0.5, so the eigenvalue of A closest to 5 is $5 + 1/(-0.5) = 5 - 2 = 3$.

41. 0.732 **43.** -0.619

47.

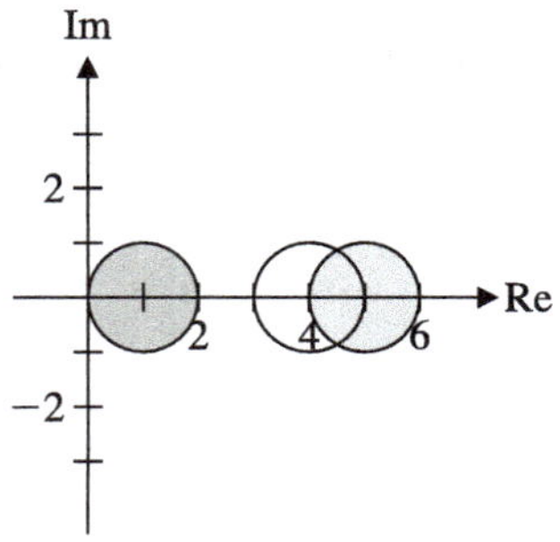

49.

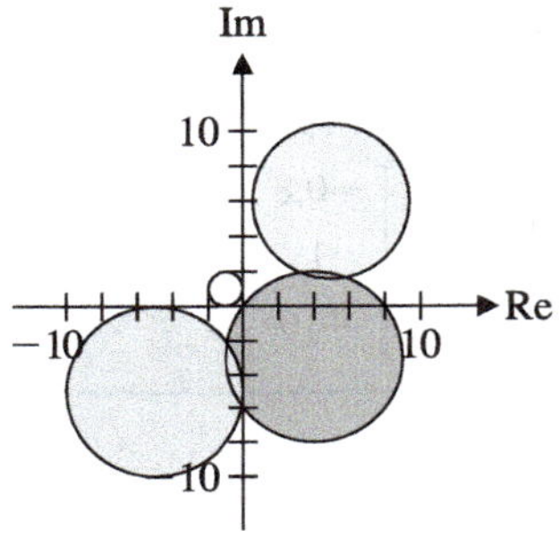

51. *Hint:* Show that 0 is not contained in any Gerschgorin disk and then apply Theorem 4.16.

53. Exercise 52 implies that $|\lambda|$ is less than or equal to all of the column sums of A for every eigenvalue λ. But for a stochastic matrix, all column sums are 1. Hence $|\lambda| \le 1$.

Exercises 4.6

1. Not regular **3.** Regular

5. Not regular **7.** $L = \begin{bmatrix}\frac{1}{5} & \frac{1}{5}\\ \frac{4}{5} & \frac{4}{5}\end{bmatrix}$

9. $L = \begin{bmatrix}0.304 & 0.304 & 0.304\\0.354 & 0.354 & 0.354\\0.342 & 0.342 & 0.342\end{bmatrix}$

11. 1, $\begin{bmatrix}2\\1\end{bmatrix}$ **13.** 2, $\begin{bmatrix}16\\4\\1\end{bmatrix}$

15. The population is increasing, decreasing, and constant, respectively.

17.

$$P^{-1}LP = \begin{bmatrix} b_1 & b_2s_1 & b_3s_1s_2 & \cdots & b_{n-1}s_1s_2\cdots s_{n-2} & b_ns_1s_2\cdots s_{n-1} \\ 1 & 0 & 0 & \cdots & 0 & 0 \\ 0 & 1 & 0 & \cdots & 0 & 0 \\ \vdots & \vdots & \vdots & \ddots & \vdots & \vdots \\ 0 & 0 & 0 & \cdots & 0 & 0 \\ 0 & 0 & 0 & \cdots & 1 & 0 \end{bmatrix}.$$

The characteristic polynomial of L is $(\lambda^n - b_1\lambda^{n-1} - b_2s_1\lambda^{n-2} - b_3s_1s_2\lambda^{n-3} - \cdots - b_ns_1s_2\cdots s_{n-1})(-1)^n$.

19. $\lambda \approx 1.746$, $\mathbf{p} \approx \begin{bmatrix} 0.660 \\ 0.264 \\ 0.076 \end{bmatrix}$

21. $\lambda \approx 1.092$, $\mathbf{p} \approx \begin{bmatrix} 0.535 \\ 0.147 \\ 0.094 \\ 0.078 \\ 0.064 \\ 0.053 \\ 0.029 \end{bmatrix}$

25. (a) $h \approx 0.082$

29. $3, \begin{bmatrix} \frac{3}{5} \\ \frac{2}{5} \end{bmatrix}$

31. $3, \begin{bmatrix} \frac{1}{2} \\ \frac{1}{4} \\ \frac{1}{4} \end{bmatrix}$

33. Reducible

35. Irreducible

43. 1, 2, 4, 8, 16

45. 0, 1, 1, 0, -1

47. $x_n = 4^n - (-1)^n$

49. $y_n = (n - \frac{1}{2})2^n$

51. $b_n = \dfrac{1}{2\sqrt{3}}[(1 + \sqrt{3})^n - (1 - \sqrt{3})^n]$

57. (a) $d_1 = 1, d_2 = 2, d_3 = 3, d_4 = 5, d_5 = 8$

(b) $d_n = d_{n-1} + d_{n-2}$

(c) $d_n = \dfrac{1}{\sqrt{5}}\left[\left(\dfrac{1 + \sqrt{5}}{2}\right)^{n+1} - \left(\dfrac{1 - \sqrt{5}}{2}\right)^{n+1}\right]$

59. The general solution is $x(t) = -3C_1e^{-t} + C_2e^{4t}$, $y(t) = 2C_1e^{-t} + C_2e^{4t}$. The specific solution is $x(t) = -3e^{-t} + 3e^{4t}$, $y(t) = 2e^{-t} + 3e^{4t}$.

61. The general solution is $x_1(t) = (1 + \sqrt{2})C_1e^{\sqrt{2}t} + (1 - \sqrt{2})C_2e^{-\sqrt{2}t}$, $x_2(t) = C_1e^{\sqrt{2}t} + C_2e^{-\sqrt{2}t}$. The specific solution is $x_1(t) = (2 + \sqrt{2})e^{\sqrt{2}t}/4 + (2 - \sqrt{2})e^{-\sqrt{2}t}/4$, $x_2(t) = \sqrt{2}e^{\sqrt{2}t}/4 - \sqrt{2}e^{-\sqrt{2}t}/4$.

63. The general solution is $x(t) = -C_1 + C_3e^{-t}$, $y(t) = C_1 + C_2e^t - C_3e^{-t}$, $z(t) = C_1 + C_2e^t$. The specific solution is $x(t) = 2 - e^{-t}$, $y(t) = -2 + e^t + e^{-t}$, $z(t) = -2 + e^t$.

65. (a) $x(t) = -120e^{8t/5} + 520e^{11t/10}$, $y(t) = 240e^{8t/5} + 260^{11t/10}$. Strain X dies out after approximately 2.93 days; strain Y continues to grow.

67. $a = 10$, $b = 20$; $x(t) = 10e^t(\cos t + \sin t) + 10$, $y(t) = 10e^t(\cos t - \sin t) + 20$. Species Y dies out when $t \approx 1.22$.

71. $x(t) = C_1e^{2t} + C_2e^{3t}$

77. (a) $\begin{bmatrix} 1 \\ 1 \end{bmatrix}, \begin{bmatrix} 3 \\ 3 \end{bmatrix}, \begin{bmatrix} 9 \\ 9 \end{bmatrix}, \begin{bmatrix} 27 \\ 27 \end{bmatrix}$ **(c)** Repeller

79. (a) $\begin{bmatrix} 1 \\ 1 \end{bmatrix}, \begin{bmatrix} 1 \\ 1 \end{bmatrix}, \begin{bmatrix} 1 \\ 1 \end{bmatrix}, \begin{bmatrix} 1 \\ 1 \end{bmatrix}$ **(c)** Neither

81. (a) $\begin{bmatrix} 1 \\ 1 \end{bmatrix}, \begin{bmatrix} 0.5 \\ -1 \end{bmatrix}, \begin{bmatrix} 1.75 \\ -0.5 \end{bmatrix}, \begin{bmatrix} 3.125 \\ -1.75 \end{bmatrix}$ **(c)** Saddle point

83. (a) $\begin{bmatrix} 1 \\ 1 \end{bmatrix}, \begin{bmatrix} 0.6 \\ 0.6 \end{bmatrix}, \begin{bmatrix} 0.36 \\ 0.36 \end{bmatrix}, \begin{bmatrix} 0.216 \\ 0.216 \end{bmatrix}$ **(c)** Attractor

85. $r = \sqrt{2}$, $\theta = 45°$, spiral repeller

87. $r = 2$, $\theta = -60°$, spiral repeller

89. $P = \begin{bmatrix} -1 & -1 \\ 1 & 0 \end{bmatrix}$, $C = \begin{bmatrix} 0.2 & -0.1 \\ 0.1 & 0.2 \end{bmatrix}$, spiral attractor

91. $P = \begin{bmatrix} 1/2 & -\sqrt{3}/2 \\ 1 & 0 \end{bmatrix}$, $C = \begin{bmatrix} 1/2 & -\sqrt{3}/2 \\ \sqrt{3}/2 & 1/2 \end{bmatrix}$, orbital center

Review Questions

1. (a) F **(c)** F **(e)** F **(g)** T **(i)** F

3. -18

5. Since $A^T = -A$, we have $\det A = \det(A^T) = \det(-A) = (-1)^n \det A = -\det A$ by Theorem 4.7 and the fact that n is odd. It follows that $\det A = 0$.

7. $A\mathbf{x} = \begin{bmatrix} 5 \\ 10 \end{bmatrix} = 5\mathbf{x}$, $\lambda = 5$

9. (a) $4 - 3\lambda^2 - \lambda^3$

(c) $E_1 = \text{span}\left(\begin{bmatrix} 1 \\ -1 \\ 0 \end{bmatrix}\right)$, $E_{-2} = \text{span}\left(\begin{bmatrix} 2 \\ -1 \\ 0 \end{bmatrix}, \begin{bmatrix} 1 \\ 0 \\ 1 \end{bmatrix}\right)$

11. $\begin{bmatrix} 162 \\ 158 \end{bmatrix}$

13. Not similar

15. Not similar

17. 0, 1, or -1

19. If $A\mathbf{x} = \lambda\mathbf{x}$, then $(A^2 - 5A + 2I)\mathbf{x} = A^2\mathbf{x} - 5A\mathbf{x} + 2\mathbf{x} = 3^2\mathbf{x} - 5(3\mathbf{x}) + 2\mathbf{x} = -4\mathbf{x}$.

Chapter 5

Exercises 5.1

1. Orthogonal **3.** Not orthogonal **5.** Orthogonal

7. $[\mathbf{w}]_{\mathcal{B}} = \begin{bmatrix} \frac{1}{2} \\ -1 \end{bmatrix}$ **9.** $[\mathbf{w}]_{\mathcal{B}} = \begin{bmatrix} 0 \\ \frac{2}{3} \\ \frac{1}{3} \end{bmatrix}$ **11.** Orthonormal

13. $\begin{bmatrix} 1/3 \\ 2/3 \\ 2/3 \end{bmatrix}, \begin{bmatrix} 2/\sqrt{5} \\ -1/\sqrt{5} \\ 0 \end{bmatrix}, \begin{bmatrix} 2/3\sqrt{5} \\ 4/3\sqrt{5} \\ -5/3\sqrt{5} \end{bmatrix}$

15. Orthonormal

17. Orthogonal, $\begin{bmatrix} 1/\sqrt{2} & -1/\sqrt{2} \\ 1/\sqrt{2} & 1/\sqrt{2} \end{bmatrix}$

19. Orthogonal, $\begin{bmatrix} \cos\theta\sin\theta & \cos^2\theta & \sin\theta \\ -\cos\theta & \sin\theta & 0 \\ -\sin^2\theta & -\cos\theta\sin\theta & \cos\theta \end{bmatrix}$

21. Not orthogonal

27. $\cos(\angle(Q\mathbf{x}, Q\mathbf{y})) = \dfrac{Q\mathbf{x}\cdot Q\mathbf{y}}{\|Q\mathbf{x}\|\|Q\mathbf{y}\|} = \dfrac{\mathbf{x}\cdot\mathbf{y}}{\|\mathbf{x}\|\|\mathbf{y}\|}$
$= \cos(\angle(\mathbf{x}, \mathbf{y}))$ by Theorem 5.6

29. Rotation, $\theta = 45°$ **31.** Reflection, $y = \sqrt{3}x$

33. (a) $A(A^T + B^T)B = AA^TB + AB^TB = IB + AI = B + A = A + B$

(b) From part (a),

$$\begin{aligned}\det(A + B) &= \det(A(A^T + B^T)B) \\ &= \det A \det(A^T + B^T)\det B \\ &= \det A \det((A + B)^T)\det B \\ &= \det A \det(A + B)\det B\end{aligned}$$

Assume that $\det A + \det B = 0$ (so that $\det B = -\det A$) but that $A + B$ is invertible. Then $\det(A + B) \neq 0$, so $1 = \det A \det B = \det A(-\det A) = -(\det A)^2$. This is impossible, so we conclude that $A + B$ cannot be invertible.

Exercises 5.2

1. $W^\perp = \left\{\begin{bmatrix} x \\ y \end{bmatrix} : x + 2y = 0\right\}, \mathcal{B}^\perp = \left\{\begin{bmatrix} -2 \\ 1 \end{bmatrix}\right\}$

3. $W^\perp = \left\{\begin{bmatrix} x \\ y \\ z \end{bmatrix} : x = t, y = t, z = -t\right\},$

$\mathcal{B}^\perp = \left\{\begin{bmatrix} 1 \\ 1 \\ -1 \end{bmatrix}\right\}$

5. $W^\perp = \left\{\begin{bmatrix} x \\ y \\ z \end{bmatrix} : x - y + 3z = 0\right\}, \mathcal{B}^\perp = \left\{\begin{bmatrix} 1 \\ 1 \\ 0 \end{bmatrix}, \begin{bmatrix} 0 \\ 3 \\ 1 \end{bmatrix}\right\}$

7. row(A): $\{[1 \ \ 0 \ \ 1], [0 \ \ 1 \ \ -2]\}$, null($A$): $\left\{\begin{bmatrix} 1 \\ 2 \\ 1 \end{bmatrix}\right\}$

9. col(A): $\left\{\begin{bmatrix} 1 \\ 5 \\ 0 \\ -1 \end{bmatrix}, \begin{bmatrix} -1 \\ 2 \\ 1 \\ -1 \end{bmatrix}\right\}$, null($A^T$):

$\left\{\begin{bmatrix} 1 \\ 0 \\ 2 \\ 1 \end{bmatrix}, \begin{bmatrix} -5 \\ 1 \\ -7 \\ 0 \end{bmatrix}\right\}$

11. $\left\{\begin{bmatrix} 1 \\ -10 \\ -4 \end{bmatrix}\right\}$ **13.** $\left\{\begin{bmatrix} -4 \\ 1 \\ 0 \\ 3 \end{bmatrix}, \begin{bmatrix} -3 \\ 0 \\ 1 \\ 0 \end{bmatrix}\right\}$

15. $\begin{bmatrix} \frac{3}{2} \\ \frac{3}{2} \end{bmatrix}$ **17.** $\begin{bmatrix} -\frac{1}{2} \\ \frac{1}{2} \\ 3 \end{bmatrix}$

19. $\mathbf{v} = \begin{bmatrix} -\frac{2}{5} \\ -\frac{6}{5} \end{bmatrix} + \begin{bmatrix} \frac{12}{5} \\ -\frac{4}{5} \end{bmatrix}$ **21.** $\mathbf{v} = \begin{bmatrix} \frac{7}{2} \\ -2 \\ \frac{7}{2} \end{bmatrix} + \begin{bmatrix} \frac{1}{2} \\ 0 \\ -\frac{1}{2} \end{bmatrix}$

25. No

Exercises 5.3

1. $\mathbf{v}_1 = \begin{bmatrix} 1 \\ 1 \end{bmatrix}, \mathbf{v}_2 = \begin{bmatrix} -1/2 \\ 1/2 \end{bmatrix}; \mathbf{q}_1 = \begin{bmatrix} 1/\sqrt{2} \\ 1/\sqrt{2} \end{bmatrix}, \mathbf{q}_2 = \begin{bmatrix} -1/\sqrt{2} \\ 1/\sqrt{2} \end{bmatrix}$

3. $\mathbf{v}_1 = \begin{bmatrix} 1 \\ -1 \\ -1 \end{bmatrix}, \mathbf{v}_2 = \begin{bmatrix} 2 \\ 1 \\ 1 \end{bmatrix}, \mathbf{v}_3 = \begin{bmatrix} 0 \\ -1 \\ 1 \end{bmatrix}; \mathbf{q}_1 = \begin{bmatrix} 1/\sqrt{3} \\ -1/\sqrt{3} \\ -1/\sqrt{3} \end{bmatrix},$

$\mathbf{q}_2 = \begin{bmatrix} 2/\sqrt{6} \\ 1/\sqrt{6} \\ 1/\sqrt{6} \end{bmatrix}, \mathbf{q}_3 = \begin{bmatrix} 0 \\ -1/\sqrt{2} \\ 1/\sqrt{2} \end{bmatrix}$

5. $\left\{\begin{bmatrix} 1 \\ 1 \\ 0 \end{bmatrix}, \begin{bmatrix} -\frac{1}{2} \\ \frac{1}{2} \\ 2 \end{bmatrix}\right\}$ **7.** $\mathbf{v} = \begin{bmatrix} -\frac{2}{9} \\ \frac{2}{9} \\ \frac{8}{9} \end{bmatrix} + \begin{bmatrix} \frac{38}{9} \\ -\frac{38}{9} \\ \frac{19}{9} \end{bmatrix}$

9. $\left\{\begin{bmatrix} 0 \\ 1 \\ 1 \end{bmatrix}, \begin{bmatrix} 1 \\ -\frac{1}{2} \\ \frac{1}{2} \end{bmatrix}, \begin{bmatrix} \frac{2}{3} \\ \frac{2}{3} \\ -\frac{2}{3} \end{bmatrix}\right\}$

11. $\left\{\begin{bmatrix}3\\1\\5\end{bmatrix}, \begin{bmatrix}-\frac{3}{35}\\ \frac{34}{35}\\ -\frac{1}{7}\end{bmatrix}, \begin{bmatrix}-\frac{15}{34}\\ 0\\ \frac{9}{34}\end{bmatrix}\right\}$

13. $Q = \begin{bmatrix}1/\sqrt{2} & 1/\sqrt{3} & 1/\sqrt{6}\\ 0 & 1/\sqrt{3} & -2/\sqrt{2}\\ -1/\sqrt{2} & 1/\sqrt{3} & 1/\sqrt{6}\end{bmatrix}$

15. $\begin{bmatrix}0 & 2/\sqrt{6} & 1/\sqrt{3}\\ 1/\sqrt{2} & -1/\sqrt{6} & 1/\sqrt{3}\\ 1/\sqrt{2} & 1/\sqrt{6} & -1/\sqrt{3}\end{bmatrix}\begin{bmatrix}\sqrt{2} & 1/\sqrt{2} & 1/\sqrt{2}\\ 0 & 3/\sqrt{6} & 1/\sqrt{6}\\ 0 & 0 & 2/\sqrt{3}\end{bmatrix}$

17. $R = \begin{bmatrix}3 & 9 & \frac{1}{3}\\ 0 & 6 & \frac{2}{3}\\ 0 & 0 & \frac{7}{3}\end{bmatrix}$

19. $A = AI$

21. $A^{-1} = (QR)^{-1} = R^{-1}Q^{-1} = R^{-1}Q^T =$

$$\begin{bmatrix}1/\sqrt{2} & -1/\sqrt{6} & -1/2\sqrt{3}\\ 0 & 2/\sqrt{6} & -1/2\sqrt{3}\\ 0 & 0 & 3/2\sqrt{3}\end{bmatrix} \cdot \begin{bmatrix}0 & 1/\sqrt{2} & 1/\sqrt{2}\\ 2/\sqrt{6} & -1/\sqrt{6} & 1/\sqrt{6}\\ 1/\sqrt{3} & 1/\sqrt{3} & -1/\sqrt{3}\end{bmatrix}$$

23. Let $R\mathbf{x} = \mathbf{0}$. Then $A\mathbf{x} = QR\mathbf{x} = Q\mathbf{0} = \mathbf{0}$. Since $A\mathbf{x}$ represents a linear combination of the columns of A (which are linearly independent), we must have $\mathbf{x} = \mathbf{0}$. Hence, R is invertible, by the Fundamental Theorem.

Exercises 5.4

1. $Q = \begin{bmatrix}1/\sqrt{2} & 1/\sqrt{2}\\ 1/\sqrt{2} & -1/\sqrt{2}\end{bmatrix}, D = \begin{bmatrix}5 & 0\\ 0 & 3\end{bmatrix}$

3. $Q = \begin{bmatrix}2/\sqrt{6} & 1/\sqrt{3}\\ 1/\sqrt{3} & -2/\sqrt{6}\end{bmatrix}, D = \begin{bmatrix}2 & 0\\ 0 & -1\end{bmatrix}$

5. $Q = \begin{bmatrix}1 & 0 & 0\\ 0 & 1/\sqrt{2} & -1/\sqrt{2}\\ 0 & 1/\sqrt{2} & 1/\sqrt{2}\end{bmatrix}, D = \begin{bmatrix}5 & 0 & 0\\ 0 & 4 & 0\\ 0 & 0 & -2\end{bmatrix}$

7. $Q = \begin{bmatrix}-1/\sqrt{2} & 0 & 1/\sqrt{2}\\ 0 & 1 & 0\\ 1/\sqrt{2} & 0 & 1/\sqrt{2}\end{bmatrix}, D = \begin{bmatrix}2 & 0 & 0\\ 0 & 1 & 0\\ 0 & 0 & 0\end{bmatrix}$

9. $Q = \begin{bmatrix}1/\sqrt{2} & 0 & 1/\sqrt{2} & 0\\ 1/\sqrt{2} & 0 & -1/\sqrt{2} & 0\\ 0 & 1/\sqrt{2} & 0 & 1/\sqrt{2}\\ 0 & 1/\sqrt{2} & 0 & -1/\sqrt{2}\end{bmatrix},$

$$D = \begin{bmatrix}2 & 0 & 0 & 0\\ 0 & 2 & 0 & 0\\ 0 & 0 & 0 & 0\\ 0 & 0 & 0 & 0\end{bmatrix}$$

11. $Q^TAQ = \begin{bmatrix}1/\sqrt{2} & 1/\sqrt{2}\\ 1/\sqrt{2} & -1/\sqrt{2}\end{bmatrix}\begin{bmatrix}a & b\\ b & a\end{bmatrix} \cdot$

$$\begin{bmatrix}1/\sqrt{2} & 1/\sqrt{2}\\ 1/\sqrt{2} & -1/\sqrt{2}\end{bmatrix} = \begin{bmatrix}a+b & 0\\ 0 & a-b\end{bmatrix} = D$$

13. (a) If A and B are orthogonally diagonalizable, then each is symmetric, by the Spectral Theorem. Therefore, $A + B$ is symmetric, by Exercise 35 in Section 3.2, and so is orthogonally diagonalizable, by the Spectral Theorem.

15. If A and B are orthogonally diagonalizable, then each is symmetric, by the Spectral Theorem. Since $AB = BA$, AB is also symmetric, by Exercise 36 in Section 3.2. Hence, AB is orthogonally diagonalizable, by the Spectral Theorem.

17. $A = \begin{bmatrix}\frac{5}{2} & \frac{5}{2}\\ \frac{5}{2} & \frac{5}{2}\end{bmatrix} + \begin{bmatrix}\frac{3}{2} & -\frac{3}{2}\\ -\frac{3}{2} & \frac{3}{2}\end{bmatrix}$

19. $A = \begin{bmatrix}5 & 0 & 0\\ 0 & 0 & 0\\ 0 & 0 & 0\end{bmatrix} + \begin{bmatrix}0 & 0 & 0\\ 0 & 2 & 2\\ 0 & 2 & 2\end{bmatrix} + \begin{bmatrix}0 & 0 & 0\\ 0 & -1 & 1\\ 0 & 1 & -1\end{bmatrix}$

21. $\begin{bmatrix}\frac{1}{2} & -\frac{3}{2}\\ -\frac{3}{2} & \frac{1}{2}\end{bmatrix}$ **23.** $\begin{bmatrix}\frac{5}{3} & -\frac{2}{3} & -\frac{1}{3}\\ -\frac{2}{3} & \frac{5}{3} & \frac{1}{3}\\ -\frac{1}{3} & \frac{1}{3} & \frac{8}{3}\end{bmatrix}$

Exercises 5.5

1. $2x^2 + 6xy + 4y^2$ **3.** 123 **5.** -5

7. $\begin{bmatrix}1 & 3\\ 3 & 2\end{bmatrix}$ **9.** $\begin{bmatrix}3 & -\frac{3}{2}\\ -\frac{3}{2} & -1\end{bmatrix}$ **11.** $\begin{bmatrix}5 & 1 & -2\\ 1 & -1 & 2\\ -2 & 2 & 2\end{bmatrix}$

13. $Q = \begin{bmatrix}2/\sqrt{5} & 1/\sqrt{5}\\ 1/\sqrt{5} & -2/\sqrt{5}\end{bmatrix}, y_1^2 + 6y_2^2$

15. $Q = \begin{bmatrix}2/\sqrt{5} & 2/3\sqrt{5} & -1/3\\ 0 & 5/3\sqrt{5} & 2/3\\ 1/\sqrt{5} & -4/3\sqrt{5} & 2/3\end{bmatrix}, 9y_1^2 + 9y_2^2 - 9y_3^2$

17. $Q = \begin{bmatrix}1/\sqrt{3} & 1/\sqrt{2} & 1/\sqrt{6}\\ -1/\sqrt{3} & 0 & 2/\sqrt{6}\\ -1/\sqrt{3} & 1/\sqrt{2} & -1/\sqrt{6}\end{bmatrix}, 2(x')^2 + (y')^2 - (z')^2$

19. Positive definite **21.** Negative definite

23. Positive definite **25.** Indefinite

27. For any vector $\mathbf{x}$, we have $\mathbf{x}^T A\mathbf{x} = \mathbf{x}^T B^T B\mathbf{x} = (B\mathbf{x})^T(B\mathbf{x}) = \|B\mathbf{x}\|^2 \geq 0$. If $\mathbf{x}^T A\mathbf{x} = 0$, then $\|B\mathbf{x}\|^2 = 0$, so $B\mathbf{x} = \mathbf{0}$. Since B is invertible, this implies that $\mathbf{x} = \mathbf{0}$. Therefore, $\mathbf{x}^T A\mathbf{x} > 0$ for all $\mathbf{x} \neq \mathbf{0}$, and hence $A = B^T B$ is positive definite.

29. **(a)** Every eigenvalue of cA is of the form $c\lambda$ for some eigenvalue λ of A. By Theorem 5.24, $\lambda > 0$, so $c\lambda > 0$, since c is positive. Hence, cA is positive definite, by Theorem 5.24.

(c) Let $\mathbf{x} \neq \mathbf{0}$. Then $\mathbf{x}^T A\mathbf{x} > 0$ and $\mathbf{x}^T B\mathbf{x} > 0$, since A and B are positive definite. But then $\mathbf{x}^T(A + B)\mathbf{x} = \mathbf{x}^T A\mathbf{x} + \mathbf{x}^T B\mathbf{x} > 0$, so $A + B$ is positive definite.

31. The maximum value of $f(\mathbf{x})$ is 2 when $\mathbf{x} = \pm\begin{bmatrix} 1/\sqrt{2} \\ -1/\sqrt{2} \end{bmatrix}$; the minimum value of $f(\mathbf{x})$ is 0 when $\mathbf{x} = \pm\begin{bmatrix} 1/\sqrt{2} \\ 1/\sqrt{2} \end{bmatrix}$.

33. The maximum value of $f(\mathbf{x})$ is 4 when $\mathbf{x} = \pm\begin{bmatrix} 1/\sqrt{3} \\ 1/\sqrt{3} \\ 1/\sqrt{3} \end{bmatrix}$; the minimum value of $f(\mathbf{x})$ is 1 when $\mathbf{x} = \pm\begin{bmatrix} 1/\sqrt{2} \\ 0 \\ -1/\sqrt{2} \end{bmatrix}$ or $\pm\begin{bmatrix} -1/\sqrt{2} \\ 1/\sqrt{2} \\ 0 \end{bmatrix}$.

35. Ellipse **37.** Parabola **39.** Hyperbola

41. Circle, $x' = x - 2$, $y' = y - 2$, $(x')^2 + (y')^2 = 4$

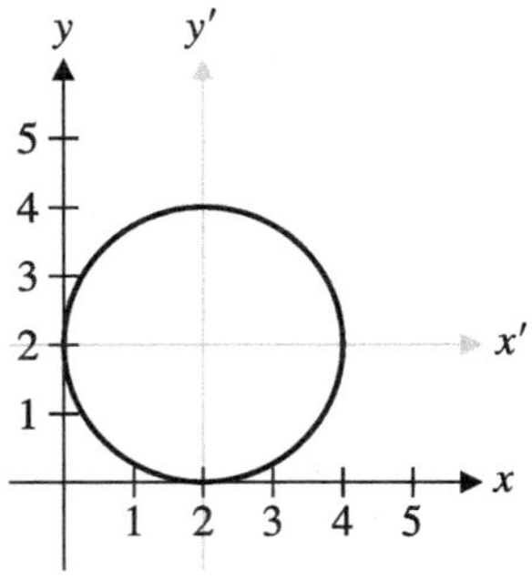

43. Hyperbola, $x' = x$, $y' = y + \frac{1}{2}$, $(x')^2/4 - (y')^2/9 = 1$

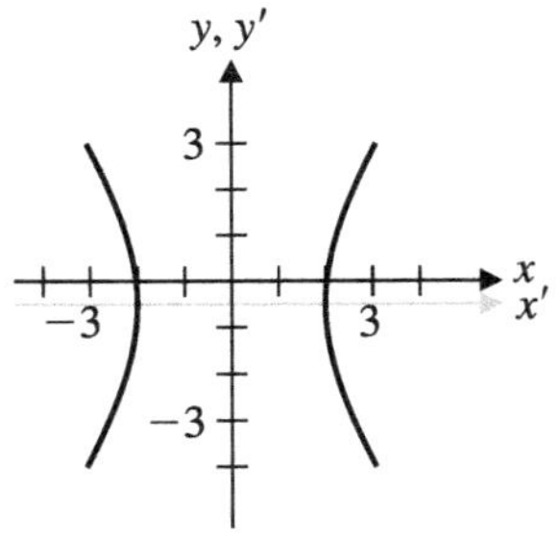

45. Parabola, $x' = x - 2$, $y' = y + 2$, $x' = -\frac{1}{2}(y')^2$

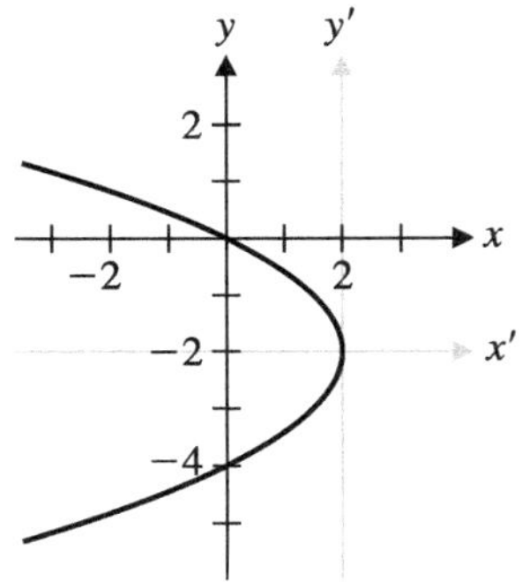

47. Ellipse, $(x')^2/4 + (y')^2/12 = 1$

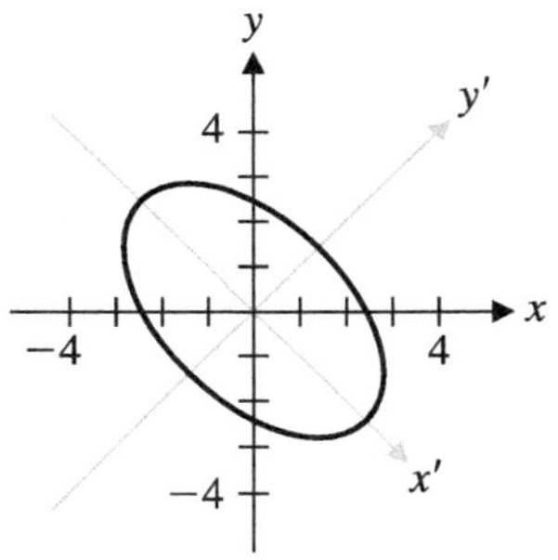

49. Hyperbola, $(x')^2 - (y')^2 = 1$

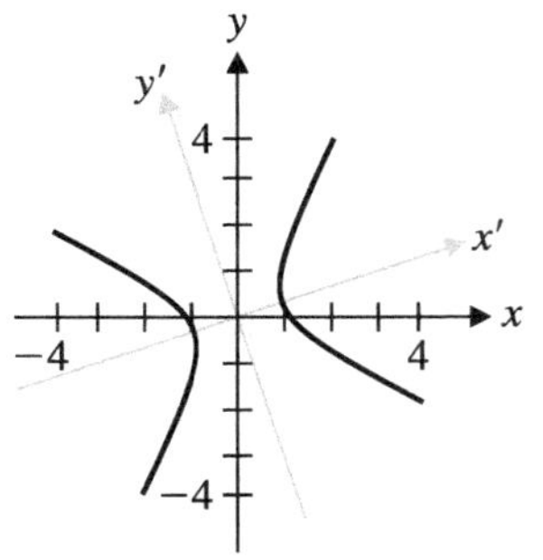

51. Ellipse, $(x'')^2/50 + (y'')^2/10 = 1$

53. Hyperbola, $(x'')^2 - (y'')^2 = 1$

55. Degenerate (two lines)

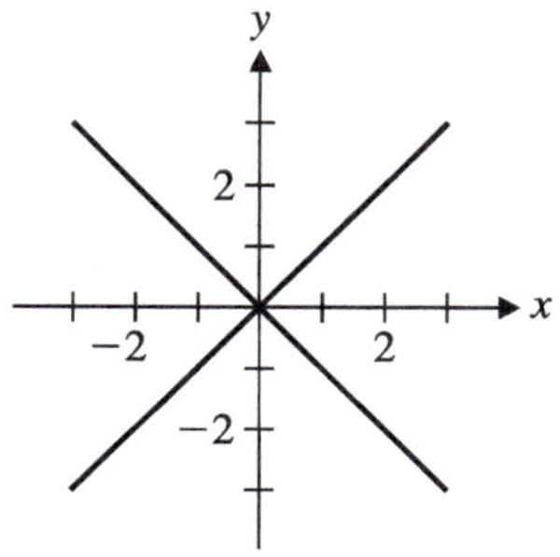

57. Degenerate (a point)

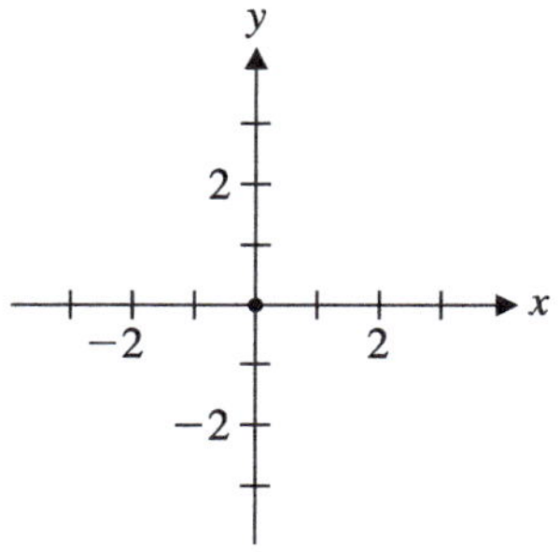

59. Degenerate (two lines)

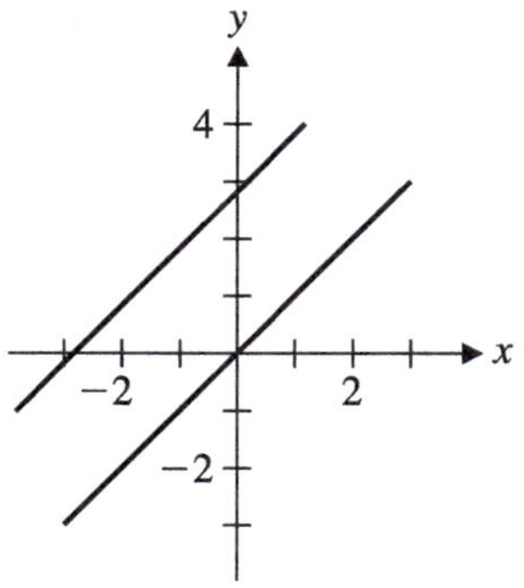

61. Hyperboloid of one sheet, $(x')^2 - (y')^2 + 3(z')^2 = 1$

63. Hyperbolic paraboloid, $z = -(x')^2 + (y')^2$

65. Hyperbolic paraboloid, $x' = -\sqrt{3}(y')^2 + \sqrt{3}(z')^2$

67. Ellipsoid, $3(x'')^2 + (y'')^2 + 2(z'')^2 = 4$

Review Questions

1. (a) T **(c)** T **(e)** F **(g)** F **(i)** F

3. $\begin{bmatrix} 9/2 \\ 2/3 \\ -11/6 \end{bmatrix}$

5. Verify that $Q^TQ = I$.

7. Theorem 5.6(c) shows that if $\mathbf{v}_i \cdot \mathbf{v}_j = 0$, then $Q\mathbf{v}_i \cdot Q\mathbf{v}_j = 0$. Theorem 5.6(b) shows that $\{Q\mathbf{v}_1, \ldots, Q\mathbf{v}_k\}$ consists of unit vectors, because $\{\mathbf{v}_1, \ldots, \mathbf{v}_k\}$ does. Hence, $\{Q\mathbf{v}_1, \ldots, Q\mathbf{v}_k\}$ is an orthonormal set.

9. $\left\{\begin{bmatrix} 2 \\ -5 \end{bmatrix}\right\}$

11. $\left\{\begin{bmatrix} -1 \\ 3 \\ 1 \end{bmatrix}\right\}$

13. row(A): $\{[1 \;\; 0 \;\; 2 \;\; 3 \;\; 4], [0 \;\; 1 \;\; 0 \;\; 2 \;\; 1]\}$

col(A): $\left\{\begin{bmatrix} 1 \\ -1 \\ 2 \\ 3 \end{bmatrix}, \begin{bmatrix} -1 \\ 2 \\ 1 \\ -5 \end{bmatrix}\right\}$

null(A): $\left\{\begin{bmatrix} -2 \\ 0 \\ 1 \\ 0 \\ 0 \end{bmatrix}, \begin{bmatrix} -3 \\ -2 \\ 0 \\ 1 \\ 0 \end{bmatrix}, \begin{bmatrix} -4 \\ -1 \\ 0 \\ 0 \\ 1 \end{bmatrix}\right\}$

null(A^T): $\left\{\begin{bmatrix} -5 \\ -3 \\ 1 \\ 0 \end{bmatrix}, \begin{bmatrix} -1 \\ 2 \\ 0 \\ 1 \end{bmatrix}\right\}$

15. (a) $\left\{\begin{bmatrix} 1 \\ 1 \\ 1 \\ 1 \end{bmatrix}, \begin{bmatrix} \frac{1}{4} \\ \frac{1}{4} \\ \frac{1}{4} \\ -\frac{3}{4} \end{bmatrix}, \begin{bmatrix} -\frac{2}{3} \\ \frac{1}{3} \\ \frac{1}{3} \\ 0 \end{bmatrix}\right\}$

17. $\left\{\begin{bmatrix} -1 \\ 1 \\ 0 \\ 0 \end{bmatrix}, \begin{bmatrix} \frac{1}{2} \\ \frac{1}{2} \\ -1 \\ 0 \end{bmatrix}, \begin{bmatrix} \frac{1}{3} \\ \frac{1}{3} \\ \frac{1}{3} \\ -1 \end{bmatrix}\right\}$

19. $\begin{bmatrix} -\frac{1}{2} & \frac{3}{2} & 0 \\ \frac{3}{2} & -\frac{1}{2} & 0 \\ 0 & 0 & 1 \end{bmatrix}$

Chapter 6

Exercises 6.1

1. Vector space

3. Not a vector space; axiom 1 fails.

5. Not a vector space; axiom 8 fails.

7. Vector space

9. Vector space

11. Vector space

15. Complex vector space

17. Not a complex vector space; axiom 6 fails.

19. Not a vector space; axioms 1, 4, and 6 fail.

21. Not a vector space; the operations of addition and multiplication are not even the same.

25. Subspace **27.** Not a subspace

29. Not a subspace **31.** Subspace

33. Subspace **35.** Subspace

37. Not a subspace **39.** Subspace

41. Subspace **43.** Not a subspace

45. Not a subspace

47. Take U to be the x-axis and W the y-axis, for example. Then $\begin{bmatrix}1\\0\end{bmatrix}$ and $\begin{bmatrix}0\\1\end{bmatrix}$ are in $U \cup W$, but $\begin{bmatrix}1\\1\end{bmatrix} = \begin{bmatrix}1\\0\end{bmatrix} + \begin{bmatrix}0\\1\end{bmatrix}$ is not.

51. No

53. Yes; $s(x) = (3 + 2t)p(x) + (1 + t)q(x) + tr(x)$ for any scalar t.

55. Yes; $h(x) = f(x) + g(x)$

57. No

59. No

61. Yes

Exercises 6.2

1. Linearly independent

3. Linearly dependent; $\begin{bmatrix}-1 & 0\\-1 & 7\end{bmatrix} = 4\begin{bmatrix}-1 & 1\\-2 & 2\end{bmatrix} + \begin{bmatrix}3 & 0\\1 & 1\end{bmatrix} - 2\begin{bmatrix}0 & 2\\-3 & 1\end{bmatrix}$

5. Linearly independent

7. Linearly dependent; $3x + 2x^2 = 7x - 2(2x - x^2)$

9. Linearly independent

11. Linearly dependent; $1 = \sin^2 x + \cos^2 x$

13. Linearly dependent; $\ln(x^2) = -2\ln 2 \cdot 1 + 2 \cdot \ln(2x)$

17. **(a)** Linearly independent

(b) Linearly dependent

19. Basis **21.** Not a basis

23. Not a basis **25.** Not a basis

27. $[A]_{\mathcal{B}} = \begin{bmatrix}-1\\-1\\-1\\4\end{bmatrix}$ **29.** $[p(x)]_{\mathcal{B}} = \begin{bmatrix}6\\-1\\3\end{bmatrix}$

35. $\dim V = 2, \mathcal{B} = \{1 - x, 1 - x^2\}$

37. $\dim V = 3, \mathcal{B} = \left\{\begin{bmatrix}1 & 0\\0 & 0\end{bmatrix}, \begin{bmatrix}0 & 1\\0 & 0\end{bmatrix}, \begin{bmatrix}0 & 0\\0 & 1\end{bmatrix}\right\}$

39. $\dim V = 2, \mathcal{B} = \left\{\begin{bmatrix}1 & 0\\0 & 1\end{bmatrix}, \begin{bmatrix}0 & 1\\0 & 0\end{bmatrix}\right\}$

41. $(n^2 - n)/2$

43. **(a)** $\dim(U \times V) = \dim U + \dim V$

(b) Show that if $\{\mathbf{w}_1, \ldots, \mathbf{w}_n\}$ is a basis for W, then $\{(\mathbf{w}_1, \mathbf{w}_1), \ldots, (\mathbf{w}_n, \mathbf{w}_n)\}$ is a basis for Δ.

45. $\{1 + x, 1 + x + x^2, 1\}$

47. $\left\{\begin{bmatrix}1 & 0\\0 & 1\end{bmatrix}, \begin{bmatrix}0 & 1\\1 & 0\end{bmatrix}, \begin{bmatrix}0 & -1\\1 & 0\end{bmatrix}, \begin{bmatrix}1 & 0\\0 & 0\end{bmatrix}\right\}$

49. $\{1, 1 + x\}$

51. $\{1 - x, x - x^2\}$

53. $\{\sin^2 x, \cos^2 x\}$

59. **(a)** $p_0(x) = \frac{1}{2}x^2 - \frac{5}{2}x + 3, p_1(x) = -x^2 + 4x - 3, p_2(x) = \frac{1}{2}x^2 - \frac{3}{2}x + 1$

61. **(c)** (i) $3x^2 - 16x + 19$ (ii) $x^2 - 4x + 5$

63. $(p^n - 1)(p^n - p)(p^n - p^2)\cdots(p^n - p^{n-1})$

Exercises 6.3

1. $[\mathbf{x}]_{\mathcal{B}} = \begin{bmatrix}2\\3\end{bmatrix}, [\mathbf{x}]_{\mathcal{C}} = \begin{bmatrix}\frac{5}{2}\\-\frac{1}{2}\end{bmatrix}, P_{\mathcal{C}\leftarrow\mathcal{B}} = \begin{bmatrix}\frac{1}{2} & \frac{1}{2}\\\frac{1}{2} & -\frac{1}{2}\end{bmatrix}, P_{\mathcal{B}\leftarrow\mathcal{C}} = \begin{bmatrix}1 & 1\\1 & -1\end{bmatrix}$

3. $[\mathbf{x}]_{\mathcal{B}} = \begin{bmatrix}1\\0\\-1\end{bmatrix}, [\mathbf{x}]_{\mathcal{C}} = \begin{bmatrix}1\\-1\\-1\end{bmatrix}, P_{\mathcal{C}\leftarrow\mathcal{B}} = \begin{bmatrix}1 & 0 & 0\\-1 & 1 & 0\\0 & -1 & 1\end{bmatrix},$

$P_{\mathcal{B}\leftarrow\mathcal{C}} = \begin{bmatrix}1 & 0 & 0\\1 & 1 & 0\\1 & 1 & 1\end{bmatrix}$

5. $[p(x)]_{\mathcal{B}} = \begin{bmatrix}2\\-1\end{bmatrix}, [p(x)]_{\mathcal{C}} = \begin{bmatrix}-3\\2\end{bmatrix}, P_{\mathcal{C}\leftarrow\mathcal{B}} = \begin{bmatrix}-1 & 1\\1 & 0\end{bmatrix},$

$P_{\mathcal{B}\leftarrow\mathcal{C}} = \begin{bmatrix}0 & 1\\1 & 1\end{bmatrix}$

7. $[p(x)]_{\mathcal{B}} = \begin{bmatrix}1\\-1\\1\end{bmatrix}, [p(x)]_{\mathcal{C}} = \begin{bmatrix}1\\0\\1\end{bmatrix}, P_{\mathcal{C}\leftarrow\mathcal{B}} = \begin{bmatrix}1 & 0 & 0\\1 & 1 & 0\\1 & 1 & 1\end{bmatrix},$

$P_{\mathcal{B}\leftarrow\mathcal{C}} = \begin{bmatrix}1 & 0 & 0\\-1 & 1 & 0\\0 & -1 & 1\end{bmatrix}$

9. $[A]_{\mathcal{B}} = \begin{bmatrix} 4 \\ 2 \\ 0 \\ -1 \end{bmatrix}, [A]_{\mathcal{C}} = \begin{bmatrix} \frac{5}{2} \\ 0 \\ -3 \\ \frac{9}{2} \end{bmatrix}, P_{\mathcal{C}\leftarrow\mathcal{B}} =$

$\begin{bmatrix} \frac{1}{2} & 0 & -1 & -\frac{1}{2} \\ 0 & 0 & 1 & 0 \\ -1 & 1 & 1 & 1 \\ \frac{3}{2} & -1 & -2 & -\frac{1}{2} \end{bmatrix}, P_{\mathcal{B}\leftarrow\mathcal{C}} = \begin{bmatrix} 1 & 2 & 1 & 1 \\ 2 & 1 & 1 & 0 \\ 0 & 1 & 0 & 0 \\ -1 & 0 & 1 & 1 \end{bmatrix}$

11. $[f(x)]_{\mathcal{B}} = \begin{bmatrix} 2 \\ -5 \end{bmatrix}, [f(x)]_{\mathcal{C}} = \begin{bmatrix} -1/2 \\ 5/2 \end{bmatrix},$

$P_{\mathcal{C}\leftarrow\mathcal{B}} = \begin{bmatrix} 1 & 1/2 \\ 0 & -1/2 \end{bmatrix}, P_{\mathcal{B}\leftarrow\mathcal{C}} = \begin{bmatrix} 1 & 1 \\ 0 & -2 \end{bmatrix}$

13. (a) $\begin{bmatrix} (3 - 2\sqrt{3})/2 \\ (-3\sqrt{3} + 2)/2 \end{bmatrix} \approx \begin{bmatrix} 3.232 \\ -1.598 \end{bmatrix}$

(b) $\begin{bmatrix} 2 + 2\sqrt{3} \\ 2\sqrt{3} - 2 \end{bmatrix} \approx \begin{bmatrix} 5.464 \\ 1.464 \end{bmatrix}$

15. $\mathcal{B} = \left\{ \begin{bmatrix} -1 \\ -1 \end{bmatrix}, \begin{bmatrix} 3 \\ 4 \end{bmatrix} \right\}$

17. $-2 - 8(x - 1) - 5(x - 1)^2$

19. $-1 + 3(x + 1) - 3(x + 1)^2 + (x + 1)^3$

Exercises 6.4

1. Linear transformation

3. Linear transformation

5. Linear transformation

7. Not a linear transformation

9. Linear transformation

11. Not a linear transformation

13. We have

$$\begin{aligned} S(p(x) + q(x)) &= S((p + q)(x)) = x((p + q)(x)) \\ &= x(p(x) + q(x)) = xp(x) + xq(x) \\ &= S(p(x)) + S(q(x)) \end{aligned}$$

and $$\begin{aligned} S(cp(x)) &= S((cp)(x)) = x((cp)(x)) \\ &= x(cp(x)) = cxp(x) = cS(p(x)) \end{aligned}$$

Therefore, S is linear. Similarly,

$$\begin{aligned} T\left(\begin{bmatrix} a \\ b \end{bmatrix} + \begin{bmatrix} c \\ d \end{bmatrix}\right) &= T\begin{bmatrix} a + c \\ b + d \end{bmatrix} \\ &= (a + c) + ((a + c) + (b + d))x \\ &= (a + (a + b)x) + (c + (c + d)x) \\ &= T\left(\begin{bmatrix} a \\ b \end{bmatrix}\right) + T\left(\begin{bmatrix} c \\ d \end{bmatrix}\right) \end{aligned}$$

and $$\begin{aligned} T\left(k\begin{bmatrix} a \\ b \end{bmatrix}\right) &= T\begin{bmatrix} ka \\ kb \end{bmatrix} = (ka) + (ka + kb)x \\ &= k(a + (a + b)x) = kT\left(\begin{bmatrix} a \\ b \end{bmatrix}\right) \end{aligned}$$

Therefore, T is linear.

15. $T\begin{bmatrix} -7 \\ 9 \end{bmatrix} = 5 - 14x - 8x^2, T\begin{bmatrix} a \\ b \end{bmatrix} = \left(\frac{a + 3b}{4}\right) - \left(\frac{a + 7b}{4}\right)x + \left(\frac{a - b}{2}\right)x^2$

17. $T(4 - x + 3x^2) = 4 + 3x + 5x^2, T(a + bx + cx^2) = a + cx + \left(\frac{3a - b - c}{2}\right)x^2$

19. *Hint:* Let $a = T(E_{11}), b = T(E_{12}), c = T(E_{21}), d = T(E_{22})$.

23. *Hint:* Consider the effect of T and D on the standard basis for $\mathscr{P}_n$.

25. $(S \circ T)\begin{bmatrix} 2 \\ 1 \end{bmatrix} = \begin{bmatrix} 4 & -1 \\ 0 & 6 \end{bmatrix}, (S \circ T)\begin{bmatrix} x \\ y \end{bmatrix} = \begin{bmatrix} 2x & -y \\ 0 & 2x + 2y \end{bmatrix}.$

$(T \circ S)\begin{bmatrix} x \\ y \end{bmatrix}$ does not make sense.

27. $(S \circ T)(p(x)) = p'(x + 1), (T \circ S)(p(x)) = (p(x + 1))' = p'(x + 1)$

29. $$(S \circ T)\begin{bmatrix} x \\ y \end{bmatrix} = S\left(T\begin{bmatrix} x \\ y \end{bmatrix}\right) = S\left(\begin{bmatrix} x - y \\ -3x + 4y \end{bmatrix}\right) = \begin{bmatrix} 4(x - y) + (-3x + 4y) \\ 3(x - y) + (-3x + 4y) \end{bmatrix} = \begin{bmatrix} x \\ y \end{bmatrix}$$

$$(T \circ S)\begin{bmatrix} x \\ y \end{bmatrix} = T\left(S\begin{bmatrix} x \\ y \end{bmatrix}\right) = T\left(\begin{bmatrix} 4x + y \\ 3x + y \end{bmatrix}\right) = \begin{bmatrix} (4x + y) - (3x + y) \\ -3(4x + y) + 4(3x + y) \end{bmatrix} = \begin{bmatrix} x \\ y \end{bmatrix}$$

Therefore, $S \circ T = I$ and $T \circ S = I$, so S and T are inverses.

Exercises 6.5

1. (a) Only (ii) is in $\ker(T)$.
(b) Only (iii) is in $\text{range}(T)$.
(c) $\ker(T) = \left\{ \begin{bmatrix} 0 & b \\ c & 0 \end{bmatrix} \right\}, \text{range}(T) = \left\{ \begin{bmatrix} a & 0 \\ 0 & d \end{bmatrix} \right\}$

3. (a) Only (iii) is in $\ker(T)$.
(b) All of them are in $\text{range}(T)$.
(c) $\ker(T) = \{a + bx + cx^2 : a = -c, b = -c\} = \{t + tx - tx^2\}, \text{range}(T) = \mathbb{R}^2$

5. A basis for ker(T) is $\left\{\begin{bmatrix}0&1\\0&0\end{bmatrix},\begin{bmatrix}0&0\\1&0\end{bmatrix}\right\}$, and a basis for range$(T)$ is $\left\{\begin{bmatrix}1&0\\0&0\end{bmatrix},\begin{bmatrix}0&0\\0&1\end{bmatrix}\right\}$; rank$(T)$ = nullity$(T) = 2$, and rank(T) + nullity$(T) = 4 = \dim M_{22}$.

7. A basis for ker(T) is $\{1 + x - x^2\}$, and a basis for range(T) is $\left\{\begin{bmatrix}1\\0\end{bmatrix},\begin{bmatrix}0\\1\end{bmatrix}\right\}$; rank$(T) = 2$, nullity$(T) = 1$, and rank$(T)$ + nullity$(T) = 3 = \dim \mathscr{P}_2$.

9. rank(T) = nullity$(T) = 2$

11. rank(T) = nullity$(T) = 2$

13. rank$(T) = 1$, nullity$(T) = 2$

15. One-to-one and onto

17. Neither one-to-one nor onto

19. One-to-one but not onto

21. Isomorphic, $T\begin{bmatrix}a&0&0\\0&b&0\\0&0&c\end{bmatrix} = \begin{bmatrix}a\\b\\c\end{bmatrix}$

23. Not isomorphic

25. Isomorphic, $T(a + bi) = \begin{bmatrix}a\\b\end{bmatrix}$

31. *Hint:* Define $T: \mathscr{C}[0, 1] \to \mathscr{C}[0, 2]$ by letting $T(f)$ be the function whose value at x is $(T(f))(x) = f(x/2)$ for x in $[0, 2]$.

33. (a) Let $\mathbf{v}_1$ and $\mathbf{v}_2$ be in V and let $(S \circ T)(\mathbf{v}_1) = (S \circ T)(\mathbf{v}_2)$. Then $S(T(\mathbf{v}_1)) = S(T(\mathbf{v}_2))$, so $T(\mathbf{v}_1) = T(\mathbf{v}_2)$, since S is one-to-one. But now $\mathbf{v}_1 = \mathbf{v}_2$, since T is one-to-one. Hence, $S \circ T$ is one-to-one.

35. (a) By the Rank Theorem, rank(T) + nullity(T) = dim V. If T is onto, then range$(T) = W$, so rank(T) = dim(range(T)) = dim W. Therefore,

$$\begin{aligned}\dim V + \text{nullity}(T) &< \dim W + \text{nullity}(T)\\ &= \text{rank}(T) + \text{nullity}(T) = \dim V\end{aligned}$$

so nullity $(T) < 0$, which is impossible. Therefore, T cannot be onto.

Exercises 6.6

1. $[T]_{\mathcal{C}\leftarrow\mathcal{B}} = \begin{bmatrix}0&1\\-1&0\end{bmatrix}$, $[T]_{\mathcal{C}\leftarrow\mathcal{B}}[4 + 2x]_{\mathcal{B}} =$

$$\begin{bmatrix}0&1\\-1&0\end{bmatrix}\begin{bmatrix}4\\2\end{bmatrix} = \begin{bmatrix}2\\-4\end{bmatrix} = [2 - 4x]_{\mathcal{C}} = [T(4 + 2x)]_{\mathcal{C}}$$

3. $[T]_{\mathcal{C}\leftarrow\mathcal{B}} = \begin{bmatrix}1&0&0\\0&1&0\\0&0&1\end{bmatrix}$, $[T]_{\mathcal{C}\leftarrow\mathcal{B}}[a + bx + cx^2]_{\mathcal{B}} =$

$$\begin{bmatrix}1&0&0\\0&1&0\\0&0&1\end{bmatrix}\begin{bmatrix}a\\b\\c\end{bmatrix} = \begin{bmatrix}a\\b\\c\end{bmatrix} = [a + b(x + 2) + c(x + 2)^2)]_{\mathcal{C}} = [T(a + bx + cx^2)]_{\mathcal{C}}$$

5. $[T]_{\mathcal{C}\leftarrow\mathcal{B}} = \begin{bmatrix}1&0&0\\1&1&1\end{bmatrix}$, $[T]_{\mathcal{C}\leftarrow\mathcal{B}}[a + bx + cx^2]_{\mathcal{B}} =$

$$\begin{bmatrix}1&0&0\\1&1&1\end{bmatrix}\begin{bmatrix}a\\b\\c\end{bmatrix} = \begin{bmatrix}a\\a+b+c\end{bmatrix} = \begin{bmatrix}a + b\cdot 0 + c\cdot 0^2\\a + b\cdot 1 + c\cdot 1^2\end{bmatrix}_{\mathcal{C}} = [T(a + bx + cx^2)]_{\mathcal{C}}$$

7. $[T]_{\mathcal{C}\leftarrow\mathcal{B}} = \begin{bmatrix}6&4\\-3&-2\\2&-1\end{bmatrix}$, $[T]_{\mathcal{C}\leftarrow\mathcal{B}}\begin{bmatrix}-7\\7\end{bmatrix}_{\mathcal{B}} =$

$$\begin{bmatrix}6&4\\-3&-2\\2&-1\end{bmatrix}\begin{bmatrix}2\\-3\end{bmatrix} = \begin{bmatrix}0\\0\\7\end{bmatrix} = \begin{bmatrix}7\\7\\7\end{bmatrix} = \begin{bmatrix}7\\7\\7\end{bmatrix}_{\mathcal{C}} = \left[T\begin{bmatrix}-7\\7\end{bmatrix}\right]_{\mathcal{C}}$$

9. $[T]_{\mathcal{C}\leftarrow\mathcal{B}} = \begin{bmatrix}1&0&0&0\\0&0&1&0\\0&1&0&0\\0&0&0&1\end{bmatrix}$, $[T]_{\mathcal{C}\leftarrow\mathcal{B}}[A]_{\mathcal{B}} =$

$$\begin{bmatrix}1&0&0&0\\0&0&1&0\\0&1&0&0\\0&0&0&1\end{bmatrix}\begin{bmatrix}a\\b\\c\\d\end{bmatrix} = \begin{bmatrix}a\\c\\b\\d\end{bmatrix} = \left[\begin{bmatrix}a&c\\b&d\end{bmatrix}\right]_{\mathcal{C}} = [T(A)]_{\mathcal{C}}$$

11. $[T]_{\mathcal{C}\leftarrow\mathcal{B}} = \begin{bmatrix}0&-1&1&0\\-1&0&0&1\\1&0&0&-1\\0&1&-1&0\end{bmatrix}$, $[T]_{\mathcal{C}\leftarrow\mathcal{B}}[A]_{\mathcal{B}} =$

$$\begin{bmatrix}0&-1&1&0\\-1&0&0&1\\1&0&0&-1\\0&1&-1&0\end{bmatrix}\begin{bmatrix}a\\b\\c\\d\end{bmatrix} = \begin{bmatrix}c-b\\d-a\\a-d\\b-c\end{bmatrix} =$$

$$\left[\begin{bmatrix}c-b&d-a\\a-d&b-c\end{bmatrix}\right]_{\mathcal{C}} = [AB - BA]_{\mathcal{C}} = [T(A)]_{\mathcal{C}}$$

13. (b) $[D]_{\mathcal{B}} = \begin{bmatrix} 0 & -1 \\ 1 & 0 \end{bmatrix}$

(c) $[D]_{\mathcal{B}}[3 \sin x - 5 \cos x]_{\mathcal{B}} = \begin{bmatrix} 0 & -1 \\ 1 & 0 \end{bmatrix}\begin{bmatrix} 3 \\ -5 \end{bmatrix} = \begin{bmatrix} 5 \\ 3 \end{bmatrix} = [3 \cos x + 5 \sin x]_{\mathcal{B}} = [D(3 \sin x - 5 \cos x)]_{\mathcal{B}}$

15. (a) $[D]_{\mathcal{B}} = \begin{bmatrix} 2 & 0 & 0 \\ 0 & 2 & 1 \\ 0 & -1 & 2 \end{bmatrix}$

17. $[S \circ T]_{\mathcal{D} \leftarrow \mathcal{B}} = \begin{bmatrix} -1 & -2 \\ 1 & -1 \end{bmatrix}$

19. Invertible, $T^{-1}(a + bx) = -b + ax$

21. Invertible, $T^{-1}(p(x)) = p(x - 2)$

23. Invertible, $T^{-1}(a + bx + cx^2) = (a - b + 2c) + (b - 2c)x + cx^2$ or $T^{-1}(p(x)) = p(x) - p'(x) + p''(x)$

25. Not invertible **27.** $-3 \sin x - \cos x + C$

29. $\frac{4}{5}e^{2x} \cos x - \frac{3}{5}e^{2x} \sin x + C$

31. $\mathcal{C} = \left\{ \begin{bmatrix} 1 \\ -1 \end{bmatrix}, \begin{bmatrix} -4 \\ 1 \end{bmatrix} \right\}$ **33.** $\mathcal{C} = \{1 - x, 2 + x\}$

35. $\mathcal{C} = \{1, x\}$

37. $[T]_{\mathcal{E}} = \begin{bmatrix} (d_1^2 - d_2^2)/(d_1^2 + d_2^2) & 2d_1d_2/(d_1^2 + d_2^2) \\ 2d_1d_2/(d_1^2 + d_2^2) & (d_2^2 - d_1^2)/(d_1^2 + d_2^2) \end{bmatrix}$

Exercises 6.7

1. $y(t) = 2e^{3t}/e^3$

3. $y(t) = ((1 - e^4)e^{3t} + (e^3 - 1)e^{4t})/(e^3 - e^4)$

5. $f(t) = \left(\dfrac{e^{(\sqrt{5}-1)/2}}{e^{\sqrt{5}} - 1}\right)[e^{(1+\sqrt{5})t/2} - e^{(1-\sqrt{5})t/2}]$

7. $y(t) = e^t - (1 - e^{-1})te^t$

9. $y(t) = ((k + 1)e^{kt} + (k - 1)e^{-kt})/2k$

11. $y(t) = e^t \cos(2t)$

13. (a) $p(t) = 100e^{\ln(16)t/3} \approx 100e^{0.924t}$

(b) 45 minutes **(c)** In 9.968 hours

15. (a) $m(t) = 50e^{-ct}$, where $c = \ln 2/1590 \approx 4.36 \times 10^{-4}$; 32.33 mg remain after 1000 years.

(b) After 3691.9 years

17. $x(t) = \dfrac{5 - 10\cos(10\sqrt{K})}{\sin(10\sqrt{K})}\sin(\sqrt{K}t) + 10\cos(\sqrt{K}t)$

19. (b) No

Review Questions

1. (a) F **(c)** T **(e)** F **(g)** F **(i)** T

3. Subspace **5.** Subspace

7. Let $c_1A + c_2B = O$. Then $c_1A - c_2B = c_1A^T + c_2B^T = (c_1A + c_2B)^T = O$. Adding, we have $2c_1A = O$, so $c_1 = 0$ because A is nonzero. Hence $c_2B = O$, and so $c_2 = 0$. Thus, $\{A, B\}$ is linearly independent.

9. $\{1, x^2, x^4\}$, $\dim W = 3$

11. Linear transformation

13. Linear transformation **15.** $n^2 - 1$

17. $\begin{bmatrix} 1 & 0 & -1 \\ 0 & 1 & -2 \\ 0 & 0 & 1 \\ 1 & 0 & -1 \end{bmatrix}$

19. $S \circ T$ is the zero transformation.

Chapter 7

Exercises 7.1

1. (a) -10 **(b)** $\sqrt{14}$ **(c)** $\sqrt{93}$

3. Any nonzero scalar multiple of $\begin{bmatrix} 3 \\ 1 \end{bmatrix}$

5. (a) 1 **(b)** $\sqrt{13}$ **(c)** $\sqrt{14}$

7. x^2 is one possibility

9. (a) π **(b)** $\sqrt{\pi}$ **(c)** $\sqrt{\pi}$

13. Axiom (4) fails: $\mathbf{u} = \begin{bmatrix} 0 \\ 1 \end{bmatrix} \neq \mathbf{0}$, but $\langle \mathbf{u}, \mathbf{u} \rangle = 0$.

15. Axiom (4) fails: $\mathbf{u} = \begin{bmatrix} 0 \\ 1 \end{bmatrix} \neq \mathbf{0}$, but $\langle \mathbf{u}, \mathbf{u} \rangle = 0$.

17. Axiom (4) fails: $p(x) = 1 - x$ is not the zero polynomial, but $\langle p(x), p(x) \rangle = 0$.

19. $A = \begin{bmatrix} 4 & 1 \\ 1 & 4 \end{bmatrix}$

21.

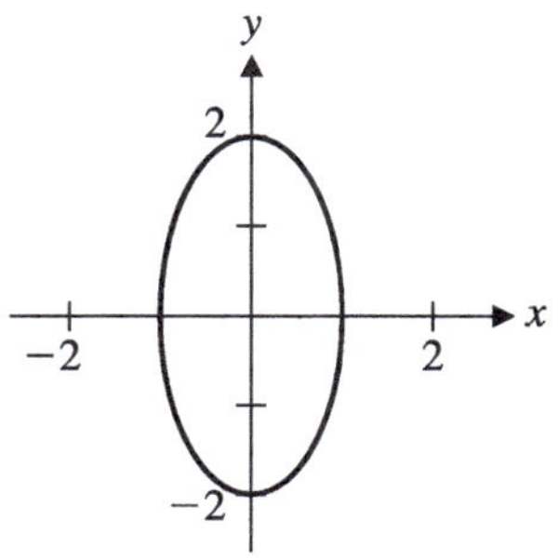

25. -8 **27.** $\sqrt{6}$

29. $\|\mathbf{u} + \mathbf{v} - \mathbf{w}\|^2 = \langle \mathbf{u} + \mathbf{v} - \mathbf{w}, \mathbf{u} + \mathbf{v} - \mathbf{w} \rangle$
$= \langle \mathbf{u}, \mathbf{u} \rangle + \langle \mathbf{v}, \mathbf{v} \rangle + \langle \mathbf{w}, \mathbf{w} \rangle + 2\langle \mathbf{u}, \mathbf{v} \rangle - 2\langle \mathbf{u}, \mathbf{w} \rangle - 2\langle \mathbf{v}, \mathbf{w} \rangle$
$= 1 + 3 + 4 + 2 - 10 - 0 = 0$

Therefore, $\|\mathbf{u} + \mathbf{v} - \mathbf{w}\| = 0$, so, by axiom (4), $\mathbf{u} + \mathbf{v} - \mathbf{w} = \mathbf{0}$ or $\mathbf{u} + \mathbf{v} = \mathbf{w}$.

31. $\langle \mathbf{u}+\mathbf{v}, \mathbf{u}-\mathbf{v}\rangle = \langle \mathbf{u},\mathbf{u}\rangle - \langle \mathbf{u},\mathbf{v}\rangle + \langle \mathbf{v},\mathbf{u}\rangle - \langle \mathbf{v},\mathbf{v}\rangle = \|\mathbf{u}\|^2 - \langle \mathbf{u},\mathbf{v}\rangle + \langle \mathbf{u},\mathbf{v}\rangle - \|\mathbf{v}\|^2 = \|\mathbf{u}\|^2 - \|\mathbf{v}\|^2$

33. Using Exercise 32 and a similar identity for $\|\mathbf{u} - \mathbf{v}\|^2$, we have

$$\begin{aligned}\|\mathbf{u}+\mathbf{v}\|^2 + \|\mathbf{u}-\mathbf{v}\|^2 &= \langle \mathbf{u}+\mathbf{v}, \mathbf{u}+\mathbf{v}\rangle + \langle \mathbf{u}-\mathbf{v}, \mathbf{u}-\mathbf{v}\rangle \\ &= \|\mathbf{u}\|^2 + 2\langle \mathbf{u},\mathbf{v}\rangle + \|\mathbf{v}\|^2 \\ &\quad + \|\mathbf{u}\|^2 - 2\langle \mathbf{u},\mathbf{v}\rangle + \|\mathbf{v}\|^2 \\ &= 2\|\mathbf{u}\|^2 + 2\|\mathbf{v}\|^2\end{aligned}$$

Dividing by 2 yields the identity we want.

35.
$$\begin{aligned}\|\mathbf{u}+\mathbf{v}\| = \|\mathbf{u}-\mathbf{v}\| &\Leftrightarrow \|\mathbf{u}+\mathbf{v}\|^2 = \|\mathbf{u}-\mathbf{v}\|^2 \\ &\Leftrightarrow \|\mathbf{u}\|^2 + 2\langle \mathbf{u},\mathbf{v}\rangle + \|\mathbf{v}\|^2 \\ &\qquad = \|\mathbf{u}\|^2 - 2\langle \mathbf{u},\mathbf{v}\rangle + \|\mathbf{v}\|^2 \\ &\Leftrightarrow 2\langle \mathbf{u},\mathbf{v}\rangle = -2\langle \mathbf{u},\mathbf{v}\rangle \Leftrightarrow \langle \mathbf{u},\mathbf{v}\rangle = 0\end{aligned}$$

37. $\left\{\begin{bmatrix}1\\0\end{bmatrix}, \begin{bmatrix}0\\1\end{bmatrix}\right\}$ **39.** $\{1, x, x^2\}$

41. (a) $1/\sqrt{2}, \sqrt{3}x/\sqrt{2}, \sqrt{5}(3x^2 - 1)/2\sqrt{2}$
(b) $\sqrt{7}(5x^3 - 3x)/2\sqrt{2}$

Exercises 7.2

1. $\|\mathbf{u}\|_E = \sqrt{42}$, $\|\mathbf{u}\|_s = 10$, $\|\mathbf{u}\|_m = 5$

3. $d_E(\mathbf{u},\mathbf{v}) = \sqrt{70}$, $d_s(\mathbf{u},\mathbf{v}) = 14$, $d_m(\mathbf{u},\mathbf{v}) = 6$

5. $\|\mathbf{u}\|_H = 4$, $\|\mathbf{v}\|_H = 5$

7. (a) At most one component of $\mathbf{v}$ is nonzero.

9. Suppose $\|\mathbf{v}\|_m = |v_k|$. Then $\|\mathbf{v}\|_E = \sqrt{v_1^2 + \cdots + v_k^2 + \cdots + v_n^2} \geq \sqrt{v_k^2} = |v_k| = \|\mathbf{v}\|_m$.

11. Suppose $\|\mathbf{v}\|_m = |v_k|$. Then $|v_i| \leq |v_k|$ for $i = 1, \ldots, n$, so

$$\begin{aligned}\|\mathbf{v}\|_s &= |v_1| + \cdots + |v_n| \leq |v_k| + \cdots + |v_k| \\ &= n|v_k| = n\|\mathbf{v}\|_m\end{aligned}$$

13.

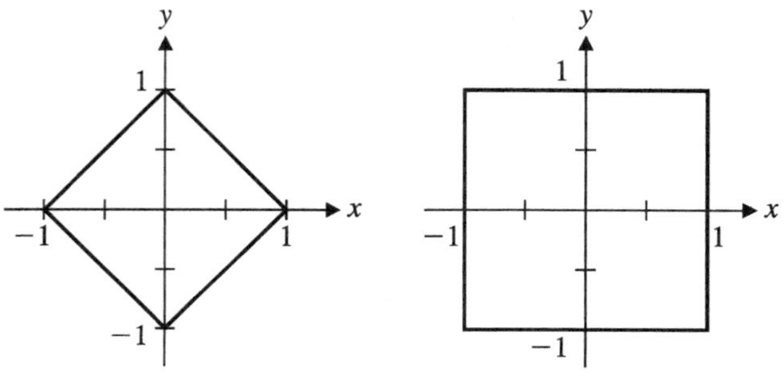

21. $\|A\|_F = \sqrt{19}$, $\|A\|_1 = 4$, $\|A\|_\infty = 6$

23. $\|A\|_F = \sqrt{31}$, $\|A\|_1 = 6$, $\|A\|_\infty = 6$

25. $\|A\|_F = 2\sqrt{11}$, $\|A\|_1 = 7$, $\|A\|_\infty = 7$

27. $\mathbf{x} = \begin{bmatrix}0\\1\end{bmatrix}, \mathbf{y} = \begin{bmatrix}-1\\1\end{bmatrix}$ **29.** $\mathbf{x} = \begin{bmatrix}0\\0\\1\end{bmatrix}, \mathbf{y} = \begin{bmatrix}1\\1\\1\end{bmatrix}$

31. $\mathbf{x} = \begin{bmatrix}1\\0\\0\end{bmatrix}, \mathbf{y} = \begin{bmatrix}1\\-1\\-1\end{bmatrix}$

33. (a) By the definition of an operator norm, $\|I\| = \max\limits_{\|\mathbf{x}\|=1}\|I\mathbf{x}\| = \max\limits_{\|\mathbf{x}\|=1}\|\mathbf{x}\| = 1$.

35. $\text{cond}_1(A) = \text{cond}_\infty(A) = 21$; well-conditioned

37. $\text{cond}_1(A) = \text{cond}_\infty(A) = 400$; ill-conditioned

39. $\text{cond}_1(A) = 77$, $\text{cond}_\infty(A) = 128$; moderately ill-conditioned

41. (a) $\text{cond}_\infty(A) = (\max\{|k| + 1, 2\}) \cdot \left(\max\left\{\left|\dfrac{k}{k-1}\right| + \left|\dfrac{1}{k-1}\right|, \left|\dfrac{2}{k-1}\right|\right\}\right)$

43. (a) $\text{cond}_\infty(A) = 40$
(b) At most 400% relative change

45. Using Exercise 33(a), we have $\text{cond}(A) = \|A\|\|A^{-1}\| \geq \|AA^{-1}\| = \|I\| = 1$.

49. $k \geq 6$ **51.** $k \geq 10$

Exercises 7.3

1. $\|\mathbf{e}\| = \sqrt{2} \approx 1.414$ **3.** $\|\mathbf{e}\| = \sqrt{6}/2 \approx 1.225$

5. $\|\mathbf{e}\| = \sqrt{7} \approx 2.646$

7. $y = -3 + \frac{5}{2}x$, $\|\mathbf{e}\| \approx 1.225$

9. $y = \frac{11}{3} - 2x$, $\|\mathbf{e}\| \approx 0.816$

11. $y = \frac{7}{10} + \frac{8}{25}x$, $\|\mathbf{e}\| \approx 0.447$

13. $y = -\frac{1}{5} + \frac{7}{5}x$, $\|\mathbf{e}\| \approx 0.632$

15. $y = 3 - \frac{18}{5}x + x^2$ **17.** $y = \frac{18}{5} - \frac{17}{10}x - \frac{1}{2}x^2$

19. $\bar{\mathbf{x}} = \begin{bmatrix}\frac{1}{5}\\ \frac{7}{15}\end{bmatrix}$ **21.** $\bar{\mathbf{x}} = \begin{bmatrix}\frac{4}{3}\\ -\frac{5}{6}\end{bmatrix}$

23. $\bar{\mathbf{x}} = \begin{bmatrix}4+t\\ -5-t\\ -5-2t\\ t\end{bmatrix}$ **25.** $\begin{bmatrix}\frac{42}{11}\\ \frac{19}{11}\\ \frac{42}{11}\end{bmatrix}$

27. $\bar{\mathbf{x}} = \begin{bmatrix} \frac{5}{3} \\ -2 \end{bmatrix}$

29. $y = 0.92 + 0.73x$

31. (a) If we let the year 1920 correspond to $t = 0$, then $y = 56.6 + 2.9t$; 79.9 years

33. (a) $p(t) = 150e^{0.131t}$

35. 139 days

37. $\begin{bmatrix} \frac{1}{2} & \frac{1}{2} \\ \frac{1}{2} & \frac{1}{2} \end{bmatrix}, \begin{bmatrix} \frac{7}{2} \\ \frac{7}{2} \end{bmatrix}$

39. $\begin{bmatrix} \frac{1}{3} & \frac{1}{3} & \frac{1}{3} \\ \frac{1}{3} & \frac{1}{3} & \frac{1}{3} \\ \frac{1}{3} & \frac{1}{3} & \frac{1}{3} \end{bmatrix}, \begin{bmatrix} 2 \\ 2 \\ 2 \end{bmatrix}$

41. $\begin{bmatrix} \frac{5}{6} & \frac{1}{3} & -\frac{1}{6} \\ \frac{1}{3} & \frac{1}{3} & \frac{1}{3} \\ -\frac{1}{6} & \frac{1}{3} & \frac{5}{6} \end{bmatrix}, \begin{bmatrix} \frac{5}{6} \\ \frac{1}{3} \\ -\frac{1}{6} \end{bmatrix}$

45. $A^+ = [\frac{1}{5} \quad \frac{2}{5}]$

47. $A^+ = \begin{bmatrix} \frac{1}{3} & -\frac{2}{3} & -\frac{1}{6} \\ \frac{1}{6} & \frac{1}{6} & \frac{1}{6} \end{bmatrix}$

49. $A^+ = \begin{bmatrix} 1 & -1 \\ 0 & 1 \end{bmatrix}$

51. $A^+ = \begin{bmatrix} \frac{2}{3} & 0 & -\frac{1}{3} & \frac{1}{3} \\ \frac{1}{3} & -1 & \frac{1}{3} & \frac{2}{3} \\ -\frac{2}{3} & 1 & \frac{1}{3} & -\frac{1}{3} \end{bmatrix}$

53. (a) If A is invertible, so is A^T, and we have $A^+ = (A^TA)^{-1}A^T = A^{-1}(A^T)^{-1}A^T = A^{-1}$.

Exercises 7.4

1. 2, 3 **3.** $\sqrt{2}$, 0 **5.** 5 **7.** 2, 3

9. $\sqrt{5}$, 2, 0

11. $A = \begin{bmatrix} 1 & 0 \\ 0 & 1 \end{bmatrix}\begin{bmatrix} \sqrt{2} & 0 \\ 0 & 0 \end{bmatrix}\begin{bmatrix} 1/\sqrt{2} & 1/\sqrt{2} \\ 1/\sqrt{2} & -1/\sqrt{2} \end{bmatrix}$

13. $A = \begin{bmatrix} 0 & 1 \\ 1 & 0 \end{bmatrix}\begin{bmatrix} 3 & 0 \\ 0 & 2 \end{bmatrix}\begin{bmatrix} -1 & 0 \\ 0 & -1 \end{bmatrix}$

15. $A = \begin{bmatrix} \frac{3}{5} & -\frac{4}{5} \\ \frac{4}{5} & \frac{3}{5} \end{bmatrix}\begin{bmatrix} 5 \\ 0 \end{bmatrix}[1]$

17. $A = \begin{bmatrix} 0 & 0 & 1 \\ 1 & 0 & 0 \\ 0 & -1 & 0 \end{bmatrix}\begin{bmatrix} 3 & 0 \\ 0 & 2 \\ 0 & 0 \end{bmatrix}\begin{bmatrix} 0 & 1 \\ 1 & 0 \end{bmatrix}$

19. $A = \begin{bmatrix} 1 & 0 \\ 0 & 1 \end{bmatrix}\begin{bmatrix} \sqrt{5} & 0 & 0 \\ 0 & 2 & 0 \end{bmatrix}\begin{bmatrix} 2/\sqrt{5} & 0 & 1/\sqrt{5} \\ 0 & 1 & 0 \\ 1/\sqrt{5} & 0 & -2/\sqrt{5} \end{bmatrix}$

21. $A = \sqrt{2}\begin{bmatrix} 1 \\ 0 \end{bmatrix}[1/\sqrt{2} \quad 1/\sqrt{2}] + 0\begin{bmatrix} 0 \\ 1 \end{bmatrix}$ $[-1/\sqrt{2} \quad 1/\sqrt{2}]$ (Exercise 3)

23. (Exercise 7) $A = 3\begin{bmatrix} 0 \\ 1 \\ 0 \end{bmatrix}[0 \quad 1] + 2\begin{bmatrix} 0 \\ 0 \\ -1 \end{bmatrix}[1 \quad 0]$

33. The line segment $[-1, 1]$

35. The solid ellipse $\dfrac{y_1^2}{5} + \dfrac{y_2^2}{4} \le 1$

37. (a) $\|A\|_2 = \sqrt{2}$ **(b)** $\text{cond}_2(A) = \infty$

39. (a) $\|A\|_2 = 1.95$ **(b)** $\text{cond}_2(A) = 38.11$

41. $A^+ = \begin{bmatrix} \frac{1}{2} & 0 \\ \frac{1}{2} & 0 \end{bmatrix}$

43. $A^+ = \begin{bmatrix} \frac{2}{5} & 0 \\ 0 & \frac{1}{2} \\ \frac{1}{5} & 0 \end{bmatrix}$

45. $A^+ = \begin{bmatrix} \frac{1}{25} & \frac{2}{25} \\ \frac{2}{25} & \frac{4}{25} \end{bmatrix}, \bar{\mathbf{x}} = \begin{bmatrix} 0.52 \\ 1.04 \end{bmatrix}$

47. $A^+ = \begin{bmatrix} \frac{1}{6} & \frac{1}{6} & \frac{1}{6} \\ \frac{1}{6} & \frac{1}{6} & \frac{1}{6} \end{bmatrix}, \bar{\mathbf{x}} = \begin{bmatrix} 1 \\ 1 \end{bmatrix}$

61. $\begin{bmatrix} \sqrt{2} & 0 \\ 0 & 0 \end{bmatrix}\begin{bmatrix} 1/\sqrt{2} & 1/\sqrt{2} \\ -1/\sqrt{2} & 1/\sqrt{2} \end{bmatrix}$

63. $\begin{bmatrix} 2 & -1 \\ -1 & 3 \end{bmatrix}\begin{bmatrix} 0 & 1 \\ -1 & 0 \end{bmatrix}$

Exercises 7.5

1. $g(x) = \frac{1}{3}$ **3.** $g(x) = \frac{3}{5}x$

5. $g(x) = \frac{3}{16} + \frac{15}{16}x^2$ **7.** $\{1, x - \frac{1}{2}\}$

9. $g(x) = x - \frac{1}{6}$

11. $g(x) = (4e - 10) + (18 - 6e)x \approx 0.87 + 1.69x$

13. $g(x) = \frac{1}{20} - \frac{3}{5}x + \frac{3}{2}x^2$

15. $g(x) = 39e - 105 + (588 - 216e)x + (210e - 570)x^2 \approx 1.01 + 0.85x + 0.84x^2$

21. $\dfrac{\pi}{2} - \dfrac{4}{\pi}\left(\cos x + \dfrac{\cos 3x}{9}\right)$

23. $a_0 = \frac{1}{2}, a_k = 0, b_k = \dfrac{1 - (-1)^k}{k\pi}$

25. $a_0 = \pi, a_k = 0, b_k = \dfrac{2(-1)^k}{k}$

Review Questions

1. (a) T **(c)** F **(e)** T **(g)** T **(i)** T

3. Inner product **5.** $\sqrt{3}$ **7.** $\left\{\begin{bmatrix} 1 \\ 1 \end{bmatrix}\right\}, \begin{bmatrix} -\frac{1}{2} \\ \frac{1}{2} \end{bmatrix}$

9. Not a norm

11. $\text{cond}_\infty(A) \approx 2432$

13. $y = 1.7x$

15. $\begin{bmatrix} \frac{7}{3} \\ \frac{2}{3} \\ \frac{5}{3} \end{bmatrix}$

17. (a) $\sqrt{2}, \sqrt{2}$

(b) $A = \begin{bmatrix} 1/\sqrt{2} & 1/\sqrt{2} & 0 \\ 0 & 0 & 1 \\ 1/\sqrt{2} & -1/\sqrt{2} & 0 \end{bmatrix} \begin{bmatrix} \sqrt{2} & 0 \\ 0 & \sqrt{2} \\ 0 & 0 \end{bmatrix} \begin{bmatrix} 1 & 0 \\ 0 & 1 \end{bmatrix}$

(c) $A^+ = \begin{bmatrix} \frac{1}{2} & 0 & \frac{1}{2} \\ \frac{1}{2} & 0 & -\frac{1}{2} \end{bmatrix}$

19. The singular values of PAQ are the square roots of the eigenvalues of $(PAQ)^T(PAQ) = Q^TA^TP^TPAQ = Q^T(A^TA)Q$. But $Q^T(A^TA)Q$ is similar to A^TA because $Q^T = Q^{-1}$, and hence it has the same eigenvalues as A^TA. Thus, PAQ and A have the same singular values.

Index

I

J

K

L

M